C0-CCP-110

Handbook of Environmental Data on Organic Chemicals

Volume 1

Handbook of Environmental Data on Organic Chemicals

Fifth Edition

Volume 1

Karel Verschueren
President
Verschueren Environmental Consultancy

WILEY
A JOHN WILEY & SONS, INC., PUBLICATION

Published by John Wiley & Sons, Inc., Hoboken, New Jersey.
Published simultaneously in Canada.

Library of Congress Cataloging-in-Publication Data is available.

Verschueren, Karel
Handbook of environmental data on organic chemicals—5th edition
ISBN: 978-0-470-17172-1

Printed in the United States of America

10 9 8 7 6 5 4 3 2 1

CONTENTS

INTRODUCTION

Since the publication of the first edition of this handbook in 1977, much more information has become available about the presence and fate of new and existing organic chemicals in the environment. These data, when given a wide distribution, will no doubt reduce the misuse of dangerous chemicals and hence their impact on the environment. The *Handbook of Environmental Data on Organic Chemicals* has now been updated for the fifth time and covers individual substances as well as mixtures and preparations.

I. ARRANGEMENT OF CATEGORIES

The information in the categories listed below is given for each product in the sequence indicated; where entries are incomplete, it may be assumed that no reliable data were provided by the references utilized.

- *Name: the commonly accepted name is the key entry.*
- Synonym: alternative names, as well as trivial names and identifiers, are indicated. Obsolete and slang names have been eliminated as far as possible.
- *Formula: the molecular and structural formulas are given.*
- CAS: the Chemical Abstracts Service number
- *Manufacturing source*
- *Use, users, and formulations*
- *Natural sources and occurrence*
- *Man-caused sources*

A. PROPERTIES The chemical and physical properties typically given are physical appearance; molecular weight (mw); melting point (mp); boiling point (bp) at 760 mm Hg unless otherwise stated; vapor pressure (vp) at different temperatures; relative vapor density (vd), the relative vapor density of air = 1; saturation concentration in air at different temperatures (sat. conc.); the maximum solubility in water at various temperatures (solub.); the liquid or solid density at room temperature; the logarithm of the octanol/water partition coefficient (log P_{oct}); the logarithm of the dimensionless constant of Henry (log H).
B. AIR POLLUTION FACTORS The following data are given: conversion factors (between volume and mass units of concentration); odor threshold values and characteristics; atmospheric reactions; natural sources (and background concentrations); man-made sources (and ground level concentrations caused by such sources); emission control methods (and results); methods of sampling and analysis.
C. WATER AND SOIL POLLUTION FACTORS Analogous to the previous category, the following data are listed: biodegradation rate and mechanisms; oxidation parameters, such as BOD, COD, and ThOD; impact on treatment processes and on the BOD test; reduction of amenities through taste, odor, and color of the water or aquatic organisms; the quality of surface water and underground water and sediment; natural sources; man-made sources; waste water treatment methods and results; methods of sampling and analysis.
D. BIOLOGICAL EFFECTS Residual concentrations, bioaccumulation values, and toxicological effects of exposing the products to ecosystems, bacteria, plants, algae, protozoans, worms, molluscs, insects, crustaceans, fishes, amphibians and birds.
The "explanatory notes" give a more detailed description of the compiled data, explain the definitions and abbreviations used throughout the book, and indicate how the data can be used to prevent or reduce environmental pollution.

II. ARRANGEMENT OF CHEMICALS

The chemicals are listed in strict alphabetical order; those that comprise two or more words are alphabetized as though they were a single word. The many prefixes used in organic chemistry are disregarded in alphabetizing because they are not considered an integral part of the name; these include *ortho-, meta-, para-, alpha-, beta-, gamma-, sec, tert, sym-, as-, uns-, cis-, trans-, d-, l-, dl-, n, and N-, as well as all numerals denoting structure. However, there are certain prefixes that are an integral part of the names (iso-, di-, tri-, tetra-, cyclo-, bio-, neo-, pseudo-), and in these cases, the name is placed in its normal alphabetical position. For example, dimethylamine appears under D and isobutane under I.*

III. ORDER OF ELEMENTS

Readers who are not acquainted with the definitions and abbreviations used throughout the book should consult the appropriate sections of this chapter. The data are given in the following sequence (each item will be discussed in detail).

A. PROPERTIES

- 1. formula
- 2. physical appearance
- 3. molecular weight (mw)
- 4. melting point (mp)
- 5. boiling point (bp)
- 6. vapor pressure (vp)
- 7. vapor density (vd)
- 8. saturation concentration (sat. conc.)
- 9. solubility (solub.)
- 10. density (d)
- 11. logarithm of the octanol/water distribution coefficient (log P_{oct})
- 12. logarithm of the dimensionless Henry's constant (log H)

B. AIR POLLUTION FACTORS

- 13. conversion factors
- 14. odor
- 15. atmospheric reactions
- 16. natural sources
- 17. man-made sources
- 18. control methods
- 19. air quality

C. WATER AND SOIL POLLUTION FACTORS

- 20. biodegradation
- 21. oxidation parameters
- 22. impact on biodegradation processes
- 23. odor and taste thresholds
- 24. water, soil, and sediment quality
- 25. natural sources
- 26. man-made sources
- 27. waste water treatment
- 28. degradation in soil
- 29. soil sorption

D. BIOLOGICAL EFFECTS

- residual concentrations
- bioaccumulation values
- toxicological effects

- 30. ecosystems
- 31. bacteria
- 32. algae
- 33. plants
- 34. worms
- 35. molluscs
- 36. insects
- 37. crustaceans
- 38. fishes
- 39. amphibians
- 40. birds

IV. EXPLANATORY ELEMENTS

A. PROPERTIES

Only the most relevant chemical and physical properties are given. Flash points, flammability limits, autoignition temperature, and the like have been omitted because they are not of direct concern to the environmentalist. These and other dangerous properties of chemicals can be found in *Dangerous Properties of Industrial Materials* by I. Sax. Chemicals are never 100% pure, but the nature and quantity of the impurities can have a significant impact on most environmental qualities. The following parameters are very sensitive to the presence of impurities: water solubility, odor characteristic and threshold values, BOD, and toxicity.
The following data (from Shell's *Chemical Guide) illustrate this point:*

- product: *diethylene glycol* $O(CH_2CH_2OH)_2$

	Normal grade	*Special grade*
distillation range:	240-255°C	242-250°C
acidity (as CH_3COOH):	max. 0.2 wt%	max. 0.002 wt%
ash content:	max. 0.05 wt%	max. 0.002 wt%
BOD_5:	0.12	0.05
COD:	1.49	1.51
goldfish 24h LD_{50}:	5,000 mg/l	5,000 mg/l

- product: *ethyleneglycol* $HOCH_2\text{-}CH_2OH$

	Normal grade	*Special grade*

distillation range:	194-205°C	max. 2°C, incl. 197.6°C
ash content:	max. 0.002 wt%	max. 0.001 wt%
BOD_5:	0.47	0.15
BOD_5 after adaptation:	0.81	0.67
COD:	1.24	1.29
goldfish 24h LD_{50}:	5,000 mg/l	5,000 mg/l

When no data are available, the distillation range can give a first indication on the presence of impurities. Therefore, in this work, whenever a distillation range (boiling range) is given, the environmental data should be interpreted carefully.

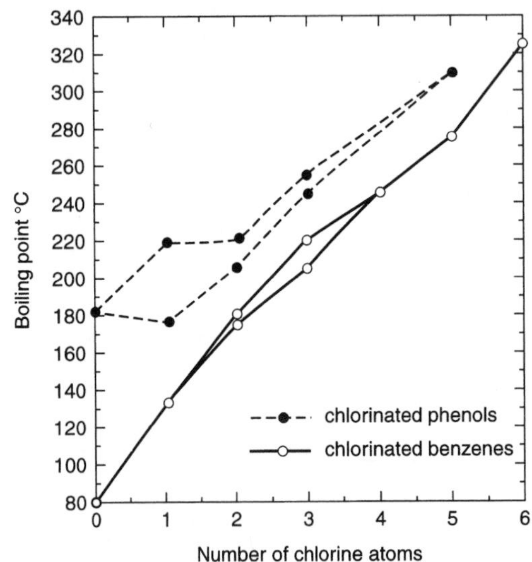

Figure 1. Relationship between boiling point and molecular weight for chlorinated benzenes and phenols.

- product: *triethanolamine* $N(CH_2CH_2OH)_3$

	Normal grade	*85%*
triethanolamine content:	min. 80 wt%	min. 85 wt%
BOD_5:	0.02	0.03
BOD_5 after adaptation:	0.17	0.90
COD:	1.50	1.50

After adaptation of the culture, the "85%" grade is much more biodegradable than the less pure "commercial" grade.

1. Boiling Points

The boiling points of the members of a given homologous series increase with increasing molecular weight. The boiling points rise in a uniform manner as shown in Figure 1.

If a hydrogen atom of one of the paraffin hydrocarbons is replaced by another atom or a group, an elevation of the boiling point results. Thus alkyl halides, alcohols, aldehydes, ketones, acids, etc. boil at higher temperatures than the hydrocarbons with the same carbon skeleton.

If the group introduced is of such a nature that it promotes association, a very marked rise in boiling point occurs. This effect is especially pronounced in the alcohols and acids, because hydrogen bonding can occur.

2. Vapor Pressure

The vapor pressure of a liquid or solid is the pressure of the gas in equilibrium with the liquid or solid at a given temperature. Volatilization, the evaporative loss of a chemical, depends on the vapor pressure of the chemical and on environmental conditions that influence diffusion from the evaporative surface. Volatilization is an important source of material for airborne transport and may lead to the distribution of a chemical over wide areas and into bodies of water (e.g., in rainfall) far from the site of release. Vapor pressure values give indications of the tendency of pure substances to vaporize in an unperturbed situation and thus provide a method for ranking the relative volatilities of chemicals. Vapor pressure data combined with solubility data permit calculations of rates of evaporation of dissolved organics from water using Henry's Law constants, as discussed by MacKay and Leinonen (1943) and Dilling (1944).

Chemicals with relatively low vapor pressures, high adsorptivity onto solids, or high solubility in water are less likely to vaporize and become airborne than chemicals with high vapor pressures or less affinity for solution in water or adsorption to solids and sediments. In addition, chemicals that are likely to be gases at ambient temperatures and that have low water solubility and low adsorptive tendencies are less likely to transport and persist in soils and water. Such chemicals are less likely to biodegrade or hydrolyze but are prime candidates for photolysis and for involvement in adverse atmospheric effects (such as smog formation and stratospheric alterations). On the other hand, nonvolatile chemicals are less frequently involved in significant atmospheric transport, so concerns regarding them should focus on soils and water.

Vapor pressures are expressed in mm Hg (abbreviated mm), in atmospheres (atm), in mbars, or in hectoPascals (hPa).

If vapor pressure data for certain compounds are not available, they can be derived graphically from the compounds' boiling points and the boiling point/vapor pressure relationship for homologous series. An example is shown in Figure 2.

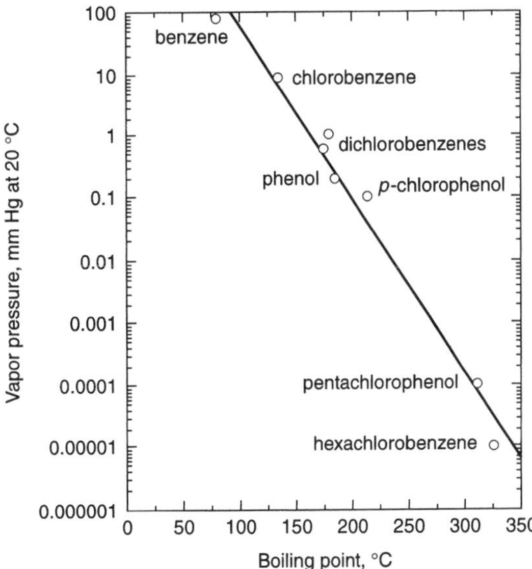

Figure 2. Relationship between boiling point and vapor pressure for homologous series of chlorinated benzenes and phenols.

3. Vapor Density

The density of a gas indicates whether it will be transported along the ground, possibly subjecting surrounding populations to high exposure, or will disperse rapidly.

The concentration term *vapor density* is often used in discussion of vapor phase systems. Vapor density is related to equilibrium vapor pressure through the equation of state for a gas:

$$PV = nRT$$

When the mass of the substance and the gram molecular weight are substituted for the number of moles n, the following equation is obtained:

$$\text{vapor density (vd)} = PM/RT$$

where

P = equilibrium vapor pressure in atmospheres

R = 0.082 liter atmospheres/mol/K

M = gram molecular weight

T = absolute temperature in kelvins (K)

In this book the relative vapor density (air = 1) is given because it indicates how the gas will behave upon release.

4. Water Solubility

4.1. Objectives. The water solubility of a chemical is an important characteristic for establishing that chemical's potential environmental movement and distribution. In general, highly soluble chemicals are more likely than poorly soluble chemicals to be distributed by the hydrological cycle.

Water solubility can also affect adsorption and desorption on soils and volatility from aquatic systems. Substances that are more soluble are more likely to desorb from soils and less likely to volatilize from water. Water solubility can also affect possible transformation by hydrolysis, photolysis, oxidation, reduction, and biodegradation in water. Finally, the design of most chemical tests and of many ecological and health tests requires precise knowledge of the water solubility of chemicals. Water solubility is an important parameter for assessment of all solid and liquid chemicals. Water solubility is generally not useful for gases, because their solubility in water is measured when the gas above the water is at a partial pressure of one atmosphere. Thus the solubility of gases does not usually apply to environmental assessment, because the actual partial pressure of a gas in the environment is extremely low.

4.2. Interpretation of Data. It is not unusual to find in the literature a wide range of solubilities for the same product. The oldest literature generally yields the highest solubility values. The reasons are twofold: First, in the years before and immediately after World War II, products were not as pure as they are today. Second, recent determinations are based on specific methods of analysis, such as gas chromatography. Nonspecific determinations do not distinguish between the dissolved product and the dissolved impurities; the latter, when they are much more soluble than the original product, move to the aqueous phase and are recorded as dissolved product. Nonspecific methods include turbidity measurement and TOD (Total Oxygen Demand).

The measurement of aqueous solubility does not usually impose excessive demands on chemical techniques, but measuring the solubility of very sparingly soluble compounds requires specialized procedures. This problem is well illustrated by the variability in the values quoted in the literature for products such as DDT and PCBs. This situation happens to be of some consequence; many of the compounds that are known to be significant environmental contaminants, such as DDT and PCBs, are those that have very low water solubilities.

4.3. Influence of the Composition of Natural Waters. The composition of natural waters can vary greatly. Environmental variables such as pH, water hardness, cations, anions, naturally occurring organic substances (e.g., humic and fulvic acids and hemicelluloses), and organic pollutants all affect the solubility of chemicals in water. Some bodies of water contain enough organic and inorganic impurities to significantly alter the solubility of poorly soluble chemicals.

The solubility of lower *n*-paraffins in salt water compared with fresh, distilled water is higher by about one order of magnitude, this difference decreasing with an increase in the molecular weight of the hydrocarbon. The increased solubility in seawater is due to simultaneous physical and chemical factors. The solubility of several higher *n*-paraffins (C_{10} and higher) has been determined in both distilled water and seawater. In all cases, the paraffins were less soluble in seawater than in distilled water. The magnitude of the salting out effect increases with increasing molar volume of the paraffins, in accordance with the McDevit-Long Theory. This theory of salt effects attributes salting in or salting out to the effect of electrolytes on the structure of water. Because the data in the literature indicate that the lower paraffins (below C_{10}) are more soluble, and the higher *n*-paraffins (C_{10} and higher) less soluble, in seawater than in distilled water, it is possible to speculate upon the geochemical fate of dissolved normal paraffins entering the ocean from rivers. If fresh water is saturated or near saturated with respect to normal paraffins (e.g., because of pollution), salting out of the higher paraffins will occur in the estuary. The salted out molecules may either adsorb on suspended minerals and on particulate organic matter or rise to the surface as slicks. In either case, they will follow a different biochemical pathway than if they had been dissolved. The salting out of dissolved organic molecules in estuaries applies not to *n*-paraffins alone, but to all natural or pollutant organic molecules whose solubilities are decreased by addition of electrolytes. Thus it is possible that regardless of the levels of dissolved organic pollutants in river water, only given amounts will enter the ocean in dissolved form because of salting out effects of estuaries. Estuaries may act to limit the amount of dissolved organic carbon entering the ocean, but they may increase the amount of particulate organic carbon entering the marine environment.

4.4. Molecular Structure-Solubility Relationship. Because water is a polar compound, it is a poor solvent for hydrocarbons. Olefinic and acetylenic linkages and benzenoid structures do not greatly affect the polarity. Hence, unsaturated or aromatic hydrocarbons are not very different from paraffins in their water solubility. The introduction of halogen atoms does not alter the polarity appreciably. It does increase the molecular weight, and for this reason the water solubility always falls off. On the other hand, salts are extremely polar. Other compounds lie between these two extremes. Here are found the alcohols, esters, ethers, acids, amines, nitriles, amides, ketones, and aldehydes-to mention a few of the classes that occur frequently.

As might be expected, acids and amines generally are more soluble than neutral compounds. The amines probably owe their abnormally high solubility to their tendency to form hydrogen-bonded complexes with water molecules. This theory is in harmony with the fact that the solubility of amines diminishes as the basicity decreases. It also explains the observation that many tertiary amines are more soluble in cold than in hot water. Apparently, at lower temperatures the solubility of the hydrate is involved, whereas at higher temperatures the hydrate is unstable and the solubility measured is that of the free amine.

Monofunctional ethers, esters, ketones, aldehydes, alcohols, nitriles, amides, acids, and amines may be considered together with respect to water solubility. As a homologous series is ascended, the hydrocarbon (nonpolar) part of the molecule continually increases while the polar function remains essentially unchanged. There follows, then, a trend toward a decrease in the solubility in polar solvents such as water.

In general, an increase in molecular weight leads to an increase in intermolecular forces in a solid. Polymers and other compounds of high molecular weight generally exhibit low solubilities in water and ether. Thus formaldehyde is readily soluble in water, whereas paraformaldehyde is insoluble:

$$\underset{\text{water soluble}}{CH_2O} \longrightarrow \underset{\text{water insoluble}}{HO(CH_2O\text{-})_x H}$$

Methyl acrylate is soluble in water, but its polymer is insoluble:

$$\underset{\text{water soluble}}{CH2{=}CHCOOCH_3} \longrightarrow \underset{\text{water insoluble}}{\left(-CH_2\underset{\substack{|\\COOCH_3}}{CH}-\right)_x}$$

Glucose is soluble in water, but its polymers-starch, glycogen, and cellulose-are insoluble. Many amino acids are soluble in water, but their condensation polymers, the proteins, are insoluble.

Lindenberg (1803) proposed a relationship between the logarithm of the solubility of a hydrocarbon in water and the molar volume of the hydrocarbon. If the logarithm of the solubilities of the hydrocarbons in water is plotted against the molar volumes of the hydrocarbons, a straight line is obtained. This relationship has been worked out further by C. McAuliffe, and solubilities as a function of molar volumes for a number of homologous series of hydrocarbons have been presented graphically.

From the given correlation between molecular structure and solubility, the following conclusions may be drawn:

Branching increases water solubility for paraffin, olefin, and acetylene hydrocarbons, but not for cycloparaffins, cyclo-olefins, and aromatic hydrocarbons.

For a given carbon number, *ring formation* increases water solubility.

Addition of a *double bond* to the molecule, ring, or chain increases water solubility. The addition of a second and third double bond to a hydrocarbon of given carbon number proportionately increases water solubility (Table 1).

A *triple bond* in a chain molecule increases water solubility to a greater extent than two double bonds.

Cary T. Chiou *et al.* (382) found a good correlation between solubilities of organic compounds and their octanol/water partition coefficients. Furthermore, functional groups such as chlorine atoms, methyl groups, hydroxyl groups, and benzene rings showed additive effects on the logarithm of the octanol/water partition coefficient (log P_{oct}) of the parent molecule.

This allowed the calculation of log P_{oct} values for many organic compounds based on the log P_{oct} value for the parent compound and the additive effects of the functional groups. Because of the correlation between solubilities of organic compounds and log P_{oct}, it is not surprising to find the same additive effects of functional groups on their water solubility. Table 2 shows this influence of functional groups on the solubility of benzene derivatives. Solubilities of homologous series of organic compounds are plotted in Figures 3, 4, and 5.

Table 1. Influence of Double Bonds on Aqueous Solubility of Cyclic Hydrocarbons (at room temperature) (242).

Hydrocarbon	Solubility, mg/l
Cyclohexane	55
Cyclohexene	213
1,4-Cyclohexadiene	700
Benzene	1,780

Table 2. Influence of Functional Groups on Solubility of Benzene Derivatives.

	Functional Group	$S_{mg/l}$ Solubility mg/l (temp., °C)		log $S_{mg/l}$	Δlog $S_{mg/l}$ log $S_{C_6H_5X}$ –log $S_{C_6H_6}$
Aniline	—NH_2	34,000	(20°)	4.53	1.28
Phenol	—OH	82,000	(15°)	4.91	1.66
Benzaldehyde	—COH	3,300		3.52	0.27
Benzoic acid	—COOH	2,900		3.46	0.21
Nitrobenzene	—NO_2	1,900		3.28	0.03
Benzene	-	1,780		3.25	0.00
Fluorobenzene	—F	1,540	(30°)	3.19	–0.06
Thiophenol	—SH	470	(15°)	2.67	–0.58
Toluene	—CH_3	515		2.71	–0.54
Chlorobenzene	—Cl	448	(30°)	2.65	–0.60
Bromobenzene	—Br	446	(30°)	2.65	–0.60
Iodobenzene	—I	340	(30°)	2.53	–0.72
Diphenylether	O—⌬	21	(25°)	1.32	–1.93
Diphenyl	—⌬	7.5	(25°)	0.88	–2.37

Effects that cannot be accounted for by this additive-constitutive character of the solubility are

- steric effects that cause shielding of an active function
- intra- and intermolecular hydrogen bonding (e.g., trihydroxyphenols)
- branching
- inductive effects of one substituent on another
- conformational effects, such as "balling up" of an aliphatic chain

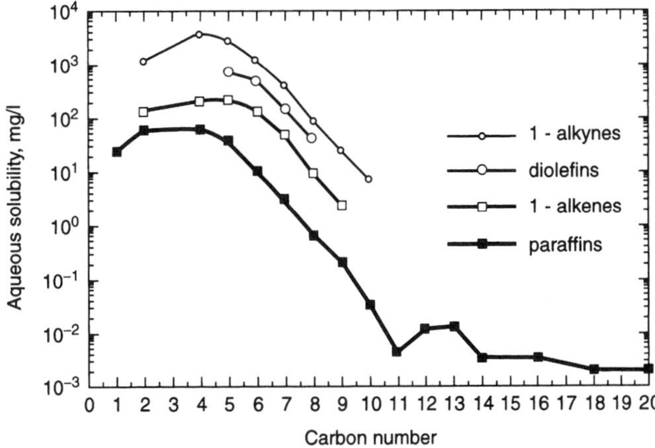

Figure 3. Relationship between aqueous solubility and molecular weight for saturated and unsaturated straight-chain hydrocarbons.

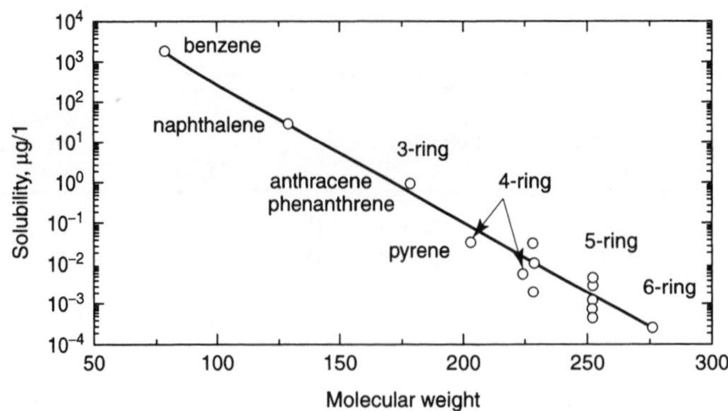

Figure 4. Relationship between aqueous solubility and molecular weight for benzene, naphthalene, and polynuclear aromatic hydrocarbons.

4.5. Solubility of Mixtures. Mixtures of compounds, whether they are natural such as oil or formulations such as many pesticides, behave differently from the single compounds when brought into contact with water. Indeed, each component of the mixture will partition between the aqueous phase and the mixture.

Components with a high aqueous solubility tend to move toward the aqueous phase while the "unsoluble" components remain in the other phase. From this, it follows that the fractional composition of the water soluble fraction (WSF) will differ from the original composition of the mixture and that concentrations of the components of the WSF are generally lower than the maximum solubilities for the individual components. Examples are shown in Tables 3 and 4.

5. Octanol/Water Partition Coefficient

The ability of some chemicals to move through the food chain, resulting in higher and higher concentrations at each trophic level, has been termed *biomagnification* or *bioconcentration.* The widespread distributions of DDT and the polychlorinated biphenyls (PCBs) have become classic examples of such movement.

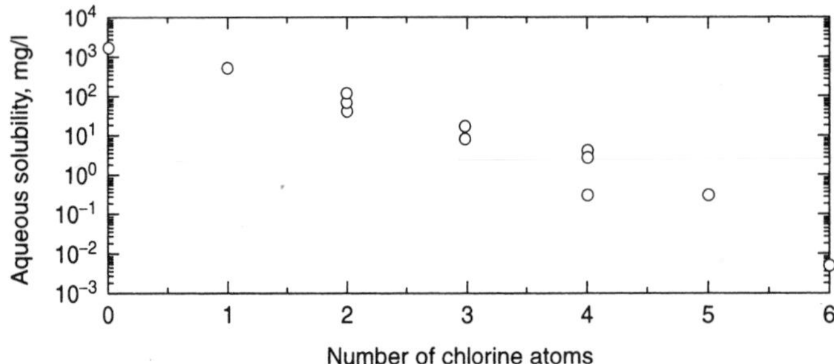

Figure 5. Relationship between aqueous solubility and molecular weight for homologous series of chlorinated benzenes.

Table 3. Comparison of Aqueous Solubility of Some PCB Isomers in the Water Soluble Phase of Aroclor 1242 with Maximum Solubility of Individual Isomers (1909).

Isomer	Solubility, µg/l, in WSF of Aroclor 1242	Max. Solubility, µg/l, for Individual Compounds	
4-	15	2,000	(calculated)
2,2′-	21	900	
2,4′-	138	637	
2,5,2′-	61	248	
2,5,2′,5′-	22	26	

The same is true for many mineral oils and petroleum products, the WSF of which consists mainly of the more soluble aromatic compounds benzene, toluene, xylene and their alkyl homologs.

From an environmental point of view, this phenomenon becomes important when the acute toxicity of the agent is low and the physiological effects go unnoticed until the chronic effects become evident. For this reason, prior knowledge of the bioconcentration potential of new or existing chemicals is desired. However, determining the bioconcentration factor of a chemical on a number of animals or in a food chain is expensive and time-consuming. If a simple relationship could be established between physico-chemical properties of a chemical and its ability to bioconcentrate, it would be of great benefit in planning the future direction of any development work on a new chemical and in directing research efforts to determine the distribution and ultimate fate of a limited number of selected chemicals.

5.1. Definition. The partition coefficient P_{oct} is defined as the ratio of the equilibrium concentrations C of a dissolved substance in a two-phase system consisting of two largely immiscible solvents, in this case *n*-octanol and water:

$$P_{oct} = \frac{C_{octanol}}{C_{water}}$$

In addition to the above, the partition coefficient is ideally dependent on only temperature and pressure. The partition coefficient P_{oct} is the quotient of two concentrations and is a constant without dimensions. It is usually given in the form of its logarithm to base ten (log P_{oct}).

The *n*-octanol/water partition coefficient has proved useful as a means of predicting soil adsorption (419), biological uptake (416), lipophilic storage (415), and biomagnification (417, 418, 339, 193).

Table 4. Composition of Aroclor 1242 and Its Water Soluble Fraction (WSF) (1909).

	Aroclor wt %	WSF wt %	WSF/Aroclor ratio of wt %
monochlorobiphenyls	3	19.4	6.5
dichlorobiphenyls	13	31.8	2.4
trichlorobiphenyls	28	31.3	1.1
tetrachlorobiphenyls	30	16.5	0.55
pentachlorobiphenyls	22	—	0.04
hexachlorobiphenyls	4		>0.02
	—		—
	100	100	

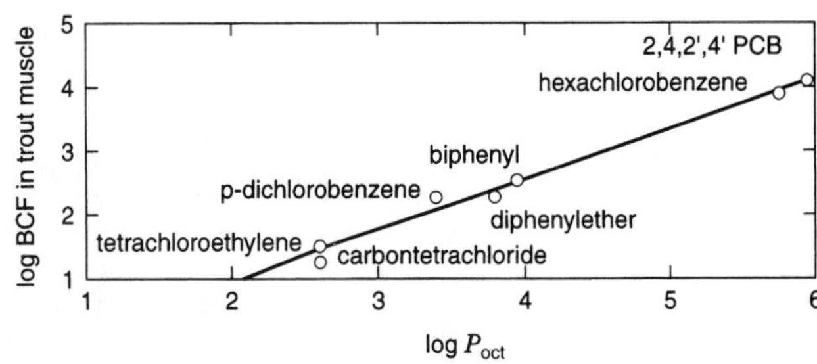

Figure 6. Relationship between octanol/water partition (P_{oct}) coefficient and bioaccumulation factor (BCF) in trout muscle (1448).

The bioconcentration of several chemicals in trout muscle was found to follow a straight-line relationship with *n-octanol/water partition coefficient* (193). Bioconcentration in this work was defined as the ratio of concentration of the chemical between trout muscles and the exposure water measured at equilibrium. The relationship was established by measuring the bioconcentration, in trout, of a variety of chemicals over a wide range of partition coefficients. An equation of the straight line of best fit was determined and used to predict the bioconcentration of other chemicals from their *n*-octanol/water partition coefficients. The predicted values agreed with the experimental values in the literature. Values are expressed as their decimal logarithms.

The linear relationship between bioconcentration factor and partition coefficient is given by

$$\log B_f = 0.542 \log P_{oct} + 0.124 \tag{1}$$

where B_f = bioconcentration factor and P_{oct} = octanol/water partition coefficient.

The relationship is shown in Figure 6.

The largest compilation of *n*-octanol/water partition coefficients has been made by Albert Leo et al. (1457).

By far the most extensive and useful partition coefficient data were obtained by the classical way of shaking a solute with two immiscible solvents and then analyzing the solute concentration in one or both phases.

Examples of physico-chemical determinations that may be appropriate are

- Photometric methods
- Gas chromatography
- HPLC
- Back-extraction of the aqueous phase and subsequent gas chromatography

5.2. Calculation of Partition Coefficients. Since partition coefficients are equilibrium constants, it should not be surprising that one finds extrathermodynamic relationships between values in different solvent systems. This relationship can be expressed by the general equation:

$$\log P_2 = a \log P_1 + b \tag{2}$$

for example:

$$\log P_{toluene} = 1.135 \log P_{oct} - 1.777$$
$$(n = 22;\ r = 0.980;\ s = 0.194) \tag{3}$$

$$\log P_{cyclohexanone} = 1.035 \log P_{oct} + 0.896 \quad (n = 10; r = 0.972; s = 0.340) \tag{4}$$

Many log P_{oct} partition coefficients in this book were calculated by A. Leo *et al.* (1457) using the above and other equations. Furthermore, it was found that the log P_{oct} of a compound could be calculated from the log P_{oct} of another compound of the same homologous series by adding or subtracting a number of times a constant value (Table 5).

Additivity was first established for a wide variety of groups in a study of the substituent constant, p, defined by the following equation:

$$p_X = \log P_X - \log P_H \tag{5}$$

where P_X is the derivative of a parent molecule P_H and thus p is the logarithm of the partition coefficient of the function X. For example p_{Cl} could be obtained as follows:

$$p_{Cl} = \log P_{chlorobenzene} - \log P_{benzene} \tag{6}$$

It has been found that p values are relatively constant from one system to another as long as there are no special steric or electronic interactions of the substituents not contained in the reference system. p Values for aliphatic and aromatic positions are shown in Table 6. Other effects that must be taken into account in the additive-constitutive character of log P_{oct} are

- steric effects, which can cause shielding of an active function by inert groups
- inductive effects of one substituent on another
- intra- and intermolecular hydrogen bonding
- branching
- conformational effects, such as "balling up" of an aliphatic chain

Because of the difficulties of estimating the influence on log P_{oct} of steric, inductive, and conformational effects, calculated log P_{oct} values of complex molecules can only be approximate and can be wrong by 1 or 2 orders of magnitude. However, for most simple molecules, calculated values are correct within 1 order of magnitude.

Table 5. Influence of Functional Groups on *n*-Octanol/Water Partition Coefficient of Benzene Derivatives.

Product	**Functional Group**	**log P_{oct}**	**D log P_{oct}** **log P_{C6H5X} – log P_{C6H6}**
benzenesulfonic acid	—SO_3H	-2.25	-4.38
benzenesulfonamide	—SO_2NH	0.31	-2.44
aniline	—NH_2	0.90	-1.23
phenol	—OH	1.46	-0.67
benzaldehyde	—COH	1.48	-0.65
benzonitrile	—CN	1.56	-0.57
benzoic acid	—COOH	1.87	-0.28
nitrobenzene	—NO_2	1.85/1.88	-0.28
benzene	–	2.13	–
fluorobenzene	—F	2.27	+0.14

thiophenol	—SH	2.52	+0.39
toluene	—CH_3	2.80	+0.67
chlorobenzene	—Cl	2.84	+0.71
bromobenzene	—Br	2.99	+0.86
iodobenzene	—I	3.25	+1.12
diphenyl	—C_6H_5	3.6	+1.47
diphenlether	—O—C_6H_5	4.21	+2.08

Table 6. Comparison of Aromatic and Aliphatic *p Values.*

Function	**Aromatic** $\log P_{C6H5} - \log P_{C6H6}$	**Aliphatic** $\log P_{RX} - \log P_{RH}$
NH_2	−1.23	−1.19
I	1.12	1.00
S—CH_3	0.61	0.45
$COCH_3$	−0.55	−0.71
$CONH_2$	−1.49	−1.71
Br	0.86	0.60
CN	−0.57	−0.84
F	0.14	−0.17
Cl	0.71	0.39
COOH	−0.28	−0.67
OCH_3	−0.02	−0.47
OC_6H_5	2.08	1.61
$N(CH_3)_2$	0.18	−0.30
OH	−0.67	−1.16
NO_2	−0.28	−0.85
CH_2	0.50	0.50

5.3. Relationship between Aqueous Solubility and Octanol/Water Partition Coefficient. Unfortunately, the partition coefficients of many components of environmental significance are not always available, despite a recent extensive compilation (1457), or cannot be easily calculated from parent molecules. Assessment of partition coefficients from a more readily available physical parameter would therefore be useful. By definition, the partition coefficient expresses the equilibrium concentration ratio of an organic chemical partitioned between an organic liquid (such as *n*-octanol) and water. This partitioning is, in essence, equivalent to partitioning an organic chemical between itself and water. Consequently, one would suspect that a correlation might exist between the partition coefficient and the aqueous solubility. Based on experimental values of aqueous solubility and *n*-octanol/water partition coefficient for various types of chemicals, the following regression equation was found (382):

$$\log P_{oct} = 5.00 - 0.670 \log S \qquad (7)$$

where S is the aqueous solubility in *m*mol/l. If the solubility is expressed in mg/l, Eq. (7) becomes

$$\log P_{oct} = 4.5 - 0.75 \log S \text{ (mg/l)} \tag{8}$$

or

$$\log P_{oct} = 7.5 - 0.75 \log S \text{ } (\mathit{m}\text{g/l})$$

This equation has been obtained empirically. This correlation covers many classes of chemicals from hydrocarbons and organic halides to aromatic acids, pesticides, and PCBs. It also spans chemicals of different polarities (from nonpolar to polar) and of different molecular states (both liquid and solid).

Cary T. Chiou *et al.* (382) found for Eq. (7) a correlation coefficient of 0.970, which allows an estimation within 1 order of magnitude of the partition coefficient of a given compound from its aqueous solubility. However, when more data points are added, the scatter increases considerably for solubilities >100 mg/l. A few products even deviate considerably from the regression Eqs. (7) and (8), as shown in the following data.

pentachlorophenol	
aqueous solubility	14 mg/l at 20°C
log P_{oct}: experimental	*5.01*
calculated Eq. (7)	3.8
Eq. (8)	3.76
l-tyrosine	
aqueous solubility	480 mg/l at 25°C
log P_{oct}: experimental	*–2.26*
calculated Eq. (7)	+2.7
Eq. (8)	+2.5

The regression equation, however, remains the same, although P_{oct} values calculated from its aqueous solubility may be wrong by more than 1 order of magnitude. Obviously, Eqs. (7) and (8) would be unlikely to apply for salts, strong acids, and bases, because the activities of these solutes in this case cannot be approximated by their concentrations. Moreover, with materials such as aliphatic acids and bases, the partition coefficient can vary drastically with changes in pH.

As previously stated, the partition coefficient is related to physical adsorption on solids, biomagnification, and lipophilic storage. Equation (9) would extend these correlations to cover compounds using their aqueous solubilities without requiring the partition coefficient data. Based on reported biomagnification data of some selected organic chemicals in rainbow trout *(Salmo gairdneri)*, the following regression equation was calculated:

$$\log(\text{BCF}) = 3.41 - 0.508 \log S \tag{9}$$

where BCF is the bioconcentration factor in rainbow trout and S is aqueous solubility in *m*mol/l. If the aqueous solubility is expressed in mg/l, Eq. (9) becomes

$$\log(\text{BCF}) = 3.04 - 0.568 \log S_{mg/l} \tag{10}$$

where $S_{mg/l}$ is the aqueous solubility in mg/l.

5.4. Ecological Magnification (E.M.). The process of bioaccumulation involves a number of fundamental events:

1. partitioning of the foreign molecule under consideration between the environment and some surface of the organism
2. diffusional transport of these molecules across cell membranes
3. transport mediated by body fluids, such as exchange between blood vessels and serum lipoproteins
4. concentration of the foreign molecule in various tissues depending on its affinity for certain biomolecules, such as nerve lipids
5. biodegradation of the foreign material

The bioaccumulation process is thus seen to be a result of both kinetic (diffusional transport and biodegradation) and equilibrium (partitioning) processes. A molecule will not bioaccumulate in an organism if its degradation rate is greater than its accumulation rate. Experience with DDT may be considered a massive experiment from which it may be concluded that degradation occurred too slowly compared to the transport and partitioning of DDT into the higher levels of the food chain, thus permitting toxic levels to result. R.L. Metcalf and co-workers (1643) have correlated the ecological magnification values for a number of organic compounds (pentachlorobiphenyl, tetrachlorobiphenyl, trichlorobiphenyl, DDE, chlorobenzene, benzoic acid, anisole, nitrobenzene, and aniline) from the fishes of model ecosystems with both water solubility and the octanol/water partition value. For the limited number of compounds included, the correlation between physical properties and biomagnification is excellent. The regression equations were:

$$\log \text{E.M.} = 4.48 \log - 0.47 \text{ S } (mg/l)$$

$$\log \text{E.M.} = 0.75 + 1.16 \log P_{oct} \qquad (11)$$

where S = aqueous solubility; P_{oct} = octanol/water partition coefficient. The correlations between ecological magnification and water solubility or octanol/water partition coefficient, as described above, are valid only for compounds that do not exhibit significant biodegradation.

Kapoor *et al.* (1937, 1938, 1939) studied in a model ecosystem the behavior of 8 DDT analogs covering a wide range of biodegradability. The basic methodology involved systematic study of the DDT molecule by replacing the environmentally stable C-Cl bonds with other groups of suitable size, shape, and polarity that could serve as degradophores by acting as substrates for the mixed-function oxidase enzymes widely distributed in living organisms. The action of the enzymes was shown to result in substantial changes in the polarity of the molecule, so that the degradation products were excreted rather than stored in lipids as was DDT and its chief degradation product DDE. A summary of model ecosystem data for a number of DDT analogs with degradaphores incorporated into aromatic and aliphatic moieties of the molecule is presented in Table 7 and Figure 7. Figure 7 shows that the difference between octanol/water partition coefficient and ecological magnification for the DDT analogs can be largely explained by the biodegradability index. The higher the biodegradability index, the larger the difference between octanol/water partition coefficient (as a predictive measure of ecological magnification) and ecological magnification itself.

Most of the correlation equations between water solubility (or octanol/water partition coefficients) and ecological magnification have indeed been established on biorefractive compounds or on homologous series of compounds with comparable biodegradability characteristics. Therefore, the equations are not universally applicable, as shown by Figure 8, which represents the relationship between water solubility of DDT, DDE, DDD, and the DDT analogs and ecological magnification in mosquito fish of a terrestrial aquatic model ecosystem. More than 30 pesticides were studied by Metcalf and Sanborn (1881), who found a highly significant correlation between log E.M. and log water insolubility. Similar correlations also exist for phenol and chlorinated derivatives and other chemical classes.

Because pesticides are now being "engineered" to be less "persistent," correlations between physico-chemical properties of compounds and ecological magnification will become less meaningful for such compounds unless degradation velocity is taken into account. For this reason, the following equation is proposed. It predicts the upper limit for ecological magnification based on aqueous solubility

$$\log \text{E.M.} = 6 - 0.66 \log S\,(mg/l) \qquad (12)$$

This correlation is only very approximate, so preference has been given to an easy-to-remember equation. The equation is therefore not based on a regression analysis.

The real E.M.s are 1 to 2 orders of magnitude smaller than the E.M.s predicted from Eq. (12). Indeed, of 43 compounds, 17 true E.M. values are smaller by less than 1 order of magnitude than the predicted values, 18 E.M. values differ by more than 1 but less than 2 orders of magnitude, and 6 E.M. values differ by more than 2 orders of magnitude.

Table 7. Biodegradability and Ecological Magnification in Fish of DDT Analog.

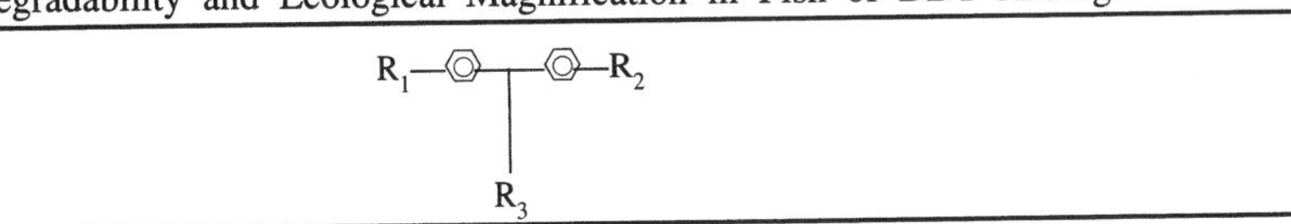

Number of DDT Analog	R_1	R_2	R_3	log P_{oct}[a]	log E.M.	log P_{oct} minus log E.M.	log (100 × B.I.)	log H_2O solub. ppb[b]
	Cl	Cl	CCl_3	6.1	4.93	1.17	0.18	0.9
	Cl	Cl	$HCCl_2$	5.4	4.92	0.48	0.73	1.8
1	CH_3O	CH_3O	CCl_3	4.7	3.19	1.51	1.97	2.7
2	CH_3	CH_3	CCl_3	6.0	2.15	3.85	2.85	1.0
3	CH_3S	CH_3S	CCl_3	5.9	0.74	5.16	3.67	1.1
4	Cl	CH_3	CCl_3	6.05	3.15	2.9	2.53	0.9
5	CH_3	C_2H_5O	CCl_3	5.9	2.60	3.3	2.08	1.1
6	CH_3O	CH_3S	CCl_3	5.3	2.49	2.81	2.44	1.9
7	CH_3O	CH_3O	$C(CH_3)_3$	4.7	3.21	1.49	1.81	2.7
8	Cl	Cl	$HC(CH_3)NO_2$	4.4	2.05	2.35	2.51	3.1

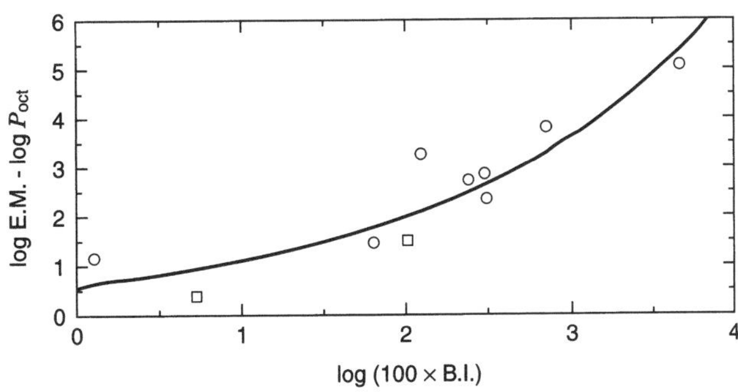

Figure 7. Influence of biodegradability of DDT analogs on experimental and predicted ecological magnification.

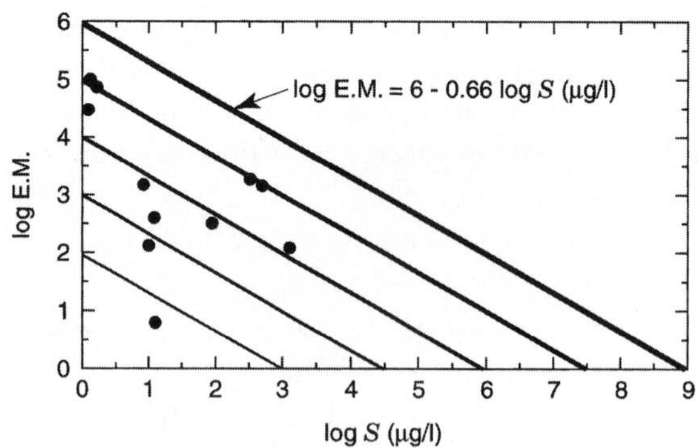

Figure 8. Relationship between water solubility of DDT, DDE, DDD, and DDT analogs and ecological magnification in mosquito fish of a terrestrial-aquatic model ecosystem.

5.5. Adsorption. The following factors influence the extent to which the adsorption of a compound from solution onto a solid occurs.

- The physical and chemical characteristics of the adsorbent (adsorbing surface)
- The actual surface area of the solid
- The nature of the binding sites on these surfaces and the actual distribution of these adsorption sites

An index of the tendency to adsorb onto a solid is the solubility of the compound. For homologous series of compounds, decreasing solubility can be interpreted as an increasing tendency to leave the water. Also, the *n*-octanol/water partition coefficient (log P_{oct}) has proved useful as a means of predicting soil adsorption, as shown in Figure 9.

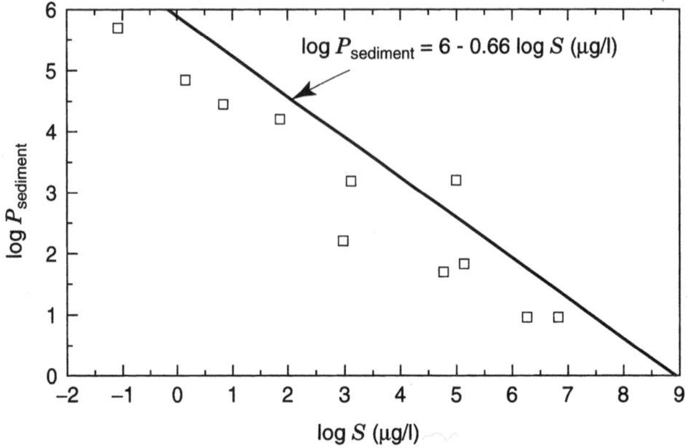

Figure 9. Plot of log water solubility and log partition coefficient (log $P_{sediment}$) for organic compounds adsorbed on natural sediment collected in Coyote Creek, California (from EPA-60017-78-074, May 1978).

5.6. Structure-Toxicity Correlations. Kopperman and co-workers (1832) noted that structure-toxicity correlations are possible only when the compounds examined have an identical mode of action. However, if toxic mechanisms of similar compounds are identical, then the internal concentration of toxicant required to elicit a specific biological response should be consistent. Because the internal concentration equals the external concentration multiplied by the partition coefficient, the external concentration is proportional to the reciprocal of the partition coefficient, and a plot of log external concentration vs. log partition coefficient should have a slope of –1.

T. Wayne Schultz and co-workers (1662) have observed excellent correlation between toxicity to the ciliate *Tetrahymena pyriformis* and partition coefficient within the following series of organic contaminants: pyridines, anilines, phenols, quinolines, and benzenes. However, no significant correlation between toxicity and partition coefficient was observed when data from all the tested contaminants were combined. Generally, an increase in alkyl substitution increases toxicity and decreases solubility. Furthermore, hetero-atom substitution into or onto the ring severely alters both the toxicity and the solubility of the compound (Figure 10). Kunio Kobayashi (1850) found that an increase of the Cl-atom number in the chlorophenols promoted an accumulation of the chlorophenols by fish and led their concentration in the fish to a lethal level, even when guppies were exposed to rather low concentrations, and consequently increased the fish-toxicity of chlorophenols. Kopperman *et al.* (1832) found approximately the same correlation for goldfish (Figure 11). The data are summarized in Table 8.

Because all the correlations obtained are valid only within homologous series and only for the test organism concerned, their predictive power is limited. Moreover, concentration values have been expressed largely in moles/liter. Although this is certainly more accurate from a scientific point of view than using mg/l, it is advisable to express concentration in mg/l or mg/l in order to make using the correlations more practicable.

Figure 12 shows that the correlation between the aqueous solubility expressed in ppm (mg/l) and LD_{50} for goldfish are still good enough.

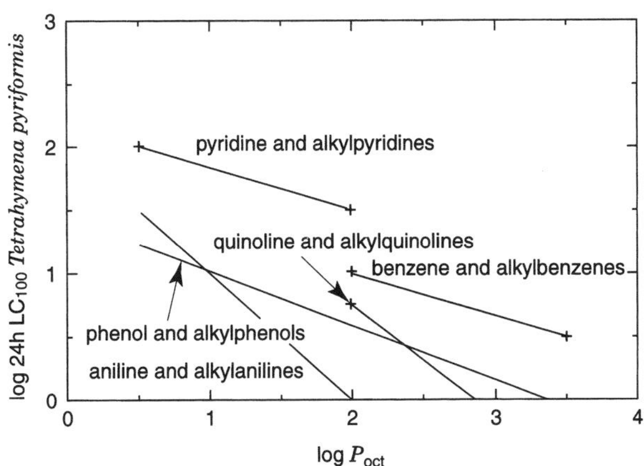

Figure 10. A least squares linear regression of log 24h LC_{100} (mMole/l) vs. log P_{oct} for the ciliate *Tetrahymena pyriformis.*

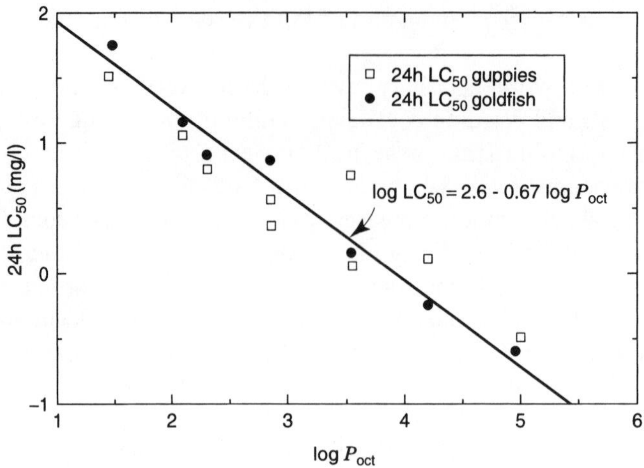

Figure 11. Toxicity versus octanol/water distribution coefficient for goldfish and guppies (after 1833 and 1850) exposed to phenol and chlorinated phenols.

Table 8. Phenol and Chlorinated Phenols: Structure–Toxicity Data.

	24h LC_{50} (ppm)[a]					
	goldfish (1850)	**guppies (1833)**	**solub.ppm**	**log P_{oct} exp.**	**log P_{oct} calculated**	**BFC goldfish**[b]
phenol	60	30	67,000 (10°C)	1.47	1.55	1.9
o-chlorophenol	16	11	28,500 (20°C)	2.17	2.27	6.4
m-chlorophenol	–	6.5	–	2.48	2.27	–
p-chlorophenol	9.0	–	27,100 (20°C)	2.41	2.27	10.0
2,4-dichlorophenol	7.8	4.2	4,600 (20°C)	–	2.93	34
3,5-dichlorophenol	–	2.7	–	–	2.93	–
2,4,6-trichlorophenol	10.0	–	800 (24°C)	3.37	3.69	20
2,4,5-trichlorophenol	1.7	–	1,190 (25°C)	3.72	3.69	62
2,3,5-trichlorophenol	–	1.6	–	–	3.69	–
2,3,6-trichlorophenol	–	5.1	–	–	3.69	–
3,4,5-trichlorophenol	–	1.1	–	–	3.69	–
2,3,4,6-tetrachlorophenol	0.75	–		–	4.42	93
2,3,4,5-tetrachlorophenol	–	0.77	–	–	4.42	–
2,3,5,6-tetrachlorophenol	–	1.37	–	–	4.42	–
pentachlorophenol	0.27	0.38	14 (20°C)	5.01	5.19	475

[a]pH 7.3

[b]Values were obtained in goldfish that died in the concentration of each chlorophenol closest to 24h LC_{50}.

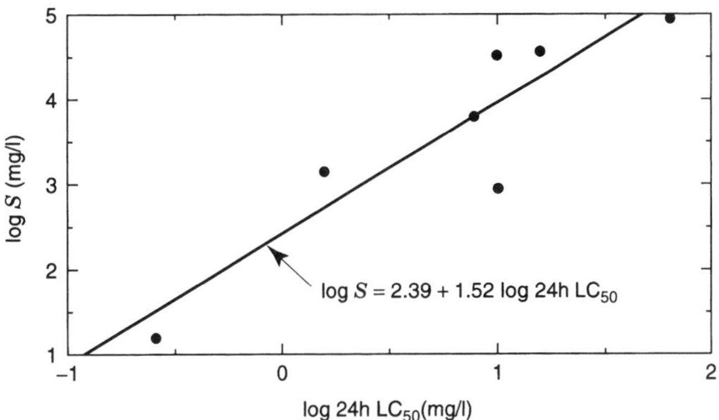

Figure 12. Toxicity and accumulation of phenols in goldfish compared to the aqueous solubility and octanol/water partition coefficients (after 1850).

6. Henry's Constant (H)

Henry's constant *(H)* is a physical property of a chemical that is a measure of its partitioning nature between the two phases in an air-water binary system. By virtue of its definition, *H* often dictates where and how a chemical tends to concentrate or "accumulate" at equilibrium. Chemicals with low *H* tend to accumulate in the aqueous phase, whereas those with high *H* partition more into the gas phase.

Because air and water are the major "compartments" of the model ecosphere, and water is considered to act as the link between all its other compartments, knowledge of *H* is very important in assessing the environmental risks associated with a chemical. *H* is also a key parameter in determination of the "cleanup" process of choice for contaminated sites and in detailed design of decontamination processes.

Henry's constant is also called the "air-to-water ratio" or the "air-water partition coefficient" and can consequently be expressed as the ratio of concentrations of a chemical in air and in water at equilibrium.

$$H = \frac{C_L}{C_W}$$

where C_L = concentration of the chemical in the air in mg/m^3

C_W = concentration of the chemical in the water in *mg*/l (= mg/m^3)

This equation represents the dimensionless *H* because the concentrations have been expressed in the same unit. This is the most convenient way of expressing *H* because it yields right away the necessary information about the partitioning of a chemical between the two phases air and water.

If a chemical behaves as an ideal gas in the atmosphere, then *H* can be calculated from the saturation concentration in the air and in the water (solubility).

The vapor pressure and a chemical's water solubility are the key physical properties that are used in the calculation of *H*.

A number of simple equations have been proposed by different researchers to calculate *H*. Henry's constant can be written in the form

$$H = \frac{P_{vp}}{S}$$

where P_{vp} is the vapor pressure in atm, and *S* is the aqueous solubility in mol/m^3. The dimension of this form is atm × m^3/mol. I must confess that with this way of expressing Henry's constant, I lose the sense of the partitioning of a chemical between air and water. Most environmental advisors share this feeling. However, the equation can easily be transformed to yield a dimensionless *H*.

$$H = \frac{P_{vp} \times mw}{0.062 \times S \times T}$$

where

P_{vp} = vapor pressure in mm Hg

mw = molecular weight

S = water solubility in mg/l

T = temperature in K

0.062 = universal gas constant

or, if the vapor pressure is expressed in pascals, then

$$H = \frac{P_{vp} \times mw}{8.3 \times S \times T}$$

where P_{vp} = vapor pressure in Pa and 8.3 = universal gas constant. In both equations, H is dimensionless.

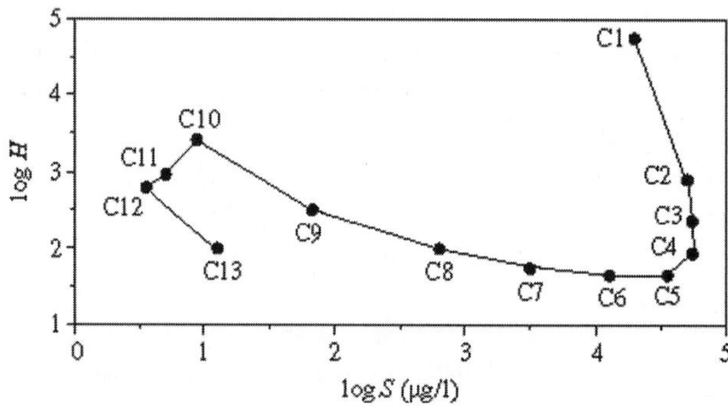

Figure 13. Relationship between Henry's constant *(H)* and the water solubility *(S)* of *n*-alkanes.

Mackay and Shiu have critically reviewed Henry's constant for 167 chemicals of environmental concern. They used vapor pressure and solubility data to calculate a "recommended" Henry's constant in the absence of experimental data and concluded that "considerable discrepancies exist in the literature even for common compounds." An important reason for these discrepancies is the lack of reliable solubility data for compounds that have a poor solubility and the interacting forces on the molecular level for polar compounds, which generally have a very high solubility. The outlier in Figure 15 (a too-low calculated *H in comparison with the curve) is probably caused by a too-low vapor pressure mentioned in the literature.*

I believe that bringing together vapor pressure, solubility, and H data for homologous series of chemicals will promote the establishment of more accurate values for all properties. Figures 12, 13, and 14 present, for a number of homologous series of chemicals, the relationship among these three parameters.

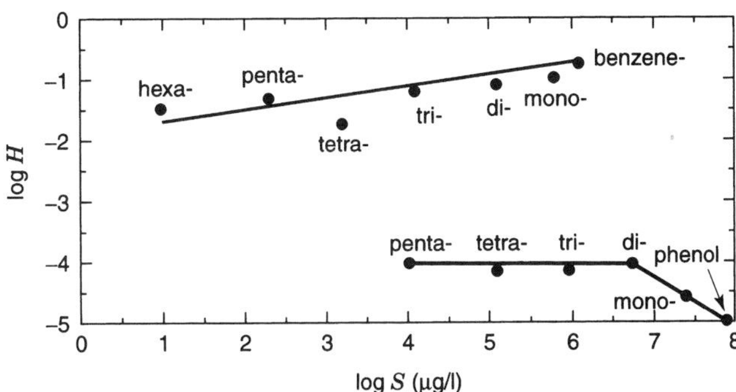

Figure 14. Relationship between the calculated Henry's constant *(H)* and water solubility (*S*) of benzene, phenol, and their chlorinated derivatives.

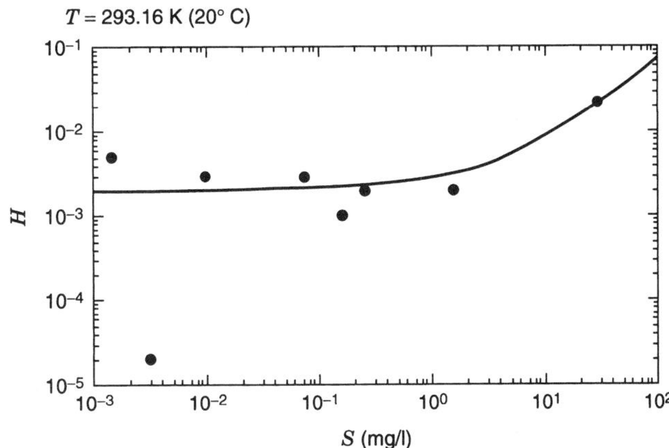

Figure 15. Relationship between the calculated Henry's constant *(H) and water solubility of polycyclic aromatics.*

In Figures 12, 13, 14, and 15, a relationship is shown between *H* and water solubility in order to complement the relationships that exist in homologous series between water solubility and (for example) toxicity toward aquatic organisms, biodegradability in soil, and soil sorption (K_{oc}).

B. AIR POLLUTION FACTORS

1. Conversion between Volume and Mass Units of Concentration

The physical state of gaseous air pollutants at atmospheric concentrations may be described generally by the ideal gas law:

$$pv = nRT \quad (1)$$

where p = absolute pressure of gas

v = volume of gas

n = number of moles of gas

T = absolute temperature (K)

R = universal gas constant

The number of moles *(n)* may be calculated from the weight of pollutant *(W)* and its molecular weight *(m)* by

$$n = \frac{W}{m} \quad (2)$$

Substituting Eq. (2) into Eq. (1) and rearranging yield

$$V = \frac{WRT}{pm} \quad (3)$$

Parts per million refers to the volume of pollutant *(v)* per million volumes of air *(V)*:

$$ppm = \frac{v}{10^6 V} \quad (4)$$

Substituting Eq. (3) into Eq. 4 yields

$$ppm = \frac{W}{V} \; \frac{RT}{pm\,10^6} \quad (5)$$

By using the appropriate values for variables in Eq. (5), a conversion from volume to mass units of concentration for methane may be derived as shown below:

T = 293.16 K (20°C)
p = 1 atm
m = 16 g/mol
R = 0.08205 1-atm/mol K

$$ppm = \frac{W(g) \times 10^3\,(mg/g)}{V(l) \times 10^{-3}\,(m^3/l)} \times \frac{0.08205\,(1\text{-atm/mol K}) \times 293.16\,(K)}{1\,(atm) \times 16\,(g/mol) \times 10^6}$$

1 ppm = 0.665 mg/m^3
1 mg/m^3 = 1.504 ppm

Whenever conversion factors are used, reference must be made to the pressure and temperature at which the conversion factors have been calculated. Unfortunately, the conditions of "Air at Normal Conditions" or "Standard Air" or "Air at STP" (STP = Standard Temperature and Pressure) often vary from country to country. ASTM D1356-60 defines "Air at Normal Conditions" as follows: "Air at 50 percent relative humidity, 70°F (21°C), and 29.92 inches of mercury (760 millimeters of mercury). These conditions are chosen in recognition of the data which have been accumulated on air-handling equipment. They are sufficiently near the 25°C and 760 millimeters of mercury commonly used for indoor air contamination work that no conversion or correction ordinarily need be applied."

For outdoor air (ambient air) pollution control, it is important to use the correct conversion factors. "Normal conditions" vary from 0°C and 760 mm Hg (dry) for Canada over 20°C and 760 mm Hg for Australia to 25°C and 760 mm Hg for Brazil.

Calculations of minimum chimney heights, in order to ensure sufficient dispersion of the waste gases before they reach ground level, are often based on the difference between Maximum Immission Concentration (M.I.C.) and the existing pollutant concentration at ground level. Significant differences in chimney heights can be obtained by not using the appropriate conversion factors.

Because of the lack of "standardization" of "standard" conditions worldwide, and because outdoor and indoor conditions are different, different conversion factors will continue to be used. The conversion factors on the data sheets should, therefore, be regarded only as approximate values. Knowing the molecular weight of the compound, one can find the correct values at 0°C and 20°C in Tables 9 and 10.

Table 9. Gaseous Air Pollutants: Conversion between Volume Units (ppm) and Mass Units (mg/m^3) of Concentration at 0°C and 760 mm Hg (m = molar weight).

m	1 ppm = mg/m^3	1 mg/m^3 = ppm	m	1 ppm = mg/m^3	1 mg/m^3 = ppm	m	1 ppm = mg/m^3	1 mg/m^3 = ppm	m	1 ppm = mg/m^3	1 mg/m^3 = ppm
16	0.714	1.401	51	2.277	0.439	86	3.839	0.260	121	5.401	0.185
17	0.759	1.318	52	2.321	0.431	87	3.884	0.257	122	5.446	0.184
18	0.804	1.244	53	2.366	0.423	88	3.929	0.255	123	5.491	0.182
19	0.848	1.179	54	2.411	0.415	89	3.973	0.252	124	5.536	0.181
20	0.892	1.121	55	2.455	0.407	90	4.018	0.249	125	5.580	0.179
21	0.937	1.067	56	2.500	0.400	91	4.062	0.246	126	5.625	0.178
22	0.982	1.018	57	2.545	0.393	92	4.107	0.243	127	5.670	0.176
23	1.027	0.974	58	2.598	0.386	93	4.152	0.241	128	5.714	0.175
24	1.071	0.934	59	2.634	0.380	94	4.196	0.238	129	5.759	0.174
25	1.116	0.896	60	2.679	0.373	95	4.241	0.236	130	5.804	0.172
26	1.161	0.861	61	2.723	0.367	96	4.286	0.233	131	5.848	0.171
27	1.205	0.830	62	2.768	0.361	97	4.330	0.231	132	5.893	0.170
28	1.250	0.800	63	2.812	0.356	98	4.375	0.229	133	5.937	0.168
29	1.295	0.772	64	2.857	0.350	99	4.420	0.226	134	5.982	0.167
30	1.339	0.747	65	2.902	0.344	100	4.464	0.244	135	6.027	0.166
31	1.384	0.722	66	2.946	0.339	101	4.509	0.222	136	6.071	0.165
32	1.429	0.700	67	2.991	0.334	102	4.554	0.217	137	6.116	0.164
33	1.473	0.679	68	3.036	0.329	103	4.598	0.217	138	6.161	0.163
34	1.518	0.659	69	3.080	0.325	104	4.643	0.215	139	6.205	0.161
35	1.562	0.640	70	3.125	0.320	105	4.687	0.213	140	6.250	0.160
36	1.607	0.622	71	3.170	0.315	106	4.732	0.211	141	6.295	0.159
37	1.652	0.605	72	3.214	0.311	107	4.777	0.209	142	6.339	0.158
38	1.696	0.590	73	3.259	0.307	108	4.821	0.207	143	6.384	0.157
39	1.741	0.574	74	3.304	0.303	109	4.866	0.206	144	6.429	0.156
40	1.786	0.560	75	3.348	0.299	110	4.911	0.204	145	6.473	0.154
41	1.830	0.546	76	3.393	0.295	111	4.955	0.202	146	6.518	0.153
42	1.875	0.533	77	3.437	0.291	112	5.000	0.200	147	6.562	0.152
43	1.920	0.521	78	3.482	0.287	113	5.045	0.198	148	6.607	0.151
44	1.964	0.509	79	3.527	0.284	114	5.089	0.197	149	6.652	0.150

45	2.009	0.498	80	3.571	0.280	115	5.134	0.195	150	6.696	0.149
46	2.054	0.487	81	3.616	0.277	116	5.179	0.173	151	6.741	0.148
47	2.098	0.477	82	3.661	0.273	117	5.223	0.192	152	6.786	0.147
48	2.143	0.467	83	3.705	0.270	118	5.268	0.190	153	6.830	0.146
49	2.187	0.457	84	3.750	0.267	119	5.312	0.188	154	6.875	0.145
50	2.232	0.448	85	3.795	0.264	120	5.357	0.187	155	6.920	0.145

Table 10. Gaseous Air Pollutants: Conversion between Volume Units (ppm) and Mass Units (mg/m^3) of Concentration at 20°C and 760 mm Hg (m = molar weight).

m	1 ppm = mg/m^3	1 mg/m^3 = ppm	m	1 ppm = mg/m^3	1 mg/m^3 = ppm	m	1 ppm = mg/m^3	1 mg/m^3 = ppm	m	1 ppm = mg/m^3	1 mg/m^3 = ppm
15	0.624	1.603	51	2.120	0.472	87	3.617	0.276	123	5.113	0.196
16	0.665	1.504	52	2.162	0.463	88	3.658	0.273	124	5.155	0.194
17	0.707	1.414	53	2.203	0.454	89	3.700	0.270	125	5.196	0.192
18	0.748	1.337	54	2.245	0.445	90	3.741	0.267	126	5.238	0.191
19	0.790	1.266	55	2.286	0.437	91	3.783	0.264	127	5.280	0.189
20	0.831	1.203	56	2.328	0.340	92	3.824	0.261	128	5.321	0.188
21	0.873	1.145	57	2.369	0.422	93	3.866	0.259	129	5.363	0.186
22	0.915	1.093	58	2.411	0.415	94	3.908	0.256	130	5.404	0.185
23	0.956	1.046	59	2.453	0.408	95	3.949	0.253	131	5.446	0.184
24	0.998	1.002	60	2.494	0.401	96	3.991	0.251	132	5.488	0.182
25	1.039	0.962	61	2.536	0.394	97	4.032	0.248	133	5.529	0.181
26	1.081	0.925	62	2.577	0.388	98	4.074	0.245	134	5.570	0.180
27	1.122	0.891	63	2.619	0.382	99	4.115	0.243	135	5.612	0.178
28	1.164	0.859	64	2.660	0.376	100	4.157	0.241	136	5.654	0.177
29	1.206	0.829	65	2.702	0.370	101	4.199	0.238	137	5.695	0.176
30	1.247	0.802	66	2.744	0.364	102	4.240	0.236	138	5.737	0.174
31	1.289	0.776	67	2.785	0.359	103	4.282	0.233	139	5.778	0.173
32	1.330	0.752	68	2.827	0.354	104	4.323	0.231	140	5.820	0.172
33	1.372	0.729	69	2.868	0.349	105	4.365	0.229	141	5.861	0.171
34	1.413	0.708	70	2.910	0.344	106	4.406	0.227	142	5.903	0.169
35	1.455	0.687	71	2.951	0.339	107	4.448	0.225	143	5.945	0.168
36	1.487	0.668	72	2.993	0.334	108	4.490	0.223	144	5.986	0.167
37	1.538	0.650	73	3.035	0.329	109	4.531	0.221	145	6.028	0.166
38	1.580	0.633	74	3.076	0.325	110	4.573	0.219	146	6.070	0.165
39	1.621	0.617	75	3.118	0.321	111	4.614	0.217	147	6.111	0.164
40	1.663	0.601	76	3.159	0.317	112	4.656	0.215	148	6.152	0.163

41	1.704	0.587	77	3.201	0.312	113	4.697	0.213	149	6.194	0.161
42	1.746	0.572	78	3.242	0.308	114	4.739	0.211	150	6.236	0.160
43	1.788	0.559	79	3.284	0.305	115	4.780	0.209	151	6.277	0.159
44	1.829	0.547	80	3.326	0.301	116	4.822	0.207	152	6.319	0.158
45	1.871	0.534	81	3.367	0.297	117	4.864	0.206	153	6.360	0.157
46	1.912	0.523	82	3.409	0.293	118	4.905	0.204	154	6.402	0.156
47	1.954	0.512	83	3.450	0.290	119	4.947	0.202	155	6.443	0.155
48	1.995	0.501	84	3.492	0.286	120	4.988	0.200	156	6.485	0.154
49	2.037	0.491	85	3.533	0.283	121	5.030	0.199	157	6.526	0.153
50	2.079	0.481	86	3.575	0.280	122	5.072	0.197	158	6.568	0.152

2. Odor

2.1. Threshold Odor Concentration (T.O.C.). A starting point in relation to quantification of odors seems to be the definition of a threshold odor concentration. At least three different odor thresholds have been determined: the absolute perception threshold, the recognition threshold, and the objectionability threshold.

At the perception threshold concentration, one is barely certain that an odor is detected, but it is too faint to identify further. Furthermore, the sense-of-smell results must be a statistical average because of biological variability. The thresholds normally used are those for 50% and for 100% of the odor panel. When the T.O.C. is given without any qualification, it is usually the 50% recognition threshold.

For the sake of clarity, a number of definitions are listed here:

- *Hedonic Tone:* the pleasure or displeasure that the odor judge associated with the odor quality being observed.
- *Absolute Odor Threshold*: the concentration at which 50% of the odor panel detected the odor.
- *50% Recognition Threshold:* the concentration at which 50% of the odor panel defined the odor as being representative of the odorant being studied.
- *100% Recognition Threshold:* the concentration at which 100% of the odor panel defined the odor as being representative of the odorant being studied.
- $P.P.T._{50}$ *(Population Perception Threshold):* the concentration at which 50% of the people who have a capable sense of smell are able to detect an odor.
- $P.I.T._{50}$ *(Population Identification Threshold):* the concentration at which 50% of the population can identify and describe the odor, or at least compare its quality with another odor.
- I.P.T. *(Individual Perception Threshold):* the lowest concentration of a particular odor at which a subject gave both an initial positive response and a repeated response when the same stimulus was given a second time.
- T.O.N. *(Threshold Odor Number):* the number of times a given volume of the sample has to be diluted with clean, odorless air to bring it to the threshold level (detected by 50% of a panel of observers). The T.O.N. is thus the value of the intensity of an odor expressed in odor units.
- O.I. *(Odor Index):* a dimensionless term that is based on vapor pressure and odor recognition threshold (100%) as follows:

$$\text{O.I.} = \frac{\text{vapor pressure (ppm)}}{\text{odor recognition treshold (100\%) (ppm)}}$$

where 1 atm = 1,000,000 ppm.

The O.I. is, in essence, a ratio between the driving force to introduce an odorant into the air and the ability of an odorant to create a recognized response. It is a concept that provides information pertaining to the potential of a particular odorant to cause odor problems under evaporative conditions. The O.I. was first proposed by T.M. Hellman and F.H. Small in 1973 as a tool to predict whether complaints are likely to arise under certain evaporative conditions. Examples of evaporative conditions include spills, leaks, and solvent evaporation processes. The O.I. takes into account the vapor pressure of a compound, which is a qualitative measure of the potential of an odorant to get into the air, as well as the odor recognition threshold, which is a measure of the detectability of an odorant in the air. The values of O.I. listed in Table 11 range from a high of 1,052,000,000 for isopropylmercaptan to a low of 0.2 for maleic anhydride.

In its present form, the O.I. does not differentiate between "good" and "bad" odor qualities. It could incorporate a quality factor that would reflect consideration of the odor quality, e.g., a "bad" odor might have a higher quality factor than a "good" odor. Because the O.I. is proposed as a general indicator of odor pollution, it would be reasonable to utilize categories of odor index values for comparison, rather than comparing individual values. These values can be separated into three categories: Category I-O.I. higher than 1,000,000 (high odor potential); Category II-O.I. between 100,000 and 1,000,000 (medium odor potential); Category III-O.I. lower than 100,000 (low potential). The odor indexes calculated by the author are based on the highest 100% recognition levels mentioned in the literature, including those mentioned by Hellman and Small. The compounds have been arranged per chemical class in Table 12.

The threshold odor concentrations for 100% recognition and the associated odor indexes are shown graphically in Figures 16, 17, and 18. For each chemical class, we observe a smooth evolution of the O.I. and of the 100% recognition level as a function of the molecular weight of the compounds.

The odor index, which is a function of the vapor pressure of the product, must be calculated at the temperature of the evaporating product. The O.I. values mentioned in the foregoing tables have all been calculated at 20°C. It is clear that certain products, in practice, will have a higher odor index than mentioned in these lists because they are handled at higher temperatures.

The 100% recognition level has been taken as the basis for calculation of the O.I. The 100% recognition level is the concentration at which all members of an odor panel recognize the odor. It shows less variation than the absolute perception level, which is much lower. For this reason, the highest recognition level mentioned in the literature has been taken as the basis for these calculations.

In general, we can say that straight-chain aliphatic molecules have the highest recognition level and that the level decreases with increasing molecular weight. For nearly every class of straight-chain molecules, the threshold odor concentration decreases with increasing molecular weight. Tables 12 and 13 show the influence of functional groups on the 100% recognition levels and on the odor indexes. The functional groups of the small molecules have dominating influence on their threshold, as shown in Table 14 for molecules with one carbon atom. Branched chains often exhibit different results, probably because of steric effects. The functional groups can intensify each other's effects on the threshold odor concentration, but they can also reduce the effect, depending on their position in the molecule. In general, a double bond reduces the threshold odor concentration. This is the case in aliphatic compounds, mercaptans, and ketones, but not in aldehydes, as shown in Table 15. The merit of the previous analysis is that the T.O.C. of products for which only the formula is known can be estimated by extrapolation.

Table 11. 100% Odor Recognition Concentration and Odor Index of Chemicals, Arranged by Chemical Class.

Chemical	Formula	Molecular Weight	Odor Index	100% Odor Recognition Concentration
BTX aromatics				
benzene	C_6H_6	78	300	300 ppm
toluene	$C_6H_5CH_3$	92	720	40 ppm
xylenes	$C_6H_4(CH_3)_2$	106	360–18,200	0.4–20 ppm
1,2,3,5-tetramethyl-benzene	$C_6H_2(CH_3)_4$	134	136,000	2 ppb
isopropylbenzene	$C_6H_5CH(CH_3)_2$	120	89,600	40 ppb
ethylesters				
ethylacetate	$CH_3COOC_2H_5$	88	1,900	50 ppm
ethylbutyrate	$C_3H_7COOC_2H_5$	116	1,982,000	7 ppb
ethyl n-valerate	$C_4H_9COOC_2H_5$	132	178,000	60 ppb
ethylhexanoate	$C_5H_{11}COOC_2H_5$	144	760,000	4 ppb
ethylpelargonate	$C_8H_{17}COOC_2H_5$	186	109,000	1 ppb
ethyldecanate	$C_9H_{19}COOC_2H_5$	200		0.17 ppb
ethylundecylate	$C_{10}H_{21}COOC_2H_5$	214		0.56 ppb
methylesters				
methylformate	$HCOOCH_3$	60	300	2,000 ppm
methylacetate	CH_3COOH_3	74	1,100	200 ppm
methylbutyrate	$C_3H_7COOH_3$	102	19,000,000	2 ppb
ketones				
acetone	CH_3COCH_3	58	720	300 ppm
methylethylketone	$CH_3COC_2H_5$	72	3,800	30 ppm
diethylketone	$C_2H_5COC_2H_5$	86	1,900	9 ppm
methylisobutylketone	$CH_3COCH_2CH(CH_3)_2$	100	1,000	8 ppm
methylisoamylketone	$CH_3COCH_2CH_2CH(CH_3)_2$	114	75,100	70 ppb
ethylisoamylketone	$C_2H_5COCH_2CH_2CH(CH_3)_2$	128	660	4 ppm
2 pentanone	$CH_3—CO—CH_2—CH_2—CH_3$	86	2,000	8 ppm
2-heptanone	$CH_3—CO—CH_2CH_2—CH_2CH_2CH_3$	114	171,000	20 ppb
2-octanone	$CH_3CO(CH_2)_5CH_3$	128	4	250 ppm
diisobutylketone	$[(CH_3)_2CHCH_2]_2CO$	142	45	50 ppm
mercaptans				

isoamylmercaptan	$(CH_3)_2CHCH_2CH_2SH$	104.2		0.2 ppb
methylmercaptan	CH_3SH	48	53,300,000	35 ppb
ethylmercaptan	CH_3CH_2SH	62	289,500,000	2 ppb
propylmercaptan	$CH_3CH_2CH_2SH$	76	263,000,000	0.7 ppb
isopropylmercaptan	$(CH_3)_2CHSH$	76	1,052,000,000	0.2 ppb
butylmercaptan	$CH_3CH_2CH_2CH_2SH$	90	49,000,000	0.8 ppb
isobutylmercaptan	$(CH_3)_2CHCHSH$	90.2		0.83 ppb
t butylmercaptan	$(CH_3)_3CSH$	90.2		0.81 ppb
phenylmercaptan	C_6H_5SH	110	940,000	0.2 ppb
o-tolylmercaptan	$C_6H_5(CH_3)SH$	125	39,000	2 ppb
sulfides				
hydrogen sulfide	H_2S	34	17,000,000	1 ppm
methylsulfide	$(CH_3)_2S$	62	2,760,000	0.1 ppm
ethylsulfide	$(CH_3—CH_2)_2S$	90	14,400,000	4 ppb
propylsulfide	$(C_3H_7)_2S$	118		19 ppb
butylsulfide	$(CH_3CH_2CH_2CH_2)_2S$	146	658,000	2 ppb
isoamylsulfide	$(CH_3)_2CHCH_2CH_2]_2S$	174	1,640,000	0.4 ppb
phenylsulfide	$(C_6H_5)_2S$	186	14,000	4 ppb
pentysulfide	$(C_5H_{11})_2S$	174		0.028 ppb
di-isopropylsulfide	$(CH_3)_2CH]_2S$	118		3.2 ppb
acrylates				
ethylacrylate	$CH_2{=}{=}CH—COOC_2H_5$	100	138,160,000	1 ppb
isobutylacrylate	$CH_2{=}{=}CH—COOCH_2—CH(CH_3)_2$	128	525,000	20 ppb
ethylhexylacrylate	$CH_2{=}{=}CH—COOC_8H_{17}$	184	7,300	150 ppb
butylrates				
methylbutyrate	$CH_3—CH_2—CH_2—COOCH_3$	102	11,000,000	3 ppb
ethylbutyrate	$CH_3—CH_2—CH_2COOC_2H_5$	116	1,982,000	7 ppb
amines				
ammonia	NH_3	17	167,300	55 ppm
methylamine	CH_3NH_2	31	940,000	3 ppm
ethylamine	$CH_3CH_2NH_2$	45	1,445,000	0.8 ppm
isopropylamine	$(CH_3)_2CHNH_2$	59	637,000	1 ppm
butylamine	$CH_3(CH_2)_3NH_2$	73	395,000	0.3 ppm
dimethylamine	$(CH_3)_2NH$	45	280,000	6 ppm
diethylamine	$(C_2H_5)_2NH$	73	880,000	0.3 ppm
dipropylamine	$(C_3H_7)_2NH$	101	395,000	0.1 ppm
di-isopopylamine	$((CH_3)_2CH]_2NH$	101	108,000	0.8 ppm
dibutylamine	$(C_4H_9)_2NH$	129	5,500	0.5 ppm

trimethylamine	$(CH_3)_3N$	59	493,500	4 ppm
tri-ethylamine	$(C_2H_5)_3N$	101	235,000	0.3 ppm
ethanolamine	$HOCH_2CH_2NH_2$	61	130	5 ppm
methylethanolamine	$HOCH_2CH_2NHCH_3$	75	400	3 ppm
dimethylethanolamine	$HOCH_2CH_2N(CH_3)_2$	89	292,000	40 ppb
alkanes				
ethane	CH_3CH_3	30	25,300	1,500 ppm
propane	C_3H_8	44	425	11,000 ppm
butane	C_4H_{10}	58	480	5,000 ppm
isobutane	C_4H_{10}	58	3,000,000	2 ppm
pentane	C_5H_{12}	72	570	900 ppm
heptane	C_7H_{16}	100	200	200 ppm
octane	C_8H_{18}	114	100	200 ppm
nonane	C_9H_{20}	128	9,800	0.4 ppm
undecane	$C_{11}H_{24}$	156	8,400	0.2 ppm
alkenes				
ethene	$CH_2{=\!=}CH_2$	28	57,100	800 ppm
propene	$CH_3CH{=\!=}CH_2$	42	14,700	80 ppm
1-butene	$CH_3CH_2CH{=\!=}CH_2$	56	43,480,000	0.07 ppm
2-butene	$CH_3CH{=\!=}CHCH_3$	56	3,330,000	0.6 ppm
isobutene	$(CH_3)_2C{=\!=}CH_2$	56	4,640	0.6 ppm
1-pentene	$CH_3CH_2CH_2CH{=\!=}CH_2$	70	376,000,000	2 ppb
1-decene	$CH_3(CH_2)_7CH{=\!=}CH_2$	140	3,900,000	20 ppb
ethers				
ethylether	$CH_3CH_2OCH_2CH_3$	74	1,939,000	0.3 ppm
isopropylether	$(CH_3)_2CHOCH(CH_3)_2$	100	3,227,000	0.06 ppm
butylether	$C_4H_9OC_4H_9$	130	13,400	0.5 ppm
phenylether	$C_6H_5OC_6H_5$	170	130	0.1 ppm
aldehydes				
formaldehyde	HCHO	30	5,000,000	1 ppm
acetaldehyde	CH_3CHO	44	4,300,000	0.3 ppm
propionaldehyde	CH_3CH_2CHO	58	3,865,000	0.08 ppm
acrylaldehyde (acroleine)	$CH_2{=\!=}CHCHO$	56	19,300	20 ppm
butyraldehyde	$CH_3CH_2CH_2CHO$	72	2,395,000	40 ppb
isobutyraldehyde	$(CH_3)_2CHCHO$	72	948,000	300 ppb
crotonaldehyde	$CH_3CH{=\!=}CHCHO$	70	125,000	0.2 ppm
methylpentaldehyde	$C_5H_{12}CHO$	101	131,500	0.15 ppm
furfuraldehyde	C_4H_3OCHO	96	5,260	0.2 ppm
benzaldehyde	C_6H_5CHO	106	22,000	5 ppb

cinnamaldehyde	$C_6H_5CH{=}{=}CHCHO$	132	131,500	2 ppb
acids				
formic acid	$HCOOH$	46	2,200	20 ppm
acetic acid	CH_3COOH	60	15,000	2 ppm
propionic acid	CH_3CH_2COOH	74	112,300	40 ppb
butyric acid	$CH_3(CH_2)_2COOH$	88	50,000	20 ppb
valeric acid	$CH_3(CH_2)_3COOH$	102	256,300	0.8 ppb
caproic acid	$CH_3(CH_2)_4COOH$	116	43,900	6 ppb
enanthic acid	$CH_3(CH_2)_5COOH$	130	7,900	20 ppb
caprylic acid	$CH_3(CH_2)_6COOH$	144	104,500	8 ppb
pelargonic acid	$CH_3(CH_2)_7COOH$	158	164,000	0.7 ppb
capric acid	$CH_3(CH_2)_8COOH$	172		1.96 ppb
lauric acid	$CH_3(CH_2)_{10}COOH$	200.3		3.4 ppb
alcohols				
methanol	CH_3OH	32	22	6,000 ppm
ethanol	CH_3CH_2OH	46	11	6,000 ppm
propanol	$CH_3CH_2CH_2OH$	60	480	45 ppm
butanol	$CH_3(CH_2)_2CH_2OH$	74	120	5,000 ppm
pentanol	$CH_3(CH_2)_3CH_2OH$	88	368	10 ppm
hexanol	$CH_3(CH_2)_4CH_2OH$	102	14,300	0.09 ppm
heptanol	$CH_3(CH_2)_5CH_2OH$	116	23,100	0.06 ppm
octanol	$CH_3(CH_2)_6CH_2OH$	130	33,000	2 ppb
decanol	$CH_3(CH_2)_8CH_2OH$	158	31,000	6 ppb
dodecanol	$CH_3(CH_2)_{10}CH_2)OH$	186	1,800	7 ppb
phenolics				
phenol	C_6H_5OH	94	16	20 ppm
cresols	$(CH_3)C_6H_4OH$	108	60–260	0.2–0.7 ppm
acetates				
methylacetate	CH_3COOH_3	74	1,100	200 ppm
ethylacetate	$CH_3COOC_2H_5$	88	1,900	50 ppm
propylacetate	$CH_3COOC_3H_7$	102	1,600	20 ppm
isopropylacetate	$CH_3COO(iC_3H_7)$	102	2,100	30 ppm
butylacetate	$CH_3COOC_4H_9$	116	1,200	15 ppm
isobutylacetate	$CH_3COO(iC_4H_9)$	116	3,300	4 ppm
amylacetate	$CH_3COOC_5H_{11}$	130	2,500	20 ppm
isoamylacetate	$CH_3COO(iC_5H_{11})$	130	526,000	20 ppm
sec. hexylacetate	$CH_3COO(sec.C_6H_{13})$	144	12,500	0.4 ppm
octylacetate	$CH_3COOC_8H_{17}$	172	2,800	300 ppb

2.2. Comparison of Techniques for Organoleptic Odor-Intensity Assessment

1. Odor room. Odor threshold can be determined by using an *odor room*. A known volume of odorous air is admitted to the room of volume V until the volume S of odorous air is found that just, and only just, allows the odor to be detected within the room. The ratio V/S at this point is the threshold dilution.
2. The syringe method. In the *ASTM syringe method* (ASTM D1391-57), variable volumes of odorous sample air are made up to 100 m^3 with clean air in a series of syringes. Variable volume samples may then be taken from these syringes and similarly made up to achieve yet higher dilutions and so on until odor threshold is reached. Assessment for odor is by injection into the nostril of panel members, one syringe being supplied to each person for each dilution tested. Hypodermic needles were sealed to produce end caps or joined together to produce transfer tubes in order to retain the sample within the syringes until required and to effect the dilution operations, respectively.
3. The dynamic dilution method. The general concept of *dynamic dilution* involves the mixing of sample air supplied at a known flowrate with dilution air at known flowrates. After each change in dilution flowrate, panel members to whom the issuing mixture is presented register the presence or absence of odor, and in this manner threshold dilutions are determined. In order to achieve a sufficiently wide range of dilutions, it is customary to use a two-stage system in which a sample of diluted odorous air from the first dilution operation is introduced to a second mixing chamber and is there diluted with flowing clean air in the second stage.

Table 12.
Classification of Chemical Classes According to Their 100% Recognition Threshold.

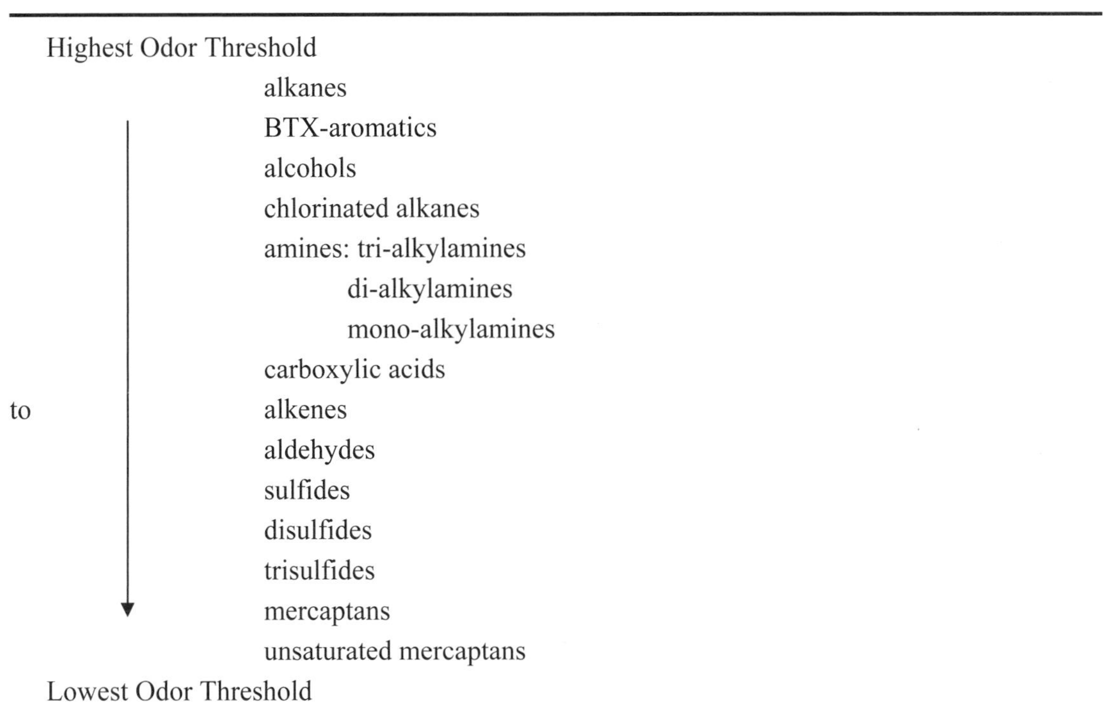

Highest Odor Threshold	
	alkanes
	BTX-aromatics
	alcohols
	chlorinated alkanes
	amines: tri-alkylamines
	di-alkylamines
	mono-alkylamines
	carboxylic acids
to	alkenes
	aldehydes
	sulfides
	disulfides
	trisulfides
	mercaptans
	unsaturated mercaptans
Lowest Odor Threshold	

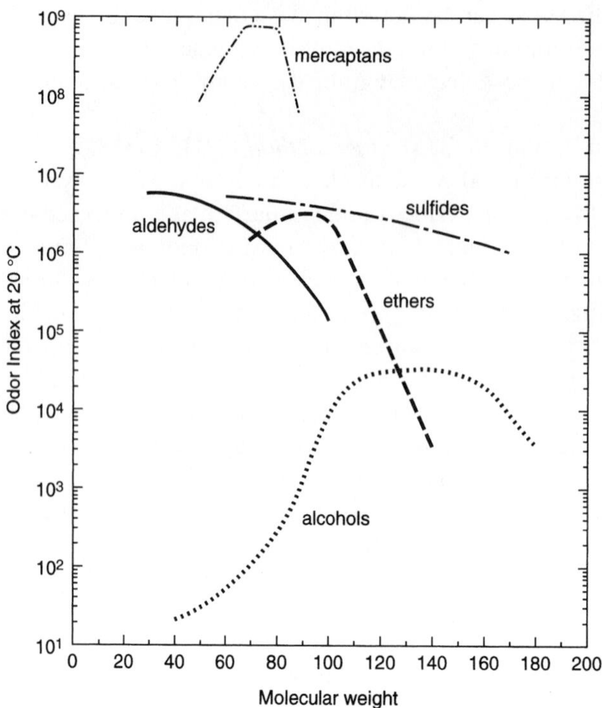

Figure 16. Odor index (at 20°C) of mercaptans, sulfides, ethers, aldehydes, and alcohols.

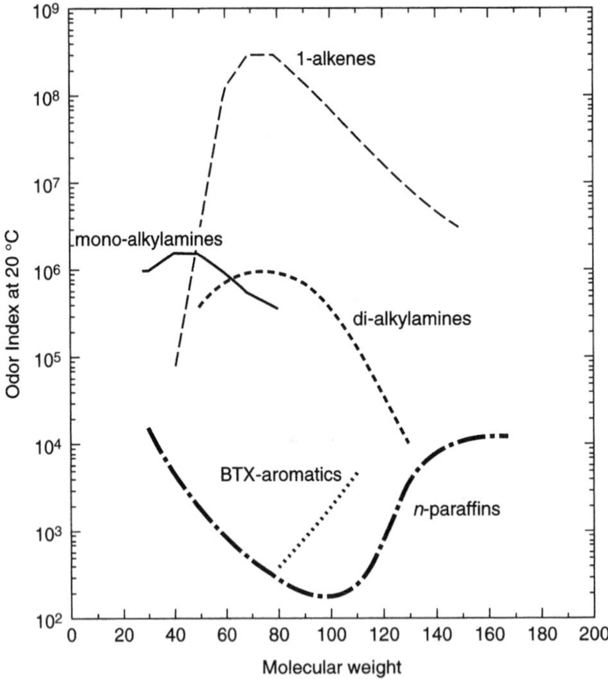

Figure 17. Odor index (at 20°C) of alkenes, mono-, di-, and tri-alkylamines, BTX-aromatics, and *n-paraffins*.

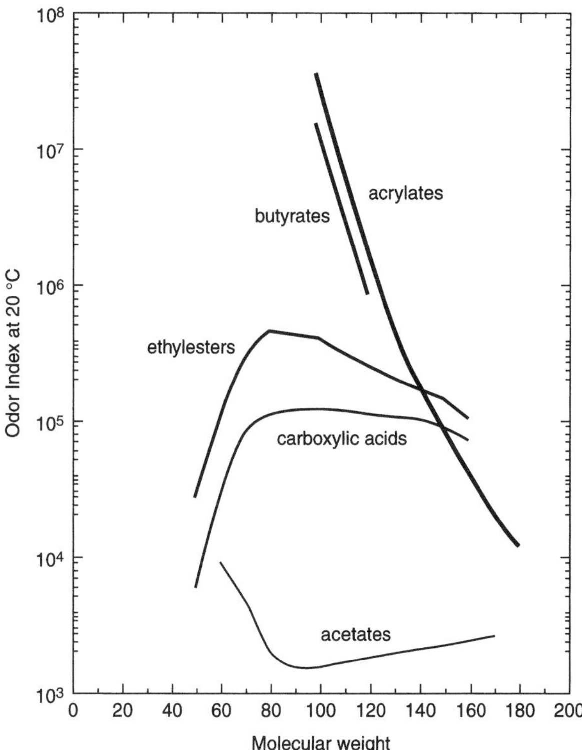

Figure 18. Odor index of butyrates, acrylates, ethylesters, carboxylic acids, and acetates.

3. Hazard Potentials of Atmospheric Pollutants

Hazard potentials of atmospheric pollutants are defined by their effects on the main components of the physical environment. These effects and related hazard potentials are as follows:

Effects	*Hazard potentials*
Global warming	Global Warming Potential (GWP)
Ozone depletion	Ozone Depletion Potential (ODP)
Photochemical smog formation	Photochemical Ozone Creation Potential
Acidification	Acidification Potential (AP)
Eutrophication	Eutrophication Potential (EP)

The atmospheric media provide an excellent matrix for photochemical or oxidative alterations of chemical contaminants. The energy provided by sunlight is able to break carbon-carbon and carbon-hydrogen bonds, cause the photodissociation of nitrogen dioxide to nitric oxide and atomic oxygen, and photolytically produce significant amounts of hydroxyl radicals. The relatively high concentration of oxygen (21 Vol %) makes it one of the most important participants in various reactions with air pollutants because the rates are concentration dependent. Similar reasoning can be used for reactions with water vapor (0.1-5 Vol %) and carbon dioxide (0.03-0.1 Vol %).

Although hydrocarbons are essential to the formation of photochemical smog, not every hydrocarbon produces the manifestations (such as eye irritation, plant damage, and visibility reduction) that are usually associated with smog. The reason for this is the differences in reactivity and chemical structure of hydrocarbons. We will define "reactivity" as the tendency of an atmospheric system containing organic products and nitrogen oxides to undergo a series of reactions in the presence of sunlight. However, because of the variety of ways in which smog is manifested, a number of reactivity scales have been developed. These scales cover

1. rates of hydrocarbon consumption (hydroc. cons.)
2. rates of photooxidation of NO to NO_2 (NO ox.)
3. eye irritation (EI)
4. chemical reactivity
5. OH rate constants

Relative reactivity scales were developed for the sake of comparison, using as a standard a fast hydrocarbon, set at a value of 1-100. In the Glasson-Tuesday Scale, sometimes referred to as the Jackson Scale, 2,3-dimethyl-2-butene is the reference hydrocarbon; it has a relative reactivity for NO oxidation (NO_2 formation) of 100. Table 16 presents an abbreviation of this scale.

Table 13. Classification of Chemical Classes According to Their Odor Index (at 20°C).

Odor index	Chemical class	
O.I. > 10^6:	mercaptans	
	alkenes	
	sulfides	
	butyrates	of low molecular weight
	acrylates	of low molecular weight
	aldehydes	of low molecular weight
	ethers	of low molecular weight
	alkylamines	of low molecular weight
O.I. between 10^4 and 10^6:	di-alkylamines	
	tri-alkylamines	
	higher ethylesters	
	carboxylic acids	
	aldehydes	of high molecular weight
	ethers	of high molecular weight
	alcohols	of high molecular weight
O.I. < 10^4:	alkanes	
	acetates	
	BTX-aromatics	
	lower alcohols	
	phenolics	

Table 14. Influence of Functional Groups on T.O.C. (100% Recognition).

C_1	T.O.C. (100% recognition)
CH_4	±10,000 ppm
CH_3OH	6,000 ppm
$CHCl_3$	300 ppm
CH_2Cl_2	214 ppm
CCl_3F	209 ppm
CCl_4	200 ppm
$CH_2(OH)_2$	60 ppm
CH_3Cl	30 ppm
HCOOH	20 ppm

HCşşN	5 ppm
CH_3NH_2	3 ppm
CCl_3NO_2	1 ppm
HCOH	1 ppm
ClCşşN	1 ppm
CH_3SH	0.035 ppm
CHI_3	0.00037 ppm

Table 15. Influence of Double Bonds on the Odor Recognition Treshold

Product	Formula	100% Recognition Threshold
propionaldehyde	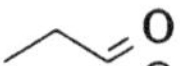 O	0.08 ppm
acroleine	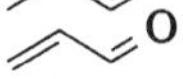 O	20 ppm
butyraldehyde	 O	0.04 ppm
crotonaldehyde	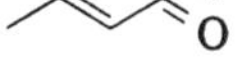 O	0.2 ppm
butane	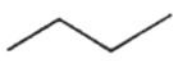	5,000 ppm
1-butene	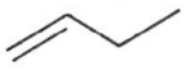	0.07 ppm
pentane		900 ppm
1-pentene		0.002 ppm
propylmercaptan	 SH	0.7 ppb
allylmercaptan	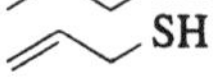 SH	0.05 ppb
butylmercaptan	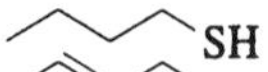 SH	0.8 ppb
crotylmercaptan	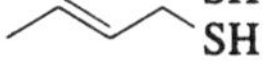 SH	0.055 ppb

The natural atmospheric environment is a complex, highly reactive system involving photochemical and secondary dark reactions to both organic and inorganic substances. To reproduce such a complex system in the laboratory, the following parameters must be controlled.

- spectral distribution and intensity of the light
- concentrations of the reactants in the test chamber
- temperature
- surface reactions (major difference between results in test reactors and in the atmosphere)

Four general laboratory procedures have been used to study atmospheric degradation: long-path infrared (LPIR) cells, plastic containers, glass flask reactors, and smog chambers. With LPIR cells, the test chemical is placed into the cell and irradiated.

Table 16. Abbreviated Glasson-Tuesday (Jackson) Reactivity Scale (Basis NO_2 Formation).

Product	Relative Reactivity
methane	0
acetylene	0

ethylene	2.9
propylene	5.9
butane	1.3
1-butene	6.0
2-butene	15
pentane	1.6
1-pentene	4.6
2-pentene	11
hexane	1.6
hexene	3.4
benzene	0.56
toluene	2.2

The rate of reaction is determined by following the disappearance of the test chemical and the formation of some products by *infrared analysis.* Fluorescent lights are usually used for irradiation, and various configurations and chamber materials (stainless steel, Teflon-coated interior, and Pyrex glass) have been used. Known amounts of nitric oxide and nitrogen dioxide can be added to simulate air pollution situations.

A popular method for studying the atmospheric chemistry of both artificial and natural atmospheric samples is the use of plastic containers made of either FEP (fluorinated ethylene-propylene), Teflon, Tedlar, or Mylar. The sample is placed in the bag, irradiated with either natural or artificial sunlight, and analyzed periodically, usually by gas chromatography. Glass flask reactors are used in a similar way, but they are usually smaller in volume than the plastic bags and are more commonly irradiated with artificial lights.

In order to study air pollution reactions, many researchers have resorted to large *photochemical smog chambers.* These chambers are extremely complicated and expensive to build and operate. Most chambers can be run in a dynamic or a static mode, although for simplicity and precision, the static procedure is more often used.

3.1. Atmospheric Lifetime. The temporal and spatial scales at which a component will be dispersed in the atmosphere, and therewith the potential environmental risk of a substance, are largely determined by its atmospheric residence time as illustrated by Table 17.
The removal processes from the atmosphere can be summarized as follows:

- dry deposition: uptake of material at the earth's surface by soil, water, or vegetation.
- wet deposition: uptake of material in cloud or rain droplets followed by droplet removal by precipitation.
- chemical reactions: strictly speaking, not a "real" removal mechanism (the species is transformed into another that may remain in the atmosphere), but chemical transformation may have a large impact on the removal rates by deposition.

The atmospheric lifetime may be defined as

$$\pi = \frac{1}{k_d \times k_w \times k_c}$$

where k_d and k_w are representative removal rates for dry and wet deposition, respectively, and k_c is the pseudo-first-order chemical transformation rate.

The dry deposition rate depends strongly on the vapor/particulate partition of the compound. In the solid phase (aerosols), the particle size determines the deposition rates. Organic compounds with a vapor pressure exceeding about 10^{-4} Pa occur in the atmosphere in general in the gaseous phase. Organic compounds having lower vapor pressures are bound to aerosols, or, depending on temperature and the available aerosol surface, an equilibrium between the gaseous and aerosol forms will be established. For most of the gaseous organic compounds, the deposition velocity will be very low.

Table 17. Correspondence between Spatial and Temporal Scales (2853).

Horizontal Transport		Time	Transport Vertical
local	0–10 km	minutes	
	0–30 km	hours	boundary layer
mesoscale	>1,000 km	day	
continental	>3,000 km	several days	troposphere
hemisphere		month	
global		year	stratosphere

Table 18. Typical Bond Dissociation Energies Related to Wavelength of Absorbed Light.

Bond	*E*, kcal/mole	*l*, nm
$CH_3CO—NH_2$	99	288
$C_2H_5—H$	98	291
$CH_3CO—OCH_3$	97	294
$C_6H_5—Cl$	97	294
$(CH_3)_2N—H$	95	300
$H—CH_2OH$	94	303
$CH_3—OH$	91	313
$C_6H_5—Br$	82	347
$C_6H_5CH_2—COOH$	68	419 (2919)

The wet removal is very efficient for particulate pollution. Again, the particle size determines the efficiency of the wet deposition process for aerosols. For gaseous compounds the solubility (Henry's constant, *H*) is the rate-determining factor.

In the atmospheric degradation of a compound, the following processes may contribute:

- direct photolysis
- reaction with hydroxyl radicals (OH radicals, OH^{o})
- reaction with ozone (O_3)

- reaction with nitrate radicals (NO_3^o) and other photochemically generated species

3.1.1. DIRECT PHOTOLYSIS. For material in the troposphere (the lowest 10-12 km of the atmosphere), the UV-B and UV-A regions (wavelengths 290-450 nm) are important. Solar radiation with wavelength shorter than 290 nm hardly reaches the troposphere because it is absorbed by ozone and oxygen in the stratosphere, and radiation with longer wavelengths is in general too "soft" to break a chemical bond.

Energetically, 290-400 nm is the most important portion of the spectrum, because the associated energies are comparable to chemical bond energies (Table 18). The energy (*E*) of each quantum, in kcal, is related to the wavelength (*l*) by

$$E = Lhc/l$$

where

L = Avogadro's number (6.0×10^{23} molecules/mole)
h = Planck's constant (6.6×10^{-27} erg sec)
c = the speed of light (3×10^{10} cm/sec)

The energy available to bring about direct photochemical transformations amounts to about 96 kcal/mole at 300 nm, 82 kcal/mole at 350 nm, and 72 kcal/mole at 400 nm. For comparison, the energy required to break the carbon-carbon bond in ethane is about 88 kcal/mole, and a carbon-hydrogen bond in the same molecule requires about 98 kcal/mole. Energy absorption is, of course, the prime requisite for a chemical reaction. Because many pesticides absorb sunlight in the range of 290-400 nm, many different chemical transformations have been observed.

Long-living species, with atmospheric lifetimes of more than one year, will be transported to the atmosphere. Here radiation with shorter wavelengths (200-290 nm) may be of importance. Examples include the fully halogenated chlorofluorocarbons (CFCs) and nitrous oxide (N_2O), which are nearly inert in the troposphere but are efficiently photolyzed in the stratosphere.

At present there are no structure-activity relations (SARs) or related procedures available from the literature to estimate the photolysis rate (2853).

3.1.2. REACTION WITH HYDROXYL RADICALS (OH RADICALS, OH^o). Reaction with OH radicals is the main oxidation pathway for the majority of organic pollutants. Various procedures are available to estimate the OH^o reaction rate constant. For a first estimate of k_{OH}, the SAR approach as proposed by Atkinson (2850, 2851) is recommended. It is simple to use and applicable to many classes of substituents. In general the method is accurate to within a factor of 2 for most aliphatics, olefins, and simple aromatics.

3.1.3. REACTION WITH OZONE (O_3). Ozonolysis, the reaction with ozone, is important only for unsaturated compounds containing double or triple bonds. Structure-activity relations have been developed by Atkinson and Carter (2852).

3.1.4. REACTION WITH NITRATE RADICALS (NO_3^o) AND OTHER PHOTOCHEMICALLY GENERATED SPECIES. In the atmosphere numerous other reactions will occur that may initiate the atmospheric degradation of organic compounds. Although most of these reactions are still poorly elucidated, it is expected that these degradation pathways will play only a minor role with respect to the OH and O_3 reactions. One exception may be the reaction between the NO_3^o radical and unsaturated hydrocarbons. NO_3^o is efficiently photolyzed, so this type of reaction is important only in night-time chemistry. Estimation procedures for k_{NO3} are not yet available.

The total pseudo-first-order phototransformation rate k_c, and therefore the phototransformation lifetime $t = 1/k_c$, is estimated by

$$k_c = J + k_{OH}[OH^o] + k_{O3}[O_3] + k_{NO3}[NO_3^o]$$

where J is the direct photolysis rate and [OH], $[O_3]$, and $[NO_3^o]$ are the atmospheric concentrations of the OH radical, of ozone, and of the nitrate radical, respectively.

As an illustration, Table 19 gives the decay times t_X for various organic compounds.

$$t_X = 1/k_X[X]$$

where k_X is the rate constant and [X] is a representative concentration of OH^o, O_3, or NO_3^o (2853).

The decay times were calculated assuming the following representative boundary layer concentrations (2854): $[O_3]$ = 40 ppb = 9.8 10^{11} mol/cm^3, a typical Northern hemisphere value $[OH^o]$ = 1.2 × 10^6 mol/cm^3, an average daytime value $[NO_3^o]$ = 10 ppt = 2.4 × 10^8 mol/cm^3, a typical night-time value for a clean atmosphere (2853)

Table 19. Atmospheric Lifetimes (in hours) for Several Organic Compounds.

Compound	t_{OHo}	t_{O3}	t_{NO3o}
methane	28,000	2.0 × 10^8	>2.9 × 10^6
ethane	840	3.0 × 10^7	1.4 × 10^5
n-pentane	57	2.8 × 10^7	1.4 × 10^4
benzene	180	3.8 × 10^6	>3.6 × 10^4
toluene	37	2.8 × 10^5	1.7 × 10^4
p-xylene	15	7.1 × 10^5	2,600
ethene	27	160	5,800
1-butene	7.4	26	93
isoprene	2.3	20	1.4
ethyne	300	2.8 × 10^4	2.3 × 10^4

3.2. Phototransformation. Phototransformations have to be taken into account for gases and compounds that occur in the gas phase in environmentally significant quantities. The probability that a compound occurs in the gas phase depends not only on its vapor pressure but also on its water solubility and adsorption/desorption behavior. Therefore, even substances that have a relatively low vapor pressure (down to 10^{-3} Pa) can be found in the atmosphere in measurable quantities.

The main transformations leading to the removal of chemicals from the atmosphere involve reactions with photochemically generated species such as the hydroxyl radical (OH^o), ozone (O_3), the hydroxyperoxyl radical (HO_2^o), singlet oxygen, and the nitrate radical (NO_3^o). Direct phototransformation-that is, all transformations resulting from direct photoexcitation of the molecule-may also be important.

It is broadly agreed that reaction with OH^o is the dominant photo-induced reaction of hydrocarbons in the atmosphere. The only class of compounds known not to react rapidly with OH^o are the fully halogenated alkanes, as can be seen from Table 20, in which half-life times are given for reactions of a variety of compounds with OH^o, O_3, and HO_2. Reactions with OH^o are generally the fastest. Only for some alkenes, alkadienes, and terpenes does the rate of removal by O_3 exceed that by OH^o, and in these cases the lifetime derived from K_{OHo} is, in any case, very short (a fraction of a day). The similarity between the reactions of OH^o and O_3 molecules with these unsaturated hydrocarbons explains these observations in that both add to the double bond, forming an additional complex that afterwards disintegrates to the reaction products. In those cases where abstraction of an H atom by OH° is the dominant pathway, the competitive reaction with O_3 is expected to be very much slower. It follows that the reaction with O_3 is usually of secondary importance in considering the fate of organic chemicals in the atmosphere. Reaction with O_3 is of interest, however, when atmospheric processes related to smog formation (where higher levels of O_3 occur) are involved.

Some evidence has been obtained that alkenes, phenols, and cresols react with NO_3^o radicals formed in photochemical smog systems. It is therefore concluded that these NO_3^o reactions are of limited importance and apply only to special environmental situations-e.g., at night in a moderately polluted atmosphere. The rate constants for reactions with singlet oxygen are very low, and this reaction can be neglected as an elimination pathway for organic molecules.

Direct phototransformation is a possible removal pathway only for those chemicals that absorb in the region of solar radiation.

Chemicals may be sorbed onto aerosol particles from the vapor phase and thence removed from the atmosphere with the aerosol. The mechanism of phototransformation of a chemical in an aerosol is not the same as that of the chemical in air because its adsorbed state, and the physiochemical properties of the aerosol substrate, influence the reactions (2716).

3.3. Global Warming. The impact of a substance on global warming depends on its IR absorption characteristics and on its atmospheric lifetime. When the substance shows absorption bands in the so-called atmospheric window (8.5-11 mm), it must be marked as a potential greenhouse gas. In this case, it is necessary to estimate the global warming potential (GWP).

The GWP is defined as the ratio of the calculated warming for each unit of mass of a gas emitted into the atmosphere to the calculated warming for a mass unit of the reference gas CFC-11 (trichlorofluoromethane). The GWP provides a measure of the cumulative effect on the radiative balance over the chemical lifetime of a mole mass emitted in the atmosphere compared to the cumulative effect of a mole mass of CFC-11.

Next to the IR absorption strength, the atmospheric lifetime is the dominant factor for the GWP. For species with lifetimes less than 1-2 years, GWP values less than 0.03 are in general expected.

Table 20. Environmental Half-lives for Reactions with OH°, O_3, and HO_2^o and for Direct Photolysis. (*t*/2 is in days.)

Compound	*t*/2 OH°	*t*/2 O_3	*t*/2 HO_2^o	t/2 Direct Photolysis
Alkanes				
methane	1,000	6,000,000		
ethane	30	7,000,000		
propane	7.3	1,000,000		
iso-butane	3.6	4,000,000		
n-butane	*3.0*	800,000	<13,000	
Haloalkane				
bromomethane		210		150,000
Aldehydes				
formaldehyde	0.8		0.6	0.23

acetaldehyde	0.5	235		1.1
Alkenes				
ethene	1.0	4.2	<13,000	
propene	0.3	0.6	<13,000	
1-butene	0.2	0.7	<13,000	
2-*cis-butene*	0.2	0.5		
2-*trans-butene*	0.1	0.03		
2-methyl-2butene	0.1	0.02		
2,3-dimethyl-2-butene	0.05	0.005	<10,000	
1-pentene	0.3	0.75		
cis-2-pentene	0.1	0.02		
1-hexene	0.25	0.7		
1-heptene	0.2	1.0		
cyclohexene	0.1	0.05		
Haloalkenes				
chloroethene	1.2	4.0		
trichloroethene	3.6	1,300		
tetrachloroethene	47	4,700		
Alkadiene				
1,3-butadiene	0.1	0.95		
Terpene				
pinene	0.3	0.05		
Alkyne				
ethyne	11	100		
Alkylbenzenes				
benzene	5.7	170,000		
toluene	1.3	29,000		
1,2-dimethylbenzene	0.6	5,000		
1,3-dimethylbenzene	0.4	6,200		
1,4-dimethylbenzene	0.8	5,000		
1,3,5-trimethylbenzene	0.2	1,100		
ethylbenzene	1.1	14,000		
2-propylbenzene	1.0	14,000		

(2716)

Table 21. Comparison of GWP Values Calculated by Different Methods.

Component	Chemical Formula	IR_{abs}, cm^{-2} atm^{-1}	Atmospheric Lifetime, years	GWP by Equation	GWP by Global Model
CFC-11*	CCl_3F	2389	60	1.0	1.0
CFC-12	CCl_2F_2	3240	120	3.1	2.8–3.4
CFC-113	CCl_2FCClF_2	3401	90	1.6	1.3–1.4
CFC-114	$CClF_2CClF_2$	4141	200	4.6	3.7–4.1
CFC-115	$CClF_2CF_3$	4678	400	11.6	7.5–7.6
HCFC-22	$CHClF_2$	2554	15	0.36	0.34–0.37
HCFC-123	CF_3CHCl_2	2859	1	0.029	0.017–0.02
HCFC-124	CF_3CHClF	4043	6	0.19	0.092–0.10
HCFC-125	CF_3CHF_2	3908	28	0.88	0.51–0.65
HCFC-134a	CF_3CH_2F	3272	15	0.48	0.25–0.29
HCFC-141b	CCl_2FCH_3	1912	7	0.12	0.087–0.097
HCFC-142b	$CClF_2CH_3$	2577	19	0.47	0.34–0.39
HCFC-143a	CF_3CH_3	3401	41	1.6	0.72–0.76
HCFC-152a	CHF_2CH_3	1648	1	0.041	0.026–0.033
tetrachloromethane	CCl_4	1195	50	0.37	0.34–0.35
methylchloroform (2853)	CCl_3CH_3	1209	6	0.055	0.022–0.026

*Reference compound

A first approxi mation of the GWP value of a substance X is obtained by

$$GWP_X = \frac{\text{atm. lifetime}_X}{\text{atm. lifetime}_{CFC\text{-}11}} \bullet \frac{mw_{CFC\text{-}11}}{mw_X} \bullet \frac{IR_{abs\ X}}{IR_{abs\ CFC\text{-}11}}$$

where IR_{abs} is the IR absorption strength in the interval 800-1200 cm^{-1} and mw is the molecular weight.

More detailed estimates of GWP can be made by means of global atmospheric models. In these models, the fate of a pollutant is estimated by taking into account atmospheric transport and dispersion and chemical and physical transformation and removal processes on a global scale (2853, 2855, 2856).

Table 21 gives a comparison between GWP values estimated using the aforementioned equation and those calculated by means of global atmospheric models for a number of CFCs and their alternatives. Provided that the atmospheric lifetimes and the IR absorption strengths are known, the former simple equation gives an approximation of GWP values within a factor of 2 compared to the GWP values calculated by means of complex atmospheric models.

3.4. Atmospheric Ozone. Concerning atmospheric ozone, it is necessary to distinguish between the possible impact of pollutants on stratospheric ozone (the ozone layer at an altitude of about 15-50 km) and the potential ozone formation in the troposphere (the lower part of the atmosphere up to about 12 km).

3.5. Stratospheric Ozone. A pollutant may have an effect on stratospheric ozone when

- it contains a Cl or Br substituent
- the atmospheric lifetime is long enough to allow for transport to the stratosphere

The potential risk of a chemical can be estimated from its ozone depletion potential (ODP). The ODP is defined as the ozone depletion in the stratosphere caused by the emission of a mass unit of a chemical relative to the ozone depletion caused by the release of a mass unit of CFC-11.

In order for a compound to have a low ODP, it must also have a very short tropospheric lifetime, so that a large fraction is destroyed before reaching the ozone layer in the stratosphere. The algorithm that has been developed for ODP has the form

$$\text{ODP} = A \bullet F_r \bullet F_s$$

where A is the normalizing constant, F_r is a reactivity factor depending on the number of chlorine atoms in the molecule, and F_s is a survival factor (the fraction of molecules surviving transport to the stratosphere). ODP values range from 0.000 for trifluoromethane to 0.23 for 1,1,1-trichloro-2,2-difluoroethane (2446).

In general, ODP values approach zero for species with atmospheric lifetimes less than one year. A first approximation of the ODP value of a chemical can be based on atmospheric lifetime and the number of Cl and Br atoms per molecule and can be obtained by

$$\text{ODP}_x = \frac{\text{atm. lifetime}_x}{\text{atm. lifetime}_{\text{CFC-11}}} \bullet \frac{\text{mw}_{\text{CFC-11}}}{\text{mw}_x} \bullet \frac{n_{\text{Cl}} + 30 n_{\text{Br}}}{3}$$

where mw is the molecular weight and n_{Cl} and n_{Br} are the numbers of Cl and Br atoms respectively, per molecule.

More detailed estimates of ODP can be made by means of global atmospheric models. Table 22 gives, for various species, a comparison between approximated ODP (see the foregoing equation) and the ODP calculated by means of global models.

Table 22. Comparison of ODP Values Calculated by Different Methods.

Component	Chemical Formula	Atmospheric Lifetime, years	ODP by Equation	ODP by Global Model
CFC-11*	CCl_3F	60	1.0	1.0
CFC-12	CCl_2F_2	120	1.5	0.87–1.0
CFC-113	CCl_2FCClF_2	90	1.1	0.76–0.83
CFC-114	$CClF_2CClF_2$	200	1.8	0.56–0.82
CFC-115	$CClF_2CF_3$	400	2.0	0.27–0.45
HCFC-22	$CHClF_2$	15	0.14	0.032–0.072
HCFC-123	CF_3CHCl_2	1.6	0.016	0.013–0.027

HCFC-124	CF_3CHClF	6.6	0.037	0.013–0.030
HCFC-125	CF_3CHF_2	28	0	0
HCFC-134a	CF_3CH_2F	15	0	0
HCFC-141b	CCl_2FCH_3	7.8	0.10	0.065–0.14
HCFC-142b	$CClF_2CH_3$	19	0.15	0.035–0.077
HCFC-143a	CF_3CH_3	41	0	0
HCFC-152a	CHF_2CH_3	1.7	0	0
tetrachloromethane	CCl_4	50	0.99	0.95–1.2
methylchloroform	CCl_3CH_3	6.3	0.11	0.092–0.20
halon-1301	$CBrF_3$	107	16	13
halon-1211	$CBrClF_2$	15	2.1	2.2
halon-1202	CBr_2F_2	1.5	0.3	0.3
halon-2402	CF_2BrCF_2Br	28	4.9	6.2
hexachlorobenzene	C_6Cl_6	1	0.02	—
pentachlorophenol (2853)	C_5Cl_5OH	>0.1	0.001	—

*Reference compound

Table 23. Comparison of Photochemical Ozone Creation Potential (POCP) Values Obtained by Derwent and Jenkin (2857) with an OH°-Reactivity Scale, Relative to Ethylene = 100.

Compound	POCP	OH°-scale	K_{OH}
methane	0.7	0.2	8.4×10^{-15}
ethane	8	3	2.7×10^{-13}
propane	42	9	1.2×10^{-12}
n-pentane	41	19	4.1×10^{-12}
benzene	19	5	1.3×10^{-12}
toluene	56	22	6.2×10^{-12}
p-xylene	89	47	1.5×10^{-11}
m-xylene	99	76	2.5×10^{-11}
o-xylene	67	46	1.5×10^{-11}
ethylene	100	100	8.5×10^{-12}
formaldehyde	42	99	9.0×10^{-12}
acetaldehyde	53	121	1.6×10^{-11}

3.6. Tropospheric Ozone. At present there is no procedure available to estimate the effect on tropospheric ozone when only the basic characteristics of a substance are known. However, a first indication of episodic ozone formation can be obtained from a reactivity scale based on the rate constant for the (OH + hydrocarbon) reaction and the molecular weight. OH-scale values can easily be estimated using structure-reactivity relations. There is a reasonable correlation between the POCP (Photochemical Ozone Creation Potential) scale and the OH-reactivity scales, and therefore the OH-scale value may, for the time being, be used as an indication for episodic ozone formation (Table 23).

The tropospheric lifetime *(T)* is the inverse of the rate constant for reaction with OH^o:

$$T = 1/k'$$

$$R\text{-H} + {}^oOH \longrightarrow R^o + H_2O$$

$$\text{Rate} = -d(\text{RH})/dt = k({}^o\text{OH})(\text{RH}) = k'(\text{RH})$$

As a molecule travels upward from the surface of the earth, it can be decomposed either in the troposphere (primarily by reaction with OH^o or, in some cases, photolysis) or in the stratosphere (by photolysis from the shorter-wavelength light). The lifetime is defined as the time it takes for the quantity of a chemical released to drop to $1/e$ or approximately 37% of its initial value. This lifetime corresponds to 1.41 half-lives.

OH-scale value is calculated as follows:

$$\text{OH-scale value} = \frac{k_X}{mw_X} \bullet \frac{mw_{ethylene}}{k_{ethylene}} \bullet 100$$

where k_X is the rate constant at $t = 298$ K for the reaction of X with the OH radical, mw_X is the molecular weight of compound X, and K_{OH} is in $mol^{-1}\ cm^3\ sec^{-1}$ (2853).

4. Natural Sources

Although the amount of manmade organic chemicals (excluding lubricants) that enter the environment may be as much as 20 million tons a year, this total is very small in comparison with the enormous tonnages of organic compounds naturally produced. Over the ages, degradation and emanation cycles have become established through which an equilibrium seems to be maintained. Although we know a great deal about the detailed mechanisms involved, the way in which many cycles operate is still obscure. Some of them are on a massive scale; for instance, it has been estimated that swamps and other natural sources emit as much as 1600 million tons of methane into the atmosphere each year. Even cattle, which emit methane equivalent to 7% of their energy intake, must contribute a world total of several million tons. It is estimated that the world's atmosphere contains 4800 million tons of methane, and it is evident that a balance exists.

Terpene-type hydrocarbons emitted into the air by forests and other vegetation amount to an estimated 170 million tons a year. It is believed that these polymerize in the air and are eventually eliminated from the air by rainfall or deposition in aerosols.

5. Manmade Sources

Emission rates for various sources, such as diesel and gasoline engines, municipal waste incinerators, and central heating furnaces, as well as ground level concentrations in residential, urban, and industrial areas, have been considered.

6. Control Methods

6.1. Incinerability Index. Controlled, high-temperature incineration, in spite of the associated high costs, is a viable organic waste reduction technology. The current performance requirement in the U.S. states that the principal organic hazardous constituents designated in each waste must be destroyed and/or removed to an efficiency of 99.99%. Organic hazardous compounds have been ranked by their heat of combustion by the U.S. Environmental Protection Agency in Appendix VIII of 40 CFR Part 261.3. This scale is based on the premise that the lower the heat of combustion, the more difficult the compound is to incinerate. Results of laboratory- and full-scale studies have indicated that this ranking is not consistent with the gas-phase thermal stability of numerous organic compounds.

It has been proposed that the temperature for 99% destruction at 2.0 sec. residence time is a viable method to determine the relative stability of organic compounds. Although destruction efficiencies are dependent upon residence time and temperature, relative destruction efficiencies are insensitive to these parameters. Consequently, it is believed that gas-phase thermal stability under oxygen-starved reaction conditions may be an effective predictor of relative destruction efficiency for organic compounds.

The thermal stability ranking of hazardous organic compounds is based on the temperature for 99% destruction at 2.0 sec. residence time under oxygen-starved reaction conditions. The ranking extends from 1 (highest thermal stability) for hydrogen cyanide to 320 (lowest thermal stability) for endosulfan (2390).

C. WATER POLLUTION FACTORS

1. Biodegradation

1.1. Objectives. The major use of data on biodegradation is for assessing the persistence of a chemical substance in a natural environment. The natural environment, for the purpose of such tests, is natural waters and various soils (including hydrosoils). If the compound does not persist, information is needed on whether the compound degrades to innocuous molecules or to relatively persistent and toxic intermediates. Secondary concerns for environmental persistence are the possibilities that (1) toxic substances may interfere with the normal operation of biological waste treatment units and that (2) toxic substances not substantially degraded within a treatment plant may be released to the natural environment.

The test procedures for biodegradation give an estimate of the relative importance of biodegradability as a persistence factor. The data are used to evaluate biodegradation rates in comparison with standard reference compounds. Most tests provide opportunities for biodegradation with relatively dense microbial populations that have been allowed to adapt to the test compound. Those compounds that degrade rapidly (in comparison with reference compounds) or extensively (as judged by such evidence as CO_2 evolution and loss of dissolved organic carbon) are likely to biodegrade rapidly in a variety of environmental situations. Even so, such compounds may persist in specialized environments and under circumstances that are poorly represented by these preliminary tests. Compounds that produce little indication of biodegradation in these tests may be relatively persistent in a wide variety of environments.

Reliable conclusions about biodegradation are not generally possible on the basis of structure alone. Biodegradation is the most important degradative mechanism for organic compounds in nature, in terms of mass of material transformed and extent of degradation. Therefore, information on biodegradability is very important in any evaluation of persistence and is generally needed on organic compounds that can be solubilized or dispersed in or on water. Highly insoluble compounds are not testable at present without the use of radioisotopes or complex analytical measurements, nor are there methods to study biodegradation with very low substrate concentrations.

These types of tests do not differentiate all chemical compounds as relatively nonbiodegradable or rapidly and extensively biodegradable. Results for many materials lead to intermediate conclusions. A more thorough understanding of the biodegradability of such compounds would result from advanced tests, such as those employing radiolabeled substrates.

Anaerobic microbial degradation of organic compounds is an important mechanism for degrading waste materials both in the natural environment and in waste treatment plants. However, there are few relatively simple state-of-the-art methods at present for evaluating the potential for anaerobic biodegradation. The types of methods most frequently cited employ microcosms such as flooded soils in flasks and require the use of radiolabeled substrate. Methane from fermentation of organic substrates is the end of a food chain that involves a wide variety of anaerobic bacteria.

The anaerobic digestion test compares the production of methane and CO_2 by anaerobic bacteria in sludge samples with and without added test material.

A desirable goal of degradation testing is to obtain some estimate of the rate at which a compound will degrade in the environment. It is relatively easy to estimate reaction rates for such degradation processes as hydrolysis, photolysis, and free-radical oxidations. Environmentally realistic reaction rates for biodegradation are much more difficult to obtain. Among the more important environmental variables that can affect the rate and the extent of biodegradation are (1) temperature, (2) pH, (3) salinity, (4) dissolved oxygen, (5) concentration of test substance, (6) concentration of viable microorganisms, (7) quantity and quality of nutrients (other than test substances), trace metals, and vitamins, (8) time, and (9) microbial species.

1.2. The Determination of Biodegradability. Biodegradability tests may be divided into two groups of tests: die-away tests in static systems and tests in flow-through systems (continuous cultures). In die-away tests, the concentration of the substance under investigation (the substrate) is determined analytically as a function of time. During the experiment, the substrate is contained in a fixed amount of test medium. In flow-through systems, a constant flow of the medium is fed into a completely mixed "reactor" in which the medium volume is also kept constant. The degradation is calculated from the difference between the substrate concentrations in the inlet and effluent streams.

1.2.1 READY BIODEGRADABILITY TESTS ACCORDING TO OECD GUIDELINES

<u>OECD 301 Tests</u>

In the Guideline adopted 17th july 1992 six methods are described that permit the screening of chemicals for ready biodegradability in an aerobic aqueous medium. They are:

301 A: DOC Die-Away
301 B: CO_2 Evolution (Modified Sturm Test)
301 C: MITI (I) (Ministry of International Trade and Industry, Japan)
301 D: Closed Bottle
301 E: Modified OECD Screening
301 F: Manometric Respirometry

In general, degradation is followed by the determination of parameters such as DOC, CO_2 production and oxygen uptake and measurements are taken at sufficient intervals to allow the identification of the beginning and end of biodegradation. Specific chemical analysis can also be used to assess primary degradation of the test substance and to determine the concentration of any intermediate substance formed. It is obligatory in the MITI method (301 C).

Normally, the test lasts for 28 days. Tests however may be ended before 28 days, i.e. as soon as the biodegradation curve has reached a plateau for at least three determinations. Tests may also be prolonged beyond 28 days when the curve shows that the biodegradation has started but that the plateau has not been reached by day 28, but in such cases the chemical would not be classed as readily biodegradable.

In order to select the most appropriate method, information on the chemical's solubility, vapor pressure and adsorption characteristics is essential. Test substances which are soluble in water to at least 100 mg/L may be assessed by all methods, provided they are non-volatile and non-adsorbing. For those chemicals which are poorly soluble in water, volatile or adsorbing, suitable methods are indicated in table 1.

Tabel 1 : Applicability of test methods

Test	Analytical method	Suitable for compounds which are:		
		Poorly soluble	Volatile	adsorbing
301 A: DOC Die-Away	DOC	-	-	+/-
301 B: CO2 Evolution	Respirometry : CO2 evolution	+	-	+
301 C: MITI (I)	Respirometry : oxygen consumption	+	+/-	+
301 D: Closed Bottle	Respirometry : dissolved oxygen	+/-	+	+
301 E: OECD Screening	DOC	-	-	+/-
301 F: Manometric Respirometry	Oxygen consumption			

The pass levels for ready biodegradability are 70% removal of DOC and 60% of ThOD or $ThCO_2$ production for respirometric methods. They are lower in the respirometric methods since, as some of the carbon from the test chemical is incorporated into new cells, the percentage of CO_2 produced is lower than the percentage of carbon being used. These pass values have to be reaced in a 10-d window within the 28-d period of the test, except where mentioned below. The 10-d window begins when the degree of biodegradation has reached 10% DOC, ThOD or $ThCO_2$ and must end before day 28 of the test. The 10-d window concept does not apply to the MITI method.

Principle of the Tests of the OECD 301 tests

301 A: DOC Die-Away :

A measured volume of inoculated mineral medium, containing a known concentration of the test substance (10-40 mg DOC/L) as the nominal source of organic carbon, is aerated in the dark or diffuse light at 22 ± 2°C. Degradation is followed by DOC analysis at frequent intervals over a 28-day period. The degree of biodegradation is calculated by expressing the concentration of DOC removed (corrected for that in the blank inoculum control) as a percentage of the concentration initially present. Primary biodegradation may also be calculated from supplemental chemical analysis for parent compound made at the beginning and end of incubation.

301 B: CO_2 Evolution (Modified Sturm Test)

A measured volume of inoculated mineral medium, containing a known concentration of the test substance (10-20 mg DOC/L or TOC/L) as the nominal source of organic carbon, is aerated by the passage of carbon dioxide-free air at a controlled rate in the dark or in diffuse light. Degradation is followed over 28 days by determining the carbon dioxide produced. The CO_2 is trapped in barium or sodium hydroxide and is measured by titration of the residual hydroxide or as inorganic carbon. The amount of carbon dioxide produced from the test substance (corrected for that derived from the blank inoculum) is expressed as a percentage of $ThCO_2$. The degree of biodegradation may also be calculated from supplemental DOC analysis made at the beginning and end of incubation.

301 C: MITI (I) (Ministry of International Trade and Industry, Japan)

The oxygen uptake by a stirred solution, or suspension, of the test substance in a mineral medium, inoculated with specially grown, unadapted micro-organisms, is measured automatically over a period of 28 days in a darkened, enclosed respirometer at 25 ± 1 °C. Evolved carbon dioxide is absorbed by soda lime. Biodegradation is expressed as the percentage oxygen uptake (corrected for blank uptake) of the theoretical uptake (ThOD). The percentage primary biodegradation is also calculated from supplemental specific chemical analysis made at the beginning and end of incubation, and optionally ultimate biodegradation by DOC analysis.

301 D: Closed Bottle Test

The solution of the test substance in mineral medium, usually at 2-5 mg/L, is inoculated with a relatively small number of micro-organisms from a mixed population and kept in completely full, closed bottles in the dark at constant temperature. Degradation is followed by analysis of dissolved oxygen over a 28-d period. The amount of oxygen taken up by the microbial population during biodegradation of the test substance, corrected for uptake by the blank inoculum run in parallel, is expressed as a percentage of ThOD or, less satisfactorily COD.

301 E: Modified OECD Screening

A measured volume of mineral medium, containing a known concentration of the test substance (10-40 mg DOC/L) as the nominal sole source of organic carbon, is inoculated with 0.5 ml effluent per litre of medium. The mixture is aerated in the dark or diffused light at 22±2°C. Degradation is followed by DOC analysis at frequent intervals over a 28 day period. The degree of biodegradation is calculated by expressing the concentration of DOC removed (corrected for that in the blank inoculum control) as a percentage of the concentration initially present. Primary biodegradation may also be calculated from supplemental chemical analysis for the parent compound at the beginning and end of incubation.

301 F: Manometric Respirometry

A measured volume of mineral medium, containing a known concentration of the test substance (100 mg test substance/L giving at least 50-100 mg ThOD/L) as the nominal sole source of organic carbon, is stirred in a closed flask at a constant temperature (±1°C or closer) for up to 28 days. The consumption of oxygen is determined either by measuring the quantity of oxygen (produced electrolytically) required to maintain constant gas volume in the respirometer flask, or from the change in volume or pressure (or a combination of the two) in the apparatus. Evolved carbon dioxide is absorbed in a solution of potassium hydroxide or another suitable absorbent. The amount of oxygen taken up by the microbial population during biodegradation of the test substance (corrected for uptake by blank inoculum, run in parallel) is expressed as a percentage of ThOD or, less satisfactorily, COD. Optionally, primary biodegradation may also be calculated from supplemental specific chemical analysis made at the beginning and end of incubation, and ultimate biodegradation by DOC analysis.

1.2.2. INHERENT BIODEGRADABILITY TESTS ACCORDING TO OECD GUIDELINES

OECD 302 Tests

Inherently Biodegradable is a classification of chemicals for which there is unequivocal evidence of biodegradation (primary or ultimate) in any test of biodegradability. Four methods are described that permit the testing of chemicals for inherent biodegradability in an aerobic aqueous medium. They are:

302 A: Modified SCAS Test
302 B: Zahn-Wellens/EMPA Test
302 C: Modified MITI Test (II)
302 D: Concawe Test

Principle of the OECD 302 tests

302 A: Modified SCAS Test

Activated sludge from a sewage treatment plant is placed in an aeration (SCAS) unit. The test compound and settled domestic sewage are added, and the mixture is aerated for 23 hours. The aeration is the stopped, the sludge allowed to settle and the supernatant liquor is removed. The sludge remaining in the aeration chambers is then mixed with a further aliquot of test compound and sewage and the cycle is repeated.

Biodegradation is established by the determination of the dissolved organic carbon content of the supernatant liquor. This value is compared with that found for the liquor obtained from a control tube dosed with settled sewage only.

302 B: Zahn-Wellens/EMPA Test

A mixture containing the test substance, mineral nutrients and a relatively large amount of activated sludge in aqueous medium is agitated and aerated at 20-25°C in the dark or in diffuse light for up to 28 days. Blank controls, containing activated sludge and mineral nutrients but no test substance, are run in parallel. The biodegradation process is monitored by determination of DOC (or COD) in filtered samples taken at daily or other time intervals. The ratio of eliminated DOC (or COD), corrected for the blank, after each time interval, to the initial DOC value is expressed as the percentage biodegradation at the sampling time. The percentage biodegradation is plotted against time to give the biodegradation curve.
Specific analysis of the test substance may be useful in cases where molecular changes, caused by biochemical reactions (primary biodegradation) are to be detected.

302 C: Modified MITI Test (II)

This test method is based on the following conditions:
– test chemicals as sole organic carbon sources
– no adaptation of micro-organisms to test chemicals
An automated closed-system oxygen consumption measuring apparatus (BOD-meter) is used. Chemicals to be tested are inoculated in the testing vessels with micro-organisms. During the test period, the biochemical oxygen demand is measured continuously by means of a BOD meter.

Biodegradability is calculated on the basis of BOD and supplemental chemical analysis, such as measurement of the dissolved organic carbon concentration, concentration of residual chemicals, etc.

302 D: Concawe Test

This method is based on International Organization for Standardization (ISO) 14593: "Carbon dioxide (CO2) Headspace Biodegradation Test" and provides a test for assessing the inherent aerobic biodegradability of organic substances. It is particularly useful for testing insoluble and/or volatile materials such as mineral oils.

The test substance is incubated in a buffered, mineral salts medium which has been inoculated with a mixed population of micro-organisms. In order to enhance the biodegradative potential of the inoculum, it is pre-exposed to the test substance using a regime based on methods described in US Environmental Protection Agency Test Guideline § 796.3100. The test is performed in sealed bottles with a headspace of air that provides a reservoir of oxygen (O2) for aerobic biodegradation. CO2 evolution from the ultimate aerobic biodegradation of the test substance is determined by measuring the inorganic carbon (IC) produced in the test bottles over that produced in
blanks which contain inoculated medium only. The ultimate aerobic biodegradation is the breakdown of an organic chemical by micro-organisms in the presence of O2, resulting in the production of CO2, water, mineral salts (i.e. mineralisation) and microbial cellular constituents (biomass). The extent of biodegradation is then expressed as a percentage of the theoretical maximum IC production (ThIC), based on the quantity of test substance (as total organic carbon) added initially. ThIC is analogous to the term ThCO2 used in OECD 301 B: CO2 evolution (modified Sturm) test.
Dissolved organic carbon (DOC) removal (water-soluble substances only) and/or the extent of primary biodegradation of the test substance can also be measured. Primary biodegradation is the structural change (transformation) of an organic chemical by micro-organisms resulting in the loss of a specific property.

1.2.3. PRINCIPLE OF THE TEST METHOD. A measured volume of inoculated medium containing a known amount of test compound as the sole source of carbon is stirred in a closed flask at a constant temperature (±0.5°C) in the range 20-25°C in the dark or in diffuse light. The microorganisms used for inoculation are not preacclimated. Activated sludge is collected from a biological waste water treatment plant or a laboratory unit receiving solely or predominantly domestic sewage.

The consumption of oxygen is determined either by measuring the quantity of oxygen required to maintain constant gas volume in the respirometer flask or from the change in volume or pressure (or a combination of the two) in the apparatus. Evolved CO_2 is absorbed in a solution of potassium hydroxide or another suitable absorbent. The degradation is followed over a 28-day period.

Table 24. Ready Biodegradability Tests, Revised in 1990. (The concentration of microorganisms in the third column corresponds to the maximum amount of effluent or activated sludge that is allowed to be used as inoculum in the test. These values are derived from the literature.)

RBT	Summary Parameter	Population Density, CFU/ml
Modified OECD Screening Test	DOC	$(0.5–2.5) \cdot 10^2$
CO_2 Evolution	CO_2	$(2–10) \cdot 10^5$
^{EEC}manometric respirometry	O_2	$(2–10) \cdot 10^5$
DOC Die-Away	DOC	$(2–10) \cdot 10^5$
Closed Bottle	O_2	$(0.5–2.5) \cdot 10^3$

MITI (I)	O_2	(2–1N) • 10^5
		(2909)

The amount of oxygen taken up by the test compound (corrected for the blank) is expressed as a percentage of the COD or the theoretical oxygen demand calculated from the formula of the compound.

The degree of biodegradability may also be calculated from supplemental chemical analyses such as dissolved organic carbon (DOC), or from specific analysis, made at the beginning and end of incubation (2790).

It may be concluded that 50% or more ThOD for a soluble substance indicates that it is readily biodegradable, provided that the % DOC removed is over 70%. The OECD recommendation of 60% or more ThOD is valid for insoluble substances.

1.2.4. MEASUREMENT OF DEGRADATION RATE. In general, the abiotic and biotic degradation of chemicals in water seems to follow a logarithmic correlation. Therefore, the degradation rate constant *(K)* is determined from the assumed first-order process in which

$$C_t = C_0 e^{-kt}$$

Accordingly,

$$\ln C_t = \ln C_0 - kt$$

where C_t is the concentration of the pesticide in culture medium after time t (days) and C_0 is the initial concentration of pesticide. The half-lives were determined from the expression

$$\frac{C_0}{C_t} = 2 = e^{kt}$$

or

$$\ln 2 = kt$$

$$t_{0.5} \text{ (days)} = 0.693/k$$

The real biodegradation rate constant (K_b) is calculated by the following equation:

$$K_b = K \text{ (metabolism or cometabolism)} - K \text{ (control)}$$

1.2.5. MINERALIZATION OF ACETATE IN METHANOGENIC RIVER SEDIMENT. In industrialized regions, sediments receive a high input of organic matter. High numbers of bacteria are present, and oxygen supplied by the overlying water will be rapidly consumed in the top layer of the sediment. The underlying part of the sediment is anoxic, and there biodegradation of natural as well as anthropogenic organic substrates is dependent on anaerobic processes. Anaerobic mineralization of organic matter by microorganisms is an essential component of the carbon cycle. Acetate is a key intermediate in the anaerobic degradation of organic matter and a major component of anaerobic waste waters.

Organic polymers like proteins, carbohydrates, and lipids are hydrolyzed to amino acids, sugars, and fatty acids. These are converted by acidogenic bacteria to produce hydrogen, acetate, propionate, butyrate, lactate, and alcohols. Acetogenic bacteria convert these products to acetate. About 70% of the methane formation in anaerobic freshwater sediments is derived from acetate. The following reaction is performed:

$$CH_3COOH \longrightarrow CH_4 + CO_2$$

The mineralization of acetate under methanogenic conditions is performed by a few specialized methanogenic bacteria, such as *Methanosarcina* sp., because most methanogens are not able to split acetate into methane and carbon dioxide. This is in contrast with the aerobic mineralization of acetate that a multitude of species can perform.

The functioning of anaerobic bacteria in river sediments is vital for the stability of freshwater ecosystems. The sensitivity of these organisms to pollutants is therefore important for the establishment of valid sediment quality criteria. Acetate is a key intermediate in the carbon cycle and was chosen as a model substrate for the activity of anaerobic bacteria. The effect of pollutants on the anaerobic mineralization of acetate is studied in sediment microcosms. A small amount of ^{14}C-labeled acetate is added to bottles with fresh anaerobic river sediment. The acetate is converted to methane and $^{14}CO_2$, with a half-life of 0.2-0.5 h. Adding a toxicant decreases the mineralization rate of acetate. The IC_{10} is defined as the toxicant concentration that decreases the mineralization rate of acetate with 10% (2693).

1.2.6. EFFECTS OF SURFACTANTS ON WATER SOLUBILITY. Aqueous surfactant solutions have the ability to solubilize compounds that are otherwise relatively water insoluble. This phenomenon is the basis for the use of surfactants as solubilizing agents for enhancing biological remediation technologies, pump-and-treat operations, and soil-washing operations. In an experimental study (2361), the linear relationship shown in Figure 19 was found between the octanol-water partition coefficient (K_{ow}) and the micelle-water partition coefficient (Km) for a number of aromatic hydrocarbons, using the anionic surfactant dodecylsulfate.

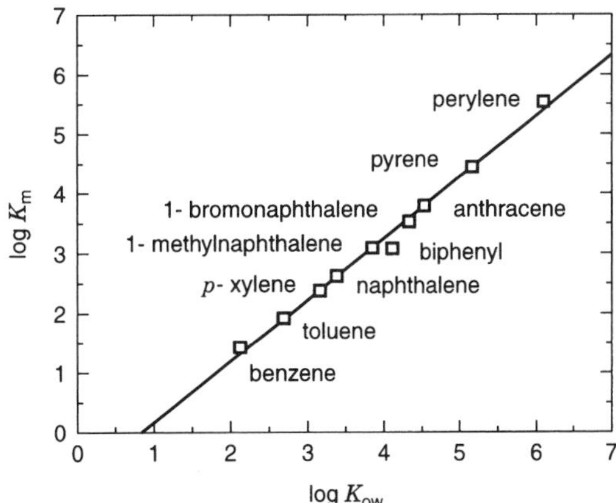

Figure 19. Relationship between micelle-water partition coefficient (K_m) and octanol-water partition coefficient (K_{ow}).

2. Oxidation Parameters

The conventional oxidation parameters, such as biological oxygen demand (BOD), chemical oxygen demand (COD) using potassium dichromate, permanganate value ($KMnO_4$), total organic carbon (TOC), total oxygen demand (TOD), and theoretical oxygen demand (ThOD) are mentioned here. The oxidation parameters are dimensionless (grams oxygen consumption per gram of product), unless stated otherwise. BOD values are assumed to be measured at 20°C, unless indicated otherwise.

2.1. Biochemical Oxygen Demand (BOD). When water containing organic matter is discharged into a river, lake, or sea, natural purification by biological action takes place. Thus biochemical oxidation is brought about by naturally occurring microorganisms that use the organic matter as a source of carbon. Dissolved oxygen in the water sustains respiration. Naturally, this is a simplified picture of a very complex set of reactions, the rates of which depend on the temperature, the type of organic matter present, the type of microorganisms, the aeration, and the amount of light available. A number of years ago an attempt was made to produce a test that would match the rate of biochemical oxidation that would occur in a river into which organic-containing water was discharged. The 5-day test has been generally adopted with the knowledge that this does not necessarily represent the time required for total oxidation of the organic matter present. In some cases, a test period of longer than 5 days is specified. It follows, therefore, that the BOD_5 test should always be considered in conjunction with other data and with a knowledge of the system being studied. Application of the test to organic waste discharges allows calculation of the effect of the discharges on the oxygen resources of the receiving water. Data from BOD tests are used for the development of engineering criteria in the design of waste water treatment plants.

The BOD test is an empirical bioassay procedure that measures the dissolved oxygen that microbial life consumes while assimilating and oxidizing the organic matter present. The sample of waste, or an appropriate dilution, is incubated for 5 days at 20°C in the dark. The reduction in dissolved oxygen concentration during the incubation period yields a measure of the biochemical oxygen demand. The standard dilution method prescribes the use of several dilutions, because faulty results can be obtained because of the toxicity of the sample (yielding a low BOD) or by depletion of the oxygen by too high a concentration of biodegradable organics. The bottles that after 5 days still contain approximately 50% of the original oxygen content are selected for the BOD calculation. In 53 laboratories, 77 analysts analyzed natural water samples plus an exact increment of biodegradable organic compounds. At a mean value of 194 mg/l BOD, the standard deviation was ±40 mg/l. There is no acceptable procedure for determining the accuracy of the BOD test.

BOD values sometimes show a wide range of variation. The main reason is that microorganisms have a tremendous capacity for adaptation. Whenever the samples are inoculated with microorganisms from a very polluted river, we may suppose that these microorganisms are already partly adapted to the specific organics. When an inoculum is taken from a polluted river or from an industrial waste water treatment plant, this should be mentioned. The foregoing is illustrated by Figure 20.

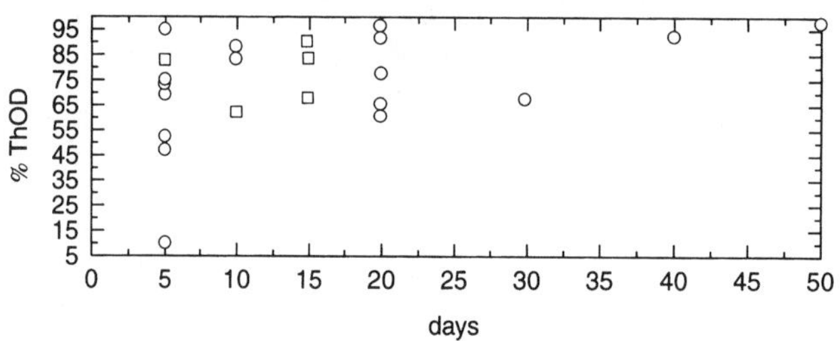

Figure 20. The variation in BOD values.

2.2. Chemical Oxygen Demand (COD). Because the BOD method takes 5 days to carry out, other methods have been sought for measuring oxygen requirements of a sample. The methods should be simple, quick, and reliable. One popular method is the chemical oxygen demand test based on the oxygen consumed from boiling acid potassium dichromate solution. This is a severe test, and a high degree of oxidation takes place. Interference from chloride ions can be a problem. Several procedures available to overcome this problem are based mainly on the complexion of the chloride ion with mercuric sulfate. This procedure is recommended in many standard methods. It has been demonstrated that high COD results are achieved when chloride ions are present and that true compensation is not made. Finally, it should be noted that the COD test is not a measure of organic carbon, although the same reactions are involved.

In this procedure, organic substances are oxidized by potassium dichromate in 50% sulfuric acid solution at reflux temperature. Silver sulfate is used as a catalyst, and, as mentioned, mercuric sulfate is added to remove chloride interference. The excess dichromate is titrated with standard ferrous ammonium sulfate, using orthophenanthroline ferrous complex as an indicator. In 58 laboratories, 89 analysts analyzed a distilled water solution containing oxidizable organic material equivalent to 270 mg/l COD. The standard deviation was ±27.5 mg/l COD, and the mean recovery was 96% of the true value.

Ideally, to measure the ultimate oxygen demand, the complete combustion of all the oxidizable elements in the sample is desired. Thus it is desired that the following reactions take place, because the end products listed represent the highest stable oxidation states of these elements (or ions) in nature.

$$C_nH_{2m} + (n + \frac{m}{2}) \quad O_2 + \longrightarrow {}_nCO_2 + {}_mH_2O$$

$$2H_2 + O_2 \longrightarrow 2H_2O$$

$$2N^{3-} + 3O_2 \longrightarrow 2NO^{3-}$$

$$S^{2-} + 2O_2 \longrightarrow SO_4^{\ 2-}$$

$$SO_3^{\ 2-} + \frac{1}{2}O_2 \longrightarrow SO_4^{\ 2-}$$

$$N_2 + O_2 \longrightarrow \text{no reaction}$$

When interpreting TOD values, keep in mind that not only hydrocarbons are oxidized but all oxidizable matter, as demonstrated by the foregoing reactions.

2.3. Total Organic Carbon (TOC). The instrument that measures the TOC of polluted water is very similar to the TOD analyzer, except for the last step. A microsample of the waste water to be analyzed is injected into a catalytic combustion tube that is enclosed by an electric furnace thermostated at 950°C. The water is vaporized, and the carbonaceous material is oxidized to carbon dioxide and steam. The carrier gas CO_2, O_2, and water vapor enter an infrared analyzer sensitized to provide a measure of CO_2. The amount of CO_2 present is directly proportional to the concentration of carbonaceous material in the injected sample.

2.4. Theoretical Oxygen Demand (ThOD). The theoretical oxygen demand is the amount of oxygen needed to oxidize hydrocarbons to carbon dioxide and water.

$$C_nH_{2m} + \left(n + \frac{m}{2}\right) O_2 \longrightarrow {}_nCO_2 + {}_mH_2O$$

The ThOD of C_nH_2m equals

$$\frac{32\left(n + \frac{m}{2}\right)g}{(12n + 2m)g} = \frac{8(2n + m)}{6n + m}$$

When the organic molecule contains other elements, such as N, S, P, etc., the ThOD depends on the final oxidation state of these elements. Most authors do not bother to define the ThOD they mention in their publications; however, ThOD can easily be calculated.

Example: Ammonium acetate: CH_3COONH_4, mw = 77

(a) If N remains as NH_4^+, then $CH_3COONH_4 + 2O_2 \longrightarrow 2CO_2 + H_2O + NH^{4+} + OH^-$

or

$$\text{ThOD} = \frac{64\text{ g}}{77\text{ g}} = 0.83$$

(b) If N is oxidized to N_2, then $2CH_3COONH_4 + 5.5O_2 \longrightarrow 4CO_2 + 7H_2O + N_2$
or

$$ThOD = \frac{5.5 \times 32}{2 \times 77} = 1.14$$

(c) If N is oxidized to NO_3^-, then $2CH_3COONH_4 + 7.5O_2 \longrightarrow 4CO_2 + 5H_2O + 2NO_3^- + 2H^+$
or

$$ThOD = \frac{7.5 \times 32}{2 \times 77} = 1.56$$

The ThOD of the substance $C_cH_hCl_{cl}N_nNa_{na}O_oP_pS_s$ of molecular weight mw can be calculated according to

$$ThOD_{NH_3} = \frac{16[2c + 1/2(h - cl - 3n) + 3s + 5/2p + 1/2\ na - o]}{mw}$$

This calculation implies that *C* is mineralized to CO_2, H to H_2O, P to P_2O_5, and Na to Na_2O. Halogen is eliminated as hydrogen halide and nitrogen as ammonia.
Example: Glucose $C_6H_{12}O_6$, mw = 180

$$ThOD = \frac{16[2 \times 6 + 1/2 \times 12 - 6]}{180} = 1.07\ mg\ O_2/mg\ glucose$$

Molecular weights of salts other than those of the alkali metals are calculated on the assumption that the salts have been hydrolyzed.
Sulfur is assumed to be oxidized to the state of 6^+.

Example: Sodium n-alkylbenzenesulfonate, $C_{18}H_{29}SO_3Na$, mw = 348

$$ThOD = \frac{16[36 + 29/2 + 3 + 1/2 - 3]}{348} = 2.34\ mg\ O_2/mg\ substance$$

In case of a nitrogen-containing substance, the nitrogen may be eliminated as ammonia, nitrite, or nitrate corresponding to different theoretical biochemical oxygen demands.

$$ThOD_{NO^{2-}} = \frac{16[2c + 1/2(h - cl) + 3s + 3/2n + 5/2p + 1/2na - o]}{m\ w}$$

$$ThOD_{NO^{3-}} = \frac{16[2c + 1/2(h - cl) + 3s + 5/2n + 5/2p + 1/2na - o]}{m\ w}$$

Suppose full nitrate formation was observed by analysis in the case of a secondary amine: $(C_{12}H_{25})_2NH$; mw = 353

$$ThOD_{NO^{3-}} = \frac{16[48 + 51/2 + 5/2]}{353} = 3.44\ mg\ O_2/mg\ substance$$

Expressing other parameters, such as BOD and COD, as a percentage of ThOD can be highly misleading. In the standard BOD_5 test, nitrification does not occur yet, and consequently the BOD_5 values of compounds containing nitrogen can be expressed as a percentage of the ThOD, under the assumption that the nitrogen remains unchanged. In determining BOD with acclimated seed tested over a longer period, nitrification may occur. This alters the oxygen demand and also the ThOD, because N is oxidized to NO_3^{-2}. When using ThOD values, one should mention the final oxidation state of the elements other than C, H, and O.

3. Impact on Biodegradation Processes
Certain compounds are very toxic to microorganisms used in waste water treatment processes and in the BOD test, and even at low concentrations, they inhibit the biodegradation processes.

Microbial activities are associated with three major biogeochemical cycles (carbon, nitrogen, and sulfur cycling) and with the decomposition of organic matter. Data obtained will provide preliminary indications of possible effects of the test chemical on the cycling of elements and nutrients in ecosystems. In addition, the test will aid in the formation of hypotheses about the ecological effects of chemicals, hypotheses that can be used in the selection of higher-level tests when appropriate.

Data on the effects of chemicals on microbial populations and functions can be obtained from laboratory studies employing nonradioisotopic analytical techniques. Studies of effects on microbial functions constitute a more direct approach and are preferred to studies of effects on populations. The activities to be observed are

- organic matter (cellulose) decomposition, by following CO_2 evolved from organically bound carbon
- nitrogen transformations, by following the release of organically bound nitrogen (in urea) and the formation of ammonia
- sulfur transformation, by following the reduction of sulfate to sulfide by *Desulfovibrio*

3.1. Nitrogen Transformation. Almost all microorganisms, higher plants, and animals require combined nitrogen. In addition, the nitrogen cycle is a major biogeochemical cycle. The main aspects of this cycle are fixation of gaseous nitrogen, ammonification of organically bound nitrogen, nitrification, and denitrification.

Ammonification is a key initial step in the reintroduction of nitrogen from protein wastes into the soil and is one of the more readily measured reactions of the nitrogen cycle. As soon as an organism dies and its organic waste returns to the soil, biological decomposition begins and fixed nitrogen is released. The breakdown of proteins and other nitrogen-containing organics in soil and the production of ammonia are the work of widespread and varied microflora. The amino groups are split off to form ammonia in a series of reactions collectively called ammonification. Urea, a waste product found in urine, is also decomposed by numerous microorganisms with the formation of ammonia. This reaction can serve as a convenient assay method for ammonification activity. There is a strong correlation between an organism's ability to degrade urea and its capacity to degrade protein. Information from such testing would be used to assess the likelihood that the test chemical interferes with the normal conversion of organically bound nitrogen into ammonia.

Consideration of at least some aspects of the nitrogen cycle is essential for assessing the effects of substances on microorganisms. An easily measured feature of the nitrogen cycle is the oxidation of nitrite to nitrate by *Nitrobacter bacteria.* This focuses on a part of the nitrogen cycle that is less important than the critical step involving the conversion of organic nitrogen into ammonia. The method uses urea as a readily obtainable, reproducible nitrogenous organic compound. Some investigators have used pieces of liver or kidney tissue, vegetable meal, dried blood, and casein hydrolysate as nitrogen sources, but these substances are not standardizable. Percolation techniques may also be used for the same purposes in the study of soluble proteins, peptones, polypeptides, and amino acids in solution. Methodology using fertile soil and urea and following the evolution of ammonia nitrogen is relatively simple to conduct and will provide meaningful results. Urea is a suitable nitrogen source not only because it is readily available and relatively easy to handle, but also because the ability to degrade urea has been associated with general proteolytic capabilities.

Nitrification is one of the most sensitive conversions in the soil. Nitrification may be inhibited by concentrations of chemicals that do not inhibit other important biochemical reactions, so there are reasons for choosing nitrification as the process that may give the most useful results. One of the arguments against its use is that ammonification is a more vital part of the nitrogen cycle than nitrification. In addition, the temporary inhibition of nitrate formation is often beneficial in that it slows down the loss of nitrogen from the root areas of plants by leaching and denitrification. Finally, nitrification by *Nitrobacter* may be too sensitive to inhibition by chemicals in laboratory studies and may give results that are not representative of natural circumstances and environments. Although the inhibition of nitrifying activity could lead to ammonium ion accumulation and serious problems such as root damage, ammonification is considered the more appropriate process to examine as a first step in looking for effects of the nitrogen cycle.

4. Waste Water Treatment

4.1. Biological Oxidation. In order to describe the basic investigation procedures and results for a wide variety of biological test methods in a compact way, the information is presented in columns, using the following column headings: methods, feed mg/l, test duration, and % product removed or % theoretical oxidation demand (ThOD). The methods column describes experimental procedures and their sequence. For example, acclimation may have been achieved in activated sludge with assimilation or oxidation measured by respirometer, BOD, or other methods.

Biological treatment performance is difficult to evaluate without information on both percent oxidation and percent removed. A characteristic balance of time, oxidation, and biosorption is necessary for effective continuous treatment. Percent oxidation shows the degradation within the test period. It rarely exceeds 60% of the influent oxygen demand within the detention period of an activated sludge or other biological treatment unit. Percent *removal* may be 90% or more because biosorption retains a significant fraction of the remaining load. The retained material is subject to further oxidation in process, so it is not likely to degrade the effluent.

4.2. Stabilization Ponds. The principal design parameter is the first-order BOD removal coefficient, *K*. The evaluation of *K* is the key to the whole design process. The method described below has been proposed by Dhandapani Thirumurthi and is based on the following definitions:

1. Standard BOD removal coefficient K_s-a constant and standard value of K_s has been chosen that corresponds to an arbitrarily selected standard environment. Under these standard environmental conditions, a stabilization pond will perform with the BOD removal coefficient K_s.

   The standard environment consists of

- a pond temperature of 20°C
- an organic load of 60 lb/day/acre (672 kg/day/ha)
- absence of industrial toxic chemicals
- minimum (visible) solar energy at the rate of 100 langleys/day
- absence of benthal load

2. Design BOD removal coefficient, *K*-design coefficient *K* corresponds to the actual environment surrounding the pond. Hence the value of *K* will be used when a pond is being designed. When the

critical environmental conditions deviate from one or more of the defined standard environmental conditions, suitable correction factors must be incorporated:

$$K = K_s C_{Te} C_o C_{Tox}$$

where

C = correction factor

Te = correction for temperature

o = correction for organic load

Tox = correction for industrial toxic chemicals

In the absence of industrial wastes, the factor C_{Tox} will equal unity. Laboratory investigations indicated that in the presence of certain industrial organic compounds, green algae could not synthesize chlorophyll pigment. Without chlorophyll production, photosynthesis cannot be sustained by algae, so oxygen production stops. The resultant decrease in dissolved oxygen (DO) concentration in the pond will result in a drastic reduction in BOD removal efficiency. Based on this observation, C_{Tox} values have been calculated for various concentrations of selected organic chemicals. Ponds designed to treat toxic industrial wastes will perform at lower efficiencies, so C_{Tox} values must be determined by laboratory investigations or by previous field experience.

4.3. The Activity of Mutant Microorganisms. In the normal biological cycle, adaptation and mutation are constantly taking place, permitting the survival of the participating microorganisms. Many of the present-day organic toxicants with which we are contaminating our environment contain a large number of carbon-chlorine bonds in addition to other substitution groups that, because of their molecular configuration and complexity, do not permit the process of adaptation to proceed at a normal rate. In an effort to overcome this obvious biochemical disability, it was decided to obtain soil and marine samples from various parts of the world to isolate and study the naturally occurring microorganisms. The numerous isolates were tested to determine their ability to degrade low levels of various synthetic organic substrates such as aryl halides, aryl and alkyl amines, halogenated phenols, inorganic and organic cyanides, halogenated insecticides, halogenated herbicides, and various synthetic surfactants. These various organic chemicals were selected for biodegradation studies because they are high on the list of toxic recalcitrant molecules with which we are defiling our environment.

After the capability of the various microorganisms to degrade selected organic molecules had been determined, tests were run to determine the maximum toxicant concentration to which they would adapt. Increasing concentrations of the challenging organic molecules were added to the growing cultures over a period of 21 days to determine the maximum level that the biomass would tolerate. Upon completion of the adaptation process, the microorganisms were exposed to programmed radiation to develop mutants with advanced biochemical capabilities. From the several thousand mutants obtained, 397 were isolated that were outstanding in their ability to degrade various types of inorganic and organic compounds. These included 180 pseudomonas, 45 nocardia, 102 streptomyces, 15 flavobacterium, 12 mycobacterium, 14 aerobacters, 14 achromobacters, 10 vibrio, and 10 micrococcus. Activated sludges were prepared from the various mutant microorganisms, and the maximum tolerated dose of toxicant, as determined by previous tests, was added to determine the maximum velocity and degree of molecular dismutation. It has been fairly well established that none of the highly halogenated organic compounds are used by microorganisms as a source of metabolic carbon. Degradation or molecular change has been found to occur most readily at levels of maximum metabolic activity. It is quite apparent that the observed changes are enzymatic and relate more to detoxification than to the use of highly halogenated organic compounds in metabolic processes. It is possible that the biodegradation observed in microbial sludges is caused essentially by extracellular enzymes that would cause desorption of the more soluble degradation intermediates.

4.4. Solvent Extraction: Distribution Coefficient. Water soluble organic compounds can be extracted from water by solvents that are much less soluble in water. The ratio of the concentration of the compound in the solvent to the concentration of the compound in the water, after a sufficient contact time, is called the *distribution coefficient.*

4.5. Stripping. Air stripping has been demonstrated to be a feasible technique for removing a portion of the organics from waste water. These operations are typically carried out in a cooling tower-type device (induced or forced draft) or in an air-sparged (bubble) vessel. Air stripping occurs to varying degrees in conventional waste water treatment techniques such as dissolved or dispersed air flotation and of course aerobic biodegradation. An analytical simulation model of desorption in aerated stabilization basins (1811) indicates that significant removals of selected industrial chemicals are occurring.

A study of ten common industrial chemicals in eleven full-scale aerated basins showed that 20%-60% removal efficiencies were possible without biochemical oxidation. Detention times ranged from 1.7 to 14 days. Laboratory observations of surface agitator desorbers support these data.

Equations have been derived to predict the evaporation rates of hydrocarbons and chlorinated hydrocarbons. These compounds have high rates of evaporation even though the vapor pressure is low. Evaporation "half-lives" of minutes and hours are due to the relatively high constant of Henry *(H)* of these components in aqueous solution.

Thus it appears that significant amounts of volatile components are being stripped and/or desorbed in conventional secondary treatment operations involving the use of air or the presence of large air-water interfaces.

4.5.1. FUNDAMENTAL DESORPTION CONCEPTS. The volatile character of dissolved constituents in waste water can be adequately quantified by the experimental determination of two parameters:

1. *Volatile fraction (Fn).* This measure denotes the maximum amount (in %) of the original organic pollutants in a water sample that can be removed by air contact. This is also the maximum efficiency of treatment that can be achieved by stripping the waste water with air.

The organic pollutants in the waste water can be expressed as BOD, COD, TOC, and other gross pollutant measures and/or as concentration of individual constituents.

2. *Relative volatilization rate (K/a).* This measure denotes the ratio of the rate of volatile removal by air contact to the rate at which water is evaporated in the same apparatus. If the experimental value of *K/a* is greater than unity, stripping with air may be a feasible treatment operation for this waste water.

If *K/a* is unity, stripping will have no effect on removing volatile constituents from this waste water, and if *K/a* is less than unity or zero, stripping will result in an increase of this constituent in the waste water. This parameter, like F_n, is dimensionless, and both are determined from a single desorption experiment.

L.J. Thibodeaux (1812) used a desorption apparatus consisting of a packed column (Figure 21) and developed a mathematical model that, if the concentration of water is much greater than the concentration of the pollutant, yields the following solution:

$$\log X_1 = \log X_0 - (k/a\ -)\log \frac{M_0}{M_t}$$

where

X_t = concentration of pollutant in the desorption apparatus at sample time t (mg/l)

X_o = concentration of pollutant in the desorption apparatus at start of experiment (mg/l)
M_t = quantity of water in the desorption apparatus at sample time t (g)
M_o = quantity of water in the desorption apparatus at start of experiment (g)
K = the specific desorption rate of the organic component in the packed column
a = the specific desorption rate of the water component in the packed column

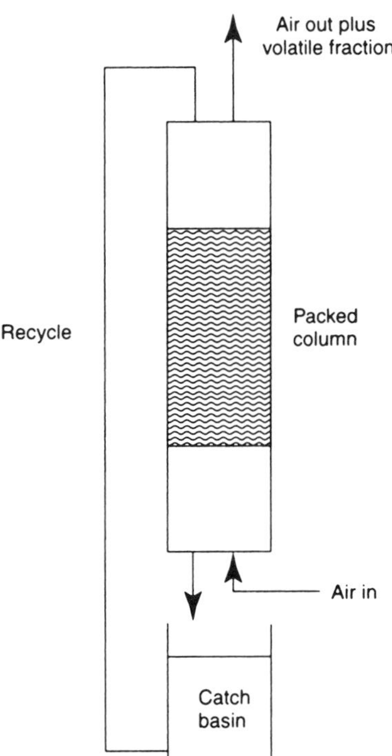

Figure 21. Volatile desorption apparatus.

Table 25. Experimental *K/a* Values Compared with Constants of Henry *(H)*.

	Experimental K/a	log H
methanol	7–13	–3.72
i-propanol	3–15	
n-propanol	12–23	–3.56
n-butanol	10–18	–3.46
formic acid	4–7	
acetic acid	1–5	–4.91
propionic acid	0–1.5	–4.74
acetaldehyde	171–207	

aceton	36–100	–3.0
benzene	107	–0.65
phenol	0.5–1.9	–4.79
furfural	10	
sucrose	0.26	

The relative volatilization rate *K/a* can be obtained graphically be preparing a log-log plot of X_t and M_o/M_t. The slope of a straight line through these raw data points yields (*K/a* – 1).

The relative volatilization rates shown in Table 25 were obtained for pure compounds at room temperature and initial concentrations of approximately 100-1000 mg/l.

From these experiments it appears that formic acid, acetic acid, propionic acid, and phenol have *K/a* values that are not significantly different from 1 and that therefore these compounds will not be removed significantly by air stripping. Acetaldehyde, acetone, and benzene show high relative volatilization rates and can thus be removed by stripping. Sucrose, on the contrary, concentrates in the solution upon stripping, as can be deduced from its *K/a* value of 0.26.

Experimental *K/a* values show a fair correlation with constants of Henry *(H)*, as illustrated by Figure 22.

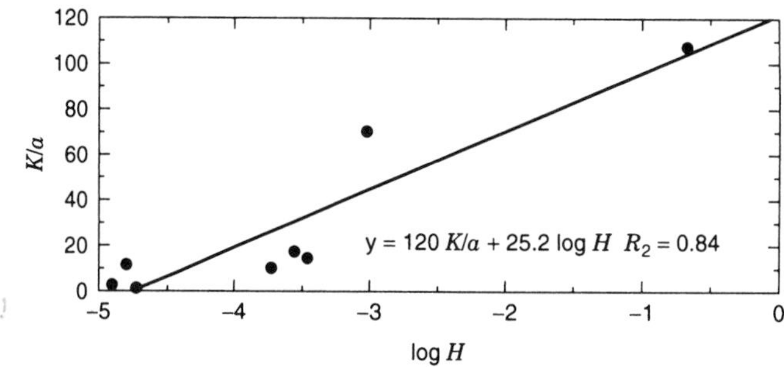

Figure 22. Relationship between experimental *K/a* values and constants of Henry *(H)* for the compounds mentioned in Table 25.

4.6. Adsorption. The affinity of a chemical substance for particulate surfaces is an important factor affecting its environmental movement and ultimate fate. Chemicals that adsorb tightly may be less subject to environmental transport in the gaseous phase or in solution. However, chemicals that adsorb tightly to soil particles may accumulate in that compartment. Substances that are not tightly adsorbed can transport through soils, aquatic systems, and the atmosphere.

The experimental information is usually expressed as an *adsorption isotherm.* The adsorption isotherm is the relationship, at a given temperature, between the amount of the substance adsorbed and its concentration in the surrounding solution.

In very dilute solutions, such as are encountered in the environment or in waste waters, a logarithmic isotherm plotting usually gives a straight line. In this connection, a useful formula is the Freundlich equation, which relates the amount of impurity in the solution to that adsorbed:

$$x/m = KC^{1/n}$$

where

x = amount of substance adsorbed

m = weight of the substrate

x/*m* = amount of substance adsorbed per unit weight of substrate

K, n = constants
C = unadsorbed concentration of substance left in solution or, in logarithmic form,

$$\log x/m = \log K + 1/n \log C$$

in which $1/n$ represents the slope of the straight-line isotherm (Figure 23). Isotherm tests also afford a convenient means of studying the effects of different adsorbents and the effects of pH and temperature.

Isotherm tests are also important in water purification by activated carbon. From an isotherm test it can be determined whether a particular purification can be effected. It will also show the approximate capacity of the carbon for the application and provide a rough estimate of the carbon dosage required.

The inherent adsorbability of a chemical in a pure-component test does not necessarily predict its degree of removal from a dynamic, multicomponent mixture. However, pure-component data do serve as a useful background for understanding why multicomponent interactions occur.

5. Alteration and Degradation Processes. The alteration and degradation processes can be divided into three categories:

1. *biodegradation*, effected by living organisms
2. *photochemical degradation*, i.e., nonmetabolic degradation requiring light energy
3. abiotic degradation *(chemical degradation)*, i.e., nonmetabolic degradation that does not require sunlight or living organisms.

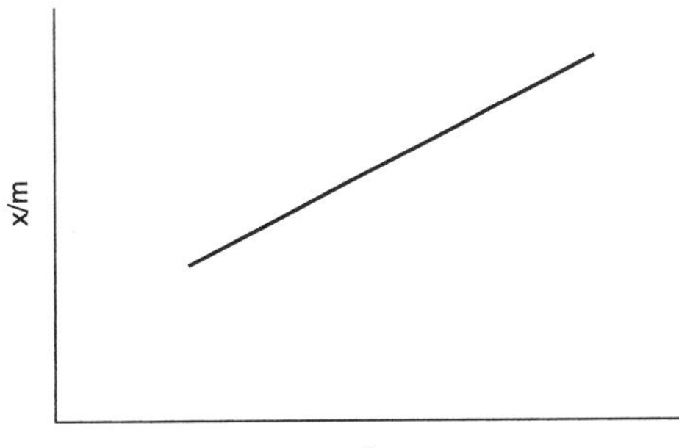

Figure 23. Freundlich isotherm.

Hydrolysis of environmental compounds has been extensively studied, and correlation between laboratory and field results is facilitated by the ease of measuring one of the more important rate-determining factors, pH, in both the laboratory and the field.

Chemical processes in soil can be studied after sterilization of the soil by autoclaving, chemical treatment, or *g*-irradiation. The sterilization processes often alter the soil to such an extent that any process observed could be artificial. Nevertheless, any biodegradation technique where the test chemical is incubated with a natural medium (such as soil or water) would allow transformations effected by chemical agents to take place; therefore, these processes would be considered during biodegradation testing.

5.1. Photochemical Degradation. Chemicals introduced into aqueous media in the environment can undergo transformation by direct photolysis in sunlight into new chemicals with different properties from those of their precursors. Data on direct photolysis rate constants and half-lives establish the importance of direct photolysis in sunlight as a dominant transformation of chemicals in aqueous media.

Although numerous papers have been published on the photolysis of chemicals in solution, rate constants for direct photolysis of chemicals in water under environmental conditions (i.e., in sunlight) have emerged only in the last few years. Zepp and Cline (1948) published a paper on photolysis in aquatic environments with equations for direct photolysis rates in sunlight. These equations translate readily obtained laboratory data into rate constants and half-lives for photolysis in sunlight. Rate constants and half-lives can be calculated as functions of season, latitude, time of day, depth in water bodies, and the ozone layer. Several published papers concerning the photolysis of chemicals in the presence of sunlight verify this method.

The soil and water media do not provide as good a matrix as the atmosphere for photochemical alterations because of the attenuation of the incident light. Nevertheless, photochemical processes on soil or vegetable surfaces and in water have been shown to be important with some compounds, so experimental procedures have been developed to study these processes. A number of different procedures can be used to study photochemical reactions. The selections that have to be made include

- light source and apparatus of irradiation
- media of reaction: solution (choice of solvent), solid, adsorbed on another solid, thin film
- effect of environmental parameters: add sensitizer, effect of oxygen, pH

The use of sunlight as the irradiation source has been a common experimental technique. However, the inherent variation of sunlight in wavelength distribution and in intensity results in poor reproducibility, inconvenience, and lack of experimental control, and therefore many researchers have resorted to artificial sources (such as medium- to low-pressure mercury lamps and fluorescent lamps).

Photochemical energy reaches the earth from the sun in the form of photons with wavelengths covering the spectrum from infrared to the far ultraviolet, including the visible. The principal energy sources of photons for environmental degradation reactions are those in the ultraviolet (UV) region. The UV radiation that reaches the earth's surface represents a wavelength range between 2,900 and 4,500 Ĺ. Shorter-wavelength UV, with high energy, is prevented from reaching the earth's surface by absorption in the upper atmosphere.

Polynuclear aromatics are prime candidates for photochemical degradation, because they have high absorptivities in the environmental UV range. Some typical data for molar absorptivities at principal bands are the following:

	Maxima, Ĺ	*Molar Absorptivity*	*Maxima, Ĺ*	*Molar Absorptivity*
anthracene	3,560	8.5×10^3	3,750	8.6×10^3
1,2-dibenzanthracene	2,875	9.0×10^4	3,410	7.1×10^3
chrysene	3,200	1.2×10^4	3,610	6.9×10^2

Monoaromatics have much lower molar absorptivities in this range, generally in the region of 10^2-10^3 liters per mole-cm. Consequently, polynuclear aromatics should more readily absorb energetic radiation to put the molecule into a higher energy state for oxidative and other environmental reactions.

Studies made on benzo(a)pyrene show it to be particularly sensitive to light. Under white fluorescent light, the photooxidation of benzo(a)pyrene absorbed onto calcite suspended in water followed a first-order rate. One would anticipate that in a natural aquatic system, photooxidation rates would be greatest in shallow systems or in the upper layers of deeper waters. A report of the Stanford Research Institute gives a photooxidation half-life for benzo(a)pyrene in a lake environment of 7.5 hours. One would expect photooxidation to be low in sediments, however, because of a lack of penetrating radiation.

5.2. Photodissociation of Nitrate Ions. The excitation of nitrate ions in aqueous solution induces the oxidation of organic substrates. The photodissociation of nitrate ions leads to formation of the hydroxyl radical and atomic oxygen according to the following processes:

$$NO_3^- + h\nu \longrightarrow \times NO_2 + O\times^-$$

$$O\times^- + H^+ \longrightarrow \times OH \text{ at pH} < 12$$

$$NO_3^- + h\nu \longrightarrow NO_2^- + O$$

These reactions can occur under daylight irradiation because the first maximum of the absorption band of NO_3^- is located at 302 nm. Hydroxyl radicals are more reactive than atomic oxygen both in gas phase and in aqueous solution, and the observed oxidations are generally attributed to hydroxyl radicals. Nitrate ions are often considered one of the main sources of hydroxyl radicals in natural waters.

Nitrate ions have some useful influence in the photodegradation of organic pollutants. However, nitrate ions can also induce the formation of nitroderivatives and of nitrosophenol and lead to the formation of mutagenic compounds (2675).

5.3. Hydrolysis. Chemicals introduced into aqueous media in the environment can undergo hydrolysis and be transformed into new chemicals with properties different from those of their precursors. In addition, processes other than nucleophilic displacement by water may occur (e.g., elimination and isomerization). The importance of these transformations of chemicals as dominant pathways in aqueous media can be determined quantitatively from data on hydrolysis rate constants and half-lives.

Hydrolysis data will generally be important in assessing risks from organic chemicals that have hydrolyzable functional groups (such as esters, amides, alkyl halides, epoxides, and phosphoric esters). Hydrolysis refers to the reaction of an organic chemical (RX) with water with the resultant net exchange of the group X for the OH group from the water at the reaction center. Therefore,

$$RX + HOH \longrightarrow ROH + HX$$

In the environment, hydrolysis of organic chemicals occurs in dilute solution. Under these conditions, water is present in a large excess, and the concentration of water is essentially constant during hydrolysis. Hence the kinetics of hydrolysis are pseudo-first-order at a fixed pH.

Processes other than nucleophilic displacement by water can take place. For example, X can be lost from RX via an elimination reaction. These elimination reactions exhibit kinetics behavior (i.e., pH independent or first-order acid or base dependent) similar to those reactions where OH substitution occurs.

The hydrolysis reaction can be catalyzed by acidic or basic species, including OH^- and H_3O^+ (H^+). The promotion of hydrolysis by H^+ or OH^- is known as *specific acid or specific base catalysis,* as contrasted to the general acid or base catalysis encountered with other cationic or anionic species. So far, the published laboratory data (1945, 1446, 1953, 1952) indicate that hydrolysis rates are the same in sterile natural freshwaters and in buffered distilled water at the same temperature and pH. Thus only specific acid or base catalysis, together with neutral water reaction, need be considered. Although other chemical species may catalyze hydrolysis reactions, the available concentrations of these species in the environment are usually too low to have an effect and are not expected to contribute significantly to the rate of hydrolysis (1954).

An extensive amount of information has been published on the hydrolysis of a wide variety of organic chemicals. However, most of the literature relating to environmental hydrolysis of chemicals pertains to pesticides. Much of this information is incomplete for the ranges of pH and temperature that are of environmental concern. Effects of buffer salts are often unrecognized.

D. BIOLOGICAL EFFECTS

1. Arrangement of Data.

The following classification has been followed in arranging the data on residues, bioaccumulation, and toxicity:

- *bacteria*
- *algae*
- *plants*
- *protozoans* (Phylum Protozoa)
- *worms* (Phyla Platyhelminthes and Aschelminthes)
- *molluscs* (Phylum Mollusca)
- *segmented worms* (Phylum Annelida)
- *arthropods* (Phylum Arthropoda)
    - *insects* (Class Insecta)
    - *crustaceans* (Class Crustacea)
    - *spiders and allies* (Class Arachnidae)
- *chordates* (Phylum Chordata)
- *vertebrates* (Subphylum Vertebrata)
    - *fishes* (Class Pisces)
    - *amphibians* (Class Amphibia)
    - *reptiles* (Class Reptilia)
    - *birds* (Class Aves)
    - *mammals* (Class Mammalia)

A more detailed classification list is given on the following pages, as well as a list of major organisms used throughout the book with their scientific and common names.

2. Classification List

protozoans	*(Phylum Protozoa)*
	the flagellates (Class Mastigophora)
	rhizopods (Class Rhizopoda)
	amoebas (Order Amoebina)
	the spore-formers (Class Spirozoa)
	the ciliates (Class Ciliata)
mesozoans	*(Phylum Mesozoa)*
sponges	*(Phylum Parazoa)*
hydroids	*jellyfishes and corals (Phylum Cnidaria or Coelenterata)*
	sea firs (hydroids), hydras and siphonopheres (Class Hydrozoa)
	jellyfishes (Class Scyphozoa)
	sea anemones and corals (Class Anthozoa)
flatworms	flukes and tapeworms (Phylum Platyhelminthes)
roundworms	*rotifers and allies (Phylum Aschelminthes)*
	roundworms (Class Nematoda)
	rotifers (Class Rotifera)

molluscs *(Phylum Mollusca)*
gastropods (Class Gastropoda)
Subclass Pulmonata
bivalves (Class Bivalvia = Pelecypoda)
Subclass Lamellibranchia
squids, cuttlefishes, octopuses (Class Cephalopoda)

segmented worms (Phylum Annelida)
polychaetes (Class Polychaeta)
nereids (Family Nereidae)
Nereis diversicolor
oligochaetes (Class Oligochaeta), including earthworms
Family Tubificidae
Tubifex tubifex

segmented worms leeches (Class Hirundinea)

arthropods *(Phylum Arthropoda)*
millipedes (Class Diplopoda)
centipedes (Class Chilopoda)
insects (Class Insecta)
exopterygote insects
mayflies (Order Ephemeroptera)
dragonflies (Order Odonata)
stoneflies (Order Plecoptera)
grasshoppers, crickets, and allies (Order Orthoptera)
cockroaches and praying mantids (Order Dictyoptera)
termites (Order Isoptera)
endopterygote insects
butterflies and moths (Order Lepidoptera)
flies (Order Diptera)
fleas (Order Siphonaptera)
ants, wasps, and bees (Order Hymenoptera)
beetles (Order Coleoptera)
crustaceans (Class Crustacea)
Subclass Branchiopoda
Order Anostraca
Artemia salina (brine shrimp)
Order Notostraca
Order Conchostraca
waterfleas (Order Cladocera)
Daphnia

copepods (Subclass Copepoda)
Order Calanoida
Calanus
Order Cyclopoida
Cyclops
malacostracans or higher crustaceans (Subclass Malacostraca)
isopods, sowbugs, and woodlice (Order Isopoda)
amphipods, scuds, and sideswimmers (Order Amphipoda)
Suborder Gammaridea
Gammarus
decapods (Order Decapoda)
Suborder Natantia
Section Caridea
Crangon vulgaris
Crangon crangon
Suborder Reptantia
Section Astacura
Homarus americanus
Homarus vulgaris
Nephrops norvegicus
Section Brachyura
Carcinus maenas (green crab)
Callinectus sapidus (blue crab)
spiders and allies (Class Arachnida)
scorpions (Order Scorpiones)
harvestmen (Order Opiliones or Phalangida)
ticks and mites (Order Acari)
spiders (Order Araneae)
starfishes, sea urchins, and allies (Phylum Echinodermata)
starfishes (Class Asteroidea)
sea urchins, sand dollars, and allies (Class Echinoidea)

chordates *(Phylum Chordata)*
vertebrates (Subphylum Vertebrata)
fishes (Class Marsipobranchii, Selachii, Bradyodonti, and Pisces)
amphibians (Class Amphibia)
newts and salamanders (Order Caudata)
frogs and toads (Order Salientia)
reptiles (Class Reptilia)
tortoises, terrapins, and turtles (Order Chelonia = Testudines)

crocodilians (Order Loricata = Crocodylia)
lizards and snakes (Order Squamata)
snakes (Suborder Ophidia = Serptentes)
birds (Class Aves)
mammals (Class Mammalia)
insect-eating mammals (Order Insectivora)
bats (Order Chiroptera)
the primates (Order Primates)
ant-eaters, sloths, armadillos (Order Edentator)
hares, rabbits, and pikas (Order Lagomorpha)
rodents (Order Rodentia)
whales (Order Cetacea)
flesh-eating animals (Order Carnivora)
dogs, cats, weasels, and bears (Suborder Fissipeda)
seals, sea lions, and walruses (Suborder Pinnipedia)
elephants (Order Proboscidea)

3. Organisms Used in Experimental Work with Polluting Substances or in Environmental Surveys

English name	**Latin name**
abalone	*Haliotis spp.*
African clawed frog	*Xenopus laevis*
Alga, blue-green	*Anabaena flos-aqua*
Alga, blue-green	*Anabaena variabilis*
Alga, blue-green	*Microcystis aeruginosa*
Alga, cryptophyte	*Rhodomonas salina*
Alga, freshwater	*N.seminulum*
Alga, green	*Scenedesmus obliquus*
Alga, green	*Scenedesmus subspicatus*
Alga, green algae	*Pseudokirchneriella subcapitata*
Alga, green fresh water	*Selenastrum capricornutum*
American char	*Salvelinus fontinalis (Mitchell)*
American toad	*Bufo americanus*
anchovy	*Stolephorus purpureus*
anole	*Anolis carolinensis*
armadillo	*Daysypus novemcinctus*
Atlantic kelp	*Laminaria digitata, L. agadhii*
Atlantic menhaden	*Brevoortia tyrannus*
Atlantic ribbed mussel	*Volsella demissa*
Atlantic salmon	*Salmo salar*

Bacterium, gram-negative rod-shaped saprophytic soil	*Pseudomonas putida*
Bacterium, gram-negative, rod-shaped	*Pseudomonas fluorescens*
Bacterium, heterotrophic gram-negative	*Vibrio fischeri*
barn owl	*Tyto alba*
barnade	*Balanus spp.*
bay anchovy	*Anchoa (Anchoiella) mitchilly*
bay mussel	*Mytilus edulis*
bay scallop	*Aequipecten (Pecten) irradians*
big brown bat	*Eptesicus fuscus*
bigmouth buffalo	*Ictiobus cyprinellus*
black bullhead	*Ameiurus melas orIctalurus melas*
Blackworm	*Lumbricus variegatus*
bleak	*Alburnus alburnus (L.)*
bloodworm	*Glycera dibranchiata*
blue crab	*Callinectes sapidus*
bluegill sunfish	*Lepomis macrochirus*
bobwhite quail	*Colinus virginianus*
box turtle	*Terrapene sp.*
Branchiopod	*Chydorus sp.*
Brine shrimp	*Artemia salina*
brook trout	*Salvelinus fontinalis*
brown bullhead	*Ameiurus nebulosus or Ictalurus nebulosus*
brown shrimp	*Crangon crangon*
brown shrimp	*Penaeus aztecus*
Brown shrimp	*Penaeus aztecus*
brown trout	*Salmo trutta*
bullfrog	*Rana catesbiana*
bullheads	*Ameiurus orIctalurus (see also catfish)*
bumble bee	*Bombus*
calico crab	*Ovalipes ocellatus*
calico scallop	*Aequipecten gibbus*
California ground squirrel	*Spermophilus beecheyi*
California sea mussel	*Mytilus californicus*
Canada goose	*Branta canadensis*
carp	*Cyprinus carpio*
cat	*Felis domestica*
catfish	*Ictalurus*
catfish (American)	*Ameiurus nebulosus (Le Sueur)*
Ciliate	*Tetrahymena pyriformis*
coalfish	*Pollachius birens*
cod	*Gadus morhua*
coho salmon	*Oncorhynchus kisutch (Walbaum)*

common cockroach	*Periplaneta americana*
common earthworm	*Lumbricus terrestris*
common frog	*Rana temporaria*
common toad	*Bufo bufo*
cone shells	*Conus spp.*
copepod	*Pseudocalanus minutus*
cotton rat	*Sigmodon hispidus*
cottontail	*Sylvilagus sp.*
crab	*Pertunus sanquinolentes*
crab	*Podophtalmus vigil*
crab	*Ranina serrata*
Crayfish	*Orconectes nais*
Crayfish	*Procambarus acutus acutus*
creek chub	*Semolitus atromaculatus*
cricket	*Gryllus sp.*
croaker	*Mycorpogon undulatus*
crucian carp (goldfish)	*Carassius carassius*
cutthroat trout	*Salmo clarki*
Cyanobacterium	*Oscillatoria sp.*
dace	*Leuciscus leuciscus (L.)*
Daphnid	*Ceriodaphnia dubia*
deer mouse	*Peromyscus maniculatus*
Diatom	*Gomphonema parvulum*
Diatom	*Navicula pelliculosa*
Diatom	*Nitzschia palea*
Diatom	*Skeletonema costatum*
dog	*Canis familiaris*
domestic chicken	*Gallus gallus*
domestic New Zealand white rabbit	*Oryctolagus cuniculus*
Duckweed	*Lemna gibba*
Duckweed	*Lemna minor*
Dungeness crab	*Cancer magister*
earthworm	*Lumbricus terrestris*
eastern chipmunk	*Tamias striatus*
eastern oyster	*Crassostrea virginica*
Edible crab	*Cancer magister*
edible frog	*Rana esculenta*
eel	*Anguilla anguilla (L.)*
eel	*Anguilla vulgaris*
egret	*Ardeola sp.*
English sole	*Parophrys vetulus*
European badger	*Meles meles*
European hares	*Lepus europaeus*

Fairy shrimp	*Streptocephalus probiscideus*
fathead minnow	*Pimephales promelas*
field cricket	*Gryllus pennsylvanicus*
flatworms	*Platyhelminthes*
Fowler's toad	*Bufo fowleri*
Freshwater amphipod	*Hyalella azteca*
Freshwater pulmonate mollusc	*Planorbis carinatus*
Freshwater rotifer	*Brachionus calyciflorus*
Freshwater shrimp	*Gammarus fasciatus*
fruit fly	*Drosophila sp.*
garter snake	*Thamnophis sirtalis*
goatfish	*Mulloidichthys spp.*
goldfish	*Carassius auratus*
green crab	*Carcinides maenas*
green frog	*Rana clamitans*
green sunfish	*Lepomis cyanellus*
guinea pig	*Cavia*
gulf menhaden	*Brevoortia partonus*
guppy	*Lebistes reticulatus*
hard clam	*Mercenaria (Venus) mercenaria*
Harelquinfish	*Rasbora heteromorpha*
hares	*Leporidea*
herring	*Lupea harengus*
honey bee	*Apis melliferra*
horned lizard	*Phrynosoma cornutum*
house sparrow	*Passer domesticus*
House sparrow	*Passer domesticus*
housefly	*Musca domestica*
channel catfish	*Ictalurus punctatus*
chicken	*Gallus domesticus*
chinook salmon	*Oncorhynchus tschawytscha*
chub	*Squalius cephalus (L.)*
Japanese eel	*Anguilla japonica*
Japanese quail	*Coturnix coturnix*
Japanese toad	*Bufo bufo japonicus*
kangaroo rat	*Dipodomys sp.*
kelp	*Macrocystis pyrifera*
killifish	*Fundulus*
king crab	*Parlithoides camtschatica*
lake trout	*Salvelinus namaycush*
land snail	*Helix sp.*
largemouth bass	*Micopterus salmoides*
leopard frog	*Rana pipiens*

little neck clam	*Protothaca staminea*
lobster	*Panulirus japonicus: P. pencillatus*
longear sunfish	*Lepomis megalotis*
mallard	*Anas platyrhynchos*
marine pin perch	*Lagodon rhomboides*
Mayfly	*Baetis sp*
meadow vole	*Microtus pennsylvanicus*
mealworm	*Tenebrio sp.*
Mexican axolotl	*Ambystoma mexicanum*
midge	*Chironomus plumosus*
Midge	*Chironomus riparius*
milkfish	*Chanos chanos*
mink	*Mustela chanos*
minnow	*Phoxinus phoxinus*
Mosquito	*Culex pipiens*
mosquito fish	*Gambusia affinis*
mosquito fly	*Aedes*
mosquito fly	*Anopheles*
mosquito fly	*Culex*
mountain bass	*Kuhlia sandvicensis*
mullet	*Mugil cephalus*
mummichog	*Fundulus heteroclitus*
Mysid shrimp	*Americamysis bahia*
Mysid shrimp	*Mysidopsis bahia*
northern lobster	*Homarus americanus*
northern pike	*Esox lucius*
nutria	*Myocaster coypus*
Oats	*Avena Sativa*
old world mouse	*Mus musculus*
old world rat	*Rattus norvegicus*
opossum	*Didelphis virginianus*
Pacific herring	*Clupea pallasii*
Pacific oyster	*Crassostrea gigas*
Pacific sardine	*Sardinops caerula*
pack rat	*Neotoma lepida*
perch	*Perca fluviatilis*
Phantom midge	*Chaoborus sp*
pheasant	*Phasianus sp.*
pickerel frog	*Rana palustris*
pigeon	*Columba livia*
pigeon	*Columba sp.*
pinfish	*Lagodon rhomboides*
pink salmon	*Oncorhynchus gorbuscha*

pink shrimp	*Penaeus duorarum*
pismo clam	*Tivela stultorum*
plaice	*Pleuronectes platessa*
pollack	*Gadus pollachius*
pompano	*Trachinotus carolinus*
pompano, jack cravally	*Caranx spp.*
Possum	*Tricosurus vulpecula*
pumpkinseed	*Lepomis gibbosus*
purple sea urchin	*Stronglo centrotus purpuratus*
rabbits	*Sylvilagus*
rainbow trout	*Salmo gairdneri*
Rainbowfish	*Melanotaenia splendida*
rainwater killifish	*Lucania parva*
rat snake	*Elaphe sp.*
razor clam	*Siliqua patula*
red algae	*Porphyra spp.*
red seaweed	*Gracileria virrucosa, G. foliifera*
red snapper	*Lutianus campechanus*
Redworm	*Eisenia foetida*
rhesus monkey	*Macaca mulatta*
ribbed limpet	*Siphonaria normalis*
Rice	*Oryza sativa*
ringnecked pheasant	*Phasianus colchicus*
roach	*Rutilus rutilus (L.)*
roundworms	*Aschelminthes*
sailfin molly	*Poecilia (Mollienisia) latipinna*
salmon (Atlantic)	*Salmo salar (L.)*
salmons	*Salmo or Oncorhynchus*
saltwater limpet	*Hecioniscus exaratus H. argentatus*
sand shrimp	*Crangon septemspinosa*
sandworm	*Nereis virens, Nereis vixillosa*
scup	*Stenotomus chrysops*
sea anemone	*Anthropleura elegantissima*
sea lamprey	*Petromyzon marinus*
sea lettuce	*Ulva spp.*
sea moss	*Chondrus crispus*
sea urchin	*Arbacid puntulata; Lytechnius spp.*
sea urchin	*Echinometra spp.*
segmented worms	*Annelida*
sheep	*Ovis sp.*
sheepshead minnow	*Cyprinodon variegatus*
shiner perch	*Cymatogaster aggregata*
shore crabs	*Hemigrapsus spp.*

shrimp	*Crangon spp.*
shrimp	*Pandalus spp.*
Silverside	*Menidia menidia*
smallmouth bass	*Micropterus dolomieui*
snapping shrimp	*Crangon spp.*
snapping turtle	*Chelydra serpentina*
soft shell clam	*Mya arenaria*
southern flounder	*Paralichthys lethostigma*
spiny lobster	*Panuliris argus*
Spot	*Leiostomus xanthurus*
sprat	*Clupea sprattus*
spring peeper	*Hyla crucifer*
Springtail	*Folsomia candida*
squirrel monkey	*Saimiri sciureus*
stable fly	*Stomoxys calcitrans*
staghorn sculpin	*Leptocottus armatus*
starling	*Sturnis vulgaris*
starry flounder	*Platichthys stellatus*
stickleback (12-spined)	*Pygosteus pingitius (L.)*
Stonefly	*Pteronarcys californicus*
striped bass	*Morone saxatilis*
striped bass	*Roccus saxatilis*
striped mullet	*Mugil cephalus*
suckers	*Catostomus or Ictiobus*
summer flounder	*Paralichthys dentatus*
sunfish (common)	*Lepomis humilis*
surf clam	*Spirula solidissima*
surgeon fish	*Acanthurus spp.*
swallows	*Hirundinidae*
swine (miniature)	*Sus scrofa*
tench	*Tinca tinca (L.)*
threespine stickleback	*Gasterosteus aculaetus*
Three-spined stickleback	*Gasterosteus aculeatus*
treefrog	*Hyla versicolor*
trout	*Salmo or Salvelinus*
turkey	*Meleagris*
Turnip	*Brassica rapa*
walleye	*Stizostedion vitreum*
water flea	*Daphnia*
water shrimp	*Gammarus pulex*
western chipmunk	*Eutamias sp.*
Western mosquitofish	*Gambusia affinis*
white shrimp	*Penaeus setiferus*

White sturgeon	*Acipenser transmontanus*
white sucker	*Catostomus commersoni*
whiting	*Gadus merlangus*
winter flounder	*Pseudopleuronectes americanus*
woodfrog	*Rana sylvatica*
yellow bullhead	*Ictalurus netalis*
Zebrafish	*Brachydanio rerio*

Latin name	**English name**
Acanthurus spp.	surgeon fish
Acipenser transmontanus	White sturgeon
Aedes	mosquito fly
Aequipecen gibbus	calico scallop
Aequipectern (Pecten) irradians	bay scallop
Alburnus alburnus	bleak
Ambystoma mexicanum	Mexican axolotl
Ameiurus melas	black bullhead
Ameiurus nebulosus	brown bullhead or American catfish
Americamysis bahia	Mysid shrimp
Anabaena flos-aqua	Alga, blue-green
Anabaena variabilis	Alga, blue-green
Anas platyrhynchos	mallard
Anguilla anguilla	eel
Anguilla japonica	Japanese eel
Anguilla vulgaris	eel
Anchoa (Anchoiella) mitchilly	bay anchovy
Annelida	segmented worms
Anolis carolinensis	anole
Anthopleura elegantissima	mosquito fly
Apis	sea anemone
Apis melliferra	honey bee
Arbacid puntulata	sea urchin
Ardeola spp.	egret
Artemia salina	Brine shrimp
Aschelminthes	roundworms
Avena Sativa	Oats
Baetis sp	Mayfly
Balanus spp.	barnacle
Bombus	bumble bee
Brachionus calyciflorus	Freshwater rotifer
Brachydanio rerio	Zebrafish

Branta canadensis	Canada goose
Brassica rapa	Turnip
Brevoortia patronus	gulf menhaden
Brevoortia tyrannus	Atlantic menhaden
Bufo americanus	American toad
Bufo bufo	common toad
Bufo bufo japonicus	Japanese toad
Bufo fowleri	Fowler's toad
Callinectes sapidus	blue crab
Cancer magister	Dungeness crab
Cancer magister	Edible crab
Canis familiaris	dog
Caranx spp.	pompano, jack cravally
Carassius auratus	goldfish
Carassius carassius	crucian carp
Carcinedes maenas	green crab
Catostomus	suckers
Catostomus commersoni	white sucker
Ceriodaphnia dubia	Daphnid
Clupea harengus	herring
Clupea pallasii	Pacific herring
Clupea sprattus	sprat
Colinus virginianus	bobwhite quail
Columba livia	pigeon
Conus spp.	cone shells
Coturnix coturnix	Japanese quail
Crangon crangon	brown shrimp
Crangon septemspinosa	sand shrimp
Crangon spp.	snapping shrimp
Crassostrea gigas	Pacific oyster
Crassostrea virginica	eastern oyster
Culex	mosquito fly
Culex pipiens	Mosquito
Cymatogaster aggregata	shiner perch
Cyprinodon variegatus	sheepshead minnow
Cyprinus carpio	carp
Daphnia	water flea
Daysypus novemcinctus	armadillo
Didelphis virginianus	opossum
Dipodomys sp.	kangaroo rat
Drosophila sp.	fruit fly
Echinometra spp.	sea urchin
Eisenia foetida	Redworm

Elaphe sp.	rat snake
Eptesicus fuscus	big brown bat
Esox lucius	northern pike
Eutamias sp.	western chipmunk
Felis domestica	cat
Folsomia candida	Springtail
Fundulus	killifish
Fundulus heteroclitus	mummichog
Gadus merlangus	whiting
Gadus morhua	cod
Gadus pollachius	pollack
Gallus gallus	domestic chicken
Gambusia affinis	mosquito fish
Gambusia affinis	Western mosquitofish
Gammarus fasciatus	Freshwater shrimp
Gammarus pulex	water shrimp
Gasterosteus aculeatus	threespine stickleback
Gasterosteus aculeatus	Three-spined stickleback
Glycera dibranchiata	bloodworm
Gomphonema parvulum	Diatom
Gracilaria verrucosa, G. foliifera	red seaweed
Gryllus pennsylvanicus	field cricket
Haliotis spp.	abalone
Helcioniscus exaratus, H. argentatus	saltwater limpet
Helix sp.	land snail
Hemigrapsus spp.	shore crabs
Hirudinidae	swallows
Homarus americanus	northern lobster
Hyalella azteca	Freshwater amphipod
Hyla crucifer	spring peeper
Hyla versicolor	treefrog
Chandrus crispus	sea moss
Chanos chanos	milkfish
Chaoborus sp	Phantom midge
Chelydra serpentina	snapping turtle
Chironomus plumosus	midge
Chironomus riparius	Midge
Chydorus sp.	Branchiopod
Ictalurus	catfish
Ictalurus melas	black bullhead
Ictalurus natalis	yellow bullhead
Ictalurus nebulosus	brown bullhead
Ictalurus punctatus	channel catfish

Ictiobus	suckers
Ictiobus cyprinellus	bigmouth buffalo
Kuhlia sandvicensis	mountain bass
Lagodon rhomboides	marine pin perch
Laminaria digibacta, L. agardhii	Atlantic kelp
Leiostomus xanthurus	Spot
Lemna gibba	Duckweed
Lemna minor	Duckweed
Lepomis cyanellus	green sunfish
Lepomis gibbosus	pumpkinseed
Lepomis humilis	common sunfish
Lepomis macrochirus	bluegill sunfish
Lepomis megalotis	longear sunfish
Leporidae	hares
Leptocottus armatus	staghorn sculpin
Lepus europaeus	European hares
Lesbistes reticulatus	guppy
Leuciscus leuciscus	dace
Lucania parva	rainwater killifish
Lumbricus terrestris	common earthworm
Lumbricus variegatus	Blackworm
Lutianus campechanus	red snapper
Lytechnius spp.	sea urchin
Macaca mulatta	rhesus monkey
Macrocystis pyrifera	kelp
Melanotaenia splendida	Rainbowfish
Meleagris	turkey
Meles meles	European badger
Menidia menidia	Silverside
Mercenaria (Venus) mercenaria	hard clam
Microcystis aeruginosa	Alga, blue-green
Micropogon undulatus	croaker
Micropterus dolomieui	smallmouth bass
Micropterus salmoides	largemouth bass
Microtus pennsylvanicus	meadow vole
Morone saxatilis	striped bass
Mugil cephalus	striped mullet
Mulloidichthys spp.	goatfish
Mus musculus	old world mouse
Musca domestica	housefly
Mustela vison	mink
Mya arenaria	soft shell clam
Myocaster coypus	nutria

Mysidopsis bahia	Mysid shrimp
Mytilus californicus	California sea mussel
Mytilus edulis	bay mussel
N.seminulum	Alga, freshwater
Navicula pelliculosa	Diatom
Neotoma lepida	pack rat
Nereis virens, Nereis vexillosa	sandworm
Nitzschia palea	Diatom
Oncorhynchus gorbuscha	pink salmon
Oncorhynchus kisuth	coho salmon
Oncorhynchus tschawytscha	chinook
Orconectes nais	Crayfish
Oryctolagus cuniculus	domestic New Zealand white rabbit
Oryza sativa	Rice
Oscillatoria sp.	Cyanobacterium
Ovalipes ocellatus	calico crab
Ovis sp.	sheep
Pandalus spp.	shrimp
Panulirus argus	spiny lobster
Panulirus japonicus, P. Pencillatus	lobster
Paralichthys dentatus	summer flounder
Paralichthys lethostigma	southern flounder
Paralithoides camtschatica	king crab
Parophrys vetulus	English sole
Passer domesticus	house sparrow
Passer domesticus	House sparrow
Penaeus aztecus	brown shrimp
Penaeus aztecus	Brown shrimp
Penaeus duorarum	pink shrimp
Penaeus setiferus	white shrimp
Perca fluviatilis	perch
Periplaneta americana	common cockroach
Peromyscus maniculatus	deer mouse
Petromyzon marinus	sea lamprey
Phasianus colchicus	ringnecked pheasant
Phoxinus phoxinus	minnow
Phrynosoma cornutum	horned lizard
Pimephales promelas	flathead minnow
Planorbis carinatus	Freshwater pulmonate mollusc
Platichthys stellatus	starry flounder
Platyhelminthes	flatworms
Pleuronectes platessa	plaice
Podophthalmus vigil	crab

Poecilia (mollienesia) latipinna	sailfin molley
Pollachins birens	coal fish
Porphyra spp.	red algae
Portunus sanquinolentus	crab
Procambarus acutus acutus	Crayfish
Protothaca staminea	little neck clam
Pseudocalanus minutus	copepod
Pseudokirchneriella subcapitata	Alga, green algae
Pseudomonas fluorescens	Bacterium, gram-negative, rod-shaped
Pseudomonas putida	Bacterium, gram-negative rod-shaped saprophytic soil
Pseudopleuronectes americanus	winter flounder
Pteronarcys californicus	Stonefly
Pygosteus pungitius	stickleback (12-spined)
Rana catesbiana	bullfrog
Rana clamitans	green frog
Rana esculenta	edible frog
Rana palustris	pickerel frog
Rana pipiens	leopard frog
Rana serrata	crab
Rana sylvatica	woodfrog
Rana temporaria	common frog
Rasbora heteromorpha	Harelquinfish
Rattus norvegicus	old world rat
Rhodomonas salina	Alga, cryptophyte
Roccus saxatilis	striped bass
Rutilus rutilus	roach
Saimiri sciureus	squirrel monkey
Salmo	trout
Salmo clarki	cutthroat trout
Salmo gairdneri	rainbow trout
Salmo salar	Atlantic salmon
Salmo trutta	brown trout
Salvelinus	trout
Salvelinus fontinalis	brown trout
Salvelinus namaycush	lake trout
Sardinops caerula	Pacific sardine
Scenedesmus obliquus	Alga, green
Scenedesmus subspicatus	Alga, green
Selenastrum capricornutum	Alga, green fresh water
Semolitus atromaculatus	creek chub
Sigmodon hispidus	cotton rat

Siliqua patula	razor clam
Siphonaria normalis	ribbed limpet
Skeletonema costatum	Diatom
Spermophilus beecheyi	California ground squirrel
Spirula solidissima	surf clam
Squalius cephalus	chub
Stenotomus chrysops	scup
Stizostedion vitreum	walleye
Stolephorus purpureus	anchovy
Stomoxys calcitrans	stable fly
Streptocephalus probiscideus	Fairy shrimp
Stronglo centrotus purpuratus	purple sea urchin
Sturnis vulgaris	starling
Sus scrofa	swine, miniature
Sylvilagus	rabbits
Sylvilagus sp.	cottontail
Tamias striatus	eastern chipmunk
Tenebrio sp.	mealworm
Terrapene sp.	box turtle
Tetrahymena pyriformis	Ciliate
Thamnophis sirtalis	garter snake
Tinca tinca	tench
Tivela stultorum	pismo clam
Trachinotus carolinus	pompano
Tricosurus vulpecula	Possum
Tyto alba	barn owl
Ulva spp.	sea lettuce
Vibrio fischeri	Bacterium, heterotrophic gram-negative
Volsella demissa	Atlantic ribbed mussel
Xenopus laevis	African clawed frog

4. Discussion of Biological Effects Tests

4.1. Ecological Effects Tests. Releases of hazardous chemical substances into the environment during manufacturing, processing, distribution, use, or disposal, whether accidental or planned, can have an adverse impact on both natural and man-modified ecosystems and their components. The societal costs may include degradation of the environment; losses in sport and commercial fisheries, shellfish populations, and wildlife resources; losses in agriculture; losses in tourism and property values; and other adverse impacts.

Testing for such effects requires the selection of indicators (i.e., indicative parameters) that provide for wide taxonomic representation and include a range of biological processes. Ideally, testing at levels of ecological organization above the individual species would provide information more directly related to ecological consequences of the release of a hazardous chemical. However, the development and standardization of such tests are difficult because of the complexity of the species interactions that characterize ecosystems. A major thrust for the future therefore will be the development of test methods to address interactions such as those that occur between predator and prey, among competitors for habitat or food, and between disease and host organisms. As methods such as microcosm studies and other laboratory model systems are developed, they may help to address these ecological testing needs.

Laboratory testing below the level of the organism is also potentially useful. It is generally rapid and readily amenable to standardization. Because many cellular and subcellular functions are common to a wide range of organisms, they have the potential of being applicable to many sets of ecological circumstances.

However, most ecological effects tests currently in use employ single-species test populations of vertebrates, invertebrates, or plants. Individual species represent an intermediate level of biological organization between subcellar functions and community/ecosystem interactions. Many single-species tests are considered state-of-the-art and have correlated well with actual ecological effects of chemicals.

The following criteria can be used to select tests:

- The test results are significant and useful for risk assessment.
- The test applies to a wide range of chemical substances or categories of chemical substances.
- The test is cost-effective in terms of personnel, time, and facilities.
- The test is adequately sensitive for detection of the subject effect.

Confidence in extrapolation from simple test to ecologically significant impacts depends not only on the appropriate kinds of tests but also on the selection of appropriate organisms to be used in those tests. Organisms useful for assessment testing should have characteristics such as the following:

- The organism is representative of an ecologically important group (in terms of taxonomy, trophic level, or realized niche).
- The organism occupies a position within a food chain leading to humans or other important species.
- The organism is widely available, is amenable to laboratory testing, is easily maintained, and is genetically stable so that uniform populations can be tested.
- There is adequate background data on the organism (its physiology, genetics, taxonomy, and role in the natural environment are well understood) so that data from these tests can be adequately interpreted in terms of actual environmental impacts.

4.2. Plant Effects Tests. All organisms require energy to perform vital functions and use the radiant energy of sunlight as their ultimate source of energy. Green plants use this energy directly through the process of photosynthesis when suitable inorganic carbon and nutrients are present. The sun's energy is converted into chemical energy, which is stored in plants in the form of sugars, starches, and other organic chemicals to be used by the plants themselves or by other organisms as energy sources.

Photosynthesis is the source of virtually all atmospheric oxygen. Plants also synthesize vitamins, amino acids, and other metabolically active compounds essential to many organisms. In this context, the maintenance of the biosphere depends on the normal functioning of green plants. Data from the tests in this section are expected to provide preliminary indications of the effects of chemical substances on the following groups of plants: blue-green algae, diatoms, green algae, monocotyledons, and dicotyledons.

Blue-green algae make up one of the two groups of organisms capable of converting atmospheric nitrogen into forms that can be utilized by all living organisms. Diatoms and unicellular green algae are responsible for most of the world's photosynthesis and are the primary food energy base for most organisms inhabiting aquatic environments. They are therefore necessary for all aquatic life and for human food taken from freshwater and marine environments (e.g., fish).

Algal inhibition. Testing for inhibition or stimulation of the growth of algae indicates the extent to which a test chemical can affect primary producers in lakes, streams, estuaries, and oceans. It can also generally indicate phytotoxicity or stimulation of plant growth. Substances that drastically inhibit growth at or near concentrations expected in the environment may reduce aquatic productivity. Even those substances that inhibit algal growth only partially, or that stimulate growth, at or near concentrations expected in the environment might shift relative algal populations so that undesirable species could increase. If diatoms or green algae grow less in the presence of the chemical than do blue-green algae, for example, a bloom of the less desirable blue-green species could develop.

Algae are often selected to represent aquatic primary productivity because they constitute the major mechanism for fixation of energy in most aquatic locations. Techniques for culturing and measuring them are simpler and less expensive than those for larger plants or for attached organisms (e.g., periphyton). The parameters recommended to measure potential effects on algae are inhibition of dry weight increase and changes in cell size. There are other potential effects that are not recommended. Lethality, for example, is commonly measured for other organisms, but it is difficult to determine for microscopic organisms and for nonmotile organisms. Inhibition of photosynthesis and/or respiration could be measured, but the balance between photosynthesis and respiration is accumulated as growth. In addition, growth represents the product that is important in the food chain and is therefore more directly relevant to assessment.

Uncertainties in using these data in assessment center on whether the selected species are adequate indicators of the potential for stimulation or toxicity to nonselected algal species, on whether the benefits of algae to the food chain can be accurately predicted from changes in dry weight and cell size, and on whether there are significant interactive effects (such as competition) that are affected at lower concentrations of the chemical substance than is any individual species.

4.3. Animal Effects Tests. The potential of a chemical to produce adverse ecological effects can be indicated by the results of preliminary testing of "representative" animals. Preliminary tests and test organisms should be selected on the basis of taxonomic, ecological, toxicological, and chemical exposure criteria. Test schemes should reflect those ecological hazards that a specific chemical substance may cause. Test responses-death, reproductive and/or behavioral dysfunction, and impairment of growth and development-are important factors for hazard assessment.

4.3.1. INVERTEBRATES. Toxicity of a chemical substance to invertebrates is an important factor in preliminary assessment of impact on ecosystems. Invertebrates have broad ecological roles and show various sensitivities to chemicals.

4.3.1.A. Aquatic Invertebrates. A number of aquatic invertebrates for acute tests are listed below:

Marine and Estuarine Invertebrates

copepods: *Acartia tonsa, Acartia clausi*
shrimp: *Penaeus setiferus, P. duorarum*
grass shrimp: *Palaemonetes pugio, P. intermedius, P. vulgaris*
sand shrimp: *Crangon septemspinosa*
mysid shrimp: *Mysidopsis bahia*
blue crab: *Callinectes sapidus*
green crab: *Carcinus maenas*
oyster: *Crassostrea virginica, C. gigas*
polychaetes: *Capitella capitata Freshwater Invertebrates*
daphnids: *Daphnia magna, D. pulex, D. pulicaria*
amphipods: *Gammarus lacustris, G. fasciatus,* or *G. pseudolimnaeus*
crayfish: *Oronectes* sp., *Cambarus* sp., *Procambarus* sp.
stoneflies: *Pteronarcys* sp.

mayflies:	*Baetis* sp. or *Ephemerella* sp.
mayflies:	*Hexagenia limbata* or *H. bilinata*
midges:	*Chironomus* sp.
snails:	*Physa integra, P. heterostropha, Amnicola limosa*
planaria:	*Dugesia tigrina*

4.3.1.B. Terrestrial Invertebrates. The ecological role and suitability for toxicity testing of a number of terrestrial invertebrates are discussed below.

Phylum Annelida

Class. Oligochaeta (Earthworms). Family Lumbricidae: *Lumbricus terrestris* (common earthworm).

Ecological Role. Earthworms occur in upper soil levels and feed on decaying organic matter. They are particularly important as soil mixers, aerators, and drainers and serve as food for many insectivores (robins, woodcock, mice, and shrews).

Suitability for Toxicity Testing. The diversity and wide distribution of worms make them desirable test species. Earthworms are particularly valuable because of their role in soil ecosystems, their part use, and their ease of maintenance.

Earthworms are important in the later stages of soil formation and in maintaining soil structure and fertility. They contribute in many ways, such as by incorporating decaying organic matter into soil, turning it over and mixing it with other soil fractions, and helping to improve soil aeration, drainage, and moisture-holding capacity. Earthworms have been reported to move as much as 250 tons of soil and organic matter per hectare annually. Certain species, particularly *Lumbricus terrestris,* pull organic matter down into the soil, fragment it, and mix it with mineral particles.

Earthworms are eaten by many vertebrates, including birds, poultry, and pigs. Ecologically they are near the bottom of the terrestrial trophic food chains and have a tendency to concentrate compounds such as organochlorine insecticides and PCBs in their tissues. These chemicals seldom harm the worms directly but can either kill vertebrates that eat the worms or be taken up into their tissues, thus indirectly affecting other animals higher in terrestrial food chains.

Earthworms have a number of characteristics that identify them as one of the most suitable soil animals to be used as a key bioindicator organism for testing for pollution by soil chemicals. In addition to their importance and key role in soil fertility, they are common in the great majority of soils and also in organic matter, they are large in size and easy to handle, they can be collected and identified readily, and they are known to be affected by, and to take up into their tissues, a number of organic and inorganic chemicals.

Earthworms are easily bred quite rapidly and in large numbers in the laboratory for toxicity testing, and their longevity makes it unlikely that many worms would die during the period of a toxicity test in untreated media. Several species are available commercially from fish bait breeders. Because of these characteristics, and because the earthworm is such a typical and important member of the soil fauna, it has been selected as a key indicator organism for the ecotoxicological testing of the toxicity of industrial chemicals by many national pesticide registration authorities and by international organizations such as the OECD, FAO, EU and others (2658).

Phylum Mollusca

Class. Gastropoda (pulmonate snails): *Helix aspersa.*

Ecological Role. Terrestrial snails and slugs are primary consumers and eat a varied diet of plant materials. They are a food source for larger insectivores.

Suitability for Toxicity Testing. Helix sp. is a very widely distributed snail and is abundant in certain moist habitats.

Phylum Arthropoda

Class Arachnida. Members of this class are spiders, mites and ticks, scorpions, and harvestmen.

Ecological Role. Mites and ticks are parasitic on plants and animals, deriving their substance directly from the fluids of their hosts. Spiders are carnivorous invertebrates whose food consists entirely of small animals, primarily insects. All arachnids are potential food sources for insectivores.

Suitability for Toxicity Testing. Mites and ticks are easily maintained under controlled conditions. Spiders are excellent test subjects because they are predators on many insects; they are relatively easy to maintain, if not to breed; and their web building provides a very useful experimental tool.

Class Insecta, order Orthoptera. This order includes many large and well-known insects-crickets, grasshoppers, roaches, locusts, and praying mantids. Toxicological research has been done on wide-ranging and easily obtainable species: *Periplaneta americana*,the common cockroach; *Gryllus pennsylvanicus*,the field cricket; *Schistocera gregaria*;and *Locusta migratoria*,locusts; *Mantis* sp., *Stagmomantis* sp., and *Tenodera* sp., mantids.

Ecological Role. Crickets and cockroaches are omnivorous insects and will feed on many kinds of organic matter. Locusts and grasshoppers are vegetarians and can occur in very large numbers, sometimes defoliating the countryside. Praying mantis are predators and feed primarily on insects. All of these species are possible food items for insectivorous invertebrates and vertebrates.

Suitability for Toxicity Testing. The insects in this order are easily maintained and very abundant. Praying mantis, being strictly carnivorous and relying heavily on insects for food, might accumulate certain chemicals or be more heavily exposed to target animals.

Orders Hemiptera, Homoptera. Hemipterans are true bugs, the homopterans are closely associated with the bugs.

Ecological Role. Hemipterans and homopterans are feeders on organic fluids, primarily plant juices. They can be destructive agricultural pests. These insect groups are food for insectivorous invertebrates and vertebrates. Aphids or plantlice are a large group of Homopterans and frequent pests on vegetation. The herbivorous aphid species are good selections for use in studies of environmental contaminants that may accumulate or deposit on vegetation.

Order Coleoptera. Beetles that have been used in research are frequently pest species, though not exclusively. Included are ground beetles *(Harpalus,Agonum,Feronia),* lady beetles (Hippodamia, Coleomegilla), and flour beetles *(Tribolium).*

Ecological Role. Some beetles are pests on agricultural crops, and others are predacious ground-dwelling species (e.g., *Harpalus*). Others feed on fungi and carrion. All beetles are potential food for insectivorous invertebrates and vertebrates.

Order Lepidoptera. Butterflies and moths are conspicuous and well-known insects.

Ecological Role. The larvae of butterflies and moths, often severe agricultural pests, are economically much more important than the adults, some of which never feed. They frequently supply food for insectivorous predators.

Order Diptera. The "flies" are a well-known group of insects and one of the larger orders. Mosquitoes, stable flies, house flies, and blow flies are pests of humans and other animals.

Ecological Role. Many adult dipterans are vectors of disease and nuisance pests of other animals. However, they also can represent staple foods for insectivorous predators (i.e., bats, swallows, frogs). Aquatic larvae are frequently major food sources for fish in quiet waters. The double association of some forms (e.g., mosquitoes) with aquatic and terrestrial systems at different times during their life cycles may make them particularly suitable subjects in experiments in which land-water transfer of a substance is to be studied.

Order Hymenoptera. The hymenopterans include ants, sawflies, ichneumons, chalcids, wasps, and bees.

Ecological Role. Many hymenopterans are important as pollinators and as parasites on other insects. They feed on pollen and plant juices, and many feed on other liquid foods.

Class Crustacea. Wen Yuh Lee (1717) recommends three laboratory-cultured crustaceans for use in marine pollution studies because they are characterized by wide distribution, a short life cycle and high reproductive potential and are representative of the plankton and benthos in coastal waters and the intertidal zone. The recommended crustaceans are *Acartia tonsa* (a planktonic copepod), *Sphaeroma quadridentatum* (an isopod), and *Amphitoc valida* (an amphipod).

The selection of specific organisms should further be based on vulnerability to marine pollutants in a critical life state (usually the larval or temporary planktonic stage), commercial or biological value, availability and ease of collection, ease of rearing and maintaining in the laboratory, and existing knowledge on ecological requirements (927). Laboratory-bred populations show several advantages in a long-term toxicity study. One of the most important is that the cultured population is able to grow and reproduce in the laboratory and is available whenever either a test is to be undertaken to determine the toxicity of products (e.g., oils) or to rank products such as dispersants in toxicity. The advantages and disadvantages of natural and laboratory cultured populations are summarized in Table 26.

Table 26. Comparison between Field- and Laboratory-Cultured Populations for Marine Pollution Studies (1717).

Field Population	Laboratory Population
A. Disadvantages	
1. Careful handling needed during collection and transportation	1. Genetic drift from wild conditions
2. Seasonal variation in mortality and in healthy condition	
3. Variations in size and life stages	
4. Seasonal changes in population abundance	
5. Various physiological states due to zonation	
B. Advantages	
1. Natural population realistic	1. Close to the normal physiological state; capable of growing and reproducing in captivity
	2. Ages known
	3. Available throughout the year
	4. Biochemical comparison possible between laboratory and suspected polluted populations
	5. Useful for chronic toxicity tests

Crustacean Life Cycle. Aquatic invertebrates are the most common food chain links between phytoplankton and desirable species of fish and shellfish. The extent to which chemical substances affect reproduction and growth of aquatic invertebrates is important because healthy stocks of fish and shellfish are dependent on adequate sources of food. A life cycle test is desirable for assessment because it gives a better estimate of total hazard than, for example, an acute toxicity test.

Freshwater Crustacean: Daphnia. One of the most widely performed and economical life cycle tests used *Daphnia,* a freshwater zooplankton. Daphnids have a planktonic existence, have a short life cycle, and can be easily cultured. They have been widely used in toxicological testing and are sensitive to toxicants. Although no invertebrate life cycle tests have been completely standardized, the *Daphnia* life cycle test has been used extensively by many researchers. Reproduction and life cycle tests on *Daphnia* begin with newborn daphnids, which are exposed to a chemical substance in culture for approximately three or four weeks. Reproductive impact of the chemical substance is evaluated by comparing the number of young produced by the organisms exposed to a chemical substance with the number produced by controls. Chronic lethal effects are evaluated by observing survival of the daphnids initially exposed throughout their life cycle.

Marine crustacean: mysid shrimp. Mysidopsis bahia is an excellent species for life cycle tests for *marine invertebrates* because of its sensitivity to known toxic chemicals, its ease of culturing, its short life cycle, and its importance in near-shore marine food webs.

Table 27. Broad Scheme of Toxicity Testing

Type of Test	Information Sought
(a) Acute	The lowest concentration having effect within a few days of continuous exposure. The effect is related to the breakdown of physiological systems, and typically death is the response sought.
(b) Subchronic	The highest concentration having no effect within perhaps one-tenth of the normal average life span. Used to determine the mode of action and functional and physiological changes.
(c) Chronic	The highest concentration having no effect over the lifetime of the animal.

4.3.2. VERTEBRATES

The term "toxicity test" covers a wide and increasing range of types of investigations. With fish, such tests can include

(a) The study of the toxic properties of fish (that is, fish toxicology).
(b) The use of fish for detecting the presence of, or determining concentrations of, metals, toxins, and hormones, for example, in solution (i.e., bioassay and tests for screening for the presence or absence of some defined response).
(c) Laboratory and field tests of selective piscicides.
(d) Basic toxicological research into the metabolism and detoxification of substances by this class of animal.
(e) Tests to compare the relative lethal toxicities of different substances, under some fixed but arbitrary set of conditions, to one or more species of fish.
(f) Tests to compare the lethal toxicity of a given substance to a single species of fish under a range of test conditions (e.g., pH, dissolved oxygen, hardness) to determine the effects of environmental conditions on toxicity.
(g) Simple, but unscientifically based, laboratory tests made under fixed conditions to assess arbitrarily for some administrative convenience the acceptability of a substance or effluent.
(h) Laboratory tests of the effects of a substance on survival, growth, reproduction, and so on, in fish.
(i) Laboratory and field tests of the effects of a waste (or of a chemical for use in agriculture) on fishes, on fish populations, and, where these exist, on fisheries.
(j) The laboratory use of fish to monitor aqueous wastes for harmful effects.
(k) The laboratory use of fish to monitor waters being abstracted from rivers for drinking, food processing,

irrigation, and so on, for harmful effects.

(l) The use of fish in cages to monitor river water for aqueous domestic and industrial wastes for harmful effects.

Toxicity studies are conveniently classified on the basis of the duration of exposure (as shown in Table 27), which of course automatically reflects the concentration of the poison. In a full investigation these tests would typically follow each other in the order given, should the substance under effect and the nature of exposure warrant it. Special studies are made for carcinogenic, teratogenic, and mutagenic effects. Rainbow trout *(Salmo gairdneri)* and bluegill sunfish *(Lepomis macrochirus)* are suggested as standard test species. These two fish, one a cold water species and the other a warm water species, have generally been the most sensitive to most previously tested chemical substances. The choice of species may depend on the geographical area of expected chemical release and on the available testing facilities. If salt water exposure is probable, testing a marine species is advised. Even though marine species generally are no more sensitive than freshwater species, toxic effects can be significantly modified by water chemistry.

4.3.2.A. Fish Embryo-Juvenile Test. Objectives: The objective of this test is to give preliminary indication of potential effects of a chemical on fish. Chemical substances can have significant chronic effects on individual fish at concentrations 2 to 500 times lower than LC_{50} values. When long-term exposure is probable, data on chronic effects contribute to assessment. Differences in sensitivities between a chronic and embryo-juvenile test are generally small or negligible, whereas differences in cost are great. The embryo-juvenile test is usually an excellent, cost-effective substitute for a chronic fish study.

Test Description. In an embryo-juvenile test, fish eggs and fry hatched from these eggs are exposed to a chemical at several concentrations for a few weeks. Effects on hatchability of eggs and on growth and survival of fry are determined by comparing responses at each concentration with the control.

4.3.2.B. Fish Bioconcentration Test. Objectives: Bioconcentration, the uptake of a compound from water into living tissue, affects the movement, distribution, and toxicity of chemical substances in the environment. A substance that bioconcentrates may affect life far removed from the initial points of environmental release and may alter ecological processes at concentrations much lower than predicted from acute and subacute results. Bioconcentration is the first step in the process of food chain biomagnification. Results of bioconcentration studies are useful in assessing risk to the environment, especially when the substance is highly lipid soluble (e.g., the octanol/water partition coefficient is greater than 1000), is poorly soluble in water, and does not undergo rapid chemical or biological transformation.

4.3.2.C. Test Description

Fish. Fish are exposed to the chemical substance in water for 28 days or until their tissue concentration reaches steady state. After steady state or 28 days, fish are placed in uncontaminated water for 7 days. During the exposure (1-28 days or steady state) and depuration (1-7 days) periods, fish are sacrificed periodically and the concentration of chemical substance in their tissues is measured. The bioconcentration factor, the relative uptake rate, and the depuration rate constants are estimated from these data. The most commonly tested species are fathead minnow *(Pimephales promelas)* and bluegill sunfish *(Lepomis macrochirus).*

Amphibians and Reptiles: Anurana. The aquatic forms of frogs and toads, the tadpoles have been widely used to study developmental biology. Common species in North America and Europe are:

Rana catesbiana: bullfrog
Rana clamitans: green frog
Rana palustris: pickerel frog
Rana pipiens: leopard frog
Rana sylvatica: woodfrog

Hyla crucifer: spring peeper
Hyla versicolor: treefrog
Bufo americanus: American toad
Bufo fowleri: Fowler's toad
Rana esculenta: edible frog
Rana temporaria: common frog
Bufo bufo: common toad

Anurans are carnivorous animals that feed on a great variety of invertebrate species, particularly insects. Many larger predators utilize them as a food source. Birds, snakes, turtles, and mammals feed on the adults, and the tadpoles provide food for predators associated with aquatic habitats.

Chelonia. Turtles are basically omnivorous reptiles and are important elements in the aquatic systems. They are predators on all types of invertebrates, and some species are avid consumers of aquatic vegetation.

Birds. Birds are a fairly large vertebrate group. They are very important in the world ecosystem. They are primary and secondary consumers, feeding on plants, invertebrates, and vertebrates alike. They in turn are food for mammalian predators (including humans), a few amphibians and reptiles, and a few species of birds. Many avian species are good indicators of environmental quality.

A toxicity test recommended by EPA (1464) is the *Quail Dietary Test.* The objective of a quail dietary test is to give preliminary indication of possible effects of a chemical substance on terrestrial birds. This test is designed to determine the concentration of a substance in food that will be lethal to 50 percent of a test population as well as to observe behavioral, neurological, and physiological effects. The test is appropriate for a chemical that might be found in or on terrestrial bird food. The bobwhite quail *(Colinus virginianus)* is an appropriate test species because it is easily and economically reared, is widely available, and is generally more sensitive to many hazardous chemicals than other common test species.

4.4. In Vitro Toxicity Assays. In vitro tests are considered useful for screening vast numbers of chemicals, detecting pollution in the environment, evaluating synergistic and antagonistic interactions between combinations of chemicals, and formulating computer-derived predictions based on quantitative structure-activity relationships (QSARs).

4.4.1. THE NEUTRAL RED ASSAY (NR ASSAY) USING GOLDFISH GF-SCALE (GFS) CELLS. *In vivo* acute lethal toxicity tests with fish are frequently used to investigate the effects of xenobiotics on aquatic biota, but their usefulness is limited by the small number of species that can be economically and conveniently studied. *In vitro* cytotoxicity assays using cultured fish cells have been developed in order to obtain toxicological data through simple, rapid, reproducible, and economical test methods, while at the same time responding to social concerns about reducing the number of test animals.

The bacterial Microtox test of *Photobacterium phosphoreum* (2681), the immobilization test of *Tetrahymena pyriformis* (2682), the bacterial growth inhibition test of *E. coli* (2683), and the cytotoxicity assays using established fish cell lines (2684, 2685, 2686, 2687) have been developed for predicting the acute toxicity of aquatic pollutants to fish. The neutral red assay (NR assay) was initially developed for use with mammalian cells and was later adapted for cytotoxicity studies with fish cells. GFS cells, a fibroblastic cell line derived from the scale of the goldfish, and their NR_{50} values were significantly correlated with *in vivo* acute toxicity to 10 aquatic species.

NR_{50} values refer to the dye "neutral red" and represent the 50% decrease of absorbance of the extracted dye neutral red from the test cultures compared to the absorbance of neutral red extracted from the control cultures. Pesticides have been conveniently classified according to their NR_{50} values as follows:

I: highly cytotoxic $NR_{50} \leq 10$ mg/l
II: moderately cytotoxic NR_{50} between 10 and 100 mg/l

III: relatively low cytotoxic $NR_{50} > 100$ mg/l (2680)

4.4.2. RUBISCO-TEST (SCHNABL). The inhibition of toxicants on the enzymatic activity of Ribulose-P2-Carboxylase in protoplasts is investigated in this microtest. This enzyme is responsible for the CO_2 uptake by the protoplasts (2698).

4.4.3. OXYGEN TEST (SCHNABL). The effect of toxicants on the oxygen consumption of protoplasts is determined in this test (2698).

GLOSSARY

10-d window The 10 days immediately following the attainment of 10% biodegradation.

acaricide (miticide) a material used primarily in the control of plant-feeding -mites (acarids), especially spider mites.

actinomycetes a group of branching filamentous bacteria, reproducing by terminal spores. They are common in the soil. Selected strains are used for production of certain antibiotics.

adjuvant an ingredient that, when added to a formulation, aids the action of the toxicant. The term includes such materials as wetting agents, spreaders, emulsifiers, dispersing agents, foaming adjuvants, foam suppressants, penetrants, and correctives.

algicide a chemical intended for the control of algae.

alkaloid a physiologically active, usually naturally occurring nitrogenous compound alkaline in reaction. Many are characteristic of specific plants, i.e., nicotine in tobacco.

anesthetic a chemical that induces insensibility to pain, such as chloroform or diethyl ether. Vinyl chloride is also an anesthetic.

anthelmintic a material used for the control of internal worms (helminths) parasitic in humans and animals.

antibiotic any of certain chemical substances that are produced by microorganisms such as bacteria and fungi (molds) and that have the capacity to inhibit the growth of, or destroy, bacteria and certain fungi that cause animal and plant diseases.

anticoagulant rodenticide a rodenticide that kills rats by inducing uncontrolled internal bleeding; an example is Warfarin.

approximate fatal concentration the geometric mean between the largest concentration allowing survival for 48 hr and the smallest concentration that was fatal in this time for practically all fish.

avicide a substance to control pest birds.

$BCF_{vegetation}$ The bioconcentration factor for vegetation is defined as the ratio of the concentration in aboveground parts (mg of compound/kg of dry plant) to the concentration in soil (mg of compound/kg of dry soil) (Travis, C. C., and Arms, A. D., "Bioconcentration of organics in beef, milk, and vegetation," *Environ. Sci. Technol.,* **22**(3), 1988. (2644).

benthic referring to aquatic organisms growing in close association with the substrate.

benthos (benthon) aquatic microorganisms capable of growing in close association with the substrate.

BFT The biotransfer factor is useful in risk assessment, because chemical exposure to cattle and cows may occur through both food and water pathways. The biotransfer factors for beef, B_b, and milk, B_m, are defined as follows:

B_b = concentration in beef (mg/kg)/daily intake of organic (mg/d)

B_m = concentration in milk (mg/kg)/daily intake of organic (mg/d) where measured concentrations of organics in beef or milk fat are converted to a fresh-meat or whole-milk base, assuming meat is 25% fat and whole milk is 3.68% fat.

bioaccumulation (bioconcentration) the process by which chemical substances are accumulated in living organisms.

- direct bio-accumulation

  1. The process by which a chemical substance accumulates in organisms by direct uptake from the ambient medium through oral, percutaneous, or respiratory routes.

  2. The increase in concentration of test material in or on test organisms (or specified tissues thereof) relative to the concentration of test material in the ambient environment (e.g., water) as a result of partitioning, sorption, or binding.

- indirect bio-accumulation: the process by which a chemical substance accumulates in living organisms through uptake via the food chain.

  bioaccumulation factor the ratio of the concentration of the test chemical in the test animal to the concentration in the test environment (e.g., water) at steady state (apparent plateau) or the ratio of the uptake rate constant (k_1) to the depuration rate constant (k_2).

bio-accumulation factor (BAF) equilibrium ratio of the organic chemical concentration resulting from the water and food routes to the water concentration; that is,

$$\text{Lipid-based BAF} = \frac{\mu\text{g chemical/kg lipid}}{\mu\text{g Chemical/l water}} = \text{l/kg}$$

bioconcentration factor (BCF) concentration resulting from the water concentration only.

biodegradability the ability of an organic substance to undergo biodegradation.

biodegradation molecular degradation of an organic substance, resulting from the complex action of living organisms.

biomagnification a process by which chemicals in organisms at one trophic level are concentrated to a level higher than in organisms at the preceding (lower) trophic level.

bird repellent a substance that drives away birds or discourages them from roosting.

bloom a concentrated growth or aggregation of plankton, sufficiently dense as to be readily visible.

blue-green algae the group Myxophyceae, characterized by simplicity of structure and reproduction, with cells in a slimy matrix and containing no starch, nucleus, or plastids and with a blue pigment in addition to the green chlorophyll.

cancer a malignant tumor anywhere in the body of a person or animal. Its origin is usually in the several types of epithelial tissue, and it invades any of the surrounding structures. The characteristic of metastasis, or seeding to other organs of the body, is positive to this diagnosis. Leukemia can be regarded as a cancer of the blood.

carbamate insecticides carbamates are esters of carbamic acid, and like the organophosphorus compounds, they inhibit chlorinesterase. Carbamic acid: H_2N-COOH.

carcinogen a highly controversial term, applied generally to any substance that produces cancer, as well as to highly specific chemicals suspected of being the cause of cancer development in any one of many target organs of the body in test animals or human beings. The words "cancer suspect agent" have been used by U.S. authorities to cover this possibility. However, NIOSH has broadened this terminology to include any agent reported in the literature to cause or to be suspected of causing *tumor* development, *malignant* or *benign* (see *oncogenic*). Examples include some mineral oils, vinyl chloride, benzene, beta-naphthylamine, hydrazine, and nickel.

chelating agents (chelates) agents that are readily soluble in water and have found wide use in many areas through their control of metal ions. These chelants (chelated metal ions) are used in the fields of textiles, water treatment, industrial cleaners, photography, pulp and paper, agriculture, and so on. In agriculture, both macronutrients and micronutrients are essential for proper plant growth. In some areas, the intensification of agricultural practices has resulted in depletion of available micronutrients. In order to achieve adequate agricultural production, it is necessary to add micronutrients to these soils. A chelated micronutrient is made by reacting a metallic salt with one of the chelating agents, forming a protective glove around the metal and retarding the normal soil chemistry reactions that tie up that metal. Thus it is more available to the plants.

chemosterilants compounds that sterilize insects to prevent reproduction.

chlorinated organic pesticides the organochlorine chemicals from one of the three principal families, including aldrin, benzene hexachloride, chlordane, DDT, endosulfan, heptachlor, lindane, methoxychlor, toxaphene, and so on.

cholinesterase a body enzyme that is necessary for proper nerve function and is destroyed or damaged by organic phosphates or carbamates taken into the body by any path of entry.

cholinesterase-inhibiting pesticides a class of pesticides having related pharmacological effects, including aldicarb, carbaryl, carbofuran, chlorpyrifos, parathion, etc.

cohort a group of individuals selected for scientific study of toxicology or epidemiology.

compatibility the ability of two or more substances to mix without objectionable changes in their physical or chemical properties.

contact herbicide a herbicide that kills primarily by contact with plant tissues rather than by translocation (systemic herbicides).

contact insecticide a chemical causing the death of an insect with which it comes in contact. Ingestion is not necessary.

controls the most important factor in any statistically meaningful experiment. The nature, number, and reproducibility of results with the controls determine the accuracy and significance of the conclusions drawn from the experimental cohort results.

coupling agent a solvent that has the ability to solubilize or to increase the solubility of one material in another.

critical range the range of concentrations in mg/l below which all fish lived for 24 hr and above which all died. Mortality is given as a fraction indicating the death rate (e.g., 3Ú4).

cyclodiene insecticides mainly aldrin, chlordane, dieldrin, endrin, heptachlor, endosulfan, and toxaphene. The cyclodienes are characterized by their endomethylene bridge structure.

defoliant a preparation intended to cause leaves to drop from crop plants such as cotton, soybeans, and tomatoes, usually to facilitate harvest.

degradation phase the time from the end of the lag period to the time when 90% of the maximum level of degradation has been reached.

desiccant a preparation intended for artificially speeding the drying of crop plant parts such as cotton leaves and potato vines.

disinfectant (1) a substance that destroys harmful bacteria, viruses and the like and makes them inactive. (2) a substance that destroys infesting organisms such as insects, mites, rats, weeds, and other organisms multicellular in nature.

dispersant a material that reduces the cohesiveness of like particles, either solid or liquid.

encapsulated pesticides pesticides enclosed in tiny capsules of such material as to control release of the chemical and extend the period of diffusion.

epidemiology discipline that attempts to evaluate the health of a defined human population and to determine cause-and-effect relationships for disease distribution. Factors such as age, sex, and ethnicity are its parameters. It is the study of the distribution and determinants of disease and injuries in human populations. Examples of epidemiological studies include those that associated a lower incidence of dental caries with fluoridated drinking water and a higher incidence of lung cancer with cigarette smoking.

eradicant fungicide a fungicide used to destroy ("burn out") fungi that have already developed and produced a disease condition.

food chain accumulation BAF/BCF is a measure of the tendency of a chemical to accumulate in an organism from both food and water exposures. BAF/BCF > 1 indicates that food chain accumulation has occurred.

fumigant a substance or mixture of substances that produces gas, vapor, fumes, or smoke intended to destroy insects, bacteria, or rodents.

fungus (fungi) all non-chlorophyll-bearing thallophytes (i.e., all non-chlorophyll-bearing plants of a lower order than mosses and liverworts) as, for example, rusts, smuts, mildews, and molds. Many cause destructive plant diseases. The simpler forms are one-celled; the higher forms are branching filaments.

green algae organisms belonging to the class *Chlorophyceae* and characterized by photosynthetic pigments similar in color to those of the higher green plants. The storage food is starch.

hematology examination of the blood.

herbicide a chemical intended for killing plants or interrupting their normal growth. Herbicides are used in five general ways:

1. pre-planting: applied after the soil has been prepared but before seeding.
2. pre-emergence (contact): nonresidual dosages are used after seeding but before emergence of the crop seedlings.
3. pre-emergence (residual): applied at time of seeding or just prior to crop emergence; it kills weed seeds and germinating seedlings.
4. post-emergence: applied after emergence of a crop.
5. sterilant (nonselective): used to effect a complete kill of all treated plant life.

histopathology microscopic examination of tissue.

Inherently Biodegradable A classification of chemicals for which there is unequivocal evidence of biodegradation (primary or ultimate) in any test of biodegradability.

inoculum the inoculum is a combination of microorganisms that, in degradability experiments, is added to the test system in order to obtain degradation of the compounds under investigation (substrate).

insecticides the various insecticides fall into six general categories according to the way they affect insects:

1. stomach: toxic quantities are ingested by the insect.
2. contact: kills upon contact with an external portion of the body.
3. residual contact: remains toxic to insects for long periods after application.
4. fumigant: possesses sufficient natural or induced vapor pressure to produce lethal concentrations.
5. repellent: does not kill but is distasteful enough to insects to keep them away from treated areas.
6. systemic: capable of being absorbed into the plant system where it makes plant parts insecticidal.

Various classes of insecticides, according to their composition, and examples of each, include:

- inorganics: calcium and lead arsenates, sodium fluoride, sulfur, and cryolite.
- botanicals: pyrethrum, nicotine, rotenone.
- chlorinated hydrocarbons: DDT, BHC, lindane, methoxychlor, aldrin, dieldrin, heptachlor, toxaphene, endrin.
- organic phosphates: parathion, diazinon, malathion, ronnel.
- carbamates: sevin, zectran

isomer a chemical the molecules of which contain the same number and kind of atoms as another chemical but arranged differently; e.g., normal (straight-chain) octylalcohol and its isomer, isooctyl alcohol. Stereoisomers are those isomers in which the same number and kind of atoms are arranged in an identical manner except for their relative position in space; e.g., endrin is a stereoisomer of dieldrin.

juvenile hormone a hormone produced by an insect in the process of its immature development that maintains its nymphal or larval form.

Lag phase Is the period from inoculation in a die-away test until the degradation percentage has increased to about 10%. The lag time is often variable and poorly reproducible.

larvicide a substance intended to kill especially the larvae of certain insect pests such as mosquitoes.

leaching downward movement of a material in solution through soil.

LC_{50} (lethal concentration fifty) a calculated concentration that, when administered by the respiratory route, is expected to kill 50% of the population of experimental animals. Ambient concentration is expressed in milligrams per liter.

LD_{50} (lethal dose fifty) a calculated dose of a chemical substance that is expected to kill 50% of a population of experimental animals exposed through a route other than respiration. Dose concentration is expressed in milligrams per kilogram of body weight.

maximum acceptable toxicant concentration (MATC) the geometric mean of the lowest concentration producing a statistically significant effect and the highest concentration producing no such effect on survival, growth, or fecundity in any life stage in a life cycle or partial life cycle, or early life stage test. It is used as a threshold for toxic effects in exposures of indefinite duration but does not correspond to any particular level or type of effect on any particular life stage.

median tolerance limit (TL_m) has been accepted by most biologists to designate the concentration of toxicant or substance at which 50% of the test organisms survive. In some cases and for certain special reasons, the LC_{10} or LC_{90} might be used. The LC_{90} might be requested by a conservation agency negotiating with an industry in an area where an important fishery exists and where the agency wants to establish waste concentrations that will definitely not harm the fish. The LC_{10} might be requested by a conservation agency that is buying toxicants designed to remove undesirable species of fish from fishing lakes.

metabolism process by which all natural and synthetic chemicals ingested or inhaled by the living body, either animal or vegetable, are continually subjected to chemical transformation in the organism into other products by myriad chemical reactions, such as synthesis and oxidative transformation in the organs of the body. Many of these primary and intermediate products find their way to body excretions through lung exhalation, urine, feces, or other expirations. The tracing of these routes is important for specific chemicals and their possible relation to disease. Isotope-tagged materials are used for these research studies. These studies are often called *pharmacokinetic and metabolism research.*

metastasis in medicine, the shifting of pathogenic cells of a disease, such as a malignant tumor, from one part or organ of the body to another unrelated to it.

mite mites are tiny organisms closely related to ticks in the group *Acarina.* They have eight legs as do spiders, except newly hatched mites, which have only six. Some mites, such as the chicken mite and the chigger, are parasitic on higher animals. A large family of mites is known as the spider mites from their habit of spinning a web on undersides of leaves where they feed.

mold any fungus, exclusive of the bacteria and yeasts, that is of concern because of its growth on foods or other products used by humans; fungus with conspicuous profuse or wooly growth (mycelium or spore masses). Occurs most commonly on damp or decaying matter and on the surfaces of plant tissues.

molluscicide a compound used to control snails that are intermediate hosts of parasites of medical importance.

mothproofer a substance that, when used to treat woolens and other materials liable to attack from fabric pests, protects the material from insect attack.

mutagenesis alteration of the genetic material of a cell in such a manner that the alteration is transmitted to subsequent generations of cells. A particular case is where a genetic change can be passed from parent to offspring.

mutation a sudden variation in some *inheritable* characteristic of an individual animal or plant, as distinguished from a variation resulting from generations of gradual change. It is an effect attributed to an action prior to conception of the embryo. It has been correlated with increased incidence of chromosome breaks in the reproductive cells, male or female.

necrosis destruction of cells.

nematocide a material, often a soil fumigant, used to control nematodes infesting roots of crop plants; an example is ethylene dibromide.

nematode a member of a large group (phylum *Nematoda*) also known as threadworms, roundworms, etc. Some larger kinds are internal parasites of humans and other animals. Nematodes injurious to plants, sometimes called eelworms, are microscopic, slender, wormlike organisms in the soil, feeding on or within plant roots or even plant stems, leaves, and flowers.

NR_{50} values values that refer to the dye "Neutral Red" and represent the 50% decrease in absorbance of the extracted dye "neutral red" from the test cultures compared to the absorbance of neutral red extracted from the control cultures. Pesticides have been conveniently classified according to their NR_{50} values as follows:

I: highly cytotoxic — NR_{50} 10 mg/l
II: moderately cytotoxic — NR_{50} between 10 and 100 mg/l
III: relatively low in cytotoxicity — NR_{50} >100 mg/l (2680)

nymph the early stage in the development of insects that have no larval stage. It is the stage between egg and adult during which growth occurs in such insects as cockroaches, grasshoppers, aphids, and termites.

organochlorine insecticides the principal pesticides included under organochlorines are the *bis*-chlorophenyls (e.g., DDT) and the cyclodienes (aldrin, etc.) with 50% chlorine content or more. These insecticides are characterized by their persistence in the environment.

organophosphorus pesticides anticholinesterase chemicals that damage or destroy cholinesterase, the enzyme required for nerve functions in the animal body. Use of some of these pesticides may involve danger for the applicator. Examples of the leading series are as follows, where R represents some organic radical:

• phosphate: dicrotophos

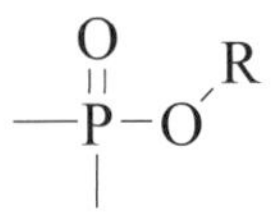

O R
—P—O

• phosphorothioate: parathion

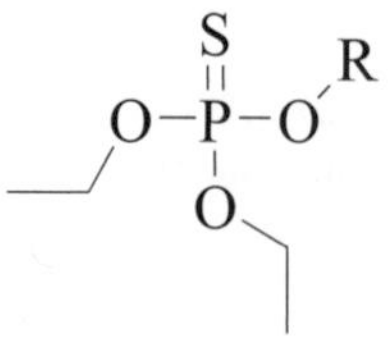

S R
O–P–O
O

• phosphorodithioate: phorate

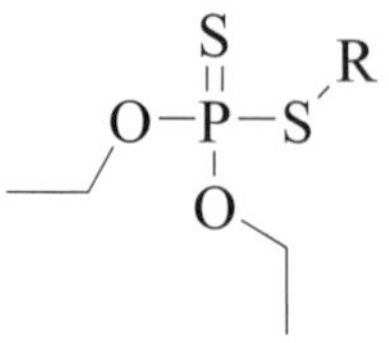

S R
O–P–S
O

organotin fungicides Several tin-based organic fungicides are commercially available, such as triphenyltin acetate, triphenyltin hydroxide, and tricyclohexyltin hydroxide.

oxidation pond an enclosure for sewage designed to promote the intensive growth of algae. These organisms release oxygen, which stimulates the transformation of the wastes into inoffensive products.

pathogen any microorganism that can cause disease. Most pathogens are parasites, but there are a few exceptions.

photosynthesis process of manufacture, by algae and other plants, of sugar and other carbohydrates from organic raw materials with the aid of light and chlorophyll.

phytoplankton plant microorganisms, such as certain algae, living unattached in the water. Contrasting term: **zooplankton**.

phytotoxicity degree to which a material is injurious (poisonous) to vegetation. It is specific for particular kinds or types of plants.

plant growth regulator a preparation that in minute amounts alters the behavior of ornamental or crop plants or the produce thereof through physiological (hormone) rather than physical action. It may act to accelerate or retard growth, to prolong or break a dormant condition, to promote rooting, or in other ways.

post-emergence herbicide a chemical applied as an herbicide to the foliage of weeds after the crop has emerged from the soil.

Primary Biodegradation The alteration in the chemical structure of a substance, brought about by biological action, resulting in the loss of a specific property of that substance.

protozoa unicellular animals, including the ciliates and nonchlorophyllous flagellates.

quaternary ammonium compounds organic nitrogen compounds in which the molecular structure includes a central nitrogen atom joined to four organic groups as well as an acid group of some sort. Nitrogen forms such pentavalent compounds, as is shown in the simplest example, ammonium chloride (NH_4Cl). When the hydrogen atoms are replaced by organic radicals, the compound is known as a quaternary ammonium compound, an example is tetramethyl ammonium chloride. These compounds are in contrast to trivalent nitrogen compounds, wherein the nitrogen combines with only three hydrogen atoms, as in ammonia, or these are replaced by one to three radicals, as in the carbamate structure.

Readily Biodegradable An arbitrary classification of chemicals which have passed certain specified screening tests for ultimate biodegradability; these tests are so stringent that it is assumed that such compounds will rapidly and completely biodegrade in aquatic environments under aerobic conditions.

red algae a class of algae *(Rhodophyceae)* most members of which are marine. They contain a red pigment in addition to chlorophyll.

reentry time the period of time immediately following the application of a pesticide to a field when unprotected workers should not enter.

rodent a member of the animal group (order *Rodentia*) to which rats, mice, gophers, and porcupines belong.

rodenticide a preparation intended for the control of rodents (rats, mice, etc.) and closely related animals (such as rabbits).

saprophytic utilizing dead organic matter as nutrients; the saprophytes include some plants and certain bacteria and molds.

sensitization an increased reaction on the second or subsequent exposure to a compound, it results from an immunological mechanism.

spray drift the movement of airborne spray particles from the intended area of application.

surface-active agent (surfactant) a substance that reduces the interfacial tension of two boundary lines. These materials are classified as nonionic, anionic, or cationic. Most emulsifying agents are of the nonionic type; they do not ionize. Wetting agents and detergents are primarily anionic; they become ionized in solution, the negative molecule exerting primary influence.

synergist a material that exhibits synergism. The joint action of different agents such that the total effect is greater than the sum of the independent effects.

systemic pesticide a pesticide that is translocated to other parts of a plant or animal than those to which the material is applied.

toxicology is the science that attempts to determine the harmful effects of materials on human populations by testing animals. It is the science that prescribes limits of safety for chemical agents. For example, the study that reported tumor development in animals when they were fed saccharin at high dosage levels was a toxicological study. Toxicology, the prospective science, warns of the potential danger to humans of a given chemical substance, and epidemiology, the retrospective science, considers a given population exposed to the chemical and determines that indeed this is or is not a hazardous case.

trademark a word, letter, device, or symbol used in connection with merchandise and pointing distinctly to the origin or ownership of the article to which it is applied. A tradename is actually a trademarked name.

translocation distribution of a chemical from the point of absorption (plant leaves or stems, sometimes roots) to other leaves, buds, and root tips. Translocation also occurs in animals treated with certain pesticides.

triazine herbicides those materials (including atrazine, simazine, and so on) that are based on a symmetrical triazine structure, where R_1, R_2, and R_3 are a variety of attached radicals:

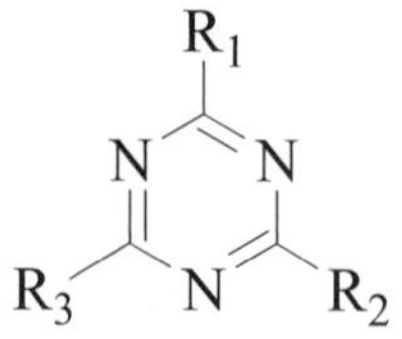

R_1 N N R_3 N R_2

Ultimate Biodegradation (aerobic) The level of degradation achieved when the test compound is totally utilised by micro-organisms resulting in the production of carbon dioxide, water, mineral salts and new microbial cellular constituents (biomass).

weed any plant that grows where it is not wanted.

wetting agent a substance that appreciably lowers the interfacial tension between a liquid and a solid and increases the tendency of the liquid to make complete contact with the surface of the solid.

wood preservative there are three main classes of wood preservatives: toxic oils (e.g., creosote) that evaporate slowly and are relatively insoluble in water; salts that are injected as water solutions into the wood; and preservatives consisting of a small percentage of highly toxic chemicals in a solvent or mixture of solvents other than water. The waterborne types are simple to apply, but they are subject to leaching, they are more or less poisonous to warm-blooded animals, and some are corrosive to iron.

zooplankton protozoa and other animal microorganisms living unattached in water.

ABBREVIATIONS

abs. perc. limit: absolute perception limit

A.C.: activated carbon

A.S.: bench scale activated sludge, fill and draw operations

ASC: activated sludge, continuous feed and effluent discharge

ASCF: activated sludge, fed slowly during aeration period

atm: atmosphere

avg: average

BCF: bioconcentration factor

BOD: biochemical oxygen demand (mg) is the amount of oxygen consumed by micro-organisms when metabolising a test compound; also expressed as mg oxygen uptake per mg test compound.

BOD_5: biological oxygen demand after 5 days at 20°C

bp: boiling point

CA: chemical analysis for the test material

cal: calorie

CAN: chemical analysis to indicate nitrogen transformation

cu ft: cubic foot

cu m: cubic meter (m^3 in equations)

°C: degree centrigrade (Celsius)

CO_2: carbon dioxide used to follow oxidation results

COD: chemical oxygen demand (mg) is the amount of oxygen consumed during oxidation of a test compound with hot, acidic dichromate; it provides a measure of the amount of oxidisable matter present; also expressed as mg oxygen consumed per mg test compound.

conc.: concentration

d: day

det. lim.: detection limit

DO: dissolved oxygen (mg/L) is the concentration of oxygen dissolved in an aqueous sample.

DOC: dissolved organic carbon is the organic carbon present in solution.

dyn. dil.: dynamic dilution

EC: electron capture

EC: effective concentration

EC_{50}: effect concentration for 50% of the organisms exposed

effl.: effluent

EIR: eye irritation reactivity

EMPA: Swiss Federal Laboratories for Materials Testing and Research.

°F: degree Fahrenheit

F: flow-through bioassay

6 f abs. app.: six-fold absorber apparatus

fp: freezing point or fusion point

FT: flow-through bioassay

g: gram

GC-EC: gas chromatography-electron capture

GC-FID: gas chromatography-flame ionization detection

geom.: geometric

glc: ground level concentration

h: hour

HC. cons.: hydrocarbon consumption

HCs: hydrocarbons

hr: hour

95%ile: 95 percentile

IC: inorganic carbon

I.D.: internal diameter

i.m.: intramuscular

infl.: influent

inh.: inhibitory or toxic action noted

inhal.: via inhalation

i.p.: intraperitoneal

I.R.: infrared

i.v.: intravenous

kcal: kilocalorie

kg: kilogram

km: kilometer

LC_{50}: lethal concentration for 50% of the organisms exposed

LD_{50}: lethal dose for 50% of the organisms exposed

liq.: liquid

m: meter

m: month

MATC: maximum acceptable toxicant concentration

max: maximum

mg: milligram

min: minute

min.: minimum

MLD: median lethal dose = LD_{50}

mm: mm Hg

mp: melting point

mph: miles per hour

***m*:** micron

mw: molecular weight

n: normality

nat: natural acclimation in surface water

NEN: Nederlandse Norm (Dutch standard test method)

NFG: nonflocculant growth activated sludge

NOEC: no observed effect concentration

NOD: nitrogenous oxygen demand

NOLC: No observed lethal concentration

NO ox: nitric oxide oxidation

n.s.i.: no specific isomer. This means that in the literature, no reference has been made to a specific isomer. It does not necessarily mean that the stated information is valid for all isomers.

O.I.: odor index

or.: oral

O.U.: odor units

P: plant treatment system for mixed wastes, including the test chemical

p.m.: particulate matter

PMS: photoionization mass spectrometer

p.p.: pour point

ppm: parts per million

ppb: parts per billion

R: renewal bioassay

R.C.R.: relative chemical reactivity

RD_{50}: concentration associated with 50% decrease in respiratory rate

Resp: special respirometer

RW: river water oxidation substrate

S: static bioassay

sat. conc.: saturation concentration in air

sat. vap.: saturated vapor

S.C.: subcutaneous

SCAS: Soap and Detergent Association **s**emi-**c**ontinuous **a**ctivated **s**ludge procedure.

scf: standard cubic foot

Sd: seed material

sel. strain: selected strain, pure culture of organisms

sew: municipal sewage

sew. dil: sewage dilution oxidation substrate

sp. gr.: specific gravity

SPM: suspended particulate matter

std. dil. sewage: the standard dilution technique has been used with normal sewage as seed material

STP: standard temperature and pressure

solub.: maximum solubility in water

$t_{1/2}$: half life

TC Total carbon, is the sum of the organic and inorganic carbon present in a sample.

TF: trickling filter

THCE: total hydrocarbon emissions

theor.: theoretical

ThCO$_2$: Theoretical carbon dioxide (mg) is the quantity of carbon dioxide calculated to be produced from the known or measured carbon content of the test compound when fully mineralized; also expressed as mg carbon dioxide evolved per mg test compound.

ThOD: theoretical oxygen demand (mg) is the total amount of oxygen required to oxidise a chemical completely; it is calculated from the molecular formula and is also expressed as mg oxygen required per mg test compound

TL$_m$: median threshold limit

TOC: total organic carbon of a sample is the sum of the organic carbon in solution and in suspension

TOD: total oxygen demand

T.O.N.: threshold odor number

TSP: total suspended particulates

UVS: spectrophotometry with ultraviolet light

vd: relative vapor density; that of air = 1

VLS: spectrophotometry with visible light

vp: vapor pressure

vs: versus

w: week

W: Warburg respirometer

y: year

yr: year

abathion *see* abate

abietan-18-oic acid

OCCURRENCE

resin acid constituent which is a major pollutant in effluents of bleached pulp and paper mills; degradation product of abietic acid

C. WATER AND SOIL POLLUTION FACTORS

Manmade sources

(7042)
In sediment of rivers and lakes downstream of a bleached kraft pulp and paper mill in New Zealand (1991/1992)

km's downstream	mg/kg dw
1.5	8.2; 18; 3.1; 2.4
11	13; 13; 1.9; 0.8
20	7.1
80	0.3
upstream river	<0.1

2 abieten-18-oic acid

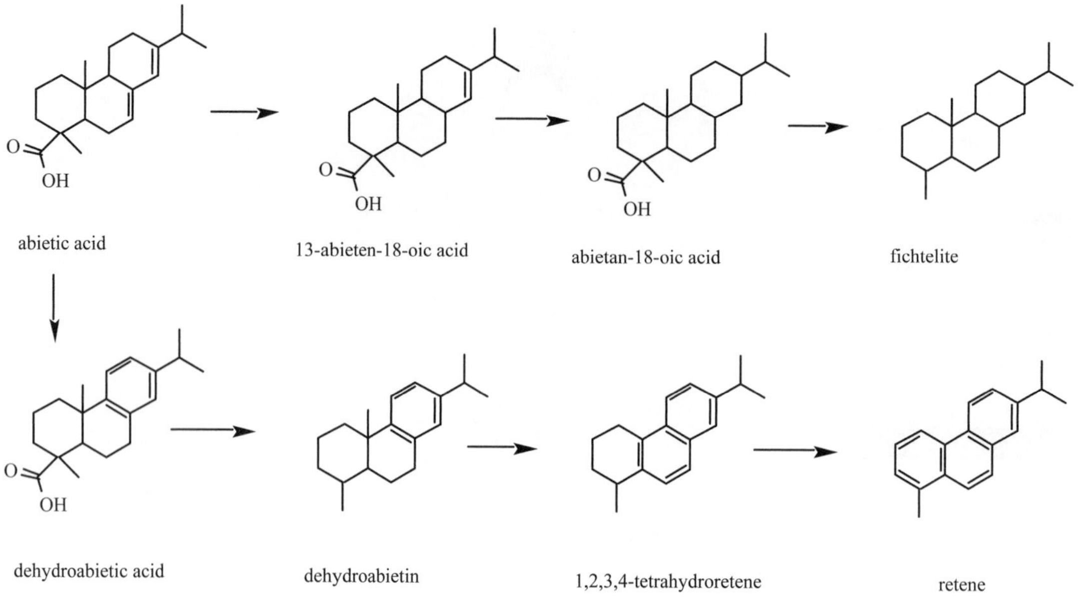

abietic acid → 13-abieten-18-oic acid → abietan-18-oic acid → fichtelite

abietic acid ↓

dehydroabietic acid → dehydroabietin → 1,2,3,4-tetrahydroretene → retene

Possible degradation pathway for abietic acid in river and lake water and sediment (7042)

Possible degradation pathway for abietic acid in river and lake water and sediment (7042)

abieten-18-oic acid

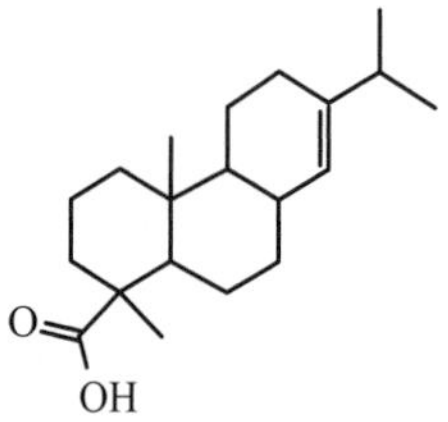

OCCURRENCE

resin acid constituent which is a major pollutant in effluents of bleached pulp and paper mills; degradation product of abietic acid

C. WATER AND SOIL POLLUTION FACTORS

Manmade sources

(7042)	km's downstream	mg/kg dw
In sediment of rivers and lakes downstream of a bleached kraft pulp and paper mill in New Zealand (1991/1992)	1.5	11; 13; 2.6; 1.2
	11	14; 9.9; 1.7; 0.3
	20	7.5
	80	0.1
	upstream river	<0.1

abietic acid → 13-abieten-18-oic acid → abietan-18-oic acid → fichtelite

dehydroabietic acid → dehydroabietin → 1,2,3,4-tetrahydroretene → retene

Possible degradation pathway for abietic acid in river and lake water and sediment (7042)

Possible degradation pathway for abietic acid in river and lake water and sediment (7042)

abietic acid (abietinic acid; sylvic acid)

COOH

$C_{19}H_{29}COOH$ (phenanthrene ring system)

$C_{20}H_{30}O_2$

CAS 514-10-3

USES

Abietates (resinates) of heavy metals as varnish driers; esters in lacquers and varnishes; fermentation industries; soaps.

OCCURRENCE

A major active ingredient of rosin, where it occurs with other resin acids. The term is often applied to these mixtures, separation of which is not achieved in technical grade material. Derived from rosin, pine resin; tall oil. Rosin acid constituent, a major pollutant in effluents of bleached pulp and paper mills. Rosin acids (in the form of crude tall oil) have been an important hydrophobe source for ethoxylate nonionic surfactants. A major source of rosin acids is a mixture with fatty acids known as *tall oil*, obtained as a by-product in the sulfate process for manufacturing paper from wood chips (*tall* is Swedish for "pine"). The rosin acids and fatty acids are separated from each other by fractional distillation. The rosin acids are tricyclic terpene derivatives of the phenanthrene series typified by abietic acid and related compounds, such as neoabietic acid, palustric acid, levopimaric acid, dehydroabietic acid, dextropimaric acid, isodextropimaric acid, isopimaric acid, dihydroabietic acid, and tetrahydroabietic acid.

A. PROPERTIES

yellowish resinous powder; molecular weight 302.44; melting point 172-175°C.

B. AIR POLLUTION FACTORS

Anthropogenic sources

emission rates released from burning wood in residential open fireplaces:

pine wood	42 mg/kg logs burned	
oak wood	nd mg/kg logs burned	(7087)

C. WATER AND SOIL POLLUTION FACTORS

Manmade sources

In sediment of rivers and lakes downstream of a bleached kraft pulp and paper mill in New Zealand (1991/1992):

K_m downstream	mg/kg dw	
1.5	15; 1.9; 4.8; 19	
11	16; 12; 2.5; 0.2	
20	6.4	
80	<0.1	
Upstream river	<0.1	(7042)

Environmental concentrations

in Fraser basin, Canada (1992-1996) (10820)

Bed sediment μg/kg dw	Suspended sediment μg/kg dw	whole water ng/L
31-2800	170-32,000	<3.5-390

Biodegradation

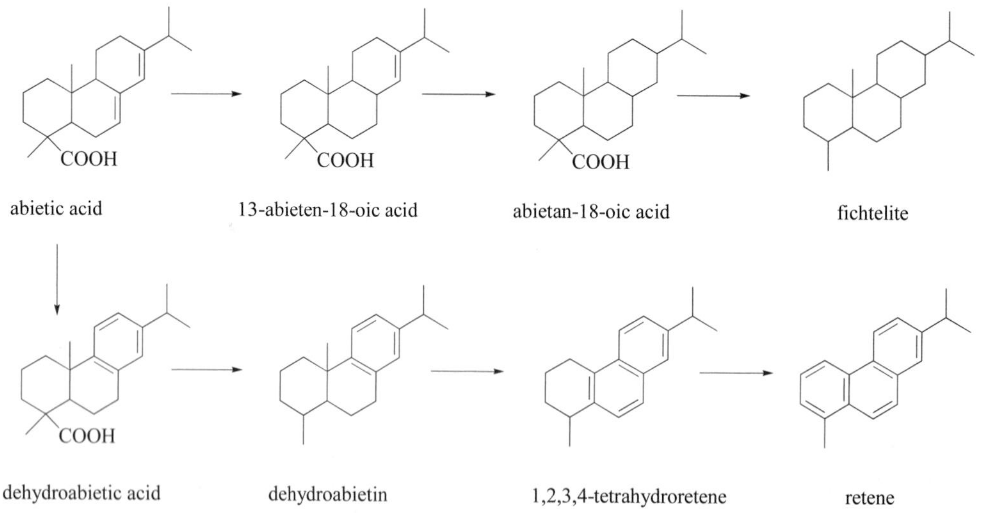

abietic acid → 13-abieten-18-oic acid → abietan-18-oic acid → fichtelite

abietic acid → dehydroabietic acid → dehydroabietin → 1,2,3,4-tetrahydroretene → retene

Possible degradation pathway for abietic acid in river and lake water and sediment. (7042)

abietic acid

palustric acid

dehydroabietic acid

7-oxo-dehydroabietic acid

Proposed pathway for abietane degradation by *Pseudomonas abietaniphila*. (7167)

Proposed pathway for abietane degradation by *Pseudomonas abietaniphila*. (7167)

D. BIOLOGICAL EFFECTS

Bioaccumulation

Species	conc. µg/L	duration	body parts	BCF	
Rainbow trout		2-30 days		2-7	
Atlantic salmon		20 days		0.12-12	(10819)

Toxicity

Crustaceae			
Daphnia magna	48h EC_{50}	19.2 mg/L	(7167)
Nitocra spinipes	96h LC_{50}	5.4-7.1 mg/L	(10819)
C. dubia	7d LC_{50}	3.1 mg/L	
	7d IC_{50}	2.6 mg/L	
	7d NOEL	1.8 mg/L	(10944)
Insecta			
Chironomus tentans	48h LC_{50}	2.8 mg/L	(10820)
Fishes			
Salmo trutta	96h LC_{50}	6.06 mg/L	(10820)
Salmo gairdneri	96h LC_{50},S	0.7 mg/L	(441)
Lepomis macrochirus	96h LC_{50}	1.9 mg/L	(10820)
Pimephales promelas	96h LC_{50}	1.9; 2.38 mg/L	(10819)
	7d IC_{50}	1.9 mg/L	
	7d NOEL	0.35 mg/L	
Oncorhynchus mykiss	96h LC_{50}	0.7; 5.45 mg/L	
Oncorhynchus kisuth Juvenile	96h LC_{50}	0.41 mg/L	(1495)
Oryzias latipes	24h LC_{50}	12 mg/L	(10944)

abietinic acid *see* abietic acid

abs *see* Teepol 715

acenaphthene (1.8-hydroacenaphthylene; ethylenenaphthalene; periethylenenaphthalene; 1,2-dihydroacenaphthylene)

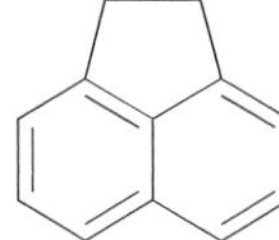

$C_{12}H_{10}$

CAS 83-32-9

USES AND FORMULATIONS

dye mfg.; plastics mfg; insecticide and fungicide mfg. (347)

MANMADE SOURCES Mean Concentrations in organic wastes and composts in Switzerland 2004 (10559)

	μg/kg dw
Organic waste	70
Organic waste compost	13
Biowaste	34
Biowaste compost	14
Green waste	170
Green waste compost	9
Input material grass	54
Foliage	203
Bark	27
Swiss compost	5
Non-Swiss compost	15

MANUFACTURING SOURCES

petroleum refining; shale oil processing; coal tar distilling. (347)

SOURCES

combustion of tobacco; constituent in asphalt; in soots generated by the combustion of aromatic fuels doped with pyridine (347; 1723); constituent of coal tar creosote: 4 wt %; constituent of diesel fuel: 100-600 mg/L.

NATURAL SOURCES (WATER AND AIR)

coal tar. (347)

A. PROPERTIES

white, yellowish, odourless solid at room temperature; molecular weight 154.21; melting point 90-

95°C; boiling point 279°C; vapor pressure 0.0027 mm at 20°C, d. 1.07; solubility 3.5-7.4 mg/L at 25°C; log P_{OW} 3.92-4.43 (measured).

B. AIR POLLUTION FACTORS

Manmade sources

emissions from open burning of scrap rubber tires: 290-2,446 mg/kg of tire. (2950)

Photodegradation

reaction with OH°: $t_{1/2}$: 7 hours (calculated) (5341)

Photodecomposition: absorbs solar radiation strongly (2903)

Environmental concentrations

monthly average gaseous concentrations in ambient air in San Francisco Bay area: june-nov 2000: < d.l. - 200 pg/m^3. (11059)

C. WATER AND SOIL POLLUTION FACTORS

Aquatic reactions

adsorption on smectite clay particles from simulated seawater at 25°C: 100 µg acenaphthene/l, 50 smectite/l; adsorption: nil. (1009)

Odor thresholds: T.O.C. in water at room temp.: 0.08 ppm, range 0.02-0.22 ppm, 14 judges. (321)

20% of population still able to detect odor at 0.026 ppm
10% of population still able to detect odor at 0.014 ppm
1% of population still able to detect odor at 0.0019 ppm
0.1% of population still able to detect odor at 0.00021 ppm

Manmade sources

constituent of diesel fuel: 100-600 mg/L (2387)
in aqueous phase of diesel fuel: 0.004-0.014 mg/L
diesel-water partition coefficient: log K_{dw} = 4.53 (2387)
constituent of coals: 0.34-0.62 mg/kg
In Canadian municipal sludges and sludge compost: September 1993- February 1994: mean values of 11 sludges ranged from 0.14 to 1.5 mg/kg dw.; mean: 0.57 mg/kg dw mean value of sludge compost: ND (7000)
in soluble fraction of leachates: <0.02-0.47 µg/L
in suspended fraction of leachates: <0.02-2.26 mg/kg (2413)

Biodegradation

Aerobic slurry bioremediation reactor with mixed aeration chamber, treating a petrochemical waste sludge at 3.5 to 7.5% solids containing approx. 25% oil and grease and 0.5 to 2.5 grams of waste material per gram of microorganisms, at 22-24°C during batch treatment:

waste residue: infl.: 5.8 mg/kg; effl. after 90 d: <2.5 mg/kg (2800)

Inoculum municipal waste water

at 5 mg/L after 7 d: 95% removal
at 10 mg/L after 28 d: 100% removal (5358)

Degradation in soil

soil type	*initial conc.*	*duration*	*removal*	
sandy loam, 0.5% oc., nonadapted	400 mg/kg	240 d	100%	(5360)
sandy loam, 0.5% oc., adapted	200 mg/kg	30 d	100%	(5361)
adapted soil	990 mg/kg	18 d	91%	(5362)
nonadapted	1.4 mg/kg	2 years	100%	(5363)
nonadapted	1 mg/L	10 d	100%	(5364)
nonadapted anaerobic Grünland soil	1 mg/L	70 d	0%	(5364)
Biodegradation $t_{1/2}$				
in nonadapted aerobic subsoil: sand Texas:	$t_{1/2}$: >161 d			(2695)

Degradation in soil system:

at initial conc. of 5 mg/kg soil:	$t_{1/2}$: 0.3 d	
500 mg/kg soil:	$t_{1/2}$: 4 d	(2903)

Degradation in sediment

aerobic sediments,			
oc. 0.7-1.6%:	5 d:	56-94%:	(5365)
Degradation in an aquifer:			
adapted microorganisms	3 d	100%	
nonadapted microorganisms	8 d	0%	(5366)

Soil sorption

	log K_{OC}	log K_S	
sandy loam	4.3	1.0	
clayish loam	4.1	2.2	(5361)
log K_{OC} (calculated):	3.3-4.3		(5355)

Environmental concentrations

average concentrations of dissolved PAH's in water of different San Francisco Estuary Segments : 1995-1999: 230-500 pg/L (11059)

Biodegradation

Oxidation of acenaphthene by *Pseudomonas aeruginosa* yielded a number of products (see table and figure) with 1-acenaphthenol (2) and 1-acenaphthenone (1) accounting for up to 65% of all recovered products. (11217)

(1) 1-acenaphthenone
(2) 1-acenaphthenol
(3a) Acenaphthylene-1,2-diol
(3b) Acenaphthylene-1,2-diol ketol tautomer
(4) cis-acenaphthene-1,2-diol
(5) trans-acenaphthene-1,2-diol
(6) Acenaphtho-1,2-quinone
(7) Naphthalene-1,8-dicarboxylic acid dimethyl ester
(8) 1,8-naphthalic anhydride

Transformation of acenaphthene and acenaphthylene by P. aeruginosa carrying Naphthalene 1,2-dioxygenase genes [11217]

Transformation of acenaphthene and acenaphthylene by *P. aeruginosa* carrying Naphthalene 1,2-dioxygenase genes	(11217)

D. BIOLOGICAL EFFECTS

Bioaccumulation

fathead minnow:	BCF: 8	(2606)
Lepomis macrochirus:	BCF: 387 after 28 d exposure to 0.009 mg/L; depuration $t_{1/2}$: < 1 d	(5367)
	BCF: 387 at 8.49 µg/L at 16 °C during 28 days	(2602)

Toxicity

Bacteria	*Anabaena flos-aquae* 14d NOEC 94% aqueous solution		(5356)
	A 48% aqueous solution caused an increase of the biomass.		(5356)
Algae			
Selenastrum capricornutum	96h EC_{50}	0.52 mg/L	(5346)
Skeletonema costatum	96h EC_{50}	0.5 mg/L	(5346)
Insects			
Paratanytarsus sp.	48h LC_{50}	0.06-2.1 mg/L	(5359)

Molluscs

Aplexa hypnorum	LC_{50}	>2.0 mg/L	(5344)
Crustaceans			
Daphnia magna	48h NOEC,S	0.6 mg/L	
	24h EC_{50},S	>280 mg/L	

	48h EC_{50},S	3.5; 41 mg/L	(5351, 5352)
Mysidopsis bahia	96h EC_{50},S	0.97 mg/L	(5346)
Fishes			
Cyprinodon variegatus	28d NOEC,F	0.52-0.97 mg/L	(5342)
	96h NOEC,S	1 mg/L	(5344)
	96h LC_{50},F	3.1 mg/L	(5342)
	96h LC_{50},S	2.2 mg/L	(5342, 5344)
	24h LC_{50},S	3.7 mg/L	(5345)
	48h LC_{50},S	2.3 mg/L	(5345)
	96h LC_{50}	3.1 mg/L	
	MATC, early life stage	0.71 mg/L	(2643)
Pimephales promelas	96h NOEC,F	0.34 mg/L	(5343)
	40-48 h NOEC,F	0.51 mg/L	(5343)
	32d NOEC,F	0.21-0.23 mg/L	(5350)
	96h LC_{50},F	0.61; 1.6; 1.7 mg/L	(5343, 5344, 5348)
Lepomis macrochirus	96h LC_{50},S	1.7 mg/L	(5342)
	24h LC_{50},S	7.2 mg/L	(5347)
Ictalurus punctatus	96h LC_{50},F	1.7 mg/L	(5344)
Oncorhynchus mykiss	24h LC_{50},F	1.6 mg/L	
	48h LC_{50},F	1.1 mg/L	
	72h LC_{50},F	0.80 mg/L	
	96h LC_{50},F	0.67 mg/L	(5344)
Salmo trutta	24h LC_{50},F	0.84 mg/L	
	48h LC_{50},F	0.65 mg/L	
	72h LC_{50},F	0.60 mg/L	
	96h LC_{50},F	0.58 mg/L	(5344)

1,2 acenaphthenedione *see* acenaphthenequinone

acenaphthenequinone (1,2 acenaphthenedione)

$C_{12}H_6O_2$

CAS 82-86-0

USES

dye synthesis.

A. PROPERTIES

yellow needles; molecular weight 182.18; melting point 249-252°C (dec.).

C. WATER AND SOIL POLLUTION FACTORS

Water quality

in Eastern Ontario drinking waters (June-Oct. 1978): n.d.-1.1 ng/L (n = 12)
in Eastern Ontario raw waters (June-Oct. 1978): 0.9-10.6 ng/L (n = 2) (1698)

acenaphthylene (cyclopenta[de]naphthalene)

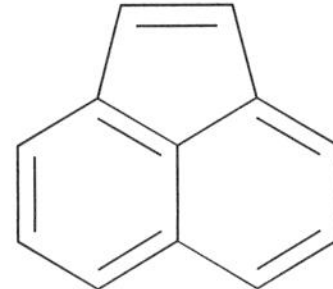

$C_{12}H_8$

CAS 208-96-8

MANMADE SOURCES

in soots generated by the combustion of aromatic hydrocarbon fuels doped with pyridine. (1723)

IMPURITIES, ADDITIVES, COMPOSITION

Technical grade typically contains 20% acenaphthene.

A. PROPERTIES

molecular weight 152.2; melting point 80-83°C; boiling point 280°C; vapor pressure 9.12 x 10^{-4} mm Hg at 25°C; sp. gr. 0.899; solubility 3.93 mg/l at 25°C in distilled water, 16 mg/L at 25°C.; LogP_{OW} 4.07.

B. AIR POLLUTION FACTORS

Photodecomposition

absorbs solar radiation strongly. (2903)

Manmade sources

in emissions from open burning of scrap rubber tires: 562- 861 mg/kg of tire. (2950)

C. WATER AND SOIL POLLUTION FACTORS

Manmade sources

In Canadian municipal sludges and sludge compost : September 1993- February 1994 : mean values of 11 sludges ranged from ND to 3.4 mg/kg dw.; mean : 0.11 mg/kg dw; mean value of sludge compost : 0.01 mg/kg dw. (7000)

Water quality

in Eastern Ontario drinking waters (June-Oct. 1978): 0.1-2.0 ng/l (*n* = 12)
in Eastern Ontario raw waters (June-Oct. 1978): 0.1-0.5 ng/l (*n* = 2) (1698)

Soil quality

typical values in Welsh surface soil (U.K.), average of 20 0-5-cm cores:
mean: 3.0 μg/kg dry wt
range: <1.0-23 μg/kg dry wt (2420)

Biodegradation

25-150 μg/L were almost totally degraded within 3 days in groundwater. (9715)
Microbial degradation of acenaphthylene in water samples was low. (9716)
In flooded soil contaminated with acenaphthylene, biodegradation occurred under aerobic and denitrifying conditions at rates of 0.53 and 0.35-0.37 mg/L/day. No significant biodegradation was seen under sulfate-reducing or methanogenic conditions. (9717)
Complete degradation after 16 months incubation in the dark at 20°C. (9718)
Degradation in sandy loam in the dark at 20°C at 700 mg/kg:
abiotic half-life: <1 month
biodegradation half-life: could not be determined due to rapid abiotic loss (2806)

Impact on biodegradation processes

Toxicity to microorganisms: A.S. respiration inhib. EC_{50}: >1,000 mg/l (2624)

D. BIOLOGICAL EFFECTS

Bioaccumulation

bullhead catfish (Black River, Ohio)	270 μg/kg	
striped bass (Potomac River, Maryland)	43 μg/kg	
Oysters	36 μg/kg	
clams	130 μg/kg	(9713, 9714)

acephate (acetylphosphoramidothioic acid ester; Acetamidophos; acetylphosphoramidothioic acid O,S-dimethyl ester; O,S-dimethyl acetylphosphoramidothioate; Orthene)

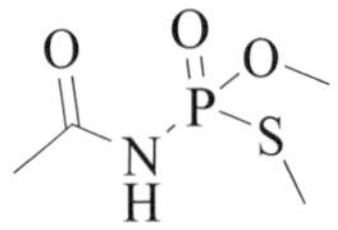

$C_4H_{10}NO_3PS$

CAS 30560-19-1

USES

Contact and systemic insecticide.

A. PROPERTIES

white crystals; molecular weigth 183.17; melting point 88-90°C; density 1.35; vapor pressure 1.72 x 10-6 mm Hg; solubility 790,000 mg/L at 20°C.

C. WATER AND SOIL POLLUTION FACTORS

Biodegradation in soil

$t_{1/2}$: 7-10 d, methamidophos was identified as a metabolite. In plants residual activity lasted for 10-15 d			(9600)
Hydrolysis			
$t_{1/2}$:	at pH 9 and 40°C	60 h	
	at pH 3 and 40°C	710 h	(9600)
Hydrolytic products formed at 37° C and varying pH, included methamidophos, O,S-dimethyl phosphorothiolate, and O-methyl acetylphosphoramidothiolate.			(9766)

D. BIOLOGICAL EFFECTS

Fishes			
	MATC life cycle	0.9 mg/L	
Menidia beryllina:	96h LC_{50}	3 mg/L	(2643)
rainbow trout	96h LC_{50}	>1,000 mg/L	
argemouth black bass	96h LC_{50}	1,725 mg/L	
bluegill sunfish	96h LC_{50}	2,050 mg/L	
channel cat fish	96h LC_{50}	2,230 mg/L	(9600)
Crustaceans			
pink shrimp	96h LC_{50}	3.8 mg/L	
mysid shrimp	96h LC_{50}	7.3 mg/L	(9765)
Mammals			
rat	oral LD_{50}	866-945 mg/kg bw	
dog	oral LD_{50}	>681 mg kg/kg bw	
mouse	oral LD_{50}	361 mg/kg bw	(9600)

acetaldehyde (ethanal; ethylaldehyde; aldehyde; acetic aldehyde)

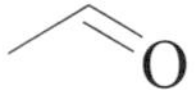

CH_3CHO

C_2H_4O

CAS 75-07-0

USES AND FORMULATIONS

organic chem. mfg.; perfumes, flavors, aniline dyes, plastics, synthetic rubbers mfg., silvering mirrors, hardening gelatin fibers. (347)

MANUFACTURING SOURCES

organic chemical mfg. (347)

SOURCES

vehicle exhaust; open burning and incineration of gas, fuel oil, and coal; evaporation of perfumes; lab use. (347)

NATURAL SOURCES (WATER AND AIR)

metabolic intermediate in higher plants; alcohol fermentation; sugar decomposition in body; by-product of most hydrocarbon oxidations. (347)

14 acetaldehyde

A. PROPERTIES

colorless liquid or gas; molecular weight 44.1; melting point 125°C; boiling point 20.2°C; vapor pressure 740 mm at 20°C; vapor density 1.52; saturation concentration in air 1811 g/m^3 at 20°C; THC 279 kcal/mole; solubility in all proportions; sp. gr. 0.783 at 18/4°C; LogP_{OW} 0.43/-0.22 (calc.).

B. AIR POLLUTION FACTORS

1 mg/m^3 = 0.55 ppm, 1 ppm = 1.831 mg/m^3.

Odor

characteristic:	
quality: sweet, green apple ripener	(73)
hedonic tone: pungent	(129)

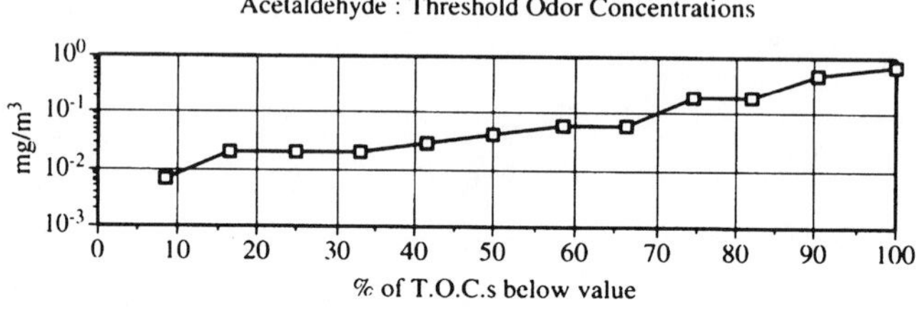

(54, 73, 291, 307, 610, 666, 710, 788, 836, 842, 871)

odor index: 5.000.000 (2)

Natural sources

glc's Pt Barrow, Alaska, Sept. 1967: n.d. 0.3 ppb (101)

Manmade sources

gasoline exhaust:	0.8-4.9 ppm	(195, 1053)
	7.2-14.3 vol. % of total exhaust aldehydes	(394, 395, 396, 397)
diesel exhaust:	3.2 ppm	(311)
emissions from cigarette smoking:	1,200 µg/cigarette	(2421)

Atmospheric half-lives

Photochemical reactions	25-30 h	(6046)
for reactions with OH°	0.5 d	
for reactions with O_3	235 d	(2716)
for direct photolysis	1.1 d	(2716)

Control methods

activated carbon:	retentivity, 7 wt% of adsorbent		(83)
wet scrubber:	water at pH 8.5:	outlet: 1,700 odor units/scf	
	$KMnO_4$:	outlet: 200 odor units/scf	(115)

Sampling and analysis

PMS: det. lim. 3-7 ppm	(200)
second derivative spectroscopy: det. lim. 400 ppb	(42)
photometry: min. full scale 2,200 ppm	(53)
nondispersive I.R.: min. full scale 275 ppm	(55)
detector tubes: UNICO: det. lim. 40 ppm	(59)

Inside and outside 6 residential houses, suburban New Jersey, summer 1992 (n = 36):

outdoor:	mean: 2.6 ppb; max: 13 ppb;	SD: 2.3 ppb	
indoor:	mean: 3.0 ppb; max: 16 ppb;	SD: 2.7 ppb	(2951)

Inside and outside 15 living rooms in Northern Italy (1983-1984)

4-7 d average	*lowest value*	*mean value*	*highest value*	
indoor	1	17	48	
outdoor	<1	5	19	(2756)

Note: all in µg/m³.

C. WATER AND SOIL POLLUTION FACTORS

BOD_5: 70; 71% ThOD (27, 41)
ThOD: 1.82

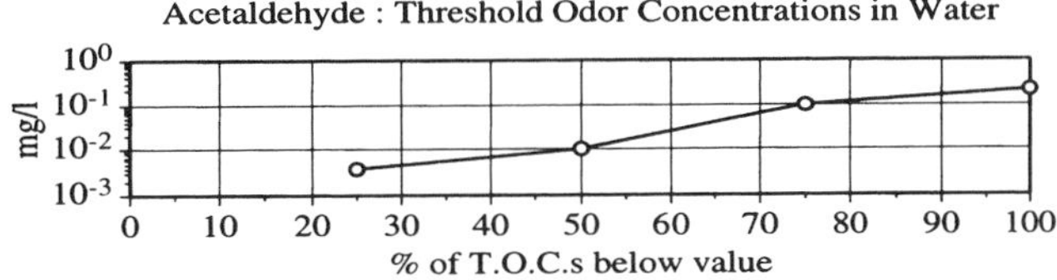

(129, 873, 889, 908)

Waste water treatment

A.S. after	6 h: 11.0% ThOD	
	12 h: 21.5% ThOD	
	24 h: 27.6% ThOD	(88)

Biodegradation

@Inoculum	*method*	*conc.*	*duration*	*elimination results*	
A.S.	Sapromat		5 d	>90% CO_2 prod.	(6045)
Ashbya gossipii		2,000 mg/L	2d	15% prod. removal	
Candida boidinii		2,000 mg/L	2d	40% prod. removal	
Candida utilis		2,000 mg/L	2d	50% prod. removal	
Dekera intermedia		2,000 mg/L	2d	75% prod. removal	
Endomycopsis fibuligera		2,000 mg/L	2d	30% prod. removal	
Hansenula glucozyma		2,000 mg/L	2d	35% prod. removal	
Pachisolen tannophilus		2,000 mg/L	2d	70% prod. removal	
Saccharomyces cerevisiae		2,000 mg/L	2d	40% prod. removal	
Sporobolomyces albo-rubescens		2,000 mg/L	2d	15% prod. removal	
Sporobolomyces singularis		2,000 mg/L	2d	100% prod. removal	
Aspergillus niger		2,000 mg/L	2d	30% prod. removal	
Fusarium lini		2,000 mg/L	2d	60% prod. removal	
Fusarium semitectum		2,000 mg/L	2d	20% prod. removal	
Nectria gliocladioides		2,000 mg/L	2d	95% prod. removal	
Neurospora crassa		2,000 mg/L	2d	20% prod. removal	
Polyporus tulipiferus		2,000 mg/L	2d	70% prod. removal	
Bacillus stearothermophilus		2,000 mg/L	2d	40% prod. removal	
Clostridium sordellii		2,000 mg/L	2d	15% prod. removal	
Lactobactillus sanfrancisco		2,000 mg/L	2d	95% prod. removal	
Pseudomonas putida		2,000 mg/L	2d	25% prod. removal	
Zymomonas mobilis		2,000 mg/L	2d	80% prod. removal	

Inhibition of biodegradation

filtrate of municipal sludge:	24h IC_0	400 mg/L	(6041)

A.C. adsorbability

0.022 g/g carbon, 11.9% reduction, infl. 1,000 mg/L, effl. 723 mg/L (32)

Anaerobic lagoon

COD/d/1,000 cu ft	infl. mg/L	effl. mg/L	
13	30	10	
22	80	35	
48	80	40	(37)

D. BIOLOGICAL EFFECTS

Bacteria			
Photobacterium phosphoreum (Microtox)	5 min EC_{50}	342 mg/L	(6042)
Pseudomonas putida	EC??	300 mg/L	(6043)
Protozoa			
Uronema parduczi Chatton-Lwoff	EC_0	57 mg/L	(1901)
Algae			
Diatomae Nitzschia linearis	5d EC_{50},S	237-249 mg/L	(6037)
Crustaceans			
Daphnia magna	48h EC_{50},S	48 mg/L	(6040)
Fishes			
pinperch	24h LC_{50}	70 mg/L	(248)
sunfish	96h LC_{50}	53 mg/L	(226)
Pimephales promelas	24h LC_{50}	60 mg/L	
	96h LC_{50}	30.8 mg/L	(2709)
Leuciscus idus	48h LC_0,S	50 mg/L	(6036)
	48h LC_0,S	117; 125 mg/L	
	48h LC_{50},S	124; 140 mg/L	
	48h LC_{100},S	156; 156 mg/L	(6036)
Lepomis macrochirus	96h LC_{50}	53 mg/L	(6037)

acetaldehydecyanohydrin *see* lactonitrile

acetaldol *see* 3-hydroxybutanal

acetamide (ethanamide; acetic acid amine)

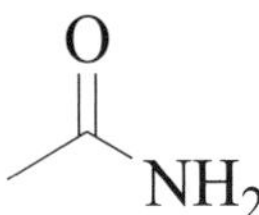

CH_3CONH_2

C_2H_5NO

CAS 60-35-5

USES

organic synthesis, general solvent, lacquers, explosives, wetting agent.

A. PROPERTIES

molecular weight 59.07; melting point 81°C; boiling point 222°C; THC 282.6 kcal/mole; solubility 975,000 mg/L at 20°C, 1,780 g/L at 60°C; sp.gr. 1.159 at 20/4°C; $LogP_{OW}$ -1.58/-1.26 (calculated).

B. AIR POLLUTION FACTORS

Odor threshold

recognition: 140-160 mg/m^3. (610)

C. WATER AND SOIL POLLUTION FACTORS

BOD_5: 69% ThOD

ThOD: 1.08 (30)

At 100 mg/L no inhibition of NH_3 oxidation by *Nitrosomonas* sp. (390)

Waste water treatment

A.S. after	6 h:	1.2% ThOD	
	12 h:	3.3% ThOD	
	24 h:	12.0% ThOD	(89)

D. BIOLOGICAL EFFECTS

Toxicity threshold

(cell multiplication inhibition test)

Bacteria			
Pseudomonas putida	16h EC_0	>10,000 mg/L	(1900)
Algae			
Microcystis aeruginosa	8d EC_0	6,200 mg/L	(329)
Green algae			
Scenedesmus quadricauda	7d EC_0	>10,000 mg/L	(1900)
Protozoa			
Entosiphon sulcatum	72h EC_0	99 mg/L	(1900)
Uronema parduczi Chatton-Lwoff	EC_0	10,000 mg/L	(1901)
Fishes			
Gambusia affinis	LC_{50} (72 h)	15,500-20,000 mg/L	(30)
mosquito fish	LC_{50} (24; 48; 96 h)	26,300; 26,300; 13,300 mg/L	(41)

acetamidoacetic acid *see* N-acetylglycine

acetaminophen (N-(4-hydroxyphenyl)acetamide; 4-acetamidophenol; N-acetyl-4-aminophenol; APAP; 4'-hydroxyacetanilide)

$C_8H_9NO_2$

CAS 103-90-2

TRADENAMES

paracetamol; Tylenol

USES

Acetaminophen is a common analgesic and antipyretic drug that is used for the relief of fever, headaches, and other minor aches and pains.

In the manufacture of azodyestuffs and photographic

A. PROPERTIES

molecular weight 151.17; density 1.26 kg/L; water solubility 14,000 mg/L at 20°C;white, crystalline powder; melting point 167-169° C

C. WATER AND SOIL POLLUTION FACTORS

Environmental concentrations

Samples from 139 streams in 30 states in the US thought to be susceptible to contamination from agricultural or urban activities during 1999-2000: detected in 17 out of 88 samples in values ranging from 0.010 – 10 µg/L (7221)

Belgium and The Netherlands 2000	ng/L	n	
in effluents of MWTP's	<100	4	
in surface waters	<100	22	
in drinking water	<100	12	(7237)

	range µg/L	n	median µg/L	90 percentile µg/L	
influent MWTP	26,000	6			(7237)
effluent MWTP (1996-2000)	<100	4			
	<200	6			(7237)
effluent MWTP (1998)	<500-6.0	49	<0.50	<0.50	(7237)
surface waters (2000)	<100	22			(7237)
drinkingwater (2000)	<100	12			(7237)
in San Francisco Estuary water (1999-2000)	1-390	5	102		(10587)

Biodegradation

The chemical is efficiently removed in municipal treatment works: maximum effluent concentratrion : 6 µg/L. (7178)

Mobility in soil

K_{OC} = 62 (10928)

D. BIOLOGICAL EFFECTS

Toxicity

Micro organisms			
Vibrio fischeri	30min EC_{50}	650 mg/L	(7237)
Protozoa			
Tetrahymena pyriformis	48h EC_{50}	112 mg/L	(7237)
Algae			
Scenedesmus subspicatus	72h EC_{50}	134 mg/L	
Crustaceae			
Daphnia magna	48h EC_{50}	9.2; 50 mg/L	
	24h EC_{50}	41; 136 mg/L	(7178)
	21d NOEC	20 mg/L	(10928)
Streptocephalus probiscideus	24h LC_{50}	29.6 mg/L	
Fish			
Brachydanio rerio	48h LC_{50}	378; 920 mg/L	(7237)
Lepomis macrochirus	96h LC_{50}	173 mg/L	(10928)
Planorbis carinatus	21d NOEC	1.02 mg/L	(10928)
Marine tests			
Microtox[TM] *(Photobacterium)* test	5 min EC_{50}	331 mg/L	
Artoxkit M *(Artemia salina)* test	24h LC_{50}	578 mg/L	
Freshwater tests			
Streptoxkit F *(Streptocephalus proboscideus)* test	24h LC_{50}	30 mg/L	
*Daphnia magna*test	24h LC_{50}	55 mg/L	
Rotoxkit F *(Brachionus calyciflorus)* test	24h LC_{50}	>5,300 mg/L	(2945)
Yeast			
Saccharomyces cerevisiae	IC_{50}	92 mg/L	(10516)
Mammals			
rat	oral LD_{50}	2,400 mg/kg	(10516, 10283)

Acetamiprid ((E)-N1-[(6-chloro-3-pyridyl)methyl]-N2-cyano-N1-methylacetamidine; (1E)-N-[(6-chloro-3-pyridinyl)methyl]-N'-cyano-N-methylethanimidamide)

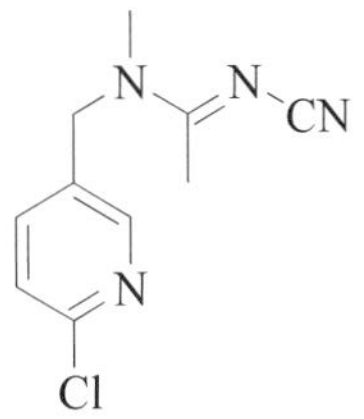

$C_{10}H_{11}ClN_4$

CAS 160430-64-8

USES

Pyridylmethylamine insecticide, neonicotinoid insecticide. Control of sucking-type insects on leafy vegetables, fruiting vegetables, cole crops, citrus fruits, pome fruits, grapes, cotton and ornamental plants and flowers.

A. PROPERTIES

White to very pale yellow fine powder; molecular weight 222.68; melting point 98.9°C; specific gravity 1.33 at 20°C; vapour pressure 1.7 x 10^{-7} Pa at 50°C; water solubility at 25°C in distilled water 4,250 mg/L, at pH 5: 3,480 mg/L, at pH 7: 2,950 mg/L at pH 9: 3,960 mg/L; log Pow 0.80 at 25°C at neutral pH;

B. AIR POLLUTION FACTORS

Photodegradation

T/2 (OH radicals) = 0.14 days

C. WATER AND SOIL POLLUTION FACTORS

Hydrolysis

Stable at pH 4, 5 and 7 at 45°C, at pH 9 at 22°C, t/2 = 812 days, at 35°C: 53 days, at 45°C: 13 days.

Photodegradation in water

T/2 = 34 days

Biodegradation in soil (10767)

	% mineralisation	% bound-residues	After
Biodegradation in soil aerobic	9.6	32	120 days
Biodegradation in soil anaerobic	0.25	12	182 days
Soil photolysis	<1	13	30 days

Biodegradation in laboratory studies (10767)

degradation in soils	%	Acetamiprid	IM-1-4	IM-1-2	IM-I-5
Aerobic at 20°C	50	0.8-5.4 days	2.7-226 days	1.1-1.6 days	388-450 days
Aerobic at 20°C	90	2.8-67 days			
Aerobic at 10°C	50	7.7 days			
Anaerobic at 20°C	50	71 days			

Dissipation studies in the field at 20°C (10767)

% dissipation	Acetamiprid	IM-1-4
50%	0.4-5.4 days	15-50 days
90%	18-31 days	50-166 days

Biodegradation in water/sediment systems (10767)

Readily biodegradable	no
DT50 water	3.6-5.8 days
DT90 water	31-37 days

Mobility in soils (10767)

	acetamiprid	IC-0	IM-I-2	IM-I-4	IM-I-5
K_{OC}	71-138	70-258	19-95	132-223	453-563

D. BIOLOGICAL EFFECTS

Toxicity

Algae			
Scenedesmus subspicatus	72h ECb_{50}	>98.3 mg/L	(10767)
Plants			
Lemna gibba	14d EC_{50}	1.0 mg/L	(10767)
Birds			

Mallard duck	Oral LD_{50}	98 mg/kg bw	(10767)
Worms			
Eisenia foetida	14d EC_{50}	9 mg/kg soil	(10767)
Crustaceae			
Daphnia magna	48h EC_{50}	49.8 mg/L	
reproduction	21d NOEC	5 mg/L	(10767)
Insecta			
Chironomus riparius	28d NOEC	0.005 mg/L	
Honey bee	Acute oral LD_{50}	14.5 μg/bee	
	Acute contact LD_{50}	8.09 μg/bee	(10767)
Fish			
Pimephales promelas	35d NOEC	19.2 mg/L	
Oncorhynchus mykiss	96h LC_{50}	>100 mg/L	(10767)
Mammals			
Rat	oral LD_{50}	213; 314; 417 mg/kg bw	(10767)

Toxicity of metabolite ***IM-I-4***			
Oncorhynchus mykiss	96h LC_{50}	98.1 mg/L	
Daphnia magna	48h EC_{50}	43.9 mg/L	
Chironomus riparius	28d NOEC	76 mg/L mg/L	
Eisenia foetida	14d EC_{50}	>1,000 mg/kg soil	(10767)

Toxicity of metabolite ***IM-I-2***			
Daphnia magna	48h EC_{50}	99.8 mg/L	
Eisenia foetida	14d EC_{50}	>1,000 mg/kg soil	(10767)

Toxicity of metabolite ***IM-I-5***			
Eisenia foetida	14d EC_{50}	>1,000 mg/kg soil	(10767)

Toxicity of metabolite ***IC-0***			
Eisenia foetida	14d EC_{50}	>1,000 mg/kg soil	(10767)

acetanilide (N-phenylacetamide; antifebrin)

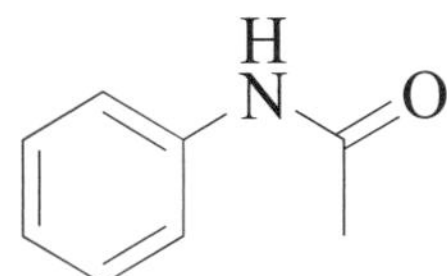

C_8H_9NO

CAS 103-84-4

USES

rubber accelerator; inhibitor in hydrogen peroxide; stabilizer for cellulose ester coatings; manufacture of intermediates (*p*-nitroaniline, *p*-nitroacetanilide; *p*-phenylenediamine); synthetic camphor; pharmaceutical chemicals; dyestuffs; precursor in penicillin manufacture; medicine (antiseptic); acetanisole.

A. PROPERTIES

molecular weight 135.16 melting point 114°C; boiling point 305°C; THC 1010 kcal/mole; solubility 5.63 g/L at 25°C, 35.0 g/L at 80°C; sp. gr. 1.21 at 4/4°C; $LogP_{OW}$ 1.16.

B. AIR POLLUTION FACTORS

Odor threshold

270 mg/m³. (735)

C. WATER AND SOIL POLLUTION FACTORS

BOD_{10}	47% ThOD	(256)
ThOD	2.55	

Biodegradation rates

adapted A.S. at 20°C-product is sole carbon source-94.5% COD removal at 14.7 mg. COD/g dry inoculum/h.

Waste water treatment

methods	*temp, °C*	*days observed*	*feed, mg/L*	*days acclim.*	*% removed*	
NFG, BOD	20	1-10	50	365 + P	51	
NFG, BOD	20	1-10	600-1,000	365 + P	inhibition	
RW, BOD	20	2-10	50	12	78	(93)

D. BIOLOGICAL EFFECTS

Fishes

Lepomis macrochirus: static bioassay in freshwater at 23°C, mild aeration applied after 24 h.

material added, mg/L	*% survival after* 24 h	*48 h*	*72 h*	*96 h*	*best fit 96h LC_{50}, mg/L*
320	40	40	40		
180	100	100	100	40	100
100	100	100	100	50	
79	100	100	100	70	(352)

Menidia beryllina: static bioassay in synthetic seawater at 23°C, mild aeration applied after 24 h.

material added, mg/L	*% survival after* 24 h	*48 h*	*72 h*	*96 h*	*best fit 96h LC_{50}, mg/L*
320	50	50	0	-	
210	100	100	100	0	
180	100	100	100	100-200 at 120 h	115 (352)

C. auratus: BCF 1.23 (1871)

acetate, sodium salt

$CH_3C(=O)ONa$

$C_2H_3O_2Na$ (a)

$C_2H_3O_2Na.3H_2O$ (b)

USES

dye and color intermediate; pharmaceuticals; cinnamic acid; soaps; photography; meat preservation; medicine; electroplating; tanning; buffer in foods.

D. BIOLOGICAL EFFECTS

Fishes

Lepomis macrochirus:	24h LC_{50}	5,000 mg/L	(1294)

Insects

Culex sp. larvae:	24h LC_{50}	7,500 mg/L	
	48h LC_{50}	7,425 mg/L	(1294)

acetic acid (ethanoic acid; methanecarboxylic acid; glacial acetic acid; vinegar acid)

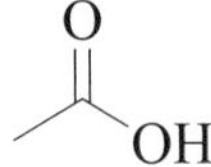

CH_3COOH

$C_2H_4O_2$

CAS 64-19-7

USES AND FORMULATIONS

food processing plants; organic chemical mfg.; nylon and fiber mfg.; dyestuff and pigments mfg.; vitamins, antibiotics, hormones mfg.; rubber mfg.; photographic chemicals mfg.; ester solvents mfg.; plastics mfg, cleaning/washing agents and disinfectants, food/foodstuff additive.

MANUFACTURING SOURCES

beetsugar mfg.; winery; vinegar mfg.; textile mills; wood distillation plants.

SOURCES

domestic use of vinegar; photographic film developing, lab use.

NATURAL SOURCES (WATER AND AIR)

both plants and animals as normal metabolite.

A. PROPERTIES

colorless liquid; molecular weight 60.05; melting point 16.7°C; boiling point 118.1°C; vapor pressure 11.4 mm at 20°C, 20 mm at 30°C, 4.7 hPa at 0°C, 15.7 hPa at 20°C, 45 hPa at 40°C, 269 hPa at 80°C, 555 hPa at 100°C; vapor density 2.07; sp. gr. 1.05 at 20/4°C; solubility 50,000 mg/L at 20°C = pH 2.5; saturation concentration in air 38 g/m^3 at 20°C, 63 g/m^3 at 30°C, $LogP_{OW}$ -0.31/-0.17.; LogH-4.91 at 25°C.

B. AIR POLLUTION FACTORS

1 mg/m³ = 0.401 ppm, 1 ppm = 2.494 mg/m³.

Odor

characteristic; quality: sour; hedonic tone: pungent.

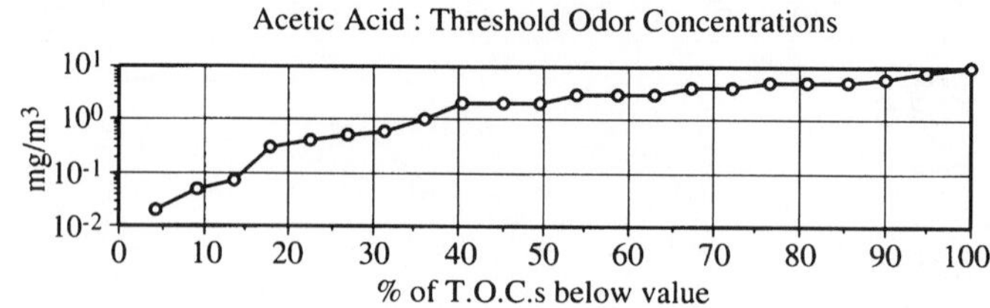

(57, 279, 610, 652, 670, 671, 679, 683, 709, 741, 753, 755, 788, 799, 825, 826, 835, 836)

odor index: 15,000. (2)

Control methods

thermal incineration: min. temp. for odor control: 743°C. (94)

Sampling and analysis

gas washing bottle: medium 200 ml water; sampling rate 0.12 cu ft/min; test conc. 520 ppm; absorption efficiency: +99%. (103)

Photodegradation

tropospheric $t_{1/2}$ based on reaction with OH°: 22 d. (3045)

C. WATER AND SOIL POLLUTION FACTORS

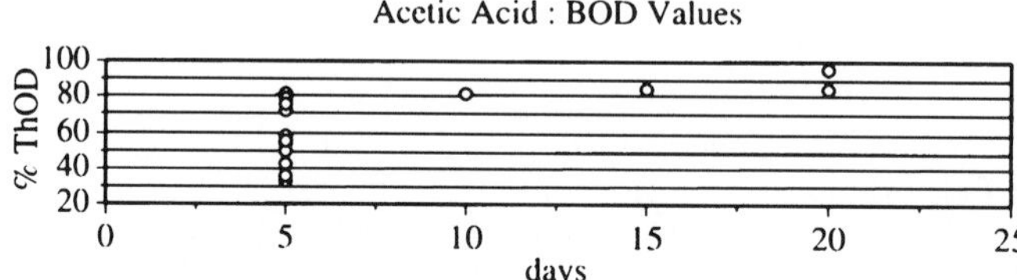

BOD_5:	32; 36; 42; 49; 50% ThOD	(220, 287, 284, 288, 281)
	55; 58; 58; 72; 79; 82% ThOD	(282, 289, 260, 259, 285, 41)
	76% ThOD in freshwater	(23)
	66% ThOD in seawater	(23)
BOD_{10}:	82% ThOD in freshwater 88% ThOD in seawater	(23)
BOD_{15}:	85% ThOD in freshwater 88% ThOD in seawater	(23)
BOD_{20}:	85% ThOD	(30)
	96% ThOD in freshwater 100% ThOD in seawater	(23)

BOD_{35}:	78% ThOD in seawater/inoculum enrichment cultures of hydrocarbon- oxidizing bacteria
ThOD:	1.07

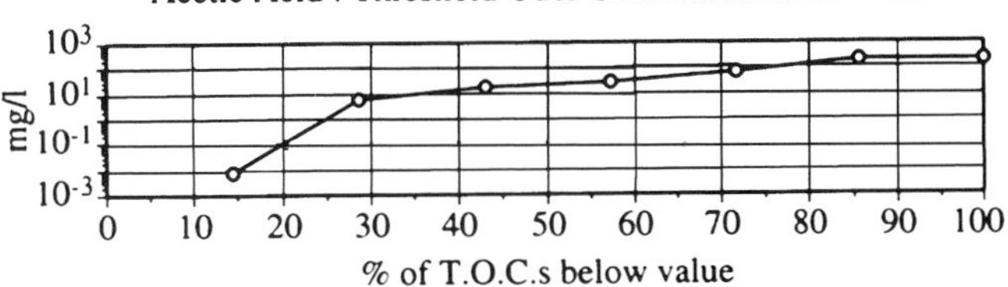

(97, 294, 295, 875, 896, 924)

Manmade sources

in domestic sewages: 2.5 to 36 mg/L	(85)
avg content of secondary sewage effluent: 0.130 mg/L	(86)

Waste water treatment

A.S.: 50% ThOD of 500 ppm by phenol acclimated sludge after 12 h aeration	(26)
A.C.: adsorbability: 0.048 g/g carbon, 24% reduction; influent: 1,000 mg/L, effluent: 760 mg/L	(32)
coagulation with 3 lb alum/1,000 gal: 8% BOD reduction	(95)
A.S.: after 6 h: 30% of ThOD; after 12 h: 35% of ThOD; after 24 h: 40% of ThOD	(89)
stabilization pond design: toxicity correction factor: 1.2 at 270 mg/L in influent	(179)
inoculum: nonadapted A.S.: at 100 mg/L at pH 7 ±1: after 140h: 100% removal	(3046)
inoculum: nonadapted industrial A.S.: after 5 d: 95% removal	(3047)

method	*temp., °C*	*d observed*	*feed, mg/L*	*d acclim.*	*theor. oxidation*	*% removed*	
A.S., BOD	20	1-5	333	15	-	99	
A.S., Resp., BOD	20	1-5	716	<1	53	99	(93)

powdered carbon: at 100 mg/L sodium salt (pH 7.5)-carbon dosage 1,000 mg/L: 1% adsorbed (520)

Impact on biodegradation processes

toxic to anaerobic bacteria at 2,500 mg/L after 24 h (3047)

The mineralization of acetate in methanogenic river sediment is affected by the following chemicals:

chemical	*EC_{50}*	
benzene	3,500 mg/kg dw	
pentachlorophenol	130 mg/kg dw	
1,2-dichloroethane	100 mg/kg dw	
chloroform	0.5 mg/kg dw	(2961)

Degradation in soil

biodegradation half-lives in nonadapted aerobic subsoil (acetate)

loam and sand:	0.36d	(acetate)	
sand, Oklahoma:	10d	(acetate)	
chalk, England:	16d	(acid)	(2695)

D. BIOLOGICAL EFFECTS

Plants

Phytotoxicity:

	EC_{50},* mg/m³	(95% conf.)
wheat	23	(5-48)
alfalfa	8	(6-10)
tobacco	41	(4-79)
soybean	20	(12-28)
corn	50	(35-65)

* EC_{50}, concentration required to cause visible injury in 50% of the leaves of the plant to population exposed for a 2-h fumigation period. (1831)

Toxicity threshold

(cell multiplication inhibition test)			
bacteria *(Pseudomonas putida)*	16h EC_0	2,850 mg/L	(1900)
algae *(Microcystis aeruginosa)*	85h EC_0	90 mg/L	(329)
green algae *(Scenedesmus quadricauda)*	16h EC_0	4,000 mg/L	(1900)
protozoa *(Entosiphon sulcatum)*	72h EC_0	78 mg/L	(1900)
protozoa (*Uronema parduczi* Chatton-Lwoff)	EC_0	1,350 mg/L	(1901)

Bacteria

Photobacterium phosphoreum	15 min EC_{50}	11 mg/L	(3057)

Algae

Euglena gracilis		15-30 min EC_{100}	720 mg/L at pH 3 (growth rate)	(3054)
Chlamydomonas dysosmos	motility	5 min NOEC	17 mg/L	
	immobil.	5 min EC_{100}	63 mg/L	(3055)
Chlamydomonas moewusii	motility	5 min NOEC	8.4 mg/L	
	immobil.	5 min EC_{100}	17 mg/L	(3055)
Chlamydomonas moewusii	motility	5 min NOEC	17 mg/L	
	immobil.	5 min EC_{100}	52 mg/L	(3055)
Chlamydomonas monoica	motility	5 min NOEC	33 mg/L	
	immobil.	5 min EC_{100}	131 mg/L	(3055)
Chlamydomonas reinhardi	motility	5 min NOEC	63 mg/L	
	immobil.	5 min EC_{100}	131 mg/L	(3055)
Chlamydomonas monoica	motility	5 min NOEC	66 mg/L	
	immobil.	5 min EC_{100}	131 mg/L	(3055)

Protozoa

Vorticella campanula:	perturbation level:	12 mg/L	(30)
Chilomonas paramecium	cell multipl. inhib.	408 mg/L	

Insects

Culex sp. larvae:	24-48h LC_{50}:	1,500 mg/L	(26)

Arthropoda

brine shrimp:	24-48h LC_{50}	42-32 mg/L	(23)
D. magna	24h LC_{50}	47 mg/L	(26)
D. magna	24h EC_0	78 mg/L	
	24h EC_{50}	95 mg/L	
	24h EC_{100}	114 mg/L	(3052)
D. magna at pH 7	24h EC_0	950 mg/L	
	24h EC_{50}	6,000 mg/L	
	24h EC_{100}	7,145 mg/L	(3052)
Gammarus pulex	perturbation level:	6 mg/L	(30)

Gammarus *(Hyale plumulosa)*	96h LC_{50}	310 mg/L	(2631)

Molluscs

Lymnaea ovata	perturbation level:	15 mg/L	(30)
Goniobasis livescens	24h LC_{50}	640 mg/L	
	48h LC_{50}	460 mg/L	(3053)
Lymnaea emarginata	24h LC_{50}	390 mg/L	
	48h LC_{50}	320 mg/L	(3053)

Fishes

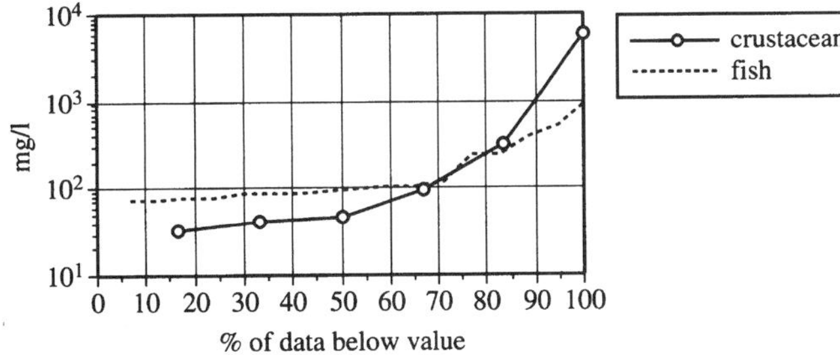

Frequency distribution curves of 24-96h EC_{50}/LC_{50} values for crustaceans (n = 6) and fishes (n = 17), based on data presented in this work

bluegill:	96h LC_{50}	75 mg/L	(23)
Lepomis macrochirus:	24h LC_{50}	100-1,000 mg/L	(26)
bluegill sunfish:	96h LC_{50}	75 mg/L	(3048)
mosquito fish:	24h-96h LC_{50}	251 mg/L	(41)
goldfish	lethal after 20h	423 mg/L	(154)
goldfish: period of survival at pH 6.8: 48 h to 4 d at 100 ppm			
goldfish: period of survival at pH 7.3: 4 d at 10 ppm			(157)
creek chub	24h LC_0	100 mg/L	(243)
creek chub	24h LC_{100}	200 mg/L	(243)
goldfish: LD_0: 10 mg/L; long-time exposure in hard water			(245)
fathead minnows: static bioassay in Lake Superior water at 18-22°C:			
	LC_{50}: (1; 24; 48; 72; 96 h):	>315; 122; 92; 88; 88 mg/L	
static bioassay in reconstituted water at 18-22°C, pH ≤ 5.9			
	LC_{50}: (1; 24; 48; 72; 96 h):	175; 106; 106; 79; 79 mg/L	(350)

Leuciscus idus melanotus	48h LC_0	368 mg/L	
	48h LC_{50}	410 mg/L	
	48h LC_{100}	452 mg/L	(3049, 3050)
Salmo gairdneri	24h LC_{100}	>315 mg/L	(3051)
red killifish *(Oryzias latipes)*	96h LC_{50}	in seawater: 530 mg/L	
		in freshwater: 94 mg/L	(2631)

acetic acid amine *see* acetamide

acetic aldehyde *see* acetaldehyde

acetic anhydride (acetic oxide; acetyl oxide; ethanoic anhydride)

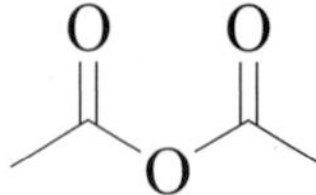

$(CH_3CO)_2O$

$C_4H_6O_3$

CAS 108-24-7

USES

acetylating agent in the chemical industry; intermediate in the mfg. of celluloseacetate; mfg. of fatty acids monoglycerides.

A. PROPERTIES

colorless liquid; molecular weight 102.09; melting point -68°C to -73°C, boiling point 139.9°C; vapor pressure 3.5 mm at 20°C, 5 mm at 25°C, 7 mm at 30°C, 4 hPa at 20°C, 13.3 hPa at 36°C, 17.3 hPa at 40°C; vapor density 3.52; sp. gr. 1.08 at 20/20°C; THC 431 kcal/mole, LHC 412 kcal/mole; saturation concentration in air 19 g/m^3 at 20°C; LogP$_{OW}$ -0.2 (measured).

B. AIR POLLUTION FACTORS

1 mg/m^3 = 0.236 ppm, 1 ppm = 4.24 mg/m^3.

Odor

sour acid, neutral to pleasant.

T.O.C.:	abs. perc. limit:	0.14 ppm
	50% recogn.	0.36 ppm
	100% recogn.	0.36 ppm
	O.I.: 14,611	

Photochemical degradation

reaction with OH°: $t_{1/2}$: 22 d (measured) (5649)

C. WATER AND SOIL POLLUTION FACTORS

Hydrolysis

At neutral pH and at 10-36 mg/L, acetic acid anhydride hydrolyses readily to acetic acid.

at 15°C	$t_{1/2}$: 8 min	
at 25°C	$t_{1/2}$: 4.4 min	(5642)

Biodegradability

inoculum/test method	*test conc.*	*test duration*	*results*	
A.S., industr., Zahn Wellens		5 d	>95% ThOD	(5644)
A.S., modif. MITI test	578 mg/L	24 hours	99% COD	(5648)

Impact on biodegradation processes

primary municip. sludge:	24h EC_0:	>2,500 mg/L	(5644)

D. BIOLOGICAL EFFECTS

Bacteria			
Pseudomonas putida	16h EC_0	1,150 mg/L	(1900)
Algae			
Chlorella pyrenoidosa:	toxic	360 mg/L	(41)
(reduction of chlorophyll content)	5d EC_{30}	100 mg/L	
	5d EC_{95}	400 mg/L	
green algae *(Scenedesmus quadricauda)*	7d EC_0	3,400 mg/L	(1900)
Microcystis aeruginosa	8d EC_0	18 mg/L	(5646)
Protozoans			
Entosiphon sulcatum	72h EC_0	30 mg/L	(1900)
Uronema parduczi at pH 7	EC_0	735 mg/L	(5650)
Chilomonas paramecium	48h EC_0	395 mg/L	(5651)
Crustaceans			
Daphnia magna	24h EC_0,S	47; 1,370 mg/L	
	24h EC_{50},S	55; 3,200 mg/L	
	24h EC_{100},S	68, 5,900 mg/L	(5645)
Fishes			
Leuciscus idus at pH 7	48h LC_0	216; 252 mg/L	
	48h LC_{50}	265; 279 mg/L	
	48h LC_{100}	324 mg/L	(5643)
	48h LC_0	>500 mg/L	(5644)

acetic ether *see* ethylacetate

acetic oxide *see* acetic anhydride

acetoacetanilide (3-oxo-N-phenylbutanamide; N-phenylacetoacetamide; N-(acetylacetyl)aniline)

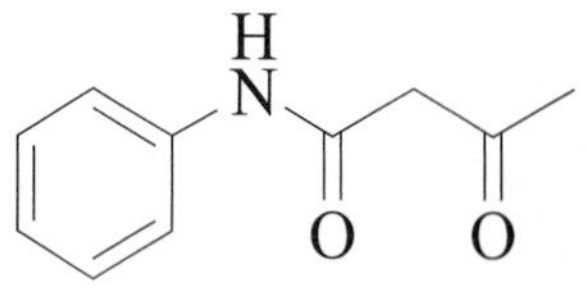

30 acetoacetic ester

$C_{10}H_{11}NO_2$

CAS 102-01-2

USES

intermediate in the production of dyes

A. PROPERTIES

molecular weight 177.2; melting point 92°C; vapor pressure 0.01 hPa at 20°C; density 1.26 at 20°C; solubility 5,000 mg/L at 20°C, 10,000 mg/L at 25°C; LogP_{OW} 0.76 (measured).

B. AIR POLLUTION FACTORS

Photochemical reactions

$t_{1/2}$: 0.22 d based on reaction with OH° (calculated) (6104)

C. WATER AND SOIL POLLUTION FACTORS

Biodegradability

Industrial A.S., Zahn-Wellens test after	3 h	Approx. 20% ThOD removal	
	5 d	85% ThOD removal	
	6 d	97% ThOD removal	(6100)

A.S., after 24 h: 18% COD and 50% TOC removal (6103)
A.S. at 100 mg/L after 24 h: 87% COD and 86% TOC removal (6102)

D. BIOLOGICAL EFFECTS

Bacteria			
Municipal sludge; A.S. respiration inhibition test:	3 h EC_0	1,250 mg/L	
	3 h EC_{50}	2,500 mg/L	(6100)
Soil bacteria in agar solution at 37°C:	48 h LC_{31}	100 mg/L	
	48 h LC_{44}	1,000 mg/L	
	48 h LC_{75}	10,000 mg/L	(6101)
Crustaceans			
Daphnia magma	24 h EC_0	17 mg/L	
	24 h EC_{50}	70- 200 mg/L	
	24 h EC_{100}	1,700 mg/L	(6100)
Fishes			
Brachydanio rerio	96 h LC_{50}	242-332 mg/L	(6098)

Cyprinus carpio: force-feeding method: no effect at 59-166 mg/kg body wt (6099)

acetoacetic ester *see* ethylacetoacetate

o-acetoacetotoluidide (butanamide,N-(2-methylphenyl)-3-oxo; AAOT; acetoacetyl-2-methylanilide)

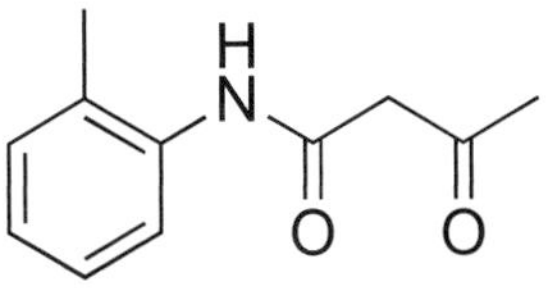

$C_{11}H_{13}NO_2$

CAS 93-68-5

USES

an intermediate in the synthesis of Pigment Yellow 9, 14, 16, 174 and Orange 1. These pigments are utilized in ink, paint, stationery goods, and coloring of resin, fiber, leather, paper, rubber, etc..

A. PROPERTIES

molecular weight 191.23; melting point 106°C; boiling point >170°C (scorched); relative density 1.3 kg/L; vapour pressure <130 Pa at 40°C; water solubility 3,000 mg/L at 25°C; log Pow 0.85 at 25°C

B. AIR POLLUTION FACTORS

Photodegradation

in sunlight: t/2 : 0.7 days (7456)

C. WATER AND SOIL POLLUTION FACTORS

Hydrolysis

t/2 > 5 days at pH 4 to 9 (7456)

Biodegradation

Inoculum	method	concentration	duration	elimination	
Activated sludge		191 mg DOC/L	1 day	>35 %	
			3 days	>65%	
			7 days	78%	(7456)
Activated sludge	Zahn-Wellens		5 days	>97% ThOD	
	Modified MITI	100 mg/L	14 days	17.6% ThOD	
			14 days	35.7 % TOC	(7456)

BOD5	0 mg/g	
BOD20	1680 mg/g	
ThOD	2280 mg/g	(7456)

D. BIOLOGICAL EFFECTS

Bioaccumulation

Species	conc. µg/L	duration	body parts	BCF	
calculated				3.2	(7456)

Toxicity

Micro organisms			
Pseudomonas putida	16h EC_{10}	ca 800 mg/L	(7456)
Algae			
Selenastrum capricornutum	72h ECb_{50}	383 mg/L	
	72h NOECb	95.3 mg/L	
	72h $ECgr_{50}$	607; 654 mg/L	
	72h NOECgr	171 mg/L	(7456)
Crustaceae			
Daphnia magna	48h EC_{50}	931 mg/L	
	48h NOEC	667 mg/L	
	96h EC_{50}	41 mg/L	
	96h EC_{100}	1000 mg/L	
reproduction	21d EC_{50}	16.5 mg/L	
	21d NOEC	10 mg/L	
	21d LOEC	20 mg/L	(7456)
Fish			
Oryzias latipes	96h LC_0-LC_{50}	>100 mg/L	
Pimephales promelas	96h LC_0	100 mg/L	
	96h LC_{50}	316 mg/L	
	96h LC_{100}	1000 mg/L	(7456)
Brachydanio rerio	96h LC_0-LC_{50}	>500 mg/L	(7456)
Mammals			
Rat	oral LD_{50}	ca 1600; 1854; 1945 ; 2500-5000; >5000 mg/kg bw	
	oral LD0	1024; 1280 mg/kg bw	
	oral $LD_{10}0$	>2000 mg/kg bw	(7456)

acetoin (3-hydroxy-2-butanone; acetylmethylcarbinol; dimethylketol)

O
OH

$CH_3COCHOHCH_3$

$C_4H_8O_2$

CAS 513-86-0

USES

preparation of flavors and essences.

A. PROPERTIES

slightly yellow liquid or crystalline solid; molecular weight 88.10; melting point 15°C; boiling point 148°C; sp. gr. 1.002 at 15/4°C; oxidizes gradually to diacetyl on exposure to air.

C. WATER AND SOIL POLLUTION FACTORS

Waste water treatment

A.S.: after 6 h: 2.6% ThOD; after 12 h: 4.3% ThOD; after 24 h: 14% ThOD.	(88)

2'-acetonaphthone (methyl-2-naphthyl ketone)

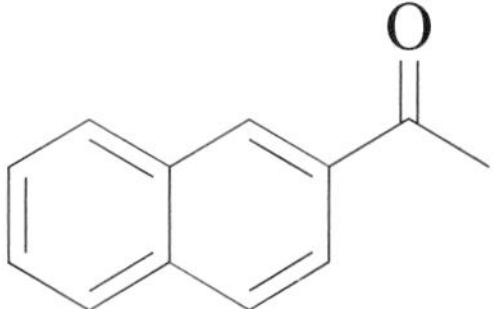

$C_{10}H_7COCH_3$

$C_{12}H_{10}O$

CAS 93-08-3

A. PROPERTIES

molecular weight 170.21; melting point 53-55°C; boiling point 300-301°C.

C. WATER AND SOIL POLLUTION FACTORS

Soil sorption

K_{OC}:	for Fullerton soil, 0.06% OC:	410	
	for Apison soil, 0.11% OC:	1,200	
	for Dormont soil, 1.2% OC:	950	(2599)

acetone (dimethylketone; 2-propanone; DMK)

CH_3COCH_3

C_3H_6O

CAS 67-64-1

USES AND FORMULATIONS

smokeless powder mfg.; paints, varnishes, lacquers mfg; organic chemical mfg.; pharmaceuticals mfg.; sealants and adhesives mfg.; solvents for cellulose acetate, nitrocellulose, acetylene. (347)

NATURAL SOURCES (WATER AND AIR)

normal microcomponent in blood and urine; minor constituent in pyroligneous acid; oxidation of alcohols and humic substances. (347)

A. PROPERTIES

clear, colorless liquid; molecular weight 58.08; melting point -95°C, boiling point 56.2°C; vapor pressure 89 mm at 5°C, 400 mm at 39.5°C, 270 mm at 30°C; vapor density 2.00; sp. gr. 0.791 at 20°C; THC 431 kcal/mole, LHC 407 kcal/mole; saturation concentration in air 553 g/m^3 at 20°C, 825 g/m^3 at 30°C; LogP_{OW} -0.24.

B. AIR POLLUTION FACTORS

1 mg/m^3 = 0.415 ppm, 1 ppm = 2.411 mg/m^3.

Odor

characteristic; quality: sweet, fruity; hedonic tone: pleasant to neutral.

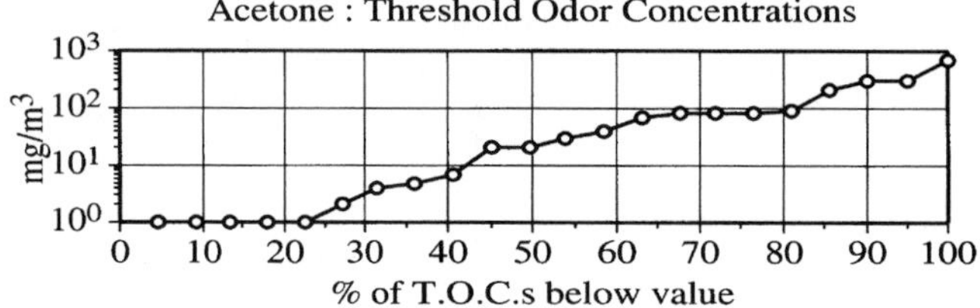

(19, 278, 279, 606, 643, 655, 675, 676, 709, 721, 722, 741, 749, 790, 804, 829, 844)

odor index: 1740 (19)

threshold for unadapted persons: 0.03% in diluent
threshold after adaption with pure odorant: 5.0% in diluent

human odor perception:	nonperception:		0.8 mg/m^3	
	perception:		1.1 mg/m^3	
	animal chronic exposure:	no effect:	0.5 mg/m^3	
	human reflex response:	no response:	0.35 mg/m^3	
		adverse response:	0.55 mg/m^3	(170)

Natural sources

glc's: Pt Barrow Alaska, Sept. 1967: 0.3 to 2.9 ppb			(101)
Manmade sources:	in cigarette smoke:	1,100 ppm	(66)
in gasoline exhaust:	2.3 to 14.0 ppm		
	(partly propionaldehyde)		(195, 1053)

Ambient air quality

Inside and outside 15 living rooms in Northern Italy (1983-84):

4-7 d averages	*lowest* *value*	*mean* *value*	*highest value* *(all in µg/m^3)*	
indoor	3	39	157	
outdoor	<2	6	16	(2756)

Manmade sources

In emissions of a municipal waste incinerator plant: 18 µg/m_3 = 1.5% of emissions of total organic carbon. (7048)

Partition coefficients

$K_{air/water}$	0.0042	
Cuticular matrix/air partition coefficient*	250 ± 20	(7077)
Cuticular matrix/water partition coefficient**	1.1	(7077)

* experimental value at 25°C studied in the cuticular membranes from mature tomato fruits (*Lycopersicon esculentum* Mill. cultivar Vendor)
** calculated from the Cuticular matrix/water partition coefficient and the air/water distribution coefficient

C. WATER AND SOIL POLLUTION FACTORS

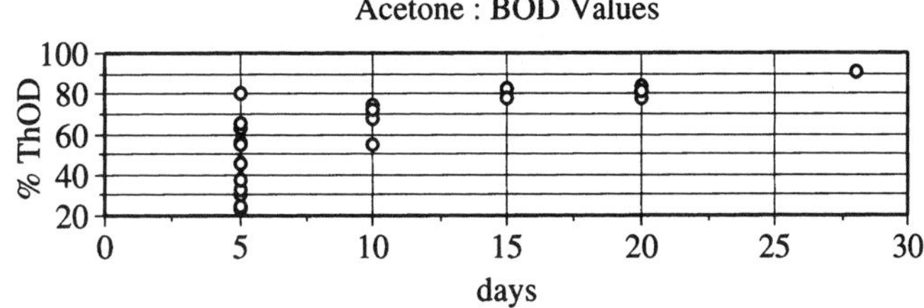

BOD_5:	23; 37; 45; 80% ThOD	(26, 27, 30)
	14-74% ThOD std. dil. acclim. sew.	(41)
	25; 31; 63% ThOD std. dil. sew.	(282, 285, 260)
	14% ThOD at 10 ppm, std. dil. sew.	(269)
	2.2-33% ThOD at 1.7-20 ppm, std. dil. sew.	(280)
	65% ThOD at 12 ppm, std. dil. sew.	(268)
	0% ThOD at 440 ppm, Sierp, 10% sewage	(280)
	56% ThOD in freshwater	
	38% ThOD in seawater	(23)
	46; 55% ThOD	(220, 79)
	38% ThOD (acclim.)	(2666)
	37% ThOD at 10 mg/L, unadapted sew., lag period: 2 d	(554)
BOD_{10}:	55% ThOD std. dil. sew.	(256)
	74% ThOD at 12 ppm, std. dil. sew.	(268)
	67% ThOD in freshwater	
	76% ThOD in seawater	(23)
	72% ThOD	(79)
	72% ThOD at 2.5 mg/L in mineralized dilution water with settled sewage seed	(405)
BOD_{15}:	83% bio. ox. in freshwater	
	69% bio. ox. in seawater	(23)
	78% ThOD	(79)
BOD_{20}	78; 81% ThOD	(26, 79)
	84% ThOD in freshwater	
	76% ThOD in seawater	(23)
	81% ThOD at 10 mg/L, unadapted sew.	(554)

[BOD values for freshwater and nonadapted inoculum are presented in the graph above.]

COD	51; 74; 87; 91; 94; 100% ThOD	(23, 27, 36, 41, 220, 277)
TOC	100% ThOD	(220)
ThOD	2.20	

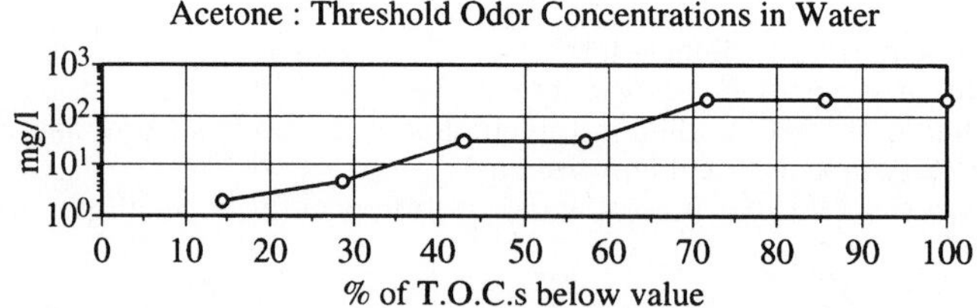

(97, 294, 296, 894, 907, 908)

Impact on biodegradation processes

BOD of sample + d at 20°C	BOD of original sample	BOD: 50 ppm acetone	
5	5 ppm	0 ppm	
10	13 ppm	1 ppm	
15	14 ppm	2 ppm	
20	15 ppm	4 ppm	
25	16 ppm	6 ppm	
30	16 ppm	6 ppm	(172)

Digestion of sludge is inhibited from 4,000 mg/L:
Nitrification of activated sludge is decreased with 75% at 840 mg/L. (30)
Slight inhibition of microbial growth follows 24 h exposure at 5 ppm. (523)
Approximately 50% inhibition of ammonia oxidation by Nitrosomonas occurs at 8,100 mg/L. (407)

Activated sludge oxygen consumption inhibition test (ISO):

municipal sludge:	EC_{50}: mean: 44 ml/L; SD: 5 ml/L	
factory sludge:	EC_{50}: mean: 27 ml/L; SD: 11 ml/L	(2802)
OECD 209 closed system inhib.:	EC_{50}: >1,000 mg/L:	
Activated sludge respiration test:	3h IC_{50}: >1,000 mg/L	
Inhibition of bacterial growth:	16h IC_{50}: >5,000 mg/L	(2676)

Manmade sources

in year-old leachate of artificial sanitary landfill: 0.6 g/L (1720)
landfill leachate from lined disposal area (Florida 1987-90): mean conc. 56 µg/L (2788)

Waste water treatment

A.C.: adsorbability: 0.043 g/g carbon; 21.8% reduction infl.: 1,000 mg/L, effl.: 782 mg/L (32)

anaerobic lagoon:

lb COD/d/1,000 cu ft	infl. mg/L	effl. mg/L	
13	150	60	
22	150	80	
48	150	70	(37)

aeration by compressed air (stripping effect): 72% removal in 8 h (30)

A.S. after	6 h	0.1% of ThOD
	12 h	0.0% of ThOD
	24 h	toxic

methods	*temp., °C*	*d observed*	*feed, mg/L*	*d acclim.*	*% removed*	
A.S., COD	20	1/3	333	30 +	86	
NFG, BOD	20	1-10	250-1,000	365 + P	47	(93)

Biodegradability

Ultimate biodegradability (CO_2 production) under aerobic conditions in the 'Sealed Vessel Test' (a modified Sturm Test) at room temp. using nonadapted secondary effluent as inoculum: 91% biodegradation after 28 d. (2596)

D. BIOLOGICAL EFFECTS

Bacteria

Pseudomonas putida	16EC_0	1,700 mg/L	(1900)

Vibrio fisheri Microtox	15min EC_{50}	85,000 mg/L	(7025)
Photobacterium phosphoreum Microtox™ test	5min EC_{20}	7,000 mg/L	
	5min EC_{50}	12,900 mg/L	
Photobacterium phosphoreum Microtox™ test	5min EC_{50}	8,600 mg/L	(7023)
Algae			
Microcystic aeruginosa	8d EC_0	530 mg/L	(329)
Scenedesmus quadricauda	7d EC_0	7,500 mg/L	(1900)
Rhaphidocellis subcapitata	72h EC_{50}	7,000 mg/L	(7025)
Protozoans			
Uronema parduczi Chatton-Lwoff	EC_0	1,710 mg/L	(1901)

Crustaceans

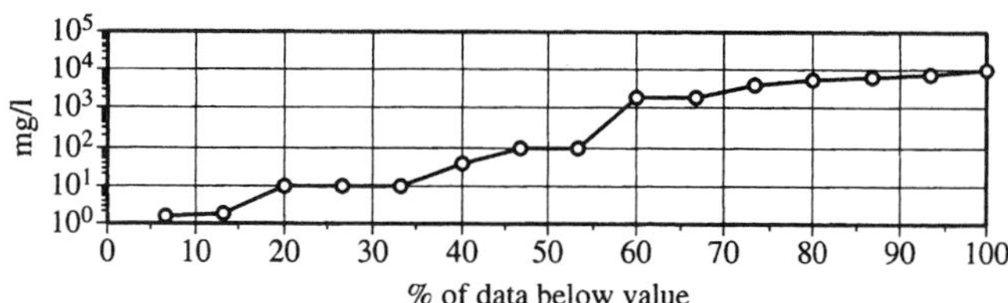

Frequency distribution of 24-96h LC_{50} values for crustaceans (n = 15) based on data from this work and from the work of Rippen (3999)

Daphnia magna	EC_{50} reproduction	4,000 mg/L	
	NOEC reproduction	3,200 mg/L	
	NOEC growth	1,000 mg/L	
	LC_{50}	7,100 mg/L	(2706)
	24-48h LC_{50}	10 mg/L	(26, 153)
	48h IC_{50}	15,800 mg/L	(7025)
brine shrimp	24h LC_{50}	2,100 mg/L	(23)
Gammarus pulex	LC_{50}	5,500 mg/L	(30)

Fishes

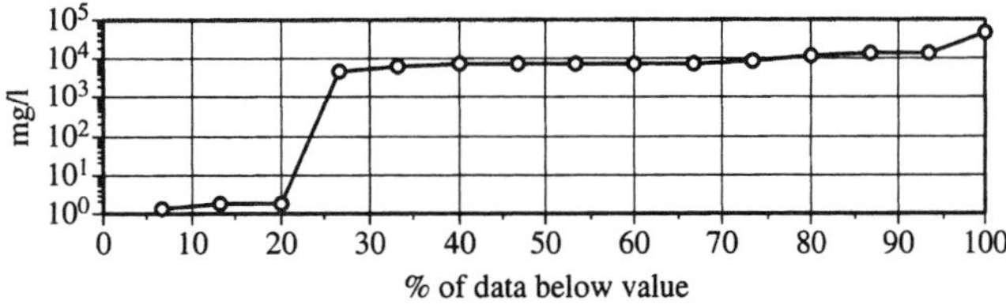

Frequency distribution of 24-96h LC_{50} values for fishes (n = 15) based on data from this work and from the work of Rippen (3999)

mosquito fish	24-96h LC_{50}	13,000 mg/L	(244)
goldfish	24h LC_{50}	5,000 mg/L	(277)
bluegill sunfish *(Lepomis macrochirus)*	96h LC_{50}	8,300 mg/L	(593)
fingerling trout	24h LC_{50},F	6,100 mg/L	(592)
guppy *(Poecilia reticulata)*	14d LC_{50}	7,032 mg/L	(1833)
Brachydanio rerio	96h LC_{50}	8,100 mg/L	(7025)

Amphibians

Mexican axolotl (3-4 w after hatching)	48h LC_{50}	20,000 mg/L	
clawed toad (3-4 w after hatching)	48h LC_{50}	24,000 mg/L	(1823)

acetonecyanohydrin (α-hydroxyisobutyronitrile; isopropylcyanohydrin; 2-hydroxy-2-methylpropanenitrile)

HO–C(CH3)2–CN

$(CH_3)_2C(OH)CN$

C_4H_7NO

CAS 75-86-5

USES

insecticides; intermediate for organic synthesis, especially methylmethacrylate.

A. PROPERTIES

Colorless liquid; molecular weight 58.10; melting point -19°C; boiling point 81°C at 23 mm, 120°C decomposes; vapor pressure 0.8 mm at 20°C; 1.1 hPa at 20°C; vapor density 2.95; sp. gr. 0.93 at 19°C; solubility miscible; $LogP_{oct}$ - 0.5/-1.0.

B. AIR POLLUTION FACTORS

1 mg/m^3 = 0.28 ppm, 1 ppm = 3.54 mg/m^3.

C. WATER AND SOIL POLLUTION FACTORS

Oxidation parameters

BOD_{20}	0.88	(7780)
COD	1.15; 1.69	(7781)

Hydrolysis

at 5 mg/L: $t_{1/2}$ = 7 d (7782)

HO–C(CH3)2–CN ⟶ HC≡N + (CH3)2C=O

Rapid hydrolysis to hydrocyanic acid and acetone (7779)

Waste water treatment

A.C.:

Influent	*Carbon dosage*	*Effluent*	*% removal*	
1,000 ppm	10 ×	400 ppm	60	
200 ppm	10 ×	110 ppm	45	
100 ppm	10 ×	70 ppm	30	(192)

Impact on digester sludge:
at 0.01-0.03 mg/L: no effect on either BOD or nitrification

at 1 mg/L: inhibition of BOD, nitrification, and growth			(7782)

D. BIOLOGICAL EFFECTS

Crustaceans			
Daphnia magna Straus	24 h EC_0	0.076 mg/L	
	24 h EC_{50}	0.19-0.38 mg/L	
	24 h EC_{100}	0.27 mg/L	(7778)
Daphnia magna Straus	48 h EC_0	0.076 mg/L	
	48 h EC_{50}	0.088-0.19 mg/L	
	48 h EC_{100}	0.13 mg/L	(7778)
Fishes			
Lepomis macrochirus	96 h LC_{50}	0.57 mg/L	(352)
Menidia beryllina	96 h LC_{50}	0.50 mg/L	(352)
Leuciscus idus	48 h LC_0	0.1; 0.8 mg/L	
	48 h LC_{100}	1.1; 1.9 mg/L	(7775)
Lepomis macrochirus	96 h LC_0	0.36 mg/L	
	96 h LC_{50}	0.32-0.58 mg/L	
	96 h LC_{100}	0.6 mg/L	
	96 h NOEC	0.079 mg/L	(7776)
Salmo gairdneri	96 h LC_0	0.078 mg/L	
	6 h LC_{50}	0.13-0.36 mg/L	(7777)

acetoneoxime *see* acetoxime

acetonitrile (ethanenitrile; methylcyanide)

$$H_3C-C\equiv N$$

CH_3CN

C_2H_3N

CAS 75-05-8

USES

solvent in hydrocarbon extraction processes, especially for butadiene; specialty solvent; intermediate; catalyst; separation of fatty acids from vegetable oils; manufacturing of synthetic pharmaceuticals.

A. PROPERTIES

colorless liquid; molecular weight 41.05; melting point -44/-41°C; boiling point 82°C; vapor pressure 74 mm at 20°C, 115 mm at 30°C; vapor density 1.42; sp. gr. 0.79 at 20/4°C; LHC 296-302 kcal/mole; saturation concentration in air 163 g/m^3 at 20°C, 249 g/m^3 at 30°C; $LogP_{OW}$ -0.34.

B. AIR POLLUTION FACTORS

1 mg/m3 = 0.586 ppm; 1 ppm = 1.706 mg/m3.

Odor

T.O.C.: 68 mg/m^3 = 39.8 ppm detection: 1,950 mg/m^3. (643)

Manmade sources

in emissions of a municipal waste incinerator plant: 14 μg/m_3 = 1.1% of emissions of total organic carbon. (7048)

Incinerability

temperature for 99% destruction at 2.0 sec. residence time under oxygen-starved reaction conditions: 1,000°C.
Thermal stability ranking of hazardous organic compounds: rank 17 on a scale of 1 (highest stability) to 320 (lowest stability). (2390)

Partition coefficients

$K_{air/water}$	0.0012	
Cuticular matrix/air partition coefficient*	380 ± 30	(7077)
Cuticular matrix/water partition coefficient**	0.44	(7077)

* experimental value at 25°C studied in the cuticular membranes from mature tomato fruits (*Lycopersicon esculentum* Mill. cultivar Vendor)
** calculated from the Cuticular matrix/water partition coefficient and the air/water distribution coefficient

C. WATER AND SOIL POLLUTION FACTORS

Waste water treatment

biodegradation by mutant microorganisms: 500 mg/L at 20°C:

% disruption:	parent:	100% in 9 h	
	mutant:	100% in 1.5 h	(152)

methods	*temp, °C*	*days observed*	*feed, mg/L*	*days acclim.*	*theor. oxidation*	*% removed*	
A.S., Resp.,BOD	20	1-5	490	<1	nil	17	
RW, CO_2,CAN	20	4	10	19	60	100	
RW, CO_2,CAN	20	8	39	25+	60	100	
RW, CO_2,CAN	5	25	39	25+ at 20°C	60	100	
ASC, C N	22-25	28	139	28	70+	98+	(93)

Impact on biodegradation processes

at 100 mg/L no inhibition of NH_3 oxidation by *Nitrosomonas* sp. (390)

D. BIOLOGICAL EFFECTS

Bacteria			
Biotox™ test*Photobacterium phosphoreum*	5min EC_{20}	10,000 mg/L	
	5min EC_{50}	12,600 mg/L	
Microtox™ test *Photobacterium phosphoreum*	5min EC_{50}	19,500 mg/L	(7023)
Pseudomonas putida	16h EC_0	680 mg/L	(1900)
Algae			
Microcystis aeruginosa	8d EC_0	520 mg/L	(329)
Scenedesmus quadricauda	7d EC_0	7,300 mg/L	(1900)

Protozoans			
Entosiphon sulcatum	72h EC_0	1,810 mg/L	(1900)
Uronema parduczi Chatton-Lwoff	EC_0	5,825 mg/L	(1901)

Fishes				
fathead minnows:	96h LC_{50}	hard water	1,020 mg/L	(41)
fathead minnows:	96h LC_{50}	soft water	1,000 mg/L	(41)
bluegill:	96h LC_{50}	soft water	1,850 mg/L	(252)
guppies:	96h LC_{50}	soft water	1,650 mg/L	(41)

3-(α-acetonylbenzene)-4-hydroxycoumarin *see* warfarin

acetophenone (methylphenylketone; hypnone; acetylbenzene; phenylmethylketone)

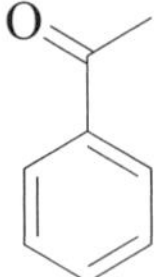

$C_6H_5COCH_3$

C_8H_8O

CAS 98-86-2

USES AND FORMULATIONS

perfume mfg.; solvent for synthesis of pharmaceuticals, rubber, chemicals, dyestuffs and corrosion inhibitors; plasticizer mfg.; tobacco flavorant; intermediate in synthesis of pharmaceuticals. (347)

MANUFACTURING SOURCES

organic chemical industry; coal processing industry. (347)

NATURAL SOURCES (WATER AND AIR)

oils of castoreum and labdanum resin; buds of balsam poplar; heavy oil fraction of coal tar. (347)

A. PROPERTIES

colorless liquid; molecular weight 120; melting point 19°C; boiling point 202°C; vapor pressure 1 mm at 15°C; vapor density 4.14; sp. gr. 1.03 at 20/4°C; solubility 5,500 mg/L; THC 989 kcal/mole; saturation concentration in air 1.96 g/m^3 at 20°C, 3.80 g/m^3 at 30°C; $LogP_{OW}$ 1.58.

B. AIR POLLUTION FACTORS

1 mg/m3 = 0.20 ppm, 1 ppm = 4.99 mg/m3.

Odor

characteristic; quality: sweet, almond hedonic tone: pleasant.

T.O.C.:	$0.01\ mg/m^3$ = 2 ppb		(57)
	average: 0.17 ppm, range: 0.0039 to 2.02 ppm, 17 panelists		(210)
	abs. perc. limit: 0.30 ppm		
	50% recogn.: 0.60 ppm		
	100% recogn.: 0.60 ppm		
	O.I.: 2183		
	human odor perception:	$0.01\ mg/m^3$	(19)
	human reflex response: no response:	$0.003\ mg/m^3$	
	adverse response:	$0.007\ mg/m^3$	
	animal chronic exposure: adverse effect:	$0.07\ mg/m^3$	(170)

Manmade sources

in gasoline exhaust: <0.1 to 0.4 ppm

Incinerability:

Temperature for 99% destruction at 2.0 sec residence time under oxygen-starved reaction conditions: 775°C

Thermal stability ranking of hazardous organic compounds

rank 85 on a scale of 1 (highest stability) to 320 (lowest stability) (2390)

C. WATER AND SOIL POLLUTION FACTORS

BOD_5:	20; 20; 32% ThOD	(30, 27, 220)
	59% ThOD (acclim.)	(2666)
BOD_{10}:	55% ThOD	(256)
COD:	100% ThOD	(220)
ThOD:	2.53	(30)

Biodegradation

(426)

Taste

in FISH: at 0.5 mg/L	(41)
T.O.C.: average: 0.17 mg/L, range 0.0039 to 2.02 mg/L	(97)
0.17 mg/L	(297)
0.065 mg/L	(297)
T.O.C. in water: 68 ppm	(326)
Water quality: in Kanawha river: 13.11.1963:	
raw: 19 ppb	
sand-filtered: not detectable: 21.11.1963:	
raw: 29 ppb	
after aeration: 17 ppb	
carbon-filtered: not detectable	(159)

Waste water treatment

A.C. adsorbability: 0.194 g/g carbon, 97.2% reduction, infl.: 1,000 mg/L effl.: 28 mg/L	(32)
Conventional municipal treatment: infl. 0.019 mg/L, effl. n.d.	(404)
Conventional + A.C.: infl. 0.021 mg/L, effl. n.d.	(404)
NFG, BOD, 20°C, 1-10 d observed, feed: 100-1,000 mg/L, acclim.: 365 + P; 14% removed	(93)

Soil sorption

K_{oc}:	for Fullerton soil, 0.06% OC:	270	
	for Apison soil, 0.11% OC:	185	
	for Dormont soil, 1.2% OC:	105	(2599)

D. BIOLOGICAL EFFECTS

Fishes

fathead minnows: static bioassay in Lake Superior water at 18-22°C:
1; 24; 48; 72; 96h LC_{50}; >200; >200; 103; 158; 155 mg/L (350)

acetothioamide *see* thioacetamide

acetoxime (2-propanoneoxime; acetoneoxime)

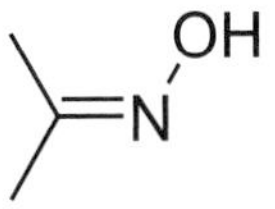

$(CH_3)_2CNOH$

C_3H_7NO

USES

organic synthesis (intermediate); solvent.

A. PROPERTIES

molecular weight 73.09; melting point 61°C; boiling point 136.3°C; sp. gr. 0.97 at 20/20°C.

C. WATER AND SOIL POLLUTION FACTORS

BOD_5:	35% ThOD	
ThOD:	2.180	(30)

D. BIOLOGICAL EFFECTS

Bacteria

Pseudomonas: toxic at 300 mg/L (30)

aceturic acid *see* N-acetylglycine

N-acetyl-2-aminoethanoic acid *see* N-acetylglycine

N-acetyl-p-phenylenediamine *see* p-aminoacetanilide

1-acetyl-2-thiourea (N-acetyl thiourea; acetamide, N-(aminothiooxomethyl))

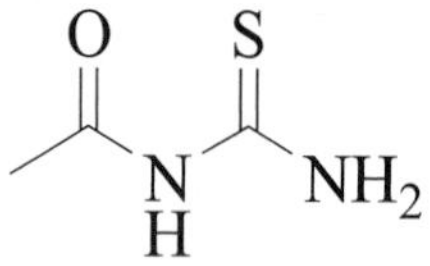

$CH_3CONHCSNH_2$

$C_3H_6N_2OS$

CAS 591-08-2

A. PROPERTIES

molecular weight 118.15; melting point 165-169°C.

B. AIR POLLUTION FACTORS

Incinerability

thermal stability ranking of hazardous organic compounds: rank 286 on a scale of 1 (highest stability) to 320 (lowest stability). (2390)

C. WATER AND SOIL POLLUTION FACTORS

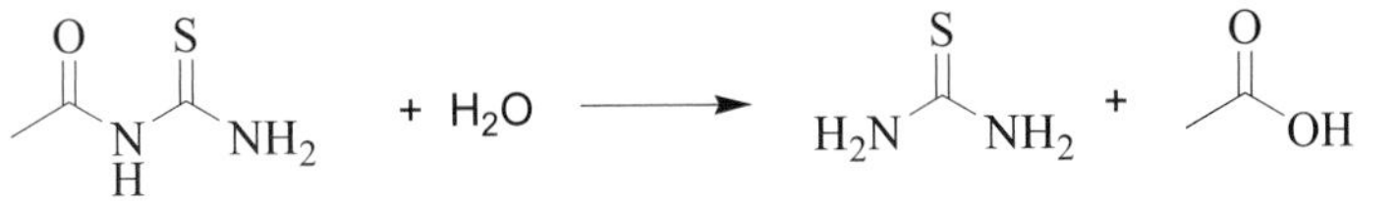

$CH_3CONHCSNH_2 + H_2O \rightarrow H_2NCSNH_2 + CH_3COOH$

thiourea acetic acid

Hydrolysis

$t_{1/2}$: 2.7 h at pH 9.6. Hydrolysis products include acetic acid and thiourea. (9770, 9771)

D. BIOLOGICAL EFFECTS

Mammals

rat and mouse	oral LDLo	50-94 mg/kg bw	(9772, 9773)

acetylacetone *see* 2,4-pentanedione

acetylaminoacetic acid *see* N-acetylglycine

9-acetylanthracene

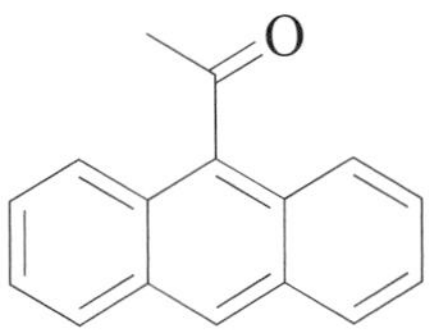

$C_{16}H_{12}O$

CAS 784-04-3

A. PROPERTIES

molecular weight 220.27; melting point 75-76°C.

C. WATER AND SOIL POLLUTION FACTORS

Soil sorption

K_{oc}:

for Fullerton soil, 0.06% OC:	1,170
for Apison soil, 0.11% OC:	8,400

for Dormont soil, 1.2% OC:	1,750	(2599)

acetylbenzene *see* acetophenone

4-acetylbiphenyl

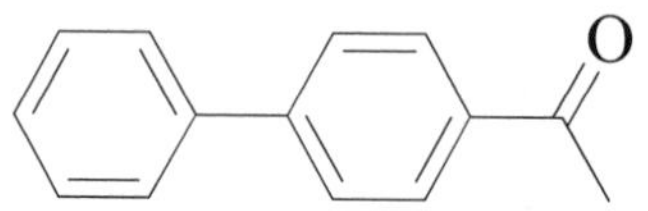

$C_{14}H_9COCH_3$

$C_{16}H_{12}O$

CAS 92-91-1

A. PROPERTIES

molecular weight 196.25; melting point 116-118°C.

C. WATER AND SOIL POLLUTION FACTORS

Soil sorption

K_{oc}:

for Fullerton soil, 0.06% OC:	710	
for Apison soil, 0.11% OC:	2,400	
for Dormont soil, 1.2% OC:	1,900	(2599)

acetylchloride (acetic chloride; ethanoyl chloride)

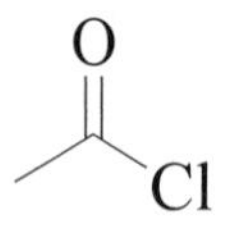

CH_3COCl

C_2H_3ClO

CAS 75-36-5

USES

chlorination of alkenes, triarylcarbonyls, inorganic compounds; catalyst for organic reactions; used in synthesis of pharmaceuticals, dyes, and pesticides.

A. PROPERTIES

molecular weight 78.5; melting point -112°C; boiling point 51°C; vapor pressure approx 320 hPa at 20°C; density 1.1 at 20°C.

B. AIR POLLUTION FACTORS

Photochemical degradation

reaction with OH°: $t_{1/2}$: 236 d (calculated) (5409)

Incinerability

Temperature for 99% destruction at 2.0 sec residence time under oxygen-starved reaction conditions: 765°C

Thermal stability ranking of hazardous organic compounds

rank 92 on a scale of 1 (highest stability) to 320 (lowest stability) (2390)

C. WATER AND SOIL POLLUTION FACTORS

Acetylchloride reacts vigorously with water to form acetic acid and HCl.

D. BIOLOGICAL EFFECTS

Fishes

Pimephales promelas	96h LC_{50}	42 mg/L	(5408)

acetylene (ethine; ethyne)

$$HC{\equiv}CH$$

C_2H_2

CAS 74-86-2

A. PROPERTIES

Colorless gas, molecular weight 26.04; melting point -81.8°C; boiling point -84°C sublimes; vapor density 0.91; sp. gr. 0.62 (liquefied); THC 312 kcal/mole, LHC 302 kcal/mole; log H -0.01 at 25°C. (67)

B. AIR POLLUTION FACTORS

1 mg/m^3 = 0.92 ppm, 1 ppm = 1.08 mg/m^3.

Odor threshold

detection: 240 mg/m^3 (637)
1,300-2,750 mg/m^3 (609)

Atmospheric reactions

reactivity; NO ox.: ranking: 0.1. (63)

Atmospheric half-lives

reactions with OH°: 11 d reactions with O_3: 100 d. (2716)

Manmade sources

diesel engine: 14.1% of emitted hydrocarbons (72)

reciprocating gasoline engine: 12-17% of emitted HCs (78, 391, 392, 393)

rotary gasoline engine: 3.3% of emitted HCs

expected glc's in U.S. urban air: 15 to 250 ppb (102)

Estimated lifetime under photochemical smog conditions in SE England: 206 h (1699, 1700)

D. BIOLOGICAL EFFECTS

Seed Plants

sweet pea: declination in seedling: 250 ppm, 3 d

tomato: epinasty in petiole: 50 ppm, 2 d (109)

Toxicity

Fishes

river trout	33h LC_{50}	200 mg/L	(30)

acetylenedichloride *see* 1,2-dichloroethylene

acetylenetetrabromide (1,1,2,2-tetrabromoethane; sym-tetrabromoethane)

Br Br
Br Br

$CHBr_2CHBr_2$

$C_2H_2Br_4$

CAS 79-27-6

USES

Solvent. Mercury substitute in gauges and balances. Separates metals by density.

A. PROPERTIES

colorless to yellow liquid; molecular weight 345.70; melting point 0.1°C; boiling point 239-242°C; vapor pressure 0.1 mm at 20°C; vapor density 11.9; sp. gr. 2.964 at 20/4°C; solubility 651 mg/l at 30°C.

B. AIR POLLUTION FACTORS

1 mg/m^3 = 0.07 ppm, 1 ppm = 14.37 mg/m^3.

Manmade sources

Downtown Los Angeles: glc's: 0.068 to 0.234 ppm C
East San Gabriel Valley: glc's: 0.050 to 0.103 ppm C (52)
Photochemical reactions: $t_{1/2}$: 1.8 months, based on reactions with OH° (calculated). (8649)

D. BIOLOGICAL EFFECTS

Mammals			
mouse	oral LD_{50}	270 mg/kg bw	
guinea pig	oral LD_{50}	400 mg/kg bw	
rabbit	oral LD_{50}	400 mg/kg bw	
rat	oral LD_{50}	1200 mg/kg bw	(9701, 9767, 9768, 9769)

acetylenetetrachloride *see* 1,1,2,2-tetrachloroethane

acetylenylcarbinol *see* propargyl alcohol

N-acetylethanolamine (N-(2-hydroxyethyl)acetamide)

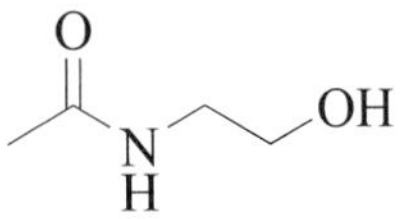

$CH_3CONHC_2H_4OH$

$C_4H_9NO_2$

CAS 142-26-7

A. PROPERTIES

molecular weight 103.12; melting point 63-65°C; boiling point 166°C at 8 mm; density 1.11 at 25/4°C.

C. WATER AND SOIL POLLUTION FACTORS

BOD_{10}: 1.10 std. dil. sew. (256)

Waste water treatment: NFG, BOD, 20°C, 1-10 d observed, feed: 200-1,000 mg/L; acclim.: 365 + P: 40% removed. (93)

N-acetylglycine (N-acetyl-2-aminoethanoic acid: aceturic acid; acetamidoacetic acid; acetylaminoacetic acid)

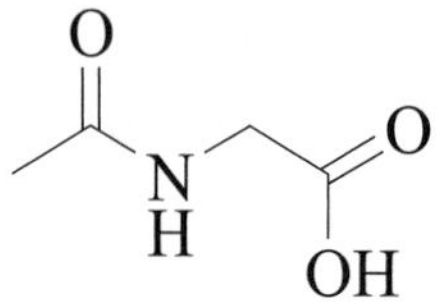

$CH_3CONHCH_2COOH$

$C_4H_7NO_3$

CAS 543-24-8

USES

medicine.

A. PROPERTIES

molecular weight 117.10; melting point 206°C; solubility 21,700 mg/L at 15°C; $LogP_{OW}$ -1.80/ -1.31 (calc.).

C. WATER AND SOIL POLLUTION FACTORS

A.S.:

after 6 h:	9.3% ThOD	
12 h:	10.0% ThOD	
24 h:	18.5% ThOD	(89)

4-acetylguaiacol

$CH_3C(O)OC_6H_3(OH)OCH_3$

$C_9H_{10}O_3$

B. AIR POLLUTION FACTORS

Manmade sources

avg. conc. in smoke from 28 woodstoves and fireplaces:

burning hardwood	3.9 g/kg carbon collected on filter	
softwood	3.4 g/kg carbon collected on filter	(2414)

acetylmethylcarbinol *see* acetoin

4-acetylmorpholine (N-acetylmorpholine)

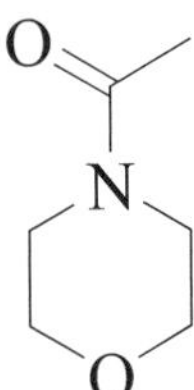

$C_6H_{11}NO_2$

CAS 1696-20-4

A. PROPERTIES

molecular weight 129.16; melting point 14°C; boiling point 152°C at 50 mm (decomposes); vapor pressure 0.02 mm at 20°C; vapor density 4.46; sp. gr. 1.12 at 20/20°C.

C. WATER AND SOIL POLLUTION FACTORS

BOD_{10}: nil. std. dil. sew.

Waste water treatment

NFG, BOD, 20°C, 1-10 d observed, feed: 200-1,000 mg/L, acclim.: 365 +P: nil% removed. (93)

acetyloxide *see* acetic anhydride

acetylsalicylic acid

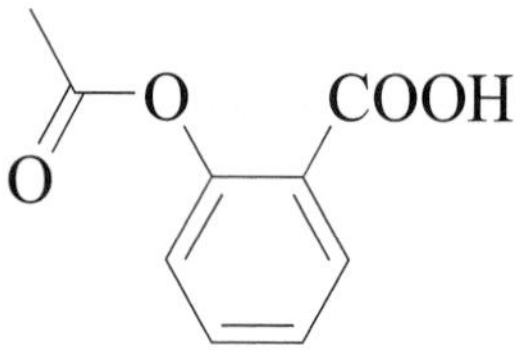

$C_9H_8O_4$

$CH_3COOC_6H_4COOH$

$C_9H_8O_4$

CAS 50-78-2

USES

medicine (analgesic, anti-inflammatory, antipyretic).

A. PROPERTIES

molecular weight 180.16; melting point 138-140°C.

C. WATER AND SOIL POLLUTION FACTORS

	range µg/L	n	median µg/L	90 percentile µg/L	
influent MWTP	3.2	6			(7237)
surface waters (1996-2000)	<0.020-0.34	43	<0.020	0.16	
	<0.020	7			
	<0.050	1			
	<0.010	19			(7237)
drinkingwater (1996)	0.290				(7237)

D. BIOLOGICAL EFFECTS

Toxicity

Crustaceae			
Daphnia magna	21d $ECrepr_{50}$	61-68 mg/L	(7237)
	24h LC_{50}	168 mg/L	

Marine tests

Microtox^{TM} *(Photobacterium)* test	5min EC_{50}	26 mg/L	
Artoxkit M *(Artemia salina)* test	24h LC_{50}	381 mg/L	
Freshwater tests			
Streptoxkit F *(Streptocephalus proboscideus)* test	24h LC_{50}	178 mg/L	
Rotoxkit F *(Brachionus calyciflorus)* test	24h LC_{50}	141 mg/L	(2945)

Acibenzolar-s-methyl (Methyl benzo[1,2,3]thiadiazole-7-carbothioate; 1,2,3-benzothiadiazole-7-carbothioic acid-S-methyl ester)

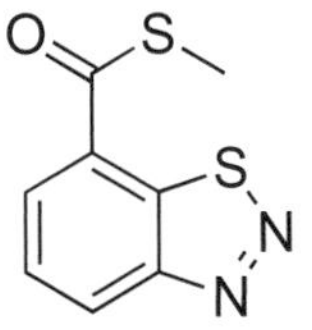

$C_8H_6N_2OS_2$

CAS 135158-54-2

USES

Plant activator, a selective systemic compound which induces host plant resistance

A. PROPERTIES

White crystalline powder; molecular weight 210.3; melting point 133°C; boiling point approx. 267°C; vapour pressure 4.6 x 10^{-4} Pa at 25°C; water solubility 7.7 mg/L at 25°C; log Pow 3.1 at 25°C; H 1.3 x 10^{-2} Pa.m^3/mol

B. AIR POLLUTION FACTORS

Photodegradation

T/2 (OH radicals)= 19040 hours

C. WATER AND SOIL POLLUTION FACTORS

Hydrolysis

T/2 : 3.8 years at pH 5, 23 weeks at pH 7, 19 hours at pH 9

Photodegradation

T/2 = 0.9 hours

Biodegradation

The main degradate is CGA-210007 = benzo[1,2,3]thiadiazole-7-carboxylic acid

1,2,3-benzothiadiazole-7-carbothioic acid -S-methyl ester

CGA-210007
benzo[1,2,3]thiadiazole-7-carboxylic acid

Degradation in water/sediment systems

	% mineralisation	% bound-residues	After
Biodegradation in soil aerobic	9.5	26-36	1 year

Biodegradation in laboratory studies (10777)

degradation in soils	%	Acibenzolar-methyl	CGA 210007
Aerobic at 20°C	50	0.2-0.5 days	16-106 days
Aerobic at 20°C	90	0.7-1.7 days	55-354 days
Aerobic at 10°C	50	1 day	70 days
Aerobic at 10°C	90	3.3 days	234 days

Dissipation studies in the field at 20°C (10777)

% dissipation	Acibenzolar-methyl	CGA 210007
50%	0.1-27 days	7-59 days

Biodegradation in water/sediment systems (10777)

	Acibenzolar-methyl	CGA 210007
DT50 water	<1 day	164-293 days
DT90 water	<3 days	544-973 days
DT50 whole system	<1 day	287-426 days
DT90 whole system	2.1-2.7 days	953-1414 days

Mobility in soils (10777)

	Acibenzolar-methyl	CGA 210007
K_{OC}	1041-1885	40-312

D. BIOLOGICAL EFFECTS

Bioaccumulation (10778)

Species	conc. µg/L	duration	body parts	BCF
Fish	16		Whole fish	118
	156			117

Toxicity

Algae			
Scenedesmus subspicatus	72h EC_{50}	0.5 mg/L	
Green algae	72h EC_{50}	3.3 mg/L	(10778)
Plants			
Lemna gibba	EC_{50}	0.31 mg/L	(10778)
Birds			
Bobwhite quail	Acute oral LD_{50}	>2,000 mg/kg bw	
Mallard duck	Acute oral LD_{50}	>2,000 mg/kg bw	(10778)
Worms			
Eisenia foetida	Acute LC_{50}	>1,000 mg/kg soil	
	NOEC	<12.3 mg/kg soil	(10778)
Crustaceae			
Daphnia magna	48h EC_{50}	2.4 mg/L	
	21d NOEC	0.044 mg/L	
Mysid shrimp	48h LC_{50}	0.88 mg/L	

Insecta			
Honeybees	Acute oral LD_{50}	>128 μg/bee	
	Acute contact LD_{50}	>100 μg/bee	(10778)
Mollusca			
Crassostrea virginica	EC_{50}	0.59 mg/L	(10778)
Fish			
Rainbow Trout	96h LC_{50}	0.4 mg/L	
	87d NOEC	0.026 mg/L	
Bluegill sunfish	96h Lc50	1.6 mg/L	
Sheephead minnow	96h LC_{50}	1.7 mg/L	(10778)
Mammals			
Rat	oral LD_{50}	>2,000 mg/kg bw	(10778)

Toxicity of metabolite ***CGA 210007*** (10778)

Rainbow Trout	96h LC_{50}	>100 mg/L
Daphnia magna	48h EC_{50}	58 mg/L
Scenedesmus subspicatus	72h $ECgr_{50}$	90 mg/L
Eisenia foetida	NOEC	2.67 mg/kg soil

Acid Blue 80

(Benzenesulfonic acid, 3,3'-[(9,10-dihydro-9,10-dioxo-1,4-anthracenediyl)diimino]bis[2,4,6-trimethyl-, disodium salt; sodium 3,3'-(9,10-dioxoanthracene-1,4-diyldiimino)bis(2,4,6-trimethylbenzenesulphonate); Acid Blue 80; Benzenesulfonic acid, 3,3'-[(9,10-dihydro-9,10-dioxo-1,4-anthracenediyl)diimino]bis[2,4,6-trimethyl-, disodium salt)

O ONa S=O HN NH O O S ONa O O

$C_{32}H_{28}N_2O_8S_2Na_2$

CAS 4474-24-2

SMILES

O=C(C3=C2C=CC=C3)C1=C(NC5=C(C)C(S(=O)([O])=O)=C(C)C=C5C)C=CC(NC4=(C)C(S(=O)([O-])=O)=C(C)C=C4C)=C1C2=O.[Na+].[Na+]

TRADENAMES

ALIZARIN E; Acid Anthraquinone Brilliant Blue; Acid Brilliant Blue Anthraquinone; Acid Brilliant Blue RAWL; Alizarine Blue BL; Alizarine Fast Blue R; Alizarine Milling Blue R; Atlantic Alizarine; Milling Blue RB; Brilliant Alizarine Milling Blue BL; C-WR Blue 10; C.I. 61585; C.I. Acid Blue 80; Coomassie Blue B; Endanil Blue B; Nylosan Blue C-L; Nylosan Blue F-L; Nylosan Blue F-L 150; Polar Brilliant Blue RAW; Polar Brilliant Blue RAWL; Sandolan Milling Blue N-BL; Sandolan Milling N-BL; Stenolana Brilliant Blue BL; Weak Acid Brilliant Blue RAW

USES

There are 4 categories of potential uses for Acid Blue 80:

1.Cleaning/washing agents; sanitation agents; colouring agents; non-agricultural pesticides and preservatives used in the following areas: general cleaning activities; industrial cleaning; specialized cleaning activities textile industry; construction; health and social work;
2.Detergent and textiles applications;
3.Disinfectants and sterilants ingredient; and

A. PROPERTIES

molecular weight 678.69; melting point 350°C; boiling point 1018°C; vapour pressure 7 x 10^{-24} Pa; water solubility 0.02 µg/L; log Pow 6.6; H 1 x 10^{-18} Pa.m^3/mol

B. AIR POLLUTION FACTORS

Photodegradation

t/2 : 0.09 days. (11172)

C. WATER AND SOIL POLLUTION FACTORS

Mobility in soils

Log K_{OC}	5.6 (modelled)	(11172)

D. BIOLOGICAL EFFECTS

Bioaccumulation

Species	BCF	
Fish	40,000-1,975,000 (modelled)	(11172)

Toxicity

Fish	14d LC_{50}	0.6-60 µg/L (modelled)	(11172)

acid butylphosphate *see* butylphosphate

acifluorfen (5-[2-chloro-4-(trifluoromethyl)phenoxy]-2-nitrobenzoic acid)

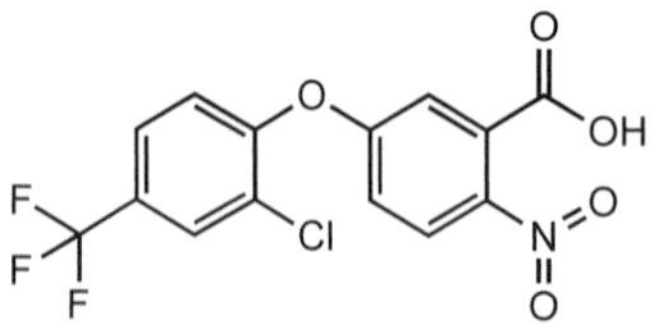

$C_{14}H_7ClF_3NO_5$

CAS 50594-66-6

USES

A diphenyl ether herbicide. A phototoxic pesticide.

OCCURRENCE

Acifluorfen is a metabolite of acifluorfen sodium and Lactofen

A. PROPERTIES

molecular weight 361.7; water solubility 120 mg/L

C. WATER AND SOIL POLLUTION FACTORS

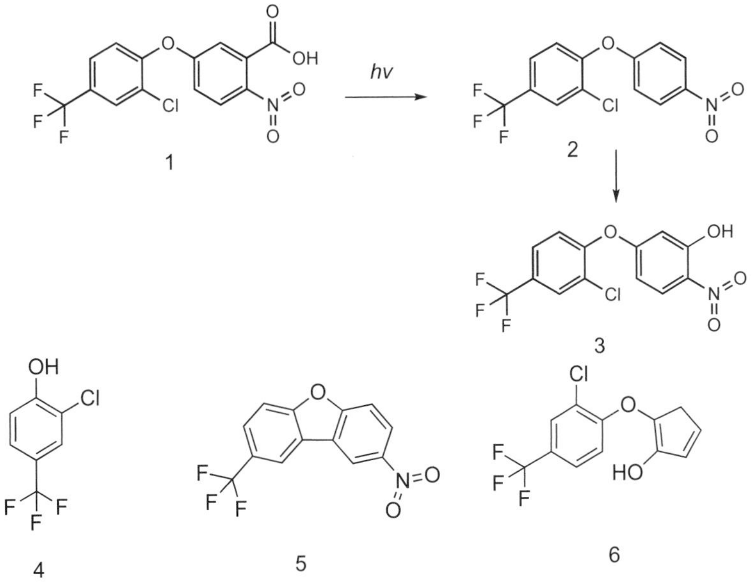

hv

1

2

3

4

5

6

The mechanism of acifluorfen photodegradation

When the reaction was stopped after 2 h of irradiation the following degradates were identified : 2-chloro-1-(4-nitrophenoxy)-4-trifluoromethylbenzene (nitrofluorfen), (Compound 2), 5-[2-chloro-4-(trifluoromethyl) phenoxy]-2-nitrophenol (hydroxy-nitrofluorfen) (Compound 3).

When the photochemical reaction was allowed to run its course, after 36 h irradiation in addition to Compounds 2 and 3, other substances were detected in the reaction mixture. These included 2-chloro-4-(trifluoromethyl)phenol (Compound 4); 5-trifluoromethyl-5'-nitrodibenzofuran, (Compound 5); and 2-[2-chloro-4-(trifluoromethyl)phenoxy]-cyclopentadien-1-ol, (Compound 6). (10730)

Biodegradation

Although acifluorfen biodegradation may largely be a cometabolic process, certain bacterial strains are capable of metabolizing the herbicide. Degradation products of acifluorfen isolated from microbial cultures include aminoacifluorfen, 5-[2-chloro-4-(trifluormethyl)phenyl]oxy)-2-aminobenzamide, and 5-([2-chloro-4-(trifluoromethyl)phenyl]oxy-2-(acetylamino)benzoic acid. Aminoacifluorfen has been recovered from soil treated with acifluorfen that was incubated 6 days. (10732)

D. BIOLOGICAL EFFECTS

For aquatic toxicity see acifluorfen sodium.

Toxicity

Mammals

Rat	oral LD_{50}	1,370-2,050 mg/kg bw	(10730)

Acifluorfen, sodium salt (Benzoic acid, 5-(2-chloro-4-(trifluoromethyl)phenoxy)-2-nitro-, sodium salt; Sodium 5-((2-chloro-alpha,alpha,alpha-trifluoro-p-tolyl)oxy)-2-nitrobenzoate; Sodium 5-(2-chloro-4-(trifluoromethyl)phenoxy)-2-nitrobenzoate)

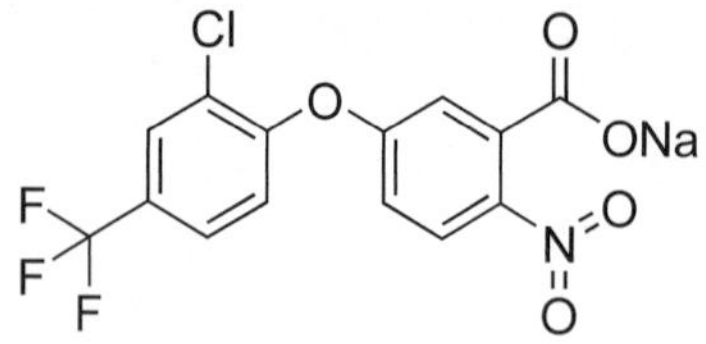

$C_{14}H_6ClF_3NO_5Na$

CAS 62476-59-9

TRADENAMES

Blazer; Blazer 2L; Blazer 2S; MC 10978; RH-6201; Tackle 2AS

USES

Sodium acifluorfen is an herbicide for post-emergent weed control on agricultural crops and for residential spot treatment. Formulations include liquid, ready-to-use and soluble concentrate. Sodium acifluorfen is sometimes formulated or packaged with other herbicides, such as bentazon, imazaquin, sethoxydim, or glyphosate. Sodium acifluorfen is applied by aircraft, boom sprayer, and other ground equipment, and by hand held sprayer and trigger bottle. Sodium acifluorfen belongs to a class of compounds known to have a phototropic mode of action in plants and animals.

A. PROPERTIES

molecular weight 361.70; melting point 172-176°C; vapour pressure 0.13 mPa at 20°C; water solubility 405 mg/L at 25°C;

C. WATER AND SOIL POLLUTION FACTORS

Photodegradation

T/2 under continuous light = 92 hours in water. (10734)

Environmental fate

Sodium acifluorfen is persistent on soils and in aquatic environments and is relatively mobile. It is stable to hydrolysis and does not break down in sunlight. Off-target transport is expected to occur initially through drift and leaching, and later through erosion and runoff. Sodium acifluorfen exists in the anion (negatively charged) form in most agricultural soils. Several factors, including soil Ph, soil organic carbon content, and soil iron content determine the extent to which acifluorfen adsorbs to soil particles. Therefore, the persistence and mobility of acifluorfen vary with different soil conditions. A prospective ground water monitoring study, which was conducted on soybeans in the central sands of Wisconsin and in which acifluorfen and two degradates were monitored; the parent only was detected at concentrations ranging from 1 to 46 ppb (average 7.33 ppb) in 56 out of 283 samples. (10733)

Biodegradation

Sodium acifluorfen's primary environmental degradate is the acifluorfen anion, which is also a degradate of another herbicide, Lactofen. Acifluorfen applied to a silt loam degraded with a half-life of 59 days. (10734)

D. BIOLOGICAL EFFECTS

Toxicity

Birds

Mallards	Acute LD_{50}	2,821 mg/kg bw

Bobwhite quail	Acute LD_{50}	325 mg/kg bw	(10733)
Crustaceae			
Fiddler crab	96h LC_{50}	>1,000 mg/L	(10733)
Mollusca			
Freshwater clams	96h LC_{50}	150 mg/L	(10733)
Fish			
Bluegill	96h LC_{50}	31 mg/L	
Rainbow trout	96h LC_{50}	54 mg/L	(10733)
Mammals			
Rat	Acute oral LD_{50}	1540 mg/kg bw	

acraldehyde *see* acrolein

acridine (2,3,5,6-dibenzopyridine)

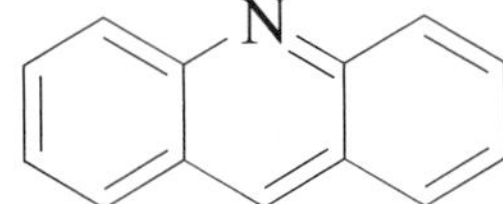

$C_{13}H_9N$

CAS 260-94-6

USES

Intermediate in manufacture of dyestuffs, alkaloids and antibacterials.

MANUFACTURING SOURCES

coal tar; constituent of coal tar creosote: 0.05 wt. %. (2386)

A. PROPERTIES

small, colorless needles; molecular weight 179.21; melting point 108°C, sublimes at 100°C; boiling point 346°C; sp. gr. 1.1 at 20/4°C; solubility 38.4 mg/L; $LogP_{OW}$ 3.4-3.6.

B. AIR POLLUTION FACTORS

Manmade sources

in airborne coal tar emissions: 8.4 mg/g of sample or 300 μg/m³ of air
in coke oven emissions: 32-173 μg/g of sample
in coal tar: 9.2 mg/g of sample
in wood preservative sludge: 3.5 g/L of raw sludge (993)

C. WATER AND SOIL POLLUTION FACTORS

Soil sorption

K_{oc}: 12,910. (2587)

D. BIOLOGICAL EFFECTS

Inhibition of photosynthesis of a freshwater, nonaxenic unialgal culture of Selenastrum capricornutum *at:*
1% saturation: 92% carbon-14 fixation (vs. controls)
10% saturation: 75% carbon-14 fixation (vs. controls)
100% saturation: 1% carbon-14 fixation (vs. controls) (1690)

DAPHNIA PULEX

bioaccumulation factor: 30 (initial conc. in water: 245 ppb) LC_{50}, 24 h: 2.9 mg/L; immobilization concentration: IC_{50}: 1.7 mg/L (1050)

Arthropoda			
Daphnia	toxic	at 0.7 mg/L	(30)
Daphnia pulex	BCF	30	(2660)

Fishes

toxic at 5 mg/L. (30)

Bioaccumulation

fathead minnow (Pimephales promelas)	*BCF*	
uptake from water:	125	
uptake via interaction with contaminated sediment:	874	
uptake via ingestion of contaminated zooplankton *(Daphnia pulex):*	30	
uptake via ingestion of benthic invertebrates *(Chironomus tentans):* living in contaminated sediments:	51	
The calculated rates of uptake of acridine via ingestion of contaminated invertebrates (0.02 μg/g/h) and ingestion of sediment (0.01 μg/g/h) were negligible compared with direct uptake from water (1.40 μg/g/h) in a hypothetical system with all compartments in equilibrium.		(1715)
Pimephales promelas	BCF: 120	(2660)
fathead minnow	BCF: 126	(2606)

acridine orange NS (N,N,N',N'-tetramethyl-3,6-acridinediamine monohydrochloride)

HCl

$C_{17}H_{19}N_3.HCl$

CAS 65-61-2

USES

selective biological stain for tumor cells, intravitam, and causes retardation of tumor growth.

D. BIOLOGICAL EFFECTS

Toxicity to microorganisms

Bacillus subtilis growth inhib. EC_{50}: 0.025 mmol/L. (2624)

acrolein (acraldehyde; acrylic aldehyde, allylaldehyde; 2-propenal; acrylaldehyde; aqualin)

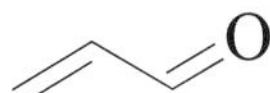

$H_2CCHCHO$

C_3H_4O

CAS 107-02-8

USES AND FORMULATIONS

stabilized with 100-200 ppm hydroquinone.

A. PROPERTIES

colorless to yellowish liquid impurities of technical grade: hydrochinon to prevent polymerization; molecular weight 56.1; melting point -87.7°C; boiling point 52.5°C; vapor pressure 220 mm at 20°C, 330 mm at 30°C; vapor density 1.94; sp. gr. 0.84 at 20/20°C; THC 393 kcal/mole, LHC 379 kcal/mole; saturation concentration in air 671 g/m^3 at 20°C, 974 g/m^3 at 30°C, solubility 20.8% at 20°C; $LogP_{OW}$ 0.90.

B. AIR POLLUTION FACTORS

1 mg/m^3 = 0.43 ppm, 1 ppm = 2.328 mg/m^3.

Odor

characteristic; quality: burnt sweet, hot fat, acrid; hedonic tone: pungent.

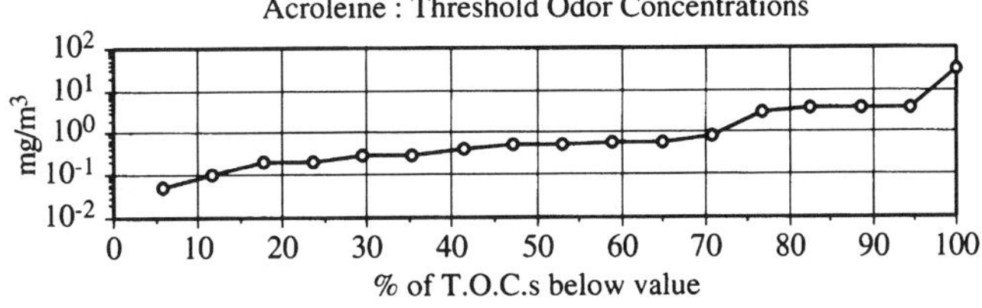

human odor perception: 0.8 mg/m^3
human reflex response: adverse response: 0.6 mg/m^3
animal chronic exposure: adverse effect: 0.15 mg/m^3 (178)

Manmade sources

in cigarette smoke: 150 ppm (66)
in gasoline exhaust: 0.2 to 5.3 ppm (195)
2.6-9.8 vol. % of total exhaust aldehydes (394, 395, 396, 397)
Emissions from cigarette smoking: 560 μg/cigarette (2421)

Control methods

activated carbon: retentivity: 15 wt% of adsorbent (83)
wet scrubber:
water at pH 8.5: outlet: 140,000 odor units/scf
$KMnO_4$ at pH 8.5: outlet: 1 odor unit/scf (115)
catalytic incineration over commercial Co_3O_4 (1.0-1.7 mm) granules, catalyst charge 13 cm^3, space velocity 45,000 h^{-1} at 100 ppm in inlet:
80% decomposition at 190°C
99% decomposition at 210°C
80% conversion to CO_2 at 190°C
95% conversion to CO_2 at 230°C

Incinerability

thermal stability ranking of hazardous organic compounds: rank 106 on a scale of 1 (highest stability) to 320 (lowest stability). (2390)

Sampling and analysis

sintered disc absorber 30 l air/30 min: VLS: lower limit: 30 μg/m^3/30 min (208)
scrubbers: liquid lift type: 15 ml of liquid, 4.5 l/min gas flow 4 min scrubbing trap method:

	% removal
H_2O, 0°C	22
NH_2OH soln, 0°C	94
H_2SO_4 (conc.), 55°C	96
open tube, -80°C	13
ethanol, -80°C	91

C. WATER AND SOIL POLLUTION FACTORS

BOD_5:	0.0% ThOD	(277)
COD:	86% ThOD	(277)
ThOD	2.0	

Odor threshold

0.11 mg/kg

Experimental concentration of 0.1 mg/L can significantly taint the flesh of rainbow trouts to make them unpalatable (1788)
A.C.: adsorbability: 0.061 g/g carbon; 30.6% reduction, infl.: 1,000 mg/L, effl.: 694 mg/L (32)

methods	*temp, °C*	*d observed*	*feed*	*d acclim.*	*% theor. oxidation*	
TF,Sd,BOD	20	10	?	?	33	(32)
RW,Sd,BOD	20	10	?	100	33	(93)

D. BIOLOGICAL EFFECTS

Protozoa

(*Uronema parduczi* Chatton-Lwoff): EC_0 0.44 mg/L (329)

Bacteria

Pseudomonas putida: 16h EC_0 0.21 mg/L (329)

Algae

Microcystis aeruginosa: 8d EC_0 0.04 mg/L (329)

Insects

mayfly nymphs (*Ephemerella walkeri*): lowest observed avoidance conc. >0.1 mg/L (1621)

Fishes

Bioaccumulation: Bluegill sunfish *(Lepomis macrochirus)* at 13.1 µg/L at 16°C during 28 d: BCF: 344; half-life in tissues: >7 d	(2602)

Toxicity

Crustaceans

Daphnia magna	48h IC_{50}	0.051 mg/L	(7025)

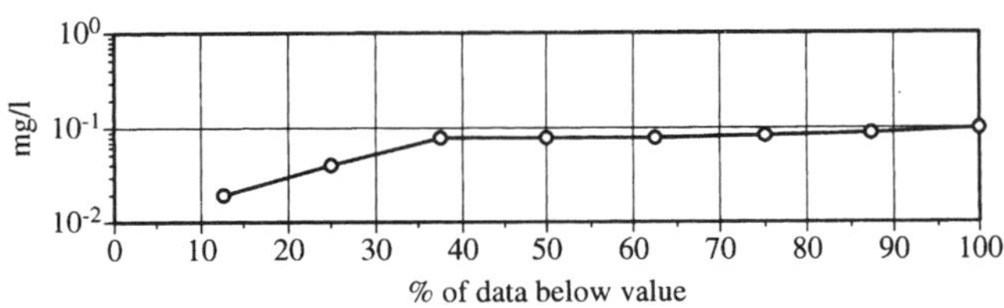

Frequency distribution of 24-96h LC_{50} values for fishes (n = 8) based on data from this work and from the work of Rippen	(3999)

goldfish: LD_{50} (24 h) <0.08 mg/L modified ASTM D 1345			(277)
rainbow trout *(Salmo gairdneri):* lowest observed avoidance conc. 0.1 mg/L			(1621)
Lepomis macrochirus	24h LC_{50}	80 µg/L	(2106)
Salmo trutta	24h LC_{50}	40 µg/L	(2108)
Lepomis macrochirus	24h LC_{50}	79 µg/L	(2108)
Fathead minnow	incipient LC_{50},F	84 µg/L	
	MATC	11 µg/L	(443)
Pimephales promelas	96h LC_{50}	20 µg/L	(2625)
Bluegill *Lepomis macrochirus*	24h LC_{50}	0.1 mg/L	
	96h LC_{50}	0.09 mg/L	(2697)
Brachydanio rerio	96h LC_{50}	0.014 mg/L	(7025)

acryl amide *see* acrylamide

acrylaldehyde *see* acrolein

acrylamide (propenamide; acrylic amide)

NH_2

O

64 acrylamide

$CH_2CHCONH_2$

C_3H_5NO

CAS 79-06-1

USES

synthesis of dyes, etc.; polymers or copolymers as plastics, adhesives; soil conditioning agents; flocculants.

A. PROPERTIES

molecular weight 71.08; melting point 84-85°C; vapor pressure 2 mm at 87°C, 10 mm at 117°C; vapor density 2.46; solubility 2,050 g/L.

B. AIR POLLUTION FACTORS

1 mg/m^3 = 0.34 ppm, 1 ppm = 2.95 mg/m^3.

Incinerability

thermal stability ranking of hazardous organic compounds: rank 60 on a scale of 1 (highest stability) to 320 (lowest stability). (2390)

C. WATER AND SOIL POLLUTION FACTORS

BOD_5:	45% ThOD std. dil./acclimated	(41)
ThOD:	2.14	

Manmade sources

in paper mill treated effluent:	0.47-1.2	µg/L	
colliery:			
coal washing effluent:	1.8	µg/L	(214)
tailings lagoon:	39-42	µg/L	(231)
in sewage effluents:	280	µg/L	

Waste water treatment

RW, COD, 20°C, 2 d observed, feed: 10 mg/L, 33 d acclim., 100% removed $KMnO_4$ oxidation: 1 mg/L at pH 5-8.5, 4h contact time: 100% removed
MnO_2 column/pH 5-7, 0.5h contact time: 17-33% removed
Ozone: 3 mg/L infl. at pH 7, 0.5h contact time: 100%

	infl. mg/L	*pH*	*contact time*	*% removed*	
Cl_2	10	1.0	4 h	100%	
	10	5.0	4 h	64%	
	10	8.5	4 h	0%	
A.C.	8	5.0	0.5 h	13%	(214)

D. BIOLOGICAL EFFECTS

Fishes

harlequin fish *(Rasbora heteromorpha)*

	mg/L				
	24 h	*48 h*	*96 h*	*3m (extrap.)*	
LC_{10} (F)	390	220	103		
LC_{50} (F)	460	250	130	10	(331)

brown trout yearlings: 48h LC_{50}: 400 mg/L (static bioassay) (939)

2-acrylamido-2-methylpropanesulphonic acid, ammonium salt

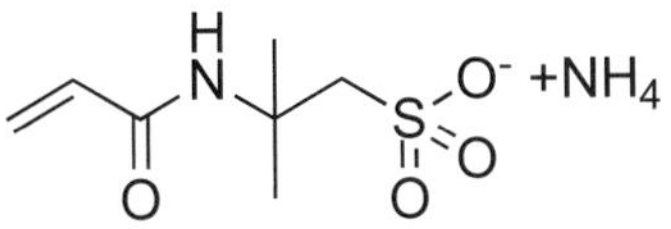

$C_7H_{12}NO_4NH_4$

CAS 58374-69-9

TRADENAMES

OS 114452; Ammonium AMPS (50% aqueous solution); LZ 2411

USES

a component of polymers used mainly in the paints and adhesive industries. Polymers containing the chemical are said to have better mechanical stability and to better stabilise the paint emulsion. In a typical polymer, the chemical will be at a concentration of approximately 5%. It would be polymerised with other monomers such as ethyl acrylate, butyl acetate, vinyl acetate a.o.. The resulting polymers will be used at a concentration of about 30% in paints or coatings. The chemical will therefore be present as a polymer component of approximately 1.5%. (7467)

IMPURITIES, ADDITIVES, COMPOSITION

Name	CAS number	weight
2-acrylamido-2-methyl-1,3-propanedisulphonic acid, ammonium salt		0.5-1.5%
2-propenamde	79-06-1	846 mg/L
2-propenenitrile	107-13-1	436 mg/L
N-(1,1-dimethylethyl)-2-propenamide	107-58-4	2138 mg/L
2-methyl-2-propene-1-sulfonic acid	3934-16-5	26 mg/L
2-methylene-1,3-propanedisulfonic acid	1561-93-9	135 mg/L
4-methoxyphenol*	150-76-5	424

* additive

A. PROPERTIES

white powder; molecular weight 224; melting point 191°C; specific gravity 1.39 kg/L at 22 °C; vapour pressure 7.4 x 10^{-12} kPa at 25 °C: water solubility >761,000 mg/L at 25°C; log Pow –3.4 at 22°C; log K_{OC} <1.8

C. WATER AND SOIL POLLUTION FACTORS

Hydrolysis

considering the chemical's structure, hydrolysis is expected to be extremely slow, ie in the order of years (7567)

Mobility in soil

the chemical is expected to be highly mobile in soils. The chemical is an ammonium salt of a sulphonic acid , and as such expected to remain highly ionised in the environment. (7467)

Biodegradation

Inoculum	method	concentration	duration	elimination	
	Modified Sturm Test		28 days	3.2% ThCO2	(7467)

D. BIOLOGICAL EFFECTS

Toxicity

Micro organisms			
Activated sludge	3h EC_{50}	>10,000 mg/L	(7567)
Algae			
Selenastrum capricornutum	96h ECb_{50}	>2,000 mg/L	
	96h NOECb	>2,000 mg/L	
	24h $ECgr_{50}$	>2,000 mg/L	(7467)
Crustaceae			
Daphnia magna	48h EC_{50}	1,200 mg/L	
	48h NOEC	640 mg/L	
	21d EC_{50}	680 mg/L	
reproduction	21d NOEC	380 mg/L	
	21d EC_{50}	>380 mg/L	(7467)
Fish			
Pimephales promelas	96h LC_{50}	1,400 mg/L	
	96h NOEC	640 mg./L	(7467)
Mammals			
Rat	oral LD_{50}	>5,000 mg/kg bw	(7467)

2-acrylamido-2-methylpropanesulphonic acid (1-propanesulphonic acid, 2-methyl-2-[(1-oxo-2-propenyl)amino]-)

$C_7H_{13}NO_4S$

CAS 15214-89-8

EINECS 239-268-0

TRADENAMES

Lubrizol 2401; Lubrizol 2404; AMPS; ATBS; TBAS-Q

USES

synthesis of acryl fibres, in closed systems. The substance is a vinyl monomer and is used with a variety of other vinyl monomers to make polymers.

A. PROPERTIES

molecular weight 207.24; melting point 181 °C; boiling point > 200°C; relative density 1.1 kg/L at 15°C; water solubility > 600,000 mg/L at 20°C; log Pow < 1 at 20°C

C. WATER AND SOIL POLLUTION FACTORS

Hydrolysis

the substance is stable over a broad range of pH. (7397)

Biodegradation

Inoculum	method	concentration	duration	elimination	
Domestic activated sludge	Modified SCAS Test	10 mg/L	28 days	< 20 %	(7397)

D. BIOLOGICAL EFFECTS

Toxicity

Crustaceae			
Daphnia magna	48h EC_{50}	280-430 mg/L	
	48h NOEC	78 mg/L	(7397)
Fish			
Lepomis macrochirus	96h LC_{50}	130 – 220 mg/L	
	96h NOEC	130 mg/L	(7397)
Mammals			
Rat	oral LD_{50}	>500-2,000; >1,000-2,000; 1830 mg/kg bw	(7397)

acrylic acid (propenoic acid; ethylenecarboxylic acid; vinylformic acid)

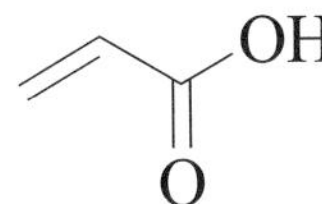

$CH_2CHCOOH$

$C_3H_4O_2$

CAS 79-10-7

USES

monomer for polyacrylic and polymethacrylic acids and other acrylic polymers. (1590)

A. PROPERTIES

molecular weight 72.06; melting point 12-14°C; boiling point 141°C; vapor pressure 3.2 mm at 20°C, 10 mm at 39°C, 3.8 hPa at 20°C, 40 hPa at 60°C; vapor density 2.50; sp. gr. 1.06 at 16°C; solubility miscible; THC 327 kcal/mole; saturation concentration in air 12.6 g/m^3 at 20°C, 22.8 g/m^3 at 30°C; LogP_{OW} 0.31; 0.43 (calc.), 0.38; 0.46 (measured).

B. AIR POLLUTION FACTORS

1 mg/m^3 = 0.33 ppm, 1 ppm = 3.00 mg/m^3.

Odor

characteristic; quality: rancid, sweet; hedonic tone: unpleasant.
T.O.C. absolute: 0.094 ppm
50% recognition: 1.04 ppm
100% recognition: 1.04 ppm

O.I.: 105,700 (19)

C. WATER AND SOIL POLLUTION FACTORS

COD:	100% ThOD
TOC:	95% ThOD
ThOD:	1.33

Natural sources

produced by marine algae such as *Phaeocystis* and *Polysiphonia lanosa;* as a result of hydrolysis of dimethyl-β-propiothetin (514)

Waste water treatment

A.C. adsorbability: 0.129 g/g carbon, 64.5% reduction, infl.: 1,000 mg/L; effl.: 355 mg/L (32)

Biodegradability

inoculum/method	*test conc.*	*test duration, d*	*removed*	
Sewage 26°C, 12 d adaptation	12 mg/L	20	35% ThOD	(93)
BOD test		5	84% ThOD	(5338)
A.S., municipal		28	>60% ThOD	(5338)
A.S., MITI test		14	>30% ThOD	(5339)
municipal waste water	10 mg/L	42	71% CO_2 prod.	(5339)
	10 mg/L	19	70% CO_2 prod.	
adapted inoculum	10 mg/L	22	81% CO_2 prod.	(5339)
anaerobic municipal A.S.	50 mg/L	56	>75% prod.	(5340)

D. BIOLOGICAL EFFECTS

Bacteria				
Pseudomonas putida		16h EC_0	41 mg/L	(1900)
Algae				
Microcystis aeruginosa		8d EC_0	0.15 mg/L	(329)
Scenedesmus quadricauda		7d EC_0	18 mg/L	(1900)
Protozoa				
Entosiphon sulcatum		72h EC_0	20 mg/L	(1900)
Uronema parduczi Chatton-Lwoff		EC_0	11 mg/L	(1901)
Crustaceans				
Daphnia magna	(neutralized)	24h EC_0	175 mg/L	
	(neutralized)	24h EC_{50}	765 mg/L	
	(neutralized)	24h EC_{100}	5,000 mg/L	(5337)
Daphnia magna		24h EC_0	51 mg/L	
		24h EC_{50}	54 mg/L	
		24h EC_{100}	91 mg/L	(5337)
Fishes				
Leuciscus idus		48h LC_0	210 mg/L	
		48h LC_{50}	315 mg/L	
		48h LC_{100}	420 mg/L	(5335)

acrylic aldehyde *see* acrolein

acrylon *see* acrylonitrile

acrylonitrile (acrylon; carbacryl; cyanoethylene; fumigrain; 2-propenenitrile; VCN; ventox; vinylcyanide)

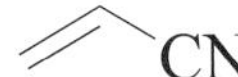

CH_2CHCN

C_3H_3N

CAS 107-13-1

USES

The major use of acrylonitrile is in the production of acrylic and modacrylic fibers by copolymerization with methylacrylate, methylmethacrylate, vinylacetate, vinylchloride, or vinylidenechloride. Other major uses include the manufacture of acrylonitrile-butadiene-styrene (ABS) and styrene acrylonitrile (SAN) resins. Acrylonitrile is also used as a fumigant.

USES AND FORMULATIONS

acritet = 34% acrylonitrile, 60% CCl_4

ventox = acritet

carbacryl: equal volumes of acrylonitrile and CCl_4

acrylofume: 3.95% acrylonitrile; 30% CCl_4; 30% chloroform; 0.5% chloropicrin

A. PROPERTIES

colorless liquid; molecular weight 53.06; melting point -83°C; boiling point 77.4°C; vapor pressure 100 mm at 23°C, 137 mm at 30°C; vapor density 1.83; sp. gr. 0.80 at 25°C; saturation concentration in air 257 g/m^3 at 20°C, 383 g/m^3 at 30°C; LogP$_{OW}$ -0.92.

B. AIR POLLUTION FACTORS

1 mg/m^3 = 0.454 ppm, 1 ppm = 2.203 mg/m^3.

Odor

characteristic; quality: onion, garlic; hedonic tone: pungent.

T.O.C.:	recogn.: 3.72-51.0 mg/m^3	
	1.7-23 ppm	
	PIT$_{50}$%: 21.4 ppm	
	PIT$_{100}$%: 21.4 ppm	(2)
	average: 18.6 ppm	
	number of panelists: 16	(210)
	41.9 mg/m^3 = 19 ppm	(279)

	45 mg/m^3 = 20.4 ppm	(291)
	detection: 3.4 mg/m^3	(819)
	recognition: 47 mg/m^3	(741)

Incinerability

temperature for 99% destruction at 2.0 sec residence time under oxygen-starved reaction conditions: 985°C. Thermal stability ranking of hazardous organic compounds: rank 20 on a scale of 1 (highest stability) to 320 (lowest stability) (2390)

Partitioning coefficients

$K_{air/water}$	0.013	
Cuticular matrix/air partition coefficient*	250 ± 10	(7077)
Cuticular matrix/water partition coefficient**	3.2	(7077)

* experimental value at 25°C studied in the cuticular membranes from mature tomato fruits (*Lycopersicon esculentum* Mill. cultivar Vendor)

** calculated from the Cuticular matrix/water partition coefficient and the air/water distribution coefficient

C. WATER AND SOIL POLLUTION FACTORS

BOD_5:	23% ThOD std. dil. acclim.	(4)
	0% ThOD	(26)
$BOD\ _5^{20}$:	0% at 10 mg/L, unadapted sew.	(36)
BOD_{10}:	22% ThOD std. dil. sew.	(256)
BOD_{30}^{20}:	38% ThOD at 10 mg/L, unadapted sew; lag period: 12 d	(554)
COD:	44% ThOD	(36)
ThOD:	3.17	(36)

Impact on biodegradation processes

BOD test is not influenced up to 1,000 mg/L (30)

at 100 mg/L no inhibition of NH_3 oxidation by *Nitrosomonas* sp. (390)

Reduction of amenities

T.O.C. average	18.6 mg/L	range: 0.0031 to 50 mg/L	(97)
	29 mg/L		(294)
T.O.C. in water:	1.86 ppm		(326)
	2.02 ppm		
	3.9 ppb		

Aquatic reactions

photooxidation by UV light in water at 50°C: 24.2% degradation to CO_2 after 24 h (1628)

Waste water treatment

biodegradation by mutant microorganisms: 500 mg/L at 20°C

% disruption:	parent:	84% in 24 h
	mutant:	100% in 4.0 h

methods	*temp., °C*	*days observed*	*feed, mg/L*	*days acclim.*	*theor. oxidation*	*% rem ove d*	
NFG, BOD	20	1-10	100-1,000	365 + P	-	25	
RW, BOD	20	5-10	50	27	-	67	
RW, BOD	20	2	10	32	-	100	
RW, CO_2, CAN	20	19	10	8	60	100	
RW, CO_2, CAN	20	9	40	30+	60	100	
RW, CO_2, CAN	5	33	40	30+ at 20°C	60	100	
ASC, C N	22-25	28	89	21	70+	95+	(93)

A.C.:	influent	carbon dosage	effluent	% reduction

1,000 mg/L	10×	490 mg/L	51	
100 mg/L	10×	72 mg/L	28	(192)

D. BIOLOGICAL EFFECTS

Bacteria

Pseudomonas putida:	16h EC_0	53 mg/L	(329)

Arthropoda

Crangon crangon in seawater at 15°C:

	EC_{50} after recovery period in unpolluted seawater		
exposure time	0 min	24 h	
1 min	>32,000 ppm	>10,000 ppm	
3 min	>32,000 ppm	<10,000 ppm	
9 min	15,000 ppm	<<10,000 ppm	(328)

Crangon crangon in seawater at 15°C (see also graph on next page)

exposure time	LC_{50}, mg/L	
3 min	>32,000	
9 min	±15,000	
27 min	4,800	
1 h	3,600	
3 h	480	
6 h	180	
24 h	25	
48 h	20	
72 h	6	
96 h	6	(328)

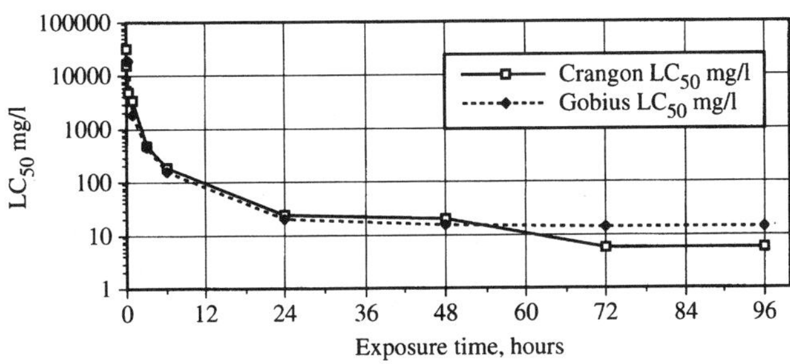

Exposure of *Crangon crangon* and *Gobius minutus* in Seawater at 15°C to Acrylonitrile

Bioaccumulation

Fishes

Bluegill sunfish *(Lepomis macrochirus)* at 9.94 μg/L at 16°C during 28 d: BCF: 48; half-life in tissues: >4 < 7 d (2602)

Toxicity

fathead minnows:	96h LC_{50}	hard water	14 mg/L	
fathead minnows:	96h LC_{50}	soft water	18 mg/L	
bluegill:	96h LC_{50}	soft water	12 mg/L	
guppies:	96h LC_{50}	soft water	33 mg/L	
pinperch:	96h LC_{50}	soft water	24 mg/L	(41)
pinperch:	24h LC_{50}	seawater	24 mg/L	(248)
bluegill sunfish:	24h LC_{50}	soft water	25 mg/L	(252)
minnows:	24h LC_{50}		37 mg/L	
	48h LC_{50}		24 mg/L	

Gobius minutus in seawater at 15°C (see also graph above): (328)

exposure time	*LC_{50}, mg/L*

3 min	±18,000
9 min	±18,000
27 min	±18,000
1 h	±1,800
3 h	435
6 h	150
24 h	20
48 h	15
72 h	14
96 h	14

Gobius minutus in seawater at 15°C:
EC_{50} after recovery period in unpolluted seawater

exposure time	0 min	24 h
1 min	±18,000 ppm	>10,000 ppm
3 min	±18,000 ppm	3,200 ppm
9 min	±18,000 ppm	>3,200 ppm

Lagodon rhomboides

24h LC_0	20 mg/L	
24h LC_{50}	24 mg/L	
24h LC_{100}	30 mg/L	(439)

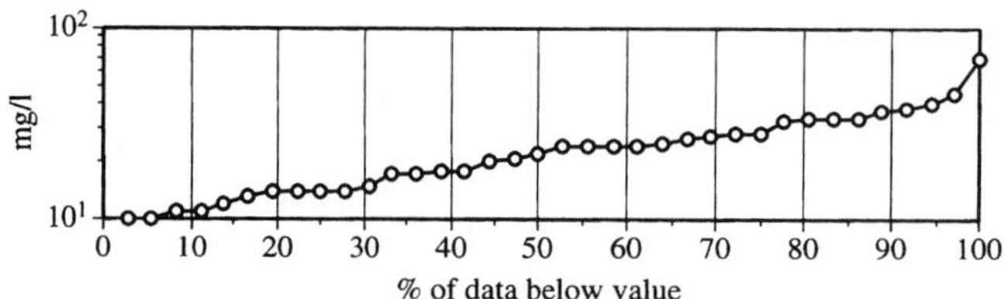

actellic (pirimiphosmethyl; PP-511; O-(4-(2-diethylamino)-6-methyl) pyrimidinyl)-phosphorothioic acid, O,O-dimethylester; Silosan)

$C_{11}H_{20}N_3O_3$

CAS 29232-93-7

USES

insecticide and acaricide.

A. PROPERTIES

molecular weight 305.34; melting point 15°C; boiling point decomposes; density 1.16 at 30°C; vapor pressure 1 x 10^{-4} mm Hg at 30°C; solubility 5 mg/L at 30°C.

D. BIOLOGICAL EFFECTS

Protozoans			
Paramecium caudatum	4h LC_{50}	< 0.58 mg/L	(9774)
Crustaceans			
Daphnia magna	48h LC_{50}	1.4 mg/L	(9600)
Fishes			
carp *(Cyprinus carpio)*	48h LC_{50}	0.005 ml/L	(1199)
rainbow trout *(Salmo gairdneri)*	48h LC_{50}	0.001 ml/L	
guppy *(Poecilia reticulata)*	48h LC_{50}	0.004 ml/L	(1199)
Mammals			
rabbit, mouse, guinea pig	oral LD_{50}	1,000-1,180 mg/kg bw	(9775)

adamantane (*sym*-tricyclodecane)

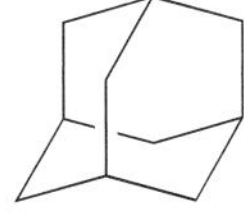

$C_{10}H_{16}$

CAS 281-23-2

USES

has unique molecular structure consisting of four fused cyclohexane rings. Derivatives (alkyl adamantanes) have potential uses in imparting heat, solvent, and chemical resistance to many basic types of plastics. Synthetic lubricants and pharmaceuticals are also based on adamantane derivatives. Adamantane diamine is used to cure epoxy resins.

A. PROPERTIES

white crystals; molecular weight 136.24; melting point 205-210°C; solub 0.027 mg/L; $LogP_{OW}$ 3.99.

D. BIOLOGICAL EFFECTS

Fishes			
Pimephales promelas	48h LC_{50}	0.325 mg/L	
	96h LC_{50}	0.285 mg/L	(2709)

2-adamantanone

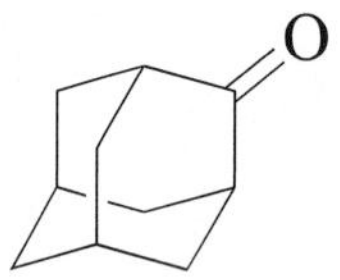

$C_{10}H_{14}O$

CAS 700-58-3

A. PROPERTIES

molecular weight 150.22; melting point 256-258°C; solub 3.32 mg/L; $LogP_{OW}$ 1.43.

D. BIOLOGICAL EFFECTS

Fishes			
Pimephales promelas	24h LC_{50}	60 mg/L	
	96h LC_{50}	60.8 mg/L	(2709)

adipic acid (hexanedioic acid; 1,4-butanedicarboxylic acid)

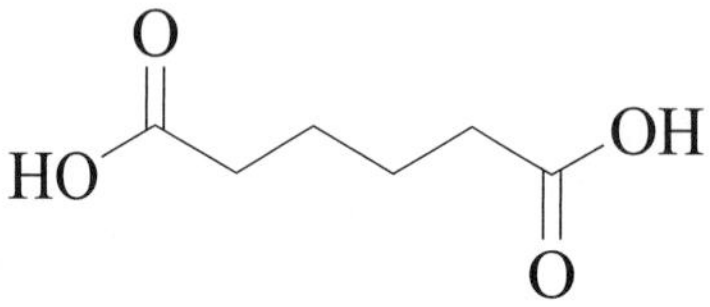

$COOH(CH_2)_4COOH$

$C_6H_{10}O_4$

CAS 124-04-9

USES

feedstock for the mfg. of polyamides (Nylon 66), polyesters (alkydresins), and polyurethanes; mfg. of dyes, pharmaceuticals.

A. PROPERTIES

molecular weight 146.14; melting point 151-153°C; boiling point 338°C; 265°C at 100 mm, 330°C at 1,013 hPa, decarboxylates at 230°C; vapor pressure 0.097 hPa at 18°C sublimes; vapor density 5.04; sp. gr. 1.37; solubility 15,000 mg/L at 15°C, 160,000 mg/L at 60°C; THC 669 kcal/mole; $LogP_{OW}$ 0.08 (calculated).

C. WATER AND SOIL POLLUTION FACTORS

BOD_5:	42% ThOD	(30)
	36% of ThOD	(220)

BOD$^{20°C}$:	79% ThOD	(3)
COD:	97% of ThOD	(220)
ThOD:	1.42	(30)

Biodegradability

inoculum/method	*test conc.*	*test duration*	*removed*	
municipal sludge			96.6% DOC	(5167)
coupled test units			99+% DOC	(5168)
A.S.			100% ThOD	(5168)
MITI test			96% DOC	(5168)
Modified Sturm test			100% CO_2 prod.	(5168)
Closed bottle test		30 d	83% ThOD	(5168)
A.S.	1 mg/L	5 d	>90% COD	(5169)
Modified OECD screening test			96% DOC	(5168)
A.S.		6 h	1.3% of ThOD	
		12 h	1.3% of ThOD	
		24 h	7.1% of ThOD	

Aquatic reactions

photooxidation by UV in aqueous medium at 90-95°C; time for the formation of CO_2 (% of theoretical) 25%: 2 h, 50%: 5 h, 75%: 32 h. (1628)

D. BIOLOGICAL EFFECTS

Bacteria			
Pseudomonas fluorescens	16h EC_0	10,000 mg/L	(5165)

Fishes			
Leuciscus idus	48h LC_0	>1,000 mg/L	(5165)
Pimephales promoxis	96h LC_{50}	10,000 mg/L (QSAR calcul.)	
Salmo gairdneri	96h LC_{50}	12,000 mg/L (QSAR calcul.)	(5166)
Brachydanio rerio	96h LC_0	>1,000 mg/L	(5165)
Macrochirus	24h LC_{50}	<300 mg/L	
fathead minnows	1-24-48-72-96h LC_{50},S	>300-172-114-97-97 mg/L	(350)

Adipic acid compound with hexane-1,6-diamine(1:1)

(adipic acid-hexamethylenediamine salt (1:1); hexamethylenediamine adipate; Nylon salt; Nylon 66 salt; 1,6-hexanediamine adipate; 1,6-hexanediamine, hexanedioate; AH salt)

HO OH · H2N NH2

$C_{12}H_{26}N_2O_4$

CAS 3323-53-3

USES

The chemical is the basic raw material for the production of nylon 66 polymers and copolymers, that are used in fibres and yarns for textiles, carpets, apparel, tire cord, and (7457) industrial applications, or in engineering resins, used for automotive parts, electrical and electronic applications, machine parts, films, wire coatings, and monfilament.

A. PROPERTIES

molecular weight 262.34; melting point 202°C; vapour pressure very low, 593 hPa at 90°C; density 1.2 kg/L; water solubility 468,000 mg/L at 21°C; log Pow −4.4

B. AIR POLLUTION FACTORS

Indirect photolysis

for the 2 components of AH salt: adipic acid: t/2 = 69 hours; 1,6-hexamethylenediamine: t/2 = 5.6 hours (7457)

C. WATER AND SOIL POLLUTION FACTORS

Hydrolysis

AH salt rapidly dissociates to form adipate and 1,6-hexanediammonium in an almost neutral aqueous solution (7457)

Photochemical degradation

for the 2 components of AH salt: adipic acid: t/2 = 67 hours; 1,6-hexamethylenediamine: t/2 = 10 days (7457)

Mobility

for the 2 components of AH salt: adipic acid: log K_{OC} = 1.33; 1,6-hexamethylenediamine: log K_{OC} = 2.46 (estimated) The soil adsorption can be only roughly estimated because of possible ionic interactions of the cations with negatively charged particles in the soil that may reduce their mobility (7457)

Biodegradation

Inoculum	method	concentration	duration	elimination	
Industrial A.S.	Modified Zahn-Wellens	400 mg DOC/L	3 days	96% ThOD	(7457)

BOD5 1.0 g/g
COD 1.7 g/g

D. BIOLOGICAL EFFECTS

Toxicity

Micro organisms			
Activated sludge	10min EC_{20}-EC_{50}	>900 mg/L	
Pseudomonas putida	17h EC_{10}-EC_{90}	>2,000 mg/L	(7457)
Algae			
Scenedesmus subspicatus	72h ECb_{50}	394.5 mg/L	
	72h ECb_{20}	269.3	
	96h ECb_{50}	291.9 mg/L	
	96h ECb_{20}	102.9 mg/L	
	96h ECb_{90}	479.1 mg/L	(7457)
Crustaceae			
Daphnia magna	48h EC_{50}	90; 98.9 mg/L	
	48h EC_{10}	55 mg/L	
	48h EC_0	62.5 mg/L	
	48h EC_{100}	250 mg/L	
	24h LC_{50}	165 mg/L	
	24h EC_0	250 mg/L	
	24h EC_{50}	353 mg/L	

	24h EC_{100}	500 mg/L	(7457)
Fish			
Leuciscus idus	96h LC_{50}	10,000 mg/L	
	96h LC_{100}	>10,000 mg/L	
	96h NOEC	2,500 mg/L	
Salmo gairdneri	96h LC_{50}	>470 mg/L	
Lepomis macrochirus	96h LC_{50}	>470 mg/L	(7457)
Mammals			
Rat	oral LD_{50}	4,900; 5,900 mg/kg bw	(7457)

adipic acid dinitrile *see* adiponitrile

adiponitrile (1,4-dicyanobutane; hexanedinitrile; adipic acid dinitrile; hexanedioic acid dinitrile; adipyldinitrile; tetramethylenedicyanide)

NC~~~CN

$NC(CH_2)_4CN$

$C_6H_8N_2$

CAS 111-69-3

USES

intermediate in the manufacture of nylon; organic synthesis.

A. PROPERTIES

colorless liquid; molecular weight 108.15; melting point 1°C; boiling point 295-306°C; vapor pressure 2 mm at 119°C; vapor density 3.73; sp. gr. 0.96 at 20/4°C; solub >100,000 mg/L; $LogP_{OW}$ -0.42.

B. AIR POLLUTION FACTORS

1 mg/m^3 = 0.22 ppm = 4.50 mg/m^3.

C. WATER AND SOIL POLLUTION FACTORS

COD:	64% ThOD	(41)
ThOD:	2.96	

Waste water treatment

methods	*temp. °C*	*d observed*	*feed, mg/L*	*d acclim.*	*theor. oxidation*	*% removed*
RW, CO_2, CAN	20	13	10	8	60	100
RW, CO_2, CAN	20	9	40	30+	60	100
RW, CO_2, CAN	5	33	40	30+ at 20°C	60	100

ASC, CAN	22-25	28	120-160	28+	80+	98+	(93)

Oxidation parameters

COD: 1.9 (41)

D. BIOLOGICAL EFFECTS

Fishes

fathead minnows	hard water	96h LC_{50}	820	mg/L	
fathead minnows	soft water	96h LC_{50}	1,250	mg/L	
bluegill	soft water	96h LC_{50}	720	mg/L	
guppies	soft water	96h LC_{50}	775	mg/L	(41)
bluegill sunfish	soft water	24h LC_{50}	1,250	mg/L	(252)
Pimephales promelas		24h LC_{50}	2,000	mg/L	
		96h LC_{50}	1,930	mg/L	(2709)

adipyldinitrile *see* adiponitrile

adronol *see* cyclohexanol

aerozine-50

CAS 8065-75-6

USES

rocket fuel.

IMPURITIES, ADDITIVES, COMPOSITION

50% hydrazine + 50% uns. dimethylhydrazine.

D. BIOLOGICAL EFFECTS

Fishes

guppy (Lebistes reticulatus) *static bioassay:*

LC_{50} 24 h: 12.0 mg/L in hard water at 22-24.5°C
48 h: 4.4 mg/L in hard water at 22-24.5°C
72 h: 2.7 mg/L in hard water at 22-24.5°C
96 h: 2.3 mg/L in hard water at 22-24.5°C

LC_{50} 24 h: 5.1 mg/L in soft water at 22-24.5°C
48 h: 2.9 mg/L in soft water at 22-24.5°C
72 h: 2.0 mg/L in soft water at 22-24.5°C
96 h: 1.2 mg/L in soft water at 22-24.5°C

comparison between previously exposed and unexposed guppies-pre-exposure period of 14 d at 1/25th 0.1 of toxicant concentration of 2.5 mg/L: mean survival time of pre-exposed fish: 43 h in soft water at 22-24.5°C; unexposed fish: 68 h in soft water at 22-24.5°C. (474)

aflatoxin B1

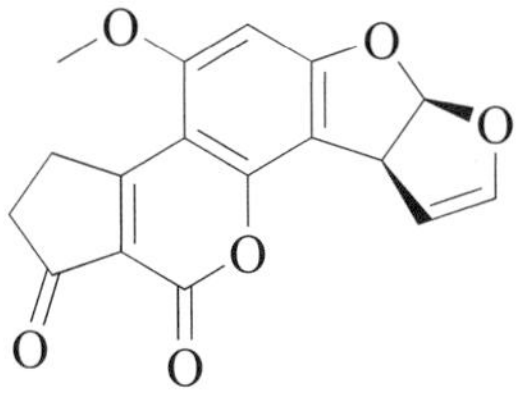

aflatoxin B1

$C_{17}H_{12}O_6$

CAS 1162-65-8

OCCURRENCE

belongs to a group of polynuclear molds (mycotoxins) produced chiefly by the fungus *Aspergillus flavus;* natural contaminants of a wide range of fruits, vegetables, and cereal grains.

A. PROPERTIES

a crystalline material; molecular weight 312.29; melting point 268°C.

B. AIR POLLUTION FACTORS

Incinerability

thermal stability ranking of hazardous organic compounds: rank 200 on a scale of 1 (highest stability) to 320 (lowest stability). (2390)

D. BIOLOGICAL EFFECTS

Confirmed human carcinogen with experimental tumorigenic, neoplastigenic, and carcinogenic data. Aflatoxin B1 is oxidized by mammals to aflatoxin M1, which is also a confirmed carcinogen with experimental tumorigenic data.

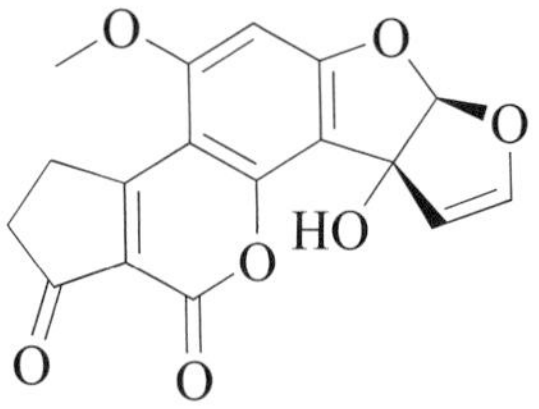

aflatoxin M1

(3998)

AGE *see* allylglycidylether

AHTN (7-acetyl-1,1,3,4,4,6-hexamethyl-1,2,3,4-tetrahydronaphtalene; 1-(5,6,7,8-tetrahydro-3,5,5,6,8,8-hexamethyl-2-naphtyl)ethan-1-one; 6-acetyl-1,1,2,4,4,7-hexamethyltetraline)

$C_{18}H_{26}O$

CAS 1506-02-1

CAS 406-02-1

CAS 21245-77-7

TRADENAMES

Tonalide; Fixolide

USES

a polycyclic musk. The polycyclic musks are used as fragrance ingredients in consumer products like cosmetics and detergents and cleaning agents. They are important ingredients in fragrances because of their typical musky scent and their fixative properties.
The substance AHTN (diluted) is mixed with other fragrance ingredients into fragrance oils or 'compounds'. A compound may consist of as many as 50 ingredients and a compounder may produce a large number of specified recipes out of the 2000 – 3000 different fragrance ingredients. These fragrance compounds are used in formulating products such as cosmetics, detergents etc. Many fragrance oils or compounds contain AHTN which is one of the most important representative polycyclic musks; when present, at a concentration of 2 to 4% in the compounds. The concentration of fragrances in detergents and soap ranges from 0.2 to 1%. (see also musk fragrances)

A. PROPERTIES

solid; molecular weight 258; melting point >54°C; boiling point 180°C at 15 hPa; vapour pressure 0.061 Pa at 25°C; water solubility 1.25 mg/L ; log Pow 5.7; log K_{OC} 3.8-4.8; H 37 Pa.m^3/mol

B. AIR POLLUTION FACTORS

Photodegradation

t/2 = 1.2 minutes (10573)

C. WATER AND SOIL POLLUTION FACTORS

Concentrations of AHTN in STP influents and effluents (10573)

Location	Influent µg/L	Effluent µg/l
Germany, Ruhr, <1994	Mean 2.2	Median 1.8; 90 perc 3.0
Germany, Berlin 1996		Mean 4.5; max. 5.8
Germany, Berlin 1996-1997		Median 2.2; 90 perc. 3.4; max 4.4
Germany Hessen 1999-2000		Median 0.4; 90 perc. 0.6
The Netherlands 1997-1998	Median 4.0; max 8.7	Median <dl.; max 0.77
The Netherlands 1995-1996		Median 0.28; max 0.42
The Netherlands 1997	03-0.4	0.3-0.6
The Netherlands 1999	2.4-3.9	1.2
The Netherlands 2001	Median 1.3; max 1.8	Median 0.7; max 1.2
Switzerland 1998		Median 1.4; 90 perc. 2.8
Switzerland 1997		Median 2,0; max 2.8
Switzerland 1997		Median 2,0; max 2.8
Switzerland 2002	Mean 1.4; 1,2-2.0; mean 1.5; 0.6	Mean 0.33; 0.2-0.5; mean 0.25
Ohia, U.S.A. 1997	Mean 10.5	Mean 1.3; max 1.7
U.S.A. 1977-1999	Mean 10.35; 7.1-33.9	Mean 1.28; 0.02-2.0
UK 1999-2000	3.7-13.2	0.6-2.7

Concentrations of AHTN in sewage sludge (10573)

	Median mg/kg dw	Mean mg/kg dw	Range mg/kg dw
The Netherlands 6 STPs 1997			
Primary		8.3	3.3-14
Activated		16	2.3-34
Digested	16		0.9-22
The Netherlands 63STPs 1997-1998			
Primary		8.2	0.3-11.7
Activated		5.3	0-13.5
Digested	12		11-13
Germany < 1997 2 STPs			
Activated sludge		8.3	4.0-12.6
Sewer slime industrial		2.1	0.1-8.9
Sewer slime rural	23.1		9.5-36.7
Germany, Hessen 9 STPs			
1996 domest activated sludge.	14.3		3.5-20.8
1996 digested sludge	15.6		14.3-20.1
1996 domestic sludge	15.0		12-20.1
1997	12.1		6.4-17.5
1998	9.1		5.8-18.4
1999	6.8		4.5-8.5
2000	4.2		2.9-6.1
Switzerland 1998 12 STP's			
Digested sludge			Up tp 5.1
UK 2001 14 STPs			
Activated sludge			
Digested sludge	4.0		0.12-16

Concentrations in sludge of Sewage water treatment plants in Nordic countries in 2002 (10586)

Country	n	Range µg/kg dw	Median µg/kg dw
Denmark	5	1,130-3.610	2.020
Finland	5	97-2,270	1,070
Iceland	4	98-553	133
Norway	5	68-3,500	2,100
Sweden	8	952-3,350	1,570

Concentrations of AHTN in surface waters (10573)

Location	n	Median µg/L	90 percentile µg/L	Max µg/L
Germany Ruhr <1994	30	0.2	0.3	
Germany, North sea 1990-1995	12	0.00019	0.0009	
Germany Elbe 1995	3	0.07		0.09

Germany Elbe, 1996-1997	25	0.05	0.07	
Germany, Berlin 1996	26	0.5	2.4	
Germany berlin 1996-1997				
Low effluent input	34	0.02	0.03	0.06
Moderate effluent input	40	0.05	0.14	0.27
High effluent input	28	0.47	0.91	1.10
Germany, Hessen 1999-2000	20	0.05	0.10	
The Netherlands, Rhine 1994-1996	32	0.05	0.10	
River Meuse 1994-1996	35	0.07	0.11	
The Netherlands 1994-1996	14	0.04	0.14	
The Netherlands 1997	5	0.027-0.354		
Switzerland, Glatt <1995	1	0.075		
Switzerland 1997	8	0.05		0.2
Switzerland 1998	20	0.025	0.045	
U.S.A. Southwest	3	0.027-0.092		

Concentrations of AHTN in sediment and suspended matter (10573)

Location	n	Median mg/kg dw	range mg/kg dw	90 percentile mg/kg dw
Dutch borders	14	0.24		
Rhine 1994-1996			0.10-0.54	
Meuse 1994-1996	14	0.84	0.06-1.2	0.96
Netherlands	24	0.12		1.0
Germany, Elbe 1996-1997	31	0.47	0.19-0.77	0.61
Germany, Berlin 1996-1997				
Low effluent input	19	0.02		0.03
Moderate effluent inpu	20	0.24		0.52
High effluent input	20	0.93		2.21
Germany, Hessen, suspended matter				
1996	11	0.29	0.09-0.84	
1997	12	0.19	0.06-0.l86	
1998	12	0.30	0.05-0.86	
1999	16	0.14	0.03-0.40	
2000	15	0.11	0.02-0.26	
Germany, Hessen, suspended matter in contaminated brooks				
1996	11	3.2	0.54-12.7	
1997	5	2.7	1.9-6.6	
1998	5	2.9	1.6-7.2	
1999	3		0.4-2.9	
2000	2		0.6-0.97	
Germany, Elbe, 1997	9	0.047	0.007-0.104	
Niedersachsen, sediment	8	<0.0005	<0.0005-0.004	

Biodegradation

Inoculum	method	concentration	duration	elimination	
Non adapted	Closed bottle test		21 dya	21% ThCO2	
Adapted	Closed bottle test		7 weeks	12% TCO2	(10573)

An extensive study was performed on the biotransformation of ^{14}C-AHTN in activated sludge at 5 to 50 µg AHTN/l. The half-life of the parent AHTN was 12-24h. The related first-order reaction rate constant was 0.029 – 0.057 h^{-1}. In a Continuous Activated Sludge (CAS) test with^{14}C-labelled AHTN at 10 µg/l and realistic STP operation conditions (addition of wastewater, sludge retention time 10d, hydraulic retention time 6h) a complete mass balance was drawn up. The study showed that in the total removal of the parent AHTN of 87.5%, half (42.5%) was caused by biotransformation and half by sorption (44.3%), whereas volatilisation played a minor role (3.3%).
The concentration decrease of AHTN was followed during two days in duplicate activated sludge samples (slurries) that were not additionally spiked. Loss due to volatilisation was also determined. The 'free' (dissolved) and total concentrations were 1.15 and 5.25 µg/l. The degradation rate constant based on total concentrations were 0.0075 h^{-1}. (10573)

Inoculum	method	concentration	duration	elimination
	Modified MITI Test		28 days	0% ThOD
	Modified OECD 301B Test			0% TOC
	Two-phase closed		7 weeks	12-21% ThOD

bottle test CO2 evolution test, OECD 301B	28 days	0% ThCO2	(7169)

Degradation in soils

In the Netherlands a variety of 64 soil samples from various locations were screened for the presence of micro-organisms able to transform AHTN into metabolites with a more polar behaviour. Several pure cultures of fungi, e.g., *Aureobasidium pullulans* and *Phanerochaete chrysosporium* have the capacity of primary biodegradation of AHTN into a series of more polar metabolites (10573)

D. BIOLOGICAL EFFECTS

Concentrations in fish (10573)

		n	Median mg/kg lipids	Max mg/kg lipids
Germany, Ruhr				
Non-eel		7	3.5	7.1
Eel		2	0.6	0.7
Non-eel fish effluent pond		8	15.3	37.2
Eel form effluent pond		5	36	57.9
Denmark fish pond				
Rainbow trout		4	0.36	
Germany, Elbe				
Eel		5	0.056	
Other fish		4	0.58	
Germany, river Stör near STP outfall		3	14 (mean)	
East sea Herring		1	0.53	
Denmark Herring		1	0.07	
Ireland Herring		1	<0.01	
Germany Berlin 1996-1997				
Eel	Low effluent input	54	<dl	<dl
	Moderate effluent input	53	0.186	0.545
	High effluent input	58	2.8	5.2
Perch	Low effluent input	19	<dl	<dl
	Moderate effluent input	19	<dl	<dl
	High effluent input	9	7.1	43.7
Common Bream	Low effluent input	37	<dl	2.7
	Moderate effluent input	37	<dl	2.7
	High effluent input	10	18.4	35.3
Roach	Low effluent input	48	<dl	2.0
	Moderate effluent input	48	<dl	2.0
	High effluent input	6	4.5	18.4
Pike	Low effluent input	12	<dl	2.2
	Moderate effluent input	12	<dl	2.2
	High effluent input	2	8-10	
Pike perch	Low effluent input	25	<dl	<dl
	Moderate effluent input	25	<dl	<dl
	High effluent input	8	10	88.3
Czech republic three rivers 1997-2000				
Chub		302	0.6-2.4	
Bream		164	0.9-3.5	
Barbel		50	0.5-11.4	
Perch		156	0.4-3.7	
Trout		117	0.3-3.1	
Norway high effluent input 1997-1998				
Thornback ray filet		1	0.089	
Haddock filet		2	0.008-0.010	
Atlantic cod filet		3	0.008	0.010
Saithe filet		1	0.093	
Thornback ray liver		1	0.003	
Haddock liver		3	0.024	0.034
Atlantic cod liver		13	0.096	0.034
Saithe liver		1	0.001	

Concentrations in fish in Nordic countries in 2002 (10586)

Sampling site	Sample type	Extractable lipid %	ng/g lipid
Norway			
Trondheim inner harbour	Thornback ray filet	0.86	89
	Thornback ray, liver	39.6	3
	Haddock filet	0.29-1.3	135-373
	Haddock liver	63.7-67.4	15-34
Tromso inner harbour	Atlantic cod, filet	0.42-1.25	13-47
	Atlantic cod, liver	15.4-28.6	6-10
	Saithe, filet	2.3	93
	Saithe, liver	37.1	1
Oslo Fjord 6 sites	Atlantic cod, liver	28.5-42.8	81-380
Larvik, inner harbour	Atlantic cod, liver	11.5	14
Grenland fjord 3 sites	Atlantic cod, liver	26.3-45.2	3-13
Sweden			
Lake Hjärtsjön	Perch	0.4-0.79	30-125
Lake Kvädöfjärden	Perch	0.61-0.95	<118-85
Lake Bysjön	Perch	0.26-0.50	<118-41
River Viskan 3 sites	Bream	0.95-3.11	<18-367
Soutern Vättern	Perch	0.69-1.15	155-301
	Arctic char	1.74-4.13	89-225
Northern Vättern	Perch	0.58-1.15	124-154
	Arctic char	2.75-3.97	42-113
Denmark			
Denmark	Rainbow trout		0.20-0.59
Fish ponds	Trout 50 samples		5.70-252

Concentrations in breast milk in Nordic countries in 2002 (10586)

Country	n	Ng/g lipid
Denmark	10	5.58-37.9
Sweden	30-44	<6-71.4

Bioaccumulation

Species	conc. μg/L	duration	body parts	BCF	
Bluegill sunfish		28 days		597	
Zebrafish		14 days		600	
Eel				200-650	
Non-eel				50-145	
Eel				250-1791	
Rudd				280	
Tench				670	
Crucian carp				400	
Eel				570	
Chrironomus riparius	5.8			50-112	
Lumbriculus variegatus	4			3000-6918	(10573)

Toxicity

Algae			
Pseudokirchneriela subcapitata	72h ECb_{50}	0.47 mg/L	
	72h NOECb	0.37 mg/L	(10573)
Worms			
Lumbriculus variegatus	5d EC_{50}	0.4 mg/L	
Eisenia foetida	4wkgNOEC	>250 mg/kg dw	(10573)
Crustaceae			
Daphnia magna	21d EC_{50}	0.24 mg/L	
reproduction	21d NOEC	0.19 mg/L	(10573)
Insecta			
Chironomus riparius	96h NOEC	>0.5 mg/L	
Springtail Folsomia candida	4 wk NOEC	45 mg/kg dw	(10573)
Copepods			
Nitorca spinipes	96h LC_{50}	0.61 mg/L	
	7d NOEC	>0.05	

Acartia tonsa	48h LC_{10}	0.45 mg/L	
	5d NOEC	0.01 mg/L	
	5d EC_{50}	0.026 mg/L	(10573)
Fish			
Lepomis macrochirus	21d NOEC	0.089 mg/L	
	21d LC_{50}	0.31 mg/L	
Pimephales promelas	32d NOEC	0.035 mg/L	
	36d LC_{50}	0.10 mg/L	
Brachydanio rerio	32d NOEC	0.035 mg/L	(10573)

Concentrations in humans

in human adipose tissue from Switzerland: median value: 6.2 µg/kg; range 1.0-23 µg/kg (7039)

alachlor (2-chloro-2',6'-diethyl-N-(methoxymethyl)acetanilide)

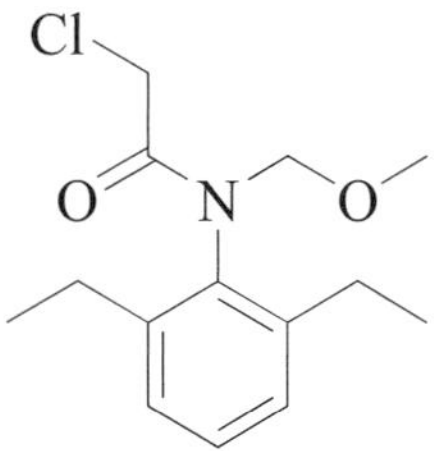

$(CH_3CH_2)_2C_6H_3N(CH_2OCH_3)COCH_2Cl$

$C_{14}H_{20}ClNO_2$

CAS 15972-60-8

USES

a selective herbicide used to control annual grasses and many broad-leaved weeds in cotton, brassicas, maize, oilseed rape, peanuts, radish, soybeans, and sugarcane.

A. PROPERTIES

colorless to yellow crystals; molecular weight 269.77; melting point 40°C; boiling point decomposes at 105°C; vapor pressure 2.9 x 10^{-5} hPa at 20°C; density 1.13 at 20°C; solubility 148 mg/L at 20°C, 240 mg/L; $logP_{OW}$ 2.8.

C. WATER AND SOIL POLLUTION FACTORS

Water treatment

initial concentration, µg/L	*treatment method*	*% removal*	
7	flocc. + sedim. + filtr. + chlorination	24	
4	chlorination + filtration	15	
1.4	chlorination	1	
1.0	chlorination	9	
3.6	flocc. + sedim. + carbonate removal	-5 (increase!)	
not reported	activated carbon	43	(2915)

In groundwater wells

in the U.S.A.-150 investigations up to 1988:		
median of the conc.'s of positive detections for all studies	0.9 µg/L	
maximum conc.	113 µg/L	(2944)

Degradation in soil (see also graph): degradation in soils (Eustis loamy sand) and the underlying vadose zone and aquifer materials: half-life ($t_{1/2}$) at 20°C:

	under aerobic conditions	*under anaerobic conditions*	
soil depth, m	*$t_{1/2}$, d*	*$t_{1/2}$, d*	
0.0-0.6	23	100	
Vadose zone			
0.6-2.4	73	144	
2.4-4.3	104	110	
4.3-6.3	116	53	
6.3-8.3	22	57	
8.3-9.6	285	148	
Aquifer			
9.6-11.6	324	337	
13.1-15.2	320	553	(2577)

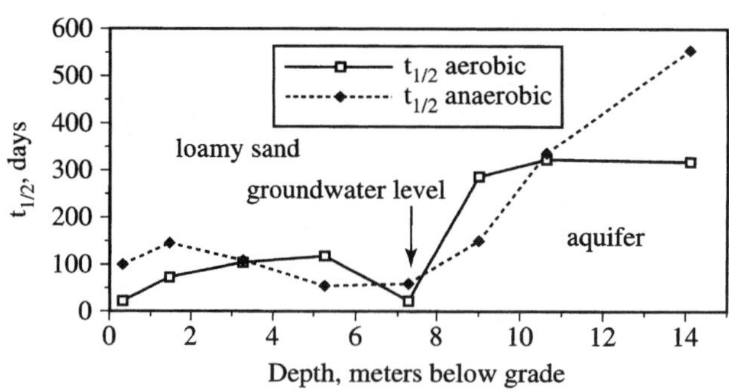

In soil and in plants, the chlorine atom is rapidly cleaved. Further breakdown to the aniline derivative occurs. After 4-5 weeks in soil, the bulk of applied alachlor is degraded. (2962)

Biodegradation half life time in soil

in greenhouse at 23-26°C at constant light at initial conc. of 5-10 mg/kg soil, experiment time 50 d

	t/2 d	
in sandy soil	63	
in organic rich orchard soil	8	
in agricultural soil	13	
in soil from volcanic area	15	
in field experiments	15 (dissipation t/2)	(7045)

Mobility

log K_{OC}: 2.2 (7045)

D. BIOLOGICAL EFFECTS

Nitrification inhibition in soil (1% organic carbon) after 14 d: 100% at 50 mg/kg. (7050)

alamine 336 (aliphatic amines, C_{12}-C_{18})

TRADENAMES

Alamine is a trademark for a series of primary, secondary, and tertiary aliphatic amines, organic substituted ammonia derivatives, chain length from C_{12} to C_{18}, with varying degrees of unsaturation. (1590)

USES

corrosion inhibitors, ore flotation agents, textile finishing agents, rubber compounding. (1590)

D. BIOLOGICAL EFFECTS

Fishes

rainbow trout: 96h LC_{50}, S: 7.5-10 mg/L (1500)

alcohol ethoxylates (EA's; linear alkyl ethoxylates; ethoxylated acohols; polyethylene glycol alkyl ethers; alkyl polyethylene glycol ethers; polyoxyethylene alcohols; polyoxyethylene alkylethers)

$CH_3(CH_2)_x(OC_2H_4)_nOH$

TRADENAMES

Heliwet; Imbentin

USES

Alcohol ethoxylates are a common nonionic surfactant employed in consumer and industrial detergents worldwide. EA's have, depending on the chemical composition and concentration, a combination of cleaning (detergent), foaming, wetting, emulsifying, solubilizing and dispersing properties. They have substituted nonylphenol ethoxylates in most applications as they have similar functions and prices but better environmental properties.

IMPURITIES, ADDITIVES, COMPOSITION

AE comprises a complex mixture of homologues that vary in alkyl chain length and degree of ethoxylation where de C number of the alkyl chain varies between 8 and 20 and the number of ethoxy groups between 0 and 18.

The following numbers have been assigned amongst others

CAS no	Alkyl chain C number	Name, synonyms
74432-13-6		Alcohol ethoxylate
26183-52-8	10	Decyl ethoxylate; decyl alcohol ethoxylate; polyethylene glycol decyl ether
61827-42-7	10	Isodecyl ethoxylate; isodecyl alcohol polyethylene glycol
67254-71-1	10-12	Alcohols(C10-12) ethoxylates; ethoxylated C10-12 alcohols
66455-15-0	10-14	Alcohols(C10-14) ethoxylates; ethoxylated C10-14 alcohols
68002-97-1	10-16	Alcohols(C10-16) ethoxylates; ethoxylated C10-16 alcohols
127036-24-2	11	Branched and linear undecyl alcohol ethoxylate
34398-01-1	11	Undecyl ethoxylate; polyethylene glycol undecyl ether
68439-54-3	11-13	Alcohols(C11-13 branched)ethoxylates
78330-21-9	11-14	Alcohols(C11-14-iso, C13-rich)ethoxylates
68131-40-8	11-15	Alcohols(C11-15 secondary)ethoxylates
9002-92-0	12	Dodecyl ethoxylate; polyoxyethylene lauryl ether; lauryl ethoxylate; Laureth
66455-14-9	12-13	Alcohols (C12-13) ethoxylates; polyethyleneglycol (C12-13) alkylether
68213-23-0	12-14	Alcohols(C12-14)ethoxylates

68439-50-9	12-14	Alcohols(C12-14)ethoxylates
84133-50-6	12-14	Alcohols(C12-14-secondary)ethoxylates
68131-39-5	12-15	Alcohols(C12-15) ethoxylates; linear primary (C12-15) alcohol ethoxylate
106232-83-1	12-15	Alcohols(C12-15-branched and linear) ethoxylates
68551-12-2	12-16	Alcohols(C12-26)ethoxylates
68213-23-0	12-18	Alcohols(C12-18)ethoxylates
68526-94-3	12-20	Alcohols(C12-20)ethoxylates
69011-36-5	13	Branched tridecyl ethoxylates, isotridecanol ethoxylates
9043-30-5	13	Isotridecylethoxylate; ethoxylated isotridecanol
24938-91-8	13	Tridecyl ethoxylate; ethoxylated tridecanol; Trideth
64425-86-1	13-15	Alcohols(C13-15) ethoxylates; ethoxylated C13-15 alcohols
27306-79-2	14	Tetradecyl ethoxylate; myristyl alcohol ethoxylate
120944-68-5	14-15	Alcohols(C14-15-branched and linear)ethoxylates
9004-95-9	16	Hexadecyl ethoxylate; polyoxyethylene cetyl ether; cetyl ethoxylate
68155-01-1	16-18	Alcohols(C16-18 unsaturated alkyl)ethoxylates
68920-66-1	16-18	Alcohols(C16-18) and C18-unsaturated ethoxylates
68439-49-6	16-18	Alcohols(C16-18)ethoxylates
68439-49-6	16-18	Alcohols(C16-18)ethoxylates; C16-18 fatty alcohols ethoxylates
61791-28-4	16-18	Tallow alcohol ethoxylate; ethoxylated tallow alcohol
69227-20-9	16-22	Alcohols(C16-22)ethoxylates
9004-98-2	18	Octadecenyl ethoxylate; polyoxyethylene oleyl ether; oleyl ethoxylate; Oleth
9005-00-9	18	Octadecyl ethoxylate; stearylalcohol ethoxylate; stearyl ethoxylate; Steareth
68439-45-2	6-12	Alcohols(C6-12)ethoxylates
71060-57-6	8-10	Alcohols(C8-10)ethoxylates
71243-46-4	8-16	Alcohols(C12-20)ethoxylates
157707-43-2	8-18	Alcohols(C8-18)ethoxylates; c8-18 fatty alcoholethoxylate
61791-13-7	8-18	Coco oil alcohol ethoxylate; ethoxylated coconut oil alcohol
69013-19-0	8-22	Alcohols(C8-22)ethoxylates
68439-46-3	9-11	Alcohols(C9-11)ethoxylates
78330-20-8	9-11	Alcohols(C9-11-iso, C10-rich)ethoxylates
97043-91-9	9-16	Alcohols(C9-16)ethoxylates

A. PROPERTIES

water solubility depend on both the length of the alkyl chain and the number of ethoxy groups.

C. WATER AND SOIL POLLUTION FACTORS

Biodegradation

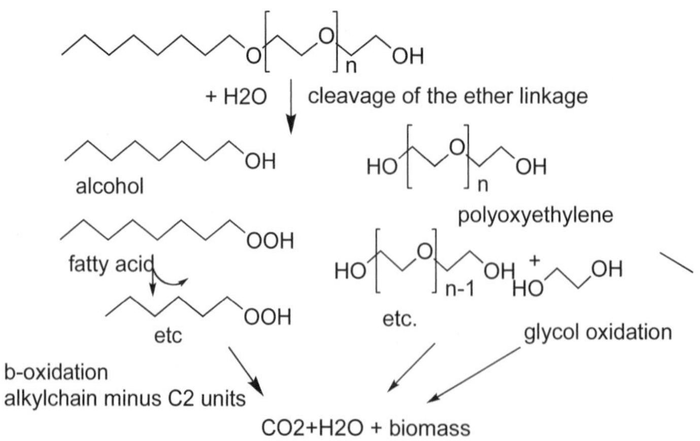

+ H2O
cleavage of the ether linkage
alcohol
polyoxyethylene
fatty acid
etc
etc.
glycol oxidation
b-oxidation
alkylchain minus C2 units
CO2+H2O + biomass

Aerobic degradation pathway

w-PEG degradation

terminal hydroxy shift

release of acetaldehyde

+ CH3CHO

fermentation of acetaldehyde

oxidation of the alcohol to the fatty acid

1/2 CH3COOH + CH_3CH_2COOH

b-oxidation
alkylchain minus C2 units

CH4 + CO2+H2O + biomass

Proposed pathway of LAE degradation by an anaerobic community of fermenting bacteria (Wagner and Schink, 1987)

D. BIOLOGICAL EFFECTS

Toxicity

			CnAEn	
Algae				
Scenedesmus subspicatus	72h ECb_{50}			
N. seminulum	5d algistatic	5-10 mg/L	C14-15AE6	
N. pellicolosa	96H EC_{50}	0.28 mg/L	C14-15AE7	
M.aeruginosa	96h EC_{50}	0.60 mg/L	C14-15AE7	
	5d algistatic	>1,000 mg/L	C14-15AE6	
Selenastrum capricornutum	96h EC_{50}	0.09 mg/L	C14-15AE7	
	5d algistatic	50 mg/L	C14-15AE6	(10589)
Crustaceae				
Daphnia magna	21d NOEC	0.24 mg/L	C14-15AE7	
Ceriodaphnia dubia	7d NOEC	0.17-0.70 mg/L	C14-15AE7	(10589)

Alcohol ethoxysulphates (AES)

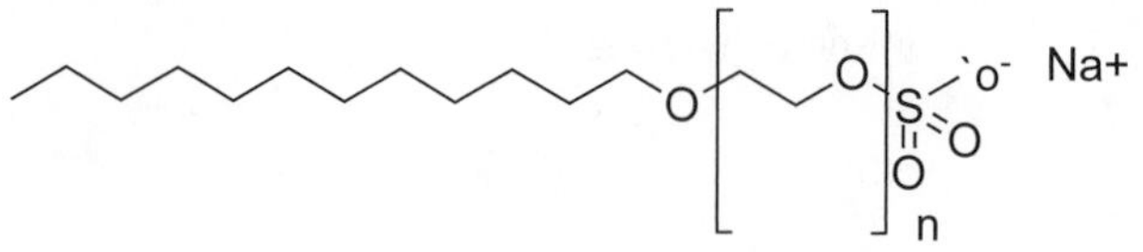

$$CH_3(CH_2)_{8-m}\underset{|\atop (CH_2)_mCH_3}{C}HCH_2O(EO)_nSO_3Na$$

USES

Alcohol ethoxysulphates (AES) are a widely used class of anionic surfactants. They are used in household cleaning products, personal care products, institutional cleaners and industrial cleaning processes, and as industrial process aids in emulsion polymerisation and as additives during plastics and paint production. Uses in household cleaning products include laundry detergents, hand dishwashing liquids, and various hard surface cleaners.

CHEMICAL IDENTITY

The principle structures are shown above, where n varies from 0-8 and m varies from 0-4, but is primarily 0. The average value of n is 2.7 for AES sold into household use and 2.4 for the total AES produced.

The alcohol ethoxysulphate family is defined to encompass commercial grades of linear-type primary alcohol ethoxysulphates containing AES components of basic structure $C_nH_{2n}(C_2H_4O)_mSO_4X$, where n=12-18 and m = 0-8 and X = sodium, ammonium or triethanolamine (TEA). Sodium salts of AES are by far the most commonly used grades.

IMPURITIES, ADDITIVES, COMPOSITION

Commercial sodium AES typically contain, approximately 2-4% of unsulphated alcohol ethoxylate, 1-2% unreacted alcohol and 15-45% alcohol sulphate, and optionally trace amounts of inorganic pH buffering agents, depending on the active matter content and the degree of ethoxylation.

The linear-type alcohols include those which are mixtures of entirely linear alkyl chains, and those which are mixtures of linear and mono-branched alkyl chains, though still with a linear backbone. Such alcohols and their blends are substantially interchangeable as feedstocks for AES used in the major applications.

The entirely-linear alcohol feedstocks include those derived from vegetable or animal sources via oleochemical processes and those derived from ethylene via Ziegler chemistry. Such alcohols contain even numbered alkyl chains only, and are produced in single carbon cuts or more usually wider cuts from C6 through C22+. C12 through C18 grades are feedstocks for AES.

The essentially-linear alcohol feedstocks, also known as linear oxo-alcohols, are derived from linear higher olefins via oxo-chemistry. The feedstock linear olefins are typically derived from ethylene or normal paraffins. Such alcohols contain mixtures of even/odd or odd numbered alkyl chains depending on the feedstock olefin, and are produced in grades ranging from C7 through C15. Typically 90-40% of the carbon chains are linear, the remainder being mono-branched 2-alkyl isomers, predominantly 2-methyl. The mono-branched isomers thus have a linear backbone. C12 through C15 grades.

Of the AES used in consumer cleaning applications in Europe, approximately 71% is derived from even carbon numbered linear alcohols (C12-14 and C16-18), with the remaining 29% derived from odd and even carbon numbered essentially-linear oxo alcohols.

Alcohol ethoxysulphates derived from alcohols shorter than C_{12} are not typically used in household cleaning products. Their uses are small and specialised.

A. PROPERTIES

All values were estimated by interpolation of values for EO2 and EO3 calculated using SRC software (10590)

Eo2.7-average

Carbon #	12	13	14	15	16	18
molecular weight (g mol^{-1})	407	422	436	450	464	492
Melting point (°C)	298	304	309	315	320	331
Boiling point (°C)	684	695	707	719	730	754
Vapour pressure at 25°C (Pa)	$1.2x10^{-13}$	$4.9x10^{-14}$	$2.1x10^{-14}$	$8.8x10^{-15}$	$3.8x10^{-15}$	$6.2x10^{-16}$
$LogP_{OW}$	0.95	1.4	1.9	2.4	2.9	3.9
Water solubility (mg/L)	425	133	41	13	4.0	0.38

C. WATER AND SOIL POLLUTION FACTORS

Biodegradation

There are 3 starting routes of AES degradation which all seem to occur: i) oxidation of the alkyl chain, ii) enzymatic cleavage of the sulphate substituent leaving an alcohol ethoxylate, iii) cleavage of an ether bond in the AES molecule producing either the alcohol (central cleavage) or an alcohol ethoxylate and an oligo(ethylene glycol) sulphate. The subsequent degradation of the resulting intermediates encompasses oxidation of the alcohol to the corresponding fatty acid (itself then degraded via ß-oxidation) or degradation of the alcohol ethoxylate (via central cleavage or degradation from either end of the molecule) or degradation of the oligo(ethylene glycol) sulphate. The ultimate biodegradability of alcohol ethoxylates is well established and glycol ether sulphates have also been shown to be fully degradable by mixed cultures forming inorganic sulphate and carbon dioxide. AES degradation will not produce any recalcitrant metabolite. (10590)

Several reviews highlight that AES are readily biodegradable, with alkyl-chain length having little effect

Removal in WWTP: The primary removal data suggest no consistent removal trend with alkyl chainlength or degree of ethoxylation. A geometric mean of 97.5% removal has been calculated (10590)

Anaerobic degradation: Based on the chemical structure of AES and the proven easy anaerobic biodegradability of the structurally related alcohol ethoxylates and alkyl sulphates, good anaerobic biodegradability of AES is likely. This is supported by the result from testing C12-14EO2S in a stringent anaerobic biodegradability screening test (ECETOC test) which showed a gas (CO_2+ methane) production of 75 % within the 41-day incubation period. AES tested in a lab digester simulating the situation in practice showed that within the 17-day incubation period 88% ultimate biodegradation (based on ^{14}C-gas formation) was found for C14[^{14}C]EO3S.
Taking these mineralisation data into account it is expected that the removal of the parent AES compound under digester conditions is appoximately 90%. (10590)

D. BIOLOGICAL EFFECTS

Different AES homologues are expected to differ in their toxicity. Dyer et al have developed QSAR for chronic toxicity to *Ceriodaphnia* using data on single AES homologues, including EO=0, ie AS. The QSAR developed was: (10591)

logNOEC (mol/l) = $0.128C^2$- 3.767C + 0.152EO + 21.182

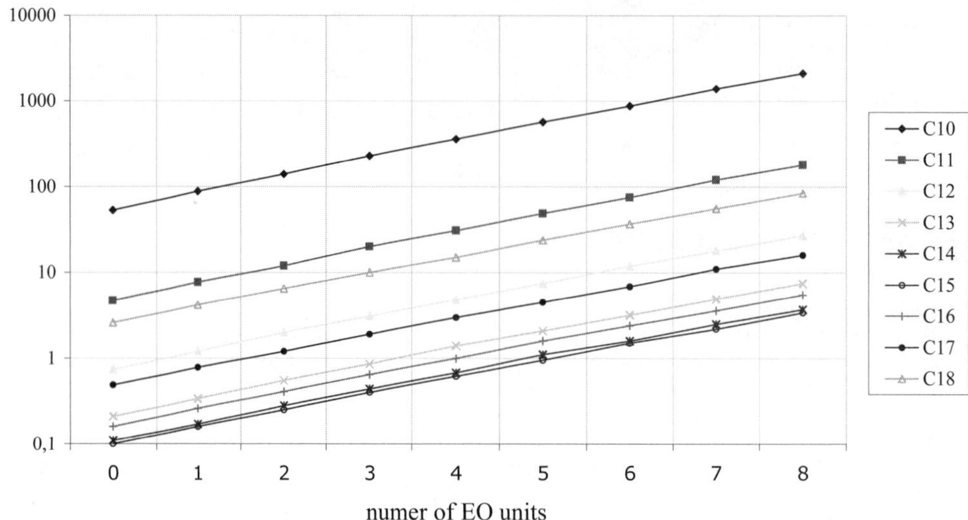

	Toxicity			
			CxEOx	
Algae				
Scenedesmus subspicatus	72h NOEC	0.72 mg/L	C12014EO2	
	96h NOEC	0.35 mg/L	C12-14EO2	
	72h NOEC	0.9 mg/L	C12-15EO3	
	72h NOEC	17 mg/L	C17.3EO0	(10590)
Selenastrum capricornutum	96h NOEC gr	12 mg/L	C12EO0	(10590)
Crustaceae				
Daphnia magna	21d NOECrepr	0.72 mg/L	C12-14EO2	
	21d NOECrepr	0.34 mg/L	C12-15EO3	
	21d NOEC	0.7 mg/L	C12EO>2	
	21d NOEC	0.27 mg/L	C13.6EO13-15	
	21d NOEC	0.18; 0.27 mg/L	C14-15EO2.25	
	2d NOEC	16.5 mg/L	C17.3EO0	(10590)
Ceriodaphnia dubia	7d NOEC	0.34 mg/L	C12EO1	
	7d NOEC	6.3 mg/L	C12EO2	
	7d NOEC	2.7 mg/L	C12EO4	
	7d NOEC	1.2 mg/L	C12EO8	
	7dNOEC	0.28 mg/L	C13EO2	
	7d NOEC	1.2 mg/L	C12EO8	
	7d NOEC	0.34 mg/L	C14EO2	
	7d NOEC	1.1 mg/L	C14EO4	
	7d NOEC	0.08 mg/L	C15EO1	
	7d NOEC	0.06 mg/L	C15EO2	
	7d NOEC	0.15 mg/L	C15EO4	
	7d NOEC	5.8 mg/L	C15EO8	
	7d NOEC	0.20 mg/L	C15EO0	
	7d NOEC	0.60 mg/L	C18EO0	(10590)
Brachionus calyciflorus	2d EC_{20}	0.97-1.1 mg/L	C12EO2	
	2d EC_{20}	2.3 mg/L	C12EO4	
	2d EC_{20}	0.97-1.1 mg/L	C12EO2	
	2d EC_{20}	0.49 mg/L	C13EO2	
	2d EC_{20}	0.13 mg/L	C14EO2	
	2d EC_{20}	0.37 mg/L	C14EO4	
	2d EC_{20}	0.22 mg/L	C15EO4	(10590)
Fish				

Pimephales promelas	30d NOEC	0.88 mg/L	C12-13EO1	
	365 NOEC	0.1 mg/L	C12-14EO2	
	45d LC_{50}	0.44; 0.63; 0.94 mg/L	C14-15EO2.25	
	45d LC_{50}	0.1 mg/L	C14-16EO2.25	
	365d NOEC	0.13 mg/L	C17EO3	
Oncorhyncus mykiss	28d NOEC	0.12 mg/L	C12-15EO3	(10590)

Alcohols, C_{12-15} ethoxylate, sulfonate, sodium salt (sodium C_{12-15}Pareth-15 Sulfonate)

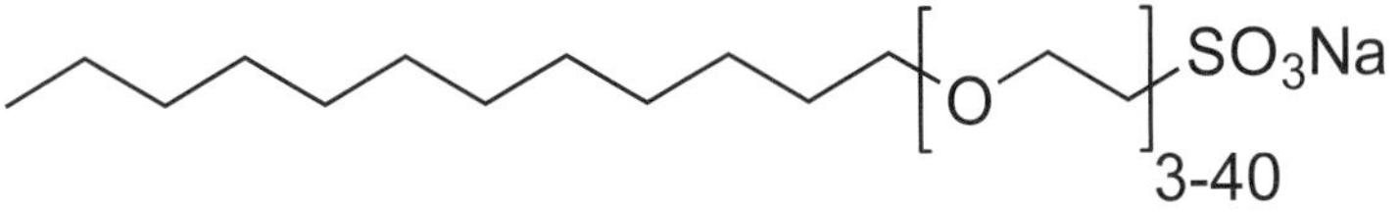

$C_{12-15}H_{24-31}(OC_2H_4)_n$ SO_3 Na where n = 3 - 40, average 15

CAS 121546-77-8

TRADENAMES

Avanel S-150; Sulfonate 300

USES

a cleaning agent in facial wash for topical rins-off applications.

A. PROPERTIES

clear yellowish liquid; molecular weight 965 (range 405-2077); boiling point > 300°C; melting point –18°C*; specific gravity 1.07*; vapour pressure 0.1 kPa at 25°C*; water solubility high

* the data refer to a 35% aqueous solution of the chemical

C. WATER AND SOIL POLLUTION FACTORS

Biodegradation

Inoculum	method	concentration	duration	elimination	
	Municipal WWTP*			99.3-99.9%	(7411)

* C_{12-15} alcohol ethoxylated $(EO_{8.2})$sulphate

D. BIOLOGICAL EFFECTS

Toxicity

Algae	72hLC_{50}	3.5-10 mg/L	(7412)
Crustaceae			
Daphnia magna	48h EC_{50}	4.2 – 72 mg/L	(7412)
Fish	96h LC_{50}	0.7-94.4 mg/L	(7412)

aldehydine *see* 2-methyl-5-ethylpyridine

aldicarb (Temik; 2-methyl-2(methylthio)propionaldehyde-O-(methylcarbomoyl)-oxime; Ambush)

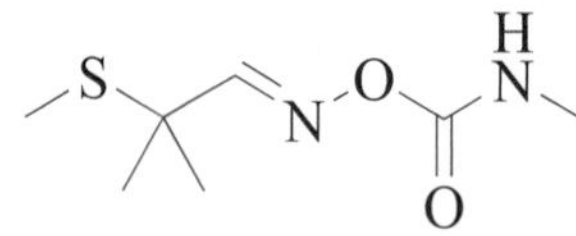

$C_7H_{14}N_2O_2S$

CAS 116-06-3

USES

systematic insecticide; acaricide and nematocide for soil use.

A. PROPERTIES

white, crystalline solid; molecular weight 190.27; melting point 100°C; boiling point decomposes above 100°C; sp. gr. 1.195 at 25/20°C; vapor pressure 3.47 × 10^{-5} mm Hg; solubility 6,000 mg/L at 25°C; K_{OC}: 4.3-6.5; 7-47; 22.

C. WATER AND SOIL POLLUTION FACTORS

Leaching and migration

Lake Hamilton site in Polk County of central Florida: 11.5 kg/ha application rate, sandy soil, shallow groundwater and Florida climate: residues can leach and contaminate upper 3 to 5 m of the saturated zone and can migrate horizontally to distances of 90 m at concentrations >10 μg/L; half-life in the saturated zone = ±8 months. (2923)

In surface waters

In superficial runoff from a potato field 49 d after treatment with Temik: up to 190 μg/L (2924)
In the St Johns River near the mouth of Deep Creek, which drains about 1,600 ha of Temik-treated potato fields: up to 1 μg/L was detected (2922)

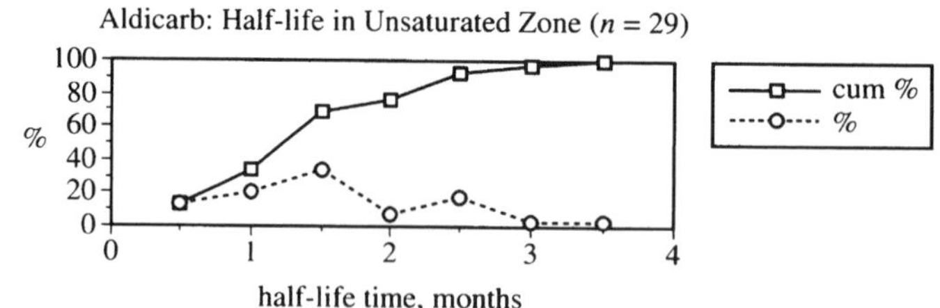

Aerobic degradation in soils

83% of the applied aldicarb was recovered as CO_2 after 63 d of incubation. (2925)

Half-lives of the sum of aldicarb and metabolites aldicarb sulfoxide and aldicarb sulfone varied with pH, temperature, and water content from 9 to 128 d (see graph). (2922)

Oxidation

in the natural environment appears to be entirely a microbial process. (2922)

Degradation in soil

/water mixtures was inhibited at pH below 6. Raising the pH above 6 restored aldicarb-degrading properties. (2922)

Pure cultures of five common soil fungi treated with AS showed that the major degradation products were aldicarb sulfoxide, aldicarb sulfoxide-oxime, and aldicarbsulfoxide-nitrile with minor amounts of the corresponding sulfones. (2927)

Incorporation into the soil

The level of unextracted [^{14}C] aldicarb from soil ranged from 3 to 16% of the applied dose after 69 d in a sandy loam soil. About 5% of the radioactivity was found in the humic fraction, 4% was found in the fulvic acid fraction, and 8% was unextracted from the sample with the highest unextractable residues. (2928)

Evaporation

Applied to rice seedlings as a solution, 5.6% of the initial aldicarb residue in the plant tissue was volatilized in a 10-d period. (2929)
Soil-catalyzed chemical hydrolysis may be an important degradation mechanism, even in the root zone of many soils. Hydrolysis is primarily a chemical, not a microbial, process. Hydrolysis is both acid and base catalyzed with base catalysis occurring more rapidly. (2922)
Applications greater than 800 mg/kg soil broke down increasingly more slowly. (2922)
Fungal growth *(Penicillium, Fusarium)* was observed in soils treated with high levels of aldicarb (250 to 2,000 mg/kg) but not in untreated soils. (2922)

Anaerobic degradation

Aldicarb rapidly degraded to aldicarb-nitrile in the presence of ground limestone or high concentrations of microorganisms under anaerobic conditions. (2930)

Groundwater

microcosms showed that ±18% of aldicarbsulfoxide was reduced to aldicarb in 71 d under anaerobic conditions. (2931)

Photolysis

Irradiation of aldicarb in solvents at 254 nm produced methylamine, dimethyl sulfide, tetramethylsuccinonitrile, and 1-(methylthio)-2,3-dimethylbutane as major products with minor amounts of N,N'-dimethylurea. (2932)

Photolysis

of aldicarb in water produced a mixture containing α-methylacronitrile, dimethyl sulfide, and 2-(methylthio)-2-methylpropanenitrile. (2932)

Photolysis

half-life in sunlight and in water: 6 d. (2933)

Degradation

Aldicarb degrades quite rapidly in soils with the evolution of CO_2.

Under field conditions a half-life of about 7 d in loam soil was found.

The following metabolites were identified: aldicarbsulfoxide [2-methyl-2-(methylsulfinyl)propionaldehyde-O-(methylcarbamoyl)oxime] and alicarbsulfone [2-methyl-2-(methylsulfonyl)-propionaldehyde-O-(methylcarbamoyl)-oxime].

Dissipation periods in soil

in the field

soil	DT_{50} (d)	DT_{90} (d)	
sand	10	67	
silt	33	110	(6020)

Groundwater

Leaching of alicarb in Houston clay and Lufkin sandy loam is insignificant, but it appeared to move more freely through columns of coarse sand.

In groundwater wells in the U.S.A. 150 investigations up to 1988

median conc. of positive detections for all studies:	9 µg/L	
maximum conc.	315 µg/L	(2944)

Representative total aldicarb concentrations-Suffolk County (U.S.A.) groundwater after 4 years of extensive use. The parent compound (aldicarb) was not detected, but the aldicarb metabolites (aldicarb sulfone and aldicarb sulfoxide) occurred as follows:

total aldicarb	= parent aldicarb	+ aldicarb sulfone	+ aldicarb sulfoxide
100%	0%	40-60%	60-40%

community	*wells sampled*	*>7 µg/L*	*1-7 µg/L*	*% below detection*	
1	222	2	18	91	
2	434	43	46	80	
3	2,161	351	345	69	
4	1,832	270	256	71	
5	3,160	359	374	77	(2943)

Degradative pathways in groundwater:

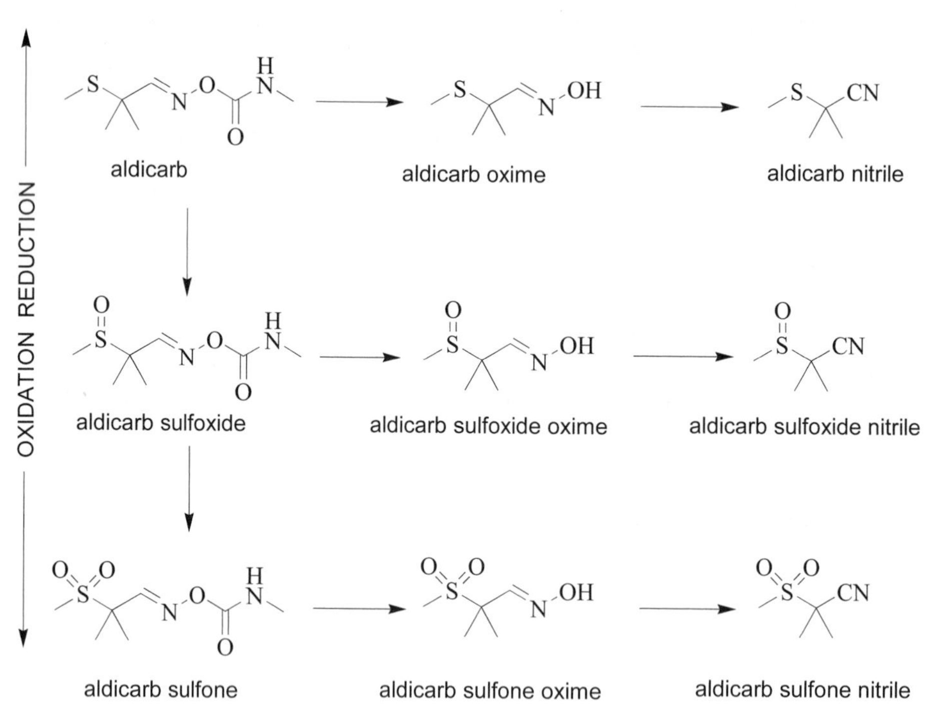

(2922)

D. BIOLOGICAL EFFECTS

Fishes			
rainbow trout	96h LC_{50}	8.8 mg/L	(2962)
bluegill	96h LC_{50}	1.5 mg/L	(2962)

aldicarb sulfone (aldoxycarb; 2-methyl-2-methylsulphonylpropionaldehyde-O-methylcarbamoyloxime)

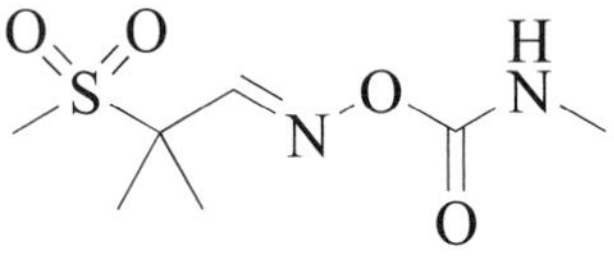

$C_7H_{14}N_2O_4S$

CAS 1646-88-4

USES

a systemic insecticide and nematicide and a potent inhibitor of cholinesterase.

OCCURRENCE

metabolite of aldicarb.

A. PROPERTIES

molecular weight 222.3; melting point 140-142°C; vapor pressure 12 mPa at 25°C; solubility 8,000 mg/L at 20°C.

C. WATER AND SOIL POLLUTION FACTORS

Soil sorption

K_{oc}: 1.7-2.2; 6-18 (2922)

Degradative pathways in groundwater:

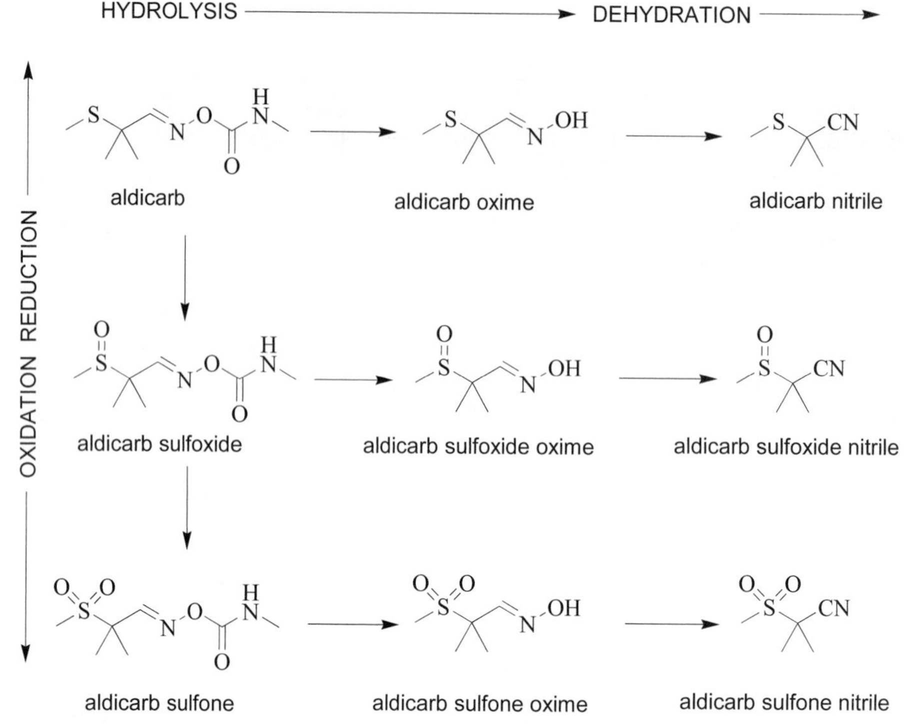

Degradation in soil

Soil degradation was fastest in a clay loam soil at pH 7.2: half-life: 18 d. (2926)

Half-lives of the sum of aldicarb and its metabolites aldicarb sulfoxide and aldicarb sulfone varied with pH, temperature, and water content from 20 to 128 d. (2922)

Soil-catalyzed chemical hydrolysis may be an important degradation mechanism, even in the root zone of many soils. Hydrolysis is both acid and base catalyzed with base catalysis occurring more rapidly. (2922)

Photolysis

half-life in sunlight and in water: 76 d (2933)

Waste water treatment

Granular activated carbon: an empty bed contact time of 10 min yielded the following results:

influent	*effluent*	*Gac usage rate lb/1,000 gal*	*service time days*	
25 µg/L	3 µg/L	0.22	110	
25 µg/L	9 µg/L	0.18	146	(2943)

Reverse osmosis

monthly averages, µg/L

month	*raw water*	*permeate*	*concentrate*	
July	31.4	1.3	79.0	
August	27.2	1.2	79.7	
September	22.8	1.0	68.4	
October	21.6	1.0	54.6	
November	20.3	<1.0	45.8	
December	19.0	<1.0	42.3	
January	17.4	1.5	42.2	
February	16.5	<1.0	42.5	
March	16.3	<1.0	43.1	
April	14.8	<1.0	42.8	
May	14.7	<1.0	42.0	
June	14.0	<1.0	39.6	(2943)

D. BIOLOGICAL EFFECTS

Fishes			
trout	96h LC_{50}	40 mg/L	
bluegill	96h LC_{50}	55 mg/L	(2963)

aldicarb sulfoxide

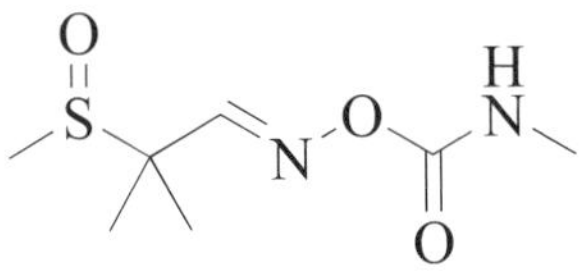

CAS 1646-87-3

OCCURRENCE

metabolite of aldicarb.

A. PROPERTIES

solubility 330 g/L.

C. WATER AND SOIL POLLUTION FACTORS

Soil sorption

K_{oc}: 0-1.7. (2922)

Degradative pathways in groundwater:

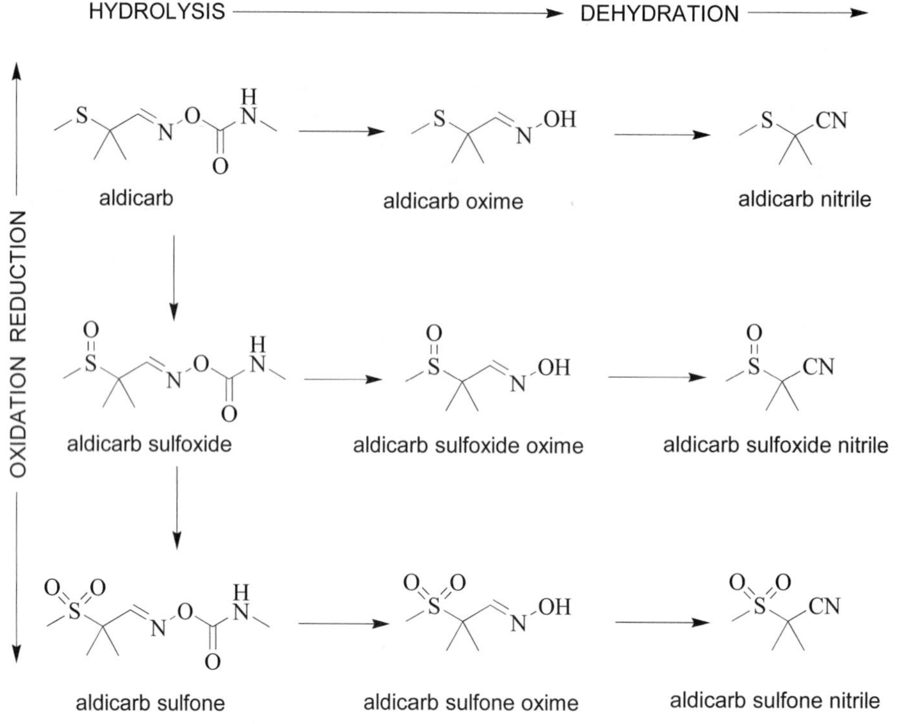

Degradation in soil

Soil degradation was fastest in a clay loam soil at pH 7.2: half-life: 14 d.	(2926)
Half-lives of the sum of aldicarb and metabolites aldicarb sulfoxide and aldicarb sulfone varied with pH, temperature and water content from 9 to 128 d.	(2922)
Soil-catalyzed chemical hydrolysis may be an important degradation mechanism, even in the root zone of many soils. Hydrolysis is both acid and base catalyzed with base catalysis occurring more rapidly.	(2922)
Oxidation in the natural environment appears to be entirely a microbial process	(2922)
Groundwater microcosms showed that ±18% of aldicarbsulfoxide was reduced to aldicarb in 71 d under anaerobic conditions.	(2931)

Photolysis

half-lives in sunlight and in water: 6 d (2933)

Waste water treatment

Granular activated carbon: an empty bed contact time of 10 min yielded the following results:

influent	*effluent*	*Gac usage rate, lb/1,000 gal*	*service time, days*	
20 µg/L	3 µg/L	0.20	126	
20 µg/L	9 µg/L	0.13	186	(2943)

Reverse osmosis

monthly averages, µg/L

month	*raw water*	*permeate*	*concentrate*	
August	20.0	<1	58.2	
September	17.2	<1	51.4	
October	17.4	<1	42.4	
November	16.2	<1	36.0	
December	14.7	<1	32.7	
January	14.0	<1	34.9	
February	13.4	<1	33.8	
March	13.0	<1	33.2	
April	13.0	<1	34.2	
May	11.6	<1	32.8	
June	11.2	<1	30.5	(2943)

aldol see 3-hydroxybutanal

aldrin (,2,3,4,10,10-hexachloro-1,4,4a,5,8,8a-hexahydro-1,4-endo,exo-5,8 dimethanonaphthalene; HHDN; aldrex; aldrite; aldrosol; drinox; octalene; seedrin liquid))

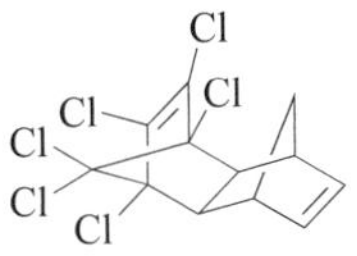

$C_{12}H_8Cl_6$

CAS 309-00-2

USES

insecticide, fumigant.

A. PROPERTIES

brown and white crystalline solid; molecular weight 364.93; melting point 104-105.5°C, vapor pressure 2.3 × 10^{-5} mm at 20°C (technical grade melting point 49-60°C), solubility 0.01 mg/L.

B. AIR POLLUTION FACTORS

Incinerability

thermal stability ranking of hazardous organic compounds: rank 162 on a scale of 1 (highest stability) to 320 (lowest stability). (2390)

Atmospheric reactions (photochemical transformations):

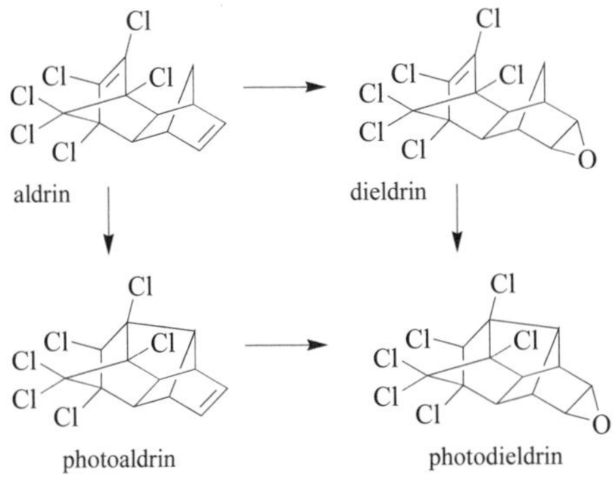

aldrin → dieldrin

aldrin → photoaldrin

dieldrin → photodieldrin

photoaldrin → photodieldrin

C. WATER AND SOIL POLLUTION FACTORS

In groundwater wells in the U.S.A.: 150 investigations up to 1988:

median of the conc.'s of positive detections for all studies	0.1 µg/L	
maximum conc.	0.1 µg/L	(2944)

Water and sediment quality

in Northern Mississippi water: avg. 0.21 ng/L; 0.01-0.49 ng/L (1082)
Hawaii: sediment: 5.5-11 µg/kg (1174)

Odor threshold

0.017 mg/kg water (326, 915)

Biodegradation

Metabolic pathway of aldrin and dieldrin under oceanic conditions:

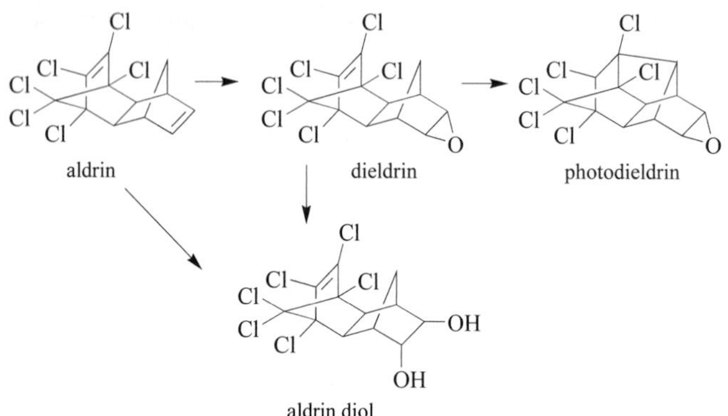

Conversion of aldrin to dieldrin was 80% complete after 8 weeks in river water kept in a sealed jar under sunlight and artificial fluorescent light-initial conc. 10 µg/L (1309)

	% of original compound found					
after	*1 h*	*1 wk*	*2 wk*	*4 wk*	*8 wk*	
	100	100	80	40	20	(1309)

Waste water treatment

Removal during primary sedimentation: range: 5-80%, mean: 31%. (7053)
Manmade sources: in sewage sludges UK 1984: range: 0.01-0.2 mg/kg; mean 0.03 mg/kg. (7054)

Photooxidation

by UV light in aqueous medium at 90-95°C, time for the formation of CO_2 (% of theoretical):

25%:	14 h	
50%:	28 h	
75%:	110 h	(1628)

75-100% disappearance from soils: 1-6 years (1815, 1816)

Evaporation from water

Calculated half-life in water at 15°C and 1 m depth, based on evaporation rate of 3.72×10^{-3} m/h: 185 h (437)

Impact on biodegradation processes

nitrification inhibition in soil (1% organic carbon) after 14 d: 3% at 50 mg/kg. (7050)

D. BIOLOGICAL EFFECTS

AQUATIC VASCULAR PLANTS

of lake Pinne, Finland (1972, 1973)

mean:	2 µg/kg dry weight, *n* =114
S.D.:	5 µg/kg dry weight
min.:	0 µg/kg dry weight

max.:	36 µg/kg dry weight	(1055)

Bioaccumulation

Algae

Chlorella fusca: BCF (wet wt.): 12,260 (2659)

Biotransfer factor in beef: log B_b: -1.07

Biotransfer factor in milk: log B_m: -1.62

Bioconcentration factor for vegetation: log BCF_v: 0.85 (2644)

Crustaceans

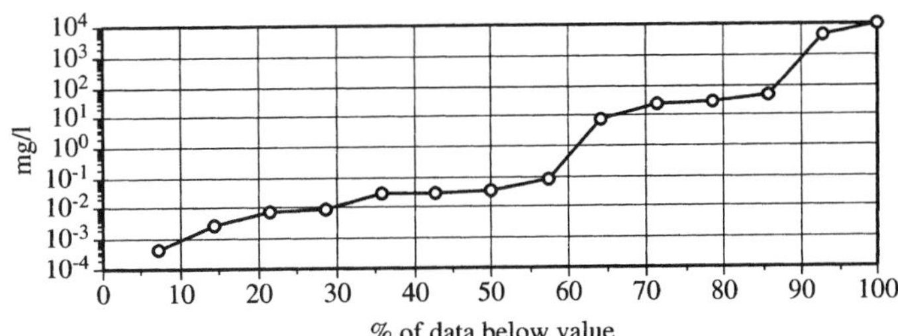

Frequency distribution of 24-96h EC_{50}/LC_{50} values for crustaceans (n = 14) based on data from this and other works (3999)

Gammarus lacustris	$96hLC_{50}$	9,800 mg/L	(2124)
Gammarus fasciatus	$96hLC_{50}$	4,300 mg/L	(2125)
Palaemonetes kadiakensis	$96hLC_{50}$	50 mg/L	(2125)
Asellus brevicaudus	$96hLC_{50}$	8 mg/L	(2125)
Daphnia pulex	$48hLC_{50}$	28 mg/L	(2127)
Simocephalus serrulatus	$48hLC_{50}$	23 mg/L	(2127)
Korean shrimp *(Palaemon macrodactylus):*	$96hLC_{50}$	0.74 µg/L	(2352)
	$96hLC_{50}$,F	3 µg/L	(2352)
Sand shrimp *(Crangon septemspinosa)*	$96hLC_{50}$,S	8 µg/L	(2327)
Grass shrimp *(Palaemonetes vulgaris)*	$96hLC_{50}$,S	9 µg/L	(2327)
Hermit crab *(Pagurus longicarpus)*	$96hLC_{50}$,S	33 µg/L	(2327)
Daphnia magna	$24hLC_{50}$	30 µg/L	
	$48hLC_{50}$	28 µg/L	(1002, 1004)
Isopod *(Asellus)*	$24hLC_{50}$	80 µg/L	(1681)

Toxicity

Toxicity ratios of aldrin (A) to photo-aldrin (PA), calculated from respective 24h LC_{50} and LT_{50} values: A/PA

Crustaceans		
Daphnia pulex (water flea)	1.43	
Gammarus spp. (amphipod)	1.83	
Asellus spp. (isopod)	2.0-2.1	
Insects		
Aedes aegypti larvae (mosquito)	5.7-6.0	
Musca domestica (house fly)	2.0-2.1	
Fishes		
Lebistes reticulatus (guppy)	2.41	
Pimephalus promelas (bass)	1.42	
Lepomis macrochirus (blue gill)	2.9-3.6	(1681, 1684, 1685)

Mollusc

Hard clam *(Mercenaria mercenaria):*			
10-d two-cell stage fertilized, 500 ppb, 37% survival			(2324)
10 d eggs introduced into test media, 1,000 ppb, 0% survival			(2324)
48 hour 50 percent of eggs develop normally, >1,000 ppb			(2324)
larvae	12d LC_{50}	410 ppb	(2324)

Bioaccumulation

BCF: 5 aquatic MOLLUSCS: 350-4,500 (1870)

Residue: mussel *(Mytilus galloprovincialis):* avg 1.02 ppb fresh wt. (n = 4) from Central Mediterranean-1976/77): range 0.4-1.7 ppb fresh wt. (1774)

Insects

Pteronarcys californica	96h LC_{50}	1.3 μg/L	(2128)
	96h LC_{50}	180 μg/L	(2128)
	30 d LC_{50}	2.5 μg/L	(2118)
Acroneuria pacifica	96h LC_{50}	200 μg/L	
	30 d LC_{50}	22 μg/L	(2118)
fourth instar larvae *Chironomus riparius*	24h LC_{50}	0.8 μg/L	(1853)
housefly (3 d old female *Musca*)	LD_{50}	14 μg/fly	(1681)
mosquito (late 3rd instar *Aedes aegypti larvae*)	24h LC_{50}	3 ppb	(1681)
Pteronarcys	48h LC_{50}	43 μg/L	(1003)

Hydropsyche larvae: significant modification of net construction after 48-h exposure to 20 μg/L (1006, 1007, 1008)

Fishes

Mummichog			
(Fundulus heteroclitus)	96h LC_{50},S	4 ppb	(2328, 3229)
Striped killifish			
(Fundulus majalis)	96h LC_{50},S	17 ppb	(2329)
Atlantic silverside			
(Menidia menidia)	96h LC_{50},S	13 ppb	(2329)
Striped mullet			
(Mugil cephalus)	96h LC_{50},S	100 ppb	(2329)
Bluehead			
(Thalassoma bifasciatum)	96h LC_{50},S	12 ppb	(2329)
Northern puffer			
(Sphaeroides macalatus)	96h LC_{50},S	36 ppb	(2329)
American eel			
(Anguilla rostrata)	96h LC_{50},S	5 ppb	(2329)
Threespine stickleback			
(Gasterosteus aculeatus)	96h LC_{50},S	27.4 ppb	(2333)
Shiner perch			
(Cymatogaster aggregata)	96h LC_{50},S	7.4 ppb	(2354)
Dwarf perch			
(Micrometus minimus)	96h LC_{50},S	18 ppb	(2354)
	96h LC_{50},F	2.03 ppb (1-4.2)	(2354)
Pimephales promelas	96h LC_{50}	28 μg/L	(2113)
Lepomis macrochirus	96h LC_{50}	13 μg/L	(2113)
Salmo gairdneri	96h LC_{50}	17.7 μg/L	(2119)
Oncorhynchus kisutch	96h LC_{50}	45.9 μg/L	(2119)
Oncorhynchus tschawytscha	96h LC_{50}	7.5 μg/L	(2119)
Striped bass			

(Morone saxatilis)	96h LC_{50},S	0.010 mg/L	
Banded killifish *(Fundulus diaphanus)*	96h LC_{50},S	0.021 mg/L	
Pumpkinseed *(Lepomis gibbosus)*	96h LC_{50},S	0.02 mg/L	
White perch *(Roccus americanus)*	96h LC_{50},S	0.042 mg/L	
American eel *(Anguilla rostrata)*	96h LC_{50},S	0.016 mg/L	
Carp	96h LC_{50},S	0.004 mg/L	
Guppy	96h LC_{50},S	0.02 mg/L	(1193)
Salmo gairdneri	96h LC_{50}	10 µg/L	(1001)
bluegill	24h LC_{50}	260 ppb	(1681)
bluegill	96h LC_{50}	0.013 ppm	
rainbow trout	96h LC_{50}	0.036 ppm	(1878)
susceptible mosquito fish	48h LC_{50}	36 ppb	
resistant mosquito fish	48h LC_{50}	2,735 ppb	(1851)

Residues

in marine animals from the Central Mediterranean (1976-1977):

fishes:		
anchovy *(Engraulis encrasicholus):*	avg. 0.26 ppb fresh wt. (*n* = 12) range 0.1-0.8 ppb fresh wt.	
striped mullet *(Mullus barbatus):*	avg. 0.59 ppb fresh wt. (*n* = 10) range 0.2-1.7 ppb fresh wt.	
tuna *(Thunnus thynnus thynnus):*	avg. 0.14 ppb fresh wt. (*n* = 5) range 0.1-0.2 ppb fresh wt.	(1774)

Barbus conchonius: histological effects after 60-120 d at 0.0466 µg/L

Heteropneustes fossilis: hematological effects after 4 d at 140 µg/L; LC_{50}, 4 d: 175 µg/L

Tilapia mossambica: hematological effects after 30 d at 100 µg/L (2625)

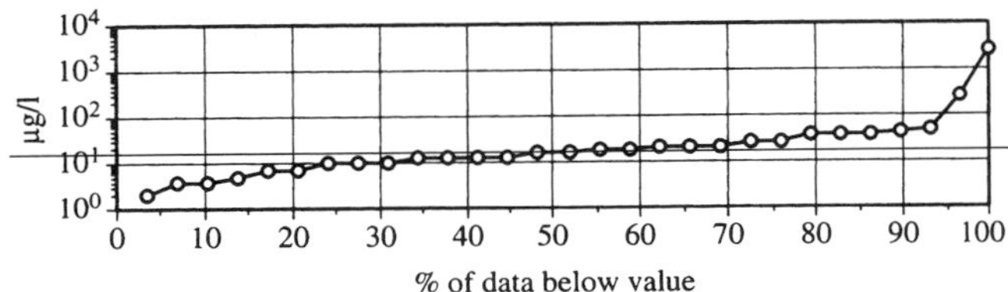

Birds

mean concentration in game bird muscle in upper Tennessee (U.S.): mg/kg fresh weight + standard error.

	grouse	*quail*	*woodcock*
Johnson County	0.29 ± 0.03 (*n* = 12)	0.28 ± 0.08 (*n* = 6)	0.18 ± 0.06 (*n* = 6)
Carter County	0.30 ± 0.04 (*n* = 9)	0.20 ± 0.06 (*n* = 7)	0.28 ± 0.08 (*n* = 6)
Washington County	0.59 ± 0.16 (*n* = 10)	0.23 ± 0.07 (*n* = 11)	0.32 ± 0.09 (*n* = 6)

algerite alba *see* hydroquinone monobenzylether

algerite powder *see* N-phenyl-α-naphthylamine

aliphatic acids C_3-C_6

C. WATER AND SOIL POLLUTION FACTORS

Göteborg (Sweden) sew. works 1989-1991: infl.: 100-500 μg/L; effl.: <0.2 μg/L (including esters). (2787)

aliphatic acids C_7-C_{18}

C. WATER AND SOIL POLLUTION FACTORS

Göteborg (Sweden) sew. works 1989-1991: infl.: 350-38,000 μg/L; effl.: 1-5 μg/L (including esters). (2787)

aliphatic aldehydes

C. WATER AND SOIL POLLUTION FACTORS

Göteborg (Sweden) sew. works 1989-1991: infl.: 5-100 μg/L; effl.: <0.1 μg/L. (2787)

aliphatic amines *see* alamine 336

alkylarylsulfonate *see* teepol 715

alkylbenzene sulfonic acid *see* dobanic acid

alkylbenzenes *see also* dobane and linear alkylbenzenes, C_{10-13}

D. BIOLOGICAL EFFECTS

Bioaccumulation factors

In muscle of Coho salmon:

	*weeks of exposure**				*1 wk of depuration (after 6 wk of exposure)*
	2	*3*	*5*	*6*	
C_2-substituted benzenes	1.1	2.4	2	1	n.d.
C_3-substituted benzenes	10	30	50	10	n.d.
C_{4-5}-substituted benzenes	150	170	550	200	n.d.

In muscle of starry flounder *(Platichthys stellatus):*

	*weeks of exposure**		*weeks of depuration (after 2 wks of exposure)*	
	1	*2*	*1*	*2*
C_2-substituted benzenes	20	4	1	n.d.
C_3-substituted benzenes	500	70	6	10
C_{4-5}-substituted benzenes	9,300	1,700	980	2,600

(n.d. = not detected) (1659)

* exposure to approx. 1 ppm of the water soluble fraction of Prudhoe Bay crude oil

alkylbenzenes, linear, C_{10}-C_{13} (LAB, alkyl(C_{20}-C_{13})benzene; benzene,C_{10}-C_{13}alkyl derivatives; Marlican; Sirene; LAB)

$C_{16}H_{26}$ to $C_{19}H_{32}$

CAS 67774-74-7

USES

The commercial mixtures of long-chain alkylbenzenes (C_{11}-C_{14}) are usually used as raw material for the manufacture of alkylbenzenesulfonates (anionic surfactants).

IMPURITIES, ADDITIVES, COMPOSITION

Composition

Marlican is a mixture of:
<1 wt% 2-, 3-, 4-, and 5-phenylnonanes
6-12 wt% 2-, 3-, 4-, and 5-phenyldecanes
35-45 wt% 2-, 3-, 4-, -5, and 6-phenylundecanes
35-45 wt% 2-, 3-, 4-, -5, and 6-phenyldodecanes
8-14 wt% 2-, 3-, 4-, -5, and 6-phenyltridecanes
<1 wt% phenyltetradecanes

108 alkylbenzenes, linear, C10-C13

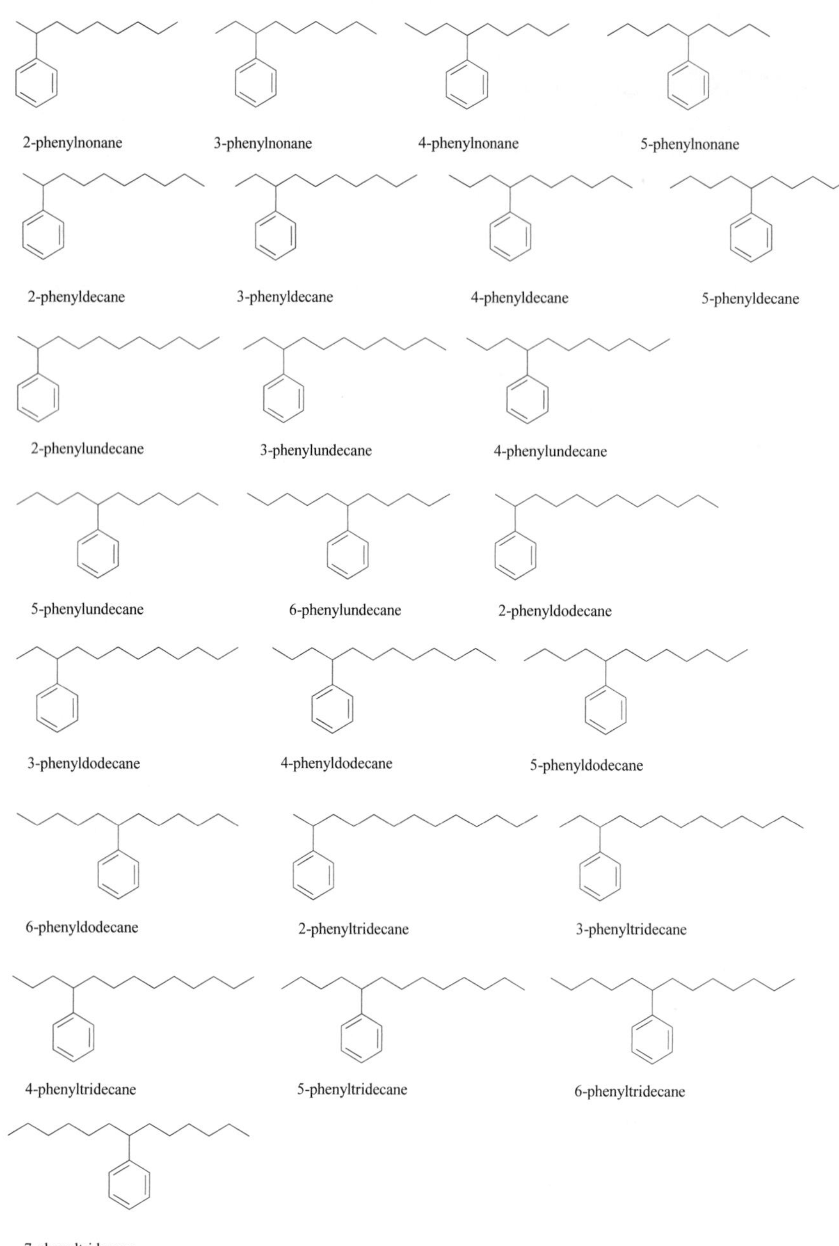

A. PROPERTIES

melting point -70°C; boiling point 280-305 °C; density 0.85-0.86 at 20°C; vapor pressure 0.00065 hPa at 25°C; solubility 0.041 mg/L at 27°C (average side chain length $C_{11.1}$); $LogP_{ow}$>5 at 23°C (measured), 8.5 (calculated for phenyldodecane); 95 hPa m^3/mol measured (C_{12} alkyl chain).

B. AIR POLLUTION FACTORS

Photochemical reactions

direct photolysis under sun light: <1% after 14 d (average $C_{11.1}$) (9446)

C. WATER AND SOIL POLLUTION FACTORS

Background concentrations

in coastal mediterranean seawater, Livorno Italy: 1.1 ng/L		(9447)
in 3 wells near Llobregat river (Barcelona Spain) in 1984 near a surfactant manufacturing plant	<500-3,000 µg/L	(9449)

IN SPANISH RIVERS NEAR BARCELONA MARCH 1985-MARCH 1986

	River Besos	River Llobregat	
Phenyldecanes	0.57 µg/L	0.14 µg/L	
Phenylundecanes	1.3 µg/L	0.87 µg/L	
Phenyldodecanes	0.21 µg/L	0.24 µg/L	
Phenyltridecanes	0.15 µg/L	0.098 µg/L	(9450, 9451)

IN RIVERS DOWNSTREAM OF SEVERAL SEWAGE TREATMENT PLANTS THROUGHOUT THE USA, 1987-1989:

Upstream:	water:	<0.1-0.5 g/L (average side chain 11.4-12.2)	
	sediment:	<0.1-0.61 mg/kg (average side chain 11.5-12.7)	
Downstream:	water:	<0.1-1.0 g/L (average side chain 11.8-12.5)	
	Sediment	<0.1-0.87 mg/kg(average side chain 11.6-12.9)	(9446)

IN SEDIMENTS OF SAN PEDRO BASIN, CALIFORNIA, USA, APRIL 1981, AT 0-2-CM DEPTH:

coastal sediments: 21 mg/kg dw (C_{10}-C_{14})

Marine sediments: 1.1 mg/kg dw. (C_{10}-C_{14}) (9452)

in sediments of rivers flowing into the Tokyo Bay: October 1982-May 1983:

	mg/kg dw	% C_{10}	% C_{11}	% C_{12}	% C_{13}	% C_{14}
River Sumidagawa	0.56-12	2-5	19-26	29-31	28-37	6-12
River Tamagawa	0.01-15	2-4	17-23	25-34	33-44	6-12

July-August 1983

	mg/kg dw	% C_{10}	% C_{11}	% C_{12}	% C_{13}	% C_{14}
River Sumidagawa	23-31	0-2	11-19	27-31	37-46	11-16
River Tamagawa	4.2-26	1-4	12-20	27-36	32-43	9-17

(9454)

in sediments of Port Phillip Bay, SE Australia, July 1984: <0.01-19 mg/kg (C_9-C_{15}) (9455)

Manmade sources

in effluents of four major municipal waste water treatment plants in southern California	(1979): 25-2,200 µg/L (C_{10}-C_{14} alkylbenzenes)	
in waste water treatment plants 1983-1986 in Tokyo, Japan:		
influents: water:	0.061 µg/L	
suspended particles:	0.026 mg/kg dw	
effluents: water:	1.97 µg/L	
suspended particles:	15 mg/kg dw.	
mean abundance of alkylbenzene homologues:	1% C_{10}, 15% C_{11}, 31% C_{12}, 40% C_{13}, 13% C_{14}	(9454)

in sewage treatment plants (n = 10 through USA)

Influents	20-31 µg/L	% removal	
Effluents			
activated sludge plants	<0.1-5.0 µg/L	>95%	
Trickling filters without tert. Treatment	2.7-11 µg/L	54-88%	
Trickling filters with additional sand filtration	1.0 µg/L	>99%	
Trickling filters with 20-min aeration of effluent	1.6 µg/L	95%	
Average side chain: influent:	11.7-12.2		
Effluent: In municipal sewage sludge in California April 1986-June 1986:	11.6-12.3		(9446)
20-430-mg/kg solids			
151-7,274 µg/L			(9456)

Biodegradation

Inoculum	*Method*	*Concentration*	*Duration*	*Elimination*

				results	
A.S. domestic adapted	Modified Sturm test	20 mg DOC/L	7 d	0% DOC	
			14 d	30% DOC	
			33 d	65% DOC	(9457)
Domestic sewage aerobic	Biodegradation test	9.3 µg/L C_{10}-C_{14}	1 d	5-20%	
	at 25-28°C		2 d	26-31%	
			3 d	41-47%	
			4 d	56-62%	
			5 d	73-74%	

The ratio of [6-C_{12}+ 5-C_{12}-alkylbenzenes] relative to [4-C_{12}+ 3-C_{12}+ 2-C_{12}-alkylbenzenes] changed from 0.74-0.75 at d 0 to 5.9-7.2 at d 6. Isomers with the phenyl group bound close to the end of the alkyl chain were degraded preferentially. (9458)

aerobic	Shake flask carbon dioxide	18 mg/L $C_{11.1}$	35 d	56-61% $ThCO_2$	(9446)
filtered river water	Die-away test	0.05-0.25 mg/L $C_{11.8}$	7 d	56-81% TS	
			14 d	79-93% TS	
			21 d	86-97% TS	
			28 d	92-99% TS	(9446)
Domestic sewage anaerobic		9.3 µg/L	6 d	8%	(9458)

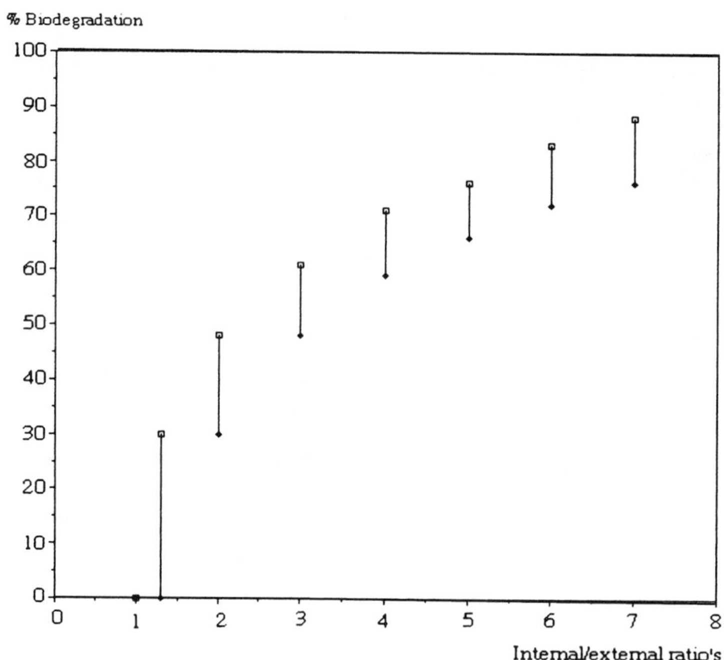

Effect of the internal/external ratio on the biodegradation of linear dodecylbenzenes

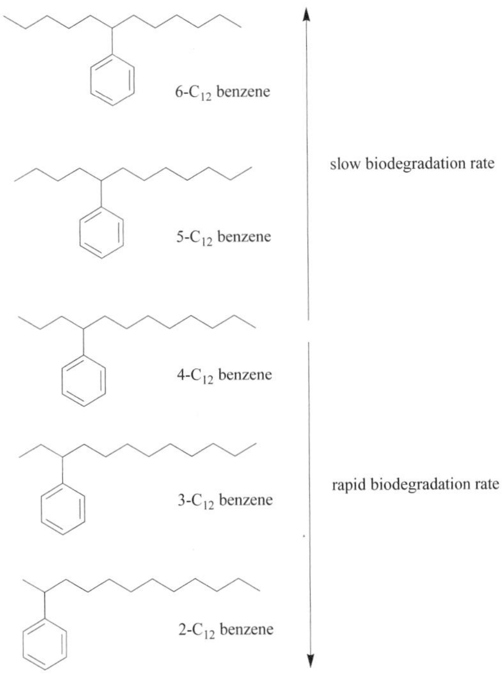

6-C_{12} benzene

5-C_{12} benzene

slow biodegradation rate

4-C_{12} benzene

3-C_{12} benzene

rapid biodegradation rate

2-C_{12} benzene

Influence of position of phenyl on the biodegradation velocity of linear C12 alkylbenzenes

(9458)

Mobility in soil: log K_{OC} 4.3 (average side chain $C_{11.1}$) (9446)

D. BIOLOGICAL EFFECTS

Bioaccumulation

	Conc, µg/L	*Exposure period*	*BCF*	*Organ/tissue*
Fishes				
Lepomis macrochirus	0.092	48 h	35*	total body
* The predicted BCF is 6,300. The discrepancy is ascribed to metabolism.				(9459)

Toxicity

Bacteria			
Pseudomonas putida	18 h EC_{10}	>10 mg/L	(9462)
	6 h EC_{10}	>8.8 mg/L	(9462)
Algae			
Selenastrum capricornutum	96 h EC_{50}	>0.041 mg/L mg/L	(9446)
Crustaceans			
Daphnia magna	48 h EC_0	>max solubility	(9463)
	24 h EC_{50}	>1,000 mg/L**	(9465)
	48 h EC_{50}	0.009 mg/L ($C_{11.8}$)	(9446)
	48 h EC_{50}	0.08 mg/L ($C_{11.1}$)	(9446)
	21 d NOEC	0.0075 mg/L ($C_{11.1}$)	
	21 d LOEC	0.015 mg/L ($C_{11.1}$)	(9446)
	21 d NOEC	0.013 mg/L ($C_{13.2}$)	
	21 d LOEC	0.023 mg/L ($C_{13.2}$)	(9446)
Artemia salina	24 h LC_{50}	25 mg/L*	(9460)
Carcinus maenas	24 h LC_{50}	500 mg/L*	(9460)

Crangon crangon	24 h LC_{50}	300 mg/L*	(9460)
Gammarus duebeni	24 h LC_{50}	5 mg/L*	(9460)

* 80% Marlican + 20% fatty acid polyglycol ester.
**+ emulsifier Marlowet ET.

Insects

Chironomus tentans	14 d NOEC	>0.125 mg/L ($C_{11.8}$)	(9446)

Fishes

Lebistes reticulatus	24 h LC_{50}	2.5 mg/L**	
	24 h LC_{50}	4 mg/L***	(9460)
Anguila anguila	24 h LC_{50}	0.01 mg/L*	
	24 h LC_{50}	0.8 mg/L**	
	24 h LC_{50}	3 mg/L***	

* 80% Marlican + 20% alkylphenyloxethylate (APO).
** 80% Marlican + 18% nonylphenyloxethylate phosphate DEA salt (NPOP) + 2% oleic acid.
*** 80% Marlican + 20% fatty acid polyglycolester (FAP).

Carassius auratus	LC_0	<0.01 mg/L	(9461)

Lepomis macrochirus, Pimephales promelas and *Salmo sp.* were not affected after 96 h, even with test substance visibly floating on the surface (average side chain $C_{11.1}$). (9446)

Mammals

rat	acute oral LD_{50}	>15,000; >21,500; <34,800 mg/kg bw

alkylbenzenesulfonate, linear *see also* Teepol 715

D. BIOLOGICAL EFFECTS

Fishes

chain length C_{10-15}: 46.7% active material (supplied by Unilever, U.K.):

	48h LC_{50}	*96h LC_{50}*	
Rasbora heteromorpha	0.9 mg/L	0.7 mg/L	
Salmo trutta	0.2-0.4 mg/L	0.1-0.5 mg/L	
Idus idus	0.8-0.9 mg/L	0.4-0.6 mg/L	
Carassius auratus	1.2 mg/L		(1905)

15.4% active material (supplied by Hüls, Germany):

	48h LC_{50}	*96h LC_{50}*	
Rasbora heteromorpha	7.6 mg/L	6.1 mg/L	
Salmo trutta	2.0-5.3 mg/L	0.9-4.6 mg/L	
Idus idus	2.1-2.9 mg/L	1.9-2.9 mg/L	
Carassius auratus	4.9 mg/L		(1905)

28.5% active material (supplied by Procter & Gamble, U.K.):

	48h LC_{50}	*96h LC_{50}*	
Rasbora heteromorpha	5.1 mg/L	4.6 mg/L	
Salmo trutta	0.7-2.3 mg/L	1.4 mg/L	
Idus idus	1.3-1.7 mg/L	1.2-1.3 mg/L	
Carassius auratus	2.4 mg/L		(1905)

alkylbenzenesulfonates, linear (LAS LAS)

$SO_3^-Na^+$

USES

Most of LAS European consumption is in household detergency (>80%). Important application products are laundry powders, laundry liquids, dishwashing products and all purpose cleaners. The remainder of the LAS (<20%) is used in Industrial and Institutional cleaners, textile processing as wetting, dispersing and cleaning agents, industrial processes as emulsifiers, polymerisation and in the formulation of crop protection agents. (10592)

IDENTIFICATION

CAS No.	EINECS No.	NAME
68411-30-3	270-115-0	Benzenesulphonic acid, C_{10-13} alkyl derivs., sodium salts
1322-98-1	215-347-5	Sodium decylbenzenesulphonate
25155-30-0	246-680-4	Benzenedodecylsulfonic acid, sodium salt
90194-45-9	290-656-6	Benzenesulphonic acid, mono-C_{10-13} alkyl derivs., sodium salt
85117-50-6	285-600-2	Benzenesulphonic acid, mono-C_{10-14} alkyl derivs., sodium salt

Linear alkylbenzene sulphonate (LAS) is an anionic surfactant. It was introduced in 1964 as the readily biodegradable replacement for highly branched alkylbenzene sulphonates (ABS). LAS is a mixture of closely related isomers and homologues, each containing an aromatic ring sulphonated at the *para* position and attached to a linear alkyl chain at any position except the terminal carbons.The linear alkyl chain has typically 10 to 13 carbon units, approximately in the following mole ratio $C_{10}:C_{11}:C_{12}:C_{13}$=13:30:33:24, an average carbon number near 11.6 and a content of the most hydrophobic 2-phenyl isomers in the 18-29% range. This commercial LAS consists of more than 20 individual components. The ratio of the various homologues and isomers, representing different alkyl chain lengths and aromatic ring positions along the linear alkyl chains, is relatively constant across the various household applications. This LAS constant ratio is unique and does not apply to the other major surfactants. (10592)

A. PROPERTIES

The data presented in Table 2 are fully described in IUCLID,1994, and SIDS,1999 and refer to the commercial $C_{11.6}$LAS or the pure C_{12} homologue.

Table 2

LAS	Protocol	Results
Molecular description	Solid organic acid sodium salt	-
molecular weight (g/M)$C_{11.6}$	$(C_{11.6}H_{24.2})C_6H_4SO_3Na$	342.4
Vapour pressure at 25°C (Pa)	Calculated as C_{12}	$(3\text{-}17) \cdot 10^{-13}$
Boiling point (°C)	Calculated as C_{12}	637
Melting point (°C)	Calculated as C_{12}	277
log P_{ow}	Calculated as $C_{11.6}$	3.32
K_{oc} (l/kg)	Calculated as $C_{11.6}$	2500
Water solubility (g/l)	Experimental	250
Relative density (kg/l)	Experimental	1.06; 0.55 (bulk)
Henry's constant (Pa m^3/mole)	Calculated as C_{12}	$6.35 \cdot 10^{-3}$

C. WATER AND SOIL POLLUTION FACTORS

Environmental concentrations

Surface water concentrations: In water of the bay of Cadiz (Spain- 1999) : LAS and metabolites sulfophenylcarboxylic acids (SPC's) (7107)

C_{13}LAS	2 -140 µg/L		
C_{12}LAS	5 - 306 µg/L		
C_{11}LAS	6 - 347 µg/L	C_{11}SPC	0.6 - 8.5 µg/L
C_{10} LAS	1.5 -118 µg/L	C_{10}SPC	5.3 - 39 µg/L
		C_8SPC	nd - 64 µg/L
		C_6SPC	nd - 3.5 µg/L

In effluents of municipal WWTP: 8-220; 2-273 µg/L
Removal in WWTP: 98-99.9% (10592)

Biodegradation

The biodegradation of C_{12}LAS was studied in two different aerobic assays resulting in a complete primary degradation of C_{12}LAS after 48h up to 8 days and the formation of sulfophenylcarboxylic acids (SPC's). C_4SPC to C_{10}SPC's could be detected as metabolites as illustrated by the following pathway. (7107) The sulpho phenyl carboxylates (SPCs), are not persistent and their toxicities is several orders of magnitude lower than that of the parent molecule. (10592)

2-dodecylbenzenesulfonic acid — SO_3^- — ω-oxidation

sulfophenyl-2-undecylcarboxylic acid C_{12}SPC — O, OH, SO_3^- — β-oxidation

O, OH, SO_3^-

O, O, OH, SO_3^-

sulfophenyl-2-nonylcarboxylic acid C_{10}SPC — OH, O, SO_3^- + CH_3-COOH

$CO_2 + H_2O + SO_4^{2-}$

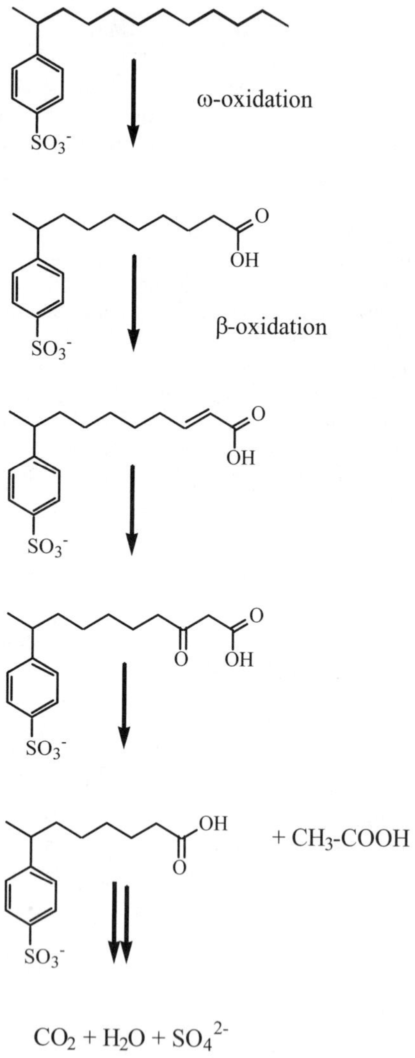

Biodegradation pathway of linear alkyl benzesulfonate (7107)

D. BIOLOGICAL EFFECTS

Fishes

bluegill		96h LC_{50},S	0.72 mg/L	
			0.89 mg/L (37% degraded)	
			1.2 mg/L (53% degraded)	
			1.6 mg/L (76% degraded)	(1190)
	homolog			
fathead minnow	C_{10}	24h LC_{50},S	48 mg/L	
	C_{10}	48h LC_{50},S	43 mg/L	
	C_{11}	24h LC_{50},S	17 mg/L	
	C_{11}	48h LC_{50},S	16 mg/L	
	C_{12}	24h LC_{50},S	4.7 mg/L	
	C_{12}	48h LC_{50},S	4.7 mg/L	
	C_{13}	24h LC_{50},S	1.7 mg/L	
	C_{13}	48h LC_{50},S	0.4 mg/L	
	C_{14}	24h LC_{50},S	0.6 mg/L	
	C_{14}	48h LC_{50},S	0.4 mg/L	(1191)

Average measured aquatic toxicity (mg/l) of LAS homologues (10593)

Alkyl chain	Invertebrate (*Daphnia magna*)		Fish (*Pimephales promelas*)	
	EC_{50}	NOEC	LC_{50}	NOEC
C_{10}	17	9.8	40	14
C_{11}	9	-	20	6.4
C_{12}	4.8	0.58	3.2	0.67
C_{13}	2.3	0.57	1.0	0.1
C_{14}	1.5	0.1	0.5	0.05

Toxicity

			Cn	
Algae				
Selenastrum capricornutum	72h EC_{50}	50-100 mg/L	C11.6	
	96h EC_{50}	29 mg/L	C11.8	
	96h EC_{50}	116 mg/L	C13	
M.aeruginosa	72h EC_{50}	10-20 mg/L	C11.6	
	96h EC_{50}	0.9 mg/L	C11.8	
	72h EC_{50}	32-56 mg/L	C12	
	96h EC_{50}	5.0 mg/L	C13	
N.fonticola	72h EC_{50}	20-50 mg/L	C11.6	
Chlorella vulgaris	72h EC_{50}	18-32 mg/L	C12	(10589)
Crustaceae				
Daphnia magna	21d NOEC	1.0; 4.2 mg/L	C11.6	(10589)
	21d NOEC	1.7-3.4 mg/L	C11.8	
	21d NOEC	1.2 mg/L	C12	
	21d NOEC	1.18 mg/L	C13	
Paratanytarsus parthenogenica	21d NOEC	3.4 mg/L	C11.6	
Cerodaphnia dubia	7d NOEC	0.32-0.89 mg/L	C11.8	
	7d NOEC	1.0 mg/L	C12	(10589)
Fish				
Bluegill	96h NOEC	1.0; 0.7-1.0 mg/L	C11.6	
Fathead minnow	28d NOEC	0.7 mg/L	C11.9	
	7d NOEC	0.3-0.9 mg/L	C11.9	
	Life cycle NOEC	0.48 mg/L	C12	(10589)

alkylether carboxylic acid, C_{6-8} (poly(oxy-1,2-ethanediyl)-α-(carboxymethyl)-ω-(C_{6-8} alkoxy))

116 alkylether carboxylic acid, C6-8

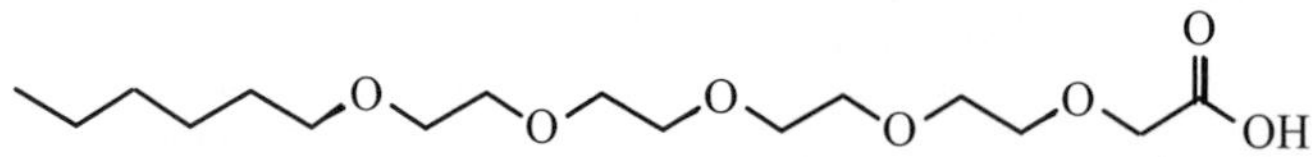

R= C_6H_{13} and n=4

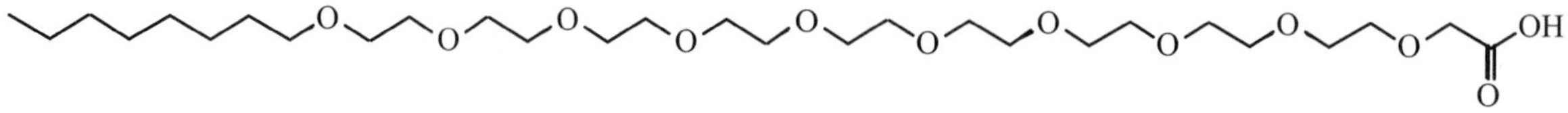

R= C_8H_{17} and n=9

$RO(CH_2CH_2O)_nCH_2COOH$

$C_{16-28}H_{32-56}O_{7-12}$

CAS 105391-15-9 C_6

CAS 107600-33-9 C_8

TRADENAMES

Akypo MB 2621; Akypo LF4

USES

a surfactant; a component (±15%) of an electroplating brightener formulation to produce a zinc cobalt alloyed deposit for automotive and building parts.

IMPURITIES, ADDITIVES, COMPOSITION

the chemical is a mixture of hexylether carboxylic acids and octylether carboxylic acids with the number of ethoxy groups varying between 4 and 9.

A. PROPERTIES

clear colourless to light yellow liquid; molecular weight 460, range 320-568; boiling point ± 100 °C; specific gravity 1.0 at 20 °C; water solubility miscible in all proportions; dissociation constant pH of a 100.000 mg/L solution at 20 °C is 1.5-3.0; $LogP_{ow}$ 2,5-3.9 calculated

C. WATER AND SOIL POLLUTION FACTORS

Hydrolysis

the chemical contains no bonds which are susceptible to hydrolysis under the environmental pH region where $4<pH<9$, and so it is expected to be stable. (7096)

Biodegradation

Inoculum	method	concentration	duration	elimination	
	OECD modified screening Test		23 days	99% BiAS	(7096)

Mobility in soil

log K_{oc} 2.7-3.5 calculated (7096)

D. BIOLOGICAL EFFECTS

Toxicity

Crustaceae

Daphnia magna	48h EC_{50}	67 mg/L

(R=C_8 and n=8)	48H NOEC	32 mg/L	(7096)
Fish			
Poecilia reticulata	96h LC_{50}	>320-560 mg/L	
(R=C_8 and n=8)	96h NOEC	<560 mg/L	
Oncorhynchus mykiss	96h LC_{50}	>100 mg/L	
(R=C_8 and n=8)	96h NOEC	>100 mg/L	(7096)
Mammals			
Rat	oral LD_{50}	>5000 mg/kg bw	(7096)

Toxicity calculated using the USEPA ASTER estimation Model (7406)

Structure	Daphnia 48h LC_{50} mg/L	Fish 96h LC_{50} mg/L
R = C_6, n = 4	45	35-66
R = C_6, n = 6	44	34-81
R = C_6, n = 9	39	29-71
R = C_7, n = 4	16	12-29
R = C_7, n = 6	16	11-28
R = C_7, n = 9	15	10-25
R = C_8, n = 4	6	4-11
R = C_8, n = 6	6	4-11
R = C_8, n = 8	6.6	4.2-11
R = C_8, n = 9	6	4-11

alkylethoxysulfate

D. BIOLOGICAL EFFECTS

Fishes
chain length C_{13}-C_{15}:

	48h LC_{50}	*96h LC_{50}*	
Rasbora heteromorpha	3.9 mg/L	-	
Salmo trutta	1.4-2.6 mg/L	1.0-2.5 mg/L	
Idus idus	3.4-7.2 mg/L	3.3-6.2 mg/L	
Carassius auratus	5.7 mg/L		(1905)

alkylnaphthalenes

D. BIOLOGICAL EFFECTS

Fishes
Bioaccumulation: BCF in muscle of Coho salmon*

	weeks of exposure				
	2	*3*	*5*	*6*	*1 week depuration after 6 wk of exp.*
C_2-substituted naphthalenes	30	40	85	40	n.d.
C_3-substituted naphthalenes	50	30	140	80	n.d.

BCF in muscle of Starry flounder*

	weeks of exposure		weeks of depuration after 2 wk of exp.	
	1	2	1	2
C_2-substituted naphthalenes	2,400	540	270	700
C_3-substituted naphthalenes	3,400	1,000	420	1,600

* exposure to approx. 1 ppm of the water soluble fraction of Prudhoe Bay crude oil (1659)

alkylolefinsulfonate

D. BIOLOGICAL EFFECTS

Fishes

chain length C_{16-18}:

	48h LC_{50}	96h LC_{50}	
Rasbora heteromorpha	0.9 mg/L	0.5 mg/L	
Salmo trutta	0.3-0.6 mg/L	0.5 mg/L	
Idus idus	1.0 mg/L	0.9 mg/L	
Carassius auratus	1.9 mg/L		(1905)

chain length C_{14-16}:

	48h LC_{50}	96h LC_{50}	
Rasbora heteromorpha	4.8 mg/L	3.3 mg/L	
Salmo trutta	2.5-5.0 mg/L	2.5-5.0 mg/L	
Idus idus	3.7-6.8 mg/L	3.4-4.9 mg/L	
Carassius auratus	5.7 mg/L		(1905)

alkylphenolethyleneoxide condensate *see* nonidet NP 50

Alkylpolyglucoside (alkylpolyglycoside; APG)

USES

surfactant, cosmetics, cleaners.

PRODUCTION, MANUFACTURING

Alkylpolyglycosides are obtained by glycosylation of fatty alcohols as raw material. Natural fatty alcohol derived for example from coconut or palm kernel oil can be used to build up the hydrophobic part of the molecule.

C. WATER AND SOIL POLLUTION FACTORS

Biodegradation

Alkylpolyglycosides show a fast biodegradation as illustrated by the comparative data in table. Results of aerobic microbial degradation assays in the laboratory test filter (7107)

Compound	Half life (h)	Complete primary degradation (h)	Detected metabolites
Alkylpolyglucosides	18	24	Not detected
Nonylphenol ethoxylates	30	95	Nonylphenol ethoxycarboxylates
Alkylbetaines	36	52	Not detected
Alkylglucamides	96	125	AG-C4-acid
LAS	120	720	sulfophenylcarboxylates

Anaerobic mineralisation

at 20 mg C/L at 35 °C in the dark

after 56 days in digested sewage sludge	>75 % (C_{12-14} APG)	
after 56 days in freshwater swamp	>75 % (C_{12-14} APG)	(7037)

Inhibition of anaerobic gas production

by unacclimatized anaerobically digested municipal sludge incubated in the dark at 35 °C:

7d EC_{20}	8.8 mg/L C_{12}APG	
7d EC_{50}	67 mg/L C_{12}APG (see also graph)	(7027)

Toxicity to methanogenic gas production (7027)

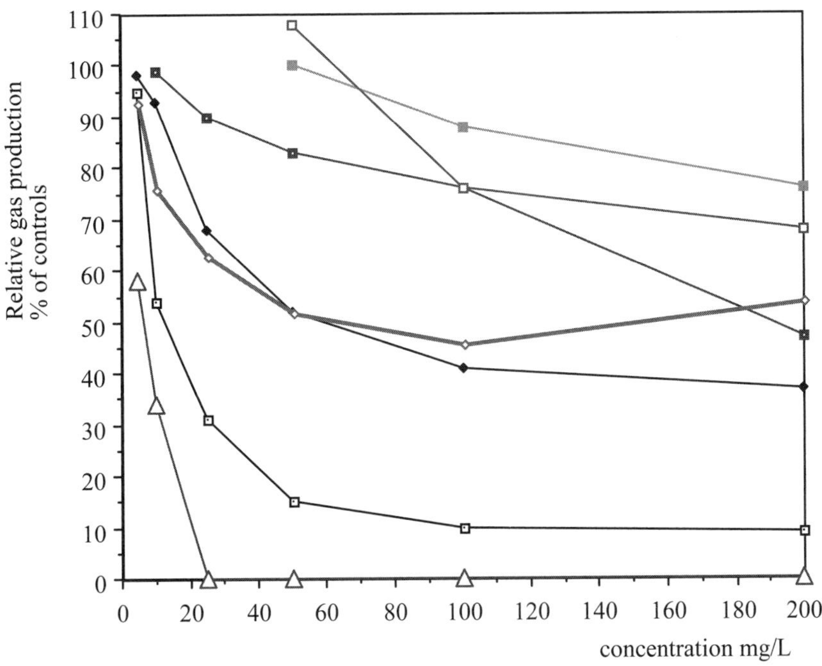

Alkylsulfates

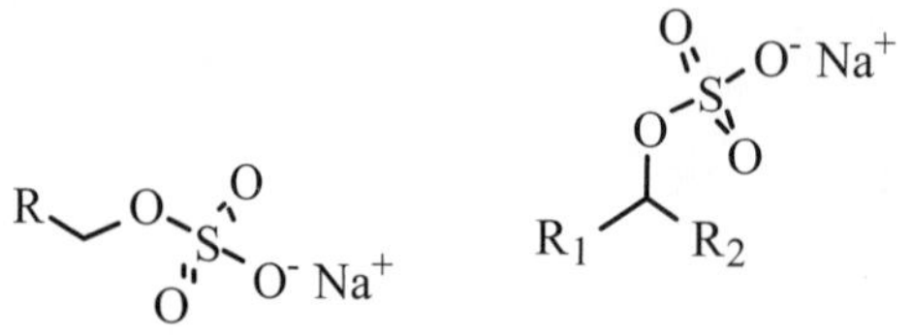

primary AS

secondary AS

USES

Alkylsulfates (AS) are used in laundry detergents, frequently in combination with other surfactants. Besides, AS are used in speciality products, including wool-washing agents, soap bars and liquid soaps, hair shampoos and tooth pastes. Most of the AS used in consumer products are linear primary AS but some linear and branced secondary AS are also used.

The hydrophobic alkyl chain (R or R1+R2) usually contains 12-18 carbon atoms. The sulfate group of secondary AS is found at all positions along the alkyl chain, exept of course at the ends. The most widely used surfactant is the sodium salt, but raw materials with various other cations like, e.g., ammonium, magnesium, mono-,di-, tri-ethanolamine and cyclohexamine, are also produced. (7169)

Alkylsulfonic acid phenyl esters

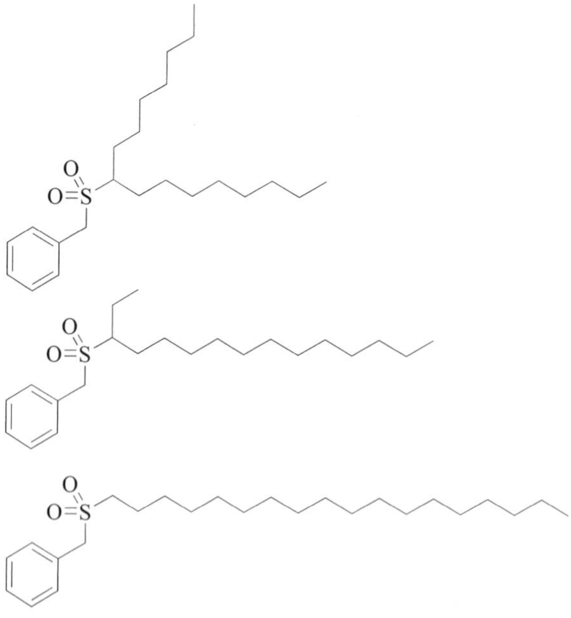

Typical alkyl sulfonic acid phenyl esters present in Mesamoll

USES

Alkylsulfonic acid phenyl esters are commercially employed as a plasticizer for PVC and similar products. Mesamoll a commercial plasticizer contains more than 30 different homologues of alkylsulfonic acid phenyl esters.

Manmade sources

Alkylsulfonic acid phenyl esters were identified as contaminants in sediment samples of the river Elbe in the Federal Republic of Germany (7105)

C. WATER AND SOIL POLLUTION FACTORS

Biodegradation

Alkyl chain homologues with substituents present at or close to the end of the aliphatic backbone (f.e. 2-tetradecylsulfonic acid phenyl ester) are degradable, while those with substituents close to the center of the alkyl chain (f.e. tetradecylsulfonic acid phenyl ester) are rather persistent.
The proposed initial reaction for the aerobic catabolism of alkylsulfonic acid phenyl esters by two strains of *Rhodococcus rhodochrous* consists of hydrolysis yielding phenol and the corresponding alkylsulfonate, the latter component is not catabolized by the two strains of *Rhodococcus rhodochrous.* They utilize the phenol for growth by channeling it into the ortho-pathway. (7105)

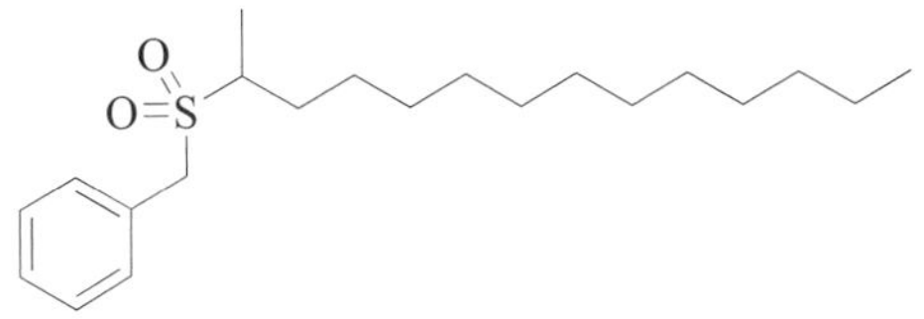

well degradable: 2-tetradecylsulfonic acid phenyl ester

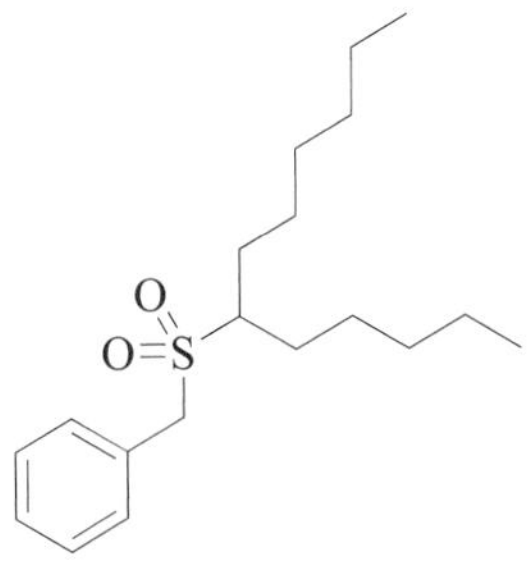

poorly degradable: 7-tetradecylsulfonic acid phenyl ester

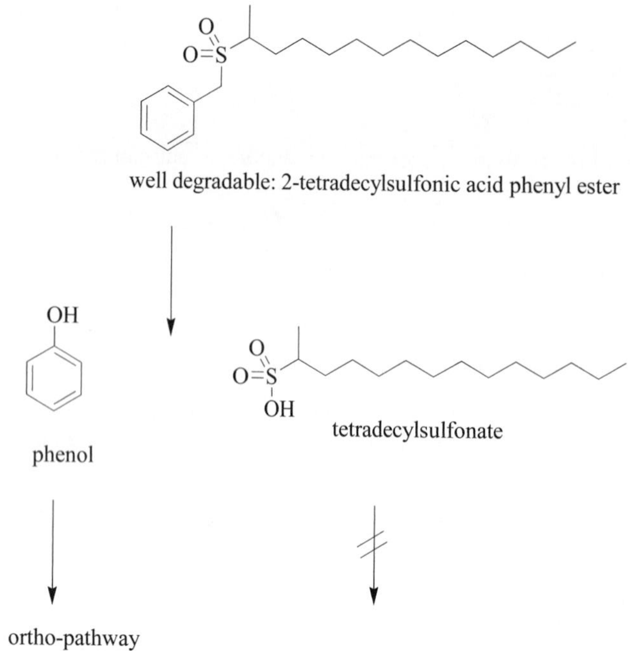

O=S(=O)
well degradable: 2-tetradecylsulfonic acid phenyl ester

OH

phenol

O=S(=O)OH

tetradecylsulfonate

ortho-pathway

---

allyl-2,3-epoxypropylether *see* allylglycidylether

2-allyl-4-hydroxy-3-methyl-2-cyclopenten-1-one ester of chrysanthemum monocarboxylic acid *see* D-*trans*-allethrin

---

allylacetate (2-propenylethanoate)

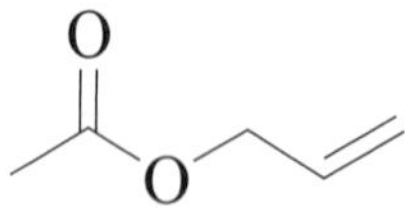

O

O

$CH_3COOCH_2CHCH_2$

$C_5H_8O_2$

CAS 591-87-7

A. PROPERTIES

molecular weight 100.11; boiling point 103-104°C; density 0.928.

B. AIR POLLUTION FACTORS

1 mg/m^3 = 0.245 ppm, 1 ppm = 4.1 mg/m^3.

Emission limits

wet scrubber: water at pH 8.5: outlet: 1,700 odor units/scf
$KMnO_4$ at pH 8.5: outlet: 25 odor units/scf (115)

allylalcohol

CH_2CHCH_2OH

C_3H_6O

CAS 107-18-6

USES

contact pesticide for weed seeds and certain fungi.

A. PROPERTIES

molecular weight 58.1; melting point -129°C; boiling point 96.9°C; vapor pressure 20 mm at 20°C, 32 mm at 30°C; vapor density 2.00; density 0.8250 at 20/4°C; THC 442 kcal/mole; saturation concentration in air 57 g/m^3 at 20°C, 98 g/m^3 at 30°C; $LogP_{OW}$ 0.17; log H -3.69 at 25°C.

B. AIR POLLUTION FACTORS

1 mg/m^3 = 0.414 ppm, 1 ppm = 2.414 mg/m^3.

Odor

characteristic: quality: alcoholic; hedonic tone: not unpleasant.

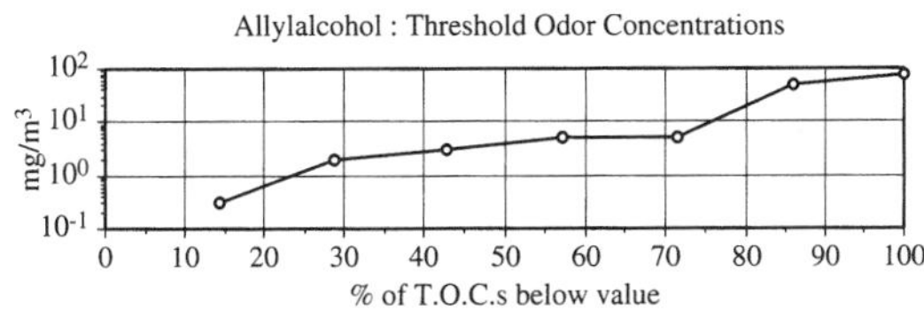

Incinerability

thermal stability ranking of hazardous organic compounds: rank 116 on a scale of 1 (highest stability) to 320 (lowest stability). (2390)

C. WATER AND SOIL POLLUTION FACTORS

BOD_5:	9.1% of ThOD	(27, 220)
	75% ThOD	(227)
BOD_{10}:	55; 73% of ThOD	(220, 256)
BOD_{15}:	78% of ThOD	(220)
BOD_{20}:	82% of ThOD	(220)
COD:	95; 96% of ThOD	(220, 227)
T.O.C:	100% of ThOD	(220)
ThOD:	2.2	

Impact on biodegradation processes: 75% inhibition of the nitrification process in activated sludge at 19.5 mg/L (43)

Aqueous reactions

photooxidation by UV light in aqueous medium at 50°C: 13.9% degradation of CO_2 after 24 h (1628)

faint odor: 0.017 mg/L water (129)

Waste water treatment

A.C.: adsorbability: 0.024 g/g carbon, 21.9% reduction, infl.: 1,010 mg/L, effl.: 789 mg/L (32)

NFG, BOD, 20°C, 1-10 d observed, feed: 200-1,000 mg/L, acclimation 365 + P, 57% removed (93)

D. BIOLOGICAL EFFECTS

Plants

highly phytocidal (1855)

Fishes			
goldfish	24h LC_{50}	1 mg/L	(277)

allylaldehyde *see* acrolein

allylamine (2-propenylamine; 3-amino-1-propene)

NH_2

$CH_2CHCH_2NH_2$

C_3H_7N

CAS 107-11-9

USES

pharmaceutical intermediate, organic synthesis.

A. PROPERTIES

molecular weight 57.07; boiling point 53°C; vapor density 2.0; density 0.76 at 20/4°C.

B. AIR POLLUTION FACTORS

1 mg/m^3 = 0.428 ppm, 1 ppm = 2.33 mg/m^3.

Odor

characteristic: similar to ammonia, irritating.

T.O.C. = 6.3 ppm; <6 mg/m^3; 14 mg/m^3. (279, 688, 710)

C. WATER AND SOIL POLLUTION FACTORS

Reduction of amenities

faint odor: 0.067 mg/L. (129)

Waste water treatment

A.C. adsorbability: 0.063 g/g carbon, 31.4% reduction, infl.: 1,000 mg/L, effl.: 686 mg/L. (32)

Degradation by Aerobacter

	200 mg/L at 30°C	
parent:	78% in 93 h	
mutant:	100% in 13 h	(152)

D. BIOLOGICAL EFFECTS

Bacteria			
Pseudomonas putida	16h EC_0	70 mg/L	(1900)
Microtox*Vibrio fisheri*	15min EC_{50}	16 mg/L	(7025)
Algae			
Rhaphidocellis subcapitata	72h EC_{50}	13 mg/L	(7025)
Microcystis aeruginosa	8d EC_0	0.35 mg/L	(329)
Scenedesmus quadricauda	7d EC_0	2.2 mg/L	(1900)
Protozoans			
Entosiphon sulcatum	72h EC_0	23 mg/L	
Uronema parduczi Chatton-Lwoff	EC_0	3,140 mg/L	(1901)
Crustaceans			
Daphnia magna	48h IC_{50}	39 mg/L	(7025)
Amphibians			
Mexican axolotl (3-4 wk after hatching)	48h LC_{50}	1.8 mg/L	
clawed toad (3-4 wk after hatching):	48h LC_{50}	5.0 mg/L	(1823)
Fishes			
Brachydanio rerio	96h LC_{50}	22 mg/L	(7025)

allylbenzene *see* 3-phenylpropene

allylchloride (3-chloro-1-propene; chloroallylene; 3-chloro-1-propylene)

CH_2CHCH_2Cl

C_3H_5Cl

CAS 107-05-1

USES

Used as a chemical intermediate in epichlorohydrin manufacture. A polymerisation monomer in the manufacture of resins, polymers, varnishes and adhesives. In the synthesis of medicinal derivatives, such as barbiturates, diuretics and cyclopropane.

A. PROPERTIES

colorless to pale yellow; molecular weight 76.53; melting point -136/-134.5°C; boiling point 44-45°C; vapor pressure 340 mm at 20°C, 440 mm at 30°C; vapor density 2.64; density 0.94 at 20/4°C; solubility 100 mg/L; saturation concentration in air 1,229 g/m^3 at 20°C, 1,772 g/m^3 at 30°C; logP_{OW} 1.45; log H -0.42 at 25°C.

B. AIR POLLUTION FACTORS

1 mg/m^3 = 0.314 ppm, 1 ppm = 3.18 mg/m^3.

Odor

characteristic: garlic-onion pungency, green.

T.O.C.:	0.660 mg/m^3	0.21 ppm	(307)
		1,500 ppm	(279)
50% recognition:		0.21 ppm	
100% recognition:		25 ppm	(211)
PIT_{50}:		0.21 ppm	
PIT_{100}:		0.47 ppm	
	1.5 mg/m^3=	0.47 ppm	(57)

Incinerability

temperature for 99% destruction at 2.0 sec residence time under oxygen-starved reaction conditions: 695°C.

Thermal stability ranking of hazardous organic compounds

rank 120 on a scale of 1 (highest stability) to 320 (lowest stability) (2390)

Photochemical reactions:

91% loss in 12 h, based on reactions with OH° (calculated) (9505)

Ozonolysis: $t_{1/2}$: 9 h (8867)

Partition coefficients

$K_{air/water}$ 0.11

Cuticular matrix/air partition coefficient* 120 ± 20 (7077)

* experimental value at 25°C studied in the cuticular membranes from mature tomato fruits (*Lycopersicon esculentum* Mill. cultivar Vendor)

Hydrolysis to allyl alcohol and hydrochloric acid

$t_{1/2}$ 7.2 d in water at 25°C

C. WATER AND SOIL POLLUTION FACTORS

BOD_5:	14% ThOD	(277)
COD:	51; 80% ThOD	(272)
ThOD:	1.77	

Impact on biodegradation processes

75% decrease of nitrification by activated sludge at 180 mg/L. (30)

Reduction of amenities

odor threshold: average: 14,700 mg/L; range: 3,660 to 29,300 mg/L (30)

Waste water treatment

evaporation rate from water at 25°C of 1 ppm solution:
50% after 27 min
90% after 89 min (313, 289)

D. BIOLOGICAL EFFECTS

Toxicity threshold (cell multiplication inhibition test)

bacteria *(Pseudomonas putida):*	16h EC_0	115 mg/L	(1900)
green algae *(Scenedesmus quadricauda):*	7d EC_0	6.3 mg/L	(1900)
protozoa *(Entosiphon sulcatum):*	72h EC_0	8.4 mg/L	(1900)
protozoa (*Uronema parduczi* Chatton-Lwoff)	EC_0	>240 mg/L	(1901)

Fishes

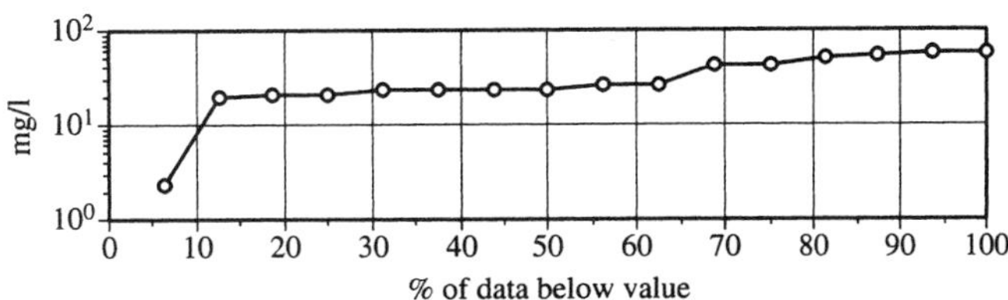

Frequency distribution of 24-96h LC_{50} values for fishes (*n* = 16) based on data from this and other works (3999)

		LC_{50}			
test fish	*dilution water*	*24 h*	*48 h*	*96 h*	
fatheads	soft	24	24	20	
fatheads	hard	26	24	24	
bluegills	soft	59	42	42	
goldfish	soft	27	21	21	
guppies	soft	58	53	52	(158)
goldfish			24h LC_{50}	10 mg/L	(277)
guppy *(Poecilia reticulata)*			14d LC_{50}	1.2 mg/L	(2696)

Mammals

rat	oral LD_{50}	700 mg/kg bw	(9696)
mouse	oral LD_{50}	425 mg/kg bw	(9697)

allylcyanide (3-butenonitrile)

H_2CCHCH_2CN

C_4H_5N

CAS 109-75-1

A. PROPERTIES

molecular weight 67.09; mp°C; boiling point 116-121°C; density 0.834; solubility >100,000 mg/L; $LogP_{OW}$ 0.12.

D. BIOLOGICAL EFFECTS

Fishes			
Pimephales promelas	24h LC_{50}	210 mg/L	
	96h LC_{50}	182 mg/L	(2709)

allylene *see* methylacetylene

allylglycidylether (AGE; allyl-2,3-epoxypropylether; 1-allyloxy-2,3-epoxypropane; 1,2-epoxy-3-allyloxypropane; glycidylallylether; [(2-propenyloxy)methyl] oxirane)

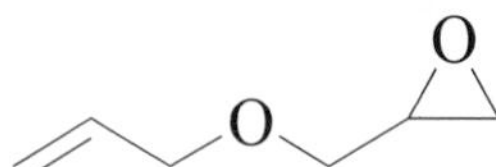

$CH_2CHCH_2OCH_2CH(O)CH_2$

$C_6H_{10}O_2$

CAS 106-92-3

USES

component of epoxy resin systems. The epoxy group of the glycidylether reacts during the curing process, and glycidylethers are therefore generally no longer present in completely cured products.

A. PROPERTIES

colorless liquid; molecular weight 114.15; melting point -100°C forms glass; boiling point 153.9°C; vapor pressure 3.6 mm at 20°C, 5.8 mm at 30°C; vapor density 3.94; density 0.97 at 20/4°C; solubility 141,000 mg/L; saturation concentration in air 22 g/m^3 at 20°C, 35 g/m^3 at 30°C.

B. AIR POLLUTION FACTORS

1 mg/m^3 = 0.21 ppm, 1 ppm = 4.74 mg/m^3.

Odor

threshold value: 47 mg/m^3. (57)

C. WATER AND SOIL POLLUTION FACTORS

BOD_5:	2.8% ThOD	(277)
COD:	95% ThOD	(277)
ThOD:	2.1	

D. BIOLOGICAL EFFECTS

Fishes			
goldfish	24h LC_{50}	78 mg/L	
	96h LC_{50}	30 mg/L	(277)

allylisocyanate (isocyanic acid, allyl ester)

CH_2CHCH_2NCO

C_4H_5NO

CAS 1476-23-9

A. PROPERTIES

molecular weight 83.09; boiling point 87 - 89°C; density 0.95.

C. WATER AND SOIL POLLUTION FACTORS

nitrification inhibition in activated sludge or trickling filter	IC_{75}	1.9 mg/L	(7050)

D. BIOLOGICAL EFFECTS

Mammals			
mouse	oral LD_{50}	18 mg/kg bw	(10546)

allylisosulfocyanate *see* allylisothiocyanate

allylisothiocyanate (mustard oil; 2-propenylisothiocyanate; allylisosulfocyanate)

CH_2CHCH_2NCS

C_4H_5NS

CAS 57-06-7

USES

fumigant; ointments and mustard plasters; military poison gas.

OCCURRENCE

Isolated from black mustard seed Brassica nigra.

A. PROPERTIES

molecular weight 99.15; melting point -100°C; boiling point 151°C; vapor pressure 1 mm at -2°C, 10 mm at 38.3°C, 40 mm at 67.4°C; density 1.01 at 15/4°C; solubility 2,000 mg/L; vapor density 3.4; $LogP_{OW}$ 2.11 (calculated).

B. AIR POLLUTION FACTORS

1 mg/m^3 = 0.243 ppm, 1 ppm = 4.120 mg/m^3.

Odor

characteristic: mustard oil, irritant.

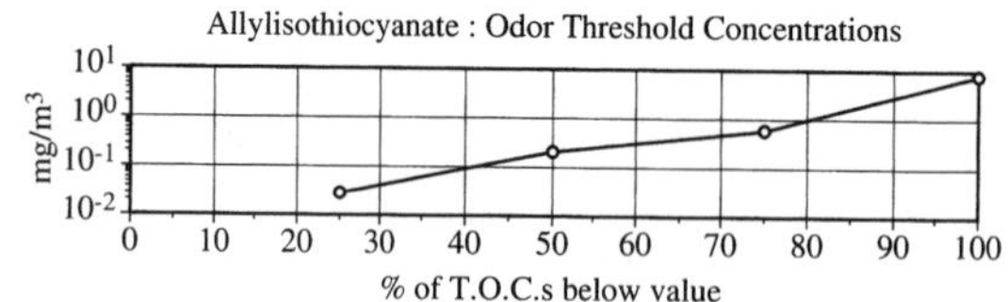

Emission limits

wet scrubber: water: pH 8: outlet: 2,500 odor units/scf; $KMnO_4$: pH 8: outlet: 1 odor units/scf (115)

C. WATER AND SOIL POLLUTION FACTORS

Impact on biodegradation processes

75% inhibition of the nitrification process in activated sludge at 1.9 mg/L (43)

Reduction of amenities

faint odor: 0.0017 mg/L	(129)

D. BIOLOGICAL EFFECTS

Mammals			
rat, mouse	oral LD_{50}	108-339 mg/kg bw	(9778, 9779)

1-allyloxy-2,3-epoxypropane *see* allylglycidylether

allylsulfocarbamide *see* allylthiourea

allylsulfourea *see* allylthiourea

allylthiourea (allylsulfocarbamide; thiosinamine; allylsulfourea)

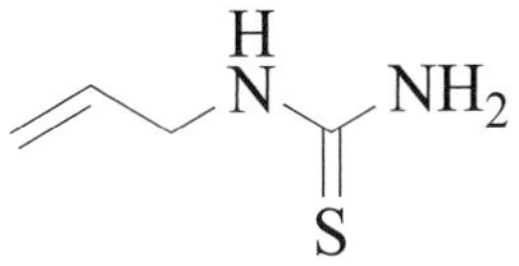

$CH_2CHCH_2NHCSNH_2$

$C_4H_8N_2S$

CAS 109-57-9

USES

medicine, corrosion inhibitor, organic synthesis.

A. PROPERTIES

white crystalline solid; slight garlic odor; bitter taste; molecular weight 116.18; melting point 78°C; density 1.22.

C. WATER AND SOIL POLLUTION FACTORS

Impact on biodegradation processes

NH_3 oxidation by *Nitrosomonas*	IC_{50}	1.2 mg/L	(407)
nitrification inhibition in activated sludge or trickling filter	IC_0	0.58 mg/L	
	IC_{16}	1.1 mg/L	
	IC_{38}	1.1 mg/L	(9698)

D. BIOLOGICAL EFFECTS

Toxicity threshold

(cell multiplication test):			
bacteria *(Pseudomonas putida):*	16h EC_0	140 mg/L	
green algae *(Scenedesmus quadricauda):*	7d EC_0	41 mg/L	
protozoa *(Entosiphon sulcatum):*	72h EC_0	13 mg/L	(1900)

Mammals			
rat	oral LD_{50}	200 mg/kg bw	(9699)

Alpha hydroxy acids (AHA's; fruit acids)

CHEMICAL IDENTITY

Alpha hydroxy acids comprise several chemicals, all of which are natural carboxylic acids with a hydroxy group at the two, or alpha position. They are mostly manufactured by chemical synthesis or fermentation, but because of their abundance in sources such as citrus fruits, apricots, apples, grapes and sugar cane, they are sometimes referred to as 'fruit acids' rather than AHA's.
AHA's are widely used in cosmetic products.

The most common AHA's are:

Glycolic acid	CH_2OH-COOH
Lactic acid	CH_3-CHOH-COOH
Malic acid	COOH-CH_2-CHOH-COOH
Tartaric acid	COOH-CHOH-CHOH-COOH
Citric acid	$(COOH\text{-}CH_2)_2$-COH-COOH

altosid-SR-10 (Isopropyl(2E-4E)-11-methoxy-3,7,11-trimethyl-dodeca-2,4-dienoate; methoprene; ZR-515)

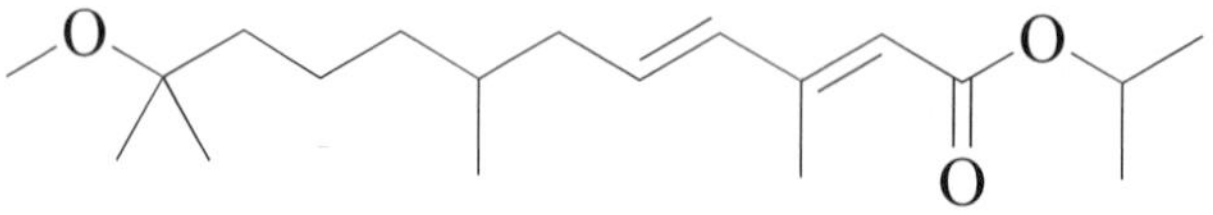

$C_{19}H_{34}O_3$

CAS 40596-69-8

USES

insect growth regulator: prevents adult emergence of mosquitoes, houseflies, stable flies, and blackflies by preventing metamorphosis of final instar larvae.

A. PROPERTIES

amber liquid; density 0.93 at 20°C; vapor pressure 2.37×10^{-5} at 25°C, 1.6×10^{-4} at 40°C; solubility 1.39 ppm.

D. BIOLOGICAL EFFECTS

Arthropods

amphipod: *Gammarus aequieaudu:*			
adult female:	96h LC_{50}	2,150 µg/L	
adult male:	96h LC_{50}	1,950 µg/L	(1128)

decapod: *Rhithropanopeus harrisii:*
adult: 1.30 mg/L, 12-15 d
Progressive inhibition of vitellogenesis and stimulation of spermatogenesis after 30-45 d (1157)

Fishes

juvenile rainbow trout:	96h LC_{50}:	106 mg/L, 95% conf. lim: 92-121 mg/L	
Coho salmon:	96h LC_{50}:	86 mg/L, 95% conf. lim: 81-91 mg/L	(1059)
bluegill:	TL_{50}:	4.6 ppm (static)	
trout:	TL_{50}:	4.39 ppm (static)	
		106 ppm (static, when aerated)	
channel catfish:	TL_{50}:	>100 ppm (static)	(1855)

ametryn (6-ethylamino-4-isoproylamino-2-methylthio-1,3,5-triazine; 2-ethylamino-4-isopropylamino-6-methylmercapto-s-triazine; gesapax; evik; G 34162)

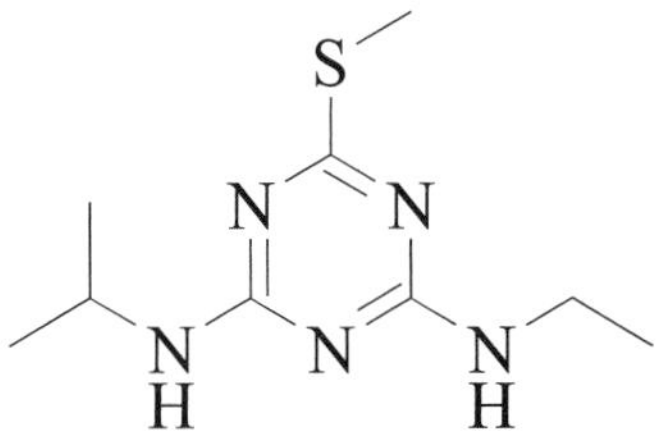

$C_9H_{17}N_5S$

CAS 834-12-8

USES

herbicide.

A. PROPERTIES

colorless crystals, molecular weight 227.33; melting point 84-85°C, solubility 185 ppm at 20°C, vapor pressure 8.4 × 10^{-7} mm at 20°C; P_{OW}: 2.6; 2.69; 2.82; 3.07.

D. BIOLOGICAL EFFECTS

Algae

Chlorococcum sp. (technical acid): 20 ppb; 50% decrease in O_2 evolution
Chlorococcum sp.: 10 ppb; 50% decrease in growth measured as ABS (525 µ) after 10 d
Dunaliella tertiolecta: 40 ppb; 50% decrease in O_2 evolution
Dunaliella tertiolecta: 40 ppb; 50% decrease in growth measured as ABS (525 µ) after 10 d
Isochrysis galbana: 10 ppb; 50% decrease in O_2 evolution
Isochrysis galbana: 10 ppb; 50% decrease in growth measured as ABS (525 µ) after 10 d
Phaeodactylum tricornutum: 10 ppb; 50% decrease in O_2 evolution
Phaeodactylum tricornutum: 20 ppb; 50% decrease in growth measured as ABS (525 µ) after 10 d

Fishes			
rainbow trout	96h LC_{50}	8.8 mg/L	
bluegill	96h LC_{50}	4.1 mg/L	
goldfish	96h LC_{50}	14 mg/L	(2962)

amidol *see* 2,4-diaminophenol hydrochloride

amidosulfuron (1-(4,6-dimethoxypyrimidin-2-yl)-3-mesyl(methyl)sulfamoylurea; N-[[[[(4,6-dimethoxy-2-pyrimidinyl)amino]carbonyl]amino]sulfonyl]-N-methylmethanesulfonamide)

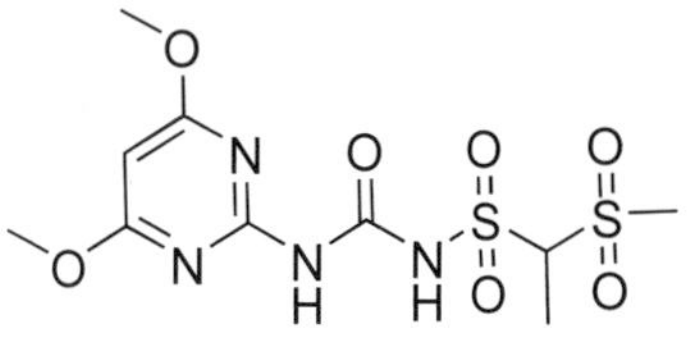

$C_9H_{15}N_5O_7S_2$

CAS 120923-37-7

TRADENAMES

Gratil

USES

A sulfonylurea herbicide for postemergence control of broadleaf weeds in cereals and other crops. Acts by inhibiting biosynthesis of the essential amino acids valine and isoleucine, hence stopping cell division and plant growth. Selectivity derives from rapid metabolism in the crop.

A. PROPERTIES

White crystalline powder; molecular weight 369.4; melting point 160-163°C; vapour pressure 2.2 x 10^{-2} mPa at 25°C; density 1.5; H 5.3 x 10^{-4} $Pa.m^3/mol$

pH	Water solubility at 20°C	Hydrolytic stability at 25°C t/2

3	3.3 mg/L	34 days (pH 5)
5.8	9.0 mg/L	365 days (pH 7)
10	13,500 mg/L	365 days (pH 9)

C. WATER AND SOIL POLLUTION FACTORS

Hydrolysis of amidosulfuron (10995)

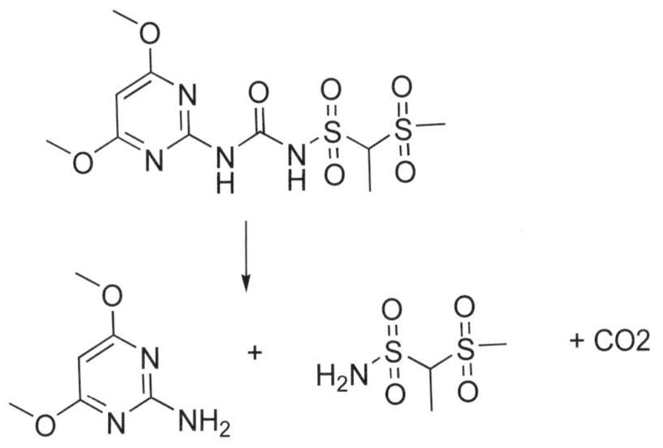

Hydrolysis of amidosulfuron

Biodegradation in laboratory studies

T/2 of degradation inaerobic soils at 20°C 3-29 days (10997)

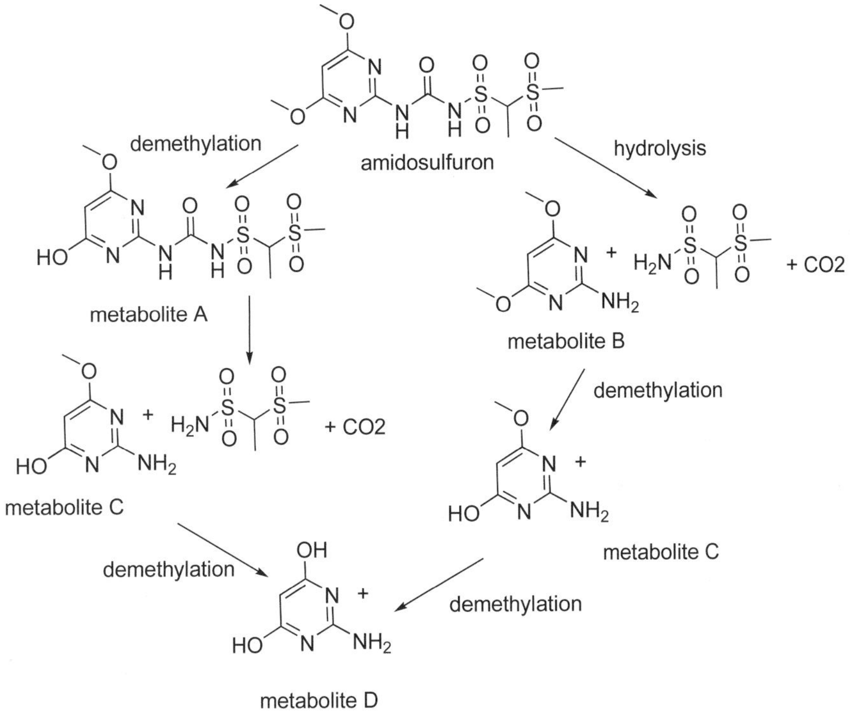

Proposed degradation pathway of amidosulfuron in soil [10995]

Proposed degradation pathway of amidosulfuron in soil (10995)

D. BIOLOGICAL EFFECTS

Toxicity

Algae			
Scenedesmus subspicatus	72h EC_{50}	47 mg/L	(10997)
Worms			
Eisenia foetida	14d EC_{50}	>1,000 mg/kg soil	(10997)
Crustaceae			
Daphnia magna	48h EC_{50}	36 mg/L	(10997)
Insecta			
Honeybees	Acute oral LD_{50}	>1,000 mg/kg bw	(10997)
Birds			
Bobwhite quail	Acute oral LD_{50}	>2,000 mg/kg bw	
Mallard duck	Acute oral LD_{50}	>2,000 mg/kg bw	(10997)
Fish			
Rainbow trout	96h LC_{50}	>320 mg/L	(10997)
Mammals			
Rat	oral LD_{50}	5,000 mg/kg	(10997)

Amidotrizoe acid

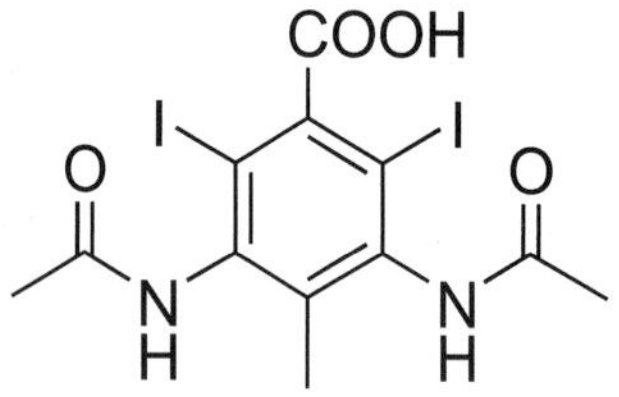

$C_{12}H_{12}I_2N_2O_4$

CAS 117-96-4

USES

Contrast fluid

A. PROPERTIES

molecular weight 502.06

C. WATER AND SOIL POLLUTION FACTORS

Environmental concentrations in Germany 2001 (10923)

	Positive samples	Min ng/L	Max ng/L	Average ng/L	Median ng/L
River Körsch upstream WWTP	0/8	<10	<10	<10	<10
River Körsch downstream WWTP	0/8	<10	<10	<10	<10
Influent WWTP Stuttgart-M	6/6	484	1101	776	817
Effluent WWTP Stuttgart-M	3/7	<10	47	13	<10
Influent WWTP Reutlingen-W	5/5	19	430	198	216
Effluent WWTP Reutlingen-W	1/5	<10	<10	<10	<10
Influent WWTP Steinlach-W	5/5	2215	5736	3817	3341
Effluent WWTP Steinlach-W	4/5	<10	279	142	82
Leachate of landfill Reutlingen-S	1/5	n.d.	14	<10	<10
Leachate of landfill Dusslingen	5/5	31	292	190	226

Amines,bis(hydrogenated tallow alkyl),oxidised (bis alkyl (C_{16-18})hydroxylamine; dialkylhydroxylamine,(C_{16-18}); TKA 40082/CGA 042)

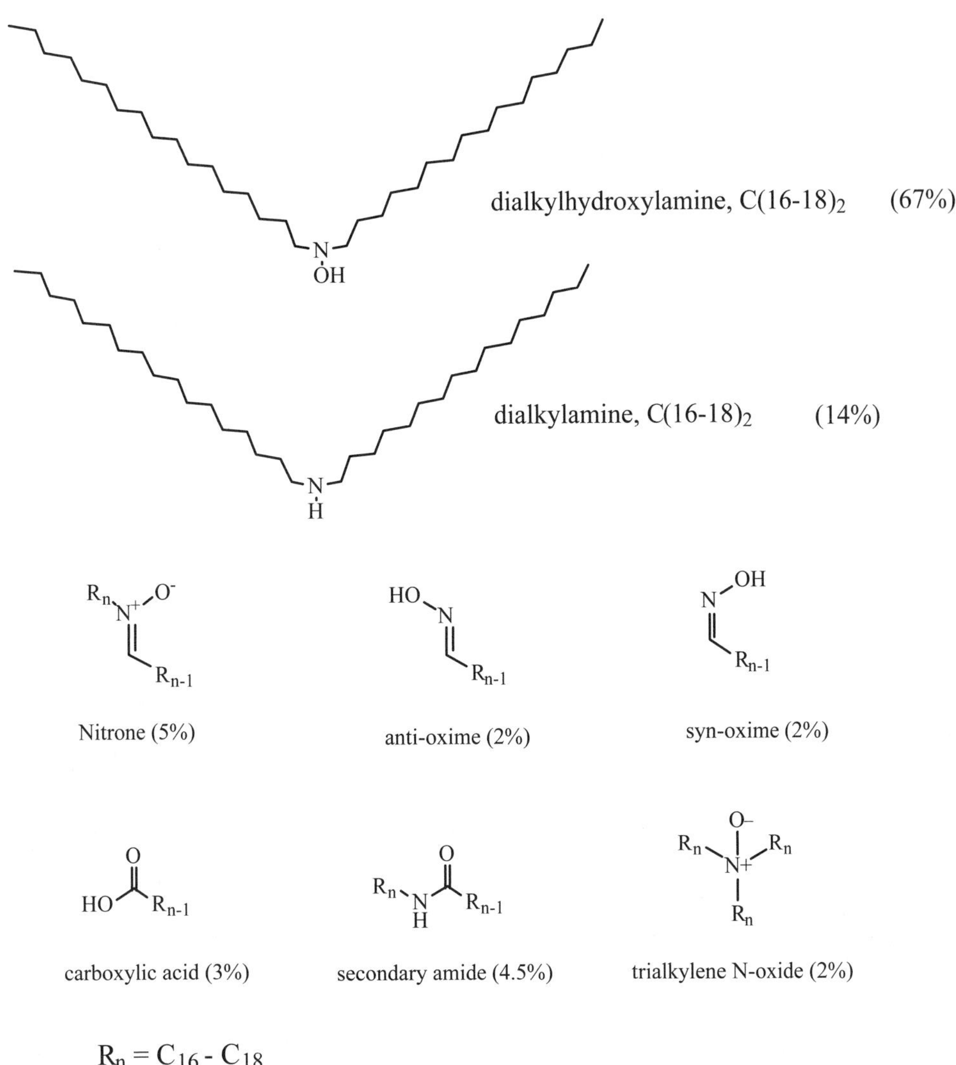

CAS 143925-92-2

USES

a component of a polymer stabiliser. Polyolefin fibres (contain less than 0.1 % of the chemical) have a wide range of applications including carpeting, carpet backing, nappies, disposable hospital gowns and packaging.

IMPURITIES, ADDITIVES, COMPOSITION

The chemical consists of a mixture of starting material and reaction products. The main component (67%) consists of bis alkylhydroxylamines with chain lengts of C_{16}-C_{18}.

A. PROPERTIES

white off white solid, no odour; molecular weight 481-538 (main component); melting point 56-92 °C; boiling point >280 °C; density 0.95 at 23 °C; vapour pressure 3 x 10^{-12} kPa at 20 °C, 1 x 10^{-11} at 25 °C; water solubility <0.5 mg/L at 20 °C; LogP_{OW} 5.5 (calculated)

C. WATER AND SOIL POLLUTION FACTORS

Hydrolysis

the major components contain no bonds which are susceptible to hydrolysis under the environmental pH region where 4<pH<9, and so it is expected to be stable. (7117)

Biodegradation

Inoculum	method	concentration	duration	elimination	
	OECD Test 301B		28 days	<10% DOC	
	Modified Sturm test		28 days	<10% product	(7117)

D. BIOLOGICAL EFFECTS

Toxicity

Micro organisms			
A.S. oxygen consumption inhibition	3h EC_{50}	> 100 mg/L*	
	3h NOEC	100 mg/L*	(7117)
Algae			
Selenastrum capricornutum	96h EC_{50}	>100 mg/L*	
	96h NOEC	100 mg/L*	(7117)
Crustaceae			
Daphnia magna	48h EC_{50}	>100 mg/L*	
	48h NOEC	>100 mg/L*	(7117)
Fish			
Brachydanio rerio	96h LC_{50}	>100 mg/L*	
	96h NOEC	100 mg/L*	(7117)
Mammals			
Rat	oral LD_{50}	>2000 mg/kg bw	(7117)

* The ecotoxicity studies were conducted at two concentrations, a nominal concentration of 100 mg/L and filtrates from solutions prepared at 100 mg/L nominal concentration. The maximum level of the chemical measured in these studies was 3 mg/L during the fish study and precipitate was noted to form in these samples with time. The maximum level measured during daphnia study was 0.6 mg/L and the substance was not detected in the filtered solutions in the algal study.

2-amino-5-chlorobenzonitrile

CN
NH_2
Cl

$C_6H_3(CN)NH_2Cl$

$C_7H_5ClN_2$

CAS 5922-60-1

A. PROPERTIES

$LogP_{OW}$: 1.83.

D. BIOLOGICAL EFFECTS

Algae

Tetrahymena pyriformis: growth inhibition at 27°C, 48 EC_{50}: 0.36 mmol/L (2704)

Fishes

Pimephales promelas: 30-35 d FT test at 25°C: log LC_{50}: 0.19 mmol/L (2704)

2-amino-p-cresol (2-amino-4-methylphenol)

$NH_2C_6H_3(CH_3)OH$

C_7H_9NO

CAS 95-84-1

A. PROPERTIES

molecular weight 123.16; melting point 135-137°C.

D. BIOLOGICAL EFFECTS

Green Algae:

Scenedesmus subspicatus: inhib. of fluorescence	IC_{10}	0.39 mg/L	
Scenedesmus subspicatus: growth inhib.	IC_{10}	0.85 mg/L	

RubisCo test

inhib. of enzyme activity of ribulose-P2-carboxylase in protoplasts	IC_{10}	1.8 mg/L	

Oxygen test

inhib. of oxygen production of protoplasts	IC_{10}	0.62 mg/L	(2698)

3-amino-2,5,dichlorobenzoic acid (chloramben; amiben; amilon-WP; vegiben)

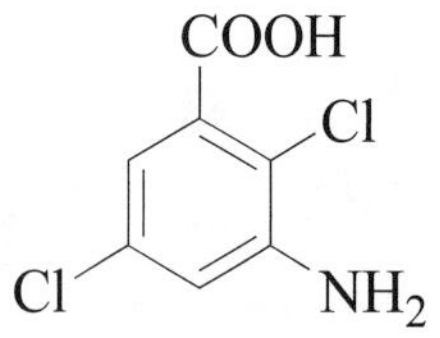

COOH, Cl, Cl, NH_2

$HOOCC_6H_2(Cl_2)NH_2$

$C_7H_5Cl_2NO_2$

CAS 133-90-4

USES

selective pre-emergence herbicide.

A. PROPERTIES

white, odorless crystalline solid; molecular weight 206.03; melting point 200-201°C; vapor pressure 7 × 10^{-3} mm at 100°C; solubility 700 mg/L at 25°C.

C. WATER AND SOIL POLLUTION FACTORS

Control methods

A.C. type BL (Pittsburgh Chem. Co.): % absorbed by 10 mg A.C. from 10^{-4} m aqueous

solution at pH 3.0:	51%	
7.0:	121%	
11.0:	5.2%	(1313)

D. BIOLOGICAL EFFECTS

Algae

Chlorcoccum sp.	10d EC_{50}	50-115 mg/L	technical acid	
		2,225-4,000 mg/L	ammonium salt	
Dunaliella tertiolecta	10d EC_{50}	50-150 mg/L	technical acid	
		2,750-4,000 mg/L	ammonium salt	
Isochrysis galbana	10d EC_{50}	15-100 mg/L	technical acid	
		1,500-3,500 mg/L	ammonium salt	
Phaeodactylum tricornutum	10d EC_{50}	25-100 mg/L	technical acid	(2348)

2-amino-3,5-diiodobenzoic acid

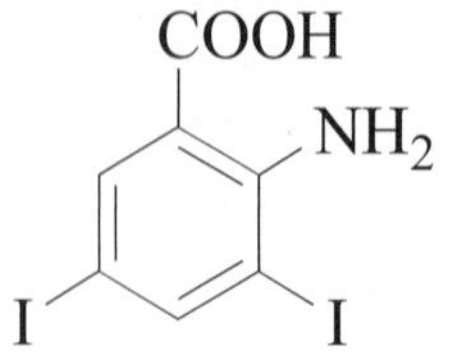

COOH, NH_2, I, I

$C_7H_5I_2N$

CAS 609-86-9

C. WATER AND SOIL POLLUTION FACTORS

Impact on biodegradation processes

at 100 mg/L no inhibition of NH_3 oxidation by *Nitrosomonas* sp. (390)

4-amino-3,5-dimethylphenol *see* 4-amino-3,5-xylenol

2-amino-4,6-dinitrophenol *see* picramic acid

2-amino-3,6-dinitrotoluene

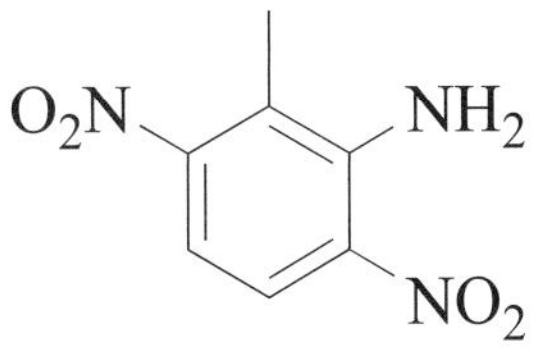

O_2N / NH_2 / NO_2

$CH_3C_6H_2(NO_2)_2NH_2$

$C_7H_7N_3O_4$

CAS 56207-39-7

C. WATER AND SOIL POLLUTION FACTORS

Manmade sources

constituent of condensate water of TNT manufacturing process.

D. BIOLOGICAL EFFECTS

Crustaceans

Water flea	48h LC_{50}	2.5 mg/L	(2627)

Fishes

Fathead minnow	96h LC_{50}	0.9 mg/L	(2627)

3-amino-2,4-dinitrotoluene

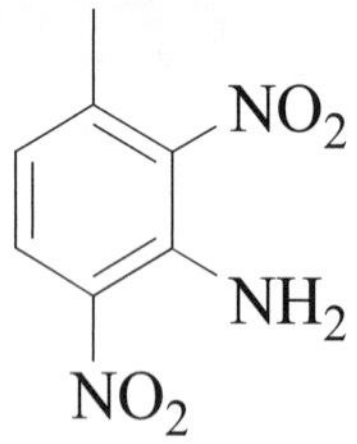

$CH_3C_6H_2(NO_2)_2NH_2$

$C_7H_7N_3O_4$

C. WATER AND SOIL POLLUTION FACTORS

Manmade sources

constituent of condensate water of TNT manufacturing process.

D. BIOLOGICAL EFFECTS

Crustaceans

Water flea	48h LC_{50}	9.6 mg/L	(2627)

Fishes

Fathead minnow	96h LC_{50}	12.1 mg/L	(2627)

2-amino-5-guanidopentanoic acid

see DL-arginine

1-amino-8-hydroxy-3,6-disulfonaphthalene, monosodium salt

(4-amino-5-hydroxy-2,7-naphthalenedisulfonic acid,monosodium salt)

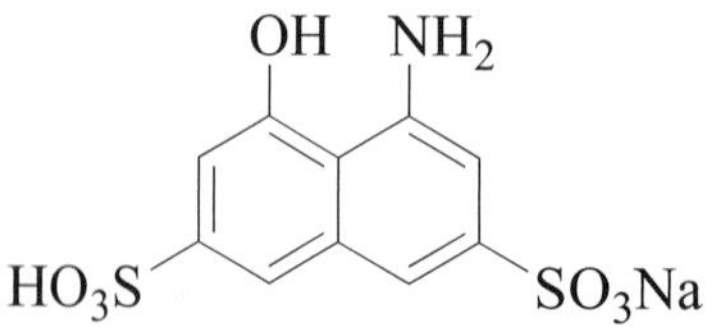

$C_{10}H_9NO_7S_2.Na$

CAS 5460-09-3

USES AND FORMULATIONS

Feedstock for chemical synthesis of dyes and pigments.

A. PROPERTIES

molecular weight 341.3; melting point >380°C; solubility 11,400 mg/L at 20°C; LogP_{OW} -2.3 (calculated for the free acid).

C. WATER AND SOIL POLLUTION FACTORS

Biodegradation

Inoculum	*Method*	*Concentration*	*Duration*	*Elimination results*	
Adapted municipal waste water	Closed bottle test	80 mg/L	20 d	<10% ThOD	(6514)
A.S.	Zahn-Wellens test	100 mg/L	7 d	3% DOC	
			14 d	8% DOC	
			21 d	10% DOC	
			29 d	8% DOC	(6514)

D. BIOLOGICAL EFFECTS

Toxicity

Bacteria			
Pseudomonas fluorescens	24 h EC_0	1,000 mg/L	(6514)
Fishes			
Leuciscus idus	96 h LC_0	>1,000 mg/L	(6514)
Salmo gairdneri	96 h LC_0	>1,000 mg/L	(6514)

6-amino-4-hydroxy-2-naphthalenesulfonic acid (gamma acid; 8-hydroxy-2-naphthylamine-6-sulfonic acid)

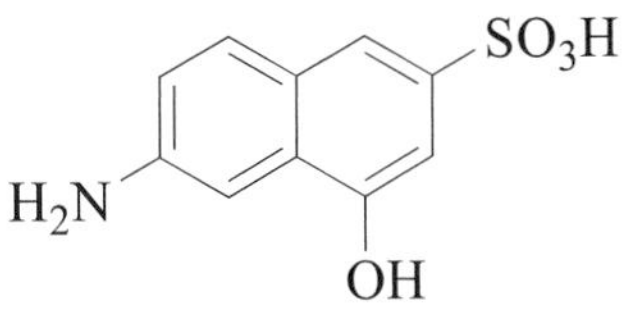

SO_3H

H_2N

OH

$C_{10}H_9NO_4S$

CAS 90-51-7

USES

Feedstock for chemical synthesis.

A. PROPERTIES

molecular weight 239.25; melting point 180-200°C decomposes; solubility 1,000 mg/L at 20°C;

$LogP_{OW}$ -0.4 (calculated).

C. WATER AND SOIL POLLUTION FACTORS

Biodegradability

Municip. Sludge, closed bottle test: at 80 mg/L after 20 d: 0% ThOD removal			(6057)
A.S., Zahn-Wellens test at 100 mg/L	after 7 d	7% DOC removal	
	after 14 d	6% DOC removal	
	after 28 d	9% DOC removal	(6057)

D. BIOLOGICAL EFFECTS

Bacteria			
Pseudomonas fluorescens	24 h EC_0	500 mg/L	(6057)
Fishes			
Leuciscus idus	48 h LC_0	>1,000 mg/L	(6057)
Oncorhynchus mykiss	24 h LC_0	>5 mg/L	(6058)
Lepomis macrochirus	24 h LC_0	5 mg/L	(6058)
Petromyzon marinus larvae	24 h LC_0	5 mg/L	(6058)

7-amino-4-hydroxy-2-naphthalenesulfonic acid (isogamma acid; 5-hydroxy-7-sulfo-2-naphthylamine)

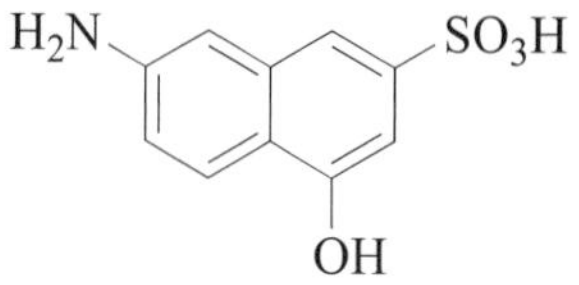

$C_{10}H_9NO_4S$

CAS 87-02-5

USES

Feedstock for chemical synthesis.

A. PROPERTIES

molecular weight 239.25; melting point 180-200°C decomposes; solubility 5,000 mg/L at 20°C, 1,500 mg/L at 60°C; $LogP_{OW}$ -0.4 (calculated).

C. WATER AND SOIL POLLUTION FACTORS

Biodegradability

Closed bottle test: at 3-30 mg/L after 30 d: 0% ThOD removal (6056)

D. BIOLOGICAL EFFECTS

Bacteria			
Pseudomonas fluorescens	24 h EC_0	1,000 mg/L	(6056)

Fishes			
Leuciscus idus	96 h LC_0	>1,000 mg/L	(6056)

2-amino-3-hydroxybutanoic acid *see* threonine

2-amino-4-hydroxybutanoic acid *see* homoserine

(2-amino-4-hydroxyphenyl)propanoic acid *see* DL-tyrosine

2-amino-3-hydroxypropanoic acid *see* DL-serine

α-amino-β-imidazolepropionic acid *see* DL-histidine

2-amino-3-indolylpropanoic acid *see* tryptophane

L-2-amino-3-mercaptopropanoic acid *see* L-cysteine

L-2-amino-3-methylpentanoic acid *see* L-isoleucine

L-2-amino-4-methylpentanoic acid *see* L-leucine

L-amino-2-methylpropane *see* isobutylamine

L-α-amino-β-methylvaleric acid *see* L-isoleucine

L-amino-2-nitrobenzene *see* *o*-nitro-aniline

L-amino-3-nitrobenzene *see* *m*-nitroaniline

L-amino-4-nitrobenzene *see* *p*-nitroaniline

2-amino-3-phenylpropionic acid *see* phenylalanine

1-amino-2-propanol *see* isopropanolamine

1-((4-amino-2-propyl-5-pyrimidiny)methyl)-2-picoliniumchloride, hydrochloride *see* amprolium

4-amino-6-tert-butyl-4,5-dyhydro-3-methylthio-1,2,4-triazin-5-one (1,2,4-Triazin-5(4H)-one, 4-amino-6-tert-butyl-3-(methylthio)-; 4-Amino-6-(1,1-dimethylethyl)-3-(methylthio)-1,2,4-triazin-5(4H)-one; 4-amino-6-tert-butyl-3-(methylthio)-as-triazin-5(4H)-one; Lexone 4L; Lexone 75DF; Lexone DF; Metribuzin; metribuzin + chlorimuron; Preview; Salute; Sencor; Sencor 4L; Sencor 75DF; Sencoral; Sencor DF; Sencorex; Sencor or metribuzin; trifuralin + metribuzin)

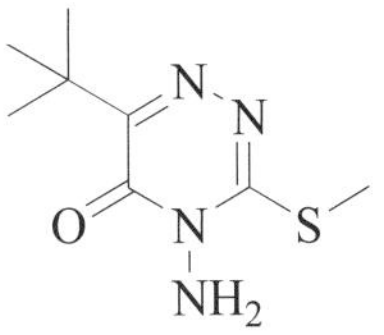

$C_8H_{14}N_4OS$

CAS 21087-64-9

A. PROPERTIES

colorless crystals soluble in alcohols; molecular weight 214.29; melting point 125; density 1.28

C. WATER AND SOIL POLLUTION FACTORS

in groundwater wells in the U.S.A.-150 investigations up to 1988:

median of the conc.'s of positive detections for all studies	0.6 μg/L
maximum conc.	6.8 μg/L (2944)

3-amino-1H-1,2,4-triazole (1-H-1,2,4-triazol-3-ylamine; 1H-1,2,4-triazol-3-amine; 3-aminotriazole; amitrol; amerol; amizol; cytrol; herbizole; weedazole)

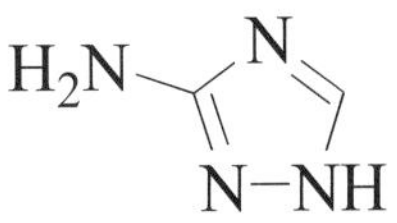

$C_2H_4N_4$

CAS 61-82-5

USES

herbicide and plant growth regulator; feedstock for chemical synthesis.

A. PROPERTIES

molecular weight 84.08; melting point 153-156°C; vapor pressure 0.00023 hPa at 60°C, 0.0075 hPa at 100°C; vapour pressure 3.3 x 10^{-5} Pa at 20°C; Relative density 1.14 at 20°C; log Pow –0.97 at pH 7 and 23°C; H 1.7 x 10^{-8} Pa.m^3/mol at 20°C;solubility 280,000 mg/L at 25°C

B. AIR POLLUTION FACTORS

Incinerability

thermal stability ranking of hazardous organic compounds: rank 208 on a scale of 1 (highest stability) to 320 (lowest stability). (2390)

Photodegradation

T/2 (OH radicals)= 4.8 hours

C. WATER AND SOIL POLLUTION FACTORS

Impact on biodegradation processes

>50% inhibition of NH_3 oxidation in *Nitrosomonas* EC_{50} 70 mg/L (n.s.i.) (407)

activated sludge, inhib. of oxygen consumption: 3h EC_{50} >10,000 mg/L (3033)

Biodegradation

inoculum: municipal waste water: at 20 mg/L after 28 d: no degradation (3033)

Hydrolysis

Insignificant hydrolysis at pH 5, 7 and 9

Photodegradation

Stable in water

Biodegradation in laboratory studies (10768)

degradation in soils	%	amitrole
Aerobic at 20°C	50	5 days
Aerobic at 20°C	90	22 days
Anaerobic at 20°C	50	>56 days

Dissipation studies in the field at 20°C (10768)

% dissipation	amitrole
50%	15-21 days
90%	50 days

Biodegradation in water/sediment systems (10768)

Readily biodegradable	
DT50 water	47; 94 days
DT90 water	156; 312 days
DT whole system	91; 95 days
DT whole system	302; 316 days

Mobility in soils

K_{OC} : 20-202 (mean 91)

D. BIOLOGICAL EFFECTS

Plants			
Duckweed *(Lemna perpusilla)*	7d EC_0	1,000 mg/L	(3035)
Algae			
Scenedesmus subspicatus	96h EC_{50}	2.3 mg/L	(3028)
	94h EC_{10}	0.16 mg/L	
	94h EC_{50}	11.4 mg/L	
	94h EC_{90}	807 mg/L	(3029)
Chlamydomonas eugametos	48h EC_{12}	84 mg/L	
	48h EC_0	8.4 mg/L	(3030)
Selenastrum capricornutum	24h NOEC	2; 5 mg/L	(3030, 3031)
	24h LOEC	5; 10 mg/L	(3030, 3031)
	6d NOEC	0.5 mg/L	(3032)
Selenastrum capricornutum in algal assay medium			
inhib. of oxygen evolution	EC_{50}	3.75 mg/L	
growth inhib.	EC_{50}	1.68 mg/L	(2624)
Crustaceans			
Gammarus fasciatus	48h NOEC	100 mg/L	(2125)
Daphnia magna	48h LC_{50}	30 mg/L	(2125)
Cypridopsis vidua	48h LC_{50}	32 mg/L	(2125)
Asellus brevicaudus	48h NOEC	100 mg/L	(2125)
Palaemonetes kadiakensis	48h NOEC	100 mg/L	(2125)
Orconectes nais	48h NOEC	100 mg/L	(2125)
Acanthocyclops vernalis nauplii	96h EC_{50}	22 mg/L	(3027)
Gammarus fasciatus	96h EC_{50}	>10 mg/L	(3026)

Molluscs

Crassostrea virginica	(2-d larvae)	12d EC_{50}	255 mg/L	
	(2-cell stage)	48h EC_{50}	734 mg/L	(303)

Fishes

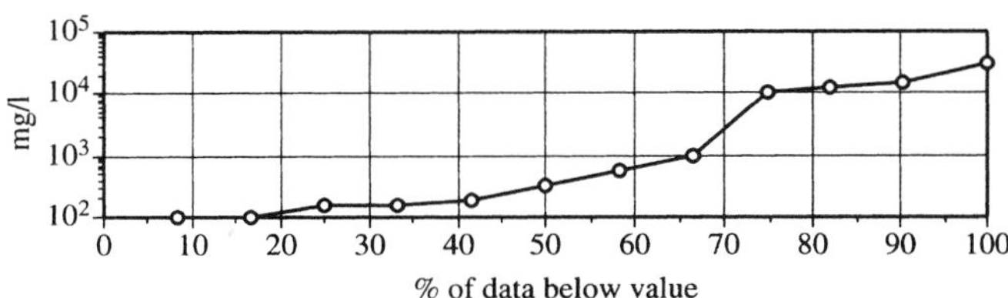

Frequency distribution of 24-96h LC_{50} values for fishes (n = 12) based on data from this and other works			(3999)
Lepomis macrochirus	48h NOEC	>100 mg/L	(2125)
	48h LC_{50}	10,000 mg/L	(3022)
fry	8d LC_0	>50 mg/L	(3024)
	12d LC_0	25 mg/L	(3024)
Lepomis cyanellus fertilized eggs	72h EC_0	50 mg/L	
Oncorhynchus kisutch	48h LC_{50}	325 mg/L	(2106)
Oncorhynchus tschawytscha fingerlings	24h LC_{50}	185 mg/L	(2106)
	48h LC_{50}	155 mg/L	(2106)
bluegill	48h LC_{50}	100 mg/L	(1878)
Poecilia reticulata	96h LC_{50}	12,500 mg/L	(3022)
	48h LC_0	600 mL	
	48h LC_{100}	1,500 mg/L	(3023)
	48h LC_{50}	567; 30,300 mg/L	(3025)

	27h LC_{50}	14,300 mg/L	(3025)
Micropterus salmoides	48h LC_{50}	>1,000 mg/L	(2106)
Cyprinus carpio	48h LC_0	>1,000 mg/L	(3023)
Salmo irideus	24h LC_0	400 mg/L	
	24h LC_{100}	1,000 mg/L	(3023)
Erimyzon sucetta fry	8d LC_0	>50 mg/L	(3024)
Pimephales promelas	96h LC_{50}	>100 mg/L	(3026)
Ictalurus punctatus	96h LC_{50}	>160 mg/L	(3026)

Toxicity

Birds			
Bobwhite quail	Acute oral LD_{50}	>2,150 mg/kg bw	(10768)
Worms			
Eisenia foetida	LC_{50}	>448 mg a.i./kg soil	(10768)
Insecta			
Honeybees	Acute oral LD_{50}	>152 µg/bee	
	Acute contact LD_{50}	>100 µg/bee	(10768)
Mammals			
Rat	oral LD_{50}	>5,000 mg/kg bw	(10768)

4-amino-1,2,4-triazole

((4-Amino-4H-1,2,4-triazole; 4-Amino-4H-1,2,4-triazole; 4H-1,2,4-Triazol-4-amine; 4H-1,2,4-Triazol-4-ylamine)

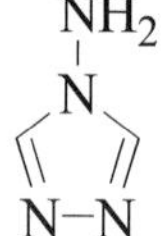

$C_2H_4N_4$

CAS 584-13-4

EINECS 209-533-5

USES

1,2,4-Triazole(triazole: a five-membered chemical ring compound with three nitrogens in the ring) and its derivatives have biological activities such as antiviral, antibacterial, antifungal and antituberculous. 4-Amino-1,2,4-triazole is used as an intermediate for the synthesis of antifungal agents such as fluconazole and other organic compounds

A. PROPERTIES

White crystals; molecular weight 84.08; melting point 84-86°C; water solubility 810,000 mg/L at 20°C

D. BIOLOGICAL EFFECTS

Toxicity

Micro organisms	nitrification inhibition in soil (1% o.c.) after 14 days : 71 % at 50 mg/kg soil		(7050)
Mammals			
Rat	oral LD_{50}	2.625	

mg/kg bw

4-amino-3,5,6-trichloropicolinic acid *see* picloram

2-amino-1,3,5-trimethylbenzene *see* mesidine

amino tris(methylenephosphonic acid) (nitrilotris(methylene)triphosphonic acid; ATMP)

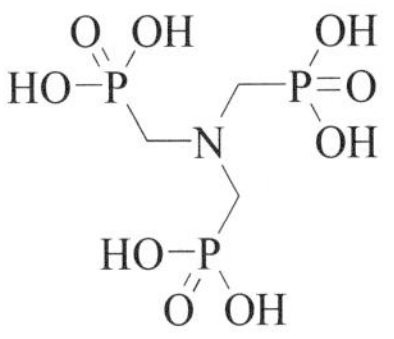

$N[CH_2P(O)(OH)_2]_3$

$C_3H_{12}NO_9P_3$

CAS 6419-19-8

CAS 2235-43-0 (penta sodium salt)

USES

ATMP is one of the major phosphonates used as corrosion and scale inhibitors in water treatment units and in boiling water applications (5-100 mg/L). Smaller amounts find use in oil field drilling, metal finishing, paper mill processes, and industrial cleaning applications. Also used in detergents.

A. PROPERTIES

molecular weight 299.05; solubility completely miscible; LogP_{OW} - 3.5; log H <- 5 (estimated)

C. WATER AND SOIL POLLUTION FACTORS

Waste water treatment

Phosphonate removal is high in sewage treatment plants, even though biological degradation may be low. The principal elimination mechanism is by adsorption on the activated sludge. (7036)

Biodegradation

Inoculum	*Method*	*Concentration*	*Duration*	*Elimination results*	
Nonacclimated	Closed bottle test		30 d	7- 20%	
Acclimated	Closed bottle test		30 d	25- 38%	(7033)

Zahn-Wellens test	28 d	23% DOC	(7034)
Modified SCAS test		0.5- 2% ThOD	(7035)
Anaerobic model digestor	4% conversion to $^{14}CO_2$ and $^{14}CH_4$		(7036)

In Meramec river water and eutrophic lake water:% $^{14}CO_2$ evolution after 60 d.

in sterile natural water:	in the dark	0.1%	
	in sunlight	6.2%	
in active natural water:	in the dark	12%	
	in sunlight	14%	(7036)

In lake water sediment at 1 mg/L phosphonate: average% $^{14}CO_2$ evolution after 50 d

sterile sediment	1.6- 8.%	
active sediment	4.7- 12%	(7036)

Thus enhancement of CO_2 evolution rates was due to microbial activity, sunlight, and their interaction. (7036)

Sewage treatment systems

ATMP removal increased significantly when the pH was buffered to about 7. Removals increased to 90% after 26 d. (7036)

Phosphonate utilizing bacteria are ubiquitous and were not only found in sludge, but in soil and peat as ell. (7036)

Drinking water purification: at 0.1- 5 mg/L: 95% removal by flocculation/precipitation using ferrous sulfate or ferric sulfate. (7036)

Degradation in soils

In two soils: mineralization to CO_2 ranged within 2- 53% over 10 weeks. (7036)

In soils supplemented with 10 mg/kg ^{14}C-labeled phosphonate:
after 148 d: average $^{14}CO_2$ evolution

sterile soil	0.3- 0.8	
active soil	0.6- 15	(7036)

Microbial degradation pathways

Arthrobacter sp. cleaves only a single phosphonate group (7036)

N(CH₂PO(OH)₂)₃ → HN(CH₂PO(OH)₂)₂ + H_3PO_4

Impact on biodegradation processes

Phosphonate levels of up to 160 mg/L had no inhibitory effect on COD or MBAS removal in the test units. (7036)

Anaerobic model digestor: no inhibition at high loading. (7036)

Quarternary phosphate removal: ATMP inhibited phosphate removal by flocculation filtration at 0.2 mg/L. Addition of $FeCl_3$ overcame this inhibition. (7036)

Soil sorption

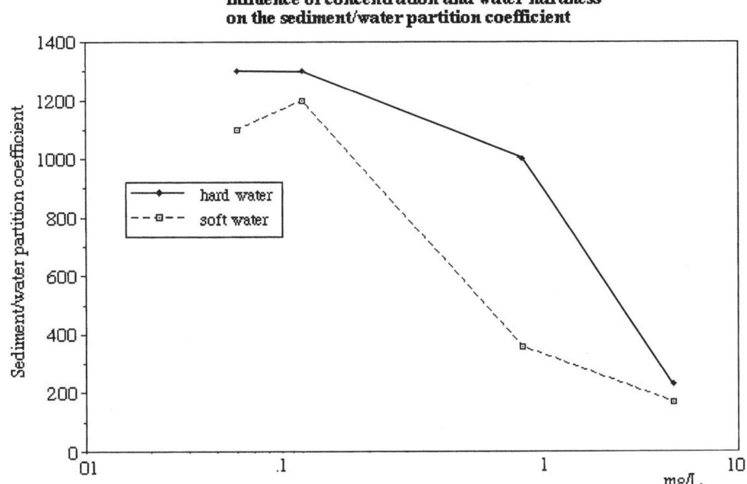

Water hardness and phosphonate concentration were found to control sorptive behavior. The sediment/water partition coefficient was measured after 24 h of mixing. Sediment properties were organic carbon 11.8% and clay 8%. (7036)

D. BIOLOGICAL EFFECTS

Bioconcentration

Zebra fish: at 1 µg/L after 4 weeks: BCF 18- 24 (7036)

Toxicity

Algae

Selenastrum capricornutum	96 h EC_{50}	20 mg/L	
	96 h NOEC	7.4 mg/L	
	14 d EC_{50}	20 mg/L	
	14 d NOEC	7.4 mg/L	(7036)

Algal growth inhibition is likely to be due to complexing essential metal nutrient by ATMP. The test result is influenced by pH; water hardness, compared with standard medium, results in increased EC_{50} values. (7033)

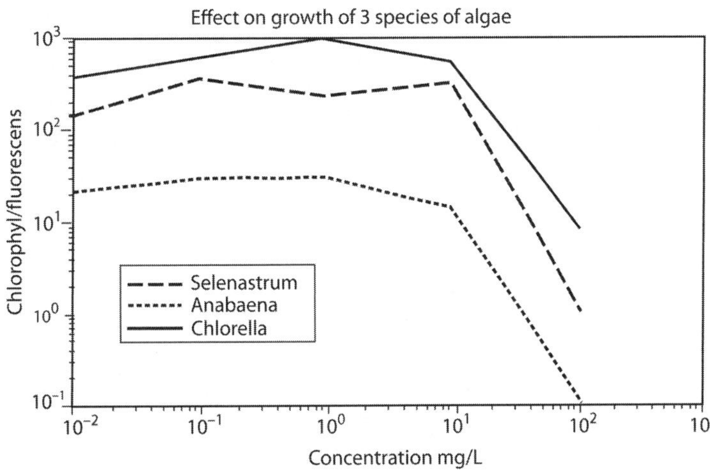

Growth measured after 8 d via chlorophyll-a fluorescence resulted in the observed growth stimulation at low concentrations (the values indicated at 10^{-2} mg/L represent the control). (7036)

Insects

Chironomus	48 h EC_{50}	11,000 mg/L	
	48 h NOEC	7,040 mg/L	(7036)
Mussels			
Eastern oyster	96 h EC_{50}	201 mg/L	
	96 h NOEC	95 mg/L	(7036)
Crustaceans			
Grass shrimp	96 h LC_{50}	7,870 mg/L	
	96 h NOEC	4,575 mg/L	
Daphnia magna	48 h EC_{50}	297 mg/L	
	48 h NOEC	125 mg/L	(7033)
	28 d NOEC	>25 mg/L	
	<54 mg/L		(7033)
Fishes			
Oncorhynchus mykiss	96 h LC_{50}	160 mg/L	
	14 d LC_{50}	150 mg/L	
	14 d NOEC	47 mg/L	(7033)
(early lifetime study)	60 d NOEC	>23 mg/L	
	60 d EC_{50}	<47 mg/L	(7033)
Bluegill sunfish	96 h LC_{50}	>330 mg/L	
	96 h NOEC	330 mg/L	
Channel catfish	96 h LC_{50}	1,212 mg/L	
	96 h NOEC	924 mg/L	
Sheephead minnow	96 h LC_{50}	8,132 mg/L	
	96 h NOEC	4,831 mg/L	(7033)

4-amino-3,5-xylenol (4-amino-3,5-dimethylphenol)

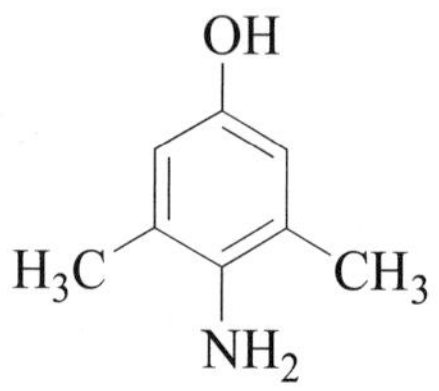

OH

H_3C CH_3

NH_2

$C_8H_{10}NO$

CAS 3096-70-6

D. BIOLOGICAL EFFECTS

Fishes			
bluegill *(Lepomis macrochirus)*	96h LC_{50}, S	0.32 mg/L (99% purity)	(1502)

p-aminoacetanilide (N-acetyl-p-phenylenediamine; 4'-aminoacetanilide)

$CH_3C(O)NHC_6H_4NH_2$

$C_8H_{10}N_2O$

CAS 122-80-5

USES

intermediates; azo dyes.

A. PROPERTIES

colorless or reddish crystals; molecular weight 150.2; melting point 162°C, boiling point 267°C.

C. WATER AND SOIL POLLUTION FACTORS

Biodegradation rates

adapted A.S. at 20°C-product is sole carbon source: 93% COD removal at 11 mg COD/g dry inoculum/h. (327)

aminoacetic acid (aminoethanoic acid; glycine; glycocoll)

H_2NCH_2COOH

$C_2H_5NO_2$

CAS 56-40-6

USES

stabilizing agent in photoprocessing.

NATURAL SOURCES

normal constituent of proteins. (211)

A. PROPERTIES

crystalline; molecular weight 75.07; melting point 233°C decomposes; boiling point 289/292°C decomposes; density 1.601; solubility 253,000 mg/L at 25°C, 575,000 mg/L at 75°C; $LogP_{ow}$ -3.03/-1.7 (calculated).

C. WATER AND SOIL POLLUTION FACTORS

BOD_5:	26% $ThOD_{NO3^-}$	(1828)
	37% $ThOD_{NO3^-}$	(30)
	86% ThOD	
COD:	64% ThOD (0.05 n Cr207) reflux COD: 99.2% recovery rapid COD: 36.1% recovery	(1828)

Poor recovery was exhibited by the rapid COD method. A possible interference was volatilization of the compound or compound oxidation products in the open digestion flask at the high test temperatures.

$ThOD_{NO3}$: 1.49 (1828)

Impact on conventional biological treatment systems

Unacclimated system at 500 mg/L: biodegradable. (1828)

Manmade sources

excreted by humans in urine: 2.3 to 18.0 mg/kg body wt/d. (203)

Waste water treatment

A.S.:	after 6 h:	4.1% of ThOD	
	after 12 h:	8.1% of ThOD	
	after 24 h:	16.9% of ThOD	(89)

methods	*temp.,°C*	*days observed*	*feed, mg/L*	*days acclim.*	*% theor. oxidation*	*% removed*	
A.S., Resp, BOD	20	1-5	720	<1	57	87	
A.S., BOD	20	1-5	333	15	-	93	(93)

[powdered carbon: at 100 mg/L, carbon dosage 1,000 mg/L: 1% absorbed (520)

p-aminoanisole *see* *p*-anisidine

1-aminoanthraquinone (9,10-anthracenedione,1-amino; Diazo fast red AL)

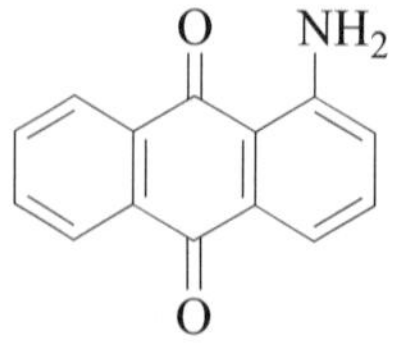

$C_{14}H_9NO_2$

CAS 82-45-1

EINECS 201-423-5

USES

Anthraquinone dye used primarily for cellulose acetate fibers; base stock for chemical synthesis. An intermediate for dyestuff and pharmaceuticals.

A. PROPERTIES

red needles; molecular weight 223.20; melting point 256-258°C; boiling point >300 °C; vapour pressure 1.2 x 10^{-4} Pa at 100°C; log Pow 3.74 at 25°C; water solubility 20 mg/L at 20°C, 32 mg/L at 25°C;

C. WATER AND SOIL POLLUTION FACTORS

Hydrolysis

the chemical is stable at pH 4 to 9 (7455)

Photodegradation

t/2 estimated at 1.4 x 10^{-2} years for direct photo degradation in water (7455)

Biodegradation

Inoculum	method	concentration	duration	elimination	
A.S. non-adapted	MITI	100 mg/L	28 days	0-1% ThOD 1-3% HPLC	(7455)
	Closed bottle test, municipal wastewater	0.8-24 mg/L	20 days	0% BOD removal	(6007)

D. BIOLOGICAL EFFECTS

Bioaccumulation

Species	conc. µg/L	duration	body parts	BCF	
carp	3,000	8 weeks		55-137	
	30,000	8 weeks		50-150	(7455)

Toxicity

Bacteria			
Pseudomonas fluorescens	24 h EC_0	1,000 mg/L	(6007)
Algae			
Selenastrum capricornutum	72h ECb_{50}	0.25 mg/L	
	72h NOECb	0.10 mg/L	(7455)
Crustaceae			
Daphnia magna	24-48h EC_{50}	>1,000 mg/L	
	21d EC_{50}	0.62 mg/L	
reproduction	21d EC_{50}	0.56 mg/L	
	21d NOEC	0.32 mg/L	(7455)
Fishes			
Oryzias latipes	24-96h LC_{50}	>1,000 mg/L	
Leuciscus idus	48h LC_0	>1,000 mg/L	(6007, 7455)
Mammals			
Rat	oral LD_{50}	>5,000 mg/kg bw	(7455)

p-aminoazobenzene *see* *p*-phenylazo-aniline

p-aminobenzenesulfonic acid *see* *p*-anilinesulfonic acid

2-aminobenzimidazole (AB)

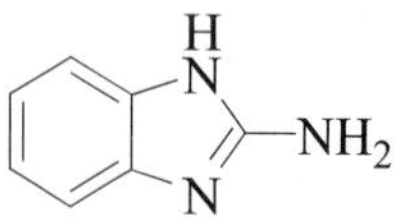

$C_7H_7N_3$

CAS 934-32-7

USES

in photographic industry as an antifoggant.

MANMADE SOURCES

degradation product (hydrolysis) of the fungicides benomyl, carbendazim, and thiophanate methyl. (1373, 1374)

A. PROPERTIES

molecular weight 133.15, melting point 229-231°C.

C. WATER AND SOIL POLLUTION FACTORS

Biodegradation

total evolution of ^{14}C in CO_2 and remaining radioactivity in soil after 218 d of incubation at 25°C with 4 ppm ^{14}C-2 aminobenzimidazole.*

	not inoculated	*inoculated with adapted soil*
^{14}C-evolution	48%	84%
^{14}C-soil residues	41%	11-18%
% recovery	89%	95-102%

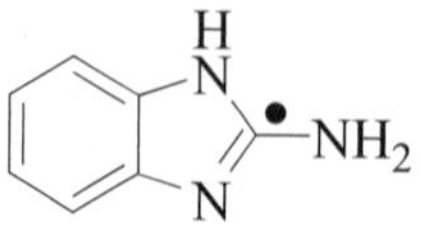

[Structure of 2-aminobenzimidazole labeled with ^{14}C as indicated by .] (1714)

Effect of AB on nitrification by *Nitrosomonas* sp.:

conc., ppm	*incubation period* 6 d	15 d

	µg nitrite recovered/ml medium:	
0	3.8	108
10	3.7	146
100	2.8	3.9
Effect of AB on nitrification by *Nitrobacteragilis:*		
	incubation period	
conc., ppm	*6 d*	*15 d*
	µg nitrite recovered/ml medium:	
0	290	0
10	352	0
100	622	661

m-aminobenzoic acid (metanilic acid)

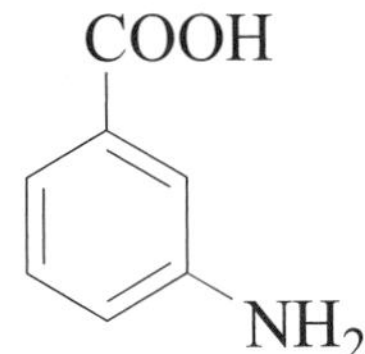

$NH_2C_6H_4COOH$

$C_7H_7NO_2$

CAS 99-05-8

A. PROPERTIES

molecular weight 137.13; melting point 174/179°C; boiling point sublimes; density 1.5 at 20/4°C; solubility 5,900 mg/L at 15°C; $LogP_{OW}$ 0.14/0.27 (calculated).

C. WATER AND SOIL POLLUTION FACTORS

Biodegradation

decomposition by a soil microflora: >64 d (176)

adapted A.S. at 20°C-product is sole carbon source: 97% COD removal at 7.0 mg COD/g dry inoculum/h (327)

o-aminobenzoic acid *see* anthranilic acid

p-aminobenzoic acid (PABA)

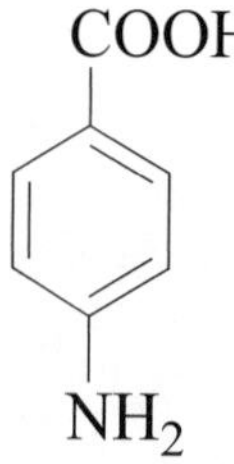

$NH_2C_6H_4COOH$

$C_7H_7NO_2$

CAS 150-13-0

A. PROPERTIES

molecular weight 137.13; melting point 187°C; density 1.47 at 20°C; solubility 3,400 mg/L at 9.6°C; $LogP_{OW}$ 0.68.

C. WATER AND SOIL POLLUTION FACTORS

Biodegradation

decomposition by a soil microflora: 8 d	(176)
Complete degradation was achieved at a concentration of 50 mg TOC/L of 4-aminobenzoic acid using anaerobic digesting sludge under methanogenic conditions.	(8441)
adapted A.S. at 20°C-product is sole carbon source: 96% COD removal at 12 mg COD/g dry inoculum/h	(327)
Lag period for degradation of 16 mg/L by wastewater or by soil at pH 7.3 and 30°C: less than 1 d	(1096)

Impact on biodegradation processes

at 100 mg/L no inhibition of NH_3 oxidation by *Nitrosomonas* sp.	(390)

D. BIOLOGICAL EFFECTS

Bacteria			
Photobacterium phosphoreum Microtox test	30 min EC_{50}	27.4 mg/L	(8899)
Mammals			
rat	oral LD_{50}	6,000 mg/kg bw	(9780)
rabbit	oral LD_{50}	1,830mg/kg bw	(9781)
mouse	oral LD_{50}	2,850 mg/kg bw	(9781)

3-aminobenzotrifluoride (3-(trifluoromethyl)benzenamine; *m*-(trifluoromethyl)aniline)

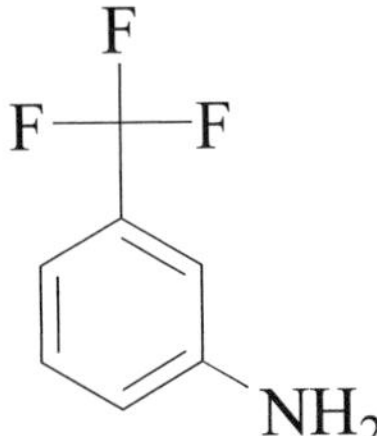

$CF_3C_6H_4NH_2$

$C_7H_6F_3N$

CAS 98-16-8

USES

pharmaceutical and pesticide intermediate.

A. PROPERTIES

colorless to oily yellow liquid, aniline-like odor; molecular weight 161.13; melting point 5-6°C; boiling point 187°C; vapor pressure 27 hPa at 85°C; vapor density 5.56; density 1.3 at 15.5/15.5°C; solubility 5,000 mg/L at 20°C, 6,000 mg/L at 30°C; $LogP_{OW}$ 2.3 (calculated).

B. AIR POLLUTION FACTORS

Atmospheric degradation

photochemical degradation: reaction with OH°: $t_{1/2}$: 10 hours (calculated). (5641)

C. WATER AND SOIL POLLUTION FACTORS

Water quality

in Waal River (Netherlands): average in 1973: 0.4 µg/L (n.s.i.). (342)

Biodegradability:

A.S., indust., nonadapted, Zahn Wellens test:

after 3 hours	14% ThOD	
after 5 d	19% ThOD	
after 10 d	20% ThOD	
after 15 d	29% ThOD	(5639)

Impact on biodegradation processes

A.S., municipal

3h EC_0	174 mg/L	
3h EC_{50}	1,400 mg/L	(5639)

D. BIOLOGICAL EFFECTS

Bacteria			
E. coli	24h IC_{50}	145 mg/L	(5640)
Crustaceans			
Scenedesmus subspicatus	2d NOEL	0.67 mg/L	(5637)
	8d IC_{10}	1.9 mg/L	(5638)
Chlorella fusca	2d NOEL	0.67 mg/L	(5637)
Fishes			
Brachydanio rerio	48-96h LC_0	25 mg/L	

48-96h LC_{50}	35 mg/L	
48-96h LC_{100}	50 mg/L	(5635)

4-aminobiphenyl (4-biphenylamine; xenylamine; p-biphenylamine; p-aminobiphenyl; p-aminodiphenyl; 4-aminodiphenyl; p-phenylaniline; (1,1'-biphenyl)-2-amine; biphenyl-4-amine)

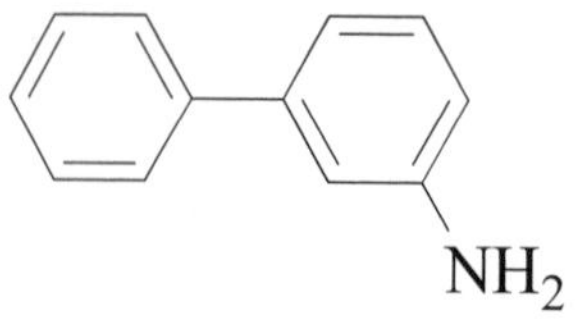

NH_2

$C_{12}H_{11}N$

$C_6H_5C_6H_4NH_2$

CAS 92-67-1

USES

In chemical analysis to detect sulfate ion. As a carcinogen in research. Formerly used as a rubber antioxidant.

A. PROPERTIES

molecular weight 169.24; melting point 53°C; boiling point 302°C; density 1.16 at 20°C; P_{ow} 2.80.

B. AIR POLLUTION FACTORS

Incinerability

thermal stability ranking of hazardous organic compounds: rank 51 on a scale of 1 (highest stability) to 320 (lowest stability). (2390)

Photochemical reactions

$t_{1/2}$: 6.9 h, based on reactions with OH° (calculated). (8481)

C. WATER AND SOIL POLLUTION FACTORS

Biodegradation

Inoculum	*method*	*conc.*	*duration*	*elimination results*	
sludge	Static biodegradation Test	2 mg/l	7 days	50%	(7730)

Soil mobility

K_{OC}: 417 (calculated) (8476)

D. BIOLOGICAL EFFECTS

Mammals

rat, rabbit, mouse	oral LD_{50}	205-690 mg/kg bw	(9782, 9783)

1-aminobutane *see* *n*-butylamine

2-aminobutane *see* sec. butylamine

2-aminobutanedioic acid *see* DL-asparatic acid

2-aminobutanedioic amide *see* L-asparagine

aminocarb (metacil; 4-(dimethylamino)-3-methylphenyl-N-methylcarbamate(ester); 4-dimethylamino-m-tolylmethylcarbamate; 4-dimethylamine-m-cresylmethylcarbamate)

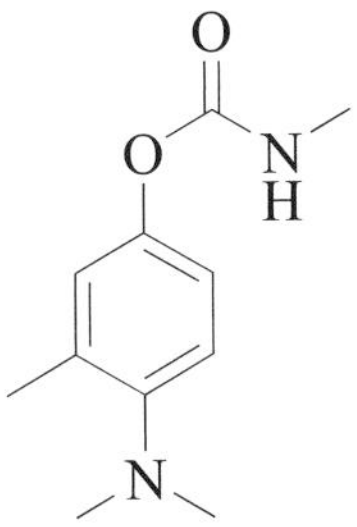

O
O
N
H
N

$C_{11}H_{16}N_2O_2$

CAS 2032-59-9

USES

nonsystemic insecticide, molluscicide.

A. PROPERTIES

white, crystalline solid; molecular weight 208.29; melting point 93-94°C.

C. WATER AND SOIL POLLUTION FACTORS

Aquatic reactions

persistence in river water in a sealed glass jar under sunlight and artificial fluorescent light-initial conc. 10 μg/L

	% of original compound found after				
1 h	*1 wk*	*2 wk*	*4 wk*	*8 wk*	
100	60	10	0	0	(1309)

D. BIOLOGICAL EFFECTS

Crustaceae

Gammarus lacustris: 96h LC_{50}; 12 μg/L (2124)

Mussels

BCF in mussel *(Mytilus edulis):* 3.8-4.9 (on wet weight) (1864)

Insects

fourth instar larval *Chironomus riparius:* 24h LC_{50}: 377 ppb (1853)

m-aminochlorobenzene *see* *m*-chloroaniline

o-aminochlorobenzene *see* *o*-chloroaniline

p-aminochlorobenzene *see* *p*-chloroaniline

aminocyclohexane *see* cyclohexylamine

2,2'-aminodiethanol *see* diethanolamine

p-aminodiethylanilinehydrochloride (N,N-diethyl-p-phenylenediamine, hydrochloride)

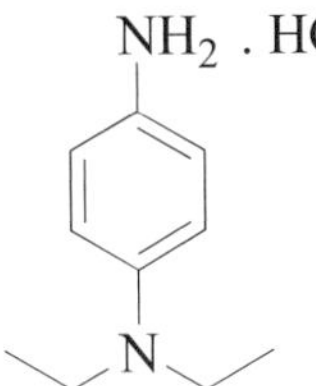

$C_{10}H_{16}N_2.HCl$

CAS 2198-58-5

USES

color photography.

A. PROPERTIES

colorless needles; molecular weight 200.74; boiling point 260-262°C.

D. BIOLOGICAL EFFECTS

GREEN ALGAE: *Microcystis aeruginosa:* LC_{100}: 1 ppm. (1094)

p-aminodimethylaniline (dimethylaminoaniline; dimethyl-*p*-phenylenediamine; N,N-dimethyl-p-phenylenediamine; N,N-dimethyl-1,4-benzenediamine; 4-aminodimethylaniline; 4-(dimethylamino)aniline; DMPD)

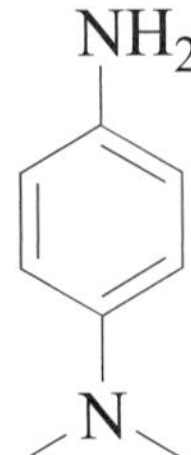

$C_8H_{12}N_2$

CAS 99-98-9

USES

base for production of methyleneblue; photodeveloper; reagent for cellulose. Manufacture of hair dyestuffs. Chemical intermediate. Analytical reagent (especially for chlorine residues in water).

A. PROPERTIES

colorless, asbestos-like needles; molecular weight 136.20; melting point 41°C; boiling point 257°C;

density 1.04 at 20°C; vapor pressure 1.9 x 10^{-3} mm Hg; solubility miscible; $logP_{OW}$ 1.11.

B. AIR POLLUTION FACTORS

Photochemical reactions

$t_{1/2}$: 97 min, based on reactions with OH° (calculated). (9786)

C. WATER AND SOIL POLLUTION FACTORS

Photooxidation

$t_{1/2}$: 19-30 h sunlight via hydroxyl and peroxy radicals in water. (9784)

Mobility in soil

K_{OC} of 10-19. (9262)

D. BIOLOGICAL EFFECTS

Bacteria			
Photobacterium phosphoreum Microtox test	30 min EC_{50}	0.84 mg/L	(8899)
Algae			
Microcystis aeruginosa	LC_{100}	2 mg/l	(1094)
Mammals			
rat	oral LDLo	50 mg/kg bw	(9785)

1-aminododecane *see* *n*-dodecyclamine

aminoethane *see* ethylamine

2-aminoethanol *see* ethanolamine

2-(2-aminoethylamino)ethanol (aminoethylethanolamine; N-hydroxyethyl-1,2-ethanediamine; β-hydroxyethyl)ethylenediamine)

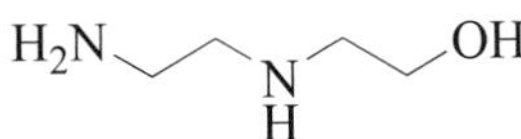

$C_4H_{12}N_2O$

CAS 111-41-1

USES

intermediate for the manufacturing of emulsifiers, corrosion inhibitors, pharmaceuticals, pesticides; additive in the textiles industry.

A. PROPERTIES

molecular weight 104.15; mp, -18°C; boiling point 140°C at 44 hPa; vapor pressure <10 hPa at 80°C; density 1.03 at 20°C; solubility miscible; $LogP_{OW}$- 1.4 (measured).

C. WATER AND SOIL POLLUTION FACTORS

Biodegradation

Inoculum	*Method*	*Concentration*	*Duration*	*Elimination results*	
	BOD test		5 d	<1% ThOD	(6193)
BASF activated sludge	Zahn-Wellens test	400 mg/L	37 d	42% DOC	(6193)

D. BIOLOGICAL EFFECTS

Bacteria			
BASF-activated sludge respiration	30 min EC_{10}	>1,000 mg/L	(6193)
Pseudomonas putida	17 h EC_{10}	82 mg/L	
(cell multiplication inhibition test)	17 h EC_{50}	135 mg/L	
	17 h EC_{90}	231 mg/L	(6192)

Algae			
Scenedesmus subspicatus	72 h EC_{20}	130 mg/L*	
(cell multiplication inhibition test)	72 h EC_{50}	210 mg/L	
	72 h EC_{90}	490 mg/L	
	72 h EC_9	500 mg/L*	(6192)

* Neutralized.

Crustaceans			
Daphnia magna Straus	24-48 h EC_0	125; 125 mg/L	
	24-48 h EC_{50}	225; 190 mg/L	
	24-48 h EC_{100}	500; 500 mg/L	(6192)

aminoform *see* hexamethylenetetramine

D-α-aminoglutaramic acid *see* L-glutamine

aminoguanidine

C. WATER AND SOIL POLLUTION FACTORS

Impact on biodegradation processes

nitrification inhib.:	EC_{50}	74 mg/L	(2624)

L-α-aminoisocaproic acid *see* L-leucine

aminomethane *see* methylamine

p-aminomethoxybenzene *see* *p*-anisidine

3-aminomethyl-3,5,5-trimethylcyclohexylamine (1,3,3-trimethyl-1-aminomethyl-5-aminocyclohexane; cyclohexanemethanamine,5-amino-1,3,3-trimethyl-)

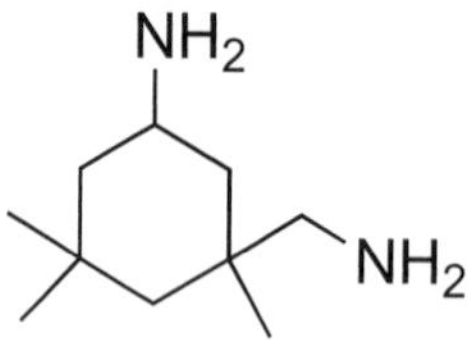

NH_2 NH_2

$C_{10}H_{22}N_2$

CAS 2855-13-2

TRADENAMES

Degamin IPDA; Isopherone diamine; Vestamin IPD

USES

a chemical intermediate to produce hardeners for epoxy resins and coatings . It is also directly used as a hardener. It had large applications in epoxy-based self-levelling and trowelable flooring systems, and various civil engineering applications such as paving, concrete protection and repair. It is further used in the production of non-crystalline speciality polyamides as a chain extender in polyurethanes and as an intermediate in dyes.

A. PROPERTIES

molecular weight 170.3; melting point 10°C; boiling point 247°C; relative density 0.92 at 20°C; vapour pressure 0.02 hPa at 20°C; water solubility miscible; log Pow 0.99 at 23°C; K_{OC} 340; H 0.0004 Pa m^3/mole

B. AIR POLLUTION FACTORS

Photodegradation

t/2 calculated at 0.2 days (7454)

C. WATER AND SOIL POLLUTION FACTORS

Hydrolysis

t/2 > 1 year at pH 4 to 9 (7454)

Biodegradation

Inoculum	method	concentration	duration	elimination	
Domestic sewage		6.9 mg DOC/L	7 days	4 % DOC	
			14 days	4 % DOC	
			28 days	8 % DOC	(7454)

D. BIOLOGICAL EFFECTS

Toxicity

Micro organisms			
Pseudomonas putida	18h EC_{10}	1120 mg/L	(7454)
Algae			
Scenedesmus subspicatus	72h $ECgr_{10}$	3.1 mg/L	
	72h $ECgr_{50}$	37 mg/L	
	72h $ECgr_{90}$	436.5 mg/L	
	72h NOECgr	1.5 mg/L	(7454)
Crustaceae			
Daphnia magna	48h EC_0	4.2 mg/L	
	48h EC_{50}	23 mg/L	
	48h EC_{100}	66.4 mg/L	
	24h EC_0	25 mg/L	
	24h EC_{50}	44 mg/L	
	24h EC_{100}	70 mg/L	(7454)
	21d NOEC	3 mg/L	
	21d LOEC	10 mg/L	(7454)
Chaetogammarus marinus	96h NOEC	100 mg/L	
	96h EC_{50}	324 mg/L	(7454)
Fish			
Leuciscus idus	96h LC_0	70 mg/L	
	96h LC_{50}	110 mg/L	
	96h LC_{100}	140 mg/L	(7454)
Mammals			
Rat	oral LD_{50}	1,030 mg/kg bw	(7454)

Mouse	Oral LD0	100 mg/kg bw	(7454)

p-aminomethylphenylether *see* *p*-anisidine

L-aminopentane *see* *n*-amylamine

2-aminopentanedioic acid *see* DL-glutamic acid

2-aminopentanedioicamide *see* L-glutamine

m-aminophenol (*m*-hydroxyaniline)

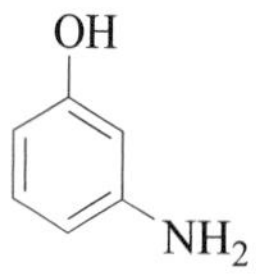

$NH_2C_6H_4OH$

C_6H_7NO

CAS 591-27-5

USES

dye intermediate.

A. PROPERTIES

molecular weight 109.12; melting point 122/123°C; boiling point 164°C at 11 mm; solubility 26,000 mg/L; $LogP_{OW}$ 0.15/0.17.

C. WATER AND SOIL POLLUTION FACTORS

$ThOD_{NH3}$: 1.91 (2790)
$ThOD_{NO3}$: 2.49 (2790)
Biodegradation: EEC respirometric method at 100 mg/L and 20°C using nonadapted sludge.

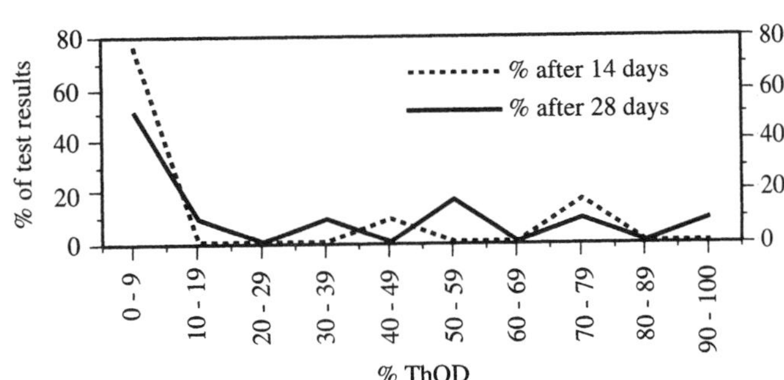

ThOD in EEC respirometric test: n = 12; median after 14 d: 0% ThOD; median after 28 d: 5% ThOD.

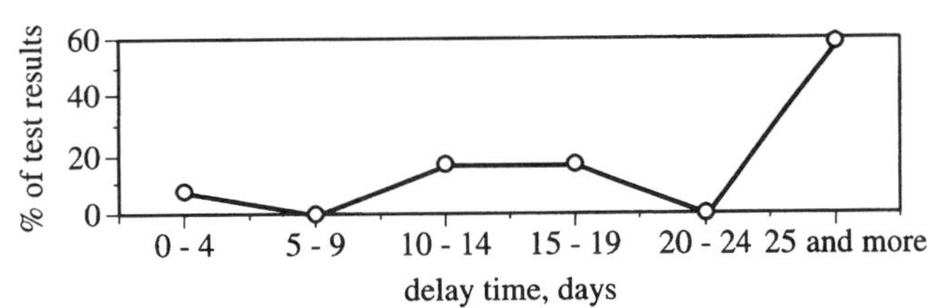

Delay times in EEC respirometric test

n = 12; median: >28 d
Oxygen uptake and DOC removed
using MITI inoculum
delay time: 7 d
% ThOD after 14 d: 1;3%
% ThOD after 28 d: 0; 1; 4; 70%
% DOC: 0; 0; 1; 99% (2790)
in repetitive Die-Away' test
delay time: 12 d
% ThOD after 28 d: 65%
% ThOD after 42 d: 77%
% DOC removal after 28 d: 98%
% DOC removal after 42 d: 98% (2790)
EEC respirometric method at 20 mg/L and 20°C: delay time: 22 d
after 14 d: 0% ThOD
after 28 d: 0-76% ThOD and 2-100% DOC removal (2790)

Biodegradation

half-lives in nonadapted aerobic subsoil:
sand Oklahoma: 10-16 d (2695)
decomposition by soil microflora: >64 d (176)

adapted A.S. at 20°C-product is sole carbon source: 90% COD removal at 11 mg COD/g dry inoculum/h (327)
lag period for degradation of 16 mg/L waste water or soil suspension at pH 7.3 and 30°C: >25 d (1096)

Impact on biodegradation processes

at 0.6 mg/L, inhibition of degradation of glucose by *Pseudomonas fluorescens* (293)
at 9 mg/L, inhibition of degradation of glucose by *E. coli*

D. BIOLOGICAL EFFECTS

Algae

Chlorella pyrenoidosa: toxic at 140 mg/L	(41)

o-aminophenol (*o*-hydroxyaniline; 2-amino-L-hydroxybenzene)

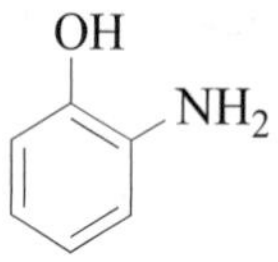

$NH_2C_6H_4OH$

C_6H_7NO

CAS 95-55-6

A. PROPERTIES

molecular weight 109.12; melting point 179/174°C; boiling point sublimes; solubility 17,000 mg/L at 0°C, $LogP_{OW}$ 0.52-0.62.

C. WATER AND SOIL POLLUTION FACTORS

Biodegradability

decomposition period by soil microflora: 4 d	(176)

Biodegradation

Inoculum	*method*	*concentration*	*duration*	*elimination results*	
adapted A.S at 20°C	product is sole carbon source	21 mg COD/g dry inoculum/h		95% COD	(327)
anaerobic digesting sludge		50 mg TOC/L		80%	(8441)

D. BIOLOGICAL EFFECTS

Bacteria			
Photobacterium phosphoreum Microtox test	5-30 min EC_{50}	134 mg/L	(9657)
Algae			
Chlorella pyrenoidosa	IC_{50}	47 mg/L	(41)
Fishes			
goldfish	48h LC_{100}	20 mg/L	(9787)
Mammals			
mouse	oral LD_{50}	1,250 mgkg bw	(9789)

p-aminophenol (*p*-hydroxyaniline; rodinal; 4-amino-1-hydroxybenzene)

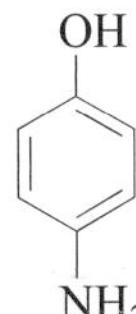

$NH_2C_6H_4OH$

C_6H_7NO

CAS 123-30-8

A. PROPERTIES

molecular weight 109.12; melting point 184°C decomposes; boiling point 284°C sublimes; solubility 11,000 mg/L at 0°C; THC 760 kcal/mol; $LogP_{OW}$ 0.04.

C. WATER AND SOIL POLLUTION FACTORS

Biodegradation

adapted A.S. at 20°C-product is sole carbon source: 87% COD removal at 17 mg COD/g dry inoculum/h. (327)

D. BIOLOGICAL EFFECTS

RubisCo test

inhib. of enzym. activity of Ribulose-P2-carboxylase in protoplasts: IC_{10}: 11 mg/L.

Oxygen test

inhib. of oxygen production of protoplasts IC_{10}: 55 mg/L. (2698)

Bacteria			
Escherichia coli:	toxic	8-10 mg/L	
Algae			
Chlorella pyrenoidosa	toxic	140 mg/L	(41)
Scenedesmus	toxic	6 mg/L	(30)
Arthropoda			
Daphnia:	toxic	0.6 mg/L	(30)
Fishes			
goldfish:	48 h	2.0 mg/L approx. fatal conc.	(226)

fathead minnows: static bioassay in Lake Superior water at 18-22°C: LC_{50} (1; 24; 48; 72; 96h): 24; 24; 24; 24; 24 mg/L (350)

4-aminophenol-N-acetate

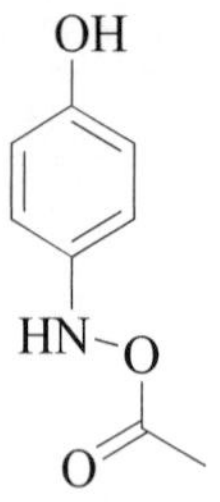

D. BIOLOGICAL EFFECTS

GREEN ALGAE: *Scenedesmus subspicatus:* inhib. of fluorescence: IC_{10}: 0.017 mmol/L.

GREEN ALGAE: *Scenedesmus subspicatus:* growth inhib.: IC_{10}: 0.17 mmol/L.

RubisCo test

inhib. of enzym. activity of Ribulose-P2-carboxylase in protoplasts: IC_{10}: 2 mmol/L.

Oxygen test

inhib. of oxygen production of protoplasts: IC_{10}: 0.25 nmol/L. (2698)

aminophenolsulfonic acid

C. WATER AND SOIL POLLUTION FACTORS

Biodegradation

adapted A.S. at 20°C-product is sole carbon source: 65% COD removal at 7 mg COD/g dry inoculum/h. (327)

N-(4-aminophenyl)aniline (4-aminodiphenylamine; N-phenyl-p-phenylenediamine; p-aminodiphenylamine)

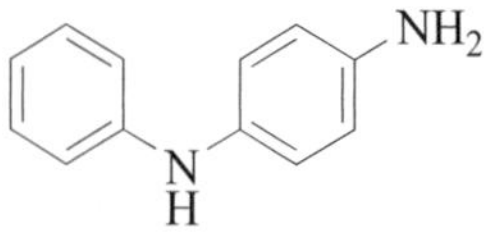

$C_{12}H_{12}N_2$

$C_6H_5NHC_6H_4NH_2$

CAS 101-54-2

USES

rubber additive

A. PROPERTIES

molecular weight 152.17; melting point 73- 75°C; boiling point 354°C; density 1.1 at 100°C; vapor pressure 7.6 x 10^{-7} hPa at 20°C, 5.7 x 10^{-5} at 50°C; solubility 500 mg/L at 20°C and pH 8.9; $LogP_{OW}$ 2.4 (calculated); H 22.8 x 10^{-5} m^3/mole at 20°C (calculated).

B. AIR POLLUTION FACTORS

1 ppm = 7.65 mg/m^3

Photochemical reactions

$t_{1/2}$: 2 h, based on reactions with OH° (calculated) (8512)

C. WATER AND SOIL POLLUTION FACTORS

Biodegradation

Inoculum	Method	Concentration	Duration	Elimination results	
A.S. adapted	DEV	2.4 mg/L	5 d	0%	
			10 d	24%	
			20 d	28%	(8513)
		24 mg/L	20 d	2%	(8513)
A.S. mixed nonadapted	Zahn-Wellens Test	100 mg DOC/L	3 h	58% DOC	
			7 d	100% DOC*	

* Elimination influenced by adsorption of the test substance at the biomass of the sludge. (8513)

Soil suspension

25 mg/L: the test substance transformed overnight to an orange-colored product characterized as chinonanilbenzene. (8514)

Mobility in soil

soil adsorption coefficient: 143 (low adsorption) (8512)

D. BIOLOGICAL EFFECTS

Toxicity

Bacteria			
Nonadapted sludge inhib. of nitrification	EC_{50}	30 mg/L	(8517)
Pseudomonas putida	30 min EC_0	31 mg/L	(8513)
Photobacterium phosphoreum	30 min EC_{50}	0.33 mg/L	(8516)
Algae			
Scenedesmus subspicatus	72 h EC_{10}	1.1; 1.8 mg/L	
(cell multiplication inhibition test)	72 h EC_{50}	2.4; 4.8 mg/L	(8513)
Crustaceans			
Daphnia magna	48 h EC_0	0.32 mg/L	
	48 h EC_{50}	0.31 mg/L	
	48 h EC_{100}	3.2 mg/L	
	48 h NOEC	0.2 mg/L	
	24 h EC_{50}	15 mg/L	(8515)
	24 h EC_0	0.32 mg/L	
	24 h EC_{50}	4.1 mg/L	
	24 h EC_{100}	31 mg/L	(8513)
	21 d NOEC	0.04 mg/L	

	21 d LOEC	0.13 mg/L	
	21 d EC_{50}	0.1 mg/L*	(8513)

* Probably conversion of test substance to p-hydroxydiphenylamine.

Fishes			
Brachydanio rerio	96 h LC_0	1.1; 7.5 mg/L	
	96 h LC_{50}	1.9 mg/L	
	96 h LC_{100}	4.4; 10 mg/L	(8513)
Mammals			
Rats	oral LD_{50}	1,000 mg/kg body wt.	
		847 mg/kg body wt.	
		720 mg/kg body wt.	

4-aminopropiophenone

4-aminopropiophenone (1-(4-aminophenyl)-1-propanone; ethyl p-aminophenyl ketone; p-aminopropiophenone; p-aminophenylpropanone)

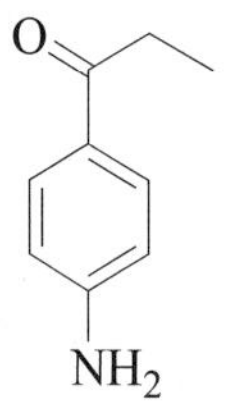

$C_9H_{11}NO$

CAS 70-69-9

SMILES C(C1=CC=C(N)C=C1)(=O)CC

USES

Chemical intermediate. Cyanide antidote.

A. PROPERTIES

molecular weight 149.19; melting point 140.1°C.

D. BIOLOGICAL EFFECTS

Bacteria			
nitrification inhibition	EC_{50}	43 mg/l.	(2624)
Nitrosomonas sp. inhibition of NH_3 oxidation	EC_{75}-EC_{100}	100 mg/L	(9790)
Mammals			
mouse, rat	oral LD_{50}	168-177 mg/kg bw	(9791, 9792)
guinea pig	oral LD_{50}	1020 mg/kg bw	(9793)

3-aminopropyldimethylamine (N,N-dimethyl-1,3-diaminopropane; dimethylamino-propylamine; 1-amino-3-(dimethylamino)propane; DMAPA)

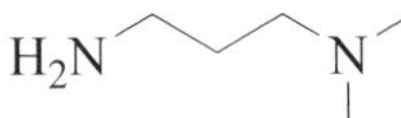

$C_5H_{14}N_2$

CAS 109-55-7

USES

as an intermediate in the production of binding agents, ion-exchange materials, flocculating agents (water treatment), cosmetic agents, washing and cleaning agents (betaines), additive for petrol and other fuels, polyurethane fibres and lubricants, dyes, agrochemicals, agents used in the photographic and textile industries, etc..DMAPA is also used directly as a hardener in epoxy resins in the plastics industry, as a cross-linking agent for cellulose fibres in the paper industry and as an anti-shrinking agent for leather.(7452)
Intermediate for the manufacture of flocculants, ion-exchange resins, betains (cosmetics), corrosion inhibitors, and additive used in the manufacture of paperware and rubber products.

A. PROPERTIES

molecular weight 102.18; melting point <60°C; boiling point 135°C; density 0.82 kg/L; vapour pressure 8 hPa at 20°C; log Pow −0.35; water solubility miscible at 20°C

B. AIR POLLUTION FACTORS

Photochemical reactions

$t_{1/2}$: 0.15 d, based on reactions with OH° (calculated). (6180)

Photodegradation

t/2 = 4.8 hours (7452)

C. WATER AND SOIL POLLUTION FACTORS

Biodegradation

Inoculum	*Method*	*Concentration*	*Duration*	*Results*	
	Closed bottle test	2-80 mg/L	20 d	0% BOD	
A.S. industrial nonadapted	Zahn-Wellens test		15 d	25% COD	(6178)
A.S. industrial nonadapted	Zahn-Wellens test	5,000 mg/L	3 h	<1% product	
			7 d	71% product	
			10 d	87% product	
			15 d	100% product	(6179)

Biodegradation

Inoculum	method	concentration	duration	elimination	
Activated sludge	Closed bottle test		20 days	65% ThOD	
Adapted A.S.	Closed Bottle Test		20 days	69% ThOD	
A.S.	Zahn-Wellens test		15 days	100% ThOD	(7452)

D. BIOLOGICAL EFFECTS

Toxicity

Micro organisms

Pseudomonas putida	17h EC_{10}	70 mg/L	
	17h EC_{50}	95 mg/L	
	17h EC_{90}	120 mg/L	(7452)
Primary municipal sludge	24 h EC_0	1,500 mg/L	(6178)
Algae			
Scenedesmus subspicatus	72h ECgr20	45.5 mg/L	
	72h $ECgr_{50}$	56 mg/L	(7452)
	72h $ECgr_{90}$	105.6 mg/L	
	96h ECgr20	37.1 mg/L	
	96h $ECgr_{50}$	57.5 mg/L	
	96h $ECgr_{90}$	110.3 mg/L	
Crustaceae			
Daphnia magna	48h EC_0	25 mg/L	
	48h EC_{50}	68.3 mg/L	
	48h EC_{100}	100 mg/L	
	24h EC_0	50 mg/L	
	24h EC_{50}	60 mg/L	
	24h EC_{100}	100 mg/L	(7452)
Fish			
Leuciscus idus	96h LC_{50}	122 mg/L	(7452)
	96h LC_0	100 mg/L	
	96h LC_{100}	140 mg/L	
Mammals			
Rat	oral LD_{50}	922; 1,870 mg/kg bw	
Mouse	Oral LD_{50}	1,500; 1,640 mg/kg bw	(7452)

3-aminopropyltriethoxysilane (1-propanamine,3-(triethoxysilyl)-; APTES; propylamine,3-(triethoxysilyl)-)

H_2N Si O O O

$C_9H_{23}NO_3Si$

CAS 919-30-2

SMILES CODE CCO[Si](CCCN)(OCC)OCC

TRADENAMES

A-1100; A-1112; AGM 9(VAN); NUCA 1100; Silane 1100; Silicone A-1100; UC-A1100

USES

The commercial uses of this material are numerous and include various applications as coupling agents and adhesive promotors in fiberglass, adhesives and sealants, foundry resins, and in pretreatment for coatings. A small percentage of this material may be found in sealants and coatings. As coupling agents and adhesion promotors, APTES is intentionally converted by hydrolysis to the trisilanols, which then bond molecularly to inorganic substrates. During hydrolysis, the ethoxygroup is liberated as ethanol. The silane modified surfaces of these inorganic substrates become incorporated within polymeric resins by a chemical reaction with the amine group. This completes the coupling process. The amino-functional silane is converted and bound within the substrate by polymer coupling. In solvent based coatings, sealants etc.., the aminofunctional silane may be present until the curing reactions take place, which is often after application. (7453)

IMPURITIES, ADDITIVES, COMPOSITION

ethylalcohol (0-1%); 2-butanone (0-2%); dibenzoyl peroxide (0-1%)

A. PROPERTIES

molecular weight 221; melting point –70°C; boiling point 223°C; relative density 0.95 at 25°C; vapour pressure 0.02 hPa at 20°C; water solubility 760,000 mg/L*; LogP_{OW} 0.31*(* these estimated values may not be applicable because the chemical is hydrolytically unstable)

B. AIR POLLUTION FACTORS

Photodegradation

the effective half life calculated at 2.4 h is even shorter due to rapid hydrolysis. The trisilanol resulting from hydrolysis in the atmosphere is not predicted to undergo direct photolysis but could react with hydroxyl radicals and ozone. (7453)

C. WATER AND SOIL POLLUTION FACTORS

Hydrolysis

The reactive nature of this material destroys the parent material in any moisture-containing environment, thus limiting environmental exposure to the silane. The parent material is rapidly hydrolyzed in a spill situation producing ethanol and trisilanols

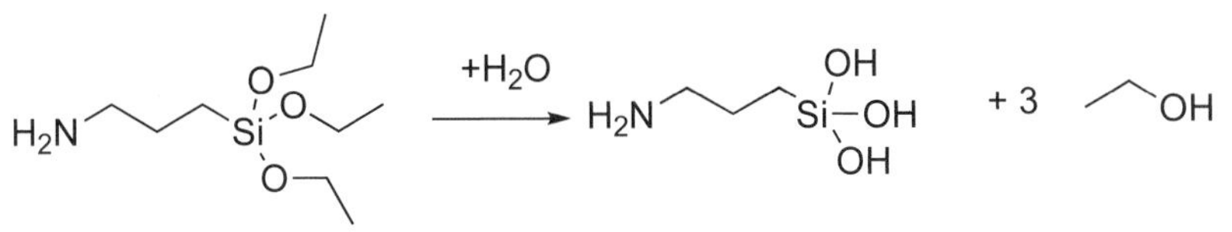

+H_2O → + 3

3-aminopropyltriethoxysilane 3-aminopropyltrisilanol ethanol

Hydrolysis

pH	half life in hours 10°C	25°C	37°C
4.7	1.0	0.4	0.2
7.0	56	8.4	3.9
9.0	0.8	0.15	0.04

The half-lives refer to the reaction to the mono-ol and the mono- and di-ol hydrolyze on a timescale similar to the silane. The Si-C bond will not further hydrolyze. The transient silanol groups will condense with other silanols to yield:

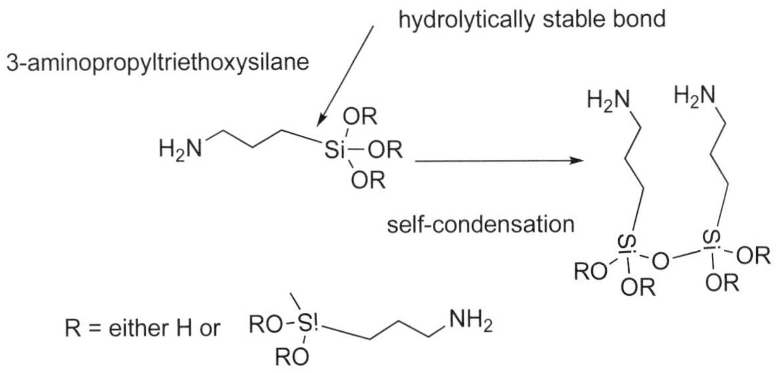

hydrolytically stable bond

3-aminopropyltriethoxysilane

self-condensation

R = either H or

In spill conditions, the concentration of the parent compound is very high. The resulting silanol concentration is also high and the silanol rapidly self-condenses to form water insoluble, resinous oligomers and polymers. The molecular weight of the resulting oligomers and polymers is predicted to be over 1000. The structure of the resulting resin is shown above. As the parent silane and the resulting silanol are diluted, it is predicted that at 1000 mg/L of a related trialkoxysilane, the equilibrium concentration will be 86% silanol monomer and 14% silanol dimer. At still lower concentrations, the silanol will exist as the uncondensed monomer. (7453)

Rapid hydrolysis generates ethanol and silanetriols.

Biodegradation

Inoculum	method	concentration	duration	elimination	
Domestic sewage	DOC Die-away test		7 days	63%	
			14 days	66%	
			28 days	67%	(7453)

D. BIOLOGICAL EFFECTS

Bioaccumulation

Bioaccumulation is not expected because the chemical is hydrolytically unstable.The water solubility of the silantriol can not be measured because of its tendency to condense at concentrations greater than 500 ppm. It is known however that the silanetriol and small condensation products will only precipitate out of water due to formation of larger, water insoluble polymeric resins. (7453)

Toxicity

Algae			
Scenedesmus subspicatus	72h ECb_{50}	603 mg/L	
	72h NOECb	1.3 mg/L	
	72h ECb_{10}	38 mg/L	(7453)
Crustaceae			
Daphnia magna	48h EC_{50}	331 mg/L	
	48h NOEC	94 mg/L	(7453)
Fish			
Brachydanio rerio	96h LC_0	>934 mg/L	(7453)
	96h NOEC		
Mammals			
Rat	oral LD_{50}	11,570-2,830; 3,980 mg/kg bw	(7453)

2-aminopyridine (α-pyridylamine; alpha-aminopyridine; o-aminopyridine; amino-2-pyridine; alpha-pyridinamine; 2-pyridinamine)

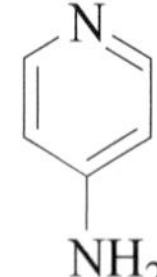

$(C_5H_4N)NH_2$

$C_5H_6N_2$

CAS 504-29-0

USES

Organic synthetic intermediate. Used in pharmaceutical manufacture, particularly antihistamines.

A. PROPERTIES

molecular weight 94.11; melting point 56°C; boiling point 204°C; vapor density 3.25; $LogP_{OW}$ -0.22.

B. AIR POLLUTION FACTORS

1 mg/m^3 = 0.26 ppm; 1 ppm = 3.91 mg/m^3.

C. WATER AND SOIL POLLUTION FACTORS

Biodegradation

adapted A.S. at 20°C,product is sole carbon source: 97% COD removal at 41 COD/g dry inoculum/h.	(327)
Of 17 mg/kg incubated in soil at pH 7 and 28°C < 1% degraded within 30 d, as evidenced via the release of inorganic nitrogen.	(9795)

D. BIOLOGICAL EFFECTS

Bacteria			
Photobacterium phosphoreum Microtox test	5-30 min EC_{50}	284 mg/L	(9657)
Algae			
Tetrahymena pyriformis	60h EC_{50}	390 mg/L	(9794)

4-aminopyridine

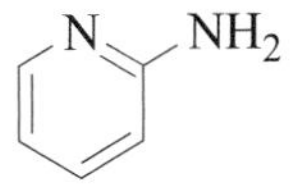

$(C_5H_4N)NH_2$

$C_5H_6N_2$

CAS 504-24-5

USES

a frightening agent for protecting grain crops from blackbirds; intermediate.

A. PROPERTIES

molecular weight 94.12; melting point 155-158°C; boiling point 273.5°C; $LogP_{OW}$ 0.28.

B. AIR POLLUTION FACTORS

Photochemical reactions

$t_{1/2}$: 8h, based on reactions with OH° (calculated)	(9795)

C. WATER AND SOIL POLLUTION FACTORS

Waste water treatment

Oxidation with ozone in aqueous solution increases with increasing pH, complete oxidation occurs at pH 9.3 in 50 min.	(9797)

Degradation in soils

degradation of 4-aminopyridine-^{14}C to $^{14}CO_2$ was negligible in soils incubated up to 2 months under anaerobic conditions; under aerobic incubation, after 3 months at 30°C and 50% moisture, evolution of $^{14}CO_2$ ranged from 0.4% for a highly acidic loam (pH 4.1) to more than 50% for a lighter-textured, alkaline, loamy sand (pH 7.8).	(1642)

D. BIOLOGICAL EFFECTS

Bacteria			
Photobacterium phosphoreum Microtox test	5-30 min EC_{50}	284 mg/L	(9657)
Algae			
Tetrahymena pyriformis	60h EC_{50}	260 mg/L	(9794)
Fishes			
bluegill	96h LC_{50},S	2.82-7.56 mg/L	
channel catfish	96h LC_{50},S	2.43-5.80 mg/L	(445)
Mammals			
rat, mouse	oral LD_{50}	20-42 mg/kg bw	(9796, 8650)

Aminopyrine (4-dimethylaminoantipyrine; dimethylaminophenazone; 1, 5-Dimethyl-4-dimethylamino-2-phenyl-3-pyrazolone)

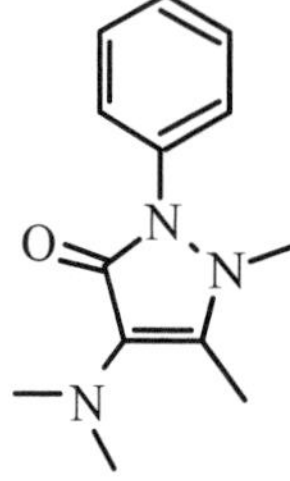

$C_{13}H_{17}N_3O$

CAS 58-15-1

TRADENAMES

Piridol; Aminopyrine

USES

Analgesic; anti-inflammatory

A. PROPERTIES

molecular weight 231.30

C. WATER AND SOIL POLLUTION FACTORS

Environmental concentrations

In influents to sewage works	removal efficiency	in effluents of sewage works
Germany: over 50 g/day	38 %	max 1000 ng/L

Environmental concentrations in Germany 2001 (10923)

	Positive samples	Min ng/L	Max ng/L	Average ng/L	Median ng/L
River Schussen	7/7	18	19	18	19
River Körsch upstream WWTP	7/8	<10	21	12	10
River Körsch downstream WWTP	8/8	14	120	38	21
Influent WWTP Stuttgart-M	6/6	23	97	58	56
Effluent WWTP Stuttgart-M	7/7	13	73	39	33
Suspended solids influent WWTP Stuttgart-M µg/kg	5/6	324	2993	1208	702
Influent WWTP Reutlingen-W	5/5	106	247	163	156
Effluent WWTP Reutlingen-W	5/5	45	243	137	137
Suspended solids influent WWTP Reutlingen-W µg/kg	5/5	434	1385	699	455
Influent WWTP Steinlach-W	5/5	23	196	111	137
Effluent WWTP Steinlach-W	5/5	54	314	129	90
Suspended solids WWTP Steinlach-W µg/kg dw	5/5	197	1266	697	751
Leachate of landfill Reutlingen-S	5/5	2137	3451	2750	2668
Leachate of landfill Dusslingen	5/5	1299	4817	3920	4764

L-α-aminosuccinamic acid *see* L-asparagine

aminothiourea *see* thiosemicarbazide

o-aminotoluene *see* *o*-toluidine

4-aminotoluene-3-sulphonic acid (benzenesulfonic acid,2-amino-5-methyl; 4-methylaniline-2-sulfonic acid; 6-amino-m-toluenesulfonic acid)

184 4-aminotoluene-3-sulphonic acid

$C_7H_9NO_3S$

CAS 88-44-8

EINECS 201-831-2

USES

The chemical is used for the synthesis of organic pigments (e.g. in Pigment Red 57 and its metal salts) in cosmetic products such as lipstick, nail polish and blush, in ink, paint and for coloring of resin, fiber, leather, paper, rubber etc.. (7449)

A. PROPERTIES

pale brown to gray solid with no distinct odour; molecular weight 187.2; melting point 312 °C; boiling point > 350°C; density 1.49 kg/L at 25°C; vapour pressure 9.5 x 10^{-10} hPa at 25°C; log Pow -0.67 at 25°C; water solubility 4,700 - 6,000 mg/L at 20°C

B. AIR POLLUTION FACTORS

Photodegradation

t/2 = 0.4 days in sunlight (7449)

C. WATER AND SOIL POLLUTION FACTORS

Hydrolysis

t/2 > says at pH 4 to 9 (7449)

Biodegradation

Inoculum	method	concentration	duration	elimination	
Activated sludge	Modified MITI Test	30-100 mg/L	14 days	0 % ThOD	(7449)

D. BIOLOGICAL EFFECTS

Bioaccumulation

Species	conc. µg/L	duration	body parts	BCF	
Cyprinus carpio	200	42 days		<4	
	2,000	42 days		<0.4	(7449)

Toxicity

Algae

Selenastrum capricornutum	72h ECb_{50}	>10 mg/L	
	72h NOECb	10 mg/L	
	72h $ECgr_{50}$	>10 mg/L	
	72h NOECgr	10 mg/L	(7449)

Crustaceae

Daphnia magna	48h EC_{50}	>10 mg/L	
	48h LC_0	>10 mg/L	
	48h NOEC	10 mg/L	(7449)
	21d NOEC	3.2 mg/L	
	21d LOEC	10 mg/L	

	21d EC_{50}	>10 mg/L	(7449)
Fish			
Oryzias latipes	96h LC_{50}	> 10 mg/L	
	96h NOEC	> 10 mg/L	(7449)
Gambusia affinis	96h NOEC	180 mg/L	
	96h LC_{50}	375 mg/L	
	24h LC_{50}	425 mg/L	
	48h LC_{50}	410 mg/L	
	6h LC_{100}	560 mg/L	(7449)
Mammals			
Rat	oral LD_{50}	>2,000; 11,700 mg/kg bw	(7449)

11-aminoundecanoic acid (11-aminoundecylic acid)

HO ... NH_2 / O

$C_{11}H_{23}NO_2$

$HOOC(CH_2)_{10}NH_2$

CAS 2432-99-7

SMILES O=C(O)CCCCCCCCCCN

EINECS 219-417-6

USES

11-aminoundecanoic acid is used exclusively as a monomer for the manufacture of polyamide 11 polymers. Polyamides 11 are used in wide-ranging applications: oil drilling pipes, brake lines for cars and heavy good vehicles, electrical cable and optical fibre sheating, medical syringues, food packaging film, sport shoe soles ets... Polyamides 11 are also used for powder anticorrosion coatings which are resistant to wear and impact.

A. PROPERTIES

White crystalline solids; molecular weight 201.31; melting point 184°C; boiling point 480°C; bulk density 550 kg/m3 at 20°C; vapour pressure 2.07 x 10^{-7} Pa at 25°C; water solubility is pH dependent, at pH 3: >20,000 mg/L, at pH: 4 3,200 mg/L, at pH 6-8: 800 mg/L at 25°C; log Pow –0.16 ; log K_{OC} 2.45 (calculated)

B. AIR POLLUTION FACTORS

1 mg/m^3 = 8.24 ppm

Photodegradation

T/2 = 4.3 hours based on reaction with OH radicals.

C. WATER AND SOIL POLLUTION FACTORS

Biodegradation

Inoculum	method	concentration	duration	elimination

Secondary effluent	OECD 301B	22.5 mg/L	1 day	1% ThCO2	
			4 days	22% ThCO2	
			7 days	45% ThCO2	
			12 days	65% ThCO2	
			19 days	77% ThCO2	(10705)

D. BIOLOGICAL EFFECTS

Toxicity

Micro organisms			
Pseudomonas putida	16h EC_{10}	0.34 mg/L	
	16h EC_{50}	1.5 mg/L	(10705)
Algae			
Pseudokirchneriella subcapitata	72h $ECgr_{50}$	53 mg/L	
	72h ECb_{50}	23 mg/L	(10705)
	72h NOECb	4.5 mg/L	
	72h NOECgr	4.5 mg/L	
Crustaceae			
Daphnia magna	48h EC_{50}	>350 mg/L	
	48h EC_{100}	>350 mg/L	(10705)
Fish			
Brachydanio rerio	96h LC_{50}	>833 mg/L	(10705)
Mammals			
Rat	oral LD_{50}	> 14,700; >15,000 mg/kg bw	(10705)

amitriptyline

amitriptyline (5-(3'-dimethylaminopropylidene)-dibenzo (a,d)(1,4) cycloheptadiene; 1-propanamine, 3-(10,11-dihydro-5H-dibenzo(a,d)cyclohepten-5-ylidene, N,N-dimethyl-; Triptisol)

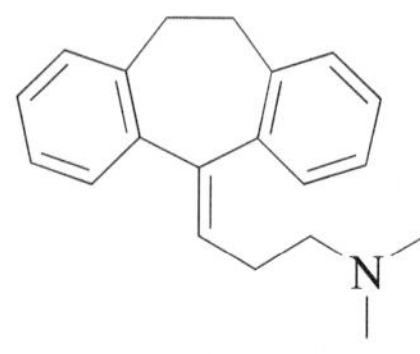

$C_{20}H_{23}N$

CAS 50-48-6

USES

Anti-depressant

A. PROPERTIES

molecular weight 277.44; melting point 196°C; log P_{OW} 4.92.

D. BIOLOGICAL EFFECTS

Toxicity

Micro organisms			
Microtox™ *(Photobacterium)* test	5 min EC_{50}	77.8 *µmol/L*	(2945)
Crustaceae			
Artoxkit M *(Artemia salina)* test	24h LC_{50}	133 *µmol/L*	(2945)

Daphnia magna	24h EC_{50}	4.93 mg/L	
	24h LC_{50}	20 µ*mol/L*	
Daphnia pulex	24h EC_{50}	1.0 mg/L	(9798)
Streptoxkit F *(Streptocephalus proboscideus)* test	24h LC_{50}	2.8 µ*mol/L*	(2945)
Rotoxkit F *(Brachionus calyciflorus)* test	24h LC_{50}	2.9 µ*mol/L*	(2945)

amitriptyline hydrochloride (3-(10,11-Dihydro-5H-dibenzo[a,d]cyclohepten-5-ylidene-N,N-dimethyl-1-propananim-hydrochloride)

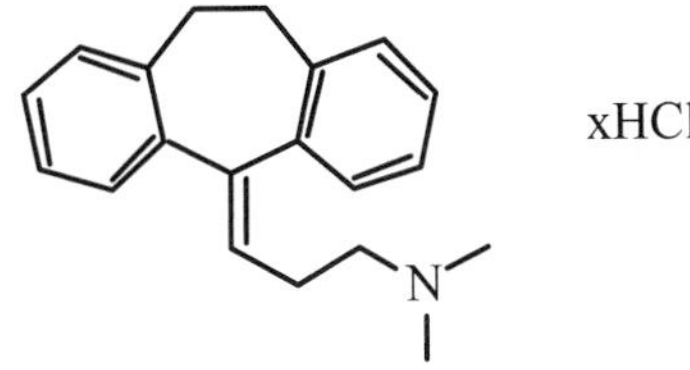

xHCl

$C_{20}H_{23}N.HCl$

CAS 5494-18-8

TRADENAMES

Adepril; Amavil; Amiprin; Amitriptyline chloride; Amyzol; Daprimen; Elavil hydrochloride; Lentizol; Novotriptyn; Rantoron; Sylvemid; Triavil; Triptizol; Trynol

Adepril; Amineurin; Domical; Elavil; Endep; Euplit; Laroxyl; Lentizol; Miketorin; Redomex; Saroten; Sarotex; Sylvemid; Tryptanol; Tryptizol

USES

Antidepressant

IMPURITIES, ADDITIVES, COMPOSITION

Product marketed as hydrochloride (7209)

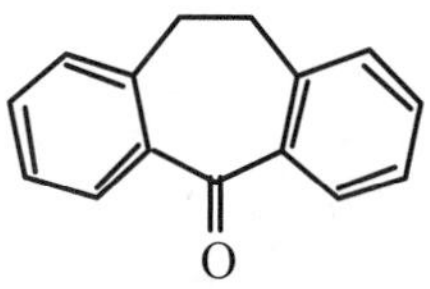

dibenzosuberone

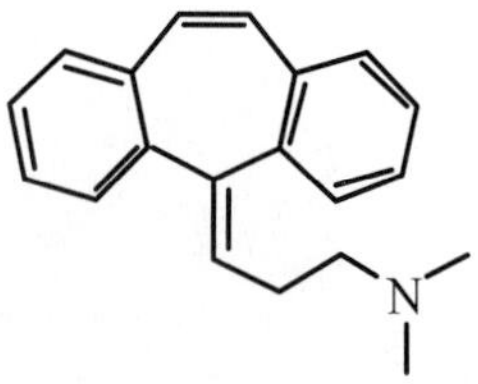

cyclobenzaprine hydrochloride

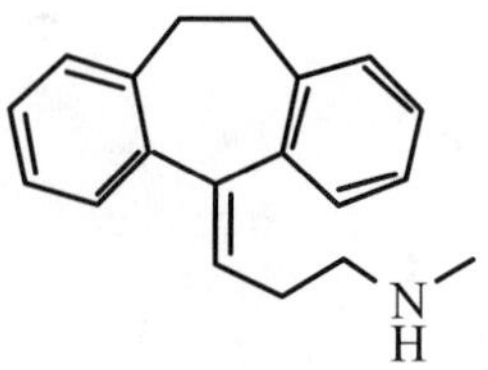

xHCl

nortryptyline

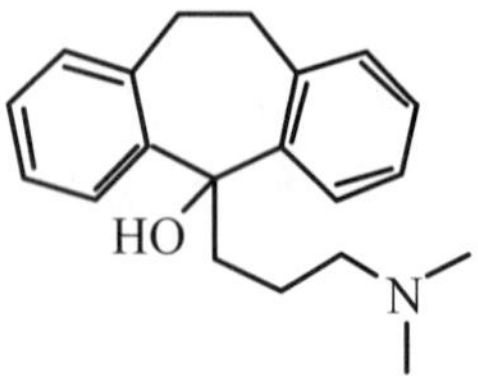

xHCl

5-(3-dimethylaminopropyl)-10-11-dihydro-5H-dibenzo[a,d]cyclohepten-5-ol

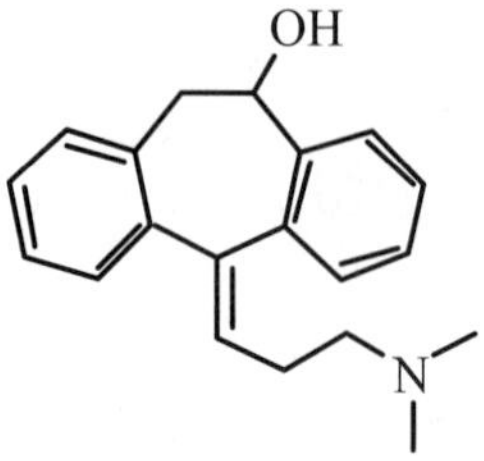

(RS)-5-(3-dimethylaminopropyliden)-10,11-dihydro-5H-dibenzo[a,d]cyclohepten-10-ol

A. PROPERTIES

Molecular weigth 313.87; melting point 196-197 °C; solubility freely

C. WATER AND SOIL POLLUTION FACTORS

Environmental concentrations

In Sante Fe River below WWTP: august 2000: 30 ng/L
In Rio Grande : august 2000: 30 ng/L (7152)

D. BIOLOGICAL EFFECTS

Micro organisms			
Vibrio fischeri	5min EC_{50}	21.5 mg/L	(7237)
Crustaceae			
Artemia salina	24h EC_{50}	36.6 mg/L	
Daphnia magna	24h EC_{50}	4.93 mg/L	
Streptocephalus probiscideus	24h LC_{50}	0.76 mg/L	
Brachionus calyciflorus	24h EC_{50}	0.80 mg/L	(7237)
Mammals			
mice and rats	oral LD_{50}	140; 240; 350; 380 mg/kg bw	(7244)

amitrole *see* 3-amino-1,2,4-triazole

ammonia

NH_3

CAS 7664-41-7

USES

intermediate, especially for the prod. of fertilizer.

A. PROPERTIES

colorless gas liquefied by compression; molecular weight 17.03; melting point -77.7°C; boiling point -33.4°C; vapor pressure 10 atm. at 25.7°C, 8.7 atm. at 20°C, 8,570 hPa at 20°C, 20,340 hPa at 50°C; vapor density 0.6; density 0.817 at 79°C; solubility 895,000 mg/L at 0°C, 531,000 mg/L at 20°C, 440,000 mg/L at 28°C; THC 91.5 kcal/mole, LHC 75.7 kcal/mole; $LogP_{OW}$ -1.14 (measured).

B. AIR POLLUTION FACTORS

1 mg/m^3 = 1.414 ppm, 1 ppm = 0.707 mg/m^3.

Odor

hedonic tone: extremely pungent.

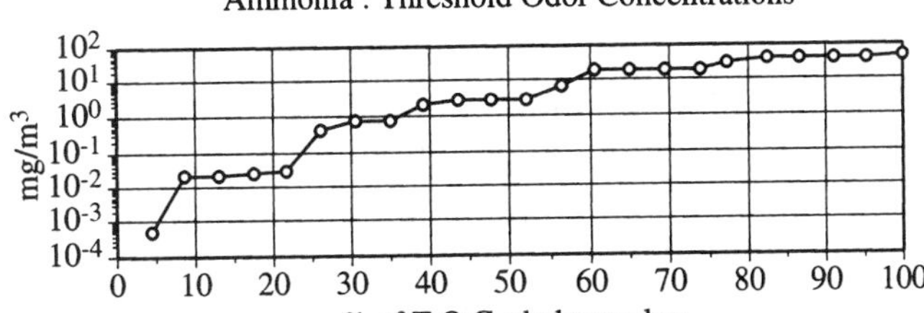

(2, 5, 10, 73, 210, 279, 629, 652, 658, 664, 670, 674, 741, 800, 816, 821, 857)

U.S.S.R.:	human odor perception:	nonperception:	0.4 mg/m^3	
		perception:	0.5 mg/m^3	
	human reflex response:	no response:	0.22 mg/m^3	
		adverse response:	0.35 mg/m^3	
	animal chronic exposure:	no effect:	0.2 mg/m^3	
		adverse effect:	2.0 mg/m^3	(170)

Manmade sources

Combustion sources:	*Amount of emission*	
coal	2 lb/ton	
fuel oil	1 lb/1,000 gal	
natural gas	0.3 to 0.56 lb/106 cu ft	
butane	1.7 lb/106 cu ft	
propane	1.3 lb/106 cu ft	
wood	2.4 lb/ton	
forest fires	0.3 lb/ton	(111)

Ammonia discharged daily in metropolitan area of 100,000 persons using each heating system:
domestic heating fuel:
coal: 2,000 lb NH_3
oil: 800 lb NH_3

gas: 0.3 lb NH_3

downtown Tokyo: Jan.-May 1969: glc's: 20 to 152 μg/m^3 Nessler's procedure, 4.0 to 25.8 μg/m^3 pyridine-pyrazolone procedure. (205)

C. WATER AND SOIL POLLUTION FACTORS

Impact on biodegradation processes

2.0 mg/L affects the self-purification of water courses. (181)

Reduction of amenities

faint odor: 0.037 mg/L. (129)

Waste water treatment

ammonia is oxidized by ozone, the reaction is first order with respect to the conc. of ammonia and is catalyzed by OH^- over the pH range 7-9.

D. BIOLOGICAL EFFECTS

Oligosaprobic and mesosaprobic organisms: toxic: 0.08 to 0.4 mg/L. (30)

Bacteria			
Photobacterium phosphoreum	toxic	5.2 mg/L (pH 9.7)	(5329)
	toxic	~3,600 mg/L (pH 7.0)	(5329)
Algae			
Phytoplankton and zooplankton	toxic	2.5-2.8 mg/L	(5327)
Crustaceans			
Daphnia magna	48h LC_{50}	24 mg/L	(5325)
	48h LC_{50},S	189 mg/L	(5325)
Daphnia pulex	48h LC_{50},S	187 mg/L	(5326)
Ceriodaphnia reticulata	48h LC_{50},S	131 mg/L	(5326)
Simochephalus vetulus	48h LC_{50}	123 mg/L	(5326)

Fishes

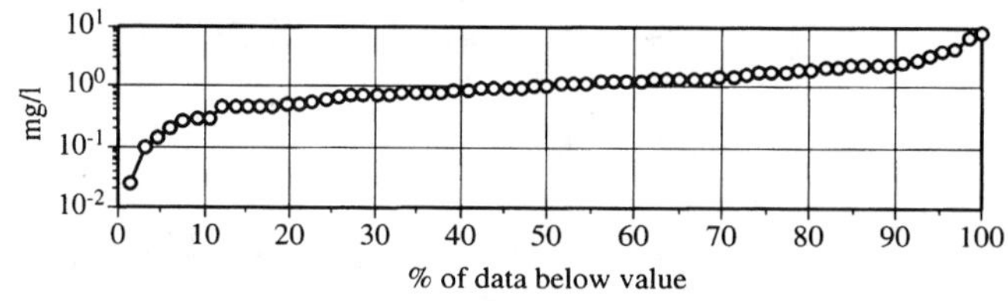

Frequency distribution of 24-96h LC_{50} values for fishes (n = 66) based on data from this and other works (3999)

fathead minnows	96h LC_{50}	8.2 mg/L (hard water)	(41)
goldfish	24-96h LC_{50}	2-2.5 mg/L	(154)
coho salmon	96h LC_{50},F	0.45 mg/L	(1505)
guppy fry	72h LC_{50},S	74 mg/L	(1503)
guppy fry	72h LC_{50},S	1.26 mg/L	(1503)
cutthroat trout *(Salmo clarki)* fry	96h LC_{50},F	0.5-0.8 mg/L	(1503)
	36d LC_{50},F	0.56 mg/L	(1504)
rainbow trout: fertilized egg	24h LC_{50},S	>3.58 mg/L	
alevins (0-50 d old)	24h LC_{50},S	>3.58 mg/L	
fry (85 d old)	24h LC_{50},S	0.068 mg/L	
adults	24h LC_{50},S	0.097 mg/L	(951)

walking catfish	48h LC_{50},S	0.28 mg/L	(948)
Salmo aguabonita	96h LC_{50}	0.76 mg/L	(5273)
Salmo trutta	18h LC_{50}	>0.15 mg/L	(5274)
	96h LC_0	0.6-0.7; 0.09 mg/L	(5274, 5276)
Oncorhynchus tschawytscha	96h LC_{50}	0.47 mg/L	(5276)
Salvelinus fontinalis	96h LC_{50}	0.96-1.05 mg/L	(5276)
Proposium williamsoni	96h LC_{50}	0.47 mg/L	(5276)
Catostomus platyrhynchos	96h LC_{50}	0.67-0.82 mg/L	(5276)
Salmo trutta	96h LC_{50}	0.47 mg/L	(5277)
Salvelinus fontinalis	1.8h LC_{50}	>3.2 mg/L	(5278)
Oncorhynchus gorbuscha (late alevins)	96h LC_{50}	0.083 mg/L	(5279)
eyed embryos	96h LC_{50}	>1.5 mg/L	(5279)
Oncorhynchus kisutch	96h LC_{50}	0.55 mg/L	(5280)
	48h LC_{50}	0.5 mg/L	(5302)
Salmo salar	96h LC_{50}	0.28 mg/L	(5281)
Pimephales promelas	96h LC_{50}	0.75-3.4; 0.73-2.3 mg/L	(5282, 5283, 5284)
Catostomus commersoni	96h LC_{50}	0.79; 1.35-1.4 mg/L	(5285, 5284)
Lepomis macrochirus	96h LC_{50}	0.26-4.6 mg/L	(5286, 5287)
	48h LC_{50}	0.024-0.93; 2.3 mg/L	(5283, 5292)
Ictalurus punctatus	48h LC_{50}	1.2-2.0; 2.3 mg/L	(5283, 5295)
	96h LC_{50}	0.21; 1.5-4.2 mg/L	(5293, 9294, 5296)
	1w LC_{50}	0.97-2 mg/L	(5296)
Micropterus dolomieui	96h LC_{50}	0.7-1.8 mg/L	(5289)
Micropterus salmoides	96h LC_{50}	>0.21; 1-1.7 mg/L	(5289, 5296)
Carassius auratus	24h LC_{50}	7.2 mg/L	(5289)
Etheostoma spectabile	96h LC_{50}	0.9-1.1 mg/L	(5296)
Notropis lutrensis	96h LC_{50}	0.9-1.1 mg/L	(5296)
Stizostedion vitreum	96h LC_{50}	0.85 mg/L	(5297)
Gambusia affinis	17h LC_{50}	1.3 mg/L	(5299)
Poecilia reticulata	96h LC_{50}	1.5; >1.5 mg/L	(5300, 5301)
Sciaenops ocellatus	96h LC_{50}	0.47 mg/L	(5303)
Mugli cephalus	96h LC_{50}	1.2-2.4 mg/L	(5304)
Monacanthus hispidus	96h LC_{50}	0.69 mg/L	(5304)
Morone americana	96h LC_{50}	0.52-2.13 mg/L	(5305)
Notropis spilopterus	96h LC_{50}	1.2-1.6; 1.35 mg/L	(5306, 5284)
Notropis whipplei	96h LC_{50}	1.25 mg/L	(5284)
Notemigonus crysoleucas	96h LC_{50}	0.72; 1.2 mg/L	(5307, 5284)
Campostoma anomalum	96h LC_{50}	1.7 mg/L	(5284)
Lepomis cyanellus	96h LC_{50}	0.6-2.1 mg/L	(5308, 5309)
Lepomis gibbosus	96h LC_{50}	0.14-0.86 mg/L	(5311)
Cottus bairdi	96h LC_{50}	1.39 mg/L	(5312)
Cyprinus carpio	96h LC_{50}	1.1 mg/L	(5314, 5315)
	18h LC_{50}	>0.24 mg/L	(5316,

			5317)
Tinca tinca	20h LC_{100}	2.5 mg/L	(5320)
Semolitus atromaculatus	24h LC_{100}	0.26; 1.2 mg/L NH_4OH solution	(5321)
Tilapia aurea	72h LC_{50}	2.85 mg/L	(5322)

ammonium salts, quaternary *see* arquad

ammonium thiolactate (propanoic acid,2-mercapto-,monoammonium salt; lactic acid,thio-monoammonium salt; ammonium-2-mercaptopropionate)

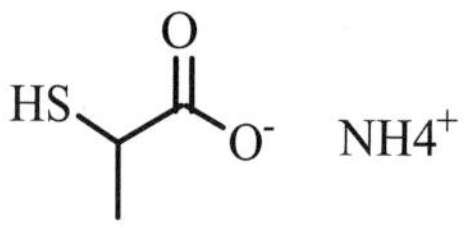

$CH_3CH(SH)C(O)ONH_4$

$C_3H_9O_2SN$

CAS 13419-67-5

USES

Cosmetics, the chemical is used as an ingredient in a permanent waving solution for human hair. At hairdressing salons, the concentrate product will be mixed with lotion to give a final concentration of <11% chemical that is applied to hair. Cosmetic concentrate products contain 35% of the chemical.

IMPURITIES, ADDITIVES, COMPOSITION

30-40% water

A. PROPERTIES

waterwhite to faint pink liquid; molecular weight 123.17; boiling point 100-105°C(water); density 1.15 kg/L; vapour pressure 1.5 kPa at 25 °C; water solubility highly soluble; $LogP_{OW}$ 1.18 (calculated)

C. WATER AND SOIL POLLUTION FACTORS

Hydrolysis

the chemical contains no bonds which are susceptible to hydrolysis under the environmental pH region where 4<pH<9, and so it is expected to be stable. (7116)

D. BIOLOGICAL EFFECTS

Toxicity

Mammals	Rat: acute oral LD_{50}: 1518 mg/kg bw	(7116)

ammoniumacetate (acetic acid ammonium salt)

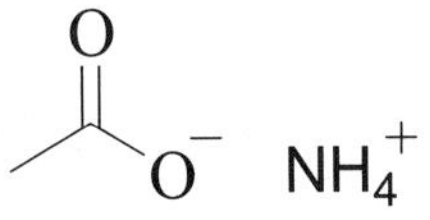

CH_3COONH_4

$C_2H_7NO_2$

CAS 631-61-8

USES

drugs; textile dyeing; foam rubbers; vinyl plastics.

A. PROPERTIES

molecular weight 77.08; melting point 114°C; density 1.07.

C. WATER AND SOIL POLLUTION FACTORS

Waste water treatment

A.S., Resp., BOD, 20°C, 1-5 d observed, feed: 1,000 mg/L; 1 d acclimation: 79% ThOD. (93)

D. BIOLOGICAL EFFECTS

Fishes

mosquito fish:	24-96h LC_{50}	238 mg/L	(41)

ammoniumbenzoate

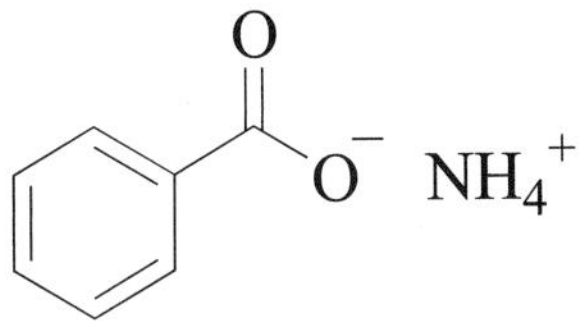

$C_6H_5CO(O)NH_4$

$C_7H_9NO_2$

CAS 1863-63-4

USES

medicine, latex preservative.

A. PROPERTIES

white crystals or powder; sublimes at 160°C; decomposes at 198°C; density 1.26; solubility 196,000 mg/L at 14°C.

C. WATER AND SOIL POLLUTION FACTORS

Waste water treatment

biodegradable dissolved organic carbon test at 21°C in continuously circulating biofilm reactor: 50% removal of DOC at: 550 mg/L DOC: 60 h; 10 mg/L DOC: 20 h (2803)

ammoniumcarbazotate *see* ammoniumpicrate

ammoniumchloride

H_4NCl

NH_4Cl

H_4ClN

CAS 12125-02-9

USES

dry batteries; soldering; manufacture of various ammonia compounds, fertilizer; electroplating.

A. PROPERTIES

white crystals; molecular weight 53.49; sublimes at 350°C; vapor pressure 3.6 x 10^{-2} mm Hg; density 1.54; solubility 283,000 mg/L at 26°C .

D. BIOLOGICAL EFFECTS

Crustaceans			
Daphnia magna	24h LC_{50}	202 mg/L	
	48h LC_{50}	161 mg/L	
	72h LC_{50}	67 mg/L	
	96h LC_{50}	50 mg/L	(153)
	100h LC_{50}	139 mg/L	(1295)
Molluscs			
Snail egg *(Lymnaea* sp.)	24h LC_{50}	241 mg/L	
	48h LC_{50}	173 mg/L	
	72h LC_{50}	73 mg/L	
	96h LC_{50}	70 mg/L	(153)

Fishes

Lepomis macrochirus	24-96h LC_{50}	725 mg/L	
Carassius carassius	24h LC_{50}	640 mg/L	(1295)

ammoniumfluoride

H_4N-F

NH_4F

H_4FN

CAS 12125-01-8

USES

fluorides; antiseptic in brewing; wood preservation.

A. PROPERTIES

white crystals; molecular weight 37.04; density 1.31; decomposed by heat; melting point 125.6°C; solubility 453,000 mg/L at 25°C.

D. BIOLOGICAL EFFECTS

	24h LC_{50}	*48h LC_{50}*	*96h LC_{50}*	
Fish				
fathead minnows*(Pimephales promelas)*	438 mg/L	417 mg/L	364 mg/L	
Crustacean				
grass shrimp*(Palaemonetes pugio)*	160 mg/L	93 mg/L	75 mg/L	(1904)

ammoniumoxalate

H_2N–O–C(=O)–C(=O)–O–NH_2

$(COONH_2)_2$

$C_2H_8N_2O_4$

CAS 6009-70-7

USES

analytical chemistry, safety explosives.

A. PROPERTIES

colorless crystals; soluble in water; density 1.5; decomposed by heat.

C. WATER AND SOIL POLLUTION FACTORS

Biodegradation

adapted A.S.-product is sole carbon source: 92% removal at 9 mg COD/g dry inoculum/h. (327)

ammoniumpicrate (ammoniumcarbazotate; ammoniumpicronitrate; trinitrophenol,ammonium salt)

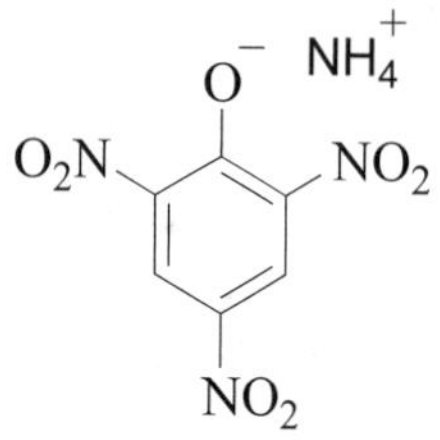

$C_6H_6N_4O_7$

CAS 131-74-8

USES

explosive, medicine.

A. PROPERTIES

yellow crystals; molecular weight 246.16; density 1.72; melting point decomposes; solubility 11,000 mg/L at 20°C.

D. BIOLOGICAL EFFECTS

Fishes

Lepomis macrochirus: static bioassay in freshwater at 23°C: 96h LC_{50}: 220 mg/L (352)

Menidia beryllina: static bioassay in synthetic seawater at 23°C: 96h LC_{50}: 66 mg/L (352)

ammoniumpicronitrate *see* ammoniumpicrate

Ammoniumpolyphosphate (APP)

$$HO\left[\begin{array}{c} O \\ \| \\ P-O \\ | \\ ONH_4 \end{array}\right]_n H$$

$(H_6NO_4P)_n$

CAS 68333-79-9

USES

This flame retardant, primarily used for polyurethane and intumescent coatings, is currently applied as liquid solution, mixed with some % of carbamide.

A. PROPERTIES

Crystalline solid; molecular weight ca 100.000 (n= ca 100); vapour pressure < 10Pa at 20°C; water solubility 10,000 mg/L

B. AIR POLLUTION FACTORS

Fire

Nitrogen oxide and ammonia in various concentration relatioships as well as phosphorous oxide are formed from APP-containing plastics in case of fire and during waste combustion.

D. BIOLOGICAL EFFECTS

Metabolism

APP is metabolised into ammonia and phosphate.

Ammoniumpropionate

$CH_3CH_2C(=O)O^- \; ^+NH_4$

$C_3H_5O_2NH_4$

CAS 17496-08-1

EINECS 241-503-7

TRADENAMES

Luprosil : 58-60% ammoniumpropionate, 15% propyleneglycol and 35-37% water.

A. PROPERTIES

molecular weight 91.11; freezing point −20°C; boiling point 115°C; density 1.04 kg/L at 20°C; vapour pressure ca 50 hPa at 50 °C; water solubility miscible

B. AIR POLLUTION FACTORS

Photolysis

not measurable (7380)

C. WATER AND SOIL POLLUTION FACTORS

Biodegradation

Ammonium propionate can be degraded by various aerobic and anaerobic micro-organisms. The aerobic degradation pathway leads to methylmalonyl-CoA and further to succinate. Another pathway runs via acrylyl-CoA and lactyl-CoA to pyruvate and further according to the citrate cyclus (7381)

Inoculum	method	concentration	duration	elimination	
	Aerobic	1068 mg COD/L	3 hours	35% COD	
			1 day	97% COD	
			2 days	99% COD	
			5 days	100% COD	(7382)
BOD_5	711 mg/L; 87 % of COD				(7380)
COD	816 mg/L				(7380)

D. BIOLOGICAL EFFECTS

Toxicity

Micro organisms			
Pseudomonas putida	17h EC_{10}	580 mg/L*	
	17h EC_{50}	890 mg/L*	
	17h EC_{90}	1200 mg/L*	(7380)
Fishes			
Leuciscus idus	96h LC_{50}	> 562 mg/L*	
Salmo gairdneri	96h LC_{50}	1500-2200 mg/L*	(7380)
Mammals			
Rat	oral LD_{50}	2000-5000 mg/kg bw*	(7380)

* Luprosil NC

ammoniumsulfate (sulfuric acid diammonium salt; diammonium sulfate)

$$O_2S(O^-)_2 \; 2NH_4^+$$

$(NH_4)_2SO_4$

$H_8N_2O_4S$

CAS 7783-20-2

USES

fertilizers; fermentation; viscose rayon; tanning; food additive.

A. PROPERTIES

brownish-gray to white crystals; molecular weight 132.13; melting point 513°C with decomposition; density 1.77; solubility 754 g/L at 20°C and pH 5; $LogP_{OW}$ -5.1 (measured).

D. BIOLOGICAL EFFECTS

Algae				
Gymnodinium splendens		3h EC,S	50; 100 µg/L	(5668)
Gymnodinium splendens + *Gonyaulax polyedra*		17d EC,S	100; 150 µg/L	(5668)
Crustaceans				
Daphnia magna:		25h LC_{50}	432 mg/L	
		50h LC_{50}	433 mg/L	
		100h LC_{50}	292 mg/L	(1295)
	pH 7.8-8.1	48h LC_{50},S	14 mg/L	(5664)
	pH 6.5-8.5	96h LC_{50},S	>100 mg/L	(5665)
Gammarus fasciatus	pH 6.5-8.5	96h LC_{50}	>100 mg/L	(5665)
Asellus intermedius	pH 6.5-8.5	96h LC_{50}	>100 mg/L	(5665)
Crangon crangon		96h IC_{50}	81-130 mg/L	(5667)

Fishes

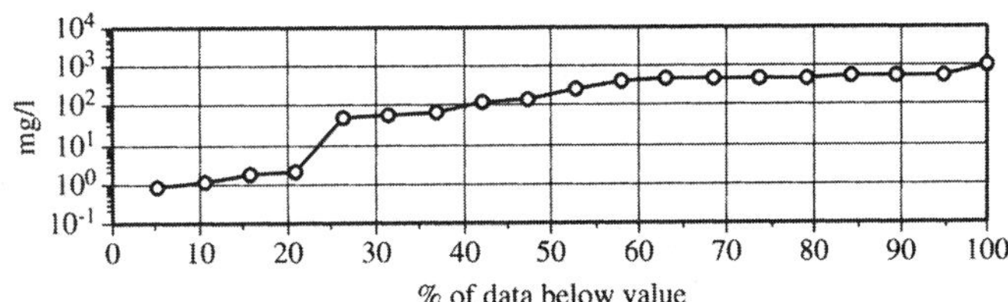

ammoniumsulfate fish 24–96h LC_{50} values (n = 19)

Poecilia reticulata		4d LC_{50}	126 mg/L	(2625)
Leuciscus idus		96h LC_{50}	460-1,000 mg/L	(5652)
Brachydanio rerio		96h LC_{50},F	250; 480 mg/L	(5653, 5654)
		24h LC_{50},F	600 mg/L	
		48h LC_{50},F	550 mg/L	(5654)
		96h LC_{50},S	420 mg/L	
		48h LC_{50},S	480 mg/L	
		24h LC_{50},S	520 mg/L	(5654)
Ictalurus punctatus	pH 6	24h LC_{50}	0.8 mg/L	(5655)
	pH 7.2	24h LC_{50}	1.2 mg/L	
	pH 8	24h LC_{50}	1.8 mg/L	
	pH 8.8	24h LC_{50}	2.2 mg/L	(5655)
Poecilia reticulata	pH 7.5-7.8	120h LC_{50}	395 mg/L	
	pH 6-6.5	120h LC_{50}	608 mg/L	(5656)
Alburnus alburnus		96h LC_{50}	592 mg/L	(5658)
Pimephales promelas		96h LC_{50}	>100 mg/L	(5659)
Labeo rohita		96h LC_{50}	67 mg/L	(5660)
Catla catla		96h LC_{50}	48 mg/L	(5660)
Cirrhinus mrigala		96h LC_{50}	62 mg/L	(5660)
Cyprinus carpio (see also graph)		96h LC_{50}	141 mg/L	(5660)
hatchling		96h LC_{50}	77 mg/L	(5660)
	egg at 20°C	96h LC_{50}	70 mg/L	(5661)
	egg at 24°C	96h LC_{50}	67 mg/L	(5661)
	egg at 28°C	96h LC_{50}	60 mg/L	(5661)

egg at 32°C	96h LC_{50}	18 mg/L	(5661)
egg at 36°C	96h LC_{50}	23 mg/L	(5661)
spawn at 20°C	96h LC_{50}	101 mg/L	(5661)
spawn at 24°C	96h LC_{50}	124 mg/L	(5661)
spawn at 28°C	96h LC_{50}	78 mg/L	(5661)
spawn at 32°C	96h LC_{50}	52 mg/L	(5661)
spawn at 36°C	96h LC_{50}	48 mg/L	(5661)
fry at 20°C	96h LC_{50}	120 mg/L	(5661)
fry at 24°C	96h LC_{50}	140 mg/L	(5661)
fry at 28°C	96h LC_{50}	93 mg/L	(5661)
fry at 32°C	96h LC_{50}	121 mg/L	(5661)
fry at 36°C	96h LC_{50}	45 mg/L	(5661)

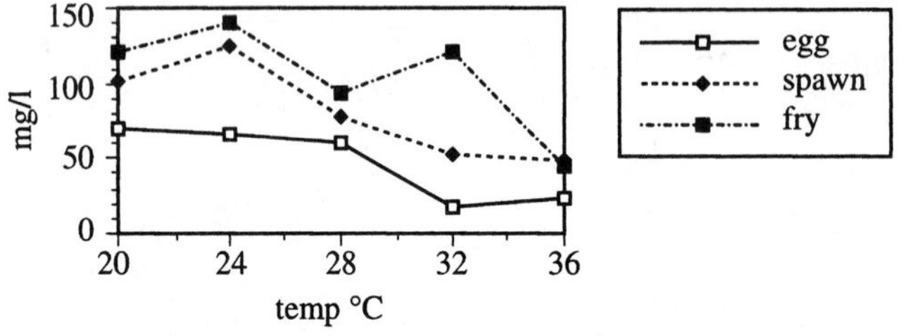

Ammonium Sulfate: Influence of Temperature on 96h LC_{50} for *Cyprinus carpio*

Barbarus ambassis	24h LC_{50}	566 mg/L	(5662)
	48h LC_{50}	546 mg/L	(5662)
Tilapia mossambica	96h LC_{50}	49; 50 mg/L	(5660)

ammoniumsulfite

$$O{=}S(O^-)(O^-)\quad NH_4^+ \quad NH_4^+$$

$(NH_4)_2SO_3 \bullet H_2O$

$H_8N_2O_3S \bullet H_2O$

CAS 10196-04-0

USES

chemical intermediate; medicine. In photography. As a reducing agent. In bricks for blast furnace linings. In lubricants for metal cold working.

A. PROPERTIES

colorless crystals; acrid; sulfurous taste; molecular weight 116.14; melting point 60-70°C decomposes; boiling point sublimes at 150°C with decomposition; density 1.41; solubility 324,000 mg/l at 0°C.

D. BIOLOGICAL EFFECTS

Crustaceans			
Daphnia magna	25h LC_{50}	299 mg/l	
	50h LC_{50}	273 mg/l	
	100h LC_{50}	203 mg/l	(1295)
Fishes			
mosquito fish	48h LC_{50}	240 mg/l	(8142)

ammoniumthiocyanate (thiocyanic acid, ammonium salt; ammonium sulfocyanide)

$N\equiv C-S^{-}\ NH_4^{+}$

CH_4N_2S

CAS 1762-95-4

USES AND FORMULATIONS

Corrosion inhibitor; stabilizer for photodevelopping fluids; additive in herbicides.

OCCURRENCE

Component of many edible plants, such as cauliflower 30- 90 mg/kg and radish 7 mg/kg.

A. PROPERTIES

molecular weight 76.12; melting point 150°C; boiling point decomposes from 70°C onward; vapor pressure <1 hPa at 20°C; density 1.3 at 20°C; solubility 1,600,000 mg/L at 20°C.

C. WATER AND SOIL POLLUTION FACTORS

Biodegradation in soil

Soil microorganisms degraded ammonium thiocyanate to thiocyanate, ammonia, nitrate, and sulfate in a few months. (8143)

D. BIOLOGICAL EFFECTS

Toxicity

Bacteria			
Pseudomonas putida (cell multiplication inhibition test)	17 h EC_{10}	8,000 mg/L	(8140)
Fishes			
Salmo trutta	LC_0	100 mg/L	
	LC_{100}	200 mg/L	(8141)
Lepomis macrochirus	1 h LC_{100}	280- 380 mg/L	(8142)

amoxycilline (α-amino-p-hydroxybenzylpenicilline; (-)-6-(2-amino-2-(p-hydroxyphenyl)acetamido)-3,3-dimethyl-7-oxo-4-thia-1-azabicyclo-(3.2.0)heptane-2-carboxylic acid)

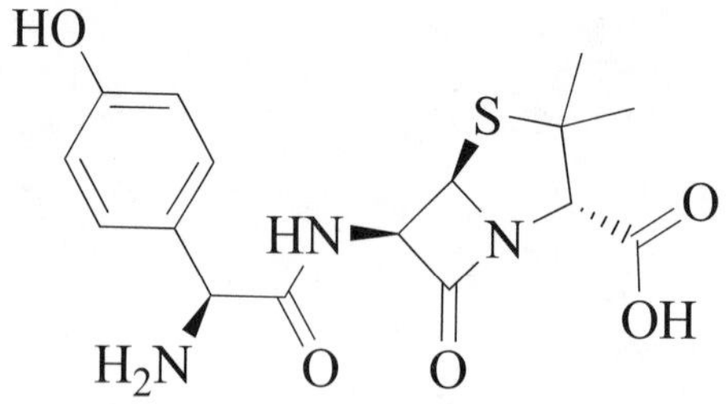

$C_{16}H_{19}N_3O_5S$

$C_{16}H_{18}N_3O_5SNa$ (amoxycilline sodium)

$C_{16}H_{19}N_3O_5S.3H_2O$ (amoxycilline trihydrate)

CAS 61336-70-7

CAS 34642-77-8(amoxycilline sodium)

CAS 26787-78-0 (amoxycilline trihydrate)

EINECS 248-003-8

TRADENAMES

Amoxil, Penamox, Moxlin;amosine, amoxa, amoxapen, amoxi, amoxibiotic, amoxicillin, amoxisol, amoxypen, ardine, biomox, bristamox, delacillin, histocillin, sumox

USES

Amoxycillin is a commonly used penicillin that works against a wide range of bacteria. It is in the group called beta-lactams. Amoxycillin (and the others in the beta-lactam group) works by stopping the making of the cell wall of susceptible bacteria, causing the bacteria to die; pharmaceutical; semisynthetic broad spectrum penicillin;

A. PROPERTIES

white or almost white powder; molecular weight: 365.4; 387.4 (sodium), 419.5 (trihydrate); water solubility 4,000 mg/L at 20°C

C. WATER AND SOIL POLLUTION FACTORS

	concentration µg/L	n	
in effluent hospital	201	estimated	(7237)

D. BIOLOGICAL EFFECTS

Bacteria			
Salmonella typhimurium	30min LOEC	20 mg/L	(7226, 10793)
Algae			
Microcystis aeruginosa	7d EC_{50}	0.0037 mg/L	
Selenastrum capricornutum	72h NOEC	250 mg/L	
Rhodomonas salina	72h EC_{50}	3108 mg/L	(7237, 10793)
Crustaceae			
Artemia salina	24h LC_{50}	477 mg/L	

	48h LC_{50}	308 mg/L
	48h LC_{50}	308 mg/L
Mammals		
rat	Oral LD_{50}	>15,000 mg/kg bw

amphetamine sulfate (α-methylphenethylamine sulfate)

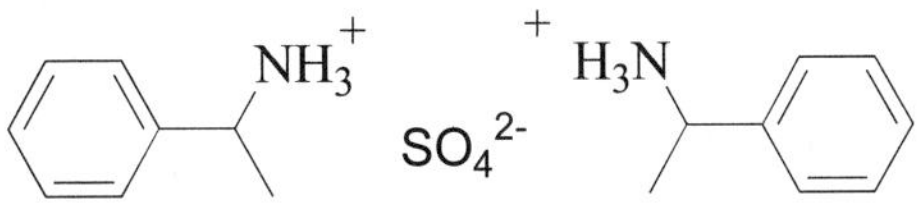

$C_{18}H_{28}N_2O_4S$

CAS 60-13-9

USES

medicine.

A. PROPERTIES

molecular weight 368.5.

D. BIOLOGICAL EFFECTS

Acute toxicity tests

		µmol/L	
Marine tests			
Microtox™ *(Photobacterium)* test	5 min EC_{50}	not tested	
Artoxkit M *(Artemia salina)* test	24h LC_{50}	4,110	
Freshwater tests			
Streptoxkit F *(Streptocephalus proboscideus)* test	24h LC_{50}	148	
Daphnia magna test	24h LC_{50}	734	
Rotoxkit F *(Brachionus calyciflorus)* test	24h LC_{50}	13.3	(2945)

ampicillin (amino-3,3-dimethyl-7-oxo-4-thia-1-azabicyclo; aminobenzylpenicillin)

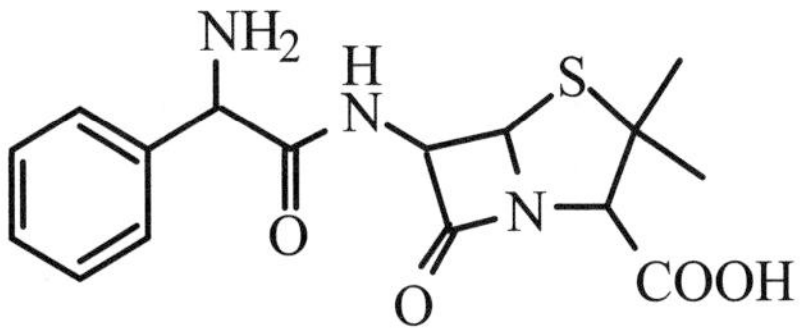

$C_{16}H_{19}N_3O_4S$

CAS 69-53-4

TRADENAMES

Amipenix; Ampicyn; Sulbactam; Omnipen; Rosampline; Synpenin; Tokiocillin; Totalciclina; Totapen; Ultrabion; Viccillin; Britacil; Copharcilin; Doktacillin; Grampenil; Guicitrina; Marsilan; Nuvapen; Pen-Bristol; Penbritin; Penbrock; Pencline; Penstabil; Pentrex; Pentrexyl; Ponecil

USES

pharmaceutical

A. PROPERTIES

molecular weight 349.4; melting point 203°C; vapor pressure 7.7 x 10^{-15} mm Hg at 35°C; water solubility 10100 mg/L at 21 °C; $LogP_{ow}$ 1.35

D. BIOLOGICAL EFFECTS

Micro organisms			
Vibrio fischeri	24h EC_{10}	90.1 mg/L	
	24h EC_{50}	163 mg/L	
	24h EC_{90}	238 mg/L	(7237)

Amprolium hydrochloride (2-Picolinium, 1-((4-amino-2-propyl-5-pyrimidinyl) methyl)-, chloride, hydrochloride; Pyridinium, 1-((4-amino-2-propyl-5-pyrimidinyl) methyl)-2-methyl-, chloride, monohydrochloride)

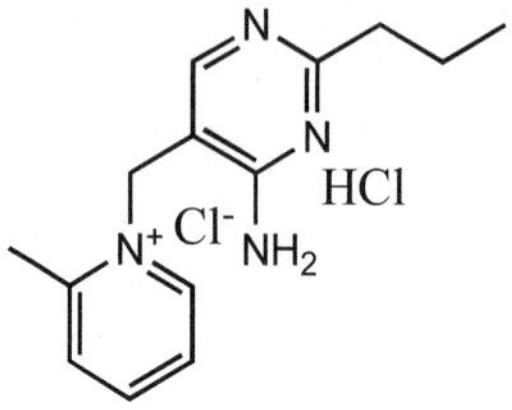

$C_{14}H_{19}ClN_4.HCl$

CAS 137-88-2

CAS 121-25-3

TRADENAMES

Amprol; Corid; Mepyrium; Thiacoccid

USES

Animal feed additive. Used in poultry feed to control coccidiosis

A. PROPERTIES

molecular weight 315.24

D. BIOLOGICAL EFFECTS

Toxicity

Algae

Chlorella sp	48h EC_{50}	160 mg/L	(10793)
Crustaceae			
Daphnia magna	48h EC_{50}	230 mg/L	
	48h LC_{50}	610 mg/L	(10793)
Fish			
Salmo gairdneri	48h LC_{50}	1,550 mg/L	
Lebistes eticulates	48h LC_{50}	270 mg/L	(10793)

t-amyl methylether (TAME; 2-methoxy-2-methylbutane; methyl-tert-pentylether; 1,1-dimethylpropylmethylether))

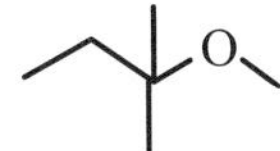

$C_6H_{14}O$

CAS 994-05-8

USES

gasoline oxygenate, additive in unleaded gasoline (± 5 %)

IMPURITIES, ADDITIVES, COMPOSITION

the chemical may contain the following impurities as a result of the manufacture process: 2-methyl-1-buten; 2-methyl-2-butene; 1,1-dimethyl-1-propanol; dimethylether.

A. PROPERTIES

colourless liquid; molecular weight 102.18; boiling point 86.3 °C; specific gravity 0.77; vapour pressure 9.0 kPa at 20°C; water solubility 10,710 mg/L at 20°C; Henry's law constant 90 Pa.m^3/mol; log Pow 1.55

B. AIR POLLUTION FACTORS

Atmospheric reactivity

half-life : 2.1 days in reaction with atmospheric hydroxyl radicals. (7268)

C. WATER AND SOIL POLLUTION FACTORS

Biodegradation

Inoculum	method	concentration	duration	elimination
	closed bottle test			<5%

TAME is not readily biodegradable (7268)

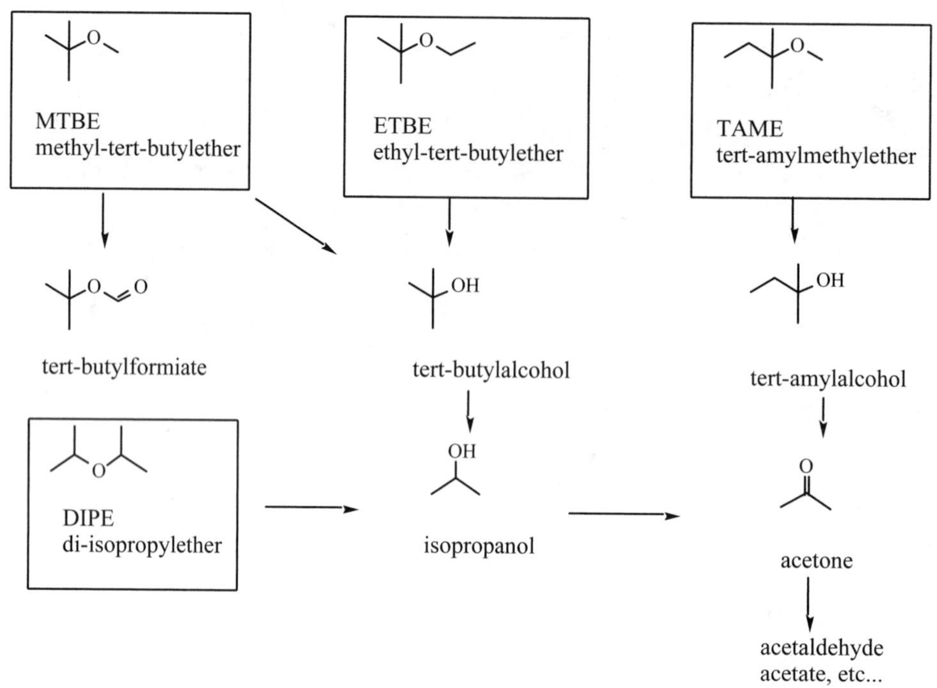

Aerobic degradation pathway for TAME and three other gasoline oxygenates (MTBE, ETBE and DIPE) by an isolate, probably a new member of the Leptothrix group (7269)

Mobility in soil

log K_{OC} 1.82 (calculated)

D. BIOLOGICAL EFFECTS

Toxicity

Crustaceae			
Daphnia magna	48h EC_{50}	100 mg/L	
	48h NOEC	83 mg/L	(7268)
Fish			
Oncorhynchus mykiss	96h LC_{50}	580 mg/L	
	496h NOEC	310 mg/L	(7268)
Mammals			
Rat	oral LD_{50}	<5000 mg/kg	(7268)
	28d repeat NOEL	125 mg/kg bw	(7268)

Metabolites

In rats, the major metabolites detected were identified as being 2-methyl-2,3-butanediol and a glucuronide of this compound and a glucuronide of t-amyl alcohol. Minor metabolites identified included 3-hydroxy-3-methylbutyric acid, t-amyl alcohol and 2-hydroxy-2-methylbutyric acid.
In humans the same metabolites were identified, tert-amylalcohol was a major metabolite in human urine, while the glucuronide of 2-methyl-2,3-butanediol was present in minor quantities. The half-lives for excretion of metabolites were in the range of 6 to 40 hours. (7268)

t-amyl peroxyoctoate (hexaneperoxoic acid,2-ethyl-,1,1-dimethylpropyl ester; t-amyl peroxy(2-ethylhexanoate)

$C_{13}H_{26}O_3$

CAS 686-31-7

TRADENAMES

Luperox 575

USES

a polymerisation initiator used at a level of 0.5-2.0% in reaction with monomers. The chemical, resin solution and final paint will be used in industry only.

ENVIRONMENTAL CONCENTRATIONS

The chemical is consumed during resin manufacture and becomes incorporated into the polymer chains. Except in the case of accident there is little likelihood of release of the chemical, but if this were to occur the chemical is expected to rapidly decompose through rupture of the peroxide bonds. The decomposition products would most likely be carbon dioxide and t-butylalcohol and 2-ethylhexanol. Rapid decomposition of organic peroxides is effectively catalysed through ferrous ions, which are invariably present in soils and natural waters. (7460)

t-butylalcohol   2-ethylhexanol

A. PROPERTIES

clear liquid; molecular weight 230.35; freezing point <-80°C; specific gravity 0.90 at 25°C; vapour pressure est. 0.88 kPa at 25°C;log Pow 4.8; log K_{OC} 3.4; log H 4.92 Pa.m^3/mole

C. WATER AND SOIL POLLUTION FACTORS

Hydrolysis

t/2 at pH 7.0= 8.3 days; t/2 at pH 8.0= 20 hours (7460)

D. BIOLOGICAL EFFECTS

Toxicity

Micro organisms			
Sewage bacteria	30min EC_{50}	64 mg/L	(7460)
Fish			
Poecilia reticulata	96h LC_{50}	8.66 mg/L	
	96h NOEC	2.1 mg/L	(7460)
Mammals			
Rat	oral LD_{50}	>5,000 mg./kg bw	(7460)

amylacetate, prim- (*n*-amylacetate; 1-pentanolacetate; amylacetic ester)

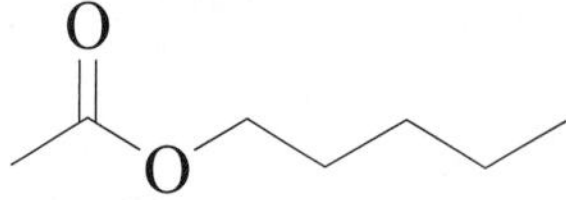

$CH_3C(O)OC_5H_{11}$

$C_7H_{14}O_2$

CAS 628-63-7

A. PROPERTIES

molecular weight 130.2; boiling point 148°C at 737 mm; density 0.879 at 20/20°C; solubility 1,800 mg/L at 20°C; THC 1042 kcal/mole; log H -1.80 at 25°C.

B. AIR POLLUTION FACTORS

1 mg/m³ = 0.188 ppm, 1 ppm = 5.32 mg/m³.

Odor

characteristic: quality: sweet, ester, banana; hedonic tone: pleasant.

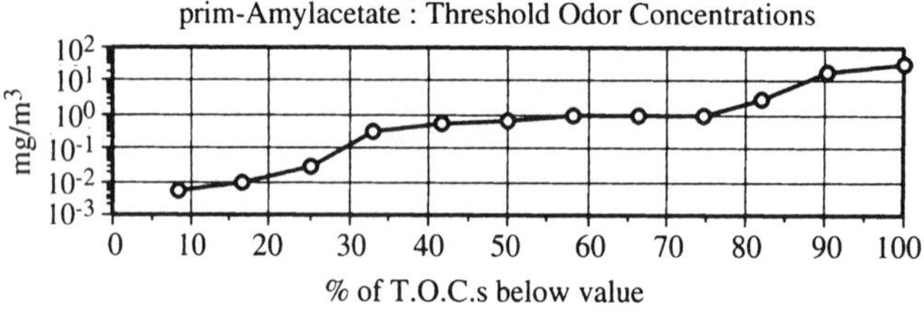

(19, 210, 307, 602, 636, 665, 671, 708, 786)

Odor

index: 25.000		(19)
threshold		
unadapted panelists:	20 ppm	
after adaptation with pure odorant:	3,000 ppm	(204)
human odor perception		
nonperception:	0.5 mg/m³	
perception:	0.6 mg/m³ = 0.11 ppm	
human reflex response		
no response:	0.12 mg/m³	
adverse response:	0.30 mg/m³	(170)

C. WATER AND SOIL POLLUTION FACTORS

BOD_5:

38% ThOD at 440 mg/L, std. dil.	(27)
13% ThOD at 1.7-20 mg/L, std. dil. sew.	(280)
37% ThOD at 440 mg/L, Sierp, sew.	(280)

ThOD: 234.

Impact on biodegradation processes

inhibition of degradation of glucose by *Pseudomonas fluorescens* at 350 mg/L; inhibition of (293) degradation of glucose by *E. coli* at: >1,000 mg/L

Odor threshold

average: 0.08 mg/L	(918, 889)
range: 0.0017 to 0.86 mg/L	(97, 294)

Waste water treatment

A.C.: adsorbability: 0.17 g/g carbon, 88% reduction, infl.: 985 mg/L, effl.: 119 mg/L (32)

D. BIOLOGICAL EFFECTS

Bacteria

Escherichia coli: no-effect level: 1 g/L (30)

Toxicity threshold (cell multiplication inhibition test)

Bacteria			
Pseudomonas putida	16h EC_0	145 mg/L	(1900)
Algae			
Microcystis aeruginosa	8d EC_0	63 mg/L	(329)
green algae			
Scenedesmus quadricauda	7d EC_0	80 mg/L	(1900)
Protozoa			
Entosiphon sulcatum	72h EC_0	226 mg/L	(1900)
Uronema parduczi Chatton-Lwoff	EC_0	550 mg/L	(1901)
Crustaceans			
Daphnia	48 h threshold toxic effect at 23°C	440 mg/L	(356)
Fishes			
mosquito fish:	24-96h LC_{50}	65 mg/L	(41)
creek chub:	24 LC_0	50 mg/L	
	24h LC_{100}	120 mg/L	(243)
creek chub:	24h critical range	350-500 mg/L	(355)
goldfish:	96h LC_{50}	10 mg/L	(354)
Lepomis macrochirus	96h LC_{50},S	650 mg/L	(352)
Menidia beryllina	96h LC_{50},S	180 mg/L	(352)

amylacetic ester *see prim*-amylacetate

amylalcohol, prim-iso- (3-methyl-1-butanol; *prim*-isobutylcarbinol; isopentanol)

OH

$(CH_3)_2CHCH_2CH_2OH$

$C_5H_{12}O$

CAS 123-51-3

USES

Solvent for fats, alkaloids and resins. In microscopy. For determining fat in milk. In the manufacture of isoamyl compounds, isovaleric acid, artificial silk, lacquers, mercury fulminate and pyroxylin.

A. PROPERTIES

molecular weight 88.15; melting point -117.2°C; boiling point 131/132°C; vapor pressure 2.3 mm at 20°C, 4.8 mm at 30°C; vapor density 3.04; density 0.81; solubility 30,000 mg/L at 20°C, 26,720 mg/L at 22°C; saturation concentration in air 11 g/m^3 at 20°C, 22 g/m^3 at 30°C; LogP$_{OW}$ 1.16.

B. AIR POLLUTION FACTORS

1 mg/m^3 = 0.27 ppm, 1 ppm = 3.66 mg/m^3.

Odor

characteristic:	quality: sweet	
	hedonic tone: pleasant	
T.O.C.:	0.027 mg/m^3 = 7.3 ppb	(307)
	recogn.: 38 mg/m^3 = 10.2 ppm	
	10 ppm = 35 mg/m^3 (n.s.i.)	(210)
	abs. perc. lim.: 0.12 ppm	
	50% recogn.: 1.0 ppm	
	100% recogn.: 1.0 ppm	
O.I.	100% recogn.: 13,000	

Manmade sources

emission from whiskey fermentation vats: 0.166 g/m^3 of grain input.

C. WATER AND SOIL POLLUTION FACTORS

BOD$_5$:	55% ThOD std. dil. sew.	(260)
	59% ThOD std. dil. sew.	(27, 30, 282)
	59% of ThOD	(274)
BOD2d	48% ThOD	
BOD5d	61% ThOD	
BOD10d	74% ThOD	
BOD30d	51% ThOD	(10548)
BOD$_5$20°:	59% ThOD (acclim.)	(2666)
COD:	77% of ThOD (0.05 $nK_2Cr_2O_7$)	(274)
ThOD:	2.740	(30)

Waste water treatment

A.S. BOD, 20°C, 1-5 d observed, feed: 333 mg/L, 30 d acclimation, 79% removed

A.S.: after	6 h:	10% of ThOD	
	12 h:	20% of ThOD	
	24 h:	30% of ThOD	(88)

Persistence

in ground water	$t_{1/2}$: 15 d	
in river water	$t_{1/2}$: 11 d	
in harbour water	$t_{1/2}$: 6 d	(10343)

D. BIOLOGICAL EFFECTS

Mammals			
rat	oral LD_{50}	1,300 mg/kg bw	(10344)

amylalcohol, sec-act. (2-pentanol; methylpropylcarbinol; 1-methyl-1-butanol)

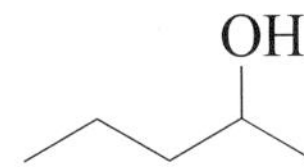

$CH_3(CH_2)_2CHOHCH_3$

$C_5H_{12}O$

CAS 6032-29-7

A. PROPERTIES

molecular weight 88.15; boiling point 119°C; vapor density 3.04; density 0.809 at 20/4°C; solubility 53,000 mg/L at 30°C; $LogP_{OW}$ 1.34 (calculated); LogH-3.22 at 25°C.

B. AIR POLLUTION FACTORS

Partition coefficients

$K_{air/water}$	0.00061	
Cuticular matrix/air partition coefficient*	4800 ± 0	(7077)
Cuticular matrix/water partition coefficient**	2.9	(7077)

* experimental value at 25°C studied in the cuticular membranes from mature tomato fruits (*Lycopersicon esculentum* Mill. cultivar Vendor)
** calculated from the Cuticular matrix/water partition coefficient and the air/water distribution coefficient

C. WATER AND SOIL POLLUTION FACTORS

Waste water treatment

A.S., BOD, 20°C, 1-5 d observed, feed: 333 mg/L, 30 d acclimation, 91% removed. (93)

D. BIOLOGICAL EFFECTS

Insects			
Chironomus riparius, 3rd instar larvae	48h LC_0	790 mg/L	
	48h LC_{50}	910 mg/L	
	48h LC_{100}	1,100 mg/L	
	48h NOLC	460 mg/L	
	48h NOEC	460 mg/L	(7073)

amylalcohol, tert- *see* 2-methyl-2-butanol

n-amylaldehyde *see* *n*-valeraldehyde

n-amylamine (pentylamine; 1-aminopentane)

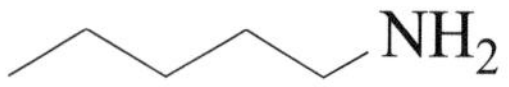

NH_2

$CH_3(CH_2)_4NH_2$

$C_5H_{13}N$

CAS 110-58-7

A. PROPERTIES

molecular weight 87.16; melting point -55°C; boiling point 104°C; vapor pressure 35 mm at 26°C; vapor density 3.01; density 0.77 at 20/4°C; $LogP_{OW}$ 1.05 (calculated).

B. AIR POLLUTION FACTORS

1 mg/m^3 = 0.281 ppm, 1 ppm = 3.56 mg/m^3.

C. WATER AND SOIL POLLUTION FACTORS

Waste water treatment

degradation by *Aerobacter:* 200 mg/L at 30°C:

parent:	100% in 25 h	
mutant:	100% in 9 h	(152)

D. BIOLOGICAL EFFECTS

Fishes

creek chub:	24h LD_0	30 mg/L	
	24h LD_{100}	50 mg/L	(243)

amylbenzene, tert- (tert-pentylbenzene)

A. PROPERTIES

log H -0.13 at 25°C.

amylcarbinol *see* *n*-hexanol

amylchloride (1-chloropentane; pentylchloride)

Cl

$CH_3(CH_2)_3CH_2Cl$

$C_5H_{11}Cl$

CAS 543-59-9

A. PROPERTIES

molecular weight 106.60; melting point -60°C; boiling point 108°C; vapor density 3.67; density 0.88 at 20/4°C; LogH -0.05 at 25°C.

C. WATER AND SOIL POLLUTION FACTORS

Biodegradation

Waste water treatment:	A.S.: after	6 h: 1.5% of ThOD	
		12 h: 1.8% of ThOD	
		24 h: 2.8% of ThOD	(88)

A thermophilic obligate methane oxidizing bacterium can degrade liquid monochloro- and dichloro-n-alkanes (C_5,C_6). Compounds are oxidised yielding their corresponding acids or chloroacids. (9801)

D. BIOLOGICAL EFFECTS

Bacteria

Photobacterium phosphoreum Microtox test	5-30 min EC_{50}	244 mg/L	(9657)
Nitrosomonas sp.	IC_{50}	99 mg/L	
methanogens	IC_{50}	150 mg/L	
aerobic heterotrophs	IC_{50}	68 mg/L	(8966)

alfa-Amylcinnamal (2-(phenylmethylene)-1-heptanal; amyl cinnamic acid aldehyde; α-amyl-β-phenylacrolein; 2-benzylidene heptanal; jasmin aldehyde; 2-pentyl cinnamaldehyde)

214 alfa-Amylcinnamal

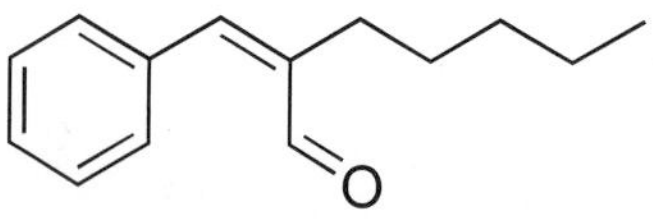

$C_{14}H_{18}O$

CAS 122-40-7

TRADENAMES

Buxine; Flomine; Flosal; Jasmine; Jasmonal

USES

a fragrance in perfumed consumer products such as dishwash, car shampoo, hand cleaner, toilet paper, nappies in concentrations of 0.009 to 0.028 wt%. Additive in chewing-gums, and tobacco (7195)

OCCURRENCE

Volatile essence of black tea

A. PROPERTIES

Pale yellow to yellow clear liquid; molecular weight 202.29 ; melting point 34°C; boiling point 287-290°C; specific gravity 0.97 at 25°C; vapour pressure 0.00013 kPa at 20°C; water solubility 8.5 mg/L; log Pow 4.3 (calculated)

B. AIR POLLUTION FACTORS

Odour

Sweet floral oily fruity herbal jasmin

C. WATER AND SOIL POLLUTION FACTORS

Biodegradation

Inoculum	method	concentration	duration	elimination	
A.S. non acclimated	OECD 301 B		28 days	65% ThCO2	
	OECD 301 B		28 days	70%	
	OECD 301F		28 days	90%	(10991)

D. BIOLOGICAL EFFECTS

Toxicity calculated (ECOSAR)

Fish	96h LC_{50}	2; 3.1 mg/L	
GREEN ALGAE	96h LC_{50}	0.6; 0.87 mg/L	
Daphnia magna	48h EC_{50}	0.42 mg/L	
Rat	oral LD_{50}	3,370 mg/kg bw	(10991)

Human metabolism of amyl cinnamaldehyde

Amyl cinnamaldehyde is readily oxidized to amyl cinnamic acid (see Figure). Human NAD+ dependent alcohol dehydrogenase (ADH) catalyzes oxidation of primary alcohols to aldehydes. Isoenzyme mixtures of NAD+ dependent aldehyde dehydrogenase catalyze oxidation of aldehydes to carboxylic acids. The urinary metabolites of cinnamyl alcohol and cinnamaldehyde are mainly derived from metabolism of cinnamic acid (see Figure).

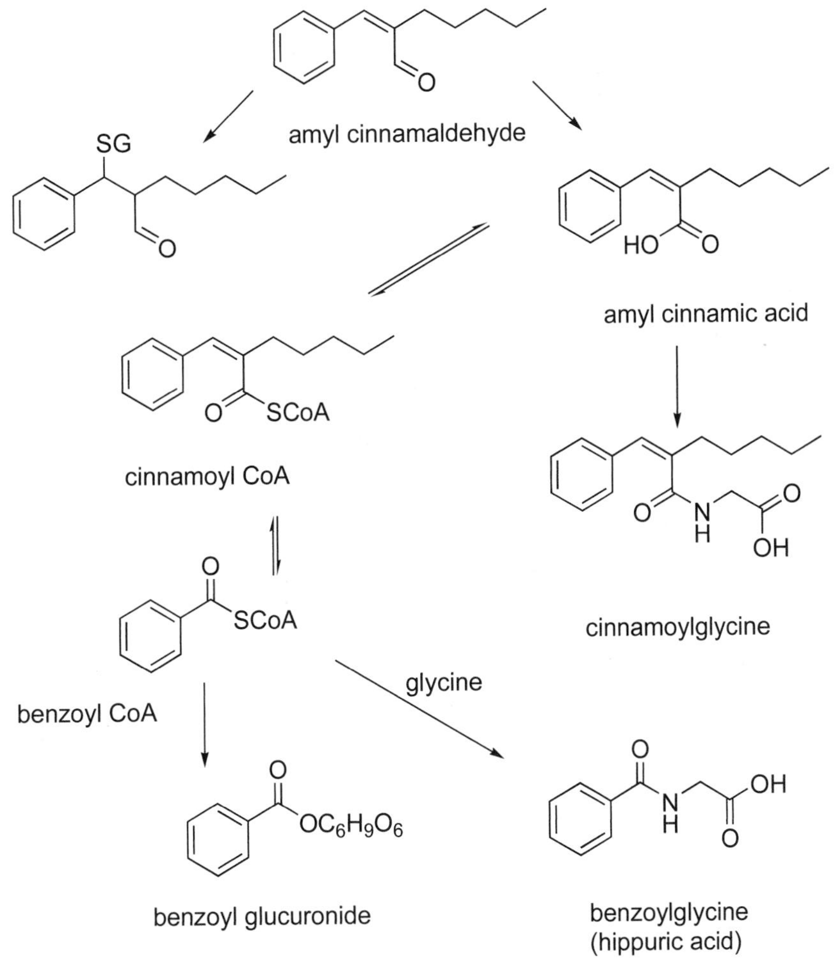

amyl cinnamaldehyde

amyl cinnamic acid

cinnamoyl CoA

cinnamoylglycine

benzoyl CoA

glycine

benzoyl glucuronide

benzoylglycine (hippuric acid)

Metabolism of amyl cinnamaldehyde [10991]

---

alfa-amylcinnamic alcohol (2-pentyl-3-phenylprop-2-en-1-ol; n-amylcinnamic alcohol; 2-benzylideneheptanol; α-pentylcinnamic alcohol)

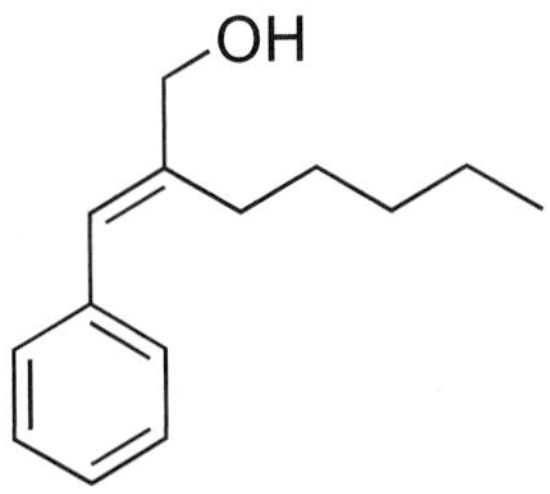

$C_{14}H_{20}O$

CAS 101-85-9

SMILES CCCCCC(=CC1=CC=CC=C1)CO

TRADENAMES

Buxinol

USES

Flavouring agent; Fragrance in parfumed cosmetic products, deodourants, tissues etc..

A. PROPERTIES

Colourless to licht yellow liquid; molecular weight 204.3; boiling point 141°C at 5 mm Hg; water solubility 30 mg/L at 25°C; log Pow 3.8-4.3

B. AIR POLLUTION FACTORS

Odour

Light floral

n-amylcyanide *see* hexanenitrile

α-n-amylene *see* 1-pentene

β-n-amylene *see* 2-pentene

amylene dimer

USES

intermediate in the manufacture of alkylated phenols.

IMPURITIES, ADDITIVES, COMPOSITION

approx. 83% C_{10} olefin isomers, 10% C_{15} olefin isomers.

A. PROPERTIES

boiling range: 150-260°C; density at 15/4°C 0.78; solubility 15 mg/L at 20°C.

C. WATER AND SOIL POLLUTION FACTORS

BOD_5:	0.14 = 4% ThOD	
	0.24 = 7% ThOD (after adaptation)	(277)
COD:	1.68 = 49% ThOD	(277)

D. BIOLOGICAL EFFECTS

Fishes

Carassius auratus: not toxic in saturated solution (16 mg/L). (277)

amylenehydrate *see* 2-methyl-2-butanol

amylmercaptan *see* pentylmercaptan

n-amylmethylketone *see* 2-heptanone

amylpropionate (n-pentyl propionate; n-pentyl propanoate)

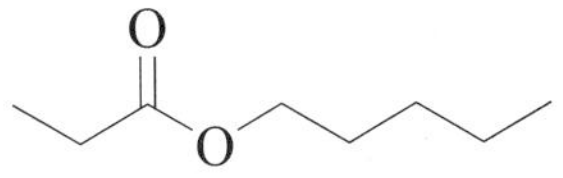

$CH_3CH_2CO_2(CH_2)_4CH_3$

$C_8H_{16}O_2$

CAS 624-54-4

OCCURRENCE

Found in flavour component of bananas.

A. PROPERTIES

molecular weight 144.21; melting point -73°C; boiling point 169°C; density 0.88 at 20°C; vapor density 5.0; LogH -1.46 at 25°C

amylxanthate, potassium (carbonodithioic acid, O-pentyl ester, potassium salt; O-pentyl carbonodothioate potassium salt; potassium pentyl xanthogenate)

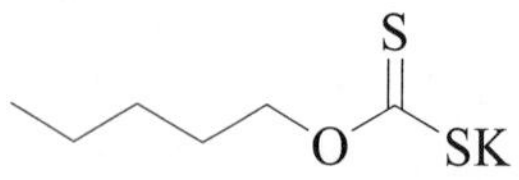

$CH_3(CH_2)_4OC(S)SK$

$C_6H_{11}KOS_2$

CAS 2720-73-2

USES

Flotation agent.

A. PROPERTIES

solid, molecular weight 204.29; melting point 255- 295°C decomposes; solubility 300,000 mg/L at 20°C.

C. WATER AND SOIL POLLUTION FACTORS

Abiotic hydrolysis

$t_{1/2}$: 7 d at pH 5.5 and 15°C	(6414)
80- 85 d (10% aqueous solution) at 20°C	(6421)
10 d (10- 25% aqueous solution) at 40°C	(6424)
13- 15 d (1- 2 mg/L aqueous solution) at 21°C	(6422)

Biodegradation

Inoculum	*Method*	*Concentration*	*Duration*	*Elimination results*	
A.S.	Respirometer test		12.5 d	52% ThOD	(6419)
A.S. not adapted		2.8 mg/L	10 d	38% ThOD	(6420)
		28 mg/L	10 d	18% ThOD	(6420)
	Zahn-Wellens test		3 h	25- 35% DOC	
			5 d	78% DOC	
			12 d	95% DOC	(6417)

D. BIOLOGICAL EFFECTS

Bacteria			
primary municipal sludge	24 h EC_0	100- 1,000 mg/L	(6417)
(facultative anaerobic bacteria)	24 h EC_0	150 mg/L	(6409)
Photobacterium phosphoreum (Microtox test)	15 min EC_{50}	0.57 mg/L	(6411)
Thiobacillus ferrooxidans	EC_{25}	250 mg/L	
(inhibition of Fe^{2+} oxidation)	EC_{83}	500 mg/L	(6418)

Algae			
Monoraphidium griffithii (cell multiplication inhibition)	72 h EC_{50}	0.1 mg/L	(6411)

Crustaceans			
Daphnia magna	24 h EC_{10}	1.7 mg/L	
	24 h EC_{50}	3 mg/L	(6416)
		0.1- 1.0 mg/L	(6410)
	24 h EC_{95}	6 mg/L	(6416)
Fishes			
Brachydanio rerio	96 h LC_{50}	10- 100 mg/L	
	96 h LC_{100}	100 mg/L	(6408)
Poecilia reticulata	48 h LC_0	25 mg/L	(6409)
Pimephales promelas	96 h LC_{50}	1.8- 18 mg/L	(6410)
		18- 180 mg/L	(6410)
Salmon salar	96 h LC_{50}	11 mg/L	(6414)
Salmo gairdneri)	96 h LC_{50},S	32- 56 mg/L (Cyanamid C 350)	
Approx.18 mg/L (Dow Chemical Z)			(1087)
		70- 80 mg/L	(6415)
	96 h LC_{100},S	56 mg/L	
	28 d LC_{100},F	1.0 mg/L	(1087)

anesthesin *see* benzocaine

aniline (aminobenzene; phenylamine)

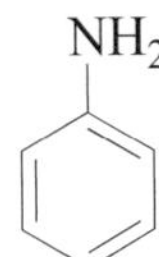

$C_6H_5NH_2$

C_6H_7N

CAS 62-53-3

USES AND FORMULATIONS

constituent of coal tar creosote: 0.05 wt %. (2386)

A. PROPERTIES

molecular weight 93.1; melting point -6°C; boiling point 184°C; vapor pressure 1 mm at 35°C, 0.3 mm at 20°C; vapor density 3.22; density 1.02; solubility 34,000 mg/L; saturation concentration in air 1.5 g/m^3 at 20°C, 3.4 g/m^3 at 30°C; LogP_{OW} 0.90/0.98.

B. AIR POLLUTION FACTORS

1 mg/m^3 = 0.259 ppm, 1 ppm = 3.87 mg/m^3.

Incinerability

thermal stability ranking of hazardous organic compounds: rank 46 on a scale of 1 (highest stability) to 320 (lowest stability). (2390)

Odor

characteristic; hedonic tone: pungent.

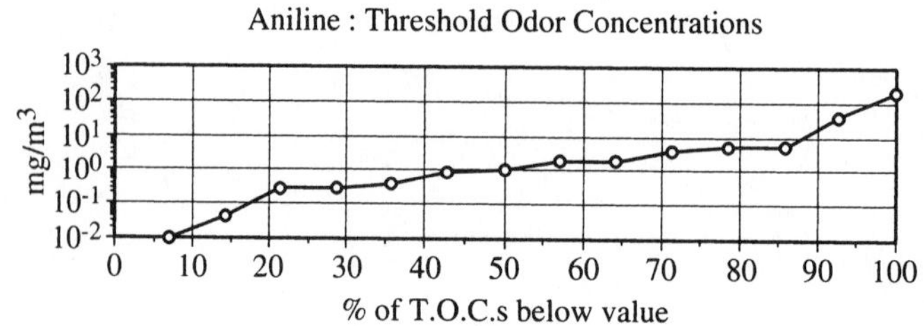

(2, 73, 279, 297, 610, 664, 695, 703, 741, 840, 845)

human odor perception:	nonperception:	0.34 mg/m^3	
	perception:	0.37 mg/m^3	
human reflex response:	adverse response:	0.07 mg/m^3	
animal chronic exposure:	adverse effect:	0.05 mg/m^3	(170)

C. WATER AND SOIL POLLUTION FACTORS

BOD_5:	1.5 std. dil. at 1.5-3 mg/L	(27)
	1.49-2.26 std. dil. sewage	(41)
	1.76	(30)
	1.42	(36)
	62% ThOD	(274)
	1.42 at 10 mg/L, unadapted sew.: lag period: 3 d	
BOD^{20}_{20}:	2.02 at 10 mg/L, unadapted sew.: lag period: 3 d	(554)
COD:		
	94% ThOD (0.05 *n* Cr_2O_7)	(274)
	2.34; 2.4	(36, 41)
$ThOD_{NH3}$:	2.41	(2790)
$ThOD_{NO3}$:	3.09	(2790)

Impact on biodegradation processes

Toxicity to microorganisms: act. sludge respiration inhib. EC_{50}: >100 mg/L

75% inhibition of nitrification in the activated sludge process at 7.7 mg/L (43)

degree of inhibition of NH_3 oxidation by Nitrosomonas *sp.*

at 100 mg/L	86% inhibition	
at <1 mg/L	±50% inhibition	(390)

inhibition of photosynthesis of a freshwater nonaxenic uni-algal culture of Selenastrum capricornutum

% C^{14} fixation vs controls

at 10 mg/L	90	
at 100 mg/L	34	
at 1,000 mg/L	3	(1690)

Effect on $BOD^{20°C}$:

d	*BOD of original sample*	*BOD after addition of 100 ppm aniline*
5	6	0
10	12	2
15	14	3

20	16	5	
25	17	6	
30	17	7	(172)

Reduction of amenities

T.O.C. 70 mg/L; range 2.0 to 128 mg/L	(294, 30)

Manmade sources

Göteborg (Sweden) sew. works 1989-1991: infl.: nd -5 µg/L; effl.: nd	(2787)

Biodegradation

Inoculum	*method*	*concentration*	*duration*	*elimination results*	
activated sludge	respirometric test	70-100 mg/L DOC	28 d	84% ThOD	(7040)
soil-water suspension	aerobic conditions at 35 °C	0.1 mg/L	5 d	17% ThCO2	
			56 d	26% ThCO2	
			56 d	87% test substance	
soil-water suspension	Anaerobic conditions at 35 °C	0.1 mg/L	56 d	12% ThCO2	
			56 d	71% test substance	(2667)
nonadapted sludge*	EEC respirometric method	100 mg/L	7 d	>40% ThOD	
			14 d	>60% ThOD	
			28 d	>70% ThOD	
			28 d	>77% DOC	(2790)
adapted A.S	product is sole carbon source	19 mg COD/g dry inoculum/h		94% COD	(327)

* delay times: in all the tests (*n* = 58) <9 d and in 86% of the tests <5 d

Biodegradation half-lives in nonadapted aerobic subsoil: Oklahoma sand: 4-12 d (2695)
decomposition period by soil microflora 4 d (176)

Aqueous reactions

photooxidation by UV light in aqueous medium at 50°C;
28% degradation to CO_2 after 24 h (1628)

Waste water treatment

A.C.: adsorbability: 0.15 g/g carbon; 75% reduction, infl.: 1,000 mg/L, effl.: 251 mg/L (32)
adsorption on Amberlite X AD-7: retention efficiency: 100%, infl.: 4.0 ppm, effl.: 0 ppm (40)
air stripping constant: *k*= 0.198 d^{-1} at 100 mg/L (82)
degradation by *Aerobacter:* 500 mg/L at 30°C:
parent: 100% ring disruption in 54 h
mutant: 100% ring disruption in 12 h (152)

D. BIOLOGICAL EFFECTS

Bacteria

E. coli:	EC_0	1,000 mg/L	(329)
Microtox *Vibrio fisheri*	15min EC_{50}	70 mg/L	(7025)
nitrification inhibition in activated sludge or trickling filter	IC_{14}	0.5 mg/L	
	IC_{60}	4 mg/L	
	IC_{75}	7.7 mg/L	
	IC_{84}	10 mg/L	(7050)
Pseudomonas putida	16h EC_0	130 mg/L	

Algae

Chlorella fusca BCF (wet wt.): 4			(2659)
Chlorella pyrenoidosa	NOEC	11mg/L	
Selenastrum capricornutum	NOEC	10 mg/L	(2907)
Tetrahymena pyriformis	24h IC_{50}	190 mg/L	
Scenedesmus	toxic	10 mg/L	
Microcystis aeruginosa:	inhibition of cell multiplication	starts at 0.16 mg/L	(329)
Microcystis aeruginosa	LD_{50}	20 ppm	(1094)
Scenedesmus subspicatus inhib. of fluorescence	IC_{10}	7 mg/L	
Scenedesmus subspicatus growth inhib.	IC_{10}	13 mg/L	(2698)
Rhaphidocellis subcapitata	72h EC_{50}	5 mg/L	(7025)
Scenedesmus quadricauda	7d EC_0	8.3 mg/L	

RubisCo Test

inhib. of enzym. activity of Ribulose-P2-carboxylase in protoplasts	IC_{10}	2.3 mg/L	

Oxygen Test

inhib. of oxygen production of protoplasts	IC_{10}	0.093 mg/L	(2698)

Worm

Tubifex	48h LC_{50}	64-100 mg/L	

Protozoans

Entosiphon sulcatum	72h EC_0	24 mg/L	(1900)
ciliate *(Tetrahymena pyriformis):*	24h LC_{100}	2,000 mg/L	(1662)
Uronema parduczi Chatton-Lwoff	EC_0	91 mg/L	(1901)

Amphibia

lethality and teratogenicity to early embryonic stages of South African clawed frog, *Xenopus laevis:*

	d								
	1		*2*		*3*		*4*		
conc., mg/L	*A/S**	*%*	*A/S*	*%*	*A/S*	*%*	*A/S*	*%*	
0	0/50	0	0/50	0	0/50	0	0/50	0	
10	0/50	0	0/47	0	4/36	11	4/36	11	
50	1/50	2	3/48	6	3/48	6	3/48	6	(1418)

* A/S = abnormals/survivors

Mexican axolotl (3-4 wk after hatching)440 mg/L			
clawed toad (3-4 wk after hatching)	48h LC_{50}	560 mg/L	(1823)
Xenopus laevis	LC_{50}	560 mg/L	
Ambystoma mexicanum	LC_{50}	440 mg/L	(2907)

Crustaceans

Daphnia magna	LC_{50}	0.64 mg/L	
	NOLC	0.34 mg/L	
Daphnia pulex	LC_{50}	0.1 mg/L	
	NOLC	0.07 mg/L	
Daphnia cucullata	LC_{50}	0.68 mg/L	(2907)
Daphnia magna: mortality reproductive effects after 21 d at 25-47 µg/L			(2625)
Daphnia	toxic	0.4 mg/L	
Daphnia magna	48h IC_{50}	0.64 mg/L	(7025)

Insects

Aedes aegypti	LC_{50}	155 mg/L	
Culex pipiens	LC_{50}	94 mg/L	(2907)

Hydroids			
Hydra oligactis	LC_{50}	406 mg/L	
	NOLC	235 mg/L	

Molluscs			
Lymnaea stagnalis	LC_{50}	800 mg/L	
	NOLC	560 mg/L	(2907)

Fishes

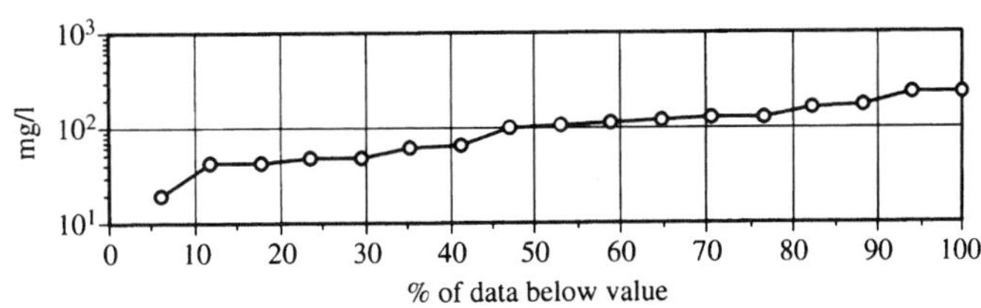

Frequency distribution for 24-96h LC_{50} values for fishes based on data from this and other works (3999)

Leuciscus idus melanotis	LC_{50}	49 mg/L	
Oryzias latipes	LC_{50}	165 mg/L	
Pimephales promelas	LC_{50}	65 mg/L	
Poecilia reticulata	LC_{50}	100 mg/L	
Salmo gairdneri	LC_{50}	43 mg/L	
Killifish *(Oryzias latipes)*	48h LC_{50}	170 mg/L	(2624)
Poecilia reticulata	14d LC_{50}	126 mg/L	(2696)
Brachydanio rerio	96h LC_{50}	32 mg/L	(7025)

aniline hydrochloride (anilinechloride)

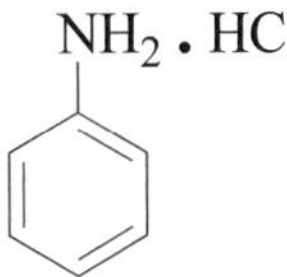

$C_6H_5NH_2.HCl$

$C_6H_7N.HCl$

CAS 142-04-1

A. PROPERTIES

white crystals, darkens in light and air; molecular weight 129.59; melting point 196-198°C; boiling point 245°C; density 1.22.

D. BIOLOGICAL EFFECTS

Fishes	48h LC_{50}	5.5 mg/L	(226)

aniline yellow *see* *p*-phenylazoaniline

anilinechloride *see* aniline hydrochloride

m-anilinesulfonic acid (3-aminobenzenesulfonic acid; 1-aminobenzene-3-sulfonic acid; m-sulfanilic acid; aniline-m-sulfonic acid; benzenesulfonic acid, 3-amino-; metanilic-acid)

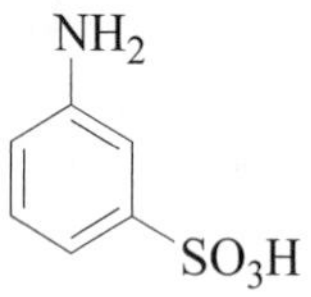

$NH_2C_6H_4SO_3H$

$C_6H_7NO_3S$

CAS 121-47-1

USES

Synthesis of azo dyes and certain sulfa drugs.

A. PROPERTIES

molecular weight 173.19; melting point decomposes without melting at approx. 288°C; density 1.69; solubility 10,800 mg/L at 20°C.

C. WATER AND SOIL POLLUTION FACTORS

Waste water treatment

decomposition period by soil microflora: >64 days. (176)

Adapted activated sludge utilises 3-aminobenzenesulfonic acid as sole carbon source yielding 95% COD removal at 4 mg COD / g dry inoculum / h at 20°C. (327)

D. BIOLOGICAL EFFECTS

Mammals			
rat	oral LD_{50}	12,000 mg/kg bw	(9802)

o-anilinesulfonic acid (orthanilic acid; 2-aminobenzenesulfonic acid; aniline-2-sulfonic acid; o-sulfanilic acid; o-aminophenylsulfonic acid; benzenesulfonic acid, 2-amino; anilino-o-sulfonic acid)

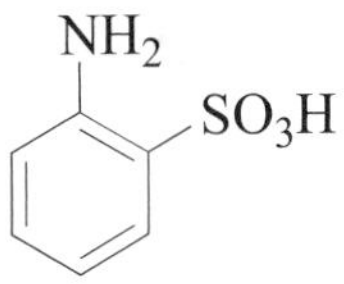

$NH_2C_6H_4SO_3H$

$C_6H_7NO_3S$

CAS 88-21-1

USES

Manufacture of azo dyestuffs. Component of water-based hydraulic fluids.

A. PROPERTIES

molecular weight 173.19; melting point >300°C.

C. WATER AND SOIL POLLUTION FACTORS

Biodegradation

decomposition period by soil microflora: >64 d. (176)

p-anilinesulfonic acid (sulfanilic acid; *p*-aminobenzenesulfonic acid; benzenesulfonic acid, 4-amino; *p*-aminophenylsulfonic acid)

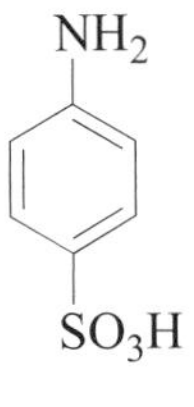

$NH_2C_6H_4SO_3H$

$C_6H_7NO_3S$

CAS 121-57-3

USES

feedstock for chemical synthesis.

A. PROPERTIES

colorless crystals; molecular weight 173.2; melting point 288°C decomposes; density 1.49 g/cm^3 at 25°C; solubility 8,000 mg/L at 20°C, 10,800 mg/L at 20°C, 66,700 mg/L at 100°C; LogP_{OW} 0.9 (calculated).

C. WATER AND SOIL POLLUTION FACTORS

Oxidation parameters

BOD_5:	1.11 std. dil. sew.	(282, 163)
$ThOD_{NH3}$:	1.29	(2790)
$ThOD_{NO3}$:	1.66	(2790)

Biodegradability

decomposition by soil microflora in >64 d (176)

adapted A.S. at 20°C-product is sole carbon source: 95.0% COD removal at 4.0 mg COD/g dry inoculum/h (327)

inoculum/method	*test conc.*	*test duration*	*removed*	
municipal sewage, adapted				
OECD screening		19 d	92% DOC	(5213)
closed-bottle test		30 d	55% ThOD	(5213)
A.S. coupled units test		14 d	19% DOC	(5213)
A.S., adapted	200 mg/L	5 d	95% COD	(5214)
A.S., coupled units test		14 d	7% DOC	(5215)
Zahn-Wellens test		14 d	0% DOC	(5215)
Original MITI test		14 d	0% DOC	(5215)
		14 d	30% ThOD	(5215)
Modified Sturm test		28 d	31% CO_2 prod.	(5215)
		28 d	66% CO_2 prod.	(5215)
		28 d	57% DOC	(5215)
		28 d	97% DOC	(5215)
Modified OECD screening test		19 d	16% DOC	(5215)
Closed-bottle test		30 d	0% ThOD	(5215)

Biodegradation

EEC respirometric method at 100 mg/L and 20°C using nonadapted sludge:

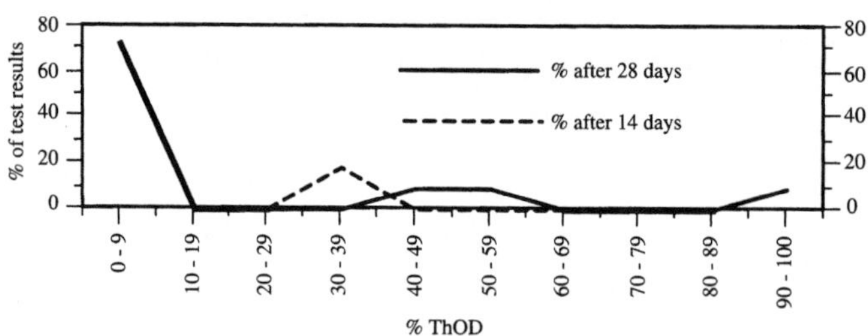

ThOD in EEC respirometric test: n = 11
median after 14 d: 0% ThOD
median after 28 d: 5% ThOD

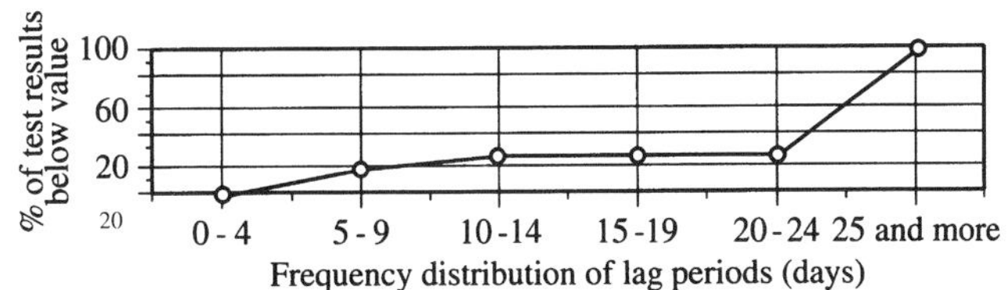

Delay times in EEC respirometric test

n = 11; median: >28 d.

Oxygen uptake and DOC removed using MITI inoculum

delay time: 6; 26 d
% ThOD after 14 d: 0; 97%
% ThOD after 28 d: 14; 103%

% DOC after 28 d: 20; 98% (2790)
in "Repetitive Die-Away" Test
delay time: 21 d
% ThOD after 28 d: 65%
% ThOD after 56 d: 75%
% DOC removal after 28 d: 88%
% DOC removal after 56 d: 99% (2790)

EEC respirometric method at 33 mg/L and 20°C

delay time: 8 d:
after 14 d: 96% ThOD
after 28 d: 112% ThOD and 105% DOC removal (2790)

D. BIOLOGICAL EFFECTS

Bacteria			
Pseudomonas fluorescens	24h EC_0	>10,000 mg/L	(5210)
Crustaceans			
Daphnia magna	24h EC_0	63 mg/L	
	24h EC_{50}	109 mg/L	
	24h EC_{100}	250 mg/L	(5212)
	48h EC_0	62 mg/L	
	48h EC_{50}	86 mg/L	
	48h EC_{100}	125 mg/L	(5212)
Fishes			
Pimephales promelas	96h LC_{50}	100 mg/L	(5211)
Leuciscus idus	48h LC_0	>1,000 mg/L	(5210)

anisaldehyde dimethyl acetal (1-(dimethoxymethyl)-4-methoxybenzene)

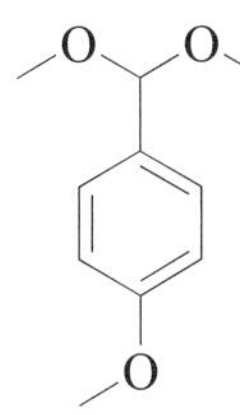

$C_{10}H_{14}O_3$

CAS 2186-92-7

USES AND FORMULATIONS

intermediate in chemical synthesis.

A. PROPERTIES

molecular weight 182.22; melting point - 20°C; boiling point 253°C; boiling point 70°C at 3 hPa; solubility low; $LogP_{OW}$2.0 (measured).

C. WATER AND SOIL POLLUTION FACTORS

Biodegradation

Inoculum	*Method*	*Concentration*	*Duration*	*Elimination results*	
	Respirometer test			0- 60% ThOD	(6495)

D. BIOLOGICAL EFFECTS

Toxicity

Bacteria			
A.S. oxygen consumption inhibition	30 min EC_{20}	90 mg/L	(6495)
Pseudomonas putida	30 min EC_{10}	580 mg/L	
(cell multiplication inhibition test)	30 min EC_{50}	1,900 mg/L	
	30 min EC_{90}	4,200 mg/L	(6494)
Algae			
Scenedesmus subspicatus	72 h EC_{10}	95 mg/L	
(cell multiplication inhibition test)	72 h EC_{50}	169 mg/L	(6494)
	96 h EC_{10}	77 mg/L	
	96 h EC_{50}	132 mg/L	(6494)
Crustaceans			
Daphnia magna	24-48 h EC_0	58 mg/L	
	24-48 h EC_{50}	123 mg/L	
	24-48 h EC_{100}	180 mg/L	(6494)
Fishes			
Leuciscus idus	96 h LC_{50}	220-460 mg/L	(6493)

m-anisic acid (3-methoxybenzoic acid)

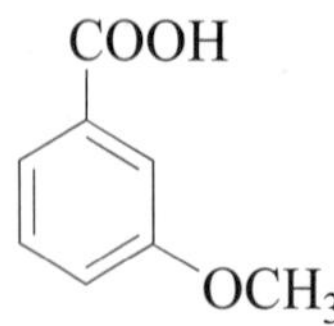

$C_8H_8O_3$

CAS 586-38-9

A. PROPERTIES

colorless needles; molecular weight 152.14; melting point 107/110°C; boiling point 170/172°C at 10 mm Hg; LogP_{OW} 2.02.

C. WATER AND SOIL POLLUTION FACTORS

Waste water treatment

decomposition period by soil microflora: 16 d. (176)

o-anisic acid (2-methoxybenzoic acid; salicylic acid methylether)

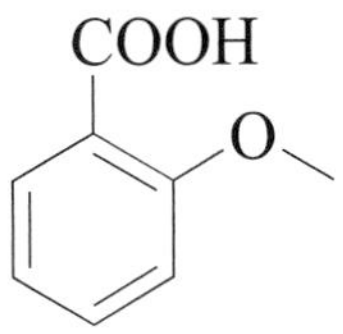

COOH

O

$CH_3OC_6H_4COOH$

$C_8H_8O_3$

CAS 579-75-9

OCCURRENCE

Major component of wastes resulting from the pressing of olives for obtaining olive oil and which is known under the generic name of "alpechin"

A. PROPERTIES

molecular weight 152.14; melting point 101°C; boiling point 200°C; density 1.2; solubility 5,000 mg/L at 30°C; log P_{OW} 0.80/2.93 (calculated).

C. WATER AND SOIL POLLUTION FACTORS

Waste water treatment

decomposition period by soil microflora: 4 d. (176)

Biodegradation

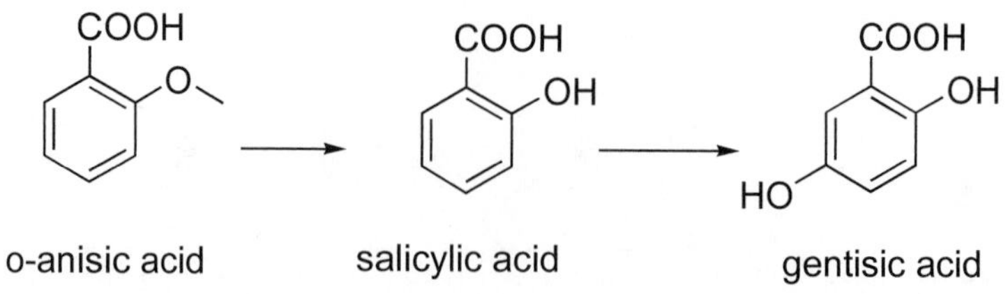

COOH, COOH, OH, COOH, OH, HO

o-anisic acid salicylic acid gentisic acid

Catabolic pathway for the mineralization of o-anisic acid by a consortium made of *Arthrobacter oxydans* and *Pantotea agglomerans.* [11139]

Catabolic pathway for the mineralization of o-anisic acid by a consortium made of *Arthrobacter oxydans*and *Pantotea agglomerans.*	(11139)

p-anisic acid (4-methoxybenzoic acid)

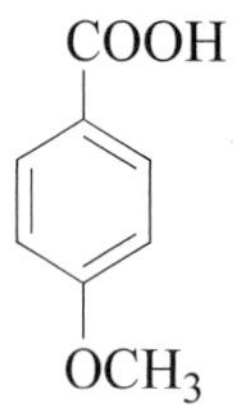

COOH

OCH_3

$CH_3OC_6H_4COOH$

$C_8H_8O_3$

CAS 100-09-4

A. PROPERTIES

molecular weight 152.14; melting point 275/280°C; density 1.4 at 4/4°C; solubility 400 mg/L at 18°C; $LogP_{OW}$ 1.96.

C. WATER AND SOIL POLLUTION FACTORS

Waste water treatment

decomposition period by soil microflora: 2 d.	(176)

m-anisidine (3-methoxyaniline)

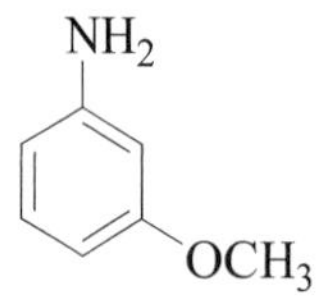

NH_2

OCH_3

$CH_3OC_6H_4NH_2$

C_7H_9NO

CAS 536-90-3

USES

manufacture of dyestuffs.

A. PROPERTIES

molecular weight 123.15; melting point <1°C; boiling point 251°C; density 1.1 at 20/4°C; $LogP_{OW}$ 0.93.

C. WATER AND SOIL POLLUTION FACTORS

Biodegradation

degradation by *Aerobacter:* 500 mg/L at 30°C

ring disruption	parent	80% in 120 h	
	mutant	100% in 24 h	(152)

decomposition period by soil microflora: >64 d (176)

o-anisidine (2-methoxyaniline; 2-methoxybenzenamine)

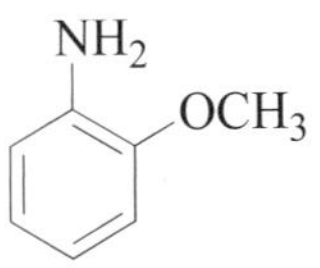

$CH_3OC_6H_4NH_2$

C_7H_9NO

CAS 90-04-0

USES

intermediate in the mfg. of dyes, pharmaceuticals, fragrances.

A. PROPERTIES

molecular weight 123.15; melting point 6.2°C; boiling point 224°C; vapor pressure <0.1 mm at 30°C, 0.02 hPa at 20°C, 2 hPa at 50°C; vapor density 4.25; density 1.09 at 20/4°C; solubility 15,000 mg/L at pH 7 and 20°C, 5,000 mg/L at pH 3 and 20°C; $LogP_{OW}$ 0.95 (calculated), 1.18 (measured).

B. AIR POLLUTION FACTORS

1 mg/m^3 = 0.20 ppm, 1 ppm = 5.12 mg/m^3.

Atmospheric degradation

reaction with OH°: $t_{1/2}$: 0.16 d (calculated). (5623)

C. WATER AND SOIL POLLUTION FACTORS

Biodegradability

inoculum/test method	*test duration*	*results*		
A.S., adapted, Zahn-Wellens	16 d	98%	ThOD	(5622)
A.S., adapted, OECD test	14 d	50%	ThOD	(5622)
A.S., Ind., Zahn-Wellens	3 hours	<10%	ThOD	(5618)
	5 d	20%	ThOD	
	10 d	55%	ThOD	(5617)

degradation by *Aerobacter:* 500 mg/L at 30°C

ring disruption	parent	92% in 120 h	
	mutant	100% in 16 h	(152)

decomposition period by soil microflora: >64 d. (176)

Impact on biodegradation processes:

prim. sew. sludge	24h EC_{50}	1,500 mg/L	
	24h EC_0	500 mg/L	(5620)
	24h EC_0	300 mg/L	(5617)
A.S., municipal	3h EC_{10}	<58 mg/L	
	3h EC_{50}	800 mg/L	
	3h EC_{80}	>1,000 mg/L	(5621)

D. BIOLOGICAL EFFECTS

Bacteria			
Pseudomonas fluorescens	24h EC_0	5,000 mg/L	(5619)
E. coli	24h EC_0	5,000 mg/L	(5619)
Algae			
Scenedesmus pannonicus	short-term EC_{50}	12 mg/L	(5618)
Crustaceans			
Daphnia magna	48h EC_{50}	6.8 mg/L	(5618)
	48h LC_{50}	12 mg/L	(5618)
Fishes			
Leuciscus idus	96h LC_0	80 mg/L	(5618)
Poecilia reticulata	14d LC_{50}	165 mg/L	(5618)
	14d EC_{50}	18 mg/L	(5618)

p-anisidine (4-methoxyaniline; *p*-aminoanisole; *p*-aminomethoxybenzene; *p*-aminomethylphenylether; 4-methoxybenzenamine)

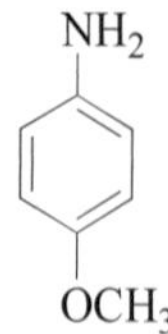

$CH_3OC_6H_4NH_2$

C_7H_9NO

CAS 104-94-9

A. PROPERTIES

molecular weight 123.15; melting point 57/59°C; boiling point 243°C; vapor pressure 0.04 hPa at 20°C; vapor density 4.25; density 1.06 at 67/4°C; solubility 21,000 mg/L at 20°C, 53,000 mg/L at 40°C; $LogP_{OW}$ 0.95.

B. AIR POLLUTION FACTORS

1 mg/m^3 = 0.20 ppm; 1 ppm = 5.12 mg/m^3.

C. WATER AND SOIL POLLUTION FACTORS

Biodegradability

degradation by *Aerobacter:* 500 mg/L at 30°C:

ring disruption:	parent:	86% in 120 h	
	mutant:	100% in 12 h	(152)

decomposition period by soil microflora: 64 d (176)

inoculum/methods	*test conc.*	*test duration*	*results*	
A.S., industr., Zahn-Wellens		3 hours	5-15% ThOD	
		10 d	52% ThOD	
		15 d	75% ThOD	
		20 d	93% ThOD	(5629)
		8 d	85% ThOD	
		13 d	96% ThOD	(5633)
soil suspension, OECD screening		8 d	100% product	(5633)
A.S., industr., Zahn-Wellens		5 d	14% ThOD	
		10 d	87% ThOD	
		15 d	96% ThOD	(5628)
A.S., Sturm test	10 mg/L	14 d	0-54% DOC	(5634)
		28 d	4-79% DOC	(5634)

A.S., adapted, Repetitive Die-Away test: $t_{1/2}$ <2 weeks (5624)

A.S., nonadapted, Repetitive Die-Away test: $t_{1/2}$: <3 weeks (5624)

Impact on biodegradation processes

primary sew. sludge	24h EC_0	800 mg/L	(5629)
	24h EC_0	1,000 mg/L	(5628)

D. BIOLOGICAL EFFECTS

RubisCo Test

nhib. of enzym. activity of Ribulose-P2-carboxylase in protoplasts:	IC_{10}	6.2 mg/L	(2698)

Bacteria

Pseudomonas putida	EC_0	1,000 mg/L	(5625)
Photobacterium phosphoreum	30 min EC_{50}	14 mg/L	(5632)

Algae

Scenedesmus pannonicus	short-term EC_{50}	14 mg/L	(5624)
Microcystis aeruginosa	24h LC_0	20 mg/L	(5631)
Scenedesmus subspicatus inhib. of fluoresc.	IC_{10}	2.5 mg/L	
growth inhib.	IC_{10}	1.2 mg/L	(2698)

Crustaceans			
Daphnia magna	24h EC_{50}	150 mg/L	(5629)
	48h EC_{50}	0.18 mg/L	
	48h LC_{50}	0.33 mg/L	(5624)
	24h EC_0	1 mg/L	
	24h EC_{50}	52 mg/L	
	24h EC_{100}	>500 mg/L	(5630)
	48h EC_0	0.24 mg/L	
	48h EC_{50}	1.9 mg/L	
	48h EC_{100}	16 mg/L	(5630)
Fishes			
Poecilia reticulata	14d LC_{50}	190 mg/L	
	14d EC_{50}	110 mg/L	(5624)
Brachydanio rerio	96h LC_0	100 mg/L	(5625)
Oryzias latipes	48h LC_{50}	40; 324 mg/L	(5626, 5627)
	24h LC_{50}	55 mg/L	(5627)
Leuciscus idus	96h LC_0	125 mg/L	(5628)

anisole *see* methylphenylether

p-anisylalcohol (4-methoxybenzyl alcohol; anisic alcohol; anise alcohol; 4-Methoxybenzenemethanol; para methoxybenzyl alcohol; Anise alcohol; Anisic alcohol)

$C_8H_{10}O_2$

CAS 105-13-5

USES

a fragrance in perfumed consumer products such as dishwash, car shampoo, hand cleaner, toilet paper, nappies in concentrations of 0.0014 wt%. (7195); Flavoring agents and adjuvants used in cigarettes; used in smokeless tobacco.

OCCURRENCE

found in: honey, tomato Industry claims used in: gelatin and puddings, baked goods, frozen dairy deserts, beverages, and soft candy.

A. PROPERTIES

white to pale yellow crystals; molecular weight 138.17; melting point 23-25°C°C; boiling point 259; density 1.1

C. WATER AND SOIL POLLUTION FACTORS

Degradation pathway of anisyl alcohol by *Penicillium simplicissimum* [10946]

D. BIOLOGICAL EFFECTS

Toxicity

Mammals		
Rat	oral LD_{50}	1200 mg/kg bw

anol *see* cyclohexanol

anone *see* cyclohexanone

ansar (disodium methanearsonic acid)

anthanthrene (dibenzo (cd, jh) pyrene; dibenzo (def-mno) chrysene)

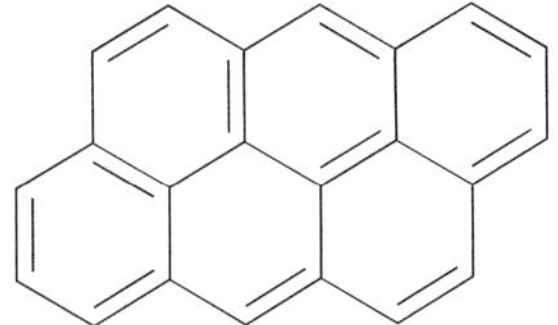

$C_{22}H_{12}$

CAS 191-26-4

MANMADE SOURCES

in coal tar: 2 g/kg. (2600)

A. PROPERTIES

molecular weight 276.34.

B. AIR POLLUTION FACTORS

Manmade sources

in tailgases of gasoline engine:	2-73 μg/m³	(340)
in tailgases of gasoline engine:	0.017-0.026 mg/L gasoline consumed	(1070)
in gasoline (high octane number):	0.028-2.1 mg/L	(1220, 380)
in bitumen:	0.04-0.30 ppm	(500)
emission from space installation burning:		
coal (underfeed stoker)	0.29 mg/10^6 Btu input	
gas	0.2 mg/10^6 Btu input	(954)

Ambient air quality

glc's in the Netherlands in the major cities		
summer 1968-1971	<1-1 ng/m³	
winter 1968-1971	<1-4 ng/m³	(1277)
in the average American urban atmosphere, 1963: 2.3 μg/g airborne particulates, or	0.26 ng/m³ air	(1293)
glc's in Birkenes (Norway): Jan.-June '77:		
	avg.: 0.03 ng/m³	
	range: n.d. -0.25 ng/m³ (n = 18)	
glc's in Rörvik (Sweden): Dec. 1976-April '77:		
	avg.: 0.04 ng/m³	
	range: n.d. -0.27 ng/m³ (n = 21)	(1236)

anthracene

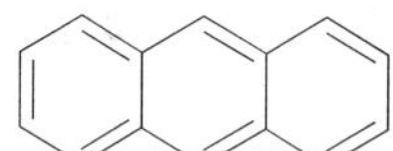

$C_{14}H_{10}$

CAS 120-12-7

USES

dyes; intermediate for anthraquinone dyestuffs.

MANMADE SOURCES

constituent of coal tar creosote: 11 wt %; constituent of diesel fuel: 100-300 mg/L; in coal tar: 12 g/kg. (2387, 2600)

A. PROPERTIES

molecular weight 178.23; melting point 216.2-216.4°C; boiling point 340°C; vapor pressure 1.96 × 10^{-4} mm Hg at 20°C; vapor density 6.15; density 1.25; log P_{OW} 4.45; solub.: 1.29 mg/L at 25°C in distilled water, 0.6 mg/L at 25°C in salt water, 0.075 mg/L at 15°C

B. AIR POLLUTION FACTORS

Manmade sources

in coke oven emissions: 46-942 µg/g of sample (960)

emissions from space heating installation burning: coal (underfeed stoker): 0.85 mg/10^6 Btu input gasoil: 3.9 mg/10^6 Btu input (954)

in gasoline: 1.55 mg/L

in exhaust condensate of gasoline engine: 0.53-0.64 mg/L gasoline consumed (1070)

emissions from typical European gasoline engine: 18-392 µg/L fuel burnt (1291)

in gasoline (high octane number): 2.6 mg/L (380)

in outlet of waterspray tower of asphalt hot-road-mix process: 1,600 ng/m^3

in outlet of asphalt air-blowing process: 220,000 ng/m^3 (1212)

emissions from open burning of scrap rubber tires: 50-56 mg/kg tire (2950)

emissions from open burning of scrap rubber tires: 0-1 mg/kg of tire (a,h)anthracene (2950)

emission rates released from burning wood in residential open fireplaces: pine wood: 0.05 mg/kg logs burned; oak wood: 0.057 mg/kg logs burned. (7087)

Ambient air quality

glc's in Birkenes (Norway) Jan.-June 1977: avg: 0.03 ng/m^3 range: n.d. -0.23 ng/m^3 (n = 18)

glc's in Rörvik (Sweden) Dec. 1976-April 1977: avg: 0.09 ng/m^3 range: n.d. -0.95 ng/m^3 (n = 21) (1236)

glc's in Budapest 1973: heavy-traffic area: winter: 63 ng/m^3 (6-20 h) summer: 55.1 ng/m^3

low-traffic area: winter: 21.5 ng/m^3 summer: 49.9 ng/m^3 (1259)

organic fraction of suspended matter: Bolivia at 5,280 m altitude (Sept.-Dec. 1975): 0.037-0.055 µg/1,000 m^3 (+ phenanthrene)

Belgium, residential area (Jan.-April 1976): 0.09-0.89 µg/1,000 m^3 (+ phenanthrene) (420)

monthly average gaseous concentrations in ambient air in San Francisco Bay area: june-nov 2000: 180-1700 pg/m^3 (11059)

Photochemical reactions

$t_{1/2}$: 1.67 d based on reactions with OH° (calculated) (8481)
$t_{1/2}$: 35 min in distilled water exposed to midday sunlight. (9726)
absorbs solar radiation strongly; $t_{1/2}$ = 35 min, midday sunlight, midsummer, 35°N latitude (2903)

C. WATER AND SOIL POLLUTION FACTORS

BOD_5	2% ThOD	(220)
	0% ThOD	(274)

BOD_{35} 25°C	35% ThOD in seawater/inoculum: enrichment cultures of hydrocarbon-oxidizing bacteria	(521)
COD	35% of ThOD	(220)
	94% of ThOD (0.05 *n* Cr_2O_7)	(274)
ThOD	3.41	(521)

Manmade sources

In storm runoff waters from M-6 motorway U.K. 1986: mean: 56 ng/L; range: 9-215 ng/L (2794)
in aqueous phase of diesel fuel: 0.0004-0.002 mg/L
In Canadian municipal sludges and sludge compost: September 1993- February 1994: mean values of 11 sludges ranged from 0.01 to 1.7 mg/kg dw.; mean: 0.2 mg/kg dw mean value of sludge compost: 0.003 mg/kg dw. (7000)
Diesel-water partition coefficient: log $K_{diesel\text{-}water}$ = 5.27 (2387)

Biodegradation pathway

of anthracene

H OH
OH
H

anthracene → anthracene-*cis*-1,2-dihydrodiol →

OH
OH
O
COOH
C
OH
OH
COOH
C
O

1,2-dihydroxyanthracene → *cis*-4-(2-hydroxynaphth-3-yl)-2-oxobut-3-enoic acid →

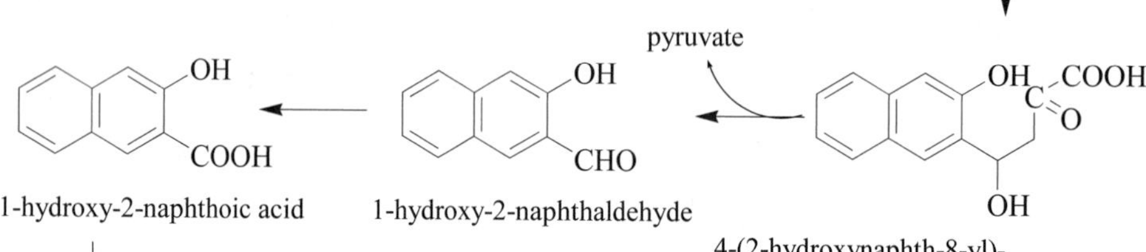

OH
OH

→ salicylate, etc.

Biodegradation

to CO_2: at 15 µg/L in October

sampling site	*incubation time, h*	*degradation rate, (mg/L/d) × 10^3*	*turnover time, d*	
Control station		0	∞	
Near oil storage tanks		70 ± 30	290	
Near oil storage tanks		0	∞	
Skidaway River	24	0	∞	
Skidaway River	72	8 ± 2	2,000	(381)

Aerobic slurry bioremediation reactor with mixed aeration chamber, treating a petrochemical waste sludge at 3.5 to 7.5% solids containing approx. 25% oil and grease and 0.5 to 2.5 grams of waste material per gram of microorganisms, at 22-24°C during batch treatment:
waste residue: infl.: 16 mg/kg; effl. after 90 d: <2.5 mg/kg (2800)

Degradation in soil

in sandy loam in the dark at 20°C at 700 mg/kg:	abiotic half-life: 12 months	
	biodegradation half-life: 6 months	(2806)
at initial conc. between 3 and 40 mg/kg soil:	half-life: 3.3; 108; 129; 138; 143; 175 d	(2903)
In soil	$t_{1/2}$: 108-175 d	(9722)
in unacclimatised sediments:	$t_{1/2}$: 57-210 d	
in oil-treated sediments:	$t_{1/2}$: 5-7 d	(9725)

Phanerochaete chrysoporium degraded anthracene to form the metabolite anthraquinone.

Biodegradation

Inoculum	*method*	*conc.*	*duration*	*elimination results*	
soil-water suspension	aerobic at 35°C	0.1 mg/L	5 d	0.1% ThCO2	
			14 d	1.3% ThCO2	
			14 d	98.4% test substance	
soil-water suspension	anaerobic at 35°C	0.1 mg/L	5 d	0.3% ThCO2	
			14 d	1.8% ThCO2	
			14 d	98.2% test substance	(2667)
sewage seed	3-weekly subcultures	5-10 mg/L	7 d	26-43%	
			28 d	51-92%	(9723)

Aquatic reactions

Adsorption on smectite clay particles from simulated seawater at 25°C-experimental conditions: 100 µg anthracene/l; 50 mg smectite/l-adsorption: 0.90 µg/mg = 46% adsorbed (1009)

In estuarine waters:	at 4 µg/L, 6% adsorbed on particles after 3 h at 15 µg/L, 22% adsorbed on particles after 3 h	(381)
After 3 h incubation in natural seawater,	11% of 15 µg/L were taken up by suspended aggregates of dead phytoplankton and bacteria	(957)

Sediment quality

in sediments in Severn estuary (U.K.): 0.1-6.4 ppm dry wt. (includes phenanthrene) (1467)

Micelle-water partition coefficient (K_{mw}) for the anionic surfactant dodecylsulfate: log K_{mw}: 3.81 (2361)

Soil quality

typical values in Welsh surface soil (U.K.), average of 20 0-5 cm cores:
mean: 7.7 µg/kg dry wt.; range: 0.6-72 µg/kg dry wt. (2420)

Mobility in soil: K_{oc}: 26,000 (9725)

Water quality

Samples from 139 streams in 30 states in the US thought to be susceptible to contamination from agricultural or urban activities during 1999-2000: detected in 2 out of 90 samples in values ranging from 0.06 – 0.11 µg/L (7221)

Waste quality

Mean Concentrations in organic wastes and composts in Switzerland 2004 (10559)

	µg/kg dw
Organic waste	144
Organic waste compost	50
Biowaste	35
Biowaste compost	52
Green waste	348
Green waste compost	35
Input material grass	129
Foliage	697

Bark	30
Swiss compost	54
Non-Swiss compost	49

D. BIOLOGICAL EFFECTS

Bioaccumulation

	Concentration µg/L	*exposure period*	*BCF*	*organ/tissue*	
Algae					
Chlorella fusca			7,770	Wet wt.	(2659)
Crustacean					
Daphnia pulex			759-912		(9721)
Fishes					
goldfish	1,000		162		
rainbow trout	$_{14}$C-anthracene	24-72h	9,000-2,000		(9720)
trout			570	(calculated)	(2917)
goldfish			160		(2661)
			4,230	(calculated)	(2917)

Algae

inhibition of photosynthesis of a freshwater, nonaxenic uni-algal culture of *Selenastrum capricornutum* at: 100% saturation: 99% ^{14}C fixation (vs controls) (1690)

Crustaceans

Daphnia pulex: BCF 760, initial conc. in water: 0.02 ppb, equilibrium reached after 4 h. Excretion of ^{14}C-after 16 h incubation with ^{14}C-anthracene and subsequent transfer to clean water resulted in a rapid release (1 h) of about 30% of the total ^{14}C, a slower elimination of roughly 60% with a half-life of 3.3 h, and a tightly bound residue of 8%.
Metabolism: the observed rate of metabolite excretion during the first 24 h of excretion was only 6% of the total ^{14}C outflux rate. (1597)

Molluscs

Uptake and depuration by oysters *(Crassostrea virginica)* from oil-treated enclosure:

time of exposure, d	*depuration time, d*	*concentration* oysters, µg/g	water, µg/L	*accumulation factor, oysters/water*
2	-	5.6	13	430
8	-	2.5	1	2,500
2	7	1.2		
8	7	0.4		
8	23	0.1		

half-life for depuration: 3 d (957)

Worms

Apparent bioconcentration factors in polychaete worms (dry wt./sediment dry wt.)

Prionospio cirrifera and *Spiochaetpoterus costarum*	BCF: 6.6	
Capitella capitata	BCF: 23.6	(2673)

Toxicity

Bacteria			
OECD 209 closed system inhibition	EC_{50}	>0.075 mg/L	(2624)
Algae			
Selenastrum capricornutum	1w EC_{50}	>>saturation	(2917)
Crustaceans			
Daphnia magna	48h LC_{50}	0.035; 3.03 mg/L	

	2h LC_{50}	0.02 mg/L	
Daphnia pulex	24h EC_{50}	0.001; >0.03 mg/L	(2917)
Artemia spp.	48h LC_{50}	>0.05 mg/L	
Artemia salina	3h LC_{50}	0.02 mg/L	(2917)
Insects			
Aedes aegypti	24h LC_{50}	<0.001; 0.15 mg/L	
	48h LC_{50}	0.027 mg/L	
Aedes taeniorhynchus	24h LC_{50}	0.26 mg/L	
Culex quinquefasciatus	24h LC_{50}	0.037 mg/L	(2917)
Culicid mosquito larvae	LC_{50}	26.8 µg/L	(9719)
Amphibians			
Rana pipiens	24h LC_{50}	0.11 mg/L	
	5h LC_{50}	0.025 mg/L	(2917)
Fishes			
Pimephales promelas	24h LC_{50}	0.36 mg/L	(2917)
bluegill sunfish	96h LC_{50}	11.9 µg/L	
trout	24h NOEC	5 mg/L	(30)

Plants
Carrots *(Daucus carota)* (U.K. 1987):
conc. through a cross section: µg/kg dry wt.
peel: 3; inner peel: 1; outer core: 0.3; core: 2
effect of food processing on conc. in whole carrot: µg/kg dry wt.
uncooked: 0.4; cooked: 2.5; tinned: 1.3; frozen: 0.5 (2674)

9,10-Anthracenedione, 1,4-bis[(4-methylphenyl)amino]-, sulfonated, potassium salts (1,4-Di-(4-methylanilino)anthraquinone, sulfonated, potassium salts; AMS)

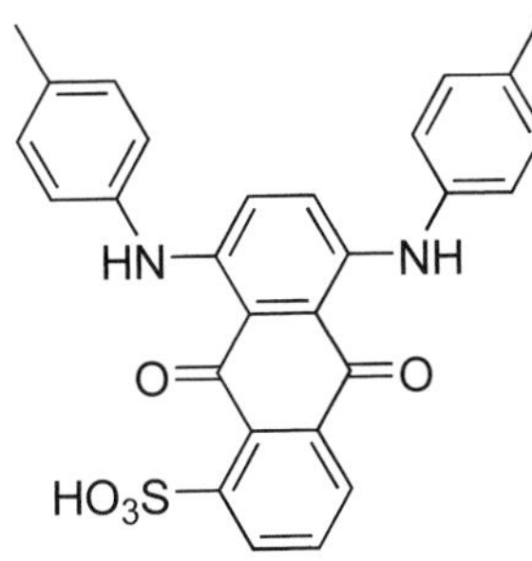

$C_{28}H_{22}N_2O_5S$

CAS 125351-99-7

SMILES
C42=C(C(=O)C3=C(C4=O)C=CC=C3S(=O)(=O)O)C(NC5=CC=C(C)C=C5)=CC=C2NC1=CC=C(C)C=C1

USES

There are 3 categories of potential uses for AMS:
Pigment for colouring printing inks;
Paint, lacquers and varnishes; and

Colorant in manufacture of rubber and plastic products.

A. PROPERTIES

(all modellled) molecular weight 498.56; melting point 321°C; boiling point 732°C; vapour pressure 5 x 10^{-19} Pa; water solubility 0.01 µg/L; log Pow 5.5; H 7 x 10^{-23} atm.m^3/mol

B. AIR POLLUTION FACTORS

Photodegradation

t/2 (OH radicals): 0.05 days. (11176)

C. WATER AND SOIL POLLUTION FACTORS

Mobility in soils

Log K_{OC} 4.5 (11176)

D. BIOLOGICAL EFFECTS

Bioaccumulation

Species	BCF	
Fish	14,000-112,000 (modelled)	(11176)

Toxicity

Fish	96h LC_{50}	0.0056-0.56 mg/L	(11176)

9-anthracenemethanol

C. WATER AND SOIL POLLUTION FACTORS

Soil sorption

K_{oc}:	for Fullerton soil, 0.06% OC:	150	
	for Apison soil, 0.11% OC:	8,000	
	for Dormont soil, 1.2% OC:	2,700	(2599)

anthranilic acid (*o*-aminobenzoic acid; vitamin L1; *o*-carboxyaniline)

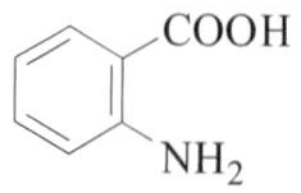

$H_2NC_6H_4COOH$

$C_7H_7NO_2$

CAS 118-92-3

USES

feedstock for dyes, drugs, perfumes, pharmaceuticals, herbicides, and saccharin.

A. PROPERTIES

molecular weight 137.13; melting point 145/147°C; boiling point sublimes; solubility 3,500 mg/L at 14°C; vapor pressure 0.001 hPa at 52°C; solubility 5,700 mg/L at 25°C; LogP_{OW} 1.21.

C. WATER AND SOIL POLLUTION FACTORS

BOD_5: 73; 74% ThOD stand. dil. sew. (281, 16, 5610)

Waste water treatment

acclimated A.S.: 50% ThOD infl.: 250 mg/L, 30 min aeration (92)

Biodegradability

decomposition period by soil microflora: 2 d (176)

adapted A.S. at 20°C-product is sole carbon source: 97% COD removal at 200 mg/L and 27 mg/COD/g dry inoculum/h
lag period for degradation of 16 mg/L by waste water or by soil at pH 7.3 and 30°C: less than 1 d

inoculum/methods	*test conc.*	*test duration*	*results*	
A.S.	400 mg/L	6 d	97% TOC	(5611)
A.S., munic.	16 mg/L	1 d	100% product	(5612)
A.C.	100 mg/L	5 d	47-50% DOC	(5614)
soil suspension	16 mg/L	1 d	100% product	(5612)
A.S., industr.	100-800 mg/L	8 d	97% DOC	(5615)
A.S., Coupled units test	12 mg/L	>1 d	100% DOC	(5616)
A.S., Zahn/Wellens test	400 mg/L	1 d	97% DOC	(5616)
A.S. MITI test	50 mg/L	14 d	86% DOC	(5616)
	50 mg/L	14 d	70% ThOD	(5616)
french AFNOR test	40 mg/L	42 d	93% DOC	(5616)
waste water, Sturm test	10 mg/L	28 d	71% CO_2	(5616)
waste water, Sturm test	10 mg/L	42 d	100% DOC	(5616)
waste water, OECD screening test	3-20 mg/L	19 d	100% DOC	(5616)
waste water	1 mg/L	30 d	90% ThOD	(5616)

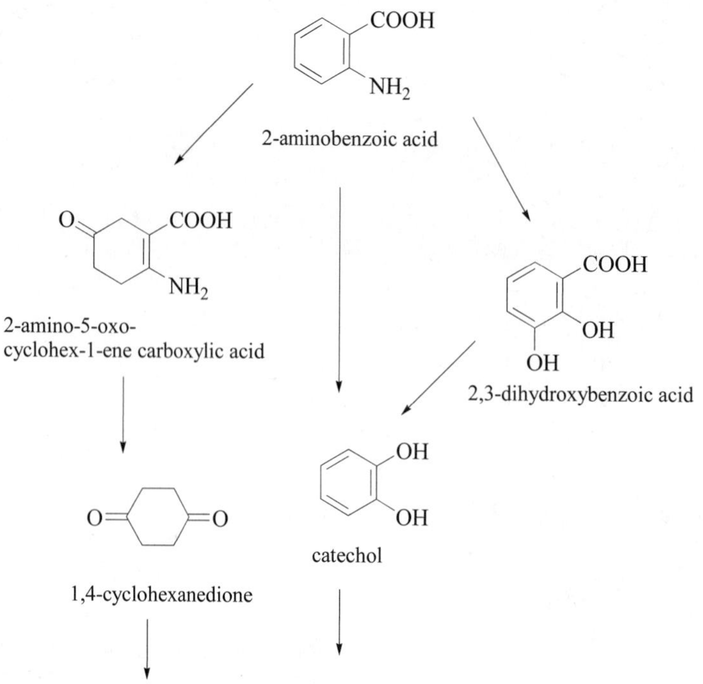
COOH
NH2
2-aminobenzoic acid
COOH
NH2
2-amino-5-oxo-
cyclohex-1-ene carboxylic acid
COOH
OH
OH
2,3-dihydroxybenzoic acid
OH
OH
catechol
1,4-cyclohexanedione

Impact on biodegradation processes

100 mg/L no inhibition of NH_3 oxidation by *Nitrosomonas* sp.	(390)

D. BIOLOGICAL EFFECTS

Bacteria			
Pseudomonas putida	17h EC_{10}	71 mg/L	
	17h EC_{50}	95 mg/L	
	17h EC_{90}	119 mg/L	(5606)
Flavobacterium devorans (Chloramphenicol-resistant mutant)	LOEC	96 mg/L	(5609)
Algae			
Scenedesmus subspicatus	72h EC_{10}	2.6 mg/L	
	72h EC_{50}	19 mg/L	
	72h EC_{90}	50 mg/L	(5606)
	96h EC_{10}	2.1 mg/L	
	96h EC_{50}	21 mg/L	
	96h EC_{90}	52 mg/L	(5606)
Chlorella vulgaris	6h EC_{50}	96-123 mg/L	(5607)
Crustaceans			
Daphnia magna Straus	24h EC_0	62 mg/L	
	24h EC_{50}	102 mg/L	
	24h EC_{100}	250 mg/L	(5606)
Daphnia magna Straus	48h EC_0	62 mg/L	
	48h EC_{50}	85 mg/L	
	48h EC_{100}	125 mg/L	(5606)
Fishes			
Leuciscus idus	1h LC_{50}	215-1,000 mg/L	(5605)
	96h NOEC	100 mg/L	

	96h LC_{50}	100-215 mg/L	(5605)

anthraquinone (9,10-dihydro-9,10-diketoanthracene)

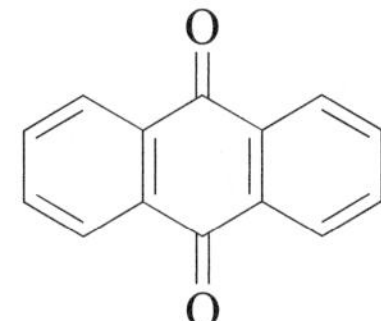

$C_{14}H_8O_2$

CAS 84-65-1

USES

intermediate for dyes and organics, bird repellent for seeds; constituent of coal tar creosote: 1 wt %. (2386)

A. PROPERTIES

yellow-green crystals; molecular weight 208.2; melting point 286°C sublimes; boiling point 379/381°C; vapor pressure 1.3 hPa at 190°C, 0.000013 hPa at 68.8°C; density 1.4 at 20/4°C; solubility 125 mg/L at 22°C; $LogP_{OW}$ 3.39 (measured), 2.7 (calculated).

B. AIR POLLUTION FACTORS

Ambient air quality

organic fraction of suspended matter:

Bolivia at 5,200-m altitude (Sept.-Dec. 1975): 0.064-0.065 µg/1,000 m^3
Belgium, residential area (Jan.-April 1976): 0.57-1.00 µg/1,000 m^3 (428)

glc's in residential area (Belgium) Oct. 1976:
in particulate sample: 1.59 ng/m^3
in gas-phase sample: 5.66 ng/m^3 (1289)

C. WATER AND SOIL POLLUTION FACTORS

Biodegradability

inoculum/method	*test conc.*	*test duration*	*results*	
munic. sew. adapted/Closed Bottle	0.8 mg/L	20 d	>70% ThOD	(5603)
A.S., munic./Sturm test	20 mg/L	24 d	75% CO_2 prod	(5604)
munic. sew./MITI modified	100 mg/L	25 d	93% ThOD	(5604)
munic. sew./Repet. Die-Away		14 d	70% DOC	(5604)
U.K.-MITI test	100 mg/L	28 d	42% ThOD	(5605)

Impact on biodegradation processes

at 2.5 mg/L inhibition of the self-purification activity of natural waters (30)

Activated sludge inhibition 3h IC_{50} 7,264 mg/L (5603)

Inhibition of photosynthesis of a freshwater nonaxenic uni-algal culture of *Selenastrum capricornutum*
at 1% sat.: 97% ^{14}C fixation (vs controls)
10% sat.: 91% ^{14}C fixation (vs controls)

100% sat.: 85% ^{14}C fixation (vs controls) (1690)

Water quality

in Eastern Ontario:
drinking waters (June-Oct. '78): 0.1-2.1 ng/L (n = 12)
raw waters (June-Oct. '78): 0.9-4.7 ng/L (n = 2) (1698)

D. BIOLOGICAL EFFECTS

Bacteria			
Pseudomonas fluorescens	24h EC_0	5,000 mg/L	(5603)
Crustaceans			
Daphnia magna	30d EC_0,S	1 mg/L	(5602)
	30d EC_{100},S	5 mg/L	(5602)
Daphnia longispina	16d EC_0,S	1 mg/L	(5602)
	30d EC_0,S	1 mg/L	(5602)
Chydorus	30d EC_0,S	1 mg/L	(5602)
Fishes			
Pimephales promelas	96h LC_{50}	2,650 mg/L	(5598)
	96h LC_0,F	0.24 mg/L	(5601)
Oncorhynchus mykiss	24h LC_0,S	>5 mg/L	(5599)
Lepomis macrochirus	24h LC_{50},S	>5 mg/L	(5599)
Petromyzon marinus larvae	24h LC_0,S	>5 mg/L	(5599)
Ptychocheilus oregonensis	13h LC_{100}	10 mg/L	(5600)
Oncorhynchus tschawytscha	9h LC_{100},S	10 mg/L	(5600)
Oncorhynchus kisutch	9h LC_{100},S	10 mg/L	(5600)

anthraquinone-α-sulfonic acid

O SO$_3$H
O

CAS 82-49-5

C. WATER AND SOIL POLLUTION FACTORS

BOD_5: 0 std. dil. sew. (n.s.i.). (281, 161)

D. BIOLOGICAL EFFECTS

Daphnia magna	24h LC_{50}:	186 mg/L (sodium salt)	
	48h LC_{50}:	186 mg/L (sodium salt)	
	72h LC_{50}:	186 mg/L (sodium salt)	
	96h LC_{50}:	50 mg/L (sodium salt)	(153)
	100h LC_{50}:	12 mg/L (sodium salt)	(1295)

Snail eggs, *Lymnaea* sp.:	24-96h LC_{50}:	186 mg/L (sodium salt)	(153)

1-anthraquinonesulfonic acid potassium salt (9,10-dihydro-9,10-dioxo-1-anthracenesulfonic acid; potassium-1-anthraquinonesulfate)

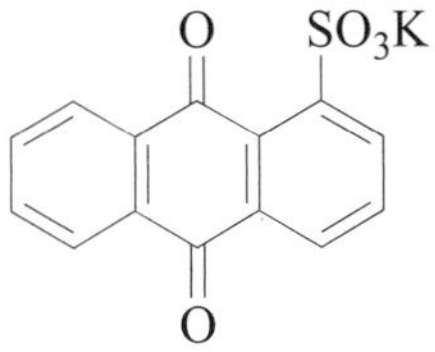

$C_{14}H_8O_5S.K$

CAS 30845-78-4

USES AND FORMULATIONS

Base stock for chemical synthesis.

A. PROPERTIES

molecular weight 327.4; melting point >250°C; solubility 10 840 mg/L at 20°C; $LogP_{OW}$ 0.9 (calculated).

C. WATER AND SOIL POLLUTION FACTORS

Biodegradation

Inoculum	*Method*	*Concentration*	*Duration*	*Elimination results*	
Municipal waste water	Closed bottle test	3 mg/L	30 d	>70% ThOD	(6540)

D. BIOLOGICAL EFFECTS

Toxicity

Bacteria			
Pseudomonas fluorescens	24 h EC_0	1,000 mg/L	(6540)
Fishes			
Leuciscus idus	48 h LC_0	20 mg/L	(6540)
	96 h LC_{50}	1.5 mg/L	
	96 h LC_{100}	2.2 mg/L	(6100)

antifebrin *see* acetanilide

antimony triacetate

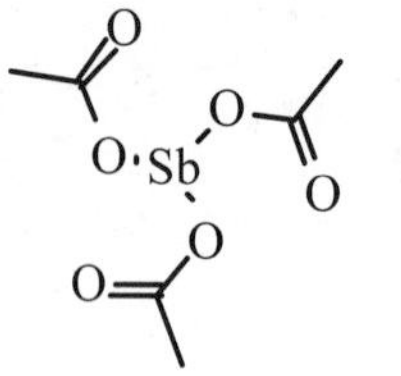

$C_6H_9O_6Sb$

$Sb(OOCCH_3)_3$

CAS 6923-52-0

TRADENAMES

catalyst S-21

USES

antimony triacetate is used as a catalyst in the manufacture of polyethylene terephthalate for the production of polyester fibre, yarn and/or chips.

A. PROPERTIES

off-white hygroscopic crystals; molecular weight 298.75; melting point 120-125 °C; decomposition commences at temperatures above 200°C to antimony trioxide and acetic acid; density 1.2 at 20°C; water solubility: decomposes in the presence of water.

B. AIR POLLUTION FACTORS

Hydrolysis

In air antimony triacetate reacts with residual moisture and decomposition is most likely diffusion controled.

C. WATER AND SOIL POLLUTION FACTORS

Hydrolysis

the chemical hydrolyses immediately with water to antimony trioxide and acetic acid : 2 $Sb(OOCCH_3)_{3(s)}$ + 3 H_2O (l) ---> $Sb_2O_{3(s)}$ + 6 $CH_3COOH_{(aq)}$
Antimony trioxide is insoluble and will settle to sediments and be immobile in soils. (7253)
Environmental fate: After conversion to textile fabrics or polyester padding, the fate of antimony triacetate is linked with the fate of the particular item. Eventually the polyester will enter the waste disposal stream for recycling or ultimately for disposal as waste in landfill. During usage, fibre abrasion will lead to particles of the chemical entering the environment. (7253)

D. BIOLOGICAL EFFECTS

Data are provided for the decomposition products antimony trioxide and acetic acid. However the pH changes *per se* - resulting from the decomposition – might lead to some detrimental effects (7253)

Toxicity

Algae

antimony trioxide	96h EC_{50}	0.61->4.2 mg/L

acetic acid	96h EC_{50}	90 mg/L	(7253)
Crustaceans			
Daphnia magna			
antimony trioxide	48h LC_{50}	423-530 mg/L	
Mysid shrimp			
antimony trioxide	48h LC_{50}	920 mg/L	
Water fleas			
acetic acid	24h LC_{50}	47 mg/L	
Brine shrimp			
acetic acid	24-48h LC_{50}	32-42 mg/L	(7253)
Worms			
Tubifex worms			
antimony trioxide	48h LC_{50}	>4.2 mg/L	(7253)
Fishes			
Pimephales promelas			
antimony trioxide	96h LC_{50}	833 mg/L	
acetic acid	96h LC_{50}	79 mg/L	(7253)
Lepomis macrochirus			
antimony trioxide	96h LC_{50}	530 mg/L	
acetic acid	96h LC_{50}	79 mg/L	(7253)
Cyprinodon variegatus			
antimony trioxide	96h LC_{50}	>6.2 mg/L	
Mammals			
Rat	acute oral LD_{50}	> 2500; < 5000; 7000 mg/kg bw	(7253)

aquathol K (7-oxabicyclo(2,2,1)heptane-2,3-dicarboxylic acid; 1,2-dicarboxy-3,6-endoxocyclohexane; 3,6-endoxohexahydrophthalic acid; endothall; accelerate; des-i-cate; hydout; hydrothol; ripenthol, triendothal; 7-oxabicyclo[2.2.1]heptane-2,3-dicarboxylic acid; 1,2-dicarboxy-3,6-endoxocyclohexane; des-i-cate; 3,6-epoxycyclohexane-1,2-dicarboxylic acid; Tri-endothal; 3,6-endoxohexahydrophthalic acid)

O

COOH

COOH

$C_8H_{10}O_5$

CAS 145-73-3

USES

herbicide, defoliant desiccant, growth regulator.

OCCURRENCE

Endothall exists as a mixture of 3 stereoisomers of which the (1R,2S,3R,4S)-isomer is the most herbicidally active

A. PROPERTIES

molecular weight 186.18 melting point 144°C; density 1.43 at 20°C; solubility 100 g/L at 20°C; $logP_{ow}$ 1.91.

B. AIR POLLUTION FACTORS

Incinerability

thermal stability ranking of hazardous organic compounds: rank 319 on a scale of 1 (highest stability) to 320 (lowest stability). (2390)

C. WATER AND SOIL POLLUTION FACTORS

Stability

Endothall is stable to oxidation, hydrolysis and photolysis. (9805)

Biodegradation

in a shake-flask study using an oligomesotrophic reservoir water: $t_{1/2}$: 8.4 d . (9803)

Incubation of $_{14}$C-ring-labelled endothall with Arthrobacter sp. revealed that $_{14}$C was incorporated into cellular amino acids, proteins, nucleic acids and lipids and was released as carbon dioxide. The major metabolite was glutamic acid. Minor metabolites included aspartic acid, citric acid, alanine and phosphate esters. (9804)

Soil mobility: Kp for lake sediment 0.4-0.9 (9806)

D. BIOLOGICAL EFFECTS

Algae

		technical acid, mg/L	*amine salt, mg/L*	
Chlorococcum sp.	10d EC_{50} O_2 evolution	100	>1,000	
	10d EC_{50} growth	40	300	
Dunaliella tertiolecta	10d EC_{50} O_2 evolution	425	>1,000	
	10d EC_{50} growth	50	45	
Isochrysis galbana	10d EC_{50} O_2 evolution	60	>1,000	
	10d EC_{50} growth	25	22	
Phaeodactylum	10d EC_{50} O_2 evolution	75	>1,000	
tricornutum	10d EC_{50} growth	15	25	(2348)

Crustaceans

Gammarus lacustris	96h NOEC	100 mg/L (dipotassium salt)	(2124)

Insects

Mayfly *(Ephemerella walkeri)*	LOEC avoidance	>10 mg/L (dipotassium salt)	

Fishes

Salmo gairdneri	LOEC avoidance	>10 mg/L (dipotassium salt)	(1621)
Pimephales promelas	96h LC_{50}	320 mg/L (dipotassium salt)	(2131)
Lepomis macrochirus	96h LC_{50}	160 mg/L (dipotassium salt)	(2131)
Pimephales notatus	96h LC_{50}	110 mg/L (disodium salt)	(2132)
Micropterus salmoides	96h LC_{50}	120 mg/L (disodium salt)	(2132)
Notropis umbratilus	96h LC_{50}	95 mg/L (disodium salt)	(2132)
Micropterus salmoides	96h LC_{50}	200 mg/L (disodium salt)	(2132)
Oncorhynchus tschawytscha	96h LC_{50}	136 mg/L (disodium salt)	(2132)

Mammals

rat	oral LD_{50}	35-51 mg/kg bw	(9807,

9808)

arachidic acid *see* eicosanoic acid

aramite (Sulfurous acid, 2-(p-t-butylphenoxy)-1-methylethyl-2-chloroethyl ester; 2-(p-tert-butylphenoxy)isopropyl-2-chloroethyl-sulfite)

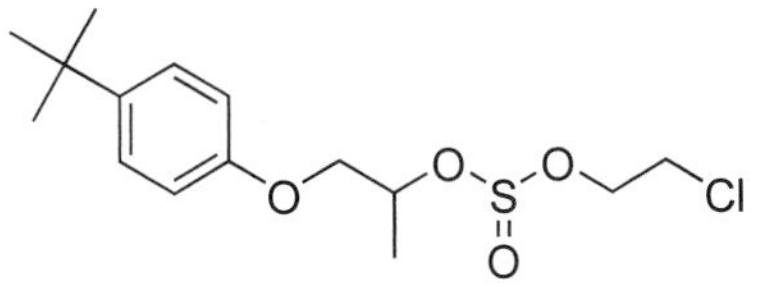

$C_{15}H_{23}ClO_4S$

CAS 140-57-8

SMILES CC(COc1ccc(cc1)C(C)(C)C)OS(=O)OCCCl

TRADENAMES

Aracide , Aramite , Aramite(r) , Armaite , ENT-16519 , Niagaramite

USES

Antimicrobial agent; miticide

A. PROPERTIES

Colorless liquid; molecular weight 334.87; melting point −31.7 °C; boiling point 175°C at 0.1 mm Hg; density 1.14 kg./L at 20°C

B. AIR POLLUTION FACTORS

Incinerability

Thermal stability ranking of hazrdous organic compounds: rank 235 on a scale of 1 (higheste stability to 320 (lowest stability) (2390)

C. WATER AND SOIL POLLUTION FACTORS

Mobility in soils

K_{OC} 15,500

D. BIOLOGICAL EFFECTS

Toxicity

Crustaceae

Daphnia magna	48h EC_{50}	0.16 mg/L
	26h LC_{50}	0.069 mg/L

Daphnia pulex	48h EC_{50}	0.16 mg/L	
Simocephalus semulatus	48h EC_{50}	0.18; 0.23 mg/L	
Gammarus fasciatus	96h LC_{50}	0.060 mg/L	
	24h LC_{50}	0.35 mg/L	
Gammarus lacustris	24h LC_{50}	0.35 mg/L	
	48h LC_{50}	0.10 mg/L	
	96h LC_{50}	0.060 mg/L	(10866)
Insecta			
Pteronarcys californicus	24-96h LC_{50}	1.0 mg/L	(10866)
Fish			
Lepomis macrochirus	96h LC_{50}	0.35 mg/L	
	24h LC_{50}	0.48 mg/L	
Onchrorhynchus mykiss	96h LC_{50}	0.32 mg/L	
	48h LC_{50}	0.39 mg/L	
	24h LC_{50}	0.72; 0.73 mg/L	
Ictalurus punctatus fingerlings	1-24h LC_{50}	100 mg/L	(10866)

aroclor 1016 *see also* polychlorobiphenyl

IMPURITIES, ADDITIVES, COMPOSITION

A mixture of polychlorobiphenyls containing on average 16 wt % of chlorine.

A. PROPERTIES

solubility 0.22-0.25 mg/L; 0.049 mg/L at 24°C.

Composition of the water soluble fraction (WSF)

monochloro isomers	111 µg/L	or	12% of WSF	
dichloro isomers	280 µg/L	or	31% of WSF	
trichloro isomers	329 µg/L	or	36% of WSF	
tetrachloro isomers	186 µg/L	or	21% of WSF	
total WSF	906 µg/L		100%	(1909)

C. WATER AND SOIL POLLUTION FACTORS

Partition coefficients to natural sediments

sediment	*% organic carbon*	*partition coefficient*
Oconee River	0.4	620
USDA Pond	0.8	1,370
Doe Run Pond	1.4	1,290
Hickory Hill Pond	2.4	1,300

D. BIOLOGICAL EFFECTS

Concentration of aroclor 1016 in fish confined to a live-cage for 14 d near Rodger's Island in the Hudson River: average conc. of aroclor 1016 in Hudson water: 0.17 µg/L.

	µg/kg whole fish	*BCF*
creek chubsucker *(Ermyzon oblongus)*	2,200	13,000
yellow perch *(Perca flavescens)*	1,800	10,600
pumpkinseed *(Lepomis gibbosus)*	2,500	14,700

brown bullhead *(Ictalurus nebulosus)*	3,800	22,300	
	2,800	17,000	(1434)

aroclor 1221 *see also* polychlorobiphenyls

CAS 11104-28-2

IMPURITIES, ADDITIVES, COMPOSITION

A mixture of polychlorobiphenyls containing on average 21 wt % of chlorine.

A. PROPERTIES

solubility 0.59 mg/L at 24°C.

Composition of the water soluble fraction (WSF)

monochloro isomers	3,241	µg/L	or	92%	of WSF	
dichloro isomers	233	µg/L	or	7%	of WSF	
trichloro isomers	31	µg/L	or	1%	of WSF	
tetrachloro isomers	11	µg/L	or	0.3%	of WSF	
total WSF	3,516	µg/L	or	100%	of WSF	(1909)

C. WATER AND SOIL POLLUTION FACTORS

Biodegradation

at 0.1 mg/L	normal sewage	adapted sewage	
after 24 h	0%	49%	
after 135 h	12%	96%	(997)

D. BIOLOGICAL EFFECTS

Fishes

harlequin fish *(Rasbora heteromorpha)*:

mg/L	*24 h*	*48 h*	*96 h*	
LC_{10}	1.1	1.05	0.98	
LC_{50}	1.3	1.15	1.05	(331)

cutthroat trout: 96h LC_{50} 1.2 mg/L (1617)

aroclor 1232

CAS 11141-16-5

D. BIOLOGICAL EFFECTS

Fishes

harlequin fish *(Rasbora heteromorpha)*

mg/L	*24 h*	*48 h*	*96 h*

LC_{10}	0.52	0.27	-	
LC_{50}	0.9	0.56	0.32	(331)

cutthroat trout: 96h LC_{50} 2.5 mg/L (1617)

aroclor 1242

CAS 53469-21-9

USES AND FORMULATIONS

dielectric liquids; thermostatic fluids; swelling agents for transmission seals; additives or bases for lubricants, oils, and greases; plasticizers for cellulosics, vinyls, and chlorinated rubbers. A mixture of polychlorobiphenyls containing an average of 42 wt % chlorine.

A. PROPERTIES

vapor pressure 50 mm at 225°C; density 1.41 at 65/15.5°C; solubility 0.10 mg/L at 24°C.

Composition

monochlorobiphenyls	3%
dichlorobiphenyls	13%
trichlorobiphenyls	28%
tetrachlorobiphenyls	30%
pentachlorobiphenyls	22%
hexachlorobiphenyls	4%

Composition of the water soluble fraction (WSF)

			%	
monochloro isomers	137 µg/L	or	19	
dichloro isomers	224 µg/L	or	32	
trichloro isomers	220 µg/L	or	31	
tetrachloro isomers	116 µg/L	or	17	
pentachloro isomers	6 µg/L	or	1	
total WSF	702.7 µg/L		100	(1909)

C. WATER AND SOIL POLLUTION FACTORS

Control methods

calculated half-life based on evaporative loss for a water depth of 1 m at 25°C: 12 h. (330)

Aquatic reactions

partition coefficient to natural sediments.

sediment	*organic carbon %*	*partition coefficient*	
Oconee River	0.4	540	
USDA Pond	0.8	1,210	
Doe Run Pond	1.4	1,090	
Hickory Hill Pond	2.4	1,250	(1068)

D. BIOLOGICAL EFFECTS

Bioaccumulation

fathead minnows (from water) 8 m BCF 32,000-274,000. (1838)

Toxicity

Crustaceans		
amphipod *(Gammarus fasciatus)*	4d LC_{50},F	0.010 mg/L
crayfish *(Orconectes nais)*	7d LC_{50},S	0.030 mg/L

Insects		
naiad of damselfly *(Ischnura fasciatus)*	4d LC_{50},F	0.40 mg/L
naiad of dragonfly (*Macromia* sp.)	7d LC_{50},S	0.80 mg/L

Fishes

LC_{50},F mg/L	days	*5*	*10*	*15*	*20*	*25*	*30*
rainbow trout		0.067	0.048	0.018	0.010	0.012	
bluegills				0.164	0.125	1.120	0.084
channel catfish				0.219	0.150	0.132	0.087
							(1617)

harlequin fish *(Rasbora heteromorpha)* *mg/L*	*24 h*	*48 h*	*96 h*	
LC_{10},F	0.63	0.275	-	
LC_{50},F	0.96	0.6	0.37	(331)

cutthroat trout: 96h LC_{50},S 5.4 mg/L.

rainbow trout: acute oral toxicity >1.5 g/kg. (1617)

aroclor 1248

CAS 12672-29-6

IMPURITIES, ADDITIVES, COMPOSITION

Composition

dichlorobiphenyls	2%	
trichlorobiphenyls	18%	
tetrachlorobiphenyls	40%	
pentachlorobiphenyls	36%	
hexachlorobiphenyls	4%	(1837)

C. WATER AND SOIL POLLUTION FACTORS

Control methods

calculated half-life based on evaporative loss for a water depth of 1 m at 25°C: 9.5 h. (330)
Complete dechlorination of aroclor 1248 was obtained with 69% nickel on Kieselguhr in the presence of sodium hydroxide and 50 atm of hydrogen at 115°C for 6 h.

D. BIOLOGICAL EFFECTS

Bioaccumulation

Pimephales promelas	8m BCF 60,000-120,000	(1919, 1838)
channel catfish:		
exposure (d)	*BCF (conc. in water 0.0058 µg/L)*	
7	3,500	
14	8,000	
28	29,000	
56	34,000	
77	56,370	(1617)

Toxicity

Crustaceans

amphipod *(Gammarus fasciatus)*	4d LC_{50}	0.052 mg/L	(1617)
Fishes			
	exposure (d)	*LC_{50} mg/L*	
rainbow trout	5	0.054	
	10	0.038	
	15	0.016	
	20	0.0064	
	25	0.0034	
bluegills	5	0.136	
	10	0.115	
	15	0.111	
	20	0.106	
	25	0.100	
	30	0.078	
fathead minnow larvae *(Pimephales promelas)*	30d	4.7 µg/L	(1919)
channel catfish	10	0.121	
	15	0.121	
	20	0.115	
	25	0.104	
	30	0.075	
Cutthroat trout	96h	5.7 mg/L	(1617)
Rainbow trout (acute oral toxicity)		>1.5 g/kg	(1617)

aroclor 1254

CAS 27323-18-8

A. PROPERTIES

solubility 0.057 mg/L at 24°C.

Composition

tetrachlorobiphenyls	11%	
pentachlorobiphenyls	49%	
hexachlorobiphenyls	34%	
heptachlorobiphenyls	6%	(1837)

Composition of the water soluble fraction (WSF)

			%	
trichloro isomers	1.8 µg/L	or	2.6	
tetrachloro isomers	28 µg/L	or	40	
penta- and hexachloro isomers	40 µg/L	or	57	
total WSF	69.8 µg/L		100	(1909)

C. WATER AND SOIL POLLUTION FACTORS

Control methods

calculated half-life based on evaporative loss for a water depth of 1 m at 25°C: 10 h. (330)

Dechlorination was achieved in 2-propanol with 2.0 mmol $NiCl_2$, 60 mmol $NaBH_4$, and 0.3 mmol aroclor 1254. Biphenyl constituted 97% of the reaction products, and monochloro- and dichlorobiphenyls the remaining products. (1551)

Biodegradation

biodegradation at 0.05 mg/L:	normal sewage	adapted sewage	
after 24 h	0%	52%	
after 135 h	0%	43%	(997)

Impact on biodegradation processes

Effect on the degradation of glucose by a mixed culture derived from A.S.:

conc., mg/L	*increase in lag period*	*respiration rate*	
1	0 hours	100%	
10	0 hours	110%	
100	0 hours	135%	
1,000	>200 hours	0%	(997)

D. BIOLOGICAL EFFECTS

Bioaccumulation

BCF for aquatic invertebrates:

	concentration in water	*exposure period, days*				
species	*µg/L*	*1*	*4*	*7*	*14*	*21*
daphnid	1.1	2,100	3,800	-	-	-
amphipod	1.6	4,400	5,200	6,000	6,300	6,200
crayfish	1.2	160	240	350	500	750
grass shrimp	1.3	1,300	1,900	2,000	2,200	2,600
stonefly	2.8	640	710	740	750	740
dobsonfly	1.1	300	1,000	1,260	1,460	1,500
mosquito	1.5	2,300	3,400	3,500	-	-
phantom midge	1.3	2,400	2,500	2,600	2,700	-

Fishes

Bioaccumulation factors: BCFs:

channel catfish at 0.0024 mg/L water:	*after x d*	*BCF*	
	7	6,500	
	14	12,000	
	28	28,000	
	56	42,500	
	77	61,190	(1617)
brook trout	118	40,000-47,000	(1531)
fathead minnows	244	109,000-238,000	(1436)
	240	46,000-307,000	(1838)
rainbow trout	30	34,000-46,000	(1437)

Toxicity

Algae			
Tetrahymena pyriformis	96h EC_{13},S	10 ppb	(2350)
Insects			
naiad of damsefly *(Ischnura verticalis)*:	4d LC_{50},F	0.20 mg/L	
naiad of dragonfly (*Macromia* sp.)	7d LC_{50},S	1.0 mg/L	(1617)
Crustaceans			
pink shrimp *(Penaeus duorarum)*	15d LC_{50},F	0.94 ppb	
	35d LC_{50},F	3.5 ppb	
spot *(Leistomus xanthurus)*	18d LC_{50},F	5 ppb	
pinfish *(Lagodon rhomboides)*	12d LC_{50},F	5 ppb	
amphipod *(Gammarus fasciatus)*	4d LC_{50},S	2.4 mg/L	
crayfish *(Orconectes nais)*	7d LC_{50},S	0.10 mg/L	
grass shrimp *(Palaemonetes kadiakensis)*	7d LC_{50},F	0.003 mg/L	
striped hermit crab *(Cibanarius vittatus)*	96h LC_0,S	30 µg/L	(1573)

Fishes

	d	LC_{50},F mg/L	
brook trout	128d LC_{50},F	6.2 µg/L	(1531)
rainbow trout	10	0.160	
	15	0.064	
	20	0.039	
	25	0.027	
bluegills	15	0.303	
	20	0.260	
	25	0.239	
	30	0.177	
channel catfish	10	0.303	
	15	0.286	
	20	0.293	
	25	0.181	
	30	0.139	
Rasbora heteromorpha	24h LC_{50},F	1.6; 6.2 mg/L	
	48h LC_{50},F	0.82; 1.45 mg/L	
	96h LC_{50},F	0.56; 1.1 mg/L	(331)
cutthroat trout	96h LC_{50}	42 mg/L	(1617)
rainbow trout	acute oral toxicity	>1.5 g/kg	
deepwater ciscoes *(Coregonus)* 22-d-old fry:			
	96h LC_{20}	10 mg/L	
	5d LC_{50}	3.2 mg/L	(1861)

Biotransfer factors

biotransfer factor in beef	log B_b	-1.28	
biotransfer factor in milk	log B_m	-1.95	
bioconcentration factor for vegetation	log BCF_v	-1.77	(2644)

aroclor 1260

CAS 11096-82-5

A. PROPERTIES

solubility 0.080 mg/L at 24°C.

Composition

pentachlorobiphenyls	12%	
hexachlorobiphenyls	38%	
heptachlorobiphenyls	41%	
octachlorobiphenyls	8%	
nonachlorobiphenyls	1%	(1837)

C. WATER AND SOIL POLLUTION FACTORS

Control methods

calculated half-life based on evaporative loss for a water depth of 1 m at 25°C: 10 h. (330)

D. BIOLOGICAL EFFECTS

Fishes

fathead minnow larvae	30d LC_{50},F:	3.3 μg/L	
	250d BCF:	270,000	(1530)
rainbow trout	30d LC_{50},F:	51 μg/L	
bluegill	30d LC_{50},F:	400 μg/L	
channel catfish	30d LC_{50},F:	433 μg/L	
cutthroat trout	96h LC_{50},S:	61 mg/L	(1617)
rainbow trout: acute oral toxicity:		>1.5 g/kg	

aroclor 1262

CAS 37324-23-5

A. PROPERTIES

Solubility 0.052 mg/L at 24°C. (1666)

D. BIOLOGICAL EFFECTS

Fish
Harlequin fish: 96h LC_{10},F not toxic below 100 mg/L. (331)

arquad (quaternary ammonium salts; trimethylstearylammoniumchloride; trimethyloctadecylammoniumchloride)

$C_{21}H_{46}N.Cl$

CAS 112-03-8

USES

trademark for a series of quaternary ammonium salts containing one or two alkyl groups ranging from C_8 to C_{18}; corrosion inhibitors; emulsifiers; germicides and sanitizing agents; textile fabric softeners.

A. PROPERTIES

molecular weight 348.13.

C. WATER AND SOIL POLLUTION FACTORS

Biodegradation
at ±18 mg/L, no degradation after 28 d exposure to nonadapted sewage. (488)

ascorbic acid (vitamin C)

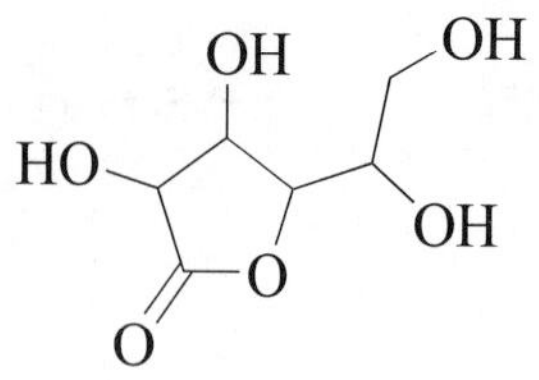

$C_6H_8O_6$

CAS 50-81-7

USES

Nutrition, color fixing, flavoring, and preservative in meats and other foods; oxidant in bread dough, abscission of citrus fruit in harvesting; reducing agent in analytical chemistry. The iron, calcium, and sodium salts are available for biochemical research.

OCCURRENCE

A dietary factor that must be present in the diet of humans to prevent scurvy. Ascorbic acid presumably acts as an oxidation-reduction catalyst in the cell. It is readily oxidized. Food sources: citrus fruits; tomatoes; potatoes; green, leafy vegetables.

A. PROPERTIES

White crystals (plates or needles); molecular weight 176.13; melting point 190-192°C decomposes; solubility 330,000 mg/L at 25°C.

C. WATER AND SOIL POLLUTION FACTORS

Photooxidation

$t_{1/2}$ in top meter of Greifensee (Switzerland): 1 d (ascorbate). (2640)

Ascorbyl phosphate, sodium (L-ascorbic acid-2-dihydrogen phosphate, trisodium salt; Vitamin C phosphate; Sodium L-ascorbic acid-2-phosphate)

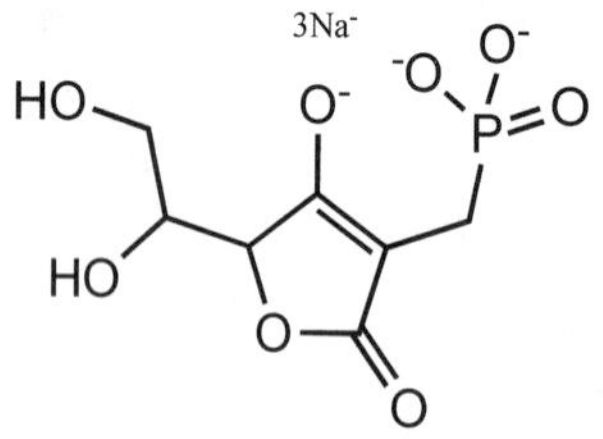

$C_6H_6O_9PNa_3$

CAS 66170-10-3

TRADENAMES

Stay-C50

USES

2 – 5% in specialised skin care products for skin lightening, 0.5 – 2 % in cosmetics as an antioxidant. (7432)

IMPURITIES, ADDITIVES, COMPOSITION

Sodium pyrophosphate (5wt%) and bis-ascorbyl phosphate (1wt%)

A. PROPERTIES

a white to slightly of-white powder with very slight odour; molecular weight 322; melting point: the chemical started to turn brown above about 245°C but remained solid up to ca. 260 °C. It chars before a melting point is reached; density 1.94 kg/L at 20°C; vapour pressure <10^{-8} kPa at 20°C; water solubility 789,000 mg/L at 20°C; log Pow <-4.0

C. WATER AND SOIL POLLUTION FACTORS

Hydrolysis

at pH 4 about 70% of the test material was hydrolysed after 5 days, whereas at pH 7 and 9, no degradation was observed. (7432)

Biodegradation

Inoculum	method	concentration	duration	elimination	
Not adapted A.S.	Manometric respirometry test	100 mg/L	6 days	20% TOC	
			10 days	20-30% TOC	
			28 days	20-30% TOC	(7432)

D. BIOLOGICAL EFFECTS

Toxicity

Micro organisms			
Pseudomonas putida	16h IC_{50}	7,700 mg/L	(7432)
Algae			
Scenedesmus subspicatus	72h $ECgr_{10}$	76.2 mg/L	
	72h $ECgr_{50}$	100 mg/L	
	72h $ECgr_{90}$	100 mg/L	
	72h LOECgr	6.25 mg/L	
	72h NOECgr	3.1 mg/L	
	72h ECb_{10}	15.7 mg/L	
	72h ECb_{50}	100 mg/L	
	72h ECb_{90}	100 mg/L	
	72h LOECb	6.25 mg/L	
	72h NOECb	3.1 mg/L	(7432)
Crustaceae			
Daphnia magna	48h EC_{50}	>100 mg/L	(7432)
Fish			
Brachydanio rerio	96h LC_{50}	5,343 mg/L	
	96h NOEC	2,150 mg/L	(7432)
Mammals			
Rat	oral LD_{50}	>5,000 mg/kg bw	(7432)

asparacemic *see* DL-aspartic acid

L-asparagine (2-aminobutanediocic amide; L-α-aminosuccinamic acid; L-β-asparagine)

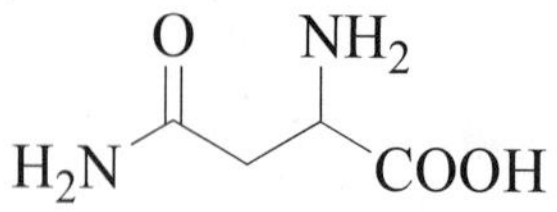

O NH$_2$

H$_2$N COOH

$NH_2COCH_2CH(NH_2)COOH$

$C_4H_8N_2O_3$

CAS 5794-13-8 (L-(+)-asparagine)

USES

biochemical research; preparation of culture media; medicine.

NATURAL SOURCES

widely distributed in plants and animals, both free and combined with proteins.

A. PROPERTIES

molecular weight 132.13; melting point 236°C decomposes; boiling point 235°C decomposes, density 1.54 at 15/4°C; solubility 24,600 mg/L at 25°C, 866,000 mg/L at 100°C.

C. WATER AND SOIL POLLUTION FACTORS

Waste water treatment

A.S.:	after 6 h	10% of ThOD	
	after 12 h	19% of ThOD	
	after 24 h	25% of ThOD	(89)

astrazon Red 6B

D. BIOLOGICAL EFFECTS

Toxicity to microorganisms

Bacillus subtilis growth inhib. EC_{50}: 0.055 mmol/L. (2624)

asulam (asulox: active ingredient: 40% w/v methyl[(4-aminophenyl)sulphonyl]carbamate (as Na salt); methylsulfanilylcarbamate; methyl-4-aminobenzenesulphonyl-carbamate)

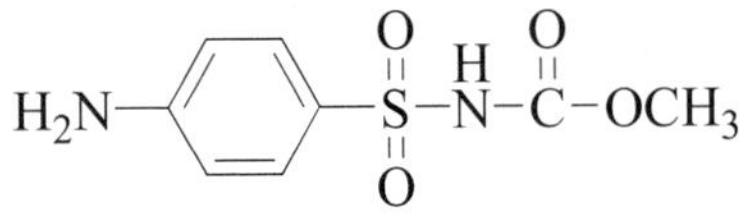

$H_2N-C_6H_4-S(=O)_2-NH-C(=O)-OCH_3$

$C_8H_{10}N_2O_4S$

CAS 3337-71-1

USES

herbicide.

A. PROPERTIES

colorless crystals; molecular weight 230.24; melting point 143- 144°C; solubility 5,000 mg/L.

C. WATER AND SOIL POLLUTION FACTORS

Impact on biodegradation processes

Cellulose decomposition, measured as weight loss of buried cotton cloth, was reduced by 8-38% in treated soil at 16 ppm, after incubation at 19°C for 8 weeks and by 0- 60% in treated soil at 160 ppm. (1825)

Experiments using pure cultures of soil-inhabiting fungi and actinomycetes, some of which were cellulolytic, showed that asulam at 10 ppm had either no effect, or only a temporary effect, on growth. (1825)

	DT_{50} (d)	DT_{90} (d)	
Dissipation periods in clay soil in laboratory experiment.	1.6- 13	17- 43	(6020)

D. BIOLOGICAL EFFECTS

Fishes			
rainbow trout	96 h LC_{50},S	>5,000 mg/L	
channel catfish *(Ictalurus punctatus)*	96 h LC_{50},S	>5,000 mg/L	
goldfish *(Carassius auratus)*	96 h LC_{50},S	>5,000 mg/L	
Bluegill	96 h LC_{50},S	>3,000 mg/L	(1108)

asulox *see* asulam

atenolol (2-[4-[2-hydroxy-3-(1-methylethylamino)propoxy]phenyl]ethanamide; 2-4-(2-Hydroxy-3-isopropylaminopropoxy)phenyl acetamide)

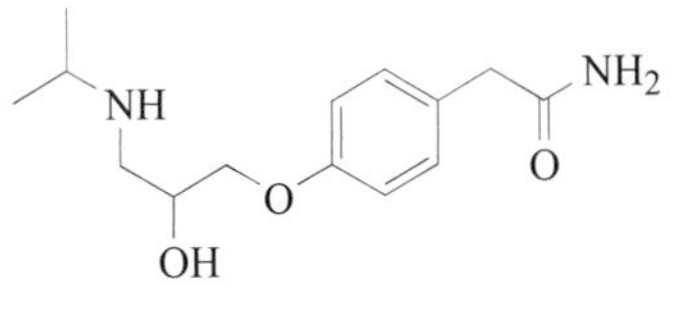

NH NH$_2$ O OH O

$C_{14}H_{22}N_2O_3$

CAS 29122-68-7

SMILES CC(C)NCC(COC1=CC=C(C=C1)CC(=O)N)O

USES

Atenolol is a drug belonging to the group of beta blockers, a class of drugs used primarily in cardiovascular diseases.

A. PROPERTIES

Odourless white powder; molecular weight 266.34; melting point 152-154°C; water solubility 13,500 mg/L; log Pow 0.23; 1.2

C. WATER AND SOIL POLLUTION FACTORS

Environmental concentrations in Germany 2001 (10923)

	Positive samples	Min ng/L	Max ng/L	Average ng/L	Median ng/L
River Körsch upstream WWTP	8/8	10	29	15	12
River Körsch downstream WWTP	8/8	39	120	71	70
Influent WWTP Stuttgart-M	6/6	73	255	149	146
Effluent WWTP Stuttgart-M	7/7	55	94	74	73
Influent WWTP Reutlingen-W	5/5	55	290	142	148
Effluent WWTP Reutlingen-W	5/5	43	78	61	66
Influent WWTP Steinlach-W	5/5	45	153	99	106
Effluent WWTP Steinlach-W	5/5	29	66	48	48
Leachate of landfill Reutlingen-S	5/5	30	35	33	34
Leachate of landfill Dusslingen	5/5	25	60	42	44

D. BIOLOGICAL EFFECTS

Toxicity

Mammals			
Rat	oral LD_{50}	>3,000; 4,960; 6,600 mg/kg bw	
Mouse	oral LD_{50}	2,000-3,000 mg/kg bw	(10927)

atraton (2-ethylamino-4-isopropylamino-6-methoxy-1,3,5-triazine)

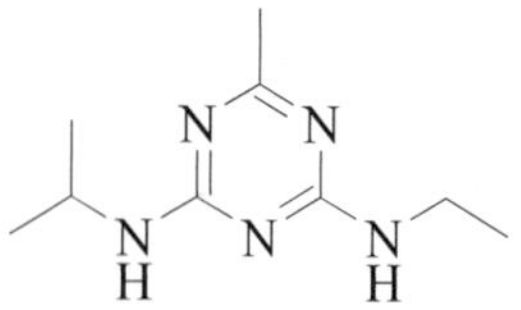

$C_9H_{17}N_5O$

CAS 1610-17-9

USES

selective herbicide used to protect sugarcane.

A. PROPERTIES

molecular weight 211.27; melting point 94-96°C; solubility 1,800 mg/L at 20°C; P_{ow} 1.98; 2.69.

C. WATER AND SOIL POLLUTION FACTORS

In groundwater wells

in the U.S.A.-150 investigations up to 1988:		
median of the conc.'s of positive detections for all studies	0.1 µg/L	
maximum conc.	0.1 µg/L	(2944)

atrazine (2-chloro-4-ethylamino-6-isopropylamino-*s*-triazine)

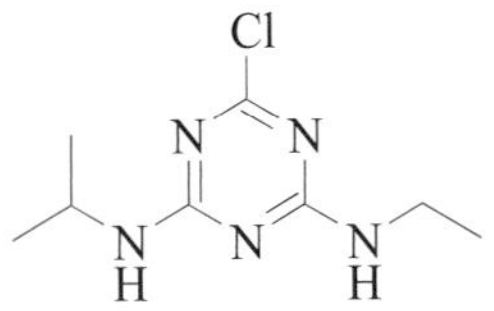

$C_8H_{14}ClN_5$

CAS 1912-24-9

USES

Most widely used chemical for pre-emergence weed control in corn. In Hawaii, it is important to the culture of sugar cane, pineapple, and macadamia nut.

A. PROPERTIES

colorless crystals; molecular weight 215.7; melting point 173-175°C, vapor pressure 3 × 10^{-7} mm at 20°C; 4 x 10^{-7} hPa at 20°C; solubility 33- 35 mg/L; 70 ppm at 25°C, density 1,2 kg/L at 20°C; LogP_{ow}: 2.64; 2.23; 2.35; 2.4; 2.61; 2.63; 2.68.

B. AIR POLLUTION FACTORS

Cuticle-water distribution coefficient (log K_{cw}) for the cuticular membrane of *Citrus aurantium:* log K_{cw}: 2.19. (2448)

Cuticle-air distribution coefficient (log K_{cw}) for the cuticular membrane of *Brassica Oleracea L:* log K_{ca}: 9.13.00

Leaf-air distribution coefficient (log K_{la}) for the whole leaf of *Brassica oleracea* L.: log K_{la}: 7.28.

C. WATER AND SOIL POLLUTION FACTORS

Soil sorption

K_{oc}: 45- 63; 100 (7045, 2576)

Water quality

In rivers in UK (1986): 139 samples from 70 river sites yielded 83 positive values (detection (2711)

limit: 0.025 μg/L): median of positive values: 0.11 μg/L; maximum: 1.12 μg/L
Reductive dehalogenation in anaerobic lake sediment suspension at 20°C in the dark: half-life: 4.2 d (2700)
In groundwater wells in the USA-150 investigations up to 1988:

median concn of positive detections for all studies	0.5 μg/L	
maximum concn	40 μg/L	(2944)

Biodegradation

Microorganisms in aerobic incubations of alluvial aquifer sediment mineralized: 9-14% of added glucose in 24 h compared with <0.1-1.5% of added atrazine (ethyl ^{14}C labeled) and <0.1% of added atrazine (U-ring ^{14}C labeled) (2572)

in submerged soils: in 90 d 0.005% of atrazine-^{14}C was recovered as $^{14}CO_2$ (from ring labeled atrizine); 48- 85% of atrazine was hydrolyzed in 30 d, depending on soil type. Chemical hydrolysis of atrazine to hydroxyatrazine is the principal pathway of detoxication in soil. Biological dealkylation without dehalogenation occurs simultaneously, leading to 2-chloro-4-amino-6-isopropylamino-*s*-triazine:

OH
N N
C_2H_5–N N N
H H
microbial degradation
chemical hydrolysis
Cl Cl
N N N N
C_2H_5–N N N H_2N N N
H H H
atrazine
microbial dealkylation

(1307)

Biodegradation in soil

in greenhouse at 23- 26°C at constant light at initial conc. of 5- 10 mg/kg soil, experiment time 50 d:	$t_{1/2}$ d		
in sandy soil	30		
in organic-rich orchard soil	40		
in agricultural soil	41		
in soil from volcanic area	20		
in field experiments	60 (dissipation $t_{1/2}$)		(7045)
Dissipation periods in sandy soil in the field	DT_{50} (d)	DT_{90} (d)	
	19- 26	133- 193	(6020)

75-100% disappearance from soils: 10 months. (1815)

The preferential dealkylation pathway of the *s*-triazine herbicides in the unsaturated zone is the removal of an ethyl side chain relative to an isopropyl side chain. It is hypothesized that deethylation reactions may proceed at 2-3 times the rate of deisopropylation reactions.

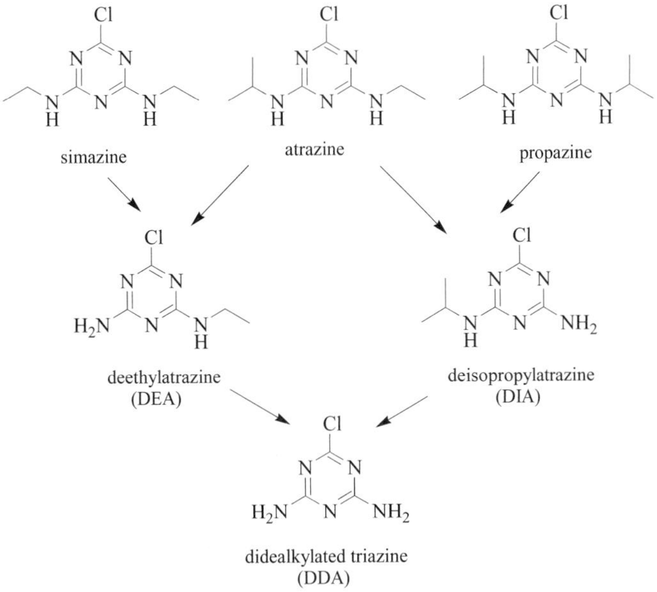

simazine

atrazine

propazine

deethylatrazine (DEA)

deisopropylatrazine (DIA)

didealkylated triazine (DDA)

Dealkylation reactions of atrazine, simazine, and propazine to deethylatrazine (DEA), deisopropylatrazine (DIA), and didealkylatrazine (DDA).	(2952)
Phototransformation under simulated sunlight in the presence of soil extract (aqueous humic acid solution). The following degradation pathway is proposed	(2914)

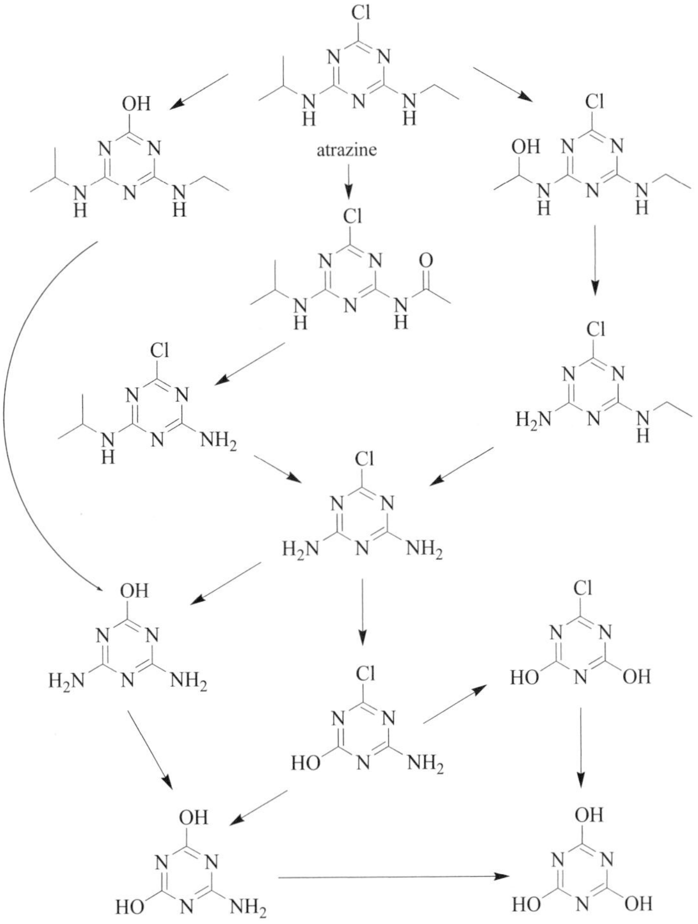

atrazine

The presence of humic substances in aqueous solution at very low concentration (10 ppm of	(2914)

organic carbon) increases the photolytic degradation of atrazine 3 times under simulated solar light. The dehalogenation path accounts for most of the initial degradation process, but dealkylated and deaminated products have also been detected in significant amounts.

Water treatment

Initial concentration, μg/L	Treatment method	% removal	
50-100	Chlorination	100	
0.0001	Chlorination	0	
not reported	river bank filtration	low	
not reported	Ozonization + flocculation	±10	
not reported	Flocculation	±5	
not reported	Activated carbon	100	
not reported	Aeration	not significant	(2915)

D. BIOLOGICAL EFFECTS

Bacteria			
Nitrification inhibition in soil (1% organic carbon)	14 d IC_{52}	50 mg/kg	(7050)
Biotox test *Photobacterium phosphoreum*	5 min EC_{10}	7.5 mg/L	
Microtox test *Photobacterium phosphoreum*	5 min EC_{10}	13 mg/L	(7023)
Pseudomonas putida	16 h EC_0	>10 mg/L	(329)

Algae			
Chlamydomonas reinhardi	Lethal, 24- 48 h	216 μg/L	
Selenastrum capricornutum:			
inhib. of oxygen evolution	EC_{50}	0.0697 mg/L; 0.854 mg/L	
growth inhib.	EC_{50}	0.0587 mg/L; 0.41 mg/L	
OECD 209 closed system	EC_{50}	>0.045 mg/L	(2624)
Selenastrum capricornutum	72 h EC_{50}	0.2-0.22 mg/L	
	72 h EC_{100}	1.0 mg/L	(7024)
Microcystis aeruginosa	8 d EC_0	0.003 mg/L	(329)
Chlorococcum sp.	10 d EC_{50}	100 ppb technical acid	
Dunaliella tertiolecta	10 d EC_{50}	300 ppb technical acid	
Isochrysis galbana	10 d EC_{50}	100 ppb technical acid	
Phaeodactylum tricornutum	10 d EC_0	100-200 ppb technical acid	

Crustaceans

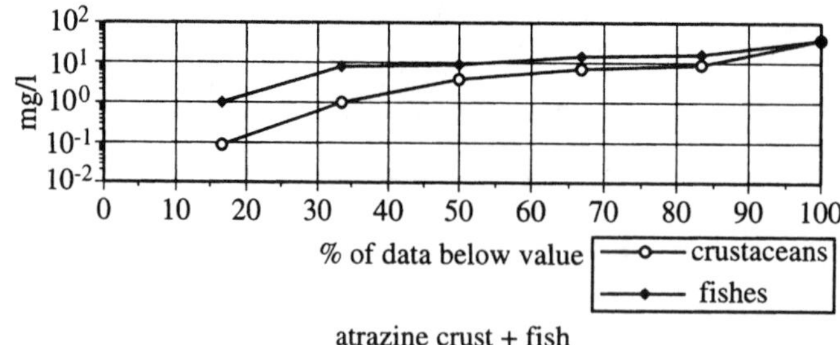

Frequency distributions for 24- 96 h LC_{50} values for crustaceans and fishes (n = 6) based on data from this and other works (3999)

Daphnia magna	48 h IC_{50}	>39 mg/L	(2625)

Bioaccumulation

BCF:	snails:	2-15

algae:		10-83		
fish:		3-10		(1891, 1892, 1893)
Chlorella fusca BCF (wet wt): 52				(2659)

Fishes				
bluegill	2 yr LC_{50},F	5.4-8.4 mg/L	at 27°C	
	MATC:	0.09 mg/L	at 27°C	
fathead minnows	1 yr LC_{50},F	11-20 mg/L	at 25°C	
	MATC:	0.21 mg/L	at 25°C	
brook trout:	1.5 yr LC_{50},F	4.0-6.0 mg/L	at 9-16°C	
	MATC:	0.06 mg/L	at 916°C	(448)
Cyprinodon variegatus	96 h LC_{50}	16 mg/L		
	MATC, early life stage	2.5 mg/L		(2643)
Menidia beryllina	96 h LC_{50}	1 mg/L		
	MATC life cycle	0.12 mg/L		(2643)
log BCF: fathead minnow: 0.90				(2606)
lowest NOEC per species	μg/L			
Cyanophyta	1.5			
Chlorophyta	15; 83			
Branchiopoda	140			
Malacostraca	60			
Diptera	110			
Pisces	65			(2906)

Plants

Bioconcentration factor for vegetation: log BCF_V: -2.00 (2644)

atropine sulfate

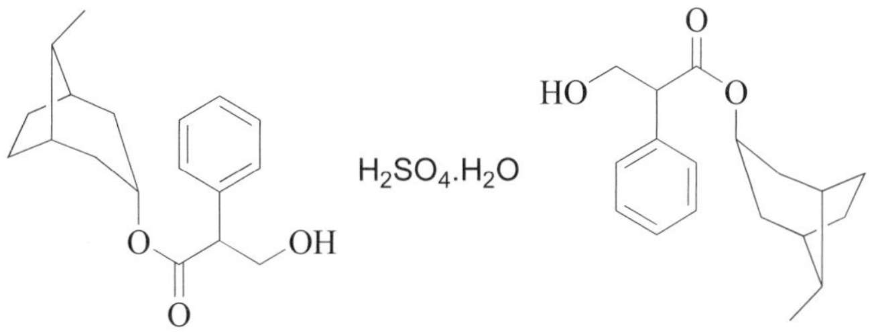

$(C_{17}H_{23}NO_3)_2.H_2SO_4.H_2O$

$C_{34}H_{46}N_2O_6.H_2SO_4.H_2O$

CAS 5908-99-6

USES

pharmaceutical; derived from atropine, parasympatholytic alkaloid isolated from *Atropa belladonna L.*, *Datura stramonium L.*, and other *Solanaceae.*

A. PROPERTIES

molecular weight 694.84; melting point 190-194 °C; water solubility 200,000 mg/L cold

D. BIOLOGICAL EFFECTS

Acute toxicity tests

Marine tests			
Microtox™ *(Photobacterium)* test	5 min EC_{50}	5,565 mg/L	
Artoxkit M *(Artemia salina)* test	24h LC_{50}	15,772 mg/L	
Freshwater tests			
Streptoxkit F *(Streptocephalus proboscideus)* test	24h LC_{50}	661 mg/L	
Daphnia magna test	24h LC_{50}	356 mg/L	
Rotoxkit F *(Brachionus calyciflorus)* test	24h LC_{50}	334 mg/L	(2945)

Micro organisms			
Vibrio fischeri	15min EC_{50}	5519 mg/l	(7237)
Crustaceae			
Artemia salina	24h EC_{50}	15,556 mg/L	
Daphnia magna	24h EC_{50}	356 mg/L	
Streptocephalus probiscideus	24h EC_{50}	664 mg/L	
Brachionus calyciflorus	24h EC_{50}	325 mg/L	(7237)
Mammals			
Rat	acute oral LD_{50}	622 mg/kg bw	(7244)

auramine (4,4'-(imidocarbonyl)bis(N,N-dimethylaniline) monohydrate)

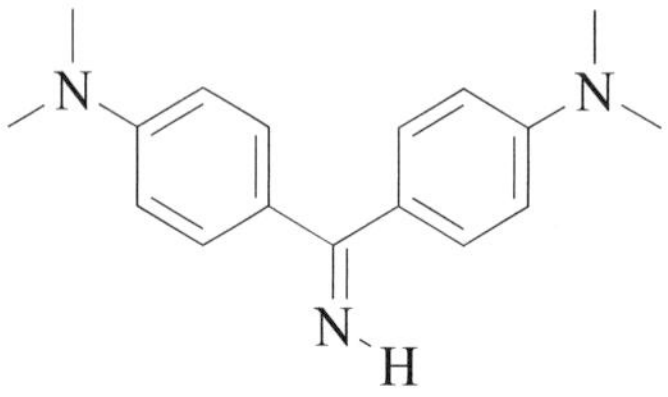

$(CH_3)_2NC_6H_4(CNH)C_6H_4N(CH_3)_2$

$C_{17}H_{21}N_3$

CAS 492-80-8

USES

yellow dye for paper, textiles, and leather; antiseptic; fungicide.

A. PROPERTIES

yellow flakes or powder.

B. AIR POLLUTION FACTORS

Incinerability

thermal stability ranking of hazardous organic compounds: rank 180 on a scale of 1 (highest stability) to 320 (lowest stability). (2390)

auramine O (basic yellow; 2; 4,4'(imidocarbonyl)bis(N,N-dimethylaniline) monohydrochloride; bis(p-dimethylaminophenyl)methyleneimine; 4,4'-carbonimidoylbis(N,N-dimethylbenzenamine); 4,4'-dimethylaminobenzophenonimide; 4,4'-(imidocarbonyl)bis(N,N-dimethylaniline); tetramethyldiaminodiphenylacetimine-; C.I. Solvent Yellow 34; aniline, 4,4-(imidocarbonyl)bis(N,N-dimethyl)-; 4,4-bisdimethylaminobenzophenoneimide; Apyonin-; Yellow-pyoctenin)

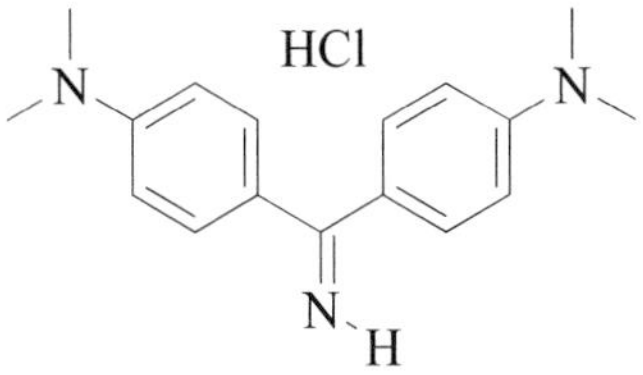

HCl

N

N

N
H

$C_{17}H_{21}N_3HCl$

CAS 2465-27-2

USES

Dyestuff for paper, textiles and leather.

A. PROPERTIES

molecular weight 303.84; melting point 136°C.

C. WATER AND SOIL POLLUTION FACTORS

Hydrolysis

$t_{1/2}$: 65 days at pH5 increasing to 74 days at pH 9. Michler's ketone has been detected as a product of hydrolysis. (8383)

D. BIOLOGICAL EFFECTS

Toxicity to microorganisms

Bacillus subtilis growth inhib. EC_{50}: 55 mg/l. (2624)

avadex *see* diallate

Avermectin B1 (5-O-demethylavermectin A1a)

272 Avermectin B1

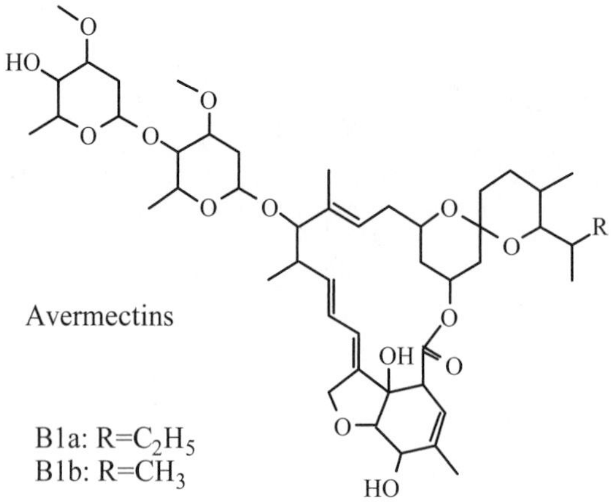

Avermectins

B1a: $R=C_2H_5$
B1b: $R=CH_3$

$C_{48}H_{72}O_{14}$

$C_{47}H_{70}O_{14}$

CAS 65195-55-3 Avermectin B1a

CAS 65195-56-4 Avermectin B1b

TRADENAMES

Abamectin; Affirm; Agri-Mek; Avid; Dynamec; Vertimec; Zephyr

USES

A macrocyclic lactone disaccharide. The Avermectins are insecticidal or anthelmintic compounds derived from the soil bacterium *Streptomyces avermitilis*. Abamectin is a natural fermentation product of this bacterium. Abamectin is used to control insect and mite pests of a range of agronomic, fruit, vegetable and ornamental crops, and it is used by homeowners for control of fire ants. (7146)

IMPURITIES, ADDITIVES, COMPOSITION of marketed product

Avermectin B1 (technical) is composed of 80% Avermectin B1a and 20% Avermectin B1b. The difference structurally between B1a and B1b is that B1a has a C_2H_5-group and the B1b has a CH_3-group attached to one of the ring structures.

A. PROPERTIES

Abamectin is a white to yellowish crystalline powder; molecular weight 873.1 (B1a), 859.1 (B1b); melting point 155-157 °C; vapor pressure 1.5 x 10^{-9} torr; water solubility 7.8 µg/L

C. WATER AND SOIL POLLUTION FACTORS

Hydrolysis

abamectin did not hydrolyse at pH 5 to 9. (7146)

Photodegradation

When Avermectin B1 is applied to plants, a plant photoproduct forms which is not present in animals. This plant photoproduct is the delta-8,9-isomer, which possesses Avermectin B1-like toxicological activity. Further photodegradation of Avermectin B1 results in polar degradates in plants but not in animals. The polar degradates do not possess Avermectin B1-like toxicological properties. (7145)
Abamectin undergoes rapid photodegradation with a half-life of 12 hours in water. (7146)

Biodegradation

In artificial pond water :	half-life: 4 days.
In pond sediment:	half-life: 2 to 4 weeks. (7146)

Dissipation periods in soil

DT_{50} (days)	DT_{90} (days)	soil
22	107	silt
17	84	silt
61	278	sand
23	183	clay
31	282	clay (6020)

Mobility

Abamectin has a strong tendency to bind to soil particles due to its very low solubility (7.8 μg/L), it is therefore immobile in soil and unlikely to leach or contaminate groundwater. Compounds produced by the degradation of abamectin are also immobile and unlikely to contaminate groundwater. (7146)

Degradation in soil

Half-lives:

at the soil surface	8 to 24 hours
Applied to the soil surface, not shaded	1 week
Under dark, aerobic conditions	2 weeks to 2 months
In fine sand loam, clay and sand	20 to 47 days

Under anaerobic conditions the degradation rate was significantly decreased. (7146)

Breakdown of chemical in vegetation

Plants do not absorb abmectin from the soil. Abamectin is subject to rapid degradation when present as a thin film, as on treated leaf surfaces. (7146)

D. BIOLOGICAL EFFECTS

Bioaccumulation

Species	Conc μg/L	duration	body parts	BCF
Bluegill sunfish	0.099 μg/L	28 days	total body	52

The residue concentration rapidly decreased after exposure indicating that abemectin does not accumulate or persist in fish. (7146)

Toxicity

Crustaceae			
Daphnia magna Straus	48h EC_{50}	0.34 μg/L	
Pink shrimp (*Panaeus duorarum*)	96h EC_{50}	1.6 μg/L	
mysid shrimp	96h EC_{50}	0.022 μg/L	
blue crab	96h EC_{50}	153 μg/L	(7146)
Insecta			
honeybees	24h contact LC_{50}	0.002 μg/bee	
	oral LD_{50}	0.009 μg/bee	(7146)

Citrus and alfalfa foliage was not toxic to bees 24 to 48 hours after treatment with abamectin.

Birds			
Bobwhite quail	Acute oral LD_{50}	2000 mg/kg	(7146)
Mollusca			
Crassostrea virginica	96h EC_{50}	430 μg/L	(7146)
Fish			
rainbow trout	96h LC_{50}	3.2 μg/L	
bluegill sunfish	96h LC_{50}	9.6 μg/L	
sheepshead minnow	96h LC_{50}	15 μg/L	
channel catfish	96h LC_{50}	24 μg/L	
carp	96h LC_{50}	42 μg/L	(7146)
Mammals			
Rat	oral LD_{50}	11 mg/kg bw	
Mouse	oral LD_{50}	14-80 mg/kg bw	(7146)

Fate in humans and mammals

Ingested Avermectin B1a by mammals is rapidly eliminated from the body within 2 days via the feces. Rats given single oral doses of radio-labeled avermectin B1a excreted most of the dose (69 to 82%) unchanged in the faces. The half-life of avermectin B1a residues in rat tissues averaged 1.2 days. Similarly, when monkeys were given a single intravenous injection of avermectin B1a, more then 90% of the dose was excreted in the feces within 7 days of thre dosing. (7146)

Azaserine (azaserin; diazoacetate (ester)-1-serine; O-diazoacetyl-1-serine; 1-serine diazoacetate)

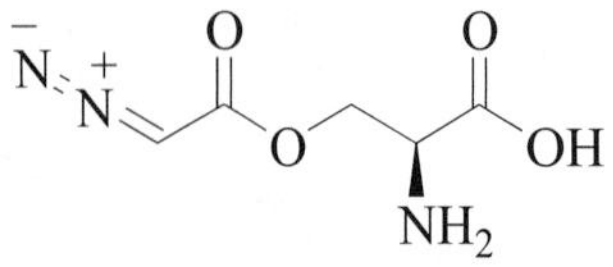

$HOOCCH(NH_2)CH_2OOCCHN_2$

$C_5H_7N_3O_4$

CAS 115-02-6

USES

Glutamine antagonist which inhibits purine biosynthesis. Antifungal and antitumour agent.

OCCURRENCE

Present in cultures of Streptomyces fragilis.

A. PROPERTIES

molecular weight 173.13; melting point 153-155°C decomposes.

B. AIR POLLUTION FACTORS

Incinerability

thermal stability ranking of hazardous organic compounds: rank 297 on a scale of 1 (highest stability) to 320 (lowest stability). (2390)
Photochemical reactions: $t_{1/2}$: 10 h, based on reactions with OH° (calculated).

C. WATER AND SOIL POLLUTION FACTORS

Hydrolysis

in aqueous solution at 25°C:	$t_{1/2}$: 2.1 h at pH 3	
	$t_{1/2}$: 111 d at pH 7	
	$t_{1/2}$: 425 d at pH 11	(8415)

D. BIOLOGICAL EFFECTS

Mammals			
rat, mouse	oral LD_{50}	150-170 mg/kg bw	(9811)

azinphosethyl (O,O-diethyl-S-(4-oxo-3-H-1,2,3-benzotriazine-3-yl)-methyldithiophosphate; S-(3,4-dihydro-4-oxobenzo(d)-(1,2,3)-triazin-3-ylmethyl)diethyl-phosphorothiolothionate; triazotion; ethylguthion, gusathion A)

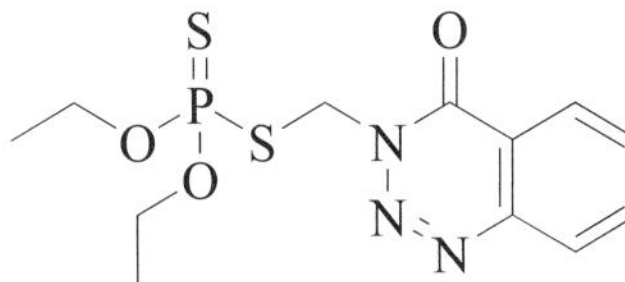

$C_{12}H_{16}N_3O_3PS_2$

CAS 2642-71-9

USES

nonsystemic insecticide and acaricide.

A. PROPERTIES

colorless crystals, molecular weight 345.36; melting point 53°C, boiling point 111°C at 0.001 mm, vapor pressure 2.2 × 10^{-7} mm at 20°C, density 1.28 at 20/4°C; solubility 4-5 mg/L; $\log P_{OW}$ 3.18.

C. WATER AND SOIL POLLUTION FACTORS

Biodegradation

In soil: $t_{1/2}$ is several weeks. Metabolites formed in soil under aerobic and anaerobic conditions are: desethyl azinphos-ethyl, sulfonmethylbenzazimide, bis(benzazimidimethyl)ether, methylthiomethylsulfoxide and methylthiomethylsulfone. (9600)

In plants, azinphos-ethyl is metabolised to azinphos-ethyl-oxon, benzazimide, and dimethyl benzazimide sulfide and disulfide. (9600)

D. BIOLOGICAL EFFECTS

Molluscs			
Penaeus monodon	96h LC_{50}	0.12 mg/L	
Crustaceans			
Daphnia pulex	48h LC_{50}	3.2 µg/L	(2127)
Daphnia	48h EC_{50}	0.0002 mg/L	(9600)
Artemia sp.	24h EC_{50}	3.3 mg/L	
Brachionus plicatilis (Rotoxkit M)	24h EC_{50}	>5.2 mg/L	
Simocephalus serrulatus	48h LC_{50}	4 µg/L	(2127)
Fishes			
golden orfe	96h LC_{50}	0.03 mg/L	
rainbow trout	96h LC_{50}	0.08 mg/L	(9600)
Salmo gairdneri	96h LC_{50}	19 µg/L	(2137)
Mammals			
rat	oral LD_{50}	12 mg/kg bw.	(9600)

azinphosmethyl (guthion; gusation M; O,O-dimethyl-S-[(4-oxo-1,2,3-benzotriazin-3(4H)-yl)methyl] phosphorodithioate; S-(3,4-dihydro-4-oxobenzo(D)-(1,2,3)-triazin-3-ylmethyl)dimethylphosphorodithioate)

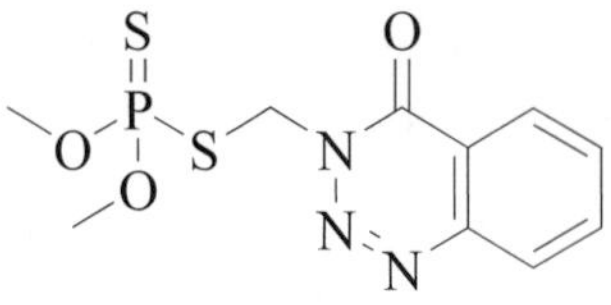

$C_{10}H_{12}N_3O_3PS_2$

CAS 86-50-0

USES

Nonsystemic insecticide and acaricide of long persistence; cholinesterase inhibitor.

A. PROPERTIES

brown waxy solid, molecular weight 317.3; melting point <73°C; vapor pressure 2.7 x 10^{-7} hPa at 20°C; solubility 29 mg/L at 25°C; density 1.44 at 20°C; $LogP_{ow}$ 2.7.

C. WATER AND SOIL POLLUTION FACTORS

Odor threshold

detection: 0.0002 mg/kg water.		(915)
Hydrolysis:		
$t_{1/2}$ in water at pH 8.6	at 6°C: 36 d	
	at 25°C: 28 d	
	at 40°C: 7.2 d	(9816)

Biodegradation

in greenhouse at 23- 26°C at constant light at initial concn of 5- 10 mg/kg soil, experiment time 50 d:	$t_{1/2}$ d	
in sandy soil	20	
in organic-rich orchard soil	4	
in agricultural soil	5	
in soil from volcanic area	12	
in field experiments	10 (dissipation $t_{1/2}$)	(7045)
Degradation involves oxidation, demethylation, and hydrolysis		(9600)
Degradation of ^{14}C-labeled compound in soil:	after 44 d: 50% degraded	
	after 197 d: 93% degraded	(9814)
The main degradation products in soil are benzazimide, thiomethylbenzazimide, bis(benzozimidylmethyl)disulfide and anthranilic acid.		(9815)
In plants, major metabolites identified include oxon, benzazimide, mercaptomethyl benzazimide, and cysteinmethyl benzazimide derivatives.		(9600)

Mobility

log K_{oc}: 3.0	(7045)

D. BIOLOGICAL EFFECTS

Molluscs

Crassostrea virginica (American oyster)	Eggs	48 h LC_{50},S	620 ppb
Mercenaria mercenaria (hard clam)	Eggs	48 h LC_{50},S	860 ppb

	Larvae	12 d LC_{50},S	860 ppb	(2324)
Crustaceans				
Gammarus lacustris		96 h LC_{50}	0.15 μg/L	(2124)
Gammarus fasciatus		96 h LC_{50}	0.10 μg/L	(2126)
Gammarus pseudolimnaeus		30 d NOEC	0.10 μg/L	(2124)
Palaemonetes kadiakensis		120 h LC_{50}	1.2 μg/L	
		20 d LC_{50}	0.16 μg/L	(2126)
Asellus brevicaudus		96 h LC_{50}	21.0 μg/L	(2126)
Insects				
Pteronarcys dorsata		96 h LC_{50}	21.1 μg/L	
		30 d LC_{50}	4.9 μg/L	(2134)
Pteronarcys californica		96 h LC_{50}	1.5 μg/L	(2128)
Acroneuria lycorias		30 d LC_{50}	1.5 μg/L	
		30 d NOEC	1.36 μg/L	(2134)
Ophiogomphus rupinsulensis		96 h LC_{50}	12.0 μg/L	
		30 d LC_{50}	2.2 μg/L	
		30 d NOEC	1.73 μg/L	(2134)
Hydropsyche bettoni		30 d LC_{50}	7.4 μg/L	
		30 d NOEC	4.94 μg/L	(2134)
Ephemerella subvaria		30 d LC_{50}	4.5 μg/L	
		30 d NOEC	2.50 μg/L	(2134)

Fishes

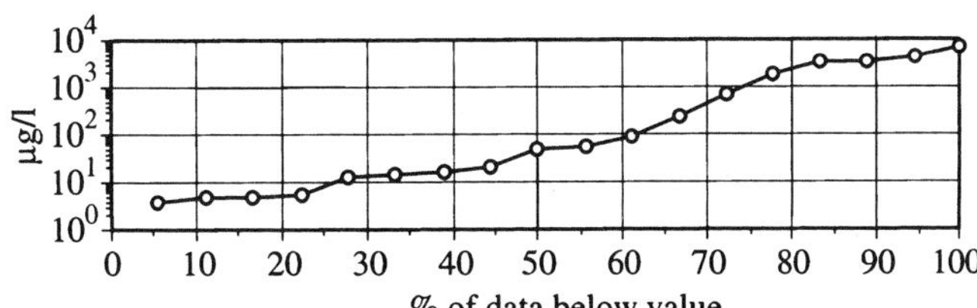

24–96h LC_{50} values for fishes based on the data in this and other works (n = 18)

Pimephales promelas	96 h LC_{50}	93 μg/L	(2119)
Lepomis macrochirus	96 h LC_{50}	5.2 μg/L	(2119)
Lepomis microlophus	96 h LC_{50}	52 μg/L	(2121)
Micropterus salmoides	96 h LC_{50}	5 μg/L	(2121)
Salmo gairdneri	96 h LC_{50}	14 μg/L	(2121)
Salmo trutta	96 h LC_{50}	4 μg/L	(2121)
Oncorhynchus kisutch	96 h LC_{50}	17 μg/L	(2121)
Perca flavescens	96 h LC_{50}	13 μg/L	(2121)
Ictalurus punctatus	96 h LC_{50}	3,290 μg/L	(2121)
Ictalurus melas	96 h LC_{50}	3,500 μg/L	(2121)
Goldfish	96 h LC_{50}	4.3 mg/L	
Minnow	96 h LC_{50}	0.24 mg/L	
Carp	96 h LC_{50}	0.70 mg/L	
Sunfish	96 h LC_{50}	0.05 mg/L	
Bluegill	96 h LC_{50}	0.02 mg/L	
Threespine stickleback (*Gasterosteus aculeatus*)	96 h LC_{50}	4.8 ppb	(2333)
rainbow trout fingerlings	96 h LC_{50},S	7.1 mg/L	(1101)
fathead minnow	96 h LC_{50},S	1.9 mg/L at 25°C	
goldfish	MATC,F	0.51 μg/L at 25°C	(455)
Cyprinodon variegatus	MATC, life cycle: 0.35		(2643)

	µg/L		
Lowest NOEC per species	µg/L		
Chlorophyta	1,800		
Branchiopoda	0.10		
Malacostraca	0.10; 0.5		
Ephemeroptera	2.0; 2.5		
Diptera	2.0		
Plecoptera	1.4		
Trichoptera	2.9		
Odonata	1.7		
Pisces	0.33		(2906)
Mammals			
Rat	oral LD_{50}	9 mg/kg bw	(9816)
Male guinea pig	oral LD_{50}	80 mg/kg bw	
Mouse	oral LD_{50}	11- 20 mg/kg bw	
Dog	oral LD_{50}	>10 mg/kg bw	(9600)

azobenzene (diphenyldiimide; benzeneazobenzene; 1,2-diphenyldiazene)

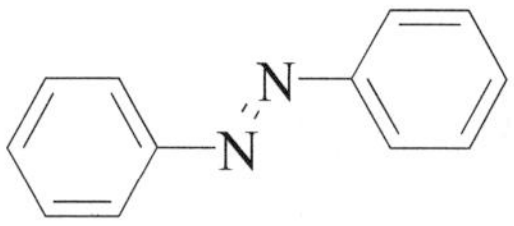

$C_6H_5NNC_6H_5$

$C_{12}H_{10}N_2$

CAS 103-33-3

USES

manufacture of dyes and rubber accelerators; fumigant, acaricide.

A. PROPERTIES

yellow or orange crystals; molecular weight 182.23; melting point 68.3°C; boiling point 293°C; density 1.1; $LogP_{OW}$ 3.82.

C. WATER AND SOIL POLLUTION FACTORS

Impact on biodegradation processes

at 100 mg/L no inhibition of NH_3 oxidation by *Nitrosomonas* sp. (390)

Manmade sources

Göteborg (Sweden) sew. works 1989-91: infl.: nd; effl.: nd. (2787)

2,2'-azobis(2-methylpropionitrile (azobisisobutyronitrile; azodiisobutyrodinitrile; AIBN; 2,2'-dicyano-2,2'-azopropane; 2,2'-azo-bis(isobutyronitrile); 2,2'-dimethyl-2,2'-azodipropionitrile)

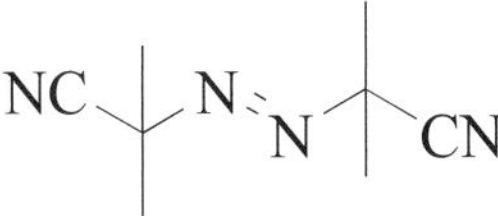

$C_8H_{12}N_4$

CAS 78-67-1

EINECS 201-132-3

USES

intermediate; a foaming agent for rubber and initiator of polymerization.

A. PROPERTIES

molecular weight ; melting point 100-103°C; vapour pressure 0.81 Pa at 25°C; water solubility 350 mg/L at 25°C; log Pow 1.10

C. WATER AND SOIL POLLUTION FACTORS

Hydrolysis

t/2 at 25°C: 263 days at pH 4,; 304 days at pH 7 and 210 days at pH 9 (7450)

Biodegradation

Inoculum	method	concentration	duration	elimination	
	OECD TG 301C		28 days	0% ThOD	
				3% TOC	
				7% HPLC	(7450)

D. BIOLOGICAL EFFECTS

Toxicity

Algae			
Selenastrum capricornutum	72h ECb_{50}	>9.4 mg/L	
	72h NOECb	4.2 mg/L	(7450)
	72h $ECgr_{50}$	6.1 mg/L	
	72h NOECgr	2.2 mg/L	(7450)
Crustaceae			
Daphnia magna	48h EC_{50}	>10; >367 mg/L	
(reproduction)	21d EC_{50}	7.5 mg/L	
	21d NOEC	2.2 mg/L	
	21d LOEC	4.6 mg/L	(7450)
Fish			
Oryzias latipes	96h LC_{50}	>10 mg/L	
Poecilia reticulata	14d LC_{50}	>10 mg/L	(7450)
Mammals			
Rat	oral LD_{50}	100 mg/kg bw	
Mouse	Oral LD_{50}	700 mg/kg bw	(7450)

azodrin (dimethylphosphate of 3-hydroxy-N-methyl-cis-crotonamide-O,O-dimethyl-O-(2-methylcarbamoyl-1-methylvinyl)phosphate; monocrotophos)

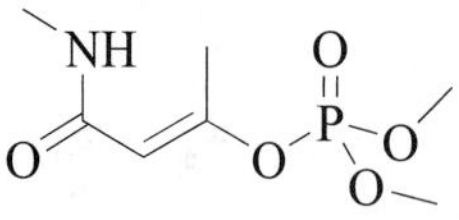

$C_7H_{14}NO_5P$

CAS 6923-22-4

USES

systemic insecticide; acaricide.

A. PROPERTIES

molecular weight 223.19; melting point 25-30°C (commercial solid product), 54-55°C (crystals); boiling point 125° C at 0.0005 mmHg; density 1.33 at 20°C; vapor pressure 7 x 10± mmHg at 20°C; solubility miscible.

C. WATER AND SOIL POLLUTION FACTORS

Photodegradation

Photodegradation was greater on soil surfaces than on glass. Photodegradation: alluvia < black < red loamy >laterite soil. Photolysis was greater on flooded moist soils than dry loam soil. (9825)

Persistence in riverwater in a sealed glass jar under sunlight and artificial fluorescent light-initial conc. = 10 µg/L: still 1 00% of original compound found after 8 weeks. (1309)

Hydrolysis

in aqueous environment at 25°C: $t_{1/2}$: 131 d at pH 3
26 d at pH 9. (9824)

Hydrolysis followed 1st-order kinetics and the major hydrolytic degradation products were N-methylacetoacetamide and O-demethylmonocrotophos.

Biodegradation

In flooded rice soil levels decreased to trace amounts after 20 d incubation. (9823)

Degradation in soils: The intermediate degradation products were N-methylacetoacetamide, N-(hydroxymethyl)monocrotophos and 3-hydroxy-N-methylbutyramide. (9824)

D. BIOLOGICAL EFFECTS

Bacteria

Inhibition of nitrogen fixation activity of *Nostoc linckia* at 5 mg/L. (9822)

Crustaceans

copecod *(Acartia tonsa):*	96h LC_{50},S	240 µg/L.	(1129)
marine edible crab	96h LC_{50}	0.58 mg/L	(9819)
shrimp	48h LC_{50}	4.46 mg/L	
	96h LC_{50}	1.59 mg/L	(9820, 9821)

Fishes

harlequin fish *(Rasbora heteromorpha)*

mg/L			
24 h	*48 h*	*96 h*	*3 m (extrapolated)*

LC_{10},F	580	580	280		
LC_{50},F	750	739	450	150	(331)

snakehead fish	96h LC_{50}	10 mg/L	(9817)
cichlid	96h LC_{100}	18.6 mg/L	(9818)
Mammals			
rat, mouse	oral LD_{50}	8-15 mg/kg bw	(9675, 9826)
rat male	oral LD_{50}	17 mg/kg bw	
rat female	oral LD_{50}	20 mg/kg bw	(9827)

azoxystrobin (Methyl (E)-2-[2[6-(2-cyanophenoxy)pyrimidin-4-yloxy] phenyl]-3-methoxyacrylate; Methyl (E)-2-[2 [6-(2-cyanophenoxy)-4-pyrimidinyl]oxy]-alpha-(methoxymethylene) benzeneacetate (9CI))

$C_{22}H_{17}N_3O_5$

CAS 131860-33-8

TRADENAMES

Quadris; Abound; Quilt; Heritage Fungicide; ICIA5504 Fungicide

USES

Broad spectrum fungicide. Azoxystrobin is the first of a new class of pesticidal compounds called ß-methoxyacrylates, which are derived from the naturally-occurring strobilurins. Their biochemical mode of action is inhibition of electron transport

A. PROPERTIES

White to pale brown crystalline powder; molecular weight 403.4; melting point 116°C; boiling point > 360°C; relative density 1.34 kg/L at 20°C; vapour pressure 1.1 x 10^{-10} Pa at 20°C; water solubility at 20°C; 6.7 mg/L at pH 5.2 and 7.0, 5.9 mg/L at pH 9.2; log Pow 2.5 at 20°C; H 7.3 x 10^{-9} Pa.m^3/mol

C. WATER AND SOIL POLLUTION FACTORS

Hydrolysis

T/2 = stable at 25°C and at pH 5 to 9; stable at 50°C at pH 5 to 7; at pH 9 : t/2 = 12 days. (10782)

Photodegradation in water

T/2 = 9-14 days at pH 7. (10782)

Photodegradation

T/2 = 11 days . Eight phosphorproducts found each in amounts less than 10%.

Degradation in soil (10782)

aerobic	After 100 days	After 360 days
Mineralization	2-2.5%	11-14%
Bound residues	9-10%	18-24%
Relevant metabolite(1)	7%	21%

(1)(E)-2-(2-[6-cyanophenoxy)-pyrimidin-4-yloxyl]-phenyl)-3-methoxyacrylic acid

anaerobic	After 100 days	After 360 days
Mineralization		50% after 231 days
Bound residues	8-10%	11-13%
Relevant metabolite(1)	15%	48-51%

Biodegradation in laboratory studies (10782)

degradation in soils	%	
Aerobic at 20°C	50	279 days
Aerobic at 5°C	50	1066 days
Anaerobic at 20°C	50	231 days

Biodegradation in water/sediment systems (10782)

Readily biodegradable	
DT50 water	34-57 days
DT50 whole system	170-294 days

Field dissipation studies (10782)

DT 50	87-407 days
DT 90	197-435 days

Mobility in soils (10782)

K_{OC}	122-594

D. BIOLOGICAL EFFECTS

Toxicity

Algae			
Selenastrum capricornutum	96h EC_{50}	0.36 mg a.i./L	(10782)
Green algae	72h EC_{50}	0.1 mg/L	
	72h NOEC	0.02 mg/L	
Marine diatom	72h EC_{50}	0.5 mg/L	
	72h NOEC	0.1 mg/L	
Freshwater diatom	72h EC_{50}	0.5 mg/L	
	72h NOEC	0.02 mg/L	
Blue-green algae	72h EC_{50}	13 mg/L	
	72h NOEC	9 mg/L	(10783)
Aquatic plants			
Duckweed	EC_{50}	3.4 mg/L	
	NOEC	0.8 mg/L	(10783)
Birds			
Bobwhite quail	Acute LD_{50}	>2,000 mg/kg bw	
Mallard duck	Acute LD_{50}	>250 mg/kg bw	(10783)
Crustaceae			
Daphnia magna	48h EC_{50}	0.259 mg a.i. /L	
Life-cycle	21d NOEC	0.044 mg a.i./L	
	21d LOEC	0.084 mg a.i./L	(10783)
Macrocyclops fuscus	48h EC_{50}	0.13 mg a.i./L	
Insecta			
Honeybees	Acute oral LD_{50}	>25 μg/bee	
	Acute contact LD_{50}	>200 μg/bee	(10783)
Worms			
Eisenia foetida	Acute LC_{50}	283 mg a.i./kg dry soil	(10783)
Fish			
Salmo gairdneri	96h LC_{50}	0.47 mg a.i./L	

Mammals		
Rat	oral LD_{50}	>5,000 mg/kg bw

B

B(a)A *see* benzo(*a*)anthracene

B(a)P *see* benzo(*a*)pyrene

B(b)F *see* benzo(*b*)fluoranthene

B(e)P *see* benzo(*e*)pyrene

B(j)F *see* benzo(*j*)fluoranthene

B(k)F *see* benzo(*k*)fluoranthene

---

balan (N-butyl-N-ethyl-2,6-dinitro-4-trifluoromethylaniline; N-butyl-N-ethyl-α, α, α-trifluoro-2,6-dinitro-*p*-toluidine; benefin; benfluralin; bethrodine; quilan)

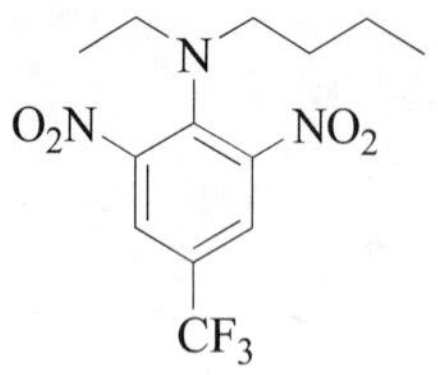

$C_{13}H_{16}F_3N_3O_4$

CAS 1861-40-1

USES

selective pre-emergence herbicidal control of annual grasses and broad-leaf weeds.

A. PROPERTIES

yellow-orange crystals; molecular weight 335.32; melting point 65-66.5°C; boiling point 148-149° C at 7 mm Hg; vapor pressure 4 × 10^{-7} mm at 25°C; density 1.28 at 20°C; solubility 70 mg/L at 25°C; <1 mg/L at 25°C; $\log P_{OW}$ 5.3 at 20°C and pH 7

C. WATER AND SOIL POLLUTION FACTORS

Biodegradation

Residual activity in soil: 4-8 months. (9600)
Degradation in soil: $t_{1/2}$: 0.4-1.8 months at 30°C. (9829)

D. BIOLOGICAL EFFECTS

Algae			
Selenastrum capricornutum specific growth	7d IC_{17}	3.68 mg/L	
terminal biomass	7d IC_{34}	3.68 mg/L	(9600)
Crustaceans			
Gammarus fasciatus	96h LC_{50}	1,100 μg/L	(2125)
Daphnia	48h EC_{50}	>0.1 mg/L	(9600)

Fishes

harlequin fish *(Rasbora heteromorpha):*

mg/L	*24 h*	*48 h*	*96 h*	*3 m (extrapolated)*	
LC_{10},F	1.0	0.95			
LC_{50},F	1.4	1.3	1.2	1.0	(331)

bluegill sunfish	96h LC_{50}	6.0 mg/L	
rainbow trout	96h LC_{50}	0.081mg/L	(9600)
Mammals			
rat	oral LD_{50}	>10,000 mg/kg bw	
mouse	oral LD_{50}	>5,000 mg/kg bw	
dog, rabbit	oral LD_{50}	>2,000 mg/kg bw	(9600)

Barium bis(2-chloro-5-(hydroxy-1-naphthyl)azotoluene-4-sulphonate

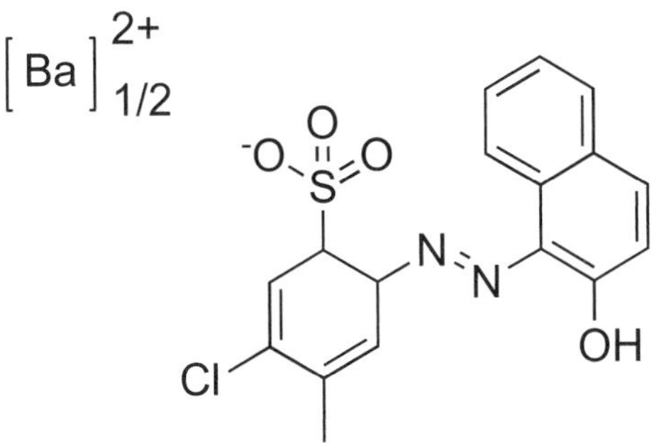

$C_{17}H_{12}ClN_2O_4S.1/2Ba$

CAS 5160-02-1

TRADENAMES

Pigment Red 53:1; D and C Red No.9

USES

Pigment red 53:1 is used particularly for short-life printed matter, for colouring plastics, PVC, polyurethane foam, natural rubber stocks and in paints. In addition, smaller amounts are also used in low-requirement products, such as crayons and water colours for the office supplies industry. (7451)

ENVIRONMENTAL CONCENTRATIONS

Releases into the environment may mainly occur during production, formulation and paper recycling.

A. PROPERTIES

melting point 330°C decomposes; vapour pressure 1.10^{-5} Pa; water solubility 2 mg/L; log Pow –0.56

C. WATER AND SOIL POLLUTION FACTORS

Biodegradation

Inoculum	method	concentration	duration	elimination	
	MITI I test		14 days	0%	
Adapted inoculum	Zahn-Wellens test		21 days	33% (1)	(7451)

(1) of which 10% was due to adsorption onto sludge

Waste water treatment: flocculation with iron sulfate : 85% elimination (7451)

D. BIOLOGICAL EFFECTS

Bioaccumulation

Species	conc. µg/L	duration	body parts	BCF	
Ozyzias latipes	700			0.9-1.8	
	70			8.5-15	(7451)

Toxicity

Crustaceae			
Daphnia magna	48h EC_0	>2 mg/L	(7451)
Fish			
Brachydanio rerio	96h LC_{50}	>500 mg/L	
Oryzias latipes	48h LC_{50}	>420 mg/L	(7451)

Mammals			
Rat	oral LD_{50}	>10,000 mg/kg bw	(7451)

Batricin

$C_{66}H_{103}N_{17}O_{16}S$

CAS 1405-87-4 Batricin

CAS 55852-84-1 Batricin methylene disalicylate

USES

Bacitracin is a polypeptide antibiotic.Bacitracin is one of the most commercially and medically important antibiotics that has been produced in large quantity. It is a metal dependent antibiotic produced by *Bacillus licheniformis* and *Bacillus subtilis* with potent antimicrobial activity directed primarily against Gram positive bacteria.This antibiotic requires a divalent metal ion such as Zn(II) for its antimicrobial activity, and has been reported to also bind several transition metal ions including Co(II), Ni(II) and Cu(II). Despite the wide use of this antibiotic, the structure and its relationship with activity of metal-bacitracin complexes have never been determined and fully understood. The different bacitracin complexes and stereoisomers are numbered A, B, C, D, etc..Commercial bacitracine is composed of at least nine bacitracines. Bacitracine A is the most important in the mixture. Complexes with organic constituents such as methylenedisalicylate are also available.

Bacitracin is used in the ration of chickens, turkeys, swine, cattle, pheasant and quail for increased rate of weight gain, improved feed conversion and control/prevention of certain diseases e.g. *Clostridial enteritis.* . Bacitracine is widely used for topical therapy such as for skin and eye infections; it is effective against gram-positive bacteria, including strains of *staphylococcus* that are resistant to penicillin . Bacitracin is toxic to humans and is no longer used internally.

A. PROPERTIES

water solubility high

C. WATER AND SOIL POLLUTION FACTORS

Half-lives of bacitracin in surface waters at 5 mg/L (7241)

ph	temperature	Farrington Lake		Farm pond	
		sterile	non-sterile	sterile	non-sterile
pH 8	4 °C	30 days	30 days	30 days	30 days

	20 °C	16 days	8 days	--	27 days
	28 °C	4 days	6 days	--	12 days
pH 7	4 °C	30 days	30 days	30 days	30 days
	20 °C	19 days	30 days	21 days	21 days
	28 °C	9 days	13 days	14 days	11 days
pH 6	4°C	30 days	30 days	30 days	30 days
	20 °C	30 days	30 days	30 days	30 days
	28 °C	24 days	27 days	21 days	28 days

Biodegradation

The primary degradation steps are deamidation, deamination and hydrolysis leading to smaller peptides and amino acids (7241)

Mobility in soils

Based on the high water solubility of bacitracin, it may be concluded that this anibiotic is mobile in soils. (7241)

D. BIOLOGICAL EFFECTS

Bioaccumulation

Complete excretion and/or partial destruction of bacitracin(s) in the gut suggests that bioaccumulation is not a problem in mammals and birds. (7241)

Toxicity

Bacteria			
Methanococcus vannielii	IC50	10 mg/L	
Halobacterium salinarium	IC50	5 mg/L	
Halobacterium halobium	IC50	12 mg/L	(7241)
Algae			
Chlorella ellipssida	IC0	> 10,000 mg/L	
Anabaena variabilis	IC50	10 mg/L	(7241)
	72h	>500; >500 mg/L	
Insects			
Rice-weevil larvae *(Sitophilus aryza)*	toxic	20,000 mg/L	
larvae of flesh-eating flies *(Agria affinis)*	toxic	50,000 mg/L in feed	(7241)
Crustaceans			
Daphnia magna	24h EC_{50}	126,4 mg/L	
	48h EC_{50}	30,5 mg/L	(7237)
Mammals			
Rat	acute oral LD_{50}	>10,000 mg/kg bw	(7241)

Metabolism in animals

It has been demonstrated in experiments that chickens, dogs and swine given bacitracin orally absorb little, if any, of the antibiotic and the portion that is excreted appears to be essentially intact in the faeces. BMD and also the bacitracin zinc salt disassociate upon ingestion releasing the active bacitracin base. (7241)

baygon (2-(1-methylethoxy)phenol methylcarbamate; propoxur; arprocarb; Bay 39007; Blattanex; Suncide; 2-isopropoxyphenylmethylcarbamate; 2-isopropoxyphenyl-N-methylcarbamate)

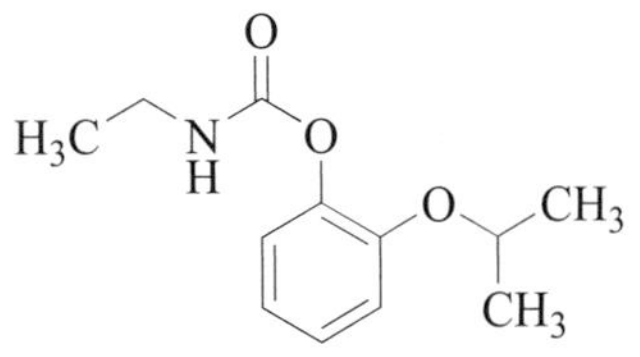

$C_{11}H_{15}NO_3$

CAS 114-26-1

USES

Insecticide.

A. PROPERTIES

white to tan crystalline solid; molecular weight 209.27; melting point 91°C; vapor pressure 1.3×10^{-5} kPa at 20°C; solubility 1,800 mg/L, <2,000 mg/L; $LogP_{OW}$ 1.5; 1.52.

C. WATER AND SOIL POLLUTION FACTORS

Aquatic reactions

persistence in river water in a sealed glass jar under sunlight and artificial fluorescent light at initial concn 10 μg/L alkaline hydrolysis half-life: 0.02 d, at pH 9 at 40°C in the dark (2690)

% of original compound found after

	1 h	*1 wk*	*2 wk*	*4 wk*	*8 wk*	
%	100	50	30	10	5	(1309)

Biodegradability

In greenhouse at 23-26°C at constant light at initial concn of 5-10 mg/kg soil, experiment time 50 d:

	$t_{1/2}$ *d*	
In sandy soil	11	
In organic-rich orchard soil	4	
In agricultural soil	5	
In soil from volcanic area	27	
In field experiments	30 (dissipation $t_{1/2}$)	(7045)

Aerobic biodegradation half-life: 0.6 d at an initial concn of 2 ppm, using a mixture of nonacclimated sludge, field soil, and river sediment as inoculum, in a stirred flask at 18-22°C, the major bacterial types in the activated sludge have been characterized as follows: *Flavobacterium* 35%; *Alcaligenes* 30%; *Corynebacterium* 20%; *Pseudomonas* 15% (2690)

Mobility

log K_{OC}: 1.5. (7045)

D. BIOLOGICAL EFFECTS

Algae

	test concn, μg/L	% reduction in O_2 evolution after 4-h exposure	
Dunaliella euchlora	1,000	25	
	100	32	
	10	27	
Phaeodactylum tricornutum	1,000	23	
	100	28	
	10	40	
Skeletonema costatum	1,000	30	
	100	23	
	10	29	
Cyclotella nana	1,000	53	(2325)

Crustaceans

Gammarus lacustris	96 h LC_{50}	34 μg/L	(2124)
Gammarus fasciatus	96 h LC_{50}	50 μg/L	(2126)

Insects			
Pteronarcys californica	96 h LC_{50}	13 µg/L	(2128)
Chironomus riparius fourth-instar larval stage	24 h LC_{50}	64 µg/L	(1853)
Anopheles quadrimaculatus	24 h LC_{50}	0.45 mg/L	(2625)
Fishes			
Pimephales promelas	4 d LC_{50}	8.8 mg/L	
Procambarus clarki	4 d LC_{50}	1.4 mg/L	(2625)

Baytex (Baycid; Entex; Hebaycid; Mercaptophos; Tiguvon; O,O-dimethyl-O-(3-methyl-4-(methylthio)phenyl)phosphorothioate; fenthion)

H_3C O
O–P
O S–(ring)–S–CH_3
CH_3 CH_3

$C_{10}H_{16}O_3PS_2$

CAS 55-38-9

USES

systemic and contact herbicide.

A. PROPERTIES

molecular weight 278.34; boiling point 87°C at 0.01 mm; vapor pressure 3 × 10^{-5} mm at 20°C; density 1.25 at 20/4°C; solubility 55 ppm at room temp.

C. WATER AND SOIL POLLUTION FACTORS

Aquatic reactions

persistence in river water in a sealed glass jar under sunlight and artificial fluorescent light at 10 µg/L initial conc.:

% of the original compound found after

	1 h	*1 wk*	*2 wk*	*4 wk*	
%	100	50	10	0	(1309)

D. BIOLOGICAL EFFECTS

Dugesia dorotocephala:	mortality: 3 d at 10 mg/L	
Dugesia tigrina:	mortality: 3 d at 10 mg/L	
Laccotrophes griseus:	reproductive effects, 10 d at 20 µg/L	(2625)

Algae

	test conc., µg/L	*% reduction in O_2 evolution after 4 h exposure*
Dunaliella euchlora	1,000	27
	100	27
	10	16

Phaeodactylum tricornutum	1,000	29	
	100	29	
	10	35	
Skeletonema costatum	1,000	19	
	100	51	
	10	26	
Cyclotella nana	1,000	50	
	100	48	(2348)

Insects			
Pteronarcys californica	96h LC_{50}	4.5 µg/L	(2128)
ricefield spider *(Oedothorax insecticeps)*	LD_{50}	500 mg/L	(1814)
insect larvae *(Chaoborus)*	48h LC_{50}	0.008 mg/L	
(Cloeon)	48h LC_{50}	0.012 mg/L	(1323)

Molluscs

gastropoda *(Lymnaea stagnalis)*	48h LC_{50}	6.4 mg/L	(1323)

Crustaceans

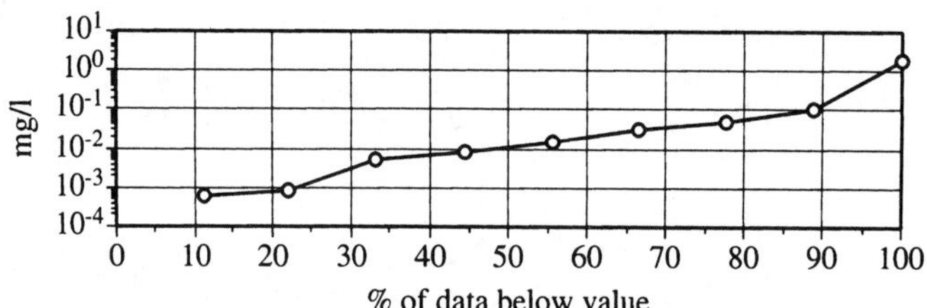

Frequency distribution of 24-96h EC_{50}/LC_{50} values for crustaceans (n = 9) based on data presented in this and in other works (3999)

Gammarus lacustris	96h LC_{50}	8.4 µg/L	(2124)
Gammarus fasciatus	96h LC_{50}	110 µg/L	(2126)
Palaemonetes kadiakensis	120h LC_{50}	5 µg/L	
	20d LC_{50}	1.5 µg/L	(2126)
Orconectes nais	96h LC_{50}	50 µg/L	(2126)
Asellus brevicaudus	96h LC_{50}	1,800 µg/L	(2126)
Simocephalus serrulatus	48h LC_{50}	0.62 µg/L	(2127)
Daphnia pulex	48h LC_{50}	0.80 µg/L	(2127)
Korean shrimp *(Palaemon macrodactylus)*	96h LC_{50}	30; 5.3 µg/L	(2348)
Gammarus pulex	48h LC_{50}	0.014 mg/L	(1323)

Fishes

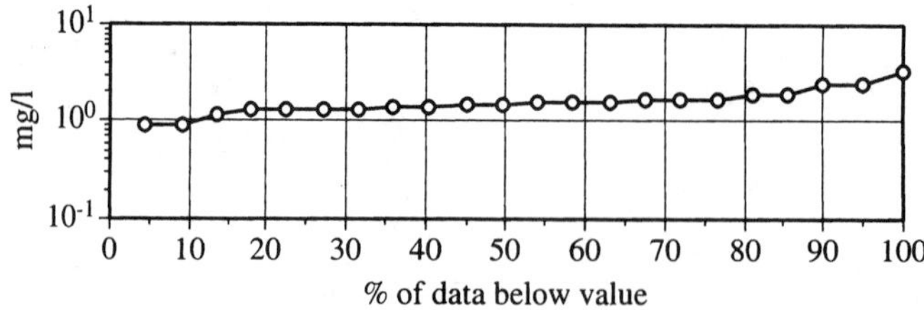

Frequency distribution of 24-96h LC_{50} values for fish (n = 22) based on data presented in this and other works (3999)

	96h LC_{50}, µg/L
Pimephales promelas	2,440

Lepomis macrochirus		
Lepomis microlophus	1,880	
Micropterus salmoides	1,540	
Salmo gairdneri	930	
Salmo trutta	1,330	
Oncorhynchus kisutch	1,320	
Perca flavesens	1,650	
Ictalurus punctatus	1,680	
Ictalurus melas	1,620	(2121)
catfish	1,700	
bullhead	1,600	
goldfish	3,400	
minnow	2,400	
carp	1,200	
sunfish	1,900	
bluegill	1,400	
bass	1,500	
rainbow	900	
brown	1,300	
coho	1,300	
perch	1,700	(1934)
Mammals		
Biotransfer factor in beef	log B_b: -4.50	
Biotransfer factor in milk	log B_m: -5.60	(2644)

BBCP *see* 1,2-dibromo-3-chloropropane

BBP *see* butylbenzylphthalate

BCME *see* dichloromethylether

2,4-D(BEE) *see* 2,4-dichlorophenoxyacetic acid, butoxyethanolester

Behenic acid (docosanoic acid; *n*-docosoic acid)

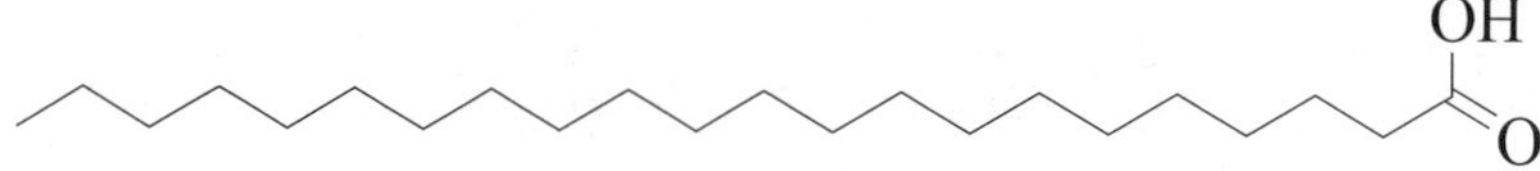

$CH_3(CH_2)_{20}COOH$

$C_{22}H_{44}O_2$

CAS 112-85-6

USES

Intermediate for the production of its metal salts, docosylamine and higher alkyl esters in the chemical industry.
Docosanoic acid occurs naturally as triglycieride in most seed fats, animal milk fats, marine animal oils and so on.
Cosmetics; waxes; plasticizers; stabilizers.

NATURAL SOURCES

a minor component of oils of the type of peanut and rapeseed. Occurs in bean oil, hydrogenated mustard oil, and rapeseed oil.

IMPURITIES, ADDITIVES, COMPOSITION

C14-20 fatty acids (ca 11%); C24 fatty acid (ca 3%)

A. PROPERTIES

molecular weight 340.58; melting point 80.2°C; boiling point 306°C at 60 mm; density 0.82 at 100/4°C; vapour pressure 6.5 x 10^{-5} Pa at 25°C (calculated); water solubility 0.016 mg/L at 25°C; log Pow >5.1 at 25°C; K_{OC} 135,000 (calculated); H 1.6 – 2.9 x 10^{-4} atm.m3/mol at 25°C (calculated)

B. AIR POLLUTION FACTORS

Organic fraction of suspended matter

Bolivia at 5,200-m altitude (Sept.-Dec. 1975): 0.29-0.69 ng/m^3
Belgium, residential area (Jan.-April 1976): 3.2-8.4 ng/m^3 (428)
glc's: Botrange (Belgium): woodland at 20-30 km from industrial area, June-July 1977: 3.2; 3.8 ng/m^3 (n = 2)
Wilrijk (Belgium): residential area: Oct.-Dec. >1976: 10; 16 ng/m^3 (n = 2) (1233)
glc's in residential area (Belgium): Oct. 1976: in particulate sample: 14 ng/m^3 (1289)

Manmade sources

emission rates released from burning wood in residential open fireplaces:

pine wood	7.98 mg/kg logs burned
pine wood	7.98 mg/kg logs burned

glc's:

Detroit freeway interchange: Oct.-Nov. 1963:	62 ng/m^3 air	
New York high-traffic location:	119 ng/m^3 air	(100)

Photodegradation

T/2 = 13.7 hours based on OH radical (10671)

C. WATER AND SOIL POLLUTION FACTORS

Hydrolysis

The chemical is stable at pH 4 to 9. (10671)

Biodegradation

Inoculum	method	concentration	duration	elimination	
A.S.	OECD 301C	100 mg/L	28 days	48-56%ThOD 67-80% GC	
A.S.	OECD 302C	30 mg/L	28 days	79-96% ThOD 94-95% GC	(10671)

D. BIOLOGICAL EFFECTS

Toxicity

Algae			
Selenastrum capricornutum	72h EC_{50}	>5.0 mg/L	
	72h NOEC	>5.0 mg/L	(10671)
Crustaceae			
Daphnia magna	48h EC_{50}	>5.0 mg/L	
reproduction	21d EC_{50}	>0.84 mg/L	
	21d NOEC	>0.84 mg/L	(10671)
Fish			
Oryzias latipes	96h LC_0	>5.0 mg/L	
	96h LC_{50}	>5.0 mg/L	
	14d LC_{50}	>4.99 mg/L	
	14d NOEC	>4.99 mg/L	(10671)
Carassius auratus			
Mammals			
Rat	oral LD_{50}	>2,000 mg/kg bw	(10671)

Behenoyl lactylate sodium

(docosanoic acid,2-(1-carboxyethoxy)-1-methyl-2-oxoethyl ester, sodium salt; sodium docosyl lactylate)

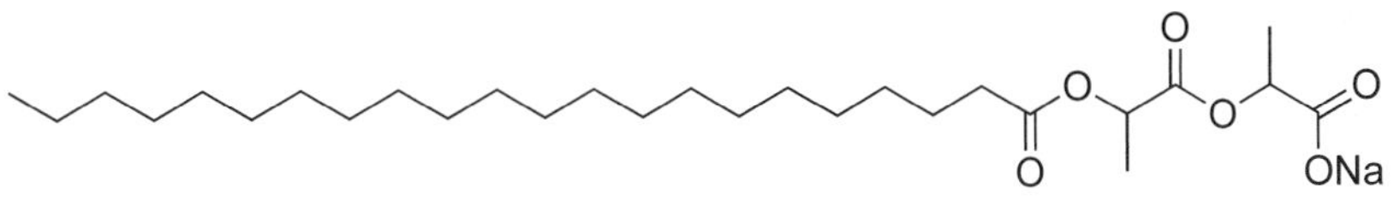

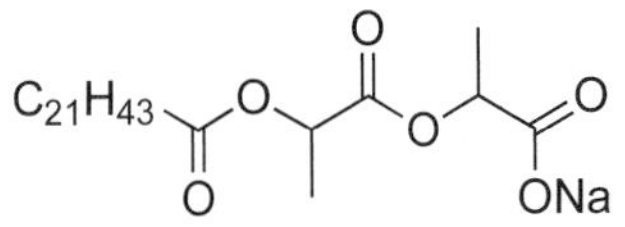

$C_{21}H_{43}$ O O ONa

$C_{28}H_{51}O_6Na$

CAS 27847-75-2

TRADENAMES

Pationic SBL

USES

component of cosmetics, an emulsifying agent in the moisterising creams at around 2.5%

IMPURITIES, ADDITIVES, COMPOSITION

the chemical contains sodium salts of lactic acid and behenic acid . The quantity of each will be less than 10% with the total not exceeding 15% (7458)

A. PROPERTIES

beige to yellow waxy solid; molecular weight 507; melting point 85-95°C; specific gravity 1.06 kg/L; water solubility 790 mg/L at 30°C; log Pow 6.0 (estimated)

C. WATER AND SOIL POLLUTION FACTORS

Hydrolysis

while it contains an ester functionality, it is not expected to hydrolyse under normal environmental conditions at pH 4 to 9 (7458)

D. BIOLOGICAL EFFECTS

Toxicity

Mammals

Rat	oral LD_{50}	est. >25,000 mg/kg bw	(7458)

benalaxyl (Methyl-N-phenylacetyl-N-2,6-xylyl-DL-alaninate; methyl-N-(2,6-dimethylphenyl)-N-(phenylacetyl)-DL-alaninate)

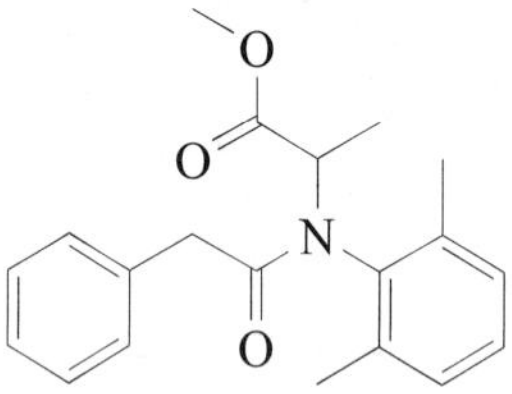

$C_{20}H_{23}NO_3$

CAS 71626-11-4

EINECS 275-728-7

TRADENAMES

Galben

USES

A racemic mixture is a broad-spectrum phenylamide fungicide that inhibits mycelial growth of fungi and germination of zoospores.

A. PROPERTIES

Whitish microcrystaline solid; molecular weight 325.4; melting point 76.8°C; decomposition at 250°C; relative density 1.18 kg/L at 20°C; vapour pressure 5.7 x 10^{-4} Pa at 20°C; water solubility 28.6 mg/L at 20°C; log Pow 3.5 at 20°C and pH 6.1; H 6.5 x 10^{-3} Pa.m^3/mol at 20°C

C. WATER AND SOIL POLLUTION FACTORS

Hydrolysis

25°C	pH 5	T/2 > 1 year
25°C	pH 7	T/2 = > 1 year
25°C	pH 9	T/2 = 86 days

Photodegradation in water

t.2 = 7 days at pH 5 and 25°C

Biodegradation in soil aerobic

% mineralisation	% bound-residues	After
25-26%	19%	100 days

Relevant metabolites: compound A, compound B and benalaxyl acid
Anaerobic degradation followed the same pattern as under aerobic conditions although much slower.(10942)

Biodegradation in laboratory studies

degradation in soils	%	Benalaxyl	Compound A	Compound B
Aerobic at 20°C	50	36-100 days	49-90 days	66-118 days
Aerobic at 10°C	50	198 days		
Anaerobic at 20°C	50	160 days		(10942)

Biodegradation in water/sediment systems

DT50 water	17-58 days	
DT90 water	57-190 days	
DT50 whole system	118-166 days	
DT90 whole system	228-309 days	(10942)

D. BIOLOGICAL EFFECTS

Bioaccumulation

Species	conc. μg/L	duration	body parts	BCF
Fish				57

Toxicity

Algae			
Selenastrum capricornutum	72h EC_{50}	2.4 mg/L	(10942)
Worms			
Dendroboena rubida	14d EC_{50}	180 mg/kg soil	(10942)
Crustaceae			
Daphnia magna	48h EC_{50}	0.59 mg/L	
	21d NOEC	0.03 mg/L	(10942)
Insecta			
Chironomus riparius	14d NOEC	3.1 mg/L	
Honeybees	Acute oral LD_{50}	>100 μg/bee	(10942)
Birds			
Bobwhite quail	Acute oral LD_{50}	3700; 4600 mg/kg bw	(10942)
Fish	96h LC_{50}	3.75 mg/L	(10942)
Mammals			
Rat	oral LD_{50}	4200 mg/kg bw	
Mouse	oral LD_{50}	680 mg/kg bw	(10942)

Toxicity of metabolite ***compound A***

Fish	96h LC_{50}	>100 mg/L	
Daphnia magna	48h EC_{50}	>100 mg/L	
Selenastrum capricornutum	72h EC_{50}	62.5 mg/L	(10942)

Toxicity of metabolite ***compound B***

Fish	96h LC_{50}	>100 mg/L	
Daphnia magna	48h EC_{50}	>100 mg/L	
Selenastrum capricornutum	72h EC_{50}	64.5 mg/L	(10942)

benazolin-ethyl (ethyl-4-chloro-2-oxobenzothiazolin-3-yl acetate)

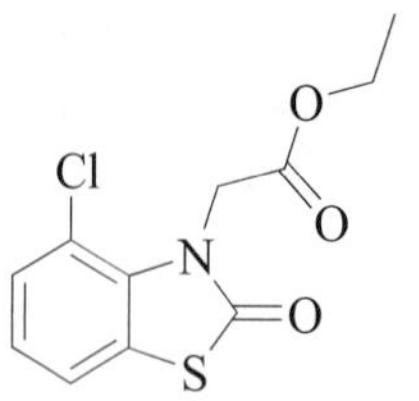

$C_9H_6ClNO_3S$

CAS 3813-05-6

USES

broad spectrum herbicide used mainly for weed control in cereals.

A. PROPERTIES

molecular weight 243.7; melting point 79°C; solubility 600 mg/L at 20°C.

C. WATER AND SOIL POLLUTION FACTORS

Degradation in soil

Degradation in soil at 25°C under aerobic conditions according to the following pathway (the phenyl ring was ^{14}C labeled):

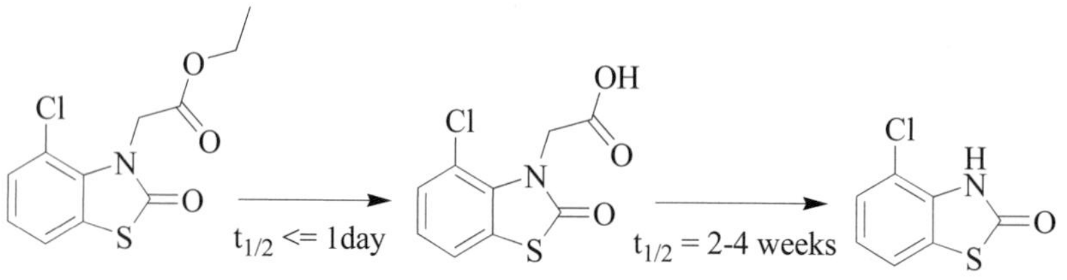

$t_{1/2}$ <= 1day

$t_{1/2}$ = 2-4 weeks

benazolin-ethyl   benazolin   4-chlorobenzothiazolin-2-o

The primary degradation route of benazolin-ethyl under aerobic conditions was by hydrolysis of the ethylester to form benazolin. This initial reaction was rapid with a half-life of less than 1 d. Degradation of benazolin occurred much more slowly with a half-life of 2-4 weeks. Further degradation via ring opening led to mineralization with up to 35% evolved as $^{14}CO_2$ in 1 year. Up to 66% of the applied radioactivity remained unextracted from the soil and was designated a bound residue.'

D. BIOLOGICAL EFFECTS

Fishes		
trout	96h LC_{50}	27 mg/L
bluegill	24h LC_{50}	204 mg/L

benefin *see* balan

benfluralin *see* balan

benomyl (methyl-1-(butylcarbamoyl)-2-benzimidazolecarbamate; benlate; tersan 1991; methyl 1-(butylcarbomyl)benzimidazol-2-ylcarbamate; methyl 1-[(butylamino)carbonyl]-1H-benzimidazol-2-ylcarbamate; methyl 1-(butylcarbamoyl)-2-benzimidazolecarbamate; benex-; fundazol)

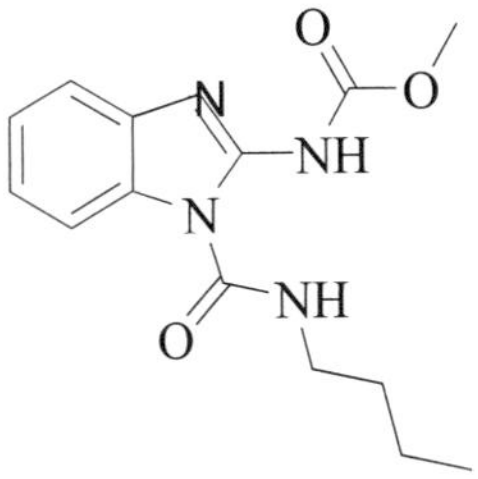

$C_{14}H_{18}N_4O_3$

CAS 17804-35-2

USES

generic name for a post-harvest fungicide for peaches, apples, etc. Also used as oxidizer in sewage treatment.

CHEMICAL STABILITY

Benomyl is easily hydrolyzed to (see Fig) methyl 2-benzimidazole carbamate (MBC) in very dilute aqueous, and in acidified methanolic solutions. MBC in turn hydrolyzes under basic conditions to give 2-aminobenzimidazole (2-AB).

By direct alkaline treatment of benomyl a triazine ring-closure takes place, forming 3-butyl-S-triazino [1.2 a] benzimidazole-2.4(1H, 3H)dione (STB) which is not stable in hot alkali. It is further converted to the stable 2-(3-butylureido)-benzimidazole (BUB). (11059)

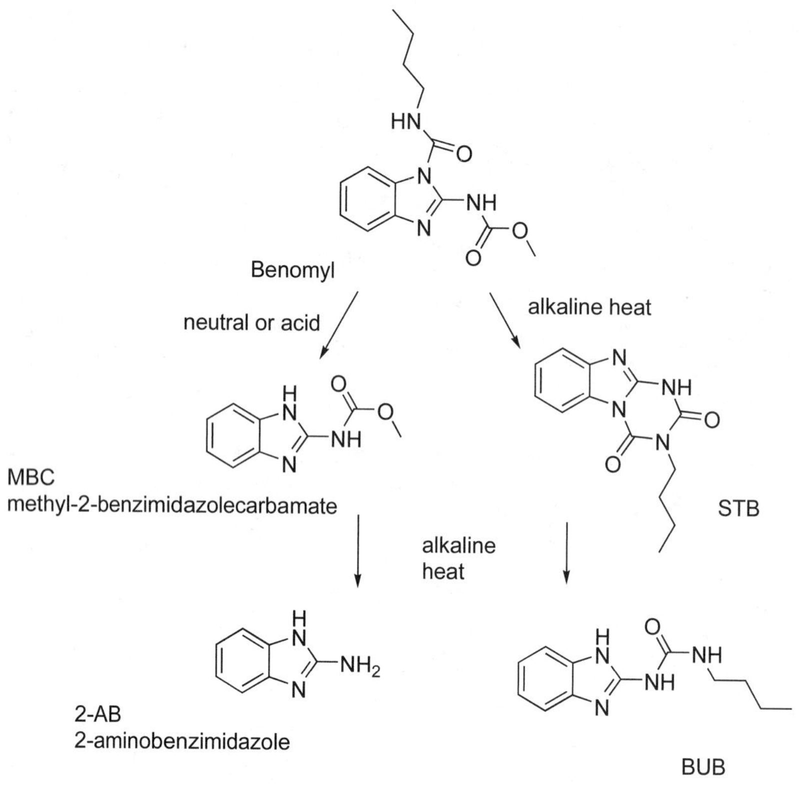

Benomyl

neutral or acid

alkaline heat

MBC
methyl-2-benzimidazolecarbamate

STB

alkaline
heat

2-AB
2-aminobenzimidazole

BUB

Chemical breakdown of Benomyl [11059]

A. PROPERTIES

molecular weight 290.62; melting point decomposes on heating, without melting; solubility 2 mg/L at 25°C.

C. WATER AND SOIL POLLUTION FACTORS

Hydrolysis

In aqueous solutions under acidic conditions, benomyl is hydrolysed to methyl 2-benzimidazole carbamate and butyl isocyanate. In water, the conversion of benomyl into methylbenzimidazole-2-yl carbamate (carbendazim) is completed within 1 wk. (9832)

Hydrolysis products detected were methyl-2-benzimidazolecarbamate and 2-aminobenzimidazole; both were less inhibitory of nitrification than benomyl (9830)

Biodegradation

Mixed bacterial cultures grew with benomyl as sole carbon source, but rates of breakdown to methyl benzimidazol-2-ylcarbamate were small and 2-aminobenzimidazole did not support growth. (9831)

$t_{1/2}$ in soil and water 2 and 19 h respectively. (9600)

Impact on biodegradation processes

effect of benomyl on nitrification:

µg nitrite recovered/ml medium:		*incubation period*
conc., ppm	*6 d*	*15 d*
by *Nitrosomonas* sp.		
0	3.8	108
10	2.0	2.1
100	1.7	2.0
by *Nitrobacter agilis*		

0	289	0	
10	444	513	
100	444	440	(1372a)

D. BIOLOGICAL EFFECTS

Bioaccumulation

	Concentration µg/L	*exposure period*	*BCF*	*organ/tissue*	
vegetation			-0.47		(2644)

Toxicity

Fishes			
goldfish	96h LC_{50}	4.2 mg/L	
rainbow trout	96h LC_{50}	0.17 mg/L	(2963)
fathead minnow	96h LC_{50}	2.2 mg/L technical material	
channel catfish	96h LC_{50}	0.029 mg/L technical material	
fathead minnow, bluegill sunfish	96h LC_{50}	1.2-1.9 mg/L wettable powder	
channel catfish	96h LC_{50}	0.028 mg/L wettable powder	(9344)
Mammals			
rat	oral LD_{50}	>9,590 mg/kg bw	(9833)

bensulide (N-(2-ethylthio)benzene sulphonamide-S,O,O-diisopropylphosphorodithioate; S-(O,O-diisopropylphosphorodithioate) of N-(2-mercaptoethyl)benzenesulfonamide; R 4461; Betasan; Prefar; Exporsan; (N-(2-ethylthio)benzene sulfonamido-S,O,O-diisopropylphosphorodithioate; S-O,O-diisopropylphosphorodithioate) of N-(2-mercaptoethyl)benzenesufonamide; R4461; Betasan-; Prefar-; Exporsan-; O,O-bis(1-methylethyl)S-[2-[(phenylsulfonyl)amino]ethyl]phosphorodithoate; phosphorodithioic acid, O,O-diisopropylester S-ester with N-(2-mercaptoethyl)benzenesulfonamide)

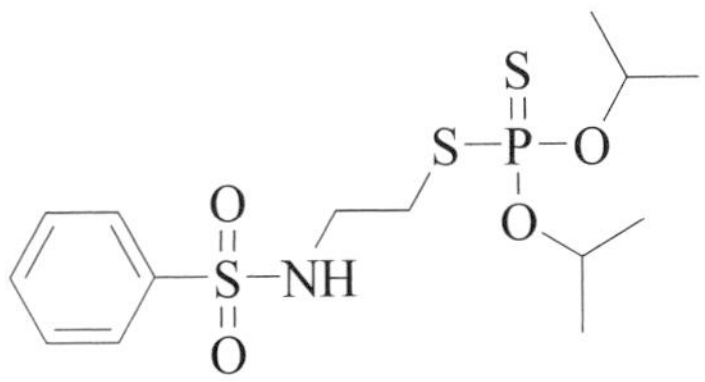

$C_{14}H_{24}NO_4PS_3$

CAS 741-58-2

USES

selective preemergence herbicide.

A. PROPERTIES

amber-colored liquid; molecular weight 397.5; melting point 34°C; boiling point decomposes on heating at 200°C; vapor pressure < 0.133 mPa at 20°C; solubility 25 mg/L; density 1.23 at 20/20°C; $LogP_{ow}$4.2.

C. WATER AND SOIL POLLUTION FACTORS

Biodegradation

Residual activity in soil 4-6 months at 21-27°C. (9600)

D. BIOLOGICAL EFFECTS

Crustaceans			
Daphnia	48h LC_{50}	0.58 mg/L	(9600)
water shrimp	96h LC_{50}	1.4 mg/L	(9344)
Fishes			
channel catfish *(Ictalurus punctatus)*	96h LC_{50},S	379 µg/L	(1202)
Carassius auratus	96h LC_{50}	1 mg/L	(2962)
rainbow trout	96h LC_{50}	1.1 mg/L	
goldfish	96h LC_{50}	1-2 mg/L	
bluegill sunfish	96h LC_{50}	0.8; 1.4 mg/L	(9344, 9600)
Mammals			
male, female rat	oral LD_{50}	360, 270 mg/kg bw	(9600, 9675)

bentazone (3-isopropyl-(1H)-benzo-2,1,3-thiadiazin-4-one-2,2-dioxyde; 3-(1-methylethyl)-(1H)-2,1,3-benzothiadiazin-4(3H)-one-2,2-dioxide; 3-isopropyl-2,1,3-benzothiadiazin-4-one-2,2-dioxide; Bentazon-; Bendioxide-; Basagran-; 1H-2,1,3-benzthiadiazin-4(3H)-one, 3-isopropyl-2,2-dioxide)

$C_{10}H_{12}N_2O_3S$

CAS 25057-89-0

TRADENAMES

Bentazon; Bendioxide; Basagran

USES

herbicide used in agriculture for selective postemergence control of many broadleaf weeds in soybeans, rice, corn, peanuts, mint, dry beans, dry peas and succulent lima beans.

A. PROPERTIES

colorless to slight brown crystalline powder; molecular weight 240.3; melting point 137-139°C; vapor pressure 3.5 x 10^{-6} mm Hg at 20°C; density 1.47; solubility 500 mg/L at 20°C, 570 mg/L at pH 7; $\log P_{OW}$ 0.35 at pH 7.

B. AIR POLLUTION FACTORS

Photolysis

In sunlight bentazone undergoes oxidation and dimerization with loss of SO_2. (9600)

C. WATER AND SOIL POLLUTION FACTORS

Aquatic reactions

very resistant to hydrolysis. In 0.1N sodium hydroxide and in 0.1N hydrochloric acid, no degradation observed after 48 h. In UV light, 50% decomposition occurs in 13 h. (2962)

Biodegradation

Aerobic $t_{1/2}$: 13.6 d at 20°C.

field $t_{1/2}$: approx. 12 d. (9600)

Rapidly metabolised in tolerant plants to extractable conjugates which are incorporated into plant components. (9666)

D. BIOLOGICAL EFFECTS

Metabolism

Parent bentazon was the major form of bentazon found in the urine of orally dosed rats. It accounted for 77-91% of the administered dose, while 8-hydroxybentazon accounted for 6.3% of the dose. Only minor amounts (2% of the dose) of the isomer 6-hydroxybentazon were found in the urine. The excretion half-life was calculated to be 4 hours.

In mice dosed orally with 14C-bentazon, the cyclic sulfonamide ring of bentazon is apparently cleaved to several metabolites. These metabolites include N-isopropylsulfamyl anthranilic acid (5.4%), anthranilic acid (6.2%) and 2-amino-N-isopropyl benzamide (6.1%). However no 6- or 8-hydroxybentazon was detected in the urine of mice. The parent compound also accounted for about 71% of the administered radioactivity.

Microorganisms

nitrification inhibition in soil (1% organic carbon) after 14 days : 77 % at 50 mg/kg (7050)

Algae			
Ankistrodesmus	72h EC_{50}	62 mg/L	(9600)
Crustaceans			
Daphnia	48h EC_{50}	125 mg/L	(9600)
Fishes			
cytotoxicity to goldfish GF-Scale cells	NR_{50}	>200 mg/L	(2680)
Cyprinus carpio	8h LC_{50}	40 mg/L	(2680)
rainbow trout	96h LC_{50}	>100 mg/L	
bluegill sunfish	96h LC_{50}	>100 mg/L	(9600)
Mammals			
rat, dog, rabbit, cat	oral LD_{50}	>1,000, >500, 750, 500 mg/kg bw	(9600)

benz(a)aceanthrylene

MANMADE SOURCES

in coal tar:	7.9 g/kg	(2600)
in carbon black:	190 mg/kg	(2600)

benz(c)acephenanthrylene

MANMADE SOURCES

in coal tar: 0.4 g/kg. (2600)

benz(a)acridine

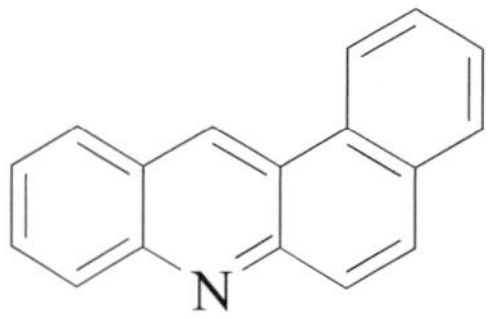

$C_{17}H_{11}N$

CAS 225-51-4

A. PROPERTIES

molecular weight 229; $LogP_{OW}$ 4.45 (calculated).

B. AIR POLLUTION FACTORS

Ambient air concentrations

in the average American urban atmosphere-1963:
2 µg/g airborne particulates or 0.2 ng/m^3 air (1293)

Cleveland (Ohio, U.S.): max. conc.: 31 ng/m^3 (144 values) (n.s.i.) (1971/1972) annual geom. mean (all sites): 1.4 ng/m^3 (n.s.i.)
annual geom. mean of TSP (all sites): 12 ppm (wt) (n.s.i.) (556)
glc's in residential area (Belgium): Oct. 1976: in particulate sample: 0.85 ng/m^3 (1239)

D. BIOLOGICAL EFFECTS

Crustaceans			
Daphnia pulex	BCF	352 (initial conc. in water: 18 ppb)	
	24h LC_{50}	0.449 mg/L	
immobilization concentration	IC_{50}	0.362 mg/L	(1050)
Daphnia pulex	BCF	320	(2660)
Fishes			
Pimephales promelas	BCF	100	(2660)

benz(c)acridine

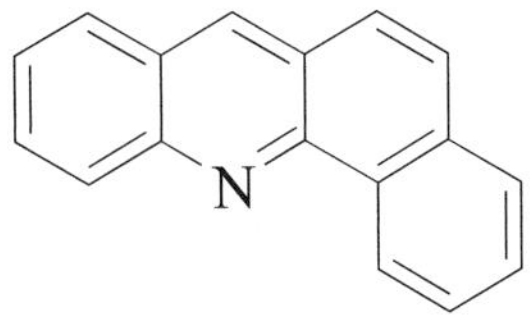

CAS 225-51-4

B. AIR POLLUTION FACTORS

Ambient air quality

in the average American urban atmosphere-1963: 4µg/g airborne particulates or 0.6 ng/cu air. (1293)

Incinerability

thermal stability ranking of hazardous organic compounds: rank 85 on a scale of 1 (highest stability) to 320 (lowest stability). (2390)

benzalchloride

(benzylidene-chloride; Dichloromethyl)benzene; (benzyl-dichloride-; alpha,alpha-dichlorotoluene; -; benzylene-chloride)

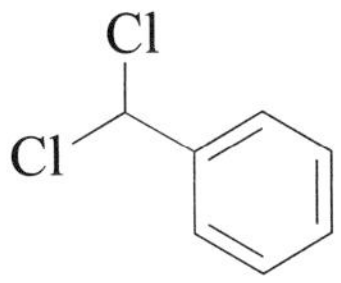

$C_7H_6Cl_2$

CAS 98-87-3

USES

In manufacture of benzaldehyde and cinnamic acid.

A. PROPERTIES

molecular weight 161.03; melting point -16°C; boiling point 205°C; density 1.26.

B. AIR POLLUTION FACTORS

Incinerability

Temperature for 99% destruction at 2.0 sec residence time under oxygen-starved reaction conditions: 625°C

Thermal stability ranking of hazardous organic compounds: rank 168 on a scale of 1 (highest stability) to 320 (lowest stability) (2390)

C. WATER AND SOIL POLLUTION FACTORS

Hydrolysis

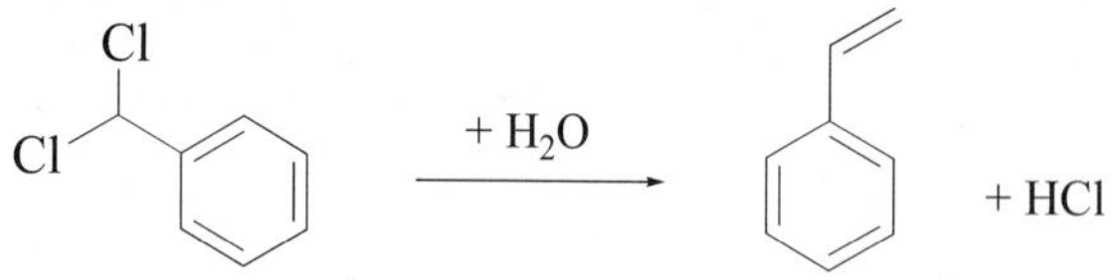

Cl, Cl, + H_2O, + HCl

benzaldehyde

Hydrolyses to benzaldehyde under both acid and alkaline conditions.	(9835)

Biodegradation

readily biodegrades in water.	(9834)

D. BIOLOGICAL EFFECTS

Bacteria			
Photobacterium phosphoreum Microtox test	30 min EC_{50}	5.9 mg/L	(9657)
Mammals			
rat	oral LD_{50}	3250 mg/kg	(9836)

benzaldehyde (benzenecarbonal; oil of bitter almonds)

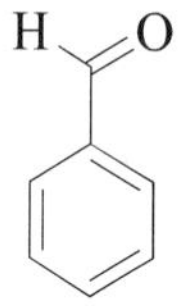

H, O

C_6H_5CHO

C_7H_6O

CAS 100-52-7

USES

chemical intermediate for dyes, flavoring materials, perfumes, and aromatic alcohols; solvent for oils, resins, some cellulose ethers, cellulose acetate and nitrate; flavoring compounds; synthetic perfumes; manufacturing of cinnamic acid, benzoic acid; pharmaceuticals; photographic chemicals.

OCCURRENCE

in oil of bitter almond.

A. PROPERTIES

molecular weight 106.1; melting point -26°C; boiling point 179°C; vapor pressure 1 mm at 26°C, 40 mm at 90°C; vapor density 3.66; density 1.05 at 15/4°C; solubility 3,300 mg/L; $LogP_{ow}$ 1.48.

B. AIR POLLUTION FACTORS

1 mg/m^3= 0.227 ppm, 1 ppm = 4.41 mg/m^3.

Odor

characteristic: quality: bitter almonds.

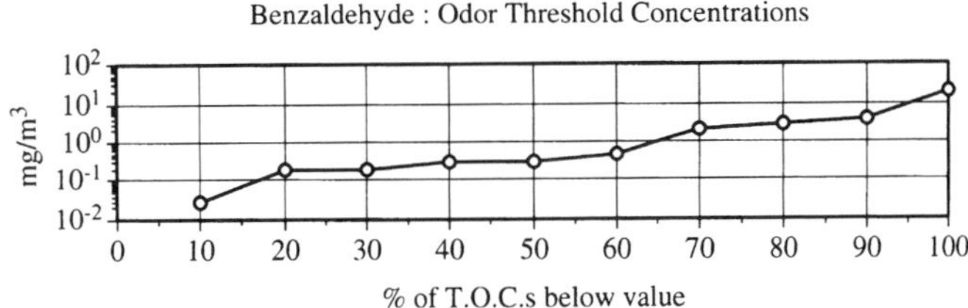

Air quality

inside and outside 6 residential houses, suburban New Jersey, summer 1992 (n = 36):

outdoor:	mean: 0.25 ppb; max: 0.9 ppb;	SD: 0.25 ppb	
indoor:	mean: 0.38 ppb; max: 1.33 ppb;	SD: 0.30 ppb	(2951)

Manmade sources

in gasoline exhaust:	<0.1-13.5 ppm	(195, 1053)
	3.2-8.5 vol % of total exhaust aldehydes	(394, 395, 396, 397)
in diesel exhaust:	0.3 ppm	(311)

Control methods

wet scrubber:	water at pH 8.5: outlet: 80 odor units/scf	
	$KMnO_4$ at pH 8.5: outlet: 1 odor unit/scf	(115)

C. WATER AND SOIL POLLUTION FACTORS

BOD_5	36; 67% of ThOD	(36, 220)
BOD_{10}	62% ThOD std. dil. sew.	(256)
BOD_{20}	67; 74% ThOD at 10 mg/L, unadapted sew.; no lag period	(554)
COD	82; 95% of ThOD	(36, 220)
ThOD	2.42	(36)

Biodegradation

adapted A.S. at 20°C-product is sole carbon source: 99% COD removal at 119 mg COD/g dry inoculum/h. (327)

Impact on biodegradation processes: at 400 mg/L standard BOD_{10} test perturbated. (30)

Reduction of amenities: faint odor: 0.003 mg/L. (129)

tentative T.O.C.: 0.002 mg/L; 0.44 ppb (27)

T.O.C. in water: 0.18; 0.44; 3.0; 3.9; 4.0; 4.2 ppb (326)

detection: 0.035 mg/kg (882)

Manmade sources

Göteborg (Sweden) sew. works 1989-1991: infl.: 0-10 µg/L; effl.: <0.1 µg/L. (2787)

Waste water treatment

A.C.: adsorbability: 0.188 g/g C, 94% reduction, infl.: 1,000 mg/L, effl.: 60 mg/L (32)

A.S.: acclimated to the following aromatics: (infl.: 250 mg/L and 30 min aeration)

phenol:	38% ThOD	
benzylalcohol:	30% ThOD	
anthranilic acid:	35% ThOD	(92)

methods	*temp, °C*	*d observed*	*feed, mg/L*	*d acclim.*	*% removed*	
NFG, BOD	20	1-10	200-400	365 + P	50	
NFG, BOD	20	1-10	600	365 + P	19	
NFG, BOD	20	1-10	800	365 + P	7	(93)

A.C.:	*influent, ppm*	*carbon dosage*	*effluent, ppm*	*% reduction*	
	1,000	10×	9	99	
	500	10×	6	99	
	100	10×	2	98	(192)

D. BIOLOGICAL EFFECTS

Toxicity threshold (cell multiplication inhibition test)

Bacteria			
Pseudomonas putida	16h EC_0	132 mg/L	(1900)
Algae			
Microcystis aeruginosa	8d EC_0	20 mg/L	(329)
Green algae			
Scenedesmus quadricauda	7d EC_0	34 mg/L	(1900)
Protozoans			
Entosiphon sulcatum	72h EC_0	0.29 mg/L	(1900)
Uronema parduczi Chatton-Lwoff	EC_0	22 mg/L	(901)

Fishes			
minnows: stop eating: 17.1 mg/L of 85% solution			(226)
Pimephales promelas	24h LC_{50}	35 mg/L	
	96h LC_{50}	7.6 mg/L	(2709)

Benzalkonium chloride (alkylbenzyldimethylammonium chloride; N-alkyl(C_{8-18})dimethylbenzylammonium chloride; cetalkonium-chloride) BAC; Benzyldimethylalkylammonium chloride))

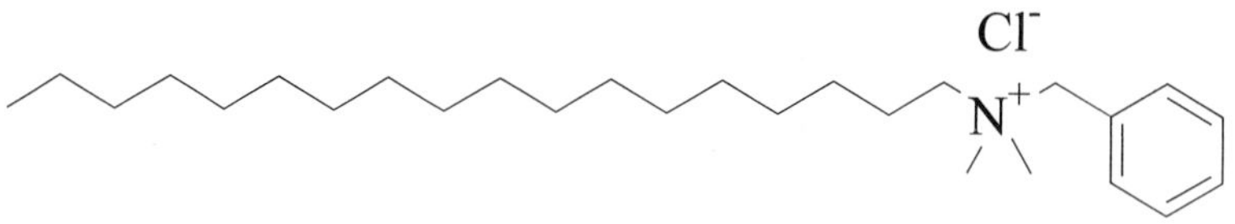

$C_6H_5CH_2N(CH_3)_2.RCl$

CAS 8001-54-5

USES

Cationic surfactant, germicide and fungicide. Used in leather processing and textile dyeing industries. As a general antibacterial agent. A yellow-white powder prepared in an aqueous solution and used as a detergent, fungicide, bactericide, and spermicide. The greatest bactericidal activity is associated with the C12-C14 alkyl derivatives. Applications are extremely wide ranging, from disinfectant formulations to microbial corrosion inhibition in the oilfield sector. It has long deemed safe for human use, and is widely used in eyewashes, hand and face washes, mouthwashes, spermicidal creams, and in various other cleaners, sanitizers, and disinfectants.Also used ascharge-control agent in electrostatographic toners and developers wherein the alkyl has 12 to 18 carbon atoms.

IMPURITIES, ADDITIVES, COMPOSITION

Benzalkonium chloride is a mixture of alkylbenzyl dimethylammonium chlorides of various alkyl chain lengths.

CAS	C8+10	C12	C14	C16	C18
61789-71-7	2.5%	61%	23%	11%	2.5%
68424-85-1		25%	60%	15%	
68424-85-1		40%	40%	20%	
68391-01-5		28%	41%	12%	19%
68989-00-4	5%	61%	23%	11%	
68424-85-1		3%	95%	2%	
122-18-9				100%	
122-19-0					100%
8001-54-5		50%	30%	17%	3%
139-07-1		100%			

A. PROPERTIES

white or yellow powder or gelatinous lumps, colorless solution;density 0.99 at 20°C

C. WATER AND SOIL POLLUTION FACTORS

Biodegradation

Biodegradation by *Aeromonas hydrophila* resulted in the follwing pathway. The first step of BAC catabolism was the cleavage of the C-alkyl-N bond resulting in the formation of benzyldimethylamine. Subsequent demethylation reactions result in the liberation of benzylmethylamine and benzylamine followed by the deamination and formation of benzaldehyde. Benzaldehyde was rapidly converted into benzoic acid, which is further degraded. However because of BAC and benzylamines toxicity, the biodegradation was not complete and its efficacy depended on the initial surfactant concentration. (10716)

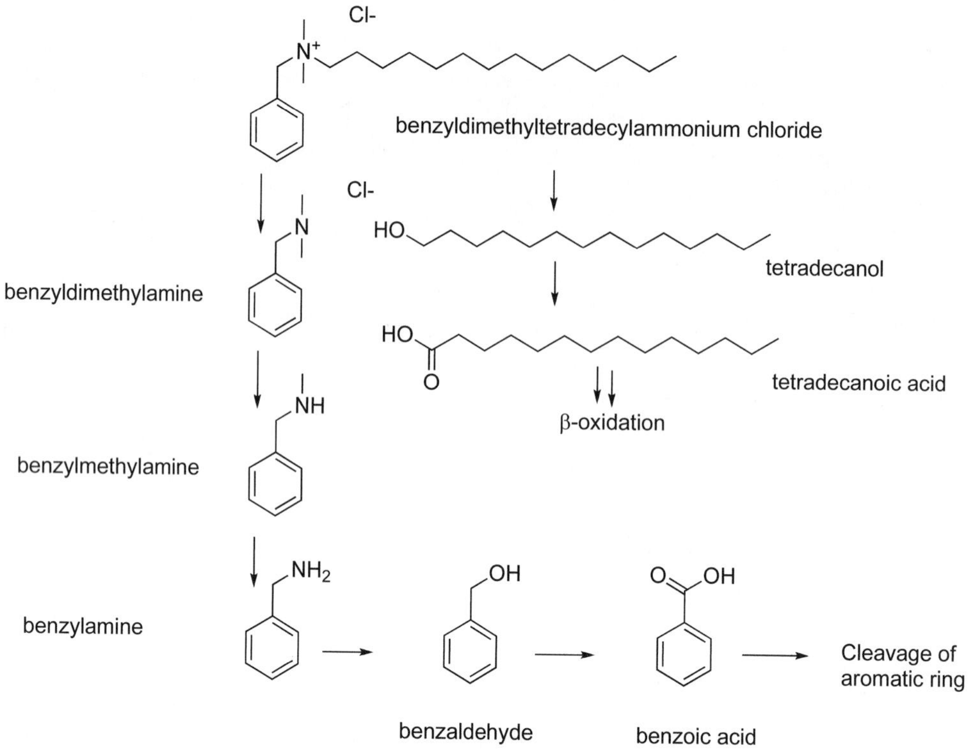

Cl-
benzyldimethyltetradecylammonium chloride
Cl-
HO
benzyldimethylamine
tetradecanol
HO
O
tetradecanoic acid
β-oxidation
benzylmethylamine
NH_2
OH
O OH
benzylamine
Cleavage of aromatic ring
benzaldehyde
benzoic acid

Proposed pathway for benzyldimethylalkylammonium chloride biodegradation by *Aeromonas hydrophila sp.K*

D. BIOLOGICAL EFFECTS

Toxicity

Bacteria			
Activated sludge respiration inhibition	EC_{50}	14 mg/L.	(2624)
Fishes			
threespine stickleback	4h LC_{100}	2 mg/L	
steelhead trout	2h LC_{100}	2 mg/L	(9837)

Micro organisms

100-200 mg/L BAC inhibited bacterial growth of *Aeromonas hydrophila sp. K* at the extent dependent on its concentration. (10716)

benzaminocetic acid *see* hippuric acid

benzanthrone

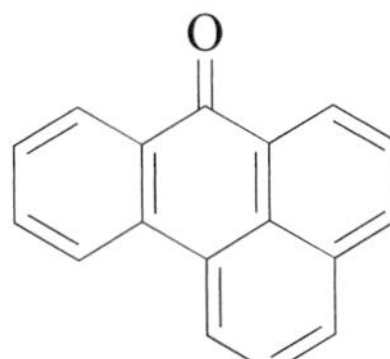

$C_{17}H_{10}O$

CAS 82-05-3

USES

dyes.

A. PROPERTIES

pale yellow needles; melting point 170°C; molecular weight 230.27.

B. AIR POLLUTION FACTORS

Ambient air quality

glc at Botrange (Belgium): woodland at 20-30 km from industrial area-June 1977: 0.18 ng/m^3. (1233)

1-benzazine *see* quinoline

benzene

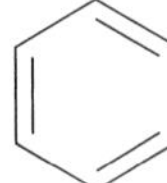

C_6H_6

CAS 71-43-2

USES AND FORMULATIONS

mfg. styrene, phenol, detergents, organic chemicals, pesticide, plastics and resins, synthetic rubber, aviation fuel, pharmaceuticals, dye, explosives, PCB gasoline, tanning, flavors and perfumes, paints and coatings; nylon intermediates; food processing; photographic chemicals. (347)

MANUFACTURING SOURCES

petroleum refinery; solvent recovery plant; coal tar distillation; coal processing; coal coking. (347)

A. PROPERTIES

colorless liquid; molecular weight 78.11; melting point 5.5°C; boiling point 80.1°C; vapor pressure 76 mm at 20°C, 60 mm at 15°C, 118 mm at 30°C; vapor density 2.77; density 0.88 at 20/4°C; solubility 1,780 mg/L at 20°C; saturation concentration in air 319 g/m^3 at 20°C, 485 g/m^3 at 30°C; log P_{ow} 2.13 at 20°C, 2.64; 2.73; log H -0.65 at 25°C.

B. AIR POLLUTION FACTORS

1 mg/m^3= 0.31 ppm, 1 ppm= 3.26 mg/m^3.

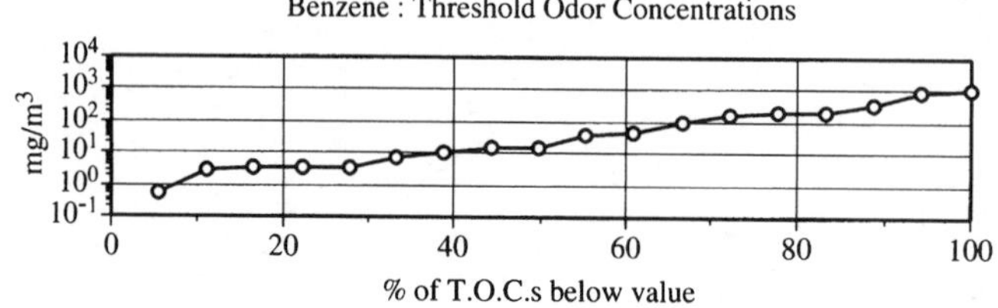

(2, 3, 4, 5, 9, 278, 291, 307)

$PIT_{50\%}$: 2.14 ppm
$PIT_{100\%}$: 4.68 ppm (2)
distinct odor: 310 mg/m^3= 90 ppm (278)
human odor perception: 3.0 mg/m^3= 1 ppm
animal chronic exposure: adverse effect: 3.2 mg/m^3 (170)

Atmospheric reactions

R.C.R.: 0.276. (49)

reactivity: HC cons.: ranking: 0.5
NO ox.: ranking: 0.04-0.15 (63)

Atmospheric half-lives

for reactions with OH°: 5.7 d
for reactions with O_3: 170,000 d (2716)

Atmospheric reactions

Only the reaction with hydroxyl radicals is important: rate constants (K_{OH}) between 1,300 and 2,350 have been reported with an average of 1,800 (ppm^{-1}, min^{-1}). Based on this latter value, an average lifetime of 5.3 d can be calculated for the Netherlands. The following reaction pathway is proposed:

+ °OH → OH + HO_2

O_2

O O

O_2 OH
endoperoxide

NO
NO_2

O
O
+ HO_2
glyoxal
+
O O
butenedial

Natural sources (2701)

volcanic eruptions, vegetation, forest fires: a benzene concentration of 35 µg/m^3 was measured in the plume of a forest fire at a distance of 6 km of the seat of the fire. (2701)

glc's Pt Barrow, Alaska, Sept. 1967: not detectable to 0.4 ppb (101)

Manmade sources

In cigarette smoke: 150-205 mg/m^3
in main stream: 0.01-0.1 mg/cigarette
in side stream: 0.05-0.49 mg/cigarette (2701)
500 µg/cigarette (2421)

In flue gases of a municipal waste incinerator plant: 15 µg/m^3 = 1.3% of total organic carbon emissions. (7048)

Atmospheric emissions in the Netherlands in 1981

	percentage	
exhaust gases from car engines	85	
production, use, and transport of benzene and derivatives	10	
remaining traffic	2	
production and transport of petrol	1	
storage and handling at gasoline stations	1	
evaporation from car tanks	1	(2701)

diesel engine: 2.4% of emitted HCs (72)

rotary gasoline engine: 1.3% of emitted HCs
reciprocating gasoline engine: 2.2% of emitted HCs (78)
expected glc's in U.S. urban air: range: 10 to 50 ppb (102)
emitted by household central heating system on gasoil:
~20 ppm at 7% CO_2
6 g/kg gasoil at 6% CO_2
2.2 g/kg gasoil at 7% CO_2 (182)
in gasoline exhaust: 0.1 to 42.6 ppm (partly methylvinylketone) (195)
in gasoline: 1.8-5% (312, 34)
in exhaust of gasoline engines: 62-car survey: 2.4 vol% of total exhaust HCs (391)
exhaust emissions from 1975 GM models equipped with bead-type converters:

% of total hydrocarbons:	avg 2.7%,	range 1-7% (1975 federal test procedure)
mg/mile:	avg 20,	range 8-75

exhaust emission from noncatalyst cars (mostly 1974 GM models)

of total hydrocarbons:	4.6% avg	
mg/mile:	113 avg	(1386)

According to an EPA source inventory published in 1989, 85% of benzene emissions in the U.S.A. are from mobile sources, of which 70% are from exhaust, 14% from evaporative emissions, and 1% from vehicle refueling. Benzene averaged 3-4% of total hydrocarbons. (2417)

Automotive emissions

precatalyst cars: 114-153 mg/mile
catalyst cars: 5-32 mg/mile
1983-87 in-use vehicles: 4.8-15 mg/mile
1983-87 in-use trucks: 9.5-14 mg/mile (2416)

Average emission by model year

model year	emission, mg/mile	
1981/1982	27	
1983/1984	13	
1985	8.6	
1986	9.6	
1987	6.6	(2416)

In 4 municipal landfill gases in Southern Finland (1989-90 data): avg 0.17-9, max 11 mg/m^3. (2605)

Ambient air quality

glc's in Netherlands:

in tunnel Amsterdam-1973: avg 6 ppb (n = 3)

in tunnel Rotterdam-1974.10.2:	avg	25 ppb (n = 12)	
	max	33 ppb	
The Hague-1974.10.11:	avg	13 ppb (n = 12)	
	max	29 ppb	
Roelofarendsveen-1974.9.11:	avg	2 ppb (n = 12)	
	max	3 ppb	(1231)

glc's in the U.S.A. during winter 1977-78:

	averages	*range (8 h average)*	
urban/suburban areas:	<0.5-3.6 ppb	<0.5-12.1 ppb	
rural areas:	<0.5-0.6 ppb	<0.5-1.8 ppb	
remote areas:	<0.5-0.7 ppb	<0.5-2.0 ppb	(1401)

glc's in Los Angeles 1966:

avg:	0.015 ppm (n = 136)	
highest value:	0.057 ppm	(1319)

Indoor/outdoor glc's winter 1981/1982 and 1982/1983 the Netherlands

µg/m^3	*median*	*maximum*	
pre-war homes	7	24	
post-war homes	7	148	
<6-year-old homes	5	53	
outdoors	3	7	(2668)

Indoors/outdoors Germany 1980: n = 15, averages of 1 to several d

µg/m^3	*mean*	*range*	
in the kitchen	15	6-27	
in the other rooms	18	6-14	
outdoors	29	4-33	(2669)

Inside and outside 15 living rooms in Northern Italy (1983-84):

4-7 d averages	*lowest*	*mean*	*highest value, µg/m^3*	
indoor	5	52	204	
outdoor	3	20	67	(2756)

Indoor concentrations

The chemical was detected in 1422 out of over 3000 air samples collected from offices, schools and homes at an average concentration of 0.024 mg/m3 and a maximum concentration of 5.2 mg/m3. (7149)

Photochemical reaction

estimated lifetime under photochemical smog conditions in SE England: 28 h. (1699, 1707)

Control methods

platinized ceramic honeycomb catalyst: ignition temp.: 180°C
inlet temp. for 90% conversion: 250-300°C (91)

Incinerability

Temperature for 99% destruction at 2.0 sec residence time under oxygen-starved reaction conditions: 1,150°C

Thermal stability ranking of hazardous organic compounds

rank 3 on a scale of 1 (highest stability) to 320 (lowest stability) (2390)

Partition coefficients

$K_{air/water}$	0.25	
Cuticular matrix/air partition coefficient*	430 ± 0	(7077)

* experimental value at 25°C studied in the cuticular membranes from mature tomato fruits (*Lycopersicon esculentum* Mill. cultivar Vendor)

C. WATER AND SOIL POLLUTION FACTORS

BOD_5:	0; 0; 10; 24; 70% ThOD	(23, 41, 27, 220, 274, 277)
	58% ThOD. (acclim.)	(23)
BOD_{10}:	67% bio.-ox. (acclim.)	(23)
	27; 39% ThOD. (nonacclim.)	(23, 256)
BOD_{15}:	76% ThOD. (acclim.)	(23)
	24% ThOD. (nonacclim.)	(23)
BOD_{20}:	80% ThOD. (acclim.)	(23)
	29% ThOD. (nonacclim.)	(23)
BOD_{35} 25°C:	61% ThOD	(62)
BOD_{35} 25°C:	51% ThOD in seawater/inoculum: enrichment cultures of hydrocarbon oxidizing bacteria	(521)
COD:	8; 45% ThOD	(27, 23)
	19% ThOD	(220)
	69% ThOD ASTM procedure	(272)
	93% ThOD	(272)
	33% ThOD (0.05 *n* $K_2Cr_2O_7$)	(274)
TOC:	40% ThOD	(220)
ThOD:	3.10	(23)

Benzene : Threshold Odor Concentrations in Water

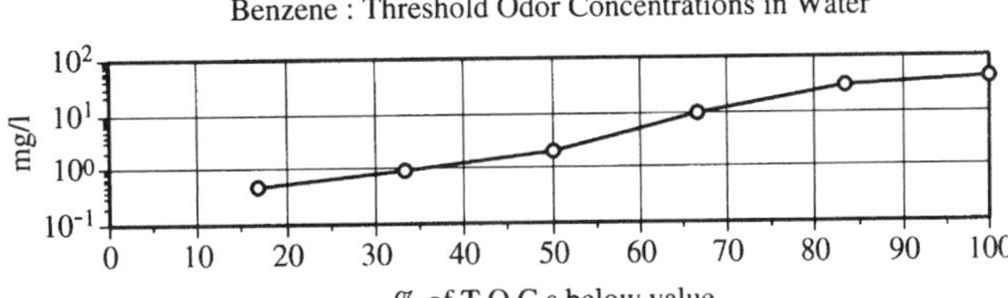

Biodegradation

CO_2 in estuarine water: in June after 24 h incubation.

Concentration (μg/L)	*Degradation rate, (μg/L/d) × 10^3*	*Turnover time, d*	
6	200±10	30	
12	260±25	46	
24	330±30	75	(381)

Incubation with natural flora in the groundwater in presence of the other components of high-octane gasoline (100 μl/L).

Biotic transformation in water

the following pathways are suggested: hydroxylation of the hydrocarbon ring:

catechol

ortho-fission of catechol:

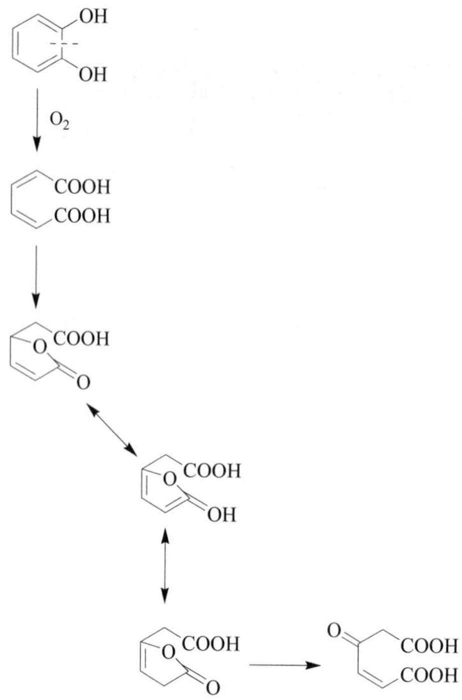

OH
OH
O_2
COOH
COOH
O COOH
O
O COOH
OH
O COOH
O
O
COOH
COOH

meta-fission of catechol:

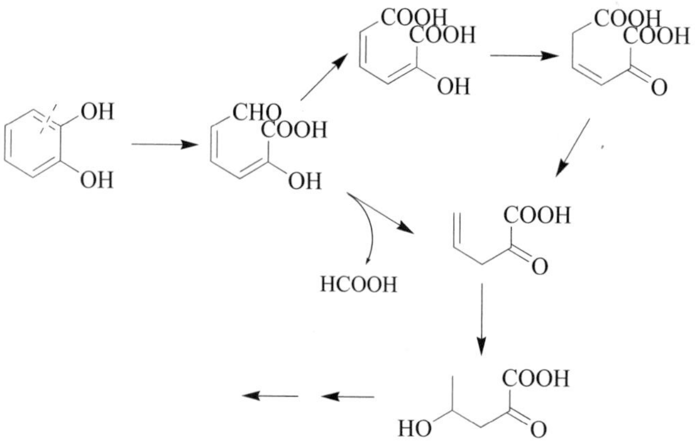

OH
OH
CHO
COOH
OH
COOH
COOH
OH
COOH
COOH
O
COOH
O
HCOOH
COOH
HO
O

(2701)

Photolysis in water

immediately below the water level, the calculated half-life is ± 80 d based on reaction with the OH radical; the half-life increases linearly with depth as a result of decreasing light intensity. (2701)

In estuaries at conc. of 6-24 µg/L, half-life: 15-38 d
In mesocosms at 5.5 m depth in the sea at 0.2-4 µg/L, half-lives:

at 3-7°C	13 d
at 8-16°C	23 d
at 20-22°C	3.1 d

At larger depths the half-lives increased proportionately. (2701)

Water treatment

Ozonation: surface water containing 80 µg/L treated with 2.8 mg ozone/L resulted in 97% removal (2701)

Activated sludge treatment: 99% removal (2701)

Reverse osmosis: 76-87% rejection was obtained by a low-pressure thin-film compos- ite polyamide membrane FT-30 in spiral wound configuration at influents of 10-25 mg/L at ambient temp (2630)

initial concentration, µg/L	*treatment method*	*% removal*	
371-750	river bank filtration	80-82	
120-416	activated carbon	17-95	(2915)

Aerobic degradation

of mixed organic wastes by a propane-fed bioreactor: >90% within 21 d at 22°C and initial conc. of 0.086 mg/L. (2432)

Micelle-water partition coefficient (K_{mw}) for the anionic surfactant dodecylsulfate: log K_{mw}: 2.88-3.06 (on mass basis), 2.11 (on vol basis). (2361)

Disappearance in soil

at 100 mg/kg and 20°C under aerobic laboratory conditions:

Half-life: >2 d (2366)

Degradation half-lives: 3.1-23 d in marine mesocosm (2367)

8.6 d in activated sludge (2368)

Waste treatment

Aerobic slurry bioremediation reactor with mixed aeration chamber, treating a petrochemical waste sludge at 3.5 to 7.5% solids containing approx. 25% oil and grease and 0.5 to 2.5 g of waste material per gram of microorganisms, at 22-24°C during batch treatment:

mixed liquor: infl.: 13 mg/L; effl. after 15 d: <1 mg/L

waste residu: infl.: 290 mg/kg; effl. after 90 d: <1 mg/kg

vapor phase: infl.: 83-156 ppm; after 4 d: 5-10 ppm; after 9 d: nd-0.3 ppm (2800)

Water quality

in river Maas at Eysden (Netherlands) in 1976:	median: 0.1 µg/L; range: n.d. to 5.7 µg/L	
Keizersveer (Netherlands) in 1976:	median: 0.1 µg/L; range: n.d. to 1 µg/L	(1368)

in Lake Zürich (Switzerland): at surface: 28 ppt; at 30 m: 22 ppt

in Zürich area: spring water: 18 ppt; ground water: 45 ppt; tap water: 36 ppt (513)

Evaporation from water

Calculated half-life in water at 25°C and 1 m depth, based on evaporation rate of 0.144 m/h: 4.81 h; based on evaporation rate of 0.137 m/h: 5.03 h. (437)

Soil adsorption

Freundlich constants for benzene sorption after 16 h incubation at concentrations 10-1,000 ppb:

adsorbent	*K*	*1/n*	
Hastings silty clay loam	2.4	0.89	
Overton silty clay loam	1.8	0.94	
A1-saturated montmorillonite	31	1.8	
Ca-saturated montmorillonite	4.4	0.99	(1862)

Waste water treatment

A.S.: 33% ThOD of 500 ppm by phenol acclimated A.S. after 12 h aeration (26)

A.C.: adsorbability: 0.08 g/g, 95% reduction, infl.: 416 mg/L, effl.: 21 mg/L (32)

anaerobic lagoon: 13 lb/d/1,000 cu ft: infl.: 10 mg/L, effl.: 5 mg/L (37)

ion exchange: adsorption on Amberlite X AD-2: retention efficiency: 100% at 100 ppm influent conc. (40)

air stripping constant: K= 1.71 d^{-1} at 100 mg/L (82)

A.S., Sd. BOD, 14 d acclim.: 2% of ThOD after 5 d at 20°C

A.S., W, 14 d acclimation: 3% of ThOD after 1/4 d at 20°C, feed: 50-200 mg/L (93)

air flotation after chemical addition: 78% removal (173)

A.C.: influent, ppm	carbon dosage	effluent, ppm	% reduction	
500	10×	27	95	
250	10×	23	91	
50	10×	20	60	(192)

Manmade sources

Göteborg (Sweden) sew. works 1989-1991: infl.: 0.1-5 μg/L; effl.: <0.5 μg/L (2787)
Landfill leachate from lined disposal area (Florida 1987-'90): mean conc. 0.10; 1.9 μg/L (2788)
In digested sewage sludge: U.K. 1993: $n = 12$

mean:	0.084 mg/kg dry wt	SD:	0.116 mg/kg dry wt	
	2.87 μg/L wet volume	SD:	3.97 μg/L wet volume	(2946)

In Canadian municipal sludges and sludge compost: September 1993- February 1994: mean values of 11 sludges ranged from 0.024 to 0.48 mg/kg dw.; mean: 0.058 mg/kg dw mean value of sludge compost: 0.0046 mg/kg dw. (7000)

Degradation in soil

biodegradation in methanogenic aquifer at 17°C and initial conc. of 613 μg/L: $t_{1/2}$ = 27 weeks. (2614)

Fate

of ^{14}C-labeled compound in a laboratory soil-plant system (7 d):

volatilization:	0.25%	
mineralization:	62.2%	
soluble metabolites in soil:	0.93%	
bound residues:	9.99%	
uptake by barley:	0.065%	
uptake by cress:	0.24%	(2616)
Biodegradation half-lives in nonadapted aerobic subsoil:		
groundwater Florida:	12 d	
river Oklahoma:	24 d	
sand Texas:	>161 d	
sand Ontario:	48 d	(2695)

Impact on biodegradation processes

Effect on the mineralization of acetate in anaerobic river sediment at 20°C:

toxic effects:	EC_{50}: 3,500 mg/kg dry sediment	
	EC_{10}: 1,200 mg/kg dry sediment	
inhibition of mineralization:	IC_{50}: 1,700 mg/kg dry sediment	
	IC_{10}: 480 mg/kg dry sediment	(2693)

Inhibition of acetoclastic methanogenesis in a granular sludge

IC_{50}: 1,500 mg/L (2694)

Toxic effect of benzene

on the mineralization of the following chemicals in methanogenic river sediment:

chemical	EC_{50}	
4-chlorophenol	2,500 mg/kg dw	
benzoate	590 mg/kg dw	
acetate	3,500 mg/kg dw	
chloroform	2,600 mg/kg dw	
methanogenesis	>10,000 mg/kg dw	(2961)

Soil sorption

K_{oc}: 38-53 at 1.0 mg/L (2588)

Anaerobic degradation

Two *Dechloromonas* strains, RCB and JJ in pure culture completely mineralize benzene to CO_2, in the absence of O_2 with nitrate as the electron acceptor. Nitrate was reduced to N_2 gas. ^{14}C-labeled benzene was oxidized to $^{14}CO_2$. Only 1.2% of the ^{14}C label was incorporated into biomass. Previous studies have demonstrated the ubiquity of the *Dechloromonas* species, and members of this genus have been identified in a broad range of environments, and even in samples collected from Antactica. (7127)

D. BIOLOGICAL EFFECTS

Bacteria

Pseudomonas putida	16h EC_0	92 mg/L	(1900)
Biotox™ test *Photobacterium phosphoreum*	5 min EC_{50}	2,600 mg/L	
Microtox™ test *Photobacterium phosphoreum*	5 min EC_{50}	531 mg/L	(7023)

Algae

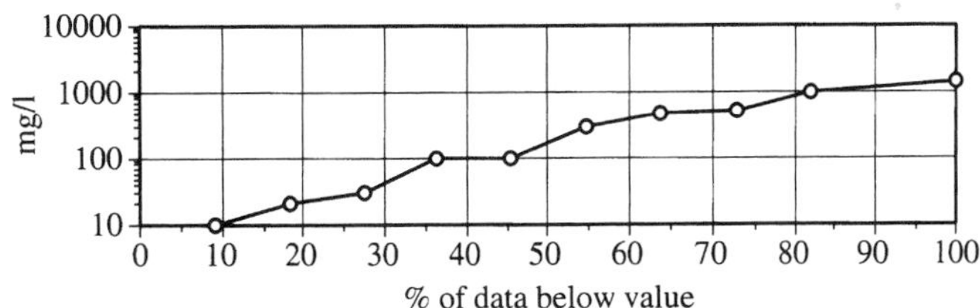

Frequency distribution of short-term EC_{50}/LC_{50} values for algae (n = 11) based on data presented in this and in other works (3999)

Microcystis aeruginosa	8d EC_0	>1,400 mg/L	(329)
Scenedesmus quadricauda	7d EC_0	>1,400 mg/L	(1900)

Chlorella vulgaris: 50% reduction of cell numbers vs controls, after 1-d incubation at 20°C: at 525 ppm

Inhibition of photosynthesis of a freshwater, nonaxenic unialgal culture of Selanastrum capricornutum
at 10 mg/L: 95% ^{14}C fixation (vs controls)
at 100 mg/L: 84% ^{14}C fixation (vs controls)
at 1,000 mg/L: 5% ^{14}C fixation (vs controls) (1690)

Tetrahymena pyriformis	24 h LC_{100}	1,000 mg/L	(1662)
Chlorella fusca	BCF	30 (wet wt)	(2659)

Protozoa

Entosiphon sulcatum	72 h EC_0	>700 mg/L	(1900)
Uronema parduczi Chatton-Lwoff	EC_0	486 mg/L	(1901)

Crustacean

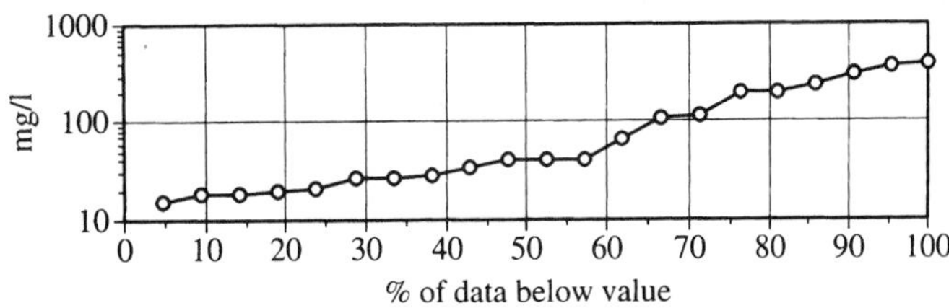

Frequency distribution of 24-96 h EC_{50}/LC_{50} values for crustaceans (n = 21) based on data presented in this and in other works (3999)

Grass shrimp *(Palaemonetes pugio)*:	96 h LC_{50}	27 ppm	(940)
Crab larvae-stage 1 *(Cancer magister)*:	96 h LC_{50}	108 ppm	(941)
Shrimp *(Crangon franciscorum)*:	96 h LC_{50}	20 ppm	(942)
brine shrimp:	24-48 h LC_{50}	66-21 mg/L	(41)

Fishes

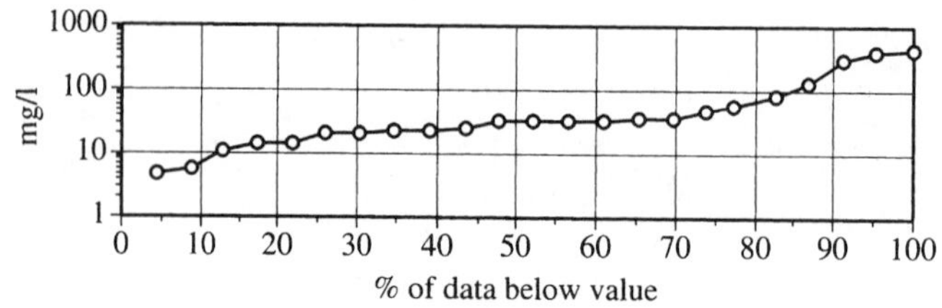

Frequency distribution of 24-96 h EC_{50}/LC_{50} values for fishes (n = 23) based on data presented in this and in other works (3999)

minnows:	6h LC_0	5-7 mg/L	(226)
bluegill sunfish:	24-48 h LC_{50}	20 mg/L	
	24 h LC_{100}	34 mg/L	
	2h LC_{100}	60 mg/L	(226)
goldfish:	24 h LC_{50}	46 mg/L	(277)
fatheads: soft water:	24-96 h LC_{50}	35 to 33 mg/L	
fatheads: hard water:	24-96 h LC_{50}	24 to 32 mg/L	
bluegills: soft water:	24-96 h LC_{50}	22 mg/L	
goldfish: soft water:	24-96 h LC_{50}	34 mg/L	
guppies: soft water:	24-96 h LC_{50}	36 mg/L	(58)
mosquito fish:	24-96 h LC_{50}	395 mg/L	
young Coho salmon:	no significant mortalities up to 10 ppm after 96 h in artificial seawater at 8°C		
mortality: 12/20 at 50 ppm after 24 up to 96 h in artificial seawater at 8°C 30/30 at 100 ppm after 24 h in artificial seawater at 8°C			(317)

guppy *(Poecilia reticulata)*:	14d LC_{50}	63 mg/L	(2696)
bass *(Morone saxatilis)*:	96 h LC_{50}	5.8-11 mg/L	(942)

herring and anchovy larvae *(Clupea pallasi;Engraulis mordex)*:	35-45 ppm caused delay in development of eggs and produced abnormal larvae 10-35 ppm caused delay in development of larvae, decrease in feeding and growth, and increase in respiration	(944)
chinook salmon *(Onchorhynchustschawytscha)*:	5-10 ppm: initial increase in respiration	
striped bass *(Morone saxatilis)*:	5-10 ppm: initial increase in respiration	(945)
Pacific herring *(Clupeaharengus pallasi)*:	BCF:	
in eggs:	10.9	
in yolk sac larvae:	6.9	
in feeding larvae:	3.9	(1918)
goldfish	4.3	(2661)
Clupea harengus	1.0-14	(2722, 2723, 2724)
Anguilla japonica	3.3-3.5	(2725)
Engraulis mordex	5-135	(2726)
Morone saxatilis	1.1	(2726)

eel: infiltration ratio: flesh/water: 0.31; eel flesh 0.14 ng/g, water: 0.45 ng/g	(412)
eel *(Anguilla japonica)*: BCF: 3.5; half-life: 0.5 d	(1926)
guppy *(Poecilia reticulata)*: 14 d, LC_{50}: 63 ppm	(1833)
brown trout yearlings: 1 h, LC_{50}: 12 mg/L (static bioassay)	(939)

Amphibians

Mexican axolotl (3-4 wk after hatching): 48 h LC_{50}: 370 mg/L	
clawed toad (3-4 wk after hatching): 48 h LC_{50}: 190 mg/L	(1823)

Plants

In higher plants, benzene administered in aqueous solution or as vapor is translocated and metabolized. The proposed sequence of benzene metabolism in plants is benzene → phenol → pyrocatechol → o-benzoquinone → muconic acid.	(2727)
The water plant *Elodea*, barley, and carrot leaves and roots exposed to aqueous solutions of at least 600 mg/L showed an avg BCF of 600 mg/L.	(2728)

Protozoa

Uronema parduczi: inhib. of cell multiplic. after 20 h at 25°C: NOEC: 490 mg/L	(2729)
Chilomonas paramecium: inhib. of cell multiplic. after 48 h at 20°C: NOEC: 490 mg/L	(2729)
Tetrahymena elliotti: survival after 24 h at 20°C: NOEC: 391 mg/L	(2730)

Algae

Chlorella sp.: photosynthesis/respiration after 12 h at 20°C: EC_{50}: 55-553 mg/L	(2731)
Amphidinium carteri: cell multiplic. inhib. after 72 h at 18°C: EC_{50}: 10-100 mg/L	(2732)

Skeletonema costatum: cell multiplic. inhib. after 72 h/10 d at 18°C: EC_{50}: 20-100 mg/L;	NOEC: <10; <50 mg/L	(2732, 2733)
Dunalliela tertiolecta: cell multiplic. inhib. after 72 h at 18°C:	NOEC: 0.1-10 mg/L	(2732)
Cricosphaera carteri: cell multiplic. inhib. after 72 h at 18°C: EC_{50}: 50-100 mg/L;	NOEC: <20 mg/L	(2732)
Macrocystis pyrifera: reduction of photosynthesis after 96 h at 18°C:	NOEC: 10 mg/L	(2732)

Algae			
Selenastrum capricornutum:	log BCF: 1.63		(2612)
	log EC_{50}, growth 8 d: 1.6 mg/L		(2612)
Chlorella fusca var. *vacuolata*	BCF 30		(2720)
Freshwater			
Rotatoria			
Dicranophorus forcipatus:	24 h LC_{100}	14 mg/L	(2733)
Hydrozoa			
Hydra oligactis	48 h LC_{50}	34 mg/L	
	48 h NOLC	24 mg/L	(2734)
Hirudinea			
Erpobdella octoculata	48 h LC_{50}	>320 mg/L	(2734)
Oligochaeta			
Tubificidae (mixture)	48 h LC_{50}	>320 mg/L	(2734)
Nematoda			
Panagrellus redivivus	96 h NOEC	78 mg/L	(2735)

Crustaceans			
Daphnia magna	24 h LC_{50}	200-250 mg/L	(2738)
	48 h LC_{50}	203; 400 mg/L	(2736, 2737)
	48 h NOLC	240 mg/L	(2736)
Daphnia pulex	96 h LC_{50}	15 mg/L	(2739)
	48 h LC_{50}	305 mg/L	(2736)
	48 h NOLC	196 mg/L	(2736)

Daphnia cucullata		48 h LC_{50}	373 mg/L	(2736)
Asellus aquaticus		48 h LC_{50}	120 mg/L	(2734)
Gammarus pulex		48 h LC_{50}	42 mg/L	(2734)
Daphnia pulex		BCF 153; 203; 225		(2721)
Insects				
Aedes aegypti:	4th instar larvae	24 h LC_{50}	59 mg/L	
		24 h NOLC	13 mg/L	(2740)
	3rd instar larvae	48 h LC_{50}	200 mg/L	
		48 h NOLC	170 mg/L	(2736)
Culex pipiens:	3rd instar larvae	48 h LC_{50}	71 mg/L	
		48 h NOLC	40 mg/L	(2736)
Chironomus gr. *thummi:*		48 h LC_{50}	100 mg/L	(2734)
Corixa punctata		48 h LC_{50}	48 mg/L	(2734)
Ischnura elegans		48 h LC_{50}	10 mg/L	(2734)
Nemoura cinerea		48 h LC_{50}	130 mg/L	(2734)
Cloeon dipterum		48 h LC_{50}	34 mg/L	(2734)
Molluscs				
Lymnaea stagnalis		48 h LC_{50}	230 mg/L	
		48 h NOLC	120 mg/L	(2736)
Degusia cf lugubris		48 h LC_{50}	74 mg/L	(2734)
Marine				
Rotatoria				
Brachionus plicatilis		48 h EC_{50}	>1.5 mg/L	(2741)
Crustaceans				
Nitroca spinipes		24 h LC_{50}	28; 112 mg/L	(2742)
Tigriopus californicus	adults	critical level	225-450 mg/L	(2743)
Palaemonetes pugio	larvae	24 h LC_{50}	75; 91 mg/L	(2742)
	adults	24 h LC_{50}	34; 41; 42 mg/L	(2742, 2744)
	adults	96 h LC_{50}	27 mg/L	(2744)
Crangon franciscorum	adults	24 h LC_{50}	19 mg/L	
		96 h LC_{50}	18 mg/L	(2745)
Cancer magister	larvae	96 h LC_{50}	108 mg/L	(2746)
Artemia salina	larvae	24 h LC_{50}	66 mg/L	(2747)
Fishes				
Barbus conchionus		24 h LC_{100}	±30 mg/L	(2748)
Cottus cognatus	juveniles	96 h LC_{50}	14 mg/L	(2749)
Gasterosteus aculeatus	adults	96 h LC_{50}	22 mg/L	(2749)
Leuciscus idus melanotus		48 h LC_{50}	132 mg/L	(2750)
Oncorhynchusgorbuscha	eggs	96 h LC_{50}	298 mg/L	
	alevins	LC_{50}	53 mg/L	
	emergent fry	LC_{50}	5 mg/L	
	fry	LC_{50}	15 mg/L	(2749)
Oncorhynchus kisutch	eggs	96 h LC_{50}	476 mg/L	
	alevins	96 h LC_{50}	35-66 mg/L	
	juveniles	96 h LC_{50}	12 mg/L	
	fry	LC_{50}	9 mg/L	(2749)
Oncorhynchus nerka	smolts	96 h LC_{50}	10 mg/L	(2749)
Oncorhynchustschawytscha	juveniles	96 h LC_{50}	10 mg/L	(2749)
Oryzias latipes	4-5 wk old	48 h LC_{50}	250 mg/L	
		48 h NOLC	126 mg/L	(2750)
Pimephales promelas		96 h LC_{50}	15 mg/L	(2751)
		48 h LC_{50}	84 mg/L	

		48 h NOLC	54 mg/L	(2750)
Poecilia reticulata		48 h LC_{50}	420 mg/L	
		48 h NOLC	265 mg/L	(2750)
		14d LC_{50}	64 mg/L	(2752)
Salmo gairdneri	adults	24 h LC_{100}	62 mg/L	
		24 h NOLC	15 mg/L	(2753)
		96 h LC_{50}	5 mg/L	(2751)
		48 h LC_{50}	56 mg/L	
		48 h NOLC	40 mg/L	(2750)
Salmo trutta	fingerlings	48 h LC_{50}	12 mg/L	(2754)
Salvelinus malma	juveniles, smolts	96 h LC_{50}	11 mg/L	(2749)
Thymallus articus	juveniles	96 h LC_{50}	13 mg/L	(2749)
Clupea harenguspallasi	eggs	96 h LC_{50}	40-50	(2755)
	2-d larvae	48 h LC_{50}	20-25 mg/L	(2755)
Engraulis mordex	1-d larvae	48 h LC_{50}	20-25 mg/L	(2755)
log BCF	rainbow trout:	1.72		
	bluegill:	1.48		(2606)

Benzene, C10-16 alkyl derivatives (dodecylbenzene and undecylbenzene; linear alkyl benzenes, LAB)

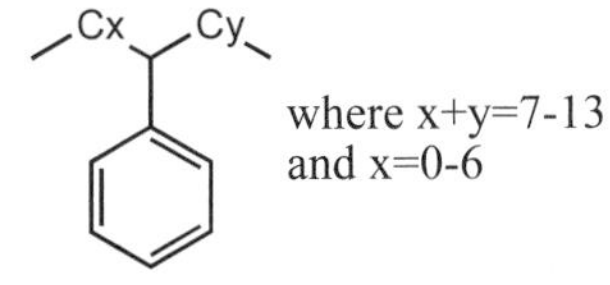

$C_6H_5C_nH_{2n+1}$ (n=10-16)

CAS 123-01-3 dodecylbenzene

CAS 6742-54-7 Undecylbenzene

CAS 68648-87-3

CAS 129813-58-7

CAS 68442-69-3

CAS 129813-59-8

CAS 129813-60-1

TRADENAMES

Alkylate 215; Alkylate 225; Alkylate 229; Alkylate 230; Nalkylene 550L; Nalkylene 600L; Nalkylene 500L; Nalkylene 580L; Nalkylene 600; Nalkyleene 575L; Detergent Alkylate

COMMENTS

The information refers to various mixtures of linear alkylbenzenes, which may contain dodecylbenzene and undecylbenzene. The chemicals are composed of mixtures of long-chain linear alkylbenzenes with the alkylgroup containing from 10 to 16 carbon atoms. (10550)

USES

most of the LAB's are used as an intermediate in the production of linear alkylbenzene sulfonate (LAS) , a detergent surfactant.LAS is used in light-duty liquid dishwashing compounds, heavy-duty liquid and powder laundry detergents, all-purpose cleaners and industrial cleaners. Consumers may be exposed to small amounts of residual LAB in the LAS contained in these products. LAB also finds minor use in the paper, flooring and functional fluid industries. (10550)

COMMERCIAL LINEAR ALKYLBENZENES TYPICAL PRODUCT COMPOSITION

	Component Percentage							
Product	**C9**	**C10**	**C11**	**C12**	**C13**	**C14**	**C15**	**C16**
Alkylate 215	<1	16	43	40	1	<1	-	-
Alkylate 225	<1	7	25	48	19	1	-	-
Alkylate 229	<1	1.1	7.6	36.4	45.2	9.6	<1	-
Alkylate 230	-	1	2	16	50	30	1	-
Nalkylene 500	1	21	39	31	7	<1	-	-
Nalkylene 500L	<1	20	44	31	5	<1	-	-
Nalkylene 550L	<1	14	30	29	20	7	<1	-
Nalkylene 575L	<1	9	17	28	30	15	<1	-
Nalkylene 580L	-	-	<1	26	74	<1	-	-
Nalkylene 600	<1	<1	1	23	50	25	<1	-
Nalkylene 600L	<1	<1	1	23	50	25	<1	-

IMPURITIES, ADDITIVES, COMPOSITION

dialklytetralins and isoalkylbenzenes

A. PROPERTIES

liquid; molecular weight ; melting point <-45°C; boiling point 251-320°C; vapour pressure 6.5x 10^{-5} kPa ; water solubility 0.04 mg/L at 25°C; log Pow 5.7 at 27°C; H 9.3 x 10^{-4} atm/m^3/mol

C. WATER AND SOIL POLLUTION FACTORS

Photodegradation

Linear alkylbenzenes do not appear to undergo direct photolysis or chemical change in the environment. Less than 1% degradation occurred when acetonitrile solutions of Alkylate 215 (<1% C9, 16% C10, 43% C11, 40% C12, 1% C13, <1% C14) were exposed to direct sunlight for 14 days. (10550)

Biodegradation

Linear alkylbenzenes undergo rapid primary and complete biodegradation in natural waters. The average half-life in waters for commercial range LAB was 4.1 days Microorganisms in sewage sludge, soil and sludge amended soil rapidly and completely biodegrade LAB. This has been confirmed with a [^{14}C] -benzene ring labeled model LAB compound. The average half lives for the conversion of LAB to CO2 were: activated sludge- 2.6 hours, soil- 3.2 to 4.5 days, and sludge amended soil- 15 to 33 days. (10550)

2-phenyl dodecane

ω-oxidation

OH

O

2-phenyl dodecanoic acid

β-oxidation

OH

O

2-phenyl butyric acid

ring oxidation and opening

biomass + CO_2 + H_2O

Inoculum	method	concentration	duration	elimination	
Aerobic		18 mg/L	35 days	56% ThCO2	Alkylate 215
Aerobic		18 mg/L	35 days	61% ThCO2	Alkylate 225
Aerobic		20 mg/L	35 days	56% ThCO2	Alkylate 230 (10550)

Sewage treatment

Sewage treatment plants remove most of the LAB that is released in sewage. Average percent removals range from 69% to >98% for trickling filter and activated sludge plants, respectively

D. BIOLOGICAL EFFECTS

Bioaccumulation

LAB possesses little potential to bioconcentrate in fish due to its rapid metabolism. Measured bioconcentration factors in fish range from 35 to 444 versus predicted values ranging from 6,300 to 3,500,000. (10550)

Species	conc. µg/L	duration	body parts	BCF	
Lepomis macrochirus	92 (1)	96 hours		35	(10550)

(1)pure C12 alkylbenzene

Toxicity

Insects			
Paratanytarsus parthenogenetica	48h EC_{50}	>water solubility	Alkylate 215
Chironomus tentans	96h LC_{50}	>water solubility	Alkylate 225
	14d MATC	125 µg/L	Alkylate 225
Algae			
Selenastrum capricornutum	96h ECb_{50}	>water solubility	Alkylate 215
Scenedesmus subspicatus	72h NOEC	0.050 mg/L	
	72h EC_{50}	>0.10 mg/L	
	72h LOEC	0.10 mg/L	European commercial LAB (C10-13)
Crustaceae			
Daphnia magna	48h EC_{50}	0.08 mg/L	Alkylate 215
	48h EC_{50}	0.009 mg/L	Alkylate 225
	48h EC_{50}	0.01 mg/L	Alkylate 230
	48h EC_{50}	>water solubility	European commercial LAB (C10-13)
	21d MATC	7.5 µg/L	Alkylate 215
	21d LC_{50}	12 mg./L	Alkylate 215
	21d MATC	13 µg/L	Alkylate 230
Gammarus fasciatus	48h LC_{50}	>water solubility	Alkylate 215
Fish			
Salmo gairdneri	24-96h LC_{50}	>water solubility	Alkylate 215
Pimephales promelas	24-96h LC_{50}	>water solubility	Alkylate 215; 225;230
Lepomis macrochirus	24-96h LC_{50}	>water solubility	Alkylate 215
Brachydanio rerio	21d NOEC	>water solubility	European commercial LAB (C10-13)
Mammals			
Rat	oral LD_{50}	>5,000 mg/kg bw	(10550)

Route	Species	Value	Type
Alkylate 215(<1% C9, 16% C10, 43% C11, 40% C12, 1% C13, <1% C14)			
oral	rat	17,000 mg/kg	LD_{50}
inhal.	rat	>1.82 mg/L	LC_{50}
dermal	rabbit	>10,200 mg/kg	LD_{50}
Alkylate 225(<1% C9, 7% C10, 25% C11, 48% C12, 19% C13, 1% C14)			
oral	rat	28,200 mg/kg	LD_{50}

dermal	rabbit	>10,200 mg/kg	LD_{50}
Alkylate 230(1% C10, 2% C11, 16% C12, 50% C13, 30% C14, 1% C15)			
oral	rat	20,800 mg/kg	LD_{50}
dermal	rabbit	>10,200 mg/kg	LD_{50}
N-alkylene 500 (1% C9, 21% C10, 39% C11, 31% C12, 7% C13, <1% C14)			
oral	rat	>34,080 mg/kg	LD_{50}
inhal.	rat	71 mg/L (nom)	LC_{50}
dermal	rabbit	~2,000 mg/kg	LD_{50}
dermal	rabbit	>5,000 mg/kg	LD_{50}
N-alkylene 600(<1% C9, <1% C10, 1% C11, 23% C12, 50% C13, 25% C14, <1% C15)			
oral	rat	>35,800 mg/kg	LD_{50}
N-alkylene 600L (<1% C9, <1% C10, 1% C11, 23% C12, 50% C13, 25% C14, <1% C15)			
oral	rat	>5,000 mg/kg	LD_{50}
dermal	rabbit	>2,000 mg/kg	LD_{50}
N-alkylene 550L (<1% C9, 14% C10, 30% C11, 29% C12, 20% C13, 7% C14, <1% C15)			
oral	rat	>5,000 mg/kg	LD_{50}
dermal	rabbit	>2,000 mg/kg	LD_{50}

(10550)

Benzene,1-chloro-2-(chloromethyl)- (o-chlorobenzyl chloride; OCBC; α,2-dichlorotoluene; α,o-dichlorotoluene)

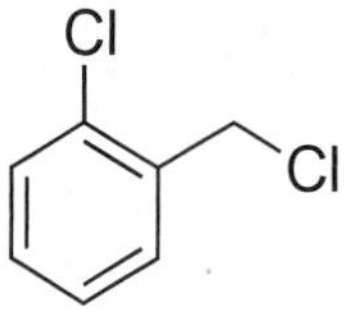

$C_7H_6Cl_2$

CAS 611-19-8

USES

OCBC is used as an intermediate for the production of the herbicides Orbencarb (CAS 34622-58-7) and Clomazone (CAS 81777-89-1).

IMPURITIES, ADDITIVES, COMPOSITION

2-chlorobenzaldehyde (0.01%);α,4-dichlorotoluene (0.2%); 1-chloro-2-(dichloromethyl)benzene (0.06%)

A. PROPERTIES

molecular weight 161.03; melting point –17°C; boiling point 217°C; relative density 1.27 kg/L at 20°C; vapour pressure 0.2 hPa at 25°C (calcul.); water solubility 100 mg/L at 25°C; log Pow 3.3; log K_{OC} 3.3 (estimated); H 157 Pa m^3/mol at 25°C

B. AIR POLLUTION FACTORS

Photodegradation

t/2 for reaction with OH radicals in air was calculated to be 103 hr. (10551)

C. WATER AND SOIL POLLUTION FACTORS

Hydrolysis

OCBC was hydrolyzed to o-chlorobenzylalcohol at 25°C with t/2 of 35, 33 and 36 hours at pH 4, 7 and 9 respectively (10551)

Biodegradation

Inoculum	method	concentration	duration	elimination	
Activated sludge	Modified MITI	100 mg/L	28 days	0% ThOD, 0% TOC (1)	
Adapted industrial A.S.	Zahn-Wellens		9 days	99% ThOD (2)	(10551)

(1) OCBC was transformed entirely to o-chlorobenzylalcohol (92%), o-chlorobenzaldehyde (2%) and o-chlorobenzoic acid (3%).The test without the activated sludge also indicated that OCBC was completely converted too-chlorobenzyl alcohol without further transformation. Based on these observations, it is concluded that OCBC is hydrolyzed in water via an abiotic process to generateo-chlorobenzyl alcohol, which is then slowly biotransformed by oxidation too-chlorobenzoic acid via *o*-chlorobenzaldehyde. Therefore OCBC and its hydrolysis products are not readily biodegradable.Degradation pathway of o-chlorobenzyl chloride is as follows: o-chlorobenzyl alcohol, o-chlorobenzaldehyde, o-chlorobenzoic acid(10551)

(2) The determination by the respirometric method showed 99% degradation of OCBC after 9 days. First 6 days were adaptation period (less than 10% degradation) and 90% degradation of OCBC was observed in the last 3 days. Thus, OCBC is inherently biodegradable with adapted industrial sludge.

abiotic

o-chlorobenzylalcohol

o-chlorobenzaldehyde

o-chlorobenzoic acid

D. BIOLOGICAL EFFECTS

Bioaccumulation

BCF 72 (calculated) (10551)

Toxicity

Algae			
Selenastrum capricornutum	72h ECb_{50}	0.78 mg/L	
	72h NOECb	0.045 mg/L	
	72h $ECgr_{50}$	1.2 mg/L	
	72h NOECgr	0.18 mg/L	(10551)
Crustaceae			
Daphnia magna	24h EiC50	0.72 mg/L	
	48h EiC50	0.38 mg/L	
	48h EiC100	1.0 mg/L	
	48h NOEC	0.1 mg/L	
	21d NOEC	0.020 mg/L	
	21d LOEC	0.041 mg/L	
	21d EC_{50}	0.23 mg/L	
	21d LC_{50}	0.39 mg/L	(10551)
Fish			
Oryzisa latipes	96h LC_{50}	0.27 mg/L	
	96h LC_{100}	0.56 mg/L	
	96h NOEC	0.18 mg/L	(10551)
Danio rerio	96h LC_{50}	0.5-0.71 mg/L	
	96h LC_{100}	1.25 mg/L	
	48h LC_{50}	1.25-1.8 mg/L	
	48h LC_0	1 mg/L	

	48h LC_{100}	1.8 mg/L	(10551)
Pimephales promelas	96h LC_{50}	0.71-0.96 mg/L	(10551)
Mammals			
Rat male	oral LD_{50}	430; 690; 880; 951; 1,700 mg/kg bw	
Rat female	Oral LD_{50}	350; 430; 533; 783; 2,200 mg/kg bw	(10551)

benzeneazobenzene *see* azobenzene

benzenecarbonal *see* benzaldehyde

benzenecarbonitrile *see* benzonitrile

benzenecarbonylchloride *see* benzoylchloride

1,2-benzenediamine *see* *o*-phenylenediamine

1,3-benzenediamine *see* *m*-phenylenediamine

1,3-benzenediamine,4,4'-[1,3-propandiylbis(oxy)]bis-, tetrahydrochloride (1,3-bis-(2,4-diaminophenoxy)-propane, tetrahydrochloride)

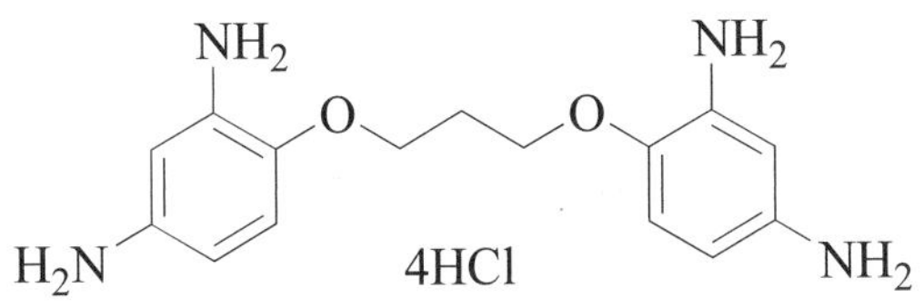

$C_{15}H_{20}N_4O_2.4HCl$

CAS 74918-21-1

TRADENAMES

Napro Vital Colors: 90 Black; Colipa A 079

USES

cosmetics, in permanent hair dye formulations (< 2%) (7273)

A. PROPERTIES

grey violet powder; molecular weight 434.2; boiling point ca 100°C; specific gravity ca 1.0; vapour pressure 2.4 kPa at 20°C; water soluble; Pow −1.54 (calcul)

C. WATER AND SOIL POLLUTION FACTORS

Soil sorption

K_{OC}: 3.46 (calculated) (7273)

D. BIOLOGICAL EFFECTS

Toxicity

Mammals			
Rat	oral LD_{50}	3570 mg/kg bw	(7274)

1,2-benzenedicarboxylic acid *see* *o*-phthalic acid

1,3-benzenedicarboxylic acid *see* isophthalic acid

1,4-benzenedicarboxylic acid *see* terephthalic acid

1,2-benzenedicarboxylic acid, bis(2-propylheptyl)ester (bis(2-propylheptyl)phthalate; di-2-propylheptyl phthalate; phthalic acid, bis(2-propylheptyl) ester)

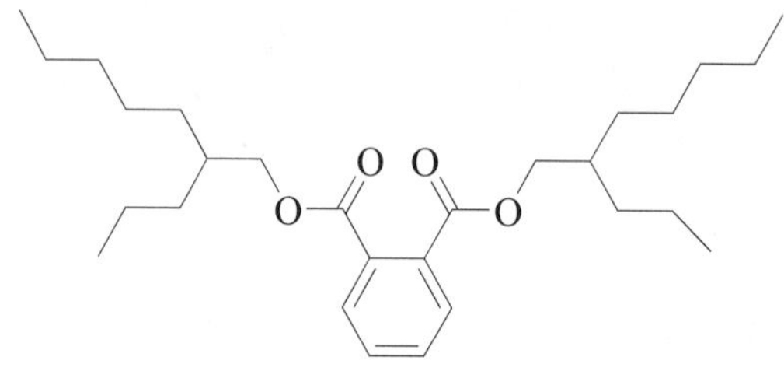

$C_{28}H_{46}O_4$

CAS 53306-54-0

TRADENAMES

Palatinol 10-P

USES

a plasticizer for PC and vinyl chloride copolymers. The end use products include automobile undercoating, building materials, wires, cables, shoes, carpet backing, pool liners and gloves. The typical concentratioon of the chemical in end-use products is 30-60% (7321)

CHEMICAL IDENTITY

The chemical is a specific isomer of di-isodecyl phthalate (CAS 68515-49-1 and 26761-40-0)

IMPURITIES, ADDITIVES, COMPOSITION

1,2-benzenedicarboxylic acid, bis(4-methyl-2-propylhexyl)ester: 2 wt%; 1,2-benzenedicarboxylic acid,4-methyl-2-propylhexyl-2-propylheptyl ester: 15 wt% (7321)

A. PROPERTIES

clear mobile liquid with a faint odour; molecular weight 446.68 ; melting point approx. –45 °C; boiling point 250-267 °C; density 0.96-0.97 kg/L; vapour pressure 3.7 x 10^{-9} kPa at 20°C; water solubility approx. 0.2 μg/L

C. WATER AND SOIL POLLUTION FACTORS

Biodegradation

Inoculum	method	duration	elimination	
Activated sludge	CO2 evolution test	10 days	5%	
		12 days	17%	
		28 days	75%	(7321)

D. BIOLOGICAL EFFECTS

Toxicity

The chemical is not toxic to Daphnia up to the limit of its water solubility in the test media . While the EC_{50} of the chemical to algae is above the limit of its water solubility, some effects were noted below this level. No adverse effects toward bacterial activity in sewage treatment plants are expected for the substance up to the limit of its solubility. (7321)

Micro organisms			
Pseudomonas putida	16h IC_{50}	>8000 mg/L	
	16h NOEC	>8000 mg/L	(7321)
Algae			
Scenedesmus subspicatus	72h ECb_{50}	> 100 mg/L	
	72h ECr_{50}	> 100 mg/L	
	72h NOEC	25 mg/L	(7321)
Crustaceae			
Daphnia magna	48h EC_{50}	>100 mg/L	
	48h NOEC	12.5 mg/L	(7321)
Mammals			
Rat	oral LD_{50}	>5000 mg/kg bw	(7321)

1,3-benzenedimethanamine,N,N,N',N'-tetrakis(oxiranylmethyl)- (N,N,N',N'-tetraglycidyl-1,3-xylylenediamine; N,N,N',N'-tetraglycodyl-metaxylilene-diamine (TGMX); polyglycidyl metaxylenediamine (PGA-X))

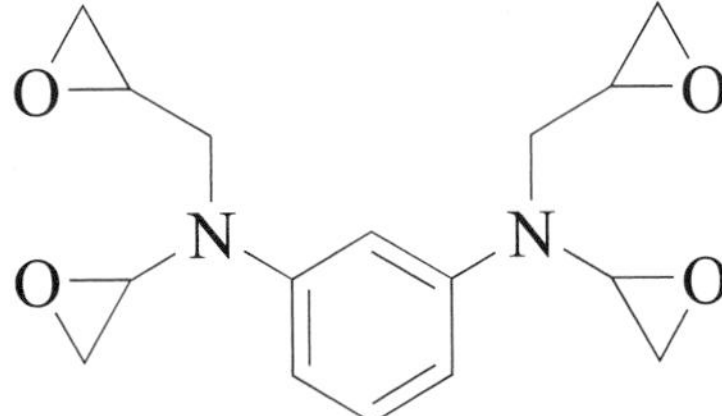

$C_{20}H_{28}N_2O_4$

CAS 63738-22-7

TRADENAMES

Tetrad-X; Antrade B15

USES

an epoxy curing agent which is used in a part of a two part epoxy coating for application to plastic surfaces.

A. PROPERTIES

pale yellow viscous liquid; molecular weight 360; boiling point >250°C; specific gravity ca. 1.15 at 25°C; vapour pressure <0.13 kPa at 25°C (estimated); water solubility : the chemical reacts with water to form a gel; rapid hydrolysis in water typical of epoxy functional groups

B. AIR POLLUTION FACTORS

Incineration

Incineration of the polymer would result in the production of water and oxides of carbon and nitrogen. (7315)

C. WATER AND SOIL POLLUTION FACTORS

Fate

The chemical will form part of a surface coating on plastic components and fixed strongly to the coated articles, thus sharing their fate. At the end of their useful lives these would be disposed of to landfill, or possibly incinerated. When disposed of to landfill the highly crosslinked nature of the material will preclude significant leaching, and the polymer would be subject to the slow biodegradation processes operative in landfill situations.

D. BIOLOGICAL EFFECTS

Bioaccumulation

the substance is not expected to bioaccumulate because it will either react with water in the environment to form a gel. (7315)

Toxicity

Mammals

Rat male	oral LD_{50}	730 mg/kg bw	(7315)
Rat female	oral LD_{50}	630 mg/kg bw	(7315)

1,2-benzenediol *see* catechol

1,3-benzenediol *see* resorcinol

1,4-benzenediol *see* hydroquinone

m-benzenedisulfonic acid

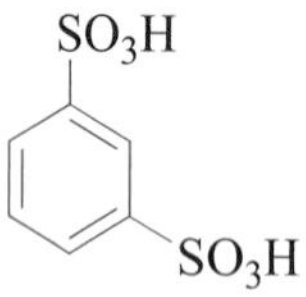

$C_6H_6O_6S_2$

CAS 831-59-4

C. WATER AND SOIL POLLUTION FACTORS

Biodegradation

adapted A.S. at 20°C-product is sole carbon source: 63% COD removal at 3.4 mg COD/g dry inoculum/h. (327)

Benzeneethanol (2-phenylethanol; (2-hydroxyethyl)benzene; β-phenylethanol; phenethylalcohol; phenethanol-; benzyl-carbinol-; beta-hydroxyethylbenzene; PEA-; 2-phenylethyl alcohol; 2-phenethyl alcohol)

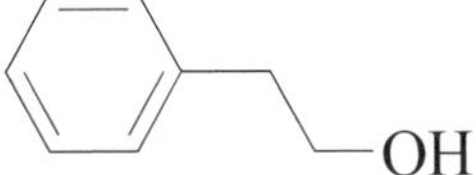

$C_6H_5CH_2CH_2OH$

$C_8H_{10}O$

CAS 60-12-8

USES

chemical intermediate, perfume and flavoring manufacture. Antibacterial activity is related to the compounds' lipophilicity.

OCCURRENCE

in etheric oils of Chapaca, Geranium, Hyacinth, Rose, Neroli, and Ylang-Ylang. L-phenylalanine is transformed by *Saccharomyces cerevisiae* into 2-phenylethanol.

A. PROPERTIES

Clear colourless liquid;molecular weight 122.17; melting point –27 °C;boiling point ±219°C; vapor pressure 0.08 hPa at 20°C; density 1.02 at 20°C; solubility 20 g/L at 20°C; log P_{OW} 1.38 exp., 1.45 calculated

B. AIR POLLUTION FACTORS

Odour

mild flowery, rose and honey

C. WATER AND SOIL POLLUTION FACTORS

Waste water treatment

inhibition of activated sludge: 30 min EC_{10}: >1,000 mg/L. (3017)

Biodegradation

Inoculum	method	concentration	duration	elimination results
adapted industrial activated			3 hours	6% TOC

sludge			
	6 d	99% TOC	
	5 d	58% COD	(3017)
	28 d	>90% DOC	(3018)

Xanthobacter

HO / OHC / HOOC

phenylacetaldehyde

phenylacetic acid

Pseudomonas

HO / HO / OH

2,3-dihydroxy-β-phenylethanol

Degraded by *Xanthobacter* to phenylacetaldehyde and phenylacetic acid.	(9839)
Pseudomonas T12 degrades benzeneethanol to the 2,3-dihydroxy derivative.	(9840)

D. BIOLOGICAL EFFECTS

Bacteria			
*Pseudomonas putida:*cell multiplic. test	17h EC_{10}	840 mg/L	
	17h EC_{50}	1,320 mg/L	
	17h EC_{90}	2,300 mg/L	(3016)
Photobacterium phosphoreum Microtox test	30 min EC_{50}	5.3 mg/L	(9657)
Algae			
Scenedesmus subspicatus	cell multiplic. inhib. test		
	72h EC_{10}	300 mg/L	
	72h EC_{50}	490 mg/L	
	72h EC_{90}	790 mg/L	(3016)
Crustaceans			
Daphnia magna Straus	24h EC_0	250 mg/L	
	24h EC_{50}	330 mg/L	
	24h EC_{100}	500 mg/L	
Daphnia magna Straus	48h EC_0	125 mg/L	
	48h EC_{50}	287 mg/L	
	48h EC_{100}	500 mg/L	(3015)
Fishes			
Leuciscus idus	6h LC_{50}	220-460 mg/L	(3014)
Mammals			
guinea pig, mouse, rat, rabbit	oral LD_{50}	400, 800, 1,790, 2,000 mg/kg bw	(9649, 9841, 9842)

Yeast

2-phenylethanol has a clear inhibitory effect above 1000 mg/L on the growth of *Sacharomyces cerevisiae* and is toxic at 4000 mg/L. (7128)

benzenehexachloride *see* hexachlorocyclohexane

benzenepropanoic acid

$C_6H_5CH_2CH_2COOH$

$C_9H_{10}O_2$

CAS 501-52-0

MANMADE SOURCES

Göteborg (Sweden) sew. works 1989-91: infl.: nd-60 µg/L; effl.: nd (including esters) (2787)

Landfill leachate from lined disposal area (Florida 1987-'90): mean conc. 1,800 µg/L (2788)

benzenesulfinic acid, *see sodium salt*

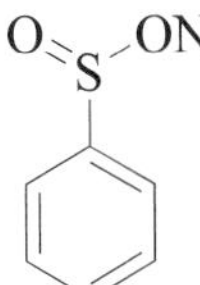

$C_6H_5SO_2Na$

CAS 873-55-2

A. PROPERTIES

molecular weight 164.15; melting point >300°C.

C. WATER AND SOIL POLLUTION FACTORS

ThOD: 1.56 (2790)

Biodegradation

EEC respirometric method at 100 mg/L and 20°C using nonadapted sludge: after 50 d: 86% ThOD and 100% DOC removal. (2790)

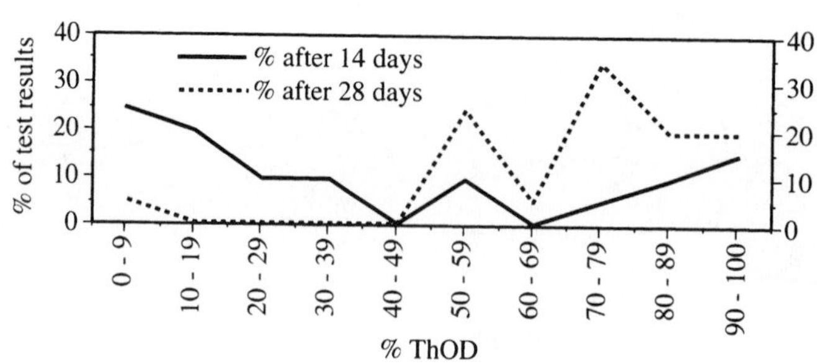

ThOD in EEC respirometric test: n = 21; median after 14 d: 26% ThOD
median after 28 d: 79% ThOD

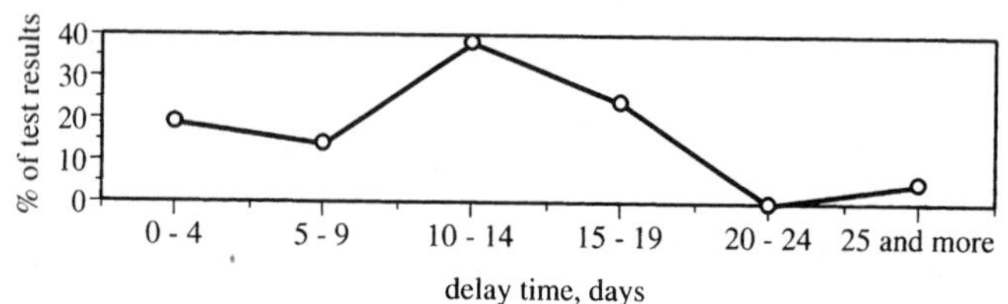

Delay times in EEC respirometric test

n = 21; median: 12 d

Oxygen uptake and DOC removed

using MITI inoculum:
delay time: 4; 5; 5; 24 d
% ThOD after 14 d: 78; 80%
% ThOD after 28 d: 24; 73; 89; 92%
% DOC after 28 d: 0; 96%

in 'Repetitive Die-Away' Test
delay time: 10; 14; 14 d
% ThOD after 28 d: 60; 82: 84%
% ThOD after 56 d: 88; 88%
% DOC removal after 28 d: 95; 96; 99%
% DOC removal after 56 d: 99; 99% (2790)

benzenesulfochloride (Benzenesulfonylchloride; benzenesulfonchloride)

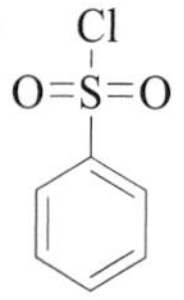

$C_6H_5ClO_2S$

CAS 98-09-9

EINECS 202-636-6

USES

Intermediate in dye, phenol and resorcinol manufacturing; accelerator in alkyl resin formulations.

A. PROPERTIES

molecular weight 176.2; melting point 14.5°C; boiling point 246°C decomposes; vapour pressure 0.00013 hPa at 20°C; 16 hPa at 50°C; density 1.38 kg/L at 23°C

B. AIR POLLUTION FACTORS

Photodegradation

T/2 (OH radicals) = 90 days

C. WATER AND SOIL POLLUTION FACTORS

Hydrolysis

Hydrolyses to benzenesulfonic acid and HCl

$$C_6H_5SO_2Cl + H_2O \longrightarrow C_6H_5SO_3H + HCl$$

Biodegradation

Inoculum	method	concentration	duration	elimination	
Activated sludge industrial	OECD 302B		5 days	62%	
			10 days	>97%	(10889)

D. BIOLOGICAL EFFECTS

Toxicity

Micro organisms			
Activated sludge, domestic	3h EC_0	>1,000 mg/L	(10889)
Fish			
Leuciscus idus	48h LC_0	10 mg/L	
Salmo trutta	48h LC_{50}	3 mg/L	(10889)
Mammals			
Rat	oral LD_{50}	1,860; 1,960 mg/kg bw	
Mouse	oral LD_{50}	828 mg/kg bw	
rabbit	oral LD_{50}	828 mg/kg bw	
Guinea pig	oral LD_{50}	828 mg/kg bw	(10889)

benzenesulfonic acid

$C_6H_5SO_3H$

$C_6H_6O_3S$

338 benzenesulfonic acid

CAS 98-11-3

A. PROPERTIES

molecular weight 158.17; melting point 1.5°C + H_2O 43/44°C, anh. boiling point decomposes $LogP_{ow}$ -2.25 (calculated).

C. WATER AND SOIL POLLUTION FACTORS

Biodegradation

Decomposition by a soil microflora: 16 d (176)

Adapted A.S. at 20°C-product is sole carbon source: 98% COD removal at 10.6 mg COD/g dry inoculum/h (327)

OH
O=S=O
O_2
OH
OH
O_2
CHO
COOH
OH
catechol
H_2SO_3

Strains of *Pseudomonas* and *Alcaligenes* isolated from environmental sources by enrichment culture were able to desulfonate benzenesulfonate early in the biodegradation process leading to catechol which was rapidly degraded via, usually, the meta pathway. (10548)

Waste water treatment

ion exchange: adsorption on Amberlite X AD-2: 31% retention effic.: infl.: 3.0 ppm effl.: 2.1 ppm. (40)

A.C.: adsorptive capacities for benzenesulfonate:

carbon 0.273-mm particle size	*benzenesulfonate, mole/g C at 30°C*
Columbia LC	131
Dareo S 51	101
Fisher	120
Norit	92
Nuchar C-190	54

D. BIOLOGICAL EFFECTS

Fishes

Cyprinus carpio: hematological effect after 75 d at 8 mg/L (sodium salt) (2625)

Benzenesulfonic acid, 2,2'-[(9,10-dihydro-5,8-dihydroxy-9,10-dioxo-1,4-anthracenediyl)diimino]bis[5-(1,1-dimethylethyl)-, disodium salt (disodium 2,2'-[(9,10-dihydro-5,8-dihydroxy-9,10-dioxo-1,4-anthrylene)diimino]bis[5-tert-butylbenzenesulphonate])

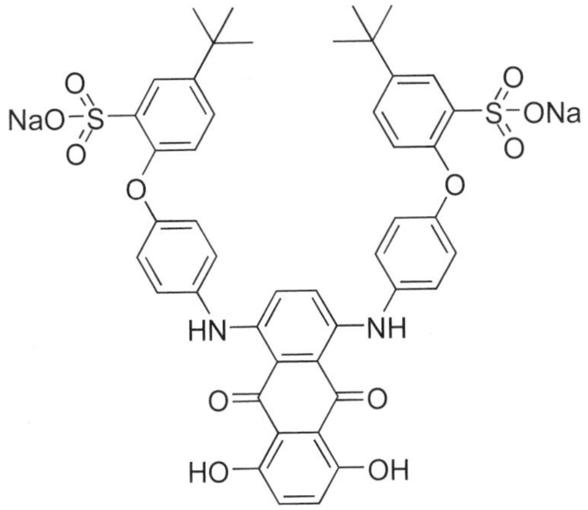

$C_{34}H_{32}N_2O_{10}S_2Na_2$

CAS 83006-67-1

SMILES
c1cc(C(C)(C)C)cc(S(=O)(=O)O[Na])c1Nc2c3C(=O)c4c(O)ccc(O)c4C(=O)c3c(cc2)Nc5c(S(=O)(=O)O[Na])cc(C(C)(C)C)cc5

USES

A textile dye

A. PROPERTIES

(all modelled) molecular weight 738.74; melting point 350°C; boiling point 115°C; vapour pressure 5 x 10^{-32} Pa; water solubility 0.16 µg/L; log Pow 5.3; H 3 x 10^{-27} $Pa.m^3/mol$

C. WATER AND SOIL POLLUTION FACTORS

Mobility in soils

log K_{OC}	7.05 (modelled)	(11175)

D. BIOLOGICAL EFFECTS

Bioaccumulation

Species	BCF	
Fish	9,000-53,000 (modelled)	(11175)

Toxicity

Fish	96h LC_{50}	0.007 mg/L (modelled)	(11175)

340 Benzenesulfonic acid, [(9,10-dihydro-9,10-dioxo-1,4-anthracenediyl)bis(imino-4,1-phenyleneoxy)]bis-, disodium salt

Benzenesulfonic acid, [(9,10-dihydro-9,10-dioxo-1,4-anthracenediyl)bis(imino-4,1-phenyleneoxy)]bis-, disodium salt (disodium [(9,10-dihydro-9,10-dioxo-1,4-anthrylene)bis(imino-4,1-phenyleneoxy)]bis-(benzenesulphonate))

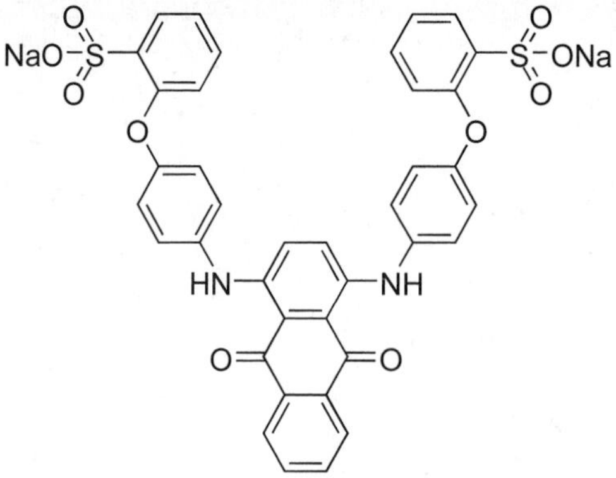

$C_{38}H_{24}N_sNa_2O_{10}S_2$

CAS 70161-19-2

SMILES
O=C1c3c(c(ccc3Nc4ccc(cc4)Oc5ccc(cc5)S(=O)(=O)O[Na])Nc6ccc(cc6)Oc7ccc(cc7)S(=O)(=O)O[Na])C(=O)c2c1cccc2

TRADENAMES

Acid Green 40:1; Lanaset Green B

USES

a textile dye

A. PROPERTIES

(all modelled) molecular weight 778.72; melting point 350°C; boiling point 1198°C; vapour pressure 4.7x 10^{-29} Pa; water solubility 0.04 µg/L; log Pow 5.0; H 1.3x 10^{-27} Pa.m^3/mol

C. WATER AND SOIL POLLUTION FACTORS

Mobility in soils

Log K_{OC}	7.3 (modelled)	(11174)

D. BIOLOGICAL EFFECTS

Bioaccumulation

Species	BCF	
Fish	5,000-18,000 (modelled)	(11174)

Toxicity

Crustaceae			
Daphnia magna	21d EC_{50}	0.028 mg/L	(11174)
Fish	96h LC_{50}	0.013 mg/L	(11174)

benzenethiol *see* thiophenol

1,2,3-benzenetriol *see* pyrogallol

benzenyltrichloride *see* benzotrichloride

benzethoniumchloride (hyamine 1622; a synthetic quaternary ammonium compound)

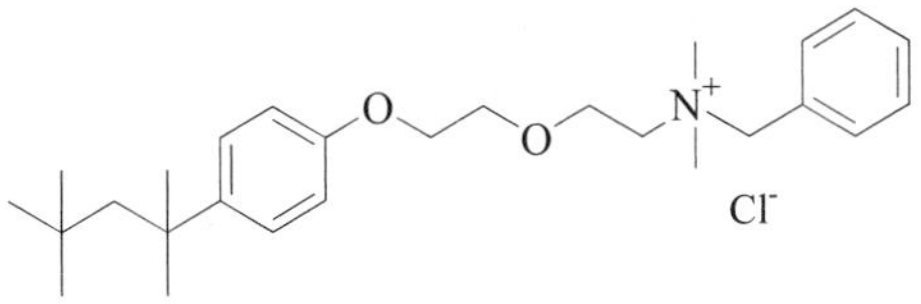

$C_{27}H_{42}ClNO_2.H_2O$

CAS 121-54-0

USES

antiseptic; cationic detergent.

A. PROPERTIES

colorless, odorless plates; very bitter taste; molecular weight 448.1; melting point 164-166°C.

C. WATER AND SOIL POLLUTION FACTORS

Biodegradation

at 18 mg/L no degradation after 28 d by nonadapted sewage. (488)

D. BIOLOGICAL EFFECTS

Bacteria

Staphylococcus aureus: at 20 mg/L bacteriolytic action after 1 h; bacteria isolated from Rhone (France) water: at 20 mg/L, no significant reduction of growth rate. (488)

Fishes			
Pimephales promelas	96h LC_{50}	1,600 µg/L	(2131)
Lepomis macrochirus	96h LC_{50}	1,400 µg/L	(2131)
Oncorhynchus kisutch	96h LC_{50}	53,000 µg/L	(2100)

benzidine (p,p'-bianiline; 4,4'-diaminobiphenyl 4,4'-biphenyldiamine; 4,4'-diphenylenediamine; biphenyl-4,4'-enediamine; 4,4'-diaminobiphenyl; p,p'-bianiline; [1,1'-biphenyl]-4,4'-diamine)

$H_2N—C_6H_4—C_6H_4—NH_2$

$NH_2C_6H_4C_6H_4NH_2$

$C_{12}H_{12}N_2$

CAS 92-87-5

USES

organic synthesis; manufacture of dyes, especially of Congo red; detection of blood stains; stain in microscopy; reagent; stiffening agent in rubber compounding.

A. PROPERTIES

grayish-yellow, white, or reddish gray crystalline powder; molecular weight 184.23; melting point 116/129°C; boiling point 402°C; vapor density 6.36; density 1.25 at 20/4°C; solubility 400 mg/L at 12°C, 9,400 mg/L at 100°C; $logP_{OW}$ 2.01

B. AIR POLLUTION FACTORS

Method of analysis

Air is drawn through a glass-fiber filter followed by a bed of silica gel to collect these substances as either particles or vapors. The compounds are extracted from the sampler and analyzed by HPLC with sensitivities in the range of 3 µg/m^3 for 48 h air samples. (1686)

Incinerability

thermal stability ranking of hazardous organic compounds: rank 60 on a scale of 1 (highest stability) to 320 (lowest stability). (2390)

C. WATER AND SOIL POLLUTION FACTORS

Biodegradation

possible bio-oxidation products scanned by GC/MS: N-hydroxybenzidine; 3-hydroxybenzidine; 4-amino-4'-nitrobiphenyl; N,N'-dihydroxybenzidine; 3,3'-dihydroxybenzidine; 4,4'-dinitrobiphenyl (419)

W, A.S. from mixed domestic/industrial treatment plant: 85-93% depletion at 20 mg/L after 6 h at 25°C (419)

Removal efficiency from water incubated with 100 units of horseradish peroxidase enzyme per liter, at pH 5.5: 99.9% (2452)

Benzidine in sludge applied to a sandy loam soil in a biological soil reactor, $t_{1/2}$: 76 d. (9845)

Benzidine has a low decomposition rate in natural soils due to its strong adsorption to clay soils and its toxicity to microorganisms at high concentrations. (9846, 9847)

Benzidine is rapidly oxidised by ferric ions and by complexing fulvic acids and clay minerals in water. (9848, 9849)

Impact on biodegradation processes

nitrification inhib. EC_{50}: 45 mg/L (benzidine dihydrochloride) (2624)

D. BIOLOGICAL EFFECTS

Algae			
Scenedesmus subspicatus inhib. of fluorescence	IC_{10}	4.4 mg/L	
Scenedesmus subspicatu s growth inhib.	IC_{10}	4.2 mg/L	
RubisCo Test: inhib. of enzym. activity of Ribulose-P2-carboxylase in protoplasts	IC_{10}	9.2 mg/	
Oxygen Test: inhib. of oxygen production of protoplasts	IC_{10}	3.7 mg/L	(2698)
Fishes			
sheepshead minnow	96h LC_{50}	64 mg/L	(9843, 9844)
red killifish	24h LC_{50}	16.5 mg/L	(9844)
Mammals			
rat, mouse	oral LD_{50}	214, 309 mg/kg bw	(9850)

benzidinedihydrochloride (benzidine hydrochloride)

HCl• H_2N—⟨C₆H₄⟩—⟨C₆H₄⟩—NH_2•HCl

$C_{12}H_{12}N_2.2HCl$

CAS 531-85-1

A. PROPERTIES

molecular weight 257.18.

C. WATER AND SOIL POLLUTION FACTORS

Control methods

concentration at various stages of water treatment works (+ chrysene):

river intake:	0.090 µg/L	
after reservoir:	0.072 µg/L	
after filtration:	0.033 µg/L	
after chlorination:	0.012 µg/L	(434)

Impact on biodegradation

NH_3 oxidation by *Nitrosomonas* sp.:

at 100 mg/L:	84% inhibition	
at 50 mg/L:	56% inhibition at 10 mg/L:	
at 10 mg/L:	12% inhibition	(390)

1,2-benzisothiazolin-3-one (IPX)

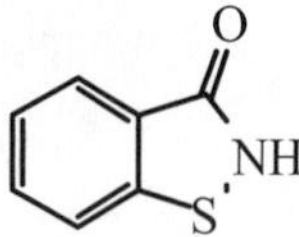

C_7H_5NOS

CAS 2634-33-5

TRADENAMES

Proxan; Proxel; Proxil

USES

an isothiazolinone preservative used in specialized cleaning agents

A. PROPERTIES

molecular weight 151.18

B. AIR POLLUTION FACTORS

Photochemical reactions

t1/2: 11 d, based on reactions with OH° (calculated)

D. BIOLOGICAL EFFECTS

Toxicity

Algae			
Green algae	72h EC_{50}	0.15 mg/L	(7169)
Crustaceans			
Daphnia magna	48h EC_{50}	1.35 mg/L	(7169)
Fish			
Salmo gairdneri	96h LC_{50}	1.6 mg/L	
Lepomis macrochirus	96h LC_{50}	5.9 mg/L	(7169)
Mammals			
Rat	acute oral LD_{50}	1020 mg/kg bw	
Mice	acute oral LD_{50}	1150 mg/kg bw	(7169)

Metabolism

1,2-benzisothiazolin-3pone is rapidly and totally metabolized in animals. Neither the substances itself nor the metabolites accumulate in the liver or adipose tissue. Excretion is mostly via the kidneys and almost completely within 24 houirs. The main metabolites are o-methylsulphonylbenzamide and o-methylsulphinylbenzamide. Rats excreted 96% of an oral dose of 1,2-benzisothiazolin-3-one. (7169)

Benzo(a)acridine (1,2-benzacridine; 7-azabenz(a)anthracene)

$C_{17}H_{11}N$

CAS 225-11-6

A. PROPERTIES

molecular weight 229.28

D. BIOLOGICAL EFFECTS

Toxicity

Insecta			
Chironomus riparius	96h LC_{50}	0.013 mg/L	(10865)
Crustaceae			
Daphnia pulex	24h LC_{50}	0.45 mg/L	(11189)

benzo(a)anthracene (1,2-benz(*a*)anthracene; B(*a*)A; tetraphene)

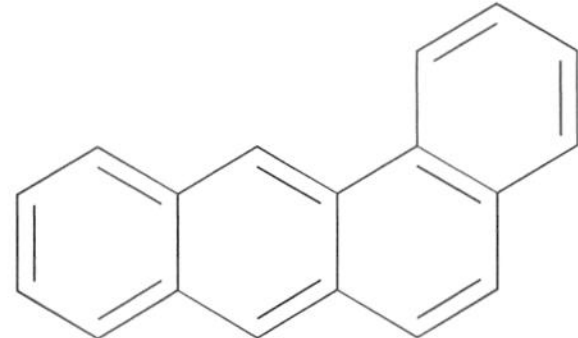

$C_{18}H_{12}$

CAS 56-55-3

MANMADE SOURCES

in gasoline	0.04-0.27 mg/L
in bitumen	0.13-0.86 mg/L
in S. Louisiana crude oil	1.7 mg/L
in Kuwait crude oil	2.3 mg/L
in no. 2 fuel oil	1.2 mg/L
in bunker C fuel oil	90 mg/L
in wood preservative sludge	5,180 mg/L of raw sludge
in compost	26 mg/kg dry wt (+ chrysene)
in horse manure	0.24 mg/kg dry wt (+ chrysene)
in coal tar	8,000 mg/L
in carbon black	10 mg/kg

A. PROPERTIES

molecular weight 228; melting point 158°C, boiling point 437°C, vapor pressure 5 × 10^{-9} mm Hg at 20°C, solubility 0.010; 0.044 mg/L at 24°C (practical grade); solubility 0.014 mg/L, LogP_{ow} 5.61.

B. AIR POLLUTION FACTORS

Manmade emissions

emissions from typical European gasoline engine:	7.3-32 µg/L fuel burnt	(1291)
in exhaust condensate of gasoline engine:	0.5-0.08 mg/L gasoline consumed	(1070)
	280 ppm	(1069)
emissions from asphalt hot-mixing plant:	5-24 ng/m^3	
in high-volume particulate matter): avg	11 ng/m^3	(491, 1379)
in coke oven airborne emissions:	105-2,740 µg/g of sample	(960)
in cigarette smoke:	0.3 µg/100 cigarettes	(1298)
in emissions from open burning of scrap rubber tires:	82-102 mg/kg of tire	(2950)

Ambient air quality

organic fraction of suspended matter:
Bolivia at 5,200 m altitude (Sept.-Dec. 1975): 0.040-0.005 µg/1,000 m^3 (+ chrysene)
Belgium, residential area (Jan.-April 1976): 2.2-13 µg/1,000 m^3 (+ chrysene) (428)

glc's in the Netherlands (ng/m^3): in the major cities:

	summer 1968-1971:	>1-3 ng/m^3	
	winter 1968-1971:	2-29 ng/m^3	(1277)

glc's in Birkenes (Norway) Jan.-June 1977:
avg: 0.34 ng/m^3

	range:	n.d.-1.86 ng/m^3 (*n* = 16)	
in Rörvik (Sweden) Dec. '76-April '77:			
	avg:	0.62 ng/m^3	
	range:	0.02-4.60 ng/m^3 (*n* = 21)	(1236)
glc's in Budapest 1973: heavy traffic area:			
	winter:	2.2 ng/m^3	
(6-20 h)	summer:	56 ng/m^3	
low-traffic area:			
	winter:	11 ng/m^3	
	summer:	85 ng/m^3	(1251)
glc's in residential area (Belgium) Oct. 1976:			
	in particulate phase:	12 ng/m^3	
	in gas phase:	3.9 ng/m^3 (+ chrysene)	(1289)
glc's in the average American urban atmosphere-1963:			
	~30 µg/g airborne particulates or ~4 ng/m^3 air		(1293)

Cleveland (16 sites) 1971/72:	1.2 ng/m^3	(556)
Los Angeles (4 sites) 1971/72:	1.2 ng/m^3	(560)
Lyon (1 site) 1972:	0.33 ng/m^3	(561)
Rome (1 site) 1970/71:	2.8 ng/m^3	(562)
Budapest (1 site) 1971/72:	8.2 ng/m^3	(563)

Cleveland (Ohio, U.S.A.) 1971/72:
max conc. 140 ng/m^3 (448 values)
annual geom. mean (all sites): 1.4 ng/m^3
annual geom. mean of TSP (all sites): 11 ppm (wt) (556)

Photodecomposition

absorbs solar radiation strongly: $t_{1/2}$= 4.8 h, winter, 35°N latitude, UV degradation creates (2903)

quinones.
Incinerability: thermal stability ranking of hazardous organic compounds: rank 9 on a scale of 1 (highest stability) to 320 (lowest stability). (2390)

C. WATER AND SOIL POLLUTION FACTORS

Biodegradation

was not observed during enrichment procedures. (1882)

Biodegradation

to CO_2 in estuarine water:

concentration, µg/L	*month*	*incubation time, h*	*degradation rate, (µg/L/d) × 10^3*	*turnover time, d*	
10	Jan.	48	0	∞	
10	June	48	0	∞	(381)

Aquatic reactions

adsorption: in estuarine water: at µg/L, 53% adsorbed on particles after 3 h. (381)
After 3 h incubation in natural seawater, 59% of 3 µg/L were taken up by suspended aggregates of dead phytoplankton cells and bacteria. (957)

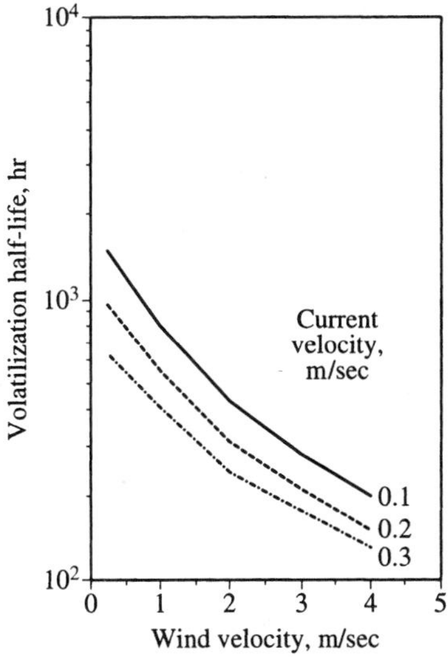

Variation in predicted volatilization rates of benzo(*a*)anthracene at 26°C under varying conditions of wind and current velocity in a stream 1.0 m in depth (1391)

Water and sediment quality

in sediment of wilderness Lake-Colin Scott-Ontario (1976): 7 µg/: (dry wt) (932)

in rapid sand filter solids from Lake Constance water: 0.2-0.3 mg/kg in river water solids:

river Rhine:	0.4 mg/kg	
river Aach at Stockach:	0.9-3.2 mg/kg	
river Argen:	0.4 mg/kg	
river Schussen:	2.6 mg/kg	(531)
in river water solids:		
river Gersprenz at Munster (Germany):	4.3-19 ng/L	
river Danube at Ulm (Germany):	11-14 ng/L	
river Main at Seligenstadt (Germany):	7.0-385 ng/L	
river Aach at Stockach (Germany):	101-199 ng/L	
river Schussen (Germany):	57 ng/L	(530)

Manmade sources

in domestic effluent:	0.19-0.32 mg/L	
in sewage (high percentage industry):	0.34-1.36 μg/L	
in sewage during dry weather:	0.025 μg/L	
in sewage during heavy rain:	10 μg/L	(531)
Göteborg (Sweden) sew. works 1989-1991: infl.: nd; effl.: nd		(2787)
In storm runoff waters from M-6 motorway U.K. 1986: mean: 447 ng/L; range: 70-1,910 ng/L (+ chrysene)		(2794)
In Canadian municipal sludges and sludge compost: September 1993- February 1994: mean values of 11 sludges ranged from 0.32 to 19.4 mg/kg dw*.; mean: 1.2 mg/kg dw*; mean value of sludge compost: 0.17 mg/kg dw.* (* + chrysene).		(7000)

Control methods

	Dec. 1965	*May 1966*	
in raw sewage	31 μg/L	1.9 μg/L	
after mechanical purification	6.0 μg/L	0.28 μg/L	
after biological purification	0.05 μg/L	0.06 μg/L	(545)

ozonation: after 1 min contact time with ozone: residual amount: 4.5% (550)
chlorination: 6 mg/L chlorine for 6 h:
initial conc. 16 ppb: 53% reduction (549)

Degradation in sandy loam in the dark at 20°C at 700 mg/kg

abiotic half-life:	5 years	
biodegradation half-life:	3 years	(2806)

D. BIOLOGICAL EFFECTS

Molluscs

OYSTERS: uptake and release

time of exposure, d	*conc. in oysters, μg/g*	*water, μg/L*	*depuration time, d*	*conc. in oysters, μg/g*	*water, μg/L*
2	4.1 (0.5)	3.8	14	0.2	0
4	1.7 (1.1)	1.1	14	0.4	0
9	1.6 (1.3)	0.3	14	0.4	0
15	0.8	0.1			

Figures in parentheses refer to compartment surrounded by a 60 μm Nitex filter to filter out large particles. (579)

Uptake and depuration by oysters (Crassostrea virginica) *from oil-treated enclosure*

time exposure, d	*depuration time, d*	*concentration oysters, μg/g*	*water, μg/L*	*accumulation factor, oysters/water*
2		2.8	5.3	530
8		1.8	0.1	18,000
2	7	1.9		
8	7	1.0		
8	23	0.3		

half-life for depuration: 9 d (957)

Apparent bioconcentration factors in polychaete worms (dry wt/sediment dry wt)

Prionospio cirrifera and *Spiochaetpoterus costarum*	BCF: 9.4	
Capitella capitata	BCF: 3.6	(2673)

Algae

Anabaena flos-aquae	2w EC_{50} growth	±0.014 mg/L
	NOEC growth	±0.003 mg/L

Selenastrum capricornutum	1w EC_{50} growth	>>sat.	
Crustaceans			
Daphnia pulex	96h LC_{50}	0.01 mg/L	(2917)
Fishes			
fish: calculated BCF 4,230			(2917)

benzo(a)chrysene (picene; 3,4-benzochrysene; dibenzo(a,i)phenanthrene)

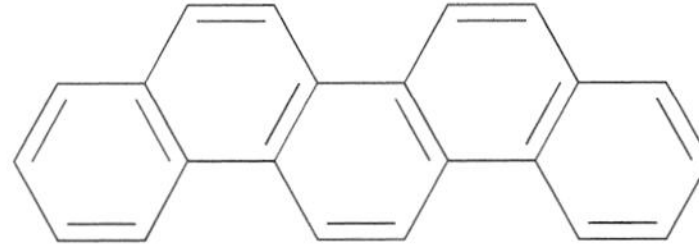

$C_{22}H_{14}$

CAS 213-46-7

A. PROPERTIES

molecular weight 278.36.

B. AIR POLLUTION FACTORS

Manmade sources

in gasoline: 1 μg/L
in exhaust condensate of gasoline engine: 1 μg/L gasoline consumed (1070)

benzo(b)chrysene (2,3-benzochrysene; 3,4-benzotetracene; 1,2: 6,7-dibenzophenanthrene)

$C_{22}H_{14}$

CAS 214-17-5

MANMADE SOURCES

in coal tar: 0.6 g/kg.

A. PROPERTIES

molecular weight 278.46.

benzo(def)phenanthrene *see* pyrene

benzo(b)fluoranthene (3,4-benzfluoranthene; B(*b*)F; benz(*e*)acephenanthrylene; 2,3-benzofluoranthene)

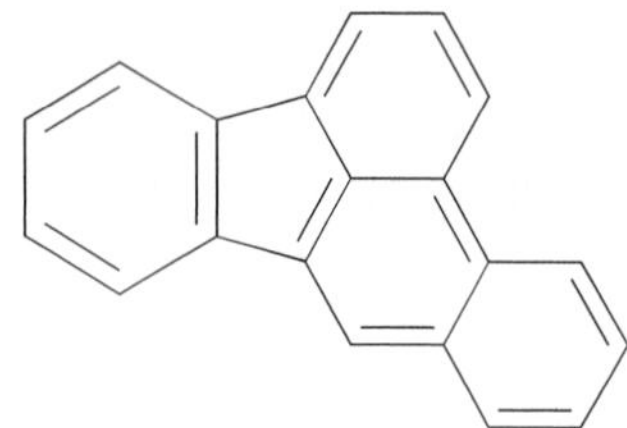

$C_{20}H_{12}$

CAS 205-99-2

MANMADE SOURCES

in Kuwait crude oil:	<1 ppm	
in South Louisiana crude oil:	<0.5 ppm	(1015)
in gasoline: low octane number:	0.16-0.49 mg/kg (*n* = 14)	(385, 1070)
in gasoline: high octane number:	0.26-1.3 mg/kg (*n* = 13)	(385)
in high-octane gasoline:	3.9 mg/kg (*b* + *j* + *k*-isomers)	(1220)
in fresh motor oil:	0.08 mg/kg (*b* + *j* + *k*-isomers)	(1220)
in used motor oil after 18 European driving cycles:	2.8-3.3 mg/kg	
in used motor oil after 5,000 km:	45-82 mg/kg	
in used motor oil after 10,000 km:	56-141.0 mg/kg	
in used motor oil after 10,000 km and 18 European driving cycles:	63-166 mg/kg (*b* + *j* + *k*-isomers)	(1220)
in lubricating motor oils:	0.04-0.30 ppm (6 samples)	(379)
in bitumen:	0.40-1.60 ppm	(506)
in compost:	24,900 µg/kg dry wt.	
in horse manure:	170 µg/kg dry wt.	(2936)

A. PROPERTIES

molecular weight 252; melting point 167°C, boiling point°C, vapor pressure 5×10^{-7} mm Hg at 20°C, solubility 0.0012 mg/L, $LogP_{OW}$ 6.57.

B. AIR POLLUTION FACTORS

Manmade sources

in emissions from the open burning of scrap rubber tires:	70-88 mg/kg of tire	(2950)
in cigarette smoke:	0.3 µg/100 cigarettes	(1298)
in exhaust condensate of gasoline engine:	0.019-0.048 mg/L gasoline consumed	(1070)
in exhaust condensate of gasoline engine:	162 µg/g	(1069)
in tail gases of gasoline engine:	16-66 µg/m^3 (n.s.i.)	(340)
in gasoline:	243 µg/kg (n.s.i.)	(340)

Ambient air quality

in air in Ontario cities (Canada):

location	*April-June '75* ng/1,000 m^3 air	µg/g p.m.	*July-Sept. '75* ng/1,000 m^3 air	µg/g p.m.	*Oct.-Dec. '75* ng/1,000 m^3 air	µg/g p.m.	*Jan.-March '76* ng/1,000 m^3 air	µg/g p.m.
Toronto	866	9.7	798	8.4	1,387	17	783	10
Toronto	890	12	693	11	1,259	21	1,829	13
Hamilton	813	5.6	2,626	19	7,841	113	2,297	27
S. Sarnia	371	6.0	243	5.2	938	18	289	11
Sudbury	255	7.8	173	4.1	417	19	650	28

(999)

Cleveland, Ohio (U.S.A. 1971-72): max. conc.: 170 ng/m^3 (448 values)
annual geom. mean (all sites): 3.6 ng/m^3
annual geom. mean of TSP (all sites): 26 ppm (wt) (556)

glc's in Birkenes (Norway)-Jan.-June '77: avg: 0.62 ng/m^3
range: n.d.-5.8 ng/m^3 (n = 18)
glc's in Rörvik (Sweden)-Dec. '76-April '77: avg: 1.2 ng/m^3
range: n.d.-9.7 ng/m^3 (n = 21) (1236)
glc's in Botrange (Belgium): Woodland at 20-30 km from industrial areas-June-July 1977: 1.6; 2.0 ng/m^3 (n = 2) (+ *b* isomer)
glc's in Wilrijk (Belgium): residential area-Oct./Dec. 1976: 15; 67 ng/m^3 (n = 2) (+ *b* isomer) (1233)

Photodecomposition

absorbs solar radiation strongly
$t_{1/2}$= 1 h midday, winter
adsorption to kaolinite clay inhibits photolysis
$t_{1/2}$= 11 h (21°C, O_2) with fluorescent light on B(*a*)P adsorbed on calcite and suspended in water
$t_{1/2}$= 5.3 h in presence of UV and without O_3
$t_{1/2}$= 0.2 h in presence of both UV and O_3 (2903)

Incinerability

thermal stability ranking of hazardous organic compounds: rank 8 on a scale of 1 (highest stability) to 320 (lowest stability). (2390)

C. WATER AND SOIL POLLUTION FACTORS

Water, sediment, and soil quality

Soil quality: typical values in Welsh surface soil (U.K.), average of 20 0-5-cm cores:
mean: 66 µg/kg dry wt
range: 8-605 µg/kg dry wt (2420)
In storm runoff waters from M-6 motorway, U.K., 1986: mean: 160 ng/L; range: 15-610 ng/L (2794)
In groundwater (Germany, 1968): 0.6-5.7 µg/m^3 (n = 10)
In tapwater (Germany, 1968): 2.6-5.4 µg/m^3 (n = 6) (955)
In dried forest soil samples:
South of Darmstadt (Germany): 30-110 ppb (*b* + *j* isomers)
Near Lake Constance: 25-35 ppb (*b* + *j* isomers) (528)
In sediment of Wilderness Lake, Colin Scott, Ontario (1976): 25 ppb (dry wt) (932)
River Gersprenz at Munster (Germany): 10-13 ng/L
River Danube at Ulm (Germany): 24 ng/L
River Main at Seligenstadt (Germany): 12-362 ng/L
River Aach at Stockach (Germany): 76-332 ng/L
River Schussen (Germany): 41 ng/L (530)

In rapid sand filter solids from Lake Constance water: 0.4 mg/kg
In river water solids: river Aach at Stockach: 2.7-6.7 mg/kg

river Argen: 1.2 mg/kg
river Schussen: 4.9 mg/kg (531)

In domestic effluent: 0.036-0.202 ppb
In sewage (high percentage industry): 0.52-0.87 ppb
In sewage during dry weather: 0.039 ppb
In sewage during heavy rain: 9.9 ppb (531)

Water treatment

initial concentration, µg/L	*treatment method*	*% removal*	
2.3-5.1	river bank filtration	17-22	
8-16	activated carbon	52-54	
4.1-12	ion exchange	-30*; 2.4	(2915)

* increase!

chlorination: 6 mg/L chlorine for 6 h: initial conc. 5.2 ppb: 34% reduction (549)

concentration at various stages of water treatment works (+ j and k isomers)

river intake:	0.15 µg/L	
after reservoir:	0.13 µg/L	
after filtration:	0.039 µg/L	
after chlorination:	0.021 µg/L	(434)

	Dec. 1965	*May 1966*	
in raw sewage	24 ppb	0.58 ppb	
after mechanical purification	4.8 ppb	0.25 ppb	
after biological purification	0.15 ppb	0.05 ppb	(545)

Aerobic slurry bioremediation reactor with mixed aeration chamber, treating a petrochemical waste sludge at 3.5 to 7.5% solids containing approx. 25% oil and grease and 0.5 to 2.5 g of waste material per gram of microorganisms, at 22-24°C during batch treatment: waste residu: infl.: 11 mg/kg; effl. after 90 d: <2.5 mg/kg. (2800)

Manmade sources

primary and digested raw sewage sludge:	0.21-0.42 ppm	
liquors from sewage sludge heat treatment plants:	0.03-0.55 ppm	
sludge cake from heat treatment plants:	0.22-0.52 ppm	
final effluent of sewage works:	0.03 ppb	(1426)

D. BIOLOGICAL EFFECTS

Worms

Apparent bioconcentration factors in polychaete worms (dry wt/sediment dry wt):

Prionospio cirrifera and *Spiochaetpoterus costarum*	BCF: 9.1	
Capitella capitata	BCF: 1.7	(2673)

Plants

Carrots *Daucus carota* (UK 1987):
conc. through a cross section: µg/kg dry wt
peel: 3; inner peel: 0.8; outer core: 1.2; core: 2.6
effect of food processing on conc. in whole carrot: µg/kg dry wt
uncooked: 17; cooked: 1; tinned: 0.8; frozen: 0.8 (2674)

benzo(g,h,i)fluoranthene

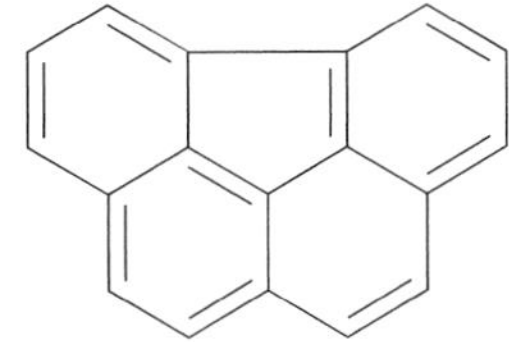

$C_{18}H_{10}$

CAS 203-12-3

MANMADE SOURCES

in lubricating motor oils:	0.16-0.96 ppm (6 samples)	(379)
in bitumen:	0.11-0.43 ppm	(506)
in Kuwait crude oil:	<1 ppm	
in South Louisiana crude oil:	1 ppm	(1015)
in gasoline:	0.003 mg/L	(1070)
in coal tar:	1.5 g/kg	(2600)
in carbon black:	60 mg/kg	(2600)

B. AIR POLLUTION FACTORS

Manmade sources

in exhaust condensate of gasoline engine:	0.11-0.24 mg/L gasoline consumed	(1070)
in airborne coal tar emissions:	3.3 mg/g of sample or 115 μg/m³ of air sampled	
in coke oven emissions:	151-677 μg/g of sample	
in coal tar:	4.4 mg/g of sample	
in wood preservative sludge:	0.91 g/L of raw sludge	(933)

C. WATER AND SOIL POLLUTION FACTORS

Sediment quality

in sediment of Wilderness Lake, Colin Scott, Ontario (1976): 14 ppb (dry wt). (932)

benzo(j)fluoranthene (10,11-benzofluoranthene; B(*j*)F; 7,8-benzofluoranthene; dibenzo (a,jk)fluorene)

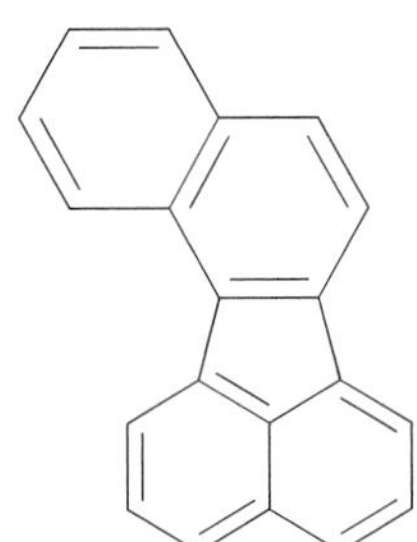

$C_{20}H_{12}$

354 benzo(j)fluoranthene

CAS 205-82-3

MANMADE SOURCES

in gasoline:	9 μg/L	(1070)
in Kuwait crude oil:	<1 ppm	
in South Louisiana crude oil:	<0.9 ppm	(1015)
in coal tar:	5.7 g/kg	(2600)
in carbon black:	20 mg/kg	(2600)

A. PROPERTIES

molecular weight 252.

B. AIR POLLUTION FACTORS

Manmade sources

in cigarette smoke: 0.6 μg/100 cigarettes (1298)
in airborne coal tar emissions: 0.12 mg/g of sample or 4 μg/m^3 of air sampled in coke oven emissions: 18.7-285.3 μg/g of sample
in coal tar: 0.73 mg/g of sample
in wood preservative sludge: 0.31 g/L of raw sludge (993)
in exhaust condensate of gasoline engine: 11-27 μg/L gasoline consumed (1070)
in exhaust condensate of gasoline engine: 94 μg/g (1069)
emissions from typical European gasoline engine: 0.3-9.2 μg/L fuel burnt (1291)

Incinerability

thermal stability ranking of hazardous organic compounds: rank 7 on a scale of 1 (highest stability) to 320 (lowest stability). (2390)

C. WATER AND SOIL POLLUTION FACTORS

Water sediment quality

river Gersprenz at Munster (Germany):	4.6-15 ng/L	
river Danube at Ulm (Germany):	10.1-23 ng/L	
river Main at Seligenstadt (Germany):	14.2-337 ng/L	
river Aach at Stockach (Germany):	144-420 ng/L	
river Schussen (Germany):	53 ng/L	(530)
in rapid sand filter solids from Lake Constance water:	0.5 mg/kg	
in river water solids: river Aach at Stockach:	0.8-4.6 mg/kg	
river Argen:	1.4 mg/kg	
river Schussen:	6.2 mg/kg	(531)
in sediment of Wilderness Lake, Colin Scott, Ontario (1976):	8 ppb (dry wt)	(932)

Manmade sources

in domestic effluent: 0.037-0.2 ppb
in sewage (high percentage industry): 1.1-1.7 ppb
in sewage during dry weather: 0.057 ppb
in sewage during heavy rain: 11 ppb (531)

Control methods

plant location	*water source*	*drinking water river water conc., ng/L*	*conc. ng/L**	*% removal transformation*
Pittsburgh, PA	Monongahela River	36	0.3	99**
Huntington, WV	Ohio River	5	0.3	94
Philadelphia, PA	Delaware River	43	n.d.	100

* treatment provided: lime, ferric sulfate or chloride; A.C., chlorination and fluoridation
** two stages A.C.: powdered and granular carbon (958)

	Dec. 1965	May 1966	
in raw sewage	30 ppb	0.56 ppb	
after mechanical purification	12 ppb	0.14 ppb	
after biological purification		0.09 ppb	(545)

D. BIOLOGICAL EFFECTS

Worms

apparent bioconcentration factors in polychaete worms (dry wt/sediment dry wt).

Prionospio cirrifera and *Spiochaetpoterus costarum*	BCF: 8.2	
Capitella capitata	BCF: 0.6	(2673)

benzo(k)fluoranthene (11,12-benzofluoranthene; B(*k*)F)

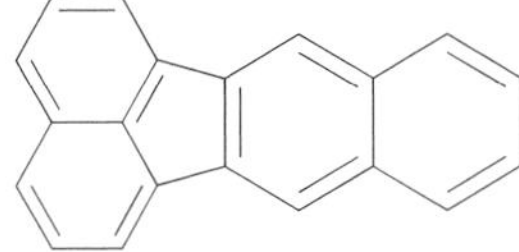

$C_{20}H_{12}$

CAS 207-08-9

MANMADE SOURCES

in bitumen:	0.34-1.4 ppm	(506)
in gasoline:	9 μg/L	(1070)
in Kuwait crude oil:	<1 ppm	
in South Louisiana crude oil:	<1.3 ppm	(1015)
in coal tar:	4.9 g/kg	(2600)
in carbon black:	40 mg/kg	(2600)

A. PROPERTIES

molecular weight 252; melting point 217°C, boiling point 480°C, vapor pressure 5 × 10^{-7} mm Hg at 20°C, solubility 0.00055 mg/L, LogP_{OW} 6.84.

B. AIR POLLUTION FACTORS

Photodecomposition

absorbs solar radiation strongly. (2903)

Manmade sources

in emissions from the open burning of scrap rubber tires: 74-99 mg/kg of tire	(2950)
in airborne coal tar emissions: 10 mg/g of sample or 350 μg/m³ of air sampled in coal tar: 32 mg/g of sample	(993)
in emissions from typical European car engine (1,608 cu cm): 0.2-19 μg/L fuel burnt	(1070, 1291)

Ambient air quality

glc's in Birkenes (Norway): Jan.-June 1977: (+ *j*-isomer)	avg.: 0.48 ng/m³ range: n.d.-4 ng/m³ (*n* = 18)

glc's in Rörvik (Sweden): Dec. '76-April '77: (+ *j*-isomer)	avg. 0.42 ng/m^3 range: n.d.-3.5 ng/m^3 (n = 21)	(1236)
glc's in residential area (Belgium)-Oct. 1976: (+ *b*-isomer)	in particulate sample: 23 ng/m^3 in gas phase sample: 2 ng/m^3	(1289)
organic fraction of suspended matter:		
Bolivia at 5,200 m altitude (Sept.-Dec. 1975):	0.036-0.055 µg/1,000 m^3 (+ benzo(*b*)fluoranthene	
Belgium, residential area (Jan.-April 1976):	2.9-30 µg/1,000 m^3 (+ benzo(*b*)fluoranthene)	(428)

in air in Ontario cities (Canada):

location	*April-June '75* ng/1,000m^3 air	µg/g p.m.	*July-Sept. '75* ng/1,000m^3 air	µg/g p.m.	*Oct.-Dec. '75* ng/1,000m^3 air	µg/g p.m.	*Jan.-March '76* ng/1,000m^3 air	µg/g p.m.
Toronto	428	4.8	571	6.0	916	11	508	6.5
Toronto	328	4.3	285	4.4	597	9.8	519	3.8
Hamilton	419	2.9	1,425	10	5,145	74	443	5.3
S. Sarnia	81	1.3	70	1.5	439	8.4	104	3.8
Sudbury	57	1.3	74	1.8	197	8.8	271	11

(999)

C. WATER AND SOIL POLLUTION FACTORS

Water and sediment quality

river Gersprenz at Munster (Germany):	4.8-9.6 ng/L	
river Danube at Ulm (Germany):	7.7-14 ng/L	
river Main at Seligenstadt (Germany):	4.2-130 ng/L	
river Aach at Stockach (Germany):	132-173 ng/L	
river Schussen (Germany):	33 ng/L	(530)
in rapid sand filter solids from Lake Constance water:	0.5-1.0 mg/kg	
in river water solids:		
river Rhine:	0.6 mg/kg	
river Aach at Stockach:	1.3-2.4 mg/kg	
river Argen:	1.4 mg/kg	
river Schussen:	3.3 mg/kg	(531)
in Thames river water:		
at Kew Bridge:	80 ng/L	
at Albert Bridge:	40 ng/L	
at Tower Bridge:	120 ng/L	(529)
in sediment of Wilderness Lake, Colin Scott, Ontario (1976):	25 ppb (dry wt)	(932)
in groundwater (Germany):	0.2-1.8 µg/m^3 (n = 10)	
in tap water (Germany-1968):	1.0-3.4 µg/m^3 (n = 6)	(955)

Manmade sources

primary and digested raw sewage sludge:	0.1-0.42 ppm	
liquors from sewage sludge heat treatment plants:	0.03-0.45 ppm	
sludge cake from heat treatment plants:	0.09-0.33 ppm	
final effluent of sewage work:	0.03 ppb	(1426)
in domestic effluent:	0.031-0.19 ppb	
in sewage (high percentage industry):	0.33-0.46 ppb	
in sewage during dry weather:	0.022 ppb	
in sewage during heavy rain:	4.2 ppb	(531)

Göteborg (Sweden) sew. works 1989-1991: infl.: nd-0.1 µg/L; effl.: nd (2787)

Control methods

plant location	*water source*	*river water conc., ng/L*	*drinking water conc., ng/L**	*% removal/ transformation*
Pittsburgh, PA	Monongahela River	19	0.2	99**

Huntington, WV	Ohio River	3.6	0.2	94	
Philadelphia, PA	Delaware River	33	n.d.	100	

* treatment provided: lime, ferric salt, A.C., chlorination and fluoridation
** two stages A.C.: powdered and granular carbon (958)

	Dec. 1965	*May 1966*	
raw sewage	8.1 ppb	0.21 ppb	
after mechanical purification	2.2 ppb	0.09 ppb	
after biological purification	0.05 ppb	0.04 ppb	(545)

chlorination: 6 mg/L chlorine for 6 h: initial conc. 69 ppb: 56% reduction (549)

Aerobic slurry bioremediation reactor with mixed aeration chamber, treating a petrochemical waste sludge at 3.5 to 7.5% solids containing approx. 25% oil and grease and 0.5 to 2.5 g of waste material per gram of microorganisms, at 22-24°C during batch treatment: waste residu: infl.: 11 mg/kg; effl. after 90 d: <2.5 mg/kg (2800)

D. BIOLOGICAL EFFECTS

Fishes

calculated BCF 8,750 (2917)

Worms

apparent bioconcentration factors in polychaete worms (dry wt/sediment dry wt):

Prionospio cirrifera and *Spiochaetpoterus costarum*	BCF: 14.1	
Capitella capitata	BCF: 1.8	(2673)

Plants

Carrots *Daucus carota* (U.K. 1987):
conc. through a cross section: µg/kg dry wt
peel: 1.2; inner peel: 0.3; outer core: 0.7; core: 1.6
effect of food processing on conc. in whole carrot: µg/kg dry wt
uncooked: 0.7; cooked: 1.8; tinned: 0.3; frozen: 0.6 (2674)

benzo(a)fluorene (1,2-benzofluorene; chrysofluorene; 11H-benzo[a]fluorene)

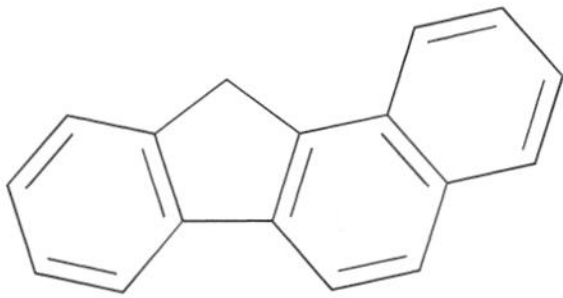

$C_{17}H_{12}$

CAS 238-84-6

CAS 30777-18-5

EINECS 205-944-9250-335-3

MANMADE SOURCES

in bitumen:	0.021-0.10 ppm	(506)
in coal tar:	20 mg/g of sample (n.s.i.)	(933)
in gasoline:	1.5 mg/L	(1070)
in coal tar:	3.7 g/kg	(2600)

A. PROPERTIES

molecular weight 216.28; melting point 187-189°C; boiling point 407°C.

B. AIR POLLUTION FACTORS

Manmade sources

in coke oven emissions: 87-971 µg/g of sample	(960)
in exhaust condensate of gasoline engine: 0.08-0.14 mg/L gasoline consumed	(1070)
in airborne coal tar emissions (n.s.i.): 24 mg/g of sample or 862 µg/m^3 of air sampled	(993)

Ambient air quality

Cleveland, Ohio (U.S.A., 1971/72): max. conc.:	7.5 ng/m^3 (400 values)	
	annual geom. mean (all sites): 0.25 ng/m^3 annual geom. mean of TSP (all sites): 2 ppm (wt)	(556)

organic fraction of suspended matter:

Bolivia at 5,200 m altitude (Sept.-Dec. 1975): 0.012-0.024 µg/1,000 m^3 (+ benzo(*c*)fluorene)	
Belgium, residential area (Jan.-April 1976): 0.38-0.54 µg/1,000 m^3 (+ benzo(*c*)fluorene)	(428)

glc's in residential area (Belgium)-Oct. 1976 (n.s.i.):

in particulate sample: 2.33 ng/m^3	
in gasphase sample: 1.87 ng/m^3	(1289)
in Birkenes (Norway), Jan.-June 1977: avg: 0.16 ng/m^3; range: n.d.-1.4 ng/m^3; (n = 18)	
in Rörvik (Sweden), Dec. 1976-April 1977: avg: 0.35 ng/m^3; range: n.d.-3.9 ng/m^3 (n = 21)	(1236)

D. BIOLOGICAL EFFECTS

Apparent bioconcentration factors in polychaete worms (polychaeta dry wt/sediment dry wt):

Prionospio cirrifera and *Spiochaetpoterus costarum*	BCF: 11	
Capitella capitata	BCF: 9.5	(2673)

benzo(b)fluorene

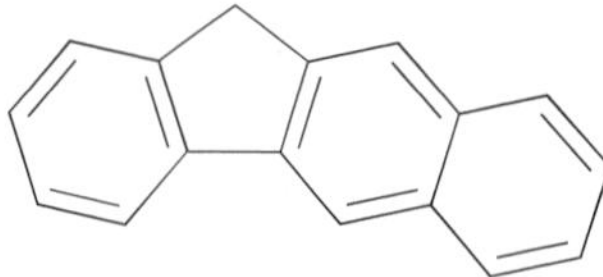

$C_{17}H_{12}$

CAS 243-17-4

EINECS 205-952-2

MANMADE SOURCES

in gasoline: 1.4 mg/L	(1070)
in bitumen: 0.02-0.09 ppm	(506)

constituent of coal tar creosote: 1 wt% (2386)
in coal tar: 3.6 g/kg (2600)

A. PROPERTIES

molecular weight 216.28; melting point 209-210.5°C.

B. AIR POLLUTION FACTORS

Manmade sources

in coke oven emissions: 17-109 µg/g of sample (960)
in exhaust condensate of gasoline engine: 0.065-0.11 mg/L gasoline consumed (1070)
emissions from typical European gasoline engine (1,608 cu cm): 37-251 µg/L fuel burnt (1291)

Ambient air quality

. glc's in Birkenes (Norway), Jan.-June 1977:
avg: 0.04 ng/m^3
range: 0.02-0.34 ng/m^3 (n = 18)
glc's in Rörvik (Sweden), Dec. 1976-April 1977:
avg: 0.2 ng/m^3
range: n.d.-2.3 ng/m^3 (n = 21) (1236)

C. WATER AND SOIL POLLUTION FACTORS

Sediment quality

in sediments in Severn estuary (U.K.): 0.2-1.2 ppm dry wt (1467)
Göteborg (Sweden) sew. works 1989-1991: infl.: nd; effl.: nd. (2787)

benzo(c)fluorene

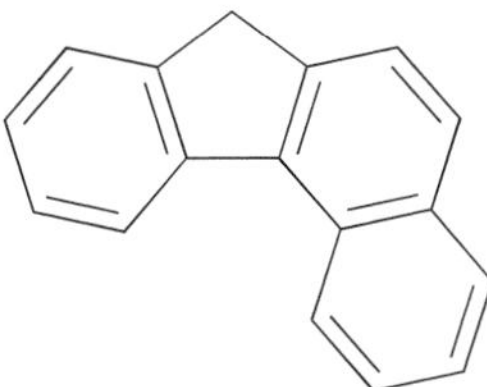

$C_{17}H_{12}$

CAS 205-12-9

B. AIR POLLUTION FACTORS

Manmade sources

in coke oven emissions: 39-627 µg/g of sample. (960)

C. WATER AND SOIL POLLUTION FACTORS

Control methods

Concentration at various stages of water treatment works (+ j and k-isomers):
river intake: 0.15 µg/L

after reservoir:	0.13 µg/L		
after filtration:	0.039 µg/L		
after chlorination:	0.021 µg/L		(434)

	Dec. 1965	*May 1966*	
in raw sewage	24 ppb	0.58 ppb	
after mechanical purification	4.8 ppb	0.25 ppb	
after biological purification	0.15 ppb	0.05 ppb	(545)
in domestic effluent:	0.036-0.20 ppb		
in sewage: high percentage industry:	0.52-0.87 ppb		
during dry weather:	0.039 ppb		
during heavy rain:	9.9 ppb		(531)

Chlorination: 6 mg/L chlorine for 6 h: initial conc. 5.2 ppb: 34% reduction (549)

benzo(ghi)cyclopenta(pqr)perylene *see* 1,12-methylenebenzo(*ghi*)perylene

benzo(ghi)perylene (1,12-benzoperylene; B(*ghi*)P)

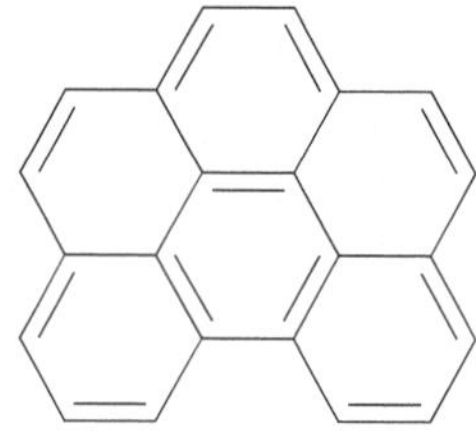

$C_{22}H_{12}$

CAS 191-24-2

MANMADE SOURCES

in compost:	5,630 µg/kg dry wt	
in horse manure:	43 µg/kg dry wt	(2936)
in coal tar:	7.0 g/kg	(2600)
in fresh motor oil:	0.12 mg/kg	
in used motor oil after 5,000 km:	108-207 mg/kg	
in used motor oil after 10,000 km:	153-289 mg/kg	(1220)
in Kuwait crude oil:	<1 ppm	
in South Louisiana crude oil:	<1.6 ppm	(1015)
in bitumen:	1.3-5.5 ppm	(506)
in gasoline: low octane number:	0.32-1.2 mg/kg (*n* = 13)	
in gasoline: high octane number:	0.42-9 mg/kg (*n* = 15)	(380, 385, 1070, 1220)

A. PROPERTIES

molecular weight 276; melting point 222°C, boiling point >500°C, vapor pressure 1 × 10^{-10} mm Hg at 20°C, solubility 0.00026 mg/L at 25°C, LogP_{OW} 7.23.

B. AIR POLLUTION FACTORS

Photodecomposition

absorbs solar radiation strongly. (2903)

Manmade sources

emissions from space heating installation burning:		
coal (underfeed stoker):	4.5 mg/10^6 Btu input	
gasoil:	0.3 mg/10^6 Btu input	
gas:	1.8 mg/10^6 Btu input	(954)
emissions from typical European car engine:	2.7-111 µg/L fuel burnt	(1291)
in exhaust condensate of gasoline engine:	0.12-0.33 mg/L gasoline consumed	(1070)
in emissions from the open burning of scrap rubber tires:	66-159 mg/kg of tire	(2950)

Ambient air quality

glc's at Botrange (Belgium): woodland at 20-30 km from industrial area, June-July 1977: 0.40; 0.70 ng/m^3 (n = 2) (1223)

glc's in Birkenes (Norway), Jan.-June 1977:
avg: 0.28 ng/m^3
range: n.d.-2.3 ng/m^3 (n = 18)
glc's in Rörvik (Sweden), Dec. 1976-April 1977:
avg: 0.42 ng/m^3
range: n.d.-2.88 ng/m^3 (n = 21) (1236)
glc's in the Netherlands: in the major cities:
summer 1968-1971: <1-4 ng/m^3
winter 1968-1971: 2-20 ng/m^3 (1277)
Cleveland, Ohio (U.S.A.): max. conc.: 98 ng/m^3 (400 values)
annual geom. mean (all sites): 1.96 ng/m^3
annual geom. mean TSP (all sites): 16 ppm (wt) (556)
glc's in air in Ontario cities (Canada):

location	*April-June 1975 ng/1,000 m^3 air*	*µg/g p.m.*	*July-Sept. 1975 ng/1,000 m^3 air*	*µg/g p.m.*	*Oct.-Dec. 1975 ng/1,000 m^3 air*	*µg/g p.m.*	*Jan.-March 1976 ng/1,000 m^3 air*	*µg/g p.m.*
Toronto	5,849	65	7,131	74	10,528	127	4,413	56
Toronto	5,077	67	3,303	50	4,693	77	9,814	71
Hamilton	5,809	39	7,183	51	7,532	108	6,418	76
S. Sarnia	1,038	16	1,049	22	2,700	51	1,158	42
Sudbury	779	23	1,104	26	2,321	104	3,009	128

(999)

glc's in the average American urban atmosphere-1963: 63 µg/g airborne particulates or 8 ng/g m^3 air (1293)

glc's in Budapest 1973 (6 A.M.-8 P.M.): heavy-traffic area:
winter: 23 ng/m^3
summer: 19 ng/m^3
low-traffic area:
winter: 5.6 ng/m^3
summer: 9.5 ng/m^3 (1259)

C. WATER AND SOIL POLLUTION FACTORS

Water, sediment, and soil quality

in Thames River water:

at Kew Bridge:	60 ng/L	
at Albert Bridge:	110 ng/L	
at Tower Bridge:	160 ng/L	(529)

in rapid sand filter solids from Lake Constance water: 0.4-1.3 mg/kg

in river water solids:

river Rhine:	1.6 mg/kg	
river Aach at Stockach:	1.4-4.8 mg/kg	
river Argen:	0.4 mg/kg	
river Schussen:	3.5 mg/kg	(531)
river Gersprenz at Munster (Germany):	1.6-13 ng/L	
river Danube at Ulm (Germany):	9.5 ng/L	
river Main at Seligenstadt (Germany):	8.0-84 ng/L	
river Aach at Stockach (Germany):	42-105 ng/L	
river Schussen (Germany):	46 ng/L	(530)
in groundwater (Germany, 1968):	0.7-6.4 μg/m^3 (n = 10)	
in tap water (Germany, 1968):	0.8-3.2 μg/m^3 (n = 6)	(955)

in wells and galleries of an aquifer: Brussels sands (sands covered with a thick loamy layer): <0.1-0.6 ng/L (1066)

in sediment of Wilderness Lake, Colin Scott, Ontario (1976): 28 ppb (dry wt) (932)

in sediments in Severn estuary (U.K.): 0.2-1.5 ppm (dry wt). (1467)

in dried forest sample:

South of Darmstadt (Germany):	10-70 ppb	
near Lake Constance:	10-20 ppb	(528)

Soil quality

typical values in Welsh surface soil (UK), average of 20 0-5-cm cores: mean: 88 μg/kg dry wt; range: 11.3-927 μg/kg dry wt. (2420)

Manmade sources

In storm runoff waters from M-6 motorway U.K. 1986: mean: 52 ng/L; range: 10-203 ng/L (2794)

Primary and digested raw sewage sludge:	0.1-0.31 ppm	
Liquors from sewage sludge heat treatment plants:	0.01-0.5 ppm	
Sludge cake from heat treatment plants:	0.09-0.44 ppm	
Final effluent of sewage work:	0.03 ppb	(1426)
In domestic effluents:	0.040-0.219 ppb	
In sewage (high percentage industry):	0.12-0.48 ppb	
In sewage during dry weather:	0.004 ppb	
In sewage during heavy rain:	3.84 ppb	(531)

Göteborg (Sweden) sew. works 1989-1991: infl.: nd-0.1 μg/L; effl.: nd (2787)

In Canadian municipal sludges and sludge compost: September 1993- February 1994: mean values of 11 sludges ranged from 0.11 to 5.2 mg/kg dw.; mean: 0.33 mg/kg dw; mean value of sludge compost: 0.06 mg/kg dw. (7000)

Water treatment

plant location	*water source*	*river water conc., ng/L*	*drinking water conc., ng/L**	*% removal/ transformation*
Pittsburgh, PA	Monongahela River	34.4	0.7	98.0**
Huntington, WV	Ohio River	10.7	2.5	76.6
Philadelphia, PA	Delaware River	48.4	4.0	91.7

* treatment provided: lime, ferric salt, A.A., chlorination, fluoridation

** two stages A.A.: powdered and granular carbon (958)

Concentrations of various stages of water treatment works

river intake:	0.072 μg/L	
after reservoir:	0.063 μg/L	
after filtration:	0.033 μg/L	
after chlorination:	0.009 μg/L	(434)

	Dec. 1965	May 1966	
raw sewage	8.7 ppb	0.26 ppb	
after mechanical purification	1.2 ppb	0.16 ppb	
after biological purification	0.03 ppb	0.12 ppb	(545)

initial concentration, µg/L	*treatment method*	*% removal*	
2	*river bank filtration*	*33*	
6.8	*activated carbon*	*68*	
2	*ion exchange*	*-35**	
*increase!			(2915)

D. BIOLOGICAL EFFECTS

Fishes

calculated BCF 26,000 (2917)

Plants

carrots *(Daucus carota)* (UK 1987):
conc. through a cross section: µg/kg dry wt.
peel: 4.5; inner peel: nd; outer core: 5; core: nd
effect of food processing on conc. in whole carrot: µg/kg dry wt
uncooked: 3; cooked: 6; tinned: 18; frozen: 21 (2674)

benzo(b)naptho(2,1-d)thiophene

MANMADE SOURCES

in carbon black: 5 mg/kg. (2600)

benzo(c)phenanthrene

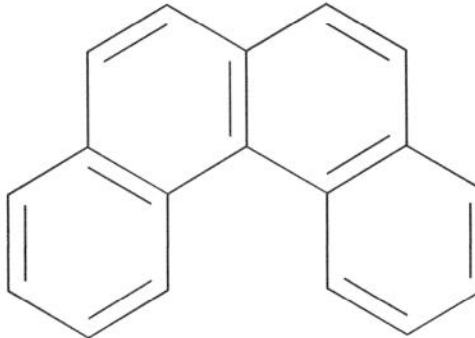

C_8H_{12}

CAS 195-19-7

CAS 46861-06-7

EINECS 205-896-9

MANMADE SOURCES

in wood preservative sludge: 0.87 g/L of raw sludge (960)
in coal tar: 1.7 g/kg (2600)

B. AIR POLLUTION FACTORS

Manmade sources

in coke oven airborne emissions: 82-2,156 μg/g of sample. (960)

Ambient air quality

. glc's in Birkenes (Norway), Jan.-June 1977:

avg:	0.10 ng/m^3
range:	n.d.-0.92 ng/m^3 (n = 18)
. glc's in Rörvik (Sweden), Dec. 1976-April 1977:	
avg:	0.45 ng/m^3
range:	n.d.-5 ng/m^3 (n = 21) (1236)

D. BIOLOGICAL EFFECTS

Worms

apparent bioconcentration factors in polychaete worms (dry wt/sediment dry wt):

Prionospio cirrifera and *Spiochaetpoterus costarum*	BCF: 15.2
Capitella capitata	BCF: 5.0 (2673)

benzo(a)pyrene (3,4-benzopyrene; B(a)P)

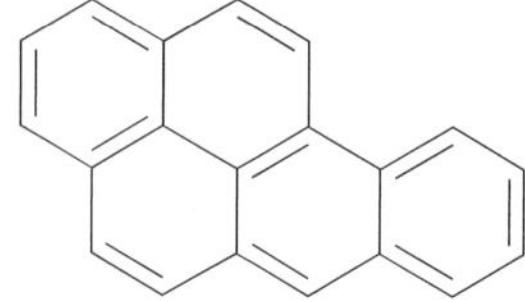

$C_{20}H_{12}$

CAS 50-32-8

MANUFACTURING SOURCES

coal tar processing; petroleum refining; shale refining; coal and coke processing, kerosene processing; heat and power generation sources. (347)

NATURAL SOURCES quantities synthesized by various bacteria:

species	*μg of B(a)P produced per kg of dry bacterial biomass*
Mycobacterium smegmatis	60
Proteus vulgaris	56
Escherichia coli (strain 1)	50
Escherichia coli (strain 2)	46
Pseudomonas fluorescens	30
Serratia marcescens	20 (931)

MAN-CAUSED SOURCES (AIR AND WATER)

combustion of tobacco, combustion of fuels; present in runoff containing greases, oils, etc.; potential roadbed and asphalt leachate. (347)

in gasoline: 0.13 mg/L; 0.14 mg/L; 0.13 mg/L; 8.28 (380, 385,

	mg/kg	1052, 1070, 1220)
	0.09-0.47 mg/kg (n = 13); 0.21-1.00 mg/kg (n = 13)	
in fresh motor oil:	0.02-0.10 mg/kg	(379, 591, 1220)
in used motor oil:	5.8 mg/L	(591)
in used motor oil after 5,000 km:	83-162 mg/kg	
in used motor oil after 10,000 km:	110-242 mg/kg	(1220)
in Kuwait crude oil:	2.8 ppm	
in South Louisiana crude oil:	0.75 ppm	(1015)
in crude oils:		
Brega (Lybia):	1.3 mg/L	
Tia-Juana (Venezuela):	1.6 mg/L	
Safania (Persian Gulf):	0.4 mg/L	(591)
in diesel oil (gasoil):	0.026 mg/L	(591)
in asphalt up to	0.0027 wt %	
in coal tar pitch up to	1.2 wt %	(1380)
constituent of coal tar creosote:	1 wt %	(2386)
in coal tar:	9.7 g/kg	(2600)
in carbon black:	250 mg/kg	(2600)
in compost:	4,410 µg/kg dry wt	
in horse manure:	44 µg/kg dry wt	(2936)

SYNTHESIZED BY ALGAE

Chlorella vulgaris. (566)

A. PROPERTIES

yellowish crystals molecular weight 252.3; melting point 179°C; boiling point 495°C, 311°C at 10 mm; vapor pressure 5 × 10^{-7} mm Hg at 20°C; solubility 0.003 mg/L; in seawater at 22°C: 0.005-0.010 mg/L; LogP_{OW} 6.04.

B. AIR POLLUTION FACTORS

Manmade sources

In emissions from the open burning of scrap rubber tires: 85-114 mg/kg of tire		(2950)
In cigarette smoke: 2.5 µg/100 cigarettes		(1298)
In emissions from typical European gasoline engine: 1.9-26 µg/L fuel burnt		(1291)
In exhaust condensate of gasoline engine:	0.05-0.08 mg/L gasoline consumed	(1070)
	340 µg/g	(1069)
In emissions from combustion of fuel oil:		
large furnaces (>1,000 hp): 0.13 mg/ton fuel		
small furnaces (<1,000 hp): 10 mg/ton fuel		(518)
In emissions from space-heating installations burning:		
coal (underfeed stoker):	10 mg/10^6 Btu input	
gasoil:	0.9 mg/10^6 Btu input	
gas:	0.2 mg/10^6 Btu input	(954)
In emissions from combustion of natural gas:		
industrial boilers: 44 × 10^6 lb/10^6 cu ft of gas		
domestic and commercial heating units: 290 × 10^6 lb/10^6 cu ft of gas		(519)

source	*g/m³ of emitted gas*	
refuse burning	11	
power station, coal	0.3	
power station, gas	0.1	
automotive diesel	5	
coke oven volatiles	35	
home furnace, coal	100	(491,

1382)

In a typical coke oven emission (benzene solubles):	0.41 wt %	(1381)
In emissions from asphalt hot-mixing plant:	3-20 ng/m^3	
(in high-volume particulate matter)	avg 11 ng/m^3	(491, 1379)
In outlet of water spray tower of asphalt hot-road-mix process: <100 ng/m^3		
In outlet of asphalt air-blowing process: <4,000 ng/m^3		(1292)

Anthropogenic emissions

emission strength of B(*a*)P (on airborn particles) from a radiant kerosene heater (11,000 Btu/h) was 30 $ng/10^3$ Btu or 0.028 ng/kJ with an emission rate of 0.32 µg/min. B(*a*)P indoor conc. in a 30 m^3 room increased from 0.78 ng/m^3 as background avg to 3.9 ng/m^3 during the heating period.

Emission strength of B(*a*)P (on airborn particulates) from an open-wall fireplace burning pine and oak firewood may be estimated to be 0.012 ng/kJ with an emission rate of 0.76 µg/h B(*a*)P. Indoor conc. in a house increased from 0.16 ng/m^3 as background avg to 3.5 ng/m^3 when fireplace was in operation. (2654)

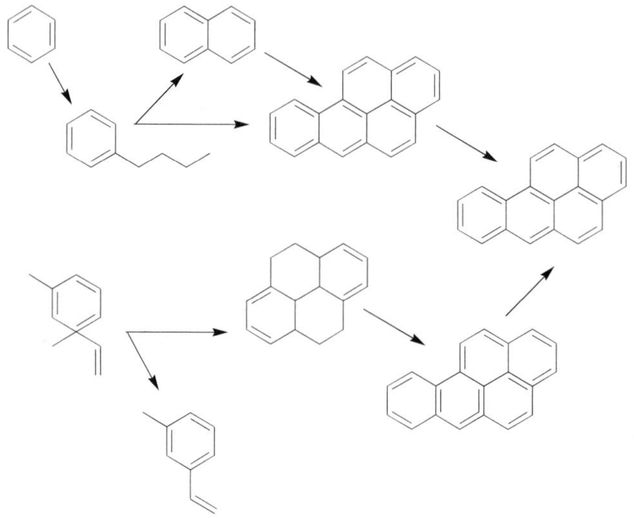

Pathway for the pyrolysis of benzo(*a*) pyrene (1287, 1288)

Ambient air quality

comparison of glc's inside and outside houses at suburban sites:

outside:	2.9 ng/m^3 or 72 µg/g SPM	
inside:	2.1 ng/m^3 or 41 µg/g SPM	(1234)

in the average American urban atmosphere-1963: 46 µg/g airborne particulates or 5.7 ng/g m^3 air (1293)

Cleveland, Ohio (U.S.A. 1971/72):
max. conc.: 126 ng/m^3 (432 values)
annual geom. mean for all sites: 1.04 ng/m^3
annual geom. mean of TSP for all sites: 8 ppm (wt) (556)

organic fraction of suspended matter: (+ benzo(*e*)pyrene + perylene)

Bolivia at 5,200 m altitude (Sept.-Dec. 1975):	0.031-0.065 µg/1,000 m^3	
Belgium, residential area (Jan.-April 1976):	3.0-23 µg/1,000 m^3	(428)

	ng/m^3	µg/g p.m.
In air of Ontario (Canada):		
April-June 1975	0.18-1.4	5.4-9.7
July-Sept. 1975	0.11-2.3	2.4-17
Oct.-Dec. 1975	0.34-3.5	11-51

Jan.-March 1976	0.19-1.9	5.9-23		(999)
	ng/m^3			
Cities in the U.S.A.				
Winter 1959:	31-74			
Summer 1959:	1.4-3.4			
Year 1960:	3.9-8.3			(1284)
	ng/m^3			
Cities in W. Europe				
Winter 1956:	15-50			
Summer 1956:	1-12			(1280)
	winter	*summer, ng/m^3*		
Industrialized W. European cities				
Liege 1958-1962	113	14		(1281)
Liverpool 1956	166	33		(1282)
Hamburg 1961-1963	183	17		(1283)
Milan 1958	198	1		(1285)
Belgium; Winter 1972: 9-64 ng/m^3				(565)
	ng/m^3			
Cities in the Netherlands				
Winter 1968-1971	>1-35			
Summer 1968-1971	>1-9			(1277)
	ng/m^3			
Cleveland				
average 1959:	24			(557)
(NASN site) 1966:	3.2		(558, 559)	
(Site 4) 1972:	0.6			(556)
Urban sites				
maximum: 1966:	11 ng/m^3		(558, 559)	
maximum: 1972:	16 ng/m^3			(556)
average: 1966:	2.8 ng/m^3		(558, 559)	
average: 1971/1972:	0.9 ng/m^3			(556)
Cleveland (16 sites):	1971/1972:	0.87 ng/m^3		(556)
Los Angeles (4 sites):	1971/1972:	1.3 ng/m^3		(560)
Lyon (1 site):	1972:	1.0 ng/m^3		(561)
Rome (1 site):	1970/1971:	1.3 ng/m^3		(562)
Budapest (1 site):	1971/1972:	2.68 ng/m^3		(563)

glc's in residential area Belgium-October 1976: B(*a*)P + B(*e*)P + perylene
in particulate sample: 20 ng/m^3
in gas phase sample: 2.7 ng/m^3 (1289)

glc's at Botrange (Belgium)-June-July 1977: woodland at 20-30 km from industrial area: 1.01; 1.22 ng/m^3 (n = 2) (+ *e*-isomer)
at Wilrijk (Belgium)-October-December 1976: residential area: 12; 62 ng/m^3 (n = 2) (+ *e*-isomer) (1233)

glc's in Birkenes (Norway), Jan.-June 1977:

avg:	0.32 ng/m^3	
range:	n.d.-2.35 ng/m^3 (n = 18)	
in Rörvik (Sweden), Dec. 1976-April 1977:		
avg:	0.54 ng/m^3	
range:	n.d.-4.32 ng/m^3 (n = 21)	(1236)

glc's in Budapest 1973 (6-20 h): heavy-traffic area:

winter:	30 ng/m^3	
summer:	31 ng/m^3	
low-traffic area:		
winter:	17 ng/m^3	
summer:	20 ng/m^3	(1259)

Ground level concentrations (average values)

country	*measuring point*	*ng B(a)P/m*^{3}	*year*	
Germany	East Berlin	4-50	1970	(1247)
Poland	Warsaw	29	1966-67	(1248)
Poland	Zabrze	117	1972	(1248)
Germany	Dusseldorf	6-37	1972	(1249)
Germany	Cologne	2-188	1970-73	(1250)
Hungary	Budapest	74	1966	(1251)
U.S.S.R.	Leningrad	19-64	1967	(1253)
W. Germany	West Berlin	8	1966-67	(1254)
U.S.A.	Washington	9	1966	(1254)
U.S.A.	32 measuring points	3	1966-'70	(1260)
U.S.A.	Cincinatti	6-9	1973	(1254)
Japan	Osaka	50	1967	(1254)

glc's in East Berlin	*ng/m*^{3}
points of dense traffic (summer):	11.5 (n = 26)
point of low traffic (summer):	4.5 (n = 26)
points of dense traffic (winter):	24.6 (n = 24)
point of low traffic (winter):	
direct heating:	7.3 (n = 15)
individual household heating:	21.1 (n = 10)
point of dense traffic (all values):	21.4 (n = 59)
point of low traffic (all values):	8.6 (n = 51)

Benzo(a)*pyrene content of urban air:*

	Benzo(a)*pyrene, ng/m*^{3}				
city	*spring*	*summer*	*fall*	*winter*	
New York					
commercial	0.5-8.1	0.7-3.9	1.5-6.0	0.5-9.4	
freeway	0.1-0.8	0.1-0.7	3.3-3.5	0.7-1.3	
residential	0.1-0.6	0.1-0.3	0.6-0.8	0.5-0.7	
					(1263)
Detroit					
central sites: commercial	7.2			5.0-17.0	
freeway		4.0-6.0	3.4-7.3	9.2-13.7	
residential		0.2		0.9-1.8	(1264)
Atlanta	2.1-3.6	1.6-4.0	12-15	2.1-9.9	
Birmingham	6.3-18	6.1-10	20-74	23-34	
Cincinatti	2.0-2.1	1.3-3.9	14-18	18-26	
Detroit	3.4-12	4.1-6.0	18-20	16-31	
Los Angeles	0.4-0.8	0.4-1.2	1.2-13	1.1-6.6	
Nashville	2.1-9.0	1.4-6.6	30-55	26	
New Orleans	2.6-5.6	2.0-4.1	3.6-3.9	2.6-6.0	
Philadelphia	2.5-3.4	3.5-19	7.1-12	6.4-8.8	
San Francisco	0.8-0.9	0.2-1.1	3.0-7.5	1.3-2.4	(1265)
Pittsburgh		0-23	2.9-37	8.2	(1266)
Hamburg, Germany	14-72	10-26	66-296	94-388	(1267)
London, England	25-48	12-21	44-122	95-147	
Sheffield, England	20-44	21-33	56-63	64-78	
Cannock, England	4-16	6-11	27-31	27-32	(1268)
London (in traffic)	20	11	57	68	
London (background)	11	1	38	42	(1269)
Milan, Italy	12	3	25	150	(1270)
Copenhagen, Denmark	6	5	14	15	(1271)
Prague,		13-36		53-145	(1272)

Czechoslovakia					
Budapest, Hungary		17-32		72-141	(1273)
Pretoria, South Africa		10		22-28	
Johannesburg, South Africa				22-49	
Durban, South Africa				5-28	(1274)
Osaka, Japan					
commercial	5.7	2.7	9.4	14	
residential	3.3	1.4	3.8	6.7	(1275)
Sydney, Australia	0.6-2.4	0.6-1.8	2.5-7.4	3.8-8.2	(1276)

Photodecomposition

absorbs solar radiation strongly. (2903)

Incinerability

thermal stability ranking of hazardous organic compounds: rank 11 on a scale of 1 (highest stability) to 320 (lowest stability). (2390)

C. WATER AND SOIL POLLUTION FACTORS

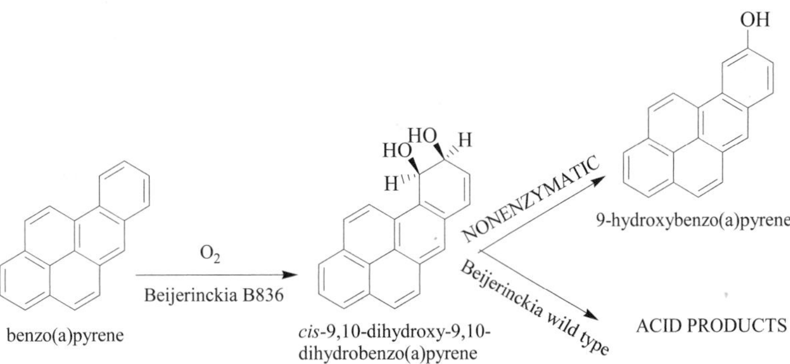

Biodegradation
Oxidation

of B(a)P by Beijerincka B836

Soil quality

typical values in Welsh surface soil (U.K.), average of 20 0-5-cm cores:
mean: 36 µg/kg dry wt
range: 3.6-285 µg/kg dry wt (2420)

Average degradation by soil bacteria after 8 d culture

	amount of extracted B(a)P, µg	*amount of B(a)P destroyed, %*
soil not inoculated with bacteria (control)	191	0
soil + N 5 bacterial strain	90	53
soil + N 13 bacterial strain	61	66
soil + N 13 bacterial strain*	33	82

* Before the experiment this strain was cultured in a medium containing B(*a*)P for 110 d (544)

Degradation in sandy loam in the dark at 20°C at 700 mg/kg

abiotic half-life:	no loss within 16 months	
biodegradation half-life:	4 years	(2806)

Initial oxidation products of soil microorganisms

Cunninghamella elegans:	trans-9,10-dihydrodiol	
	trans-7,8-dihydrodiol	
	B(*a*)P-1,6-quinone	
	B(*a*)P-3,6-quinone	
	9-hydroxy-B(*a*)P	
	3-hydroxy-B(*a*)P	(2904)

Degradation in soil system

at initial conc. between 0.048 and 32 mg/kg soil: $t_{1/2}$: 30, 37, 50, 54, 147, 175, 264, 294, 406, 420, 694 d. (2903)

Microbial degradation

to CO_2 in seawater at 12°C in the dark after 48 h incubation at 16 µg/L (n.s.i.): 0 µg/L/d; after addition of water extract of fuel oil 2, after 24 h incubation: 0.01 µg/L/d-turnover time: 1,400 d. (477)

Biodegradation

to CO_2 in estuarine water:

concentration, µg/L	*month*	*incubation time, h*	*degradation rate, (µg/L/d) × 10³*	*turnover time, d*	
5	Jan.	24	0		
5	June	24	0		
5	May	96	2 ± 1	3,500	(381)

Degradation

in seawater by oil-oxidizing microorganisms (in presence of 0.36 mg/L pyrene and 0.35 mg/L fluorene at 10°C): initial conc. 190 µg/L; after 12 d: 90 µg/L: 53% decrease. (1237)

Transformation by soil microorganisms

(934)

	% transformed			
strain	*2 d*	*3 d*	*4 d*	*8 d*
N-13	13	43	53	69
N-2/II	11	23	33	-
N-5	0	0	0	55
N-2/I	0	0	0	-
N-13 cultivated on B(*a*)P				84

Manmade sources

B(*a*)P levels in industrial effluents:

industry	*conc., ppb*	
shale oil: after dephenolization	2; 312	(536, 537)
coke by-products: after biochemical treatment	12-16	(538)
gasworks:		
after filtration through coke bed	20	(539)
after dephenolization	130-290	(540)
oil refineries:		
coking residue after direct distillation of oil	0.48-5.0	(541)
catalytic cracking of: kerosene	0.14	
catalytic cracking of gasoil fraction at 450°C	0.11-0.19	
	0.05-0.29	
catalytic cracking	0.07-0.11	
thermal cracking 700-800°C	0.09-0.23	
	0.10-3.0	
pyrolysis of ethane-ethylene fractions, 700-800°C	3.6	(541)

after settling of various refineries	up to 0.22	(542)
in wood preservative sludge:	3.39 g/L of raw sludge	(960)
in domestic effluent:	0.038-0.074 ppb	
in sewage (high percentage industry):	0.1-0.36 ppb	
in sewage during dry weather:	0.001 ppb	
in sewage during heavy rain:	1.84 ppb	
primary and digested raw sewage sludge:	0.27-0.57 ppm	
liquors from sewage sludge heat-treatment plants:	0.03-0.84 ppm	
sludge cake from heat-treatment plants:	0.31-0.52 ppm	
final effluent of sewage work:	0.03 ppb	(1426)
Göteborg (Sweden) sew. works 1989-1991: infl.: nd-0.1 μg/L; effl.: nd		(2787)
In storm runoff waters from M-6 motorway U.K. 1986: mean: 13 ng/L; range: 14-680 ng/L		(2794)

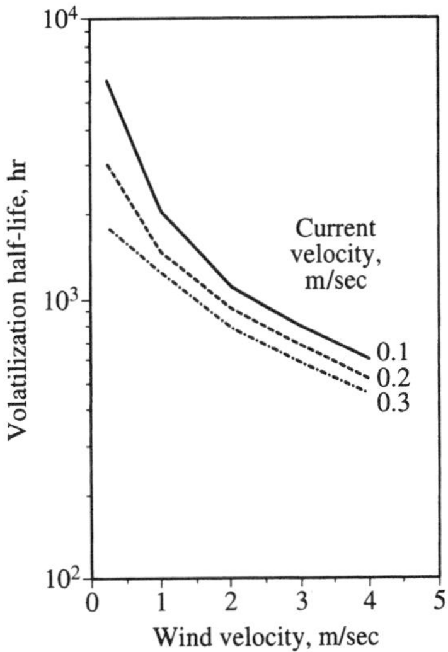

Aquatic reactions

evaporation:

Variation in predicted volatilization rates of benzo(*a*)pyrene at 25°C under varying conditions of wind and current velocity in a stream 1.0 m in depth (1391)

Adsorption

In estuarine water: at 3 μg/L, 71% adsorbed on particles after 3 h (381)

After 3 h incubation in natural seawater, 75% of 2 μg/L were taken up by suspended aggregates of dead phytoplankton cells and bacteria (957)

Water, sediment, and soil quality

in Thames River water:		
at Kew Bridge:	130 ng/L	
at Albert Bridge:	160 ng/L	
at Tower Bridge:	350 ng/L	(529)
river Gersprenz at Munster (Germany):	9.6 ng/L	
river Danube at Ulm (Germany):	0.6 ng/L	
river Main at Seligenstadt (Germany):	1.1-43 ng/L	
river Aach at Stockach (Germany):	4-16 ng/L	
river Schüssen (Germany):	10 ng/L	(530)

in groundwater (Germany-1968):	0.4-3.8 µg/m³ (*n* = 10)	
in tap water (Germany-1968):	0.5-4.0 µg/m³ (*n* = 6)	(955)
in wells and galeries of an aquifer: Brussels sands (sands covered with a thick loamy layer):	<0.1-0.4 ng/L	(1066)
in rapid sand filter solids from Lake Constance water: 0.05-0.2 mg/kg		
in river water solids:		
river Rhine:	0.3 mg/kg	
river Aach at Stockach:	0.5-2.0 mg/kg	
river Argen:	0.1 mg/kg	
river Schussen:	0.4 mg/kg	(531)

Typical levels in marine sediments

location	*depth, m*	*conc., ppb dry wt*	
Greenland	0.2	5	(532)
Italy (Bay of Naples)	15-45	1,000-3,000	
Highly industrialized areas	13	7.5	
	2-65	10-530	
Near volcanic pollution	55	260-960	
	120	1.4	
Island affected by pollution	-	100-560	(533)
French Mediterranean coast	14	400	
	16	1,500	
	48	75	
	58	traces	
	82	400	
estuary	-	34	
	1	20	
	4	15	
	5	25	
	102	not detected	(534)

In sediment of Wilderness Lake, Colin Scott, Ontario (1976): 13 ppb (dry wt) (932)

In forest soil: South of Darmstadt (W. Germany): 1.5-4.0 ppb of dried oil sample near Lake Constance: 1.5-2.5 ppb of dried soil sample (528)

In plant-bearing soil approximately 10 µg/kg is found; 1.0 µg/m³ is dissolved in the water (566)

Water treatment methods

plant location	*water source*	*river water conc., µg/L*	*drinking water conc., µg/L**	*% removal/ transformation*
Pittsburgh, PA	Monongahela River	42	0.4	99.0**
Huntington, WV	Ohio River	5.6	0.5	91.1
Philadelphia, PA	Delaware River	41.1	0.3	99.3

* treatment provided: lime, ferric salt, A.C., chlorination, fluoridation
** two stages A.C.: powdered and granular carbon (958)

chlorination: 6 mg/L chlorine for 6 h: initial conc.: 53 ppb: 98% reduction (549)

oxidation by ClO_2 produces dichloro-3,4-benzopyrenes (590)

Effect of chlorination on levels of B(a)P in water:

initial chlorine dose, mg/L	*initial B(a)P conc., ppb*	*time, h*	*% reduction*	
0.3	5	3	50	(547)
0.5	5	2	50	
		22	20	
0.5	2	2	50	
		13	100	(547)
0.5	1	0.5	81	(548)
		2	94	
0.3	1	0.5	82	
		2	92	(548)

Ozonation

after 1 min contact time with ozone; residual amount: 39%			(550)
	Dec. 1965	*May 1966*	
raw sewage	34 ppb	0.43 ppb	
after mechanical purification	1.4 ppb	0.15 ppb	
after biological purification	0.02 ppb	0.07 ppb	(545)

Concentration at various stages of water treatment works

(+ benzo (e)*pyrene):*		
after reservoir:	0.051 µg/L	
after filtration:	0.030 µg/L	
after chlorination:	0.009 µg/L	(434)

initial concentration, µg/L	*treatment method*	*% removal*	
2	river bank filtration	25	
6.6	activated carbon	61	
1.8	ion exchange	0	(2915)

Manmade sources

In Canadian municipal sludges and sludge compost: September 1993- February 1994: mean values of 11 sludges ranged from 0.2 to 6.8 mg/kg dw.; mean: 0.35 mg/kg dw; mean value of sludge compost: 0.05 mg/kg dw. (7000)

D. BIOLOGICAL EFFECTS

Residues

Content of foodstuffs:

	benzo(a)pyrene, ppb	*Ref.*
Meat		
Raw		
hamburger	ND(0.1)*	(1022)
Cooked		
sausage	0.11; 0.15	(1023, 1024)
hamburger	2.5	(1022)
Smoked		
ham	0.17; 0.7; 0.7; 3.2	(1023, 1025, 1026)
mean of 47 products	0.225	(1023)
fat bacon	0.22	(1023)
chicken	0.7	(1026)
Barbecued		
ribs	6-10	(1027)
beef in sauce	3.3	(1026)
pork	5.0	(1026)
Charcoal broiled		
beefsteak	8; 0.6-0.8	(1027, 1028)
bratwurst	8-12	(1028)
pork loin	4-8	(1028)
Vegetable oils		
margarine	0.4; 0.6; 0.4-3.7	(1038, 1039)
vegetable	ND(0.2)	(49, 1038)
sunflower	10.6; 5.2; ND(3)	(1040, 1038)
peanut	4; 0.6; 1.9	(1025, 1040, 1041)
peanut (fried)	3	(1041)
soybean	1.4; 1.7	(1025, 1040)
cottonseed	0.4; 1.4	(1025, 1040)
Miscellaneous		
butter	ND(0.21; 0.5)	(1038)
coffee	0.2-0.31; 0.2-6.0	(1042, 1043)
instant coffee	0.02-0.06	(1044)
coffee substitute	0.1-0.3	(1042)
coffee beans (aqueous infusions)	0.003	(1044)

tea	0.0-16**	(1045)
barley, rye, wheat	0.2; 0.2; 0.3; 0.5; 0.3, 1.1; 0.2; 0.3	(1046, 1047)
Fish		
Smoked		
herring	2.6; 0.3; 1.3	(1023)
whitefish	4.3; 6.9	(1026)
mackerel	1.0; 0.53; 0.76	(1029, 1023)
Vegetables		
lettuce	12; 2.9-12.8; 0.2	(1030, 1031, 1032)
	1.3; 0.7; 1.1; 150***	(1022, 1033)
leek	12.6-24.5	(1031)
spinach	3.3; 20; 50	(1031, 1034)
cabbage	7.4	(1031)
endive	6.7	(1030)
carrots, peeled	0.07-0.14	(1032)
Fruits		
pruned (dried edible portion)	0-6.2	(1035)
citric acid (7 brands)	ND(0.1)	(1036)
apples, bananas, pineapple, mandarin oranges	mean = 0.03	(1037)

* Not detected at the concentration shown in parentheses.
** <33% is extracted by hot water.
*** Dry weight basis.

Carrots (Daucus carota) *U.K. 1987:*
conc. through a cross section: µg/kg dry wt
peel: 3.5; inner peel: 1.2; outer core: 1; core: 1.7
effect of food processing on conc. in whole carrot: µg/kg dry wt
uncooked: 2; cooked: 4; tinned: 3; frozen: 3 (2674)

Bioconcentration factor for vegetation

log BCF_V : *-1.25* (2644)

marine tissue	*source*	*benzo(a)pyrene ppb*	
Mussel	Toulon Roads, France	0.2-3.0	(1016)
	Falmouth, MA	ND(0.5)	(1017)
	Vancouver, B.C., Canada		
	Remote area	0.0-0.2	
	Outer area	2.0 ± 0.3	
	Around wharf, marina, docks	18	
	Inner harbor	42	(1018)
oyster	Norfolk Harbor, VA	2-6	(1019)
	Long Island Sound	2	(1017)
	Chincoteague, VA	0.2	(1017)
	French Coast	0.1-7.0	(1020)
codfish	West Coventry, Greenland	1.5	(1021)
	Atlantic Ocean, off NJ	ND	(1017)
clam	Chinocoteague, VA	0.3	(1017)
	Darien, CT.: Scott's Cove	ND	(1017)
	Fish market, Linden, NJ	ND	(1017)
crab	Chesapeake Bay, MD	ND	(1017)
	Raritan Bay	3	(1017)
menhaden	Raritan Bay	1.5	(1017)
shrimp	Palacios, TX	ND	(1017)
lake trout	Lake Maskinonge, ON	<1	(1017)

Typical B(a)P levels in marine fauna and flora

Plankton

Greenland:	5 ppb	(532)
Italy:	6-21 ppb	(533)
French Channel Coast:	400 ppb	(535)
Algae		
Greenland (sample at 40 m):	60 ppb	(532)
Greenland (sample at seabed):	60 ppb	(532)
Italy:	2 ppb	(533)
Mollusc		
(Greenland):	60 ppb	
(Greenland) shell:	18 ppb	
body	55 ppb	(532)
(Italy) shell:	11 ppb	
body	130; 540 ppb	(533)
(Italy):	2 ppb	
Sardine (Italy):	65 ppb	(533)

Mussel *(Mytilus edulis)*: conc. in tissue: 0.05 µg/g after chronic pollution; depuration half-life: 16 d (580)

In *Mytilus edulis* from 13 sites in Yaguina Bay (Oregon) 1977-1978:
10%: 0.4 µg/kg
50%: 3.2 µg/kg
90%: 15 µg/kg (n = 150) (1671)

Bioaccumulation

uptake and depuration by oysters *(Crassostrea virginica)* from oil-treated enclosure:

time of exposure, d	*depuration, d*	*concentration oysters, µg/g*	*water, µg/L*	*accumulation factor oysters/water*
2		0.36	1.9	190
8		0.30	0.1	3,000
2	7	0.40		
8	7	0.20		
8	23	0.12		

half-life for depuration: 18 d (957)

Fishes

calculated BCF 8,750 (2917)

Apparent bioconcentration factors in polychaete worms (polychaeta dry wt/sediment dry wt)

Prionospio cirrifera and *Spiochaetpoterus costarum*	BCF: 13.8	
Capitella capitata	BCF: 0.7	(2673)

Toxicity

Algae

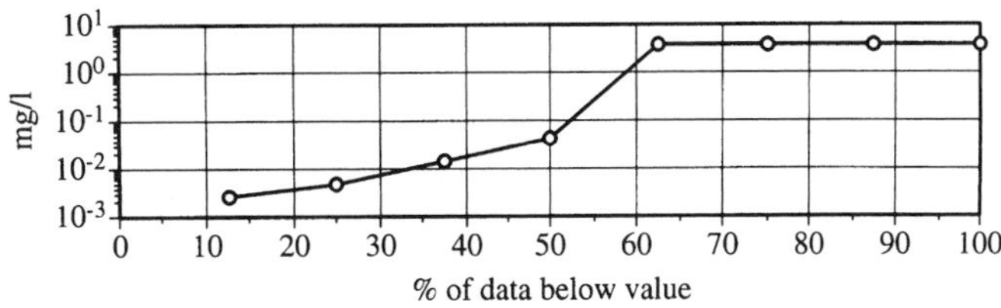

Frequency distribution of short-term EC_{50}/LC_{50} values for algae (n = 8) based on data in this and other works (3999)

Selenastrum capricornutum	1w NOEC growth	12 mg/L	
	1w EC_{50} growth	0.0025 0.044 mg/L	(2917)
Anabaena flos-aquae:	3d EC_{50} growth	>4 mg/L	
Chlamydomonas reinhardi:	3d EC_{50} growth	>4 mg/L	
Euglena gracilis:	3d EC_{50} growth	>4 mg/L	

Poeciliopsis monacha:	enzyme effect after 1 d	800-1,000 μg/L	
Poeciliopsis sp.:	lethal effect after 1 d	3.75 mg/L	
Poteriochromonas malhamensis:	3d EC_{50} growth	>4 mg/L	
Scenedesmus obliquus:	3d EC_{50} growth	5 μg/L	
Selenastrum capricornutum:	3d EC_{50} growth	15 μg/L	(2625)

Annelids			
Neanthes arenaceodentata	96h LC_{50}	>1.0 mg/L	
Crustaceans			
Daphnia pulex	96h LC_{50}	0.005 mg/L	
Insects			
Aedes aegypti	± 12h LC_{50}	0.008 mg/L	
	± 36h LC_{50}	0.002 mg/L	
Amphibians			
Rana pipiens	24h LC_{50}	>6.7 mg/L	
Pleurodeles waltl stage 53 larvae	lw NOLC	0.1 mg/L	(2917)
Fishes			
Poeciliopsis lucida	24h LC_{50}	1.2-3.7 mg/L	(2917)
Atlantic salmon *(Salmo salar)* eggs:	168 h BCF	71 (static test)	(1507)

benzo(e)pyrene (1,2-benzpyrene; B(*e*)P)

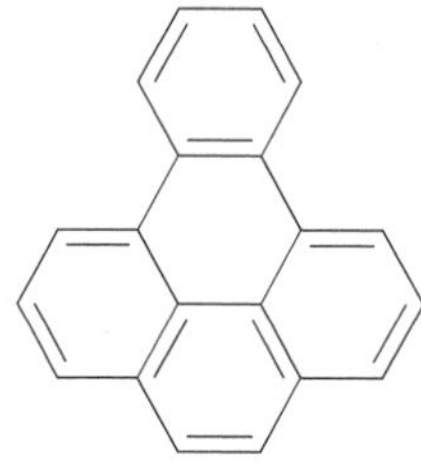

$C_{20}H_{12}$

CAS 192-97-2

MANMADE SOURCES

in gasoline:		
low octane number:	0.18-0.87 mg/kg (*n* = 13)	(385)
high octane number:	0.45-1.82 mg/kg (*n* = 13)	(385)
	6.6 mg/kg	(1220)
in gasoline:	0.30; 0.13; 0.32 mg/L	(380, 1051, 1070)
in bitumen:	1.6-6.5 ppm	(506)
in lubricating motor oils:	0.07-0.49 ppm (6 samples)	(379, 1220)
in used motor oil after 5,000 km:	92-182 mg/kg	
in used motor oil after 10,000 km:	140-278 mg/kg	(1220)
in asphalts:	up to 0.0052 wt %	
in coal tar pitches:	up to 0.70 wt %	(1380)
in Kuwait crude oil:	0.5 ppm	

in South Louisiana crude oil:	2.5 ppm	(1015)
in coal tar:	6.1 g/kg	(2600)
in carbon black:	190 mg/kg	(2600)
in compost:	9,800 µg/kg dry wt	
in horse manure:	75 µg/kg dry wt	(2936)

A. PROPERTIES

molecular weight 252; solubility 0.004 mg/L at 25°C.

B. AIR POLLUTION FACTORS

Manmade sources

in cigarette smoke: 0.3 µg/100 cigarettes		(1298)
in exhaust condensate of gasoline engine: 0.037-0.059 mg/L gasoline consumed		(1070)
emissions from typical European gasoline engine: 3.6-38.9 µg/L fuel burnt		(1291)
emissions from space heating installations burning:		
coal (underfeed stoker):	7.9 mg/10^6 Btu input	
gas:	0.49 mg/10^6 Btu input	(954)
in coke oven airborne emissions: 103 µg/g of sample		(960)
emissions from asphalt hot-mixing plant: 14-40 ng/m^3 (in high-volume particulate matter): avg. 26 ng/m^3		(1379)

Ambient air quality

Cleveland (16 sites) 1971/72:	1.2 ng/m^3	(556)
Los Angeles (4 sites) 1971/72:	2.0 ng/m^3	(560)
Lyon (1 site) 1972:	1.8 ng/m^3	(561)
Rome (1 site) 1970/71:	0.9 ng/m^3	(562)
Budapest (1 site) 1971/72:	1.5 ng/m^3	(563)
glc's in the Netherlands (ng/m^3): in the major cities		
summer 1968-1971:	1-6 ng/m^3	
winter 1968-1971:	9-31 ng/m^3	(1277)
glc's in Birkenes (Norway), Jan.-June 1977:		
avg:	0.38 ng/m^3	
range:	n.d.-2.8 ng/m^3 (n = 18)	
in Rörvik (Sweden), Dec. '76-April '77:		
avg:	0.51 ng/m^3	
range:	n.d.-3.30 ng/m^3 (n = 12)	(1236)
in Budapest 1973:		
heavy-traffic area:		
winter 6-20 h:	8.4 ng/m^3	
summer:	9.7 ng/m^3	
low-traffic area:		
winter:	19 ng/m^3	
summer:	4.7 ng/m^3	(1259)

In the average urban atmosphere, 1963: 42 µg/g airborne particulates or 5 ng/m^3 air (1293)

comparison of glc's inside and outside houses at suburban sites:		
outside:	2.2 ng/m^3 or 46 µg/g SPM	
inside:	2.3 ng/m^3 or 41 µg/g SPM	(1234)

in air of Ontario cities (Canada):

location	*April-June 1975 ng/1,000 m^3 air*	*µg/g p.m.*	*July-Sept. 1975 ng/1,000 m^3 air*	*µg/g p.m.*	*Oct.-Dec. 1975 ng/1,000 m^3 air*	*µg/g p.m.*	*Jan.-March 1976 ng/1,000 m^3 air*	*µg/g p.m.*
Toronto	440	4.9	519	5.4	1,294	15.6	781	10.0
Toronto	478	6.3	375	5.7	400	6.6	791	5.8

Hamilton	606	4.2	1,407	10.1	3,771	54.4	1,607	19.2	
S. Sarnia	118	1.9	52	1.1	603	11.5	64	2.4	
Sudbury	23	0.7	45	1.1	255	11.4	317	13.6	
									(999)

Cleveland, Ohio (U.S.A.): max. conc,: 78 ng/m^3 (464 values) (1971/1972): annual geom. mean (all sites): 1.45 ng/m^3; annual geom. mean of TSP (all sites): 11 ppm (wt) (556)

C. WATER AND SOIL POLLUTION FACTORS

Manmade sources

in wood preservative sludge: 2.48 g/L of raw sludge. (960)

Sediment quality

in sediment of Wilderness Lake-Colin Scott-Ontario (1976): 28 ppb (dry wt) (932)
Göteborg (Sweden) sew. works 1989-1991: infl: nd-0.3 µg/L; effl.: nd (2787)

D. BIOLOGICAL EFFECTS

Worms

apparent bioconcentration factors in polychaete worms (dry wt/sediment dry wt).

Prionospio cirrifera and *Spiochaetpoterus costarum*	BCF: 11.6	
Capitella capitata	BCF: 1.5	(2673)

benzo(b)pyridine *see* quinoline

benzo(b)pyrrole *see* indole

benzo(f)quinoline (5,6-benzoquinoline; 1-azaphenanthrene; β-naphtoquinone)

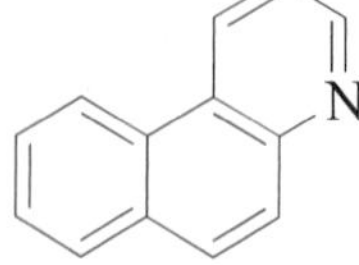

$C_{13}H_9N$

CAS 85-02-9

MANMADE SOURCES

in wood preservative sludge: 7.1 g/L of raw sludge (n.s.i.). (993)

A. PROPERTIES

molecular weight 179.22; melting point 90-93°C; boiling point 349°C/721 mm.

B. AIR POLLUTION FACTORS

Manmade sources

in coke oven emissions: 8.3-79 µg/g of sample (n.s.i.). (993)

Ambient air quality

in the average American urban atmosphere-1963: 2 µg/g airborne particulates or 0.2 ng/m^3 air. (1293)

benzo(h)quinoline (7,8-benzoquinoline)

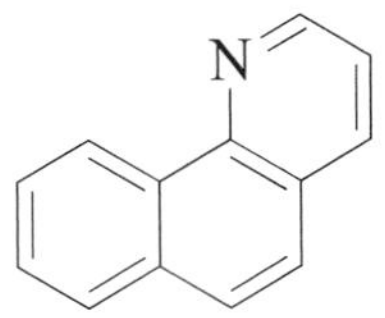

$C_{13}H_9N$

CAS 230-27-3

A. PROPERTIES

molecular weight 179.22; melting point 49-51°C; boiling point 338°C/719 mm.

B. AIR POLLUTION FACTORS

Ambient air quality

in the average American urban atmosphere-1963: 3 µg/g airborne particulates or 0.3 ng/m^3 air.

benzo(b)thiophene (isothionaphthene; isobenzothiofuran)

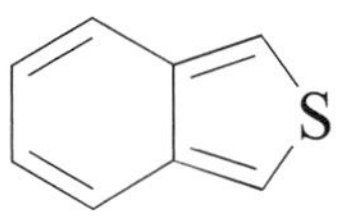

C_8H_6S

CAS 95-15-8

MANMADE SOURCES

constituent of coal tar creosote: 0.1 wt %. (2386)

A. PROPERTIES

molecular weight 134.2; density 1.15.

benzocaine (ethyl-*p*-aminobenzoate; benzoic acid,4-amino,ethylester)

$C_9H_{11}NO_2$

CAS 94-09-7

USES

medicine; suntan preparations.

A. PROPERTIES

white, crystalline, odorless, tasteless powder; molecular weight 165.19; melting point 88-90°C; boiling point approx. 310°C; solubility 800 mg/L at 20°C; $LogP_{OW}$ 1.86 (measured).

C. WATER AND SOIL POLLUTION FACTORS

Biodegradability

inoculum/method	*test duration*	*removed*	
A.S., industrial, Zahn-Wellens test	5 d	>90% ThOD	(5406)

Impact on biodegradation

NH_3 oxidation by *Nitrosomonas* sp.:
at 100 mg/L: 30% inhibition
at 50 mg/L: 27% inhibition
at 10 mg/L: 0% inhibition (390)
primary municipal effluent: inhibition concentration approx. 40 mg/L (5406)

D. BIOLOGICAL EFFECTS

Bacteria

Staphylococcus aureus	3h EC_{50}	3,400 mg/L	(5407)

Fishes

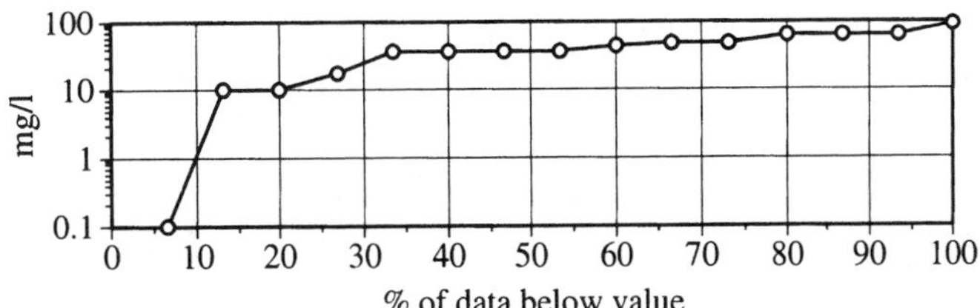

Frequency distribution of 24-96h LC_{50} values for fishes (n=15) based on data from this and other works (3999)

Salmo gairdneri	24h LC_{50},S	10; 88 mg/L	(5402, 5405)
	96h LC_{50},S	0.1; 10 mg/L	(5405)
	96h LC_{30},S	1 mg/L	(5405)
	5d LC_{50},S	47 mg/L	(5402)
	10d LC_{50}	42 mg/L	(5402)

% mortality after exposure				
conc., mg/L	*0.5 h*	*1 h*	*3 h*	
100	0%	0%	0%	
200	0%	0%	0%	
500	0%	0%	100%	(5402)

Salmo trutta	24h LC_{50},S	62 mg/L	(5402)
	5d LC_{50},S	18 mg/L	(5402)
	10d LC_{50}	14 mg/L	(5402)
Salvelinus namaycush, eggs	24h LC_{50},S	65 mg/L	(5402)
	5d LC_{50}	36 mg/L	(5402)
	10d LC_{50}	29 mg/L	(5402)
Oncorhynchus kisutch, eggs	24h LC_{50},S	43 mg/L	(5402)
	5d LC_{50},S	38 mg/L	(5402)
	10d LC_{50}	32 mg/L	(5402)
Oncorhynchus tschawytscha	24h LC_{50},S	64 mg/L	(5402)
	5d LC_{50},S	46 mg/L	(5402)
	10d LC_{50}	44 mg/L	(5402)
Pimephales promelas	24h LC_{50},F	35 mg/L	(5403)
	48h LC_{50},F	35 mg/L	(5403)
	72h LC_{50},F	35 mg/L	(5403)
Lepomis macrochirus	96h NOEC,S	10 mg/L	(5405)

2,3-benzofuran (coumarone)

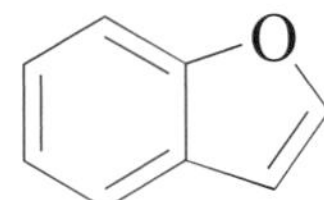

C_8H_6O

CAS 271-89-6

USES

manufacture of coumarone and indene resins.

A. PROPERTIES

molecular weight 118.13; melting point < -18°C; boiling point 173-175°C; density 1.1; LogP_{OW} 2.67.

B. AIR POLLUTION FACTORS

Manmade sources

in gasoline exhaust: <0.1-2.8 ppm. (195)

D. BIOLOGICAL EFFECTS

Fishes

Pimephales promelas:	4d LC_{50}	14 mg/L	(2625)

Benzoic acid

C_6H_5COOH

$C_7H_6O_2$

CAS 65-85-0

USES AND FORMULATIONS

food preservative; pharmaceutical and cosmetic preparations; mfg. of alkyl resins; intermediate in the synthesis of dyestuffs and pharmaceuticals; production of phenol and caprolactam; plasticizer mfg. (to modify resins-PVC, PV acetate, phenol-formaldehyde).

NATURAL SOURCES

cranberries, prunes, ripe cloves, bark of wild black cherry tree, scent glands of beavers, and oil of anise seeds.

A. PROPERTIES

white powder; molecular weight 122.1; melting point 121.7°C; boiling point 249°C; vapor density 4.21; density 1.27; solubility 2,900 mg/L; 2,700 mg/L at 18°C; log P_{OW} 1.87 at 20°C; vapour pressure 0.005 hPa at 20°C; H 0.004-0.02 Pa x m^3/mol (calculated)

B. AIR POLLUTION FACTORS

Manmade sources

in flue gases of a municipal waste incinerator plant: 101 μg/m_3 = 8.4% of total organic carbon emissions. (7048)

C. WATER AND SOIL POLLUTION FACTORS

Biodegradation

Biodegradation half-lives in nonadapted aerobic subsoil: loam and sand: 0.30 d. (2695)
Decomposition period by a soil microflora: 1 d. (176)
Adapted A.S. at 20°C-product is sole carbon source: 99% COD removal at 88 mg COD/g dry inoculum/h.

Bacterial degradation pathway

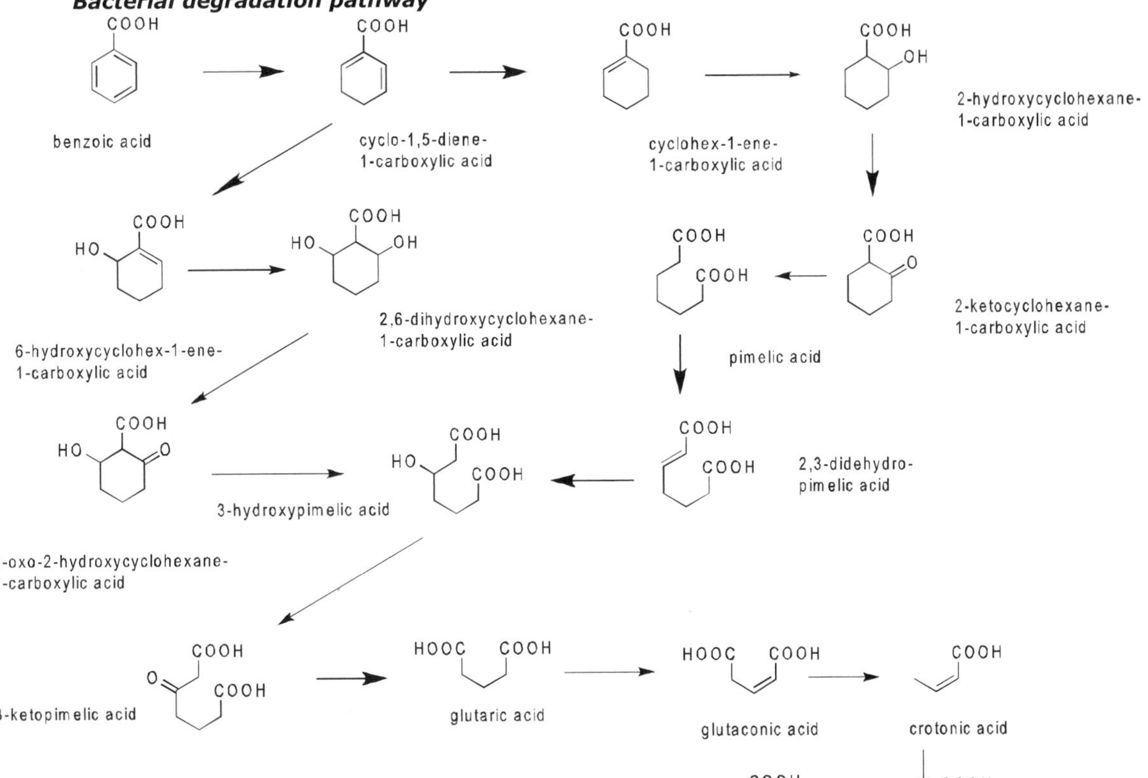

(7129)

Inoculum	method	concentration	duration	elimination
A.S.	Modified Zahn-Wellens	508 mg/L	2 days	>90% COD
Anaerobic sludge	At 35°C in the dark	73 mg/L	7 days	96-100%
Anaerobic sludge	At 35°C in the dark	50 µg/L DOC	60 days	>75% ThCH4
Industrial A.S.	Respirometric		5 days	86.9% ThCO2
Domestic A.S.	Respirometric	10 mg/L	5 days	74% ThOD
			20 days	78% ThOD
Domestic A.S.	Respirometric	100 mg/L	4 days	61-69% ThOD
Domestic A.S.	Respirometric	500 mg/L	1 days	42% ThOD
A.S. adapted		102 mg/L	5 days	99%
A.S. domestic	30°C	16 mg/L	1 days	100%

A.S. domestic	29°C	0.059 mg/L	7 days	99.5% C^{14}	
A.S. industrial	Semi-continuous	150 mg/L	1 days	86%	
A.S. domestic		1000 mg/L COD	2 days	97%	
Anaerobic sludge	35°C	300 mg/L	18 days	91%	
Anaerobic sludge	35°C	50 mg/L	21 days	110%	
Anaerobic sludge	35°C	50 mg/L	35 days	89%	
Anaerobic sludge adapted	37°C	24 mg/L	23 days	86-93%	
A.S.	Closed bottle test	0.8 mg/L	5 days	>71%	
Domestic A.S.	Respirometric test	700 mg/L	5 days	76% ThOD	
Soil loamy sand			35 days	50%	
Soil microorganims	Anaerobic	1 mg/kg soil	18 days	50% ThCO2	
Sea water		2 mg/L	5 days	75%	(10711)

(1219)

Lag period for degradation of 16 mg/L by waste water or by soil at pH: 7.3 and 30°C: less than 1 d. (1096)

Anaerobic mineralisation at 50 mg C/L at 35°C in the dark (benzoate):

after 56 d in digested sewage sludge	>75% of theoretical mineralization	
after 56 d in freshwater swamp	>75% of theoretical mineralization	
after 96 d in marine sediment	30-75% of theoretical mineralization	(7037)

Biodegradation of benzoate in anoxic habitats in the dark

% substrate disappearance after 3 months:

in sewage sludge at 37°C:	100%	
in pond sediment at 20°C:	100%	
in methanogenic aquifer at 20°C:	100%	
in sulfate reducing aquifer at 20°C:	100%	(2621)

BOD_5	47; 65; 69-73; 75% ThOD	(275, 281, 282, 220, 30, 36)
	70% ThOD std. dil. acclimated	(41)
BOD_{10}	73% ThOD std. dil. sew.	(256, 166)
BOD_{20}	75; 79% ThOD at 10 mg/L, unadapted sew.: no lag period	
COD	97; 100; 101% ThOD	(36, 41, 220)
ThOD	1.96	(36)

Manmade sources

Göteborg (Sweden) sew. works 1989-91; infl.: nd-40 µg/L; effl.: nd (including esters) (2787)
Landfill leachate from lined disposal area (Florida 1987-'90): mean conc. 1,170 µg/L (2788)
in primary domestic sewage plant effluent: 0.003 mg/L. (517)

Photooxidation

$t_{1/2}$ in top meter of Greifensee (Switzerland): >250 d (benzoate) (2640)

Waste water treatment

A.C.: adsorbability: 0.183 g/g C, 91% reduction at infl.: 1,000 mg/L (32)
A.C. (Pittsburgh Chemical Comp.-type BC): % adsorbed by 10 mg A.C. from 10^{-4} M aqueous solution:

at pH 3.0:	50%	
at pH 7.0:	11%	
at pH 11.0:	2.5%	(1313)

coagulation with 3 lb alum/1,000 gal: 8% BOD reduction (95)
ion exchange:

adsorption on Amberlite X AD-2:
23% retention effic., infl.: 10 ppm, effl.: 0.8 ppm;
at pH 3.2: 100% retention effic. at infl.: 1.0 ppm (40)

methods	*temp, °C*	*d observed*	*feed, mg/L*	*d acclim.*	*% theor. oxidation*	*% removed*
A.S., BOD	20	1-5	333	15		99
NFG, BOD	20	1-10	400-1,000	365 + P		46

select. strain	30	1/12	none	66	(93)

Impact on biodegradation processes

The mineralization of benzoate in methanogenic river sediment is affected by the following chemicals:

chemical	EC_{50}	
benzene	590 mg/kg dry wt	
pentachlorophenol	27 mg/kg dry wt	
1,2-dichloroethane	520 mg/kg dry wt	
chloroform	0.4 mg/kg dry wt	(2961)

Distribution coefficient

Observed Cuticle/water partition coefficient log Log K_{CW} for isolated cuticles from leaves of bitter orange (*Citrus aurantium L.*), rubber plant (*Ficus elastica*) fruits of tomato (*Lycopersicon esculentum* Mill.) and green pepper (*Capiscum annuum L.)* (7110)

Log K_{CW}: 1.58

Soil sorption

Kd at 50 µg/L : loamy sand : 1.92; sand: 0.62 (10711)

D. BIOLOGICAL EFFECTS

Algae

Chlorella fusca	BCF (wet wt): 3	(2659)

Toxicity threshold (cell multiplication inhibition test)

Bacteria			
Pseudomonas putida	16h EC_0	480 mg/L	(1900)
Algae			
Microcystis aeruginosa	8d EC_0	55 mg/L	(329)
green algae			
Scenedesmus quadricauda	7d EC_0	1,630 mg/L	(1900)
Protozoa			
Entosiphon sulcatum	72h EC_0	218 mg/L	(1900)
Uronema parduczi Chatton-Lwoff	EC_0	31 mg/L	(1901)

Arthropods

Daphnia magna: immobilization at 146 mg/L; prolonged exposure.			(226)
Daphnia magna	48h IC_{50}	1,540 mg/L	(7025)

Fishes

mosquito fish:	24h LC_{50}	240 mg/L	
	48h LC_{50}	255 mg/L	
	96h LC_{50}	180 mg/L	(41)
goldfish:	lethal in 7 to 96 h:	200 mg/L	(226)
orange spotted sunfish:	lethal in 1 h:	550 to 570 mg/L	(226)

Biomagnification

in model ecosystem at 0.01 – 0.1 mg/L radiolabelled after 48 hours (10711)

Species	BCF
Gambusia affinis (mosquito fish)	21
Daphnia magna	1,772
Oedogonium cardiacum(green algae)	102
Culex quinquifasciatus (midge, larvae)	138
Physa (snail)	2,786

Bioaccumulation

Species	conc. µg/L	duration	body parts	BCF
Leuciscus idus		3 days	Fresh weight	<10

Chlorella fusca	1 day	Fresh weight	<10	
Activated sludge	5 days	Dry weight	1,300	(10711)

Toxicity

Micro organisms			
Pseudomonas Stamm Berlin	1h EC_{10}	50 mg/L	
Pseudomonas fluorescens	24h EC_0	1,000 mg/L	
Activated sludge	3h EC_{50}	>1,000 mg/L	
Photobacterium phosphoreum	30min EC_{50}	16.8 mg/L	
Domestic sewage	24h EC_{50}	500 mg/L	(10711)
Algae			
Chlorella pyrenoidosa	3h EC_{50}	60 mg/L	
	14d EC_{50}	>10 mg/L	
Scenedesmus quadricauda	3h EC_{50}	75 mg/L	
	14d EC_{50}	>10 mg/L	
Anabaena variabilis	14d EC_{50}	>10 mg/L	
	3h EC_{50}	55 mg/L	
Anabaena cylindrical	3h EC_{50}	60 mg/L	
	14d EC_{50}	>10 mg/L	
Anabaena inequalis	14d EC_{50}	9 mg/L	
	3h EC_{50}	5 mg/L	
Selenastrum capricornutum	72h EC_{50}		(10711)
Crustaceae			
Daphnia magna	24h EC_0	260; 540 mg/L	
	24h EC_{50}	300; 500; 1,540 mg/L	
	24h EC_{100}	1,000 mg/L	
	48h NOEC	100 mg/L	
	48h EC_{50}	>100 mg/L	(10711)
Fish			
Salmo gairdneri	96h LC_{50}	47.3 mg/L	
	96h NOEC	10 mg/L	
Leuciscus idus	48h LC_0	400 mg/L	
	48h LC_{50}	460 mg/L	
	48h LC_{100}	600 mg/L	
Lepomis macrochirus	96h LC_{50}	44.6 mg/L	
	96h NOEC	10 mg/L	
	96h LC_0	180 mg/L	
Lepomis humilis	1h LC_{100}	550-570 mg/L	(10711)
Mammals			
Rat	oral LD_{50}	1,700; 2,530; 2,565; 3,040 mg/kg bw	
Mouse	oral LD_{50}	1,940; 2,250; 2,370 mg/kg bw	(10711)

Benzoic acid, potassium salt (potassium benzoate)

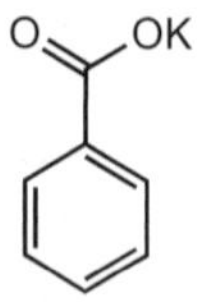

$C_7H_6O_2K$

CAS 582-25-2

EINECS 209-481-3

A. PROPERTIES

molecular weight melting point 330.6°C; boling point 465°C; vapour pressure 4.9 x 10^{-9} hPa at 25°C; water solubility 556,000 mg/L at 20°C; log Pow −2.3

C. WATER AND SOIL POLLUTION FACTORS

Biodegradation

Inoculum	method	concentration	duration	elimination	
Domestic A.S.	OECD 301B	50 mg/L	7 days	90% ThCO2	(10711)

D. BIOLOGICAL EFFECTS

Bioaccumulation

BCF:3.16 (calculated)

Toxicity

Mammals

Rat	oral LD_{50}	>10,000 mg/kg bw	
Guinea pig	oral LD_{50}	>10,000 mg/kg bw	(10711)

Benzoic acid, sodium salt (sodium benzoate)

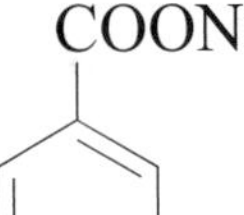

$C_7H_6O_2$.Na

CAS 532-32-1

EINECS 208-534-8

A. PROPERTIES

Melting point 410-430 °C; boiling point 465 °C; relative density 1.44 kg/L; bulk density 0.35-0.65 kg/L; vapour pressure 4.9 x 10-9 hPa at 25°C; water solubility 630,000 mg/L at pH 7 and 20°C; log Pow −2.2

C. WATER AND SOIL POLLUTION FACTORS

Biodegradation

Inoculum	Method	Concentration	Duration	Elimination results

A.S. nonadapted	Modified MITI test	100 mg/L	10 d	84 (64-98) % ThOD	
			28 d	75-111% ThOD	(8436)
A.S. domestic	Modified Sturm test	50 mg/L	7 d	90% $ThCO_2$	(8437)
soil suspension	Closed bottle test	5 mg/L	30 d	75-111% ThOD	
+ effluent sewage plant					(8438)
aerobic seawater	Modified OECD screening test	34 mg/L	12 d	95% DOC	
			28 d	100% DOC	(8484)
aerobic seawater shake	flask test	20 mg/L	5 d	57% DOC	
	lag time 4 d		20 d	82% DOC	
			61 d	96% DOC	
		40 mg/L	5 d	31% DOC	
	lag time 3 d		20 d	73% DOC	
			61 d	98% DOC	(8445)

Anaerobic sewage domestic anaerobic degrad. 50-90 mg/L at 35°C 47 d 61% (gas production)
49 d 83% (gas production)
28 d 74% (gas production) (8440)

Anaerobic laboratory sewage adapted	300 mg/L at 35°C	4 d	98% (gas production)	(8439)
Anaerobic sewage	50 mg/L at 35°C	61 d	50% (gas production)	
	60 mg/L at 35°C	35 d	95% (gas production)	
	60 mg/L at 35°C	56 d	97% (gas production)	(8440)
anaerobic sewage	85 mg/L at 35°C	7 d	93% (gas production)	(8441)
methanogenic sewage	3,000 mg/L at 35°C and pH 6.8	5 d	99% (gas production)	(8442)

anaerobic sewage, industrial, benzoate adapted	307 mg/L	2 d	99% (gas production)	(8442)
anaerobic enreichment culture adapted	2,306 mg/L at 39°C, pH 6.7	4 d	100% (gas production)	(8444)
Sediment of eutrophic lake, anaerob. Degrad.	724 mg/L	20 d	100% (gas production)	(8446)

Anaerobic degradation in laboratory aquifer column

sediment/slate soil debris	28 mg/L	retention time: 10 h	>95%	(8447)

Biodegradation

Inoculum	method	concentration	duration	elimination	
Anaerobic A.S.		85 mg/L	7 days	93% ThCH4	
Marine bacteria	Modified Sturm	10 mg/L DOC	2 days	20% ThCO2	
			6 days	55% ThCO2	
			8 days	70% ThCO2	
			20 days	85% ThCO2	
Anaerobic sewage	At 35°C	50-90 mg/L	47 days	60.5 % ThCH4	
			49 days	83% ThCH4	
			28 days	74% ThCH4	
Domestic sewage	Modified MITI test		14 days	84%	
			28 days	92%	
Seawater	Ready DOC Die away	11.6 mg/L DOC	5 days	75%	
			10 days	73%	
			30 days	83%	
			50 days	92%	
			61 days	96%	
A.S.	OECD 301D		28 days	100%	
A.S.	Modified MITI test	100 mg/L	10 days	64-98%	
			28 days	75-111%	
A.S.	OECD 301A		28 days	88%	
			60 days	95%	(10711)

D. BIOLOGICAL EFFECTS

Bacteria			
Achromobacter liquefaciens	24 h EC_{50}	>3,000 mg/L	
	7 d EC_0	>3,000 mg/L	(8451)
Micrococcus flavus	24 h EC_{50}	>500 mg/L	
	7 d EC_0	>3,000 mg/L	(8451)
Micrococcus luteus	24 h EC_{50}	500 mg/L	
	7 d EC_0	500 mg/L	(8451)
Sarcina flava	24 h EC_{50}	<100 mg/L	
	7 d EC_0	>3,000 mg/L	(8451)
Sarcina lutea	24 h EC_{50}	<100 mg/L	
	7d EC_0	1,000 mg/L	(8451)

Fungi and Yeats

Talaromyces flavus	35 d MIC*	100 mg/L at pH 3.5				
		>600 mg/L at pH 5.4				(8453)
Byssochlamys fulva	16 d MIC*	100 mg/L at pH 3.5				(8454)
	PH		2.6	5.0	7.0	
Saccharomyces cerevisiae	MIC*	mg/L	200	2,000	30,000	
Willia anomala	MIC*	mg/L	120	1,000	20,000	
Penicillium glaucum	MIC*	mg/L	600	4,000	60,000	(8455)
Saccharmyces ellipsoides	MIC*	500 mg/L at pH 3.5				
5,000 mg/L at pH 5.0						
>25,000 mg/L at pH 6.5						(8456)

*MIC Minimum inhibitory concentration: Determination of MIC is done by visually judging growth (turbidity of culture tubes). MIC is reached when no colonies occur.

Worms and Snails			
Asellus intermedius	96 h LC_{50}	>100 mg/L	(8449)
Lumbricus variegatus	96 h LC_{50}	>100 mg/L	(8449)
Dugesia tigrina	96 h LC_{50}	>100 mg/L	(8449)
Helisoma trivolvis	96 h LC_{50}	>100 mg/L	(8449)

Crustaceans			
Daphnia magna Straus	48 h EC_{50}	<650 mg/L	(8450)
	96 h EC_{50}	>100 mg/L	(8449)
Gammarus fasciatus	96 h EC_{50}	>100 mg/L	(8449)
Asellus intermedius	96 h EC_{50}	>100 mg/L	(10711)

Fishes			
Pimephales promelas	96 h EC_{50}	484 mg/L	(8448)
	96h LC_{50}	>100 mg/L	(8449)

Mammals			
Rat	oral LD_{50}	2,100; 3,140; 3,450; 4,070 mg/kg bw	(10711)

benzoic trichloride *see* benzotrichloride

benzonitrile (benzenecarbonitrile; phenylcyanide; phenylnitrile)

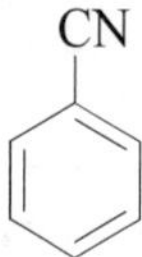

C_6H_5CN

C_7H_5N

CAS 100-47-0

A. PROPERTIES

molecular weight 103.12; mp: -13°C; boiling point 190.7°C; density 1.01 at 15/15°C; solubility 10,000 mg/L at 100°C; $LogP_{OW}$ 1.56.

C. WATER AND SOIL POLLUTION FACTORS

Waste water treatment

	temp, °C	*d observed*	*feed, mg/L*	*d acclim.*	*% ThOD*	*% removed*
RW, CO_2, CAN	20	4	10	28	60	100
RW, CO_2, CAN	20	7	51	30+	60	100
RW, CO_2, CAN	5	18	51	30+	60	100
ASC, CAN	22-25	28	134-179	28	75+	99+

(93)

D. BIOLOGICAL EFFECTS

Toxicity threshold (cell multiplication inhibition test)

bacteria *(Pseudomonas putida)*	16h EC_0	11 mg/L	(1900)
algae *(Microcystis aeruginosa)*	8d EC_0	3.4 mg/L	(329)
green algae *(Scenedesmus quadricauda)*	7d EC_0	75 mg/L	(1900)
protozoa *(Entosiphon sulcatum)*	72h EC_0	30 mg/L	(1900)
protozoa (*Uronema parduczi* Chatton-Lwoff)	EC_0	119 mg/L	(1901)

Fishes

fathead minnow	hard water	96h LC_{50}	78 mg/L	(41)
fathead minnow	soft water	96h LC_{50}	135 mg/L	(41)
bluegill	soft water	96h LC_{50}	78 mg/L	(252)
guppies	soft water	96h LC_{50}	400 mg/L	(41)
adult bluegills	no organic influence at		35 mg/L	(226)

1,2-benzophenanthrene *see* chrysene

benzophenone (diphenylmethanone; diphenyl-ketone-; benzoylbenzene-; diphenylmethanone)

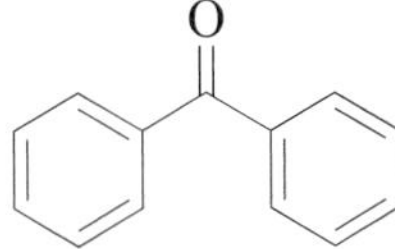

$(C_6H_5)_2CO$

$C_{13}H_{10}O$

CAS 119-61-9

USES

Used in the manufacture of antihistamines, hypnotics and insecticides. Used as a fixative for heavy perfumes, especially when used in soaps.

OCCURRENCE

: In Baltic Sea shale tar.

A. PROPERTIES

molecular weight 182.22; melting point 49-51°C; boiling point 305°C; density 1.11 at 18°C

C. WATER AND SOIL POLLUTION FACTORS

Soil adsorption

K_{OC}:	for Fullerton soil,	0.06% OC:	530	
	for Apison soil,	0.11% OC:	580	
	for Dormont soil,	1.2% OC:	440	(2599)

D. BIOLOGICAL EFFECTS

Bacteria			
Photobacterium phosphoreum Microtox test	30 min EC_{50}	8.9 mg/L	(9657)
Fishes			
fathead minnow	96h LC_{50}	15.3 mg/L	(9851)
Mammals			
mouse	oral LD_{50}	2,900 mg/kg bw	(9852)

benzophenone-o-carboxylic acid *see* *o*-benzoylbenzoic acid

1,2-benzopyrone *see* coumarin

benzoquinhydrone *see* quinhydrone

p-benzoquinone (2,5-cyclohexadien-1,4-dione; 1,4-benzoquinone; quinone)

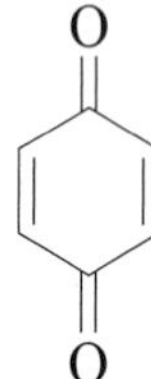

$C_6H_4O_2$

CAS 106-51-4

USES

manufacture of dyes and hydroquinone.

A. PROPERTIES

yellow crystals; molecular weight 108.09; melting point 115/124°C decomposes; boiling point sublimes; vapor pressure 0.09 mm at 20°C; density 1.31 at 20°C; $LogP_{OW}$ 0.20.

B. AIR POLLUTION FACTORS

1 mg/m³ = 0.226 ppm; 1 ppm = 4.49 mg/m³.

Odor

characteristic:	pungent	
absolute perception limit:	0.1 ppm	
100% recogn.:	0.15 ppm	(54)
O.I. at 20°C:	790	(316)

Incinerability

thermal stability ranking of hazardous organic compounds: rank 89 on a scale of 1 (highest stability) to 320 (lowest stability). (2390)

C. WATER AND SOIL POLLUTION FACTORS

Reduction of amenities

taste: average: 0.71 mg/L; range: 0.016 to 4.3 mg/L (30)
tainting of fish flesh: 0.5 mg/L (81)

Impact on biodegradation processes

at 0.2 mg/L: inhibition of degradation of glucose by *Pseudomonas fluorescens*
at 55 mg/L: inhibition of degradation of glucose by *E. coli* (293)

D. BIOLOGICAL EFFECTS

Bacteria

E. coli: toxic: 55 mg/L (30)

Algae

blue algae: toxic: <1 mg/L
Scenedesmus: toxic: 6 mg/L (30)

Inhibition of photosynthesis of a fresh water nonaxenic uni-algal culture of Selenastrum capricornutum:

at 0.1 mg/L:	37%	^{14}C fixation (vs controls)	
at 1 mg/L:	17%	^{14}C fixation (vs controls)	
at 10 mg/L:	7-13%	^{14}C fixation (vs controls)	
at 100 mg/L:	1%	^{14}C fixation (vs controls)	
at 1,000 mg/L:	2%	^{14}C fixation (vs controls)	(1690)

Arthropods

Daphnia: toxic: 0.4 mg/L

Fishes

fathead minnows *(Pimephales promelas):* probable toxic conc.: <0.1 mg/L (after 120 h)

benzothiazol-2-yl sulphide, sodium

(2(3H)-Benzothiazolethione, sodium salt; 2-mercaptobenzothiazole, sodium salt; NaMBT)

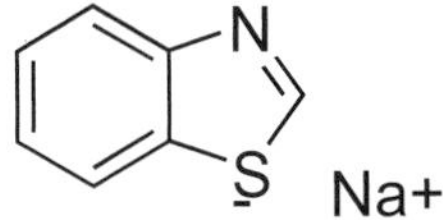

$C_7H_5NS_2Na$

CAS 2492-26-4

SMILES [Na]c12nc(S)sc1cccc2

EINECS 219-660-8

TRADENAMES

Nacap®;

USES

Rubber and Plastic Additive.Vulcanizing agent, corrosion initiators. The substance is only supplied as aqueous solutions (18-50%)

A. PROPERTIES

Yellow to brown oily liquid; molecular weight 189.23; melting point -6°C*; boiling point 103°C*; density 1.25 kg/L* at 25°C; vapour pressure 34 hPa at 30°C*; water solubility > 500,000 mg/L at 25°C log Pow -0.46 at 25°C ; H 0.0064 atm.m^3/mole
* 47-50% aqueous solution NaMBT

B. AIR POLLUTION FACTORS

Photodegradation

t/2 (OH radicals) : 2.8 hours.

C. WATER AND SOIL POLLUTION FACTORS

Hydrolysis

Below pH 7, NaMBT will be protonated to form insoluble MBT. If iron is present, NaMBT will be reduced to benzothiazole. In aqueous solution, NaMBT is not oxidized, even at temperatures of 100°C, nor is it readily hydrolyzed. In weak alkaline or neutral solutions, the mercaptobenzothiazole (MBT) anion can readily complex with various metal ions and form insoluble, relatively undissociable salts. (11062)

Biodegradation

Inoculum	method	concentration	duration	elimination	
Activated sludge	EPA OTS 796.3100	23.8 mg/L	28 days	<1%	(11062)

D. BIOLOGICAL EFFECTS

Bioaccumulation

BCF : 1 (calculated)

Toxicity

Micro organisms			
Activated sludge	3h EC_{50}	857 mg/L	(11062)
Algae			
Selenastrum capricornutum	96h EC_{50}	0.3; 0.4 mg/L	
	24h EC_{50}	2 mg/L	(11062)
Crustaceae			
Daphnia magna	48h EC_{50}	19 mg/L	
	48h NOEC	10 mg/L	(11062)
Fish			
Salmo gairdneri	96h LC_{50}	1.8 mg/L	
	96h NOEC	1.4 mg/L	
	96h LOEC	1.8 mg/L	
Leuciscus idus	48h LC_{50}	>5 mg/L	
Oncorhynchus mykiss	96h LC_{50}	1.8; 2.6-3.2 mg/L	
	96h NOEC	1.99 mg/L	
	96h LOEC	2.6 mg/L	
	4h LC_{100}	10 mg/L	
Lepomis macrochirus	96h NOEC	2.1 mg/L	
	96h LC_{50}	3.8 ; 12-15 mg/L	
	96h LOEC	2.8 mg/L	
Ptychocheilus oregonensis	11h LC_{100}	10 mg/L	(11062)
Mammals			
Rat	oral LD_{50}	750; 1,476; 2,160; 3,120; 3,968; 4,350; 5,200; 9,500	(11062)

Benzothiazole (benzosulfonazole-; 1-thia-3-azaindene)

C_7H_5NS

CAS 95-16-9

USES

In organic synthesis.

OCCURRENCE

In flowers and fruit. Provides fungal resistance to cedar, red pine and beech trees.

A. PROPERTIES

molecular weight 135.18; boiling point 231°C; melting point >112°C; density 1.24.

C. WATER AND SOIL POLLUTION FACTORS

Biodegradation

Benzothiazole is oxidised by activated sludge. The sulfur and nitrogen moieties are converted into sulfate and ammonium ions . (9853, 9657)

D. BIOLOGICAL EFFECTS

Bacteria			
act. sludge respiration inhibition	EC_{50}	650 mg/L	(2624)
Photobacterium phosphoreum Microtox test	5-30 min EC_{50}	1.8 mg/L	(9657)
Mammals			
rat	oral LD_{50}	479 mg/kg bw	(9855)
mouse	oral LD_{50}	900 mg/kg bw	(9856)

N-(2-benzothiazolyl)-N'-methylurea (benzthiazuron; gatnon; bay 60618)

$C_9H_9N_3OS$

CAS 1929-88-0

A. PROPERTIES

white powder that decomposes at 287°C; molecular weight 207.3; solubility 12 ppm at 20°C; vapor pressure 1 × 10^{-5} mm at 90°C.

D. BIOLOGICAL EFFECTS

Fishes

harlequin fish *(Rasbora hetermorpha):*

mg/L	*24 h*	*48 h*	*96 h*	*3 m*	
LC_{10},F	850	700	200		
LC_{50},F	1,300	920	400	100	(331)

4-(2-benzothiazolylthio)morpholine (2-morpholinothiobenzothiazol)

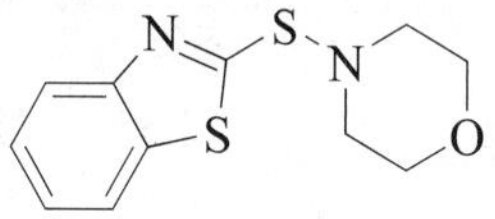

$C_{11}H_{12}N_2OS_2$

CAS 102-77-2

USES

Accelerator of vulcanization of rubber.

A. PROPERTIES

molecular weight 252.4; melting point >78°C; boiling point decomposes during distillation; vapor pressure 2.3 10-6 hPa at 20°C, 4.5 10-6 hPa at 25°C; density 1.35 at 20°C; solubility 60 mg/L at 25°C.

C. WATER AND SOIL POLLUTION FACTORS

Biodegradability

respirometer test	at 100 mg/L after 28 d	0% ThOD removal	(6109)

D. BIOLOGICAL EFFECTS

Bacteria

test for inhibition of oxygen consumption by A.S.	3 h EC_{50}	>10,000 mg/L	(6109)

Fishes

Brachydanio rerio	96 h LC_0	1; 8 mg/L	
	96 h LC_{100}	5; 30 mg/L	(6109)
Ptychocheilus (northern squawfish)	24 h LC_0	>10 mg/L	(6110)
Oncorhynchus tschawytscha	24 h LC_0	>10 mg/L	(6110)

benzotrichloride *see* trichloromethylbenzene

benzoximate (3-chloro-α-ethoxyimino-2,6-dimethoxybenzyl benzoate; Benzomate-; Citrazon-; Azomate-; Artaban-; Acarmate-; ethylO-benzol-3-chloro-2,6-dimethoxybenzohydroximate)

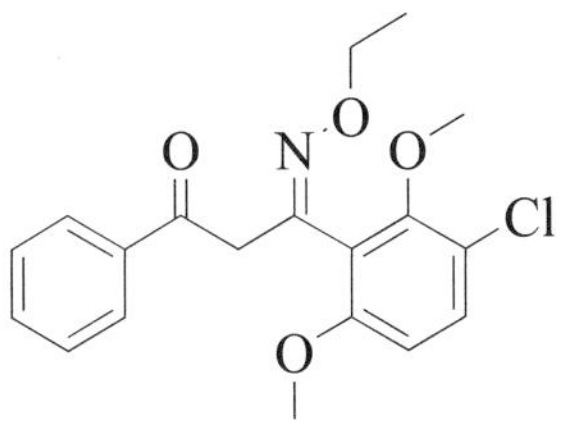

$C_{18}H_{18}ClNO_5$

CAS 29104-30-1

USES

nonsystemic acaricide.

A. PROPERTIES

colorless crystals; molecular weight 363.8; melting point 73°C; vapor pressure 3.4 x 10-6 mm Hg; solubility 30 mg/L at 25°C; $LogP_{OW}$ 2.4

D. BIOLOGICAL EFFECTS

Fishes			
cytotoxicity to goldfish GF-Scale cells:	NR_{50}	21 mg/L	(2680)
Cyprinus carpio	48h LC_{50}	12 mg/L	(2680)
carp	48h LC_{50}	1.75 mg/L	(9600)
Mammals			
rat	oral LD_{50}	>10,000 mg/kg bw	(9600)

Benzoyl peroxide

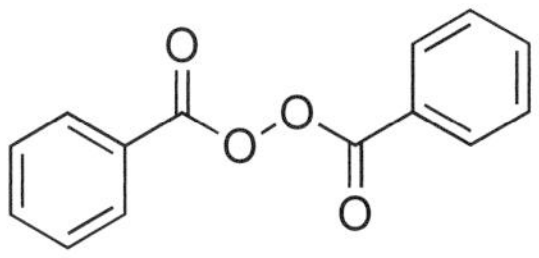

$C_{14}H_{10}O_4$

CAS 94-36-0

EINECS 202-327-6; Smiles O=C(OOC(=0)c(cccc1)c(cccc2)c2

TRADENAMES

Lucidol

USES

initiator of polymerisation; bactericidal agent, 2.5-10% in consumer products; a powerful oxidizer (10562)

A. PROPERTIES

molecular weight 242.24; melting point 104-106°C; boiling point : decomposes explosively above 105°C; density 1.33 kg/L at 25°C; vapour pressure 0.0093 Pa at 25°C; water solubility 9.1 mg/L at 25°C; log Pow 3.43 at 25°C; K_{OC} 1,800; H 0.36 Pa m^3/mol (estimated)

B. AIR POLLUTION FACTORS

Photodegradation

t/2 = 3 days (10562)

C. WATER AND SOIL POLLUTION FACTORS

Hydrolysis

t/2 at pH 4=11.9 hours, at pH 7=5.2 hours. From the chemical structure benzoyl acid is expected to hydrolyse to benzoic acid. (10562)

Biodegradation

Inoculum	method	concentration	duration	elimination	
Activated sludge	Modified MITI	100 mg/L	7 days	47% ThOD	
			14 days	83% ThOD	
			21 days	83% ThOD	
			21 days	100% HPLC analysis	(10562)
Adapted secondary A.S.	Closed Bottle		28 days	56% ThOD	(10562)

D. BIOLOGICAL EFFECTS

Bioaccumulation

BCF 92-250 (calculated) (10562)

Toxicity

Micro organisms			
Secondary activated sludge	30min EC_{50}	35 mg/L	(10562)
Algae			
Selenastrum capricornutum	72h ECb_{50}	0.07 mg/L	
	72h $ECgr_{50}$	0.44 mg/L	(10562)
Pseudokirchneriella subcapitata	72h ECb_{50}	0.44 mg/L	
	72h $ECgr_{50}$	0.83 mg/L	(10562)
Worms			
Eisenia foetida	14d LC_{50}	>1,000 mg/kg soil	(10562)
Crustaceae			
Daphnia magna	48h EC_{50}	0.07; 2.9 mg/L	(10562)
Fish			
Oryzias latipes	96h LC_{50}	0.24; 3.9 mg/L	
Poecilia reticulata	96h LC_{50}	2.0 mg/L	(10562)
Mammals			
Rat albino	oral LD_{50}	>5,000 mg/kg bw	
Mouse	oral LD_{50}	>2,000; 5,700; 7,710 mg/kg	(10562)

bw

benzoylaminoacetic acid *see* hippuric acid

o-benzoylbenzoic acid (benzophenone-*o*-carboxylic acid)

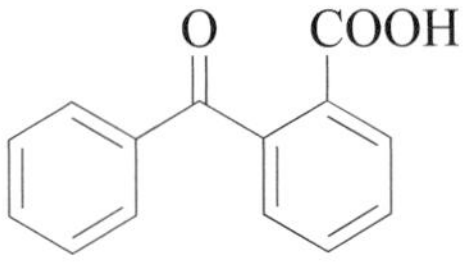

$C_6H_5COC_6H_4COOH$

$C_{14}H_{10}O_3$

CAS 85-82-9

A. PROPERTIES

molecular weight 226.22; melting point 93°C + H_2O, 127°C anh.; boiling point 257-265°C.

C. WATER AND SOIL POLLUTION FACTORS

BOD_5: 0.001 std. dil. sew. (n.s.i.) (258, 165)

benzoylchloride (benzenecarbonylchloride)

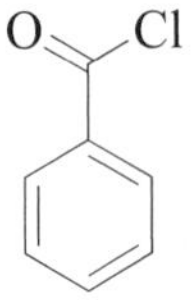

C_6H_5COCl

C_7H_5ClO

CAS 98-88-4

A. PROPERTIES

molecular weight 140.57; melting point -1°C; boiling point 197°C; vapor pressure 0.4 mm at 20°C, 1 mm at 32°C, 0.5 hPa at 20°C, 1.1 hPa at 30°C, 3.7 hPa at 50°C; vapor density 4.88; density 1.22 at 15/15°C; solubility decomposes; saturation concentration in air 3.1 g/m³ at 20°C, 5.2 g/m³ at 30°C; $LogP_{OW}$ decomposes.

B. AIR POLLUTION FACTORS

1 mg/m³ = 0.17 ppm, 1 ppm = 5.84 mg/m³.

C. WATER AND SOIL POLLUTION FACTORS

Biodegradability

inoculum: municipal waste water: closed bottle test: at 2.4 mg/L after 20 d: 95% ThOD removal. (5609)

Impact on biodegradation processes

activated sludge respiration inhibition test: 3h EC_{50}: >100 mg/L. (5609)

D. BIOLOGICAL EFFECTS

Bacteria			
Photobacterium phosphoreum	30 min EC_{50}	12 mg/L	(5610)
Crustaceans			
grass shrimp *(Palaemonetes pugio)*	96h LC_{50}	180 mg/L	(1904)
Fishes			
fathead minnows *(Pimephales promelas)*	24h LC_{50}	43 mg/L	
	48h LC_{50}	35 mg/L	
	96h LC_{50}	35 mg/L	(1904)
Leuciscus idus	72h LC_0	200 mg/L	
	72h LC_{100}	500 mg/L	(5609)
Brachydanio rerio	96h LC_0	7.5 mg/L	(5609)

benzoylglycin *see* hippuric acid

benzoylglycocoll *see* hippuric acid

benzoylprop-ethyl (ethyl-N-benzoyl-N-(3,4-dichlorophenyl)-DL-alaninate; ethyl-N-benzoyl-N-(3,4-dichlorophenyl)-2-aminopropionate; L-alanine, N-benzoyl-N-(3,4-dichlorophenyl)ethyl ester; N-benzoyl-N-(3,4-dichlorophenyl)-L-alanine ethyl ester; propionic acid, 2-(N-benzoyl-N-(3,4-dichlorophenyl)amino ethyl ester; ethyl N-benzoyl-N-(3,4-dichlorophenyl)-2-aminopropionate; Enaven-; Suffix)

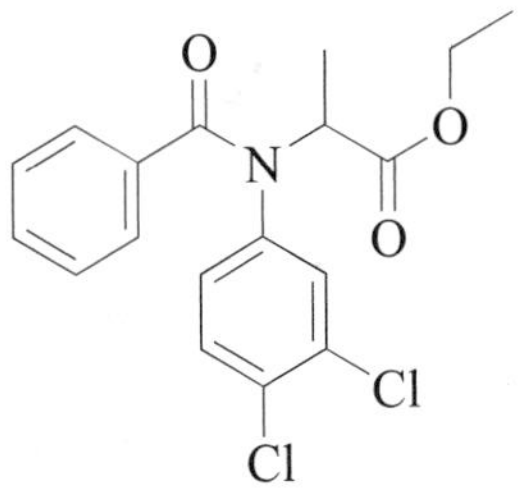

$C_{18}H_{17}Cl_2NO_3$

CAS 22212-55-1

USES

selective postemergence herbicide.

A. PROPERTIES

molecular weight 366.25; melting point 55-72°C; vapor pressure 4.7 × 10^{-8} mbar at 20°C; solubility 20 mg/L at 20°C.

C. WATER AND SOIL POLLUTION FACTORS

Stability

Photochemically and hydrolytically stable range pH 3-6. (9662)

Biodegradation

In plants, metabolism involves hydrolysis of the ester group to give the benzoyl moiety which is ultimately converted into a biologically inactive conjugate. (9861)

D. BIOLOGICAL EFFECTS

Fishes			
rainbow trout	96h LC_{50}	2.2 mg/L	(2963)
harlequin fish	96h LC_{50}	5 mg/L	(9662)

Mammals			
Biotransfer factor in beef:	log B_b	-4.81	
Biotransfer factor in milk:	log B_m	-4.72	(2644)
rat, guinea pig, rabbit	oral LD_{50}	1,000-1,550 mg/kg bw	(9859, 9775)
mouse	oral LD_{50}	716 mg/kg bw	(9860)

4-benzoylpyridine (phenyl-4-pyridyl ketone)

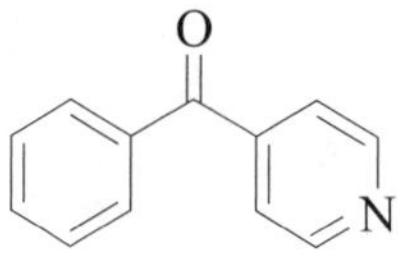

$C_6H_5CO(C_5H_4N)$

$C_{12}H_9NO$

CAS 14548-46-0

A. PROPERTIES

molecular weight 183.21; melting point 69-71°C; boiling point 215°C; solubility 1,120 mg/L; $LogP_{OW}$ 1.98; 2.07.

D. BIOLOGICAL EFFECTS

Algae			
Tetrahymena pyriformis	48h EC_{50}, growth	220 mg/L	(2704)
Fishes			
Pimephales promelas	30-35 d log LC_{50},F	103 mg/L	(2704)
	24h LC_{50}	105 mg/L	
	96h LC_{50}	103 mg/L	(2709)

benzthiazuron *see* N-(2-benzothiazolyl)-N'-methylurea

(5-benzyl-3-furyl)methyl-cis,trans-(+)-2,2-dimethyl-3-(2,2-dimethyl-3-(2-methyl-propenyl)cyclopropane-carboxylate *see* SBP-1382

(5-benzyl-3-furyl)methyl-α-trans-(+)-3-(cyclopentylidene-methyl)-2,2-dimethylcyclo-propanecarboxylate *see* RU-11679

Benzyl-3-isobutyryloxy-1-isopropyl-2,2-dimethylpropylphthalate

$C_{27}H_{34}O_6$

CAS 16883-83-3

EINECS 240-920-1

TRADENAMES

Santicizer; Texanol benzyl phthalate

A. PROPERTIES

molecular weight 455; boiling point 243 °C at 13 hPa; density 1.09 kg/L at 25°C; vapour pressure 0.66 hPa at 200°C, 19.8 hPa at 250°C; water solubility 0.81 mg/L at 22 °C; log Pow > 6 at 20°C (calculated)

C. WATER AND SOIL POLLUTION FACTORS

Biodegradation

Inoculum	method	concentration	duration	elimination
Adapted AS	Modified Sturm test	19 mg/L	35 days	94 % ThCO2
Activated Sludge	SCAS Test	3 mg/L	24 hours	44 – 68 % (7387)

D. BIOLOGICAL EFFECTS

Toxicity

Algae			
Selenastrum capricornutum	96h EC_{50}	> 5 mg/L	(7387)
Crustaceae			
Daphnia magna	48h EC_{50}	> 5 mg/L	
	48h NOEC	3.2 mg/L	(7387)
Fish			
Oncorhynchus mykiss	96h LC_{50}	> 5 mg/L	
Pimephales promelas	96h LC_{50}	> 5 mg/L	(7387)
Mammals			
Rat	oral LD_{50}	> 15,800 mg/kg bw	(7387)
Rabbit	oral LD_{50}	> 10,000 mg.kg bw	(7387)

Benzyl salicylate (2-hydroxybenzoic acid, benzyl ester; benzoic acid, 2-hydroxy-, phenylmethyl ester; benzyl-2-hydroxybenzoate)

$C_{14}H_{12}O_3$

CAS 118-58-1

EINECS 204-262-9

USES

a fragrance in perfumed consumer products such as dishwash, car shampoo, hand cleaner, toilet paper, nappies in concentrations of 0.007 to 0.059 wt%. (7195)

A. PROPERTIES

molecular weight 228.3; melting point 18-20°C; boiling point 300°C; vapour pressure 0.00045 Pa; specific gravity 1.17 kg/L; water solubility 24.6 mg/L; log Pow 4.3

C. WATER AND SOIL POLLUTION FACTORS

Environmental concentrations

Concentrations and removal in WWTP near Cincinnati (USA) october 1997

	Influent ng./L	Effluent ng/L	% removal
Activated sludge WWTP	8960	117	98.7
Trickling filter WWTP	4920	186	97.5

D. BIOLOGICAL EFFECTS

Toxicity

Mammals

Rat	oral LD_{50}	2,227; 3,000 mg/kg bw

Benzylacetic acid (benzylacetate)

$CH_3COOCH_2C_6H_5$

$C_9H_{10}O_2$

CAS 140-11-4

USES

In perfumery. In the manufacture of lacquers, polishes, printing ink and varnish removers. Solvent for cellulose acetate and cellulose nitrate.	(9801)
Fragrance	

OCCURRENCE

In jasmine, gardenia and other essential oils

A. PROPERTIES

vapour pressure 21.9 Pa;vapour pressure 1.0 x 10^{-6} mm Hg at 45°C; water solubility 1265 mg/L; log Pow 2.1;molecular weight 150.18; melting point -51°C; boiling point 206°C; density 1.04; vapour density 5.2; solubility < 1,000 mg/L at 25°C

C. WATER AND SOIL POLLUTION FACTORS

Ultimate biodegradability

(CO_2 production) under aerobic conditions in the "Sealed Vessel Test" (a modified Sturm Test) at room temp. using nonadapted secondary effluent as inoculum: 100% biodegradation after 28 d (acetate). (2596)

Environmental concentrations

WWTP, september 1997 (USA) (10844)

	Influent ng/L	Effluent ng/L	% removal
Activated sludge WWTP	1,810	93	95
Trickling filter WWTP	1,880	361	81

D. BIOLOGICAL EFFECTS

Bacteria			
Photobacterium phosphoreum Microtox test	30 min EC_{50}	4.5 mg/L	(9657)
Mammals			
mouse	oral LD_{50}	830 mg/kg bw	(9862)
rat, rabbit, guinea pig	oral LD_{50}	2,200-2,490 mg/kg bw	(9863, 9864)

benzyl acetate → benzyl alcohol → benzaldehyde → benzoate : R= Na, K or H

80% → hippuric acid (urine)

20% → benzoyl glucuronide (urine)

Metabolism in mammals	(10711)

Benzylacetate is very rapidly hydrolyzed by esterases in several species including man to benzyl alcohol and acetic acid. The benzyl alcohol is then very rapidly metabolized to benzoate and subsequent conjugation to hippuric acid en benzoyl glucuronide in the urine. No unchanged benzyl acetate was found (see figure).	(10711)

benzylalcohol (phenylcarbinol; α-hydroxytoluene; phenylmethanol; benzoylmethanol; phenylmethylalcohol)

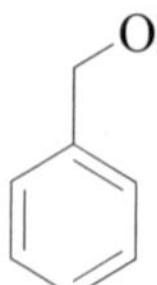

$C_6H_5CH_2OH$

C_7H_8O

CAS 100-51-6

USES

perfumes and flavors; solvent; intermediate for chemical synthesis; inks; surfactant.

A. PROPERTIES

colorless liquid; molecular weight 108.13; melting point -15°C; boiling point 206°C; vapor pressure 1 mm at 58°C, 0.03 hPa at 20°C, 0.09 hPa at 30°C, 0.67 hPa at 50°C; vapor density 3.72; density 1.05 at 15/15°C; solubility 40,000 mg/L at 17°C, 35,000 mg/L at 20°C, 44,000 mg/L at 50°C; $LogP_{OW}$ 1.1 (measured).

B. AIR POLLUTION FACTORS

1 mg/m^3 = 0.226 ppm, 1 ppm = 4.42 mg/m^3.

Manmade sources

in gasoline exhaust: <0.1 to 0.7 ppm.	(1053, 195)

C. WATER AND SOIL POLLUTION FACTORS

BOD_5:	33; 61; 61; 62; 63; 89% ThOD	(1828, 30, 220, 274, 282, 30, 27, 5520)
BOD_{20}:	77% ThOD	(30)
COD:	96% of ThOD	(220)
	95% of ThOD (0.05 n $K_2Cr_2O_7$)	(274)
	Reflux COD = 98.4% recovery	
	Rapid COD = 94.8% recovery	(1828)
TOC:	100% of ThOD	(220)
ThOD:	2.52	(30)

Impact on biodegradation processes

inhibition of degradation of glucose by *Pseudomonas fluorescens* at		350 mg/L	
inhibition of degradation of glucose by *E. coli* at:		>1,000 mg/L	(293)
inhibition of NH_3 oxidation by *Nitrosomonas:*	24h IC_{50}	390 mg/L	(5516)
A.S.	48h IC_{50}	2,100 mg/L	(5516)

(The degradation activity was not affected in a pilot unit (Ascomat) at 100 mg/L.) (5517)

Waste water treatment

reverse osmosis 0.6% benzylalcohol in product water, negative rejection. (221)

Biodegradability

A.S., W, 20°C, 1/2 d observed, feed: 250 mg/L, 25 d acclimation: 29% of ThOD (93)

Activated sludge acclimated to the following aromatics (influent: 250 mg/L benzylalcohol, aeration: 30 min):

benzylalcohol: 29% of ThOD
mandelic acid: 31% of ThOD (92)

inoculum/methods	*test conc.*	*test duration*	*removed*	
A.S., municipal, closed bottle test		30 d	<90% ThOD	(5517)
		5 d	62% ThOD	(5518)
		20 d	77% ThOD	(5518)
A.S., anaerobic		1 week	100% CH_4 prod.	(5521)

Soil sorption

K_{oc}:		
for Fullerton soil, 0.06% OC:	<5	
for Apison soil, 0.11% OC:	<5	
for Dormont soil, 1.2% OC:	<5	(2599)

D. BIOLOGICAL EFFECTS

Bacteria

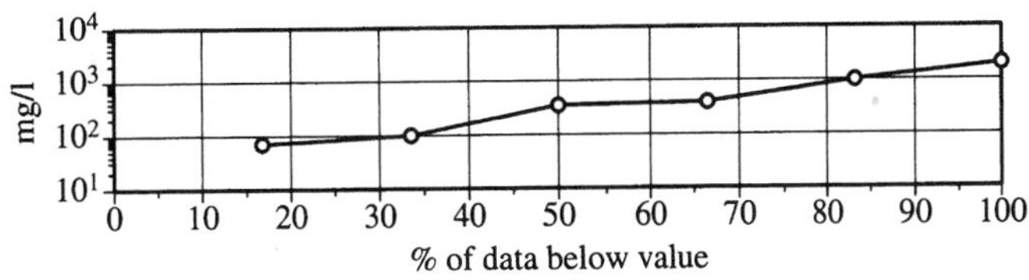

Frequency distribution of short-term EC_{50}/LC_{50} values for bacteria based on data from this and other works (2999)

E. coli	48h EC_0	1,000 mg/L	(356)
Pseudomonas putida	16h EC_{10}	658 mg/L	(5506)
Photobacterium phosphoreum	30 min EC_{50}	71 mg/L	(5514)
	5 min EC_{50}	50 mg/L	(5515)

Algae

Scenedesmus quadricauda	96h LC_0	640 ppm	(356)
	3h EC_{50}	79 mg/L	(5513)
Haematococcus pluvialis	4h EC_{50}	2,600 mg/L	(5506)
Anabaena inaequalis	3h EC_0	30 mg/L	(5513)

Anabaena cylindrica	3h EC_{50}	90 mg/L	(5513)
Anabaena variabilis	3h EC_{50}	35 mg/L	(5513)
Chlorella pyrenoidosa	3h EC_{50}	95 mg/L	(5513)
Crustaceans			
Daphnia	48h LC_0	369 ppm	(356)
	24h EC_0	26; 300 mg/L	
	24h LC_{50}	55; 400 mg/L	(5506, 5512)
	24h LC_{100}	100 mg/L	(5512)

Fishes

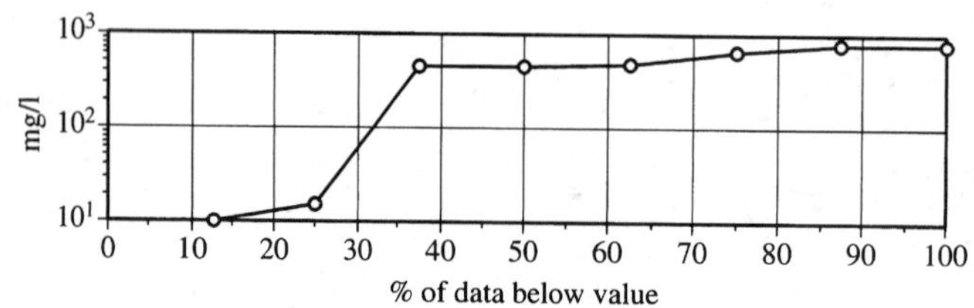

Frequency distribution of 24-96h LC_{50} values for fishes based on data from this and other works			(2999)
Lepomis macrochirus	96h LC_{50},S	10 mg/L	(352)
	24h LC_0	>5 mg/L	(5511)
Menidia berryllina	96h LC_{50},S	15 mg/L	(352)
fathead minnows	1h LC_{50},S	770 mg/L	
	24h LC_{50},S	770 mg/L	
	48h LC_{50},S	770 mg/L	
	72h LC_{50},S	480 mg/L	
	96h LC_{50},S	460 mg/L	(350)
Leuciscus idus	48h LC_0	630 mg/L	
	48h LC_{50}	646 mg/L	(5506)
Pimephales promelas	96h LC_{50}	460 mg/L	(5508)
Petromyzon marinus larvae	24h LC_{50}	>5 mg/L	(5510)
Carassius auratus	24h LC_0	>5 mg/L	(5511)
Salmo trutta	24h LC_0	>5 mg/L	(5511)

benzylamine (α-aminotoluene; benzenemethanamine-; Moringine; phenylmethylamine)

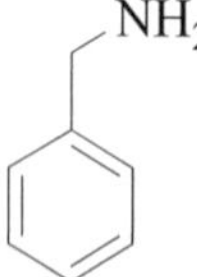

$C_6H_5CH_2NH_2$

C_7H_9N

CAS 100-46-9

USES

chemical intermediate for dyes, pharmaceuticals, polymers.

A. PROPERTIES

molecular weight 107.15; melting point 10°C; boiling point 185°C; density 0.98 at 19/4°C; LogP_{OW} 1.09.

C. WATER AND SOIL POLLUTION FACTORS

Impact on biodegradation processes

inhibition of degradation of glucose by *Pseudomonas fluorescens* at 400 mg/L

inhibition of degradation of glucose by *E. coli*	at >1,000 mg/L	(293)
NH_3oxidation by *Nitrosomonas* sp.:	at 100 mg/L: 26% inhibition	
	at 50 mg/L: 10% inhibition	
	at 10 mg/L: 0% inhibition	(390)

Toxicity to microorganisms

nitrification inhib. EC_{50}: >100 mg/L (2624)

D. BIOLOGICAL EFFECTS

Bacteria			
E. coli	EC_0	1,000 mg/L	
Photobacterium phosphoreum Microtox test	30 min EC_{50}	17 mg/L	(9657)
Algae			
Scenedesmus:	toxic:	6 mg/L	
Arthropods			
Daphnia:	toxic:	60 mg/L	(30)
Fishes			
Steelhead trout	6h LC_{100}	6 mg/L	(9837)

benzylchloride (α-chlorotoluene)

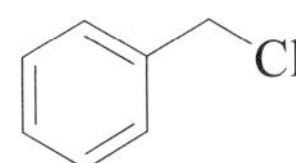

$C_6H_5CH_2Cl$

C_7H_7Cl

CAS 100-44-7

USES

Dyes, intermediate.

410 benzylchloride

A. PROPERTIES

molecular weight 126.58; melting point –41/–43°C; boiling point 179°C; vapor pressure 1 mm at 22°C, 1.7 mm at 30°C; water solubility 1,200 mg/L; vapor density 4.36; density 1.1 at 18/4°C; saturation concentration in air 6.2 g/m^3 at 20°C, 11 g/m^3 at 30°C; P_{ow} 2.30

B. AIR POLLUTION FACTORS

1 mg/m^3 = 0.19 ppm, 1 ppm = 5.262 mg/m^3.

Odour

characteristic: lacrimator aromatic. (129)

threshold values: 0.25 mg/m^3 = 0.047 ppm; 0.04 ppm (37, 279)
PIT 50%; 0.01 ppm
PIT 100%: 0.047 ppm (2)

Incinerability

Temperature for 99% destruction at 2.0-sec residence time under oxygen-starved reaction conditions: 685°C.
Thermal stability ranking of hazardous organic compounds: rank 127 on a scale of 1 (highest stability) to 320 (lowest stability). (2390)

C. WATER AND SOIL POLLUTION FACTORS

Hydrolysis

Benzyl chloride is rapidly hydrolysed to benzyl alcohol which is readily biodegradable. T/2 : 6 days at 0°C, 3 days at 15°C, 14 hours at 25°C, 6 hours at 30°C. (10617)

Reduction of amenities

faint odour: 0.0016 mg/L. (129)

D. BIOLOGICAL EFFECTS

Toxicity

Algae			
Selenastrum capricornutum	72h $ECgr_{50}$	19.3 mg/L	
	72h NOEC	10.0 mg/L	(10617)
Crustaceae			
Daphnia magna	24h EC_{50}	4.2 mg/L	
	48h EC_{50}	3.2 mg/L	
reproduction	21d NOEC	0.10 mg/L	
Penaeus setiferus	24h LC_{50}	7.1 mg/L	
	48h LC_{50}	4.4 mg/L	
	96h LC_{50}	3.9 mg/L	(10617)
Nematoda			
Panogrellus redivivus	96h LC60	126 mg/L	(10617)
Fish			
Oryzias latipes	24h LC_{50}	7.5 mg/L	
	48h LC_{50}	4.2 mg/L	
	72h LC_{50}	2.4 mg/L	
	96h LC_{50}	5.0 mg/L	
Brachydanio rerio	96h LC_{50}	4.0 mg/L	(10617)
Mammals			
Rat	oral LD_{50}	1,231 mg/kg bw	
Mouse	oral LD_{50}	1,500 mg/kg bw	(10617)

Toxicity threshold			
(cell multiplication inhibition test):			
bacteria *(Pseudomonas putida):*	16 h EC_0	4.8 mg/L	(1900)
algae *(Microcystis aeruginosa):*	8 d EC_0	30 mg/L	(329)
green algae *(Scenedesmus quadricauda):*	7 d EC_0	50 mg/L	(1900)
protozoa *(Entosiphon sulcatum):*	72 h EC_0	25 mg/L	(1900)
protozoa (*Uronema parduczi* Chatton-Lwoff):	EC_0	50 mg/L	(1901)

Protozoa			
Vorticella campanula:	toxic	11 mg/L	
Paramecium caudatum:	toxic	800 mg/L	(30)

Fishes			
Trutta iridea:	paralysis	10 mg/L	
Cyprinus carpio:	paralysis	17 mg/L	(30)

	24 h LC_{50}	*48 h LC_{50}*	*96 h LC_{50}*	
fathead minnows *(Pimephales promelas)*	11.6 mg/L	7.3 mg/L	6 mg/L	
white shrimp *(Penaeus setiferus)*	7.1 mg/L	4.4 mg/L	3.9 mg/L	(1904)
guppy *(Poecilia reticulata)*	14 d LC_{50}	0.39 mg/L		(2696)

benzylcyanide (α-tolunitrile; phenylacetonitrile; benzeneacetonitrile)

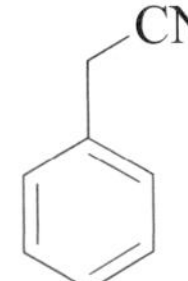

$C_6H_5CH_2CN$

C_8H_7N

CAS 140-29-4

USES

chemical synthesis.

A. PROPERTIES

molecular weight 117.14; melting point -23/-26°C; boiling point 234°C; vapor pressure 0.1 mm at 20°C, 0.1 mm at 30°C, 0.2 hPa at 20°C, 1 hPa at 60°C; vapor density 4.05; density 1.01 at 18°C; solubility 1,740 mg/L at 20°C; saturation concentration in air 0.06 g/m^3 at 30°C; LogP_{OW} 1.56 (measured).

B. AIR POLLUTION FACTORS

1 mg/m^3 = 0.21 ppm, 1 ppm = 4.87 mg/m^3.

C. WATER AND SOIL POLLUTION FACTORS

Biodegradability

Biodegradation by a mutant microorganism: 500 mg/L at 20°C:

Parent:	84% disruption in 48 h	
Mutant:	100% disruption in 12 h	(152)
Inoculum: munic. waste water, at 5 mg/L after 20 d: >60% ThOD removal		(5611)

D. BIOLOGICAL EFFECTS

Bacteria			
Pseudomonas fluorescens	24h EC_0	100 mg/L	(5611)
Photobacterium phosphoreum	30 min EC_{50}	1.3 mg/L	(5614)
Fishes			
Leuciscus idus	48h LC_0	50 mg/L	(5611)
Petromyzon marinus larvae	24h LC_0	>5 mg/L	(5612)
Cyprinus carpio oral	44h LC_0	66-134 mg/kg	(5613)

benzylhydroquinone *see* hydroquinonemonobenzylether

3-benzyloxy-aniline

$C_6H_5CH_2OC_6H_4NH_2$

$C_{13}H_{13}NO$

CAS 1484-26-0

A. PROPERTIES

$LogP_{OW}$: 2.79.

D. BIOLOGICAL EFFECTS

Algae

Tetrahymena pyriformis: growth inhibition at 27°C, 48 EC_{50}: 26 mg/L (2704)

Fishes

Pimephales promelas: 30-35 d FT test at 25°C: LC_{50}: 9 mg/L (2704)

p-benzyloxyphenol *see* hydroquinonemonobenzylether

1-benzylpyridinium-3-sulfonate

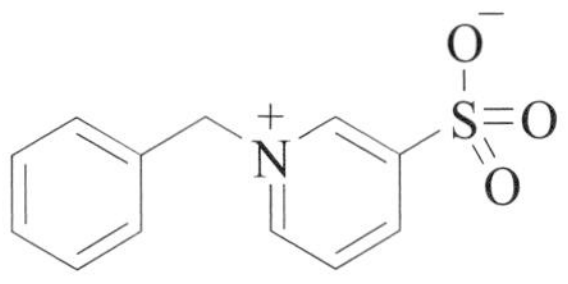

$C_{12}H_{11}NO_3S$

CAS 69723-94-0

A. PROPERTIES

molecular weight 249.28; melting point 205-207°C.

D. BIOLOGICAL EFFECTS

Algae			
Tetrahymena pyriformis: growth	48 IC_{50}	5,700 mg/L	(2704)
Fishes			
Pimephales promelas:	30-35d LC_{50},F	2,400 mg/L	(2704)

benzyltrichloride *see* benzotrichloride

Benzyltrimethylammonium chloride (BTMAC; N,N,N-trimethyl-benzenemethanaminium chloride)

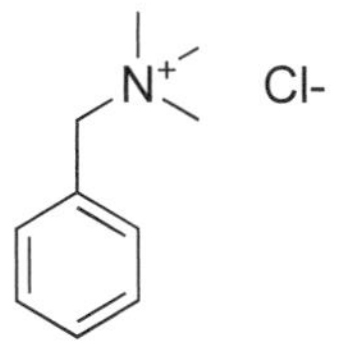

$C_{10}H_{16}N.Cl$

414 Benzyltrimethylammonium chloride

CAS 56-93-9

SMILES CN(C)(C)(Cc1ccccc1)Cl

USES

Quaternary ammonium compound combines disinfecting, wetting, frothing, anti-corrosion and waterproofing properties. BTMAC is used as a solvent for cellulose, a gelling inhibitor for polyester resins, an intermediate, a dye assistant for acrylics and a phase transfer agent.

A. PROPERTIES

White to light yellow molecular weight 185.70; melting point <-50°C; boiling point >135°C; relative density 1.07 at 20°C; vapour pressure 3.1 x 10-8 hPa at 25°C; solubility high; log Pow −2.17 at 25°C

B. AIR POLLUTION FACTORS

Photodegradation

T/2 (OH radicals)= 7.4 hours. (10715)

Odour

Slight almond

C. WATER AND SOIL POLLUTION FACTORS

Biodegradation

Inoculum	method	concentration	duration	elimination	
A.S.	OECD 301C	100 mg/L	28 days	1%	
A.S.	Aerobic	100 mg/L	10 days	0%	(10715)

D. BIOLOGICAL EFFECTS

Bioaccumulation

BCF <1.5; <2; 3.16 (calculations)

Toxicity

Micro organisms			
Anabaena variabilis	14d NOEC	1,857 mg/L	
Oscillatoria sp.	14d NOEC	1,857 mg/L	(10715)
Crustaceae			
Daphnia pulex	48h EC_{50}	11.9 mg/L	(10715)
Fish	48h LC_{50}	>1,000 mg/L (ECOSAR)	(10715)
Mammals			
Rat	oral LD_{50}	180; 250 mg/kg bw	(10715)

Betaxolol (1-[4-[2-(cyclopropylmethoxy)ethylphenoxy]-3-[(1-methyl-ethyl)amino]-2-propanol)

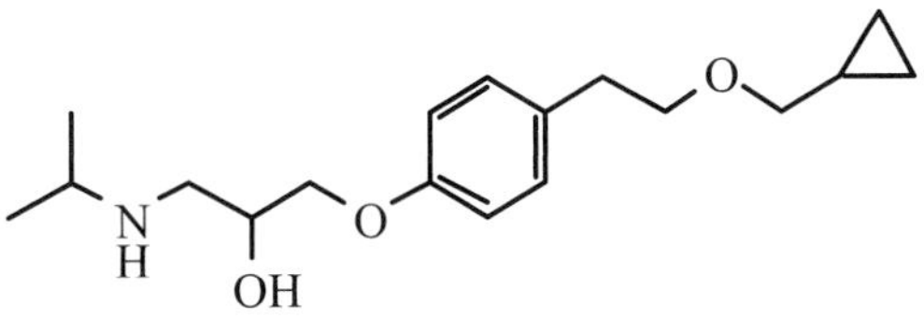

$C_{18}H_{29}NO_3$

CAS 63659-18-7

TRADENAMES

Betoptic, Betaxolol

USES

Beta-blocker, antihypertensive, antiglaucoma

A. PROPERTIES

molecular weight 307.43

C. WATER AND SOIL POLLUTION FACTORS

Environmental concentrations

	range µg/L	n	median µg/L	90 percentile µg/L	
effluent MWTP (1996-2000)	<0.025-0.188	25	0.063	0.103	
	<0.025-0.19	29	0.057	0.10	(7237)
		1	0.19		(7178)
surface waters (1996)	<0.003-0.028	24	0.006	0.009	(7237)
drinkingwater (1996)	<0.003	16			
	<0.010	1			(7237)

bezafibrate (2-[4-[2-[(4-chlorobenzoyl)amino]ethyl]phenoxy]-2-methyl-propanoic acid; 2-(p-(2-(p-Chlorobenzamido)ethyl)phenoxy)-2-methylpropionic acid)

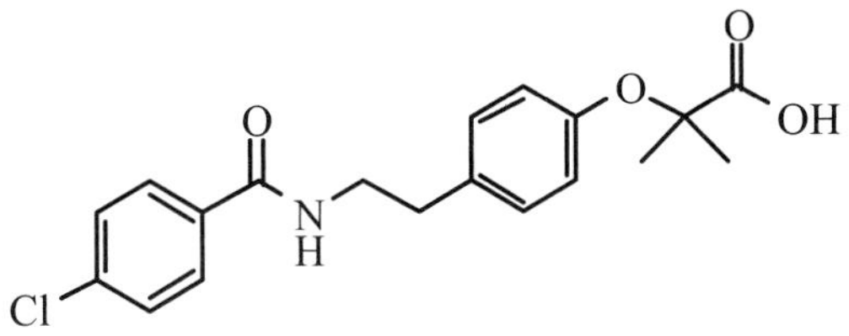

$C_{19}H_{20}ClNO_4$

CAS 41859-67-0

TRADENAMES

Bezafibrate; Bezatol; Cedur; Prestwick 724, Befizal

USES

Antilipemic agent that lowers cholesterol and triglycerides. It decreases low density lipoproteins and increases high density lipoproteins

IMPURITIES, ADDITIVES, COMPOSITION

pharmaceutical impurities (7209)

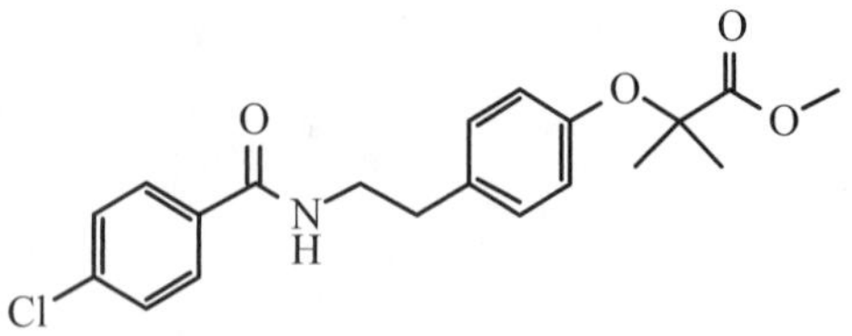

methyl-2-[4-[2-[(4-chlorobenzoyl)amino]ethyl]phenoxy]-2-methylpropionate

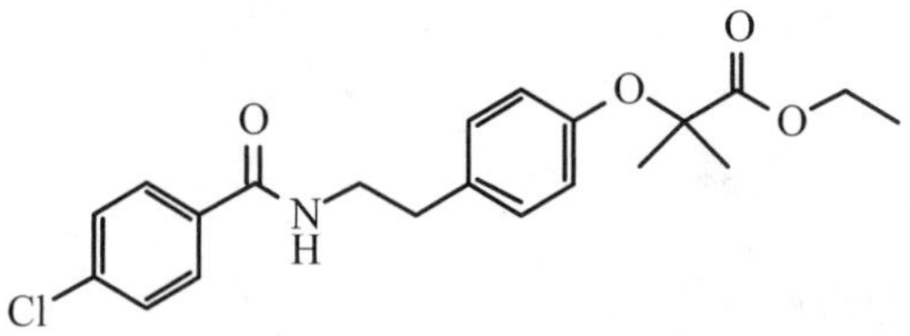

ethyl-2-[4-[2-[(4-chlorobenzoyl)amino]ethyl]phenoxy]-2-methylpropionate

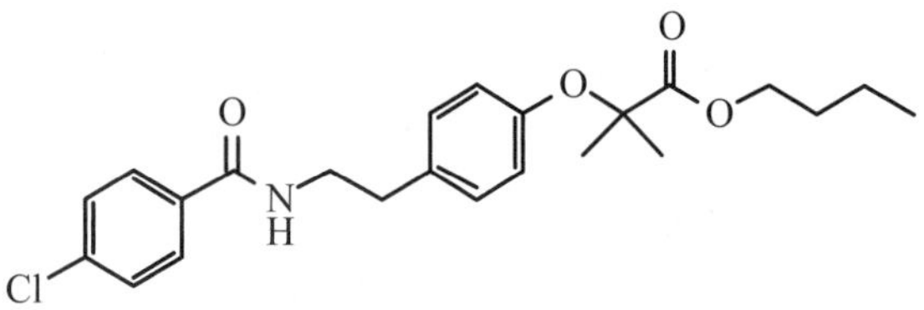

propyl-2-[4-[2-[(4-chlorobenzoyl)amino]ethyl]phenoxy]-2-methylpropionate

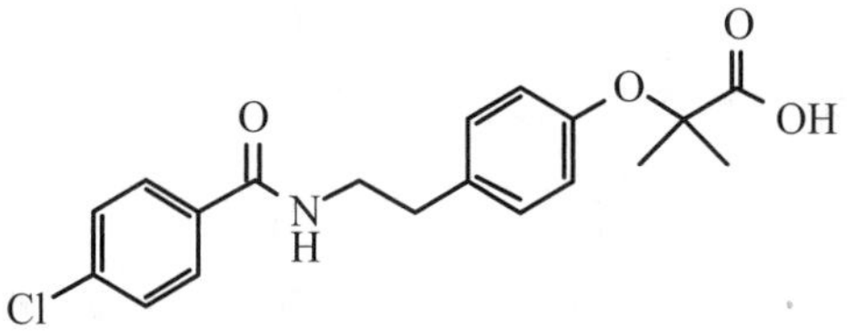

N,O-bis-(4-chlorobenzoyl)tyramine

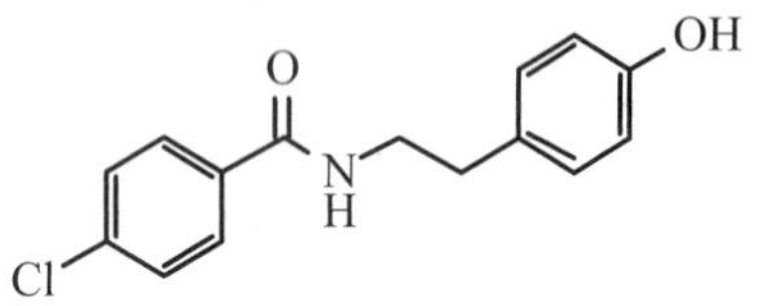

N-(4-chlorobenzoyl)tyramine

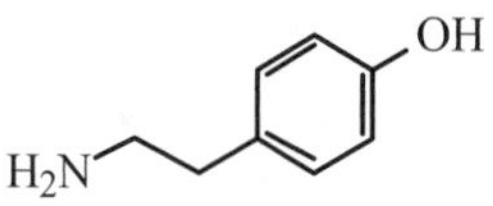

tyramine

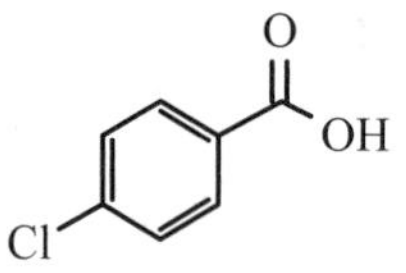

4-chlorobenzoic acid

A. PROPERTIES

molecular weight 361.82

C. WATER AND SOIL POLLUTION FACTORS

Environmental occurrence

In influents to sewage works	removal efficiency	in effluents of sewage works
Germany: 300 g/day	83 %	max 4600 ng/L
Brazil: 1200 ng/L	27 - 50 %	(7178)

	range μg/L	n	median μg/L	90 percentile μg/L	
drinkingwater (1996)	<0.010	12			(7237)
influent MWTP	up to 4.4	11			
	1.2	1			(7237)
effluent MWTP (1996-2000)	<0.250-4.560	39	2.61	3.49	
	3.32	1			
	0.85	1			
	0.60	1			
	<0.010-0.020	4			(7237)

waste waters (1996-1997)	3.32				(7237)
surface waters (1996-2000)	<0.010-0.040	22			
	<0.025-0.295	8			
	0.156-0.380	11			
	<0.025-3.10	43	0.350	1.20	
	<0.10-0.20	7			
	up to 0.315	14			
	<0.010-0.210			0.049	
	<0.010-0.075	35		0.059	
	0.38	1			
	<0.025	8			(7237)

BGE *see* *n*-butylglycidylether

BHC *see* hexachlorocyclohexane

BHT *see* 2,6-di-tert-butyl-4-methylphenol

p,p'-bianiline *see* benzidine

bianisidine *see* *o*-dianisidine

bibenzyl (dibenzyl; 1,2-diphenylethane)

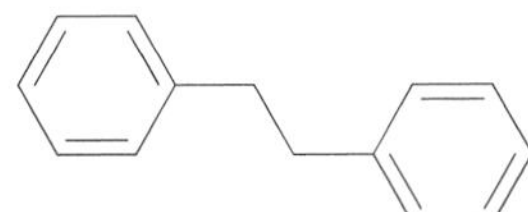

$C_6H_5CH_2CH_2C_6H_5$

$C_{14}H_{14}$

CAS 103-29-7

USES

organic synthesis.

A. PROPERTIES

molecular weight 182.27; melting point 50-53°C; boiling point 284°C; density 1.0; $LogP_{OW}$ 4.79/4.82.

C. WATER AND SOIL POLLUTION FACTORS

Water quality

in Eastern Ontario drinking waters (June-Oct. 1978): n.d.-0.8 ng/L (*n* = 12)
in Eastern Ontario raw waters (June-Oct. 1978): 0.3-1.0 ng/L (*n* = 12) (1698)

bichloromethylether *see* dichloromethylether

bidisin (2-chloro-3(4-chlorophenyl)-methylpropionate; chlorfenpropmethyl; Bay 70533)

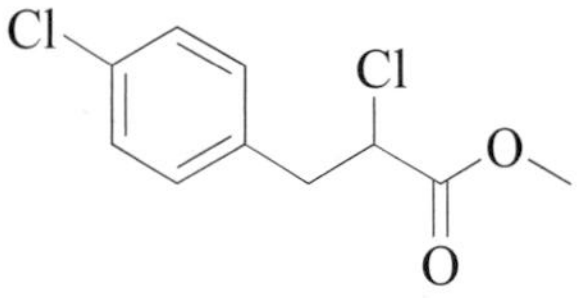

$C_{10}H_{10}Cl_2O_2$

CAS 14437-17-3

USES

a postemergence herbicide.

A. PROPERTIES

colorless to light-brown liquid; molecular weight 233.1; melting point > -20°C; vapor pressure 10^{-4} mbar at 50°C; density 1.3 at 20°/4°C; solubility 400 mg/L at 20°C.

C. WATER AND SOIL POLLUTION FACTORS

Degradation in soil

hydrolysis of ester, degradation to 2-chloro-3-(4-chlorophenyl)propionic acid, 4- (2962)

chlorobenzoic acid and sequential products.

D. BIOLOGICAL EFFECTS

Fishes

harlequin fish *(Rasbora heteromorpha)*:

mg/L	*24 h*	*48 h*	*96 h*	*3 m (extrap.)*	
LC_{10} (F)	1.2	1.3	0.8		
LC_{50} (F)	1.85	1.7	1.1	0.6	(331)
goldfish	96h LC_{50}	1-10 mg/L			(2963)

biisopropyl *see* 2,3-dimethylbutane

binapacryl (2-(1-methyl-*n*-propyl)-4,6-dinitrophenyl-2-methylcrotonate; 2-*sec*-butyl-4,6-dinitrophenyl-3-methyl-2-butenoate; 2(2-butyl-4,6-dinitrophenyl)-3,3-dimethylacrylate; Acricid; Ambox; Dinoseb methacrylate; Endosan; Dapacril; Morrocid; BP 855, 736)

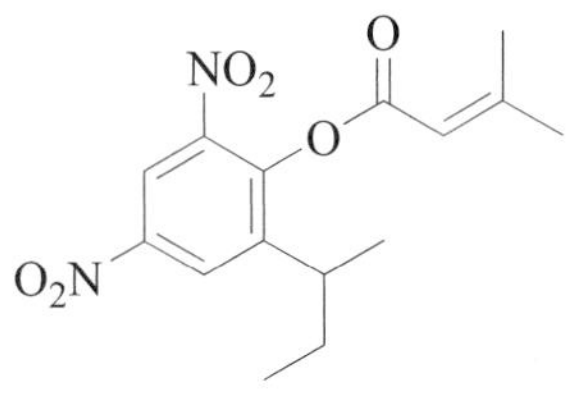

NO_2 O O O_2N

$C_{15}H_{18}N_2O_6$

CAS 485-31-4

USES

contact miticide, fungicide.

A. PROPERTIES

crystalline solid; melting point 65-69°C; molecular weight 322; vapor pressure 0.42×10^{-6} mbar at 20°C; density 1.16; solubility 1 mg/L at 20°C

C. WATER AND SOIL POLLUTION FACTORS

Biodegradation

Residual activity in soil 15-25 d. Degraded in the environment to the amine and carboxylic acid . (9662)

D. BIOLOGICAL EFFECTS

Insects

Asellus brevicaudus	96h LC_{50}	0.029 mg/L (technical grade)	(9344)

Fishes			
guppies:	LC_{100}	0.5 mg/L	(2962)
channel catfish, bluegill sunfish, rainbow trout	96h LC_{50}	0.015-0.050 mg/L technical grade	(9344)
Mammals			
rat, guinea pig, rabbit, dog	oral LD_{50}	150-640 mg/kg bw	
mice	oral LD_{50}	1,600-3,200 mg/kg bw	(9344)

bioctyl *see* *n*-hexadecane

biphenyl *see* diphenyl

4-biphenylmethanol (4-phenylbenzylalcohol)

OH

$C_6H_5C_6H_4CH_2OH$

$C_{13}H_{12}O$

CAS 3597-91-9

A. PROPERTIES

molecular weight 184.24; melting point 99-100°C.

C. WATER AND SOIL POLLUTION FACTORS

Soil sorption

K_{oc}:		
for Fullerton soil, 0.06% OC:	105	
for Apison soil, 0.11% OC:	700	
for Dormont soil, 1.2% OC:	495	(2599)

p-biphenylphenylether

$C_{18}H_{14}O$

CAS 3933-94-6

EINECS 223-504-4

A. PROPERTIES

$LogP_{ow}$ 5.55

D. BIOLOGICAL EFFECTS

Rainbow trout: log BCF: 3.22. (2606)

4-biphenylsulfonic acid, sodium salt (1,1'-biphenyl)

$C_{12}H_{10}O_3S.Na$

CAS 2217-82-5

USES AND FORMULATIONS

product is manufactured as a 25% aqueous solution/suspension; feedstock for chemical synthesis.

A. PROPERTIES

molecular weight 257.3; melting point >300°C; solubility approx. 10,000 m g/L at 15°C; $LogP_{ow}$ 2.2

C. WATER AND SOIL POLLUTION FACTORS

Biodegradation

Inoculum	*Method*	*Concentration*	*Duration*	*Elimination results*	
	Modified OECD screening test	20 mg/L	28 d	12% DOC	(6496)

D. BIOLOGICAL EFFECTS

Toxicity

Bacteria			
A.S. oxygen consumption inhibition	3 h EC_5	5,164 mg/L	
	3 h EC_{50}	9,131 mg/L	
	3 h EC_{95}	16,146 mg/L	(6498)
Fishes			
Salmo gardneri	24 h LC_0	>5 mg/L	(6497)
Lepomis macrochirus	24 h LC_0	>5 mg/L	(6497)
Petromyzon marinus larvae	24 h LC_0	>5 mg/L	(6497)
Oncorhynchus kisuth	24 h LC_0	>10 mg/L	(6497)
Oncorhynchus tshawytscha	24 h LC_0	>10 mg/L	(6498)
Ptychocheilus oregonensis	24 h LC_0	>10 mg/L	(6498)

2,2'-bipyridine (2,2'-dipyridyl)

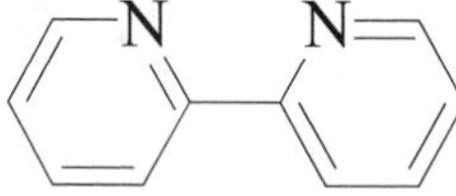

$(C_5H_4N)(C_5H_4N)$

$C_{10}H_8N_2$

CAS 366-18-7

USES

iron-chelating agent.

A. PROPERTIES

molecular weight 156.19; melting point 70-73°C; boiling point 273°C.

C. WATER AND SOIL POLLUTION FACTORS

Impact on biodegradation processes

NH_3 oxidation by *Nitrosomonas:*	at 100 mg/L	91% inhibition	
	at 50 mg/L:	81% inhibition	
	at 10 mg/L:	23% inhibition	(390)
Nitrification	EC_{50}: 23 mg/L		(2624)

2,2'-biquinoline

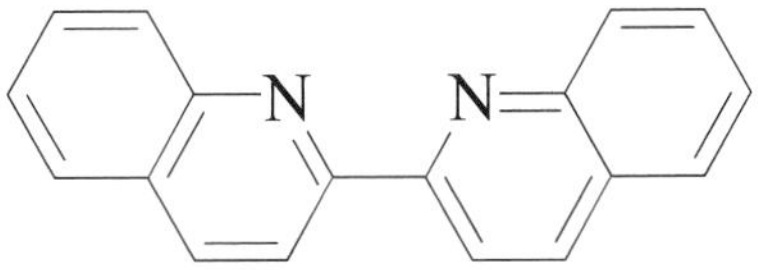

$C_{18}H_{12}N_2$

CAS 119-91-5

A. PROPERTIES

molecular weight 256.31; melting point 193-196°C; solubility 1.02 mg/L; $LogP_{OW}$ 4.3.

C. WATER AND SOIL POLLUTION FACTORS

Soil sorption

K_{oc}: 10,404. (2587)

bis-2-aminoethylamine *see* diethylenetriamine

1,3-bis(aminomethyl)benzene (1,3-benzenedimethanamine; m-phenylenebis(methylamine); m-xylylenediamine; MXDA)

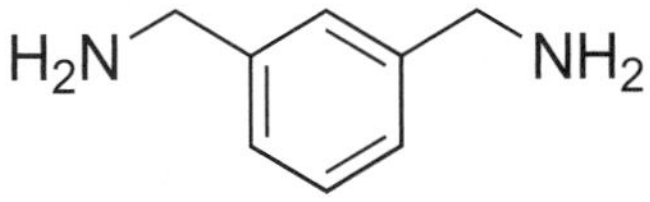

$C_8H_{12}N_2$

CAS 1477-55-0

USES

The chemical is an intermediate in the production of epoxy curing agents, polyamides, polyurethanes, lacquers and plastics.

IMPURITIES, ADDITIVES, COMPOSITION

3-methylbenzylamine; 3,5-dimethylbenzyl alcohol; 3,4-dimethylbenzylalcohol; 2,4-dimethyl benzylalcohol. All impurities at ca 0.08% or less.

A. PROPERTIES

molecular weight 136.22; melting point 14.1°C; boiling point 273°C; density 1.05 at 20°C; vapour pressure 0.04 hPa at 25°C; water solubility >100,000 mg/L at 25°C; log Pow 0.18 at 25°C

B. AIR POLLUTION FACTORS

Photodegradation

T/2 = 5.4 hours based on reaction with OH radical. (10666)

C. WATER AND SOIL POLLUTION FACTORS

Hydrolysis

The chemical is stable at pH 4 to 9.

Biodegradation

Inoculum	method	concentration	duration	elimination	
A.S.	OECD 301B	17 mg/L	28 days	49%	
A.S.	OECD 301C	100 mg/L	14 days	4% ThOD	
A.S.	OECD 302C		28 days	22% ThOD	
				6% TOC	(10666)

D. BIOLOGICAL EFFECTS

Bioaccumulation

Species	conc. µg/L	duration	body parts	BCF	
Cyprinus carpio	2,000	42 days		<0.3	
	200	42 days		<2.7	(10666)

Toxicity

Algae			
Selenastrum Capricornutum	72h ECb_{50}	20.3 mg/L	
	72h NOECb	10.5 mg/L	
	72h $ECgr_{50}$	33 mg/L	
	72h NOECgr	22.9 mg/L	
Scenedesmus subspicatus	72h ECb_{50}	12 mg/L	
	24h $ECgr_{50}$	14 mg/L	(10666)
Crustaceae			
Daphnia magna	24h EC_{50}	35.1 mg/L	
	24h EC_{100}	50 mg/L	
	24h NOEC	16 mg/L	
	48h EC_{50}	15.2; 16 mg/L	
	48h EC_{100}	28 mg/L	
	48h NOEC	8.9 mg/L	
reproduction	21d EC_{50}	6.8 mg/L	
	21d NOEC	4.7 mg/L	
	21d LOEC	15 mg/L	
	21d LC_{50}	8.4 mg/L	(10666)
Fish			
Oryzias latipes	96h LC_0	56 mg/L	
	96h LC_{50}	87.6 mg/L	
	96h LC_{100}	>100 mg/L	
Golden Orfe	96h LC_{50}	75 mg/L	
Rainbow trout	96h LC_{50}	>100 mg/L	(10666)
Mammals			
Rat	oral LD_{50}	980; 1,090; 1,180 mg/kg bw	(10666)

1,3-bis(aminomethyl)cyclohexane (1,3-cyclohexanedimethanamine; 1,3-BAC)

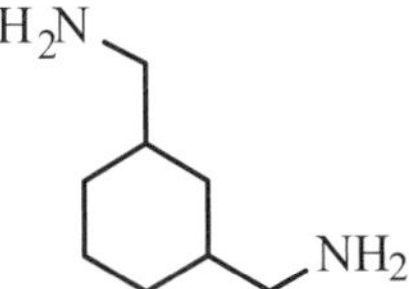

$C_8H_{18}N_2$

CAS 2579-20-6

USES

the chemical is a component of hand dishwashing liquid preparations, which is used directly by householders.

IMPURITIES, ADDITIVES, COMPOSITION

0.01% of 1,3-benzenedimethanamine

A. PROPERTIES

colourless liquid with slight ammonia odour; molecular weight 142.2; boiling point 254-255°C; density 0.94; vapour pressure $1.24x10^{-3}$ kPa at 20°C; water solubility >1,000,000 mg/L; $LogP_{OW}$ 0.44

C. WATER AND SOIL POLLUTION FACTORS

Hydrolysis

hydrolysis is unlikely at environmental temperatures and pH (7124)

Biodegradation

Inoculum	method	concentration	duration	elimination
Activated sewage sludge	OECD 301B	14.8 mg/l	28 days	29% ThCO2
Activated sewage sludge	OECD 303A		28 days	92% COD 96% DOC
Activated sewage sludge	Modified SCAS		36 days	86% (1)

(1)of which 12% adsorption and 74% mineralisation
The chemical is not readily but inherent biodegradable. (7124)

Soil sorption

log K_{OC} 2.1-2.9 (7124)

D. BIOLOGICAL EFFECTS

Bioaccumulation

The chemical is unlikely to bioaccumulate due to the highly water solubility, low $LogP_{OW}$ and inherent biodegradability. (7124)

Toxicity

Micro organisms			
Pseudomonas putida	acute EC_{10}	90 mg/L	
(cell multiplication inhibition test)	acute EC_{50}	330 mg/L	(7124)
Algae			
Selenastrum capricornutum	96h EC_{50}	>100 mg/L	(7124)
terrestrialPLANTS			
Luctuca sativa growth	NOEC	320 mg/kg soil	
Emergence, survival	NOEC	>1000 mg/kg soil	

Avena sativa growth, emergence, survival	NOEC	>1000 mg/kg soil	
Lycopersicum esculentum growth, emergence, survival	NOEC	>1000 mg/kg soil	(7124)
Worms			
Eisenia foetida	acute EC_{50}	>1000 mg/kg soil	(7124)
Crustaceae			
Daphnia magna	48h EC_{50}	0.32; 2.4 mg/L	(7124)
Fish			
Brachydanio rerio	96h LC_{50}	2.4 mg/L	
	96h NOEC	1.8 mg/L	(7124)
Mammals			
Mouse	oral LD_{50}	200-2000 mg/kg bw	(7124)

bis-2-chloro-1-methylethylether *see* dichloroisopropylether

1,2-bis(2-chloroethoxy)methane (triethyleneglycol dichloride; triglycol dichloride)

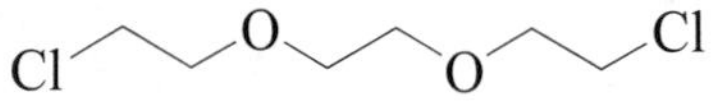

Cl–CH₂CH₂–O–CH₂CH₂–O–CH₂CH₂–Cl

$(CH_2OCH_2CH_2Cl)_2$

$C_6H_{12}Cl_2O_2$

CAS 112-26-5

A. PROPERTIES

molecular weight 187.07; boiling point 235°C; density 1.2.

B. AIR POLLUTION FACTORS

Incinerability

thermal stability ranking of hazardous organic compounds: rank 189 on a scale of 1 (highest stability) to 320 (lowest stability). (2390)

bis(2-chloroethyl)ether *see* β,β'-dichloroethylether

N,N-bis (2-chloroethyl)-2-napthylamine (2-bis (2-chloroethyl)- aminonaphthalene; chlornaphazine; Chlornaftina-; dichloroethyl-beta-napthylamine; Erysan-; 2-naphthylamine mustard))

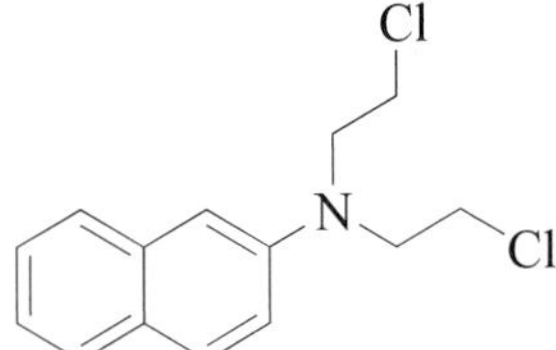

$C_{14}H_{15}Cl_2N$

CAS 494-03-1

USES

Antineoplastic agent, used in treatment of leukaemia and Hodgkin's disease.

A. PROPERTIES

molecular weight 268.20; melting point 54-56°C; boiling point 210°C at 5 mmHg; solubility < 1,000 mg/L at 22°C

B. AIR POLLUTION FACTORS

Incinerability

thermal stability ranking of hazardous organic compounds: rank 132 on a scale of 1 (highest stability) to 320 (lowest stability). (2390)

bis(2-chloroethyl)sulfide (sulfur mustard gas; di-2-chloroethylsulfide; mustardgas; ethane, 1,1'-thiobis[2-chloroethane]; bis(beta-chloroethyl) sulfide; Yperite)

$S(CH_2CH_2Cl)_2$

$C_4H_8Cl_2S$

CAS 505-60-2

USES

warfare.

A. PROPERTIES

molecular weight 159.08; melting point 13-14°C; boiling point 215-217°C; density 1.27 at 20°C; vapor pressure 9 x 10^{-12} mm Hg at 30°C; vapor density 5.4; solubility 680 mg/L at 25°C

B. AIR POLLUTION FACTORS

Incinerability

thermal stability ranking of hazardous organic compounds: rank 132 on a scale of 1 (highest stability) to 320 (lowest stability). (2390)

bis (β-chloroisopropyl)ether *see* dichloroisopropylether

bis(chloromethyl)ether *see* dichloromethylether

bis(chloromethyl)naphthalene

$C_{12}H_{10}Cl_2$

A. PROPERTIES

molecular weight 225.12; vapor density 7.78.

B. AIR POLLUTION FACTORS

1 mg/m^3 = 0.11 ppm, 1 ppm = 9.36 mg/m^3.

C. WATER AND SOIL POLLUTION FACTORS

Aquatic reactions

evaporation: measured half-life for evaporation from 1 mg/L aqueous solution at 25°C, still air and an average depth of 6.5 cm: 45 min. (369)

2,2-bis(p-chlorophenyl)-1,1-dichloroethane *see* DDD

1,1-bis(4'-chlorophenyl)-2,2,2-trichloroethenol *see* kelthane

2,2-bis(3,5-dichloro-4-hydroxyphenyl)propane

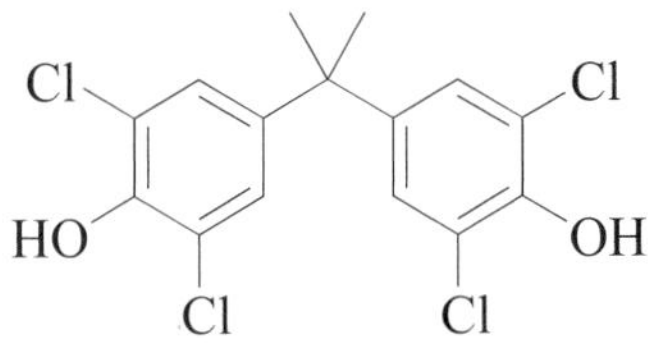

D. BIOLOGICAL EFFECTS

Fishes			
Pimephales promelas	96h LC_{50}	1.3 mg/L	(2625)

Bis(2,4-dicumylphyenyl)pentaerythritol diphosphite (2,4,8,10-tetraoxa-3,9-diphosphaspiro[5.5]undecane,3,9-bis[2,4-bis(1-methyl-1-phenylethyl)phenoxy]-)

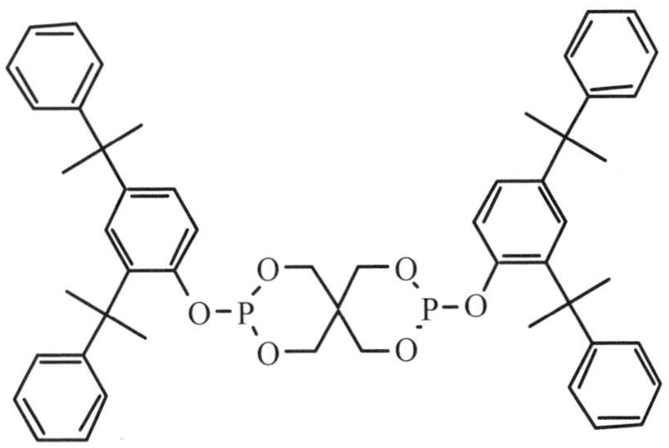

$C_{53}H_{58}O_6P_2$

CAS 154862-43-8

USES

The chemical is used as an antioxidant/stabilizer for thermoplastic engineering polymers, including polyethylene, polyamides, polyesters, polyethers and polycarbonates. Products made from these plastics include containers for food, clothing and other items, food wrap, household appliances and automotive parts. The chemical will be present in the finished polymer blend at 0.05 – 0.15 wt%. (7157)

IMPURITIES, ADDITIVES, COMPOSITION and additives of marketed product

<2%: 2,4-bis(1-methyl-1-phenylethyl)phenol

A. PROPERTIES

off white, free flowing powder or compacted pellets; molecular weight 852; melting point 221-230°C;

density 1.26; vapour pressure <10^{-6} kPa at 25°C; water solubility <0.05 mg/L at 20°C; log P_{OW} >6 at 22 °C

C. WATER AND SOIL POLLUTION FACTORS

Hydrolysis

Hydrolysis of the chemical could not be determined due to its low water solubility. Hydrolysis of the phosphorus-oxygen bond is however likely to occur under environmental conditions. (7157)

Mobility in soils

log K_{OC} >4.6 (calculated). This high value indicates that the chemical will bind strongly to the organic component of soils and sediments. (7157)

Biodegradation

Inoculum	method	concentration	duration	elimination	
	BOD Test		28 days	0.2-2 % ThOD	(7157)

D. BIOLOGICAL EFFECTS

Toxicity

Micro organisms			
sewage bateria		not inhibitory at maximum solubility	(7157)
Algae			
Scenedesmus subspicatus	72h EC_{50}	> 0.49 mg/L	(7157)
Crustaceae			
Daphnia magna	48h EC_{50}	> 0.20 mg/L	
	48h NOEC	0.06 mg/L	(7157)
Fish			
Brachydanio rerio	96h LC_{50}	>0.22 mg/L	
	96h NOEC	>0.22 mg/L	(7157)
Mammals			
Rat	oral LD_{50}	>5000 mg/kg bw	(7157)

bis(dimethylthiocarbamoyl)disulfide *see* thiram

S-(1,2-bis (ethoxycarbonyl)ethyl)O,O-dimethylphosphorodithioate *see* malathion

bis-2-ethoxyethylether *see* diethyleneglycoldimethylether

2,2-bis-hydroxymethyl-1,3-propanediol *see* pentaerythritol

3,4-bis (p-hydroxyphenyl)-3-hexene *see* diethylstilbestrol

Bis(isopropyl)naphthalene (DIPN; di(isopropyl)naphthalene)

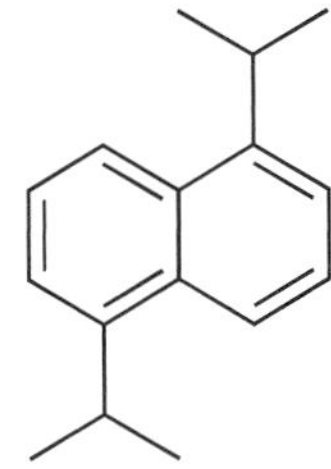

$C_{16}H_{20}$

CAS 38640-62-9

EINECS 254-052-6

USES

Diisopropyl naphthalene is used as a solvent. It can also be used as raw material for further synthesis. It is reported to be used as a substitute for PCB, which is in capacitors and transformers, but can also be contained in chemical products like for instance in lubricants for spinning textile fibres. Diisopropyl naphthalene is not chlorinated like PCB and can thus not be used in applications where there is a risk of fire as diisopropyl naphthalene burns well.

OCCURRENCE

The substance is a constituent of crude oil and coal tar and is also formed, like other aromatics, in different refinery processes to be one of the hydrocarbons in petroleum products like diesel and lighter fuel oils. It is also formed by combustions like in forest fires and in gasoline and diesel engines

A. PROPERTIES

a clear, yellow-brown liquid with faint sweet odour;molecular weight 212.34; boiling point 290-299 °C; density 0.96 at 15°C; vapour pressure 0.003 hPa at 20°C, 0.6 hPa at 60°C; log Pow >4 at 25°C; water solubility <0.02 - 0.44 mg/L at 20°C;

B. AIR POLLUTION FACTORS

Photodegradation

under high pressure mercury lamp: at 1 mmol/L t/2 = 6.4-16 h (7353)

C. WATER AND SOIL POLLUTION FACTORS

Biodegradation

Inoculum	method	concentration	duration	elimination

Activated sludge	^{14}C-DIPN	13 mg/L DOC	7 days	> 90%
	^{14}C-DIPN	27 mg/L DOC	4 days	> 86 %
Activated sludge	^{14}C-DIPN	4 mg/L	28 days	83. 9% (7354, 7355)
COD: 0.01 g/g				
BOD5/COD: 0.56	(7356)			

D. BIOLOGICAL EFFECTS

Bioaccumulation

Species	conc. µg/L	duration	body parts	BCF	
Cyprinus carpio	5	28 days		203	(7357)
Goldfish	2500	70 days		<100	(7358)

Toxicity

Micro organisms			
Photobacterium phosphoreum	EC_{50}	> 0.1 mg/L	
Pseudomonas putida	EC_{10}	> 0.16 mg/L	(7362)
Algae			
Inhibition of growth	48h EC_0	> 0.19 mg/L	(7362)
Crustaceae			
Daphnia magna	24h EC_{50}	2.3 mg/L	
	24h EC_{100}	14 mg/L	(7361)
	21d NOEC	0.013 mg/L	
	21d LOEC	0.025 mg/L	(7362)
Fish			
Leuciscus idus	96h LC_0	> 0.5 mg/L	(7359)
Cyprinus carpio	96h LC_{50}	> 1000 mg/L	(7360)
Oryzias latipes	96h LC_{50}	> 1000 mg/L	(7360)
Mammals			
Rat	oral LD_{50}	15,182 mg/kg bw	(7363)
Rat	oral LD_{50}	4300 – 5600 ; 3900-4500 mg/kg bw	(7364, 7365)
Mouse	oral LD_{50}	5100 – 5400 mg/kg bw	(7366)

2,2-bis (p-methoxyphenyl)-1,1,1-trichloroethane *see* methoxychlor

bis (methylpropyl)amine *see* di-*s*-butylamine

bis (β-methylpropyl)amine *see* diisobutylamine

bis(1-octyloxy-2,2,6-tetramethyl-4-piperidyl)sebacate (Tinuvin 123; TK12382; GL 123)

Major product

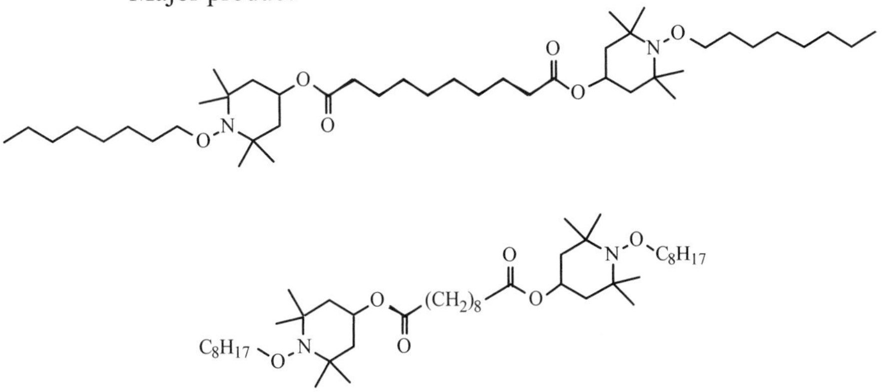

Minor product

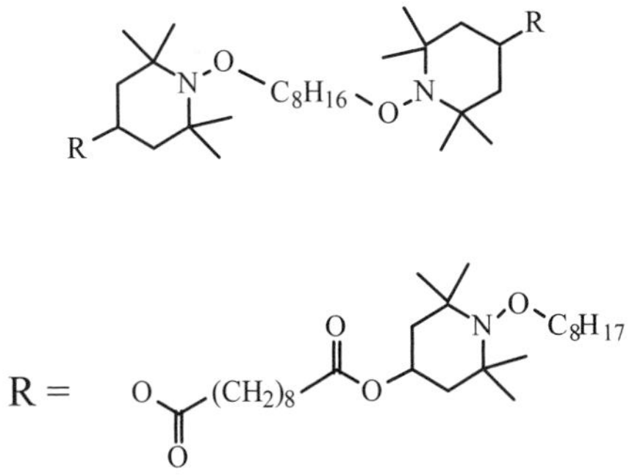

$C_{44}H_{84}N_2O_6$

CAS 129757-67-1

USES

Tinuvin 123 is a N-substituted hindered amine. These compounds find widespread use as heat and light stabilisers, oxidants and radical scavengers in polymer applications. As such, they are used in the manufacture of surface coatings, plastics and printing inks to counteract the effects of long term degradation to weather and sunlight. The final concentration of Tinuvin 123 in paint is 5000 to 10000 mg/L paint (0.5-1.0%). (7113)

IMPURITIES, ADDITIVES, COMPOSITION

Tinuvin 23 is a mixture of a number of chemical entities of which there are two principal components : $C_{44}H_{84}$ N_2O_6 (major product) and $C_{80}H_{150}$ N_4O_{12} (minor product). Main component (monomer) 65-85%, minor component (dimer) 5-20%, other by-products < 3.5% per compound. (7113)

A. PROPERTIES

yellow-amber coloured viscous liquid; molecular weight ; specific gravity 0.97 at 20 °C; vapour pressure 0.00036 Pa at 25 °C (extrapolated); water solubility <6 mg/L at 20 °C; LogP_{OW} <10.

C. WATER AND SOIL POLLUTION FACTORS

Biodegradation

Tinuvin is not readily biodegradable (7113)

D. BIOLOGICAL EFFECTS

Toxicity

Micro organisms			
A.S. respiration inhibition	3h EC_{20}	> 100 mg/L	(7113)
Crustaceae			
Daphnia magna	24h EC_{50}	83 mg/L*	(7113)
Fish			
Brachydanio rerio	96h LC_{50}	>58 mg/L*	(7113)
Mammals			
Rat	oral LD_{50}	>2000 mg/kg bw	(7113)

* nominal initial concentration of > 500 mg/L solubilised with Tween 20

Bis (pentabromophenyl) ethane (1,2-bis(pentabromophenyl)ethane)

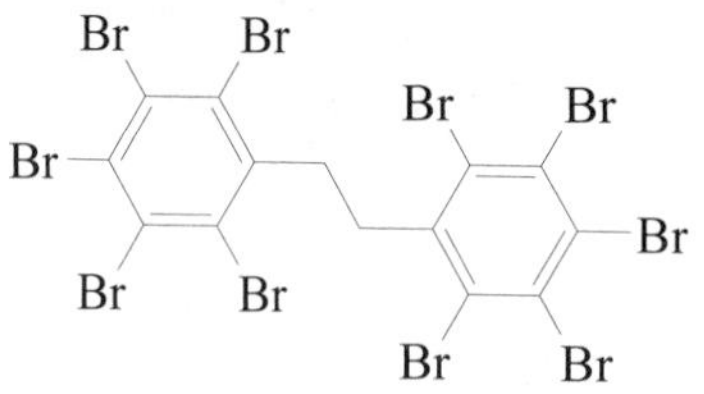

$C_{14}H_4Br_{10}$

CAS 84852-53-9

TRADENAMES

Saytex 8010

USES

The product which is mainly used as additive flame retardant for electric parts (Polystyrene, Polyamide, Polyolefin) is, at a minimum 98,5 % pure. The principal impurity is Nonabromodiphenylethane.

A. PROPERTIES

molecular weight ; vapour pressure <1 x 10^{-4} Pa at 20°C; water solubility 0.72 µg/L; log Pow 3.2

B. AIR POLLUTION FACTORS

Fire

In case of fire or during thermic influence > 320°C hydrobromic acid, bromine and carbon oxides are formed from Decabromodiphenyl oxide. Whilst on one hand 2,3,7,8 - substituted dioxins and furans lay below their respective detection limit of 0.1 - 100 ppb, other examinations illustrated a furan concentration of up to 47.5 µg 2,3,4,7,8, - PBDF/kg. (11031)

D. BIOLOGICAL EFFECTS

Toxicity		
Mammals		
Rat	oral LD_{50}	>5,000 mg/kg bw

1,2-bis(4-pyridyl)ethane (4,4'-ethylene dipyridine)

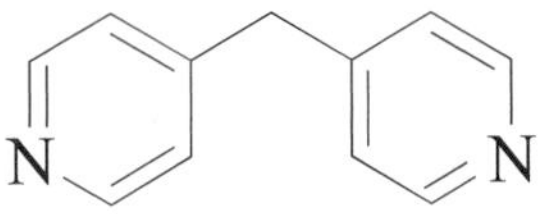

$(C_5H_4N)CH_2CH_2(C_5H_4N)$

$C_{12}H_{12}N_2$

CAS 4916-57-8

A. PROPERTIES

molecular weight 184.24; melting point 107-110°C; $LogP_{OW}$: 1.93.

D. BIOLOGICAL EFFECTS

Algae

Tetrahymena pyriformis: growth inhibition at 27°C, 48 EC_{50}: 203 mg/L (2704)

Fishes

Pimephales promelas: 30-35 d FT test at 25°C: log LC_{50}: 1,217 mg/L (2704)

Bis(2,2,6,6-tetramethyl-4-piperidyl) sebacate (Sebacic acid bis(2,2,6,6-tetramethyl-4-piperidyl) ester; Decanedioic acid bis(2,2,6,6-tetramethyl-4-piperidinyl)ester)

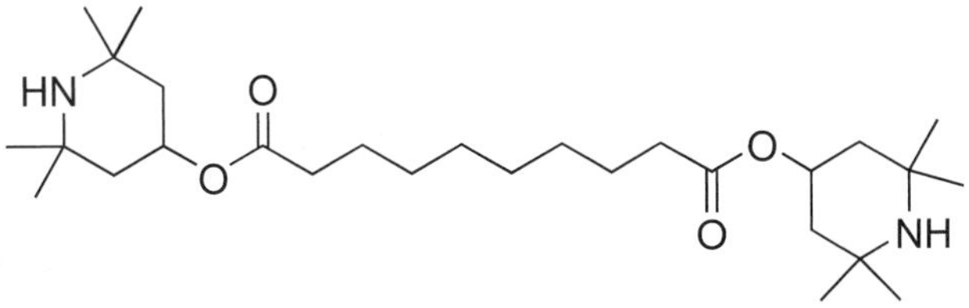

$C_{28}H_{52}N_2O_4$

CAS 52829-07-9

EINECS 258-07-9

TRADENAMES

Photo-stabilizer HS-770; Sunsorb 770; Chisorb 770; Tinuvin 770

USES

UV absorber and light stabilizer for polyolefins

A. PROPERTIES

White crystalline powder; molecular weight 480.73; melting point 82-85°C;

D. BIOLOGICAL EFFECTS

Toxicity

Algae	72h EC_{50}	0.39; 1.1 mg/L	(11221)
	72h NOEC	0.050; 0.093 mg/L	(11221)
Daphnids	48h EC_{50}	8.6 mg/L	(11221)
	21d EC_{50}	0.96 mg/L	(11221)
	21d NOEC	0.23 mg/L	(11221)
Fish	96h LC_{50}	5.3 mg/L	(11221)

bis(tri-n-butyltin)oxide (hexabutyldistannoxane; TBTO; biometTBTO; bis(tributylstannyl) oxide; BTO; butinox)

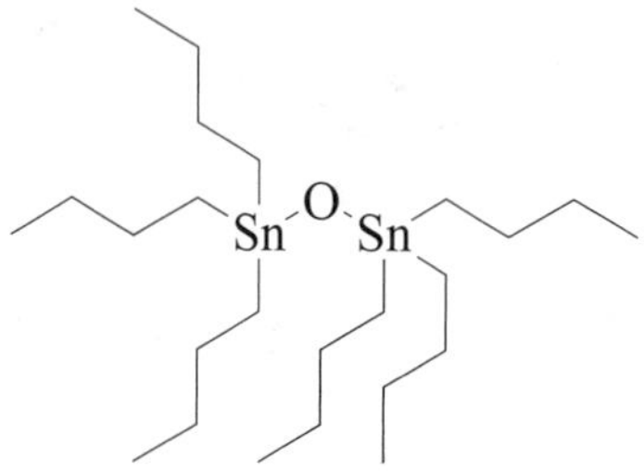

$[CH_3(CH_2)_3]_3SnOSn[(CH_2)_3$

$[(C_4H_9)_3Sn]_2O$

$C_{24}H_{54}OSn_2$

CAS 56-35-9

USES

algicide in antifouling paints, conservation of wood, molluscicide, prevention of Bilharzia. Polymerisation catalyst.

A. PROPERTIES

colorless liquid at ambient temp.; molecular weight 596.1; melting point -45°C; boiling point 210-214°C at 10 mm Hg; density 1.17; solubility 18-19 mg/L at 20°C in dist. water; 8-10 mg/L at 22°C in seawater, 100 mg/L; $LogP_{OW}$2.2; LogH -4.3 at 25°C (calculated).

C. WATER AND SOIL POLLUTION FACTORS

Biodegradation

$t_{1/2}$: 4 months in water/sediment mixture derived from Toronto Harbour. (6155)

Impact on biodegradation processes

act. sludge respiration inhib. EC_{50}: 26 mg/L. (2624)

Mobility in soils

K_{OC}: 90,800 for Toronto Harbour sediments. (6156)

D. BIOLOGICAL EFFECTS

Bioaccumulation

	Concentration µg/L	*exposure period*	*BCF*	*organ/tissue*	
salt water goby	0.21-2.1	12 weeks	2,000-12,000		(6151)
goldfish	1	12 d	2800	muscle*	(6154)

* depuration time $t_{1/2}$: 6 d

Bacteria

Photobacterium phosphoreum Microtox test	30min EC_{50}	1.1 µg/L	(8899)

Algae

Skeletonema costatum, Thalassiosira pseudonana growth inhibition	IC_{50}	0.33-1.1 µg/L	(6153)
Rhaphidocellis subcapitata	72h EC_{50}	0.016 mg/L	(7025)

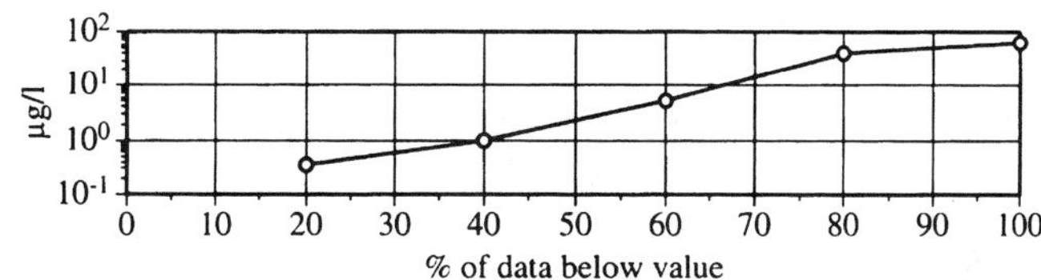

Frequency distribution of EC_{50}/LC_{50} values for algae (n = 5) based on data from this and other works (3999)

µg/L	*96h LC_{50}*	*96h LC_{50}*	*96h NOEC*	
Algae				
Chlorella pyrenoidosa	n.d.	42	18	
Scenedesmus pannonicus	n.d.	64	32	
Crustaceans				
Daphnia magna (48 h)	4.7	0.75	0.56	
Molluscs				
Lymnaea stagnalis	42	2.4	1.8	
FISHES:				
Gasterosteus aculeatus (1-2 d)	19	7.5	5.6	
Gasterosteus aculeatus (4-5 wk)	13	4.2	3.2	(2707)
Poecilia reticulata	21	7.5	5.6	
Oryzias latipes	17	7.5	5.6	

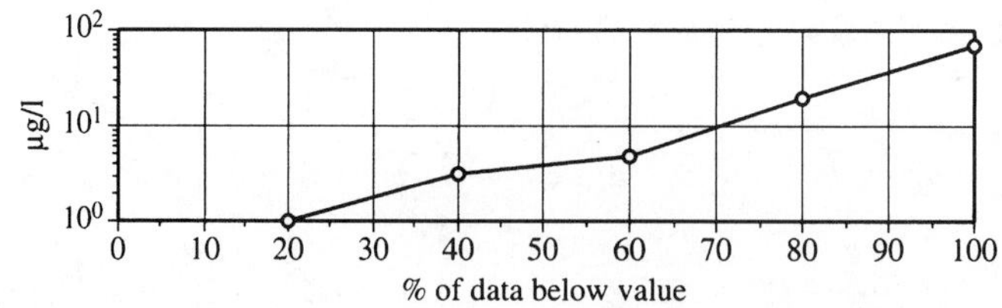

Frequency distribution of EC_{50}/LC_{50} values for crustaceans (n = 5) based on data from this and other works (3999)

Crustaceans			
Daphnia magna	48h IC_{50}	0.0017 mg/L	(7025)

Crustacean

Daphnia magna: μg/L	*LC_{50}*	*EC_{50}*	*NOLC*	*NOEC*	
week 1	3.2	2.4	1.8	1.8	
week 2	2.4	1.4	1.8	1.0	
d 20	1.8	1.4	1.0	0.56	(2707)

Mollusc

Lymnaea stagnalis: μg/L	*LC_{50}*	*EC_{50}*	*NOLC*	*NOEC*	
d 12	>5.6	0.56	3.2	0.32	
d 19	3.4	0.54	3.2	0.32	
d 26	2.6	0.50	1.0	0.32	
d 33	1.5	0.38	1.0	0.32	(2707)

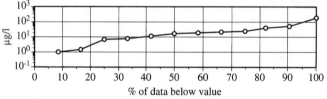

Frequency distribution of LC_{50} values for fishes (n = 12) based on data from this and other works (3999)

Fish

Poecilia reticulata: μg/L	*LC_{50}*	*EC_{50}*	*NOLC*	*NOEC*	
week 1	14	5.6	3.2	3.2	
week 2	18	5.8	3.2	3.2	
week 3	17	5.8	3.2	3.2	
week 4	16	5.8	3.2	1.0	
week 8	13	5.8	3.2	<3.2	
d 99	12	1.8	3.2	0.32	(2707)

Fish

Oryzias latipes: μg/L	*LC_{50}*	*EC_{50}*	*NOLC*	*NOEC*	
week 1	18	1.8	10	1.0	
week 2	16	1.8	3.2	1.0	
week 3	13	1.8	3.2	1.0	
week 4	11	1.8	3.2	1.0	
week 8	11	5.6	3.2	3.2	
d 104	7.5	5.6	3.2	3.2	(2707)

Fish			
Cyprinodon variegatus	96h LC_{50}	0.96 µg/L	
	MATC, early life stage	3.7 µg/L	(2643)
Brachydanio rerio	96h LC_{50}	0.0027 mg/L	(7025)
Three-spined stickleback in sea water	2m LC_{80}	10 µg/L	(6150)
goby, rainbow trout, fathead minnow in salt water	24-96h LC_{50}	3-31 µg/L	(6151, 6152)
bleak	96h LC_{50}	15 µ/L	(6152, 7694)
tilapia *(Tilapia rendalli)*	24h EC_{50}	53 µg/L	(1543)

Mammals			
rabbit, mouse, rat	oral LD_{50}	50, 55, 190 mg/kg bw respectively	(9801)

Bis(2,4,4-trimethylpentyl)phosphinic acid (Cyanex 272 Extractant)

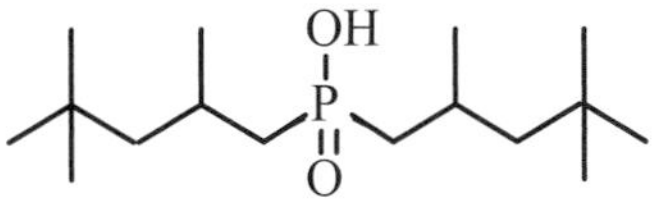

$C_{16}H_{35}O_2P$

CAS 83411-71-6

USES

The chemical is used in the mineral processing industry where it will be used as an extractant in mining operations. A 10% solution with an anorganic liquid detergent is used to separate zinc and iron from cobalt solutions.

IMPURITIES, ADDITIVES, COMPOSITION

tris(2,4,4-trimethylpentyl)phosphine oxide 10-12 wt %. (7097)

A. PROPERTIES

colourless to light amber liquid; molecular weigth 290; boilng point > 300 °C; specific gravity 0.92 at 24 °C; water solubility 16 mg/L at pH 2.6, 38 mg/L at pH 3.7, with increasing pH solubility is expected to increase; $LogP_{OW}$ > 2.7 measured.

C. WATER AND SOIL POLLUTION FACTORS

Hydrolysis

the chemical is stable at pH 5 and 50 °C for 4 weeks. Under environmental conditions hydrolysis is not expected. (7097)

D. BIOLOGICAL EFFECTS

Toxicity

Crustaceae			
Daphnia magna	48h EC_{50}	>9.9 mg/L	
	48h NOEC	2.9 mg/L	(7097)

Fish			
Rainbow trout	96h LC_{50}	22 mg/L	
	96h NOEC	7.7 mg/L	(7097)
Bluegill	96h LC_{50}	46 mg/L	
	96h NOEC	22 mg/L	(7097)
Mammals			
Rat	oral LD_{50}	>3500 mg/kg bw	(7097)

Bisopropol (1-[4-[[2-(1-methylethoxy)-ethoxy]methyl]phenoxy]-3-[(methylethyl)amino]-2-propanol)

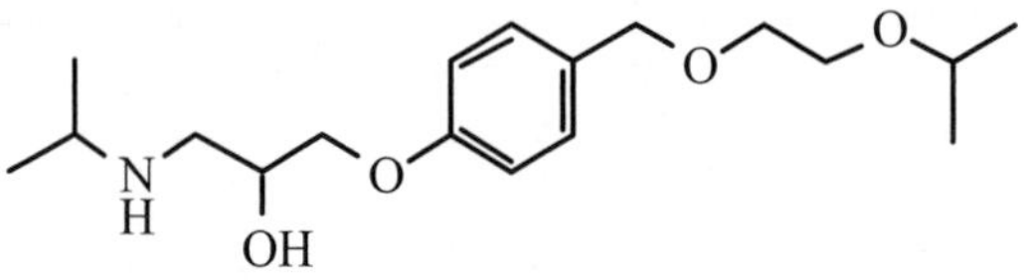

$C_{18}H_{31}NO_4$

CAS 66722-44-9

TRADENAMES

Concor

USES

Beta-blocker; antihypertensive

IMPURITIES, ADDITIVES, COMPOSITION

(7209)

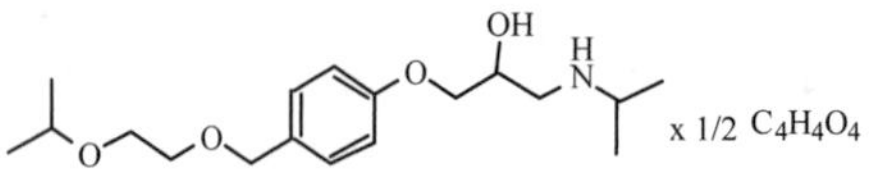

x 1/2 $C_4H_4O_4$

bisoprolol fumarate

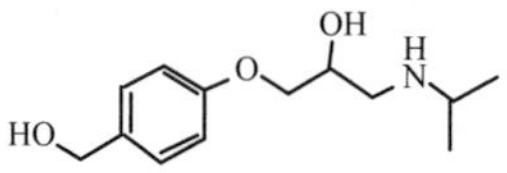

1-[4-(hydroxymethyl)phenoxy]-3-isopropylamino-2-propanol

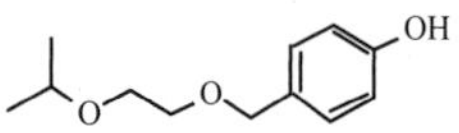

4-(2-isopropoxy-ethoxymethyl)phenol

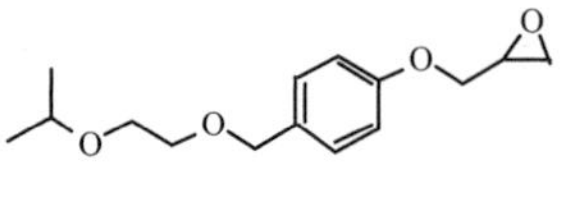

2-[4-(2-isopropoxyethoxymethyl)phenoxymethyl]oxirane

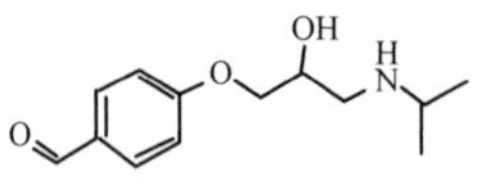

4-[(2RS)-2-hydroxy-3-(isopropylamino)-propoxy]benzaldehyde

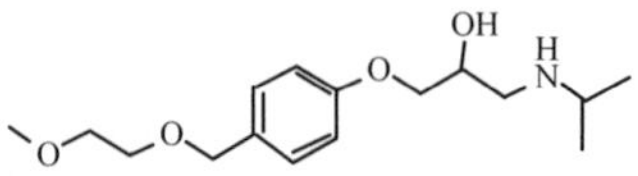

1-[4-(2-methoxyethoxymethyl)phenoxy]-3-isopropylamino-2-propanol

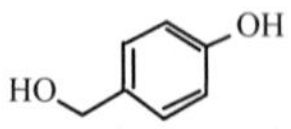

4-hydroxybenzylalcohol

A. PROPERTIES

molecular weight 298.38

C. WATER AND SOIL POLLUTION FACTORS

Environmental concentrations

	range µg/L	n	median µg/L	90 percentile µg/L	
drinkingwater (1996)	<0.003	16			(7237)
effluent MWTP	<0.025-0.370	25	0.057	0.176	(7237)
surface waters		1	0.110		(7178)

Bisphenol A (4,4'-isopropylidenediphenol; 4,4'-dihydroxydiphenyldimethylmethane; bis(4-hydroxyphenyl)dimethylmethane; BPA; 2,2-*bis*-4-hydroxyphenylpropane)

$C_{15}H_{16}O_2$

$(CH_3)_2C(C_6H_4OH)_2$

CAS 80-05-7

EINECS 201-245-8

USES

Fungicide. Manufacture of epoxy resins and polycarbonates. Bisphenol A is a plastic monomer which is used extensively in the production of polycarbonate, epoxy and other plastics. It can be found in a diverse range of products including adhesives, reinforced pipes, interior coatings of tins and drums, flooring, artificial teeth, electronic goods, powder paints, headlights, lenses, helmets and paper coatings, such as that used in thermal fax paper.

OCCURRENCE

A constituent leached from polycarbonate flasks during autoclaving.

A. PROPERTIES

white powder; molecular weight 228,29 ; melting point 153-157 °C; boiling point 220 °C at 4 mm Hg, 398 °C at 760 mm Hg; vapor pressure 0.2 mm Hg at 170°C; specific gravity 1.06-1.19 at 25°C; water solubility 120-300 mg/L; $LogP_{OW}$ 2.20(measured), 3,84 (calculated), 3.32 at pH 3 (measured); H 10^{-10} atm.m^3/mol (calculated)

B. AIR POLLUTION FACTORS

Photochemical reactions

$t_{1/2}$: 4 h, based on reactions with OH° (calculated) (8481)

HO–C6H4–C(CH3)2–C6H4–OH → 4-isopropylphenol + phenol + a semiquinone

Photodecomposition

products of the vapour are phenol; 4-isopropylphenol; and a semiquinone derivative (10353)

C. WATER AND SOIL POLLUTION FACTORS

Hydrolysis

It is not expected to undergo chemical hydrolysis under environmental conditions, since it contains no hydrolysable functional groups (8383)

Inhibition of degradation processes

Activated sludge: inhibition of nitrification	IC_0	50 mg/L	(10176)

Anthropogenic sources

In effluent of sewage treatment plants in Germany in 2000: 0.1-0.27; 0.80-29 µg/L (7172)

Waste water treatment

Partitioning between particle and water phase in 24 hours samples of influent of sewage treatment plant: ca. 3 %
Elimination in sewage treatment plant: 70-94 %.
Elimination in sewage treatment plant with reference to influent: biodegraded: 46%, sorbed on dried composite sludge: 32%, in effluent 22%. (7172)

Mobility in soil

log K_{OC} : 2.6 – 3.2 (calculated) (7177)

Biodegradation

Inoculum	method	concentration	duration	elimination	
	U.S.EPA Test		28 days	83.6% $ThCO_2$	
	OECD 301F Test		28 days	81-93% ThOD 76-91% $ThCO_2$	
	BOD		20 days	71% ThOD	
	Closed Bottle Test		20 days	0% ThOD	(7172)
A.S. acclimated		105 mg/L	24 hours	72% COD	(8030)
		105 mg/L	24 hours	57% TOC	(8030)

Metabolic Pathway for BPA biodegradation by Strain MV1 (7177)

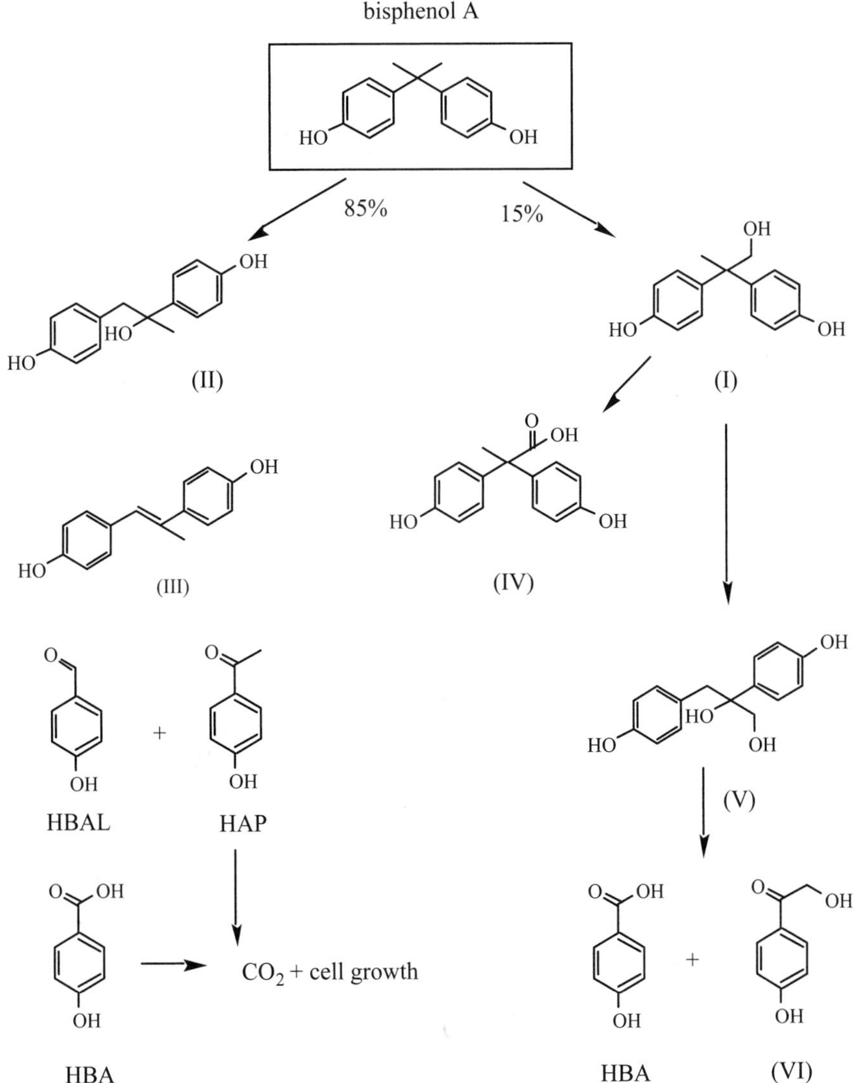

(I) = 2,2-bis(4-hydroxyphenyl)-1-propanol
(II) = 1,2-bis(4-hydroxyphenyl)-2-propanol
(III) = 4,4-dihydroxy-a-methylstilbene
(IV) = 2,2-bis(hydroxyphenyl)propanoic acid
(V) = 2,3-bis(4-hydroxyphenyl)-1,2-propanediol
(VI) = 4-hydroxyphenacyl alcohol
BPA = bisphenol A
HBAL = 4-hydroxybenzaldehyde
HBA = 4-hydroxybenzoic acid
HAP = 4-hydroxyacetophenone

D. BIOLOGICAL EFFECTS

Toxicity

Algae			
Skeletonema costatum	96h EC_{50}	1.0 mg/L	(10352)
Selenastrum capricornutum	4d EC_{50}	2.7 mg/L	(2625)

Crustaceans			
Mysidopsis bahia	96h LC_{50}	1.1 mg/L	(10352)
Daphnia magna	2d IC_{50}	10 mg/L	(2625)
Fishes			
Atlantic silverside	96h LC_{50}	9.4 mg/L	(10352)
Pimephales promelas	4d LC_{50}	4.6-4.7 mg/L	(2625)
Mammals			
rat	oral LD_{50}	3,250 mg/kg bw	
mouse	oral LD_{50}	2,500 mg/kg bw	
rabbit	oral LD_{50}	2,230 mg/kg bw	(10354)

Bioaccumulation

BCF : 42 – 196 (calculated) (7177)

Bisphenol A-diglycidylether

(BADGE; 2,2-bis(4-hydroxyphenyl)propane- bis(2,3-epoxypropyl)ether; 2,2-bis(p-2,3-epoxypropoxy) phenyl)propane; 4,4'-Isopropylidenediphenol diglycidyl ether; DGEBPA; 2,2'-[(1-methylethylidene)bis(4,1-phenyleneoxymethylene)]bisoxirane))

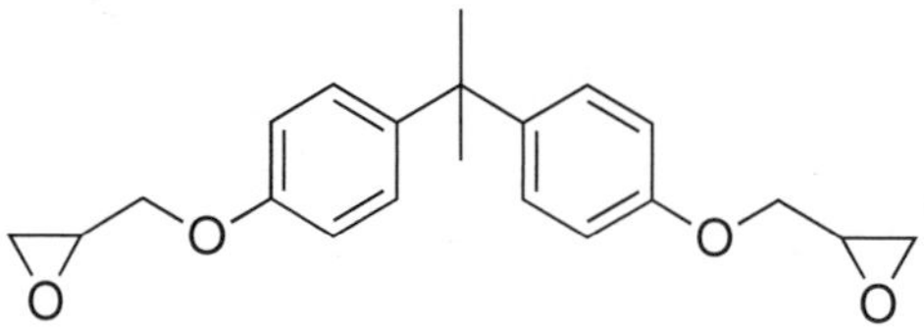

$C_{21}H_{24}O_4$

CAS 1675-54-3

EINECS 216-823-5

USES

For roughly 40 years, bisphenolA diglycidyl ether (BADGE) has been used commercially as a raw material. BADGE is the primary chemical building block for the broad spectrum of materials referred to generally as epoxy resins (se structural formulae below).

Depending on the end-use application, BADGE is used without additives, or in solutions including various solvents. Epoxy resins vary widely in some physical properties such as molecular weight, solvent type and specific composition, but they rely on the same basic chemistry. Major uses epoxy resins are adhesives, potting and encapsulating media and coatings. Some BADGE-based materials, when properly formulated and cured, are approved for food contact applications such as can coatings.

Epoxy surface coatings are among the most widely used industrial finishes and provide superior adhesion, flexibility and corrosion resistance when applied to metallic substrates. BADGE products are used in the following markets and applications: Adhesives, Civil Engineering, Composites, Electrical Laminates, Electrical Castings / Potting and Encapsulation / Tooling, Industrial Coatings, Automotive, Can and coil, Marine and protective, UV / photocure Powder.

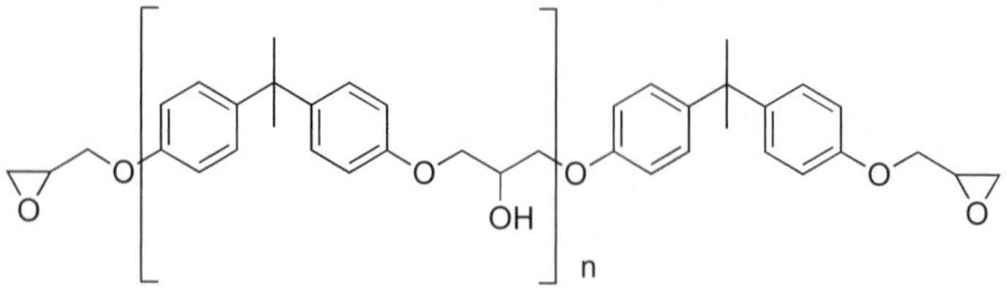

Epoxy prepolymer

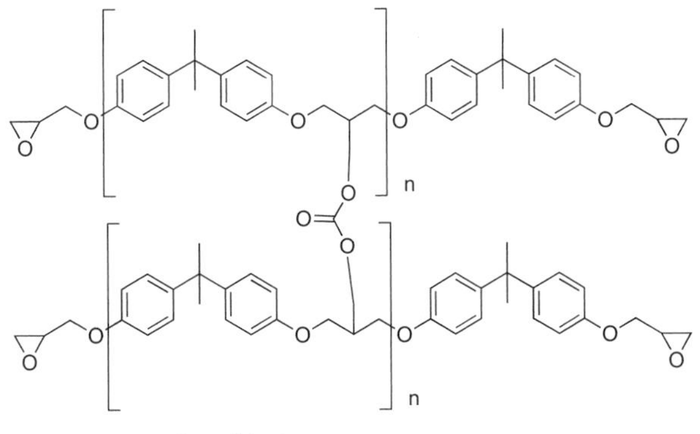

Crosslinked epoxy

Crosslinked epoxypolymer after curing

PRODUCTION, MANUFACTURING

BADGE is produced by reacting epichlorohydrin with bisphenolA as shown below

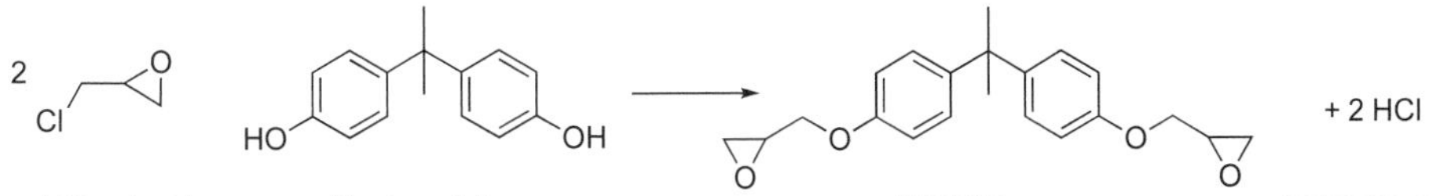

2 epichlorohydrin + Bisphenol A → BADGE + 2 HCl hydrogen chloride

A. PROPERTIES

molecular weight 340.4

C. WATER AND SOIL POLLUTION FACTORS

Hydrolysis

Hydrolysis t/2 at 40°C: 43 hours (3% acetic acid).
In contact with water or with aqueous products the following hydrolysis-and hydrochlorinated derivatives are produced : (7468)

BADGE.H_2O	2-[4-(2,3-epoxypropoxy)phenyl]-2-[4-(2,3-dihydroxy)phenyl]propane
BADGE.2H_2O	2,2-bis-[4-(2,3-dihydroxypropoxy)phenyl]propane
BADGE.HCl.H_2O	2-[4-(3-chloro-2-hydroxypropoxy)phenyl]-2-[4-(2,3-dihydroxypropoxy)phenyl]propane

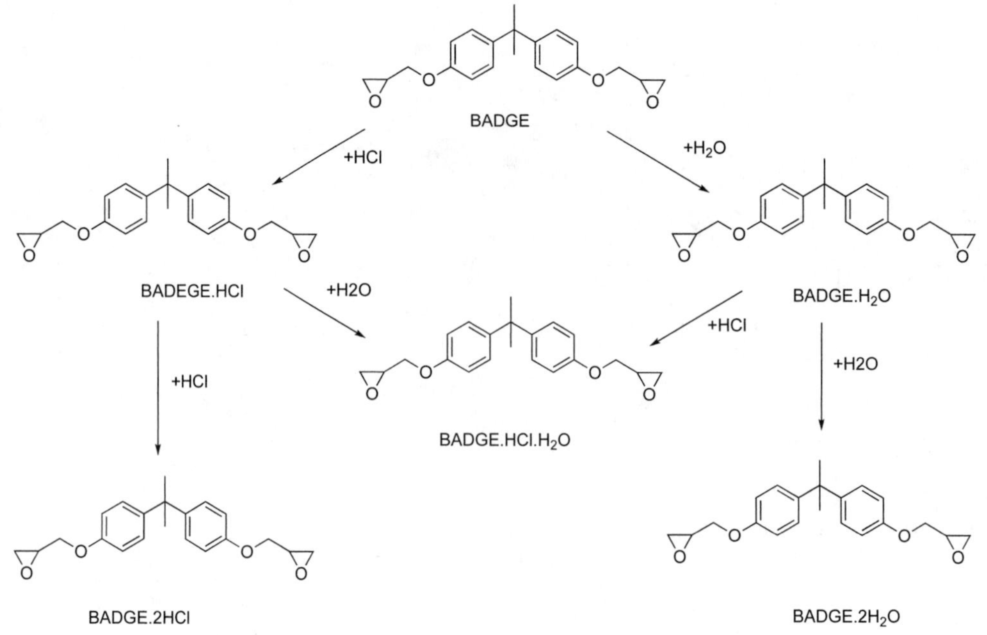

BADGE

+HCl

$+H_2O$

BADEGE.HCl

+H2O

$BADGE.H_2O$

+HCl

+HCl

+H2O

$BADGE.HCl.H_2O$

BADGE.2HCl

$BADGE.2H_2O$

---

bitertanol (1-(biphenyl)-4-yloxy)-3,3-dimethyl-1-(1H-1,2,4-triazol-1-yl)butan-2-ol; β-([1,1'-biphenyl]-4-yloxy)-α-(1,1-dimethylethyl)-1H-1,2,4-triazole-1-ethanol)

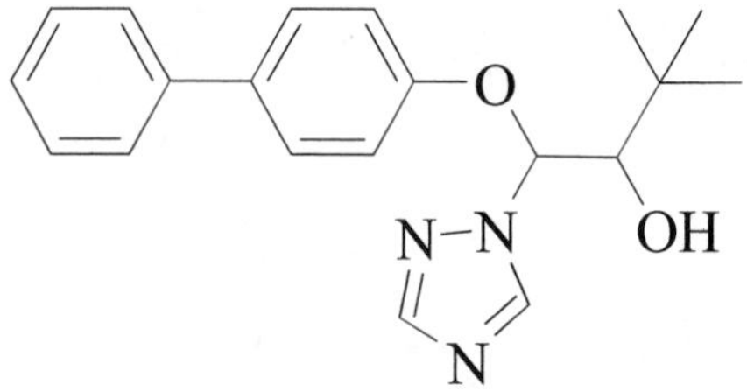

O

N–N OH

N

$C_{20}H_{23}N_3O_2$

CAS 55179-31-2

TRADENAMES

Baycor; Sibutol; Bitertanol

USES

agricultural triazole fungicide

A. PROPERTIES

colorless crystals; molecular weight 337.4; melting point 125-129°C; vapor pressure 10^{-5}mbar at 20°C; water solubility 5 mg/L at 20°C

C. WATER AND SOIL POLLUTION FACTORS

Adsorption

Observed Cuticle/water partition coefficient log Log K_{CW} for isolated cuticles from leaves of bitter orange (*Citrus aurantium L.*), rubber plant (*Ficus elastica*) fruits of tomato (*Lycopersicon esculentum* Mill.) and green pepper (*Capiscum annuum L.)* Log K_{CW}: 3.77-3.85	(7110)

D. BIOLOGICAL EFFECTS

Uptake by PLANTS

cuticle-water partition coefficient for conifer needles: Log K_{CW} = 4.05.	(2436)

Fishes			
rainbow trout	96h LC_{50}	2.2-2.7 mg/L	
carp	48h LC_{50}	2.5 mg/L	(2963)
Mammals			
Rats and mice	acute oral LD_{50}	4202; 4488; >5000 mg/kg bw	(7244)

biviny*l* *see* 1,3-butadiene

bleomycine sulfate

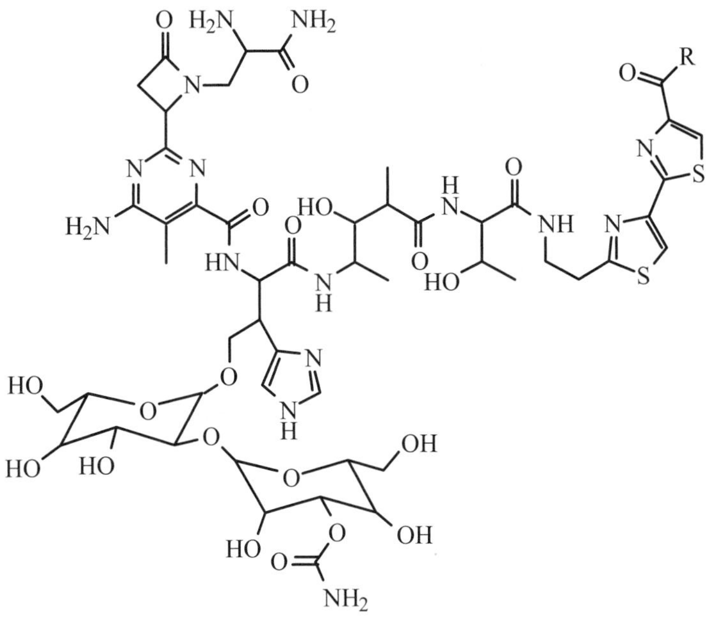

R=terminal amine (variable)

$C_{55}H_{85}N_{17}O_{21}S_3$

$C_{55}H_{83}N_{17}O_{21}S_3$(Bleomycine A2)

$C_{55}H_{83}N_{20}O_{21}S_2$(Bleomycine B2)

CAS 11056-06-7

TRADENAMES

Bleocin; Blenoxane; Bleo; BLM; B

USES

pharmaceutical; a glycopeptide

A. PROPERTIES

colorless to yellowish powder;melting point 71 °C; water solubility 14,000 mg/L

C. WATER AND SOIL POLLUTION FACTORS

Environmental concentrations

	range µg/L	n	median µg/L	90 percentile µg/L	
effluent MWTP (1990)	0.011-0.019				(7237)
in effluent hospital	0.02	estimated			(7237)
surface waters (1990)	<0.005-0.017				(7237)
drinkingwater (1990)	0.0005-0.013	9			(7237)

D. BIOLOGICAL EFFECTS

Mammals

Rat	acute oral LD_{50}	>200 mg/kg bw

borane-tert-butylaminecomplex *see* *t*-butylamine borane

bornylalcohol *see* DL-borneol

BPMC *see* *o-sec*-butylphenyl-N-methylcarbamate

bravo *see* chlorothalonil

bromacil (5-bromo-6-methyl-3-(1-methylpropyl)uracil; 5-bromo-3-*sec* -butyl-6-methyluracil; Borea; Hyvar X; Hyvar X-L)

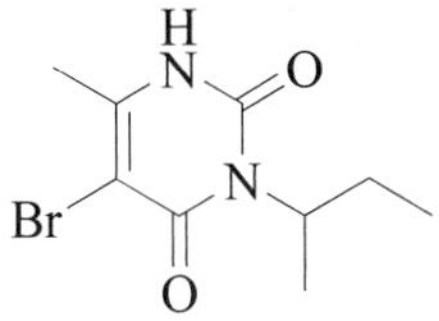

$C_9H_{13}BrN_2O_2$

CAS 314-40-9

USES

herbicide used in amounts up to 12 kg per acre to control a wide range of grasses and broad-leaf weeds.

A. PROPERTIES

molecular weight 261.1; melting point 158-159°C; vapor pressure 10^{-3} mbar at 100°C; density 1.55 at 25°C; solubility 815 mg/L at 25°C.

C. WATER AND SOIL POLLUTION FACTORS

Aqueous reactions

photodecomposition: the action of four months sunlight on dilute (1-10 ppm) aqueous solution of bromacil resulted in the formation of only one detectable photoproduct, 5-bromo-6-methyluracil in very low yield. The N-dealkylated photoproduct proved to be much less stable toward sunlight wavelengths, forming principally 6-methyluracil.

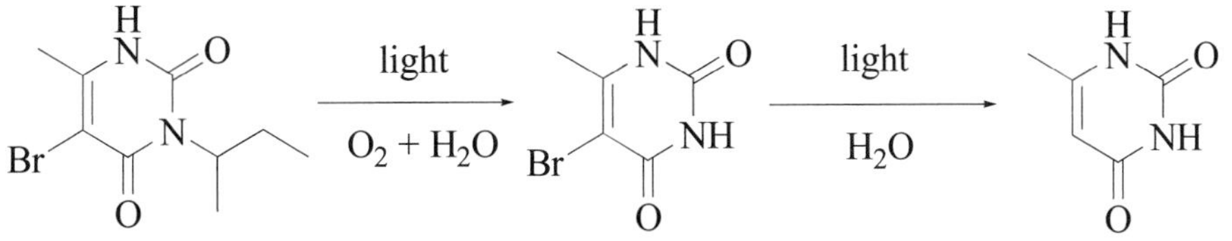

light, $O_2 + H_2O$ — light, H_2O

bromacil — 5-bromo-6-methyluracil — 6-methyluracil

(1638)

In groundwater wells in the U.S.A.-150 investigations up to 1988: median of the conc.'s of positive detections for all studies: 9 µg/L; maximum conc.: 22 µg/L. (2944)

D. BIOLOGICAL EFFECTS

Fishes			
Pimephales promelas	4d LC_{50}	186 mg/L	(2625)
bluegill	48h LC_{50}	71 mg/L	
rainbow trout	48h LC_{50}	75 mg/L	
carp	48h LC_{50}	164 mg/L	(2963)

Brominated polystyrene (Benzene,ethenyl-, homopolymer brominated)

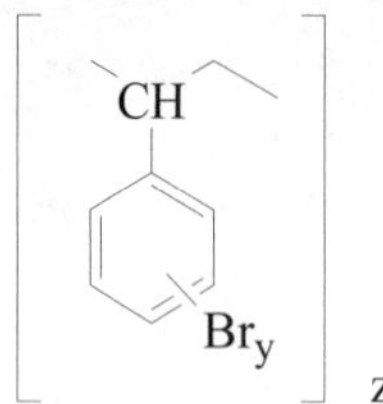

$(C_8H_xBr_y)_z$ (x = 5-6, y=2-3, z=4-100)

CAS 88497-56-7

TRADENAMES

Pyro-Chek; JM-631

USES

Used as a flame retardant, melt-blendable (at approx. 180°C) additive to polymeric products including acrylonitrile-butadiene-styrene products, polyethylene (high density), polyamide, polybutylene terephthalate, and others, and typically in conjunction with antimony oxide

A. PROPERTIES

molecular weight 1,000-200,000; typically 750-1,500; vapour pressure 2 x 10^{-5} Pa; relative density 2.1;

D. BIOLOGICAL EFFECTS

Toxicity

Mammals

Rat	oral LD_{50}	>5,000 mg/kg bw	(11030)

brominecyanide *see* cyanogenbromide

4-bromo-1,2-difluorobenzene (1-bromo-3,4-difluorobenzene)

$C_6H_3BrF_2$

CAS 348-61-8

EINECS 2064815

A. PROPERTIES

molecular weight 192.99; boiling point 150-151°C; density 1.7 kg/L at 25°C

D. BIOLOGICAL EFFECTS

Toxicity

Mouse	oral LD_{50}		

Toxicity

Algae	72h EC_{50}	10; 14 mg/L	(11221)
	72h NOEC	3.4; 3.9 mg/L	(11221)
Daphnids	48h EC_{50}	6.3 mg/L	(11221)
	21d EC_{50}	2.9 mg/L	(11221)
	21d NOEC	0.91 mg/L	(11221)
Fish	96h LC_{50}	7.8 mg/L	(11221)

2-bromo-2',5'-dimethoxyacetophenone

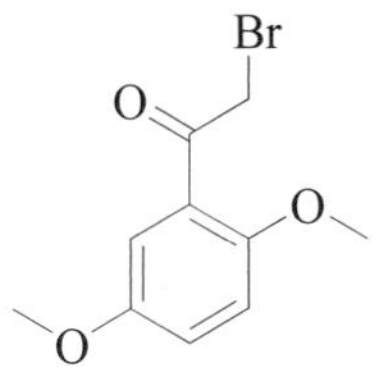

$(CH_3O)_2C_6H_3COCH_2Br$

$C_{10}H_{11}BrO_3$

CAS 1204-21-3

A. PROPERTIES

molecular weight 259.10; melting point 83-85°C.

D. BIOLOGICAL EFFECTS

Fishes

Pimephales promelas	4d LC_{50}	64 µg/L	(2625)

1-bromo-2-ethylbenzene

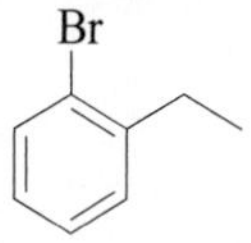

C_8H_9Br

CAS 1973-22-4

A. PROPERTIES

LogH -0.87 at 25°C.

3-bromo-4-hydroxybenzaldehyde

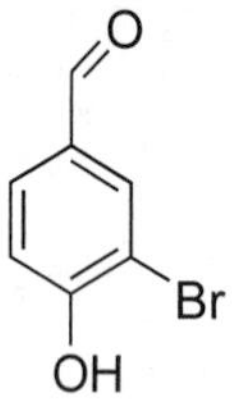

$C_7H_5BrO_2$

CAS 2973-78-6

OCCURRENCE

*Odonthaliafloccosa*is a marinered alga which, along withmanyothers,contains2,3-dibromo-4,5-dihydroxybenzylalcohol(lanosol). Cell-free homogenatesof*O.floccosa* have beenshowntoproduce 3-bromo-4-hydroxybenzaldehydefromtyrosine. The followingpathway wasproposed: L-tyrosine -->4-hydroxyphenylpyruvicacid-->4-hydroxyphenylaceticacid-->4-hydroxymandelicacid-->4-hydroxybenzaldehyde
-->3-bromo-4-hydroxybenzaldehyde.(see figure)Asimilarpathway, resultingin the formationof3-bromo-4-hydroxybenzoic acid,hasrecentlybeendescribed(16)for the diatom *Navicula incerta*and thehaptophyceanalga*Isochrysisgalbana.*(11187)

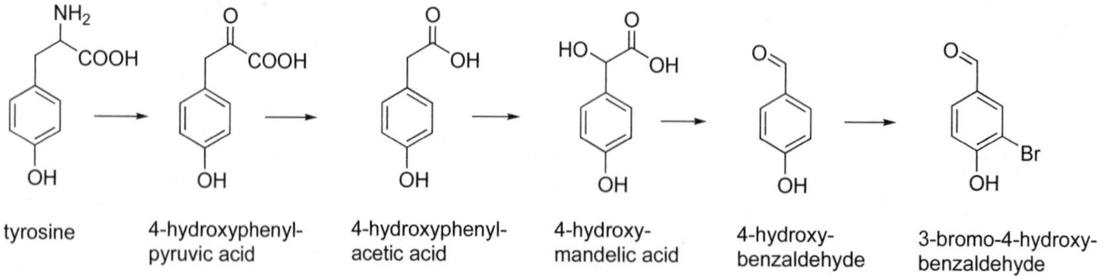

tyrosine → 4-hydroxyphenyl-pyruvic acid → 4-hydroxyphenyl-acetic acid → 4-hydroxy-mandelic acid → 4-hydroxy-benzaldehyde → 3-bromo-4-hydroxy-benzaldehyde

Production of 3-bromo-4-hydroxybenzaldehyde from the essential aminoacid tyrosine in the marine red alga *Odonthalia floccosa* [11187]

Production of 3-bromo-4-hydroxybenzaldehyde from the essential aminoacid tyrosine in the marine red alga *Odonthalia floccosa* (11187)

A. PROPERTIES

molecular weight 201.02

C. WATER AND SOIL POLLUTION FACTORS

3-bromo-4-hydroxybenzaldehyde → 3-bromo-4-hydroxybenzoic acid → phenol

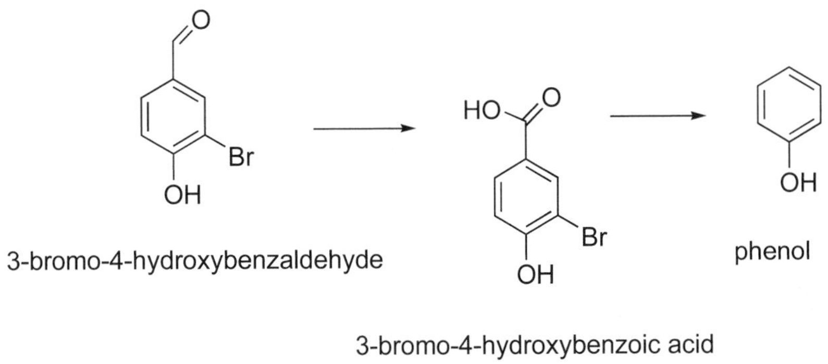

Transformation pathway of 3-bromo-4-hydroxybenzaldehyde by metabolically stable anaerobic enrichment cultures obtained from sediments from the Gulf of Bothnia and the Baltic sea (10821)

2-bromo-3-hydroxypyridine

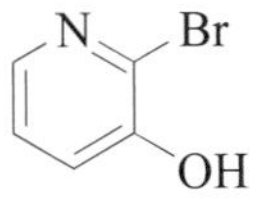

C_5H_4BrNO

CAS 6602-32-0

C. WATER AND SOIL POLLUTION FACTORS

Reductive dehalogenation in anaerobic lake sediment suspension at 20°C in the dark

half-life: 43 d, 3-hydroxypyridine detected as metabolite. (2700)

1-bromo-2-methylbutane

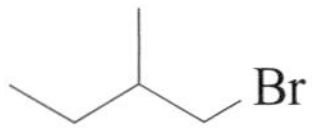

$C_5H_{11}Br$

CAS 10422-35-2

CAS 534-00-9 (S-)

A. PROPERTIES

LogH -0.02 at 25°C.

1-bromo-3-methylpentane

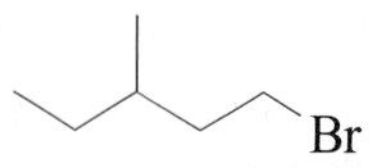

$C_6H_{13}Br$

CAS 51116-73-5

A. PROPERTIES

LogH 0.15 at 25°C.

5-bromo-5-nitro-1,3-dioxane

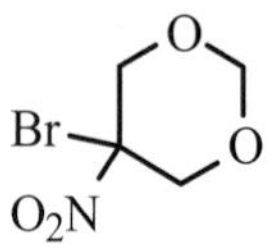

CAS 30007-47-7

USES

a preservative in cosmetic products. liquid soaps and cleaning agents. It may react with amines and amides to form nitrosamines or nitrosamides which are considered as carcinogenic substances. (7169)

D. BIOLOGICAL EFFECTS

Toxicity

Mammals

Rat (ingestion/inhalation)	oral LD_{50}	455-590	(7169)

2-bromo-4-nitrophenol

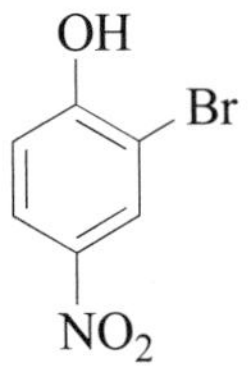

$C_6H_4BrNO_3$

CAS 5847-59-6

D. BIOLOGICAL EFFECTS

Fishes			
larvae of a sea lamprey:	LD_{100}	5 mg/L	(226)
rainbow trout:	LD_{10}	13 mg/L	(226)
brown trout:	LD_{10}	11 mg/L	(226)

3-bromo-4-nitrophenol

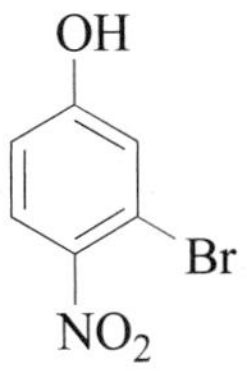

$C_6H_4BrNO_3$

D. BIOLOGICAL EFFECTS

Fishes			
larvae of sea lamprey:	LD_{100}	3 mg/L	(226)
rainbow trout:	LD_{10}	5 mg/L	(226)
brown trout:	LD_{10}	5 mg/L	(226)

2-bromo-2-nitropropane-1,3-diol (BNPD; 2-Bromo-2-nitro-1,3-propanediol)

$C_3H_6BrNO_4$

CAS 52-51-7

TRADENAMES

Bronopol

USES

A broad spectrum preservative with a wide range of antimicrobial properties.BNPD is used in cosmetic products, liquid soaps and cleaning agents. It is active against gram positive and gram negative bacteria, fungi and yeast. BNPD is a formaldehyde-releasing compound, also called a formaldehyde donor. In alkaline solution and at increasing temperature, it dissociates to form formaldehyde, bromide and nitrite.

A. PROPERTIES

White to slightly yellow crystals; molecular weight 199.99; melting point 130-133°C; vapour pressure 1.26 x 10^{-5} mm Hg; water solubility 250,000 mg/L; log Pow −0.64; H 1.3 x 10^{-11} atm.m^3/mol; K_{OC} 5;molecular weight 195.96

C. WATER AND SOIL POLLUTION FACTORS

Hydrolysis

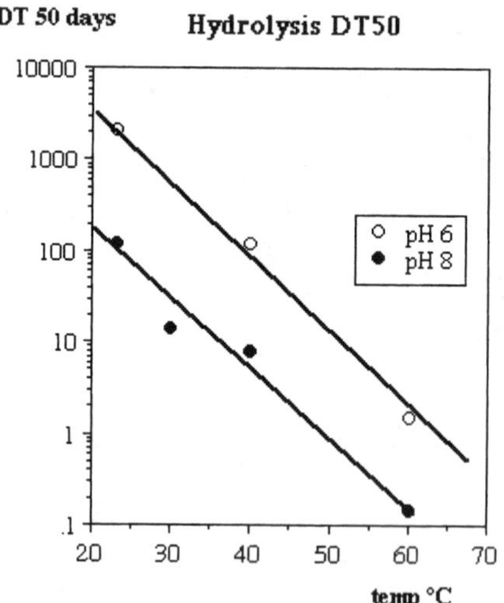

Biodegradation pathways

(7073)

$HOC(Br)(NO_2)CH_2OH$ $\xrightarrow[\text{light}]{H_2O}$ HO–C(OH)(NO₂)–OH

tris(hydroxy)nitromethane

Br–CH(NO₂)–CH₂OH + CH_3OH + H_2CO

bromonitroethanol

HCOOH

formic acid

Br^- + NOx + HNO_2 + HO–CH₂–COOH → $CO_2 + H_2O$

glycolic acid

$H_3C{-}NO_2$

nitromethane

Biodegradation (7169)

Inoculum	method	concentration	duration	elimination
activated sludge		1 mg/L	17 days	40% ThCO2
			21 days	80% ThCO2 + biomass

D. BIOLOGICAL EFFECTS

Algae			
Selenastrum capricornutum	72h EC_{50}	0.37 mg/L	
Chlorella vulgaris	72h EC_{50}	1.87 mg/L	
Chlorella	72h EC_{50}	0.02 mg/L	(7169)
	MIC	6 - 50 mg/L	(7073)
Scenedesmus subspicatus	72h EC_{50}	>1.0 mg/L	(7169)
Crustaceans			
Daphnia magna	48h EC_{50}	1.4 mg/L	
	48h Max NLD	0.56 mg/L	(7073)
Mysid shrimp	48h EC_{50}	5.9 mg/L	
	48h Max NLD	3.3 mg/L	(7073)
Mollusca			
Crassostrea giga	48h LC_{50}	0.78 mg/L	
Mysidopsis bahia	96h LC_{50}	0.59 mg/L	(7169)
Oyster larvae	48h NOEC	0.32 mg/L	
	48h Max NLD	0.56 mg/L	(7073)
Fishes			
Rainbow trout	96h LC_{50}	96h LC_{50}	
	96h NOEC	32 mg/L	
	96h Max NLD	32 mg/L	(7073)
Bluegill sunfish	96h LC_{50}	35.7 mg/L	
	96h NOEC	18 mg/L	
	96h Max NLD	32 mg/L	(7073)
Sheepshead minnow	96h LC_{50}	57.6 mg/L	
	96h NOEC	18 mg/L	
	96h Max NLD	18 mg/L	(7073)
Mammals			
Rat	oral LD_{50}	180-400 mg/kg bw	

Mouse	oral LD_{50}	250-500; 374 mg/kg bw	
Dog	oral LD_{50}	250 mg/kg bw	(7169)

Metabolism

BNDP and its breakdown products administered intravenously to rats and rabbits were excreted in the urine and expired air. BNPD did not accumulate in the organism. Metabolic breakdown products included 2-nitropropane-1,3-diol, which may be further metabolized to glycerol and CO_2. (7169)

β-bromo-β-nitrostyrene

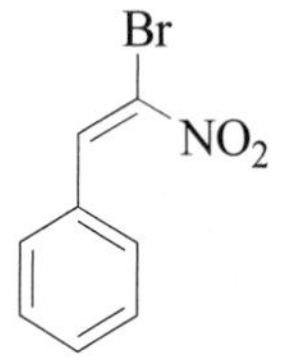

β-bromo-β-nitrostyrene

$C_8H_6BrNO_2$

CAS 7166-19-0

USES

biocide in recirculating cooling towers.

A. PROPERTIES

molecular weight 228.05.

C. WATER AND SOIL POLLUTION FACTORS

Biodegradation pathway

(7073)

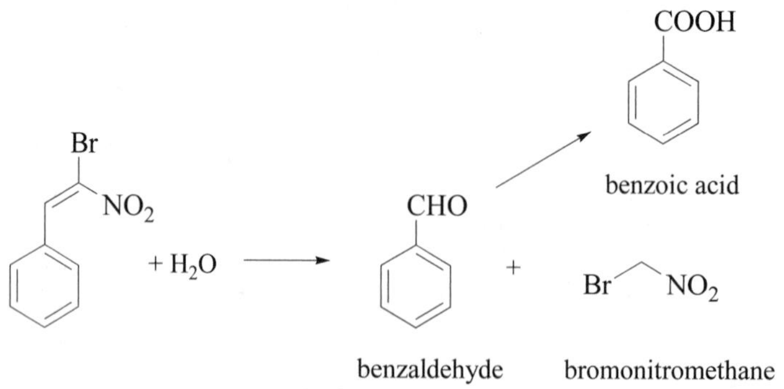

β-bromo-β-nitrostyrene + H_2O → benzaldehyde + bromonitromethane; benzaldehyde → benzoic acid

Detoxification reaction using sodium bisulfate. (7073)

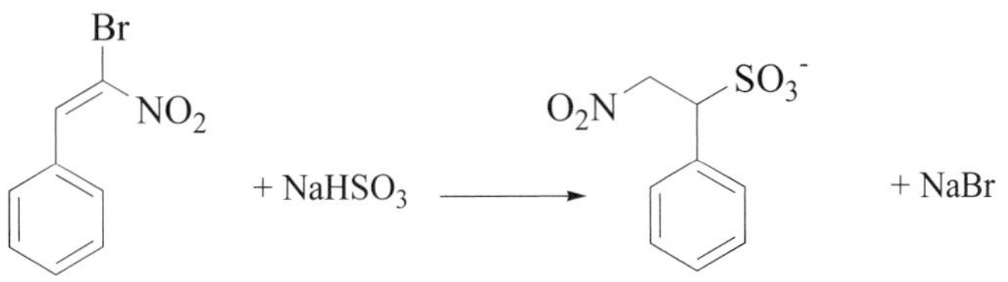

$$C_6H_5CH{=}C(Br)NO_2 + NaHSO_3 \longrightarrow O_2NCH_2CH(C_6H_5)SO_3^- + NaBr$$

D. BIOLOGICAL EFFECTS

Crustaceans			
Daphnia magna	48h EC_{50}	0.024 mg/L	
Mysid shrimp	96h LC_{50}	0.051 mg/L	(7073)
Fishes			
Rainbow trout	96h LC_{50}	0.027 mg/L	
Bluegill sunfish	96h LC_{50}	0.017 mg/L	
Sheepshead minnow	96h LC_{50}	0.057 mg/L	(7073)

1-bromo-3-phenylpropane ((3-bromopropyl)benzene; 3-phenylpropylbromide)

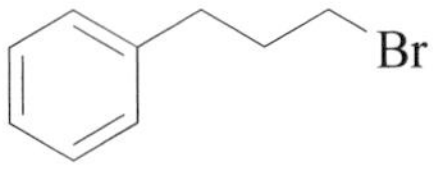

$C_6H_5(CH_2)_2CH_2Br$

$C_9H_{11}Br$

CAS 637-59-2

A. PROPERTIES

molecular weight 199.09; boiling point 219-220°C; density 1.31.

C. WATER AND SOIL POLLUTION FACTORS

In aqueous solution hydrolysis $t_{1/2}$ at 25°C and pH 7: 290 d. (2662)

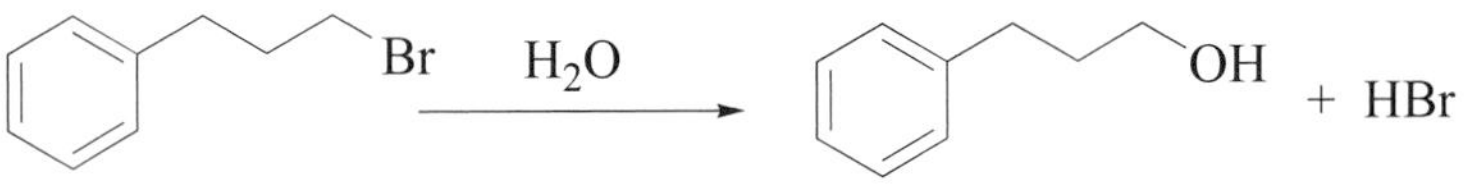

$$C_6H_5(CH_2)_3Br \xrightarrow{H_2O} C_6H_5(CH_2)_3OH + HBr$$

1-bromo-2-propanone (bromoacetone)

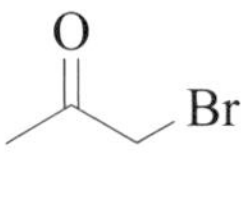

$$CH_3C(=O)CH_2Br$$

$BrH_2CC(O)CH_3$

C_3H_5BrO

CAS 598-31-2

A. PROPERTIES

molecular weight 136.99.

B. AIR POLLUTION FACTORS

Incinerability

thermal stability ranking of hazardous organic compounds: rank 136 on a scale of 1 (highest stability) to 320 (lowest stability). (2390)

5-bromo-3-sec-butyl-6-methyluracil *see* bromacil

2-bromoaniline

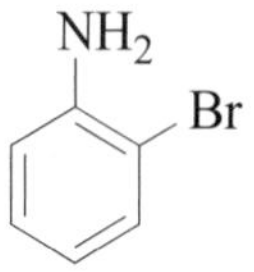

$C_6H_4NH_2Br$

C_6H_6BrN

CAS 615-36-1

A. PROPERTIES

molecular weight 172.02; melting point 29-31°C; boiling point 229°C.

D. BIOLOGICAL EFFECTS

RubisCo Test

inhib. of enzym. activity of Ribulose-P2-carboxylase in protoplasts: IC_{10}: 1.3 mmol/L.

Oxygen Test

inhib. of oxygen prod. of protoplasts: IC_{10}: 10 mmol/L. (2698)

GREEN ALGAE:

Scenedesmus subspicatus inhib. of fluorescence	IC_{10}: 0.032 mmol/L
Scenedesmus subspicatus growth inhib.	IC_{10}: 0.049 mmol/L

4-bromoaniline (p-bromoaniline; 4-bromobenzenamine; p-bromophenylamine)

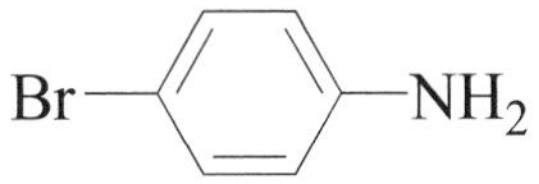

Br—C_6H_4—NH_2

$C_6H_4NH_2Br$

C_6H_6BrN

CAS 106-40-1

USES

Synthesis of azo dyestuffs and dihydroquinazolines.

A. PROPERTIES

molecular weight 172.02; melting point 60-64°C; density 1.5 at 99.6° C; $LogP_{OW}$ 2.08.

B. AIR POLLUTION FACTORS

Photochemical reactions

$t_{1/2}$: 2 d, based on reactions with OH° (calculated). (9870)

Significantly absorbs UV light above 290 nm in alcohol solution indicating a potential for direct photolysis in the environment . (9870)

C. WATER AND SOIL POLLUTION FACTORS

Waste water treatment

removal efficiency from water incubated with 1.000 units of horseradish peroxidase enzyme per liter, at pH 5.5: 84.5%. (2452)

Biodegradation

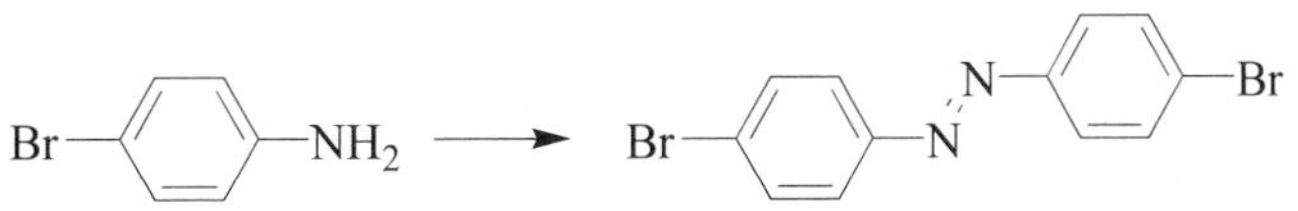

Br—C_6H_4—NH_2 → Br—C_6H_4—N=N—C_6H_4—Br

4,4'-dibromoazobenzene

Purified enzymes of the soil fungus *Geotrichum candidum* biotransformed 4-bromoaniline to 4, 4'-dibromoazobenzene. (9868)

A strain of Moraxella sp. used 4-bromoaniline as sole source of carbon and nitrogen. (9869)

Undergoes rapid and reversible covalent bonding with humic materials in aqueous solution. (9872)

Mobility in soils

K_{OC}: 7 on silt loam soils (9871)

D. BIOLOGICAL EFFECTS

Fishes			
Pimephales promelas	96h LC_{50}	48 mg/L	(2679)
Mammals			
mouse, rat	oral LD_{50}	289, 456 mg/kg bw	(9873, 9874)

m-bromobenzamide

O NH_2

Br

$C_6H_4C(O)NH_2Br$

C_7H_6BrNO

CAS 22726-00-7

A. PROPERTIES

molecular weight 200.04; melting point 155.3°C; solub 15.8 mg/L; $LogP_{OW}$ 2.48.

D. BIOLOGICAL EFFECTS

Pimephales promelas	24h LC_{50} 96h LC_{50}	100 mg/L 92.7 mg/L	(2709)

bromobenzene (phenylbromide)

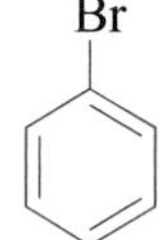

C_6H_5Br

CAS 108-86-1

USES AND FORMULATIONS

solvent (fats, waxes, or resins); intermediates in synthesis of specialty organic chemicals; additive to motor oil and fuels.

SOURCES

general lab use; use as solvent; discharge of waste motor oils to water; road surface runoff.

A. PROPERTIES

molecular weight 157.02; melting point -31°C; boiling point 156°C; vapor pressure 3.3 mm at 20°C; vapor density 5.4; density 1.50 at 15/15°C; solubility 500 mg/L at 20°C, 446 mg/L at 30°C, $LogP_{OW}$ 2.99 at 20°C; LogH -1.07 at 25°C.

B. AIR POLLUTION FACTORS

1 mg/m^3 = 0.15 ppm, 1 ppm = 6.53 mg/m^3.

Odor thresholds

30 mg/m^3; recognition 1.7-2.1 mg/m^3.	(748, 610)

D. BIOLOGICAL EFFECTS

Bacteria			
Photobacterium phosphoreum Microtox test	30 min EC_{50}	9.5 mg/L	(9657)
Crustaceans			
Ceriodaphnia dubia	24h LC_{50}	5.8 mg/L	(9875)
Fishes			
fathead minnow	96h LC_{50}	35.7 mg/L	(9875)
Mammals			
guinea pig, rabbit	oral LD_{50}	1,700, 3,300 mg/kg	(9876)
mouse, rat	oral LD_{50}	2,700, 2,699 mg/kg	(9877, 9878)

p-bromobenzoic acid

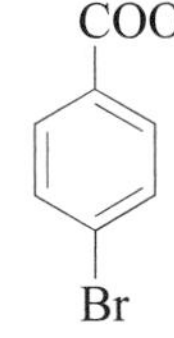

BrC_6H_4COOH

$C_7H_5BrO_2$

CAS 586-76-5

A. PROPERTIES

molecular weight 201.02.

D. BIOLOGICAL EFFECTS

Algae

Chlorella fusca BCF (wet wt): 25 (2659)

1-bromobutane (*n*-butylbromide)

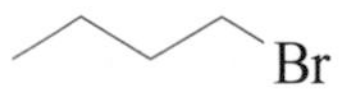

$CH_3CH_2CH_2CH_2Br$

C_4H_9Br

CAS 109-65-9

USES

In preparation of drugs.

A. PROPERTIES

molecular weight 137.02; melting point -112°C; boiling point 100-104°C; vapor density 4.7; density 1.28; $LogP_{OW}$2.75; LogH -0.30 at 25°C.

D. BIOLOGICAL EFFECTS

Fishes

Pimephales promelas	4d LC_{50}	36.7 mg/L	(2625)

bromochloromethane (chlorobromomethane; methylene-chlorobromide-; Fluorocarbon 1011; Halon 1011)

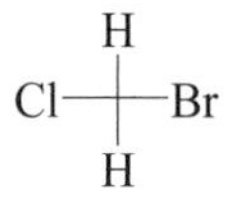

CH_2BrCl

CAS 74-97-5

USES

Solvent. Component in fire extinguishers. Diesel fuel additive. Nail varnish remover.

A. PROPERTIES

molecular weight 129.38; melting point -88°C; boiling point 68°C; density 1.99; solubility 16,700 mg/L at 25°C; $LogP_{OW}$1.4.

B. AIR POLLUTION FACTORS

Photochemical reactions

Does not absorb ultraviolet light at >290 nm which suggests direct photochemical degradation in the atmosphere or water is unlikely. (9882)
$t_{1/2}$: 160 d, based on reactions with OH° (calculated). (8415)

glc's at Tsukuba land site location (Japan):	0.90-1.4 ppt	
at Otake beach (Japan):	0.52-0.64 ppt	(2447)

C. WATER AND SOIL POLLUTION FACTORS

Calculated half-life in nonpolluted anaerobic aquifer at varying organic carbon contents

organic carbon content	1%	0.1%	0.01%	0.001%	
average half-life	0 d	1 d	15 d	278 years	(2695)

Hydrolysis

in water: $t_{1/2}$ is 44 yr. (9883)

Biodegradation

Inoculum	*method*	*conc.*	*duration*	*elimination results*	
domestic waste water	screening test	5-10 mg/L	7 d	100% product	(9879)

Reported to undergo microbial degradation by soil bacteria under anoxic conditions. (9881)

D. BIOLOGICAL EFFECTS

Mammals

mouse, rat	oral LD_{50}	4,300, 5,000 mg/kg bw	(9884, 9885)

o-bromocumene

A. PROPERTIES

LogH -0.62 at 25°C.

1-bromodecane (decylbromide)

Br

$CH_3(CH_2)_8CH_2Br$

$C_{10}H_{21}Br$

CAS 112-29-8

A. PROPERTIES

molecular weight 221.18; melting point -29.6°C; boiling point 238°C; density 1.07.

D. BIOLOGICAL EFFECTS

Crustaceans

Daphnia pulex	48h EC_{50}	284 µg/L	(2625)

bromodichloromethane (dichlorobromomethane)

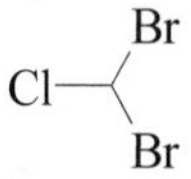

$CHBrCl_2$

CAS 75-27-4

USES AND FORMULATIONS

fire extinguisher fluid ingredient; solvent (fats, waxes, resins); synthesis intermediate; heavy liquid for mineral and salt separations. (347)

SOURCES

results from chlorination of finished water; use of fire extinguishers, lab use. (347)

A. PROPERTIES

colorless liquid; molecular weight 163.8, boiling point 90°C; melting point -55°C; vapor pressure 50 mmHg at 20°C; density 1.97 at 25/25°C.

C. WATER AND SOIL POLLUTION FACTORS

Hydrolysis

$t_{1/2}$: 137 year (calculated) at 25°C and pH 7. (9367)

Water quality

in N.W. England tap waters (1974):	1-27 ppb	(933)
in drinking water (Frankfurt, Germany, 1977): average:	7.4 µg/L; max. 10 µg/L	(2960)

Water treatment

Experimental water reclamation plant: sand filter effluent: 63 ng/L; after chlorination: 82 ng/L. (928)

Methanogenic laboratory column reactor at 35°C in the dark and a liquid detention time of 2 d, at influent conc. of 30 µg/L and 0 weeks acclimation: >99% removal at steady state. (2959)

Degradation in soil

Biodegradation half-lives in nonadapted aerobic subsoil: sandy clay Oklahoma: >242 d. (2695)

Calculated half-life in nonpolluted anaerobic aquifer at varying organic carbon contents

organic carbon content	1%	0.1%	0.01%	0.001%	
average half-life	0 d	0 d	5 d	87 years	(2695)

Biodegradation

Inoculum	*method*	*conc.*	*duration*	*elimination results*	
anaerobic			8 weeks	50%	(9886)
mixed methanogenic anaerobic			2 weeks	100%	(9887)
	static flask		28 d	51-59%	(8580)

D. BIOLOGICAL EFFECTS

Bacteria

methanogenic bacteria at 35°C	IC_{50}	1.6 mg/L	(8966)

Mammals

rat	oral LD_{50}	450 mg/kg bw	(9888)

bromoethane (ethylbromide; bromic-ether; Halon 2001; hydrobromic-ether)

Br

CH_3CH_2Br

C_2H_5Br

CAS 74-96-4

USES

Ethylating agent in organic synthesis. Refrigerant and extraction solvent. Investigated as a possible substitute for chlorofluorocarbons in compression heat pumps.

A. PROPERTIES

molecular weight 108.97; melting point -119°C; boiling point 37-40°C; vapor pressure 40 mm Hg at 21°C; vapor density 3.8; density 1.46; solubility 10,670 mg/L at 0°C; $\log P_{OW}$ 1.61; LogH -0.51 at 25°C.

C. WATER AND SOIL POLLUTION FACTORS

Hydrolysis

In aqueous solution hydrolysis to ethanol at 25°C and pH 7: half-life: 30 d. (2662)

Br $\xrightarrow{+H_2O}$ OH + HBr

Biodegradation

Biodegraded by *Acinetobacter sp.* strain GJ70. (9889)

D. BIOLOGICAL EFFECTS

Mammals			
rat	oral LD_{50}	1,350 mg/kg bw	(9701)

2-bromoethanol (ethylenebromohydrin)

$BrCH_2CH_2OH$

C_2H_5BrO

CAS 540-51-2

A. PROPERTIES

molecular weight 124.97; boiling point 56-57°C at 20 mm; density 1.76.

C. WATER AND SOIL POLLUTION FACTORS

Biodegradation

Inoculum	*method*	*concentration*	*duration*	*elimination results*	
activated sludge	respirometric test	70-100 mg/L DOC	28 days	17 % ThOD	(7040)

Bromoform (tribromomethane)

H Br / Br Br

$CHBr_3$

CAS 75-25-2

USES AND FORMULATIONS

pharmaceutical mfg.; ingredient in fire-resistant chemicals; gage fluid; heavy liquid in solid separations based on differences in specific gravity; geological assaying; solvent for waxes, greases, and oils. May be stabilized with 1-3% ethanol.

A. PROPERTIES

colorless liquid; molecular weight 252.77; melting point 6/7°C; boiling point 149°C; vapor pressure 5.6 mm at 25°C; vapor density 8.7; density 2.9 at 20/4°C; solubility 3,190 mg/L at 30°C.

B. AIR POLLUTION FACTORS

Odor

characteristic: chloroform-like, sweetish. (211)

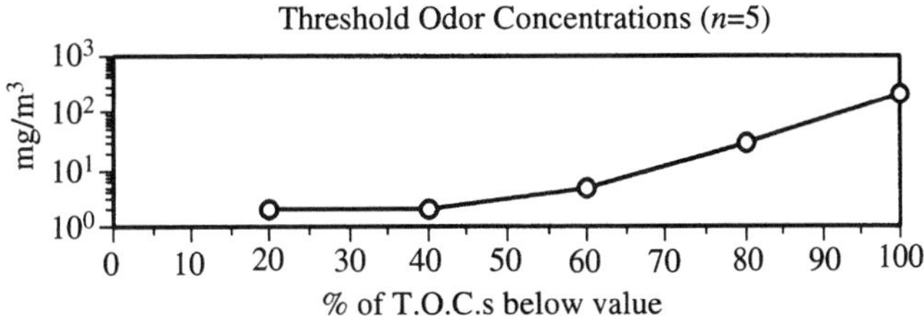

glc's at Tsukuba land site location (Japan): 0.79-1.71 ppt
at Otake beach (Japan): 0.91-2.18 ppt (2447)

Incinerability

Temperature for 99% destruction at 2.0 sec residence time under oxygen-starved reaction conditions: 585°C.
Thermal stability ranking of hazardous organic compounds: rank 202 on a scale of 1 (highest stability) to 320 (lowest stability). (2390)

1 ppm = 10.34 mg/m^3, 1 mg/m^3 = 0.0966.

C. WATER AND SOIL POLLUTION FACTORS

Odor threshold

detection: 0.3 mg/kg. (894)

Water quality

In N.W. England tap waters (1974): <0.01-2.5 ppb. (933)
Bromoform in tap water is caused by the chlorination of bromides in the raw water. (2915)
In tap water in Germany, 1980: average of 100 cities: 0.86 µg/L. (2915)
In drinking water (Frankfurt, Germany, 1977): average: 1.1 µg/L; max. 1.7 µg/L. (2960)

Water treatment
Experimental water reclamation plant

	sample 1	*sample 2*	
sand filter effluent	211 ng/L	154 ng/L	
after chlorination	1,723 ng/L	3,711 ng/L	
final water after A.C.	175 ng/L	135 ng/L	(928)

Methanogenic laboratory column reactor at 35°C in the dark and a liquid detention time of 2 d, at influent conc. of 34 µg/L and 0 weeks acclimation: >99% removal at steady state. (2959)

Degradation in soil and aquifer

biodegradation $t_{1/2}$ in nonadapted aerobic subsoil: sand: 400 d (2695)

Calculated half-life in nonpolluted anaerobic aquifer at varying organic carbon contents

organic carbon content	1%	0.1%	0.01%	0.001%	
average half-life	0 d	1 d	5 d	93 years	(2695)

D. BIOLOGICAL EFFECTS

Toxicity

Molluscs

Larvae of eastern oyster *(Crassostrea viginica)*	48h LC_{50},S	1 mg/L (after 48 h only approx. 30% of original	(1545)

		conc. was still present).	
Fishes			
Cyprinodon variegatus	96h LC_{50}	71 mg/L	
	MATC, early life stage	6.4 mg/L	(2643)
Bluegill *Lepomis macrochirus*	24h LC_{50},S	33 mg/L	
	96h LC_{50},S	29 mg/L	(2697)

Toxicity

Algae			
Pseudokirchneriella subcapitata	96h EC_{50}	38.6; 40.1 mg/L	
	96h NOEC	10 mg/L	
Skeletonema costatum	96h EC_{50}	12.3 mg/L	(10810)
Crustaceae			
Daphnia magna	24h LC_{50}	46 mg/L	
Daphnia pulex	96h EC_{50}	44 mg/L	
Americamysis bahia	96h LC_{50}	24.4 mg/L	
Penaeus aztecus	96h LC_{50}	26 mg/L	(10810)
Mollusca			
Crassostrea virginica	48h LC_{50}	1.5 mg/L	(10810)
Fish			
Cyprinodon Variegates	96h LC_{50}	7.1; 18 mg/L	
	28d NOEC	4.8 mg/L	
Cyprinus carpio, eggs	3-5d EC_{50}	52; 76; 80 mg/L	
Brevoortia tyrannus	96h LC_{50}	12 mg/L	(10810)

3-bromofuran

C_4H_3BrO

CAS 22037-18-1

A. PROPERTIES

molecular weight 146.97; boiling point 103°C; density 1.63.

C. WATER AND SOIL POLLUTION FACTORS

Reductive dehalogenation in anaerobic lake sediment suspension at 20°C in the dark

half-life: 3.7 d. (2700)

1-bromoheptane (heptylbromide)

$CH_3(CH_2)_5CH_2Br$

$C_7H_{15}Br$

CAS 629-04-9

A. PROPERTIES

molecular weight 179.10; melting point -58°C; boiling point 180°C; density 1.14.

C. WATER AND SOIL POLLUTION FACTORS

In aqueous solution hydrolysis to 1-heptanol at ambient temp. and pH 7. (2662)

1-bromoheptane → 1-heptanol

D. BIOLOGICAL EFFECTS

Fishes			
Pimephales promelas	4d LC_{50}	1.47 mg/L	(2625)

1-bromohexane (hexylbromide)

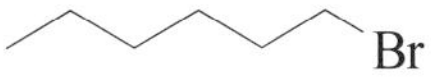

$CH_3(CH_2)_4CH_2Br$

$C_6H_{13}Br$

CAS 111-25-1

A. PROPERTIES

molecular weight 165.07; melting point -85°C; boiling point 154-158°C.

D. BIOLOGICAL EFFECTS

Fishes			
Pimephales promelas	4d LC_{50}	3.45 mg/L	(2625)

5-bromoindole

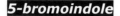

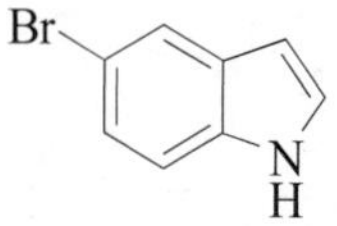

C_8H_5BrN

CAS 10075-50-0

A. PROPERTIES

molecular weight 196.05; melting point 90-92°C; $LogP_{OW}$: 2.97.

D. BIOLOGICAL EFFECTS

Fathead minnow: log BCF: 1.15. (2606)

bromomethane

see methylbromide

3-(bromomethyl)cyclohexene

D. BIOLOGICAL EFFECTS

Crustaceans			
Daphnia pulex	48h EC_{50}	12 mg/L	(2625)

2-(bromomethyl)tetrahydro-2H-pyran

D. BIOLOGICAL EFFECTS

Fishes			
Pimephales promelas	4d LC_{50}	205 mg/L	(2625)

1-bromonaphthalene

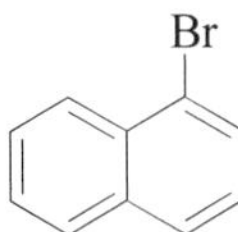

$C_{10}H_7Br$

CAS 90-11-9

A. PROPERTIES

molecular weight 207.07; melting point -1°C; boiling point 279-281°C; density 1.49; $LogP_{OW}$: 4.35.

C. WATER AND SOIL POLLUTION FACTORS

Micelle-water partition coefficient (K_{mw}) for the anionic surfactant dodecylsulfate: log K_{mw}: 3.53. (2361)

1-bromooctane (octylbromide)

$CH_3(CH_2)_6CH_2Br$

$C_8H_{17}Br$

CAS 111-83-1

A. PROPERTIES

molecular weight 193.13; melting point -55°C; boiling point 201°C; density 1.12.

D. BIOLOGICAL EFFECTS

Fishes

Pimephales promelas	4d LC_{50}	838 µg/L	(2625)

m-bromophenol (3-BP; 3-bromophenol)

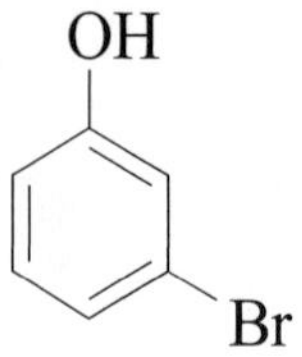

BrC_6H_4OH

C_6H_5BrO

CAS 591-20-8

A. PROPERTIES

molecular weight 173.02; melting point 33°C; boiling point 236°C; log P_{ow} 2.63.

B. AIR POLLUTION FACTORS

Odor threshold

recognition: 0.000007 mg/m^3. (712)

Manmade sources

In flue gas from incinerating chlorinated and brominated solvents: 0.024; 0.031-0.23 µg/m^3 (3- + 4-bromophenol) (11083)

C. WATER AND SOIL POLLUTION FACTORS

Biodegradation
Decomposition

rate in suspended soils: >72 d for complete disappearance		(175)
Degradation by *Pseudomonas:*	200 mg/L at 30°C:	
Ring disruption: parent:	51% in 96 h	
Ring disruption: mutant:	100% in 25 h	(152)

Environmental concentrations

Untreated sewage	Essex U.K. 1995	0.01 µg/L	
Treated effluent with peracetic acid	Essex U.K. 1995	0.001-0.007 µg/L	(11083)

D. BIOLOGICAL EFFECTS

Algae

Chlorella pyrenoidosa: 36 mg/L: toxic (41)

o-bromophenol

BrC_6H_4OH

C_6H_5BrO

CAS 95-56-7

A. PROPERTIES

molecular weight 173.02; melting point 5.6°C; boiling point 195°C; density 1.49 at 20/4°C; log P_{OW} 2.35.

B. AIR POLLUTION FACTORS

Manmade sources

In flue gas from hazardous waste incinerator incinerating chlorinated and brominated solvents : 0.036; 0.016-0.11 µg/m^3. (11083)

C. WATER AND SOIL POLLUTION FACTORS

Waste water treatment

degradation by *Pseudomonas:*	200 mg/L at 30°C	
ring disruption: parent:	100% in 85 h	
ring disruption: mutant:	100% in 14 h	(152)

Manmade sources

In untreated sewage effluents (U.K. 1995) : 0.003 µg/L (11083)

Biodegradation

In 3-day biodegradation tests, 2-BP (1 mg/litre) was degraded by 2% and 3% in river water and seawater, respectively, whereas 2,4,6-TBP (10 mg/litre) was degraded by 82% in river water and by 9% in seawater. The thermophilic bacterium Bacillus sp. transformed 2-BP to 3-bromocatechol (365 µmol/litre after 8 h) and 3-BP to 3- and 4-bromocatechol (21 µmol/litreafter 8 h).
The anaerobic biodegradation of 2-BP to phenol and the subsequent utilization of phenol by microorganisms enriched from marine and estuarine sediments from pristine and polluted sites were determined under iron-reducing, sulfidogenic, and methanogenic conditions. 2-BP was debrominated with the subsequent utilization of phenol under all three reducing conditions. Debromination of 3-BP and 4-BP was also observed under sulfidogenic and methanogenic conditions, but not under iron-reducing conditions. The production of phenol as a transient intermediate demonstrated that reductive dehalogenation is the initial step in the biodegradation of bromophenols under iron- and sulfate-reducing conditions. In the presence of added sulfate, 2-BP and phenol were completely degraded by a sulfate-reducing consortium of bacteria. In the absence of sulfate, 2-BP was dehalogenated, and phenol accumulated.
Debromination of 2,4,6-TBP and 2,6-DBP to 2-BP was more rapid than the debromination of the monobrominated phenols.(11083)

D. BIOLOGICAL EFFECTS

Concentrations in aquatic organisms

Location	year	species	conc µg/kg
Exmouth Gulf, W.Australia	1990	Brown and red macroalga	0.36-17
	1981	Bryozoa	1.3-2.4

	1990	Hydroid	2.2	
	1990	sponge	0.2-5.8	
Turimetta Head, Sydney	1998	*Polysiphonia sphaerocarpa*	0.4-1.0	
	1998	*Ulva lactuca*	0.1-3	
Hong Kong	2000	Molluscs	0.2-17.2	
		Crustacea	1.1-1.6	
	2000	Marine fish	0.5-30.8	
Anchor Point, USA		Marine fish	1.4-1.6	
Eastern coast Australia	1992	Marine fish	0.1-5.2	(11083)

Toxicity

Algae			
Scenedesmus subspicatus	48h EC_{50}	110 mg/L	
Tetrahymena pyriformis	60h EC_{50}	54.2 mg/L	(11083)
Chlorella pyrenoidosa	toxic	78 mg/L	(41)
Scenedesmus subspicatus (inhib. of fluorescence)	IC_{10}	9.8 mg/L	
Scenedesmus subspicatus (growth inhibition)	IC_{10}	17 mg/L	
Crustaceae			
Daphnia magna	24h EC_{50}	1.6; 13 mg/L	
	48h EC_{50}	0.9 mg/L	
	21d NOEC	0.2 mg/L	(11083)
Fish			
Mammals			
Rat	oral LD_{50}	652 mg/kg bw	(11083)

Biodegradation

decomposition rate in soil suspensions: 14 d for complete disappearance. (175)

RubisCo Test

inhibition of enzym activity of Ribulose-P2-carboxylase in protoplasts: IC_{10} 43 mg/L.

Oxygen Test

inhibition of oxygen production of protoplasts: IC_{10} 0.87 mg/L. (2698)

p-bromophenol (4-bromophenol; 4-BP; p-Bromohydroxybenzene; p-Bromophenic acid)

HO, Br

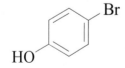

C_6H_5BrO

BrC_6H_4OH

C_6H_5BrO

CAS 106-41-2

OCCURRENCE

Metabolite of bromobenzene by rat and guinea pig.

A. PROPERTIES

molecular weight 173.0091; melting point 63.5°C; boiling point 238°C; density 1.84 at 15°C; solubility 14,200 mg/L at 15°C; log P_{OW} 2.95; log H -5.21 at 25°C.

C. WATER AND SOIL POLLUTION FACTORS

Biodegradation

Decomposition rate in soil suspensions: 16 d for complete disappearance (175)

Degradation by *Pseudomonas:* 200 mg/L at 30°C:

Ring disruption	parent	87%	in 85 h	
	mutant	100%	in 22 h	(152)

Dissociation

pH 5	0%
pH 7	1%
pH 8	6%
pH 9	40%

The degree of dissociation increase the solubility of the compound. (11083)

Biodegradation

4-BP was found to be a readily utilizable substrate for bacteria in estuarine sediments. Similar rates of 4-BP degradation at bromophenol-containing and non-bromophenol-containing locations show that adaptation of sediment bacteria by prior exposure to bromophenols is not required for degradation of these compounds.

Sulfidogenic consortia enriched from an estuarine sediment were maintained on monochlorophenols as the only source of carbon and energy for over 5 years. The culture was capable of degrading 4-BP (100 µmol/litre) within 6 days. Utilization of 4-BP yielded stoichiometric release of bromide. To verify that 4-BP was mineralized under sulfate-reducing conditions, the evolution of $^{14}CO_2$ from [^{14}C]4-BP was examined. A 4-BP concentration of 275 µmol/litre was depleted within 30 days, with concomitant release of bromide at 228 µmol/litre; this demonstrated that [^{14}C]4-BP was mineralized, with over 90% of the radiolabel recovered as carbon dioxide.

Isolated anaerobic bacteria from estuarine sediments of the Arthur Kill in the New York/New Jersey harbour, USA, twere capable of reductively dehalogenating 2,4,6-TBP to phenol. The organism was found to debrominate 2-BP, 4-BP, 2,4-DBP, 2,6-DBP, and 2,4,6-TBP, but not 3-BP or 2,3-DBP.(11083)

Environmental concentrations

River water	India 1988-89	Not detected in 4 polluted rivers	
Untreated effluent	Essex UK	0.0003 µg/L	
Treated effluent	Essex UK	0.04 µg/L	(11083)

Fate of p-bromophenol in wastewater treatment plants

96% in effluent	3.4% removed	0% biodegraded	(11083)

Mobility in soils

Log K_{OC} 2.4; 2.6 (11083)

D. BIOLOGICAL EFFECTS

Concentrations in aquatic organisms

Location	year	species	conc. µg/kg
Exmouth Gulf, W.Australia	1990	Brown and red macroalgae	0.1-13
	1981	Bryozoa	2.3-18
	1990	Hydroid	4.9
	1990	Sponge	0.4-62
Turimetta head, Sydney	1998	*Polysiphonia sphaerocarpa*	1-8
	1998	*Ulva lactuca*	0.2-70
Hong Kong	2000	Molluscs	n.d.-55.6
		Crustaceae	n.d.-47.9

	2000	Marine fish gut	206	
Eastern coast Australia	1992	Marine fish gut	n.d.- 100	(11083)

Bioaccumulation

BCF : 20 (calculated)

Toxicity

Protozoa			
Tetrahymena pyriformis	60h EC_{50}	36.1 mg/L	
Crustaceae			
Daphnia magna	48h EC_{50}	5.95 mg/L	(11088)
Mammals			
Mouse	oral LD_{50}	523 mg/kg bw	(11088)
Algae	oral LD_{50}	523 mg/kg bw	(11088)
Chlorella pyrenoidosa	toxic	36 mg/L	(41)

4-bromophenyl-3-pyridylketone

CAS 14548-45-9

A. PROPERTIES

$LogP_{OW}$: 2.97.

D. BIOLOGICAL EFFECTS

Algae			
Tetrahymena pyriformis: growth inhibition at 27°C	48 EC_{50}	0.15 mmol/L	(2704)
Fishes			
Pimephales promelas: 30-35 d FT test at 25°C	log LC_{50}	0.078 mmol/L	(2704)

1-bromopropane (propylbromide)

Br

$CH_3CH_2CH_2Br$

C_3H_7Br

CAS 106-94-5

A. PROPERTIES

molecular weight 122.99; melting point -110°C; boiling point 71°C; density 1.35; LogH -0.41 at 25°C.

C. WATER AND SOIL POLLUTION FACTORS

Br $\xrightarrow{+ H_2O}$ OH + HBr

1-bromopropane 1-propanol

in aqueous solution hydrolysis to 1-propanol at 25°C and pH 7, $t_{1/2}$: 26 d	(2662)

D. BIOLOGICAL EFFECTS

Fishes			
Pimephales promelas	4d LC_{50}	67.3 mg/L	(2625)

2-bromopropane (isopropylbromide)

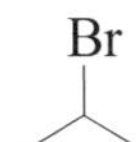

$(CH_3)_2CHBr$

C_3H_7Br

CAS 75-26-3

A. PROPERTIES

molecular weight 122.99; melting point -89°C; boiling point 59°C; density 1.31; LogH -0.35 at 25°C.

C. WATER AND SOIL POLLUTION FACTORS

In aqueous solution at 25°C and pH 7, $t_{1/2}$: 2.1 d.	(2662)

2-bromopyridine

$(C_5H_4N)Br$

C_5H_4BrN

CAS 109-04-6

A. PROPERTIES

molecular weight 158; boiling point 192-194°C; density 1.66.

C. WATER AND SOIL POLLUTION FACTORS

Reductive dehalogenation in anaerobic lake sediment suspension at 20°C in the dark

half-life: 12.4 d. (2700)

5-bromopyrimidine

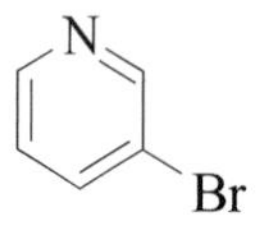

$(C_5H_3N_2)Br$

$C_5H_3BrN_2$

CAS 4595-59-9

A. PROPERTIES

molecular weight 158.99; melting point 71-73°C.

C. WATER AND SOIL POLLUTION FACTORS

Reductive dehalogenation in anaerobic lake sediment suspension at 20°C in the dark

half-life: 24 d. (2700)

2-bromothiazole

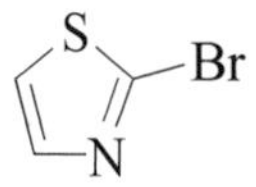

C_3H_2BrNS

CAS 3034-53-5

A. PROPERTIES

molecular weight 164.02; boiling point 171°C; density 1.82.

C. WATER AND SOIL POLLUTION FACTORS

Reductive dehalogenation in anaerobic lake sediment suspension at 20°C in the dark

half-life: 5.2 d, thiazole detected as metabolite. (2700)

3-bromothiophene

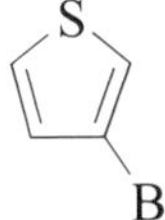

C_4H_3BrS

CAS 872-31-1

A. PROPERTIES

molecular weight 163.03; boiling point 150°C; density 1.74.

C. WATER AND SOIL POLLUTION FACTORS

Reductive dehalogenation in anaerobic lake sediment suspension at 20°C in the dark

half-life: 4.7 d, thiophene detected as metabolite. (2700)

D. BIOLOGICAL EFFECTS

Fishes

Pimephales promelas	4d LC_{50}	6.19 mg/L	(2625)

p-bromotoluene

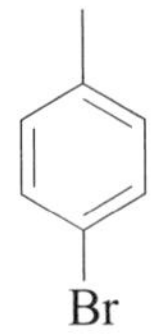

C_7H_7Br

CAS 75-63-13

A. PROPERTIES

LogH -1.11 at 25°C.

bromotrifluoromethane (bromofluoroform; F-13B1; Freon 13B1; Halon 1301; monobromotrifluoro methane)

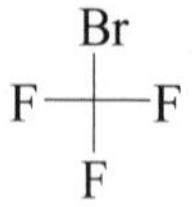

$CBrF_3$

CAS 75-63-8

A. PROPERTIES

molecular weight 148.92; LogH1.31 at 25°C.

5-bromovanillin (3-bromo-4-hydroxy-5-methoxy-benzaldehyde; Benzaldehyde, 3-bromo-4-hydroxy-5-methoxy)

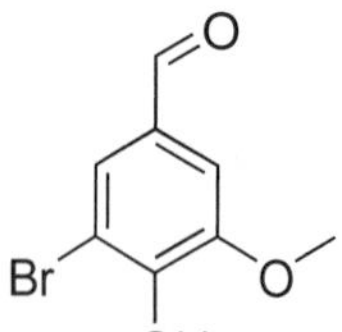

5-bromovanillin

$C_8H_7BrO_3$

CAS 2973-76-4

EINECS 21-016-6

A. PROPERTIES

molecular weight 231.06; melting point 161°C; log Pow 2.09

C. WATER AND SOIL POLLUTION FACTORS

Biodegradation

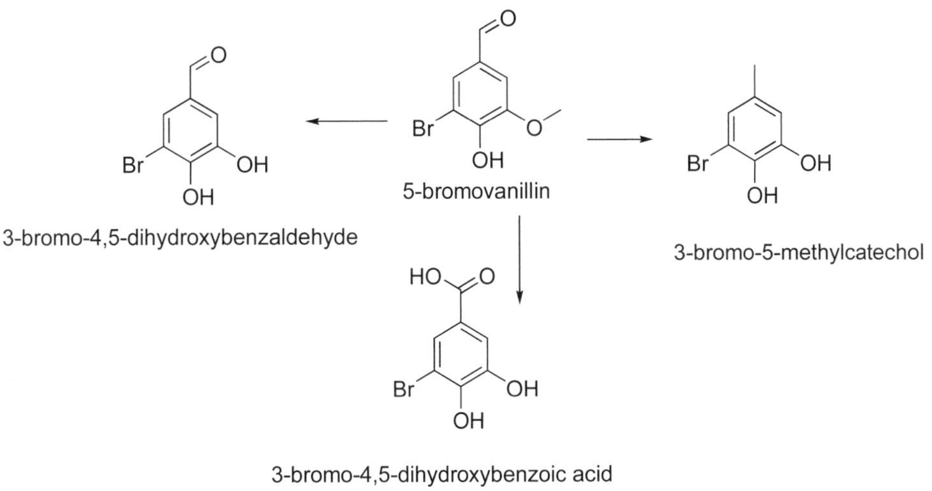

3-bromo-4,5-dihydroxybenzaldehyde

5-bromovanillin

3-bromo-5-methylcatechol

3-bromo-4,5-dihydroxybenzoic acid

Transformation pathway of 5-bromovanillin by metabolically stable anaerobic enrichment cultures obtained from sediments from the Gulf of Bothnia and the Baltic sea	(10821)

D. BIOLOGICAL EFFECTS

Toxicity

Fish	96h LC_{50}	59.7 mg/L (QSAR calculated)	(11152)

6-bromovanillin (2-bromo-4-hydroxy-5-methoxybenzaldehyde)

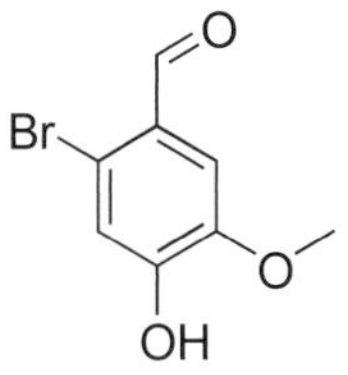

6-bromovanillin

$C_8H_7BrO_3$

CAS 60632-40-8

A. PROPERTIES

molecular weight 231.06

C. WATER AND SOIL POLLUTION FACTORS

Biodegradation

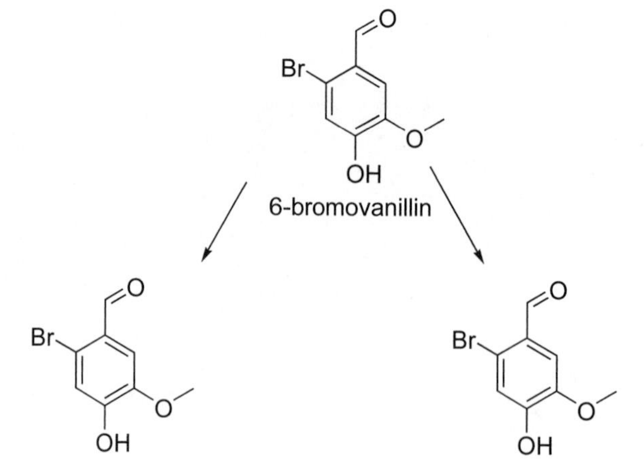

Transformation pathway of 6-bromovanillin by metabolically stable anaerobic enrichment cultures obtained from sediments from the Gulf of Bothnia and the Baltic sea (10821)

bromoxynil (3,5-dibromo-4-hydroxybenzonitrile; 3,5-dibromo-4-hydroxyphenyl cyanide; 2,6-dibromo-4-cyanophenol; Broxynil; Brominil; Buctril)

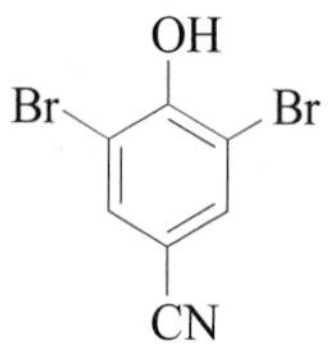

$C_7H_3Br_2NO$

$C_7H_3Br_2NO$

CAS 1689-84-5

EINECS 216-882-7

USES

Bromoxynil is a nitrile herbicide that is used for post-emergent control of annual broadleaved weeds. It is especially effective in the control of weeds in cereal, corn, sorghum, onions, flax, mint, turf, and on non-cropland. The compound works by inhibiting photosynthesis in the target plants.

A. PROPERTIES

White crystalline powder; molecular weight 276.9; melting point 188.7°C;melting point 194-195°C; 45-46°C (octanoate); 360°C (sodium salt); about 360°C (potassium salt); boiling point sublimes at 135°C at 0.2 mbar; boiling point 318.7°C; relative density 1.6 kg/L; vapour pressure 1.7 x 10^{-4} Pa at 25°C; H 5.3 x 10^{-4} Pa.m^3/mol;solubility 130 mg/L at 20-25°C

pH	Water solubility	$LogP_{OW}$
5	539 mg/L	1.3 (pH 2)
7	90 mg/L	1.04
9	>3100 mg/L	

B. AIR POLLUTION FACTORS

Photodegradation

T/2 (OH radicals): 12-51 days.

C. WATER AND SOIL POLLUTION FACTORS

Hydrolysis

No hydrolysis at pH 5, 7 and 9

Photodegradation in water

t.2 < 10 hours. (2 major degradation products)

Environmental concentrations

Traces (0.01 µg/L) of Bromoxynil were
 detected in two of 48 municipal water samples in Manitoba (Canada) and in one of 149 private wells in Ontario (detection limit 0.1 µg/L) (10953)

Biodegradation

In soil $t_{1/2}$: ca. 10 d. Degraded by hydrolysis and debromination to less toxic substances such as hydroxybenzoic acid. In plants, the ester and nitrile groups are hydrolysed, and debromination also occurs. (9600)

Flexibacterium rapidly degraded bromoxynil. After 5 weeks only 5% remained. Benzamide and benzoic acid metabolites were identified. (9893)

Inhibition of degradation

Inhibits nitrification in soil at 50 mg/kg. (9892)

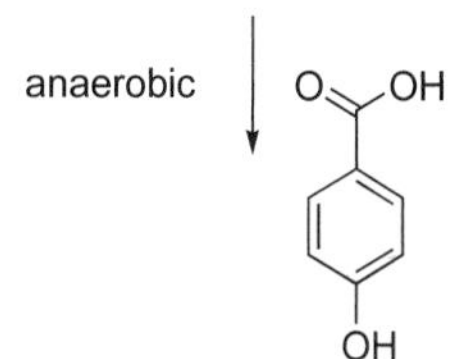

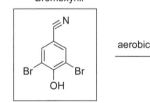

Degradation pathways of Bromoxynil under aerobic conditions by *Klebsiella pneumoniae* and under anaerobic conditions by *desulfitobacterium chlororespirans*

Degradation pathways of Bromoxynil under aerobic by *Klebsiella pneumoniae* and under anaerobic conditions by *Desulfitobacterium chlororespirans* (10948, 10950)

Biodegradation

	% mineralisation	% bound-residues	After
Biodegradation in soil aerobic	27-34%	73-74%	28 days

Relevant metabolites: 3,5-diBr-4-OH-benzamide; 3,5-diBr-4-OH benzoic acid,(10949)

Biodegradation in laboratory studies

degradation in soils	%	bromoxynil	3,5-diBr-4-OH-benzamide	3,5-diBr-4-OH-benzoic acid
Aerobic at 20°C	50	<1 day	1.5-5.2 days	<0.5 day
Aerobic at 20°C	90		<6.7 days	<5.3 days
Aerobic at 10°C	50	<2.5 days		

Biodegradation in water/sediment systems

DT50 water	9.6-16 days	
DT90 water	32-53 days	
DT50 whole system	9.6-16 days	
DT90 whole system	32-53 days	(10949, 10949)

Mobility in soils (10949)

	bromoxynil	3,5-diBr-4-OH-benzamide	3,5-diBr-4-OH-benzoic acid
K_{OC}	108-239	32-330	284-639

D. BIOLOGICAL EFFECTS

Toxicity

Algae			
Navicula pelliculosa	72h EC_{50}	0.12 mg/L	(10949)
Aquatic plants			
Lemna gibba	14d EC_{50}	0.033 mg/L	(10949)
Worms			
Eisenia foetida	14d EC_{50}	45 mg/kg soil	(10949)
Crustaceae			
Daphnia magna	48h EC_{50}	12.5 mg/L	
	21d NOEC	3.1 mg/L	(10949)
Insecta			
Honeybees	Acute oral LD_{50}	5 µg/bee	
	Acute contact LD_{50}	150µg/bee	(10949)
Birds			
Bobwhite quail	Acute oral LD_{50}	217 mg/kg bw	(10949)
Fish			
Oncorhynchus mykiss	21d NOEC	2 mg/L	
Lepomis macrochirus	96h LC_{50}	29.2 mg/L	(10949)
Pimephales promelas	4d LC_{50}	14 mg/L	(2625)
harlequin fish	48h LC_{50}	5.0 mg/L(potassium)	
rainbow trout	48h LC_{50}	0.15 mg/L (octanoate)	
goldfish	48h LC_{50}	0.46 mg/L	
catfish	48h LC_{50}	0.063 mg/L	(2963)
Mammals			
rat	oral LD_{50}	81 mg/kg bw	(10949)
	oral LD_{50}	190 mg/kg bw	
mouse	oral LD_{50}	110 mg/kg bw	
rabbit	oral LD_{50}	260 mg/kg bw	
dog	oral LD_{50}	100 mg/kg bw	(9600)

Toxicity of metabolite ***4-hydroxybenzonitrile 4-cyanophenol***

Oncorhynchus mykiss	96h LC_{50}	8.8 mg/L	
Daphnia magna	48h EC_{50}	33 mg/L	(10949)

Bromoxynil heptanoate (2,6-dibromo-4-cyanophenyl heptanoate; heptanoic acid, 2,6-dibromo-4-cyanophenyl ester)

$C_{14}H_{15}Br_2NO_2$

CAS 56634-95-8

USES

Bromoxynil heptanoate is a nitrile herbicide that is used for post-emergent control of annual broadleaved weeds. It is especially effective in the control of weeds in cereal, corn, sorghum, onions, flax, mint, turf, and on non-cropland. Bromoxynil heptanoate is readily hydrolysed under alkaline conditions to the the parent phenol Bromoxynil which is considered to be the active species. Bromoxynil works by inhibiting photosynthesis in the target plants.

A. PROPERTIES

White fine powder; molecular weight 389.1; melting point 441.°C: no boiling point up to 180°C; relative density 1.63 kg/L; vapour pressure $<10^{-7}$ Pa at 25°C;

pH	Water solubility	$LogP_{OW}$	Hydrolysis t/2
5	0.17 mg/L		11.7 days
7	0.08 mg/L	5.9	5.3 days
9	0.15 mg/L		4.1 days

B. AIR POLLUTION FACTORS

Photodegradation

T/2 (OH radicals) : 4 days. (10949)

C. WATER AND SOIL POLLUTION FACTORS

Photodegradation in water

T/2 = 18 hours . Major degradation products are bromoxynil and 2-Br-4-cyanophenyl heptanoate. The environmental behaviour of bromoxynil heptanoate is similar to bromoxynil octanoate.

bromoxynil heptanoate — hydrolysis + H_2O → bromoxynil + heptanoic acid

bromoxynil → 3,5-dibromo-4-hydroxybenzamide; bromoxynil → 3,5-dibromo-4-hydroxybenzoate

3,5-dibromo-4-hydroxybenzamide — anaerobic → 4-cyanophenol

3,5-dibromo-4-hydroxybenzamide — aerobic → 3,5-dibromo-4-hydroxybenzoate

3,5-dibromo-4-hydroxybenzoate — anaerobic → 4-hydroxybenzoate

Degradation pathways of Bromoxynil heptanoate under aerobic conditions by *Klebsiella pneumoniae* and under anaerobic conditions by *desulfitobacterium chlororespirans.* (according to the degradation pathway of bromoxynil octanoate)

Degradation pathways of Bromoxynil heptanoate under aerobic conditions by *Klebsiella* (10949, *pneumoniae* and under anaerobic conditions by *Desulfitobacterium chlororespirans.* (according to the degradation pathway of bromoxynil octanoate) 10950)

D. BIOLOGICAL EFFECTS

Toxicity

Algae			
Selenastrum capricornutum	120h EC_{50}	0.083 mg/L	(10949)
Aquatic plants			
Lemna gibba	14d EC_{50}	0.21 mg/L	(10949)
Worms			
Eisenia foetida	15d EC_{50}	29 mg/kg soil	(10949)
Crustaceae			
Daphnia magna	48h EC_{50}	0.031 mg/L	(10949)
Birds			
Bobwhite quail	Acute oral LD_{50}	379 mg/kg bw	(10949)
Fish			
Lepomis macrochirus	96h LC_{50}	0.029 mg/L	(10949)
Mammals			
Rat	oral LD_{50}	291 mg/kg bw	(10949)

Bromoxynil octanoate

(2,6-dibromo-4-cyanophenyl octanoate; octanoic acid, 2,6-dibromo-4-cyanophenyl ester)

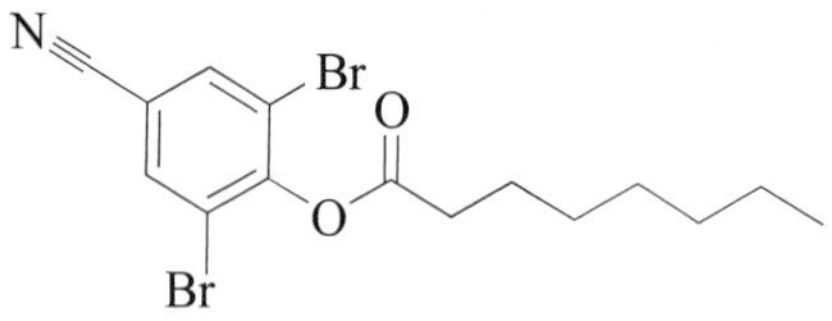

$C_{15}H_{17}Br_2NO_2$

CAS 1689-99-2

EINECS 216-885-3

USES

Bromoxynil octanoate is a nitrile herbicide that is used for post-emergent control of annual broadleaved weeds. It is especially effective in the control of weeds in cereal, corn, sorghum, onions, flax, mint, turf, and on non-cropland. Bromoxynil octanoate is readily hydrolysed under alkaline conditions to the the parent phenol Bromoxynil which is considered to be the active species. Bromoxynil works by inhibiting photosynthesis in the target plants.

A. PROPERTIES

White fine powder; molecular weight 403.0; melting point 45.3 °C; boiling point : no boiling poinr uo to 185°C; relative density 1.64 kg/L; vapour pressure $<10^{-7}$ Pa at 25°C;

pH	Water solubility	$LogP_{OW}$	Hydrolysis t/2
5	0.04 mg/L		34 days
7	0.03 mg/L	5.9	11 days
9	0.03 mg/L		1.7 days

B. AIR POLLUTION FACTORS

Photodegradation

T/2 (OH radicals)= 3.5 days

C. WATER AND SOIL POLLUTION FACTORS

Photodegradation in water

At pH 7.5: t.2 = 4.5 hours. Major degradation products are 2-Br-4-cyanophenyl octanoate and bromoxynil
At pH 5: t/2 = 18 hours.

Biodegradation in soil aerobic

% mineralisation	% bound-residues	After
64%	14%	90 days

Relevant metabolites: bromoxynil, 3,5-diBr-4-OH-benzamide and 3,5-diBr-4-OH-benzoic acid.(10949)

Biodegradation in laboratory studies

degradation in soils	%	Bromoxynil-octanoate	bromoxynil	3,5-diBr-4-OH-benzamide	3,5-diBr-4-OH-benzoic acid
Aerobic at 20°C	50	<1 day	<1 day	1.5-5.2 days	<0.5 day
Aerobic at 20°C	90			<6.7 days	<5.3 days
Aerobic at 10°C	50	<1.5 days	<2.5 days		
Anaerobic at 20°C	50	3.7 days (1)	<7 days (1)		(10949)

(1) water-sediment study

Biodegradation in water/sediment systems

	Bromoxynil octanoate	bromoxynil	
DT50 water	< 1 hours	9.6-16 days	
DT90 water	< 6 hours	32-53 days	
DT50 whole system	< 4 hours	9.6-16 days	
DT90 whole system	< 3 days	32-53 days	(10949, 10950)

Degradation pathways of Bromoxynil octanoate under aerobic conditions by *Klebsiella pneumoniae* and under anaerobic conditions by *desulfitobacterium chlororespirans*

Degradation pathways of Bromoxynil octanoate under aerobic conditions *by Klebsiella pneumoniae* and under anaerobic conditions *by Desulfitobacterium chlororespirans*. (10949, 10950)

Mobility in soils (10949)

	Bromoxynil-octanoate	bromoxynil	3,5-diBr-4-OH-benzamide	3,5-diBr-4-OH-benzoic acid
K_{OC}	4847 (calculated)	108-239	32-330	284-639

D. BIOLOGICAL EFFECTS

Bioaccumulation

Species	conc. µg/L	duration	BCF	
Lepomis macrochirus			230	(10949)

Toxicity

Algae			
Navicula pelliculosa	120h EC_{50}	0.043 mg/L	(10949)
Aquatic plants			
Lemna gibba	14d EC_{50}	>0.073 mg/L	(10949)
Crustaceae			
Daphnia magna	48h EC_{50}	0.046 mg/L	
	48h EC_{50}	>0.13 mg/L (1)	
	48h NOEC	0.09 mg/L (1)	
(1) with sediment	21d NOEC	0.0025 mg/L	(10949)
Insecta			
Chironomus riparius	22d NOEC	0.1 mg/L	
Honeybees	Acute oral LD_{50}	>119.8 µg/bee	
	Acute contact LD_{50}	>100 µg/bee	(10949)
Birds			
Bobwhite quail	Acute oral LD_{50}	170 mg/kg bw	(10949)
Fish			
Pimephales promelas	35d NOEC	0.0034 mg/L	
Oncorhynchus mykiss	96h LC_{50}	0.041 mg/L	
	96h LC_{50}	0.18 mg/L (1)	
	96h NOEC	0.093 mg/L (1)	(10949)
(1) with sediment			
Lepomis macrochirus	96h LC_{50}		
Mammals			
Rat	oral LD_{50}	238 mg/kg bw	(10949)

brucine (dimethyoxystrychnine)

$C_{23}H_{26}O_4N_2.2H_2O$ or $4H_2O$

CAS 357-57-3

USES

denaturing alcohol, lubricant additive, separation of racemic mixtures.

OCCURRENCE

in *Nux vomica* or *ignatica* seeds.

A. PROPERTIES

white crystalline alkaloid; very bitter taste; molecular weight 394.45; melting point 178°C.

B. AIR POLLUTION FACTORS

Incinerability

thermal stability ranking of hazardous organic compounds: rank 245 on a scale of 1 (highest stability) to 320 (lowest stability). (2390)

C. WATER AND SOIL POLLUTION FACTORS

Degradation in soil

In soil, the nitrile group is converted to the acid amide, then to the carbonic acid. The benzene ring is debrominated and hydroxylated. $t_{1/2}$ in heavy clay is approximately 14 d. (2962)

Impact on biodegradation processes

at 100 mg/L no inhibition of NH_3 oxidation by *Nitrosomonas* sp.

D. BIOLOGICAL EFFECTS

Fishes

Lepomis macrochirus: static bioassay in freshwater at 23°C, mild aeration applied after 24 h

material added	*% survival after*				*best fit 96h LC_{50}*	
mg/L	*24 h*	*48 h*	*72 h*	*96 h*	*mg/L*	
63	0				36	
40	40	30	20	20		
32	90	80	60	40		
25	100	100	100	100		(352)

Menidia beryllina: static bioassay in synthetic seawater at 23°C, mild aeration applied after 24 h

material added	*% survival after*				*best fit 96h LC_{50}*	
mg/L	*24 h*	*48 h*	*72 h*	*96 h*	*mg/L*	
32	80	60	40	20		
18	100	100	80	70	20	
10	100	100	100	90		(352)

busan 25

USES

microbicide; active ingredients: 2-(thiocyanomethylthio)-benzothiazol (13%) and 2-hydroxypropylmethanethiolsulphonate (11.7%).

D. BIOLOGICAL EFFECTS

Fishes

harlequin fish *(Rasbora heteromorpha):*

mg/L	*24 h*	*48 h*	*96 h*	*3 m (extrap.)*	
LC_{10},F	0.6	0.43	0.34		
LC_{50},F	1.0	0.57	0.42	0.07	(331)

busan 70

USES

microbicide; active ingredients: butanethiol sulphonate.

D. BIOLOGICAL EFFECTS

Fishes

harlequin fish *(Rasbora heteromorpha)*:

mg/L	*24 h*	*48 h*	*96 h*	*3 m (extrap.)*	
LC_{10},F	0.48	0.37	0.36		
LC_{50},F	0.76	0.47	0.43	0.3	(331)

busan 72

USES

microbicide; active ingredient: 2-(thiocyanomethylthio)-benzothiazol (60%).

D. BIOLOGICAL EFFECTS

Fishes

harlequin fish *(Rasbora heteromorpha)*:

mg/L	*24 h*	*48 h*	*96 h*	*3 m (extrap.)*	
LC_{10},F	0.08	0.044	0.031		
LC_{50},F	0.13	0.075	0.036	0.006	(331)
active ingredient: 15%					
LC_{10},F	1.4	0.64	0.34		
LC_{50},F	1.7	0.88	0.46	0.1	(331)

busan 74

USES

microbicide; active ingredients: 2-(thiocyanomethylthio)-benzothiazol (40%) and 2-hydroxypropylmethanethiolsulphonate (35%).

D. BIOLOGICAL EFFECTS

Fishes

harlequin fish *(Rasbora heteromorpha)*:

mg/L	*24 h*	*48 h*	*96 h*	*3 m (extrap.)*	
LC_{10},F	0.12	0.052	0.035		
LC_{50},F	0.21	0.084	0.045	0.001	(331)

busan 76

USES

microbicide; active ingredient: β-cyanoethyl-2,3-dibromopropionate (60%).

D. BIOLOGICAL EFFECTS

Fishes

harlequin fish *(Rasbora heteromorpha)*:

mg/L	*24 h*	*48 h*	*96 h*	*3 m (extrap.)*	
LC_{10},F	0.31	0.29	0.29		
LC_{50},F	0.47	0.35	0.31	0.26	(331)

busan 77

USES

microbicide; active ingredient: quaternary ammonium compound.

D. BIOLOGICAL EFFECTS

Fishes

harlequin fish *(Rasbora heteromorpha)*:

mg/L	*24 h*	*48 h*	*96 h*	*3 m (extrap.)*	
LC_{10},F	0.47	0.32	-		
LC_{50},F	0.66	0.39	0.17	0.01	(331)

butachlor (butoxymethyl)

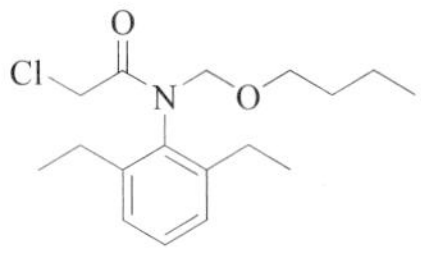

$C_{17}H_{26}ClNO_2$

CAS 23184-66-9

USES

butachlor is a member of the α-chloroacetanilide class of herbicides. It is used as a soil-applied preemergence herbicide in India for use in both direct seeded and transplanted rice fields.

A. PROPERTIES

molecular weight 311.9; melting point -5°C; boiling point 156°C at 0.66 mbar (decomposes at 165°C); density 1.07; solubility 20 mg/L at 20°C.

C. WATER AND SOIL POLLUTION FACTORS

Degradation

Two soil fungi, *Fusarium solani* and *Fusarium oxysporum,* degraded effectively more than 50% of butachlor within 3 d at 28°C in the dark. Degradation yielded after 4 d at least 32 metabolites that could be detected by GC-MS. The structure of 20 components could not be identified. The main degradative pathways involved dechlorination, hydroxylation, dehydrogenation, debutoxymethylation, C-dealkylation, N-dealkylation, O-dealkylation, and cyclization according to the following proposed partial metabolic pathways of butachlor by *Fusarium solani.*

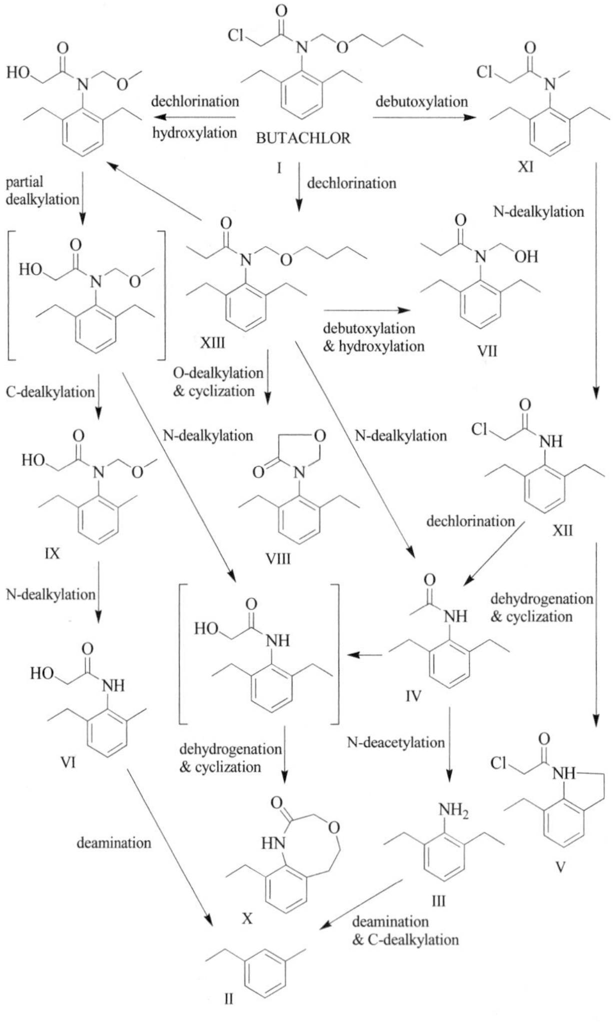

Proposed partial metabolic pathways of butachlor by *F. solani* (2689)

I	= butachlor
II	= m-ethyltoluene
III	= 2,6-diethylaniline
IV	= 2',6'-diethyl-acetanilide
V	= N-chloroacetyl-7-ethyl-2,3-dihydroindole
VI	= 2-hydroxy-2'-ethyl-6'-methylacetanilide

VII = 2',6'-diethyl-N-(hydroxymethyl)acetanilide
VIII = N-2',6'-diethylphenyl-2,6-dihydrooxazolone
IX = 2-hydroxy-2'-ethyl-6'-methyl-N-(methoxymethyl)acetanilide
X = m/z 205
XI = N-methyl-2-chloro-2',6'-diethylacetanilide
XII = 2-chloro-2',6'-diethylacetanilide
XIII = 2',6'-diethyl-N-(butoxymethyl)acetanilide

D. BIOLOGICAL EFFECTS

Anacystic sp.:	growth effect at ≤2 mg/L	
Nostoc muscorum:	lethal effect at 6 mg/L	
	photosynthesis effect at 1 mg/L	(2625)

Fishes			
rainbow trout	96h LC_{50}	0.52 mg/L	
bluegill	96h LC_{50}	0.44; 0.47 mg/L	
carp	96h LC_{50}	0.32; 0.76 mg/L	(2962, 2963)

1,3-butadiene (vinylethylene; divinyl; bivinyl; pyrrolylene; biethylene; erythrene)

$CH_2CHCHCH_2$

C_4H_6

CAS 106-99-0

USES

Principally in styrene-butadiene rubber; in latex paints; resins; organic intermediate.

A. PROPERTIES

colorless gas; molecular weight 54.09; melting point -108.9°C; boiling point -4.41°C; vapor pressure 2.5 mm at 20°C; 3.3 mm at 30°C; vapor density 1.87; density 0.62 at 20°C liquefied; solubility 500-550, 735 mg/L at 20°C; THC 618 kcal/mole, LHC 587 kcal/mole; LogP_{OW}1.85 (measured), 1.92- 1.99 (calculated); LogH 0.41 at 25°C.

B. AIR POLLUTION FACTORS

1 mg/m^3 = 0.45 ppm, 1 ppm = 2.25 mg/m^3.

Odor

characteristic: quality: undefined; hedonic tone: unpleasant to neutral.

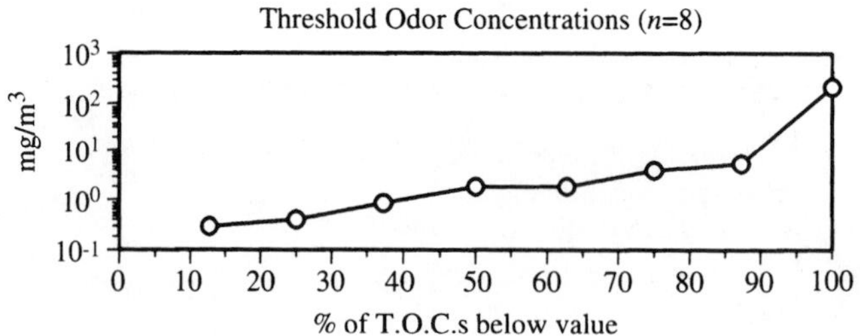

(19, 279, 307, 637, 737, 761, 794)

Odor

index (100% recognition): 770,000 (19)

Atmospheric reactions

R.C.R.: 4.31 (49)
reactivity: HC consumption: ranking: 1 (63)
estimated lifetime under photochemical smog conditions in S.E. England: 0.48 h (1699, 1700)

Atmospheric half-lives

for reactions with OH°: 0.1 d; 5.8 h-1.9 d (9078, 9079, 9080, 9081, 9082, 9083)
for reactions with O_3: 0.95 d (2716)

Manmade sources

emissions from cigarette smoking: 400 µg/cigarette: (2421)
Glc's downtown Los Angeles 1967: 10%ile: 1 ppb; average: 2 ppb; 90%ile: 5 ppb (64)
Expected glc's in U.S. urban air: range: 0.5- 10 ppb (102)
Sampling and analysis: photometry: min. full scale: 67 ppm (53)
Scrubbers: liquid lift type; 15 ml of liquid, 4.5 L/min gasflow, 4-min scrubbing

Trap method:	*% removal*	
H_2O, 0°C	0	
NH_2OH solution, 0°C	0	
H_2SO_4 (concn), 55°C	29	
H_2SO_4, 4% Ag_2SO_4 (concn), 55°C	87	
Open tube, -80°C	0	
Ethanol, -80°C	37	(311)

C. WATER AND SOIL POLLUTION FACTORS

Biodegradation

The overall catabolic pathway of butadiene as sole carbon and energy source by *Nocardia* sp. 249 is believed to be the following.

O_2 OH OH O COOH CO_2 COOH OH COOH O COOH CO_2 COOH

(427)

D. BIOLOGICAL EFFECTS

Bioaccumulation

	Concentration (µg/L)	Exposure period	BCF	Organ/tissue	
Fishes: *Pimephales promelas*			13	total body	(9084)

Crustaceans			
Daphnia magna	96 h EC_{50}	25 mg/L	(9084)
Fishes			
Lepomis macrochirus	96 h LC_{50}	38 mg/L	
Pimephales promelas	96 h LC_{50}	50 mg/L	
	32 d MATC*	7.3 mg/L	(9084)
Ictalurus punctatus	96 h LC_{50}	21 mg/L	
Oncorhynchus mykiss	96 h LC_{50}	22 mg/L	(9084)
* MATC=SQRT(NOEC*LOEC).			
Mammals			
Rat	oral LD_{50}	5,480 mg/kg body wt	

1-butanal *see* butyraldehyde

butanamide *see* butyramide

n-butane

$CH_3CH_2CH_2CH_3$

C_4H_{10}

CAS 106-97-8

A. PROPERTIES

colorless gas; molecular weight 58.14; melting point -135/138°C; boiling point -1°C; vapor pressure 1,823 mm at 25°C, 2.9 atm at 30°C; vapor density 2; density 0.60 liquefied; solubility 61 mg/L at 20°C, 30 mg/L at 15°C, 21 mg/L at 38°C; LHC 636 kcal/mole; LogH 1.58 at 25°C.

B. AIR POLLUTION FACTORS

1 mg/m^3 = 0.41 ppm, 1 ppm = 2.42 mg/m^3.

Odor

T.O.C.:
12,100 mg/m^3 = 4,960 ppm
not detectable <5,000 ppm (211)
recognition: 6,160 mg/m^3 (761)
3,000 mg/m^3 (737)
5.5 ppm (279)

Atmospheric reactions

R.C.R.: 0.79 (49)
reactivity: NO ox.: ranking: 0.1 (63)
Estimated lifetime under photochemical smog conditions in SE England: 15 h (1699, 1705)

Atmospheric half-lives

for reactions with OH°: 3 d
for reactions with O_3: 800,000 d (2716)
for reactions with hydroperoxyl radical HO^0_2: <13,000 d

Manmade sources

glc's at Pt Barrow, Alaska, Sept. 1967: 0.03 to 0.19 ppb (101)
glc's downtown Los Angeles 1967:
0.13 to 0.30 ppm C (52)
10% ile: 20 ppb
average: 46 ppb
90% ile: 80 ppb (64)
glc's East Gabriel Valley 1967: 0.098 to 0.175 ppm C (52)
expected glc's in U.S.A. urban air: range: 0.05 to 0.45 ppm (102)
in flue gas of municipal incinerator: <0.4 ppm (196)
exhaust gas of diesel engine: 5.3% of emitted HCs (72)
combustion gas of household central heating
appr. 50 ppm at 7% CO_2 system on gas oil:
3.3 g/kg gas oil at 6% CO_2
1.6 g/kg gas oil at 7% CO_2 (182)
in gasoline: 4.3-5 vol % (312)
in auto exhaust-gasoline engine:
62-car survey: 5.3 vol % of total exhaust HCs (391)
15-fuel study: 4 vol % of total exhaust HCs (392)
engine variable study: 2.3 vol % of total exhaust HCs (393)
evaporation from gasoline fuel tank: 16-48 vol % of total evaporated HCs
evaporation from carburator: 9-23 vol % of total evaporated HCs (398, 399, 400, 401, 402)

C. WATER AND SOIL POLLUTION FACTORS

Incubation with natural flora in the groundwater-in presence of the other components of high-octane gasoline (100 µl/L): biodegradation: 0% after 192 h at 13°C and initial conc. 63 µl/L. (956)

butane dioxime *see* dimethylglyoxime

1,4-butanedicarboxylic acid *see* adipic acid

butanedinitrile *see* succinonitrile

butanedioic acid *see* succinic acid

1,4-butanediol (tetramethyleneglycol; 1,4-dihydroxybutane; 1,4-butyleneglycol)

HO‿‿‿OH

$(CH_2CH_2OH)_2$

$C_4H_{10}O_2$

CAS 110-63-4

EINECS 203-786-5

USES

intermediate for the mfg. of polyurethanes, polyesters. Plasticizer, solvent, textile additive, moisturizer.

A. PROPERTIES

molecular weight 90.1222;melting point 16°C; boiling point 230°C; density 1.020 at 20/4°C; vapor pressure <1 hPa at 20°C; water solubility 100,000 mg/L at 25°C;log P_{OW} 0.88 (measured).

B. AIR POLLUTION FACTORS

Photodegradation

T/2 = 24 hours based on reaction with OH radical.

C. WATER AND SOIL POLLUTION FACTORS

Biodegradation

adapted A.S.-product as sole carbon source at 200 mg/L-98.7% COD removal at 40.0 mg (327)

COD/g dry inoculum/h.

Waste water treatment

reverse osmosis: 65.9% rejection from 0.01 M solution. (221)

Biodegradability

inoculum	*test conc.*	*test duration*	*removed*	
not mentioned	400 mg/L	10 d	95% TOC	(5246)
A.S., industrial	430 mg/L	3 hours	6% DOC	
		7 d	97% DOC	(5247)
BOD:		5 d	5% COD	(5247)

Biodegradation

Inoculum	method	concentration	duration	elimination	
A.S.	OECD 301C		14 days	83% ThOD	
				94% TOC	
				100% GC	(10690)

D. BIOLOGICAL EFFECTS

Bacteria			
Pseudomonas putida	EC_{10}	10,000 mg/L	(5245)
Crustaceans			
Daphnia magna Straus	24-48h EC_0	500 mg/L	
	24-48h EC_{100}	>500 mg/L	(5244)
Fishes			
Leuciscus idus	96h LC_{50}	>10,000 mg/L	(5240)
bluegill sunfish, goldfish, rainbow trout, trout		not toxic after 24 h exposure at 5 mg/L	(5241)

Toxicity

Algae			
Selenastrum capricornutum	72h ECb_{50}	>1,000 mg/L	
	72h NOECb	>1,000 mg/L	(10690)
Crustaceae			
Daphnia magna	48h EC_{50}	>1,000 mg/L	
reproduction	21d EC_{50}	>85 mg/L	
	21d NOEC	>85 mg/L	(10690)
Fish			
Oryzias latipes	96h LC_{50}	>100 mg/L	
	14d LC_{50}	>100 mg/L	(10690)
Mammals			
Rat	oral LD_{50}	1,525; 1,500-2,500 mg/kg bw	
Mouse	oral LD_{50}	2,062 mg/kg bw	
Rabbit	oral LD_{50}	2,531 mg/kg bw	
Guinea pig	oral LD_{50}	1,200 mg/kg bw	(10690)

Metabolism

1,4-butanediol is rapidly absorbed and metabolized to γ-hydroxybutyric acid by the enzyme alcohol dehydrogenase in animals as well as humans

butanenitrile *see* butyronitrile

butanethiol *see* *n*-butylmercaptan

butanethiolsulfonate *see* busan 70

n-butanol (*n*-butylalcohol; propylcarbinol; 1-butanol)

$CH_3CH_2CH_2CH_2OH$

$C_4H_{10}O$

CAS 71-36-3

A. PROPERTIES

colorless liquid; molecular weight 74.12; melting point -89.9°C; boiling point 117.7°C; vapor pressure 4.4 m at 20°C, 6.5 mm at 25°C, 10 mm at 30°C; vapor density 2.55; density 0.810 at 20/4°C; solubility 77,000 mg/L; THC 638.2 kcal/mole, LHC 597.0 kcal/mole; saturation concentration in air 20 g/m^3 at 20°C, 39 g/m^3 at 30°C; log P_{OW} 0.88; log H -3.46 at 25°C.

B. AIR POLLUTION FACTORS

1 mg/m^3 = 0.33 ppm, 1 ppm = 3.03 mg/m^3.

Odor

characteristic: quality: rancid, sweet; hedonic tone: neutral to pleasant.

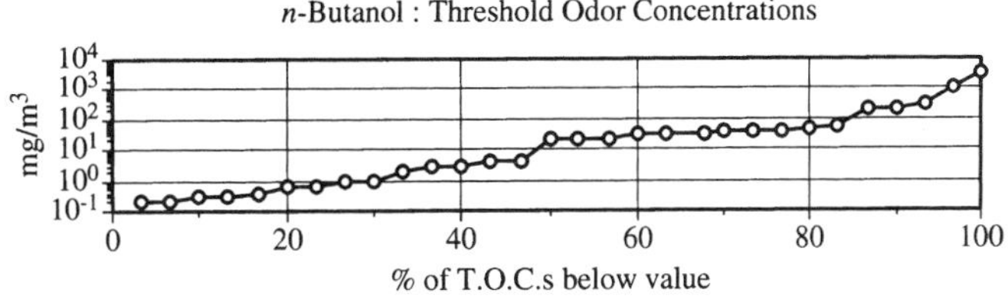

(73, 210, 278, 291, 307, 610, 627, 634, 641, 663, 704, 706, 707, 708, 709, 715, 737, 749, 756, 776, 785, 788, 804)

Odor

index:
(100% detection): 2,600 (19)
threshold for unadapted panelists: 50 ppm
threshold after adaption with pure odorant: 10,000 ppm (204)
distinct odor: 48 mg/m^3 = 16 ppm (278)

Manmade sources

glc's at Pt Barrow, Alaska, Sept. 1967: 51 to 126 ppb (101)
concentrations of 5 to 100 ppm have been reported in workrooms (211)

Control methods

Comparison of catalyst performance at space velocity of 80,000/h:

catalyst	*feed conc., ppm*	*reactor inlet temp., °C*	*% odor removal ASTM*	*% conversion GC*	
0.5% Pt on γ-Al_2O_3	1,405	225	96	100	
0.1% Pt on γ-Al_2O_3	1,405	236	95	100	
0.5% Pd on γ-Al_2O_3	1,405	376	96	100	
10% Cu on γ-Al_2O_3	1,405	411	94	100	(1221)

C. WATER AND SOIL POLLUTION FACTORS

Biodegradability

of higher alcohols in the activated sludge process and in BOD 5 d test decreases with increasing number of carbon atoms as illustrated by the figure above (only highest values found have been reported).

BOD2d	43% ThOD	(10548)
BOD_5	33; 58; 55; 42-79; 58-77% ThOD	(10548, 220, 30, 41, 271, 280, 27)
$BOD_5$20°C	61% ThOD (acclim.)	(2666)
BOD20°C	73% ThOD	(30)
BOD_{10d}	61% ThOD	(10548)
BOD_{30d}	54% ThOD	(10548)
COD	73; 92% of ThOD	(220, 27)
TOC	97% of ThOD	(220)
ThOD	2.59	(30)

n-Butanol : Threshold Odor Concentrations in Water

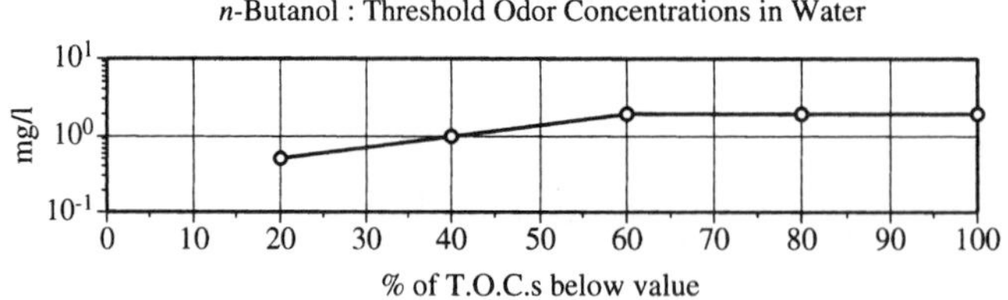

(97, 294, 326, 875, 883, 889, 894, 906, 907)

20% of the population still able to detect odor at 1.5 ppm
10% of the population still able to detect odor at 1.2 ppm
1% of the population still able to detect odor at 0.44 ppm (321)

Manmade sources

in year-old leachate of artificial sanitary landfill: 0.21 g/L (1720)

Waste water treatment

A.C.: adsorbability: 0.107 g/g C, 53% reduction, infl.: 1,000 mg/L, effl.: 466 mg/L (32)
Reverse osmosis: 41.3% rejection from a 0.01 M solution (221)
Stabilization pond design: toxicity correction factor: 2.0 at 4,000 mg/L influent (179)
Anaerobic lagoon: 22 lb COD/d/1,000 cu ft: infl.: 170 mg/L, effl.: 75 mg/L
48 lb COD/d/1,000 cu ft: infl.: 170 mg/L, effl.: 80 mg/L (37)
A.S.:
after 6 h: 16% of ThOD
after 12 h: 31% of ThOD
after 24 h: 36% of ThOD (88)

methods	*temp, °C*	*d observed*	*feed, mg/L*	*d acclim.*	*% removed*	
A.S., W	20	1	500	24	44% ThOD	
A.S., BOD	20	1/3-5	333	30	96% product	(93)

	influent ppm	carbon dosage	effluent ppm	% reduction	
A.C.:	1,000	10×	249	75	
(n.s.i.):	500	10×	163	67	
	100	10×	52	48	(192)

Adapted A.S.: product as sole carbon source: 98.8% COD removal at 84.0 mg COD/g dry inoculum/h. (327)

Impact on biodegradation processes

50% inhibition of NH_3 oxidation in *Nitrosomonas* at 8,200 mg/L. (407)

Partition coefficients

$K_{air/water}$	0.00038	
Cuticular matrix/air partition coefficient	5000 ± 0	(7077)
Cuticular matrix/water partition coefficient**	1.9	(7077)

* experimental value at 25°C studied in the cuticular membranes from mature tomato fruits (*Lycopersicon esculentum* Mill. cultivar Vendor)
** calculated from the Cuticular matrix/water partition coefficient and the air/water distribution coefficient

D. BIOLOGICAL EFFECTS

Toxicity

Bacteria			
Pseudomonas putida	16h EC_0	650 mg/L	(1900)
Bacteria Inhibition of nitrification	EC_{50}	8,200 mg/L.	(2624)
Microtox*Vibrio fisheri*	15min EC_{50}	910 mg/L	(7025)
Biotox™ test *Photobacterium phosphoreum*	5min EC_{50}	1,560 mg/L	
Microtox™ test *Photobacterium phosphoreum*	5min EC_{50}	3,170 mg/L	(7023)
Protozoans			
protozoa *(Entosiphon sulcatum):*	72h EC_0	55 mg/L	(1900)
protozoa (*Uronema parduczi* Chatton-Lwoff):	EC_0	8.0 mg/L	(1901)
Algae			
Microcystis aeruginosa	8d EC_0	100 mg/L	(329)
Scenedesmus quadricauda):	7d EC_0	875 mg/L	(1900)
RubisCo Test: inhib. of enzym. activity of Ribulose-P2-carboxylase in protoplasts	IC_{10}	2,594 mg/L.	
inhib. of oxygen production of protoplasts	IC_{10}	44 mg/L	(2698)
Chlorella pyrenoidosa	toxic	8,500 mg/L	(41)

Scenedesmus subspicatus inhib. of fluorescence	IC_{10}	889 mg/L	
Scenedesmus subspicatus growth inhib.	IC_{10}	222 mg/L	
Crustaceans			
Daphnia magna	48h IC_{50}	1,880 mg/L	(7025)
Fishes			
creek chub	24h LC_0	1,000 mg/L	(243)
	24h LC_{100}	1,400 mg/L	(243)
fathead minnow	LC_{50},S (1; 24; 48; 72; 96h):	1,950; 1,950; 1,950; 1,950; 1,910 mg/L	
	LC_{50},S (1; 24; 48; 72; 96h):	1,940 mg/L	(350)
Bleak *(Alburnus alburnus)*	96h LC_{50}	2,300 mg/L	
Nitocra spinipes:	96h LC_{50}	2,100 mg/L	(2841)
Brachydanio rerio	96h LC_{50}	1,730 mg/L	(7025)

Toxicity

Micro organisms			
Vibrio fischeri	5min EC_{50}	3,300; 3,388 mg/L	
	15min EC_{50}	2,800; 2,938 mg/L	(10810)
Protozoa			
Spirostomum ambiguum	48h EC_{50}	875 mg/L	
	48h LC_{50}	1,097 mg/L	
Tetrahymena pyriformis	48h EC_{50}	2,466 mg/L	
Chilomonas paramaecium	48h NOEC	28 mg/L	(10810)
Algae			
Chlorococcales mixed culture	24h EC_{50}	>1,000 mg/L	(10810)
Crustaceae			
Daphnia magna	48h EC_{50}	1,983 mg/L	
	24h EC_{50}	1,855 mg/L	
Artemia salina	24h LC_{50}	2,950 mg/L	
Nitocra spinipes	96h LC_{50}	2,100 mg/L	(10810)
Amphibia			
Xenopus laevis	48h LC_{50}	1,200 mg/L	(10810)
Fish			
Leuciscus idus	48h LC_{50}	1,200; 1,770 mg/L	
Lepomis macrochirus	96h LC_{50}	100-500 mg/L	
Carassius auratus	24h Lc50	1,900 mg/L	
Oryzias latipes	48h LC_{50}	500; >1,000 mg/L	
Pimephales promelas	96h LC_{50}	1,730 mg/L	
Semotilus atromaculatus	24h LC_{50}	1,000-1,400 mg/L	
Alburnus alburnus	96h LC_{50}	2,300; 2,250-2,400 mg/L	(10810)
Mammals			
rat	acute oral LD_{50}	790; 2020; 2510;4360 mg/kg bw	(7166)

t-butanol (3-butanol; 2-methyl-2-propanol; trimethylcarbinol; *tert*-butylalcohol; trimethylmethanol)

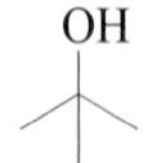

$(CH_3)_3COH$

$C_4H_{10}O$

CAS 75-65-0

USES

as blending agent up to 7% to increase the octane rating of unleaded gasoline; alkylating agent; solvent; anti-icing additive.

A. PROPERTIES

molecular weight 74.1; melting point 25°C; boiling point 83°C; vapor pressure 31 mm at 20°C, 42 mm at 25°C, 56 mm at 30°C, 41 hPa at 20°C; vapor density 2.55; density 0.78 at 20/4°C; solubility miscible at 20°C at pH 7; saturation concentration in air 121 g/m^3 at 20°C, 219 g/m^3 at 30°C; LogP$_{OW}$ 0.37 (measured); LogH -3.31 at 25°C.

B. AIR POLLUTION FACTORS

1 mg/m^3 = 0.330 ppm, 1 ppm = 3.03 mg/m^3.

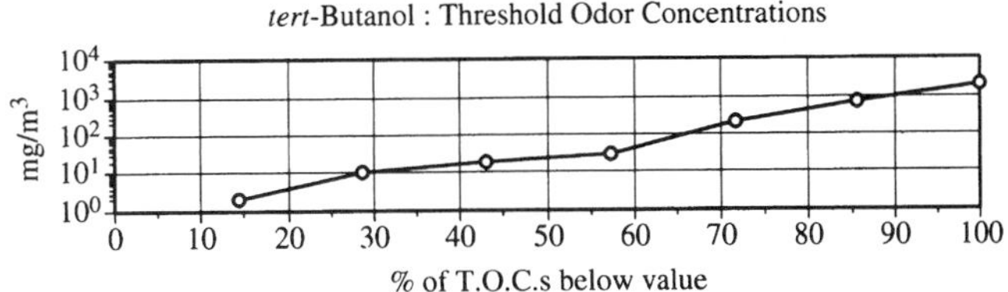

(279, 307, 610, 643, 708, 776)

Photochemical degradation

reaction with OH°: $t_{1/2}$: 11 d. (5632)

The following reactions occur:

$(CH_3)_3COH + NO + OH° \rightarrow H_2CO + (CH_3)_2CO + HO_2 + NO_2$

formaldehyde + acetone (5632)

Partitioning coefficients

$K_{air/water}$ 0.00050

Cuticular matrix/air partition coefficient* 820 ± 0 (7077)

* experimental value at 25°C studied in the cuticular membranes from mature tomato fruits (*Lycopersicon esculentum* Mill. cultivar Vendor)

C. WATER AND SOIL POLLUTION FACTORS

BOD$_{2d}$	4% ThOD	(10548)
BOD$_5$	0% ThOD	(10548, 36, 220)
	0% ThOD at 10 mg/L, unadapted sew.	(554)
BOD$_{20}$	0% ThOD at 10 mg/L, unadapted sew.	(554)
BOD$_{10d}$	3% ThOD	(10548)
BOD$_{30d}$	0% ThOD	(10548)
COD	80; 84% of ThOD	(36, 220)
TOC	2.59	(36)
ThOD	2.59	

Biodegradation

Adapted A.S.-product as sole carbon source-98.5% COD removal at 30.0 mg COD/g dry (327)

inoculum/h.
Slight inhibition of microbial growth after 24 h exposure at 100 ppm. (523)

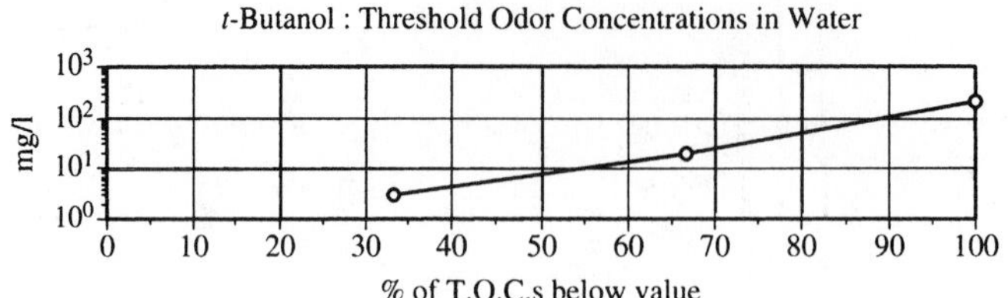

(874, 907, 908)

EEC respirometric method at 100 mg/L and 20°C using nonadapted sludge:

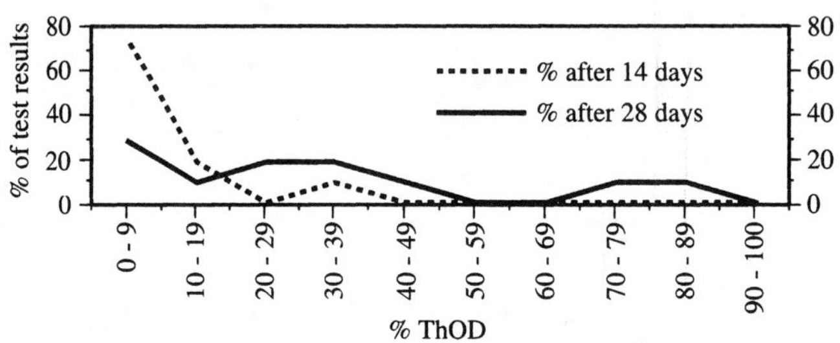

ThOD in EEC respirometric test: n = *11:*
median after 14 d: 0% ThOD
median after 28 d: 27% ThOD

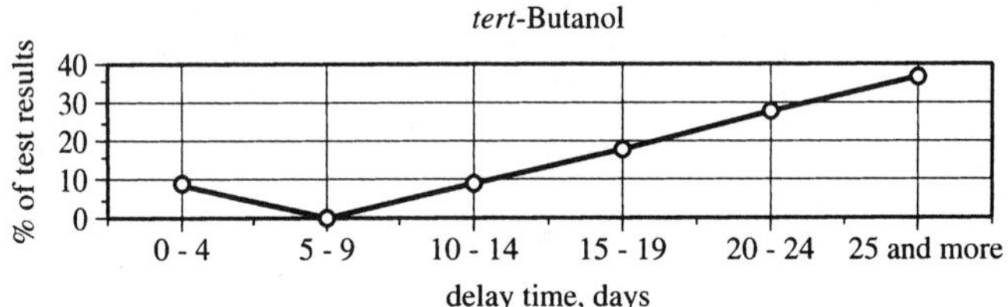

Delay times in EEC respirometric test

n = 11; median: 23 d.

Oxygen uptake and DOC removed using MITI inoculum

delay time: 17-21 d
% ThOD after 14 d: 0; 2%
% ThOD after 28 d: 38; 64%
% DOC after 28 d 10; 49; 76% (2790)

In 'Repetitive Die-Away' Test
delay time: 18 d
% ThOD after 28 d: 58%
% ThOD after 49 d: 70%
% DOC removal after 28 d: 87%
% DOC removal after 49 d: 99% (2790)

inoculum/test method	*test conc.*	*test duration*	*removed*	
A.S.: Coupled Units test	12 mg/L	42 d	33% DOC	(5621)
A.S.: nonadapted,	400 mg/L	6 d	96% DOC	(5621)

Zahn-Wellens				
A.S.: MITI	50 mg/L	14 d	0% DOC	
			7% ThOD	(5621)
waste water, adapted, Sturm	10 mg/L	42 d	32% DOC	
			0% CO_2 prod.	(5621)
waste water, modif. OECD Screen	20 mg/L	19 d	29% DOC	(5621)
waste water, Closed Bottle	2 mg/L	30 d	0% ThOD	(5621)
A.S., adapted	200 mg/L	5 d	98.5% COD	(5622)
A.S., nonadapted, Warburg	500 mg/L	1 d	8% ThOD	(5623)
not specified	10 mg/L	20 d	0% ThOD	(5625)
A.S., nonadapted, Zahn-Wellens	300 mg/L	19 d	>99.9% prod.	(5615)
A.S., Coupled Units	10 mg/L	3 hours	46% DOC	(5615)
munic. waste water, OECD screen	20 mg/L	28 d	43% DOC	(5615)
A.S., adapted, BOD test	330 mg/L	8 hours	0% ThOD	(5627)
A.S., industrial	100 mg/L	15 d	64% TOC	(5628)
Alcaligenes faecalis, Warburg	500 mg/L	6 d	0% ThOD	(5629)
anaerobe soil microorganisms	6 mg/L	100 d	30% product	(5630)
anaerobe soil microorganisms	100 mg/L	25 d	5% product	
		50 d	18% product	
		100 d	54% product	(5630)

Waste water treatment

A.C.: adsorbability: 0.059 g/g C; 29% reduction at infl.: 1,000 mg/L. (32)

A.S.: after	6 h: 0.5% of ThOD	
	12 h: 0.7% of ThOD	
	24 h: 0.8% of ThOD	(88)

Reverse osmosis

90% rejection from a 0.01 M solution. (221)

methods	*temp, °C*	*d observed*	*feed, mg/L*	*d acclim.*	*% removed*	
A.S., W	20	1	500	34	21% ThOD	
A.S., COD	20	1/3	333	30	30% product	(93)

Partitioning coefficients

$K_{air/water}$	0.00050	
Cuticular matrix/water partition coefficient**	0.41	(7077)

** calculated from the Cuticular matrix/water partition coefficient and the air/water distribution coefficient

D. BIOLOGICAL EFFECTS

Bacteria			
E. coli	EC_{50}	11,263 mg/L	(5620)
	EC_0	13,560 mg/L	(5620)
M. smegmatis	EC_0	9,707 mg/L	(5620)
Pseudomonas putida	18h EC_{10}	2,050 mg/L	(5615)

Algae			
Chlorella pyrenoidosa	toxic	24,200 mg/L	(41)
Scenedesmus subspicatus inhib. of	IC_{10}	697 mg/L	

fluorescence

Scenedesmus subspicatus growth inhib.	IC_{10}	1,186 mg/L	(2698)

RubisCo Test: inhib. of enzym. activity of Ribulose-P2-carboxylase in protoplasts: IC_{10}: 8,892 mg/L.
Oxygen Test: inhib. of oxygen production of protoplasts: IC_{10}: 1,482 mg/L. (2698)

Crustaceans			
Daphnia magna	24-48h EC_0	2,915 mg/L	
	24-48h EC_{50}	5,504 mg/L	
	24-48h EC_{100}	8,000 mg/L	(5619)
Insects			
Aedes aegypti larvae	4h LC_{50}	18,600 mg/L	(5635)
Chironomus riparius, 3rd instar larvae	48h LC_0	5,000 mg/L	
	48h LC_{50}	5,800 mg/L	
	48h LC_{100}	6,700 mg/L	
	48h NOLC	2,800 mg/L	
	48h NOEC	1,600 mg/L	(7073)
Amphibians			
Xenopus laevis	48h LC_{50}	2,450 mg/L	(5636)
Worms			
Tubifex tubifex	2 min EC_{100}	31,196 mg/L	(5638)
Fishes			
creek chub:	24h LD_0	3,000 mg/L	
	24h LD_{100}	6,000 mg/L	(243)
guppy *(Poecilia reticulata)*	7d LC_{50}	3,550 ppm	(1833)
Leuciscus idus	48h LC_0	>1,000 mg/L	(5615)
	96h LC_{50}	6,410 mg/L	(5615)
Carassius auratus	24h LC_{50}	>5,000 mg/L	(5617)
Semolitus atromaculatus	24h LC_0,S	<3,000 mg/L	
	24h LC_{100},S	>6,000 mg/L	(5618)

butanol, sec- (2-butanol; methylethylcarbinol; 2-butylalcohol; SBA)

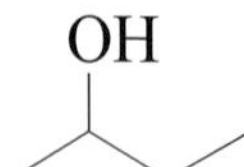

$CH_3CHOHCH_2CH_3$

$C_4H_{10}O$

CAS 78-92-2

A. PROPERTIES

colorless liquid; molecular weight 74.12; melting point -89/-108°C; boiling point 99.5/107.7°C; vapor pressure 12 mm at 20°C, 24 mm at 30°C; vapor density 2.55; density 0.81 at 20/4°C; solubility 125,000 mg/L at 20°C, 201,000 mg/L at 20°C; saturation concentration in air 52 g/m^3 at 20°C, 94 g/m^3 at 30°C; log P_{OW} 0.61; log H -3.31 at 25°C.

B. AIR POLLUTION FACTORS

Odor

characteristic: quality: sweet; hedonic tone: pleasant to neutral.

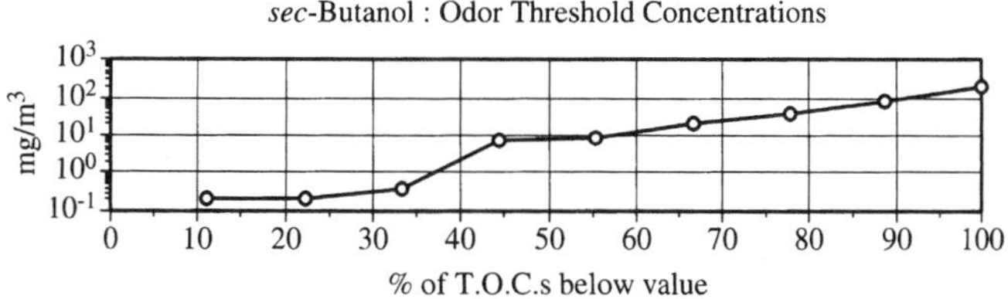

(57, 279, 307, 708, 709, 737)

Odor

index (100% recognition): 28,000 (19)

Partition coefficients

$K_{air/water}$	0.00049	(7077)
Cuticular matrix/air partition coefficient*	2,000 ± 0	

* experimental value at 25°C studied in the cuticular membranes from mature tomato fruits (*Lycopersicon esculentum* Mill. cultivar Vendor)

1 mg/m^3 = 0.330 ppm, 1 ppm = 3.03 mg/m^3.

C. WATER AND SOIL POLLUTION FACTORS

BOD_5:	0; 33; 72% of ThOD	(79, 220, 270)
10 d:	44.2% of ThOD	(79)
15 d:	69.2% of ThOD	(79)
20 d:	72.3% of ThOD	(79)
30 d:	73.2% of ThOD	(79)
50 d:	77.9% of ThOD	(79)
COD:	91; 95% of ThOD	(220)
TOC:	96% of ThOD	(220)
ThOD:	2.59	

Biodegradation

adapted A.S.-product as sole carbon source-98.5% COD removal at 55.0 mg COD/g dry inoculum/h. (327)

Waste water treatment

A.S.: after	6 h:	4% of ThOD	
	12 h:	6% of ThOD	
	24 h:	9% of ThOD	(88)

reverse osmosis: 58% rejection from a 0.01 M solution (221)

methods	*temp, °C*	*d observed*	*feed, mg/L*	*d acclim.*	*% removed*	
A.S., W	20	1	500	24	58% ThOD	
A.S., BOD	20	1/3 to 5	333	30	96% product	(93)

Partition coefficients

$K_{air/water}$	0.00049	
Cuticular matrix/water partition coefficient**	0.95	(7077)

** calculated from the Cuticular matrix/water partition coefficient and the air/water distribution coefficient

D. BIOLOGICAL EFFECTS

RubisCo Test

inhib. of enzym. activity of Ribulose-P2-carboxylase in protoplasts: IC_{10}: 5,188 mg/L.

Oxygen Test

inhib. of oxygen production of protoplasts: IC_{10}: 0.7 mg/L. (2698)

Toxicity

Bacteria			
Pseudomonas putida	16h EC_0	500 mg/L	(1900)
Protozoa			
Chilomonas paramaecium	48h NOEC	745 mg/L	(10810)
Entosiphon sulcatum	72h EC_0	1,280 mg/L	(1900)
Uronema parduczi Chatton-Lwoff	EC_0	1,416 mg/L	(1901)
Algae			
Chlorococcales mixed culture	24h EC_{50}	3,400 mg/L	(10810)
Scenedesmus quadricauda	8d NOEC	95 mg/L	(10810)
Microcystis aeruginosa	8d EC_0	312 mg/L	(329)
Chlorella pyrenoidosa	toxic	8,900 mg/L	(41)
Scenedesmus subspicatus inhib. of fluorescence	IC_{10}	312 mg/L	(329)
Scenedesmus subspicatus growth inhib.	IC_{10}	1,111 mg/L	
GREEN ALGAE			
Scenedesmus quadricauda	7d EC_0	95 mg/L	(1900)
Crustaceae			
Daphnia magna	48h EC_{50}	4,227 mg/L	
	24h EC_{50}	2,300; 3,750 mg/L	(10810)
Amphibia			
Xenopus laevis	48h LC_{50}	1,530 mg/L	(10810)
Fish			
Leuciscus idus	48h LC_{50}	3,520; 3,540 mg/L	(10810)
Pimephales promelas	96h LC_{50}	3,670 mg/L	(10810)
goldfish	24h LC_{50}	4,300 mg/L	(277)
Mammals			
rat	acute oral LD_{50}	2640-3100; 2650-3100 mg/kg bw	
mouse	acute oral LD_{50}	3500 mg/kg bw	(7166)

2-butanone *see* methylethylketone

2-butanone peroxide (methyl-ethyl-ketone-peroxide; Lupersol-; MEK-peroxide; methyl-ethyl-ketone-hydroperoxide; Quickset-extra; Thermacure)

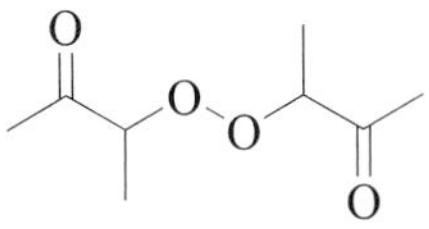

$CH_3COCH(CH_3)OOCH(CH_3)COCH_3$

$C_8H_{16}O_4$

CAS 1338-23-4

USES

Cross-linking agent for thermosetting resins. Strong oxidising agent and contact with organic materials can create a fire hazard.

IMPURITIES, ADDITIVES, COMPOSITION

Usually supplied as a 50% solution in dimethyl phthalate. (9897)

A. PROPERTIES

molecular weight 240.21; boiling point 118°C decomposes; density 1.17.

B. AIR POLLUTION FACTORS

Incinerability

thermal stability ranking of hazardous organic compounds: rank 279 on a scale of 1 (highest stability) to 320 (lowest stability). (2390)

Manmade sources

Contaminant in air samples collected in the vicinity of a fibreglassing plant. (9898)

D. BIOLOGICAL EFFECTS

Mammals			
mouse, rat	oral LD_{50}	470-484 mg/kg bw	(9895, 9896)

2-butanoneoxime (ethylmethylketoxime; methylethylketoxime; methyl-ethyl-ketone-oxime; MEK-oxime; 2-oximinobutane; ethyl-methyl-ketoxime))

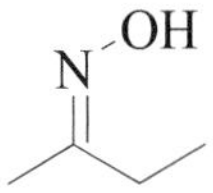

$CH_3C(NOH)C_2H_5$

C_4H_9NO

CAS 96-29-7

USES

Blocking agent for polymerisation. Catalyst. Oxygen scavenger in steam generators. Extraction of silver.

A. PROPERTIES

molecular weight 87.12; melting point -29.5°C; boiling point 152°C; density 0.923 at 20/4°C; solubility 100,000 mg/L; $LogP_{OW}$ 0.36.

D. BIOLOGICAL EFFECTS

Bacteria			
Pseudomonas	toxic	630 mg/L	
Photobacterium phosphoreum Microtox test	5 min EC_{50}	955 mg/L	(9657)
Algae			
Scenedesmus:	still toxic at 2.5 g/L		
Protozoa			
Colpoda:	still toxic at 2.5 g/L		(30)
Fishes			
Pimephales promelas	24h LC_{50}	1,500 mg/L	
	96h LC_{50}	843 mg/L	(2709)
fathead minnow	96h LC_{50}	9.8 mg/L	(7773)
Mammals			
rat	oral LD_{50}	930 mg/kg bw	(9894)

2-buten-1-ol (crotyl-alcohol; crotonyl-alcohol-; 1-hydroxy-2-butene; 3-methylallyl alcohol)

OH

$CH_2OHCHCHCH_3$

C_4H_8O

CAS 6117-91-5

USES

Catalyst for polymerisation of ketones

OCCURRENCE

Natural occurrence in Victoria plums *Prunus domesticus* , lemon juice, colza oil and wood-alcohol oil. (9801)

A. PROPERTIES

molecular weight 72.11; melting point < -30°C; boiling point 118-122°C; density 0.85 at 20°C; solubility 166,000 mg/L at 20°C; $logP_{OW}$ 0.54.

B. AIR POLLUTION FACTORS

Manmade sources

in gasoline exhaust: 0.1 to 3.6 ppm (+C_5H_8O).

C. WATER AND SOIL POLLUTION FACTORS

Waste water treatment

reverse osmosis: 18% rejection from a 0.01 M solution. (221)

D. BIOLOGICAL EFFECTS

Mammals			
rat	oral LD_{50}	790 mg/kg bw	(9900)

2-butenal *see* crotonaldehyde

1-butene *see* α-butylene

3-butenoic acid (vinylacetic acid; β-butenoic acid)

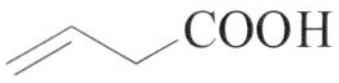

COOH

CH_2CHCH_2COOH

$C_4H_6O_2$

CAS 625-38-7

A. PROPERTIES

molecular weight 86.09; melting point -39°C; boiling point 163°C; density 1.01 at 15/15°C.

D. BIOLOGICAL EFFECTS

Algae

Chlorella pyrenoidosa: toxic: 280 mg/L (n.s.i.) (41)

1-butoxy-2,3-epoxypropane *see* *n*-butylglycidylether

3-butoxy-1,2-epoxypropane *see* *n*-butylglycidylether

2-butoxy-1-propanol

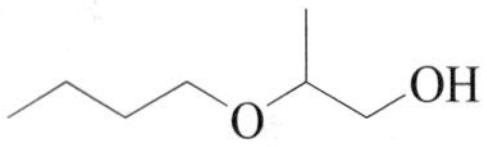

OH

O

$C_7H_{16}O_2$

C. WATER AND SOIL POLLUTION FACTORS

Göteborg (Sweden) sew. works 1989-1991: infl.: n.d.-200 µg/L; effl.: n.d.-3.0 µg/L. (2787)

1-butoxybutane *see* *n*-butylether

2-butoxyethanol *see* butylcellosolve

2(β-butoxyethoxy) ethanol *see* diethyleneglycolmonobutylether

α-(2-(2-butoxyethoxy)ethoxy)-4,5-methylene-2-propyl-toluene *see* piperonylbutoxide

(butoxymethyl)-oxirane *see* *n*-butylglycidylether

butter yellow *see* 4-dimethylaminoazobenzene

butyl-m-cresol, 6-tert- ((2-tert-butyl-5-methylphenol; 6-tert-butyl-3-methylphenol; 3-methyl-6-tert-butylphenol; 1-tert-butyl-2-hydroxy-4-methylbenzene; 2-(1,1-dimethylethyl)-5-methylphenol)

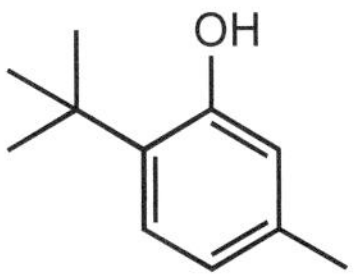

$C_{11}H_{16}O$

CAS 88-60-8

EINECS 201-842-3

TRADENAMES

3M6B; MBMC

USES

Flavoring agent in or on foodstuffs; intermediate to synthesize a.o. the following antioxidants added to polymers and rubbers: 4'butylidene bis(6-tert-butyl-3-methylphenol) (CAS 85-60-9; 4,4'-thio bis(6-tert-butyl-3-methylphenol) (CAS 96-69-5); 1,1,3-tris(5-tert-butyl-4-hydroxyphenyl)butane (CAS 1843-03-4)

IMPURITIES, ADDITIVES, COMPOSITION

2-tert-butyl-4-methylphenol (<2%)

A. PROPERTIES

molecular weight164.25 ; melting point 21.3°C; boiling point 244°C; vapour pressure 3.3 Pa at 25°C; water solubility 420 mg/L at 25°C; log Pow 4.11 at 25°C; H 1.3 Pa.m^3/mol

B. AIR POLLUTION FACTORS

Photodegradation

T/2 (OH radicals) = 1.2 hours

C. WATER AND SOIL POLLUTION FACTORS

Hydrolysis

No hydrolysis at pH 4, 7 and 9 at 50°C for 5 days.

Biodegradation

Inoculum	method	concentration	duration	elimination	
A.S.	OECD 301C	100 mg/L	28 days	1% ThOD	
				1% HPLC	(10718)

D. BIOLOGICAL EFFECTS

Bioaccumulation

Species	conc. µg/L	duration	body parts	BCF	
Cyprinus carpio	1	33 days		39-93	
	10	33 days		41-92	(10718)

Toxicity

Algae			
Selenastrum capricornutum	48h $ECgr_{50}$	1.8 mg/L	
	48h ECb_{50}	0.90 mg/L	
	48h NOECgr	0.62 mg/L	
	48h NOECb	0.25 mg/L	(10718)
Crustaceae			
Daphnia magna	48h EC_{50}	2.8 mg/L	
	48h EC_0	1.2 mg/L	
	48h EC_{100}	3.9 mg/L	
	21d LC_{50}	0.87 mg/L	
reproduction	21d EC_{50}	0.57 mg/L	
	21d NOEC	0.24 mg/L	
	21d LOEC	0.49 mg/L	(10718)
Fish			
Oryzias latipes	96h LC_{50}	2.7 mg/L	
	96h LC_0	2.1 mg/L	
	96h LC_{100}	3.7 mg/L	
Lepomis macrochirus	96h Lc50	3.4 mg/L	(10718)
Mammals			
Rat	oral LD_{50}	130-320; 320-800 mg/kg bw	
Mouse	oral LD_{50}	580; 740 mg/kg bw	(10718)

butyl-1,2-dihydroxybenzene, 4-tert- *see* *p-tert*-butylcatechol

2(2-butyl-4,6-dinitrophenyl)-3,3-dimethylacrylate *see* binapacryl

butyl-4,6-dinitrophenylacetate, 2-tert- (dinoterbacetate; 2-(1,1-dimethylethyl)-4,6-dinitrophenolacetate)

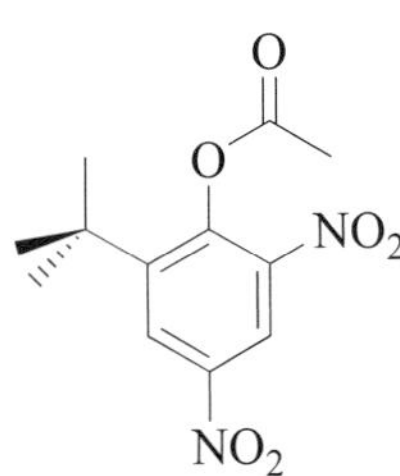

$C_{12}H_{14}N_2O_6$

CAS 3204-27-1

USES

herbicide.

A. PROPERTIES

pale yellow crystals; molecular weight 282.28; melting point 133-134.5°C.

D. BIOLOGICAL EFFECTS

Harlequin fish (Rasbora heteromorpha):

mg/L	*24 h*	*48 h*	*96 h*	*3m (extrapolated)*	
LC_{10} (F)	0.045	0.038	0.031		
LC_{50} (F)	0.068	0.051	0.039	0.03	(331)

butyl-2,3-epoxypropylether *see* *n*-butylglycidylether

N,butyl-N-ethyl-2,6-dinitro-4-trifluoromethylaniline *see* balan

N,butyl-N-ethyl-α,α,α-trifluoro-2,6-dinitro-p-toluidine *see* balan

butyl-O-(2-ethylhexyl)monoperoxycarbonate, OO-ter- (carbonperoxoic acid, OO-(1,1-dimethylethyl)-O-(2-ethylhexyl)ester)

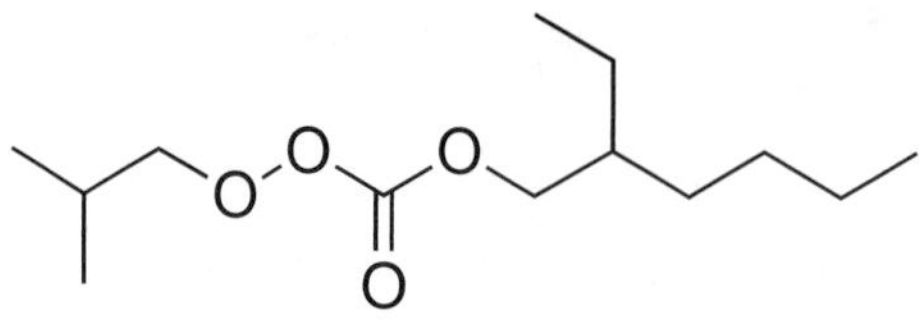

$C_{13}H_{26}O_4$

CAS 34443-12-4

TRADENAMES

Luperox TBEC, Lupersol TBEC

USES

The chemical is an organic peroxide which decomposes under appropriate conditions to free radicals such as. tert.butoxy radical [$(CH_3)_3CO.$]. The free radicals are effective initiators for the polymerisation of olefins, for example in the production of polyethylene and polystyrene. The radicals are consumed in the reactions and become incorporated into the polymer, probably at the ends of the polymer chains.

IMPURITIES, ADDITIVES, COMPOSITION

Chemical	Weight %
2-ethylhexanol	1-3
Isobutanol	<0.1
t-butyl hydroperoxide	>0.5%
Unidentified impurities	1-2%

A. PROPERTIES

light yellow liquid with fruity odour; molecular weight 246.3; boiling point <-60°C; specific gravity 0.93 at 25°C; vapour pressure 0.13 kPa (estimated); water solubility 2.6 mg/L at 25°C(estimated); log Pow 4.7 (estimated); log K_{OC} 3.6 (estimated)

C. WATER AND SOIL POLLUTION FACTORS

Hydrolysis

Organic radicals are inherently thermodynamically unstable and their rapid decomposition (sometimes explosively) is initiated by a variety of compounds, including ferrous ions, which are ubiquitous in industrial and natural environments. Decomposition products are 1-ethylhexanol, tert-butanol and CO_2. (7447)

D. BIOLOGICAL EFFECTS

Toxicity

Crustaceae			
Daphnia magna	48h EC_{50} (calculated) 48h LC_5	0.76 mg/L	(7447)
Fish	96h LC_{50} (calculated)	0.84 mg/L	(7447)
Mammals			
Rat	oral LD_{50}	>5,000 mg/kg bw	(7447)

butyl hydroperoxide, tert- (TBHP; 2-hydroxyperoxy-2-methylpropane; (1,1-)dimethylethylhydroxyperoxide; tert-butyl hydrogen peroxide)

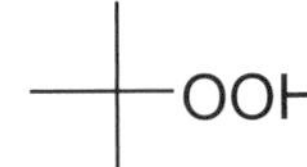

$C_4H_{10}O_2$

CAS 75-91-2

EINECS 200-915-7

TRADENAMES

TBHP-70 (T-hydro); Cadox TBH; Trigonox AW 70; perbutyl H

USES

Tert-butyl hydroperoxide (TBHP) is primarily used in the chemical industry. TBHP is used as starting material (or intermediate) and as a reactive ingredient (catalyst, initiator or curing agent). Applications are:
-the epoxidation of propylene to propylene oxide (intermediate);
-free radical initiator for polymerisations, copolymerisations, graft polymerisations and curing of polymers (plastic industry);
-free radical initiator to polymerise unsaturated monomers, usually to high polymers. Mainly used by manufacturers of synthetic lattices or water borne dispersions. Also used as a component of catalysts systems for unsaturated polyester resins;
-the synthesis of other organic peroxy molecules (as a precursor of initiators) such as perester, persulphate, dialkyl peroxide and perketal derivatives;
-the preparation of speciality chemicals required by fine chemical and performance chemical industries, such as pharmaceuticals and agrochemicals (fungicide).
-the use as an ingredient of hardeners for plastics. These products contain 5 - 20 % TBHP. Hardeners for plastics are also used in the plastic industry.
TBHP is used in several products. Product types are paint, lacquer and varnishes, adhesives and binding agents. (11046)

A. PROPERTIES

molecular weight 90.1; melting point –8 to 5.4 °C; boiling point 96-160°C; density 0.79-0.90 kg/L at 20°C; vapour pressure 0.7-3 kPa at 20°C; water solubility ca 700,000 mg/L;

B. AIR POLLUTION FACTORS

Conversion factor

1 ppm = 3.75 mg/m^3

Photodegradtion

T/2 (OH radicals) = 3 days

C. WATER AND SOIL POLLUTION FACTORS

Hydrolysis

An hydrolysis study with TBHP-70 did not show an appreciable degradation of TBHP during the 5-d test period at a temperature of 50 °C and pH values of 4 to 9.

Biodegradation

Inoculum	method	concentration	duration	elimination
A.S.	Modified Sturm Test	30 mg/L	28 days	0% ThCO2

A.S.	Closed Bottle Test	2 mg/L	96 days	0% ThOD
A.S.	14C-radiolabeled TBHP	25 mg/L	1 hour	85% degradation (1)
Raw sewage	Coupled unit test	19 mg/L	HRT: 9 h	86%

(1)THBP was not mineralised during the 1-h tests, but mainly degraded to the primary metabolite tertiary butyl alcohol (TBA); in addition, one unidentified metabolite was found.
(2)The overall removal of radioactivity averaged 86%. Of the remaining 14% in effluent, around two-third (9%) was CO2 and one-third (5%) was the primary degradation product tertiary butyl alcohol (TBA) plus two other, unidentified metabolites. (11046)

D. BIOLOGICAL EFFECTS

Toxicity

Micro organisms			
Activated sludge	30 min EC_{50}	17 mg/L	(11048)
Algae			
Selenastrum capricornutum	72h EC_{50}	0.84; 1.5 mg/L	
	72h NOEC	0.22 mg/L	(11048)
Crustaceae			
Daphnia magna	48h EC_{50}	14 mg/L	
	48h NOEC	7 mg/L	(11048)
Fish			
Pimephales promelas	96h LC_{50}	29 mg/L	
	96h NOEC	22 mg/L	
Poecilia reticulates	96h LC_{50}	57 mg/L	
	96h NOEC	30 mg/L	(11048)
Mammals			
Rat	oral LD_{50}	406; 560 mg/kg bw (70% TBHP)	(11046)
Mouse	oral LD_{50}	800 mg/kg bw	

The absorbed TBHP is rapidly converted to 2-methylpropan-2-ol and distributed over the body. 2-Methylpropan-2-ol is either excreted in exhaled air, conjugated and eliminated in the urine or oxidised to and excreted in the urine as 2-methyl-1,2-propanediol and 2-hydroxyisobutyric acid (see Figure). 2-hydroisobutyric acid was the main metabolite in all tissues at 12 hours after treatment. (11046)

Proposed metabolic pathway of TBHP in rats [11046]

Proposed metabolic pathway of TBHP in rats (11046)

Butyl-p-hydroxybenzoate (butylparaben ; 4-hydroxybutylbenzoate; 4-hydroxybenzoic acid butyl ester)

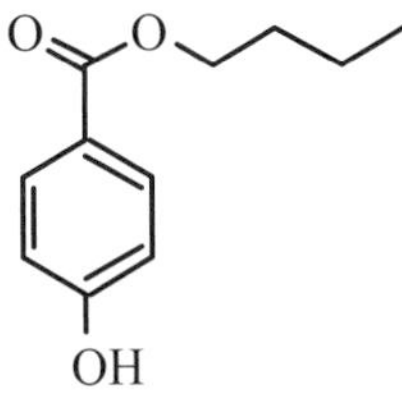

$C_{11}H_{14}O_3$

$HOC_6H_4COOCH_2CH_2CH_2CH_3$

CAS 94-26-8

TRADENAMES

Butoben, parasept, butyl chemosept, solbrol B, butylparaben, tegosept butyl, butyl butex

USES

Butylparaben is used as a preservative in foods, beverages and cosmetics.

A. PROPERTIES

odourless white crystalline powder; molecular weight 194.23; melting point 68-69 °C ; $LogP_{OW}$ 3.57 (calculated)

C. WATER AND SOIL POLLUTION FACTORS

Hydrolysis

Parabens are stable in acidic solutions. Hydrolysis occurs above pH 7. In strong alkaline solutions parabens hydrolyzes to benzoic acid and the corresponding butylalcohol. (7169)

D. BIOLOGICAL EFFECTS

Mammals

Mouse	oral LD_{50}	13200 mg/kg bw	(7170)

butyl-2-methacrylate (2-methyl-butylacrylate; 2-methyl-2-propenoic acid, n-butyl ester)

$C_8H_{14}O_2$

$CHC(CH_3)C(O)OCH_2CH_2CH_2CH_3$

CAS 97-88-1

USES AND FORMULATIONS

The product may contain the following impurities:

methacrylic acid: approx. 30 mg/L

methyl methacrylate: 0- 3,000 mg/L
hydroquinone: 10- 700 mg/L (polymerization inhibitor)
p-methoxyphenol: 10- 100 mg/L (polymerization inhibitor)
dimethyl-6-tert-butylphenol: 10- 100 mg/L (polymerization inhibitor)

A. PROPERTIES

molecular weight 142.22; melting point - 25°C; boiling point 163°C; density 0.90 at 20°C; vapor pressure 2.7- 6.5 hPa at 20°C; solubility 250- 882 mg/L at 20°C; LogP_{OW} 2.26; 2.6; 2.88; 3.01 (measured values); H 1.1 x 10^{-4} atm x m^3/mol.

B. AIR POLLUTION FACTORS

1 ppm = 5.81 mg/m^3 1 mg/m^3 = 0.17 ppm

Odor threshold

- 0.06 ppm; >0.15 mg/m^3

Photochemical reactions

$t_{1/2}$: 7.5 h, based on reactions with OH° (calculated). (8990)

C. WATER AND SOIL POLLUTION FACTORS

Evaporation

into the atmosphere from a model river (1-m deep flowing 1 m/s with wind speed of 3 m/s): $t_{1/2}$: 13 h. (8991)

Biodegradation

Inoculum	*Method*	*Concentration*	*Duration*	*Elimination results*	
	Modified MITI test	100 mg/L	1 d	0.6% ThOD	
			2 d	17% ThOD	
			5 d	30% ThOD	
			10 d	38% ThOD	
			28 d	38% ThOD	(8992)

Mobility in soil

K_{OC}: 878 (calculated).

D. BIOLOGICAL EFFECTS

Toxicity

Bacteria			
Pseudomonas putida	18 h EC_0	32 mg/L	
(cell multiplication inhibition test)	18 h EC_{50}	>254 mg/L	(8999)
Luminescence inhibition assay	5 min EC_{50}	37 mg/L	
	10 min EC_{50}	49 mg/L	
	15 min EC_{50}	55 mg/L	(9000)
Algae			
Selenastrum capricornutum	96 h EC_{50}	57;130 mg/L	
(biomass and growth rate)	96 h NOEC	26; 26 mg/L	
	96 h LOEC	54; 56 mg/L	(8998)
Crustaceans			
Daphnia magna	48 h NOEC	23 mg/L	
	48 h EC_{50}	32 mg/L	
	48 h EC_{100}	75mg/L	(8997)

Fishes			
Pimephales promelas	96 h LC_{50}	11 mg/L	
	96 h LC_{100}	20 mg/L	
	96 h NOEC	7 mg/L	(8993)
Carassius auratus	72 h LC_{50}	112; 124 mg/L	(8994, 8996)
Lepomis macrochirus	24 h LC_0	>5 mg/L	(8995)
Oncorhynchus mykiss	24 h LC_0	>5 mg/L	(8995)
Petromyzon marinus	24 h LC_0	>5 mg/L	(8995)
Salmo gairdneri	24 h LC_0	>5 mg/L	(8995)
Mammals			
rat	acute oral LD_{50}	>2,000; 16,000; 22,600; 18,020; 18,561; >20,300 mg/kg bw	

butyl-α-methylhydrocinnamaldehyde, p-tert- (3-(p-t-butylphenyl)-2-methylpropanal; 4-(1,1-Dimethylethyl)-alpha-methyl-Benzenepropanal; para-tert-Butyl-alpha-methyl dihydrocinnamic aldehyde; Beta-(4-tert-butylphenyl)- alpha-methyl-Propionaldehyde; 2-Methyl-3-(4-tert-butylphenyl) propanal; lilialdehyde; 2-methyl-3-(4-tert-butylbenzyl)propanal)

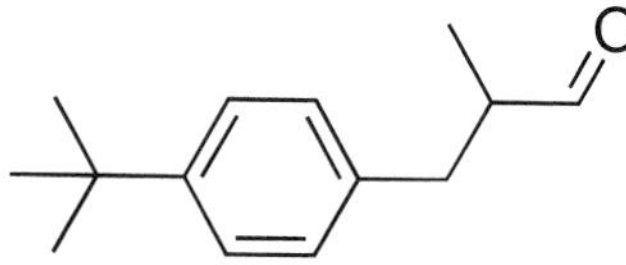

$C_{14}H_{20}O$

CAS 80-54-6

EINECS 201-289-8

TRADENAMES

Lilestralis; Lilial; Lilyal; Lysmeral

USES

Lilialdehyde is one of common ingredients in fragrances. It's end applications include soap,detergent, beauty care product, household product. a fragrance in perfumed consumer products such as dishwash, car shampoo, hand cleaner, toilet paper, nappies in concentrations of 0.0009 to 0.05 wt%. (7195).Lilialdehyde is used as an intermediate for the synthesis of agrochemicals (fenpropimorph). (10991)

A. PROPERTIES

Clear to slightly yellow liquid; molecular weight 204.30; melting point <-20°C; boiling point 279°C; specific gravity 0.95; Vapour pressure 1 hPa at 50°C; water solubility 33 mg/L; log Pow 4.2 at 24°C

B. AIR POLLUTION FACTORS

Odour

Lilial is a fragrance with an aroma similar to that of lily flowers. Floral-muguet, fresh, powerful

Photodegradation

T/2 (OH radicals) = 11.5 hours

C. WATER AND SOIL POLLUTION FACTORS

Hydrolysis

Stable at pH 4 to 9. (10991)

Manmade sources

Concentrations and removal % in municipal WWTP's in september 1997 (USA) (10844)

	Influent ng/L	Effluent ng/L	% removal
Activated sludge WWTP	2300	24	99.0
Trickling Filter WWTP	2580	144	94.4

Biodegradation

Inoculum	method	concentration	duration	elimination	
A.S. domestic	OECD 301F	100 mg/L	7 days	7%	
			14 days	27%	
			21 days	60%	
			28 days	68%	
A.S. domestic	OECD 301F	50 mg/L	7 days	6%	
			14 days	68%	
			21 days	81%	
			28 days	84%	
A.S.	OECD 302 C	100 mg/L	28 days	8%	
A.S. non-adapted		61 mg/L	1 day	69%	(10992)

Mobility in soils

K_{OC} : 1285 (calculated) (10991)

D. BIOLOGICAL EFFECTS

Toxicity

Crustaceae			
Daphnia magna	24h EC_{50}	41.6 mg/L	
	24h EC_0	12.5 mg/L	
	24h EC_{100}	100 mg/L	
	48h EC_0	6.2 mg/L	
	48h EC_{50}	10.7 mg/L	
	48h EC_{100}	25 mg/L	(10992)
	21d NOEC		
Fish			
Brachydanio rerio	96h LC_{50}	>2.2-4.6 mg/L	
	96h NOEC	2.15 mg/L	
Leuciscus idus	96h LC_{50}	>4.6-10 mg/L	
	96h NOEC	4.6 mg/L	(10992)
Mammals			
Rat	oral LD_{50}	1390; 2880; 3700 mg/kg bw	(10992)

butyl sulfide, tert- (di-*tert*-butylsulfide)

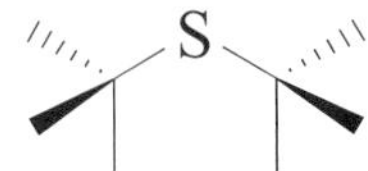

$(CH_3)_3CSC(CH_3)_3$

$C_8H_{18}S$

CAS 107-47-1

A. PROPERTIES

molecular weight 146.29; melting point -11°C; boiling point 147-151°C; density 0.82.

D. BIOLOGICAL EFFECTS

Fishes			
Pimephales promelas	4d LC_{50}	29.1 mg/L	(2625)

n-(n-butyl)thiophosphoric triamide (NBPT)

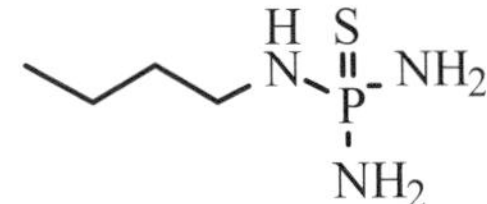

$C_4H_{14}N_3PS$

CAS 94317-64-3

TRADENAMES

Agrotrain (the chemical is 25% of the formulation)

USES

a urea fertilizer additive that temporarily retards the enzymatic breakdown of urea by inhibition of urease. It is incorporated into granular urea. (7208)

IMPURITIES, ADDITIVES, COMPOSITION

tetrahydrofuran (0-2%); triethylamine (0-2%); N,N-di-(n-butyl)thiophosphoric triamide (0-3%); N,N,N-tri-(n-butyl)thiophosphoric triamide (0-1%); thiophosphoric triamide (0-3%); other (eg dimers and more complex materials((0-10%).

A. PROPERTIES

white crystalline solid; molecular weight ;melting point 58-60 °C; vapour pressure 1.1 kPa at 40°C; density 1.04; water solubility 4300 mg/L at 25 °C; log Pow 0.42

C. WATER AND SOIL POLLUTION FACTORS

Hydrolysis

t/2 at pH 3.0	58 minutes at 25 °C	
t/2 at pH 7.0	92 days at 25 °C	
t/2 at pH 11.0	16 days at 25 °C	(7208)

Biodegradation

In soils in the dark at 22 °C: at an initial concentration of 9.5 mg/kg soil: 52-55% mineralization to CO_2 after 16 days (7208)

D. BIOLOGICAL EFFECTS

Toxicity

Algae			
Selenastrum capricornutum	96h EC_{50}	280 mg/L	(7208)
Crustaceae			
Daphnia magna	48h EC_{50}	290 mg/L	(7208)
Fish			
Lepomis macrochirus	96h LC_{50}	1140 mg/L	(7208)
Mammals			
Rat	oral LD_{50}	> 4200 mg/kg bw	(7208)

butyl-2,4-xylenol, 6-tert- ((2-(1,1-dimethylethyl)-4,6-dimethylphenol)

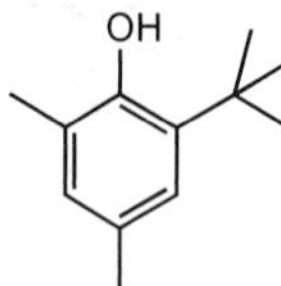

$C_{12}H_{18}O$

CAS 1879-09-0

EINECS 217-533-1

USES

Antioxidant, rubber processing agent

A. PROPERTIES

molecular weight 178.3; melting point 21-22°C; boiling point 248°C; vapour pressure 1.7 Pa at 25°C; water solubility 150 mg/L at 25°C; log Pow 4.8

C. WATER AND SOIL POLLUTION FACTORS

Hydrolysis

The chemical is stable in water at pH 4, 7 and 9.

Photodegradation

T/2 = 2.16 years , estimated by direct photodegradation.

Biodegradation

Inoculum	method	concentration	duration	elimination	
A.S.	OECD 301 C	100 mg/L	28 days	3-5% ThOD	
				0-4% GC	(10704)

D. BIOLOGICAL EFFECTS

Toxicity

Algae			
Selenastrum capricornutum	72h ECb_{50}	3.6 mg/L	
	72h NOECb	1.7 mg/L	(10704)
Crustaceae			
Daphnia magna	24h EC_{50}	5.6 mg/L	
	48h EC_{50}	3.5 mg/L	
	21d EC_{50}	2.5 mg/L	
reproduction	21d EC_{50}	0.60 mg/L	
	21d NOEC	0.32 mg/L	
	21d LOEC	0.56 mg/L	(10704)
Fish			
Oryzias latipes	24h LC_{50}	6.0 mg/L	
	72h LC_{50}	5.0 mg/L	
	96h LC_{50}	4.4 mg/L	
Carassius auratus			(10704)
Mammals			
Rat	oral LD_{50}	1,400 mg/kg bw	(10704)

n-butylacetate (butylethanoate; acetulacid, butyl ester)

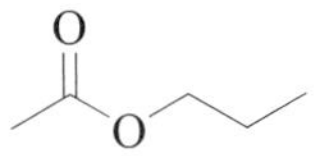

$CH_3COO(CH_2)_3CH_3$

$C_6H_{12}O_2$

CAS 123-86-4

USES

solvents in production of lacquers, perfumes, natural gums, and synthetic resins.

A. PROPERTIES

molecular weight 116.2; melting point -76.8°C; boiling point 124/127°C; vapor pressure 10 mm at 20°C, 15 mm at 25°C, vapor pressure 13 hPa at 20°C; vapor density 4.0; density 0.88 at 20/4°C; solubility 14,000 mg/L at 20°C; 5,000 mg/L at 25°C; $LogP_{OW}$ 1.81 (measured); LogH -1.87 at 25°C.

B. AIR POLLUTION FACTORS

1 mg/m^3 = 0.211 ppm, 1 ppm = 4.75 mg/m^3.

Odor

characteristic: quality: sweet, ester; hedonic tone: pleasant.

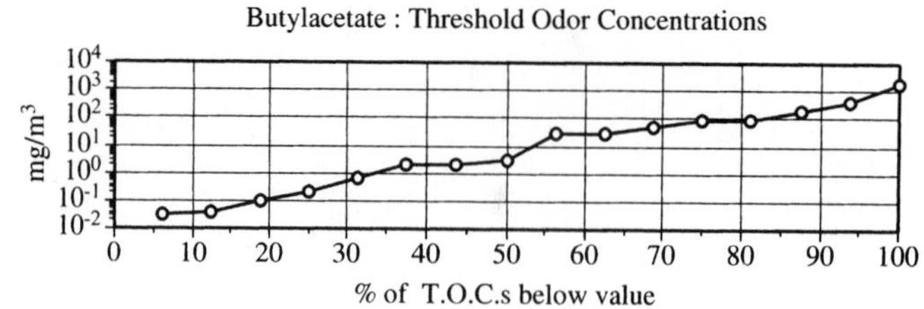

(19, 73, 210, 298, 307, 610, 643, 665, 709, 727, 749, 786, 804)

Odor

index (100% recognition): 284,000. (19)

human odor perception:	nonperception:	0.5 mg/m^3	
	perception:	0.6 mg/m^3	
human reflex response:	no response:	0.1 mg/m^3	
	adverse response:	0.13 mg/m^3	
animal chronic exposure:	no effect:	0.1 mg/m^3	
	adverse effect:	20 mg/m^3	(170)

distinct odor: 55 mg/m^3 = 11 ppm (278)

Photochemical degradation

reaction with OH°: $t_{1/2}$: 45 hours (calculated) (5205)

in smog room in the presence of 1 ppm NO_X and UV light: after 5 hours: 26% degraded (5206)

Manmade sources

in flue gase of a municipal waste incinerator plant: 5.7 μg/m_3 = 0.5% of emissions of total organic carbon. (7048)

Partition coefficients

$K_{air/water}$ 0.015

Cuticular matrix/air partition coefficient* 1900 ± 0 (7077)

* experimental value at 25°C studied in the cuticular membranes from mature tomato fruits (*Lycopersicon esculentum* Mill. cultivar Vendor)

C. WATER AND SOIL POLLUTION FACTORS

BOD_5	7; 24; 46% of ThOD	(30, 274, 282, 255)
BOD_{20}	60% of ThOD	(30)
COD	78% of ThOD (0.05 $nK_2Cr_2O_7$)	(274)
ThOD	2.207	(30)

Impact on biodegradation processes

0.1 mg/L affects the self-purification of surface waters. (181)

Odor threshold

detection: 0.066 mg/kg; 0.043 mg/kg. (889, 911)

Waste water treatment

A.C.: adsorbability: 0.17 g/g C; 84.6% reduction, infl.: 1,000 mg/L. (32)

Biodegradability

inoculum/method	*test conc.*	*test duration*	*removed*	
municipal sludge, nonadapted	3-10 mg/L	5 d	58% ThOD	
		10 d	68% ThOD	
		15 d	79% ThOD	
		20 d	83% ThOD	(5201)
saltwater m.o. nonadapted	3-10 mg/L	5 d	40% ThOD	
		10 d	52% ThOD	
		15 d	56% ThOD	
		20 d	61% ThOD	(5201)
municipal sludge, closed bottle test	2 mg/L	28 d	98% ThOD	(5196)
waste water	2.3 mg/L	5 d	23% ThOD	
		10 d	51% ThOD	
		20 d	57% ThOD	(5202)
Alcaligenes faecalis	500 mg/L	6 d	64% ThOD	(5203)
saltwater m.o.	3 mg/L	5 d	52% ThOD	(5204)

Partition coefficients

$K_{air/water}$	0.015	
Cuticular matrix/water partition coefficient**	28	(7077)

** calculated from the Cuticular matrix/water partition coefficient and the air/water distribution coefficient

D. BIOLOGICAL EFFECTS

Bacteria			
Pseudomonas putida	18h EC_{50}	960 mg/L	(5196)
E. coli	no toxic effect	1,000 mg/L	
Pseudomonas putida	16h EC_0	115 mg/L	(1900)

Algae			
Scenedesmus subspicatus	72h EC_{50}	675 mg/L	(5196)
Scenedesmus	96h LC_{50}	320 mg/L	(356)
Microcystis aeruginosa	8d EC_0	280 mg/L	(329)
Scenedesmus quadricauda	7d EC_0	21 mg/L	(1900)

Protozoans			
Chilomonas paramecium	48h EC_0	670 mg/L	(5207)
Uronema parduczi	20h EC_0,S	574 mg/L	(5208)
Entosiphon sulcatum	72h EC_0,S	321 mg/L	(5209)
Uronema parduczi Chatton Lwoff	EC_0	574 mg/L	(1901)

Crustaceans			
Daphnia magna	24h EC_{50},S	24; 73 mg/L	(5196, 5199)
	24h EC_0,S	93 mg/L	
	24h EC_{50},S	205 mg/L	
	24h EC_{100},S	500 mg/L	(5200)
Daphnia	48h LC_{50}	44 mg/L	(356)
Artemia salina	24h EC_{50},S	150 mg/L	
	24h EC_{50},S	32 mg/L	(5201)

Fishes

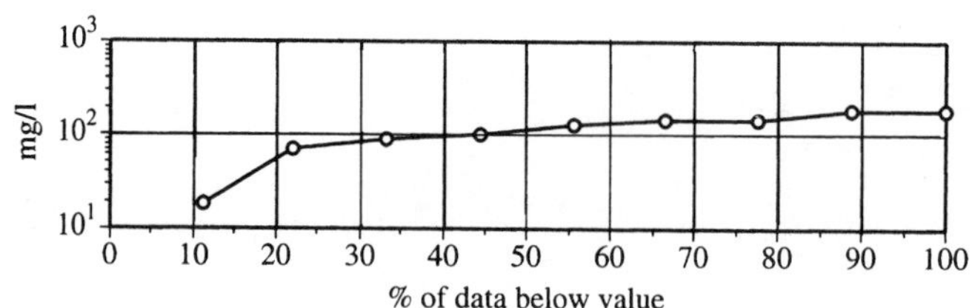

Frequency distribution of 24-96h LC_{50} values for fishes, based on data from this and other works (3999)

Leuciscus idus	48h LC_0	44; 70; 80 mg/L	
	48h LC_{50}	70; 90; 141 mg/L	
	48h LC_{100}	72; 176 mg/L	(5196, 5197)
Pimephales promelas	96h LC_{50},F	18 mg/L	(5198)

Lepomis macrochirus: static bioassay in freshwater at 23°C, mild aeration applied after 24 h

material added	*% survival after*				*best fit 96h LC_{50}*	
mg/L	*24 h*	*48 h*	*72 h*	*96 h*	*mg/L*	
250	0					
180	0				100	
125	100	90	50	0		
100	100	100	90	50		
79	100	100	100	100		(352)

Menidia beryllina: static bioassay in synthetic seawater at 23°C: mild aeration applied after 24 h

material added	*% survival after*				*best fit 96h LC_{50}*	
mg/L	*24 h*	*48 h*	*72 h*	*96 h*	*mg/L*	
320	0					
240	0					
180	64	60	50	50		(185)
132	100	80	80	80		
100	100	100	100	100		(352)

butylacetate, tert- (acetic acid, tert-butyl ester; acetic acid, 1,1-dimethylethyl ester)

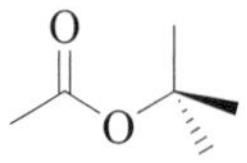

$CH_3COOC(CH_3)_3$

$C_6H_{12}O_2$

CAS 540-88-5

USES

Gasoline additive. Used in the synthesis of tert-butyl esters, N-protected amino acids. Solvent for paints, varnishes and printing inks

A. PROPERTIES

colorless liquid; molecular weight 116.16; boiling point 96°C; density 0.90 at 20°C.

B. AIR POLLUTION FACTORS

Photochemical reactions

$t_{1/2}$: 26 d, based on reactions with OH° (calculated). (8481)

Manmade sources

emissions from printing ink, paints and varnishes and automobile industry. (9801)

D. BIOLOGICAL EFFECTS

Toxicity threshold (cell multiplication inhibition test)

Bacteria			
Pseudomonas putida	16h EC_0	78 mg/L	(1900)
Algae			
Microcystis aeruginosa	8d EC_0	420 mg/L	(329)
Green algae			
Scenedesmus quadricauda	7d EC_0	3,700 mg/L	(1900)
Protozoa			
Entosiphon sulcatum	72h EC_0	970 mg/L	(1900)
Uronema parduczi Chatton Lwoff	EC_0	1,850 mg/L	(1901)

n-butylacrylate (2-propenoic acid, butyl ester; butyl-2-propenoate

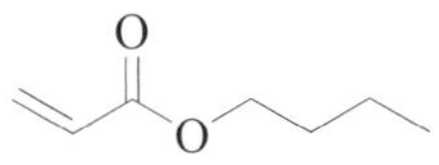

$CH_2CHCOOC_4H_9$

$C_7H_{12}O_2$

CAS 141-32-2

SMILES O=C(OCCCC)C=C

EINECS 205-480-7

USES

monomer for the mfg. of polyacrylates.

IMPURITIES, ADDITIVES, COMPOSITION

Butyl propionate <0.3%), water (<0.1%), acrylic acid (<0.01%) and 4-methoxyphenol as an additive.

A. PROPERTIES

molecular weight 128.2; melting point -64°C; boiling point 145°C; vapor pressure 4 mm at 20°C, 10 mm at 36°C, vapor pressure 4.3 hPa at 20°C, 25 hPa at 50°C; density 0.9 kg/L at 20°C;vapor density

4.42; density 0.90 at 20/4°C, solubility 1,600 mg/L at 20°C; log P_{OW} 2.38 (measured); K_{OC} : 40-148; H 46.6 Pa m^3/mol at 25°C

B. AIR POLLUTION FACTORS

1 mg/m^3 = 0.19 ppm, 1 ppm = 5.33 mg/m^3.

C. WATER AND SOIL POLLUTION FACTORS

BOD_{15}: >60% ThOD. (5162, 5164)

Reduction of amenities

organoleptic limit: 0.015 mg/L.

Waste water treatment

A.C.: adsorbability: 0.19 g/g C; 96% reduction, infl.: 1,000 mg/L. (32)

Hydrolysis

At pH 4 to 7 : t/2 = >1000 days at 25°Ç, at pH 11 : t/2 = ca 4 hours at 25°C (10650)

Biodegradation

Inoculum	method	concentration	duration	elimination	
A.S.	OECD 301C	100 mg/L	14 days	61% ThOD	
A.S.	Closed bottle test		28 days	57.8% ThOD	
A.S.	DEV H5 DIN 28409		15 days	>60%	
A.S.	Original MITI test	100 mg/L	14 days	>30%	
River water			3 days	98-100%	
			3 days	39-100%	(10650)

BOD5 = 700 mg/g substance
BOD5/COD = 0.497; 0.37
COD = 1874 mg/g substance

D. BIOLOGICAL EFFECTS

Bacteria			
Pseudomonas putida	16h EC_0	80 mg/L	(5163)
Algae			
Microcystis aeruginosa	8d EC_0	1.3 mg/L	(5161)
Scenedesmus quadricauda	8d EC_0	9.3 mg/L	(5161)
Protozoans			
Uronema parduczi Chatton Lwoff	EC_0	21 mg/L	(1901)
Crustaceans			
Daphnia magna Straus	24h EC_0	7.8 mg/L	
	24h EC_{50}	42 mg/L	
	24h EC_{100}	125 mg/L	(5159)
Daphnia magna	24h EC_0	16 mg/L	
	24h EC_{50}	230 mg/L	
	24h EC_{100}	500 mg/L	(5160)
Fishes			
Leuciscus idus	48h LC_{50}	23 mg/L	(5157)
Carassius auratus	72h LC_{50}	5 mg/L	(5158)

Bioaccumulation

BCF : 13 (calculated) (10650)

Toxicity			
Micro organisms			
Activated sludge	72h EC_0	>150 mg/L	
A.S. domestic	3 min EC_{20}-80	>1,000 mg/L	(10650)
Algae			
Selenastrum capricornutum	96h NOEC	<3.8 mg/L	
	96h EC_{10}	<3.8 mg/L	
	96h EC_{50}	5.2 mg/L	
Desmodesmus subspicatus	72h EC_{10}	1.3 mg/L	
	72h EC_{50}	3.2 mg/L	
	72h EC_{90}	6.4 mg/L	(10650)
Protozoans			
Chilomonas paramaecium	48h EC_{10}	3.5 mg/L	
Entosiphon sulcatum	72h EC_{10}	50 mg/L	(10650)
Crustaceae			
Daphnia magna	48h NOEC	2.4 mg/L	
	48h EC_0	2.4; 3.1; 12.5 mg/L	
	48h EC_{10}	4.6 mg/L	
	48h EC_{50}	8.2; 9.7; 19.8 mg/L	
	48h EC_{100}	17; 25; 50 mg/L	(10650)
Birds			
Redwinged blackbirds	18h LC_{50}	>103 mg/kg bw	(10650)
Fish			
Salmo gairdneri	96h NOEC	3.8 mg/L	
	96h LC_0	3.8 mg/L	
	96h LC_{50}	5.2 mg/L	
	96h LC_{100}	7.2 mg/L	
Pimephales promelas	96h LC_{50}	2.09; 10-20 mg/L	
	96h NOEC	10 mg/L	
Cyprinodon variegates	96h NOEC	<1.3 mg/L	
	96h LC_0	1.3 mg/L	
	96h LC_{50}	2.1 mg/L	
	96h LC_{100}	3.5 mg/L	(10650)
Mammals			
Rat	oral LD_{50}	900; 3,140; 3,143; 3,730; 64,920; ,220; 8,125; 9,050 mg./kg bw	
mouse	oral LD_{50}	756; 7,550 mg/kg bw	
Rabbit	oral LD_{50}	900-3,600 mg/kg bw	
Cat	oral LD_{50}	450-900 mg/kg bw	(10650)

butylacrylate, tert- (2-propenoic acid, 1,1-dimethylethyl ester; tert-butyl-2-propenoate; 2-propenoic acid, tert-butyl ester)

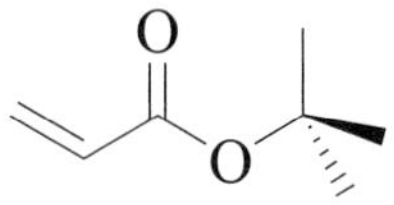

$C_7H_{12}O_2$

CAS 1663-39-4

USES

Monomer for the of polyacrylates

A. PROPERTIES

molecular weight 128.17; melting point -69°C; boiling point 121°C; vapor pressure 16 hPa at 20°C, 85 hPa at 50°C; density 0.88 at 20°C; solubility 2,000 mg/L at 20°C; LogP_{OW} 2.3 (measured), 1.9 (calculated).

D. BIOLOGICAL EFFECTS

Bacteria			
A.S.	EC_0	>500 mg/L	(6363)
Pseudomonas putida	30 min EC_{10}	8,800 mg/L	
(oxygen consumption inhibition test)	30 min EC_{50}	>10,000 mg/L	
	30 min EC_{90}	>10,000mg/L	(6361)
Algae			
Scenedesmus subspicatus	72 h EC_{10}	60 mg/L	
(cell multiplication inhibition test)	72 h EC_{50}	280 mg/L	
	72 h EC_{90}	1,300 mg/L	(6362)
Crustaceans			
Daphnia magna Straus	24 h EC_0	62 mg/L	
	24 h EC_{50}	151 mg/L	
	24 h EC_{100}	500 mg/L	(6361)
	48 h EC_0	31 mg/L	
	48 h EC_{50}	57 mg/L	
	48 h EC_{100}	125 mg/L	(6361)
Fishes			
Leuciscus idus	96 h LC_{50}	46-68 mg/L	(6360)

2-butylalcohol *see* sec-butanol

n-butylalcohol *see* *n*-butanol

butylaldehyde *see* butyraldehyde

n-butylamine (1-aminobutane)

NH_2

$CH_3(CH_2)_3NH_2$

$C_4H_{11}N$

CAS 109-73-9

USES

intermediate for emulsifying agents; pharmaceuticals; insecticides; dyes; tanning agents.

A. PROPERTIES

molecular weight 73.1; melting point -50°C; boiling point 78°C; vapor pressure 72 mm at 20°C, vapor pressure 93 hPa at 20°C; density 0.74; solubility miscible: $LogP_{OW}$ 0.81 (measured), 0.68/0.88 (calculated).

B. AIR POLLUTION FACTORS

1 mg/m^3 = 0.334 ppm, 1 ppm = 2.99 mg/m^3.

Odor

characteristic: quality: sour, ammoniacal; hedonic tone: unpleasant to pleasant.

T.O.C.:		
abs. perc. limit:	0.08 ppm	
50% recogn.:	0.24 ppm	
100% recogn.:	0.24 ppm	
O.I.: 100% recogn.:	449,166	(19)
O.I. at 20°C:	395,000	(316)

C. WATER AND SOIL POLLUTION FACTORS

BOD_5	26% of ThOD	
10 d	48% of ThOD	
15 d	50% of ThOD	
20 d	48% of ThOD	
30 d	48% of ThOD	
50 d	52% of ThOD	(79)

Biodegradability

inoculum/method	*test conc.*	*test duration*	*removed*	
BOD		5 d	>60% COD	(5254)
A.S., nonadapted	1,000 mg/L	2 d	>90% COD	(5256)

Waste water treatment

A.C.: adsorbability: 0.103 g/g C; 52.0% reduction; infl.: 1,000 mg/L; effl.: 480 mg/L. (32)

degradation by	*Aerobacter:*	200 mg/L at 30°C:	
	parent:	100% in 22 h	
	mutant:	100% in 7 h	(152)

Reverse osmosis: 39.2% rejection from a solution of 0.01 M (221)

D. BIOLOGICAL EFFECTS

Toxicity threshold (cell multiplication inhibition test)

Algae			
Microcystis aeruginosa	*8d EC_0*	0.14-0.10 mg/L	(329)
green algae			
Scenedesmus quadricauda	7d EC_0	0.53 mg/L	(1900)
Protozoa			
Entosiphon sulcatum	72h EC_0	9 mg/L	(1900)
Uronema parduczi Chatton-Lwoff	EC_0	1,752 mg/L	(1901)
Bacteria			
Pseudomonas putida	16h EC_0	65 mg/L (not neutralized)	(5253)
Pseudomonas putida	16h EC_0	800 mg/L (neutralized)	(1900)
Pseudomonas fluorescens	EC_0	65 mg/L (not neutralized)	(5255)
Crustaceans			
Daphnia magna Straus	24h EC_0	19 mg/L	
	24h EC_{50}	43 mg/L	
	24h EC_{100}	71 mg/L	(5251)
Daphnia magna	24h EC_0	19 mg/L	
	24h EC_{50}	75 mg/L	
	24h EC_{100}	150 mg/L	(5252)
brine shrimp:	24h LC_{50},S	30-70 ppm	(355)

Fishes

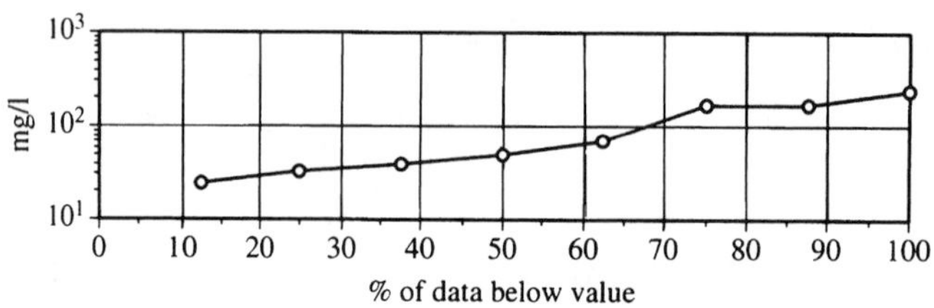

Frequency distribution of 24-96h LC_{50} values for fishes based on data from this work and the work of Rippen (3999)

Creek chub: critical range: 30-70 mg/L: 24 h (226)

Lepomis macrochirus: static bioassay in freshwater at 23°C, mild aeration applied after 24 h

material added	*% survival after*				*best fit 96h LC_{50}*	
ppm	*24 h*	*48 h*	*72 h*	*96 h*	*ppm*	
79	80	20	10	10		
50	80	70	50	20	32	
32	80	50	60	50		
10	100	100	100	100		(352)

Menidia beryllina: static bioassay in synthetic seawater at 23°C: mild aeration applied after 24 h

material added	*% survival after*				*best fit 96h LC_{50}*	
ppm	*24 h*	*48 h*	*72 h*	*96 h*	*ppm*	
100	100	100	0	-		
50	95	60	0	-		
32	100	33	33	33	24	
18	100	100	75	65		
10	100	100	100	100		(352)

Leuciscus idus	48h LC_{50}	171-236 mg/L	(5249)
	96h LC_{50}	171 mg/L	(5250)

t-butylamine borane (borane-*tert*-butylamine complex)

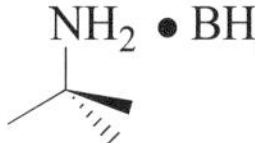

$(CH_3)_3CNH_2.BH_3$

$C_4H_{14}BN$

CAS 7337-45-3

USES

fogging agent in photoprocessing.

A. PROPERTIES

molecular weight 86.97; melting point 98°C (decomposes).

C. WATER AND SOIL POLLUTION FACTORS

theoretical	*analytical*
ThOD = 2.95	Reflux COD = 38.7% recovery
COD = 2.21	Rapid COD = 24.6% recovery
NOD = 0.74	TKN = 92.3% recovery
	BOD_5 = 0.39
	BOD_5/COD = 0.176
	BOD_5 (acclimated) = 0.31

Both COD analytical methods exhibited poor recovery of the theoretical COD. The compound exhibited a moderate response to the BOD_5 analysis. (1828)

Impact on conventional biological treatment systems

	chemical conc., mg/L	*effect*	
unacclimated system	50	inhibitory	
	500	inhibitory	
acclimated system	134	biodegradable	(1828)

D. BIOLOGICAL EFFECTS

Algae			
Selenastrum capricornutum	no effect	0.1 mg/L	(1828)
	inhibitory	1.0 mg/L	(1828)
Arthropods			
Daphnia magna	LC_{50}	0.7 mg/L	(1828)
Fishes			
Pimephales promelas:	LC_{50}	10 - 18 mg/L	(1828)

butylamine, sec- (2-aminobutane; 2-butanamine; 2-butylamine)

$CH_3CH_2CH(CH_3)NH_2$

$C_4H_{11}N$

CAS 13952-84-6

USES

intermediate for pesticides and rubber additive.

A. PROPERTIES

molecular weight 73.1; melting point -104°C; boiling point 63/63°C; vapor density 2.52; density 0.72; vapor pressure 182 hPa at 20°C; solubility miscible at 20°C; LogP_{OW} 0.6 (measured).

B. AIR POLLUTION FACTORS

1 mg/m^3 = 0.334 ppm, 1 ppm = 2.99 mg/m^3.

C. WATER AND SOIL POLLUTION FACTORS

Impact on biodegradation processes

A.S., industrial	30 min EC_{10}	>1,995 mg/L	(5122)
Nitrosomonas at 100 mg/L no inhibition of NH_3 oxidation			(390)

Biodegradability

inoculum	*test conc.*	*test duration*	*removed*	
A.S., industrial	400 mg/L	13 d	100% TOC	(5123)

D. BIOLOGICAL EFFECTS

Bacteria			
Pseudomonas putida	17h EC_{10}	70 mg/L	
	17h EC_{50}	100 mg/L	
	17h EC_{90}	160 mg/L	(5121)
Algae			
Scenedesmus subspicatus	72h EC_{50}	0.54 mg/L	
	72h EC_{20}	0.22 mg/L	
	72h EC_{90}	2.4 mg/L	(5121)
	96h EC_{50}	0.2 mg/L	
	96h EC_{20}	0.07 mg/L	
	96h EC_{90}	1.1 mg/L	(5121)
Crustaceans			
Daphnia magna Straus	24h EC_0	31 mg/L	
	24h EC_{50}	45 mg/L	

	24h EC_{100}	125 mg/L	(5121)
	48h EC_0	8 mg/L	
	48h EC_{50}	40 mg/L	
	48h EC_{100}	125 mg/L	(5121)
Fishes			
Leuciscus idus	96h LC_{50}	46-68 mg/L	(5120)
at pH 7.5-7.7	96h LC_0	100 mg/L	(5120)
creek chub	24h critical range	20-60 mg/L	(226)

4-n-butylaniline

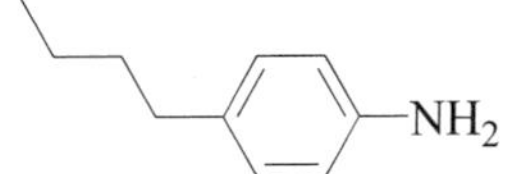

$C_4H_9C_6H_4NH_2$

$C_{10}H_{15}N$

CAS 104-13-2

A. PROPERTIES

molecular weight 149; $LogP_{OW}$ 3.15.

D. BIOLOGICAL EFFECTS

Algae			
Tetrahymena pyriformis: growth inhibition at 27°C,	48h EC_{50}	13 mg/L	(2704)
Fishes			
Pimephales promelas:	30-35d LC_{50},F	10 mg/L	(2704)

butylated hydroxyanisole (3-BHA; tert-butyl-4-hydroxyanisole; phenol, (1,1-dimethylethyl)-4-methoxy-; phenol, tert-butyl-4-methoxy-tert-butyl-4-hydroxyanisole; Antracine 12; Embanox)

3-BHA 2-BHA

$C_{11}H_{16}O_2$

CAS 25013-16-5

USES

BHA is one of the extensively used food antioxidants. Commercial BHA is a mixture of two isomers, 2-tert-butyl-4-hydroxyanisole (2-BHA) and 3-tert-butyl-4-hydroxyanisole (3-BHA) and contains 90% of the 3-isomer. Levels used in food products vary from 10 to 1000 mg/kg. Used in cosmetics and essential oils. Antimicrobial properties. (6030)

A. PROPERTIES

white waxy tablets with phenolic odor; molecular weight 180; melting point 48 - 65°C; boiling point 264 - 270°C.

D. BIOLOGICAL EFFECTS

Mammals

rat	oral LD_{50}	2000 mg/kg bw	(9711)
rabbit	oral LD_{50}	2100 mg/kg bw	(9712)

butylated hydroxytoluene *see* 2,6-di-*tert*-butyl-4-methylphenol

n-butylbenzene (1-phenylbutane)

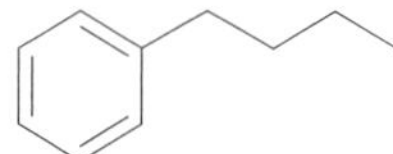

$C_{10}H_{14}$

CAS 104-51-8

USES AND FORMULATIONS

organic synthesis; pesticide mfg.; solvent for coating compositions; plasticizer; surface active agents; polymer linking agent; asphalt component; naphtha constituent. (347)

MANUFACTURING SOURCES

petroleum refining. (347)

A. PROPERTIES

molecular weight 134.21; melting point -81°C; boiling point 183°C; vapor pressure 1 mm at 23°C; vapor density 4.62; density 0.86 at 20°C; LogH -0.29 at 25°C.

B. AIR POLLUTION FACTORS

Atmospheric reactions

R.C.R.: 1.03. (49)

Manmade sources

in gasoline (high octane number): 0.08 wt %. (387)

Ambient air concentrations

indoor/outdoor glc's winter 1981-1982 and 1982-1983 the Netherlands:

μg/m³	*median*	*maximum*	
pre-war homes	0.8	30	
post-war homes	0.9	40	
homes <6 years old	1	20	
outdoors	<0.3	0.6	(2668)

C. WATER AND SOIL POLLUTION FACTORS

$BOD^{25°C}_{35\ d}$: 61% ThOD (n.s.i.) in seawater/inoculum: enrichment cultures of hydrocarbon-oxidizing bacteria. (521)
ThOD: 3.22. (521)

Reduction of amenities

organoleptic limit: 0.1 mg/L (n.s.i.). (181)

Waste water treatment

methods	*temp, °C*	*d observed*	*feed, mg/L*	*d acclim.*	*% theor. oxidation*	
A.S., Sd, BOD	20	5		14	14	
A.S., W	20	1/4	50-100	14	6	(93)

Manmade sources

Göteborg (Sweden) sew. works 1989-1991: infl.: 0.1-5 μg/L; effl.: n.d. (sum of isomers). (2787)

butylbenzene, sec- (2-phenylbutane)

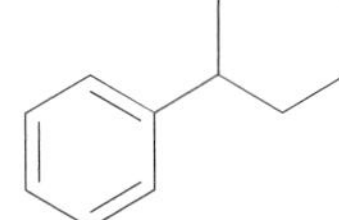

$C_6H_5CH(CH_3)C_2H_5$

$C_{10}H_{14}$

CAS 135-98-8

A. PROPERTIES

molecular weight 134.21; melting point -83°C; boiling point 173°C; vapor pressure 1.1 mm at 20°C; vapor density 4.62; density 0.86 at 20°C; LogH -0.33 at 25°C.

B. AIR POLLUTION FACTORS

Atmospheric reactions

R.C.R.: 1.31. (49)

butylbenzene, tert- (2-methyl-2-phenylpropane)

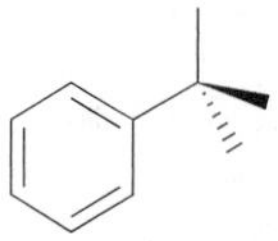

$C_6H_5(CH_3)_3$

$C_{10}H_{14}$

CAS 98-06-6

A. PROPERTIES

molecular weight 134.21; melting point -58°C; boiling point 169°C; vapor pressure 1.5 mm at 20°C; vapor density 4.62; density 0.87 at 20°C; $LogP_{OW}$ 4.11; LogH -0.32 at 25°C.

B. AIR POLLUTION FACTORS

Atmospheric reactions

R.C.R.: 0.66. (49)

Ambient air quality

Los Angeles 1966: glc's: avg.: 0.002 ppm (*n* = 136); highest value: 0.006 ppm. (1319)

C. WATER AND SOIL POLLUTION FACTORS

Reduction of amenities

T.O.C. = 0.05 mg/L. (295)

Waste water treatment

methods	*temp, °C*	*d observed*	*feed, mg/L*	*d acclim.*	*% theor. oxidation*	
A.S., W	20	5	50-200	14	1	
A.S., Sd, BOD	20	5		14	1	(93)

butylbenzoic acid, p-tert- (PTBBA; 4-(1,1-dimethylethyl)benzoic acid)

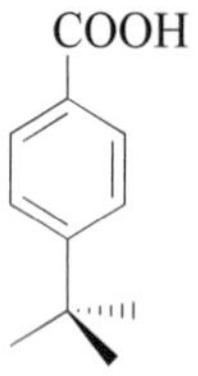

COOH

$(CH_3)_3CC_6H_4COOH$

$C_{11}H_{14}O_2$

CAS 98-73-7

USES

intermediate for mfg of resins.

A. PROPERTIES

colorless crystalline powder; molecular weight 178.23; melting point 166°C; vapor pressure <0.01 hPa at 20°C; density 1.14 at 20/4°C; solubility 300 mg/L at 20°C; LogP_{OW} 3.86 (calculated).

C. WATER AND SOIL POLLUTION FACTORS

BOD_5:	11% ThOD.	(277)
COD:	100% ThOD.	(277)
ThOD:	2.37.	

Biodegradability

Inoculum/test method	*test conc.*	*test duration*	*removed*	
river water, adapted, BOD test	0.8-3.2 mg/L	5 d	4% ThOD	
		10 d	26% ThOD	
		15 d	33% ThOD	
		20 d	40% ThOD	(5641)
groundwater, adapted, BOD test	0.8-3.2 mg/L	5 d	4% ThOD	
		10 d	4% ThOD	
		15 d	4% ThOD	
		20 d	4% ThOD	(5641)
harbor water, adapted, BOD test	0.8-3.2 mg/L	5 d	11% ThOD	
		10 d	40% ThOD	
		15 d	64% ThOD	
		20 d	67% ThOD	(5641)
harbor water, BOD test	0.8-3.2 mg/L	5 d	31% ThOD	
harbor water, BOD test	0.8-3.2 mg/L	5 d	33% ThOD	
harbor water, BOD test	0.8-3.2 mg/L	5 d	8% ThOD	
harbor water + nutrients, BOD test	0.8-3.2 mg/L	5 d	27% ThOD	
harbor water, adapted + nutrients, BOD test	0.8-3.2 mg/L	5 d	37% ThOD	(5642)

D. BIOLOGICAL EFFECTS

Bacteria

Pseudomonas fluorescens	48h EC_{50}	1,000 mg/L	(5640)

Fishes

goldfish	pH 5	24-96h LD_{50}	4 mg/L	
	pH 7	24-96h LD_{50}	33 mg/L	(277)
Poecilia reticulata	pH 7	48h LC_0	200 mg/L	(5639)

butylbenzothiazole-2-sulphenamide, N-tert- ((N-(1,1-dimethylethyl)-benzothiazolesulfenamide))

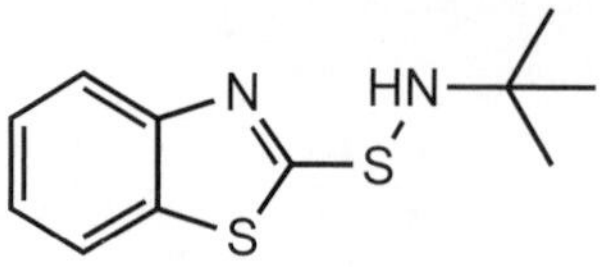

$C_{11}H_{14}N_2S_2$

CAS 95-31-8

TRADENAMES

Cure-rite BBTS; Accel TBS-R; Nocceler NS-P; Pennac Tbbs; Perkacit NS; Sanceler NS-G; Santocure NS; Santocure NS vulcanization accelerator; TBBS; Vanax NS; Vulkacit NZ

USES

An accelerator in the vulcanisation of rubber. The chemical and related substances produced during synthesis may be released from the final products, for example tyres, during use.

IMPURITIES, ADDITIVES, COMPOSITION :

Dibenzthiazylsulfide (0.8%)

A. PROPERTIES

molecular weight 238.39; melting point 105°C; relative density 1.28 kg/L at 25°C; vapour pressure 2.1 x 10^{-6} hPa at 20°C; water solubility 0.345 mg/L at 20°C; log Pow 3.9 at 20°C

B. AIR POLLUTION FACTORS

Photodegradation

T/2 = 4.6 days based on reaction with OH radicals . (10663)

C. WATER AND SOIL POLLUTION FACTORS

Hydrolysis

7/2 = 1.7 hours at pH 4; 1.8 hours at pH 7 and 21 hours at pH 9 at 25°C. The identified hydrolysis products are mercaptobenzothiazole, di(benzothiazoyl-2)disulfide, tert-butylamine and benzothazole (10663)

Biodegradation

Inoculum	method	concentration	duration	elimination	
A.S.	Modified MITI test	100 mg/L	28 days	0% ThOD(1)	
AdaptedA.S.		29.4 mg/L	32 days	63.5% ThCO2	(10663)

(1) under the test conditions, no biodegradation was observed. However, the chemical did hydrolyse to produce mercaptobenzothiazole, di(benzothiazoyl-2)disulfide, tert-butylamine, 2-sulfo(sulfino)benzothazole and benzothazole as the degradation products.

D. BIOLOGICAL EFFECTS

Bioaccumulation

The hydrolysis products have been tested and shown to haved low potential for bioaccumulation. (10663)

Toxicity

Micro organisms			
A.S.	3h EC_{50}	>10,000 mg/L	(10663)
Algae			
Selenastrum capricornutum	72h NOECgr	0.023 mg/L	
	72h $ECgr_{50}$	0.071 mg/L	
	96h ECb_{50}	>0.3 mg/L	(10663)
Crustaceae			
Daphnia magna	48h EC_{50}	>0.3; 1.3; 6 mg/L	
	48h EC_{100}	1.7 mg/L	
	48h NOEC	>0.3; 0.95 mg/L	
	24-48h LC_{50}	>100 mg/L	
	48h NOEC	100 mg/L	
reproduction	21d NOEC	>0.16 mg/L	
	14d NOEC	0.042 mg/L	
	14d LC_{50}	>0.16 mg/L	(10663)
Fish			
Oryzias latipes	96h LC_0	0.15; 0.52 mg/L	
	96h LC_{50}	1.02; 1.4 mg/L	
	96h LC_{100}	1.3; 4.0 mg/L	
	7d LC_{50}	1.3 mg/L	
	14d LOEL	0.44 mg/L	
Pimephales promelas	14d LC_{50}	>0.3 mg/L	
	96h LC_{50}	21 mg/L	
	96h NOEC	5.6 mg/L	
	24h LC_{50}	24 mg/L	
	48h LC_{50}	21 mg/L	
Brachydanio rerio	96h LC_{50}	>0.5 mg/L	
Lepomis macrochirus	24h LC_{50}	3.6 mg/L	
	48h LC_{50}	1.5 mg/L	
	96h LC_{50}	1.2 mg/L	
Oncorhynchus mykiss	24-96h LC_{50}	1.6 mg/L	(10663)
Mammals			
Rat	oral LD_{50}	>2,000; 6,850 mg/kg bw	(10663)

butylbenzylphthalate (santicizer 160; BBP; butyl phenylmethyl)

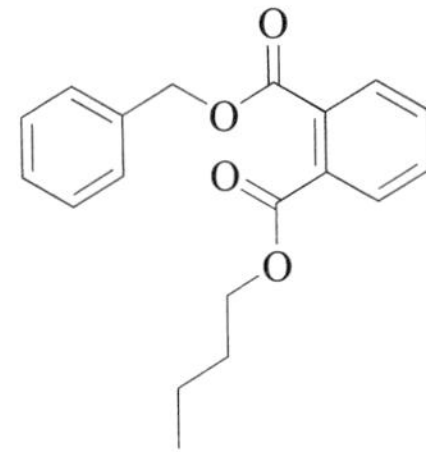

$C_{19}H_{20}O_4$

CAS 85-68-7

USES

to plasticize or flexibilize synthetic resins, chiefly polyvinylchloride.

A. PROPERTIES

clear, oily liquid; molecular weight 312.4; mp. < -35°C; boiling point 370°C; density 1.1 at 25/25°C; vapor pressure 8.6 × 10^{-6} mm at 20°C, 1.9 mm at 200°C; vapor density 10.8; solubility 2.9 ± 1.2 mg/L (in deionized water); $LogP_{OW}$ 4.91; 4.78; 4.05.

B. AIR POLLUTION FACTORS

Incinerability

Temperature for 99% destruction at 2.0 sec residence time under oxygen-starved reaction conditions: 415°C.

Thermal stability ranking of hazardous organic compounds

rank 253 on a scale of 1 (highest stability) to 320 (lowest stability). (2390)

C. WATER AND SOIL POLLUTION FACTORS

Manmade sources

In Canadian municipal sludges and sludge compost: September 1993- February 1994: mean values of 11 sludges ranged from 0.3 to 10.1 mg/kg dw.; mean: 2.1 mg/kg dw; mean value of sludge compost: 0.05 mg/kg dw. (7000)

Water, soil and sediment quality

soil sorption coefficient: 68-350 (1830)

residue in natural water (U.S. rivers):
avg: approx. 0.35 µg/L
range: n.d.-4.1 µg/L (n = 53)
residue in sediments of natural waters:
BBP detected in 7 out of 28 samples: avg 136 ng/g
range: n.d.-567 ng/g (1830)

residue in Delaware river (U.S.A.): conc. range:
winter: 0.4-1 ppb
summer: 0.3 ppm (1051)

river Rhine at Dutch-German border (Lobith, October 1986):
in water: avg 0.04; range 0.01-0.10 µg/L
in suspended particulate matter avg 0.3; range 0.1-0.9 mg/kg dry wt (2718)

Surface waters

In Rhine River near Lobith (Dutch-German border) Nov. 1986: <0.01-0.09 µg/L and 0.1-0.9 mg/kg SPM
In Lake Yssel (the Netherlands) Oct. 1986: <0.01 µg/L and <0.01-0.05 mg/kg SPM (2655)

Degradation

	*% degradation primary**	*ultimate***	*time, d*	*half life, d*	
biodegradation					
activated sludge	93-99		1		
CO_2 evolution, aerobic		96	28		
gas production, anaerobic		<10	28		
river water	100		9	2	
lake water microcosm	>95		7	<4	
lake water microcosm		51-65	28		(1830)
photodegradation	<5		28	>100	(1830)
chemical degradation (hydrolysis)	<5		28	>100	(1830)

* disappearance of BBP as measured by gas chromatography

** mineralization under aerobic conditions to CO_2, under anaerobic conditions to H_2, CH_4, and CO_2

Biodegradation in river water

100% loss in 7 d at 3 µg/L at 20°C; no significant loss in 10 d at 3 µg/L at 4°C. (2655)

Water treatment methods

domestic activated sludge plant:
inlet: 8.0 µg/L
outlet: 1.3 µg/L
outlet aerated lagoon: 1.0 µg/L (1830)

Biodegradation

A.S. 48 h: 99% (1840)
continuous A.S.: 96% (1841)
in river water, 1 week: 80% (1841)
Göteborg (Sweden) sew. works 1989-1991: infl.: 1-10 µg/L; effl.: nd (2787)
Aerobic slurry bioremediation reactor with mixed aeration chamber, treating a petrochemical waste sludge at 3.5 to 7.5% solids containing approx. 25% oil and grease and 0.5 to 2.5 grams of waste material per gram of microorganisms, at 22-24°C during batch treatment: waste residue: infl.: 9.8 mg/kg; effl. after 90 d: <2.5 mg/kg. (2800)

Impact on biodegradation processes

toxicity to microorganisms: OECD 209 closed system inhib. EC_{50}: >2.9 mg/L. (2624)

D. BIOLOGICAL EFFECTS

Acute lethality, measured in static tests:

Algae

species	*96h EC_{50} or LC_{50} mg/L*	*NOEC mg/L*
Microcystis	1,000	560
Dunaliella	1.0	0.3
Navicula	0.6	0.3
Skeletonema	0.6	0.1
Selenastrum	0.4	0.1

Crustaceans

species	*96h EC_{50} or LC_{50} mg/L*	*NOEC mg/L*	
*Daphnia magna**	3.7	1.0	
mysid shrimp	0.9	0.4	
Mysid shrimp *(Mysidopsis bahia):*	9.6 mg/L		
Daphnia magna:	92 mg/L		(2807)

* 48h EC_{50}

Fishes

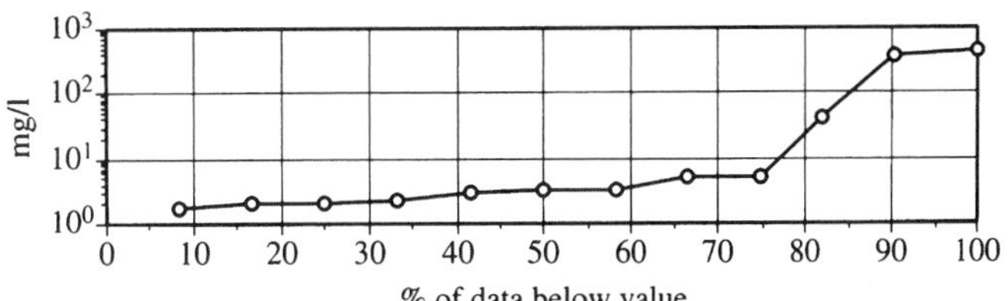

Frequency distribution of 24-96h LC_{50} values for fishes based on data from this work and from the work of G. Rippen (3999)

species	*96h EC_{50} or LC_{50} mg/L*	*NOEC mg/L*

fathead minnow	2.1-5.3	1.0-2.2	
bluegill	1.7	0.38	
rainbow trout	3.3	<0.36	
sheepshead minnow	3.0	1.0	(1830)
Bluegill *(Lepomis macrochirus):*	43 mg/L		
Fathead minnows *(Pimephales promelas):*	2.1; 5.3 mg/L		
Rainbow trout *(Salmo gairdneri):*	96h LC_{50}	3.3 mg/L	
Sheepshead minnow *(Cyprinodon variegatus):*	96h LC_{50}	360; 445 mg/L	(2807)

Chronic Toxicity

Fishes

fathead minnows:	4d LC_{50},F	2.3 mg/L	
	14d LC_{50},F	2.2 mg/L	(1830)
fathead minnow post-hatch embryo-larval stage: MATC: 0.14-0.36 mg/L			(1830)

Crustaceans

Daphnia magna:	MATC	0.26-0.76 mg/L	(1830)

Bioconcentration

Bluegill: BCF: 663 (based on ^{14}C determinations); depuration half-life: <2 d. (1830)

Bluegill sunfish *(Lepomis macrochirus)* at 9.73 µg/L at 16°C during 21 d: BCF: 772; half-life in tissues: >1 d < 2 d. (2602)

n-butylbromide *see* 1-bromobutane

n-butylcarbamate (carbamic acid, butyl ester)

$C_5H_{11}NO_2$

CAS 592-35-8

USES

Intermediate in the manufacturing of resins.

A. PROPERTIES

molecular weight 117.15; melting point 54°C; boiling point 203°C; density 0.97 at 55°C; solubility 25,800 mg/L at 37°C; 1,000 mg/L at 20°C; $LogP_{OW}$0.88 (calculated), 0.85 (measured).

B. AIR POLLUTION FACTORS

Photochemical reactions

$t_{1/2}$: 3.7 d, based on reactions with OH° (calculated) (6296)

C. WATER AND SOIL POLLUTION FACTORS

Biodegradation

Inoculum	*Method*	*Concentration*	*Duration*	*Elimination results*	
	Zahn-Wellens test	200 mg/L	3 h	30% DOC	
			5 d	30% DOC	
			10 d	71% DOC	
			15 d	97-100% DOC	(6264)

D. BIOLOGICAL EFFECTS

Bacteria			
primary municipal sludge	24 h EC_0	100-1,000 mg/L	(6294)
Bacillus subtilis (cell multiplication inhibition test)	18 h EC_{100}	8,500 mg/L	(6295)
Crustaceans			
Daphnia magna	24 h EC_0	>1,000 mg/L	(6294)
Fishes			
Brachydanio rerio	96 h LC_0	250 mg/L	
	96 h LC_{50}	350-500 mg/L	
	96 h LC_{100}	500 mg/L	(6293)

n-butylcarbinol *see* *n*-pentanol

butylcarbitol *see* diethyleneglycolmonobutylether

butylcarbitolacetate (butyldiglycolacetate; 2-(2-butoxyethoxy)ethylacetate; diethyleneglycol monobutyletheracetate)

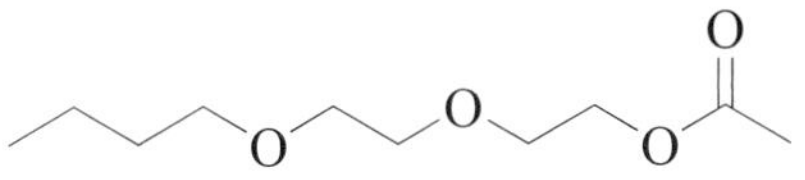

$C_4H_9OCH_2CH_2OCH_2CH_2O$

550 butylcarbitolacetate

$C_{10}H_{20}O_4$

CAS 124-17-4

USES

mfg. of paints, lacquers and varnishes, solvent.

IMPURITIES, ADDITIVES, COMPOSITION

Impurities: 0.1% ethylenediacetate, 0.1% 2-ethylacetate, 0.3% 2-(2-butoxyethoxyethanol), 0.6% oxydiethylene di(acetate). (5191)

A. PROPERTIES

molecular weight 204.27; melting point -32°C; boiling point 246°C; vapor pressure 0.04 mm at 20°C; vapor density 7.02; density 0.98 at 20°C; solubility 65,000 mg/L at 20°C; $LogP_{OW}$ 1.77 at 25°C (calculated).

B. AIR POLLUTION FACTORS

Photochemical degradation

reaction with OH°: $t_{1/2}$: 13 hours (calculated). (5185)

C. WATER AND SOIL POLLUTION FACTORS

BOD_5: 13% of ThOD
10 d: 18% of ThOD
15 d: 24% of ThOD
20 d: 67% of ThOD (79)

Biodegradability

inoculum/method	*test conc.,*	*test duration*	*removed*	
A.S., municipal nonadapted	500 mg/L	60 hours	35% CO_2 prod.	(5186)
A.S., industrial nonadapted		5 d	30% ThOD	
		10 d	69% ThOD	
		15 d	100% ThOD	(5187)
A.S., municipal nonadapted		5 d	14% ThOD	
		10 d	54% ThOD	
		20 d	73% ThOD	(5188, 5189)
A.S., industrial nonadapted	500 mg/L	14 d	>90% COD	(5190)

Impact on biodegradation processes

A.S.	16h EC_{50}	>5,000 mg/L	(5185)
anaerobic bacteria from municipal sludge	24h EC_0	2,500 mg/L	(5185)
	24h EC_0	1,000 mg/L	(5187)

D. BIOLOGICAL EFFECTS

Crustaceans

Daphnia magna	24h EC_0	186 mg/L	
	24h EC_{50}	430 mg/L	(5193)
	48h LC_{50}	665 mg/L	(5194)

Fishes

Brachydanio rerio	96h LC_0	50 mg/L	
	96h LC_{50}	50-70 mg/L	
	96h LC_{100}	100 mg/L	(5191)
Leuciscus idus	48h LC_0	100 mg/L	(5187)

Pimephales promelas	24h LC_{50}	>100 mg/L	(5192)
	48h LC_{50}	83 mg/L	
	96h LC_{50}	77 mg/L	(5192)
Poecilia reticulata	96h LC_0	100 mg/L	(5186)

butylcatechol, p-tert- (4-*tert*-butylpyrocatechol; 4-*tert*-butyl-1,2-dihydroxybenzene)

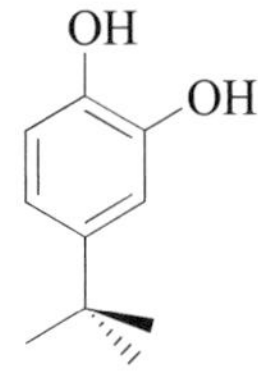

$(CH_3)_3CC_6H_3(OH)_2$

$C_{10}H_{14}O_2$

CAS 98-29-3

USES

polymerization inhibitor for styrene-butadiene and other olefins.

A. PROPERTIES

white, crystalline solid; molecular weight 166.22; melting point 56-58°C; boiling point 285°C; density 1.05 at 60/25°C; vapor pressure 0.0028 mm at 25°C; solubility 2,000 ppm at 25°C.

C. WATER AND SOIL POLLUTION FACTORS

Odor threshold

detection: 1.0 mg/L. (998)

butylcellosolve (butylglycolether; glycol monobutylether; butylglycol; 2-butoxyethanol; ethyleneglycolmono-n-butylether; butyl "Oxitol")

$C_4H_9OCH_2CH_2OH$

$C_6H_{14}O_2$

CAS 111-76-2

TRADENAMES

Butyl Cellosolve, Butyl Icinol, Butyl Oxitol, Dowanol EB, Ektasolve, Gafcol EB, Glycol ethert, Jeffersol EB, Poly-Solv EB.

USES

The main use is in paints and surface coatings, followed by its use in cleaning products and then in inks.The main types of cleaning products are: surface cleaners, floor strippers, glass/window cleaners, carpet cleaners, laundrY detergents, rust removers, oven cleaners, ink/resin remover. Concentrations of the chemical ranges between 1 and 94% in the cleaning products. Other products which contain 2-butoxyethanol include acrylic resin formulations, asphalt release agents, firefighting foam, leather protectors, oil spill dispersants and photographic strip solutions. The chemical is also used as an ingredient in agricultural chemicals, cosmetics and brake oils, and as a raw material in the production of acetate esters and phthalate and stearate plasticisers. (7156)

IMPURITIES, ADDITIVES, COMPOSITION

a stabiliser, 2,6-bis(1,1-dimethylethyl)-4-methylphenol, can be added at ca. 0.01% to prevent formation of peroxides.

A. PROPERTIES

colorless liquid: molecular weight 118.17; melting point -70°C; boiling point 171°C; vapour density 4.9 g/L at 20°C;vapor pressure 0.6 mm at 20°C, 0.89 hPa at 20°C, 3.7 hPa at 40°C; vapor density 4.07; density 0.90 at 20°C; solubility miscible at 20°C; log P_{oct} 0.74; 0.81 (calculated).

B. AIR POLLUTION FACTORS

1 mg/m^3 = 0.204 ppm; 1 ppm = 4.90 mg/m^3.

Odor

characteristic; quality: sweet, ester; hedonic tone: pleasant.

T.O.C.:	abs. Perc. limit:	0.10 ppm	
	50% recogn.:	0.35 ppm	
	100% recogn.:	0.48 ppm	
O.I.	100% recogn.:	2,800	(19)
O.I.	at 20°C:	1,600	(316)

C. WATER AND SOIL POLLUTION FACTORS

BOD:	31% ThOD	
	73% ThOD adapted sew.	(277)
COD:	96% ThOD	(277)
ThOD	2.3	

Manmade sources

Göteborg (Sweden) sewage works 1989-1991: influent.: nd -250 µg/L; effluent.: nd -3.0 µg/L. (2787)

Waste water treatment

A.C.: adsorbability: 0.11 g/g carbon, 55.9% reduction, infl.: 1,000 mg/L. (32)

Biodegradation

Inoculum	*Method*	*Concentration*	*Duration days*	*Elimination results*	
	BOD test		5	8; 52% ThOD	(8062, 8063)
A.S.	MITI test	100 mg/L	14	>30%	(8064)
industrial A.S.	Zahn-Wellens test	400 mg/L	7	92% ThOD	
			8	98% ThOD	(8065)
			8	98% ThOD	(8065)
Soil and sewage	Closed Bottle Test		28	88% ThOD	
Domestic sewage	20day BOD Test		20	75% ThOD	(7156)

Half-lives in surface water range from 7 days to 4 weeks.	(7156)
2-butoxyethanol is oxidised to 2-butoxyacetic acid	(7156)

2-butoxyethanol → 2-butoxyacetic acid

Hydrolysis

the chemical is unlikely to hydrolyse as alcohols and ethers are generally resistant to hydrolysis.	(7156)

Soil sorption

a K_{OC} of 67 (calculated) indicates that 2-butoxyethanol will not partition into organic matter contained in sediments and suspended solids, and should be highly mobile in soil.

D. BIOLOGICAL EFFECTS

Bacteria			
Pseudomonas putida	16 h EC_0	700 mg/L	(1900)
Industrial A.S. respiration inhibition	30 min EC_{10}	>1,995 mg/L	(8065)
Pseudomonas fluorescens	16 h EC_{10}	700 mg/L	(8072)

Algae			
Microcystis aeruginosa	8 d EC_0	35 mg/L	(329)
Scenedesmus quadricauda	7 d EC_0	900 mg/L	(1900)

Protozoans			
Entosiphon sulcatum	72 h EC_0	91 mg/L	(1900)
Uronema parduczi Chatton-Lwoff	EC_0	463 mg/L	(1901)

Crustaceans			
arthropods: brown shrimp *(Crangon crangon)*			
48 h LC_{50}	avg	800 mg/L	
	range	600-1,000 mg/L	
96 h LC_{50}	avg	775 mg/L	
	range	550-950 mg/L	(310)
Daphnia magna	24 h EC_0	1,140; 1,283 mg/L	
	24 h EC_{50}	1,720; 1,815 mg/L	
	24 h EC_{100}	2,500; 2,500 mg/L	(8070, 8071)

Fishes

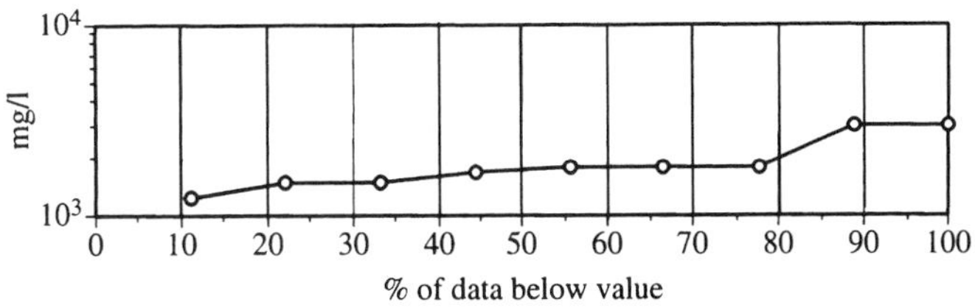

Frequency distribution of 24-96 h LC_{50} values for fishes based on data from this work and from the work of Rippen.	(3999)

goldfish:	24 h LC_{50}	1,650 mg/L	(277)
guppy *(Poecilia reticulata):*	7 d LC_{50}	983 mg/L	
Leuciscus idus	48 h EC_0	1,170-1,350 mg/L	
	48 h EC_{50}	1,395-1,575 mg/L	
	48 h EC_{100}	1,490-1,620 mg/L	(8068)
Lepomis macrochirus	96 h LC_{50}	2,950 mg/L	(8069)

Lepomis macrochirus: static bioassay in fresh water at 23°C, mild aeration applied after 24 h.

material added,	*% survival after*				*Best fit 96 h LC_{50},*	
ppm	*24 h*	*48 h*	*72 h*	*96 h*	*ppm*	
2,400	40	40	20	0		
1,800	50	50	50	30		
1,000	100	100	90	80	1,490	
790	100	100	100	100		
320	100	100	100	100		(352)

Menidia beryllina: static bioassay in synthetic seawater at 23°C, mild aeration applied after 24 h.

material added,	*% survival after*				*best fit 96 h LC_{50},*	
ppm	*24 h*	*48 h*	*72 h*	*96 h*	*ppm*	
1,800	90	70	30	20		
1,320	100	100	90	30	1,250	
1,000	100	100	100	70		(352)
bluegill *(Lepomis macrochirus)*	24h LC_{50},S	2,950 mg/L		96h LC_{50},S	2,950 mg/L	(2697)

Bioaccumulation

The calculated BCF of 2 indicates that the chemical is unlikely to accumulate in aquatic organisms. (7156)

Toxicity

Micro organisms			
bacteria from sewage	16h EC_{50}	> 1000 mg/L	(7156)
Algae			
Selenastrum capricornutum	7d EC_{50}	>1000 mg/L	(7156)
Crustaceae			
Daphnia magna Straus	48h LC_{50}	835 mg/L	
Paenaeus setiferus	4d LC_{50}	130 mg/L	(7156)
Mollusca			
Crassostrea virginica	4d LC_{50}	89.4 mg/L	(7156)
Fish			
Salmo gairdneri	48h LC_{50}		
*Lebistes _variegates_*	48h LC_{50}		
Leuciscus idus			
Pimephales promelas	28d NOEC		
Oncorhynchus mykiss	96h LC_{50}		
Mammals			
Rat	oral LD_{50}	530-3000 mg/kg bw	
Mouse	oral LD_{50}	1230 mg/kg bw	
Guinea pig	oral LD_{50}	950-1400 mg/kg bw	
Rabbit	oral LD_{50}	320-370 mg/kg bw	(7156)

butylcellosolveacetate (ethyleneglycolmonobutyletheracetate; 2-butoxy-ethanolacetate; 2-butoxyethylacetate; butylglycol acetate)

$C_4H_9OCH_2CH_2OOCCH_3$

$C_8H_{16}O_3$

CAS 112-07-2

USES

high-boiling solvent for nitrocellulose lacquers, epoxy resins, dyes, and printing inks.

A. PROPERTIES

colorless liquid; fruity odor; molecular weight 160.21; melting point -64°C; boiling point 184-195°C at 1,013 hPa; vapor pressure 0.4 hPa at 20°C; 1.4 hPa at 40°C; density 0.94 at 20/20°C; solubility 15,000 mg/L at 20°C; LogP_{OW} 1.51 (measured).

B. AIR POLLUTION FACTORS

1 mg/m^3 = 0.150 ppm; 1 ppm = 6.64 mg/m^3.

Odor

characteristic; quality: sweet, ester; hedonic tone: pleasant.

T.O.C.:	abs. perc. limit:	0.10 ppm	
	50% recogn.:	0.35 ppm	
	100% recogn.:	0.48 ppm	(19)
	O.I.: 2700		

C. WATER AND SOIL POLLUTION FACTORS

Impact on biodegradation processes

A.S., municipal	30 min EC_{20}	900 mg/L	(5237)
	180 min EC_{20}	>1,000 mg/L	(5237)

Biodegradability

inoculum/method	*test conc.*	*test duration*	*removed*	
A.S., municipal		28 d	88% CO_2 prod.	(5237)
A.S., nonadapted, Zahn- Wellens test	1,000 mg/L	6.5 d	>90% COD	(5238)
Effluent treatment plant, ISO test 7827	20 mg/L	14 d	96% DOC	(5239)

D. BIOLOGICAL EFFECTS

Bacteria

Pseudomonas putida	17h EC_{10}	720 mg/L	
	17h EC_{50}	960 mg/L	
	17h EC_{90}	1,200 mg/L	(5236)

Algae

Scenedesmus subspicatus	72h EC_{50}	>500 mg/L	(5235)

Crustaceans

Daphnia magna Straus	24h EC_0	58 mg/L	
	24h EC_{50}	150 mg/L	
	24h EC_{100}	320 mg/L	(5235)
	48h EC_0	10 mg/L	
	48h EC_{50}	37 mg/L	
	48h EC_{100}	320 mg/L	(5235)

n-butylchloride (1-chlorobutane)

$CH_3(CH_2)_2CH_2Cl$

C_4H_9Cl

CAS 109-69-3

USES

alkylation agent in organic synthesis.

IMPURITIES, ADDITIVES, COMPOSITION

Isobutylchloride; 2-chlorobutane; butanol

A. PROPERTIES

molecular weight 92.57; melting point -123°C; boiling point 78°C; vapor pressure 80.1 mm at 20°C, vapor pressure 110 hPa at 20°C; 267 hPa at 41°C; vapor density 3.20; density 0.88; water solubility 370 mg/L at 25°C; solubility 660 mg/L at 12°C; LogP_{OW} 2.39 (calculated), 2.66 (measured); LogH -0.10 at 25°C.

B. AIR POLLUTION FACTORS

Odor

characteristic: quality: pungent; hedonic tone: unpleasant.

T.O.C.:	abs. perc. limit:	8.8 ppm	
	50% recogn.:	13. ppm	
	100% recogn.:	17 ppm	(19)
	O.I.: 6,000		

C. WATER AND SOIL POLLUTION FACTORS

Biodegradability

inoculum/method	*test conc.*	*test duration*	*removed*	
A.S., nonadapted	73 mg/L	28 d	70% product	(5248)
A.S.		6 hours	0.9% of ThOD	
		12 hours	1.3% of ThOD	
		24 hours	2.6% of ThOD	(88)

Biodegradation

Inoculum	*Method*	*Concentration*	*Duration days*	*Elimination results*	
A.S.	OECD 301D	5.2 mg/L	28	0%	(10675)

Photolysis in water

T/2 = 9.6 years

D. BIOLOGICAL EFFECTS

Bacteria			
Pseudomonas putida	18h EC_{10}	330 mg/L	(5248)
Fishes			
Leuciscus idus	48h LC_0	200 mg/L	
	48h LC_{50}	245 mg/L	(5248)
guppy *(Poecilia reticulata)*	7d LC_{50}	97 mg/L	(1833)

Bioaccumulation

Species	conc. µg/L	duration	body parts	BCF	
Carp	360	6 weeks		90-110	
	36	6 weeks		300-450	(10675)

Toxicity

Algae			
Selenastrum capricornutum	72h ECb_{50}	>1,000 mg/L	(10675)
Crustaceae			
Daphnia magna	24h EC_{50}	330; 380 mg/L	
	48h LC_{50}	190 mg/L	
	96h LC_{50}	110 mg/L	
	7d LC_{50}	110 mg/L	
	14d LC_{50}	77 mg/L	
	21d LC_{50}	60 mg/L	
reproduction	21d EC_{50}	40 mg/L	
	21d NOEC	14 mg/L	(10675)
Fish			
Oryzias latipes	24-96h LC_{50}	120 mg/L	(10675)
Mammals			
Rat	oral LD_{50}	2,670 mg/kg bw	(10675)

butylchloroacetate

O
O Cl

$C_4H_9OOCCH_2Cl$

$C_6H_{11}ClO_2$

CAS 590-02-3

C. WATER AND SOIL POLLUTION FACTORS

Manmade sources

60 ml/min pure water passed through 25 ft, 1/2 inch I.D. tube of general chemical grade (430)

PVC contained 0.66 ppb butylchloroacetate, which constituted 6.14% of total contaminant concentration.

n-butylcyanide *see* valeronitrile

butylcyclohexanol, 4-tert- (4-(1,1-dimethylethyl)-cyclohexanol)

$C_{10}H_{20}O$

CAS 98-52-2

USES

Feedstock for peroxides and fragrances.

A. PROPERTIES

molecular weight 156.3; melting point approx. 64°C; boiling point approx. 226°C; vapor pressure <0.1 hPa at 20°C; density 0.87; solubility <100 mg/L; LogP_{OW} 3.23 (measured).

C. WATER AND SOIL POLLUTION FACTORS

Biodegradability

Modified OECD screening test: at 10 mg/L after 21 d: 89% DOC removal. (6073)

D. BIOLOGICAL EFFECTS

Bacteria			
Pseudomonas putida	6 h EC_{10}	221 mg/L	
(inhib. of oxygen consumption)			(6073)
Fishes			
Leuciscus idus	48 h LC_{50}	17 mg/L	(6073)

butyldiglycol *see* diethyleneglycolmonobutylether

butyldigol *see* diethyleneglycolmonobutylether

---

butyldigolacetate *see* butylcarbitolacetate

---

butyldioxitol *see* diethyleneglycolmonobutylether

---

t-butyldisulfide

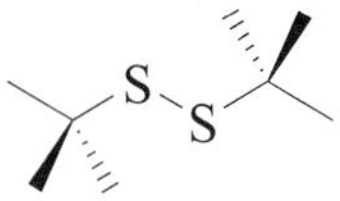

$(CH_3)_3CSSC(CH_3)_3$

$C_8H_{18}S_2$

CAS 110-06-5

A. PROPERTIES

molecular weight 178.36; boiling point 198-204°C; density 1.49; solubility 2 mg/L; $LogP_{OW}$ 4.02.

D. BIOLOGICAL EFFECTS

Fishes			
Pimephales promelas	48h LC_{50}	1.5 mg/L	
	96h LC_{50}	1.37 mg/L	(2709)

α-butylene (1-butene; ethylethylene)

560 α-butylene

$CH_3CH_2CHCH_2$

C_4H_8

CAS 106-98-9

A. PROPERTIES

molecular weight 56.10; melting point -130°C; boiling point -6°C; vapor pressure 400 mm at -21.7°C, 760 mm at -6.3°C; vapor density 1.94; density 0.67 liquefied; LogH 1.01 at 25°C.

B. AIR POLLUTION FACTORS

1 ppm = 2.33 mg/m^3; 1 mg/m^3 = 0.43 ppm.

Odor

characteristic: quality: gaseous odor.
T.O.C.: 0.160 mg/m^3 = 69 ppb (307)
faint odor: 50-59 mg/m^3 (129)
O.I. at 20°C: 43,000,000 (316)

T.O.C.: recognition:	39 mg/m^3	(761)
	2.1 mg/m^3	(710)
	1.2 mg/m^3	(724)

Atmospheric reactions

R.C.R.: 2.26 (n.s.i.) (49)

Atmospheric half-lives

for reactions with OH°: 0.2 d
for reactions with O_3: 0.7 d (2716)
for reactions with hydroperoxyl radical HO_2°: <13,000 d (2716)

Manmade sources

in diesel engine exhaust gas: 1.8% of emitted HCs (n.s.i.) (72)
expected glc's in U.S. urban air: range: 1-20 ppb (n.s.i.) (102)
in exhaust of gasoline engines:
62-car survey: 1.8 vol % of total exhaust HCs (391)
15-fuel study: 3 vol % (+ isobutylene) of total exhaust HCs (392)
engine-variable study: 6.0 vol % (+ isobutylene) total exhaust HCs (393)
evaporation from gasoline fuel tank: 4.6 vol % of total evaporated HCs
evaporation from carburator: 0-0.3 vol % of total evaporated HCs (398, 399, 400, 401, 402)

C. WATER AND SOIL POLLUTION FACTORS

Reduction of amenities

organoleptic limit: 0.2 mg/L. (181)

D. BIOLOGICAL EFFECTS

Plants

tomato: epinasty in petiole: 50,000 ppm, 2 d. (109)

β-butylene *see* *trans*-2-butene

γ-butylene *see* isobutene

1,2-butyleneoixide (1-butene oxide; 1,2-epoxybutane; ethylethyleneoxide; ethyloxirane; 1,2-butane oxide)

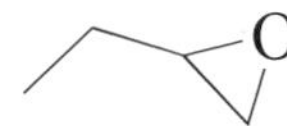

C_4H_8O

CAS 106-88-7

USES

Fuel or oil products; intermediate for various polymers, stabilizers for chlorinated solvents.

A. PROPERTIES

molecular weight 72.11; melting point 130°C; boiling point 62– 65°C; density 0.83 at 20°C; vapor pressure 188 hPa at 20°C, 227 hPa at 24°C, 654 hPa at 51°C; vapor density 2.49; solubility 59,000 mg/L at 20°C, 82,400 mg/L at 25°C, 63,200 mg/L at 60°C; log P_{oct} 0.68 at 25°C (measured), 0.42 (calculated): H 7.6 x 10^{-4} Pa x m^3/mol at 25°C.

B. AIR POLLUTION FACTORS

1 mg/m^3 = 0.340 ppm, 1 ppm = 2.94 mg/m^3

Odor

characteristic: quality: sweet, alcohol; hedonic tone: pleasant.

T.O.C.:	abs. perc. limit: 0.07 ppm	
	50% recogn.: 0.71 ppm	
	100% recogn.: 0.71 ppm	
Odor index	260,000	(19)

Photolysis

Indirect photolysis: simulated atmospheric photodecomposition:	$t_{1/2}$	
in the presence of ozone	0.8 h	
in the presence of OH	7.6 d	
in the presence of NO	16 h	
in the presence of NO_2	15 h	(8796, 8797, 8798, 8799)

C. WATER AND SOIL POLLUTION FACTORS

COD	2.42	
ThOD	2.0; 2.44	(8807)

Hydrolysis: $t_{1/2}$: 6.5- 13 d at pH 7. (8800, 8801)

Biodegradation

Inoculum	Method	Concentration	Duration	Eli mination results	
A.S. municipal	Modified MITI test	50- 200 mg/L	28 d	20- 30% ThOD	
				84% DOC*	(8803)
A.S. industrial	Modified Zahn-Wellens test	400 mg DOC/L	3 h	1% DOC	
			6 d	100% DOC*	(8804)
			5 d	6% ThOD	
			10 d	14% ThOD	
			15 d	16% ThOD	
			20 d	16% ThOD	(8805)
municipal effluent	Closed bottle test	2 mg/L	29 d	17% ThOD	
	Closed bottle test		5 d	0.08; 0.2% ThOD	(8807)
A.S.	DOC Die Away test	28 mg/L DOC	3 d	0%	
			7 d	17%	
			10 d	20%	
			14 d	34%	
			21 d	86%	
A.S.	Closed bottle test	4 mg/L	28 d	80-90%	(10683)

* Mainly by stripping.

Stability in soil

DT_{50}: 7- 12 d. (8801)

D. BIOLOGICAL EFFECTS

Toxicity

Bacteria			
Mycobacter sp. inactivation of alkene oxidation	EC_{50}	7,211 mg/L	
Activated sludge industrial, respiration	30 min EC_{10}	>1,000 mg/L	
Activated sludge municipal, oxygen consumption	30 min EC_{20}	120 mg/L	
Pseudomonas putida	17 h EC_{10}	2,920 mg/L	
(cell multiplication inhibition test)	17 h EC_{50}	4,840 mg/L	
	17 h EC_{90}	7,350 mg/L	(8811)
Algae			
Scenedesmus subspicatus	72 h EC_{20}	130 mg/L	
(cell multiplication inhibition test)	72 h EC_{50}	>500 mg/L	
	72 h EC_{90}	>500 mg/L	(8811)
Crustaceans			
Daphnia magna	24 h EC_{0}	62 mg/L	
	24 h EC_{50}	160 mg/L	
	24 h EC_{100}	250 mg/L	
	48 h EC_{0}	31 mg/L	
	48 h EC_{50}	70 mg/L	
	48 h EC_{100}	125 mg/L	(8810)
Fishes			
Poecilia reticulata	14 d LC_{50}	33 mg/L	
	96 h LC_{50}	100– 200 mg/L	
	96 h NOEC	46 mg/L	(8809)
Mammals			
Rat	oral LD_{50}	900; 500-1,000; mg/kg bw	(10683)

butylethanoate *see* *n*-butylacetate

n-butylether (di-*n*-butylether; 1-butoxybutane; butyloxide; dibutyloxide; 1,1'-oxybis(butane); 5-oxanonane)

$C_4H_9OC_4H_9$

$C_8H_{18}O$

CAS 142-96-1

USES

solvent for hydrocarbons, fatty materials; extracting agent. Alkylating agent. Catalyst and catalyst activator.

A. PROPERTIES

molecular weight 130.2; melting point -95°C; boiling point 141°C; vapor pressure 4.8 mm at 20°C; vapor density 4.5; density 0.77 at 20/20°C; solubility 300 mg/L at 20°C; $LogP_{OW}$ 3.21

B. AIR POLLUTION FACTORS

1 mg/m^3 = 0.188 ppm; 1 ppm = 5.33 mg/m^3.

Odor

characteristic: quality: fruity, sweet; hedonic tone: pleasant.

T.O.C.:	abs. perc. limit: 0.07 ppm	
	50% recogn.: 0.24 ppm	
	100% recogn.: 0.47 ppm	
O.I.:	14,000	(19)
detection:	8 mg/m^3	(691)

C. WATER AND SOIL POLLUTION FACTORS

Waste water treatment

A.C.: adsorbability: 0.039 g/g C, 100% reduction, infl.: 1,000 mg/L. (32)

D. BIOLOGICAL EFFECTS

Bacteria			
Photobacterium phosphoreum Microtox test	5 min EC_{50}	62 mg/L	(9657)
Protozoans			
Uronema parduczi.Chatton-Lwoff cell multiplication inhibition	IC10	>40 mg/L	(1901)

Crustaceans			
Daphnia magna	EC_{50}	>150 mg/L	(7747)
Fishes			
Pimephales promelas:	4 d LC_{50}	32 mg/L	(2625)
fathead minnow	96h LC_{50}	52 mg/L	(7643)
Mammals			
rat	oral LD_{50}	7,400 mg/kg bw	(9902)

butylethylene *see* 1-hexene

butylethylketone *see* 3-heptanone

n-butylformate (formic acid butylester)

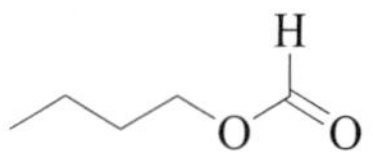

$HCOOC_4H_9$

$C_5H_{10}O_2$

CAS 592-84-7

USES

Resin hardner. Solvent.

A. PROPERTIES

molecular weight 102.13; melting point -90°C; boiling point 106.8°C; vapor pressure 30 mm at 25°C; vapor density 3.5; density 0.89 at 20/4°C.

B. AIR POLLUTION FACTORS

1 mg/m^3 = 0.240 ppm, 1 ppm = 4.17 mg/m^3.

Odor

detection: 6.0 mg/kg	(894)
T.O.C.: 17 ppm = 70 mg/m^3	(210)
distinct odor: 60 mg/m^3 = 20 ppm	(278)

C. WATER AND SOIL POLLUTION FACTORS

Reduction of amenities

T.O.C. = 6.0 mg/L. (299)

D. BIOLOGICAL EFFECTS

Mammals

rabbit	oral LD_{50}	2,700 mg/kg bw	(9903)

n-butylglycidylether (BGE, 1-butoxy-2,3-epoxypropane; 3-butoxy-1,2-epoxypropane; 2,3-epoxypropylbutylether; ERL 0810; butyl-2,3-epoxypropylether; glycidylbutyl-ether; (butoxymethyl)oxirane)

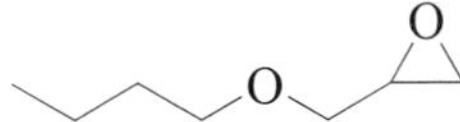

$C_4H_9OCH(O)CH_2$

$C_7H_{14}O_2$

CAS 2426-08-6

USES

component of epoxy resin systems. The epoxy group of the glycidylether reacts during the curing process, so glycidylethers are generally no longer present in completely cured products.

A. PROPERTIES

colorless liquid; molecular weight 130.21; boiling point 164/168°C; vapor pressure 3.2 mm at 25°C; vapor density 3.78/4.50; density 0.91 at 25/4°C; solubility 20,000 mg/L at 20°C.

B. AIR POLLUTION FACTORS

1 mg/m^3 = 0.188 ppm, 1 ppm = 5.32 mg/m^3.

Odor

characteristic: hedonic tone: irritating, not unpleasant.

butylglycol *see* butylcellosolve

n-butylglycolate (hydroxyacetic acid butyl ester; butyl hydroxyacetate)

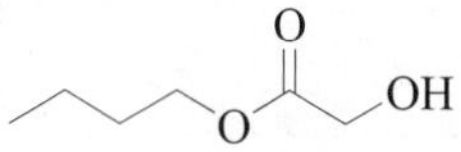

$CH_2OHCOO(CH_2)_3CH_3$

$C_6H_{12}O_3$

CAS 7397-62-8

USES

solvent for lacquers.

A. PROPERTIES

colorless liquid; molecular weight 132.16; melting point -26°C; boiling point 186-200°C; vapor pressure 1.3 hPa at 20°C; density 1.0 at 20°C; solubility 80,000 mg/L at 20°C; $\log P_{OW}$ 1.1 (calculated).

B. AIR POLLUTION FACTORS

Photochemical reactions

$t_{1/2}$: 3.6 d, based on reactions with OH° (calculated). (6440)

C. WATER AND SOIL POLLUTION FACTORS

Biodegradation

Inoculum	*method*	*conc.*	*duration*	*elimination results*	
Industrial A.S.	Zahn-Wellens Test		5 days	80 % DOC	(6437)

D. BIOLOGICAL EFFECTS

Bacteria			
primary municipal sludge	24h EC_0	100 - 1000 mg/L	(6438)
Pseudomonas putida	18h EC_{10}	454 mg/L	
	18h EC_{50}	2320 mg/L	(6439)
(cell multiplication inhibition test)	3h EC_{10}	347 mg/L	
(reduction of adenosine-5'-triphosphate concentration)	3h EC_{50}	2240 mg/L	(6439)
Crustaceans			
Daphnia magna	24h EC_{10}	approx. 200 mg/L	
	24h EC_{50}	280 mg/L	
	24h EC_{90}	400 mg/L	(6438)
Fishes			
Leuciscus idus	48h LC_0	50 mg/L	(6437)

butylglycolether *see* butylcellosolve

butylhydroquinone, tert- (TBHQ; 1,4-benzenediol, 2-(1,1-dimethylethyl)-; MTBHQ)

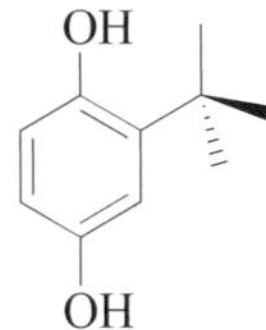

$C_{10}H_{14}O_2$

CAS 1948-33-0

USES

TBHQ is used as food antioxidant and is very effective in stabilizing fats and oils, especially polyunsaturated crude vegetable oils. (6030)

A. PROPERTIES

white to light tan crystals; molecular weight 166.22; melting point 126.5-128.5°C; solubility < 10,000 mg/L.

D. BIOLOGICAL EFFECTS

Mammals			
rat	oral LD_{50}	70 mg/kg bw	
mouse	oral LD_{50}	1000 mg/kg bw	(9704, 9705)

n-butylmercaptan (1-butanethiol; n-butyl thioalcohol; thiobutyl-alcohol)

SH

C_4H_9SH

$C_4H_{10}S$

CAS 109-79-5

USES

Solvent. Intermediate in organic synthesis. Guided missile propellant and oxidiser. Odour agent.

OCCURRENCE

Residues detected in the environment after commercial spraying with the defoliant DEF.

A. PROPERTIES

molecular weight 90.18; melting point -116°C; boiling point 98°C; vapor density 3.1; density 0.84 at 20°C; solubility 590 mg/L at 22°C; LogP_{OW} 2.28.

B. AIR POLLUTION FACTORS

1 mg/m^3 = 0.27 ppm, 1 ppm = 3.75 mg/m^3.

Odor

characteristic: strong, unpleasant.

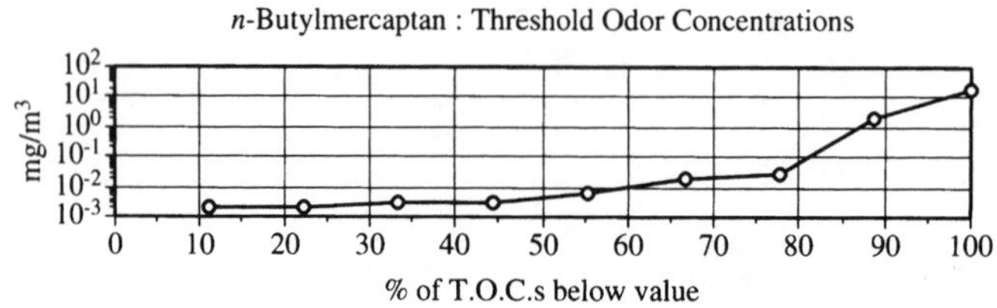

(71, 129, 279, 307, 602, 622, 637, 710, 723)

O.I. at 20°C: 49,000,000 (316)

Control methods

wet scrubber:

water:	effluent:	200,000 odor units/scf	
$KMnO_4$:	effluent:	33 odor units/scf	(115)

Photochemical reactions

$t_{1/2}$: 38h, (calculated). (9801)

Biodegradation

Alcaligenes faecalis , a microorganism in activated sludge, oxidised n-butylmercaptan. (9904)

C. WATER AND SOIL POLLUTION FACTORS

Reduction of amenities

T.O.C.:

average:	0.006 mg/L	
range:	0.001 to 0.06 mg/L	(294, 97)

D. BIOLOGICAL EFFECTS

Fishes

Lepomis macrochirus	24 h LC_{50}	7.4 mg/L	
	48 h LC_{50}	5.5 mg/L	

Mammals

rat	oral LD_{50}	1,500 mg/kg bw	(9905)

butylmethylcarbinol *see* 2-hexanol

butylmethylketone (2-hexanone; methylbutylketone; MBK)

$C_4H_9COCH_3$

$C_6H_{12}O$

CAS 591-78-6

USES

Solvent.

A. PROPERTIES

colorless liquid; molecular weight 100.2; melting point -57°C; boiling point 128°C; vapor density 3.45; density 0.83 at 0/4°C; solubility 35,000 at 20°C; vapor pressure 2 mm at 20°C; $LogP_{OW}$ 1.38.

B. AIR POLLUTION FACTORS

Odor threshold

0.28-0.35 mg/m³. (610)

C. WATER AND SOIL POLLUTION FACTORS

$BOD_5^{20°C}$: 61% ThOD (acclim.). (2666)

Waste water treatment

A.C.: adsorbability: 0.16 g/g C, 80.7% reduction, infl.: 998 mg/L. (32)

D. BIOLOGICAL EFFECTS

Bacteria			
mixed microbial culture	IC_{50}	5,500 mg/L	(8213)
methanogenic bacterial culture	IC_{50}	6,100 mg/L	(9906)
Fishes			
fathead minnow	LC_{50}	430 mg/L	(9801)
Mammals			
rat, mouse	oral LD_{50}	2,400-2,600 mg/kg bw	(9907, 9908)

butyloctylfumarate

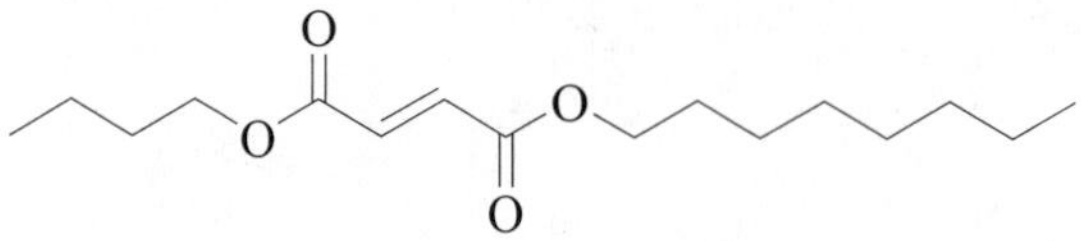

$C_4H_9OOCCHCHCOOC_8H_{17}$

$C_{16}H_{28}O_4$

USES

plasticizer.

C. WATER AND SOIL POLLUTION FACTORS

60 ml/min pure water passed through 25 ft, 1/2 in. I.D. tube of general chemical grade PVC contained 1.4 ppb butyloctylfumarate, which constituted 13% of total contaminant concentration. (430)

butyloxitol *see* butylcellosolve

butylphenol, 2-sec-

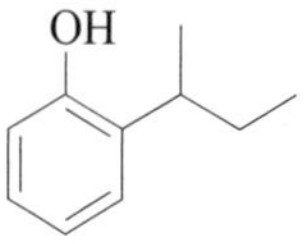

$C_2H_5CH(CH_3)C_6H_4OH$

$C_{10}H_{14}O$

CAS 89-72-5

USES

feedstock for chemical synthesis.

A. PROPERTIES

molecular weight 150.22; melting point 12-16°C; boiling point 226-228°C at 1,013 hPa; density 0.98 at 25°C; vapor pressure 0.1 hPa at 20°C, 100 hPa at 150°C; solubility 2,000 mg/L at 20°C and pH 6.4; $LogP_{OW}$ 3.6 at 25°C (calculated).

B. AIR POLLUTION FACTORS

Photodegradation

reaction with OH°: $t_{1/2}$: 4 d (calculated). (5398)

C. WATER AND SOIL POLLUTION FACTORS

Biodegradability

inoculum/method	*test duration*	*removed*	
A.S., aerobic, sapromat	4 d	0% ThOD	(5399)
aerobic, Coupled unit test		74%	(5398)

Impact on biodegradation processes

anaerobic bacteria from domestic water treatment plant: 24h threshold limit: 150 mg/L. (5399)

D. BIOLOGICAL EFFECTS

Bacteria			
Staphylococcus aureus	LOEC	60 mg/L	(5401)
Crustaceans			
Crangon septemspinosa	96h LC_{50},S	1.3 mg/L	(5400)
Fishes			
Poecilia reticulata	96h LC_0	5 mg/L	(5399)

butylphenol, m-tert- (3-tert-butylphenol; phenol,3-(1,1-dimethylethyl)-)

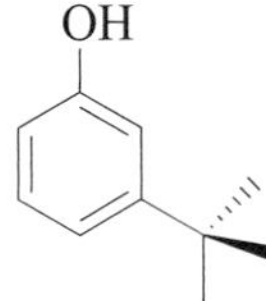

$C_{10}H_{14}O$

CAS 585-34-2

EINECS 209-553-4

A. PROPERTIES

molecular weight 150.22; melting point 40-43°C; boiling point 125-130°C at 20 mm Hg;

C. WATER AND SOIL POLLUTION FACTORS

Environmental concentrations

Landfill leachate from lined disposal area (Florida 1987-1990): mean conc.: 28 µg/L (2788)

D. BIOLOGICAL EFFECTS

Toxicity

Micro organisms			
Vibrio fischeri	15 min EC_{50}	0.1 mg/L	(11051)

butylphenol, p-tert- (1-hydroxy-4-tert-butylbenzene; 2-(p-hydroxyphenyl)-2-methylpropane; 4-hydroxy-1-tert-butylbenzene; 4-tert-butylphenol; 1-hydroxy-4-tert-butylbenzene; phenol, 4-(1,1-dimethylethyl); 4-(α,α-dimethylethylphenol))

OH

$(CH_3)_2CHCH_2C_6H_4OH$

$C_{10}H_{14}O$

CAS 98-54-4

USES

A chemical intermediate for the production of vulcanization agents and phenolic resins. Ingredient in de-emulsifiers for oil field use. In motor oil additives. Intermediate in the manufacture of varnish and lacquer resins. Soap antioxidant.

A. PROPERTIES

White aromatic flake; molecular weight 150.21; melting point 99°C; boiling point 236°C; density 0.91 at 114/4°C; solubility 700 mg/L; log P_{OW} 3.31; log H -4.34 at 25°C; vp 0.3 hPa at 50 °C, 1 mm Hg at 70°C; vapor density 5.1; solubility 500 - 800 mg/L at 20 °C; log P_{oct}2.4; 3.3 (measured);

B. AIR POLLUTION FACTORS

Photodegradation

Photodegradation is expected because p-tert-butylphenol has an absorption band at the UV region.t/2 = 8.9 hours based on reaction with OH radicals. (10608)

C. WATER AND SOIL POLLUTION FACTORS

Reduction of amenities

odor threshold: detection: 0.8 mg/L (998)
approx. concn causing adverse taste in fish: 0.03 mg/L (41, 998)

Impact on biodegradation processes

inhibition of degradation of glucose by *Pseudomonas fluorescens* at 25 mg/L
inhibition of degradation of glucose by *Escherichia coli* at >100 mg/L (293)

Photodegradation

Photomineralisation test : U.V. lamp : after 17h : 47% mineralisation to CO_2 (8631)

Hydrolysis

the chemical is stable at environmental conditions.

Monitoring data

River Rhein (FRG) : p-tert-butylphenol was identified in 5 of 16 samples			
River Main (FRG) : p-tert-butylphenol was identified in 5 of 16 samples			(8632)
USA (1975-1976) (8634)	effluent from chemical plant	1 - 150 µg/L	
	in receiving surface water	max 3 µg/L	
	in related river sediment	0.2 - 7 mg/kg	

Biodegradation

Inoculum	method	concentration	duration	elimination	
	Die away test		28 days	98% DOC	(10608)
A.S.	Biodegradation Test	0.05 mg/l	5 days	<0.1- 0.2% $ThCO_2$	(8635, 8631)
A.S. domestic non-adapted	DOC Die away Test	10 mg DOC/l	28 days	98% DOC	(8636)

D. BIOLOGICAL EFFECTS

Bioaccumulation

Species	conc. µg/L	Exposure period	BCF	
Leuciscus idus melanotus	46	3 days	120	
Chlorella fusca	50	24 hours	34	(10608)
activated sludge	50	5 days	240	(8631)

Toxicity

Micro organisms			
Pseudomonas putida	6h EC_{10}	145 mg/L	(10608)
primary municipal sludge	24h EC_0	100 - 1000 mg/L	(6100)
A.S.	3h EC_{50}	> 10,000 mg/L	(6000)
Photobacterium phosphoreum	5min EC_{50}	0.21 mg/L	(8645)
Staphylococcus aureus	18h MIC	98 mg/L	(8646)
	EC_{100}	<100 mg/L	(8647)
Tetrahymena pyriformis	60h EC_{50}	18 mg/L	(8648, 8649)
Bacillus subtilis	EC_{100}	<100 mg/L	(8647)
E.coli	EC_{100}	<100 mg/L	(8647)
	LD_0	>100 mg/L	
Algae			
Selenastrum capricornutum	72h ECb_{50}	22.7 mg/L	
	72h NOECb	9.5 mg/L	
	72h LOECb	17.2 mg/L	(10608)
Scenedesmus subspicatus	72h EC_{50}	11 mg/L	(8648)
	LD_0	10 mg/L	(30)
Chlorella vulgaris	6h EC_{50}	22-34 mg/L	(8644)
Protozoa			
Tetrahymena pyriformis	60h IC_{50}	18.4 mg/L	(10608)
Crustaceae			
Daphnia magna	24h EC_{50}	3.4; 7.3; 4.2; 4.8 mg/L	
	48h EC_0	0.34; 2.6 mg/L	
	48h EC_{50}	3.4; 3.9; 6.7 mg/L	
	48h EC_{100}	7.1 mg/L	(8643)
reproduction	21d NOEC	0.73 mg/L	

	21d EC_{50}	2.0 mg/L	
	21d LOEC	2.3 mg/L	
Crangon septemspinosa	96h LC_{50}	1.9 mg/L	(10608)
Fish			
Pimephales promelas	24h LC_{50}	6.2 mg/L	
	48h LC_{50}	5.7 mg/L	
	72h LC_{50}	5.3 mg/L	
	96h LC_{50}	5.1 mg/L	
Oryzias latipes	24-96h LC_{50}	5.1 mg/L	(10608)
	48h LC_{50}	23 mg./L	(8641)
Lepomis macrochirus	3h LC_{100}	>5 mg/L	(8639)
Leuciscus idus	48h LC_{50}	1.6 mg/L	(8648)
Petromyzon marinus	24h NOEC	>5 mg/L	(8639)
Salmo gairdneri	3h LC_{100}	>5 mg/L	(8639)
Cyprinus carpio	50h NOEL	<200 mg/kg bw.	(8640)
juvenile Atlantic salmon	96h LC_{50}	0.74 mg/L	(9909)
Arthropods			
Daphnia	LD_0	8 mg/L	(30)
Mammals			
Rat	oral LD_{50}	2,990; 3,500; 3,620; 4,000; 5,360 mg/kg bw	
Guinea pig	oral LD_{50}	400 mg/kg bw	(10608)

butylphenyl-N-methylcarbamate, o-sec- (BPMC; baycarb; 2-(1-methylpropyl) phenylmethylcarbamate)

HN—
O
O

$C_{12}H_{17}NO_2$

CAS 3766-81-2

USES

BPMC is one of several carbamate insecticides applied in large quantities in Japan to control planthoppers and leafhoppers on rice plants.

A. PROPERTIES

molecular weight 207.3; melting point 31-32°C; vapor pressure 48 mPa at 20°C; density 1.03; solubility 610 mg/L at 20°C; $LogP_{OW}$ 3.18.

C. WATER AND SOIL POLLUTION FACTORS

Degradation in soil

disappearance in Saga soil at 30°C humidity:

at 0.2 ppm initial concentration, 55% BPMC remained in paddy soil after 50 d
at 1.0 ppm initial concentration, 45% BPMC remained in paddy soil after 50 d
at 10 ppm initial concentration, 5% BPMC remained in paddy soil after 50 d

Because the disappearance rate of BPMC in soils was retarded by addition of sodium azide, it was suggested that soil microorganisms participated in the degradation of BPMC. (1324)

Aerobic biodegradation half-life

0.5 d. At an initial conc. of 2 ppm, using a mixture of nonacclimated sludge, field soil and river sediment as inoculum, in a stirred flask at 18-22°C, the major bacterial types in the activated sludge have been characterized as follows: *Flavobacterium* 35%; *Alcaligenes* 30%; *Corynebacterium* 20%; *Pseudomonas* 15%. (2690)

Alkaline hydrolysis

$t_{1/2}$: 0.01 d, at pH 9 at 40°C in the dark (2690)

D. BIOLOGICAL EFFECTS

Fishes			
carp	48h LC_{50}	13 mg/L	(2963)

butylphenyldiphenylphosphate, tert- (BPDP)

$C_{22}H_{24}O_4P$

CAS 28108-99-8

USES

Triaryl phosphate esters (TAPs) are used as fire-resistant hydraulic fluids and lubricant additives and are present as plasticizers in many consumer products.

A. PROPERTIES

solubility 3.2 mg/L at 20°C; $LogP_{OW}$ 5.12.

C. WATER AND SOIL POLLUTION FACTORS

Biodegradation

by activated sludge:

addition rate, mg/L/24h	*% ThOD*	*test duration, weeks*	
3	93	9	
13	84	8	(2558)

in river die-away test: % CO_2 evolution of theory for elapsed d of test

	7 d	*28 d*	*48 d*	
at 20 mg/L	43%	90%	92%	(2558)

Biodegradation

The fungal metabolism of tert-butylphenyl diphenyl phosphate (BPDP) was studied. *Cunninghamella elegans* metabolized BPDP predominantly at the tert-butyl moiety to form the carboxylic acid 4-(2-carboxy-2-propyl)triphenyl phosphate. In addition, 4-hydroxy-4'-(2-carboxy-2-propyl)triphenyl phosphate, triphenyl phosphate, diphenyl phosphate, 4-(2-carboxy-2-propyl)diphenyl phosphate, 2-(4-hydroxyphenyl)-2-methyl propionic acid, and phenol were detected. Similar metabolites were found in the 28 fungal cultures which were examined for their ability to metabolize BPDP. (see structures of identified metabolites) (11054)

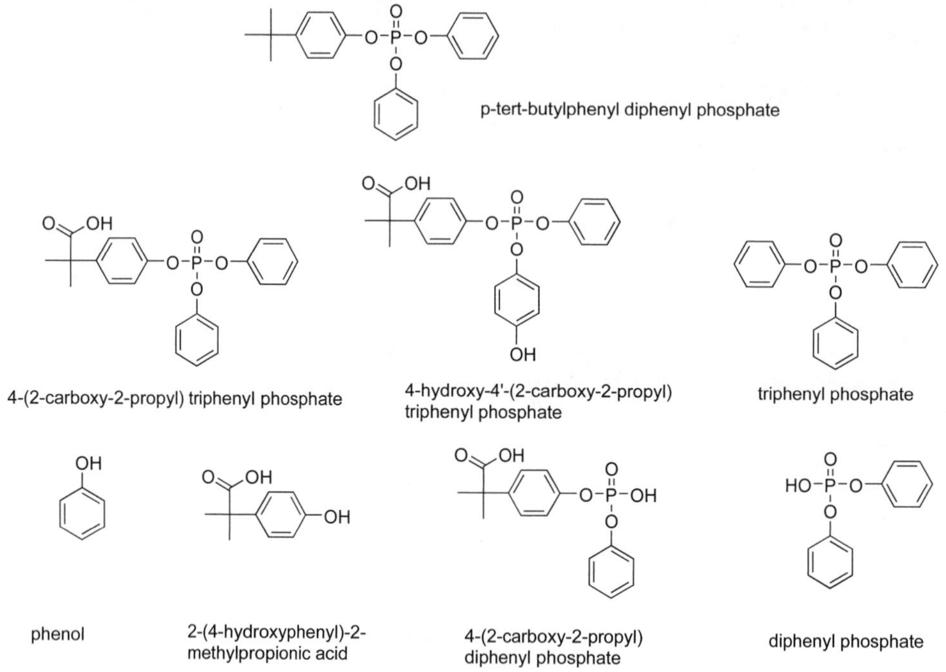

p-tert-butylphenyl diphenyl phosphate

4-(2-carboxy-2-propyl) triphenyl phosphate

4-hydroxy-4'-(2-carboxy-2-propyl) triphenyl phosphate

triphenyl phosphate

phenol

2-(4-hydroxyphenyl)-2-methylpropionic acid

4-(2-carboxy-2-propyl) diphenyl phosphate

diphenyl phosphate

Structures of identified metabolites from fungal metabolism of BPDP. [11054]

Structures of identified metabolites from fungal metabolism of BPDP (11054)

butylphosphate (*n*-butylphosphoric acid; acid butylphosphate)

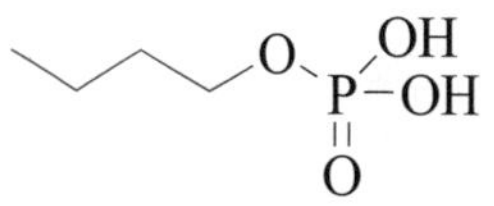

$CH_3(CH_2)_3H_2PO_4$

$C_4H_{11}O_4P$

CAS 12788-93-1

USES

Antifoaming agent for drilling muds. Catalyst. Flame retardant.

A. PROPERTIES

water white liquid; molecular weight 153.11; density 1.12 at 25/4°C; $LogP_{OW}$ 0.28.

C. WATER AND SOIL POLLUTION FACTORS

Anthropogenic sources

: Contaminant in natural and drinking water supplies from River Po at Turin, Ferrara and Como in Northern Italy

D. BIOLOGICAL EFFECTS

Bacteria			
Pseudomonas putida	16h EC_0	>100 mg/l	(329)
Algae			
Microcystis aeruginosa	8d EC_0	4.1 mg/l	(329)

n-butylphosphoric acid *see* butylphosphate

butylphthalate *see* di-*n*-butylphthalate

butylphthalate, di-sec-

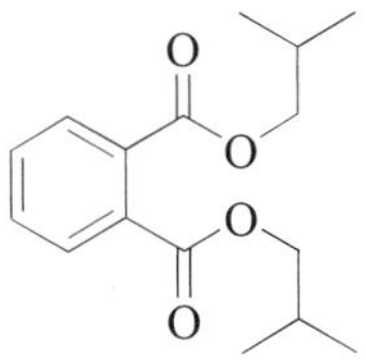

$C_{16}H_{22}O_4$

B. AIR POLLUTION FACTORS

Organic fraction of suspended matter

Bolivia at 5,200-m altitude (Sept.-Dec. 1975): 0.32-0.40 µg/1,000 m^3
Belgium, residential area (Jan.-April 1976): 2.4-7.2 µg/1,000 m^3 (428)

glc's in residential area (Belgium), Oct. 1976:
in particulate sample: - ng/m^3
in gas phase sample: 60 ng/m^3 (428)

butylpyrocatechol, 4-tert- *see* *p-tert*-butylcatechol

n-butylsulfide (dibutylsulfide; butylthiobutane)

$(CH_3CH_2CH_2CH_2)_2S$

$C_8H_{18}S$

CAS 544-40-1

A. PROPERTIES

molecular weight 146.29; melting point -79.7°C; boiling point 182°C; density 0.84 at 16/0°C; vapor pressure 1 mm at 21.7°C, 10 mm at 66.4°C, 40 mm at 96.0°C; solubility 34.3 mg/L; $LogP_{OW}$ 4.02.

B. AIR POLLUTION FACTORS

Odor

hedonic tone: unpleasant
T.O.C.: 0.012 mg/m^3 = 2 ppb (307)
O.I. at 20°C: 658,000 (316)

C. WATER AND SOIL POLLUTION FACTORS

Reduction of amenities

faint odor: at 0.0011 mg/L. (129)

D. BIOLOGICAL EFFECTS

Fishes			
Pimephales promelas	48h LC_{50}	3.5 mg/L	
	96h LC_{50}	3.6 mg/L	(2709)

butylsulfonate, sodium

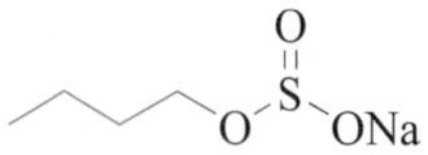

$C_4H_9SO_3Na$

D. BIOLOGICAL EFFECTS

Fishes

Daphnia magna:	24h LC_{50}	8,000 mg/L	
	48h LC_{50}	8,000 mg/L	
	72h LC_{50}	5,400 mg/L	
	96h LC_{50}	2,700 mg/L	(153)

2,4-D(butylthio) *see* 2,4-dichlorophenoxyacetic acid, butylthio

butylthiobutane *see* *n*-butylsulfide

butyltin

C. WATER AND SOIL POLLUTION FACTORS

Manmade sources

in effluents of 5 municipal sewage treatment plants: 9-19 ng Sn/L
in effluent of organotin production plant: 2,130 ng Sn/L (2419)

Water and sediment quality

In the water of the Schwarzbach, a tributary of the Rhine near Mainz (Germany): 7 ng/L
In the sediment of the Schwarzbach, a tributary of the Rhine near Mainz (Germany): 547,000 ng/kg dry wt (2419)
Harbors in Mainz/Wiesbaden (Germany):
in the water: 0.9-6 ng Sn/L
in sediment: 27,000-34,000 ng Sn/kg dry wt
In wastewater treatment plant of Zürich (Switzerland):

in raw wastewater:	245 ± 162 ng/L
in primary effluent:	181 ± 48 ng/L
in secondary effluent:	69 ± 7 ng/L
in tertiary effluent:	9 ± 6 ng/L
in sludge:	0.10-0.97 mg/kg dry weight

Adsorption into sludge is the most important removal process in sewage treatment. Aerobic and anaerobic degradation are of minor importance. (2545)

In surficial sediment in 12 sites in 3 lowland rivers in Norfolk broads in S.E. England after a retail ban on organotin-based antifouling paints in 1987:

μg/kg as Sn	*1989*	*1992*	*% reduction*	
average	15.4	10.5	32%	
range	< 1-45	< 1-56		(2954)

Biodegradability

Most of the tributylin (TBT) concentrations in a marine enclosure with near-natural water column and benthos, at an initial concentration of 590 ng/L at 20°C, was lost from the water column through biodegradation, which occurred at a rate of 0.08/d. Two-thirds of the degradation proceeded through debutylation to dibutyltin (DBT), which in turn degraded to monobutyltin (MBT) at 0.04/d. There was no evidence for degradation of MBT in the water. (2373)

D. BIOLOGICAL EFFECTS

In the presence of eelgrass (*Zostera marina* L.): $t_{1/2}$: 2-14 d in filtered seawater. (2568)

In bivalves from US coastal estuaries

Mussel *(Mytilus edulis):* avg 221, range <5-1,240 ng Sn/g dry animal tissue, representing 25% of total butyltins.
Oysters *(Crassostrea virginica* and *Ostrea sandwichensis):* avg 58, range ≤5-3,760 ng Sn/g dry animal tissue, representing 12% of total butyltins. (2649)

butyltoluene, p-tert- (1-methyl-4-tert-butylbenzene; 8-methylparacymene; PTBT; 1-(1,1-dimethylethyl)-4-methylbenzene; p-methyl-tert-butylbenzene)

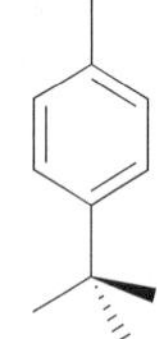

$(CH_3)_3CC_6H_4CH_3$

$C_{11}H_{16}$

CAS 98-51-1

USES AND FORMULATIONS

Feedstock for the synthesis of fragrances and agricultural chemicals. Solvent in preparation of resins. Oil additive. Perfume component. Intermediate in organic synthesis.

A. PROPERTIES

molecular weight 148.25; melting point -62.53°C; boiling point 192.8°C; vapor pressure 0.8 hPa at 20°C; vapor density 4.6; density 0.857 at 20/20°C; solubility 600 mg/L at 20°C; $LogP_{OW}$4.4 (measured).

B. AIR POLLUTION FACTORS

1 mg/m^3 = 0.16 ppm, 1 ppm = 6.05 mg/m^3.

Odor

immediate recognition at 5 ppm. (211)
Photochemical reactions: $t_{1/2}$: 11 d, based on reactions with OH° (calculated). (6100)

C. WATER AND SOIL POLLUTION FACTORS

BOD_5:	2% ThOD	
	6% ThOD adapted sew.	(277)
COD:	77% ThOD	(277)
ThOD	3.24	

Biodegradation

Inoculum	*Method*	*Concentration*	*Duration*	*Elimination results*	
municipal waste water	Closed bottle test	2 mg/L	28 d	7% product	(7677)

D. BIOLOGICAL EFFECTS

Bacteria			
Pseudomonas putida (inhibition of oxygen consumption test)	6 h EC_{10}	>8.8 mg/L	(7677)
Fishes			
goldfish	24 h LD_{50}	3 mg/L	(277)
Leuciscus idus	48 h LC_{50}	9.9 mg/L	(7677)
Mammals			
rat, rabbit	oral LD_{50}	1,500, 2,000 mg/kg bw	(9912)

butylxanthate, sec- (butylxanthic acid; butyldithiocarbonic acid)

D. BIOLOGICAL EFFECTS

Fishes		
rainbow trout *(Salmo gairdneri)*	96h LC_{50},S	(1087)
Cyanamid C 301-sodium salt:	100-166 mg/L	(1087)
Dow Chemical Z 12-sodium salt:	320 mg/L	(1087)

2-butyn-1-ol

—C≡C—
OH

CH_3CCCH_2OH

C_4H_6O

CAS 764-01-2

A. PROPERTIES

molecular weight 70.09; melting point -2.2°C; boiling point 142-143°C; density 0.94.

D. BIOLOGICAL EFFECTS

Fishes			
Pimephales promelas	96h LC_{50}	10.1 mg/L	(2625)

1-butyne (ethylacetylene; ethylethyne)

CH_3CH_2CCH

C_4H_6

CAS 107-00-6

OCCURRENCE

Identified in exhaust gases from motor vehicles.

A. PROPERTIES

molecular weight 54.09; melting point -130°C; boiling point 8.3°C; density 0.67 at 0°C; LogH -0.12 at 25°C.

D. BIOLOGICAL EFFECTS

Bacteria			
Inhibition of nitrification in soils at 30°C	at partial pressure mm Hg	% inhibition	
	7.5×10^{-2}	90-97%	
	7.5×10^{-3}	36-80%	
	7.5×10^{-4}	0-6%	(9913)

2-butyne-1,4-diol (1,4-dihydroxy-2-butyne)

$CH_2OHCCCH_2OH$

$C_4H_6O_2$

CAS 110-65-6

USES

intermediate in chemical synthesis; additive for corrosion inhibition and galvanic applications.

A. PROPERTIES

molecular weight 86.09; melting point 52-54°C; boiling point 238°C; vapor pressure 1.33 hPa at 102°C; solubility >600,000 mg/L; $LogP_{OW}$ 0.73 (measured).

C. WATER AND SOIL POLLUTION FACTORS

Biodegradability

inoculum/method	*test conc.*	*test duration*	*removed*	
A.S., adapted	372 mg/L	15 d	88% TOC	(5417)

D. BIOLOGICAL EFFECTS

Bacteria			
Pseudomonas putida	17h EC	1,990 mg/L	
	17h EC_{50}	3,940 mg/L	
	17h EC_{90}	6,900 mg/L	(5415)
Algae			
Scenedesmus subspicatus	72h EC_{20}	240 mg/L	
	72h EC_{50}	480 mg/L	
	72h EC_{90}	>500 mg/L	(5415)
	96h EC_{20}	220 mg/L	
	96h EC_{50}	430 mg/L	
	96h EC_{90}	>500 mg/L	(5415)
Crustaceans			
Daphnia magna	24h EC_{0}	25 mg/L	
	24h EC_{50}	43 mg/L	
	24h EC_{100}	100 mg/L	(5415)
	48h EC_{0}	12; 62 mg/L	
	48h EC_{50}	27; 142 mg/L	
	48h EC_{100}	100; 250 mg/L	(5414, 5416)
Fishes			
Leuciscus idus	96h LC_{50}	46-100 mg/L	(5411)
Pimephales promelas	96h LC_{50}	54 mg/L	(2625)
Fathead minnow	96h LC_{50}	54 mg/L	(5412)
Salmo trutta	24h NOEC,S	5 mg/L	(5413)
Lepomis macrochirus	24h NOEC,S	5 mg/L	(5413)
Perca flavescens	24h NOEC,S	5 mg/L	(5413)
Carassius auratus	24h NOEC,S	5 mg/L	(5413)
Salmo gairdneri	24h NOEC	5 mg/L	(5413)

butyraldehyde (1-butanal; butylaldehyde; butyric aldehyde)

O

H

$CH_3CH_2CH_2CHO$

C_4H_8O

CAS 123-72-8

USES

Added to sulfite in German wines, it could be detected in wines in concentrations of up to 2 mg/L. Wines spoiled by lactic bacteria contained 1.4 and 0.4 mg/L butyraldehyde.	(9288)

A. PROPERTIES

molecular weight 72.1; melting point -97/-99°C; boiling point 75/76°C; vapor pressure 71 mm at 20°C; vapor density 2.48; density 0.82 at 20/4°C; solubility 37,000 mg/L, 71,000 mg/L, 63,500 mg/L, 69,900 mg/L; $LogP_{OW}$ 0.83; 1.18; log H: -2.33 at 25°C.

B. AIR POLLUTION FACTORS

1 ppm = 2.9 mg/m^3, 1 mg/m^3 = 0.340 ppm.

Odor

characteristic: quality: sweet, rancid; hedonic tone: unpleasant, animal rendering odor.

T.O.C.:		
abs. perc. limit:	0.0046 ppm	
50% recogn.:	0.0092 ppm	
100% recogn.:	0.039 ppm	
O.I.:	3,000,000	(19)
T.O.C.:		
recognition:	0.013-0.014 mg/m^3	(610)
	15 mg/m^3	(788)
	0.042 mg/m^3	(842)

Photochemical reactions

$t_{1/2}$: 12-25 h, based on reactions with OH° (calculated).	(9255, 9256)

Manmade sources

in municipal sewer air: 10-100 ppb	(212)
In diesel exhaust: 0.3 ppm	(311)
In exhaust of gasoline engine: 0.4-4 vol% of total exhaust aldehydes	(394, 395, 396, 397)
in condensate of municipal dump gas: 78 µg C/N m^3 gas	(9263)
in combustion gases of Bunsen burner using natural gas: 0.01-0.1 mg/m^3	(9265)
in hospital air: <0.001 mg/m^3	(9265)
in natural gas: >2 µg/m^3	(9265)
cigarette smoke: 269-452 µg/cigarette	(9273)

In exhaust of a 1970 Ford Maverick gasoline engine: 4-8 ppm (incl. Unknown compound)

Control methods

catalytic incineration over commercial Co_3O_4 (1.0-1.7 mm) granules, catalyst charge 13 cm^3, space velocity 45,000 h^{-1} at 100 ppm in inlet:

80% decomposition at	200°C	
99% decomposition at	225°C	
80% conversion to CO_2 at	210°C	
95% conversion to CO_2 at	245°C	(346)

Comparison of catalyst performance at space velocity of 80,000 h^{-1}:

	feed concn	Reactor inlet	% odor removal	%

catalyst	Ppm	temp., °C	(ASTM)	conversion (GC)	
0.5% Pt on μ-$Al2O_3$	1,537	160	69	64	
0.1% Pt on μ-$Al2O_3$	1,537	193	93	94	
0.5% Pd on μ-$Al2O_3$	1,537	290	99	100	
10% Cu on μ-$Al2O_3$	1,537	381	97	100	(1221)

Outdoor background concentrations

in pine forest in Germany (near Berlin) July 18, 1991 at 3:00 a.m.: 0.17 μg/m^3	(9264)
in pine forest in central Italy Feb 27, 1992 at noon: 0.84 μg/m^3	(9264)
in cloud water (Henniger Flats, CA, USA): 0-0.52 mg/L (mean 0.07 mg/L)	(9268)
in fog water collected from Pasadena, CA, USA: 0-0.052 mg/L	(9268)
Los Angeles, CA, USA, air during photochemical pollution episode: 0-7 ppb (mean 1.5 ppb)	(9269)
Los Angeles, CA, USA, Fall 1981: 0-5 ppb	(9271)
Raleigh, NC, USA, near highway, May 1983: 2.88-7.29 ppb	(9272)
Clarement, CA, USA, September 1985, no smog: 0.2-0.8 ppb	(9270)

Ambient air

Inside and outside 6 residential houses, suburban New Jersey, Summer 1992 (n = 36):

outdoor:	mean: 0.5 ppb; max: 1.2 ppb;	SD: 0.36 ppb	
indoor:	mean: 0.7 ppb; max: 2.4 ppb;	SD: 0.6 ppb	(2951)

Inside and outside 15 living rooms in Northern Italy (1983-1984)

4-7-d averages	*lowest*	*Mean*	*highest value (all in μg/m^3)*	
Indoor	<1	3	34	
Outdoor	<1	<2	10	(2756)

in new and recently renovated buildings in Switzerland: range: 5-84 μg/m^3; 90%ile: 71 μg/m^3 (9267)

C. WATER AND SOIL POLLUTION FACTORS

BOD_5:	28; 43; 66% ThOD	(79, 255, 27, 220)
BOD_{10}:	60% ThOD	(79)
15 d:	61% ThOD	(79)
20 d:	66% ThOD	(79)
BOD_{30}:	64% ThOD	(79)
40 d:	72% ThOD	(79)
50 d:	69% ThOD	(79)
COD:	99% of ThOD	(220)
ThOD:	2.44	
Reduction of amenities:	T.O.C.: 0.009 mg/L	(305)
	detection: 0.0373 mg/kg	(874)

Background concentrations

Straits of Florida, in seawater: 0-0.048 mg/L up to 518-m depth. (9274)

Volatilization half-lives of 9 h and 4.1 d have been estimated for a model river (1 m deep) and an environmental pond, respectively. (9257, 9258)

Volatilization

Volatilization half-lives of 9 h and 4.1 d have been estimated for a model river (1 m deep) and an environmental pond, respectively. (9257, 9258)

Persistency

Half-lives	in surface waters: 24-168 h	
estimation based on aerobic biodegradation half-lives	in groundwater: 48-336 h	(9259)

Waste water treatment

A.C.: adsorbability: 0.106 g/g carbon; % reduction = 52, infl.: 1,000 mg/L.			(32)
	influent	*effluent*	
anaerobic lagoon:			
22 lb COD/d/1,000 cu ft:	190 mg/L	50 mg/L	
48 lb COD/d/1,000 cu ft:	190 mg/L	35 mg/L	(32)
Aeration by compressed air (stripping effect): 85% removal in 8 h			(30)
Reverse osmosis: 72% rejection from a 0.01 M solution			(221)

Biodegradation

Inoculum	*Method*	*Concentration*	*Duration*	*Elimination results*	
A.S			6 h	14% of ThOD	
			12 h	21% of ThOD	
			24 h	22% of ThOD	(88)
soil suspension	+ inorganic nutrients	100 mg/L	24 h	88; >90% test substance	
		100 mg/L	96 h	93; >97% test substance	
		200 mg/L	48 h	60; 73% test substance	
		200 mg/L	96 h	87; >95% test substance	(9260)
A.S. industrial nonadapted	Modified Zahn-Wellens test		5 d	>95% DOC	(9275)
unacclimated m.o., aerobic			15 d	71% ThOD	
			20 d	73% ThOD	(9276)
A.S.	BOD test	100 mg/L	14 d	100% ThOD	(9277)
sewage	BOD test		5 d	>100% ThOD	(9278)
A.S.	BOD test	2.5 mg/L	5 d	43% ThOD	
			10 d	60% ThOD	
			15 d	62% ThOD	
			20 d	66% ThOD	
			30 d	64% ThOD	
			40 d	72% ThOD	
			50 d	68% ThOD	(9279)
sewage	BOD test		5 d	28% ThOD	(9280)
A.S.	BOD test	500 mg/L	1 d	23% ThOD	(9281)
A.S.	BOD test		5 d	>60% ThOD	(9282)
sewage	BOD test	2.5 mg/L	5 d	43% ThOD	
			10 d	60% ThOD	
			15 d	62% ThOD	
			20 d	66% ThOD	(9283)
sewage	OECD 301A	10 mg DOC/L	7 d	93% DOC	(9284)
Anaerobic upflow filter adapted		270 mg/L	2-10 d	82% CH4 prod.	(9287)

butylaldehyde → butyric acid

In a biofilter system to clean malodorous air streams, bacteria like *Pseudomonas putida* metabolized butyraldehyde via the production of butyric acid. (9308)

Mobility in soil

K_{OC}: 9-71 (estimated). (9261, 9262)

D. BIOLOGICAL EFFECTS

Chemotaxis

At a threshold of 7.2 mg/L butyraldehyde attracts zoospores of *Phytophthora palmivora* and leads to positives chemotaxis. The fungus is plant pathogenic, especially in the tropics on cocoa.

Bacteria			
Pseudomonas putida	16 h EC_0	100 mg/L	(1900)
primary municipal sludge	24 h EC_{10}	150 mg/L	(9275)
Photobacterium phosphoreum	5 min EC_{50}	16; 154 mg/L	(9303, 9304)
Escherichia coli: 72 mg/L inhibited the growth and profileration of *E. coli* for about 1.5 h at 37°C. After 2 h, the growth rate was equal to that of the control.			(9306)
Protozoans			
Entosiphon sulcatum	72 h EC_0	4.2 mg/L	(1900)
Uronema parduczi Chatton-Lwoff	20 h EC_{10}	98 mg/L	(1901)
Chilomonas paramaecium	48 h EC_{10}	44 mg/L	(9301)
Algae			
Microcystis aeruginosa	8 d EC_0	19 mg/L	(329)
Scenedesmus quadricauda (cell multiplication inhibition test)	7 d EC_0	83 mg/L	(1900)
Crustaceans			
Daphnia magna Straus	24 h EC_0	100; 120 mg/L	
	24 h EC_{50}	195; 340 mg/L	
	24 h EC_{100}	383; 740 mg/L	(9296, 9298)
	96 h EC_{50}	27; 340 mg/L	(9295, 9297)
Insects			
Aedes aegypti larvae	4 h LC_{50}	2,000 mg/L	(9299)
Fishes			
Pimephales promelas	12 h LC_{50}	>80 mg/L	
	24 h LC_{50}	20 mg/L	
	96 h LC_{50}	16; 15; 26 mg/L	(2709, 9289, 9294)
Poecilia reticulata	14 d LC_{50}	14 mg/L	(9290)
Leuciscus idus	48 h LC_0	41; 70 mg/L	(9292, 9293)
	48 h LC_{50}	50; 57; 114 mg/L	(9291, 9292, 9293)
	8 h LC_{100}	66; 158 mg/L	(9292, 9293)
Mammals			
Rat	oral LD_{50}	5,890 mg/kg	

butyramide (butanamide; butyric amide)

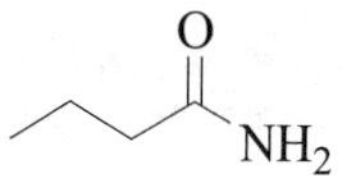

$CH_3CH_2CH_2CONH_2$

C_4H_9NO

CAS 541-35-5

A. PROPERTIES

molecular weight 87.12; melting point 116°C; boiling point 216°C; density 1.032 at 20/4°C; solubility 162,800 mg/L at 15°C; $LogP_{OW}$ -0.21.

C. WATER AND SOIL POLLUTION FACTORS

Waste water treatment

A.S.:
after 6 h: 1.3% of ThOD
after 12 h: 3.8% of ThOD
after 24 h: 6.4% of ThOD (89)
reverse osmosis: 40% rejection from a 0.01 M solution (221)

n-butyric acid (*n*-butanoic acid; ethylacetic acid)

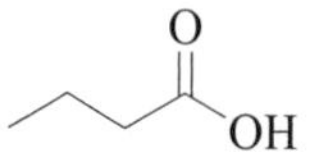

$CH_3(CH_2)_2COOH$

$C_4H_8O_2$

CAS 107-92-6

A. PROPERTIES

colorless liquid; molecular weight 88.1; melting point -5.5/-8°C; boiling point 163.7°C at 757 mm; vapor pressure 0.43 mm at 20°C, 1.4 mm at 30°C; vapor density 3.04; density 0.96 at 20/4°C; solubility 56,200 mg/L at -1.1°C; saturation concentration in air 2.9 g/m^3 at 20°C, 5.6 g/m^3 at 30°C; $LogP_{OW}$ 0.79; 0.24; LogH -4.66 at 25°C.

B. AIR POLLUTION FACTORS

1 mg/m^3 = 0.27 ppm, 1 ppm = 3.66 mg/m^3.

Odor

quality: sour.

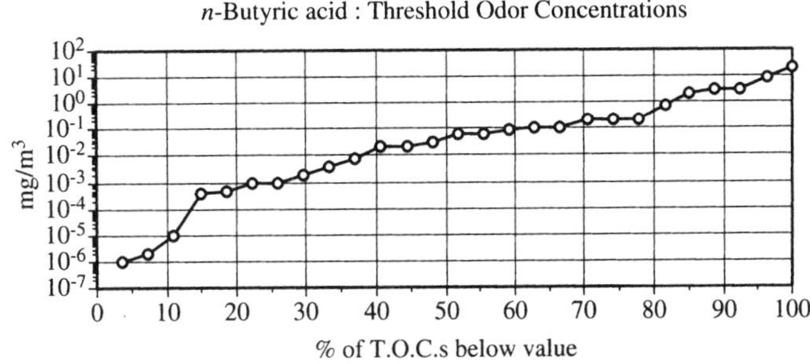

(57, 151, 210, 279, 307, 602, 610, 623, 635, 667, 683, 684, 706, 707, 708, 713, 753, 762, 765, 778, 779, 831, 837)

Control methods

A.C.: retentivity: 35 wt% of adsorbent (83)
thermal incineration for odor control: min. temp.: 1426°F (94)

catalytic incineration over commercial Co_3O_4 (1.0-1.7 mm) granules, catalyst charge 13 cm^3, space velocity 45,000 h^{-1} at 140 ppm inlet conc.:

80% decomposition at	210°C	
90% decomposition at	230°C	
99% decomposition at	275°C	
80% conversion to CO_2 at	270°C	
90% conversion to CO_2 at	360°C	(346)

Comparison of catalyst performance at space velocity of 80,000/h

catalyst	*% odor removal feed conc., ppm*	*reactor inlet temp.,°C*	*conversion (ASTM)*	*(GC)*	
0.5% Pt on γ-Al2O3	1,473	225	100	98	
0.1% Pt on γ-Al2O3	1,473	240	75	81	
0.5% Pd on γ-Al2O3	1,473	251	97	100	
10% Cu on γ-Al2O3	1,473	400	96	100	(1221)

C. WATER AND SOIL POLLUTION FACTORS

BOD_5:	19; 49; 49% ThOD	(41, 282, 284)
	19% ThOD std. dil./spec. culture	(283)
	64% ThOD std. dil./acclimated	(41)
BOD_{10}:	41% ThOD std. dil./spec. culture	(283)
BOD_{20}:	80% ThOD	(30)
BOD_{25}:	76% ThOD in seawater/inoculum: enrichment cultures of hydrocarbon-oxidizing bacteria	(521)
COD:	91; 96% ThOD	(27, 41)
ThOD:	1.82	(30)

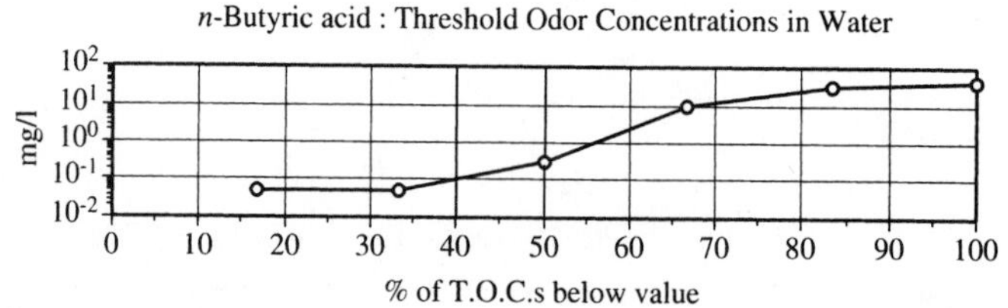

(295, 296, 898, 896, 907, 924)

Manmade sources

Contents in domestic sewages:	0.4-17 mg/L	(85)
Average content in secondary sewage effluents:	30 μg/L	(86)
In domestic sewage effluent:	5 μg/L	(227)
In year-old leachate of artificial sanitary landfill:	49 mg/L	(1720)
Landfill leachate from lined disposal area (Florida, July 1990): mean conc. 10.7 mg/L		(2788)

Waste water treatment

A.C.: adsorbability: 0.119 g/g carbon; 60% reduction; infl.: 1,000 mg/L (32)
A.S.:
after 6 h: 17% of ThOD
after 12 h: 25% of ThOD
after 24 h: 27% of ThOD (88)
Reverse osmosis: 16% rejection from a 0.01 M solution (221)
Powdered carbon: at 100 mg/L sodium salt (pH 7.5)-carbon dosage 1,000 mg/L: 4% absorbed (520)

Biodegradation in soil

biodegradation half-lives in nonadapted aerobic subsoil:
chalk England: 16 d. (2695)

D. BIOLOGICAL EFFECTS

Toxicity threshold (cell multiplication inhibition test)

Bacteria			
Pseudomonas putida	16h EC_0	875 mg/L	(1900)
Algae			
Microcystis aeruginosa	8d EC_0	318 mg/L	(329)
GREEN ALGAE			
Scenedesmus quadricauda	7d EC_0	2,600 mg/L	(1900)
Protozoa			
Entosiphon sulcatum)	72h EC_0	26 mg/L	(1900)
Uronema parduczi Chatton-Lwoff	EC_0	129 mg/L	(1901)

Algae			
Chlorella pyrenoidosa	toxic	340 mg/L	(41)
Scenedesmus	toxic	200 mg/L	(30)
Protozoa			
Vorticella campanula	toxic	10 mg/L	(30)
Paramecium caudatum	toxic	250 mg/L	(30)
Arthropoda			
Daphnia	toxic	60 mg/L	(30)
D. magna	48h LC_{50}	61 mg/L	(153)
Gammarus *(Hyale plumulosa)*	96h LC_{50}	250 mg/L	(2631)

Mollusc			
Limnaea ovata	toxic	50 mg/L	(30)
Fishes			
L. macrochirus:	24h LC_{50}	200 mg/L	(153)
(sodium salt)		5,000 mg/L	(1294)
Salmo irideus:	LC_{50}	400 mg/L	(30)
trout:	LC_{50}	20-40 mg/L	(30)
red killifish *(Oryzias latipes):*			
in seawater	96h LC_{50}	230 mg/L	
in freshwater	96h LC_{50}	90 mg/L	(2631)

butyric aldehyde *see* butyraldehyde

butyric amide *see* butyramide

butyrolactam *see* 2-pyrrolidone

γ-butyrolactone (4,5-dihydro-2(3H)-furanone; 4-deoxytetronic acid; 2-oxotetrahydrofuran; 4-butanolide; 4-hydroxybutanoic acid lactone)

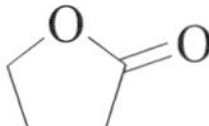

$C_4H_6O_2$

CAS 96-48-0

USES AND FORMULATIONS

Intermediate in the manufacturing of pharmaceuticals, herbicides, pesticides, and photchemicals; solvent.

A. PROPERTIES

molecular weight 86.09; melting point - 43°C; boiling point 204- 206°C; vapor pressure 0.34 hPa at

20°C, 3 hPa at 50°C; solubility miscible; $LogP_{OW}$0.57 (measured), 0.30 (calculated).

C. WATER AND SOIL POLLUTION FACTORS

Biodegradation

Inoculum	*Method*	*Concentration*	*Duration*	*Elimination results*	
BASF A.S.			13 d	97% TOC	(7660)
	Respirometric test		12 d	48% ThOD	(7660)

D. BIOLOGICAL EFFECTS

Bacteria			
Pseudomonas putida	17 h EC_{10-90}	>10,000 mg/L	
(cell multiplication inhibition test)			(7659)
Algae			
Scenedesmus subspicatus	72 h EC_{20}	14 mg/L	
(cell multiplication inhibition test)	72 h EC_{50}	360 mg/L	
	72 h EC_{90}	>500 mg/L	(7659)
	96 h EC_{20}	20 mg/L	
(cell multiplication inhibition test)	96 h EC_{50}	79 mg/L	
	96 h EC_{90}	>500 mg/L	(7659)
Crustaceans			
Daphnia magna Straus	24- 48 h EC_0	500 mg/L	
	24- 48 h EC_{50}	>500 mg/L	
	24- 48 h EC_{100}	>500 mg/L	(7659)
Fishes			
Oncorhynchus mykiss	24 h EC_0	>5 mg/L	(7656)
Ptychocheilus oregonensis	24 h EC_0	>10 mg/L	(7657)
Lepomis macrochirus	24 h EC_0	>5 mg/L	(7656)
Petromyzon marinus	24 h EC_0	>5 mg/L	(7656)
Leuciscus idus	96 h EC_{50}	220- 460 mg/L	
	96 h EC_{100}	460 mg/L	
	96 h NOEC	100 mg/L	(7658)

butyrone *see* 4-heptanone

butyronitrile (butanenitrile; *n*-propylcyanide; cyanopropane)

CN

$CH_3CH_2CH_2CN$

C_4H_7N

CAS 109-74-0

USES

Polymer synthesis. Intermediate in organic synthesis.

A. PROPERTIES

colorless liquid; molecular weight 69.10; melting point -112.6°C; boiling point 118°C; vapor pressure 10 mm at 15°C, 40 mm at 38°C; vapor density 2.4; density 0.80 at 20°C; $LogP_{OW}$ 0.60

C. WATER AND SOIL POLLUTION FACTORS

Waste water treatment

A.S.: after 6 h: 1.2% of ThOD
after 12 h: 1.5% of ThOD
after 24 h: 1.7% of ThOD (89)

Biodegradation

by a mutant microorganism: 500 mg/L at 20°C:
parent: 100% disruption in 13 h
mutant: 100% disruption in 4 h (152)

$CH_3CH_2CH_2CN \rightarrow CH_3CH_2CH_2CONH_2 + NH_3$

butyramide

Klebsiella pneumoniae adapted to benzonitrile as the sole source of C and N metabolised butyronitrile to butyramide and NH_3 . (9914)

Can be used as sole nitrogen source by the soil microorganisms *Candida fabianii, C. guilliermondii and Williopsis saturnus* (9915)

D. BIOLOGICAL EFFECTS

Mammals			
mouse, rat	oral LD_{50}	28; 140 mg/kg bw	(9916, 9917)

C

cabacryl *see* acrylonitrile

cacodylhydride *see* dimethylarsine

cacodylic acid (hydroxydimethylarsine oxide; dimethylarsinic acid; trade names: Phytar; Ansar)

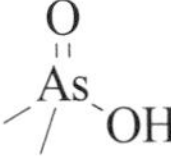

$(CH_3)_2AsOOH$

$C_2H_7AsO_2$

CAS 75-60-5

USES

contact herbicide; cotton defoliant; nonselective contact herbicide on noncrop areas; Phytar 138 contains 65.6% cacodylic acid; Fisher purified cacodylic acid contains 95.5% cacodylic acid.

A. PROPERTIES

colorless, odorless crystals; molecular weight 138.01; melting point 192-198°C; solubility 2,000 g/L at 25°C.

C. WATER AND SOIL POLLUTION FACTORS

Biodegradation

the degradation of cacodylic acid in soils proceeds by two mechanisms:

Under anaerobic conditions, 61% was converted to a volatile organoarsenical within a 24-week period and was lost from the soil system.

Under aerobic conditions, 35% was converted to a volatile organoarsenical compound and 41% to CO_2 and AsO_4^{3-} within the same 24-week period.

The ultimate environmental fate of the arsenic from cacodylic acid appears to be that it is metabolized to inorganic arsenate that is bound as insoluble compounds in the soil. (1299)

D. BIOLOGICAL EFFECTS

Residues in growth media and some natural waters (salt)

Diatoms	1.3-2.4 nmol/L	
coccolithophorids:	n.d.-1.9 nmol/L	
dinoflagellates:	0.06-0.1 nmol/L	
green algae:	1.9 nmol/L	
sterile medium:	0.02 nmol/L	
surface seawater:	1.6 nmol/L	
Salton Sea, surface:	67.5 nmol/L	(1933)

Bioaccumulation in model ecosystem

(multi-organism experiment) after 32 d of exposure:

conc. of cacodylic acid in water:	1st d:	10	ppb
	32nd d:	6.1	ppb
bioaccumulation ratio after 32 d:	algae:	1,635	
	snails:	419	
		110	(after 16 d in clean water)
	Daphnia:	1,658	
	fish:	21	(after 3 d of exposure)

bioaccumulation ratio at:	*conc. of cacodylic acid in water* 0.1 ppm	*1.0 ppm*	*10.0 ppm*	
one-organism experiment:				
algae: after 2 d	45	17	7	
Daphnia: after 2 d	39	42	25	
fishes: after 2 d	1.4	0.9	1.1	
snails: after 7 d	20	68	6.8	(1300)

Caffeine

Caffeine (1,3,7-trimethyl-2,6-dioxo-1,2,3,6-tetrahydropurine; 1,3,7-Trimethyl-2,6-dioxopurine; 1,3,7-Trimethylxanthine; 1-methyltheobromine; 3,7-dihydro-1,3,7-trimethyl-1H-Purine-2,6-dione; 7-Methyltheophylline; Alert-Pep; Cafeina; caffenium; Cafipel; eldiatric c; Guaranine; Koffein; Mateina; methyltheobromide; Methyltheobromine; No-Doz; organex; Refresh'n; Stim; Theine)

O N N N N O

$C_8H_{10}N_4O_2.H_2O$

CAS 58-08-2

TRADENAMES

Neocaf; Anacin; Vivarian, Dexatrim

USES

Caffeine is often combined with analgesics or with ergot alkaloids for the treatment of migraine and other types of headache.Beverages, medicine. A central nervous system stimulant and diuretic. Used in soft drinks and some foods as a flavouring ingredient.

OCCURRENCE

In mate leaves, guarana paste, cola nuts, coffee beans, tea leaves and kola nuts.
Caffeine is found in a wide rang of food products such as coke, Pepsi, Mountain Dew , milk chocolate etc.. The chief source of pure caffeine is the process of decaffeinating coffee and tea.

A. PROPERTIES

white crystalline powder that tastes very bitter;molecular weight 212.11; melting point 236.8; boiling point 178°C (sublimes); density 1.23; solubility 13,500 mg/L at 16°C, 455,500 mg/L at 65°C; log P_{ow} -0.07; vapour pressure 4.7 x 10^{-6} Pa at 25°C; H 1.9 x 10^{-19} atm x m^3/mol

B. AIR POLLUTION FACTORS

Manmade sources

concentrations in particulate organic matter in New York City, Jan.-March 1975: 0.7-7.0 µg/1,000 m^3 (equivalent concentration); (n = 8), emissions from coffee-roasting plants are implicated as contributors to ambient air levels. (429)

Photodegradation

T/2 (OH radicals) = 20 hours

C. WATER AND SOIL POLLUTION FACTORS

Manmade sources

in primary domestic sewage plant effluent: 0.010-0.046 mg/L. (517)
Göteborg (Sweden) sew. works 1989-1991: infl.: 6-30 µg/L; effl.: 0.1-0.5 µg/L. (2787)

Biodegradation

Bacteria isolates metabolized caffeine as a nitrogen source rather than carbon source. The strains produced xanthine derivatives as degradation intermediates. (9918)
Pseudomonas sp. isolated from soil under coffee cultivation were able to utilize high concentrations of caffeine (50,000 mg/L) as the sole source of carbon and nitrogen. (9919)

Hydrolysis

Stable

Environmental concentrations

In Rio Grande : august 2000	200 ng/L	
In treated sewage in USA, august 2000	1000 ng/L	(7152)
Samples from 139 streams in 30 states in the US thought to be susceptible to contamination from agricultural or urban activities during 1999-2000	detected in 42 out of 90 samples in values ranging from 14 – 5700 ng/L	(7221)

Degradation in soil

Bacteria isolates metabolized caffeine as a nitrogen source rather than carbon source. The strains produced xanthine derivatives as degradation intermediates. (9918)
Pseudomonas sp. isolated from soil under coffee cultivation were able to utilize high concentrations of caffeine (50,000 mg/L) as the sole source of carbon and nitrogen. (9919)
A strain of *Serratia marcescens* was isolated from soil under coffee cultivation. Microbial growth was only observed with xanthines methylated at the 7 position (caffeine (1,3,7-trimethylxanthine); paraxine (1,7-dimethylxanthine); theobromine (3,7-dimethylxanthine) and 7-methylxanthine. Paraxine and theobromine were released in liquid medium when caffeine was used as the sole source of carbon and nitrogen. When paraxine or theobromine were used, 3-methylxanthine, 7-methyl-xanthine and xanthine were detected in the liquid medium. *Serratia marescens* did not grow with theophilline, 1-methylxanthine and 3-methylxanthine and poor growth was observed with xanthine. Methyluric acid formation could not be confirmed. (7223)

caffeine
(1,3,7-trimethylxanthine)

paraxine (PX)
(1,7-dimethylxanthine)

theobromine (TB)
3,7-dimethylxanthine

theophylline (TP)
1,3-dimethylxanthine

3-monomethylxanthine
(3-MX)

7-monomethylxanthine
(7-MX)

xanthine

Degradation pathway

D. BIOLOGICAL EFFECTS

Acute toxicity tests

Marine tests			
Microtox^{TM} *(Photobacterium)* test	5 min EC_{50}	733 mgl	
Artoxkit M *(Artemia salina)* test	24h LC_{50}	3,773 mg/L	
Freshwater tests			
Streptoxkit F *(Streptocephalus proboscideus)* test	24 h LC_{50}	447 mg/L	
Daphnia magna test	24h LC_{50}	174 mg/L	
Rotoxkit F *(Brachionus calyciflorus)* test	24h LC_{50}	5,088 mg/L	(2945)
Pimephales promelas	4d LC_{50}	151 mg/L	(2625)
Mammals			
rat, mouse, guinea pig	oral LD_{50}	127-230 mg/kg bw	(9920, 9921, 9922)

Bioaccumulation: BCF : 0.52 – 2.35 (calculated) (10712)

Toxicity

Micro organisms		
Pseudomonas putida	17h EC_{50}	3,490 mg/L

	17h EC_{10}	1,530 mg/L	
	17h EC_{90}	5,240 mg/L	(10712)
Algae			
Scenedesmus subspicatus	72h EC_{50}	>100 mg//L	
	72h NOEC	6.25 mg/L	(10712)
Plants			
Oryza sativa	IC growth	1060-2120 mg/L	(10712)
Crustaceae			
Daphnia magna	48h EC_{50}	182 mg/L	
	48h EC_{0}	3.9 mg/L	
	48h EC_{100}	>500 mg/L	(10712)
Birds			
Red-winged blackbird	LD_{50}	316 mg/kg bw	
Starling	LD_{50}	>500 mg/kg	(10712)
Insects			
Red flour beetle *tribolium castaneum*	20d LD_{50}	288 mg/kg bw (adults0	
	20d LD_{50}	251 mg/kg bw (larvae)	(10712)
Amphibians			
Xenopus laevis	LC_{50}	130; 190 mg/L	
	LOEC	80 mg/L	(10712)
Fish			
Salmo gairdneri	48h LC_{50}		
Lebistes reticulates	48h LC_{50}		
Pimephales promelas	5d LC_{50}	720 mg/L	
Leuciscus idus	96h LC_{50}	87 mg/L	
	96h NOEC	46 mg/L	
	96h LC_{100}	215 mg/L	(10712)
Mammals			
Rat	oral LD_{50}	192; 233; 261-283; 200-400; 355; 247; 344; 421; 450; 483; 700 mg/kg bw	
Mouse	oral LD_{50}	127; 185; 200 mg/kg bw	
Rabbit	oral LD_{50}	224; 246 mg/kg bw	
dog	oral LD_{50}	140 mg/kg bw	
Guinea pig	oral LD_{50}	230 mg/kg bw	(10712)

Metabolism in humans (7222)

The primary metabolic pathways that contribute to caffeine (1,3,7-trimethylxanthine) metabolism in adults are:
- N-demethylation to form paraxanthine (PX; 1,7-dimethylxanthine), theobromine (TB; 3,7-dimethylxanthine) and theophylline (TP; 1,3-dimethylxanthine) account, on average for 80%, 11% and 4% respectively of caffeine metabolism in vivo;
- formation of the hydroxylated metabolite 1,3,7-trimethyluric acid (1,3,7-TMU)
- the C-N bond scission product 6-amino-5-(N-formylmethylamino)-1,3-dimethyluracil (1,3,7-TAU).

The remainder (ca 6%) is cleared from the body as unchanged drug.
Once formed, PX, TP and TB are subject to extensive metabolism. Each dimethylxanthine can undergo two separate N-monodemethylations to form the corresponding monomethylxanthines (ie 1-, 3- and 7-methylxanthine; 1-, 3- and 7-MX).
Further oxidation by way of 8-hydroxylation, of 1-MX and 7-MX, but not 3-MX, produces 1- and 7-methylmercuric acid (1- and 7-MU) respectively.
PX, TB and TP similarly undergo hydroxylation giving rise to the respective dimethyluric acids (1,7-, 3,7- and 1,3-DMU).
C-N bond scission, to form a dimethylaminouracil (DAU), only 1,7-DAU and 3,7-DAU.
It is believed that 5-acetylamino-6-formylamino-3-methyluracil (AFMU) forms together with 1-MX during demethylation pf PX. Internal rearrangement or acetylation of a putative unstable ring-opened intermediate results in the formation of 1-MX and AFMU, respectively.
Deformylation of AFMU produces 5-acetyl-6-amino-3-methyluracil (AAMU).

Caffeine excretion: As would be expected, those compounds derived from PX, whose formation accounts for approximately 80% of caffeine clearance, are the most abundant metabolites. In adult males for example, demethylation products of PX (ie 1-MX, 1-MU, AFMU) account for approximately 60% of all metabolites formed from caffeine. (7222)

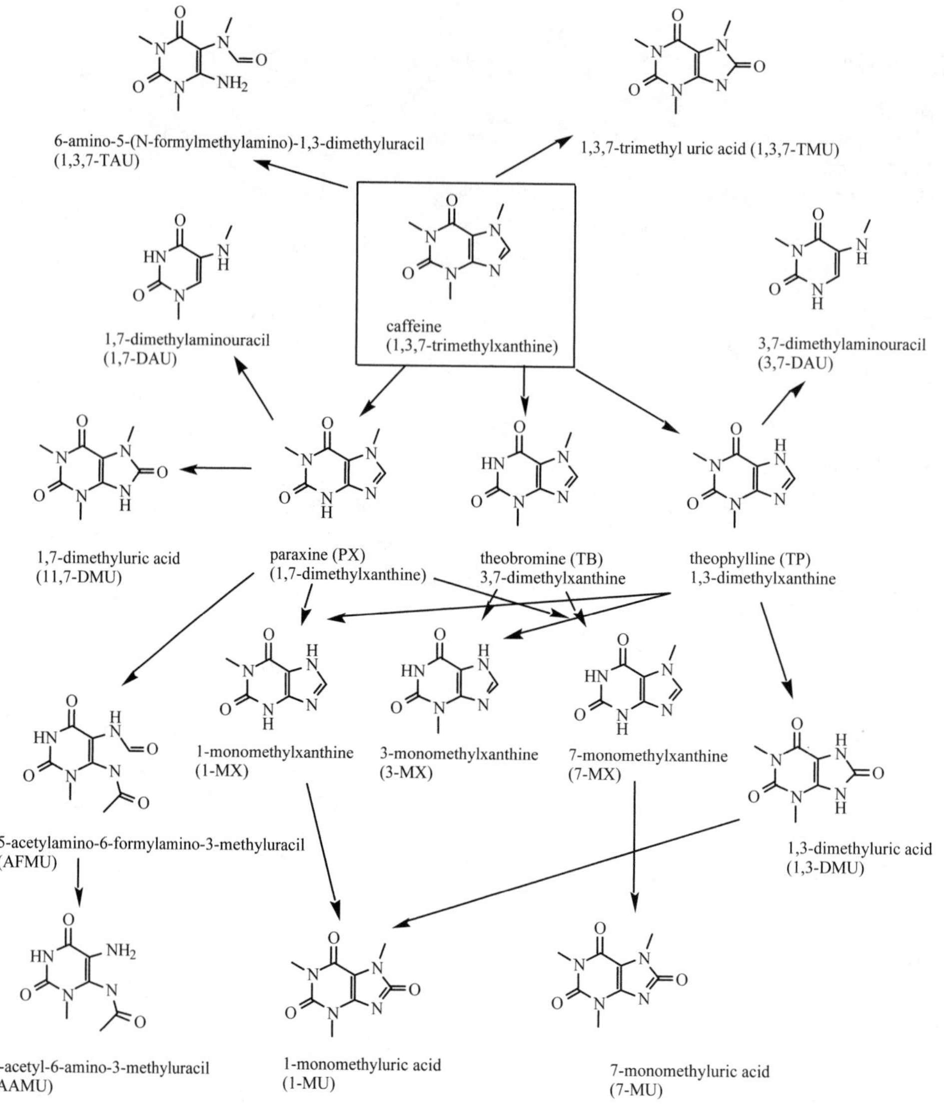

Pathways of caffeine metabolism in humans. (7222)

Callitris intratropica Oil (Northern Cypress Pine Oil; Blue Cypress Essential Oil; Callitris intratropica Essential Oil)

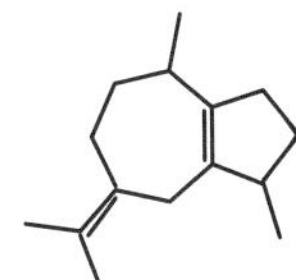

guaiol

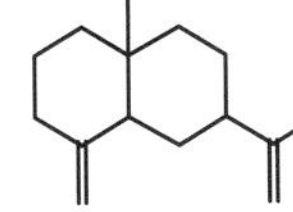

5-azulene methanol derivative

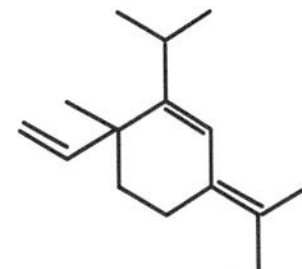

cyclohexane methanol derivative

2-naphthalene methanol derivatives

guaiene

decahydro-4a-methyl-1-naphthalene

elemene

CAS 187348-13-6

USES

The oil is used in perfumery (e.g. after shave).

IMPURITIES, ADDITIVES, COMPOSITION

The oil is steam distilled from wood chips from *Callitris intratropica* pine. It consists of a mixture of around 190 constituents. The identified constituents of the oil are mainly sesquiterpenes (C15) and aryl alcohols. The oil contains 8 major constituents (>1%) which are closely related in structure. These can be further devided into two subgroups; the alcohols (56% of the oil) and the alkenes (10% of the oil). (7115)

Chemical name	Weight %	CAS No.
guaiol	30.0	489-86-1
2-naphthalene methanol derivatives	17.0	1209-71-8
		473-15-4
		473-16-5
5-azulene methanol derivative	8.0	22541-73-6
decahydro-4a-methyl-1-naphthalene	5.0	17066-67-0
guaiene	4.0	88-84-6
cyclohexane methanol derovative	1.5	639-99-6
elemene	1.0	11029-06-4
menthol	<1.0	1490-04-6
terpinol	<1.0	98-55-5
linalyl ester	<1.0	7149-26-0
azulene	<1.0	275-51-4

A. PROPERTIES

blue-green viscous oil with an aromatic, woody odour; boiling range 122-288 °C; vapour pressure 10^{-4}

to 10^{-3} mm Hg (estimated); specific gravity 0.89-0.92 ; water solubility 0.01 mg/L (estimated); $LogP_{OW}$ 5.0-6.0 (estimated).

C. WATER AND SOIL POLLUTION FACTORS

Hydrolysis

the chemical contains no bonds which are susceptible to hydrolysis under the environmental pH region where 4<pH<9, and so it is expected to be stable. (7115)

D. BIOLOGICAL EFFECTS

Toxicity

Crustaceae			
Daphnia magna	24h EC_{50}	1.3 mg/L*	(7115)
Fish			
Lepomis macrochirus	96h EC_{50}	1.8 mg/L*	
Pimephales promelas	96h EC_{50}	2.1mg/L*	
Ictalurus punctatus	96h EC_{50}	1.8 mg/L*	
Oncorhynchus mykiss	96h EC_{50}	1.8 mg/L*	(7115)
Mammals			
Rat	oral LD_{50}	>2000 mg/kg bw	(7115)

*QSAR calculation for guaiol

campesterol

$C_{28}H_{48}O$

CAS 474-62-4

OCCURRENCE

small amounts are found in rape-seed oil in soybean oil and in wheat germ oil.

A. PROPERTIES

molecular weight 400.7; melting point 157-158°C

C. WATER AND SOIL POLLUTION FACTORS

Manmade sources

In tobacco smoke in 30 m^3 teflon chambers from smoking 1 to 4 Kentucky reference cigarettes: 1.5±0.58 μmol/g particulate sample (7108)

2-camphanol *see* DL-borneol

2-camphanone *see* camphor

Camphene (2,2-dimethyl-3-methylene-bicyclo(2,2,1)heptane; 2,2-dimethyl-3-methylenenorbornane)

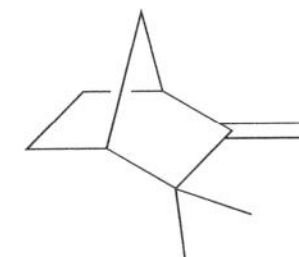

$C_{10}H_{16}$

CAS 79-92-5

A. PROPERTIES

molecular weight 136.24; melting point 36-38°C; boiling point 159-160°C; vapor pressure 2.4 hPa at 20°C, 12.5 hPa at 50°C; density 0.85; solubility 4.2 mg/L at 20°C; log P_{OW} 4.1.

B. AIR POLLUTION FACTORS

Natural sources

Emitted from the leaves of *Cupressus sempervirens*: 3.2% of total terpene emissions which varied between 3-35 µg/g dw.h. The major component was limonene which constituted 83% of total terpene emissions. (7049)

Calculated tropospheric lifetimes

reactant	*lifetime*	
OH°	3.5 h	
ozone	18 d	
NO_3^0	18 h	(2451)

Photochemical atmospheric degradation

reaction with OH°: $t_{1/2}$: 0.27 d (calculated). (5648)

Uses

Off-gas treatment,mfg. of camphor and camphor derivatives, isobornylesters, terpenphenol derivatives; fragrance; plasticizer for resins and lacquers.

Biofiltration of emissions of wood kiln-drying (lab scale biofilter with bark) (7135)

Degradation efficiencies with bark inoculated with bacterial cultures and a fungus: (figures between brackets are for uninoculated bark)

monoterpene	White rot fungus	Bacterial culture BB22	Bacterial culture LSC

α-pinene	99.5 (94.3)	98.6 (92.8)	98.2 (97.0)
Camphene	99.5 (91.9)	96.6 (94.1)	94.1 (97.4)
β-pinene	99.8 (96.1)	95.8 (85.7)	98.5 (97.6)
Limonene	99.8 (98.9)	94.9 (76.7)	96.6 (96.3)

Sources

Thermal degradation products formed in the presence of air at a temperature of 120 °C, mimicing the conditions of wood drying such as occurring during drying of wood flakes, wood veneers or lumber. The degradation products resulted from the following reactions:
·Dehydrogenation to form aromatic systems;
·Oxidative cleavage of the carbon-carbon bonds;
·Epoxide formation;
·Allylic oxidation (oxidation of the carbon adjacent to the carbon-carbon double bond) to form alcohols, ketones and aldehydes;
·Carbon skeleton rearrangements of the parent or oxidised compounds. (7088)
The following thermal degradation products of camphene were identified:

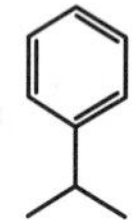
(1-methylethyl)benzene
[cumene]

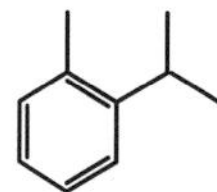
1-methyl-2-(1-methylethyl)benzene
[o-cymene]

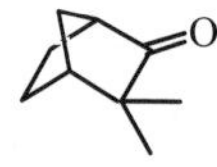

3,3-dimethylbicyclo[2.2.1]heptan-2-one
[camphenilone]

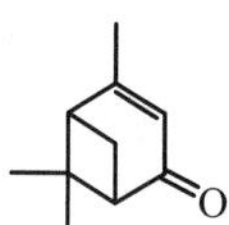

4,6,6-trimethylbicyclo[3.1.1]hept-3-en-2-one
[verbenone]

C. WATER AND SOIL POLLUTION FACTORS

Manmade sources

landfill leachate from lined disposal area (Florida 1987-1990): mean conc. 147 μg/L. (2788)

Biodegradability

Inoculum/test method	*test duration*	*removed*	
A.S., municip., Sapromat	10 d	5% ThOD	(5646)
A.S., municip., modif. MITI	28 d	<20% ThOD	(5647)
A.S., municip., Zahn-Wellens	28 d	0% ThOD	(5647)
A.S., industr. nonadapted, Zahn-Wellens	5 d	<95% product	(5646)

Impact on biodegradation processes

primary municipal effluent	24h EC_0	>2,500 mg/L	(5646)
A.S., municipal	3h EC_{10}	490 mg/L	
	3h EC_{50}	>1,000 mg/L	
	3h EC_{50}	>1,000 mg/L	(5647)

D. BIOLOGICAL EFFECTS

Toxicity			
Crustaceans			
Daphnia magna	24h EC_{50},S	46 mg/L	
	48h EC_{50},S	22 mg/L	(5645)
Fishes			
Brachydanio rerio	48-96h LC_0	125 mg/L	
	48-96h LC_{50}	150 mg/L	
	48-96h LC_{100}	180 mg/L	(5643)
Cyprinodon variegatus	24h LC_{50},S	1.8 mg/L	(5644)
	48-72h LC_{50},S	2 mg/L	(5644)
	96h LC_{50},S	1.9 mg/L	
	96h NOEC,S	1 mg/L	(5644)

camphor (gum camphor; 2-camphanone; bornan-2-one; 1,7,7-trimethylbicyclo(2,2,1)heptan-2-one; root bark oil; spirit of camphor; 1,7,7-trimethylnorcamphor; 2-bornanone)

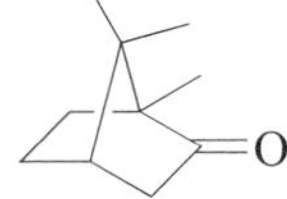

$C_{10}H_{16}O$

CAS 21368-68-3 (DL-camphor)

CAS 76-22-2 (2-bornanone)

CAS 464-49-3 (D-camphor)

CAS 464-48-2 (L-camphor)

CAS 21368-68-3 (DL-camphor)

USES AND FORMULATIONS

odorant/flavorant in household, pharmaceutical, and industrial products; plasticizer for cellulose esters and ethers; insect repellant and incense mfg.; lacquers and varnishes; explosives; embalming fluid; plastics mfg.; chemical intermediate. (347)

MANUFACTURING SOURCES

Organic chemical industry; wood processing industry. (347)

IMPURITIES, ADDITIVES, COMPOSITION

terpenes (0.3- 1.3%) mostly camphene

Terpenealcohols (0.5- 3.7%) such as isofenchol (1.7%) and borneol (1%)

Other terpeneketones (2- 4%) such as fenchol (2.3%) and pseudocamphor (0.8%)

NATURAL SOURCES (WATER AND AIR)

major component of pine oil (leaves, twigs, stems of camphor tree of China, Formosa, Japan); present in forest runoff. (347)

A. PROPERTIES

colorless or white crystals; molecular weight 152.24; melting point 174- 179°C; boiling point 204°C; vapor pressure 1 mm at 41.5°C; 400 mm at 182°C, 700 mm at 209.2°C, 0.85 hPa at 20°C, 5 hPa at 65°C; density 0.99; solubility 1,500 mg/L at 20°C; $LogP_{OW}$1.6 at 25°C

B. AIR POLLUTION FACTORS

Odor

quality: penetrating aromatic odor. O.I. at 20°C: 40.

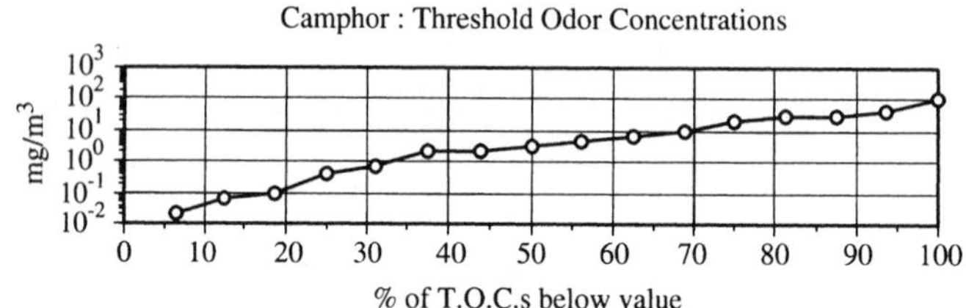

(210, 279, 307, 610, 672, 690, 753, 755, 771, 774, 775, 837, 871)

Photochemical reactions

$t_{1/2}$: 1.3 d, based on reactions with OH° (calculated). (7783)

C. WATER AND SOIL POLLUTION FACTORS

Reduction of amenities

T.O.C. in water at room temp.: 1.29 ppm, range 0.25-3.83.
20% of the population still able to detect odor at 0.33 ppm
10% of the population still able to detect odor at 0.041 ppm
1% of the population still able to detect odor at 0.0092 ppm
0.1% of the population still able to detect odor at 0.021 ppb (321)

Manmade sources

in Zürich Lake: at surface 12 ppt; at 30 m depth 2 ppt
in Zürich area: spring water 2 ppt; tapwater: 2 ppt (513)

Biodegradation

Inoculum	*Method*	*Concentration*	*Duration*	*Elimination results*	
A.S	Closed bottle test	519 mg DOC/L	7 d	73% DOC	
			14 d	92% DOC	
			21 d	87% DOC	
			28 d	92% DOC	(7785)
industrial nonadapted A.S.	Zahn-Wellens test		5 d	90%	
	* of which >50% by stripping		10 d	95%*	(7784)

Several *Pseudomonas* strains are able to degrade camphor as sole carbon source. The following metabolites werde identified: 1,2-champolid; 2,5-dioxobornane; 5-exo-hydrocamphor; 5-endo-hydrocamphor. (7786)

Diphtheria-like soil bacteria oxidized D-camphor to 6-endo-hydroxycamphor; 2,6-diketobornane, isohydroxycamphoric acid and 3,4,4-trimethyl-oxo-trans-2-hexanoic acid. (7787)

D. BIOLOGICAL EFFECTS

Bacteria

anaerobic bacteria from municipal	24 h EC_{10}	300 mg/L	(7784)

waste water treatment plant (fermentation tube method)			
Pseudomonas putida	16 h EC_{0-50}	>1,000 mg/L	(7792)
Fungi			
Aspergillus niger	EC_{50}	400 mg/L	(7791)
Saccharomyces cerevisiae	EC_{40}	400 mg/L	(7793)
	48 h EC_{100}	1,520 mg/L	(7795)
Protomyces inundatus	EC_{100}	400 mg/L	(7797)
Aspergillus niger	EC_{50}	400 mg/L	(7791)
Saprolegnia	24 h EC_7	100 mg/L	
Aspergillus niger	24 h EC_{40}	1,000 mg/L agar plate	(7800)
Fusarium oxysporum	24 h EC_{76}	1,000 mg/L agar plate	(7800)
Penicillium digitatum	24 h EC_{50}	1,000 mg/L agar plate	(7801)
Rhizopus stolonifer	24 h EC_{75}	1,000 mg/L agar plate	(7800)
Arthropods			
Amitermes evuncifer	24 h EC_{53-87}	500 mg/kg soil	(7804)
Bruchus pisorum	24 h LC_{100}	96,000 mg/m^3 air	(7805)
Bruchus rufimanus	24 h LC_{100}	48,000 mg/m^3 air	(7805)
Callosobruchus chinensis	24 d LC_{100}	12,000 mg/m^3 air	(7806)
Fishes			
Fathead minnows	1 h LC_{50},S	145 mg/L	
	24 h LC_{50},S	112 mg/L	
	48 h LC_{50},S	111 mg/L	
	72 h LC_{50},S	110 mg/L	
	96 h LC_{50},S	110 mg/L	(350)
Brachydanio rerio	48- 96 h LC_0	25 mg/L	
	48- 96 h LC_{50}	35- 50 mg/L	
	48- 96 h LC_{100}	50 mg/L	(7788)
Carassius auratus	96 h NOEC	1 mg/L	(7789)
Cyprinus carpio	96 h NOEC	1 mg/L	
Lepomis macrochirus	96 h NOEC	1 mg/L	
Salmo gairdneri	96 h NOEC	1 mg/L	
Catostomus commersoni	96 h NOEC	1 mg/L	
Ictalurus melas	96 h NOEC	1 mg/L	
Lepomis cyanellus	96 h NOEC	1 mg/L	
Perca flavescens	96 h NOEC	1 mg/L	(7789)
Petromyzon marinus	24 h EC_0	5 mg/L	(7790)

n-capric acid (decanoic acid; *n*-decoic acid; *n*-decylic acid)

$CH_3(CH_2)_8COOH$

$C_{10}H_{20}O_2$

CAS 334-48-5

USES

esters for perfumes and fruit flavors, base for wetting agents; intermediate; plasticizer.

A. PROPERTIES

molecular weight 172.26; melting point 31.5°C; boiling point 268-270°C; vapor pressure 1 mm at 125°C; sp. gr. 0.89 at 40/4°C; log P_{OW} 1.88.

B. AIR POLLUTION FACTORS

Odor

threshold odor conc.

0.014 mg/m^3 = 1.96 ppb	(307)
detection 0.05 mg/m^3	(778, 779)
recognition 0.08-0.09 mg/m^3	(610)

Anthropogenic sources

emission rates released from burning wood in residential open fireplaces:

pine wood: burned	0.095 mg/kg logs	
oak wood	0.39 mg/kg logs burned	(7087)

C. WATER AND SOIL POLLUTION FACTORS

BOD$_5$:	9% of ThOD
COD:	85% of ThOD

Waste water treatment

A.S.: after	6 h: 11% of ThOD	
	12 h: 19% of ThOD	
	24 h: 23% of ThOD	(89)

Odor threshold

detection: 10 mg/kg. (886)

Biodegradation

Inoculum	method	concentration	duration	elimination	
STW effluent	Closed bottle		30 days	71-110%	
	Warburg respirometer	500 mg/L	1 day	23.4% ThOD	
Adapted dom.sewage			5 days	61%	(10788)

D. BIOLOGICAL EFFECTS

Toxicity

Micro organisms			
Photobacterium phosphoreum	25min microtox	0.47-0.57 mg/L	
Bacillus subtilis	60min EC$_{50}$	43 mg/L	
Pseudomonas putida	30min EC$_{10}$	1000 mg/L	
Methanotrix	26h EC$_{50}$	1016 mg/L	(10788)
Algae			
Nitzschia closterium	72h EC$_{50}$	0.3 mg/L	(10788)
Crustaceae			
Daphnia magna	24h EC$_{50}$	65 mg/L	
Artemia salina	16h EC$_{50}$	36 mg/L	(10788)
Fish			

Oryzias latipes	96h LC_{50}	20; 54 mg/L	
Leuciscus idus	48h LC_{50}	95 mg/L	
	48h LC_0	30 mg/L	
	48h LC_{100}	300 mg/L	(10788)
Gammarus *(Hyale plumulosa)*	96h LC_{50}	41 mg/L	(2631)
Lepomis macrochirus	chemical is too insoluble to be toxic		(1294)
Red killifish *(Oryzias latipes)*	96h LC_{50}	in seawater: 31 mg/L	
	96h LC_{50}	in freshwater: 20 mg/L	
	96h LC_{50}	54 mg/L (sodium caprate)	(2631)
Mammals			
Rat	oral LD_{50}	3301 mg/kg bw	(7166)

caproaldehyde *see* *n*-hexaldehyde

Caproic acid (hexanoic acid; *n*-hexoic acid)

$CH_3(CH_2)_4COOH$

$C_6H_{12}O_2$

CAS 142-62-1

OCCURRENCE

hexanoic acid is a degradation product of octane, 1-octanol, octanal and octanoic acid.

A. PROPERTIES

oily liquid; molecular weight 116.2; melting point -6/-2°C; boiling point 204/208°C; vapor pressure 0.2 mm at 20°C, 1 mm at 70°C; vapor density 4.01; sp. gr. 0.945 at 0/0°C; solubility 11,000 mg/L, LHC 831.0 kcal/mole; log P_{OW} 1.88/1.92.

B. AIR POLLUTION FACTORS

Odor

characteristic: like limburger cheese. (211)

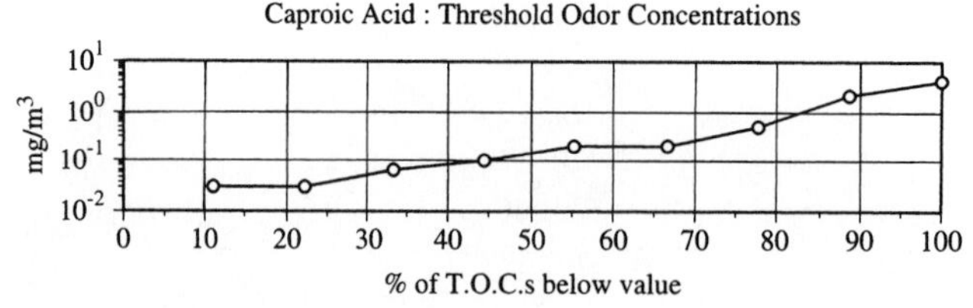

(307, 610, 647, 683, 684, 778, 779, 840, 872)

Control methods

thermal incineration: min. temp. for odor control: 774°C.	(94)

1 mg/m^3 = 0.21 ppm; 1 ppm = 4.83 mg/m^3.

C. WATER AND SOIL POLLUTION FACTORS

BOD_5	44% ThOD	(220)
$BOD_{2d}^{25°C}$	80% ThOD (substrate conc.: 3.5 mg/L; inoculum: soil microorganisms)	
BOD_5	95% ThOD	
BOD_{10}	95% ThOD	(1304)
COD	100% of ThOD	

Manmade sources

average content of secondary effluents: 48 µg/L.	(86)
landfill leachate from lined disposal area (Florida 1990): mean conc. 62 mg/L.	(2788)

Odor threshold

detection: 3.0 mg/kg.	(886)

Waste water treatment

A.S.:	after 6 h:	13% of ThOD	
	after 12 h:	22% of ThOD	
	after 24 h:	39% of ThOD	(89)

A.S.: BOD, 20°C, 1-5 d observed, feed: 333 mg/L; 15 d acclimation: 99% removed.	(93)
A.C.: adsorbability: 0.194 g/g carbon, 97.0% reduction, infl.: 1,000 mg/L.	(32)
Stabilization pond design: toxic correction factor: 1.3 at 200 mg/L in pond influent; 5.0 at 300 mg/L in pond influent.	(179)

Biodegradation pathway

hexanoic acid enters the beta-oxidation cycle and is used as both a carbon and energy source	(7132)

hexanoic acid + CH_3-COOH (acetic acid) ⟶ $CO_2 + H_2O$

β-oxidation ↓

butanoic acid

↓

+ CH_3-COOH (acetic acid) ⟶ $CO_2 + H_2O$

D. BIOLOGICAL EFFECTS

Arthropoda			
Daphnia magna:	24h LC_{50}	22 mg/L	(246)
Gammarus *(Hyale plumulosa):*	96h LC_{50}	235 mg/L	(2631)

Fishes

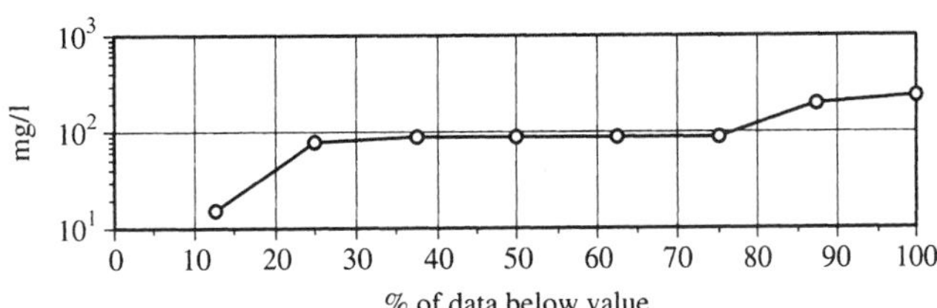

Frequency distribution of 24-96h LC_{50} values for fishes based on data from this work and from the work of Rippen	(3999)

L. macrochirus	24h LC_{50}	15-200 mg/L	(153)
Fathead minnows	LC_{50},S	(1; 24; 48; 72; 96 h): 140; 88; 88; 88; 88 mg/L	(350)
Red killifish *(Oryzias latipes)*	96h LC_{50}	235 mg/L (in seawater)	
	96h LC_{50}	80 mg/L (in freshwater)	(2631)

caproic nitrile *see* hexanenitrile

Caprolactam (perhydroazepinone; 2-oxohexamethylenimine; 2-ketohexamethylenimine; 6-hexanelactam; aminocaproic lactam; hexahydro-2H-azepi-2-one; a-caprolactam; e-caprolactam; w-caprolactam; 1-aza-2-cycloheptanone; 6-caprolactam; 6-hexanelactam; hexanoic acid,6-amino-,cyclic lactam); cyclohexanone iso oxime)

O N H

$C_6H_{11}NO$

CAS 105-60-2

USES

Nylon mfg. and processing; mfg. of plastics, bristles, film coatings, synthesis leather plasticizers, and paint vehicles; cross-linking agent for curing polyurethanes; synthesis of amino acid lysine.

A. PROPERTIES

molecular weight 113.16; melting point 69.2°C; boiling point 271°C; vapor pressure 0.0014 hPa at 20°C, 0.089 hPa at 60°C; density 1.01 at 80°C; vapor density 3.91; solubility 456 000 mg/L at 20°C, 5,250 000 mg/L at 25°C; sat concn 0.006 g/m3 at 20°C, 0.021 g/m³ at 30°C; log P_{ow} 0.12 (measured); K_{OC} 12 (estimated); H 3.3 x 10^{-6} Pa/mol

B. AIR POLLUTION FACTORS

1 mg/m³ = 0.21 ppm, 1 ppm = 4.70 mg/m³.

Odor

T.O.C. = 0.3 mg/m3 = 63 ppb (730, 57)

Photodegradation

T/2 = 4.9 hours

C. WATER AND SOIL POLLUTION FACTORS

Reduction of amenities

T.O.C. in water at room temp.: 59 mg/L, range 36- 100 mg/L, 8 judges.
20% of population still able to detect odor at 25 mg/L (321)
10% of population still able to detect odor at 16 mg/L (321)
1% of population still able to detect odor at 3.8 mg/L (321)
0.1% of population still able to detect odor at 0.92 mg/L (321)

Biodegradation

Inoculum	*Method*	*Conc.*	*Duration*	*Elimination results*	
A.S. adapted		200 mg/L		94% COD	(7634)
Zahn-Wellens test		400 mg/L	12 d	98% TOC	(7633)
A.S. nonadapted	Zahn-Wellens test	1,000 mg/L	6 d	>90% TOC	(7635)
BOD test			5 d	57% ThOD	(7636)
BASF A.S.	Respirometer test	185 mg/L	21 d	78% ThOD	(7637)
A.S. adapted	Product as sole carbon source			94% COD	(327)

Impact on biodegradation processes

1.0 mg/L affects the self-purification capacity of river water	(181)
nitrification in river water decreases from 100 mg/L onward	(30)

Hydrolysis

Stable at pH 4 to 9.

D. BIOLOGICAL EFFECTS

Toxicity

Bacteria			
Pseudomonas putida	17 h EC_{10}	1,740 mg/L	
(cell multiplication inhibition test)	17 h EC_{50}	4,200 mg/L	
	17 h EC_{90}	8,400 mg/L	(7630)
	17 h EC_{10}	74 mg/L	(7632)
Crustaceans			
Daphnia magna Straus	24-48 h EC_0	500 mg/L	
	24-48 h EC_{50}	>500 mg/L	
	24-48 h EC_{100}	>500 mg/L	(7630)
	24 h EC_0	1,992; 4,000 mg/L	
	24 h EC_{50}	4,380; 7,000 mg/L	
	24 h EC_{100}	7,653; 8,000 mg/L	(7631, 7632)
Algae			
Selenastrum capricornutum	72h EC_{50}	4,550 mg/L	
	72h NOEC	1,250 mg/L	(10681)
Scenedesmus subspicatus	72 h EC_{20}	34 mg/L	
(cell multiplication inhibition test)	72 h EC_{50}	130 mg/L	
	72 h EC_{90}	>500 mg/L	(7630)
	96 h EC_{20}	62 mg/L	
	96 h EC_{50}	160 mg/L	
	96 h EC_{90}	>500 mg/L	(7630)
Fishes			
Lepomis macrochirus	96 h LC_{50}	930 mg/L	(7628)
Pimephales promelas	96 h LC_{50}	1,400 mg/L	(7628)
Ictalurus punkt atus	96 h LC_{50}	1,000 mg/L	(7628)
Salmo gairdneri	96 h LC_{50}	500– 1,000 mg/L	
	96 h LC_{100}	1,000 mg/L	
	96 h NOEC	500 mg/L	(7628)
Mammals			
Rat	oral LD_{50}	1,475-1,876 mg/kg bw	(10681)

capronitrile *see* hexanenitrile

caprylalcohol *see* 2-octanol

caprylaldehyde *see* octanal

Caprylic acid (octanoic acid; *n*-octoic acid; *n*-octylic acid)

$CH_3(CH_2)_6COOH$

$C_8H_{16}O_2$

CAS 124-07-2

OCCURRENCE

octanoic acid is a degradation product of octane, 1-octanol and octanal

A. PROPERTIES

colorless liquid or solid; molecular weight 144.21; melting point 16°C; boiling point 237°C; vapor pressure 1 mm at 92°C; vapor density 5.00; sp. gr. 0.910 at 20/4°C; solubility 2,500 mg/L at 100°C, 620 mg/L at 15°C; log P_{ow} 2.92 (calculated), log P_{ow}: 1.06.

B. AIR POLLUTION FACTORS

Odor

characteristic: unpleasant, irritating.

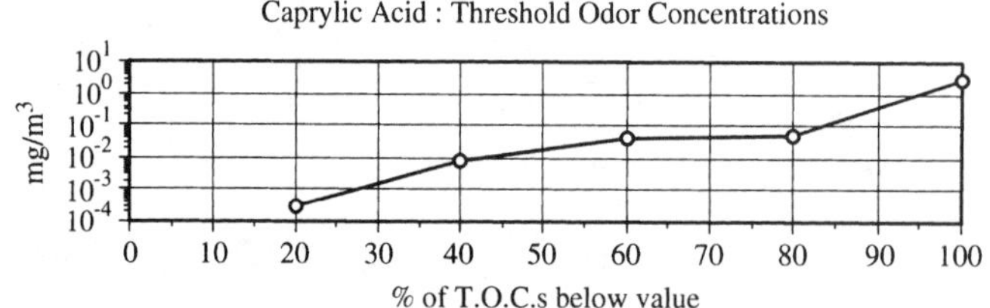

(211, 307, 610, 708, 778, 779)

1 ppm = 5.994 mg/m^3, 1 mg/m^3 = 0.167 ppm.

C. WATER AND SOIL POLLUTION FACTORS

$BOD^{25°C}_{2\ d}$:	44% ThOD (substrate conc.: 2.9 mg/L; inoculum: soil microorganisms)	
$BOD_{5\ d}$:	52% ThOD	
$BOD_{10\ d}$:	73% ThOD	
$BOD_{20\ d}$:	100% ThOD	(1304)
ThOD	2.44	

Odor thresholds

T.O.C. detection: 3.0 mg/kg.	(886)

Manmade sources

landfill leachate from lined disposal area (Florida 1987-1990): mean conc. 87 µg/L.	(2788)

Biodegradability

inoculum/method:	*test conc.*	*test duration*	*removed*	
municipal sludge, closed bottle test	2 mg/L	30 d	>90%	(5184)
A.S.		6 hours	10% of ThOD	
		12 hours	20% of ThOD	
		24 hours	33% of ThOD	(89)

Treatment methods

powdered carbon: at 100 mg/L sodium salt (pH 7.5)-carbon dosage 1,000 mg/L; 54% adsorbed.	(520)

Biodegradation pathway

Octanoic acid enters the beta-oxidation cycle and is used as both a carbon and energy source	(7132)

octanoic acid

β-oxidation

hexanoic acid + CH_3-COOH → $CO_2 + H_2O$

acetic acid

β-oxidation

butanoic acid

+ CH_3-COOH → $CO_2 + H_2O$

acetic acid

D. BIOLOGICAL EFFECTS

Bacteria			
Pseudomonas putida	16h EC_{10}	100 mg/L	
	16h $EC_{>10}$	30 mg/L	(5173)
	30 min EC_{10}	30 mg/L	
	30 min $EC_{>10}$	10 mg/L	(5170)
Bacillus subtilis	30 min EC_{50}	260 mg/L	(5170)
(inhib. initial germ. rate)	EC_{50}	7,210 mg/L	(5178)

Bacillus megaterium	24h MIC	288 mg/L	(5179)
Pseudomonas phaseolicola	24h MIC	>424 mg/L	(5179)
Methanothrix sp.	24h MIC	973 mg/L	
	24h EC_{50}	>1,440 mg/L	(5180)
heterogeneous culture	EC_{50}	1,615 mg/L (sodium salt)	(5181)
mixed aerobic culture	EC_{50}	15,719 mg/L (potassium salt)	(5182)
Vibrio parahaemolyticus	9h MIC	100 mg/L	(5183)
Algae			
Nitzschia closterium	72h EC_{50}	144 mg/L	(5176)
Crustaceans			
Daphnia magna	24h EC_0	300 mg/L	
	24h EC_{50}	550 mg/L	
	24h EC_{100}	990 mg/L	(5173)
	24h EC_{50}	900 mg/L (sodium salt)	(5174)
Artemia salina	16h EC_{50}	240 mg/L	(5175)
Hyale plumulosa	48h EC_{50}	128 mg/L (salinity 25 ppt)	(5172)
Fishes			
Leuciscus idus	48h LC_0	100 mg/L	
	48h LC_{50}	173 mg/L	
	48h LC_{100}	300 mg/L	(5170)
Brachydanio rerio	96h LC_0,SS	90 mg/L	
	96h LC_{50},SS	110 mg/L	
	96h LC_{100},SS	130 mg/L	(5171)
Oryzias latipes	48h LC_{50}	150 mg/L (salinity 30 ppt)	(5172)
	96h LC_{50}	310 mg/L (sodium salt)	(5172)

Lepomis macrochirus: chemical is too insoluble in water to be toxic (1294)

Red killifish *(Oryzias latipes):* 96h LC_{50}:
in seawater: 105 mg/L
in freshwater: 57 mg/L
310 mg/L (sodium caprylate) (2631)

captafol (1,2,3,6-tetrahydro-N-(1,1,2,2-tetrachloroethylthio)phthalimide; 1H-isoindole-1,3(2H)-dione, 3a,4,7,7a-tetrahydro-2-[(1,1,2,2-tetrachloroethyl)thio]; N-1,1,2,2-tetrachloroethylmercapto-4-cyclohexene-1,2-carboximide; Captofol; Difolatan; Difosan; Folcid; Sulfonimide)

$C_{10}H_9Cl_4NO_2S$

CAS 2425-06-1

USES

Fungicide. Wood preservative in timber industry.

A. PROPERTIES

pale yellow crystals with a slight mercaptan odor; molecular weight 349.1; melting point 162°C; vapor pressure 1.3×10^{-9} mbar at 20°C; solubility 1.4 mg/L at 20°C.

C. WATER AND SOIL POLLUTION FACTORS

Hydrolysis

Slow hydrolytic cleavage in aqueous emulsion or suspension. Rapidly hydrolyzed in acidic and alkaline media. At 50°C, 50% decomposition occurs in about 3 h at pH 6. (2962)

D. BIOLOGICAL EFFECTS

Bacteria			
Photobacterium phosphoreum Microtox test	5 min EC_{50}	7.0 mg/L	(9657)
Crustaceans			
Gammarus lacustrus	96h LC_{50}	0.8 mg/L	(9923)
Insects			
Pternarcys californica	96h LC_{50}	0.4 mg/L	(9924)
Fishes			
cytotoxicity to goldfish GF-Scale cells: NR_{50}*: 6.8 mg/L = 0.019 mmol/L*			(2680)
Cyprinus carpio	48h LC_{50}	0.12 mg/L	(2680)
rainbow trout	96h LC_{50}	0.5 mg/L	
goldfish	96h LC_{50}	3.0 mg/L	
bluegill	96h LC_{50}	2.8 mg/L	(2962)
Mammals			
rat	oral LD_{50}	2,500 mg/kg bw	(9736)

captan (N-trichloromethylthiotetrahydrophthalimide; *cis*-N-((trichloromethyl) thio)-4-cyclohexene-1,2-dicarboximide; Merpan; Orthocide; Voncaptan)

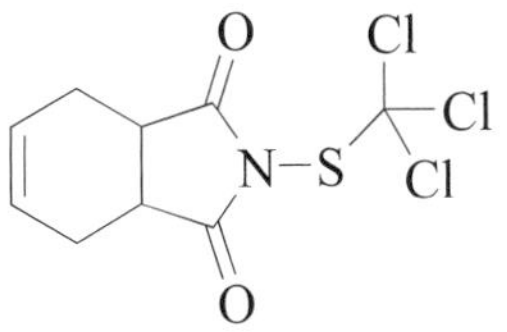

$C_9H_8Cl_3NO_2S$

CAS 133-06-2

USES

protectant-eradicant fungicide. Bacteriostat in soap.

A. PROPERTIES

pure chemical white solid; molecular weight 300.6; melting point 175°C; boiling point decomposes above 175°C; vapor pressure <1.3 × 10^{-7} mbar at 25°C; solubility <0.5 mg/L at 20°C; $LogP_{ow}$ 2.35.

C. WATER AND SOIL POLLUTION FACTORS

Hydrolysis at 20°C

at pH 7: $t_{1/2}$: 32 h
at pH 10: $t_{1/2}$: < 2 min (9600)

Biodegradation

Degradation of captan in soil incubated under laboratory conditions started early, continued at an exponential rate for 35 d and then stopped. After 60 d of incubation 57-64% of the original captan was recovered. (9927)

D. BIOLOGICAL EFFECTS

Crustaceans			
Daphnia pulex	48h EC_{50}	1.5 mg/L	(9925)

Worms

earthworm upon contact	48h LC_{50}	73-80 mg/kg soil	(9926)

Fishes

harlequin fish (Rasbora heteromorpha) *(89% active ingredient):*

mg/L	*24 h*	*48 h*	*96 h*	*3 m (extrap.)*	
LC_{10}(F)	0.23	0.14	-		
LC_{50}(F)	0.46	0.33	0.3	0.2	(331)

cytotoxicity to goldfish GF-Scale cells	NR_{50}	6.5 mg/L = 0.022 mmol/L	(2680)
Cyprinus carpio	48h LC_{50}	0.25 mg/L	(2680)
brook trout, bluegill sunfish	96h LC_{50}	0.034-0.072 mg/L	(9600)

Mammals			
rat	oral LD_{50}	9,000 mg/kg bw	(9928)
rabbit	oral LD_{50}	740 mg/kg bw	(9929)

captax *see* 2-mercaptobenzothiazole

Carazolol (1-(9H-carbazol-4-yloxyl)-3-[(1-methylethyl)amino]-2-propanol))

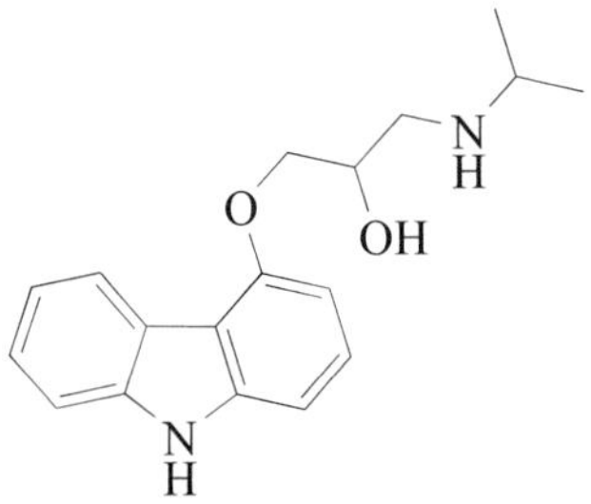

$C_{18}H_{22}N_2O_2$

CAS 57775-29-8

TRADENAMES

Conducton, Carazolol

USES

Veterinary drug. Beta-blocker, antihypertensive, antianginal, antiarrhythmic. Carazolol is a ß-adrenoceptor blocking agent primarily used in pigs to prevent sudden death due to stress during transport, thus preventing death losses and deterioration of meat quality.

A. PROPERTIES

molecular weight 298.38

C. WATER AND SOIL POLLUTION FACTORS

Environmental concentrations

	range μg/L	n	median μg/L	90 percentile μg/L	
surface waters (1996-2000)	<0.003-0.124	24	<0.003	0.008	
	<0.010-0.110	45	<0.010	0.100	(7237)
drinkingwater (1996)	<0.003	16			(7237)
effluent MWTP	<0.025-0.117	25	<0.025	0.088	(7237)
	max 0.120				(7178)

D. BIOLOGICAL EFFECTS

Biotransformation

One male beagle dog was given 10 mg side-chain labelled ^{14}C-carazolol . Urine was collected for 24 hours and used for metabolite investigations. Within 24 hours about 31% of the dose was excreted in urine. Apart from carazolol 6 metabolites were isolated and identified by mass spectrometry as carazolol with a shortened side-chain, a diastereomer pair of carazolol-O-glucuronides, an O-glucuronide of 6(7)-hydroxy carazolol and a diastereomer pair of carazolol-bis-O-glucuronide.
Unchanged carazolol, carazolol mono-and bis-glucuronide and carazolol lactate were identified in the urine of pigs intramuscularly injected with 1 mg carazolol/100 kg b.w. Two human volunteers each ingested 5 mg carazolol. Urine was collected during the following 4.5 hours; the parent compound, carazolol monoglucuronide, carazolol lactate and carazolol acetate were found. (see metabolic pathway). (11032)

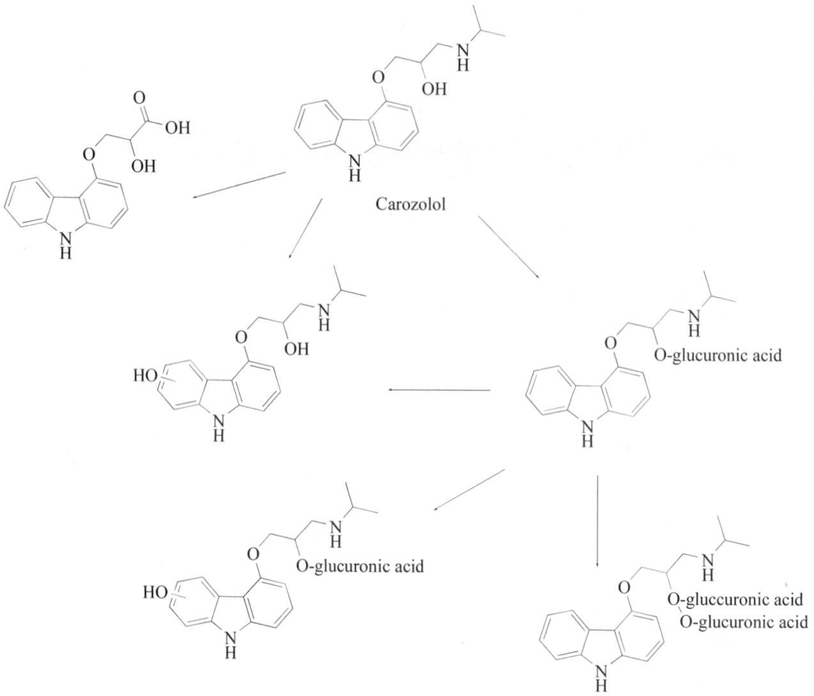

Metabolic pathway of carozolol in the dog

Metabolic pathway of carazolol in the dog			(11032)
Mammals			
Rat	Acute oral LD_{50}	80; 88 mg/kg bw	
Mouse	Acute oral LD_{50}	132; 160 mg/kg bw	(11032)

Carbamazepine (5H-dibenz[b,f]azepine-5-carboxamide)

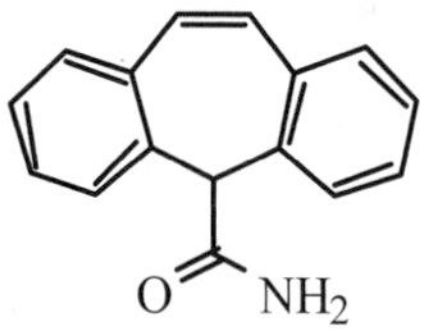

$C_{15}H_{12}N_2O$

CAS 298-46-4

TRADENAMES

Tegratal

USES

Analgesic; antiepileptic

A. PROPERTIES

molecular weight 236.27

C. WATER AND SOIL POLLUTION FACTORS

Waste water treatment

In influents to sewage works Germany: over 100 g/day	removal efficiency 7 %	in effluents of sewage works max 6300 ng/L	(7178)

	ng/L	n	
Belgium and The Netherlands 2000			
in effluents of MWTP's	580-870	2	
in surface waters	<10-310	11	
in drinking water	<10	4	(7237)

	range µg/L	n	median µg/L	90 percentile µg/L	
effluent MWTP (1996-2000)	0.58-0.87	2			
	up to 6.3	30	2.1	3.7	(7237)
in effluent farming industry	2500	1			(7237)
influent MWTP	0.15-1.7	8	1.3		(7237)
waste waters (1996-1997)	5.0-46.0				
	0.50-2.0				(7237)
surface waters (1996-2000)	<0.010-0.23	11			
	<0.030-1.10	26	0.25	0.82	
	<0.80				
	<0.020-2.10	161		0.69	
	<0.020-0.17	35		0.042	(7237)
drinkingwater (2000)	<0.010	6			(7237)

Environmental concentrations in Germany 2001 (10923)

	Positive samples	Min ng/L	Max ng/L	Average ng/L	Median ng/L
River Schussen	7/7	101	272	170	153
River Körsch upstream WWTP	8/8	35	58	45	45
River Körsch downstream WWTP	8/8	185	1448	803	655
Influent WWTP Stuttgart-M	6/6	151	1472	761	768
Effluent WWTP Stuttgart-M	7/7	545	1282	912	744
Suspended solids influent WWTP Stuttgart-M µg/kg	0/6	n.d.	n.d.	n.d.	n.d.
Influent WWTP Reutlingen-W	5/5	657	2724	1337	948
Effluent WWTP Reutlingen-W	5/5	232	1269	783	926
Suspended solids influent WWTP Reutlingen-W µg/kg	5/5	39	73	53	51
Influent WWTP Steinlach-W	5/5	62	1298	579	329
Effluent WWTP Steinlach-W	5/5	161	1251	612	605
Suspended solids WWTP Steinlach-W µg/kg dw	5/5	68	393	211	214
Leachate of landfill Reutlingen-S	5/5	64	268	182	202
Leachate of landfill Dusslingen	5/5	664	2118	1465	1415

D. BIOLOGICAL EFFECTS

Mammalian metabolism

only 1-2% excreted free; 10,11-epoxycarbamazepine is major metabolite; also excreted as glucuronides. (7178)

carbamide *see* urea

carbanil *see* phenylisocyanate

carbaryl (1-naphthyl-N-methylcarbamate; sevin; methylcarbamate-1-naphthol)

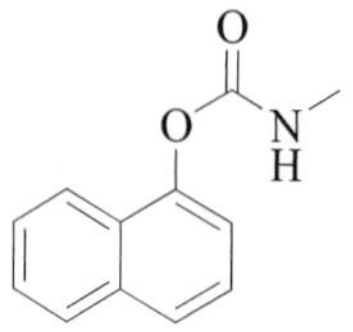

$C_{12}H_{11}NO_2$

CAS 63-25-2

USES

contact insecticide.

A. PROPERTIES

white crystalline solid; molecular weight 201; melting point 142°C; sp. gr. 1.23 at 20/20°C; vapor pressure <0.005 mm at 26 °C, 5.3 x 10^{-5} hPa at 20 °C; solubility 40; 120 mg/L at 30°C; $logP_{OW}$ 2.3

C. WATER AND SOIL POLLUTION FACTORS

Aquatic reactions

comparison of calculated hydrolysis and photolysis under given conditions:

*hydrolysis**	*direct photolysis*	
pH	*half-life, d*	*half-life, d*
5	1,500	6.6
7	15	6.6
9	0.15	-

* Calculation based on neutral and alkaline hydrolysis assuming pseudo-first-order kinetics. (1070)

Biolysis by bacteria

$t_{1/2}$: >30,000 d (minimum value assuming a bacterial population of 0.1 mg/L. (1071)

Mechanism for alkaline hydrolysis:

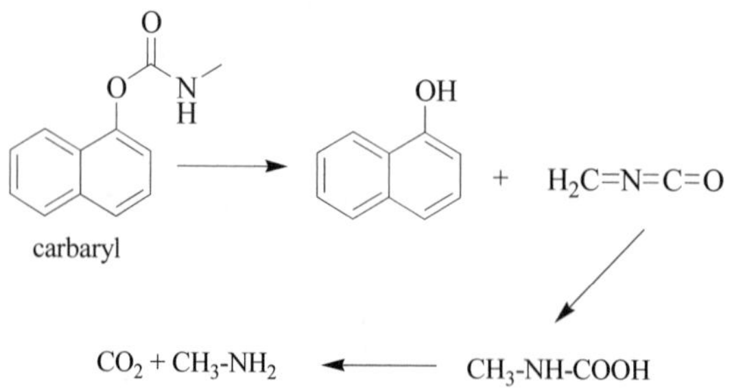

carbaryl → 1-naphthol + $H_2C=N=C=O$

$H_2C=N=C=O$ → CH_3-NH-COOH → $CO_2 + CH_3$-NH_2

Persistence in river water

in a sealed glass jar under sunlight and artificial fluorescent light-initial conc. 10 μg/L:

% of original compound found after					
1 h	*1 wk*	*2 wk*	*4 wk*	*8 wk*	
90	5	0	0	0	(1309)

Persistence

in vitro, in capped bottles at 21°C in darkness: % remaining after 8 weeks at initial conc. of 5 ppm:

in sterilized natural water:	0%	
in sterilized distilled water:	5%	
in natural water:	0%	
in distilled water:	18%	(2559)

Biodegradation in soil

aerobic biodegradation half-life: 0.34 d, at an initial conc. of 20 ppm, using a mixture of nonacclimated sludge, field soil, and river sediment as inoculum, in a stirred flask at 18-22°C, the major bacterial types in the activated sludge have been characterized as follows: *Flavobacterium* 35%; *Alcaligenes* 30%; *Corynebacterium* 20%; *Pseudomonas* 15%. (2690)

Alkaline hydrolysis

half-life: 0.013 d, at pH 9 at 40°C in the dark. (2690)

Mobility

log K_{OC}: 2.5 (7045)

Biodegradation in soil

in greenhouse at 23-26°C at constant light at initial conc. of 5-10 mg/kg soil

experiment time 50 d:	t/2 d	
in sandy soil	4	
in organic rich orchard soil	4	
in agricultural soil	4	
in soil from volcanic area	20	
in field experiments	8 (dissipation t/2)	(7045)

D. BIOLOGICAL EFFECTS

Bacteria			
Pseudomonas putida	16h EC_0	>50 mg/L	(329)
Algae			
Microcystis aeruginosa	8d EC_0	0.03 mg/L	(329)
Dunaliella euchlora	10d EC_{35}	1 ppm	
Phaeodactylum tricornutum	10d EC_{100}	0.1 ppm	
Monochrysis lutheri	10d EC_{100}	1 ppm	
Chlorella sp.	10d EC_{20}	1 ppm	
Chlorella sp.	10d EC_{100}	10 ppm	
Protococcus sp.	10d EC_{26}	1 ppm	
Protococcus sp.	10d EC_{100}	10 ppm	(2347)

Crustaceans

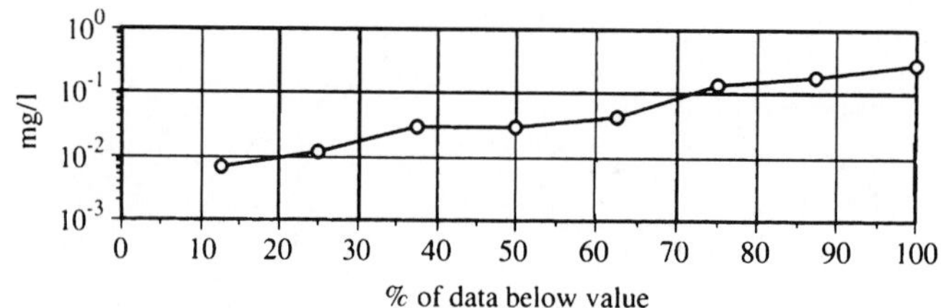

Frequency distribution of 24-96h EC_{50}/LC_{50} values for crustaceans (n = 8) based on the data in this work and in the work of Rippen (3999)

Gammarus lacustris	96h NOEC	16 µg/L	(2124)
Gammarus fasciatus	96h NOEC	26 µg/L	(2126)
Palaemonetes kadiakensis	96h NOEC	5.6 µg/L	(2126)
Orconectes nais	96h NOEC	8.6 µg/L	(2126)
Asellus brevicaudus	96h NOEC	240 µg/L	(2126)
Simocephalus serrulatus	48h NOEC	7.6 µg/L	(2127)
Daphnia pulex	48h NOEC	6.4 µg/L	(2127)
Daphnia magna	63d NOEC	5.0 µg/L	(2135)
Gammarus pulex	96h LC_{50}	0.029 ppm	(1323)
Cancer magister			
egg/prezoeal, prevention of hatching and molting	24h EC,S	6 ppb	
zoea, prevention of molting and death	96h EC,S	10 ppb	
juvenile, death or paralysis	96h EC,S	280 ppb	
adult, death or paralysis	96h EC,S	180 ppb	(2321)
juvenile male, death, paralysis, loss of equilibrium	24h EC_{50},S	600 ppb	(2346)
Hemigrapsus oregonensis			
adult female, death, paralysis, loss of equilibrium	24h EC_{50},S	270 ppb	(2346)
Palaemon macrodactylus	96h LC_{50},S	12 ppb	
	96h LC_{50},F	7 ppb	(2352)
Upogebia pugettensis	48h LC_{50},S	40 ppb	(2346)
Callianassa californiensis	48h LC_{50},S	30 ppb	(2346)
adult	48h LC_{50},S	130 ppb	(2346)

Mollusc

Crassostrea gigas larvae, prevention of development	48 EC_{50},S	2,200 ppb	(2346)
Crassostrea virginica, eggs	48h LC_{50},S	3,000 ppb	(2324)
larvae	14d EC_{50},S	3,000 ppb	(2324)
Mercenaria mercenaria, eggs	14h EC_{50},S	3,820 ppb	(2324)
larvae	14d EC_{50},S	>2,500 ppb	(2324)
Clinocardium nuttalli, adults	24h LC_{50},S	7,300 ppb	(2346)
juvenile	96h LC_{50},S	3,850 ppb	(2346)
Mytilus edulis, larvae	96h EC_{50}	2,300 ppb	(2346)

Gastropod			
Lymnaea stagnalis	48h LC_{50}	21 ppm	(1323)
	Growth rate decreases from 1 mg/L onwards.		(1910)

Insects			
Pteronarcys californica	96h NOEC	4.8 µg/L	(2128)
Pteronarcys dorsata	30d LC_{50}	23 µg/L	
	30d NOEC	11 µg/L	(2134)

Pteronarcella badia	96h NOEC	1.7 µg/L	(2128)
Claassenia sabulosa	96h NOEC	5.6 µg/L	(2128)
Acroneuria lycorias	30d LC_{50}	2.2 µg/L	
	30d NOEC	1.3 µg/L	(2134)
Hydropsyche bettoni	30d LC_{50}	2.7 µg/L	
	30d NOEC	1.8 µg/L	(2134)
insect larvae: *Chaoborus*	48h LC_{50}	0.30 ppm	(1323)
Cloeon	48h LC_{50}	0.48 ppm	(1323)
Chironomus riparius, fourth instar larval	24h LC_{50}	104 ppb	(1853)
rice-field spider *(Oedothorax insecticeps)*	LD_{50}	840 ppm	(1814)
bees	48h LC_{50}	3.8-4.5 ppm in food	(1683)

Fishes

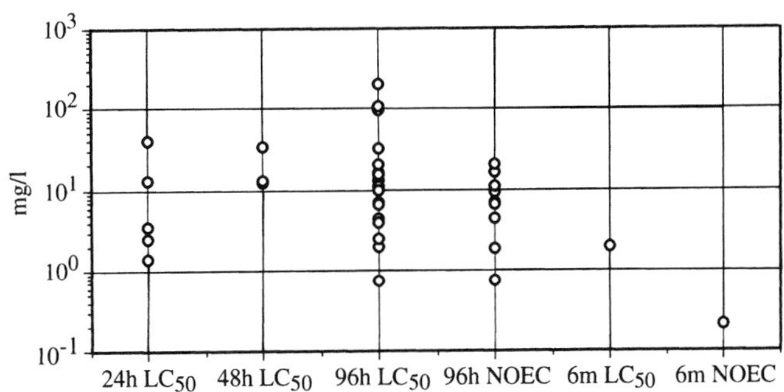

Summary of fish toxicities

the data reflected in this graph are presented below

Pimephales promelas	96h NOEC	9,000 µg/L	
(decline survival and reproduction 6 months)		680 µg/L	
	6m NOEC	210 µg/L	(2136)
Lepomis macrochirus	96h NOEC	6,760 µg/L	(2121)
Lepomis microlophus	96h NOEC	11,200 µg/L	(2121)
Micropterus salmoides	96h NOEC	6,400 µg/L	(2121)
Salmo gairdneri	96h NOEC	4,340 µg/L	(2121)
Salmo trutta	96h NOEC	1,950 µg/L	(2121)
Oncorhynchus kisutch	96h NOEC	764 µg/L	(2121)
Perca flavescens	96h NOEC	745 µg/L	(2121)
Ictalurus punctatus	96h NOEC	15,800 µg/L	(2121)
Ictalurus melas	96h NOEC	20,000 µg/L	(2121)
	96h LC_{50}	20,000 µg/L	(2121)

	96h LC_{50} mg/L		*96h LC_{50} mg/L*	
catfish	15.8	bluegill	6.8	
bullhead	20.0	bass	6.4	
goldfish	13.2	rainbow	4.3	
minnow	14.6	brown	2.0	
carp	5.3	Coho	0.76	
sunfish	11.2	perch	0.75	(1934)

Channa punctatus	180d LC_{50}	2.0 mg/L	(1508)
Mosquitofish *(Gambusia affinis)*	24h LC_{50},S	40 mg/L	(1104)
Mosquitofish	48h LC_{50},S	35 mg/L	(1104)
Mosquitofish	96h LC_{50},S	32 mg/L	(1104)
Carp *(Cyprinus carpio)*	24h LC_{50},S	13 mg/L	(1106)
Carp	48h LC_{50},S	12 mg/L	(1106)

Carp	72h LC_{50},S	10 mg/L	(1106)
Parophyrs vetulus, juvenile	24h LC_{50},S	4,100 ppb	(2346)
Cymatogaster aggregata, juvenile	24h LC_{50},S	3,900 ppb	(2346)
Gasterosteus aculeatus, juvenile	24h LC_{50},S	6,700 ppb	(2346)
Gasterosteus aculeatus	96h LC_{50},S	3,990 ppb	(2333)
Leiostomus xanthurus	5m LC_{35},F	100 ppb	(2339)
Oncorhynchus keta, juvenile	96h LC_{50},S	2,500 ppb	(2343)
bluegill	24h LC_{50}	3.4 ppm	
rainbow trout	24h LC_{50}	3.5 ppm	(1878)
cytotoxicity to goldfish GF-Scale cells	NR_{50}	>200 mg/L	(2680)
Cyprinus carpio	48h LC_{50}	13.0 mg/L	
Barbus conchonius: histological effect after 15 d	0.19 mg/L		
Clarius batrachus	96h LC_{50}	47-108 mg/L	
Gambusia affinis	96h LC_{50}	204 mg/L	
Melanopsis dufouri	96h LC_{50}	10-15 mg/L	
Pimephales promelas	96h LC_{50}	6.7-10 mg/L	
Salmo gairdneri	96h LC_{50}	1.4 mg/L	(2625)

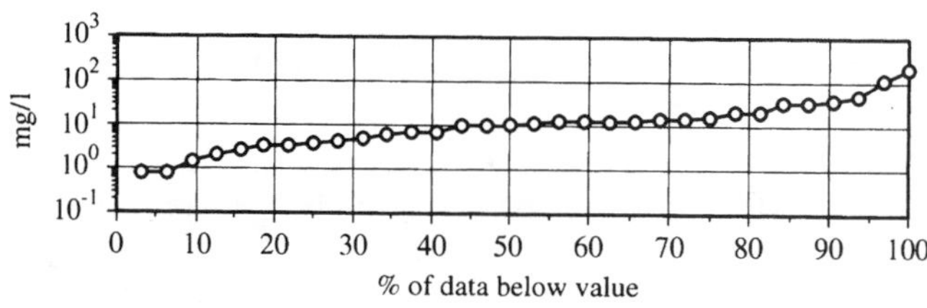

Frequency distribution of 24-96h LC_{50} values for fish (n = 32) based on data from this work and from the work of Rippen (3999)

Bioaccumulation

BCF:	algae:	4,000	
	duckweed:	3,600	
	snails:	300	
	catfish:	140	
	crayfish:	260	(1891, 1892, 1893)

Algae: *Chlorella fusca* BCF (wet wt): 73 (2659)

carbathione *see* methyldithiocarbamate

carbazole (dibenzopyrrole; 9-azafluorene; diphenyleneimine)

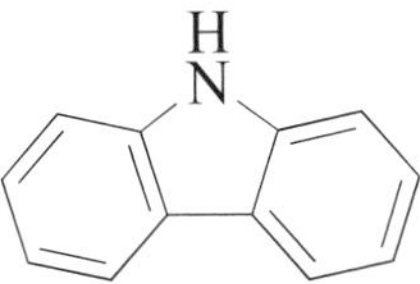

$C_{12}H_9N$

CAS 86-74-8

USES

feedstock for the synthesis of dyes, insecticides, and vinylcarbazole for polymer production; as an odour inhibitor in detergents; in the maufacture of lubricants; a key raw material for the synthesis of pigment Violet-23 which is used in many products ranging from printing inks tot photographic plates. Synthesis of the insecticide Nirosan (1,3,6,8-tetranitrocarbazole). An intermediate in the bulk drug manufacture of the following drugs : carvediol: an antihypertensive drug, Carprofen (6-chloro-α-methylcarbazole-2-acetic acid): an anti-inflammatory drug , Carnazolyl alanines : anticancer drugs.

carprofen
6-chloro-α-methylcarbazole-2-acetic acid
CAS 53716-49-7

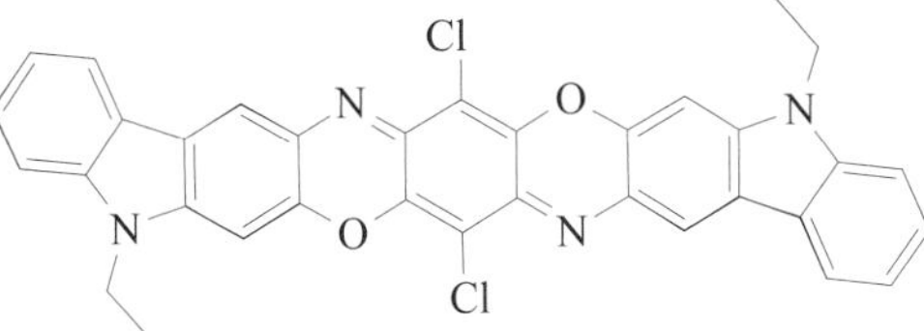

Pigment Violet 23
CAS 6458-30-1

insecticide Nirosan
1,3,6,8-tetranitrocarbazole

OCCURRENCE

Carbazole is naturally contained in coal, crude oil and peat.In foods as a result of char-broiling.

MANMADE SOURCES

constituent of coal tar creosote: 0.1 wt %.

A. PROPERTIES

molecular weight 167.21; melting point 240-248°C; boiling point 354-355°C; d. 1.1 at 18/4°C; vapor pressure 533 hPa at 323°C; solubility 0.1-1.9 mg/L at 20°C; log P_{OW} 3.84-3.84 (measured).

B. AIR POLLUTION FACTORS

Manmade sources

in coal tar pitch fumes: 9.6 wt % (also methylphenanthrene); combustion of rubber, petroleum, coal and wood; waste incineration; tobacco smoke and aluminum manufacturing. Carbazole has been identified in cigarette smoke at a concentration of 1 μg/cigarette.

C. WATER AND SOIL POLLUTION FACTORS

Photochemical degradation in water

$t_{1/2}$: 2.9 hours at 23-28°C. (5377)

Soil sorption

K_{oc}: 1,200 (calculated). (5376)

Biodegradation

inoculum/method	*test conc.*	*test duration*	*removed*	
mixed bacteria, adapted	1 mg/L	14 d at 28°C	24% $^{14}CO_2$ prod.	(5379)
Pseudomonas sp. *HL 7b*	1 mg/L	16 d at 28°C	8.5% $^{14}CO_2$ prod.	(5380)
soil microorganisms	2.9 mg/L	24 hours	28% product	
		72 hours	59%	
		8 d	72%	
		14 d	66%	(5381)

Aerobic mixed culture from oligotrophic and eutrophic lakes: $t_{1/2}$: 14 hours at 15-25°C. (5384)

Methanogenic anaerobic mixed culture at 24 mg/L, after 203 d at 36°C: 100% product removal. (5389)

The fungus *Planerochaete chrysosporium,* anaerobic, at 0.9 mg/L, after 30 d at 39°C: 98% product removal. (5390)

The fungus *Cunninghamella elegans* degraded carbazole to 2-hydroxydibenzopyrrole. (5388)

Degradation in soil system

at initial conc. of 5 mg/kg soil: half-life: 10.5 d
500 mg/kg soil: half-life: 3 d (2903)

Biodegradation

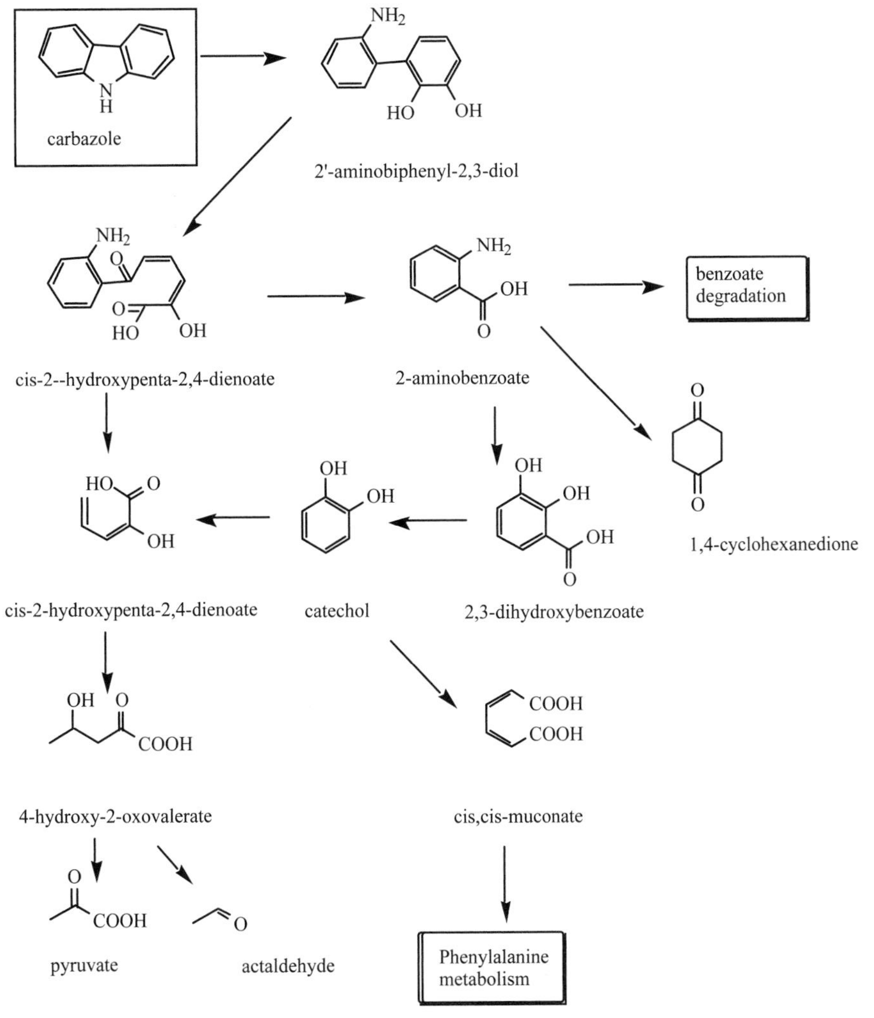

Carbazole degradation pathway (7196)

D. BIOLOGICAL EFFECTS

Bioaccumulation

Daphnia pulex	*BCF*	115 (wet wt)	(1570)
Daphnia magna	*BCF*	108	(5382)

Toxicity

Algae			
Tetrahymena pyriformis	60h IC_{50},S	6.7 mg/L	(5391)
Insects			
Culex larvae	24h LC_7	0.1 mg/L (in acetone)	(5385)
	24h LC_9	5 mg/L (in ethanol)	(5386)
	24h LC_0	0.05 mg/L (in ethanol)	(5387)
Fishes			
Mugil cephalus	48h LC_{50},S	1 mg/L	

carbetamex (*d*-N-ethyllactamide carbanilate; carbetamide; N-ethyl-2-[(phenylcarbamoyl)oxy]propionamide)

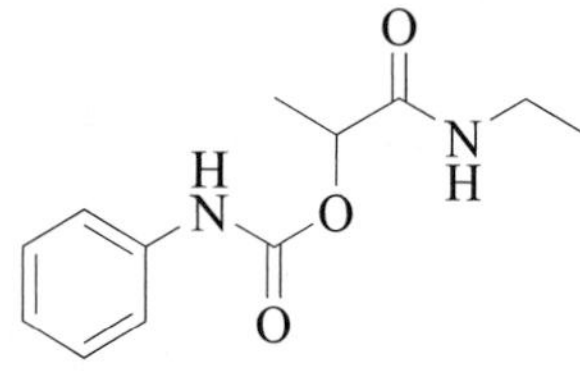

$C_{12}H_{16}N_2O_3$

CAS 16118-49-3

USES

selective herbicide absorbed through the roots and leaves.

A. PROPERTIES

molecular weight 236.27; melting point 118-110°C; boiling point decomposes on distillation; vapor pressure <10^{-5} mbar at 20°C; solubility 3,500 mg/L at 20°C.

D. BIOLOGICAL EFFECTS

mg/L	*24 h*	*48 h*	*96 h*	*3 m (extrap.)*	
LC_{10}(F)	170	150	125		
LC_{50}(F)	220	190	165	100	(331)

carbetamide *see* carbetamex

carbitol *see* diethyleneglycolmonoethylether

carbitolacetate *see* diethyleneglycolmonoethyletheracetate

carbofuran (furadan; NIA 10,242; ENT 27,164; 2,3-dihydro-2,2-dimethyl-7-benzofuranylmethylcarbamate; 2,2-dimethyl-2,3-dihydrobenzofuranyl-7N-methylcarbamate-N-acetylfuradan)

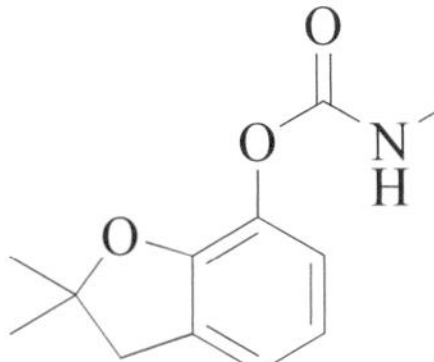

$C_{12}H_{15}NO_3$

CAS 1563-66-2

USES

systemic insecticide widely used to control corn rootworms.

A. PROPERTIES

odorless, white crystalline solid; molecular weight 221.3; melting point 150-152°C; sp. gr. 1.18 at 20/20°C; vapor pressure 2×10^{-5} mm at 33°C; solubility 250-700 mg/L at 25°C.

C. WATER AND SOIL POLLUTION FACTORS

In groundwater wells in the U.S.A. 150 investigations up to 1988

median of the conc.'s of positive detections for all studies	5.3 μg/L	
maximum conc.	176 μg/L	(2944)

Persistence

in vitro, *in capped bottles at 21°C in darkness: % remaining after 8 weeks at initial conc. of 5 ppm:*

in sterilized natural water:		5%	
in sterilized distilled water:	10%		
in natural water:	16%		
in distilled water:		18%	(2559)

Water treatment

initial concentration μg/L	*treatment method*	*% removal*	
0.13	chlorination	24	
0.79	flocc. + filtr. + chlorin. + filtr. + sed.	54	
1.62	filtr. + sed. + carbonate removal	100	
0.49	carbonate removal	100	(2915)

Reverse osmosis

monthly averages μg/L

month	*raw water*	*permeate*	*concentrate*	
July	9.8	<1	24.5	
August	8.0	<1	21.5	
September	8.1	<1	22.5	
October	7.8	<1	19.5	
November	6.2	<1	13.0	
December	5.8	<1	13.8	
January	5.9	<1	14.0	
February	5.7	<1	13.4	
March	5.3	<1	12.8	
April	4.9	<1	12.6	
May	4.7	<1	13.3	
June	4.3	<1	12.7	(2943)

Degradation in soil

In 4 soils with known insecticide use: the technical carbofuran had a calculated half-life of 11-13 d (pH 6.5), and the granular formulation had a half-life of 60-75 d (pH 6.5). (1663)

Time for 95% disappearance from the soil 3.1-5.6 kg/ha: 145 to 434 d. (1650)

Volatilization

at 25°C from soils in the laboratory:
sandy loam: 0.5% after 60 d
sand: 36% after 60 d (1650)
half-lives: 21 to 227 d depending on soil type and incubation conditions (2938)
shortest half-lives obtained in three soil incubation treatments: 4.7 to 5.3 d (2938)

Degradation

The dominant fate of the carbonyl group of carbofuran added to the soil is hydrolysis of the carbamate bond followed by rapid mineralization of the methylamine. Production of $^{14}CO_2$ accounted for 95% of the added ^{14}C-carbonyl-labeled carbofuran over a 13-d period. The $^{14}CO_2$ production from the ring-labeled carbofuran exhibited linear kinetics, and only 12% of the added label was released as $^{14}CO_2$ during the 13-d incubation period. (2938)

D. BIOLOGICAL EFFECTS

BLUE-GREEN ALGAE

Nostoc muscorum at 25 mg/L enhanced survival; growth and nitrogen fixation; gradual inhibition from 50-1,000 mg/L; at 1,200 mg/L algicidal. (1352)

Fishes			
Mosquitofish *(Gambusia affinis)*	72h LC_{50},S	0.52 mg/L	
Green sunfish *(Lepomis cyanellus)*	72h LC_{50},S	0.16 mg/L	(1203)
Channa punctatus	histological effect after 30 d	at 5 mg/L	(2625)
Cyprinodon variegatus	96h LC_{50},F	386 µg/L	
MATC, early life stage		19 µg/L	(2643)
cytotoxicity to goldfish GF-Scale cells	NR_{50}	>400 mg/L = >1.8 mmol/L	(2680)
Cyprinus carpio	48h LC_{50}	1.0 mg/L	(2680)

carbondisulfide (carbonbisulfide)

S=C=S

CS_2

CAS 75-15-0

USES AND FORMULATIONS

mfg. rayon, cellophane, carbon tetrachloride, rubber chemicals and flotation chemicals, soil disinfectants, electronic vacuum tubes; solvent (phosphorus, sulfur, bromine, iodine, selenium, fats, resins, rubbers); mfg. grain fumigants, soil conditioners, herbicides; paper mfg.; pharmaceutical mfg. (347) Solvent for the manufacturing of viscose fibres and cellophane (as xanthogenate). Feedstock for the manufacture of carbontetrachloride, several thiurams, and dithiocarbamates. many applications.

A. PROPERTIES

colorless liquid; molecular weight 76.14; melting point - 108.6°C/- 116.6°C; boiling point 46.3°C; vapor pressure 260 mm at 20°C, 430 mm at 30°C; 200 mm at 10°C, 264 hPa at 10°C, 398 hPa at 20°C; 580 hPa at 30°C; vapor density 2.64; density 1.263 at 20/4°C; solubility 2,100 mg/L at 20°C; 2,300 mg/L at 22°C; $LogP_{OW}$ 1.84- 2.16 (calculated).

B. AIR POLLUTION FACTORS

1 mg/m^3 = 0.315 ppm, 1 ppm = 3.17 mg/m^3.

Odor

characteristic: quality: vegetable sulfide, aromatic; hedonic tone: slightly pungent.

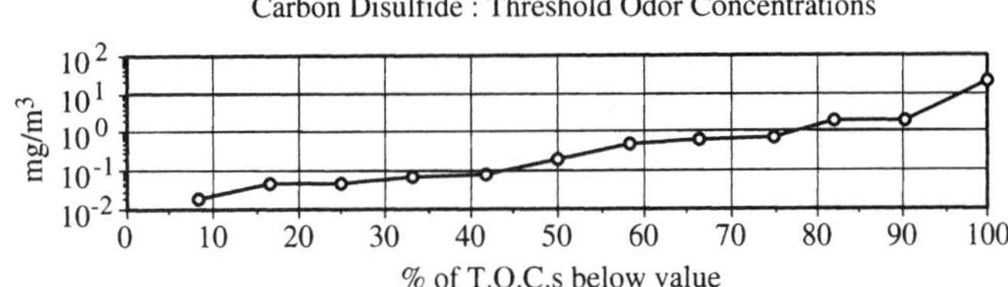

(73, 210, 307, 612, 637, 662, 686, 741)

O.I. 100% recognition: 1,600,000		(2)
human odor perception:		
nonperception:	0.04 mg/m^3	
perception:	0.05 mg/m^3 = 16 ppb	
human reflex response:		
no response:	0.03 mg/m^3	
adverse response:	0.04 mg/m^3	(170)

Natural sources

biogenic carbon disulfide emissions from soils in USA:

avg sulfur flux				
sampling sites	*soil orders*		*g S/m^2/yr*	
Wadesville, IN	Alfisol	Sept.-Oct. 1977	0.002	
Philo, OH	Inceptisol	Oct. 1977	0.0012	
Dismal Swamp, NC	Histosol	Oct. 1977	0.0001	
Ceder Island, NC	freshly clipped marsh		1.13	
Cox's Landing, NC	freshly clipped marsh		0.021	(1385)

Control methods

catalytic combustion; platinized ceramic honeycomb catalyst; ignition temp.: 350°C, inlet temp. for 90% conversion: 375- 400°C. (91)

Manmade sources: atmospheric emissions from the viscose production: 200- 300 g CS_2/kg viscose for fibre manufacturing.

C. WATER AND SOIL POLLUTION FACTORS

Impact on biodegradation processes

75% reduction of the nitrification process in activated sludge at 35 mg/L. (30)

Reduction of amenities

faint odor: 0.0026 mg/L; organoleptic limit: 1.0 mg/L. (129)

D. BIOLOGICAL EFFECTS

Bacteria

mixed population	24 h EC_0	4.6 mg/L	
	24 h EC_{50}	13 mg/L	
	24 h EC_{100}	100 mg/L	(7770)
Photobacterium phosphoreum	15 min EC_{50}	341 mg/L	(2712)
Nitrosomonas/Nitrobacter	3 h MIC	28 mg/L	(2712)
Algae			
Chlorella pyrenoidosa	96 h EC_{50}	21 mg/L	(2712)
Crustaceans			
Poecilia reticulata	96 h LC_{50}	4.0 mg/L	(2712)
Daphnia magna	48 h LC_{50}	2.1 mg/L	(2712)
Fishes			
Leuciscus idus	96 h LC_0	25- 164 mg/L	
	96 h LC_{50}	95- 265 mg/L	
	96 h LC_{100}	126- 290 mg/L	(7767)
Alburnus alburnus	96 h LC_{50}	65 mg/L	(7768)
Gambusia affinis	96h LC_{50}	135 mg/L	(7769)
Mosquitofish:	24-96 h LC_{50}	135-162 mg/L	(41)

carbonhexachloride *see* hexachloroethane

carbonoxysulfide *see* carbonylsulfide

Carbontetrachloride (tetrachloromethane)

CCl_4

CAS 56-23-5

USES AND FORMULATIONS

fire extinguisher mfg.; dry cleaning operations; mfg. of refrigerants, aerosols, and propellants; mfg. of chlorofluoromethanes; extractant; solvent; veterinary medicine; metal degreasing; fumigant; chlorinating organic compounds.

A. PROPERTIES

Henry's constant

temp, °C	*H*	
10	0.57	
18	0.9	
25	1.2	
35	1.8	(2603)

colorless liquid; molecular weight 153.82; melting point -23°C; boiling point 76.7°C; vapor pressure 90 mm at 20°C, 56 mm at 10°C, 113 mm at 25°C, 137 mm at 30°C; vapor density 5.5; sp. gr. 1.59 at 20°C; solubility 1,160 mg/L at 25°C, 800 mg/L at 20°C; saturation concentration in air 754 g/m^3 at 20°C, 1,109 g/m^3 at 30°C; log P_{OW} 2.64 at 20°C (calculated).

B. AIR POLLUTION FACTORS

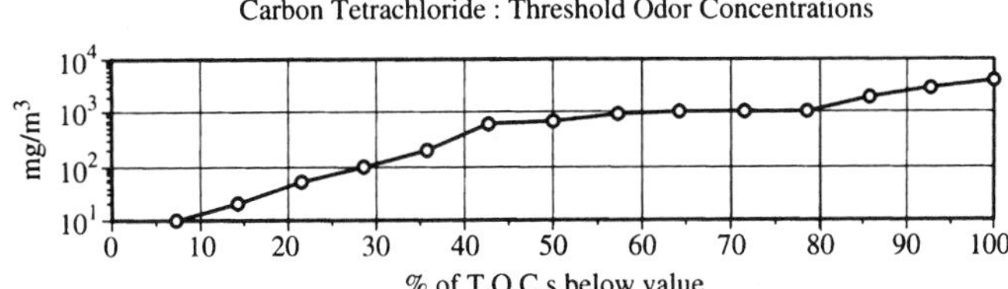

(57, 73, 210, 602, 617, 643, 740, 741, 749, 767)

Reduction of amenities

threshold odor concentrations:

chlorination of CS_2:	50% response:	10 ppm	
	100% response:	21 ppm	
chlorination of CH_4:	50% response:	47 ppm	
	100% response:	100 ppm	(2)
distinct odor:		1,600 mg/m^3 = 250 ppm	(278)

Photochemical reactions

$t_{1/2}$: >400 years, based on reactions with OH° (calculated).
atmospheric lifetime based on experiments and emission estimations: 36-50 years.
CCl_4 passes into the stratosphere where chlorine formation by photolytic decomposition contributes significantly to ozone depletion. (8518, 8519, 8520)

Manmade sources

in 4 municipal landfill gases in Southern Finland (1989-1990 data): avg 1.7-34.3, max. 88 mg/m^3. (2605)

Control methods

A.C.: retentivity: 45 wt % of adsorbent. (83)

Incinerability

temperature for 99% destruction at 2.0 sec. residence time under oxygen-starved reaction conditions: 645°C. Thermal stability ranking of hazardous organic compounds: rank 148 on a scale of 1 (highest stability) to 320 (lowest stability). (2390)

Ambient air concentrations

Background concentrations: over the sea: 0.26-0.95 μg/m$_3$ (1976-1985). (8522, 8523, 8524)

glc's in Los Angeles county: 0.12-1.63 ppb (46)
glc's in rural Washington: Dec. 1974-Feb. 1975: 120 ppt (315)

Inside and outside 15 living rooms in Northern Italy (1983-1984)

4-7 d averages	*lowest*	*mean*	*highest value (all in μg/m³)*	
indoor	<1	6	12	
outdoor	<1	7	22	(2756)

indoor/outdoor glc's winter 1981/1982 and 1982/1983 the Netherlands:			
μg/m³	*median*	*maximum*	
pre-war homes	<4	25	
post-war homes	<4	6	
homes <6 years old	<4	25	
outdoors	<4	20	(2668)

sampling *sampling date*	*measured* *sampling location*	*height, m*	*conc., ppt*	
December 1971	northern lat. 0-50°N	sea level	71 ± 6.8	(374)
January 1974	Greenland Sea and Arctic Ocean	30-5,500	173 ± 39	(376)
June/July 1974	W. Ireland	sea level	111 ± 11	(375)
October 1973	North Atlantic	sea level	138 ± 15	(375)
Dec. 1974/Feb. 1975	Pullman, WA	not reported	120 ± 15	(315)
March 1975	Farnborough Aberporth, U.K.	2,000-6,100	53 ± 8	(371)
March 1975	Boscombe Down Exeter, U.K.	3,000-6,100	66 ± 4	(371)
January 1976	Off SW Wales	900-7,300	66 ± 6	(371)
February 1976	Off NE England	400-5,200	64 ± 2	(371)

Partition coefficients

Cuticular matrix/air partition coefficient* 270 ± 20 (7077)
* experimental value at 25 °C studied in the cuticular membranes from mature tomato fruits (*Lycopersicon esculentum* Mill. cultivar Vendor)

1 mg/m³ = 0.16 ppm, 1 ppm = 6.39 mg/m³.

C. WATER AND SOIL POLLUTION FACTORS

Hydrolytic degradation

$t_{1/2}$: ca. 1,000 years (calculated) (8521)

BOD_5	0% ThOD std. dil. sew.	(275, 27)
ThOD	0.21.	(27)

Reduction of amenities

odor threshold: 50 mg/L. (84)

Manmade sources

In digested sewage sludge: U.K. 1993: *n* = 12
mean:
0.019 mg/kg dry wt
0.53 μg/L wet volume
SD:
0.028 mg/kg dry wt
0.73 μg/L wet volume (2946)

Water quality

waters from upland reservoirs in NW England (1974):
during dry cloudy weather: 1-2 ppb
during prolonged heavy rain: 13-24 ppb (933)

in river Maas at Eysden (Netherlands) in 1976: median: 1.2 μg/L; range: 0.1 to 3.7 μg/L

in river Maas at Keizersveer (Netherlands) in 1976: median: 1.3 µg/L; range: n.d. to 4.2 µg/L	(1368)
in drinking water (Frankfurt, Germany, 1977): average: 0.1 µg/L; max. 0.3 µg/L	(2960)
In German cities in 1977: average 0.035 µg/L (range <0.0001 to 0.5 µg/L), occasionally up to 4.1 µg/L.	(8526, 8527)

in drinking water treatment plants where pure chlorine is used for disinfection in the Netherlands: jan-march 1994 (ng/L)

Pumpstation	Enschede	Andijk	Berenplaat	Kralingen	
raw water (intake)	6	2.4	1	1.1	
after rapid filtration	< 0.6	3.3	1	0.9	
after activated carbon		2.1		< 0.6	
rein water (effluent)	< 0.6	2.1	1	< 0.6	(7060)

Background concentrations in soils

In a crop farming area (Vorharz, FRG):	at 0-20 cm depth	0.26 mg/kg soil dw.	
	at 50 cm depth	0.40 mg/kg soil dw.	
	at 60-100 cm depth	0.035 mg/kg dw.	(8525)

Background concentrations in river sediment

In Elbe river FRG	3.3 to 13 µg/kg dw.	(8525)

Background concentrations in food-stuffs

In many food-stuffs in the FRG: the highest values have been reported for	dairy products	<1-30 µg/kg	
	plant fats	5-16 µg/kg	
	fruits and vegetables	<1-29 µg/kg	
	wheat flour and grits	<1-62 µg/kg	(8529)

Aquatic reactions

measured evaporation from 1 ppm aqueous solution, still air and an average depth of 6.5 cm: at 25°C: 50% after 29 min, 90% after 97 min	(313, 369)
environmental hydrolysis $t_{1/2}$ at 25°C and pH 7: 40 years	(2434)
reductive dehalogenation in anoxic sediment with 6% organic carbon: $t_{1/2}$: 4.4 d at 22°C	(2842)

Waste water treatment

experimental water reclamation plant:

	sample 1	*sample 2*	
sand filter effluent	122 ng/L	19 ng/L	
after chlorination	685 ng/L	308 ng/L	
final water after A.C.	152 ng/L	20 ng/L	(928)

Dimensionless distribution coefficient at initial conc. = 1 mg/L:

on mixed liquor:	500	
on primary wastewater sludge:	340	
on anaerobically digested sludge:	200	(2395)

initial concentration, µg/L	*treatment method*	*% removal*	
3.3-12	river bank filtration	76-99	(2915)

Methanogenic laboratory column reactor at 35°C in the dark and a liquid detention time of 2 d, at influent conc. of 17 µg/L and 0 weeks acclimation: >99% removal at steady state.	(2959)

Impact on biodegradation processes

act. sludge respiration inhib.	EC_{50}	>100 mg/L	
OECD 209 closed system inhib.	EC_{50}	204 mg/L	(2624)

Degradation in soil

Disappearance in soil at 100 mg/kg and 20°C under aerobic laboratory conditions: $t_{1/2}$: 5.0 d; confidence limits: 2.6-55.6 d	(2366)

Degradation

$t_{1/2}$: 32.9 d in activated sludge	(2368)

Calculated half-life in nonpolluted anaerobic aquifer at varying organic carbon contents

organic carbon content	1%	0.1%	0.01%	0.001%	
average half-life	4 d	13 d	112 d	5 years	(2695)

Biodegradation

half-lives in nonadapted aerobic subsoil: sand Canada: >2,426 d

Biodegradation

Inoculum	*method*	*conc.*	*duration*	*elimination results*	
A.S. adapted	model sewer plant	19 mg/L		99% test substance*	(8530)
adapted methanogenic bacteria	flow through system	0.15 mg/L		>99% $ThCO_2$	(8531)
primary sewage effluent	denitrifying bacteria in static system	0.054 mg/L	14 d	60% test substance	
			21 d	97% test substance	
			28 d	99% test substance	
			42 d	99% test substance	
			56 d	90% test substance	(8532)

Biodegradation

Reductive dehalogenation

1. Sequential reductive dehalogenation
2. Substitutive transformation
3. Oxidative transformation
4. CO_2 assimilation by the acetyl-CoA pathway

Type 1 reactions were observed in *Desulfobacterium autotrophicum*. Type 1, 2 and 4 reactions were observed in *Acetobacterium Woodii* which in addition may catalyze type 3 reactions.

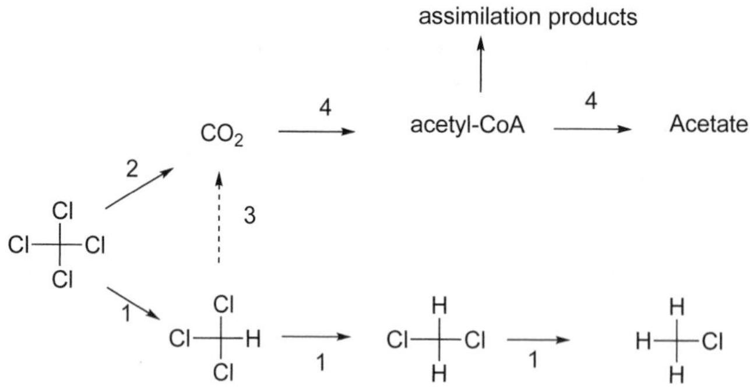

Proposed pathway of CCl_4 degradation in bacteria which utilize the acetyl-CoA pathway.

Proposed pathway of CCl_4 degradation in bacteria which utilize acetyl-CoA pathway

Acetobacterium Woodii converted 92% of the added $14CCl_4$ (40µM) to nonhalogenated products. Much of the initial radioactivity (67%) was recovered as CO_2, acetate, pyruvate, and cell material; the remainder comprised an unknown, hydrophobic material and dichloromethane (CH_2Cl_2).	(7427)

Carbon tetrachloride can be degraded under anaerobic conditions via three different pathways. No enzyme is involved in these pathways; rather, reactions are catalyzed by cofactors present in microorganisms. Carbon tetrachloride is degraded to chloroform, dichoromethane, chloromethane and ultimately methane by hydrogenolytic dechlorinations. Many anaerobic bacteria can catalyze the first two reactions of this pathway.

Carbon tetrachloride can also be converted to carbon monoxide in a reduction process in which chloroform is only a minor product. Carbon tetrachloride, in the right-most pathway, can be degraded via sulfur and oxygen substitution in one-electron reduction reactions (see figure).

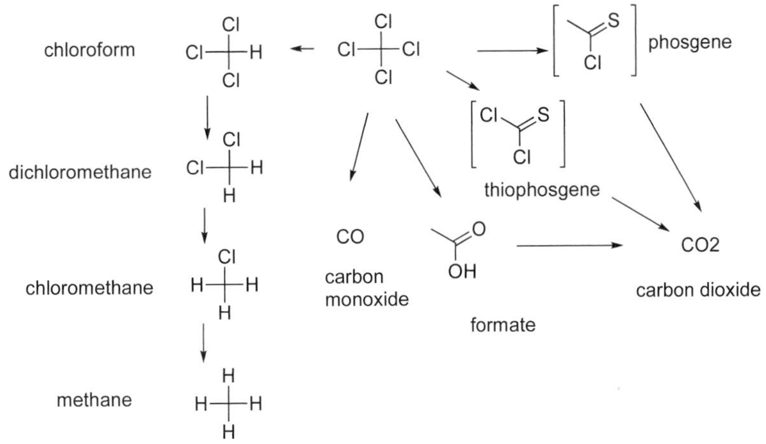

Anaerobic degradation pathway of carbon tetrachloride.

D. BIOLOGICAL EFFECTS

Residues

concentrations in various organs of molluscs and fishes collected from the relatively clean waters of the Irish Sea in the vicinity of Port Erin, Isle of Man (only organs with highest and lowest concentrations are mentioned-concentrations on dry weight basis):

Mollusc			
Baccinum undatum	muscle:	5 ng/g	
	digestive gland:	8 ng/g	
Modiolus modiolus	digestive tissue:	20 ng/g	
	mantle:	114 ng/g	
Pecten maximus	mantle, testis:	2-3 ng/g	
	ovary, gill:	14-16 ng/g	
Fishes			
Conger conger (eel)	gill:	3 ng/g	
	liver:	51 ng/g	
Gadus morhua (cod)	stomach, skeletal tissue, muscle, liver:	4-7 ng/g	
	brain:	29 ng/g	
Pollachius birens (coalfish)	muscle:	7 ng/g	
	gill: alimentary canal:	32-35 ng/g	
Scylliorhinus canicula (dogfish)	spleen:	3 ng/g	
	gill:	55 ng/g	
Trisopterus luscus (bib)	gut, liver:	16-18 ng/g	
	gill:	209 ng/g	(1092)

Bioconcentration

Fishes

Salmo gairdneri:	log BCF	1.24	(193)

Bluegill sunfish *(Lepomis macrochirus)* at 52.3 µg/L at 16 °C during 21 d: BCF: 30; $t_{1/2}$ in tissues: <1 d (2602)

Algae			
Chlorella fusca:	BCF (wet wt)	300	(2659)

Toxicity

Bacteria			
Anaerobic sludge inhibition of methanogenesis	EC_{100}	16 mg/L	(8542)
Activated Sludge Oxygen consumption inhibition	5d EC_{50}	>1,000 mg/L	(7047)
Photobacterium phosphoreum (Microtox test)	15min EC_{50}	5 mg/L	(7047)
Biotox™ test *Photobacterium phosphoreum*	5min EC_{20}	190 mg/L	
	5min EC_{50}	563 mg/L	
Microtox™ test *Photobacterium phosphoreum*	5min EC_{50}	729 mg/L	(7023)
Pseudomonas cepacia, Aeromonas hydrophila and Bacillus subtilis	MIC	750 mg/L	(8533)
Microtox™*(Photobacterium)* test:	5 min EC_{50}	997 mg/L	
Pseudomonas putida	16h EC_0	30 mg/L	(1900)
Protozoans			
Entosiphon sulcatum	72h EC_0	770 mg/L	(1900)
Uronema parduczi Chatton-Lwoff	EC_0	616 mg/L	(1901)
Algae			
Microcystis aeruginosa	8d EC_0	105 mg/L	(1329)
Scenedesmus quadricauda	7d EC_0	>600 mg/L	(1900)
Tetrahymena pyriformis	24h EC_{50}	830 mg/L	(2624)
Scenedesmus subspicata	72h EC_{50}	21 mg/L	(7047)
Haematococcus pluvialis	4h EC_{10}	>136 mg/L	(8537)
Crustaceans			
Artoxkit M™*(Artemia salina)* test:	24h LC_{50}	2,153 mg/L	
Streptoxkit F test *(Streptocephalus proboscideus)*	24h LC_{50}	6,429 mg/L	
Daphnia magna test:	24h LC_{50}	20,763 mg/L	
Rotoxkit F *(Brachionus calyciflorus)* test:	24h LC_{50}	5,798 mg/L	(2945)
Daphnia magna (in open glass beaker)	24h EC_{50}	>770 mg/L	(8538)
(in covered beaker)	48h NOEC	7.7 mg/L	
	48h EC_{50}	29 mg/L	
	24h EC_0	9 mg/L	
	24h EC_{50}	20; 28 mg/L	
	24h EC_{100}	159 mg/L	(8539)
Amphibians			
Bufo fowleri embryos	3d LC_{50}	>92 mg/L	
Bufo fowleri larvae	4d LC_{50}	2.8 mg/L	
Rana catebeiana embryos	4d LC_{50}	1.5 mg/L	
Rana catebeiana larvae	4d LC_{01}	0.024 mg/L	
	4d LC_{10}	0.11 mg/L	
	4d LC_{50}	0.9 mg/L	
Rana palustris embryos	4d LC_{50}	3.6 mg/L	
Rana palustris larvae	4d LC_{01}	0.11 mg/L	
	4d LC_{10}	0.44 mg/L	
	4d LC_{50}	2.4 mg/L	(8533)

Worms

Eisenia foetida	48h LC_{50}	0.16 mg/cm2 filter paper	(8543)
Tubifex tubifex	24h NOEC	100 mg/L	
	24h EC_{50}	ca. 600 mg/L	(8537)
Tubifex	48h LC_{50}	1,100-2,000 mg/L	(2624)

Fishes

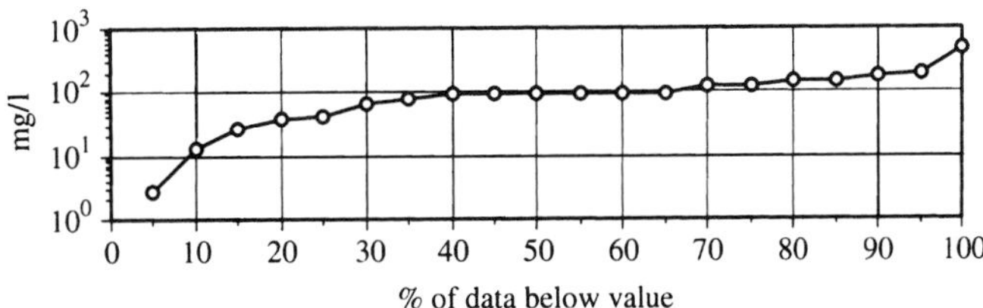

Frequency distribution of 24-96h LC_{50} values for fishes (n = 20) based on data from this work and from the work of Rippen (3999)

Guppy *(Poecilia reticulata)* 14d LC_{50}: 67 ppm (1833)

Lepomis macrochirus: static bioassay in fresh water at 23°C, mild aeration applied after 24 h:

material added, ppm	*% survival after 24h*	*48h*	*72h*	*96h*	*best fit 96h LC_{50}, ppm*
320	0	-	-	-	
200	0	-	-	-	
(narcosis) 125	70	60	60	50	125
(narcosis) 100	30	20	20	20	
75	100	100	100	100	(352)

Menidia beryllina: static bioassay in synthetic seawater at 23°C, mild aeration applied after 24 h:

material added, ppm	*% survival after 24h*	*48h*	*72h*	*96h*	*best fit 96h LC_{50}, ppm*
320	0	-	-	-	
180	100	80	60	50	
100	100	60	60	60	150
75	100	100	100	90	
					(352)

Leuciscus idus	48h LC_0	5; 16; 40; 272 mg/L	
	48h LC_{50}	13; 47; 95; 472 mg/L	
	48h LC_{100}	33; 143; 672 mg/L	(8534, 8536, 8537)
Limanda limanda	96h LC_{100}	50 mg/L	(8533)
Pimephales promelas	LC_{50}	43 mg/L	(8533)
Salmo gairdneri	14d LC_0	>80 mg/L	(8533)
Lepomis macrochirus	24h LC_{50}	38 mg/L	(8534)
	96h LC_{50}	27 mg/L	(8534)
Menidia beryllina	96h LC_{50}	150 mg/L	(8533)
Salmo trutta	14d LC_0	24 mg/L	
	14d LC_{100}	56 mg/L	(8533)
Killifish *Oryzias latipes*	48h LC_{50}	93 mg/L	(2624)
Bluegill *Lepomis macrochirus*	24h LC_{50},S	38 mg/L	
	96h LC_{50},S	27 mg/L	(2697)

carbonylchloride *see* phosgene

carbonylsulfide (carbonoxysulfide)

S=C=O

COS

CAS 463-58-1

OCCURRENCE

Volatile organic sulfur compounds, including carbonyl sulfide, are formed in marine and freshwater lake environments.

MANMADE SOURCES

Toxic gas encountered during petroleum refining or destructive distillation of coal. May liberate highly toxic hydrogen sulfide upon decomposition.

NATURAL SOURCES

Emitted as a reduction product of sulfur gases from vegetation and soils.

A. PROPERTIES

molecular weight 60.07; melting point -138°C; boiling point -50.2°C; vapor density 2.1; density 1.24 at -87°C, 1.03 at 17° C; solubility 1,000 cm^3/l.

B. AIR POLLUTION FACTORS

Odor

characteristic: typical sulfide odor except when pure.

Natural sources

biogenic carbonylsulfide emissions from soils in U.S.A.:

sampling sites	*soil orders*	*sampling dates*	*avg sulfur flux, g/S/m²/yr*
Wadesville, IN	Alfisol	9/20-10/3 1977	0.002
Philo, OH	Inceptisol	10/7-10/10 1977	0.0022
Cedar Island, NC	Saline swamp	10/19-10/28 1977	0.0016
Cox´s Landing, NC	Saline marsh	11/1-11/9 1977	6.36
Clarkedale, AR	Alluvial clay	11/16-11/20 1977	0.0014
Cedar Island, NC	freshly clipped marsh		0.013
Cox´s Landing, NC	freshly clipped marsh		0.0005

(1385)

C. WATER AND SOIL POLLUTION FACTORS

Hydrolysis

Hydrolysed in water to carbon dioxide and hydrogen sulfide. (9934)

Biodegradation

High concentrations of carbonyl sulfide were found in the hypolimnion of Lake Ciso, Spain. The concentrations dropped rapidly in the metalimnion dominated by plates of purple sulfur bacteria and *C. phaseolus* indicating a rapid microbial degradation of this compound, most likely by chemotrophic bacteria. (9933)

carbophenothion (S-((p-chlorophenylthio)methyl)O,O-diethylphosphorodithioate; S- (4-chlorophenylthiomethyl)diethylphosphorothiolothionate; trithion)

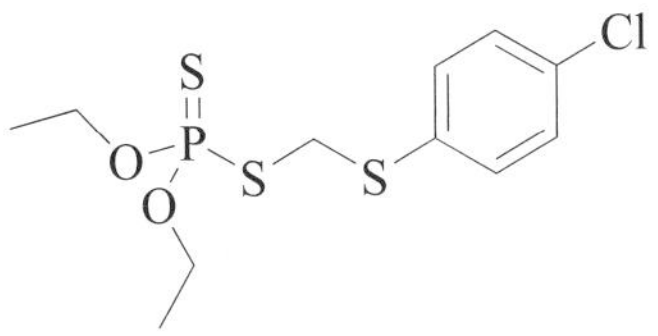

$C_{11}H_{16}ClO_2PS_3$

CAS 786-19-6

USES

insecticide, acaricide.

A. PROPERTIES

off-white to amber liquid with a mild, mercaptan-like odor; sp. gr. 1.275-1.3 at 20/20°C; boiling point 82°C at 0.01 mm; vapor pressure 3×10^{-7} mm at 20°C; solubility <40 mg/L.

C. WATER AND SOIL POLLUTION FACTORS

Degradation

50% degradation in soil occurs in 100 d or longer, depending on soil type. (1855)

Persistence in river water

in a sealed glass jar under sunlight and artificial fluorescent light-initial conc. = *10 μg/L:*

% of original compound found

after	*1 h*	*1 wk*	*2 wk*	*4 wk*	*8 wk*	
	90	25	10	0	0	(1309)

D. BIOLOGICAL EFFECTS

Crustacean

Gammarus lacustris	96h, LC_{50}	5.2 μg/L	(2124)
Palaemonetes kadiakensis	96h, LC_{50}	1.2 μg/L	(2126)
Asellus brevicaudus	96h, LC_{50}	1,100 μg/L	(2126)

Fishes			
Cyprinodon variegatus	96h LC_{50}	2.8 µg/L	
	MATC, early life stage	1.9 µg/L	(2643)
Menidia beryllina	96h LC_{50}	3 µg/L	
	MATC life cycle	0.76 µg/L	
Palaemonetes pugio	96h LC_{50}	2.9 µg/L	
	MATC life cycle	0.28 µg/L	(2643)

o-carboxybenzenesulfonic acid *see* 2-sulfobenzoic acid

p-carboxybenzenesulfonic acid *see* 4-sulfobenzoic acid

1-(carboxymethyl)pyridinium chloride

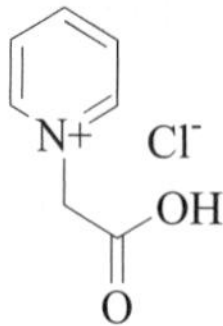

CAS 6266-23-5

EINECS 228-434-8

A. PROPERTIES

molecular weight 175.6

D. BIOLOGICAL EFFECTS

Fishes

Pimephales promelas	96h LC_{50}	162 mg/L	(2625)

carboxymethyltartronate (CMT; trisodium-2-oxa-1,1,3-propanetricarboxylate, trisodiumcarboxymethyltartronate)

COONa
NaOOC
O
COONa

$C_5H_3Na_3O_7$

IMPURITIES, ADDITIVES, COMPOSITION commercial-grade CMT contains:

78% trisodium carboxymethyltartronate
4% tetrasodium ditartronate
8% disodium diglycolate
10% water

C. WATER AND SOIL POLLUTION FACTORS

Screening tests for CMT biodegradability

screening test	*CMT conc., mg/L*	*acclim. time, weeks*	*% biodegradation*	
semi-continuous A.S.	50	4-8	>95	
river water	5	4-6	>95	
CO_2 evolution	20	-	15-40*	
	20	-	65-90**	(1072)

* raw sewage employed as inoculum
** acclimated activated sludge mixed liquor or soil suspension (0.5%) used as inoculum

6-carboxyyuracil *see* orotic acid

2-carene (3,7,7-trimethylbicyclo[4.1.0]hept-2-ene)

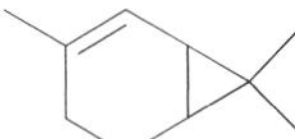

$C_{10}H_{16}$

CAS 4497-92-1

A. PROPERTIES

molecular weight 136.23

B. AIR POLLUTION FACTORS

Calculated tropospheric lifetimes

reactant	*lifetime*	
OH°	2.3 h	
ozone	1.7 h	
NO_3^0	36 min	(2451)

3-Carene (3,7,7-trimethylbicyclo[4.1.0]hept-3-ene)

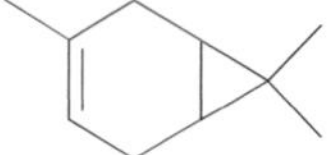

$C_{10}H_{16}$

CAS 498-15-7

CAS 13466-78-9

A. PROPERTIES

molecular weight 136.24; melting point 5°C; boiling point 170-172°C; density 0.865.

B. AIR POLLUTION FACTORS

Natural sources

emitted from the leaves of *Cupressus sempervirens* : 0.24 % of total terpene emissions which varied between 3-35 µg/g dw.h. The major component of the emitted terpenes was limonene which constituted 83% of total terpene emissions. (7049)

Calculated tropospheric lifetimes

reactant	*lifetime*	
OH°	2.1 hr	
ozone	10 hr	
NO_3^0	1.1 hr	(2451)

Pyrolysis

Thermal degradation products formed in the presence of air at a temperature of 120 °C, mimicing the conditions of wood drying such as occurring during drying of wood flakes, wood veneers or lumber. The degradation products resulted from the following reactions:
·Dehydrogenation to form aromatic systems;
·Oxidative cleavage of the carbon-carbon bonds;
·Epoxide formation;
·Allylic oxidation (oxidation of the carbon adjacent to the carbon-carbon double bond) to form alcohols, ketones and aldehydes;
·Carbon skeleton rearrangements of the parent or oxidised compounds. (7088)

The following thermal degradation products of Δ-carene were identified:

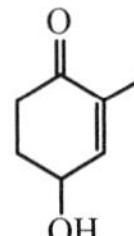

4-hydroxy-2-methyl-2-cyclohexenone

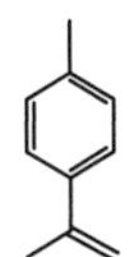

4-methyl-1-(1-methylethenyl)benzene
[p-cymene]

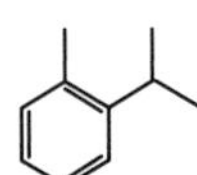

1-methyl-2-(1-methylethenyl)benzene
[o-cymene]

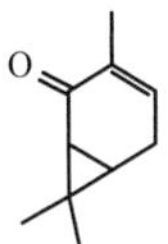

3,7,7-trimethylbicyclo[4.1.0]hept-3-en-2-one
[3-caren-2-one]

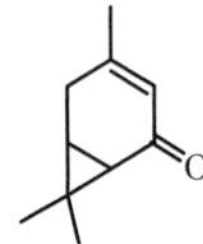

4,7,7-trimethylbicyclo[4.1.0]hept-3-en-2-one
[3-caren-5-one]

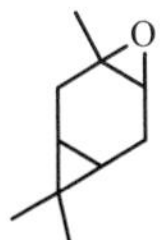

3,4-epoxy-3,7,7-trimethylbicyclo[4,1,0]hept-3-en-2,5-dione
[3-carene-2,5-dione]

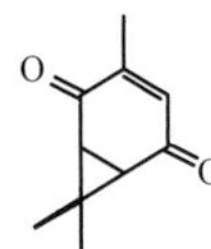

3,7,7-trimethylbicyclo[4.1.0]hept-3-en-2,5-dione
[3-caren-2,5-dione]

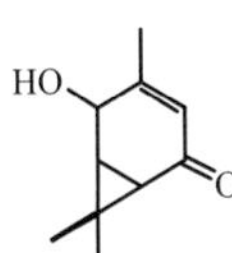

trans-2hydroxy-3-caren-5-one

C. WATER AND SOIL POLLUTION FACTORS

Manmade sources

Göteborg (Sweden) sew. works 1989-1991: infl.: 0-5 µg/l; effl.: n.d. (2787)

4-carene (4,7,7-trimethylbicyclo[4.1.0]hept-2-ene)

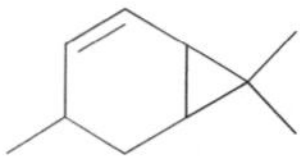

$C_{10}H_{16}$

CAS 5208-50-4

A. PROPERTIES

molecular weight 136.23

C. WATER AND SOIL POLLUTION FACTORS

Manmade sources

Göteborg (Sweden) sew. works 1989- 1991: infl.: 0-5 µg/L; effl.: n.d. (2787)

Carfentrazone-ethyl (Ethyl (RS)-2-chloro-3-[2-chloro-5-(4-difluoromethyl-4,5-dihydro-3-methyl-5oxo-1*H*1,2,4-triazol-1-yl)-4-fluorophenyl] propionate; Ethyla,2-dichloro-5-[4-(difluoromethyl)-4,5-dihydro-3-methyl-5-oxo-1*H*-1,2,4-triazol-1-yl]-4-fluorobenzenepropanoate)

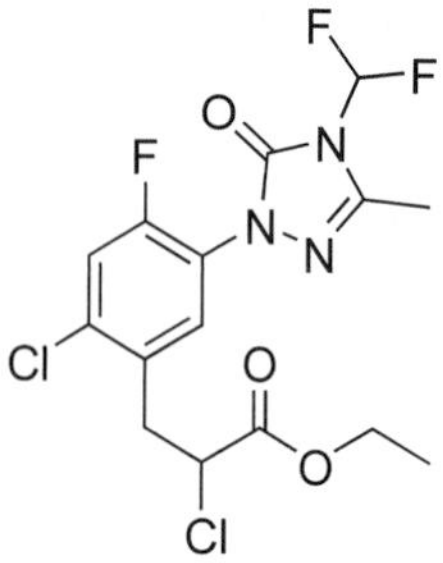

$C_{15}H_{14}Cl_2N_3O_3F_3$

CAS 128639-02-1

TRADENAMES

Aim; Affinity 400 DF Herbicide

USES

An aryl triazolinone. Carfentrazone-ethyl is a post-emergence herbicide and is registerd in the USA for use on wheat, corn and soybeans.

A. PROPERTIES

molecular weight 412.19; melting point –22.1°C; boiling point 350-355°C; density 1.46 kg/L at 20°C; vapour pressure 7.2 x 10^{-6} Pa at 20°C; water solubility at pH 7 : at 20°C: 12 mg/L; at 25°C: 22 mg/L; at 30°C: 23 mg/L; log Pow 3.36 at 20°C; H 2.5 x 10^{-4} Pa.m^3/mol

C. WATER AND SOIL POLLUTION FACTORS

Hydrolysis

T/2 at 20°C: at pH 7 : 13.7 hours; at pH 9: 5.1 hours; at pH 5 : stable. Relevant metabolite = F8426-chloropropionic acid which is hydrolytically stable. (10885)

Photodegradation in water

Light t/2 = 8.3 days. Relevant metabolite : 5-hydroxy-derivates of carfentrazone. (10885)

Biodegradation (10885)

	% mineralisation	% bound-residues	After
Biodegradation in soil aerobic	<3%	Max 15%	110 days

Analyses of metabolites indicated that under aerobic test conditions, carfentrazone-ethyl hydrolysed to F8426-chloropropionic acid and transformations then followed on the acid side- chain to form F8426-propionic acid, F8426-cinnamic acid and F8426-benzoic acid. Hydroxylation of the methyl group on the triazolinone ring also occurred. Under anaerobic conditions, F8426-chloropropionic acid and F8426-propionic acid were formed, but further changes to the acid side-chain occurred very slowly and hydroxylation of the methyl group on the triazolinone ring was absent. No ring separation metabolites were identified and mineralisation of the molecule to CO_2 occurred very slowly (<3% of applied after 12 months). (10888)

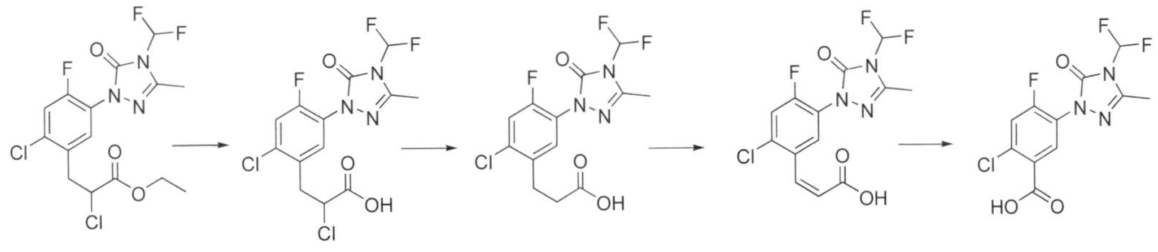

Carfentrazone-ethyl (F8426) → F8426-chloropropionic acid → F8426-propionic acid → F8426-cinnamic acid → F8426-benzoic acid

Aerobic degradation pathway of Carfentrazone-ethyl in soil [10888]

(10888)

Biodegradation in laboratory studies (10885)

degradation in soils	%	Carfentrazon-ethyl	F-8426-Chloropropionic acid
Aerobic at 20°C	50	<0.1-<1.3 days	11-86 days
Aerobic at 20°C	90	<0.5-299 days	
Aerobic at 10°C	50	0.1 day	92 days
Anerobic at 20°C	50	<1 day	>100 days

Biodegradation in water/sediment systems (10885)

DT50 water	<0.4 days
DT90 water	<1.2 days
DT50 whole system	<0.4 days
DT90 whole system	<1.2 days

Dissipation studies in the field at 20°C (10885)

	DT 50	DT 90
F8426-chloropropionic acid	3-14 days	11-47 days
F8426-cinnamic acid	5-29 days	17-97 days
F8426-benzoic acid	11-31 days	37-104 days
F8426-propionic acid	no residue after 57 days	

Mobility in soils (10885)

	Carfentrazon-ethyl	Chloropropionic acid	Benzoic acid	Cinnamic acid	Propionic acid
K_{OC}	Not applicable	7.4-46	4-41	44-333	27-260

D. BIOLOGICAL EFFECTS

Bioaccumulation

Species	conc. µg/L	duration	body parts	BCF	
Trout			Whole fish	176	
			Edible parts	34	
			Non-edible parts	379	(10885)

Toxicity

Micro organisms			
Nitrogen mineralization	61d IC15	>0.52 mg/kg soil	
Carbon mineralization	28d IC15	>0.52 mg/kg soil	(10885)
Algae			
Anabaena flos-aquae	72h EC_{50}	0.012 mg/L	(10885)
Aquatic plants			
Lemna gibba	14d EC_{50}	0.0057 mg/L	(10885)
Worms			
Eisenia foetida	15d EC_{50}	>820 mg/kg soil	(10885)
Crustaceae			
Daphnia magna	48h EC_{50}	>9.8 mg/L	
	21d NOEC	0.22 mg/L	(10885)
Mysid shrimp Americamysis bahia	96h LC_{50}	1.16 mg/L	(10886)
	48h EC_{50}	>9.8 mg/L	
	96h NOEC	0.62 mg/L	
Insecta			
Chironomus riparius	21d NOEC	7.4 mg/L	
Honeybees	Acute oral LD_{50}	>200 µg/bee	
	Acute contact LD_{50}	>200 µg/bee	(10885)
Birds			
Bobwhite quail	Acute oral LD_{50}	>2,250 mg/kg bw	(10885)
Mollusca			
Crassostrea virginica	96h LC_{50}	2.05 mg/L	(10885)
	96h NOEC	0.62 mg/L	
Fish			
Trout	96h LC_{50}	1.6 mg/L	
	96h NOEC	1.2 mg/L	
	28d NOEC	0.11 mg/L	(10885)
Bluegill sunfish	96h LC_{50}	2.0 mg/L	(10886)
	96h NOEC	1.5 mg/L	
Tidewater silverside fish	96h LC_{50}	1.14 mg/L	(10886)
	96h NOEC	0.44 mg/L	
Oncorhynchus mykiss	96h LC_{50}	62.8 mg/L	(10887)
Mammals			
Rat	oral LD_{50}	>5,000 mg/kg bw	(10885)

Toxicokinetics and Metabolism

After oral dosing, carfentrazone-ethyl appeared rapidly in both plasma and red blood cells of rats and mice. Carfentrazone-ethyl was eliminated rapidly from plasma and red blood cells. Most of an orally administered dose was excreted in the urine rather than the faeces. Two major metabolites were identified (carfentrazone-ethyl-chloropropionic acid and 3-hydroxymethyl- carfentrazone-ethyl-chloropropionic acid) in the plasma of mice and rats. Six metabolites were identified in excreta; carfentrazone-ethyl-chloropropionic acid, 3-hydroxymethyl- carfentrazone-ethyl-chloropropionic acid, 3-hydroxymethyl-carfentrazone-ethyl-propionic acid, carfentrazone-ethyl-cinnamic acid, 3-hydroxymethyl-carfentrazone-ethyl-cinnamic acid, and carfentrazone-ethyl-propionic acid in descending order. A maximum of 2.8% of an oral dose was excreted unchanged. (10888)

Carpropamid ((1R,3S)-2,2-dichloro-N-[(R)-1-(4-chlorophenyl)ethyl]-1-ethyl-3-methylcyclopropane-carboxamide, (1S,3R)-2,2-dichloro-; 2,2-dichloro-*N*-[1-(4-chlorophenyl)ethyl]-1-ethyl-3-methylcyclopropanecarboxamide)

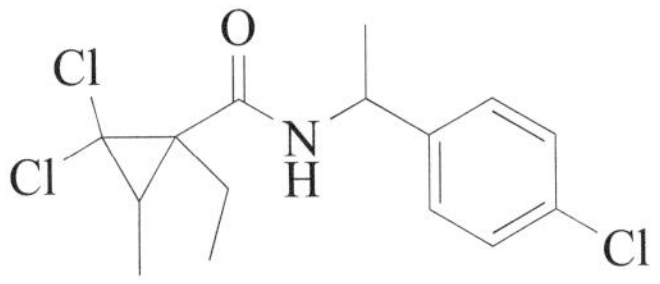

$C_{15}H_{18}Cl_3NO$

CAS 104030-54-8

TRADENAMES

Win®

USES

A rice fungicide for controlling rice blast, a dangerous disease caused by the fungus *Pyricularia oryzae*.

A. PROPERTIES

Melting point 152°C; solubility very low; log Pow 4.25

B. AIR POLLUTION FACTORS

Photodegradation

T/2 : >150 days

C. WATER AND SOIL POLLUTION FACTORS

Hydrolysis

T/2 : > 1 year

Photodegradation in water

t.2 = 7 days at pH 5 and 25°C

Biodegradation

The half-life (DT50) was found to be 120 days in the Toyama soil and 220 days in the Tochigi soil. As in the metabolic studies in plants and animals, the essential metabolites were products of oxidative transformation, among which the only one identified, other than carbon dioxide, was the carboxylic acid. The intermediate metabolites alcohol and phenol are obviously degraded to the carboxylic acid or CO_2 faster than in plants or animals.(10939)

Proposed pathway of degradation of carpropamid in paddy soils (10939)

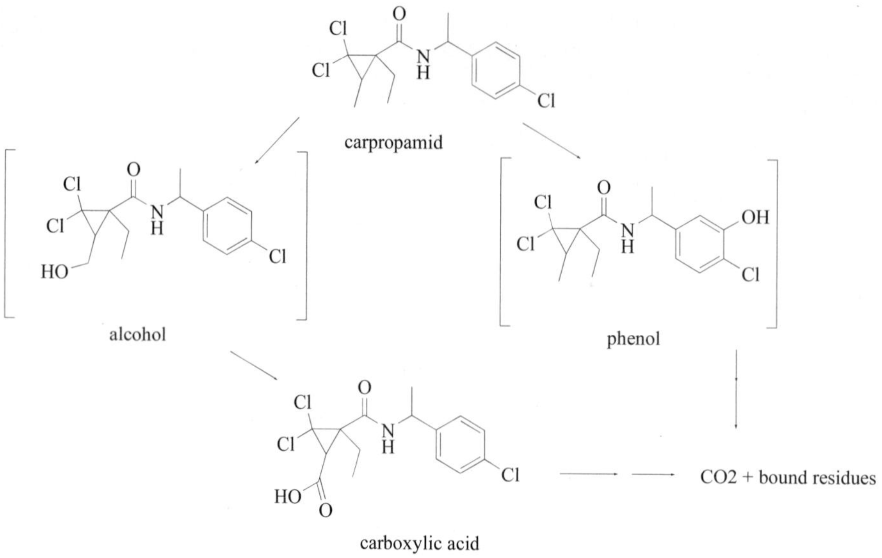

Proposed pathway of degradation of carpropamid in paddy soils

Mobility in soils

K_{OC}	574-1412	(10939)

Proposed pathway of degradation of carpropamid in paddy soils (11033)

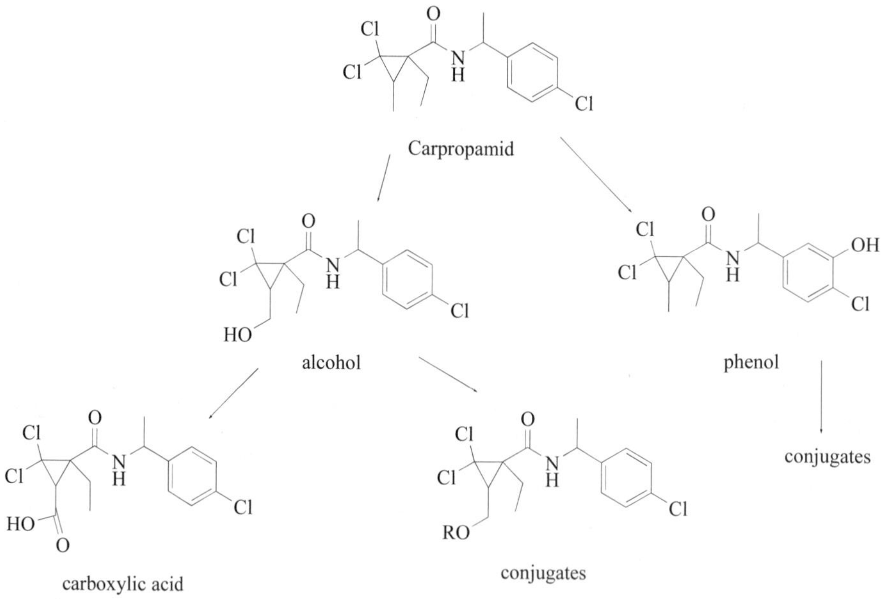

Poposed degradation pathway of carpropamid in rice [11033]

cartap (S,S'-2-dimethylaminotrimethylene bis-thiocarbamate hydrocic acid-5,5-(2-(dimethylamino)trimethylene)ester hydrochloride; S,S'-(2-Dimethylaminotrimethylene)bis(thiocarbamate))

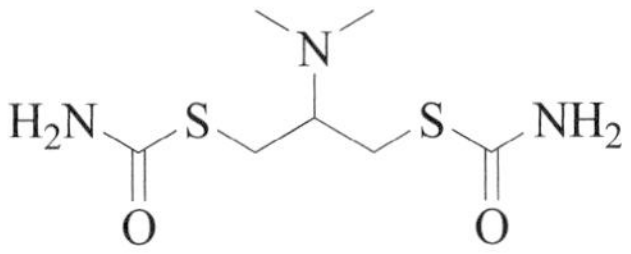

$C_7H_{15}N_3O_2S_2$.HCl

CAS 22042-59-7

A. PROPERTIES

molecular weight 237.35; melting point 179-181°C; solubility 200,000 mg/L at 25°C.

C. WATER AND SOIL POLLUTION FACTORS

Degradation in soil

aerobic biodegradation half-life: 0.22 d, at an initial conc. of 2 ppm, using a mixture of nonacclimated sludge, field soil, and river sediment as inoculum, in a stirred flask at 18-22°C, the major bacterial types in the activated sludge have been characterized as follows: *Flavobacterium* 35%; *Alcaligenes* 30%; *Corynebacterium* 20%; *Pseudomonas* 15%. (2690)

Aquatic reactions

alkaline hydrolysis half-life: 0.25 d, at pH 9 at 40°C in the dark. (2690)

It is stable under acidic conditions but hydrolyzes in neutral or alkaline media.

D. BIOLOGICAL EFFECTS

Fishes

carp	48h LC_{50}	1.3 mg/L	(2962)

Cashmeran (1,2,3,5,6,7,-hexahydro-1,1,2,3,3,-pentamethyl-4,H-inden-4-one; 6,7-dihydro-1,1,2,3,3-pentamethyl-4(5H)indanone; DPMI)

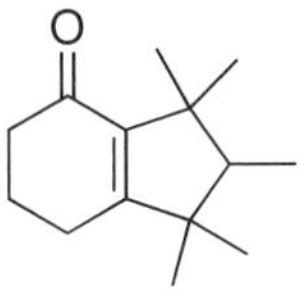

CAS 33704-61-9

EINECS 251-649-3

USES

a polycyclic musk. The polycyclic musks are used as fragrance ingredients in consumer products like cosmetics and detergents and cleaning agents. They are important ingredients in fragrances because

of their typical musky scent and their fixative properties.
The substance is mixed with other fragrance ingredients into fragrance oils or 'compounds'. A compound may consist of as many as 50 ingredients and a compounder may produce a large number of specified recipes out of the 2000 – 3000 different fragrance ingredients. These fragrance compounds are used in formulating products such as cosmetics, detergents etc. The concentration of fragrances in detergents and soap ranges from 0.2 to 1%. (see also musk fragrances)

A. PROPERTIES

molecular weight 206.3; vapour pressure 5.2 Pa; water solubility 0.17 mg/L; log Pow 4.9; H 9.9 Pa m^3/mol

B. AIR POLLUTION FACTORS

Odour

diffusive, spicy, musky-like, odour with strong floral reinforcement. powdery, velvet nuance.

C. WATER AND SOIL POLLUTION FACTORS

Concentrations in sludge of Sewage water treatment plants in Nordic countries in 2002

(10586)

Country	n	Range µg/kg dw	Median µg/kg dw
Denmark	5	0-54	35
Finland	5	<12-59	29
Iceland	4	n.d.	n.d.
Norway	5	0-128	n.d.
Sweden	8	0-17	<12

D. BIOLOGICAL EFFECTS

Concentrations in fish in Nordic countries in 2002 (10586)

Sampling site	Sample type	Extractable lipid %	ng/g lipid
Denmark Fish ponds	Trout 50 samples	2.23	0-9523

Concentrations in breast milk in Nordic countries in 2002 (10586)

Country	n	ng/g lipid
Denmark	10	n.d
Sweden	30-44	n.d

Cassiffix (3-cyclohexene-1-methanol.3(or 4)-methyl-1-(2,2,3-trimethyl-3-cyclopenten-1-yl)-, acid-isomerised)

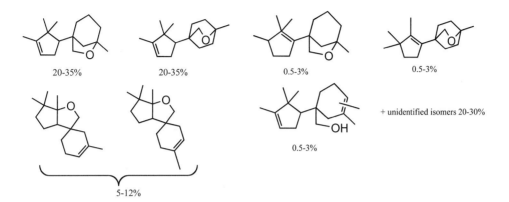

20-35% 20-35% 0.5-3% 0.5-3%

+ unidentified isomers 20-30%

0.5-3%

5-12%

$C_{16}H_{26}O$

CAS 426218-78-2

TRADENAMES

Cassiffix

USES

A fragrance in a variety of consumer products at 0.01-0.1% such as alcoholic perfumery, cosmetics, toiletries, household products, soaps and detergents.

EXPOSURE

The chemical is used in the formulation of numerous products, which will be available to the general public. Public exposure will be widespread and will result through the use of consumer products containing up to 0.1% Cassiffix. Members of the public will make dermal contact and possibly accidental ocular and/or inhalation exposure with products containing Cassiffix.

A. PROPERTIES

Clear colorless liquid; molecular weight 234; freezing point <-25°C; boiling point 301-309°C; relative density 0.99 kg/L at 20°C; vapour pressure 0.015 kPa at 25°C; water solubility 11 mg/L at 20°C; log Pow >3.7

C. WATER AND SOIL POLLUTION FACTORS

Hydrolysis

At 25°C: t/2 = 710 hours at pH 4; 520 hours at pH 7 and 830 hours at pH 9

Biodegradation

Inoculum	method	concentration	duration	elimination	
A.S.	OECD 301D		15 days	0%	
			28 days	3%	(10863)

D. BIOLOGICAL EFFECTS

Toxicity

Micro organisms

Activated sludge respiration	3h IC_{50}	>100 mg/L

	3h NOEC	100 mg/L	(10863)
Algae			
Selenastrum capricornutum	72h ECb_{50}	8.6 mg/L	
	72h NOEC	2.6 mg/L	(10863)
Fish			
Oncorhynchus mykiss	96h LC_{50}	3.8 mg/L	
	96h NOEC	0.52 mg/L	(10863)
Mammals			
Rat	oral LD_{50}	>2,000 mg/kg bw	
	dermal LD_{50}	>2,000 mg/kg bw	(10863)

Castor oil

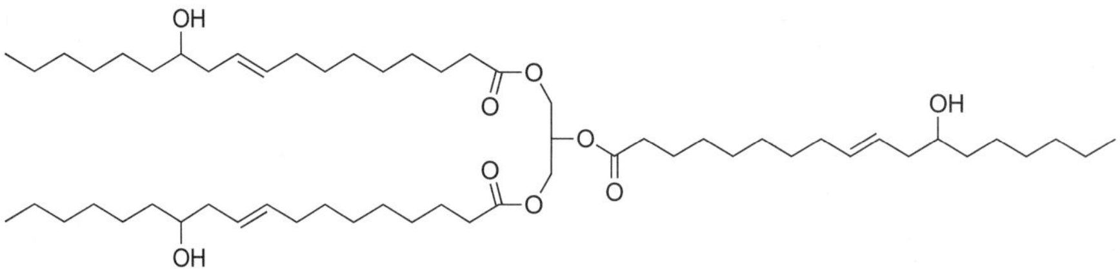

OH OH OH

CAS 8001-79-4

USES

It was used as a laxative or purgative or a vermifuge (to expel intestinal worms). Historically, it was used for lighting by means of torches and candles.
Today castor oil and its derivatives have many uses such as: Polyamide 11 (Nylon 11) engineering plastic, lubricating grease, coatings, inks, sealant, aircraft lubricants, surfactants emulsifiers , encapsulants, plastic films, plasticizer for coatings, and components for shatterproof safety glass. Castor oil has even made its way into cosmetics and related products. The ester linkages, double bonds and the hydroxyl groups in castor oil provide reaction sites for the preparation of many useful derivatives. Castor oil is unique among all fats and oils in that it is the only source of an 18-carbon hydroxylated fatty acid with one double bond (Ricinoleic acid).
Castor oil, owing to its chemical structure can be used as a bio-fuel in place of petrol-based fuels

SOURCES

oil of the seeds of the plant *Ricinus communis L.* native from Central Africa. The plant is cultivated for the production of industrial oils.

IMPURITIES, ADDITIVES, COMPOSITION

The composition of castor fatty acids is shown below

90% Ricinoleic acid

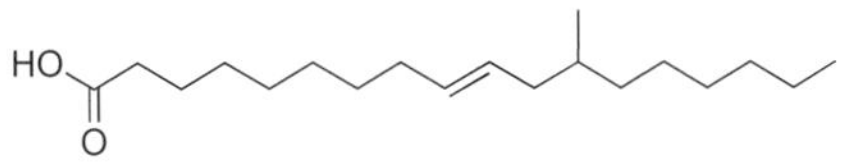

4.2% Linoleic acid

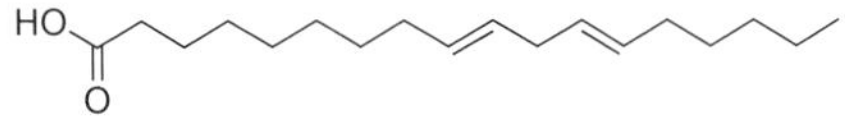

3% Oleic acid

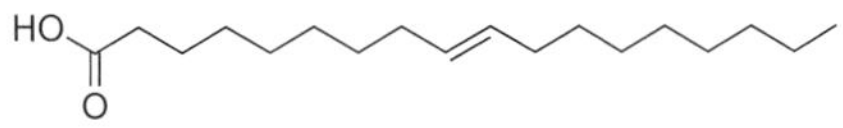

1% Stearic acid

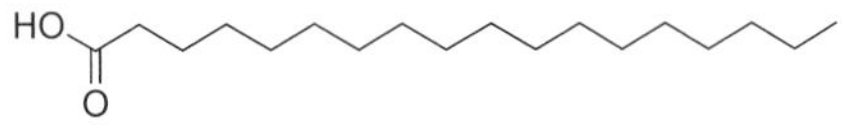

1% Palmitic acid

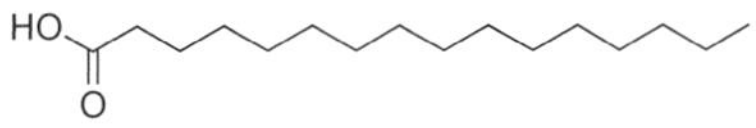

0.7% dihydroxystearic acid

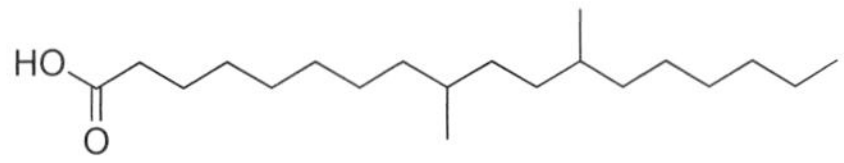

0.3% Linolenic acid

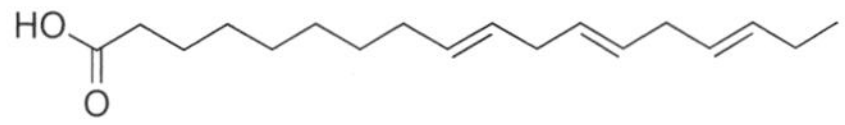

0.3% Eicosanoic acid

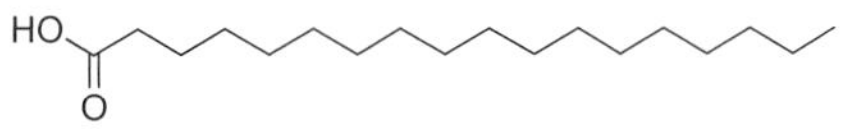

A. PROPERTIES

Castor oil consists almost entirely of the triglycerides ricinoleic acid. It is ranges in colour from colourless to greenish. It is a viscous liquid. It has a faint but characteristic odour. It has a slightly acrid taste and leaves a nauseating after taste.

Castor oil monomaleate (castoryl maleate; ricinoleyl manomaleate triglyceride; maleated castor oil)

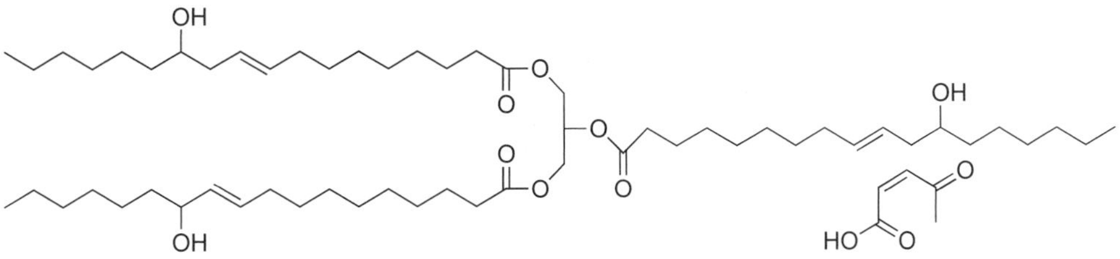

CAS 241153-84-4

TRADENAMES

Ceraphyl MTE; Ceraphyl RMT

USES

a skin conditioning aid to be used in rins-off personal care consumer products. The finished product contains up to 2.5% of the chemical.

658 Castor oil monomaleate

ENVIRONMENTAL FATE

As the product will be rinsed off upon use, most of the chemical will be disposed of via domestic wastewater treatment plant.

IMPURITIES, ADDITIVES, COMPOSITION

Castor Oil, also known as ricinus oil, is a triglyceride of fatty acids which occurs in the seed of the castor plant, *Ricinus communis* (India, Brazil).
The composition of castor oil fatty acids is shown below:

Castor Oil is unique among all fats and oils in that:

- it is the only source of an 18-carbon hydroxylated fatty acid with one double bond
- ricinoleic acid (12-Hydroxyoleic Acid) comprises approximately 90% of the fatty acid composition

The ester linkages, double bonds and hydroxyl groups in castor oil provide reaction sites for the preparation of many useful derivatives.

89.5% Ricinoleic acid

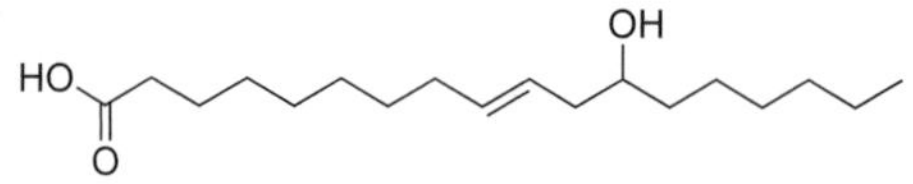

4.2% Linoleic acid

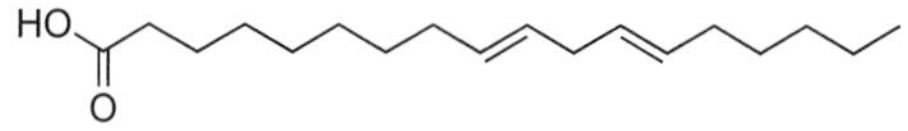

3.0% Oleic acid

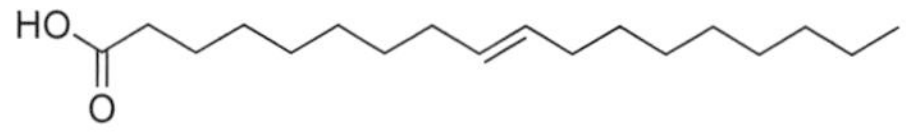

1.0% Stearic acid

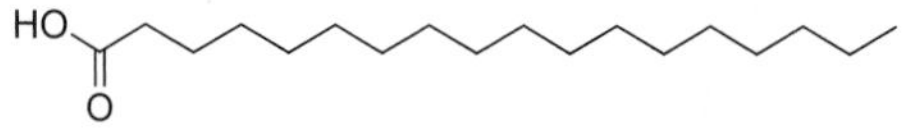

1.0% Palmitic acid

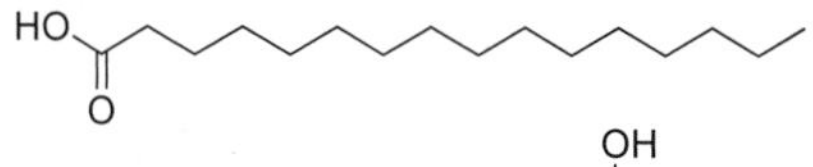

0.7% Dihydrostearic acid

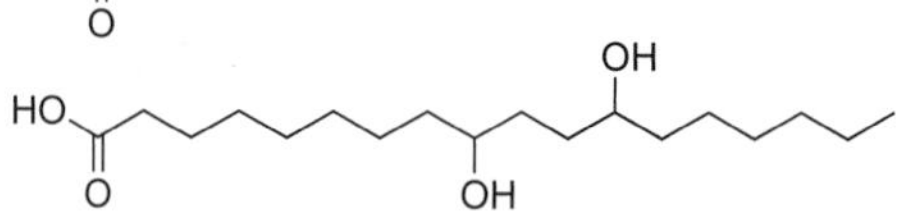

0.3% Linolenic acid

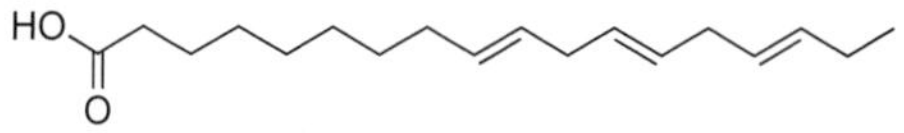

0.3% Eicosanoic acid

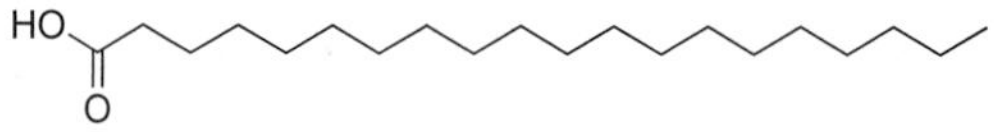

A. PROPERTIES

yellow oily viscous liquid; molecular weight ; melting point <-50°C; boiling point 337.7 °C; density 0.99 at °C; vapour pressure 2.3 x 10^{-12} kPa at 25°C; water solubility 6.2 mg/L at 20°C; $LogP_{OW}$ >3 (estimation)

C. WATER AND SOIL POLLUTION FACTORS

Biodegradation

Inoculum	method	concentration	duration	elimination	
Activated sludge	CO2 evolution test	15 mg OC/L	28 days	71.5 %ThCO2	(7418)

the chemical is readily biodegradable (see also graph). (7418)

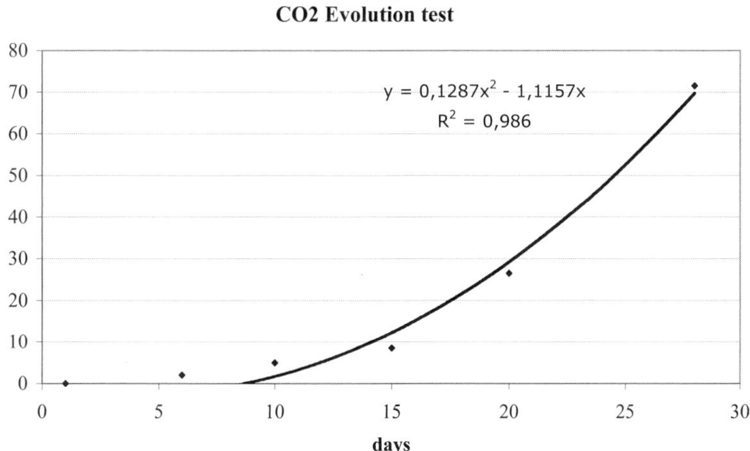

D. BIOLOGICAL EFFECTS

Toxicity

Micro organisms			
Activated sludge	3h IC_{50}	>1,000 mg/L	
	3h NOEC	1,000 mg/L	(7418)
Algae			
Selenastrum capricornutum	72h ECb_{50}	>5.63 mg/L	
	72h NOECb	>5.63 mg/L	
	72h $ECgr_{50}$	>5.63 mg/L	
	72h NOECgr	>5.63 mg/L	(7418)
Crustaceae			
Daphnia magna	48h EC_{50}	>10.2; >12 mg/L	
	48h LOEC	10.2; 12 mg/L	(7418)
Fish			
Oncorhynchus mykiss	96h LC_{50}	>6 mg/L	
	96h NOEC	6 mg/L	(7418)
Mammals			
Rat	oral LD_{50}	>5,000 mg/kg bw	(7418)

catechol (1,2-dihydroxybenzene; 1,2-benzenediol; pyrocatechol; pyrocatechin)

OH
OH

$C_6H_4(OH)_2$

$C_6H_6O_2$

CAS 120-80-9

USES

Important uses include the preparation of dyes and pharmaceuticals and the production of antioxidants for rubber and lubricatory oils. It is also used in photography, in rubber, in fur dyeing, and in specialty inks as an agent for oxygen removal.

A. PROPERTIES

colorless leaflets; molecular weight 110.11; melting point 105°C; boiling point 240°C decomposes; vapor density 3.79; sp. gr. 1.371 at 15°C; solubility 451,000 mg/L at 20°C; $LogP_{OW}$ 0.88/1.01.

B. AIR POLLUTION FACTORS

1 mg/m^3 = 0.22 ppm, 1 ppm = 4.50 mg/m^3.

Manmade emissions

avg conc. in smoke from 28 woodstoves and fireplaces:

burning hardwood:	36 g/kg carbon collected on filter	
softwood:	12 g/kg carbon collected on filter	(2414)

C. WATER AND SOIL POLLUTION FACTORS

Biodegradation rates

adapted A.S. at 20°C-product is sole carbon source: 96% COD removal at 55 mg COD/g dry inoculum/h. (327)

Biodegradation

decomposition by a soil microflora: 1 d. (176)

catechol
'ortho'cleavage
'meta'cleavage
cis,cis-muconate
(+)-muconolactone
3-oxoadipate-enol-lactone
3-oxoadipate
$CoAS\text{-}CO\text{-}CH_2\text{-}CO\text{-}CH_2\text{-}CH_2\text{-}COOH$
3-oxoadipyl CoA
$CH_3\text{-}COO\text{-}S\text{-}CoA$ + $HOOC\text{-}CH_2\text{-}CH_2\text{-}COOH$
acetyl-CoA succinate
2-hydroxymuconic semialdehyde
H_2O → H-COOH
$CH_2{=}CH\text{-}CH_2\text{-}CO\text{-}COOH$
2-oxo-pent-4-enoate
$CH_3\text{-}CH(OH)\text{-}CH_2\text{-}CO\text{-}COOH$
4-hydroxy-2-oxovalerate
$CH_3\text{-}CHO$ + $CH_3\text{-}CO\text{-}COOH$
acetaldehyde pyruvate

Degradation in soil system

at initial conc. of 500 mg/kg soil: $t_{1/2}$: 0.2 d (2903)

COD: 100% ThOD. (27)

ThOD: 1.89. (27)

Reduction of amenities

Taste in fish (carp): at 2.5 mg/L (41)
Tainting of fish flesh: 2-5 mg/L (81)
Odor threshold (detection): 8.0 mg/L (998)

Waste water treatment

A.S.: % ThOD at 250 mg/L catechol after 30 min, acclimated to the following aromatics:

phenol:	13% ThOD
o-cresol:	34% ThOD
m-cresol:	22% ThOD
p-cresol:	30% ThOD
benzylalcohol:	22% ThOD
mandelic acid:	36% ThOD
anthranilic acid:	10% ThOD

A.S. W, 20 C, 1/2 d observed, feed: 500 mg/L, 25 d acclimation: 26% ThOD. (93)

Ozonation

catechol + $O_3 \rightarrow$ *o*-quinone.

$$o\text{-quinone} + O_3 + H_2O \longrightarrow \text{(COOH, COOH diacid)} + O_2 + {}_4O$$
$$\downarrow$$
$$4O_2 + 2\ \begin{matrix}COOH\\|\\COOH\end{matrix} \xleftarrow{+4O_3} \text{(COOH, COOH)} + \begin{matrix}COOH\\|\\COOH\end{matrix}$$

(251)

Autooxidation at 25°C

$t_{1/2}$: 447 h at pH 7.0, 412 h at pH 9.0. (1908)

Manmade sources

in secondary domestic sewage plant effluent: 0.001 mg/L. (517)

D. BIOLOGICAL EFFECTS

Bacteria			
E. coli	LD_0	90 mg/L	(30)
Biotox™ test *Photobacterium phosphoreum*	5min EC_{20}	305 mg/L	
	5min EC_{50}	591 mg/L	
	15min EC_{20}	403 mg/L	
Microtox™ test *Photobacterium phosphoreum*	5min EC_{20}	19 mg/L	
	5min EC_{50}	67 mg/L	
	15min EC_{20}	18 mg/L	
	15min EC_{50}	46 mg/L	(7022)
Algae			
Scenedesmus	LD_0	6 mg/L	(30)
Protozoa			
Vorticella campanula	LD_0	1.6 mg/L	(30)
Paramecium caudatum	LD_0	35 mg/L	(30)
Arthropoda			

Daphnia	LD_0	4 mg/L	(30)
Fishes			
Trutta iridea	perturbation level	3 mg/L	(30)
Cyprinus carpio	perturbation level	2.8 mg/L	(30)
Goldfish	48h LC_0	14 mg/L	(226)

CDT *see* 1,5,9-cyclododecatriene

cefuroxime (4-(carbamoyloxymethyl)-8- [2-(2-furyl)-2- methoxyimino-acetyl]amino -7-oxo- 2-thia-6-azabicyclo[4.2.0]oct -4-ene-5-carboxylic acid)

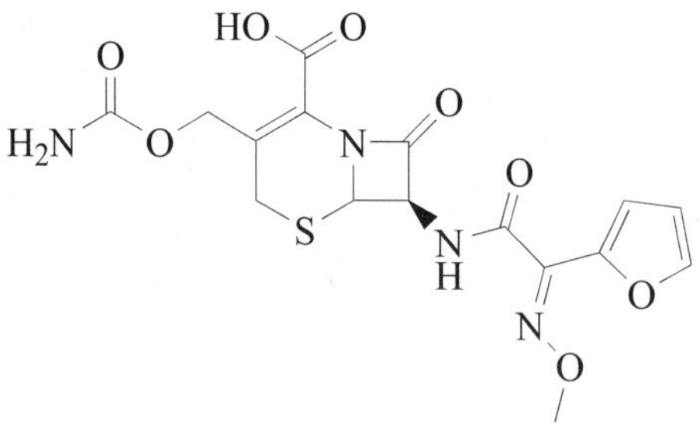

$C_{16}H_{16}N_4O_8S$

CAS 55268-75-2

TRADENAMES

Ceftin

USES

Cefuroxime is a second-generation cephalosporin antibiotic

A. PROPERTIES

molecular weight 424.39

D. BIOLOGICAL EFFECTS

Toxicity

Micro organisms			
Activated sludge respiration inhibition	IC_{50}	>100 mg/L	(10928)
Algae			
Selenastrum capricornutum	72h EC_{50}	91 mg/L	(10928)
Crustaceae			
Daphnia magna	48h EC_{50}	>1,000 mg/L	(10928)
Fish			
Oncorhynchus mykiss	96h LC_{50}	120 mg/L	(10928)

Celestolide (4-acetyl-1,1-dimethyl-6-tert-butyldihydroindene; ADBI; Musk TDT; 4-Acetyl-6-t-butyl-1,1-dimethylindan; 6-tert-Butyl-1,1-dimethylindan-4-yl methyl ketone)

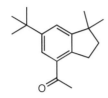

$C_{17}H_{24}O$

CAS 13171-00-1

EINECS 236-114-4

TRADENAMES

Celestolide, Crysolide

USES

a polycyclic musk. The polycyclic musks are used as fragrance ingredients in consumer products like cosmetics and detergents and cleaning agents. They are important ingredients in fragrances because of their typical musky scent and their fixative properties.
The substance is mixed with other fragrance ingredients into fragrance oils or 'compounds'. A compound may consist of as many as 50 ingredients and a compounder may produce a large number of specified recipes out of the 2000 – 3000 different fragrance ingredients. These fragrance compounds are used in formulating products such as cosmetics, detergents etc. The concentration of fragrances in detergents and soap ranges from 0.2 to 1%. (see also musk fragrances)

A. PROPERTIES

molecular weight 244.4; melting point 77-79°C; boiling point 319°C; vapour pressure 0.020 Pa; water solubility 0.015-0.22 mg/L; log Pow 5.4-6.6; H 1801 Pa m^3/mol

C. WATER AND SOIL POLLUTION FACTORS

Environmental concentrations

In River Glatt in Switzerland 1995: 0.0032 µg/L (7039)
In surface waters downstream from a tertiary STP In the USA 2001: 0.02 – 2.1 ng/L (7200)
in sludge of Sewage water treatment plants in Nordic countries in 2002 (10586)

Country	n	Range µg/kg dw	Median µg/kg dw
Denmark	5	63-294	115
Finland	5	0-100	12
Iceland	4	0-175	83
Norway	5	0-170	139
Sweden	8	52-153	68

D. BIOLOGICAL EFFECTS

Concentrations in fish in Nordic countries in 2002 (10586)

Sampling site	Sample type	Extractable lipid %	ng/g lipid
Sweden			
Lake Hjärtsjön	Perch	0.4-0.79	n.d
Lake Kvädöfjärden	Perch	0.61-0.95	n.d

Lake Bysjön	Perch	0.26-0.50	n.d
River Viskan 3 sites	Bream	0.95-3.11	0-149
Soutern Vättern	Perch	0.69-1.15	n.d
Soutern Vättern	Arctic char	1.74-4.13	n.d
Northern Vättern	Perch	0.58-1.15	n.d
Northern Vättern	Arctic char	2.75-3.97	n.d
Denmark			
Fish ponds	Trout 50 samples		0-204

Concentrations in breast milk in Nordic countries in 2002 (10586)

Country	n	Ng/g lipid
Denmark	10	0-11.2
Sweden	30-44	<3-9.3

Mammals: in human adipose tissue from Switzerland:

Median: 0.57 μg/kg, Range: 0.12-3.5 μg/kg (7039)

cellosolve *see* ethyleneglycolmonoethylether

cellosolveacetate *see* ethyleneglycolmonoethyletheracetate

cerinic acid *see* *n*-hexacosanoic acid

cerotic acid *see* *n*-hexacosanoic acid

cetab *see* cetyltrimethylammoniumbromide

cetane *see* *n*-hexadecane

cetylalcohol *see* 1-hexadecanol

cetylpyridiniumbromide (hexadecylpyridiniumbromide)

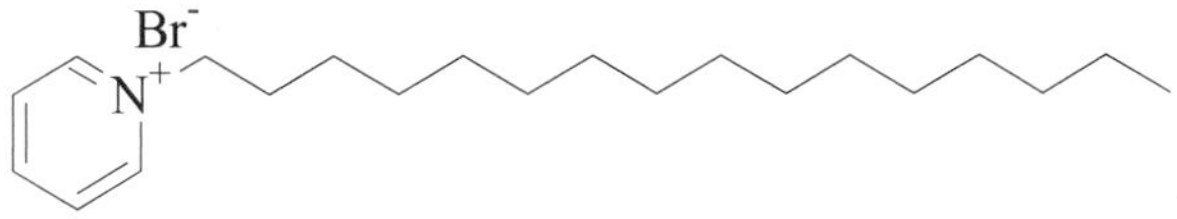

$C_6H_5N^+(CH_2)_{15}CH_3Br^-$

$C_{22}H_{38}BrN$

CAS 140-72-7

USES

surface active agent; germicide.

A. PROPERTIES

cream-colored waxy solid.

C. WATER AND SOIL POLLUTION FACTORS

Biodegradation

At 18 mg/L (inoculum = sew.) the lag period = 6 d; after 17 d there is complete degradation. (488)

D. BIOLOGICAL EFFECTS

Staphylococcus aureus at 20 mg/L, bacteriolytic action after 5 h.
Escherichia coli at 20 mg/L, slight reduction of growth rate;at 5 mg/L, no significant reduction of growth rate. (488)

cetyltrimethylammonium bromide (hexadecyltrimethylammoniumbromide; cetab; Acetoquat-CTAB; Bromat; Kissolamine-; trimethylhexadecylammoniumbromide-; N,N,N-trimethyl-1-hexadecanaminium bromide)

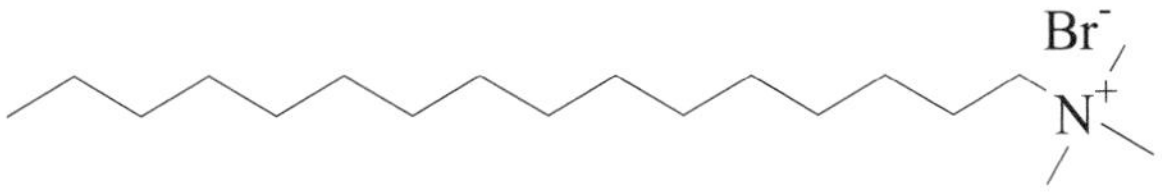

666 cetyltrimethylammonium bromide

$CH_3(CH_2)_{15}N(CH_3)_3Br$

$C_{19}H_{42}BrN$

CAS 57-09-0

TRADENAMES

bromat; centimide; cetarol; cetrimonium bromide; cirrasol-od; cycloton; softex kw

USES

an antiseptic with detergent properties and variable activity against different type of bacteria and some fungi. It is also used as an ingredient of shampoos for treating seborrhoea and psoriasis.

A. PROPERTIES

white powder; molecular weight 364.46; melting point 219 °C; boiling point ca 235 decomposes; vapour density 13;solubility 100,000 mg/L

C. WATER AND SOIL POLLUTION FACTORS

Biodegradation

at 15 mg/L: nonadapted sew.: lag time: 2 d, complete degradation after 4 d, adapted sew.: complete degradation after 2 d. (488)

Impact on biodegradation processes

BOD inhib.	EC_{50}	2.2; 4.9; 14.5 mg/L	
OECD 209 closed system inhib.	EC_{50}	7.5; 23; 37 mg/L	
biodegradation inhib.	EC_{50}	12 mg/L	
nitrification inhib.	EC_{50}	16.5 mg/L	(2624)

D. BIOLOGICAL EFFECTS

Bacteria			
Escherichia coli growth	EC_0	20 mg/L	(488)
microtox	5 min EC_{50}	0.4 mg/L	(2624)
Photobacterium phosphoreum Microtox test	5 min EC_{50}	9.8 mg/L	(9657)
Poteriochromonas malhamensis	3d LC_{10}	4.4 mg/L	(9935)
Vibrio fischeri	24h EC_{10}	1.62 mg/L	
	24h EC_{50}	2.21 mg/L	
	24h EC_{90}	2.51 mg/L	(7237)

Algae			
Selenastrum capricornutum	96h EC_{50}	0.09 mg/L	
Microcystis aeruginosa	96h EC_{50}	0.03 mg/L	(9936)

Mammals			
rat	oral LD_{50}	410 mg/kg bw	(9937)

chandor

USES

herbicide; active ingredients: trifluralin 24%, linuron 12%.

D. BIOLOGICAL EFFECTS

Harlequin fish (Rasbora heteromorpha):

mg/L	*24 h*	*48 h*	*96 h*	*3 m (extrap.)*	
LC_{10},F	0.87	0.58	0.58		
LC_{50},F	1.1	0.74	0.6	0.3	(331)

chinone *see* *p*-benzoquinone

chloracetamide (2-chloroacetamide; 2-chloroethanamide; Mergal-AF; Microcide)

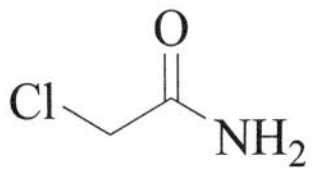

$ClCH_2CONH_2$

C_2H_4ClNO

CAS 79-07-2

A. PROPERTIES

molecular weight 93.52; melting point 116-118°C; boiling point 225° C (decomp.); solubility 1,700,000 mg/L at 40°C.

D. BIOLOGICAL EFFECTS

Worms

artificial soil toxicity to *Eisenia foetida*	LC_{50}	38.5 mg/kg soil	
Artisol test with *Eisenia foetida*	LC_{50}	99 mg/kg	(2658)

Mammals

rat	oral LD_{50}	70 mg/kg	(9938)

chloral hydrate (trichloroethanal hydrate; trichloroacetaldehyde hydrate)

$C_2H_3Cl_3O_2$

CAS 302-17-0

USES

manufacture of DDT: organic synthesis.

A. PROPERTIES

colorless liquid; molecular weight 165.40; melting point -57.5°C; boiling point 98°C; vapor pressure 35 mm at 20°C; vapor density 5.1; sp. gr. 1.5 at 20/4°C; miscible hygroscopic.

B. AIR POLLUTION FACTORS

1 mg/m^3 = 0.166 ppm, 1 ppm = 6.0 mg/m^3.

Odor

characteristic: sweet.

T.O.C.:		
$PIT_{50\%}$:	0.047 ppm	
$PIT_{100\%}$:	0.047 ppm	(2)
O.I. at 20°C:	980,000	(316)
recognition:	0.035-0.050 mg/m^3 (hydrate)	(610)

Incinerability

thermal stability ranking of hazardous organic compounds: rank 189 on a scale of 1 (highest stability) to 320 (lowest stability). (2390)

D. BIOLOGICAL EFFECTS

Toxicity threshold

(cell multiplication inhibition test) (chloral hydrate):

Bacteria			
Pseudomonas putida	16h EC_0	1.6 mg/L	(1900)
Algae			
Microcystis aeruginosa	8d EC_0	78 mg/L	(329)
green algae			
Scenedesmus quadricauda	7d EC_0	2.8 mg/L	(1900)
Protozoa			
Entosiphon sulcatum):	72h EC_0	79 mg/L	(1900)
Uronema parduczi Chatton-Lwoff	EC_0	86 mg/L	(1901)

chloramben *see* 3-amino-2,5-dichlorobenzoic acid

chlorambucil (4-[bis(2-chloroethyl)amino]benzenebutanoic acid; 4-[p-[bis(2-chloroethyl)amino]phenyl]butyric acid; Chormanimophene; Chlorbutium; Leukeran)

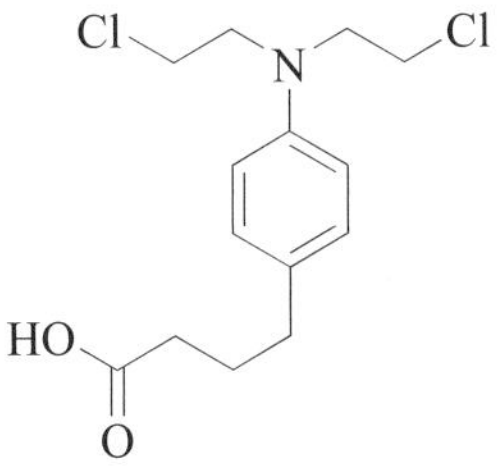

$C_{14}H_{19}Cl_2NO_2$

CAS 305-03-3

USES

Chemotherapy drug.

A. PROPERTIES

moleculare weight 304.22; melting point 64-66°C; solubility < 1000 mg/L at 20°C.

B. AIR POLLUTION FACTORS

Incinerability

thermal stability ranking of hazardous organic compounds: rank 141 on a scale of 1 (highest stability) to 320 (lowest stability). (2390)

D. BIOLOGICAL EFFECTS

Mammals			
rat, mouse	oral LD_{50}	76; 100 mg/kg bw	(9939, 9940)

chloramine T trihydrate sodium (N-chloro-*p*-toluenesulfonamide, sodium salt)

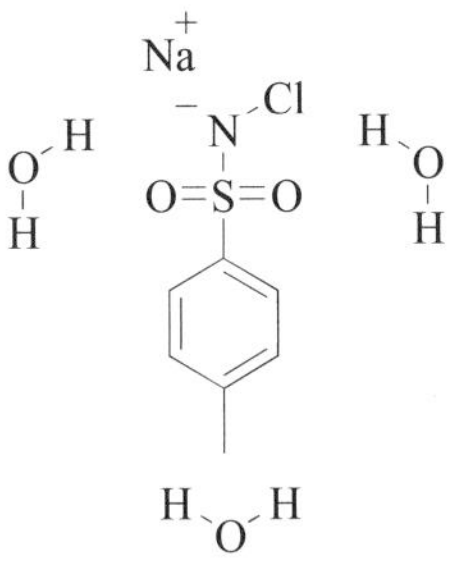

$CH_3C_6H_4SO_2N(Cl)Na.3H_2O$

$C_7H_7ClNaNO_2S.3H_2O$

CAS 7080-50-4 (trihydrate)

CAS 127-65-1

A. PROPERTIES

molecular weight 281.69; melting point 170-177°C decomposes.

C. WATER AND SOIL POLLUTION FACTORS

Impact on biodegradation processes

inhib.	EC_{50}: 6.3 mg/L; 12.5 mg/L; 13 mg/L	
OECD 209 closed system	EC_{50}: 124 mg/L; 250 mg/L	
biodegradation inhib.	EC_{50}: >43.5 mg/L	
inhib. of nitrification	EC_{50}: 290 mg/L	(2624)

D. BIOLOGICAL EFFECTS

Toxicity to microorganisms

5 min microtox inhib.	EC_{50}	<1.0 mg/L.	(2624)

chloramphenicol (chloramfenicol; chloromycetine; D-(-)-threo-2-dichloroacetamido-1-(4-nitrophenyl)-1,3-propanediol; D-(-)-threo-2,2-dichloro-N-(β-hydroxy-α-(hydroxy-methyl-*p*-nitrophenethyl)acetamide)

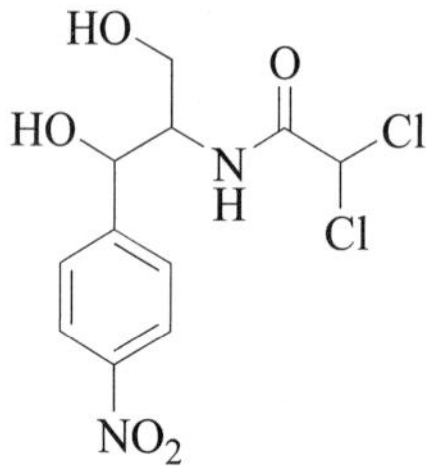

$C_{11}H_{12}Cl_2N_2O_2$

$Cl_2CHCONHCH(CH_2OH)CH(OH)C_6H_4NO_2$

$C_{11}H_{12}Cl_2N_2O_5$

CAS 56-75-7

USES

an antibiotic derived from *Streptomyces venezuelae* or by organic synthesis; antifungal agent.

A. PROPERTIES

molecular weight 323.1;melting point 148-150°C.

C. WATER AND SOIL POLLUTION FACTORS

Biodegradation rates

adapted A.S. at 20°C-product is sole carbon source: 86% COD removal at 3.3 mg COD/g dry inoculum/h. (327)

Environmental concentrations

	range µg/L	n	median µg/L	90 percentile µg/L	
effluent MWTP (1999)	up to 0.56	10			
surface waters (1999)	up to 0.060	52			(7237)

D. BIOLOGICAL EFFECTS

Acute toxicity tests

Marine tests			
Microtox™*(Photobacterium)* test	5 min EC_{50}	1,715 mg/L	
Artoxkit M *(Artemia salina)* test	24h LC_{50}	2,041 mg/L	
Freshwater tests			
Streptoxkit F *(Streptocephalus proboscideus)* test	24h LC_{50}	305 mg/L	
Daphnia magna test	24h LC_{50}	1,085 mg/L	
Rotoxkit F *(Brachionus calyciflorus)* test	24h LC_{50}	2,073 mg/L	(2945)

Micro organisms			
Vibrio fischeri	24h EC_{10}	0.0187 mg/L	
	24h EC_{50}	0.0643 mg/L	
	24h EC_{90}	0.129 mg/L	
	5min EC_{50}	1696 mg/L	(7237)
Crustaceae			
Artemia spp.	24h LC_{50}	2039 mg/L	
Daphnia magna	24h EC_{50}	1095 mg/L	
Streptocephalus probiscideus	24h LC_{50}	302 mg/L	
Brachionus calyciflorus	24h EC_{50}	2086 mg/L	(7237)

chloraniformethansee *see* imugan

chloranil (tetrachloroquinone; tetrachloro-*p*-benzoquinone; spergon)

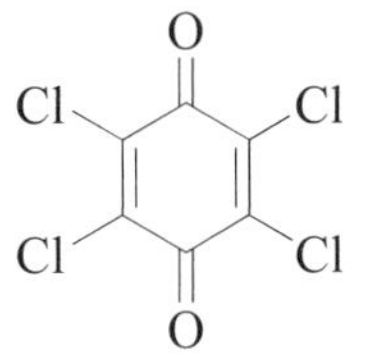

$C_6Cl_4(O_2)$

$C_6Cl_4O_2$

CAS 118-75-2

USES

agricultural fungicide; dye intermediate; electrodes for pH measurements; vulcanizing agent.

A. PROPERTIES

yellow leaflets; molecular weight 245.88; melting point 290°C; sp. gr. 1.97; molecular weight 245.9; solubility 250 ppm at room temp.

D. BIOLOGICAL EFFECTS

Fishes			
fathead minnows *(Pimephales promelas):*	96h LC_{50},S	0.01-1 mg/L	(935)

chlorbromuron (3-(4-bromo-3-chlorophenyl)-1-methoxy-1-methylurea)

$C_9H_{10}BrClN_2O_2$

CAS 13360-45-7

USES

selective herbicide, absorbed by roots and leaves; inhibits photosynthesis.

A. PROPERTIES

whitish powder; molecular weight 293.5; melting point 95-97°C; vapor pressure 53 μPa at 20°C; density 1.69 at 20°C; solubility 35 mg/L at 20°C; $LogP_{OW}$ 3.05.

C. WATER AND SOIL POLLUTION FACTORS

Aquatic reactions

slowly hydrolyzes in neutral, slightly acidic, and slightly alkaline media.

Degradation in soil

$t_{1/2}$: ca. 45 d. (2962)

D. BIOLOGICAL EFFECTS

Toxicity

Fishes			
Salmo gairdneri	*96h LC_{50}*	1.5 mg/L	
Ictalurus punctatus	*96h LC_{50}*	10 mg/L	
Lepomis macrochirus	*96h LC_{50}*	5.0 mg/L	
Poecilia reticulata	*96h LC_{50}*	7.9 mg/L	(2679)
rainbow trout	*96h LC_{50}*	5 mg/L	
bluegill	*96h LC_{50}*	5 mg/L	(2962)

chlordane (1,2,4,5,6,7,8,8-octachloro-4,7-methano-3a,4,7,7a-tetrahydroindane)

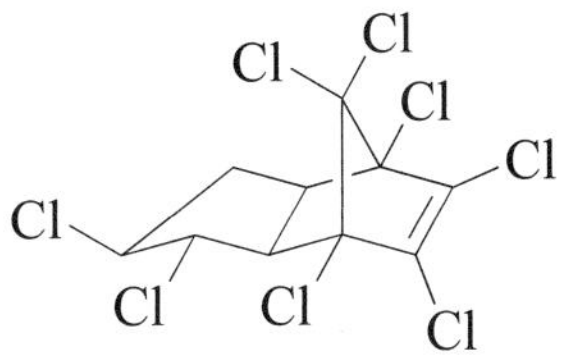

$C_{10}H_6Cl_8$

CAS 12789-03-6 (technical)

USES

insecticide, nonsystemic, termite control.

A. PROPERTIES

colorless to amber, odorless, viscous liquid; molecular weight 409.8; boiling point 175°C; vapor pressure at 25°C 1 × 10^{-5} mm Hg; solubility 0.056 mg/L; *H* 3.9 × 10^{-4}; $LogP_{OW}$ 6.0.

Technical chlordane consists of 60 to 75% isomers of chlordane and 25 to 40% of related compounds including two isomers of heptachlor and one each of enneachloro- and decachlorodicyclopentadiene.

Two isomers of octachlorodicyclopentadiene have been isolated from chlordane, of which α-chlordane is the endo-*cis* and β-chlordane is the endo-*trans* isomer.

The commercial product known as γ-chlordane is substantially the α-isomer.
Technical chlordane is a mixture of 26 organochlorine compounds whose aqueous solubility has been reported as 9 µg/L. (1393)

cis:trans *(75:25) chlordane:* water solubility: 0.056 ppm. (1396)

approx. composition of technical chlordane:

fraction	*% present*
cis(α)chlordane ($C_{10}H_6Cl_8$)	19 ± 3
trans(γ)chlordane ($C_{10}H_6Cl_8$)	24 ± 2
chlordene (4 isomers) ($C_{10}H_6Cl_6$)	21.5 ± 5
heptachlor ($C_{10}H_5Cl_7$)	10 ± 3
nonachlor ($C_{10}H_5Cl_9$)	7 ± 3
($C_{10}H_{7-8}Cl_{6-7}$)	8.5 ± 2
hexachlorocyclopentadiene (C_5Cl_6)	>1

octachlorocyclopentadiene (C_5Cl_8)	1 ± 1	
Diels-Alder adduct of cyclopentadiene and pentachlorocyclopentadiene ($C_{10}H_6Cl_5$)	2 ± 1	
others	6 ± 5	(1661)

B. AIR POLLUTION FACTORS

Incinerability

thermal stability ranking of hazardous organic compounds: rank 221 on a scale of 1 (highest stability) to 320 (lowest stability) (α-)(γ-). (2390)

Ambient air concentrations

Bloomington, IN, U.S.A. 1985-1986:

	γ-chlordane	*α-chlordane*	
inside 12 homes	0.5-29 ng/m^3	0.3-20 ng/m^3	
outside	0.3-0.6 ng/m^3	0.2-0.9 ng/m^3	(2642)

C. WATER AND SOIL POLLUTION FACTORS

Reduction of amenities

T.O.C. in water:	2.5 ppb	(326)
detection:	0.5 ppb	(915)

Aquatic reactions

persistence in river water in a sealed glass jar under sunlight and artificial fluorescent light-initial conc. = 10 µg/L:

	% of original compound found					
after	*1 h*	*1 wk*	*2 wk*	*4 wk*	*8 wk*	
	100	90	85	85	85	(1309)

Biodegradation

75-100% disappearance from soils: 3-5 years. (1815, 1816)

Removal efficiency in the activated rotating biological contactor ranged from 75 to 96% at initial conc. of 0.1 to 4.6 mg/L with organic supplement levels of 500 and 175 mg/L of BOD. 58 to 89% of the influent chlordane was not accounted for in the effluent, the system, or waste solids. Biological transformation was the primary removal mechanism. (2574)

Impact on biodegradation processes

Toxicity to microorganisms: act. sludge respiration inhib. EC_{50}: >2,000 mg/L. (2624)

Water and sediment quality

Hawaii:	α-chlordane: sediments: 400-5,270 ppt	
	γ-chlordane: sediments: 1,330-5,120 ppt	(1174)

in groundwater wells in the U.S.A.-150 investigations up to 1988:

median of the conc.'s of positive detections for all studies	1.7 µg/L	
maximum conc.	1.8 µg/L	(2944)

D. BIOLOGICAL EFFECTS

Bacteria

oral *Viridans streptococci:* total inhibition at 3.0 ppm (1661)

Algae

Scenedesmus quadricauda	0.1-100 µg/L: stimulation of growth	(1395)
Chlamydomonas sp. (soil alga)	0.1-50 µg/L: stimulation of growth 100 µg/L: inhibition of cell division	(1395)

Oedogonium (filamentous green algae): BCF 98,000 ^{14}C-*cis:trans* (75:25) (1396)

Scenedesmus quadricauda: BCF 6,000 to 15,000 for *cis*(α) and *trans*(γ)chlordane at treatment levels of 0.1 to 100 µg/L water. Bioconcentration was rapid, occurring within the first 24 h. (1392)

Arthropoda			
pelecypod *(Crassostrea virginica)*	96h EC_{50},F	6.2 µg/L	
decapod *(Penaeus duorarum)*	96h LC_{50},F	0.4 µg/L	
decapod *(Palaemonetes pugio)*	96h LC_{50},F	4.8 µg/L	(1132)

Crustaceans

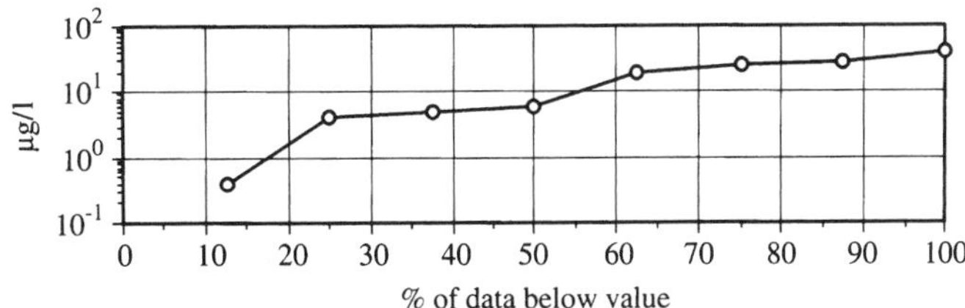

Frequency distribution of 24-96h EC_{50}/LC_{50} values for crustaceans (*n* = 8) based on data from this work and from the work of Rippen (3999)

Gammarus lacustris	96h LC_{50}	26 µg/L	(2124)
Gammarus fasciatus	96h LC_{50}	40 µg/L	(2126)
Palaemonetes kadiakensis	96h LC_{50}	4.0 µg/L	
	120h LC_{50}	2.5 µg/L	(2126)
Simocephalus serrulatus	48h LC_{50}	20 µg/L	(2127)
Daphnia pulex	48h LC_{50}	29 µg/L	(2127)

lobster *(Homarus americanus):* Canadian East Coast, lipid: 0.078-0.100 µg/g (74, 1479)

INSECT:			
Pteronarcys californica	96h LC_{50}	15 µg/L	(2128)

Ni_2B-catalyzed dechlorination of technical grade chlordane yielded a mixture of partially dechlorinated products: the major one was 4,5,6,7,8-pentachloro-2,3,3a,4,7,7a-hexahydro-8-antihydromethanoindene, which contains five chlorine atoms. (1413)

Acute toxicity of chlorinated and dechlorinated chlordane:

		LC_{50} (µg/L)[a]		
species	*formulation*	*chlorinated*	*dechlorinated*	*detoxification factor*[b]
bluegill	technical	41	582	14
	72% EC[c]	62	800	13
Daphnia	technical	97	813	8
	72% EC	156	1174	8

[a]96h LC_{50} for bluegill, 48h EC_{50} for *Daphnia;* calculated on basis of insecticide present in formulation.

[b]factor: LC_{50} dechlorinated/LC_{50} chlorinated.

[c]EC = emulsifiable concentrate. (1413)

Molluscs

BCF: oyster: 7,300 (1924)

Fishes

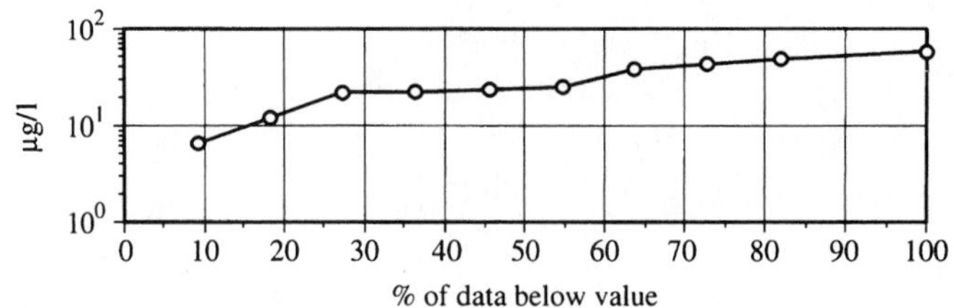

Frequency distribution of 24-96h LC_{50} values for fishes (*n* = 11) based on data from this work and from the work of Rippen (3999)

fathead minnow *(Pimephales promelas):*	96h LC_{50},FT	36.9 µg/L	
brook trout:	96h LC_{50},F	47 µg/L	
	LOEC,F	0.32 µg/L	
bluegill:	96h LC_{50},F	59 µg/L	
	MATC,F	0.54 µg/L	(1204)
bluegill:	96h LC_{50}	22 µg/L	
rainbow trout:	96h LC_{50}	22 µg/L	(1878)
Lagodon rhomboides	96h LC_{50},FT	6.4 µg/L	
Cyprinodon variegatus	96h LC_{50},FT	24.5 µg/L	(1132)
Saccobranchus fossilis	96h LC_{50},S	42 µg/L	(1509)
Cyprinodon variegatus	96h LC_{50},FT	12 µg/L	(1472)
	MATC, early life stage	11 µg/L	
	MATC, life cycle	0.6 µg/L	(2643)
Lepomis macrochirus	24h LC_{50},S	0.11 mg/L	
	96h LC_{50},S	0.025 mg/L	(2697)

goby fish *(Acanthogobius flavimanus)* collected at the seashore of Keihinjima along Tokyo Bay-Aug. 1978: residue level: 6 ppb *(cis)*, 9 ppb *(trans)* (1721)

herring *(Clupea harengus):* Canadian East Coast, lipid: 0.039-0.114 µg/g. (1479)

Carassius auratus: adsorption of ^{14}C-*cis*-chlordane from water by goldfish is very rapid.

Over 99% of the radioactivity recovered from the fish on d 10 and d 15 post-treatment times was unchanged *cis*-chlordane, indicating its inert storage in body tissues. (1569)

BCF following exposure to 5 ppb *(cis-)* in a static system:

	BCF	*time for maximum absorption*	
frogs *(Xenopus laevis)*	108	96 h	
bluegills *(Lepomis macrochirus)*	322	24 h	
goldfish *(Carassius auratus)*	990	16 h	(1839)

Elimination half-life after maximum absorption in a static system *(cis-):*

goldfish	4.4 weeks	
tropical fish *(Cichlasoma)*	20 weeks	
bluegill *(Lepomis macrochirus)*	16 weeks	
frog *(Xenopus)*	3.2 weeks	(1839)
Biotransfer factor in beef:	log B_b: -2.13	
Biotransfer factor in milk	log B_m: -3.43	
Bioconcentration factor for vegetation	log BCF_v: -1.81	(2644)

Worms

artificial soil toxicity to *Eisenia foetida*	LC_{50}	75 mg/kg soil (techn. chlordane)	
Artisol test with *Eisenia foetida*	LC_{50}	301 mg/kg	(2658)

α-chlordane (cis-chlordane)

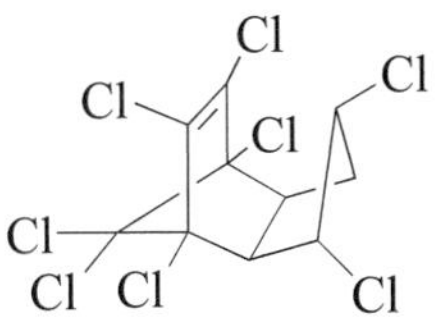

$C_{10}H_6Cl_8$

CAS 5103-71-9

A. PROPERTIES

molecular weight 409.76.

D. BIOLOGICAL EFFECTS

Rainbow trout	*BCF*	*conc. ng/L water*	
laboratory BCF (d 96):	16,000-22,000	at 1,2-13	(2399, 2400)
field BCF in Lake Ontario (U.S.A.):	1,400,000	0.03	(2399)
fathead minnow:	38,000		(2606)
BAF: rainbow trout			
Lake Ontario (U.S.A.):	17,400,000 l/kg(lp)		(2440)

γ-chlordane (trans-chlordane)

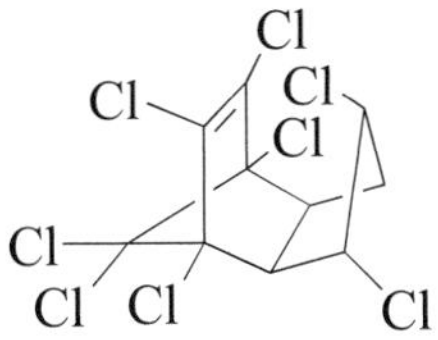

$C_{10}H_6Cl_8$

CAS 57-74-9

A. PROPERTIES

molecular weight 409.76.

D. BIOLOGICAL EFFECTS

Rainbow trout:	*BCF*	*conc. ng/L water*	
laboratory BCF (d 96):	15,000-20,000	at 1.4-17	(2399, 2400)
field BCF in Lake Ontario (U.S.A.):	76,000	0.02	(2399)
BAF: rainbow trout Lake Ontario (U.S.A.):	955,000 l/kg(lp)		(2440)

Biouptake

by Oligochaete worms Tubifex tubifex *and* Limnodrilus hoffmeisteri:

worm/sediment accumulation factor after 79 d at 8°C and 0.36 mg/kg sediment: 4.7 $t_{1/2}$ in worm: 63 d.

BCF from water at 0.15 µg/L: 25,000. (2618)

chlordecone *see* kepone

chlordene

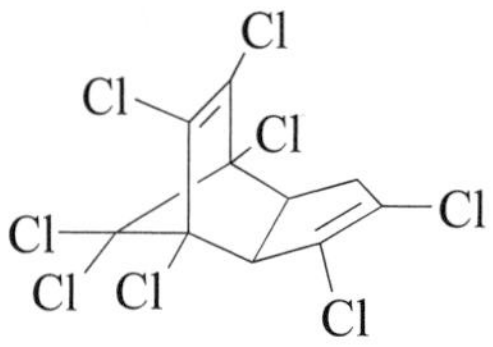

$C_{10}H_4Cl_8$

CAS 3734-48-3

USES

intermediate in the manufacturing of chlordane.

D. BIOLOGICAL EFFECTS

Toxicity ratios of chlordene (C) to photochlordene (PC), calculated from respective 24h LC_{50} and LT_{50} values:

	C/PC
Daphnia pulex (waterflea)	0.91
Musca domestica (housefly)	0.78 (0.70)*
bluegill:	0.72 (0.63)*

		(1684, 1685)
bluegill	24h LC_{50}: 218 ppb	
mosquito (late 3rd instar *Aedes aegypti* larvae)	24h LC_{50}: 130 ppb	
housefly (3-d-old female *Musca*)	LD_{50}: 158 µg/fly	(1681)

chlordimeform *see* chlorphenamidine

chlorfenpropmethyl *see* bidisin

chlorfenvinphos (2-chloro-1-(2,4-dichlorophenyl)vinyldiethylphosphate; O,O-diethyl-O-1-(2',4'-dichlorophenyl)-2-chlorovinylphosphate; Birlane; phosphoric acid, 2-chloro-1-(2,4-dichlorophenyl)ethenyl diethyl ester; O,O-diethyl O-[2-chloro-1-(2,4-dichlorophenyl)vinyl] phosphate; 2,4-dichloro-alpha-(chloromethylene)benzyl alcohol, diethyl phosphate; Sapecron; Steladone; Supona)

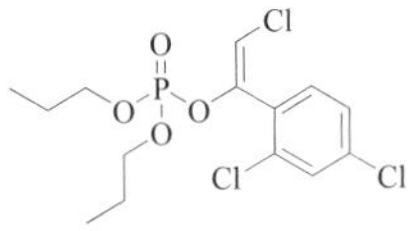

$C_{12}H_{14}Cl_3O_4P$

CAS 470-90-6

USES

insecticide.

A. PROPERTIES

amber-colored liquid; molecular weigth; 359.57; melting point -16/-22°C; boiling point 168-170°C at 0.5 mm; vapor pressure 1.7×10^{-7} mm at 25°C; sp. gr. 1.36 at 15.5/15.5°C; solubility 145 mg/L at 23°C; $\log P_{oct}$ 3.85 (Z-isomer), 4.22 (E-isomer).

C. WATER AND SOIL POLLUTION FACTORS

Slowly hydrolyzes in aqueous, alkaline, and acidic solutions.		(2962)
Persistence in aqueous solution:	at -5°C: $t_{1/2}$: 86-103 d	
	at 35°C: $t_{1/2}$: 40-55 d	(9946)

Biodegradation

Persistence in soil at 10 ppm initial concentration:

weeks incubation to

	50% remaining	5% remaining	
sterile sandy loam	>24		
sterile organic soil	>24		
nonsterile sandy loam	<1	5	
nonsterile organic soil	1	9	(1433)

Persistence

In soil:	DT50: 2 months; 12-28 d	(9943, 9945)
	DT90: 12 months	(9943)
On a peaty soil:	70% of the applied dose remaining after 21 wk 30% after nearly 12 months, 3-15% remaining after 4 months	(9944)

2,4-dichlorophenacyl chloride 2,4-dichlorobenzoic acid

2-hydroxy-4-chlorobenzoic acid 2,4-dihydroxybenzoic acid

Major metabolites detected were 2,4-dichlorophenacyl chloride, 2,4-dichlorobenzoic acid, 2-hydroxy-4-chlorobenzoic acid and 2, 4-dihydroxybenzoic acid. (9945)

Inhibition of biodegradation

process: In its active form chlorfenvinphos is an inhibitor of 'B' esterases and can effect the activity of these enzymes in soil microorganisms.

D. BIOLOGICAL EFFECTS

Protozoans			
Paramecium caudatum	48h LC_{50}	48 mg/L	(9941)
Crustaceans			
Daphnia magna	48h EC_{50}	18.5 mg/L	(9941)
Molluscs			
Mytilus galloprovincialis	48h LC_{50}	26.3 mg/L	(9942)
Fishes			
harlequin fish	24h LC_{50}	0.36 mg/L	(2962)
Mammals			
rat	oral LD_{50}	9-14 mg/kg bw	(9947, 9948, 9949)

chlorhexidine *see* 1,6-di(4-chlorophenyldiguanido)hexane

chloridazon (1-phenyl-4-amino-5-chloropyridaz-6-one; 5-amino-4-chloro-2-phenylpyridazin-3(2H)-one; burex; Phenazon; Pyramin; Pyrazone; Suzon)

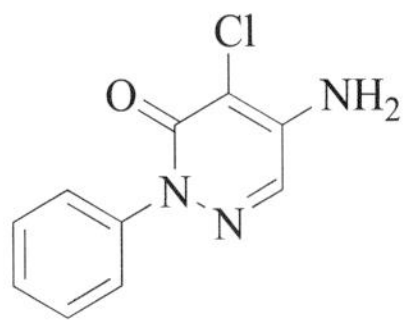

$C_{10}H_8ClN_3O$

CAS 1698-60-8

USES

Herbicide

A. PROPERTIES

molecular weight 221.65; melting point 205-206°C decomposes; vapor pressure < 7.5 x 10^{-8} mm Hg at 20°C; solubility 400 mg/L at 20°C; logP_{OW} 2.2.

C. WATER AND SOIL POLLUTION FACTORS

Biodegradation

In soil microbial degradation involves cleavage of the phenyl group to give 5-amino-4-chloropyridazin-3(2H)-one, which is not active herbicidally. Persists in soil for 6-8 wk in sufficiently moist conditions. (9662, 9757)

Inhibition of biodegradation

nitrification inhibition in soil (1% organic carbon) after 1 day : 53 % at 25 mg/kg. (7050)

D. BIOLOGICAL EFFECTS

Fishes			
trout	96h LC_{50}	27 mg/L (65% wettable powder formulation)	(9662)

Mammals			
rat	oral LD_{50}	647 mg/kg bw	
guinea pig	oral LD_{50}	760 mg/kg bw	
mouse	oral LD_{50}	1000 mg/kg bw	
rabbit	oral LD_{50}	2000 mg/kg bw	(9662, 9758, 9759, 9600)

chlorimuron ethyl (ethyl-2-[[[[(4-chloro-6-methoxypyrimidin-2-yl)amino]carbonyl]amino]sulfonyl]benzoate))

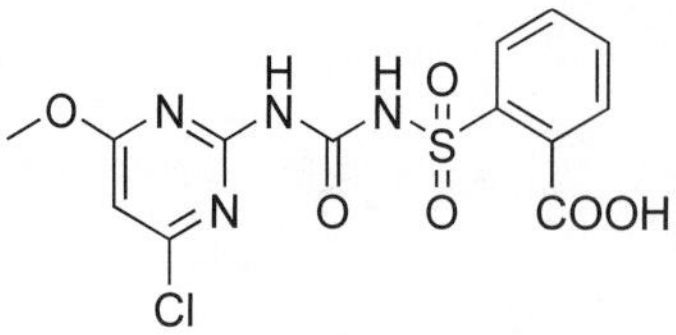

$C_{13}H_{11}ClN_4O_6S$

CAS 90982-32-4

USES

a sulfenylurea herbicide and pesticide.

A. PROPERTIES

off-white to pale yellow solid; molecular weight 414.83; melting point 181 °C; density 1.5; solubility in water is pH dependent 4.1 mg/L at pH 4.2, 9.0 mg/L at pH 5.0, 99 mg/L at pH 5.8, 450 mg/L at pH 6.5, 1200 mg/L at pH 7.0

C. WATER AND SOIL POLLUTION FACTORS

Dissipation periods in soil

DT_{50} (days)	DT_{90} (days)	soil	
17	122	sand	
1.9	15	Silt	(6020)

D. BIOLOGICAL EFFECTS

Toxicity

Daphnia magna	48h EC_{50}	>10 mg/L	(7162)
Birds			
Mallard duck	Acute oral LD_{50}	>2510 mg/kg bw	(7162)
Fish			
Rainbow trout	96h LC_{50}	>12 mg/L	
Bluegill sunfish	96h LC_{50}	>10 mg/L	(7162)
Mammals			
Rat	oral LD_{50}	>5000 mg/kg bw	(7162)

chlorinated camphene *see* toxaphene

chlorinated paraffins (CP)

USES

secundary plasticizer for polyvinyl chloride, chloro rubber paints, and acrylic resins; extreme pressure lubricant additives in the metal working industry; flame retardants; additives in PUR coatings; modifier in fat liquors for improving the suppleness of leather.

USES AND FORMULATIONS

in chloro rubber paints: <50%; in extreme pressure lubricants: 5-70%. (6001)

CHEMICAL IDENTITY

Chlorinated paraffins are characterized by the carbon-chain length range of their *n*- alkanes and by their chlorine content. A general classification of chlorinated paraffins is presented below.

carbon-chain length	*average % chlorine*	
C_{10-13}	50-70	
C_{14-17}	45-60	
C_{20-30}	40-70	(6000)

MANMADE SOURCES

Chlorinated paraffins have been found in human foodstuffs in the U.K.

	average conc., mg/kg C_{10-20}	C_{20-30}	
dairy products	0.3		
vegetable oils + derivatives	0.15		
fruit and vegetables	0.025		
beverages	<0.05		
cheese		0.19 (1 sample)	
potato crisps		0.025 (1 sample)	
peach fruit		0.025 (1 sample)	
Mussels	3.2 (range 0.1-12)	<0.1	
sheep grazing near a CP plant	<0.05-0.2	<0.05	(6004, 6008)

B. AIR POLLUTION FACTORS

Photochemical degradation

$t_{1/2}$ *(calculated)*			
reaction with OH°:	C_{10-13}	1.2-1.8 d	
	C_{14-17}	0.8-0.85 d	
	C_{15-30}	0.5-0.8 d	
	not specified	0.85-7.2 d	(6008)

C. WATER AND SOIL POLLUTION FACTORS

In surface waters

source	*location*	*conc., µg/L* C_{10-20}	C_{20-30}	
fresh and nonmarine waters remote from industry	rivers	<0.5-0.5	<0.5-0.5	
	lakes	<0.5-1.0	<0.5-0.5	
	estuaries	not measured	not measured	
close to industry	rivers	<0.5-6.0	<0.5	
	lakes	not measured	not measured	
	estuaries	<0.5-1.5	<0.5-0.5	
marine waters	harbors	0.5	<0.5	
	bays	<0.5	<0.5	
	seas and sounds	<0.5-4.0	<0.5-2.0	(6002)

U.K. 1985-1986 (averages)	*Trent*	*Humber*	
C_{10-13}	2.0 µg/L	0.56 µg/L	
C_{14-17}	7.6 µg/L	1.6 µg/L	(6003)

In sediments

		conc., mg/kg (w/w)		
		C_{10-20}	C_{20-30}	
fresh and nonmarine sediments				
remote from industry	rivers	<0.05-1.0	<0.05-0.05	
	lakes	not measured	not measured	
	estuaries	not measured	not measured	
Close to industry	rivers	0.15-15	0.05-3	
	lakes	not measured	not measured	
	estuaries	0.05	<0.05	
Marine sediments	harbors	0.5	<0.5	
	bays	<0.05	<0.05	
	seas and sounds	<0.05-0.1	<0.05-0.6	(6004)

In sewage sludge in the U.K.: conc., mg/kg (w/w)			
	C_{10-20}	C_{20-30}	
in Liverpool	4-10	>0.05	
in Manchester	>0.05	>0.05	
in Zürich (Switzerland)	>0.05	not measured	(6005, 6006, 6007)

Biodegradation

expected to occur in sediment and soil, the biodegradation rate being higher under anaerobic conditions than under aerobic conditions. The degradation rate has been shown to decrease with increasing carbon-chain length and chlorine content. (6017)

D. BIOLOGICAL EFFECTS

Bioaccumulation

	grade	*exposure conc., µg/L*	*exposure time*	*BCF*	
Mytilus edulis	C_{12}; 69% Cl	0.003	21 d	95,000	
		0.13	28 d	140,000	
	C_{16}; 34% Cl	0.13	28 d	7,000	(6009)
Alburnus alburnus	C_{10-13}; 49% Cl	125	14 d	800	
	C_{10-13}; 59% Cl	125	14 d	750	
	C_{10-13}; 71% Cl	125	14 d	150	(6010)
	C_{11}; 70% Cl	100	14 d	100-150	(6011)
	C_{14-17}; 50% Cl	125	14 d	30	
	C_{18-26}; 49% Cl	125	14 d	15	(6010)

Toxicity

Bacteria				
microtox test	C_{10-13}; 49% Cl	15 min EC_{50}	1-1.5 mg/L	(6012, 6013)
Crustaceans				
N. spinipes	C_{10-13}; 49% Cl	96h LC_{50}	0.06; 0.1 mg/L	
	C_{10-13}; 70% Cl	96h LC_{50}	>0.3; <5 mg/L	
	C_{14-17}; 45% Cl	96h LC_{50}	9 mg/L	
	C_{14-17}; 52% Cl	96h LC_{50}	>1,000 mg/L	
	C_{22-26}; 42% Cl	96h LC_{50}	>1,000 mg/L	

$C_{22\text{-}26}$; 49% Cl	96h LC_{50}	>1,000 mg/L	(6012, 6013, 6014)
Daphnia magna same values as for *N. spinipes*			
L. leander	LC_0	1,000 mg/L	(6012)
Fishes			
Salmo gairdneri and *L. macrochirus*			
$C_{10\text{-}13}$; 58% Cl	96h LC_{50}	>300 mg/L	
C_{20}; 34% Cl	96h LC_{50}	>300 mg/L	
C_{23}; 40% Cl	96h LC_{50}	>300 mg/L	
C_{24}; 48% Cl	96h LC_{50}	>300 mg/L	
$C_{20\text{-}30}$	96h LC_{50}	>300 mg/L	(6015)
Alburnus alburnus			
$C_{10\text{-}13}$, 49-71% Cl	LC_{50}	>5,000 mg/L	
	14d NOLC	>125 mg/L	
C_{11}, 70% Cl	LC_{50}	>5,000 mg/L	
$C_{14\text{-}17}$, 50-52% Cl	LC_{50}	>5,000 mg/L	
	14d NOLC	>125 mg/L	
$C_{22\text{-}26}$, 42% Cl	LC_{50}	>5,000 mg/L	(6010, 6016)

chlorinated paraffins, short chain

(SCCPs; chloroalkanes; chlorinated n-paraffins; chlorinated wax; paraffin waxes, chlorinated; chloroparaffin waxes)

$C_{10\text{-}13}H_{17\text{-}23}Cl_{3\text{-}8}$

CAS 85535-84-8

USES

Chlorinated paraffins were first produced commercially in the 1930s. They are manufactured by the chlorination of specified normal paraffin fractions obtained from petroleum refining. Substances of this type were used extensively for the first time during the 1940-44 World War when they were aplied to cotton fabrics employed by the military as flame retardants.

The major use of SCCPs nowadays is in the formulation of products used as extreme pressure lubricants, where they find application in metal cutting and metal working such as pressing, drawing and forming. SCCPs are blended with mineral oil, water and stabilizers to produce stable solvent-based emulsions containing approximately 50% SCCPs. Water-based emulsions with 2 to 15% SCCPs formed by mixing with water, mineral oil, antioxidants and dispersants are other types of products used in metal cutting and working fluids. These products are marketed for subsequent dilution and use.

SCCPs are used as flame retardants in products used in building materials such as sealers or fillers and adhesives. Sealers or fillers contain up to 30% SCCPs where these are waterbased emulsions, or 50% in the case of solvent-based adhesives. Chlorinated waxes and antimony trioxide are also incorporated into adhesives together with SCCPs. Based on the plasticising ability, small amounts of SCCPs are also used as plasticizers in producing chlorinated rubber coating materials and sealants used in the building industry. The concentration of SCCPs in the final products ranges from 1 to 10%.

SCCPs are also used in dispersion of pigments prior to incorporation of these substances mainly into polyurethane foams and to a lesser extent into other polymer formulations such as industrial paints. Pigments which may be hard to disperse into the final polymer are first blended with SCCPs, together with other additives such as fillers and surfactants, to form pastes or viscous liquids, depending on customer requirements. Formulated products are dilute dispersions of various types of pigment – diazo, azo condensation, phthalocyanine, copper phthalocyanine, or carbon black – in the $C_{10\text{-}13}$ chlorinated paraffin. The concentration of SCCPs in the products ranges from 80 to 90%. These dispersions are marketed for addition to the ultimate products by end users.

Flame retardancy is the major reason for their use in rubbers. SCCPs are blended at moderate temperatures with other flame retardants such as hydrated alumina and antimony oxide into rubbers which are used to make drive and conveyor belting and other flexible materials for use in a wide range of industries. These rubber products typically contain 5 to 15% of SCCPs for use by end users.

A traditional use of SCCPs is in formulation of leather treatment products in which the SCCPs are blended with animal fats, vegetable and mineral oils and surfactants. The final products contain 15 to 40% SCCPs and sell to footwear manufacturers. They are applied at the tannery to finished products (shoes) or sheets of leather ' to impart a waxy feel'.

Some of the final products containing SCCPs such as adhesives, paints and sealants may be available to the general public. (7123)

CHEMICAL IDENTITY

Table 1: of SCCPs

Chemical name	CAS	EINECS	Trade names
Alkanes,C_{6-18},chloro	68920-70-7	272-924-4	N/A
Alkanes,C_{10-13},chloro	85535-84-8	287-476-5	Cereclor 50IV PCA60 PCA 70 Witachlor 149 Witachlor 171P
Alkanes,C_{12-13},chloro	71011-12-6	N/A	N/A
Alkanes,C_{12-14},chloro	85536-22-7	287-504-6	N/A
Alkanes,C_{10-14},chloro	85681-73-8	288-211-6	N/A
Alkanes,C_{10-12},chloro	108171-26-2	N/A	N/A
Paraffin waxes and hydrocarbon waxes, chlorinated	63449-39-8	264-150-0	A70(wax) Adekacizer E 410 ADK Cizer E450 Aquamix 108 Arubren CP Cereclor Cereclor 48 Cereclor 51L Chlorocosane Clorez HMP Chlorez 760 Chloroflo 35 Chlorowax Chlorowax 70 CP 52 (wax) CPW 7 CW 35 Diablo 700X Enpara L50 Paroil 140 Paroil 150 HVH Paroil 170HV Platichlor Toyoparax 40 Unichlor

ENVIRONMENTAL CONCENTRATIONS

in blubber of beluga whales:
in the Arctic: 0.23 mg/kg (7123)
in the St Lawrence Estuary in Canada: 0.53 mg/kg

IMPURITIES, ADDITIVES, COMPOSITION

Ordinary commercial chlorinated paraffins are mixtures which contain several homologous n-alkanes corresponding to their manufacture from n-paraffin fractions with several different degrees of chlorination. (7123)

A. PROPERTIES

SCCPs with 10 to 13 carbons are liquids with a faint, sweetish characteristic odour. The physical properties depend on the number of carbons and the degree of chlorination as evidenced by the data in tables 2 and 3.

Table 2

Physical properties of SCCPs (7123)

Percentage chlorine (%)	50	56	70
Vapour pressure at 20°C (Pa)	0.16	N/A	N/A
Specific gravity at 20°C	1.19	1.3	1.51 (50°C)
Pour point (°C)	-30	-10	20
$LogK_{ow}$	5.9-7.1	N/A	5.7-8.7
Aqueous solubility		0.15-0.47 mg/L	

Gas chromatography permitted the separation of a commercial mixture (C_{10-13},50% chlorine) into nine components, each comprised of mixtures of unstated numbers of congeners (table 3).

Table 3

The $LogK_{ow}$ values for components of C_{10-13}, 50% chlorine (7123)

Congener group	Molecular formula	Log K_{ow}
1	$C_{10}H_{19}Cl_3$	5.85
2	$C_{10}H_{18}Cl_4$	5.93
	$C_{11}H_{20}Cl_4$	
3	$C_{10}H_{17}Cl_5$	6.04
	$C_{11}H_{19}Cl_5$	
4	$C_{10}H_{17}Cl_5$	6.20
	$C_{11}H_{19}Cl_5$	
5	$C_{11}H_{19}Cl_5$	6.40
	$C_{11}H_{18}Cl_6$	
	$C_{12}H_{20}Cl_6$	
6	$C_{12}H_{20}Cl_6$	6.61
	$C_{13}H_{23}Cl_5$	
7	$C_{12}H_{20}Cl_6$	6.77
	$C_{13}H_{22}Cl_6$	
8	$C_{12}H_{19}Cl_7$	7.00
	$C_{13}H_{22}Cl_6$	
	$C_{12}H_{18}Cl_8$	
9	$C_{13}H_{21}Cl_7$	7.14

The $LogK_{ow}$ values show a general trend for increasing partitioning to lipids with increasing degree of chlorination. (7123)

% chlorine	$LogK_{ow}$
49	4.4-6.9
60	4.5-7.4
63	5.5-7.3
70	5.7-8.7
71	5.4-8.0

Henry's Law Constant for SCCPs show a tendency of decreasing with increasing degree of chlorination. H = 17.1 Pa.m^3/mole (calculated). (7123)

Fugacity modeling

As can be seen from the results of the fugacity modelling exercise (table 4) , once released into the environment, SCCPs are expected to distribute mainly onto the soil and sediment phases

Table 4: results of fugacity modelling (7123)

Compartiment	Release : 100% to air	Release: 100% to water	Release: 100% to soil
Air	0.11%	0.05%	<0.001%
Water	0.02%	1.16%	0.005%
Sediment	0.8%	53.5%	0.23%
Soil	99.0%	45.3%	99.8%

B. AIR POLLUTION FACTORS

Photochemical reactions

t1/2: 1.9 tot 7.2 days, based on reactions with OH° (calculated) for C $_{10-13}$, 49 to 71% wt Cl (7123)

Thermal degradation

SCCPs are easily degraded thermally, either through prolonged heating at temperatures as low as 70 °C. The major mode of thermal decomposition is the elimination of hydrochloride (HCl) to form unsaturated materials which are themselves polymerised and/or further degraded. (7123)

C. WATER AND SOIL POLLUTION FACTORS

Hydrolysis

a partial and slow hydrolysis may be observed (7123)

Soil sorption

LogK_{ow} 4.4-8.0 (calculated from K_{ow} values). (7123)

Biodegradation

method	concentration	duration	elimination	
Modified MITI I	20-100 mg/l2% products C_{10-12}		58% Cl	
Modified Zahn-Wellens	69-134 mg/l	28 days	7-16% $ThCO_2$ C_{10-12}	
Coupled units test	10 mg/L		58% Cl 93% removal mainly by adsorption onto the sludge	(7123)

It can be concluded from the biodegradation results that SCCPs with low chlorine contents (e.g.<50% wt Cl) may biodegrade slowly in the environment, particularly in the presence of adapted micro-organisms. Certain bacteria have also been shown to dechlorinate SCCPs with high chlorine contents in a co-metabolic process and so under certain conditions, biodegradation of these compounds might also be expected to occur in the environment. (7123)

D. BIOLOGICAL EFFECTS

Bioaccumulation

Species	Conc µg/L	duration	body parts	BCF	
Oncorhynchus mykiss	33-3,050*	60 days	total body	574-7273	
Oncorhynchus mykiss	*3.1-14.3*	168 days	total body	3600-5300	
Alburnus alburnus	125 µg/L**	14 days	total body	800-1000 (49-59%Cl)	
			total body	200 (71%Cl)	
Mytilus edulis	2.3-10.1	147 days	total body	24,800-40,900	
Mytilus edulis	13-930	60 days	total body	5,785-25,952	(7123)

* C_{10-12},58%Cl

** C_{10-13},49%Cl, C_{10-13},59%Cl, C_{10-13},71%Cl

Toxicity

Micro organisms			
Anaerobic activated sludge inhibition of gas production			
C_{10-12}, 58% wt Cl	24h $EC_{>10}$	>32,000 mg/L	
Anaerobic bacteria from a domestic wastewater treatment plant			
C_{10-13}, 52% wt Cl	24h $EC_{>10}$	5,000 mg/L	
C_{10-13}, 56% wt Cl	24h $EC_{>10}$	1,700 mg/L	
C_{10-13}, 58% wt Cl	24h $EC_{>10}$	2,500 mg/L	
C_{10-13}, 62% wt Cl	24h $EC_{>10}$	2,000 mg/L	
C_{10-13}, 70% wt Cl	24h $EC_{>10}$	600 mg/L	(7123)
Algae			
Selenastrum capricornutum	96h EC_{50}	3.7 mg/L	
	7d EC_{50}	1.6 mg/L	
	10d EC_{50}	1.3 mg/L	
	10d NOEC	0.39 mg/L	(7123)
Skeletonema costatum	96h EC_{50}	0.043; 0.056 mg/L	
C_{10-12}, 58% wt Cl	48h EC_{50}	0.032 mg/L	
	96h NOEC	0.012 mg/L	(7123)
Crustaceae			
Daphnia magna			
C_{10-13}, 20% wt Cl	21d EC_{50}	0.228 mg/L	
	21d NOEC	0.05 mg/L	(7123)
C_{10-13}, 56% wt Cl	24h EC_{50}	0.44; 0.45; 0.55; 0.7; 0.82; 11; 11.1 mg/L	
	24h NOEC	<0.1; 0.1; 0.1; 0.13; 0.13; <0.3; 2 mg/L	
	21d EC_{50}	0.137 mg/L	
	21d NOEC	0.05 mg/L	(7123)
C_{10-12}, 58% wt Cl	24h EC_{50}	1.9; 1.9 mg/L	
	24h NOEC	0.5; 0.5 mg/L	
	48h EC_{50}	0.53 mg/L	
	72h EC_{50}	0.024 mg/L	
	96h EC_{50}	0.018 mg/L	
	5d EC_{50}	0.014 mg/L	
	21d EC_{50}	0.124 mg/L	
C_{10-13}, 60% wt Cl	24h EC_{50}	0.51; 0.7; 0.95; 4.0 mg/L	
	24h NOEC	0.06; 0.1; 0.5; 1.0 mg/L	
	21d EC_{50}	0.101 mg/L	
	21d NOEC	<0.05 mg/L	(7123)
C_{10-13}, 61% wt Cl	24h EC_{50}	0.3; 0.51; 1.02; 3.0 mg/	
	24h NOEC	<0.1; 0.1; 0.1; <0.3 mg/L	
	21d EC_{50}	0.104 mg/L	
	21d NOEC	0.02 mg/L	(7123)
Mysid shrimp *Mysidopsis bahia*			
C_{10-12}, 58% wt Cl	96h NOEC	0.014 mg/L	
	28d NOEC	0.007 mg/L	(7123)

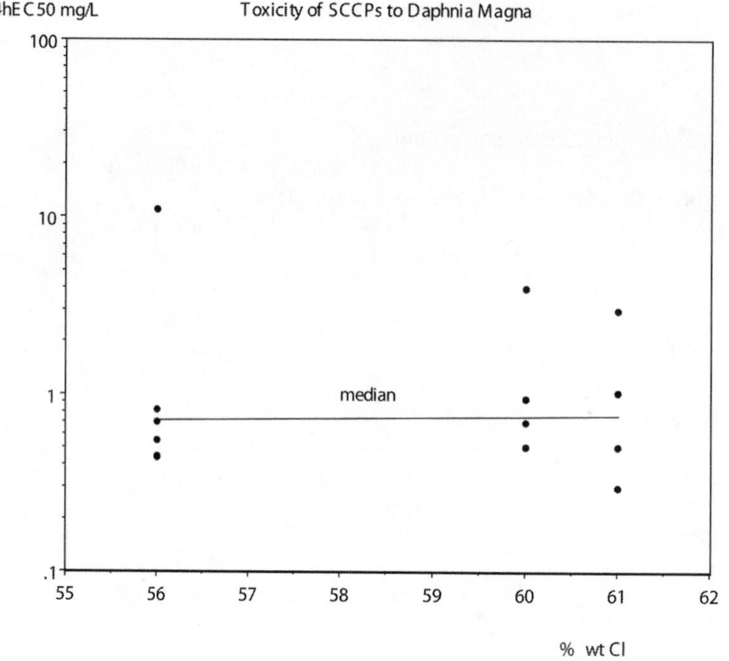

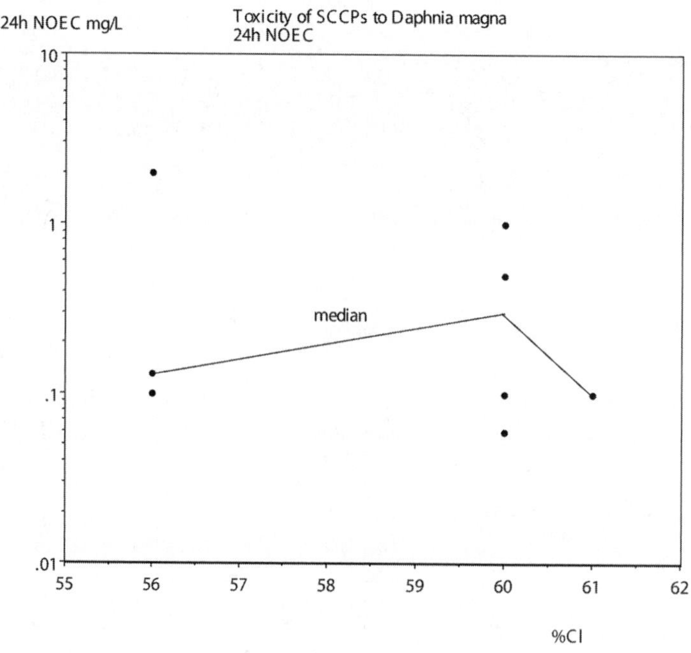

Insecta			
midge *Cihronomus tentans*			
C_{10-12}, 58% wt Cl	48h NOEC	>0.162 mg/L	
	49d NOEC	0.061 mg/L	(7123)
Mollusca			
Mytilus edulis			
C_{10-12}, 58% wt Cl	60d LC_{50}	0.074 mg/L	
	12w NOEC	<0.0093 mg/L	(7123)
Fish			
Ictalurus punctatus (C_{10-12}, 58%Cl)	96h LC_{50}	>300 mg/	

Leuciscus idus (C_{10-13}, 52-70%Cl)	48h LC_{50}	>500 mg/L	
Pimephales promelas (C_{10-12}, 58%Cl)	96h LC_{50}	>100 mg/L	
Oncorhynchus mykiss (C_{10-12}, 58%Cl)	96h LC_{50}	>300 mg/L	
	60d LC_{50}	0.34 mg/L	
	20d NOEC	<0.040 mg/L	
Cyprinodon variegatus (C_{10-12}, 58%Cl)	32d NOEC	0.28 mg/L	(7123)
Mammals			
Rat	oral LD_{50}	>4000 mg/kg bw	(7123)

Restrictions

Oslo and Paris Convention for the Protection of the Marine Environment of the North-East Atlantic **(OSPAR):** phase out use of SCCPs as metal working fluids, plasticizers in paint, coating and sealants and flame retardants in rubber, plastics and textiles.

chlorine cyanide *see* cyanogenchloride

chlormephos (S-chlormethyl-O,O-diethylphosphorothiolothionate)

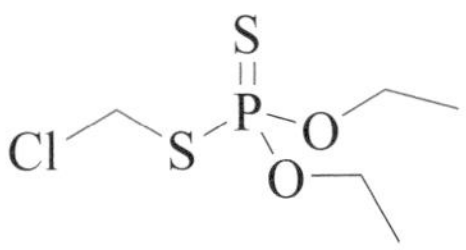

$C_5H_{12}ClO_2PS_2$

CAS 24934-91-6

USES

insecticide.

A. PROPERTIES

pale-colored liquid; molecular weight 234.71; boiling point 81-85°C at 0.1 mbar; vapor pressure 0.47 × 10^{-2} mbar at 20°C; sp. gr. 1.26; solubility 60 mg/L at 20°C.

C. WATER AND SOIL POLLUTION FACTORS

Aquatic reactions

rapidly hydrolyzes in alkaline media. (2962)

Degradation in soil

is transformed to ethion.

D. BIOLOGICAL EFFECTS

Harlequin fish (Rasbora heteromorpha):

mg/L	24 h	48 h	96 h	3 m (extrap.)	
LC_{10},F	3.5	2.8	-		
LC_{50},F	4.8	3.5	2.5	1.5	(331)

S-chlormethyl-O,O-diethylphosphorothiolothionate *see* chlormephos

chlornitrofen (2,4,6-trichlorophenyl-4'-nitrophenylether)

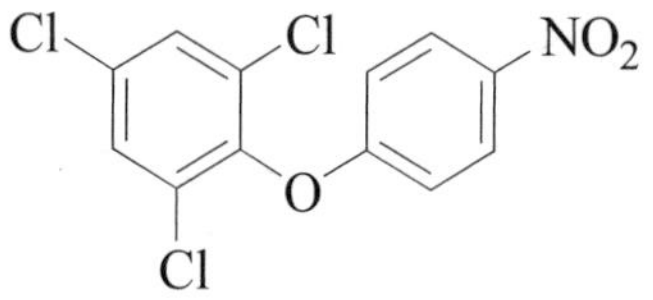

$C_{13}H_6Cl_3O_3$

CAS 1836-77-7

A. PROPERTIES

molecular weight 318.6; $LogP_{OW}$ 3.67.

C. WATER AND SOIL POLLUTION FACTORS

Degradation in soil

Aerobic biodegradation half-life: 0.5 d, at an initial conc. of 0.1 ppm, using a mixture of nonacclimated sludge, field soil, and river sediment as inoculum, in a stirred flask at 18-22°C, the major bacterial types in the activated sludge have been characterized as follows: *Flavobacterium* 35%; *Alcaligenes* 30%; *Corynebacterium* 20%; *Pseudomonas* 15%. Metabolites were amino-chlornitrofen and 4-aminophenyl-2,6-dichlorophenylether. (2690)

Aquatic reactions

alkaline hydrolysis half-life: 1.4 d, at pH 9 at 40°C in the dark. (2690)

D. BIOLOGICAL EFFECTS

Fishes

cytotoxicity to goldfish GF-Scale cells	NR_{50}	3.6 mg/L = 0.011 mmol/L	(2680)
Cyprinus carpio	48h LC_{50}	>40 mg/L	(2680)

2-chloro-4,6-bis(ethylamino)-s-triazine (princep; simazine; tafazine)

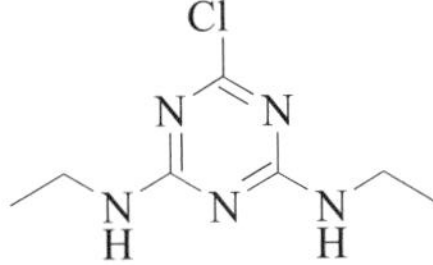

$C_7H_{12}ClN_5$

CAS 122-34-9

USES

herbicide.

A. PROPERTIES

molecular weight 201.7; melting point 225-227°C; vapor pressure 8.1 × 10^{-7} Pa at 20°C; solubility 5 ppm at 20°C; $LogP_{OW}$ 1.5, 1.9, 1.95, 2.26, 2.34.

C. WATER AND SOIL POLLUTION FACTORS

Degradation in soil

75-100% disappearance from soils: 12 months. (1815)

The preferential dealkylation pathway of the *s*-triazine herbicides in the unsaturated zone is the removal of an ethyl side chain relative to an isopropyl side chain. It is hypothesized that deethylation reactions may proceed at 2-3 times the rate of deisopropylation reactions.

simazine atrazine propazine

deethylatrazine deisopropylatrazine

didealkylated triazine

Dealkylation reactions of atrazine, simazine, and propazine to desethylatrazine (DEA), deisopropylatrazine (DIA), and didealkylatrazine (DDA) (2952)

Groundwater quality

groundwater wells in the U.S.A.-150 investigations up to 1988:

median of the conc.'s of positive detections for all studies	0.3 µg/L	
maximum conc.	9 µg/L	(2944)

Surface water quality

In rivers in U.K. (1986): 139 samples from 70 river sites yielded 74 positive values (det. (2711)

limit: 0.025 µg/L): median of positive values: 0.12 µg/L; maximum: 0.82 µg/L.

In surface waters

in Germany 1985/1987: up to 0.35 µg/L.
In groundwater wells in Germany 1984/1986: up to 0.2 µg/L. (2915)

Water treatment

initial concentration, µg/L	*treatment method*	*% removal*	
not reported	activated carbon	>90	
not reported	flocculation	<5	
not reported	river bank filtration	low	(2915)

D. BIOLOGICAL EFFECTS

Lowest NOEC's per species, µg/L:

Chlorophyta	5.0; 10; 10; 10; 32; 100; 100; 200	
Cyanophyta	1.0; 1.0; 1.0; 5.0; 10; 10; 100	
Branchiopoda	100	(2906)

Bioconcentration factor for vegetation

log BCF_V: 0.22. (2644)

Algae				
Selenastrum capricornutum in algal assay medium				
inhib. of oxygen evolution	EC_{50}	0.0022 mg/L		
growth inhib.	EC_{50}	0.0006 mg/L		(2624)
Selenastrum capricornutum	72h EC_{50}	0.2 mg/L		
	72h EC_{100}	3.2 mg/L		(7024)
Chlorococcum sp.	10d EC_{50}	2-2.5 mg/L	technical acid	
Dunaliella tertiolecta	10d EC_{50}	4-5 mg/L	technical acid	
Isochrysis galbana	10d EC_{50}	0.5-0.6 mg/L	technical acid	
Phaeodactylum tricornutum	10d EC_{50}	0.5-0.6 mg/L	technical acid	(2348)

Crustaceans			
Gammarus lacustris	96h LC_{50}	13 mg/L	(2124)
Gammarus fasciatus	48h NOEC	100 mg/L	(2125)
Daphnia magna	48h LC_{50}	1 mg/L	
Cypridopsis vidua	48h LC_{50}	3 mg/L	
Asellus brevicaudus	48h NOEC	100 mg/L	(2125)
Palaemonetes kadiakensis	48h NOEC	100 mg/L	(2125)
Orconectes nais	48h NOEC	100 mg/L	(2125)
Daphnia magna	2d EC_{50}	>3.5 mg/L	
Lymnaea stagnalis	20d EC_{50}	<0.20 mg/L	(2625)

Fishes			
Oncorhynchus kisutch	48h LC_{50}	6.6 mg/L	(2106)
bluegill	48h LC_{50}	130 mg/L	
rainbow trout	48h LC_{50}	85 ppm	(1878)

1-chloro-2-bromoethane (1-bromo-2-chloroethane; ethylene-chlorobromide)

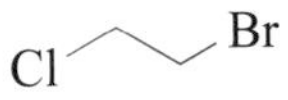

$BrCH_2CH_2Cl$

C_2H_4BrCl

CAS 107-04-0

A. PROPERTIES

molecular weight 143.41; melting point -16.6°C; boiling point 106-107°C; density 1.74 at 20°C; solubility 6830 mg/L; LogH -1.43 at 25°C.

B. AIR POLLUTION FACTORS

Photochemical reactions

$t_{1/2}$: 49 d, based on reactions with OH° (calculated). (8374)

Volatilisation from a model river

$t_{1/2}$: 4.7 h (depth: 1 m; current : 1 m/sec; windspeed : 3 m/sec). (8383)

D. BIOLOGICAL EFFECTS

Mammals

rat	oral LD_{50}	64 mg/kg bw	(9966)

2-chloro-1,3-butadiene *see* chloroprene

1-chloro-2-(β-chloroethoxy)ethane *see* β,β'-dichloroethylether

1-chloro-2-(β-chloroisopropoxy)propane *see* dichloroisopropylether

chloro(chloromethoxy)methane *see* dichloromethylether

2-chloro-3(4-chlorophenyl)methylpropionate *see* bidisin

p-chloro-m-cresol (6-chloro-3-hydroxytoluene; 4-chloro-3-hydroxytoluene; PCMC; 4-chloro-3-methylphenol; 4-chloro-1-hydroxy-3-methylbenzene)

OH
CH3
Cl

C_7H_7ClO

CAS 59-50-7

USES

external germicide; preservative for glues, gums, inks, textiles, and leather goods.

A. PROPERTIES

odorless crystals (when pure); molecular weight 142.6; melting point 66°C; boiling point 235°C; density 1.37 at 20°C; vapor pressure 0.08 hPa at 20°C, 7 hPa at 100 °C; solubility 3,600-3,900 mg/L at 20°C, 4,000 mg/L at 25°C at pH 5.1; pH 5.6 in saturated aqueous solution; $LogP_{ow}$3.0; 3.02; 3.1 (measured); H 0.28 Pa x m^3/mol.

B. AIR POLLUTION FACTORS

Incinerability

thermal stability ranking of hazardous organic compounds: rank 116 on a scale of 1 (highest stability) to 320 (lowest stability). (2390)

C. WATER AND SOIL POLLUTION FACTORS

Odor threshold

detection: 0.1 mg/kg. (892)

COD: 1.85

Photolysis

Indirect photolysis in water at 2.7 mg/L in the presence of humic acid : $t_{1/2}$: 3.3-46 hours depending on exposure to artificial light or to sunlight (8559)

Manmade sources

effluent of municipal waste water treatment plant after chlorination:	0.5-1.5 µg/L	(8566, 8567)
effluent of digested sludge of 6 municipal waste water treatment plants:	effluent: <0.12-0.4 µg/L digested sludge: <0.12-0.4 µg/L	(8569)
effluent of waste water treatment plant of paper mill:	1.8 µg/L	(8568)
Environmental concentrations: in river Ruhr mean 0.1 µg/L (n=27), max 0.4 µg/L.		(8565)

Biodegradation

Inoculum	*method*	*concentration*	*duration*	*elimination results*	
river water	Closed Bottle Test	100 mg/L 200 mg/L	7-14 d 28 d	100% product 100% product	(8560)
A.S. domestic	Modified MITI Test	100 mg/L	4 d 6 d 8 d 10 d 14 d	0% ThOD 62% ThOD 68% ThOD 72% ThOD 87% ThOD	(8572)
A.S. adapted	Closed Bottle test	2.4 mg/L	21days 28 d	0% ThOD 61% ThOD	(8573)
A.S. adapted	Closed Bottle Test	2.4 mg/L	5 d 10 d 5 d 10 d 20 d	58% ThOD 79% ThOD 62% COD 84% COD 89% COD	(8573)
A.S. industrial		5 mg/L	2 d	>99% product	(8573)
non-adapted domestic sewage		50 mg/L	3 d	10-20% product	
			3.5 d	100% product	(8575, 8576)
Pseudomonas putida		14 mg/L	14 hours	100% product	
	in co-metabolism with phenol	17 mg/L	4.8 d	100% product	(8577)
A.S. *	OECD-confirmaatory test	20 mg/L	21 d	100% product	(8578)
soil extract, non-adapted	OECD method	20 mg/L	21 d	30% product	(8578)
A.S. non-adapted	Closed Bottle Test	4.5 mg/L	5 d	4% ThOD	
			15 d 28 d	78% ThOD 84% ThOD	(8573)
A.S.different origins	Modified MITI Test	100 mg/L	28 d	0% ThOD	(8579)
domestic sewage	static culture flask screening	10 mg/L	7 d	76% TOC	
			14 d	100% TOC	(8580)
		0.013 mg/L	16 d	90% product	(8581)
A.S.	pilot plant			59-67% product	(8583)
soil extract in synthetic sewage	anaerobic conditions	20 mg/L	21 d	0% product	(8578)

* in a complex organic medium

Pseudomonas putida is able to metabolise 4-chloro-m-cresol as the sole carbon and energy source. (9972)
Nitrification inhibition in sewage occurs at a threshold concentration 4.0 mg /L. (9971)

Stability in soil

at 25 °C, humidity at field capacity	Dissipation after	acid silty clay	neutral sandy clay	
at 10 mg/kg dw	10 d	41%	42%	
	20 d	58%	60%	
	40 d	80%	81%	
1,000 mg/L	10 d	33%	39%	
	20 d	52%	52%	
	40 d	72%	75%	(8563)
aerobic not adapted sandy silt loam at 20 mg/kg and 20 °C	DT_{50}		1.4 d	
aerobic not adapted in sandy loam at 10 mg/kg and 20 °C	DT_{50}		4.2 d	(8564)

D. BIOLOGICAL EFFECTS

Bioaccumulation

	Concentration µg/L	*exposure period*	*BCF*	*organ/tissue*	
Fishes: *Cyprinus carpio*	2	42 d	5-11	total body	
	20	42 d	7-13	total body	(8579)

Bacteria

A.S. oxygen consumption inhibition	30min EC_{10}	32 mg/L	
	3h EC_5	4.3 mg/L at pH 9	
	3h EC_{50}	60 mg/L at pH 9	
	3h EC_{95}	855 mg/L at pH 9	(8573)
	70h EC_{50}	200 mg/L at pH 7.5	
	70h EC_{90}	300-500 mg/L at pH 7.5	(8597)
	5d EC_{15}	100 mg/L	
	5d EC_{100}	500 mg/L	(8575)
Bacillus sp. inhibition of spore germination	6h EC2-82	63 mg/L	(8581)
	24h EC_{100}	1,000 mg/L	(8599)
Bacillus subtilis	24h EC_{25}	28 mg/L	(8578)
Pseudomonas aeruginosa	7d EC_{100}	>1,000 mg/L	(8600)
	40min EC_{100}	657 mg/L	(8601)
Pseudomonas putida	16h EC_0	70 mg/L	(8602)
(oxygen consumption inhibition)	30min NOEC	250 mg/L	(8573)
anaerobic microorganisms	8d EC_{10}	1 mg/L	
	8d EC_{16}	100 mg/L	(8604)
	5h EC_{21}	1 mg/L	
	5h EC_{31}	100 mg/L	(8604)
Desulfovibrio desulfuricans (anaerobic conditions)	MIC	35 mg/L	
Bacillus subtilis (on agar plates)	MIC	150 mg/L	
Bacillus punctatum (on agar plates)	MIC	200 mg/L	
Leuconostoc mesenteroides (on agar plates)	MIC	200 mg/L	
Proteus vulgaris (on agar plates)	MIC	200 mg/L	
Staphylococcus aureus (on agar plates)	MIC	200 mg/L	
Escherichia coli (on agar plates)	MIC	250 mg/L	
Pseudomonas aeruginosa (on agar plates)	MIC	800 mg/L	
Pseudomonas fluorescens (on agar plates)	MIC	800 mg/L	(8606)

Fungi

Aspergillus niger	EC_{29}	25 mg/L	(8605)
Aureobasidium pollulans (on agar plates)	14d MIC	30 mg/L	
Torula rubra (on agar plates)	14d MIC	50 mg/L	

Chaetomium globosum (on agar plates)	14d MIC	80 mg/L	
Aspergillus flavus (on agar plates)	14d MIC	100 mg/L	
Aspergillus niger (on agar plates)	14d MIC	100 mg/L	
Coniophora puteana (on agar plates)	14d MIC	100 mg/L	
Penicillium citrinum (on agar plates)	14d MIC	100 mg/L	
Penicillium glaucum (on agar plates)	14d MIC	100 mg/L	
Rhizopus nigricans (on agar plates)	14d MIC	100 mg/L	
Sclerophoma pityophila (on agar plates)	14d MIC	100 mg/L	
Stachybotrys atra Corda (on agar plates)	14d MIC	100 mg/L	
Trichophyton pedis (on agar plates)	14d MIC	100 mg/L	
Trichoderma viride (on agar plates)	14d MIC	140 mg/L	
Alternaria tenuis (on agar plates)	14d MIC	200 mg/L	
Candida albicans (on agar plates)	14d MIC	200 mg/L	
Cladosporium herbarum (on agar plates	14d MIC	200 mg/L	
Paecilomyces variotii (on agar plates)	14d MIC	200 mg/L	
Lentinus tigrinus (on agar plates)	14d MIC	3,500 mg/L	
Polyporus versicolor (on agar plates)	14d MIC	5,000 mg/L	(8606)
Algae			
Scenedesmus subspicatus	96h EC_{10}	5.2 mg/L	
(biomass)	96h EC_{50}	>10 mg/L	(8596)
	72h EC_{10}	1.8 mg/L	
	72h EC_{50}	4.2 mg/L	(8573)
(cell multiplication inhibition test)	48h EC_{10}	5.7 mg/L	
	72h EC_{10}	4.7 mg/L	(8596)
Plants			
Lactuca sativa Ravel R2 (growth)	14d EC_{50}	>32 <100 mg/kg dry soil	
	16-21d EC_{50}	2.3 mg/L	(8607)
Crustaceans			
Daphnia magna Straus	24h EC_0	1.9-2.0 mg/L	
	24h EC_{50}	3.5-10 mg/L	
	24h EC_{100}	9 mg/L	(8592, 8593)
	48h EC_{50}	2 mg/L	(8594)
	21d EC_{60}	2.5 mg/L	
	21d NOEC	1.2 mg/L	(8593)
Daphnia pulex	96h EC_{50}	3.1 mg/L	(8595)
Daphnia magna reproductive rate	NOEL	1.3 mg/L	(9970)
Fishes			
fathead minnows *(Pimephales promelas)*	96h LC_{50},S	0.1-0.01 mg/L	(935)
	96h LC_{50}	4-7.4 mg/L	(8584, 8585)
	96h LC_{50}	7.6 mg/L	
	24h LC_{50}	13 mg/L	
	48h LC_{50}	11 mg/L	
	72h LC_{50}	9.2 mg/L	(8586)
Brachydanio rerio	5d LC_{100}	3.2 mg/L	(8573)
	14d NOEC	1 mg/L	(8573)
	24h LC_{50}	1-3.5 mg/L	(8588)
	2-6h LC_{50}	10-35 mg/L	(8588)
Oncorhynchus mykiss	96h LC_{100}	0.97 mg/L	
	96h NOEC	0.37 mg/L	(8587)
Leuciscus idus	48h LC_0	0.5; 2.0 mg/L	
	48h LC_{50}	1.2; 2.4 mg/L	

	48h LC_{100}	2.0; 3.0 mg/L	(8573, 8589)
Poecilia reticulata	24h LC_{100}	2.2 mg/L	(8590, 8591)
Salmo trutta	24h LC_{50}	50 mg/L	(8581)
Oryzias latipes	48h LC_{50}	4.6 mg/L	(8579)
Mammals			
rat	oral LD_{50}	1,830 mg/kg bw	(9973)

p-chloro-o-cresol (4-chloro-2-methylphenol; PCOC; 5-chloro-2-hydroxytoluene)

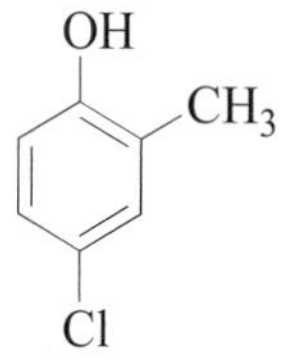

C_7H_7ClO

CAS 1570-64-5

SMILES Cl-c(ccc1O)cc1C

EINECS 216-381-3

USES

feedstock for the synthesis of pesticides; Additive tracer for resist plasma etching. Disinfectant. Manufacture of the herbicide 4-chloro-2-methylphenoxyacetic acid (MCPA). Exhibits fungicidal activity.

OCCURRENCE

PCOC is a photolysis degradation product of the herbicides MCPA (4-chloro-2-methoxyphenoxy acetic acid) and possibly MCPP (2-(4-chloro-2-methylphenoxy)-propionic acid. PCOC is an impurity in phenoxyherbicides such as MCPA and MCPP. Microbial metabolite of MCPA, impurity in technical grade MCPA (up to 4%). (1463)

IMPURITIES, ADDITIVES, COMPOSITION

2-chloro-6-methylphenol (<1%); 2-methylphenol (<1%); 2,4-dichloro-6-methylphenol (<1%); 4-chloro-2,6-dimethylphenol (<0.5%); 4-chlorophenol (<0.5%); 5-chloro-2-methylphenol (<0.2%)

A. PROPERTIES

molecular weight 142.59; melting point 46°C; boiling point 150-155°C; density 1.2 at 50°C; vapour pressure 26.7 pa at 20°C;vapor pressure 1.6 hPa at 70 °C; solubility 7,600 mg/L at 25°C; log P_{oct} 3.1 (measured), 2.7 (calculated); H = 0.11 Pa x m^3/mol (calculated)

B. AIR POLLUTION FACTORS

Photochemical reactions

$t_{1/2}$: 25 d, based on reactions with OH° (calculated). (9786)

C. WATER AND SOIL POLLUTION FACTORS

Biodegradation

Inoculum	method	concentration	duration	elimination results	
	ISO degradation Test 7827			<20% DOC	(6354)
	Closed flask			0%	(6355)

Persistence in soil

in sandy clay and silty clay soils: $t_{1/2}$: 19-23 d. (8563)

Impact on biodegradation processes

BOD inhibition*	EC_{50}	39; 120 mg/L	
OECD 209 closed system inhibition*	EC_{50}	59; 76; 115 mg/L	
biodegradation inhibition*	EC_{50}	>25 mg/L	
nitrification inhibition*	EC_{50}	0.5 mg/L	(2624)

* *o-chlorocresol*

Hydrolysis

The chemical is stable

Photodegradation

PCOC is photodegraded to o-cresol and methylhydroquinon. Photodecomposition results in substitution of the chlorine by an hydroxyl group or by dehalogenation. Other suggestions are photodecomposition to chlorosalicylaldehyde and o-cresol. (10703)

dedhalogenation

cresol

substitution of Cl by OH

methylhydroquinone

oxidation

chlorosalicylaldehyde

oxidation

chlorosalicylic acid

Photodegradation products

Environmental concentrations

River at outfall of production site	<0.2-3.0 µg/L	(10703)
sewage works in agricultural region	infl.:2400 µg/L; effl.: 4 µg/L; removal: 99.8 %	(7055)

Biodegradation

Inoculum	method	concentration	duration	elimination
Sea water	Shake flask die away	18 µg/L	3 days	50%
Sandy clay soil	aerobic	10 mg/kg soil	18 days	50%
		1000 mg/kg soil	24 days	50%
			49 days	69%

Soil	anaerobic		32 days	0%	
Sewage sludge	Treatment plant	10 µg/L		0%	
Sandy loam		10 mg/kg	14 days	50%	
			56 days	90%	(10703)

Soil sorption

for a New zealand acidic soil (pH 5.2 and 3.6% OC) K_{OC} was found to be only 0.22
for non-acidic soils a K_{OC} of 400 has been estimated. (10703)

D. BIOLOGICAL EFFECTS

Bacteria			
Microtox	5 min EC_{50}	1.1 mg/L	
growth inhib.	16h EC_{50}	32 mg/L	(2624)
adapted A.S. oxygen consumption inhibition	30min EC_{20}	30 mg/L	(6354)
A.S. oxygen consumption inhibition	30min EC_{50}	55 mg/L	(6355)
Pseudomonas putida	17h EC_{10}	37 mg/L	
(cell multiplication inhibition test)	17h EC_{50}	110 mg/L	
	17h EC_{90}	150 mg/L	(6353)
** o-chlorocresol			
Algae			
Scenedesmus subspicatus	72h EC_{10}	0.97 mg/L	
(cell multiplication inhibition test)	72h EC_{50}	15 mg/L	(6353)
	96h EC_{10}	0.89 mg/L	
	96h EC_{50}	8.2 mg/L	(6353)
Crustaceans			
Daphnia magna	24h EC_{50},S	1.9 mg/L	
	48h EC_{50},S	0.29 mg/L	
	NOEC,S	0.028 mg/L	(6352)
Daphnia magna Straus	24h EC_{0}	1.0 mg/L	
	24h EC_{50}	1.8 mg/L	
	24h EC_{100}	5.8 mg/L	(6353)
	48h EC_{0}	0.32 mg/L	
	48h EC_{50}	1.0 mg/L	
	48h EC_{100}	1.8 mg/L	(6353)
Fishes			
trout	24h LC_{50},S	2.1 mg/L	(1463)
	24h LC_{50}	2.5 mg/L	(6350)
Brachydanio rerio	96h LC_{0}	1.0 mg/L	
	96h LC_{50}		
	96h LC_{100}		
Bluegill	96h LC_{50}	2.3 mg/L	(6349)

Bioaccumulation

Species	conc. µg/L	duration	body parts	BCF	
Oryzias latipes	2	42 days		6.4-14	
	20	42 days		8.2-28	
Salmo trutta	500	28 days		6.6	
	1,500	28 days		4.3	(10703)
trout	500-1500	3 weeks	wet weight	4-5	(1463)

Toxicity

Plants			
Lemna minor	48h EC_{50}	93 mg/L	(10703)

Fish			
Salmo trutta	21d NOEC	0.55 mg/L	
Oryzias latipes	96h LC_{50}	6.3 mg/L	
Lepomis macrochirus	96h LC_{50}	2.3 mg/L	
	24h LC_{50}	3.8 mg/L	
Brachydanio rerio	96h LC_{50}	3-6 mg/L	(10703)
Mammals			
Rat	oral LD_{50}	1,190; 2,650; 2,700; 3,195 mg/kg bw	
Mouse	oral LD_{50}	1,330 mg/kg bw	(10703)
Rat mouse	oral LD_{50}	1.190, 1.320 mg/kg bw	(9968, 9969)

5-chloro-2(2,4-dichlorophenoxy)phenol (triclosan; 2,4,4'-trichloro-2'-hydroxydiphenyl ether)

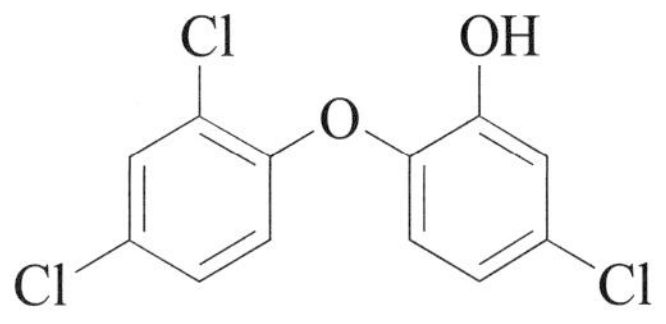

$C_{12}H_7Cl_3O_2$

CAS 3380-34-5

TRADENAMES

Aquasept; Gamophen; Irgasan; Sapoderm; Ster-Zac; Triclosan

USES

Triclosan is a broad-spectrum antibacterial/anti-microbial agent. It is manufactured in the U.S. by Ciba-Geigy, under their trade name Irgasan DP300, and by several other manufacturers outside of the U.S. As a result of its bacteriostatic activity against a wide range of both gram-negative and gram-positive bacteria it has found increasing and recent popular use in personal care products, i.e. toothpaste,deodorant soaps, deodorants, antiperspirants, body washes, detergents, dish washing liquids, cosmetics, anti-microbial creams, lotions, and hand soaps. It is also used as an additive in plastics, polymers, and textiles to give these materials antibacterial properties.

IMPURITIES, ADDITIVES, COMPOSITION

Since triclosan is by its chemical structure a polychloro phenyoxy phenol it is possible that several polychlorodibenzo-p-dioxins (dioxins) and polychloro-dibenzofurans are found in varying low level amounts, as synthesis impurities in triclosan.

A. PROPERTIES

white to off-white crystalline powder; molecular weight 289.5; melting point 54-57°C; vapor pressure 4 x 10-6 mm Hg at 20°C; log Pow 4.76

Environmental concentrations

Samples from 139 streams in 30 states in the US thought to be susceptible to contamination from agricultural or urban activities during 1999-2000: detected in 49 out of 90 samples in values ranging from 0.04 – 2.3 µg/L	(7221)
Residues of methyl triclosan (4-chloro-1(2,4-dichlorophenoxy)-2-methoxybenzen) have been reported in rivers, industrial wastewater and aquatic biota: - 1-38 µg/kg bw in the freshwaterfish topmouth gudgeon in Tama river - 1-2 µg/kg bw in the goby fish *(A. flavimanus)*	(7169)

- 3-20 µg/kg bw in clam, oyster and mussels in Tokyo bay (1984)

Biodegradation

Inoculum	method	concentration	duration	elimination	
	standard readibility test	30-100 mg/L	4 weeks	0 %ThOD	(7169)

C. WATER AND SOIL POLLUTION FACTORS

Biodegradation

after 3 weeks of adaptation at 1-5 mg/L at 22°C:

aerobic degradation:	product is sole carbon source:	0% degradation
	+ synthetic sewage:	50% degradation
anaerobic degradation:	product is sole carbon source:	0% degradation
	+ synthetic sewage:	50% degradation

D. BIOLOGICAL EFFECTS

Bioaccumulation

Species	conc. µg/L	duration	body parts	BCF	
Fish		8 weeks		2.7-90	(7169)

Toxicity

Crustaceae			
Daphnia magna	48h EC_{50}	0.39 mg/L	(7169)
Fish			
Pimephales promelas	96h LC_{50}	0.25 mg/L	(7169)
Mammals			
Rat	oral LD_{50}	3700 mg/kg bw	(7169)

2-chloro-1-(2,4-dichlorophenyl)vinyldiethylphosphate *see* chlorfenvinphos

2-chloro-α-(((diethoxyphosphinothioyl)oxy)imino)-benzeneacetonitrile *see* chlorphoxim

1-chloro-1,2-difluoroethane

H Cl
F F

$ClFCHCH_2F$

$C_2H_3ClF_2$

CAS 338-64-7

B. AIR POLLUTION FACTORS

Tropospheric lifetime (estd.): 0.49 year.
Ozone depletion potential (estd.): <0.001 per molecule. (2446)

1-chloro-2,2-difluoroethane

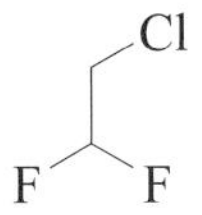

ClH_2CCHF_2

$C_2H_3ClF_2$

CAS 338-65-8

B. AIR POLLUTION FACTORS

Tropospheric lifetime (estd.): 0.76 year.
Ozone depletion potential (estd.): <0.001 per molecule. (2446)

1-chloro-1,1-difluoroethane

(Freon 142; difluoromonochloroethane; 1,1-difluoro-1-chloroethane; alpha-chloroethylidene fluoride; HCFC 142b)

$ClF_2C\text{-}CH_3$

$C_2H_3ClF_2$

CAS 75-68-3

USES

Aerosol propellant. Blowing agent for polymers. Solvent. Working fluid for heat pumps. Refrigerant.

A. PROPERTIES

molecular weight 100.50; melting point -131°C; boiling point -10°C; vapor pressure 1103 mm Hg at 0°C; vapour pressure 339.000 Pa at 20°C; water solubility 1.900 mg/L at 25°C; log Pow 1.6-2.0

(calculated)

B. AIR POLLUTION FACTORS

1 ppm = 4.1 mg/m^3; 1 mg/m^3 = 0.24 ppm

Tropospheric lifetime (estd.): 13.9 years.

Ozone depletion potential (estd.): 0.038 per molecule. (2446)

Odour

slightly ethereal

C. WATER AND SOIL POLLUTION FACTORS

Biodegradation

Inoculum	method	concentration	duration	elimination
Activated sludge	OECD 301B		28 days	5% ThOD

Waste water treatment

the chemical will be mainly eliminated by stripping to the air.

D. BIOLOGICAL EFFECTS

Bioaccumulation

Due to its relatively low log Pow of 1.6-2.0, the chemical is not expected to bio-accumulate to any significant degree. (10598)

Toxicity

Algae			
Green algae	96h EC_{50}	45 mg/L (ECOSAR)	(10598)
Crustaceae			
Daphnia magna	48h EC_{50}	160; >190 mg/L	(10598)
Fish			
Poecilia reticulata	96h LC_{50}	220 mg/L	(10598)

3-chloro-4,5-dihydroxybenzaldehyde

O

HO Cl

OH

C7H5O3Cl

OCCURRENCE

Degradation product of 5-chlorovanillin

A. PROPERTIES

molecular weight 172.57; melting point 182-186°C

C. WATER AND SOIL POLLUTION FACTORS

Biodegradation

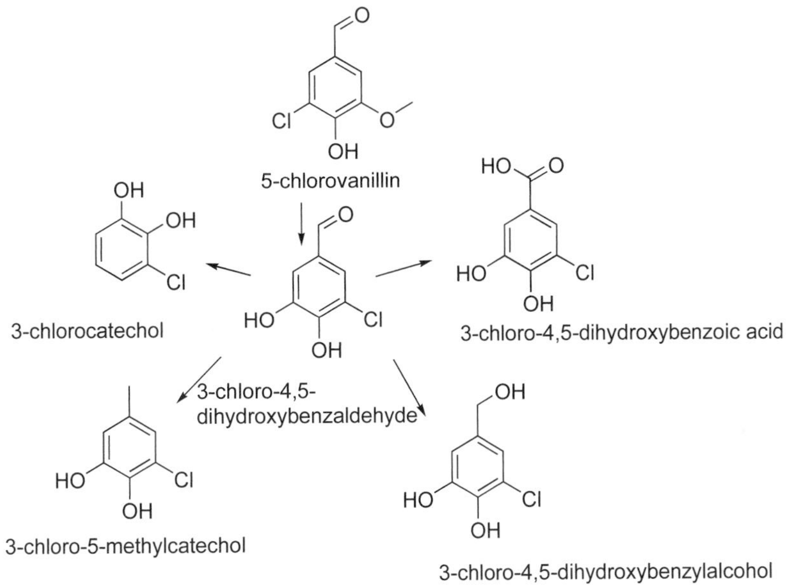

5-chlorovanillin

3-chlorocatechol

3-chloro-4,5-dihydroxybenzaldehyde

3-chloro-4,5-dihydroxybenzoic acid

3-chloro-5-methylcatechol

3-chloro-4,5-dihydroxybenzylalcohol

Transformation pathway by metabolically stable anaerobic enrichment cultures obtained from sediments from the Gulf of Bothnia and the Baltic sea (10821)

4-chloro-2,5-dihydroxydiphenylsulfone

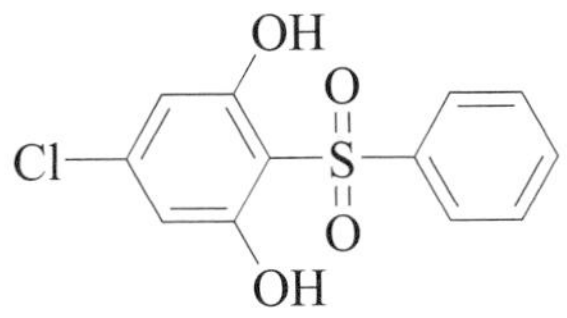

$C_{12}H_9ClO_4S$

D. BIOLOGICAL EFFECTS

Fishes

goldfish: approx. fatal conc.: 35 mg/L, 48 h (226)

4-chloro-2,5-dimethoxyaniline (4-chloro-2,5-dimethoxybenzeneamine; aminochlorohydroquinonedimethylether)

$C_8H_{10}ClNO_2$

CAS 6358-64-1

USES AND FORMULATIONS

Intermediate in the chemical industry, manufacture of pigments and dyestuffs.

A. PROPERTIES

molecular weight 187.63; melting point 117.5°C; vapor pressure 1 x 10^{-5} hPa at 20°C, 0.093 hPa at 90°C; solubility 4,000 mg/L at 20°C; $LogP_{OW}$ 1.8 (calculated).

B. AIR POLLUTION FACTORS

Photochemical reactions

$t_{1/2}$: 0.08 d, based on reactions with OH° (calculated) (6518)

C. WATER AND SOIL POLLUTION FACTORS

Biodegradation

Inoculum	*Method*	*Concentration*	*Duration*	*Elimination results*	
A.S.	Closed bottle test		7 d	0% ThOD	
			14 d	3% ThOD	
			21 d	10% ThOD	
			28 d	4% ThOD	(6517)
Industrial A.S.	Zahn-Wellens test	81 mg/L	3 h	20% COD	
			5 d	24% COD	
			10 d	20% COD	
			20 d	89- 95% COD	(6517)

D. BIOLOGICAL EFFECTS

Toxicity

Bacteria			
primary municipal sludge	24 h EC_0	1- 10 mg/L	(6517)
Crustaceans			
Daphnia magna QSAR	EC_0	5.0 mg/L	(6516)
Fishes			
Brachydanio rerio	96 h LC_0	35 mg/L	
	96 h LC_{50}	111 mg/L	
	96 h LC_{100}	180 mg/L	(6515)
	48 h LC_0	35 mg/L	
	48 h LC_{50}	125- 180 mg/L	

	48 h LC_{100}	180 mg/L	(6515)
Pimephales promelas QSAR	96 h LC_{50}	5.4 mg/L	(6516)
Salmo gairneri QSAR	96 h LC_{50}	4.2 mg/L	(6516)
lepomis macrochirus QSAR	96 h LC_{50}	4.2 mg/L	(6516)

5-chloro-2,4-dimethoxyaniline

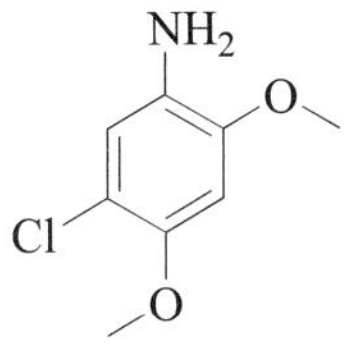

$ClC_6H_2(OCH_3)_2NH_2$

$C_8H_{10}ClNO_2$

CAS 97-50-7

A. PROPERTIES

molecular weight 187.63; melting point 90-92°C.

D. BIOLOGICAL EFFECTS

RubisCo test

inhib. of enzym. activity of Ribulose-P2-carboxylase in protoplasts:	IC_{10}	19 mg/L	

Oxygen test

inhib. of oxygen production of protoplasts:	IC_{10}	56 mg/L	(2698)

Algae			
Scenedesmus subspicatus inhib. of fluorescence	IC_{10}	0.21 mg/L	
Scenedesmus subspicatus growth inhib.	IC_{10}	2.8 mg/L	(2698)

4-chloro-2,5-dimethoxynitrobenzene (1-chloro-2,5-dimethoxy-4-nitrobenzene; nitrochlorohydroquinonedimethylether)

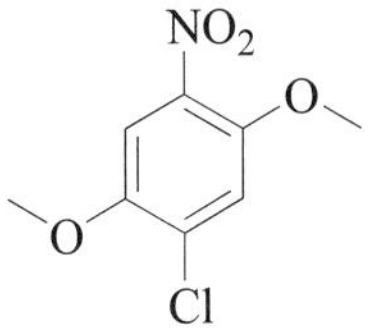

$C_8H_8ClNO_4$

CAS 6940-53-0

A. PROPERTIES

molecular weight 217.61; melting point 140°C; boiling point 315°C; vapor pressure 3.4 x 10^{-6} hPa at 20°C; density 1.4 at 1°C; solubility 55 mg/L at 20°C; $logP_{OW}$ 2.5.

B. AIR POLLUTION FACTORS

Photochemical reactions

$t_{1/2}$: 4.6 d, based on reactions with OH° (calculated). (6521)

C. WATER AND SOIL POLLUTION FACTORS

Biodegradation

Inoculum	*method*	*conc.*	*duration*	*elimination results*	
Municipal A.S.	Closed Bottle Test	1.8 mg/l	28 days	<20% ThOD	(6520)

D. BIOLOGICAL EFFECTS

Bacteria			
primary municipal sludge	24h EC_0	2.3 mg/L	
	24h EC_{50}	6.9 mg/L	
	24h EC_{100}	>18 mg/L	(6520)
Crustaceans			
Daphnia magna	24h EC_0	7.2 mg/L	
	24h EC_{50}	>14 mg/L	(6520)
Fishes			
Brachydanio rerio	48-96h LC_0	7.7 mg/L	
	48-96h LC_{50}	11 mg/L	
	48-96h LC_{100}	15 mg/L	(6519)

1-chloro-2,4-dinitrobenzene (2,4-dinitrochlorobenzene)

Cl

NO_2

NO_2

$ClC_6H_3(NO_2)_2$

$C_6H_3ClN_2O_4$

CAS 97-00-7

USES

intermediate for the mfg. of dyes, pesticides, photochemicals, and fragrances.

A. PROPERTIES

molecular weight 202.55; melting point 49-52°C; boiling point 315°C; vapor pressure 0.0001 hPa at 20°C; density 1.52 g/cm^3; solubility 360 mg/L at 20°C; 160,000 mg/L at 100°C; $LogP_{OW}$ 2.1 (calculated).

B. AIR POLLUTION FACTORS

Photochemical degradation

$t_{1/2}$: 740 d. (5428)

C. WATER AND SOIL POLLUTION FACTORS

Biodegradability

inoculum/method	*test conc.*	*test duration*	*removed*	
industrial, Zahn-Wellens test	350 mg/L	25 d	0%	(5421)
A.S., municipal adapted	50 mg/L		99% product	(5424)

Aerobic degradation

by *Pseudomonas* sp. CBS3 yielded the following metabolites:
4-chloro-3-nitroaniline
4-chloro-1,3-diaminobenzene
N-acetyl-4-chloro-1,3-diaminobenzene (5427)

Impact on biodegradation processes

primary municipal effluent:	24h EC_0	approx. 40 mg/L	(5421)

D. BIOLOGICAL EFFECTS

Rotifers			
Brachionus calyciflorus	24h LC_{50}	1.3 mg/L	
Brachionus plicatilis	24h LC_{50}	2.0 mg/L	(5426)
Crustaceans			
Daphnia magna	24h EC_0	0.5; 0.66; 0.7 mg/L	
	24h EC_{50}	0.7; 0.85; 0.7-0.9 mg/L	
	24h EC_{100}	2; approx. 12 mg/L	(5421, 5422, 5423)
	48h EC_{50}	0.49; 0.59 mg/L	(5423)
Fishes			
guppy *(Poecilia reticulata)*	14d LC_{50}	0.31 mg/L	(2696)
Leuciscus idus	48-96h LC_{50}	0.75 mg/L	
	48-96h LC_{100}	1 mg/L	
	48-96h NOLC	0.315 mg/L	(5419)
Brachydanio rerio	48-96h LC_0	0.5 mg/L	
(technical grade)	48-96h LC_{50}	0.71 mg/L	
	48-96h LC_{100}	1 mg/L	(5420)

1-chloro-2,3-epoxypropane *see* epichlorohydrin

2-chloro-4-ethylamino-6-isopropylamino-s-triazine *see* atrazine

2-(4-chloro-6-ethylamino)-s-triazine-2-ylamino-2-propionitrile *see* cyanazine

1-chloro-1-fluoroethane

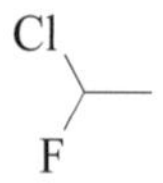

$ClFCHCH_3$

C_2H_4ClF

CAS 1615-75-4

B. AIR POLLUTION FACTORS

Tropospheric lifetime (estd.): 0.11 year.
Ozone depletion potential (estd.): <0.001 per molecule. (2446)

1-chloro-2-fluoroethane

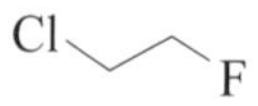

ClH_2CCH_2F

C_2H_4ClF

CAS 762-50-5

A. PROPERTIES

molecular weight 82.51.

B. AIR POLLUTION FACTORS

Tropospheric lifetime (estd.): 0.19 year.
Ozone depletion potential (estd.): <0.001 per molecule. (2446)

4-chloro-1-hydroxy-3-methylbenzene *see p*-chloro-*m*-cresol

3-chloro-4-hydroxybenzaldehyde

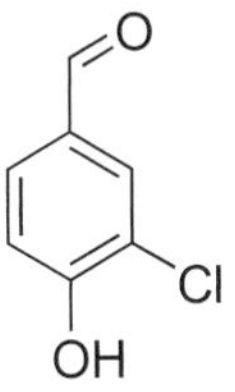

$C_7H_5ClO_2$

CAS 2420-16-8

OCCURRENCE

In effluents of bleaching of kraft cellulosic pulp using chlorine-containing bleaching agents.

A. PROPERTIES

molecular weight 156.57

C. WATER AND SOIL POLLUTION FACTORS

Biodegradation

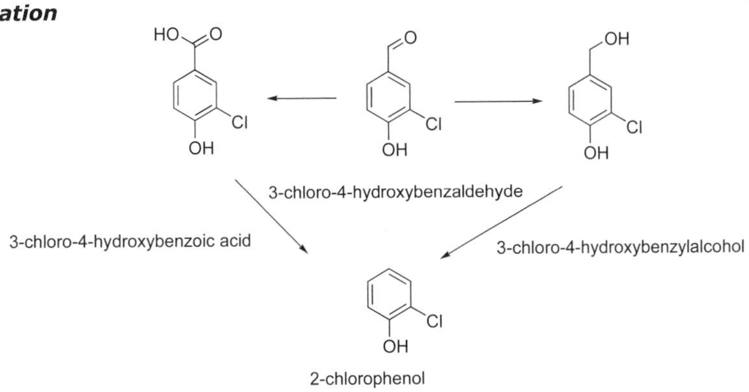

3-chloro-4-hydroxybenzaldehyde

3-chloro-4-hydroxybenzoic acid

3-chloro-4-hydroxybenzylalcohol

2-chlorophenol

Transformation pathway of 3,5-dichloro-4-hydroxybenzaldehyde by metabolically stable anaerobic enrichment cultures obtained from sediments from the Gulf of Bothnia and the Baltic sea	(10821)

1-chloro-2-hydroxybenzene *see* *o*-chlorophenol

1-chloro-3-hydroxybenzene *see* *m*-chlorophenol

1-chloro-4-hydroxybenzene *see* *p*-chlorophenol

2-chloro-3-hydroxypyridine (2-chloro-3-pyridinol)

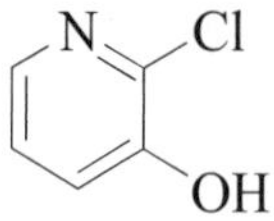

$Cl(C_5H_3N)OH$

C_5H_4ClNO

CAS 6636-78-8

A. PROPERTIES

molecular weight 129.55; melting point 170-172°C.

C. WATER AND SOIL POLLUTION FACTORS

Reductive dehalogenation

in anaerobic lake sediment suspension at 20°C in the dark: half-life: >70 d.	(2700)

4-chloro-3-hydroxytoluene *see* *p*-chloro-*m*-cresol

6-chloro-3-hydroxytoluene *see* *p*-chloro-*m*-cresol

3-chloro-isobutene *see* methallylchloride

4-chloro-4'-isopropylbiphenyl

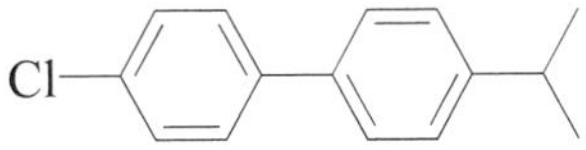

Cl

$C_{15}H_{15}Cl$

CAS 610-78-6

C. WATER AND SOIL POLLUTION FACTORS

Biodegradation

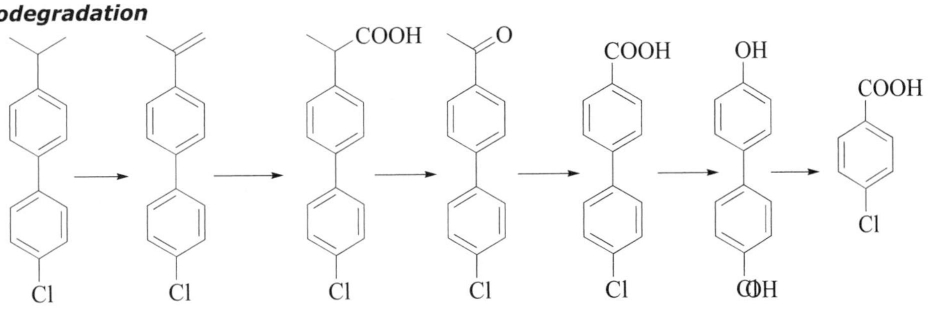

COOH O COOH OH COOH

Cl Cl Cl Cl Cl Cl COH

Metabolism by a mixture culture of aerobic bacteria from activated sludge

(1215)

N'-(3-chloro-4-methoxyphenyl)-N,N-dimethylurea *see* dosanex

3-chloro-4-methyl-7-coumarinyldiethylphosphorothionate (O,O-diethyl-O-(3-chloro-4-methyl-2-oxo(2H)-1-benzopyran-7-yl)-phosphorothioate)

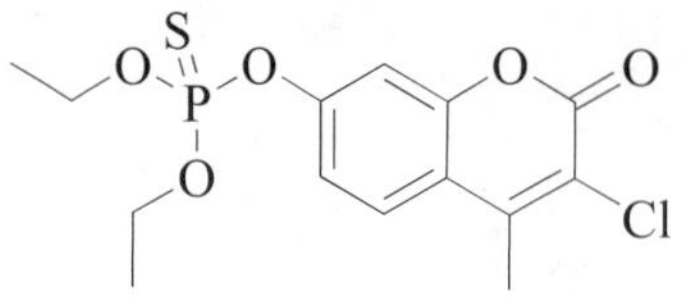

$C_{14}H_{16}ClO_5PS$

CAS 56-72-4

TRADENAMES

co-ral; Muscatox; Resistox; Coumaphos; Bay 21/199; Asuntol; Baymix; Meldane

USES

livestock insecticide. Nematodicide. Therapeutically as (veterinary) antihelmintic.

A. PROPERTIES

tan, crystalline solid; molecular weight 362.77; melting point 90-92°C, hydrolyzes slowly under alkaline conditions; vapor pressure 1 × 10^{-7} mm at 20°C; density 1.47 at 20/4°C; solubility 1.5 mg/L at 20°C.

C. WATER AND SOIL POLLUTION FACTORS

Biodegradation

Bacteriological degradation was shown to be responsible for loss of activity in cattle-dips. (10039)

D. BIOLOGICAL EFFECTS

Crustaceans

Gammarus lacustris	96h LC_{50}	0.07 μg/l	(2124)
Gammarus fasciatus	96h LC_{50}	0.15 μg/l	(2126)
Daphnia magna	48h LC_{50}	1.0 μg/l	

Molluscs

eastern oyster *(Crassostrea virginica)*	egg	48h LC_{50}	110 μg/l	
	larvae	14d LC_{50}	>1,000 μg/l	(2324)
hard clam *(Mercenaria mercenaria)*	egg	48h LC_{50}	9,120 μg/l	
	larvae	12d LC_{50}	1,470 μg/l	(2324)
threespine stickleback *(Gasterosteus aculeatus)*		96h LC_{50}	1,470 μg/l	(2333)

Insects

Hydropsyche sp.	24h LC_{50}	5 μg/l	(2110)
Hexagenia sp.	24h LC_{50}	430 μg/l	(2110)

Fishes

Pimephales promelas	96h LC_{50}	18,000 μg/l	(2119)
Lepomis macrochirus	96h LC_{50}	180 μg/l	(2119)
Salmo gairdneri	96h LC_{50}	1,500 μg/l	(2119)
Oncorhynchus kisuth	96h LC_{50}	15,000 μg/l	(2119)

Mammals			
rat	oral LD_{50}	16-41 mg/kg bw	(9827)

5-chloro-2-methyl-4-isothiazolin-3-one (CMI)

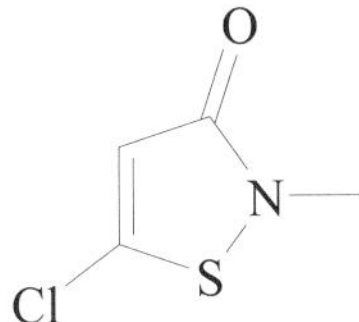

C_4H_4ClNOS

CAS 26172-55-4

EINECS 247-500-7

TRADENAMES

Kathon (a mixture of 2-methyl-4-isothiazolin-3-one (MI) and 5-chloro-2-methyl-4-isothiazolin-3-one (CMI))

USES

an isothiazolinone preservative used in household detergents and cosmetic detergent products.

A. PROPERTIES

Solid; molecular weight 149.6; melting point 52°C

C. WATER AND SOIL POLLUTION FACTORS

Half-life times

		hours	
Hydrolysis	at pH 5	>720	
	at pH 7	>720	
	at pH 9	528	
Photolysis		158	
Aerobic aquatic microcosm		17	
Anaerobic aquatic microcosm		5	
Aerobic terrestrial microcosm	at 25°C	5	
Sterile microcosm control		>1536	(7255)

Environmental partitioning

Adsorption equilibrium K_{OC}	30-310	
Desorption eqquilibrium	40-420	
Metabolite adsorption equilibrium	>3000	
log Pow	0.4	(7255)

Biodegradation (7255)

Inoculum	method	concentration	duration	elimination
Activated sludge	OECD protocol	0.3 mg/L	28 days	39-55 %ThCO2
		0.03 mg/L	28 days	62 % ThCO2
river sediment		1 mg/L	1 days	70% parent compound
river sediment	anoxic		30 days	17% ThOD
			365 days	56% ThOD

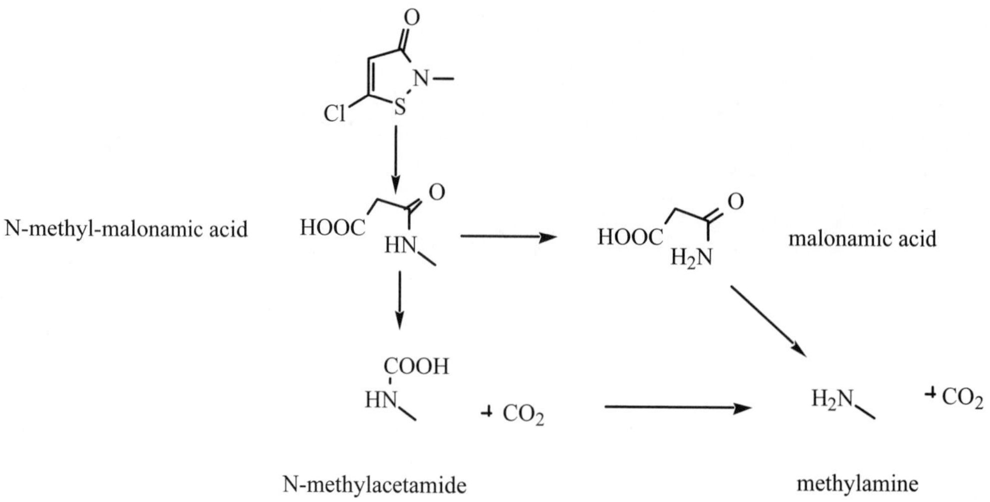

N-methyl-malonamic acid

malonamic acid

N-methylacetamide

methylamine

Biodegradation pathway

Proposed pathway for aerobic biodegradation: The isothiazolone ring is cleaved between theN-S bond yielding N-methyl-malonamic acid. Oxidation continues resulting in the formation of malonamic acid, N-methyl-acetamide, CO2.and the respective amine. On the basis of the observed mineralization and the fate of ^{14}C residuals it has been proposed that the anaerobioc degradation of CMI leads to the same type of metabolites as proposed for the aerobic degradation.(7255)

D. BIOLOGICAL EFFECTS

Bioaccumulation (7255)

	conc. µg/L	exposure period	BCF	organ/tissue
Bluegill sunfish			2	total body

Toxicity for Kathon (MI:CMI, 1:3)

Algae			
Selenatrum capricornutum	72h EC_{50}	0.003 mg/L	
Crustaceans			
Daphnia magna	48h EC_{50}	0.16 mg/L	(7169)
Mussels			
Crassostrea virginica	48h EC_{50}	0.028 mg/L	
Bay mussel (embryo/larvae)	48h EC_{50}	0.014 mg/L	(7169)
Fish			
Salmo gairdneri	96h LC_{50}	0.19 mg/L	
Cyprinodon variegatus	96h LC_{50}	0.3 mg/L	
Lepomis macrochirus	96h LC_{50}	0.28 mg/L	(7169)
Mammals			
Rat and mouse (CMI)	acute oral LD_{50}	53-60 mg/kg bw	(7169)

5-chloro-2-methyl-4-isothiazoline-3-onecalciumchloride

D. BIOLOGICAL EFFECTS

Fishes

bluegill	49d BCF	22-27 (whole fish minus viscera) F

49d BCF	157-300 (viscera) F	
43d BCF	30 (whole fish minus viscera) F	
43d BCF	203 (viscera) F	(1511)

3-chloro-2-methyl-1-propene *see* methallylchloride

3-chloro-4-methylaniline (3-chloro-p-toluidine; 4-amino-2-chlorotoluene; 1-amino-3-chloro-4-methylbenzene; 3-chloro-4-toluidine; 3-chloro-4-methylaniline; 3-chloro-4-methylbenzenamine; m-chloro-p-toluidine)

$ClC_6H_3(CH_3)NH_2$

C_7H_8ClN

CAS 95-74-9

USES

Dyestuff industry. Avicide.

OCCURRENCE

: Metabolite of the herbicide chlorotoluron, found in winter wheat.

A. PROPERTIES

molecular weight 141.60; melting point 25°C; boiling point 237- 238°C; density 1.17; vapor pressure 0.04 hPa at 20°C, 6.1 hPa at 100°C; 62 hPa at 150°C; solubility 1,000 mg/L, 2,650 mg/L at 20°C, 1,500 mg/L at - 60°C; $LogP_{OW}$2.6 (calculated); pH value ca. 7.3 in saturated aqueous solution.

C. WATER AND SOIL POLLUTION FACTORS

ThOD: 2.035 g/g

Biodegradation

Inoculum	*Method*	*Concentration*	*Duration*	*Elimination results*	
Domestic sewage, adapted	Closed bottle test	2- 80 mg/L	20 d	0% ThOD	(8353)
Industrial A.S.	Modified Zahn-Wellens test		1 d	4% DOC	
			7 d	89% DOC	
			14 d	100% DOC	

	28 d	100% DOC	(8353)

Pseudomonas cepacia isolated from soil, used 3-chloro-4-toluidine in concentrations of up to 200 mg/L as the sole source of carbon, nitrogen, and energy.	(9979)
Degraded by soil after 72 h to 3,3',4,4'-dichlorodimethylazobenzene and 3-chloro-4-methyl-6-(3-chloro-4-methylphenylimino)cyclohexa-2,4-dienone.	(9978)

D. BIOLOGICAL EFFECTS

Bioaccumulation

	Concentration µg/L	*Exposure period*	*BCF*	*Organ/tissue*	
Bluegill sunfish*	100	28 d	88	total body	(9977)

* N-Acetyl-3-chloro-4-toluidine was shown to be a metabolite

Bacteria			
Pseudomonas putida (oxygen consumption inhibition)	30 min EC_0	125 mg/L	(8353)

Algae			
Scenedesmus subspicatus inhib. of fluorescence	IC_{10}	1.8 mg/L	(2698)
Scenedesmus subspicatus	7 d EC_{10}	2.1 mg/L	
(cell multiplication inhibition test)	7 d EC_{50}	22 mg/L	(8353)
RubisCo test: inhib. of enzym. activity of Ribulose-P2-carboxylase in protoplasts:	IC_{10}	28 mg/L	(2698)
Oxygen test: inhib. of oxygen prod. of protoplasts	IC_{10}	7,080 mg/L	(2698)

Crustaceans			
Daphnia magna Straus	24 h EC_0	0.23 mg/L	
	24 h EC_{50}	0.82 mg/L	
	24h EC_{100}	30 mg/L	(8353)
	48h EC_0	0.12 mg/L	
	48 h EC_{50}	0.62 mg/L	
	48 h EC_{100}	30 mg/L	(8353)

Fishes			
Brachydanio rerio	96 h LC_0	18 mg/L	
	96 h LC_{100}	42 mg/L	(8353)
	48 h LC_0	25 mg/L	
	48 h LC_{100}	30 mg/L	(8353)

Mammals			
rat	oral LD_{50}	1,500 mg/kg bw	(9980)
mouse	oral LD_{50}	316 mg/kg bw	(9981)

4-chloro-2-methylaniline (4-chloro-*o*-toluidine; azogene fast red TR; 2-amino-5-chlorotoluene; 4-chloro-2-methylbenzenamine; 4-chloro-2-toluidine; p-chloro-o-toluidine)

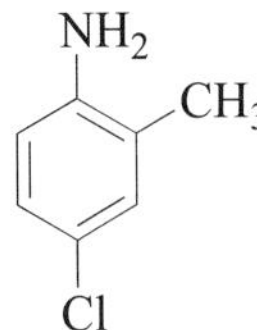

$ClC_6H_3(CH_3)NH_2$

C_7H_8ClN

CAS 95-69-2

USES

Manufacture of dyestuffs and pesticides.

A. PROPERTIES

molecular weight 141.6; melting point 27°C; boiling point 241°C.

Biodegradation

product of chlorphenamidine. (1547)

D. BIOLOGICAL EFFECTS

Mammals			
rat	oral LD_{50}	1,058 mg/kg bw	(9976)

3-chloro-4-methylbenzenaminehydrochloride (3-chloro-*p*-toluidine hydrochloride; CPTH; 3-chloro-4-methylaniline hydrochloride)

$C_7H_5ClN \cdot HCl$

USES

avian toxicant currently used to control starlings at cattle and poultry feed lots.

D. BIOLOGICAL EFFECTS

Crustaceans

shrimp

19% pink shrimp *(Penaeus duorarum)* and 81% white shrimp *(Penaeus setiferus):*

	96h LC_{50},S	10.8 ppm	
blue crab *(Callinectes sapidus):*	96h LC_{50},S	16.0 ppm	(1400)

2-chloro-1-methylbenzene *see* *o*-chlorotoluene

5-chloro-3-methylcatechol

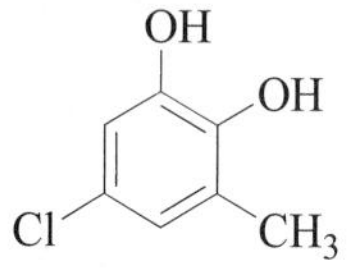

OCCURRENCE

microbial metabolite of MCPA.

4-chloro-3-methylphenol *see* *p*-chloro-*m*-cresol

2-chloro-1-methylpyridinium iodide

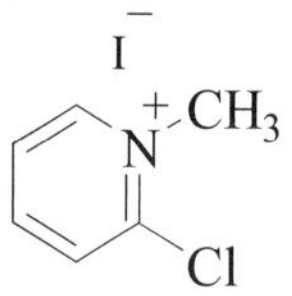

C_6H_7ClIN

CAS 14338-32-0

A. PROPERTIES

molecular weight 255.49; melting point 200°C decomposes.

D. BIOLOGICAL EFFECTS

Fishes			
Pimephales promelas	96h LC_{50}	199 mg/L	(2625)

2-chloro-4-nitroaniline (2-chloro-4-nitrobenzenamine)

$C_6H_5ClN_2O_2$

CAS 121-87-9

USES

intermediate for the mfg. of dyes, pesticides.

A. PROPERTIES

molecular weight 172.57; melting point 107-109°C; boiling point >200°C; vapor pressure <0.0000046 hPa at 25°C; density 1.38 g/cm^3; solubility 230 mg/L at 20°C; LogP_{OW} 2.17, 2.2 (calculated).

B. AIR POLLUTION FACTORS

Photochemical degradation

reaction with OH°: $t_{1/2}$: 0.66 d (calculated). (5439)

C. WATER AND SOIL POLLUTION FACTORS

Biodegradability

inoculum/method	*test conc.*	*test duration*	*removed*	
A.S., industrial, nonadapted	300 mg/L	3 hours	10-20% COD	
		28 d	<10% COD*	
A.S., nonadapted, closed bottle test		20 d	10% ThOD	(5429)
A.S., adapted		>>4 weeks	50%	(5430)

*removal probably caused by adsorption (5438)

Impact on biodegradation processes

primary municipal effluent:	24h EC_0	approx. 100 mg/L	(5438)

D. BIOLOGICAL EFFECTS

Bacteria			
Pseudomonas putida	EC_0	125 mg/L	(5429)
Algae			
Tetrahymena pyriformis	48h EC_{50}, growth	31 mg/L	(2704)
Scenedesmus pannonicus	96h EC_{50}	12 mg/L	(5429)
Crustaceans			
Daphnia magna	48h EC_{50}	1.8 mg/L	
	48h LC_{50}	5.6 mg/L	(5430)
Chaetogammarus marinus	96h EC_{50}	7.1 mg/L	(5437)

Fishes			
Pimephales promelas	30-35d LC_{50}	222 mg/L	(2704)
	96h LC_{50}	20 mg/L	(5431)
Brachydanio rerio	96h LC_0	100 mg/L	(5429)
	48h LC_0	10 mg/L	
	48h LC_{50}	12 mg/L	
	48h LC_{100}	18 mg/L	(5435)
Poecilia reticulata	96h LC_{50}	24 mg/L	
	96h EC_{50}	2.4 mg/L	(5430)
Oncorhynchus tschawytscha	24h LLC*	<1 mg/L	(5432)
Oncorhynchus kisutch	24h LLC	<1 mg/L	(5432)
Ptychocheilus oregonensis	24h LLC	<1 mg/L	(5433)

*LLC = lowest lethal concentration

4-chloro-2-nitroaniline (4-chloro-2-nitrobenzenamine; p-chloro-o-nitroaniline)

NH2 NO2 Cl

$ClC_6H_3(NO_2)NH_2$

$C_6H_5ClN_2O_2$

CAS 89-63-4

USES

Preparation of dyestuffs.

A. PROPERTIES

molecular weight 172.57; melting point 117-119°C.

B. AIR POLLUTION FACTORS

Photochemical reactions

Photolysis: $t_{1/2}$: 18 h, based on reactions with OH° (calculated) (9986)

C. WATER AND SOIL POLLUTION FACTORS

Biodegradation

Inoculum	*method*	*concentration*	*duration*	*elimination results*	
sewage non-acclimated	MITI Test	2-80 mg/L	2 weeks	< 30% ThOD	(8064)

D. BIOLOGICAL EFFECTS

Bacteria				
Photobacterium phosphoreum	Microtox	30 min EC_{50}	20 mg/L	(9657)

Algae			
RubisCo test: inhib. of enzym. activity of Ribulose-P2-carboxylase in protoplasts:	IC_{10}	6.9 mg/L	
Oxygen test: inhib. of oxygen prod. of protoplasts:	IC_{10}	1.3 mg/L	(2698)
Scenedesmus subspicatus inhib. of fluorescence	IC_{10}	2.1 mg/L	
Scenedesmus subspicatus growth inhib.	IC_{10}	0.15 mg/L	(2698)
Crustaceans			
Daphnia magna	24h EC_{50}	3.7 mg/L	
	48h EC_{50}	3.2 mg/L	(8064)
Mammals			
rat	oral LD_{50}	400 mg/kg bw	
mouse	oral LD_{50}	800 mg/kg bw	(9987)

4-chloro-3-nitrobenzenesulphonate, sodium

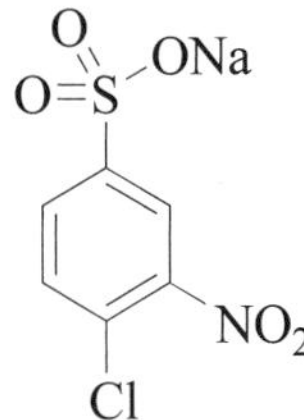

$C_6H_3ClNO_5SNa$

CAS 17691-19-9

EINECS 241-680-0

A. PROPERTIES

molecular weight 259.6; decomposition at 210-220°C; density 0.6 kg/L; water solubility 95,000 mg/L at 20°C

C. WATER AND SOIL POLLUTION FACTORS

Biodegradation

Inoculum	method	duration	elimination	
Adapated activated sludge	Closed Bottle Test	20 days	0 %	(7373)

COD	0.42 g/g	(7373)

D. BIOLOGICAL EFFECTS

Toxicity

Micro organisms			
Pseudomonas putida	30min EC_0	250 mg/L	(7373)

Fish			
Brachydanio rerio	96h LC_0	>100 mg/L	(7373)
Mammals			
Rat	oral LD_{50}	> 5000 mg/kg bw	(7374)

2-chloro-4-nitrophenol

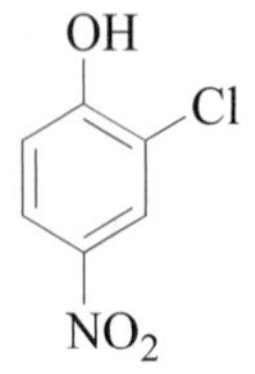

$ClC_6H_3(NO_2)OH$

$C_6H_4ClNO_3$

CAS 619-08-9

USES

Fungicide.

A. PROPERTIES

molecular weight 173.56; melting point 111°C.

C. WATER AND SOIL POLLUTION FACTORS

Odor threshold

detection: 5 mg/L. (998)

Biodegradation rates

adapted A.S. at 20°C -product is sole carbon source: 71% COD removal at 5.3 mg COD/g dry inoculum/h. (327)

Inhibition of biodegradation inhibition EC_{50}: 245 mg/L. (2624)

D. BIOLOGICAL EFFECTS

Fishes			
Rainbow trout, white sucker	24h LC_0	2 mg/L	(9999)
Threespine stickleback, rainbow trout	24 h LC_0	10 mg/L	(10000)
Mammals			
rat	oral LD_{50}	900 mg/kg bw	(9884)

3-chloro-4-nitrophenol (2-chloro-4-hydroxynitrobenzene; 4-nitro-3-chlorophenol)

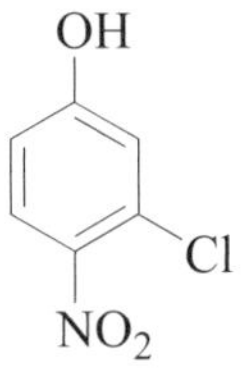

$ClC_6H_3(NO_2)OH$

$C_6H_4ClNO_3$

CAS 491-11-2

A. PROPERTIES

molecular weight 173.56; melting point 121-122°C

C. WATER AND SOIL POLLUTION FACTORS

Inhibition. of biodegradation EC_{50}: 85 mg/L. (2624)

D. BIOLOGICAL EFFECTS

Fishes			
Rainbow trout, white sucker	24h LC_0	3 mg/L	(9999)
Threespine stickleback,rainbow trout	24h LC_0	5 mg/L	(10000)

3-chloro-6-nitrophenol *see* 5-chloro-2-nitrophenol

4-chloro-2-nitrophenol

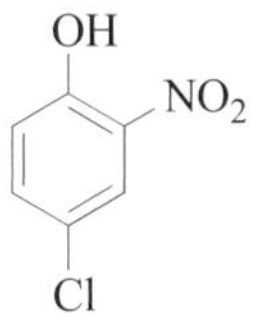

$ClC_6H_3(NO_2)OH$

$C_6H_4ClNO_3$

CAS 89-64-5

USES

Increases shelf life of thermosetting vinyl ester resins.

A. PROPERTIES

molecular weight 173.56; melting point 85-87°C; boiling point 235-236°C; density 1.5 at 20°C; $logP_{OW}$ 2.5.

C. WATER AND SOIL POLLUTION FACTORS

Biodegradation

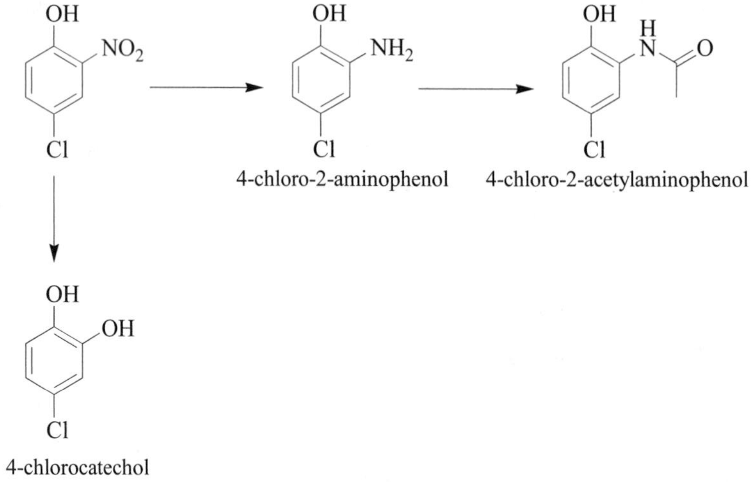

4-chloro-2-aminophenol 4-chloro-2-acetylaminophenol

4-chlorocatechol

Enterobacter cloacae (50 h incubation) converted 4-chloro-2-nitrophenol into 4-chloro-2-aminophenol (8.1%) and 4-chloro-2-acetoaminophenol (16%) under anaerobic conditions. *Alcaligenes sp.* further degraded the reduction product 4-chloro-2-aminophenol under aerobic conditions.	(10001)
Pseudomonas sp. used 4-chloro-2-nitrophenol as a source of nitrogen, eliminating nitrite and accumulating 4-chlorocatechol. Other strains utilising 4-chloro-2-nitrophenol as a sole source of carbon or nitrogen included *Alcaligenes eutrophus.*	(10002)

Inhibition of biodegradation processes

biodegradation inhibition EC_{50}: 164 mg/l. (2624)

D. BIOLOGICAL EFFECTS

Fishes			
rainbow trout ,white sucker	24 h LC_0	4 mg/L	(9999)
rhreespine stickleback	9h LC_{100}	10 mg/L	
rainbow trout	24h LC_0	10 mg/L	(10000)

4-chloro-3-nitrophenol

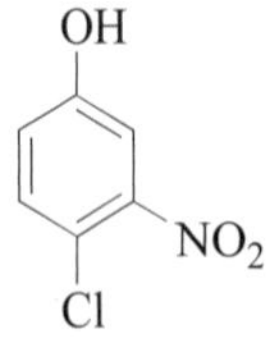

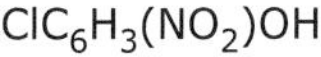

$ClC_6H_3(NO_2)OH$

$C_6H_4ClNO_3$

A. PROPERTIES

molecular weight 173.56.

C. WATER AND SOIL POLLUTION FACTORS

Toxicity to microorganisms

biodegradation inhib. EC_{50}: 52 mg/L.	(2624)

5-chloro-2-nitrophenol (3-chloro-6-nitrophenol; 2-nitro-5-chlorophenol)

$ClC_6H_3(NO_2)OH$

$C_6H_4ClNO_3$

CAS 611-07-4

A. PROPERTIES

molecular weight 173.56; melting point 38.9°C; boiling point sublimes.

D. BIOLOGICAL EFFECTS

Fishes			
larvae of sea lamprey	LC_{100}	3 mg/l	(30)
rainbow trout, white sucker	24h LC_0	0.5 mg/L	
rhreespine stickleback , rainbow trout	24h LC_0	10 mg/L	(9999)

O-(2-chloro-4-nitrophenyl)O,O-dimethylphosphorothioate *see* dicapthon

1-chloro-1-nitropropane (1-nitro-1-chloropropane)

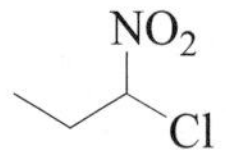

$C_2H_5CH(NO_2)Cl$

$C_3H_6ClNO_2$

CAS 600-25-9

USES

Manufacture of adhesives and pesticides.

A. PROPERTIES

molecular weight 123.5; boiling point 139-143°C; vapor pressure 5.8 mm at 25°C; vapor density 4.26; density 1.21 at 20/20°C; solubility 6 mg/L at 20°C.

B. AIR POLLUTION FACTORS

1 mg/m^3 = 0.198 ppm, 1 ppm = 5.05 mg/m^3.

D. BIOLOGICAL EFFECTS

Mammals			
mouse	oral LD_{50}	510 mg/kg bw	(10003)

2-chloro-4-nitrotoluene (2-chloro-1-methyl-4-nitrobenzene)

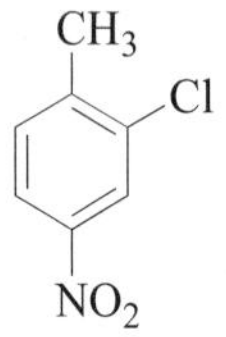

$C_7H_6ClNO_2$

CAS 121-86-8

A. PROPERTIES

molecular weight 171.58; melting point 63°C; boiling point 260°C; vapor pressure 4 hPa at 100°C; density 1.28 at 70°C; solubility 49 mg/L at 20°C; $LogP_{OW}$ 3.2 (calculated).

B. AIR POLLUTION FACTORS

Photochemical reactions

$t_{1/2}$: 84 d, based on reactions with OH° (calculated) (6215)

C. WATER AND SOIL POLLUTION FACTORS

Biodegradation

Inoculum	*Method*	*Concentration*	*Duration*	*Elimination results*	
	Closed bottle test		5 d	0% BOD	
			14 d	40% BOD	
			28 d	75% BOD	(6213)

D. BIOLOGICAL EFFECTS

Bacteria			
Primary municipal sludge	24 h EC_0	60 mg/L	(6213)
Pseudomonas putida (cell multiplication inhibition test)	30 min EC_{10}	>56 mg/L	(6214)
Algae			
Haematococcus pluvialis (inhibition of oxygen production)	4 h EC_{50}	9 mg/L	(6214)
Crustaceans			
Daphnia magna Straus	24 h EC_0	2.0; 4.0 mg/L	
	24 h EC_{50}	10; 6 mg/L	
	24 h EC_{100}	60; 10 mg/L	(6213)
Fishes			
Brachydanio rerio	96 h LC_0	34 mg/L	
	96 h LC_{100}	35 mg/L	(6212)

α-chloro-4-nitrotoluene

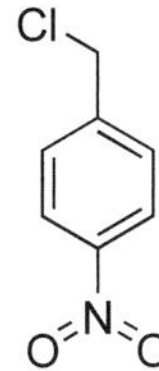

$C_7H_6ClNO_2$

CAS 100-14-1

EINECS 202-822-7

A. PROPERTIES

molecular weight 171.58; melting point 71-74°C; boiling point 112°C;

D. BIOLOGICAL EFFECTS

Toxicity

Algae	72h EC_{50}	0.037; 0.038 mg/L	(11221)
	72h NOEC	0.012; 0.017 mg/L	(11221)
Daphnids	48h EC_{50}	1.5 mg/L	(11221)
	21d EC_{50}	0.53 mg/L	(11221)
	21d NOEC	0.24 mg/L	(11221)
Fish	96h LC_{50}	0.61 mg/L	(11221)

6-chloro-2-picoline (6-chloro-2-methylpyridine)

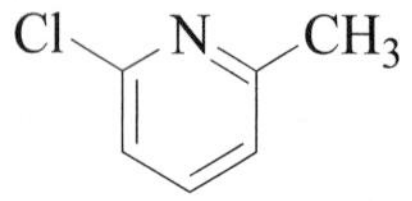

$ClC_5H_3NCH_3$

C_6H_6ClN

CAS 18368-63-3

A. PROPERTIES

$LogP_{OW}$ 2.09.

D. BIOLOGICAL EFFECTS

Algae
Tetrahymena pyriformis: growth inhibition at 27°C, 48h EC_{50}: 2.6 mmol/L (2704)

Fishes
Pimephales promelas: 30-35 d FT test at 25°C: log LC_{50}: 1.8 mmol/L (2704)

3-chloro-1-propanol (1-chloro-3-hydroxypropane; trimethylenechlorohydrin)

$CH_2ClCH_2CH_2OH$

C_3H_7ClO

CAS 627-30-5

USES

Chemical intermediate.

A. PROPERTIES

molecular weight 94.54; melting point -20°C; boiling point 160-162°C; density 1.13; solubility 50,000-100,000 mg/L at 23°C; $logP_{OW}$ 0.007.

C. WATER AND SOIL POLLUTION FACTORS

$BOD_5^{20°C}$:	42% ThOD (acclim.).	(2666)
ThOD:	1.43	

D. BIOLOGICAL EFFECTS

Fishes			
goldfish	24h LC_{50}	170 mg/L	(7813)
Mammals			
mouse	oral LD_{50}	2300 mg/kg bw	(9897)

1-chloro-1-propene (propenylchloride)

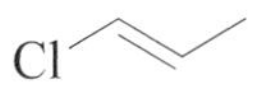

Cl

$CHClCHCH_3$

C_3H_5Cl

CAS 590-21-6

A. PROPERTIES

molecular weight 76.53.

C. WATER AND SOIL POLLUTION FACTORS

Waste water treatment

evaporation from water at 25°C of 1 ppm solution:

50% after 16 min
90% after 59 min (313)

2-chloro-1-propene (2-chloro-1-propylene)

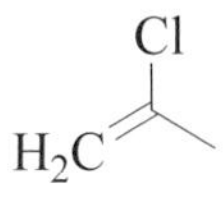

Cl

H_2C

CH_2CClCH_3

C_3H_5Cl

CAS 557-98-2

USES

Refrigerant. Chemical intermediate. Solvent.

A. PROPERTIES

molecular weight 76.53; melting point -139°C; boiling point 22.5°C; density 0.9 at 20°C

C. WATER AND SOIL POLLUTION FACTORS

Waste water treatment

evaporation from water at 25°C of 1 ppm solution:

50% after 29 min
90% after 110 min (313)

Evaporation

measured half-life for evaporation from 1 ppm aqueous solution at 25°C, still air, and an average depth of 6.5 cm: 33 min. (369)

3-chloro-1-propene *see* allylchloride

2-chloro-1-propylene *see* 2-chloro-1-propene

3-chloro-1-propylene *see* allylchloride

6-chloro-2-pyridinol (6-chloro-2-hydroxypyridine)

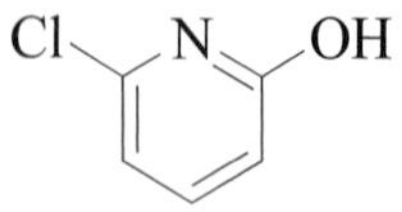

Cl N OH

$Cl(C_5H_3N)OH$

C_5H_4ClNO

CAS 16879-02-0

A. PROPERTIES

molecular weight 129.55; melting point 128-130°C; $LogP_{OW}$ 1.78.

D. BIOLOGICAL EFFECTS

Algae			
Tetrahymena pyriformis: growth inhibition at	27°C, 48h EC_{50}	23 mg/L	(2704)
Fishes			
Pimephales promelas	30-35d LC_{50},F	167 mg/L	(2704)

1-chloro-1,1,2,2,-tetrafluorethane

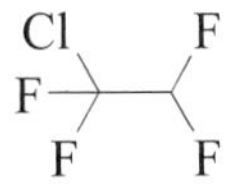

CF_2ClCHF_2

C_2HClF_4

CAS 63938-10-3

A. PROPERTIES

molecular weight 136.48.

B. AIR POLLUTION FACTORS

Tropospheric lifetime (estd.): 57 years.

Ozone depletion potential (estd.): 0.047 per molecule. (2446)

1-chloro-1,2,2,2-tetrafluorethane

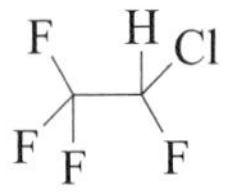

$CHClFCF_3$

C_2HClF_4

CAS 2837-89-0

A. PROPERTIES

molecular weight 136.48.

B. AIR POLLUTION FACTORS

Tropospheric lifetime (estd.): 11.2 years.

Ozone depletion potential (estd.): 0.035 per molecule. (2446)

5-chloro-2-toluidine (5-chloro-2-methylaniline; 2-amino-4-chlorotoluene; 5-chloro-2-methylbenzenamine; 5-chloro-o-toluidine)

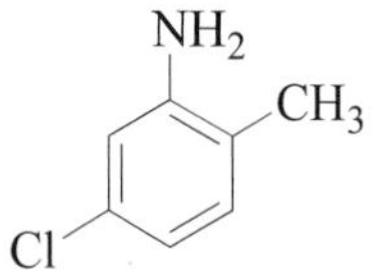

$ClC_6H_3(CH_3)NH_2$

C_7H_8ClN

CAS 95-79-4

USES

Intermediate in the manufacture of dyestuffs and in organic synthesis.

A. PROPERTIES

molecular weight 141.60; melting point 22°C; boiling point 237°C; solubility < 1,000 mg/L at 22°C.

C. WATER AND SOIL POLLUTION FACTORS

Manmade sources

landfill leachate from lined disposal area (Florida 1987-1990): mean conc. 2 µg/L. (2788)

D. BIOLOGICAL EFFECTS

Mammals			
rat	oral LD_{50}	464 mg/kg bw	(9981)

3-chloro-p-toluidinehydrochloride *see* 3-chloro-4-methylbenzenaminehydrochloride

N'-(4-chloro-o-tolyl)-N,N-dimethylformamidine *see* chlorphenamidine

2-chloro-6-(trichloromethyl)pyridine (nitrapyrin; N-Serve; N-Serve 24E; DOWCO-163)

$C_6H_3Cl_4N$

CAS 1929-82-4

TRADENAMES

N-Serve; Stay-N 2000; Nitrapyrin

USES

Fertilizer additive to prevent loss of soil nitrogen and to control nitrification, a potent inhibitor of nitrification now in use with ammonium fertilizers.

IMPURITIES, ADDITIVES, COMPOSITION

Technical grade contains approximately 10% related chlorinated pyridines.

A. PROPERTIES

white crystalline solid with a mildly sweet odour; molecular weight 230.9; melting point 62-63°C; vapour pressure 370 mPa at 23°C; water solubility 40 mg/L at 22°C; 92 mg/L at 25°C; log Pow 3.3; log K_{oc} 2.2-2.8; H 0.93 Pa m^3/mol; boiling point 136-137.5°C at 11 mm Hg; vapor pressure 0.0028 mm at 20°C

C. WATER AND SOIL POLLUTION FACTORS

Biodegradation in soil

Halflife in soils : 5; 12; 25; 27; 30 days under aerobic conditions. Its main metabolite and residue in soil is 6-chloropicolinic acid. (10575)

nitrapyrin → 6-chloropicolinic acid

Environmental fate and transport

Nitrapyrin hydrolyzes and photodegrades rapidly and hence should not persist in most environments. In aerobic mineral soils half lives ranged from 11 to 17.9 days, and in anaerobic aquatic environments, nitrapyrin had a half-life of less than 3 hours. 6-chloropicolinic acid (6-CPA) was identified as the major degradate in both hydrolysis and photolysis. 6-CPA appears to degrade through hydroxylation (breaking the pyridine ring) and microbial mineralization.

Biodegradation

6-chloropicolinic acid is the sole detectable metabolite, other than carbon dioxide in soil. 5lb active ingredient acre-1 applied to soil prior to planting with strawberries. No nitrapyrin residues were detected in strawberries (>= 0.04 mg/kg). The metabolite 6-chloropicolinic acid was detected in strawberries at 0.09 mg/kg.

Aquatic reactions

Hydrolysis: in buffered, distilled water followed simple first-order kinetics over the concentration range $6.2 \times 10^{-7}$to 8.7×10^{-5} M.
Half-life times ranged from 1.7 to 4.0 d at 35°C depending on concentration.
Photolysis: in a natural water followed simple first-order kinetics over the concentration range 7.1×10^{-6} to 7.5×10^{-6} M.
Half-life under these conditions was 0.5 d.
The products of this reaction were 6-chloropicolinic acid, 6-hydroxypicolinic acid, and unidentified polar material. (1654)

Mobility in soil

Field studies with Drummer silty clay loam and Cisne silt loam found that nitrapyrin did not move beyond 7.5 cm of the point of application in the soil. The highest concentrations were found within 2.5 cm of the point of application and a concentration gradient existed out to 7.5 cm, movement was less in silty clay loam soil. There was no indication that soil accumulation would occur from a once a year application . (10023)

Hydrolysis

Hydrolysis rate constant : 0.09 /day at 25 °C and pH 6.0

Field dissipation

Dissipation half life : 5 – 58 days.

Mobility in soils

Nitrapyrin was shown to be mobile to moderately mobile in several soils, according to available mobility studies. The adsorption coefficient (K_d) for nitrapyrin ranged from 0.947 to 19.9 with K_{oc} values ranging from 254 to 360, respectively. The major degradate 6-CPA is mobile in mineral soils and high organic matter soils, with approximate K_d values ranging from 0.387 (mineral soils) to 1.02 (high organic matter soils). Nitrapyrin also has a high vapor pressure and hence is prone to volatilize from the application site. Nitrapyrin volatilization from soil appears to be dependent on the depth of incorporation as well as air-flow rates and soil temperatures. Hence, Nitrapyrin could move off site through leaching and volatilization. (10576)

D. BIOLOGICAL EFFECTS

Bioaccumulation

Species	conc. µg/L	duration	body parts	BCF	
Bluegill sunfish		21 days		303	(10576)

Bacteria			
Nitrosomonas inhibition of NH_3 oxidation	IC_{50}	11 mg/L	(407)
Microtox	30 min EC_{50}	1.45 mg/L	(2624)
A.S. inhibition of NH_3oxidation	IC_{75}	100 mg/L	(9660)
recirculating system inhibition of nitrification	IC_{50}	50 mg/L	(10022)
Crustaceans			
Daphnia	LC_0	10 mg/L	(9662)
Snails			
Ramshorn snail	LC_0	10 mg/L	(9662)

Fishes			
channel catfish	LC_{50}	5.8 mg/L	(9662)

Mammals			
rabbit, mouse, rat	oral LD_{50}	500-940 mg/kg bw	(9810, 10024, 9675)

Bioaccumulation

Species	conc. µg/L	duration	body parts	BCF	
Bluegill sunfish		21 days		303	(10576)

Toxicity

Nitrification inhibition in soil (1% organic carbon) after 14 days : 68 % at 50 mg/kg (7050)
Nitrapyrin inhibited growth, CH_4 oxidation, and NH_4 oxidation, but not the oxidation of CH_3OH, HCHO, or HCOONa, by *Methylosinus trichosporium OB3b*, suggesting that nitrapyrin acts against the methane monooxygenase enzyme system. The inhibition of methane oxidation could be reversed by repeated washing of nitrapyrin-inhibited cells, indicating that its effect is bacteriostatic. The addition of Cu did not release the inhibition. Methane oxidation was also inhibited by 6-chloro-2-picoline.

Crustaceae			
Daphnia magna	48h EC_{50}	2.2 mg/L	
	48h NOAEC	1.5 mg/L	
Grass shrimp	96h LC_{50}	3.1 mg/L	
	96h NOAEC	2.2 mg/L	(10567)
Mollusca			
Easter oyster shell deposition	96h EC_{50}	0.41; 1.5 mg/L	
Fish	96h NOAEC	0.16; 0.7 mg/L	(10567)
Bluegill sunfish	96h LC_{50}	3.4 mg/L	
	96h NOAEC	1.5 mg/L	
Silverside minnow	96h LC_{50}	4.28 mg/L	
	96h NOAEC	<1.26 mg/L	(10576)
Mammals			
Rat	oral LD_{50}	1070; 1230 mg/kg bw	(10576)
Birds			
Beltsville small white turkey poults	Acute single oral LD_{50}	118 mg/kg bw	

1-chloro-1,1,2-trifluoroethane

CF_2ClCH_2F

$C_2H_2ClF_3$

B. AIR POLLUTION FACTORS

Tropospheric lifetime (estd.): 25 years.
Ozone depletion potential (estd.): 0.043 per molecule. (2446)

1-chloro-1,2,2-trifluoroethane

$CHClFCHF_2$

$C_2H_2ClF_3$

B. AIR POLLUTION FACTORS

Tropospheric lifetime (estd.): 2.1 years.

Ozone depletion potential (estd.): 0.008 per molecule. (2446)

1-chloro-2,2,2-trifluoroethane

(2-chloro-1,1,1-trifluoroethane; 2,2,2-trifluorochloroethane; 1,1,1-trifluoro-2-chloroethane; 1,1,1-trifluorethyl chloride; CFC 133a; FC 133a; HCFC-133a; R 133a)

CH_2ClCF_3

$C_2H_2ClF_3$

CAS 75-88-7

USES

Blowing agent. Refrigerant. Intermediate in synthesis of halothane.

OCCURRENCE

Chemical intermediate in the production of the anaesthetic halothane. Human exposure results from the presence as a low level impurity in, and as a metabolite of, halothane.

A. PROPERTIES

molecular weight 118.49; melting point -105°C; boiling point 6.9°C; density 1.39 at 0°C; log H 0.04 at 25°C.

B. AIR POLLUTION FACTORS

Tropospheric lifetime (estd.): 4.8 years.

Ozone depletion potential (estd.): 0.023 per molecule. (2446)
Photochemical reactions: $t_{1/2}$: 5 year, based on reactions with OH° (calculated). (9975)

2-chloro-1,1,1-trifluoroethane

see 1-chloro-2,2,2-trifluoroethane

1-chloro-1,2,2-trifluoroethene

(CTFE; chlorotrifluoroethylene; fluorothene; trifluorovinylchloride; trifluorochloroethylene; Trithene)

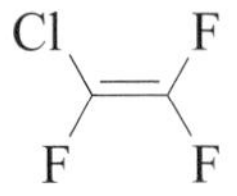

Cl F F F

$CClFCF_2$

C_2ClF_3

CAS 79-38-9

USES

Chemical intermediate. Monomer for resins which are used as high performance lubricants, plastics and elastomers.

A. PROPERTIES

molecular weight 116.47; melting point -157°C; boiling point -28°C; vapor density 4.1;density 1.54 at -60°C;

C. WATER AND SOIL POLLUTION FACTORS

Degradation in microcosms at 20°C and initial conc. 1.5 mg/L:

	half-life ± SD d	
control buffer	17 ± 1	
redox buffer	0.2 ± 0.01	
redox-hematin 1 mg/L	0.5 ± 0.03	
methanogenic landfill leachate	43 ± 26	(2598)

Degradation pathway

see 1,1,2-trichloro-1,2,2-trifluoroethane. (2598)

4-chloro-3,5-xylenol

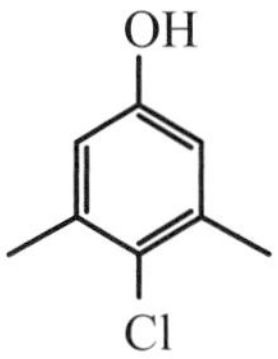

OH Cl

C_8H_9ClO

CAS 88-04-0

TRADENAMES

Chloroxylenol, Benzytol

USES

Antiseptic

A. PROPERTIES

molecular weight 156.61

C. WATER AND SOIL POLLUTION FACTORS

Environmental concentrations

in municipal sewage works: occasionally found in both influents and effluents (< 100 ng/L) in Germany. (7178)

chloroacetaldehyde (chloroacetaldehyde-monomer-; 2-chloroethanol; monochloroacetaldehyde)

Cl

CH_2ClCOH

C_2H_3ClO

CAS 107-20-0

USES

Intermediate in the manufacture of 2-aminothiazole. Used to remove bark from tree trunks.

OCCURRENCE

Metabolite of vinyl chloride and of chloroethanol. Rearrangement product of chloroethylene oxide.

A. PROPERTIES

molecular weight 78.5; melting point -16°C; boiling point 85°C; vapor pressure 139 hPa at 25°C; solubility 443,000 mg/L at 20°C.

B. AIR POLLUTION FACTORS

Incinerability

thermal stability ranking of hazardous organic compounds: rank 166 on a scale of 1 (highest stability) to 320 (lowest stability). (2390)

D. BIOLOGICAL EFFECTS

Algae			
Tetrahymena pyriformis	IC_{50}	10 mg/L (flask technique), 3 mg/L (microplate technique)	(9950)

Mammals

mouse	oral LD_{50}	69 mg/kg bw	(9951)

Chloroacetate, sodium (monochloroacetic acid sodium salt; SMA; SMCA)

$C_2H_2ClO_2.Na$

CAS 3926-62-3

USES

Intermediate for the production of SMCA, thioglycolic acid, ethyl and methyl chloroacetate, herbicides. Raw material for pigments, dyes, printing ink, surfactant, carboxymethyl cellulose, paints, pharmaceuticals, lacquers and varnishes.

A. PROPERTIES

molecular weight 116.48; melting point 120°C; vapour pressure <0.0087 kPa (<0.065 mm Hg) at 25°C; water solubility 820,000 mg/L at 20°C; log Pow <0.2 (calculated)

C. WATER AND SOIL POLLUTION FACTORS

Biodegradation

Inoculum	method	concentration	duration	elimination	
A.S.	OECD 301C		9 days	91%	
A.S. adapted	OECD 301C		5.5 days	91%	
A.S.	OECD 301B	1000 mg/L	5 days	80%	
			6 days	90%	
Methanogenic bacteria	Anaerobic	348 mg/L		(1)	
Methanogenic bacteria	Anaerobic	9 mg/L	2 days	86% (2)	
		14 mg/L	2 days	90% (2)	(10641)

(1)degradation product was glycolate
(2)degradation products were methane, CO_2 aand chloride ions

D. BIOLOGICAL EFFECTS

Toxicity

Algae			
Scenedesmus subspicatus	72h ECb_0	0.0058 mg/L	
	72h ECb_{10}	0.006; 0.007 mg/L	
	72h ECb_{50}	0.025; 0.033 mg/L	(10641)
Snails			
Planorbarius corneus	6h LOEC	15,000 mg/L	
	16.5h LOEC	7,000 mg/L	(10641)
Crustaceae			
Daphnia magna	24h LC_0	<100 mg/L	
	24h LC_{50}	800 mg/L	
	24h LC_{100}	2,000 mg/L	
	24h EC_{50}	427 mg/L	(10641)

Fish			
Rasbora heteromorpha	6.5h LC_{50}	17,000 mg/L	
	8h LC_{50}	7,000 mg/L	
	24h LC_{50}	2,600 mg/L	
	96h LC_{50}	1,400 mg/L	
Oncorhynchus mykiss	24h LC_{50}	2,000 mg/L	
	48h LC_{50}	900 mg/L	(10641)
Mammals			
Rat	oral LD_{50}	76; 165; 225-339; 335; 474; 487; 580 mg/kg bw	
Rabbit	oral LD_{50}	256 mg/kg bw	
Guinea pig	oral LD_{50}	79 mg/kg bw	
Golden hamster	oral LD_{50}	245 mg/kg bw	(10641)

chloroacetic acid (chloroethanoic acid)

Cl COOH

$CH_2ClCOOH$

$C_2H_3ClO_2$

CAS 79-11-8

USES

mfg. of carboxymethylcellulose, intermediate for pesticides and other chemicals.

A. PROPERTIES

colorless crystals; molecular weight 94.5; melting point α: 63°C, β: 55-56°C, γ: 50°C; boiling point 189°C; vapor pressure 1 mm at 43°C, 0.2 hPa at 20°C, 2 hPa at 50°C, 43 hPa at 100°C, 190 hPa at 140°C, 400 hPa at 160°C; vapor density 3.25; sp. gr. 1.58 at 20/20°C; LogP_{OW} 0.2; pH at 500 mg/L = 3.8.

B. AIR POLLUTION FACTORS

Odor

T.O.C.: 0.045 ppm. (279)

O.I. at 20°C: 1,460 (316)

Photochemical degradation

based on reaction with OH°: $t_{1/2}$: 22 d. (3218)

C. WATER AND SOIL POLLUTION FACTORS

In drinking water supply U.S.A., 1988-1989: up to 1.2 µg/L. (3219)

Waste water treatment

photolysis (UV light, *l* = 253 nm) produced Cl^-, CO_2, glycol acid, acetic acid, methane, and formaldehyde. (3220, 3221, 3222)

Biodegradability

Industrial A.S.:	after	5 d: 87% COD removal 10 d: 98% COD removal		(3208)
at 1,140 mg/L	after	*d*	*% product removal*	
		0.12	10-20	
		3	27	
		6	71	
		8	89	
		10	98	(3212)

A.S. nonadapted, at 1,000 mg/L, after 5.5 d:	90% COD removal	(3215)
A.S. nonadapted, at 4.5 mg/L, after 7 d:	73% DOC removal	(3217)
A.S. nonadapted, at 9 mg/L, after 7 d:	14-24% DOC removal	(3217)

A.S. nonadapted at 570 mg/L, after	3 hours	>10% ThOD removal	
	3 d	35% ThOD removal	
	6 d	87% ThOD removal	
	8 d	100% ThOD removal	(3212)

A.S. adapted, at 5 mg/L, after 28 d: 100% removal of product. (3213)

Methanogenic bacteria: the following metabolites have been identified: glycolate, carbonate, methane, and carbon dioxide. (3216)

Methanogenic bacteria, at 11 mg/L (1-^{14}C), after 2 d: 90% removal followed by the production of methane, CO_2, and Cl^-. (3216)

Methanogenic bacteria, at 5 mg/L (2-^{14}C), after 2 d: 86% removal followed by the production of methane, CO_2, and Cl^-. (3216)

Pseudomonas putida is able to degrade chloroacetic acid as sole carbon and energy source. (3214)

Inhibition of biodegradation processes

municipal sewage:	24h EC_0	60; 10-100 mg/L	(3208, 3212)

D. BIOLOGICAL EFFECTS

Bacteria			
Pseudomonas putida	3h LC_0	>1,000 mg/L	(3213)
	growth inhibition	1,930 mg/L	(3214)
Algae			
Scenedesmus subspicatus	48h EC_{10}	0.007-0.014 mg/L	
	48h EC_{50}	0.028-0.07 mg/L	(3211)
Worms			
Tubifex tubifex	perturbation conc.	150 mg/L	(30)
Protozoa			
Vorticella campanula	perturbation conc.	9 mg/L	(30)
Paramecium caudatum	toxic	150 mg/L	(30)
Crustaceans			
Gammarus pulex	perturbation conc.	30 mg/L	(30)
Daphnia magna	48h EC_0	55 mg/L	

	48h EC_{50}	77 mg/L	
	48h EC_{100}	107 mg/L	(3209)
	21d NOEC	32 mg/L	(3210)
Insects			
Chironomus plumosus	perturbation conc.	140 mg/L	(30)
Fishes			
Trutta iridea	perturbation conc.	20 mg/L	(30)
Cyprinus carpio	perturbation conc.	14 mg/L	(30)
Leuciscus idus	96h LC_{50}	100-500 mg/L	(3207)
	LC_0	5 mg/L	(3208)

chloroacetone (chloro-2-propanone; acetonyl-chloride-; 1-chloro-2-oxopropane; 1-chloro-2-ketopropane; monochloroacetone)

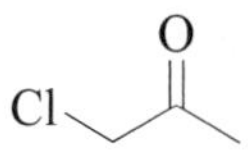

$CH_2ClCOCH_3$

C_3H_5ClO

CAS 78-95-5

USES

Tear gas component for police and military use. Manufacture of couplers for colour photography. Enzyme inactivator. Intermediate in the manufacture of perfumes, antioxidants and drugs. Insecticide formulations. Photopolymerisation of vinyl compounds. Catalyst in tetraethyllead production. Selective solvent for separating diolefins.

A. PROPERTIES

molecular weight 92.52; melting point -44.5°C; boiling point 119°C; density 1.16; solubility 100,000 mg/L.

C. WATER AND SOIL POLLUTION FACTORS

Hydrolysis

Chloroacetone undergoes hydrolysis, releasing hydrochloric acid, which is responsible for its lachrymatory effect. (9952)

D. BIOLOGICAL EFFECTS

Fishes			
guppy *Poecilia reticulata*	14d LC_{50}	0.70 mg/L	(2696)
Mammals			
rat, mouse	oral LD_{50}	100-127 mg/kg bw	(9952)

chloroacetonitrile (chloromethylcyanide)

Cl CN

H_2ClCN

C_2H_2ClN

CAS 107-14-2

USES

Analytical reagent. Intermediate in synthetic chemistry. Has been used as a fumigant.

A. PROPERTIES

molecular weight 75.50; boiling point 123-124°C; vapor pressure 8.0 mm Hg at 20°C; vapor density 3.0; density 1.20; solubility >100,000 mg/L; $LogP_{OW}$ 0.219; 0.45.

D. BIOLOGICAL EFFECTS

Fishes			
Pimephales promelas	36h LC_{50}	4 mg/L	
	96h LC_{50}	1.35 mg/L	(2709)
Mammals			
mouse, rat	oral LD_{50}	136, 220 mg/kg bw	(9953)

α-chloroacetophenone (phenacylchloride; 2-chloro-1-phenylethanone; alpha-chloroacetophenone; MACE-; phenyl-chloromethyl-ketone)

O Cl

$C_6H_5COCH_2Cl$

C_8H_7ClO

CAS 532-27-4

USES

Tear gas preparations. Soil pesticide.

OCCURRENCE

metabolite of the insecticide chlorfenvinphos.

A. PROPERTIES

molecular weight 154.59; melting point 59-60°C; boiling point 244-247°C; vapor pressure 0.004 mm at 20°C, 0.014 mm at 30°C; vapor density 5.32; sp. gr. 1.32 at 15/4°C; saturation concentration in air 0.034 g/m^3 at 20°C, 0.11 g/m^3 at 30°C.

B. AIR POLLUTION FACTORS

1 mg/m^3 = 0.16 ppm, 1 ppm = 6.43 mg/m^3.

Odor

characteristic: quality: apple blossom odor
hedonic tone: strong lacrimator
T.O.C.: 0.1 mg/m^3 = 0.016 ppm; 0.1-0.7 mg/m^3 (57, 710)
O.I. at 20°C = 330 (316)

C. WATER AND SOIL POLLUTION FACTORS

Reduction of amenities

faint odor: 0.0085 mg/L. (129)

D. BIOLOGICAL EFFECTS

Mammals

rat, guinea pig	oral LD_{50}	127, 158 mg/kg bw	(9954)

N-chloroalanine

Cl–NH–CH(CH_3)–COOH

$C_3H_6ClNO_2$

USES

a rapidly formed chlorination product of alanine. N-chloroalanine decomposes rapidly in water to form acetaldehyde, ammonia, carbon dioxide, and chloride ion or, depending on pH, pyruvic acid, ammonia, and chloride ion. The half-life of the reaction in the 5-9 pH range is 46 min at 25°C. The rate constant shows a marked temperature dependence at all pH values, changing by a factor of more than 3 for each 10°C temperature change. (1906)

chloroallylene *see* allylchloride

m-chloroaniline (3-chlorophenylamine)

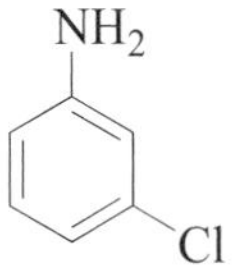

$ClC_6H_4NH_2$

C_6H_6ClN

CAS 108-42-9

USES

intermediate for azo dyes and pigments; pharmaceuticals; insecticides; agricultural chemicals.

A. PROPERTIES

molecular weight 127.57; melting point -10.4°C; boiling point 229.8°C; vapor pressure <0.1 mm at 30°C; vapor density 4.41; sp. gr. 1.2 at 20/4°C; $LogP_{ow}$ 1.88.

B. AIR POLLUTION FACTORS

1 mg/m^3 = 0.19 ppm, 1 ppm = 5.39 mg/m^3.

C. WATER AND SOIL POLLUTION FACTORS

Manmade sources

Göteborg (Sweden) sew. works 1989-1991: infl.: n.d.-3 µg/L; effl.: n.d. (2787)

Waste water treatment

Biodegradation rates: adapted A.S. at 20°C-product is sole carbon source: 97% COD removal at 6.2 mg COD/g dry inoculum/h. (327)

Degradation by Aerobacter: *500 mg/L at 30°C:*

parent:	100% ring disruption in 68 h	
mutant:	100% ring disruption in 16 h	(152)
decomposition period by a soil microflora	>64 d	(176)

W, A.S. from mixed domestic/industrial treatment plant: 14-25% depletion at 20 mg/L after 6 h at 25°C (419)

D. BIOLOGICAL EFFECTS

RubisCo test

inhib. of enzym. activity of Ribulose-P2-carboxylase in protoplasts	IC_{10}	19 mg/L	

Oxygen test

inhib. of O_2 production of protoplasts	IC_{10}	0.26 mg/L	(2698)

Algae

Scenedesmus subspicatus inhib. of fluorescence	IC_{10}	1.8 mg/L	
Scenedesmus subspicatus growth inhib.	IC_{10}	4.1 mg/L	(2698)

Earthworms

Eisenia andrei	14d LC_{50}	568-725 µg/L (estd. conc. in the soil pore water phase)	(2789)
Lumbricus rubellus	14d LC_{50}	388-469 µg/L (estd. conc. in the soil pore water phase)	(2789)
Fishes			
Poecilia reticulata	14d LC_{50}	13 mg/L	(2696)

o-chloroaniline (2-chlorophenylamine; *o*-aminochlorobenzene)

$ClC_6H_4NH_2$

C_6H_6ClN

CAS 95-51-2

A. PROPERTIES

molecular weight 127.57; melting point α:14°C; β:.5°C; boiling point 208.8°C; vapor pressure 0.13 hPa at 20°C, 0.36 hPa at 30°C; 1.7 hPa at 50°C; sp. gr. 1.21 at 20/4°C; solubility 5130; 5,600 mg/L at 20°C; $LogP_{OW}$ 1.90.

C. WATER AND SOIL POLLUTION FACTORS

COD	1.74
ThOD	1.86

Manmade sources

Göteborg (Sweden) sew. works 1989-1991: infl.: n.d.-1 µg/L; effl.: n.d. (2787)

Biodegradation rates

Degradation by *Aerobacter:* 500 mg/L at 30°C

parent:	100% ring disruption in 60 h	
mutant:	100% ring disruption in 18 h	(152)

decomposition by a soil microflora: >64 d	(176)
Adapted A.S. at 20°C-product is sole carbon source: 98% COD removal at 17 mg COD/g dry inoculum/h	(327)
W, A.S. from mixed domestic/industrial treatment plant: 22-41% depletion at 20 mg/L after 6 h at 25°C	(419)

Biodegradation

Inoculum	*Method*	*Concentration*	*Duration*	*elimination results*	
Industrial	Modified Zahn-Wellens	700 mg DOC/L	28 d	94% DOC	(8981)

sewage	test				
A.S. adapted	Zahn-Wellens test	35 mg DOC/L	19 d	16% DOC	(8981)
A.S.	Modified MITI test	100 mg/L	14 d	2.7% ThOD	(8982)

Impact on biodegradation processes

BOD EC_{50}: 5->100 mg/L

Biodegradation inhibition EC_{50}: >27 mg/L

Nitrification inhibition EC_{50}: 0.8 mg/L

Act. sludge respiration inhibition EC_{50}: >1,000 mg/L (2624)

D. BIOLOGICAL EFFECTS

Bioaccumulation

	Concentration µg/L	*Exposure period*	*BCF*	*Organ/tissue*	
Fishes: *Cyprinus carpio*	100	56 d	5.4- 9	whole body	
	10	56 d	<14- 32	whole body	(8981)
Fishes: *Brachydanio rerio*	0.2	24 h	15	whole body	(8982)

Bacteria			
microtox	30 min EC_{50}	16.9 mg/L	
microtox	5 min EC_{50}	18.8 mg/L	
OECD 209 closed system	EC_{50}	42- 593 mg/L	(2624)
A.S. oxygen consumption inhibition	3 h EC_{50}	>1,000 mg/L	(8981)
Pseudomonas putida	16 h EC_0	55 mg/L	(8981)

Algae			
RubisCo test: inhib. of enzym. activity of Ribulose-P2-carboxylase in protoplasts	IC_{10}	153 mg/L	
Oxygen test: inhib. of O_2 production of protoplasts	IC_{10}	0.0096 mg/L	(2698)
Tetrahymena pyriformis	24 h EC_{50}	200 mg/L	
Scenedesmus subspicatus inhib. of fluorescence	IC_{10}	1.7 mg/L	
Scenedesmus subspicatus growth inhib.	IC_{10}	4.0 mg/L	(2698)
Scenedesmus subspicatus	72 h EC_{20}	6; 25 mg/L	
(cell multiplication inhibition test)	72 h EC_{50}	32; 40; 150 mg/L	
	7 d EC_{50}	58 mg/L	(8981)

Worms

Tubifex	48 h LC_{50}	130- 220 mg/L

Crustaceans			
Daphnia magna Straus	24 h EC_0	1.2; 1.4; 4.4 mg/L	
	24 h EC_{50}	4.2; 6; 11 mg/L	
	24 h EC_{100}	35; 36 mg/L	
	21 d NOEC	0.03 mg/L	
	48 h EC_0	0.3 mg/L	
	48 h EC_{50}	0.46; 1.5; 1.8 mg/L	
	8 h EC_{100}	4.7 mg/L	(8981)

Fishes			
Killifish *(Oryzias latipes)*	48 h LC_{50}	6.4 mg/L	(2624)
Guppy *(Poecilia reticulata)*	14 d LC_{50}	6.2 mg/L	(2696)
Brachydanio rerio	96 h LC_{50}	5.2 mg/L	(8983)
	96 h LC_0	2.0 mg/L	(8981)
Leuciscus idus	48 h LC_0	2.0 mg/L	

	96 h LC_{40}	5.0 mg/L	
	96 h LC_{90}	10 mg/L	(8981)
Poecilia reticulata	14 d LC_{50}	6.3 mg/L	(8981)
Oryzias latipes	48 h LC_{50}	6.3 mg/L	(8982)

p-chloroaniline (4-chlorophenylamine)

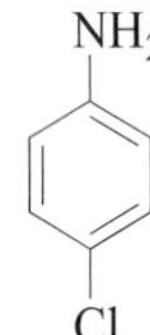

$ClC_6H_4NH_2$

C_6H_6ClN

CAS 106-47-8

USES

dye intermediate; pharmaceuticals; agricultural chemicals.

OCCURRENCE

phenylurea metabolite.

A. PROPERTIES

rhombic prisms; molecular weight 127.57; melting point 70/72°C; boiling point 231/232°C; vapor pressure 0.015 mm at 20°C, 0.05 mm at 30°C; vapor density 4.41; sp. gr. 1.43 at 19/4°C; saturation concentration in air 0.01 g/m^3 at 20°C, 0.34 g/m^3 at 30°C; $LogP_{OW}$ 1.83.

B. AIR POLLUTION FACTORS

1 mg/m^3 = 0.19 ppm, 1 ppm = 5.30 mg/m^3.

Odor

hedonic tone: sweet.

Incinerability

thermal stability ranking of hazardous organic compounds: rank 37 on a scale of 1 (highest stability) to 320 (lowest stability) (n.s.i.). (2390)

C. WATER AND SOIL POLLUTION FACTORS

Manmade sources

Göteborg (Sweden) sew. works 1989-1991: infl.: n.d.; effl.: n.d. (2787)

Biodegradation rates

adapted A.S. at 20°C-product is sole carbon source: 96% COD removal at 5.7 mg COD/g dry inoculum/h. (327)

Waste water treatment

***degradation by* Aerobacter*:** 500 mg/L at 30°C:*

parent: 100% ring disruption in 59 h

mutant: 100% ring disruption in 12 h (152)

Decomposition

period by a soil microflora: >64 d (176)

W, A.S. from mixed domestic/industrial treatment plant: 9-34% depletion at 20 mg/L after 6 h at 25°C. (419)

Removal efficiency from water incubated with 1,000 units of horseradish peroxidase enzyme per liter, at pH 5.5: 62.5%. (2452)

Transformation

4-chloroaniline reacted with the humic acid monomer catechol, in the presence of an enzyme (tyrosinase), to form 4,5-*bis*(4-chlorophenylamino)3,5-cyclohexadiene-1,2-dione according to the proposed reaction scheme:

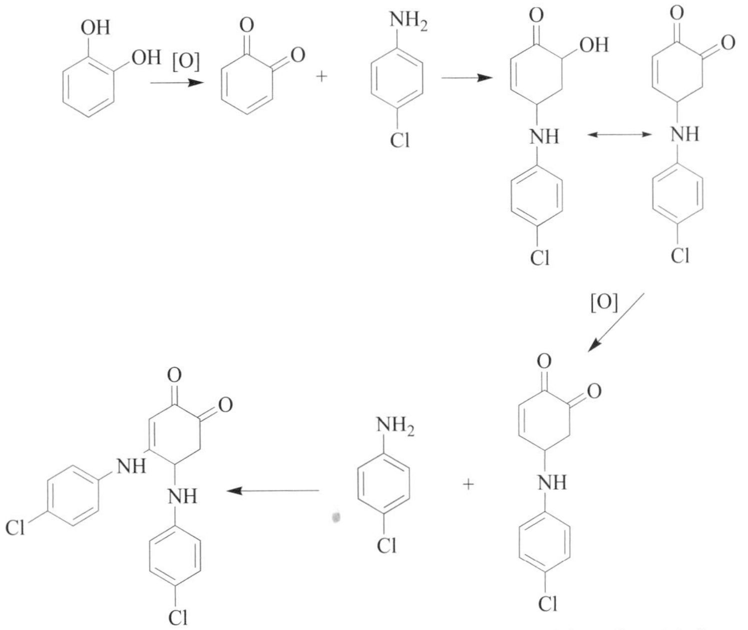

4,5-bis-(4-chlorophenylamino)-3,5-cyclohexadiene-1,2-dione

(2634)

Biodegradation

in a soil-water suspension at 35°C at initial conc. of approx. 0.1 mg/L:

	% mineralization to CO_2	
under aerobic conditions:	1.5% after 5 d, 3% after 56 d 15% of initial conc. remaining in the water after 56 d	
under anaerobic conditions:	2% after 56 d 22% of the initial conc. remaining in the water after 56 d	(2667)

D. BIOLOGICAL EFFECTS

RubisCo test

inhib. of enzym. activity of Ribulose-P2-carboxylase in protoplasts

IC_{10} 13 mg/L

Oxygen test

inhib. of O_2 production of protoplasts	IC_{10}	0.1 mg/L	(2698)

Algae			
Chlorella fusca BCF: 260 (wet wt)			(2659)
Scenedesmus subspicatus inhib. of fluorescence	IC_{10}	0.38 mg/L	
Scenedesmus subspicatus growth inhib.	IC_{10}	0.026 mg/L	(2698)

Fishes

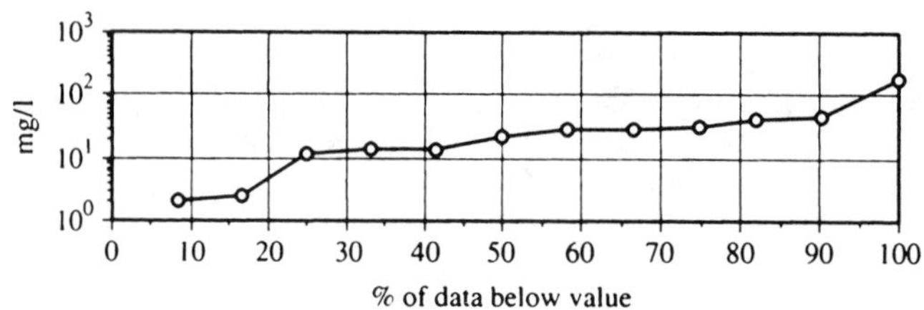

Frequency distribution of 24-96h LC_{50} values for fishes (*n* = 12) based on data from this work and from the work of Rippen (3999)

rainbow trout:	96h LC_{50},S	14 mg/L	
fathead minnow:	96h LC_{50},S	12 mg/L	
channel catfish:	96h LC_{50},S	23 mg/L	
bluegill:	96h LC_{50},S	2 mg/L	(1510)
Pimephales promelas	96h LC_{50}	31 mg/L	(2625)
Salmo gairdneri	96h LC_{50}	14 mg/L	
Pimephales promelas	96h LC_{50}	29 mg/L	
Lepomis macrochirus	96h LC_{50}	2.4 mg/L	(2679)
guppy: *Poecilia reticulata*	14d LC_{50}	26 mg/L	(2696)

2-chloroanthraquinone (2-chloro-9,10-anthracenedione)

$C_{14}H_7ClO_2$

CAS 131-09-9

USES

Intermediate in the manufacturing of dyestuffs.

A. PROPERTIES

molecular weight 242.66; melting point 206- 208; boiling point decomposes >340°C; vapor pressure

0.003 hPa at 110°C; $LogP_{OW}$4.9 (measured).

C. WATER AND SOIL POLLUTION FACTORS

Biodegradation

Inoculum	*Method*	*Concentration*	*Duration*	*Elimination results*	
				<20%	(6254)

D. BIOLOGICAL EFFECTS

Bacteria			
Pseudomonas putida (cell multiplication inhibition test)	30 min EC_{10-90}	>10,000 mg/L	(6252)
Algae			
Scenedesmus subspicatus			
(cell multiplication inhibition test)	72 h EC_{10-50}	>500 mg/L	(6252)
Gyrodinium sp. (inhibition of mitose)	12 d EC	0.015 mg/L (salt water)	(6253)
Crustaceans			
Daphnia magna Straus	24 h EC_0	400 mg/L	
	24 h EC_{50}	>500 mg/L	
	24 h EC_{100}	>500 mg/L	(6252)
	48 h EC_0	200 mg/L	
	48 h EC_{50}	420 mg/L	
	48 h EC_{100}	>500 mg/L	(6252)
Fishes			
Leuciscus idus	96 h LC_{50}	>500 mg/L	(6251)

o-chlorobenzaldehyde (2-chlorobenzaldehyde; alpha-chlorobenzaldehyde; o-chlorobenzenecarboxaldehyde)

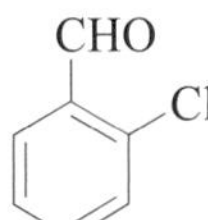

ClC_6H_4CHO

C_7H_5Cl

CAS 89-98-5

USES

feedstock for chemical synthesis.

A. PROPERTIES

molecular weight 140.57; melting point 10-11.5°C; boiling point 209-215°C; sp. gr. 1.25; vapor pressure 3 hPa at 50°C; solubility 1,117 mg/L; $LogP_{OW}$ 2.33 (measured).

C. WATER AND SOIL POLLUTION FACTORS

hydrolysis product of *o*-chlorobenzylidene-malononitrile.

Biodegradability

inoculum/test method	*test conc.*	*test duration*	*removed*	
municipal waste water, Closed Bottle	12 mg/L	20 d	60-70% ThOD	(5649)

CHO
Cl
OH
Cl

o-chlorobenzylalcohol

Euglena gracilis Z and *Dunaliella tertiolecta* transform 2-chlorobenzaldehyde to 2-chlorobenzyl alcohol. (9955, 9956)

D. BIOLOGICAL EFFECTS

Toxicity

Bacteria			
Pseudomonas fluorescens	24h EC_0	500 mg/L	(5649)
Fishes			
rainbow trout *(Salmo gairdneri)*	12h LC_{50}	5.2 mg/L	
	24h LC_{50}	3.6 mg/L	
	48h LC_{50}	2.8 mg/L	
	96h LC_{50}	2.5 mg/L	(1913)
Leuciscus idus	48h LC_0	20 mg/L	
	48h LC_{100}	50 mg/L	(5649)
Gambusia affinis	24h LC_{50},S	8.9 mg/L	(5650)

chlorobenzene (phenylchloride)

Cl

C_6H_6Cl

CAS 108-90-7

USES AND FORMULATIONS

solvent recovery plants; intermediate in dyestuffs mfg.; mfg. aniline, insecticide, phenol, chloronitrobenzene. (347)

A. PROPERTIES

colorless liquid; molecular weight 112.56; melting point -45 °C; boiling point 132 °C; vapor pressure

8.8 mm at 20 °C, 11.8 mm at 25 °C, 15 mm at 30 °C, 12 hPa at 20 °C, 20 hPa at 30 °C, 53 hPa at 50 °C; vapor density 3.88; density 1.11 at 20/4°C; solubility 500 mg/L at 20°C, 488 mg/L at 30°C; saturation concentration in air 54 g/m^3 at 20°C, 89 g/m^3 at 30°C; LogP_{OW} 2.84 at 20°C; LogH -0.74 at 25°C.

B. AIR POLLUTION FACTORS

1 mg/m^3 = 0.217 ppm, 1 ppm = 4.678 mg/m^3.

Odor

characteristic: quality: chlorinated mothballs, aromatic.

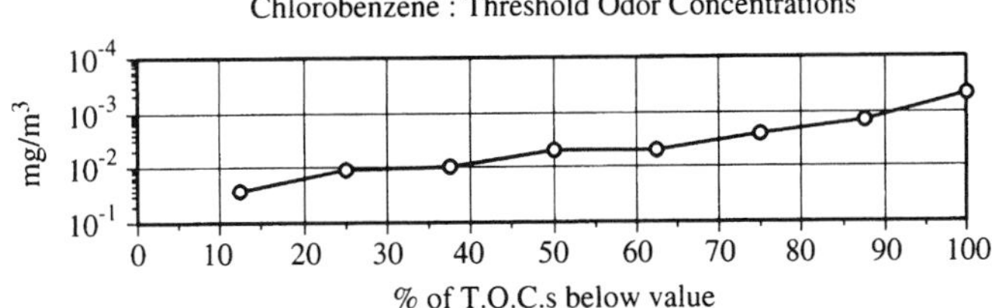

(57, 73, 279, 298, 307, 610, 748, 815, 838)

human odor perception:	0.4 mg/m^3 = 0.09 ppm	
human reflex response:	no response: 0.1 mg/m^3	
	adverse response: 0.2 mg/m^3	
animal chronic exposure:	no effect: 0.1 mg/m^3	
	adverse effect: 1.0 mg/m^3	(170)
O.I. at 20°C: 52,600		(316)

Incinerability

Temperature for 99% destruction at 2.0-sec residence time under oxygen-starved reaction conditions: 990°C.

Thermal stability ranking of hazardous organic compounds

rank 19 on a scale of 1 (highest stability) to 320 (lowest stability). (2390)

Manmade sources

in 4 municipal landfill gases in Southern Finland (1989-1990 data): avg 0.01-0.19, max. 0.33 mg/m^3. (2605)

Ambient air quality

Indoor/outdoor glc's winter 1981-1982 and 1982-1983, the Netherlands:

	median	*maximum (all in μg/m^3)*	
pre-war homes	<0.4	3	
post-war homes	<0.4	<0.4	
homes <6 years old	<0.4	27	
outdoors	<0.4	<0.4	(2668)

Partition coefficients

$K_{air/water}$	0.15	
Cuticular matrix/air partition coefficient*	3300 ± 0	(7077)

* experimental value at 25 °C studied in the cuticular membranes from mature tomato fruits (*Lycopersicon esculentum* Mill. cultivar Vendor)

C. WATER AND SOIL POLLUTION FACTORS

BOD_5:	1.5% ThOD std. dil. sew.	(41, 27, 298)
COD:	20% ThOD.	(27)

ThOD: 2.06. (27)

Impact on biodegradation processes: at 100 mg/L, no inhibition on NH_3 oxidation by *Nitrosomonas* sp. (390)

Reduction of amenities: T.O.C.: 0.1 mg/L. (296, 903)

Water quality

in Zürich Lake: at surface: 3 ppt; at 30 m depth: 12 ppt
in Zürich area: in groundwater: 14 ppt; in tap water: 6 ppt (513)

in Delaware River (U.S.A.): conc. range: winter: 7.0 ppb; summer: n.d. (1051)

in river Mass at Eysden (Netherlands) 1976: median: n.d.: range: n.d.-1.9 μg/L
in river Maas at Keizersveer (Netherlands) 1976: median: n.d.: range: n.d.-9.6 μg/L (1368)
in rivers in U.K. (1986): 139 samples from 70 river sites yielded 24 positive values (detection limit: 0.1 μg/L):median of positive values: 0.23 μg/L; maximum: 7.6μg/L (2711)

in Bayou d'Inde (Louisiana, U.S.A.) near industrial outfall:
bottom sediments: 1.5 mg/kg of organic carbon
suspended sediments: 0.22 mg/kg of organic carbon
water: 18 ng/L (2646)

Manmade sources

Occurrence in sewage sludges, NW England 1989:

urban	mean	57,300 μg/kg dry wt	
	SD	104,000 μg/kg dry wt	
urban/industrial	mean	7,010 μg/kg dry wt	(2953)

In Canadian municipal sludges and sludge compost: September 1993- February 1994: mean values of 11 sludges ranged from 0.003 to 0.20 mg/kg dw.; mean: 0.062 mg/kg dw mean value of sludge compost: ND. (7000)

Waste water treatment

air stripping constant: K = 0.969 d^{-1} at 100 mg/L (82)

ring disruption by *Pseudomonas* at 200 mg/L:
parent: 100% in 58 h
mutant: 100% in 14 h (152)

Aerobic slurry bioremediation reactor with mixed aeration chamber, treating a petrochemical waste sludge at 3.5 to 7.5% solids containing approx. 25% oil and grease and 0.5 to 2.5 grams of waste material per gram of microorganisms, at 22-24°C during batch treatment: waste residue: infl.: 4.4 mg/kg; effl. after 90 d: <1 mg/kg. (2800)

Dimensionless distribution coefficient at initial conc. = 1 mg/L:

on mixed liquor:	290	
on primary waste water sludge:	480	
on anaerobically digested sludge:	290	(2395)

Degradation in soil and in sediment

Disappearance in soil at 100 mg/kg and 20 °C under aerobic laboratory conditions: half-life: 2.1 d; confidence limits: 1.7-3.0 d (2366)

Reductive dechlorination in sediment under sulfate reducing conditions at 25 °C to benzene: half life period assuming peudo first order kinetics: 46 d. (7041)

Degradation half-lives:	75 d in sediment	(2369)
	4.6-21 d in marine mesocosm	(2367)
	7.9 d in activated sludge	(2368)

Biodegradation half-lives in nonadapted aerobic subsoil:

sand Oklahoma:	>539 d	
coarse sand Oklahoma:	81; 242 d	(2695)

Biodegradation

Inoculum	*method*	*conc.*	*duration*	*elimination results*	
domestic sewage	Closed Bottle Test		20 d	50-60% ThOD	(9147)
A.S. industrial non adapted	Respirometric Test		5 d	30% ThOD	
			10 d	70% ThOD	
			15 d	>90% ThOD	(9149)
A.S.	Modified MITI Test		28 d	15% ThOD	(9146)
different sludges	Modified MITI Test	100 mg/L	28 d	0% ThOD	(9148)

Partition coefficients

Cuticular matrix/water partition coefficient**	500	(7077)
log K_{OC}	2.60	(2597)

** calculated from the Cuticular matrix/water partition coefficient and the air/water distribution coefficient

D. BIOLOGICAL EFFECTS

Bioaccumulation

	Concentration µg/L	*exposure period*	*BCF*	*organ/tissue*	
Fishes: *Cyprinus carpio*	150	8 weeks	4.3-40	whole body	
	15	8 weeks	3.9-23	whole body	(9148)
fathead minnow			447		(2606)

Bacteria			
Pseudomonas putida	16h EC_0	17 mg/L	(1900)
A.S. respiration inhibition	30 min EC_{50}	140 mg/L	(2624)
A.S. oxygen consumption test	30 min EC_{50}	2,950 mg/L	(9146)
A.S. domestic ETAD Test	24d EC_{50}	17 mg/L	(9147)
Photobacterium phosphoreum	5 min IC_{50}	9.4 mg/L	(9153)
	10 in IC_{50}	20 mg/L	(9147)
	30 min IC_{50}	11 mg/L	(9147)
Aerobic heterotrophs	15h IC_{50}	310 mg/L	(9153)
Methanogens	48h IC_{50}	270 mg/L	(9153)
Nitrosomonas inhibition of N oxidation	24h IC_{50}	0.71 mg/L	(9153)

Algae			
Scenedesmus subspicatus	48h EC_{10}	38; 50 mg/L	
(growth rate + biomass)	48h EC_{50}	110; 220 mg/L	(9152)
Selenastrum capricornutum	96h EC_{50}	12 mg/L	(9147)
	5d EC_{50}	280 mg/L	(9150)
Skeletonema costatum	5d EC_{50}	203 mg/L	(9150)

algae *(Microcystis aeruginosa)*:	8d EC_0	120 mg/L	(329)
green algae *(Scenedesmus quadricauda)*:	7d EC_0	>390 mg/L	(1900)
protozoa *(Entosiphon sulcatum)*:	72h EC_0	>390 mg/L	(1900)
protozoa (*Uronema parduczi* Chatton-Lwoff)	EC_0	>392 mg/L	(1901)

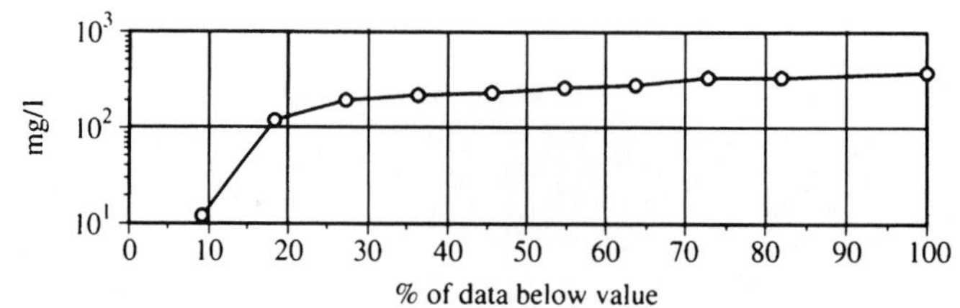

Frequency distribution of short-term EC_{50}/LC_{50} values for algae (n = 11) based on data from this work and from the work of Rippen	(3999)

Plants			
Lemna gibba	7d EC_{50}	593 mg/L	
	7d NOEL	294 mg/L	(9154)
Lemna minor	7d EC_{50}	353-545 mg/L	
	7d NOEL	294 mg/L	(9154)
Lactuca sativa Ravel R2	7d EC_{50}	1,000 mg/kg soil	
	14d EC_{50}	>1,000 mg/kg soil	(9156)
(soilless culture)	16-21 EC_{50}	9.3 mg/L	(9156)

Earthworms

Eisenia andrei: 14d LC_{50}: 797-1,453 µg/L (estim. conc. in the soil pore water phase)	(2789)
Lumbricus rubellus: 14d LC_{50}: 2,243-4,281 µg/L (estim. conc. in the soil pore water phase)	(2789)

Crustaceans

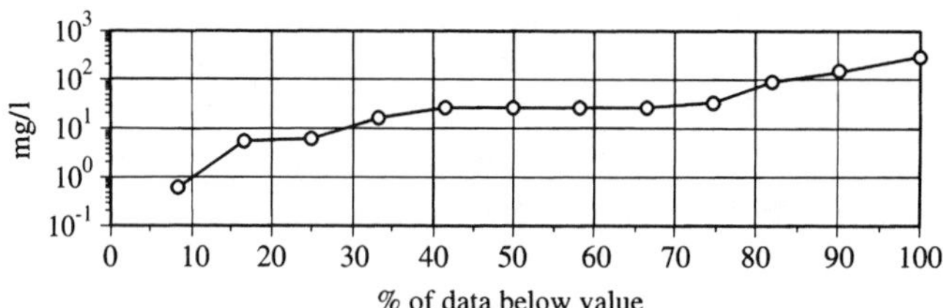

Frequency distribution of 24-96h EC_{50}/LC_{50} values for crustaceans (n = 12) based on data from this work and from the work of Rippen	(3999)

Daphnia magna	EC_{50} reproduction:	1.9 mg/L	
	NOEC reproduction:	1 mg/L	
	NOEC growth:	0.32 mg/L	
	LC_{50}:	5.6 mg/L	(2706)
Daphnia magna	24h EC_0	110 mg/L	
	24h EC_{50}	310 mg/L	
	24h EC_{100}	390 mg	(9147)
	24h EC_{50}	16; 34; 140 mg/L	(9147, 9152)
	21d NOEC	2.5 mg/L	(9152)
	16d EC_{50}	3.4 mg/L	
	16d NOEC	0.32 mg/L	(9147)
	48h EC_{50}	31 mg/L	(9150)
Ceriodaphnia dubia	48h EC_{50}	47 mg/L	(9150)

Fishes

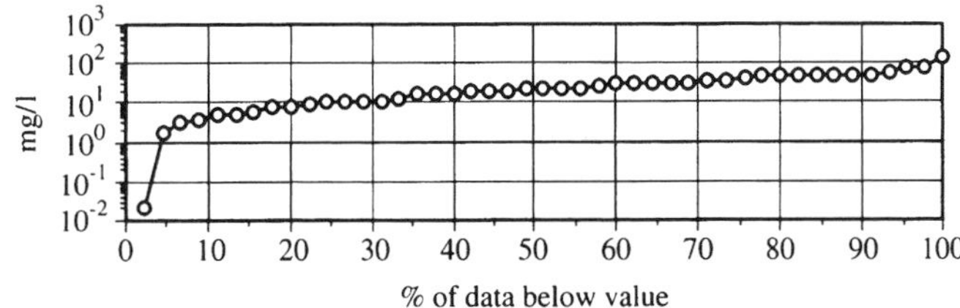

Frequency distribution of 24-96h LC_{50} values for fishes (n = 45) based on data from this work and from the work of Rippen (3999)

fatheads:	24-96h LC_{50}	29-39 mg/L	
bluegills:	2496h LC_{50}	24 mg/L	(41)
bluegill *(Lepomis macrochirus):*	24h LC_{50},S	17 mg/L	
	96h LC_{50},S	16 mg/L	(2697)
goldfish:	24-96h LC_{50}	51-73 mg/L	
guppies:	24-96h LC_{50}	45 mg/L	(41)
guppy *(Poecilia reticulata):*	14d log LC_{50}	2.23 µmol/L	(2696)
rainbow trout:	24h LC_{50}	1.8 ml/kg	(1528)
guppy *(Poecilia reticulata):*	14d LC_{50}	19 mg/L	(1833)
Brachydanio rerio	96h LC_0	18 mg/L	
	96h LC_{50}	91 mg/L	
	96h LC_{100}	178 mg/L	(9147)
Lepomis macrochirus	96h LC_{50}	4.5; 7.4 mg/L	(9147)
Pimephales promelas	96h LC_{50}	19; 26 mg/L	(9147, 9150)
Poecilia reticulata	14d LC_{50}	19 mg/L	(9147)
Cyprinodon variegatus	96h LC_{50}	10 mg/L	
	96h NOEC	6.2 mg/L	(9147)
Leuciscus idus	96h LC_{50}	42 mg/L	(9147)
Salmo gairdneri	48h LC_{50}	4.1 mg/L	(9147)
Oryzias latipes	48h LC_{50}	17 mg/L	(9148)

in catfish: Bayou d'Inde industrial outfall: not detect. mg/kg of lipid
junction of Bayou d'Inde and Calcasieu River: 0.05 mg/kg of lipid
Lake Charles: not detect. mg/kg of lipid (2646)

Mammals

rat	acute oral LD_{50}	1,427; 2,455; 1,110; 1,760; 3,400; 2,300; 2,910; 2,390 mg/kg bw

p-chlorobenzenesulfonic acid

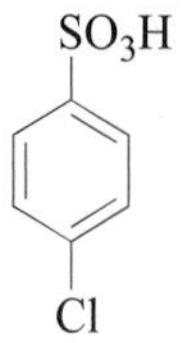

$C_6H_5ClO_3S$

CAS 98-66-8

A. PROPERTIES

molecular weight 192.62; melting point 68°C; boiling point 147-148°C at 25 mm.

C. WATER AND SOIL POLLUTION FACTORS

Biodegradation

decomposition period by a soil microflora: 16 d. (176)

D. BIOLOGICAL EFFECTS

Crustaceans			
Daphnia magna	24h LC_{50}	8,600 mg/L	
(sodium salt)	48h LC_{50}	7,659 mg/L	
	72h LC_{50}	3,964 mg/L	
	96h LC_{50}	2,150 mg/L	(153)
	100h LC_{50}	2,394 mg/L	(1295)
Mollusc			
Snail eggs *Lymnaea* sp.	24h LC_{50}	8,600 mg/L	
(sodium salt)	48h LC_{50}	7,633 mg/L	
	72h LC_{50}	6,343 mg/L	
	96h LC_{50}	5,053 mg/L	(153)
Fishes			
Lepomis macrochirus (sodium salt)	24h LC	>3,219 mg/L	(1295)

chlorobenzilate

(4,4'-dichlorobenzilic acid, ethylester; ethyl-2-hydroxy-2,2-*bis*(4-chlorophenyl)acetate; chlorobenzylate-; ethyl 4,4'-dichlorobenzilate; ethyl p,p-dichlorobenzilate)

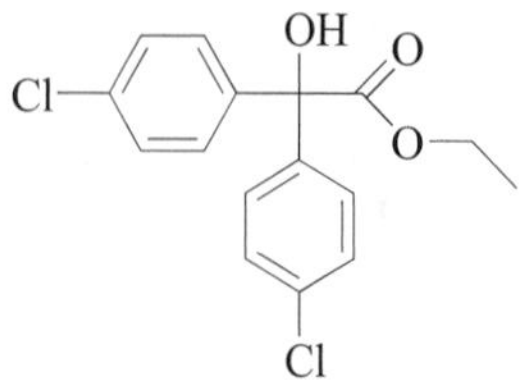

$C_{16}H_{14}Cl_2O_3$

CAS 510-15-6

USES

Acaricide.

A. PROPERTIES

molecular weight 325.2; melting point 35-37°C; boiling point 156-158°C; density 1.28 at 20°C; vapor pressure 2.2 x 10_{-6} mm Hg at 20°C.

B. AIR POLLUTION FACTORS

Incinerability

thermal stability ranking of hazardous organic compounds: rank 204 on a scale of 1 (highest stability) to 320 (lowest stability). (2390)

C. WATER AND SOIL POLLUTION FACTORS

No inhibition of nitrification capacity of soils at 0.25-1.0 mg/kg for 16 weeks. (9957)

D. BIOLOGICAL EFFECTS

Fishes			
cytotoxicity to goldfish GF-Scale cells	NR_{50}	5.7 mg/L	(2680)
Cyprinus carpio	48h LC_{50}	1.5 mg/L	(2680)
rainbow trout	96h LC_{50}	0.6 mg/L	
bluegill sunfish	96h LC_{50}	1.8 mg/L	(9600)

Mammals			
rat	oral LD_{50}	1,040-1,220 mg/kg bw	(9958)
hamster	oral LD_{50}	700 mg/kg bw	(9959)

m-chlorobenzoic acid

COOH

Cl

ClC_6H_4COOH

$C_7H_5ClO_2$

CAS 535-80-8

A. PROPERTIES

molecular weight 156.57; melting point 158°C; boiling point sublimes; density 1.5 at 25/4°C; solubility 400 mg/L at 0°C; log P_{OW} 2.68.

C. WATER AND SOIL POLLUTION FACTORS

Biodegradation

Decomposition period by a soil microflora: 32 d. (176)

Lag period for degradation of 16 mg/L by waste water or by soil at pH 7.3 and 30°C: 7-14 d.

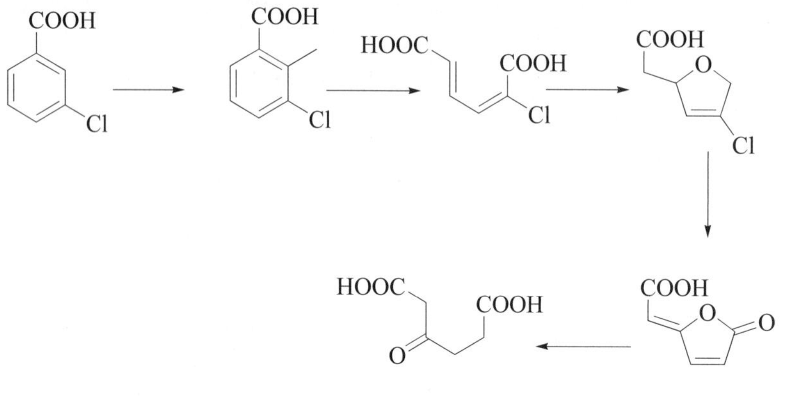

COOH Cl → COOH Cl → HOOC COOH Cl → COOH O Cl → COOH O O → HOOC COOH O

Bacterial degradation pathway

for m-chlorobenzoic acid (1219)

Biodegradation of 3-chlorobenzoate in anoxic habitats in the dark

% substrate disappearance after 3 months

in sewage sludge at 37°C:	2
in pond sediment at 20°C:	93
in methanogenic aquifer at 20°C:	100
in sulfate-reducing aquifer at 20°C:	3

Benzoate was detected as metabolite in methanogenic aquifer. (2621)

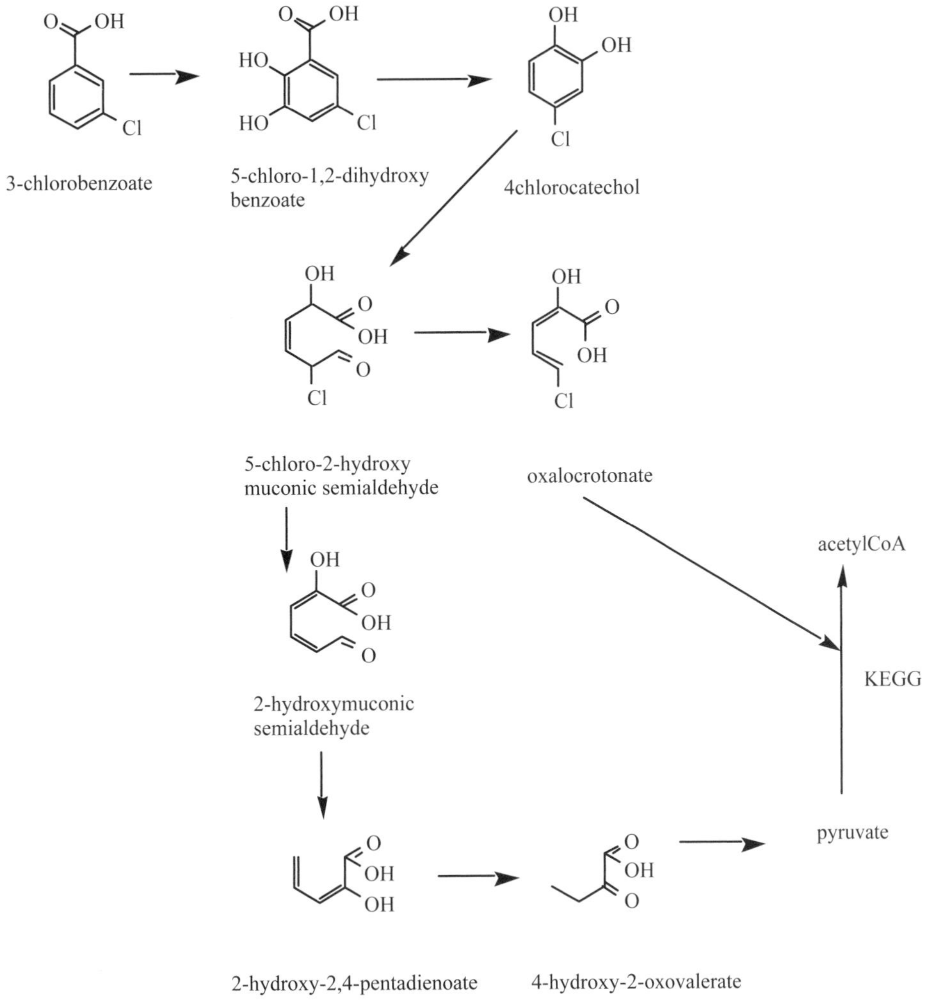

Biodegradation pathway with meta cleavage of 4-chlorocatechol (7176)

o-chlorobenzoic acid (OCBA; 2-chlorobenzoic acid; 2-CBA)

ClC_6H_4COOH

$C_7H_5ClO_2$

CAS 118-91-2

USES

Preservative for glues, paints. Intermediate in the preparation of fungicide and dyestuff.

OCCURRENCE

2-Chlorobenzoic acid is found bound to marine and riverine humic substances. (9965)

A. PROPERTIES

molecular weight 156.57; melting point 142°C; boiling point 285°C sublimes; density 1.54 at 20/4°C; solubility 2,100 mg/L at 25°C; LogP_{OW} 1.98.

C. WATER AND SOIL POLLUTION FACTORS

Biodegradation

Decomposition period by a soil microflora: >64 d. (176)
Lag period for degradation by 16 mg/L:
by waste water at pH 7.3 and 30°C: more than 25 d
by soil suspension at pH 7.3 and 30°C: 7-14 d (1096)
2-Chlorobenzoic acid is utilized by anaerobic denitrifying cultures from river sediments under denitrifying conditions. (9960)
A strain of *Pseudomonas cepacia* isolated from soil caused complete degradation with the formation of chlorine.

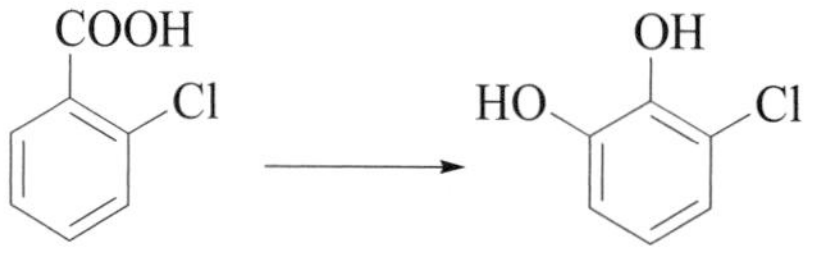

COOH, Cl → OH, HO, Cl

2-chlorocatechol

Pseudomonas stutzeri oxidizes 2-chlorobenzoic acid to 2-chlorocatechol . (9961)
2-Chlorobenzoic acid disappeared from soil columns inoculated with Pseudomonas stutzeri within 8 d. (9962)

Inhibition of degradation

The concentration of 2-chlorobenzoic acid not substantially inhibiting *Pseudomonas stutzeri* (able to utilise the acid as sole carbon and energy source) was 250-500 mg/L. (9963)

D. BIOLOGICAL EFFECTS

Mammals

rat	oral LD_{50}	6,460 mg/kg bw	(9964)

p-chlorobenzoic acid (4-CBA)

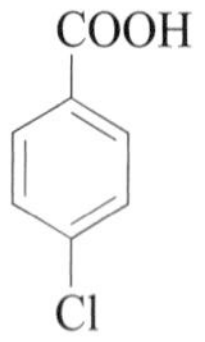

COOH, Cl

ClC_6H_4COOH

$C_7H_5ClO_2$

CAS 74-11-3

OCCURRENCE

Chlorobenzoates are the main intermediates that accumulate in the bacterial co-metabolism of polychlorinated biphenyls. (2792)

A. PROPERTIES

molecular weight 156.57; melting point 243°C; boiling point sublimes; density 1.54 at 24/4°C; solubility 77 mg/L at 25°C; $LogP_{OW}$ 2.65.

C. WATER AND SOIL POLLUTION FACTORS

Biodegradation
Decomposition

period by a soil microflora: 64 d. (176)

Lag period for degradation of 16 mg/L by waste water or soil suspension at pH 7.3 and 30°C: more than 25 d. (1096)

Acinetobacter sp., strain ST-1, could completely mineralize 4-CBA in pure culture using 4-CBA as the sole carbon and energy source under aerobic conditions. Under these conditions the 4-CBA was hydrolytically dehalogenated to 4-hydrobenzoic acid (4-HBA). The conversion of this metabolite into protocatechuic acid (PCA) and further to *β*-carboxy-*cis,cis*-muconic acid proceeded so fast that neither 4-HBA nor PCA could be detected.

Under anaerobic conditions, the conversion of 4-CBA into 4-HBA occurred with a yield of >80%.

Proposed total biodegradation pathway of 4-CBA by strain ST-1 of *Acinetobacter* sp.:

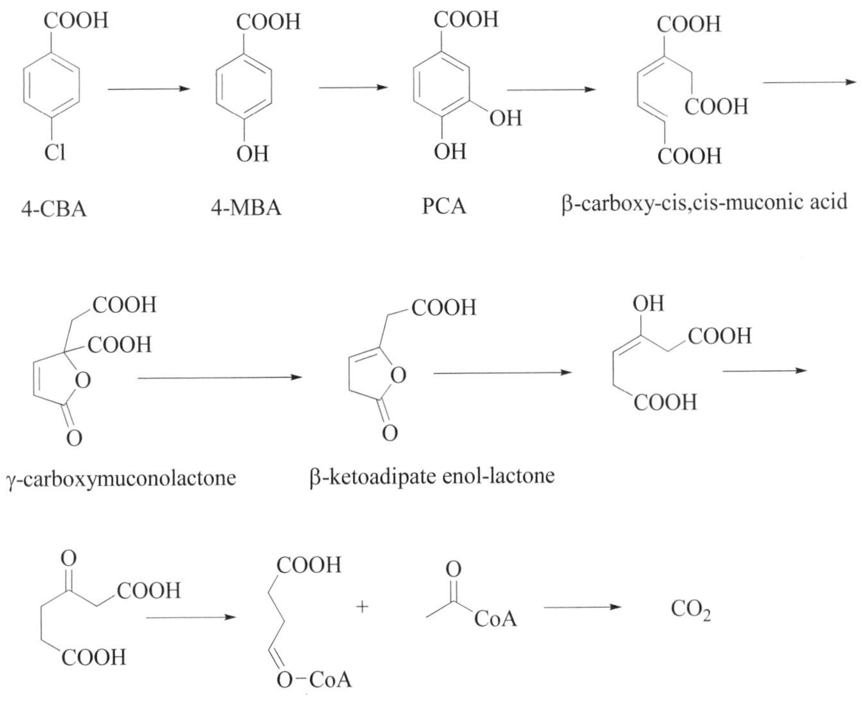

(2791)

Biodegradation of 4-chlorobenzoate in anoxic habitats in the dark

% substrate disappearance after 3 months

in sewage sludge at 37°C:	17
in pond sediment at 20°C:	6

in methanogenic aquifer at 20°C:	14	
in sulfate-reducing aquifer at 20°C:	11	
No evidence of degradation in any of the incubations.		(2621)

D. BIOLOGICAL EFFECTS

Algae
Chlorella fusca BCF (wet wt): 63 (2659)

2-(4-chlorobenzoyl)benzoic acid (4-chlorobenzophenone-2'-carboxylic acid)

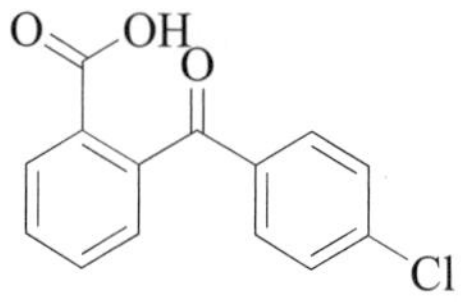

$C_{14}H_9ClO_3$

CAS 85-56-3

USES

Intermediate for dyes and pharmaceuticals.

A. PROPERTIES

molecular weight 260.68; melting point 147-148°C; solubility approx. 1,000 mg/L at 20°C; $LogP_{ow}$ -0.2 (measured).

C. WATER AND SOIL POLLUTION FACTORS

Biodegradability

Zahn-Wellens test: >70% DOC removal (6055)

Inhibition of biodegradation

A.S. inhib. of oxygen consumption:	30 min EC_{20}	approx. 750 mg/L	(6055)

D. BIOLOGICAL EFFECTS

Bacteria

Pseudomonas putida			
Cell multiplication inhib.	17 h EC_{10-90}	>100 mg/L	
Inhib. of oxygen consumption	30 min EC_{10}	60 mg/L	
	30 min EC_{50}	140 mg/L	
	30 min EC_{90}	200 mg/L	(6054)

Algae

Scenedesmus subspicatus	72 h EC_{10}	8.3 mg/L	
	72 h EC_{50}	88 mg/L	
	96 h EC_{10}	6.7 mg/L	
	96 h EC_{50}	40 mg/L	(6054)

Crustaceans			
Daphnia magna	24- 48 h EC_0	125 mg/L	
	24- 48 h EC_{50}	180 mg/L	
	24- 48 h EC_{100}	250 mg/L	(6054)
Fishes			
Leuciscus idus	96 h LC_{50}	237 mg/L	(6053)

o-chlorobenzylidene malononitrile (OCBM; *β,β*-dicyano-*o*-chlorostyrene)

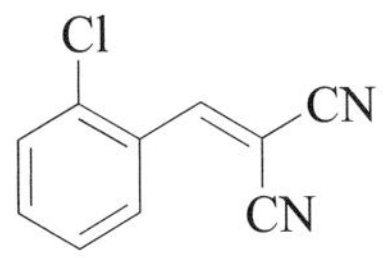

$C_{10}H_5ClN_2$

CAS 2698-41-1

A. PROPERTIES

molecular weight 188.62; melting point 95°C; boiling point 310/315°C; vapor density 6.52.

B. AIR POLLUTION FACTORS

1 mg/m^3 = 0.13 ppm, 1 ppm = 7.84 mg/m^3.

C. WATER AND SOIL POLLUTION FACTORS

Hydrolysis

in water:

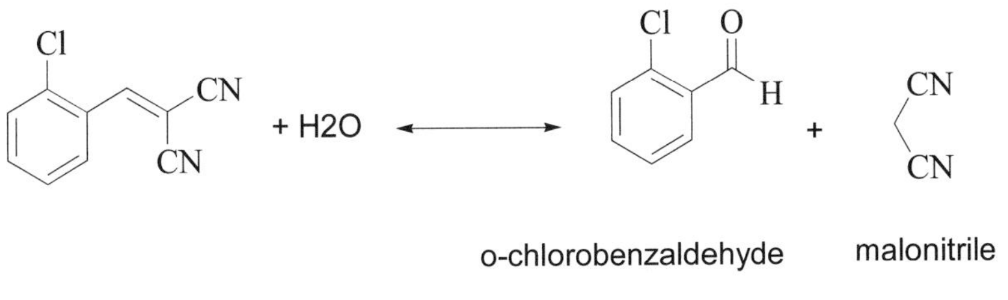

+ H2O ⟷ o-chlorobenzaldehyde + malonitrile

D. BIOLOGICAL EFFECTS

Fishes			
rainbow trout:	LC_{50}	12 h: 1.28 mg/L	
		24 h: 0.45 mg/L	
		48 h: 0.42 mg/L	
		96 h: 0.22 mg/L	(1913)

rainbow trout exposed to equimolar mixture of *o*-chlorobenzaldehyde and malononitrile (products of hydrolysis):

LC_{50}	12 h: 4.7 mg/L	
	24 h: 2.0 mg/L	

24 h: 1.2 mg/L	
96 h: 1.1 mg/L	(1913)

2-chlorobiphenyl

$C_{12}H_9Cl$

CAS 2051-60-7

A. PROPERTIES

solubility 5.8 mg/L.

C. WATER AND SOIL POLLUTION FACTORS

Biodegradation

100% degradation after 1 h by *Alcaligenes* Y42 (cell number 2 × 10^9/ml) and *Acinetobacter* P6 (cell number 4.4 × 108/ml) at 9.3 mg/L initial conc.; trimethylsilyl derivative of monochlorobenzoic acid was detected in the metabolite. (1086)

D. BIOLOGICAL EFFECTS

Marine yeast			
Rhodotorula rubra:	BCF:	737 in whole cells ±37,000 in their lipid portion	(1566)

3-chlorobiphenyl

$C_{12}H_9Cl$

CAS 2051-61-8

A. PROPERTIES

solubility 3.3 mg/L.

C. WATER AND SOIL POLLUTION FACTORS

Biodegradation

100% degradation after 1 h by *Alcaligenes* Y42 (cell number 2 × 10^9/ml) and *Acinetobacter* P6 (cell number 4.4 × 10^8/ml) at 9.3 mg/L initial conc.; trimethylsilyl derivative of monochlorobenzoic acid was detected in the metabolite. (1086)

D. BIOLOGICAL EFFECTS

MARINE YEAST

Rhodotorula rubra:	BCF	1,180 in whole cells ±59,000 in their lipid portion	(1566)

4-chlorobiphenyl

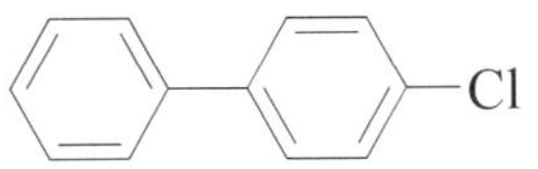

$C_{12}H_9Cl$

CAS 2051-62-9

A. PROPERTIES

molecular weight 188.66; melting point 76-78°C; boiling point 282°C; solubility 0.8 mg/L.

C. WATER AND SOIL POLLUTION FACTORS

Biodegradation

Bacterial degradation

Eighteen lichens from a variety of habitats were treated with 4-CB. All were shown to partially convert 4-CB to 4-chloro-4'-hydroxybiphenyl. It took between 6 and 22 hours for the hydroxy derivative to appear. Only one species *(Pseudocyphellaria crocata)* produced a further metabolite: 4-chloro-4'-methoxybiphenyl. (1651)

100% degradation after 1 hour by *Alcaligenes* Y42 (cell number 2 × 10^9/ml) and *Acinetobacter* P6 (cell number 4.4 × 10^8/ml) at 9.3 mg/L initial conc.; trimethylsilyl derivative of monochlorobenzoic acid was detected in the metabolite. (1086)

D. BIOLOGICAL EFFECTS

MARINE YEAST

Rhodotorula rubra: BCF	1,550 in whole cells ±77,500 in their lipid portion	(1566)

chlorobromomethane *see* bromochloromethane

1-chlorobutane *see* n-butylchloride

4-chlorocatechol

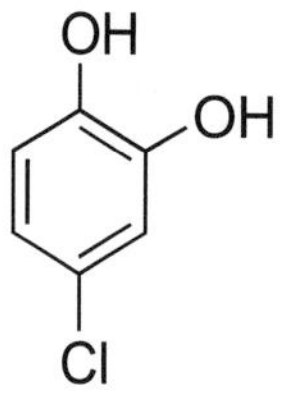

$C_6H_5ClO_2$

CAS 2138-22-9

OCCURRENCE

Degradation product of 4-chloroguaiacol, 4- and 5-chlorosalicylates

A. PROPERTIES

molecular weight 144.56

C. WATER AND SOIL POLLUTION FACTORS

Anthropogenic sources

In effluents from pulp mills using 100% chlorine dioxide (1990-1997): maximum 0,3 µg/L (7192)

Biodegradation pathways

Pseudomonas sp. strain MT1 is capable of degrading 4- and 5-chlorosalicylates via 4-chlorocatechol, 3-chloromuconate, and maleylacetate by a novel pathway. 3-Chloromuconate is transformed by muconate cycloisomerase into protoanemonin, a dominant reaction product, as previously shown for other muconate cycloisomerases. (11211)

OH
OH
Cl
4-chlorocatechol
-OOC COO⁻ Cl
3-chloro-cis,cis-muconate
A
O O
protoanemonin
COO⁻
O O
cis-dienelactone
B
COO⁻ O -OOC
maleylacetate
COO⁻ O -OOC
3-oxoadipate

Degradation of 4-chlorocatechol by enzymes of the 3-oxoadipate pathway (A) or the chlorocatechol pathway (B) [11211].

Degradation of 4-chlorocatechol by enzymes of the 3-oxoadipate pathway (A) or the chlorocatechol pathway (B) (11211)

4-chlorocatechol

5-chloro-2-hydroxy muconic semialdehyde

oxalocrotonate

acetylCoA

2-hydroxymuconic semialdehyde

KEGG

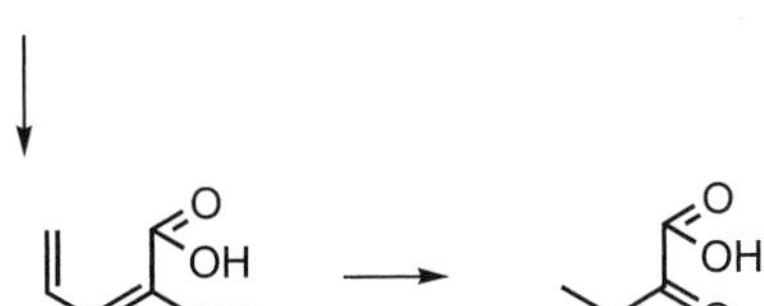

pyruvate

2-hydroxy-2,4-pentadienoate

4-hydroxy-2-oxovalerate

meta-cleavage of 4-chlorocatechol [7176]

Meta-cleavage of 4-chlorocatechol (7176)

D. BIOLOGICAL EFFECTS

Toxicity

Micro organisms

Nitrification inhibition in soil (1% organic carbon) after 14 days : 14 % at 10 mg/kg (7050)

Fish

Zebra fish egg/larvae	NOEC	0.70 mg/L	(7161)

1-chlorodecane (*prim-n*-decylchloride)

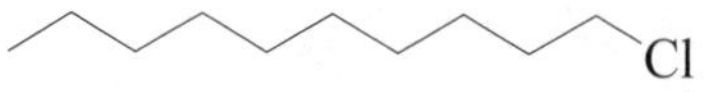

$CH_3(CH_2)_8CH_2Cl$

$C_{10}H_{21}Cl$

CAS 1002-69-3

A. PROPERTIES

molecular weight 176.73; melting point -34°C; boiling point 223°C; density 0.87.

B. AIR POLLUTION FACTORS

Odor threshold

recognition: 22 mg/m^3. (761)

C. WATER AND SOIL POLLUTION FACTORS

Waste water treatment

A.S.: after 6 h: 1.6% of ThOD
after 12 h: 3.2% of ThOD
after 24 h: 5.9% of ThOD

12,14-chlorodehydroabietic acid

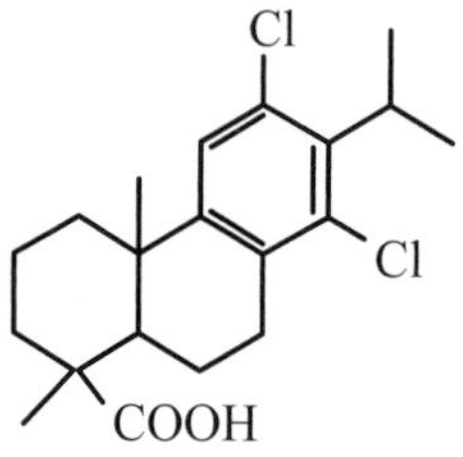

CAS 57055-38-6

OCCURRENCE

chlorinated resin acid found in effluent of pulp and paper mills.

D. BIOLOGICAL EFFECTS

Crustaceae			
Daphnia magna	24h EC_{50}	1.5 mg/L	(7167)
Fish			
Fathead minnow	96h LC_{50}	0.67 mg/L	(7167)

chlorodifluoromethane (F 22; difluorochloromethane; Algeon 22; CFC 22; Freon 22; monochlorodifluoromethane; Refrigerant 22)

$CHClF_2$

CAS 75-45-6

USES

Refrigerant. Aerosol propellant. Low-temperature solvent. Component of fluorocarbon resins such as tetrafluoroethylene polymers.

A. PROPERTIES

molecular weight 86.47; melting point -160°C; boiling point -41°C; density 1.5 at -69°C, 1.21 at 20°C; solubility 3,000 mg/L at 20°C; $LogP_{OW}$ 1.1 (calculated), 1.08 (measured); LogH 0.08 at 25°C.

B. AIR POLLUTION FACTORS

Incinerability

Temperature for 99% destruction at 2.0-s residence time under oxygen-starved reaction conditions: 645°C
Thermal stability ranking of hazardous organic compounds: rank 148 on a scale of 1 (highest stability) to 320 (lowest stability) (2390)
Tropospheric lifetime (estd.): 7.5 yr
Ozone depletion potential (estd.): 0.030 per molecule (2446)
Photodegradation: reaction with OH°: 50% after 530 d (6047)

Global Warming Potential (GWP) and Ozone Depletion Potential (ODP)

Name	CAS	GWP	ODP	
CFC 11	75-69-4	1.0	1.0	
CFC 12	75-71-8	3.0	1.0	
CFC 113	76-13-1	1.4	0.8	
CFC 114	76-14-2	3.9	1.0	
CFC 115	76-15-3	7.5	0.6	
HCFC 22	75-45-6	0.36	0.055	(6048)

C. WATER AND SOIL POLLUTION FACTORS

Impact on biodegradation

Anaerobic bacteria from municipal waste water treatment plant (ETAD fermentation tube method)	24 h IC_0	>400 mg/L	(6049)

D. BIOLOGICAL EFFECTS

Fishes

Poecilia reticulata	24 h LC_0	approx. 180 mg/L	(6049)

1-chlorododecane (*prim-n*-dodecylchloride)

$CH_3(CH_2)_{10}CH_2Cl$

$C_{12}H_{25}Cl$

CAS 112-52-7

A. PROPERTIES

molecular weight 204.79; melting point -9.3°C; boiling point 260°C; density 0.87.

C. WATER AND SOIL POLLUTION FACTORS

Waste water treatment

A.S.: after	6 h:	3.5% of ThOD	
	12 h:	6.2% of ThOD	
	24 h:	12% of ThOD	(88)

chloroethane *see* ethylchloride

chloroethanoic acid *see* chloroacetic acid

2-chloroethanol (*β*-chloroethylalcohol; ethylenechlorohydrin; glycolchlorohydrin)

CH_2ClCH_2OH

C_2H_5ClO

CAS 107-07-3

A. PROPERTIES

colorless liquid; molecular weight 80.52; melting point -89°C; boiling point 128.8°C; vapor pressure 4.9 mm at 20°C, 10 mm at 30°C; vapor density 2.78; density 1.1 at 20/4°C; saturation concentration in air 24 g/m^3 at 20°C, 42 g/m^3 at 30°C.

B. AIR POLLUTION FACTORS

1 mg/m³ = 0.304 ppm, 1 ppm = 3.29 mg/m³.

C. WATER AND SOIL POLLUTION FACTORS

BOD_5:	0; 46% of ThOD	(41, 79)
BOD_{10}:	16; 46% of ThOD	(79, 256)
BOD_{15}:	74% of ThOD	(79)
BOD_{20}:	88% of ThOD	(79)
ThOD:	1.09	

Waste water treatment

methods	temp, °C	d observed	feed, mg/L	d acclim.	% removed	
NFG, BOD	20	1-10	200	365 + P	nil	
NFG, BOD	20	1-10	400	365 + P	nil	
NFG, BOD	20	1-10	1,000	365 + P	3	(93)

Biodegradation

Inoculum	method	concentration	duration	elimination results	
activated sludge	respirometric test	70-100 mg/L DOC	28 d	93% ThOD	(7040)

D. BIOLOGICAL EFFECTS

Scenedesmus subspicatus inhib. of fluorescence	IC_{10}	2.0 mg/L	
Scenedesmus subspicatus growth inhib.	IC_{10}	35 mg/L	

RubisCo test

inhib. of enzym. activity of Ribulose-P2-carboxylase in protoplasts:	IC_{10}	4,830 mg/L	

Oxygen test

inhib. of O_2 production of protoplasts:	IC_{10}	1.4 mg/L	(2698)

chloroethene *see* vinylchloride

2-chloroethylacetate

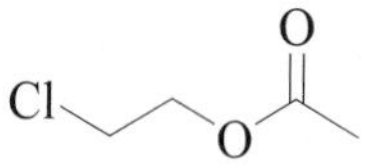

$CH_3CHClCOOCH_3$

$C_4H_7ClO_2$

CAS 542-58-5

D. BIOLOGICAL EFFECTS

Toxicity to microorganisms

30 min microtox EC_{50}: 0.69 mg/L. (2624)

4-chloroethylacetate

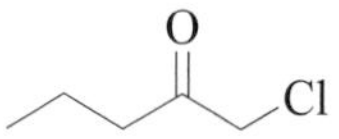

$CH_3CH_2COOCH_2Cl$

$C_4H_7ClO_2$

D. BIOLOGICAL EFFECTS

Toxicity to microorganisms

30 min microtox inhib. EC_{50}: 0.51 mg/L. (2624)

β-chloroethylalcohol *see* 2-chloroethanol

4-chloroethylbutyrate

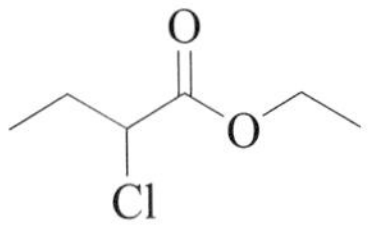

$CH_3CH_2CHClCOOCH_2CH_3$

$C_6H_{11}ClO_2$

D. BIOLOGICAL EFFECTS

Toxicity to microorganisms

30 min microtox inhib. EC_{50}: 0.51 mg/L. (2624)

chloroethylene *see* vinylchloride

2-chloroethylphosphonic acid (Mono-2-chloroethyl ester, phosphonic acid; MEPHA)

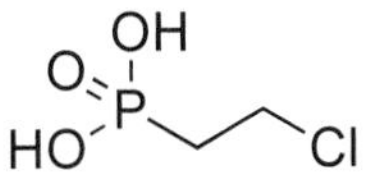

$C_2H_6ClO_3P$

CAS 16672-87-0

EINECS 240-718-3

TRADENAMES

Ethephon; Chlorethephon; Bromeflor; Ethrel; Florel; Cerone; Prep; Flordimex

USES

Plant growth regulator that promotes fruit ripening , abscission, flower induction, and other responses by releasing ethylene gas, a natural plant hormone.

A. PROPERTIES

molecular weight 144.5; melting point 73°C; boiling point : decomposes; vapour pressure <1.0 x 10^{-3} Pa at 80°C; relative density 1.65 kg/L at 20°C; water solubility at pH 4 : 800,000 mg/L, above pH 5 : decomposition; log Pow at room temperature: −0.6 at pH 2, -1.9 at pH 7 and −1.8 at pH 10; H <1.4 x 10^{-7} $Pa.m^3/mol$.

B. AIR POLLUTION FACTORS

Photodegradation

T/2 due to irradiation : 139 days. Only degradation product is ethylene.

C. WATER AND SOIL POLLUTION FACTORS

Hydrolysis

T/2 = 73 days at pH 5, 2.4 days at pH 7, 1 day at pH 9.

Aerobic biodegradation in plants and soils

Radio-tracer studies show that the compound is metabolized in plants to ethylene and 2-hydroxy ethyl phosphonic acid and chloride ions. When applied to the soil ethephon releases ethylene gas. (10774)

OH O P HO Cl ⟶ OH O P HO OH ⟶ = + $PO4^{---}$

ethephon 2-chloroethylphosphonic acid | 2-hydroxy ethyl phosphonic acid | ethylene | phosphate

Biodegradation pathway of Ethephon

Anaerobic aquatic metabolism

Ethephon degrades fairly rapidly to ethylene gas and 2-hydroxy ethyl phosphonic acid in flooded sediment under anaerobic conditions. (10774)

Degradation in soil (10773)

	% mineralisation	% bound-residues	% ethylene	After
aerobic	22%	27%	25%	44 days
	<1%	35%	0-62%	180 days
anaerobic	0.03%	2.1%	94%	30 days
Soil photolysis				

Biodegradation in laboratory studies (10773)

degradation in soils	%	
Aerobic at 20°C	50	2.7-38 days
Aerobic at 20°C	90	12-173 days
Aerobic at 10°C	50	51 days
Anaerobic at 20°C	50	2.2 days

Biodegradation in water/sediment systems (10773)

Readily biodegradable	
DT50 water	2.2; 2.6 days
DT90 water	7.2; 8.5 days
DT whole system	2.7; 3.0 days
DT whole system	8.6; 9.9 days

Mobility in soils

	2-chloroethylphosphonic acid	2-hydroxyethylphosphonic acid
K_{OC}	608-4,078	1,464-12,055

D. BIOLOGICAL EFFECTS

Toxicity

Algae

Chlorella vulgaris	72h EC_{50}	20.9 mg a.i./L

Selenastrum capricornutum	120h EC_{50}	>1.4 mg a.i./L	
Anabaena flos aquae	120h EC_{50}	>1.8 mg a.i./L	
Navicula pelliculosa	120h EC_{50}	>1.5 mg a.i./L	
Pseudokircchneriella subcapitata	72h EC_{50}	7.1 mg a.i./L	(10773)
Plants			
Lemna gibba	14d EC_{50}	>1.6 mg a.i./L	
	14d NOEC	<0.10 mg a.i./L	(10773)
Birds			
Bobwhite quail	Oral LD_{50}	764 mg/kg bw	
Mallard duck	Oral LD_{50}	1,425 mg/kg bw	(10773)
Crustaceae			
Daphnia magna	48h EC_{50}	32; 54 mg a.i./L	
	21d NOEC	67 mg a.i./L	
reproduction	21d LOEC	160 mg a.i./L	
	21d EC_{50}	>160 mg a.i/L	(10773)
Gammarus fasciatus	Acute EC_{50}	92.5 mg/L	(10774)
Insecta			
Chironomus teatans	EC_{50}	165 mg/L	(10774)
Fishes			
Pimephales promelas	34d LOEC	86 mg a.i./L	
Cyprinus carpio	96h LC_{50}	>100 mg a.i./L	
Fathead minnow	34d NOEC	43 mg a.i./L	(10773)
	96h LC_{50}	88 mg a.i./L	
Bluegill sunfish	96h LC_{50}	222 mg a.i./L	
Rainbow trout	96h LC_{50}	254.5 mg a.i./L	(10774)
Mammals			
Rat	oral LD_{50}	1,564 mg/kg bw	(10773)

Metabolism in animals

Ethephon is rapidly eliminated by rats and dogs.After 72 hours the rats and dogs retained < 1% of the ingested radioactivity. In both species 40-68% of the radioactivity was excreted as ethephon, 29-35% as ethylene, and less than 0.1 to 0.5% as carbon dioxide. This metabolism pattern is considered similar to that found in plants. (10774)

2-chloroethylvinylether ((2-chloroethoxy)ethene; 2-vinyloxyethyl chloride)

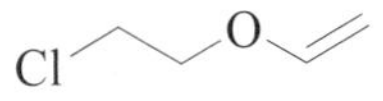

$ClCH_2CH_2OCHCH_2$

C_4H_7ClO

CAS 110-75-8

USES

Manufacturing intermediate for anaesthetics, sedatives, cellulose ethers and esters. Polymer and copolymer intermediate. Plant regulator.

A. PROPERTIES

molecular weight 106.55; melting point -70.3°C; boiling point 109°C; density 1.05; solubility 3 mg/L at 25°C; $logP_{OW}$ 1.28.

B. AIR POLLUTION FACTORS

Incinerability

Temperature for 99% destruction at 2.0-sec residence time under oxygen-starved reaction conditions: 565°C.
Thermal stability ranking of hazardous organic compounds: rank 211 on a scale of 1 (highest stability) to 320 (lowest stability). (2390)

C. WATER AND SOIL POLLUTION FACTORS

Biodegradation

Inoculum	*method*	*concentration*	*duration*	*elimination results*	
	Closed Bottle Test	5 mg/L	7 d	75% ThOD	(9552)

D. BIOLOGICAL EFFECTS

Worms

Eisenia foetida	14d LC_{50}	740 mg/kg soil	(9552)

Fishes

bluegill sunfish	24h LC_{50}	452 mg/L	
	96h LC_{50}	350 mg/L	(8102, 9974)

Mammals

rat	oral LD_{50}	250 mg/kg bw	(9552)

chlorofluoromethane (fluorochloromethane; HCF3 31; Freon 31; R31; FC31)

F Cl

CH_2ClF

CAS 593-70-4

USES

Blowing agent. Refrigerant. Alkylating agent.

A. PROPERTIES

molecular weight 68.48; melting point -133°C; boiling point -9.1°C; LogH -0.57 at 25°C.

B. AIR POLLUTION FACTORS

Tropospheric lifetime (estd.): 2.83 years.
Ozone depletion potential (estd.): 0.013 per molecule. (2446)

Photochemical reactions

$t_{1/2}$: 1.6 year, based on reactions with OH radical (calculated). (9975)

chloroform (trichloromethane)

$CHCl_3$

CAS 67-66-3

USES AND FORMULATIONS

mfg. fluorocarbon refrigerants and propellants and plastics; mfg. anesthetics and pharmaceuticals, primary source for chlorodifluoromethane; fumigant; solvent; sweetener; fire extinguisher mfg.; electronic circuitry mfg.; analytical chemistry; insecticide. May contain 0.2% ethanol as stabilizer. (347)

MANUFACTURING SOURCES

organic chemical industry. (347)

A. PROPERTIES

colorless liquid; molecular weight 119.38; melting point -64°C; boiling point 62°C; vapor pressure 160 mm at 20°C, 245 mm at 30°C, 211 hPa at 20°C; vapor density 4.12; density 1.49 at 20°C; solubility 8,000 mg/L at 20°C, 9,300 mg/L at 25°C, 10,000 mg/L at 15°C; saturation concentration in air 1027 g/m^3 at 20°C, 1,540 g/m^3 at 30°C; $LogP_{OW}$ 1.97 at 20°C.

Henry's constant

temp, °C	H	
10	0.06	
18	0.1	
25	0.15	
35	0.22	(2603)

B. AIR POLLUTION FACTORS

1 mg/m^3 = 0.20 ppm, 1 ppm = 4.96 mg/m^3.

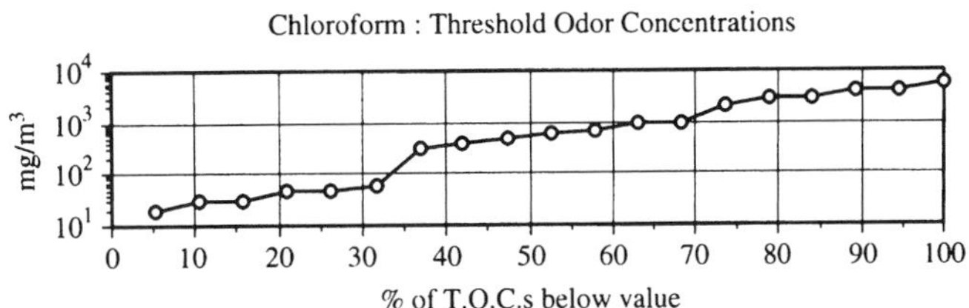

(57, 211, 602, 610, 643, 671, 704, 740, 753, 777, 804, 805, 840)

O.I. at 20°C: 70 (316)

Atmospheric reactions

Tropospheric lifetime (estd.): 0.17 year.
Ozone depletion potential (estd.): <0.001 per molecule. (2446)

reaction with OH°:	$t_{1/2}$: 160 d (calculated)	
reaction with OH° + NO_3:	$t_{1/2}$: 110 d (calculated)	
reaction with NO_3:	$t_{1/2}$: 310 d (calculated)	(5671)

Manmade sources

in 4 municipal landfill gases in S. Finland (1989-1990 data): avg 3.2-54, max 70 mg/m^3. (2605)

Ambient air quality

inside and outside 15 living rooms in Northern Italy (1983-1984):

4-7 d averages	*lowest*	*mean*	*highest value (all in μg/m³)*	
indoor	<1	2	15	
outdoor	<1	<1	3	(2756)

in rural Washington: Dec. 1974-Feb. 1975: 20 ppt. (315)

Incinerability

Temperature for 99% destruction at 2.0-sec residence time under oxygen-starved reaction conditions: 625°C.
Thermal stability ranking of hazardous organic compounds: rank 158 on a scale of 1 (highest stability) to 320 (lowest stability). (2390)

Partition coefficients

$K_{air/water}$	0.12	
Cuticular matrix/air partition coefficient*	430 ± 30	(7077)

* experimental value at 25 °C studied in the cuticular membranes from mature tomato fruits (*Lycopersicon esculentum* Mill. cultivar Vendor)

C. WATER AND SOIL POLLUTION FACTORS

BOD_5: 0-0.2; 6% ThOD (41, 275)

ThOD: 0.33 (27)

Reduction of amenities

odor thresholds: 20 mg/L; 0.1 mg/kg. (84, 894)

Manmade sources

(7000)
In Canadian municipal sludges and sludge compost: September 1993- February 1994: mean values of 11 sludges ranged from 0.072 to 0.22 mg/kg dw.; mean: 0.12 mg/kg dw; mean value of sludge compost: 0.0072 mg/kg dw.
in digested sewage sludge: U.K. 1993: n = 12

mean:	0.006 mg/kg dry wt	SD:	0.013 mg/kg dry wt	
	0.30 μg/L wet volume		0.75 μg/L wet volume	(2946)

Drinking water treatment plants in the Netherlands where pure chlorine is used for disinfection: januari - march 1994 (ng/L).

Pumpstation	Enschede	Andijk	Berenplaat	Kralingen	
raw water (intake)	30	13	54	13	
after rapid filtration	41	7509	3150	23	
after activated carbon		2699		349	
rein water (effluent)	8064	2912	4421	454	(7060)

Aquatic reactions

measured half-life for evaporation from 1 ppm aqueous solution, still air, and an average depth of 6.5 cm:
at 1-2°C: 34 min
at 25°C: 18-25 min (369)
Environmental hydrolysis half-life at 25°C and pH = 7: 1850 years. (2434)

Water quality

in NW England tap waters (1974): 0.7-38 ppb (933)
waters from upland reservoirs in NW England (1974):
during dry, cloudy weather: <0.1-0.1 ppb

during prolonged heavy rain: 11-21 ppb (933)

in river Maas at Eysden (Netherlands) in 1976: median: 1.2 μg/L, range: n.d.-5.6 μg/L
in river Maas at Keizersveer (Netherlands) in 1976: median: 0.75 μg/L, range: 0.1-2.1 μg/L (1368)
in drinking water (Frankfurt, Germany, 1977): average: 37 μg/L; max 55 μg/L (2960)

Waste water treatment

experimental water reclamation plant:

	sample 1	*sample 2*	
sand filter effluent	0.75 μg/L	0.56 μg/L	
after chlorination	33.6 μg/L	40.6 μg/L	
final water after A.C.	4.1 μg/L	0.57 μg/L	(928)

Removal in an activated sludge aeration basin: 7.4% stripped into the air
73% biotransformed (2793)

initial conc., μg/L	*treatment method*	*% removal*	
0.75-4.0	river bank filtration	51-65	
0.4-3.8	activated carbon	10-60	
0.8-2.2	ozonization	0-29	(2815)

dimensionless distribution coefficient at initial conc. of 1 mg/L:

on mixed liquor:	90	
on primary waste water sludge:	270	
on anaerobically digested sludge:	50	(2395)

Reverse osmosis

49-89% rejection was obtained by a low-pressure thin-film composite polyamide membrane FT-30 in spiral wound configuration at influents of 1-9 mg/L at ambient temp. (2630)

Biodegradability

inoculum/test method	*test conc.*	*test duration*	*removed*	
Bottle test, Bunch/Chambers	5 mg/L volatility loss: 0% at 5°C, 24% at 25°C	7 d	49% DOC	(5672)
A.S., municip., adapted, Bottle test	5 mg/L	7 d	100% DOC	(5672)
municip. sew., Closed bottle test	0.1 mg/L	175 d	0%	(5674)
anaerob. methanogenic bacteria (Closed Bottle test)	0.2 mg/L	14 d	25%	
		28 d	43%	
		56 d	50%	
		84 d	59%	
		112 d	78%	(5674)
anaerob. methanogenic bacteria (Closed Bottle test)	0.03 mg/L	14 d	15%	
		28 d	97%	
		56 d	70%	
		84 d	85%	
		112 d	100%	(5674)
anaerob. methanogenic bacteria (Closed Bottle test)	0.016 mg/L	14 d	81%	
		28 d	88%	
		56 d	89%	
		84 d	89%	
		112 d	100%	(5674)

A microbial aerobic heterotroph population-adapted to chloroform-reduced chloroform under aerobic conditions in a respirometer test: at an initial conc. of 570 mg/L after 27 d approx. 67% chloroform reduction was obtained. (2958)

Soil degradation

disappearance in soil at 100 mg/kg and 20°C under aerobic laboratory conditions: $t_{1/2}$: 4.1 d; confidence limits: 2.5-11.5 d.	(2366)

calculated half-life in nonpolluted anaerobic aquifer at varying organic carbon contents:

organic carbon content	1%	0.1%	0.01%	0.001%	
average $t_{1/2}$:	5 d	25 d	333 d	16 years	(2695)

chloroform-contaminated soil: at 500 mg/L in respirometer test: 68% removal in 27 d.	(5673)

Degradation in Lincoln fine sands at 25°C in sealed tubes shaken in the dark:
without prior methane enrichment of the soil: 26% removed after 6 d
with prior methane enrichment of the soil: 63% removed after 6 d
The addition of methane to this soil resulted in a microbial community that was capable of removing several halocarbons at rates greatly exceeding the rate in the nonamended soil. (2956)

Impact on biodegradation processes

Effect on the mineralization of acetate in anaerobic river sediment at 20°C:

toxic effects:	EC_{50}: 0.5 mg/kg dry sediment	
	EC_{10}: 0.2 mg/kg dry sediment	
inhibition of mineralization:	IC_{50}: 0.16 mg/kg dry sediment	
	IC_{10}: 0.04 mg/kg dry sediment	(2693)

Nitrosomonas, inhib. of NH_3 oxidation	24h EC_{50}	0.48 mg/L	(5671)
A.S.	24h EC_{50}	640 mg/L	(5671)
A.S., domestic	3h EC_{50}	1,010 mg/L	(5672)
	30 min EC_{50}	840 mg/L	(5672)
methanogenic bacteria	48h EC_{50}	0.9 mg/L	(5671)

Toxic effect of chloroform on the mineralization of the following chemicals in methanogenic river sediment:

chemical	EC_{50}	
4-chlorophenol	1.5 mg/kg dw	
benzoate	0.4 mg/kg dw	
acetate	0.5 mg/kg dw	
methanogenesis	7 mg/kg dw	(2961)
benzene	2,600 mg/kg dw	
pentachlorophenol	350 mg/kg dw	
1,2-dichloroethane	0.7 mg/kg dw	(2961)

Partition coefficients

Cuticular matrix/water partition coefficient**	54	(7077)

** calculated from the Cuticular matrix/water partition coefficient and the air/water distribution coefficient

D. BIOLOGICAL EFFECTS

Bacteria

OECD 209 closed system inhib	EC_{50}	500 mg/L	
microtox inhib.	5 min EC_{50}	520 mg/L	
microtox inhib.	15 min EC_{50}	670 mg/L	
polytox inhib.	30 min EC_{50}	670 mg/L	
microtox inhib.	30 min EC_{50}	1,360; 1,550 mg/L	(2624)
Microtox^{TM}*(Photobacterium)* test	5 min EC_{50}	1,549 mg/L	(2945)
Pseudomonas putida	16h EC_0	125 mg/L	(1900)

Aeromonas hydrophila	EC_0	815 mg/L	(5689)
Bacillus subtilis	EC_5	4,077 mg/L	(5689)
BiotoxÔ test *Photobacterium phosphoreum*	5min EC_{20}	360 mg/L	
	5min EC_{50}	658 mg/L	
MicrotoxÔ test *Photobacterium phosphoreum*	5min EC_{50}	736 mg/L	(7023)
	15min EC_{50}	512 mg/L	(7047)
A.S. Oxygen consumption inhibition	5d EC_{50}	890 mg/L	(7047)

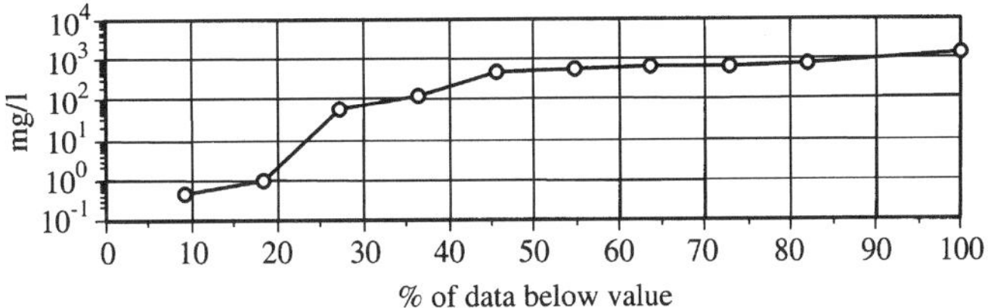

Frequency distribution of short-term EC_{50} values for bacteria (n = 11) based on data from this and other works (3999)

Algae			
Microcystis aeruginosa	8d EC_0	185 mg/L	(329)
Scenedesmus quadricauda	7d EC_0	110 mg/L	(1900)
Haematococcus pluvialis	4h EC_{10}	440 mg/L	(5677)
Scenedesmus subspicatus	48h EC_{10}	225; 360 mg/L	
	48h EC_{50}	560; 950 mg/L	(5686)
Skeletonema costatum	7d $EC_{25\text{-}50}$	>32 mg/L	(5687)
Thalassiosira pseudonana	>32 mg/L		(5687)
Protozoa			
Entosiphon sulcatum	72h EC_0	>6,560 mg/L	(1900)
Uronema parduczi Chatton-Lwoff	EC_0	>6,560 mg/L	(1901)
Chilomonas paramecium	48h EC_0	>3,200 mg/L	(5670)
Glenodinium halli	7d $EC_{25\text{-}50}$	>32 mg/L	(5687)
Isochrysis galbana	7d $EC_{25\text{-}50}$	>32 mg/L	(5687)
Crustaceans			
Daphnia magna	EC_{50} reproduction	35 mg/L	
	NOEC reproduction	14 mg/L	
	NOEC growth	27 mg/L	
	LC_{50}	62 mg/L	(2706)
	24-48h NOEC	>7.8 mg/L	
	24-48h LC_{50}	29 mg/L	(5681)
	24h EC_0	48; 62 mg/L	
	24h EC_{50}	79; 290 mg/L	
	24h EC_{100}	500 mg/L	(5677)
	24h EC_{50}	65 mg/L	(5684)
	48h EC_{50}	66; 79 mg/L	(5682, 5684)
	48h LC_{50}	353 mg/L	(5683)
	16d NOEC	15 mg/L	
	16d EC_{50}	60 mg/L	(5685)
reproduction rate	9d LC_{50}	290 mg/L	
reproduction rate	9d NOEC	120 mg/L	(5683)
reproduction rate	21d NOEC	13 mg/L	(5688)
	24h EC_{50}	116 mg/L	(7047)
Ceriodaphnia dubia	9d LC_{50}	235 mg/L	

reproduction rate	9d NOEC	3.4 mg/L	(5683)
Artoxkit M *(Artemia salina)* test	24h LC_{50}	564 mg/L	
Streptoxkit F *(Streptocephalus proboscideus)* test	24h LC_{50}	771 mg/L	
Daphnia magna test	24h LC_{50}	316 mg/L	
Rotoxkit F *(Brachionus calyciflorus)* test	24h LC_{50}	7,783 mg/L	(2945)

Molluscs

larvae of eastern oyster *(Crassostrea virginica)*: 48h, LD_{50},S: 1 mg/L initial conc. (after 48h only approx. 15% of original conc. was still present). (1545)

Fishes

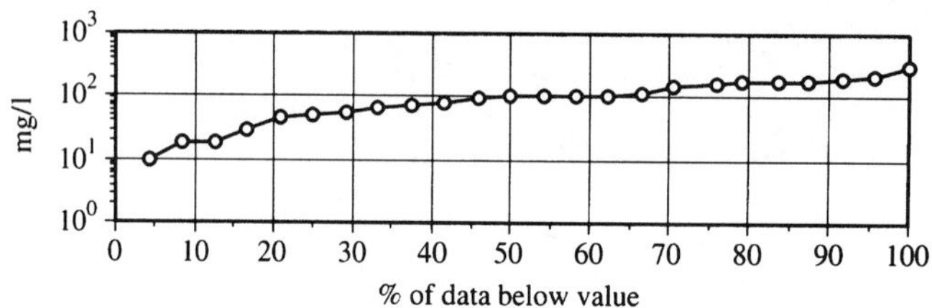

Frequency distribution of 24-96h LC_{50} values for fishes (*n* = 24) based on data from this and other works (3999)

Poecilia reticulata (guppy)	14d LC_{50}	102 ppm	(1833)
	14d LC_{50}	101 mg/L	(2696)
	96h LC_{50}	300 mg/L	(5680)
Brachydanio rerio	48h LC_{50}	100 mg/L	(5675)
Pimephales promelas	96h LC_{50},F	71 mg/L	(5676)
	96h LC_{50}	103; 171 mg/L	(5679)
Leuciscus idus	48h LC_0,S	51 mg/L	
	48h LC_{50},S	92 mg/L	
	48h LC_{100},S	151 mg/L	(5677)
Leuciscus idus melanotus	48h LC_0	118; 147 mg/L	
	48h LC_{50}	162; 191 mg/L	
	48h LC_{50}	176; 220 mg/L	(5678)

Residual concentrations: Concentrations in various organs of molluscs and fishes collected from the relatively clean water of the Irish Sea in the vicinity of Port Erin, Isle of Man (only organs with highest and lowest concentrations are mentioned- concentrations on dry weight basis):

Mollusc

	Baccinum undatum	digestive gland:	117 ng/g
		muscle:	129 ng/g
	Modiolus modiolus	digestive tissue:	56 ng/g
		mantle:	438 ng/g
	Pecten maximus	mantle:	224 ng/g
		gill:	1040 ng/g
Fishes	*Conger conger* (eel)	gut:	43 ng/g
		liver:	474 ng/g
	Gadus morhua (cod)	stomach:	7 ng/g
		muscle, brain:	167 ng/g
	Pollachius birens (coal fish)	alimentary canal:	51 ng/g
		liver:	851 ng/g
	Scylliorhinus carnicula (dogfish)	liver, spleen:	76-80 ng/g

	gill:	755 ng/g	
Trisopterus luscus (bib)	liver, skeletal tissue:	48-50 ng/g	
	gill:	212 ng/g	(1092)

Bioaccumulation

bluegill sunfish *(Lepomis macrochirus)* at 110 µg/L at 16°C: 14d BCF: 6; $t_{1/2}$ in tissues: <1 d. (2602)

chloroformylchloride *see* phosgene

chlorofos *see* dimethyl-(2,2,2-trichloro-1-hydroxyethyl)phosphonate

4-chloroguaiacol (4-chloro-2-methoxyphenol)

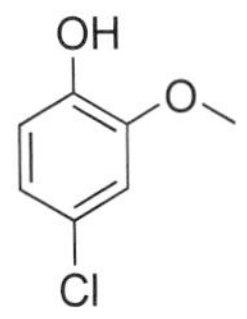

OH, O, Cl

$C_7H_7ClO_2$

CAS 16766-30-6

OCCURRENCE

In bleached kraft mill effluents

A. PROPERTIES

molecular weight 158.58

C. WATER AND SOIL POLLUTION FACTORS

Anthropogenic sources

In effluents from pulp mills using 100% chlorine dioxide (1990-1997): maximum 1,2 µg/L (7192)

Biodegradation

A strain,identified as *Acinetobacter junii* 5ga, is able toinitiate metabolism of specific chlorinatedguaiacols by 0-demethylation. Transient formation of chlorocatechols resulting from incubation of cells with 4-chloroguaiacol or 4,5-dichloroguaiacol was suggested by UV spectroscopy. (11208)

The metabolism of chloroguaiacols by a soil bacterium was studied. The strain was identified as a *Rhodococcus ruber* CA16. None of seven chlorinated guaiacols supported bacterial growth. However resting cells grown on guaiacol degraded completely 4-chloroguaiacol, 5-chloroguaiacol, and 6-chloroguaiacol and, to a lesser extent, 4,5-dichloroguaiacol. Gas chromatographic analysis suggested microbial formation of 4-chlorocatechol and 4,5-dichlorocatechol from 4-chloroguaiacol and 4,5-dichloroguaiacol, respectively. Although mono- and dichloroguaiacols did not affect the strain's ability to grow on guaiacol, chlorocatechols completely arrested growth. (11209)

OH O Cl → OH OH Cl → dechlorination

Metabolism of 4-chloroguaiacol by Acinetobacter junii strain [11208]

Metabolism of 4-chloroguaiacol by *Acinetobacter junii* strain (11208)

D. BIOLOGICAL EFFECTS

Toxicity

Fish			
Zebra fish egg/larvae	NOEC	0.70 mg/L	(7161)

1-chloroheptane (heptylchloride)

Cl

$CH_3(CH_2)_5CH_2Cl$

$C_7H_{15}Cl$

CAS 629-06-1

A. PROPERTIES

molecular weight 134.65; melting point -69.5°C; boiling point 159.5°C; density 0.87 at 20/0°C.

B. AIR POLLUTION FACTORS

Odor

T.O.C.: 0.33 mg/m³ =	59 ppb	(307)
	0.06 ppm	(279)
recognition:	157 mg/m³	(761)

C. WATER AND SOIL POLLUTION FACTORS

Waste water treatment

A.S.: after	6 h: 0.9% of ThOD	
	12 h: 4.3% of ThOD	
	24 h: 4.4% of ThOD	(88)

1-chlorohexane (prom-*n*-hexylchloride)

$CH_3(CH_2)_4CH_2Cl$

$C_6H_{13}Cl$

CAS 544-10-5

A. PROPERTIES

molecular weight 120.62; melting point -83°C; boiling point 132°C; vapor density 4.2; density 0.90 at 20°C.

C. WATER AND SOIL POLLUTION FACTORS

Waste water treatment

A.S.: after	6 h: 3.9% of ThOD	
	12 h: 4.8% of ThOD	
	24 h: 6.4% of ThOD	(88)

chloromethane *see* methylchloride

chloromethoxymethane *see* chloromethylether

chloromethylcyanide *see* chloroacetonitrile

chloromethylether *see* chloromethylmethylether

chloromethylmethylether (chloromethoxymethane; monochlorodimethylether; chloromethylether; CMME; dimethylchloroether)

Cl⌒O⌒

CH_2ClOCH_3

C_2H_5ClO

CAS 107-30-2

USES AND FORMULATIONS

mfg. irritant gases (lacrymators); chloromethylating agent; chemical; intermediate. (347)

Can be contaminated with 1 to 8% dichloromethylether. (1349, 1350)

A. PROPERTIES

colorless liquid; molecular weight 80.52; melting point -103.5°C; boiling point 59.5°C; density 1.0625 at 10/4°C; solubility decomposes.

B. AIR POLLUTION FACTORS

1 mg/m^3 = 0.304 ppm, 1 ppm = 3.29 mg/m^3.

Incinerability

thermal stability ranking of hazardous organic compounds: rank 218 on a scale of 1 (highest stability) to 320 (lowest stability). (2390)

C. WATER AND SOIL POLLUTION FACTORS

Aquatic reactions

hydrolysis:
CMME in aqueous solution is hydrolyzed very fast with a half-life on the order of <1 sec; extrapolated to pure water. (1364)

The hydrolytic reaction can be depicted as follows:
$CH_2ClOCH_3 + H_2O \rightarrow CH_3OH + HCl + CH_2O$ (1365)

N-chloromethylsulfo-pentachloro-2-aminodiphenylether, sodium salt *see* eulan

chloromycetin *see* chloramphenicol

1-chloronaphthalene (alpha-naphthyl chloride)

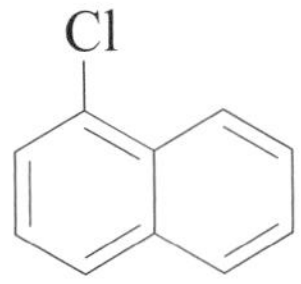

$C_{10}H_7Cl$

CAS 90-13-1

USES

Monochloronaphthalenes and mixtures of mono- and dichloronaphthalenes have been used for chemical-resistant gauge fluids and instrument seals, as heat exchange fluids, as high boiling speciality solvents, for colour dispersions, as engine crankcase additives, and as ingredients in motor tune-up compounds. Monochloronaphthalenes have also been used as a raw material for dyes and as a wood preservative with fungicidal and insecticidal properties.

Immersion liquid in the microscopic determination of refractive index of crystals. Solvent for oils, fats, DDT. The major component of Halowax 1031.

USES AND FORMULATIONS

contains 2-chloronaphthalene, up to 10%.

A. PROPERTIES

molecular weight 162.61; boiling point 250-280°C; melting point -20°C; density 1.19 at 20/4°C; log P_{ow} 3.90.; vapour pressure 0.003 kPa; water solubility 2.9 mg/L

B. AIR POLLUTION FACTORS

Incinerability

thermal stability ranking of hazardous organic compounds: rank 21 on a scale of 1 (highest stability) to 320 (lowest stability). (2390)

C. WATER AND SOIL POLLUTION FACTORS

Manmade sources

Göteborg (Sweden) sew. works 1989-1991: infl.: n.d.-1.0 µg/L; effl.: n.d. (2787)

Biodegradation

metabolic pathway:

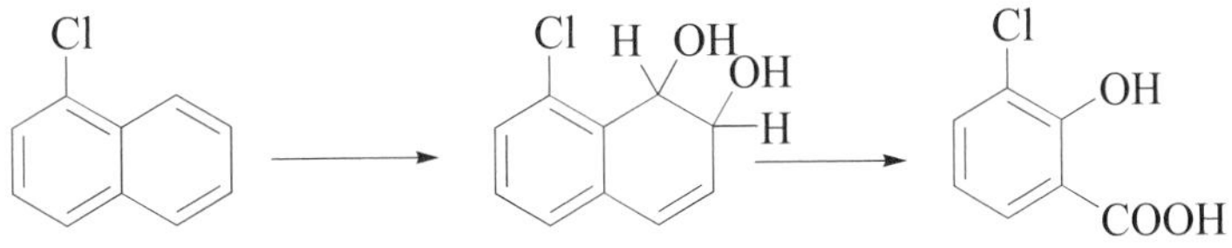

d-8-chlor-1,2-dihydroxynaphthalene →3-chlorosalicylic acid (177)

D. BIOLOGICAL EFFECTS

Bioaccumulation

Species	conc. µg/L	duration	body parts	BCF	
Cyprinus carpio				191	(11112)

Toxicity

Amphibians			
frog, *Rana esculenta*	oral LD_{50}	900 mg/kg bw	(9985)
Algae			
Tetrahymena pyriformis	60h EC_{50}	25 mg/L	(11112)
Crustaceae			
Daphnia magna	48h LC_{50}	0.82; 1.6 mg/L	
	48h EC_0	<0.17 mg/L	
Artemia salina	24h LC_{50}	0.91; 1.84 mg/L	
Mysidopsis bahia	96h LC_{50}	0.37 mg/L	(11112)
Fish			
Lepomis macrochirus	96h LC_{50}	2.3 mg/L	
Cyprinodon variegatus	96h LC_{50}	0.69; 2.4 mg/L	
	96h LC_0	1.2 mg/L	(11112)
	MATC, early life stage	555 µg/L	(2643)
Mammals			
mouse	oral LD_{50}	1,090 mg/kg bw	
rat	oral LD_{50}	1,540 mg/kg bw	(9982, 9983)
guinea pig	oral LD_{50}	2,000 mg/kg bw	(9984)

2-chloronaphthalene (beta-naphthyl chloride)

$C_{10}H_7Cl$

CAS 91-58-7

USES

A component of Halowax 1031.
Monochloronaphthalenes and mixtures of mono- and dichloronaphthalenes have been used for chemical-resistant gauge fluids and instrument seals, as heat exchange fluids, as high boiling speciality solvents, for colour dispersions, as engine crankcase additives, and as ingredients in motor tune-up compounds. Monochloronaphthalenes have also been used as a raw material for dyes and as a wood preservative with fungicidal and insecticidal properties.

OCCURRENCE

as contaminant in 1-chloronapthalene: up to 10%.

A. PROPERTIES

molecular weight 162.62; melting point 59°C; boiling point 256°C; log P_{OW} 4.0; vapour pressure 0.0011 kPa; water solubility 0.92 mg/L

B. AIR POLLUTION FACTORS

Incinerability

temperature for 99% destruction at 2.0-sec residence time under oxygen-starved reaction conditions: 975°C. (2390)

C. WATER AND SOIL POLLUTION FACTORS

Disappearance in soil

at 100 mg/kg and 20°C under aerobic laboratory conditions: half-life: 11 d; confidence limits: 8.6-16 d. (2366)

D. BIOLOGICAL EFFECTS

Bioaccumulation

Species	conc. µg/L	duration	body parts	BCF	
Poecilia reticulata	100-1000			4,266	(11112)

Toxicity

Crustaceae			
Daphnia magna	48h EC_{50}	1.99 mg/L	
Artemia salina	24h LC_{50}	2.82 mg/L	(11112)
Mammals			
rat	oral LD_{50}	2,080 mg/kg bw	
mouse	oral LD_{50}	886 mg/kg bw	(9709)

m-chloronitrobenzene (m-nitrochlorobenzene; 1-chloro-3-nitrobenzene)

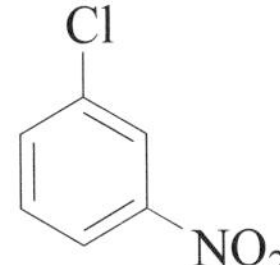

$C_6H_4ClNO_2$

$C_6H_4ClNO_2$

CAS 121-73-3

USES AND FORMULATIONS

Dyestuffs mfg.; intermediate in organic chemical synthesis.

SOURCES

Formed in small quantities during chlorination, when nitrobenzene is present. (347)

A. PROPERTIES

pale yellow crystals; molecular weight 157.56; melting point unst. 23.7°C, stb. 44.4°C; boiling point 235- 236°C; vapor pressure 0.011 hPa at 20°C, 0.019 at 25°C; density 1.53 at 20/4°C; solubility 390 mg/L at 20°C; $LogP_{OW}$2.4 (measured), $2.4_{1/2}.46$.

B. AIR POLLUTION FACTORS

Odor

Odor threshold: 0.02 mg/m^3. (605)

Photochemical reactions

Photolysis: $t_{1/2}$: 11 d, based on reactions with OH° (calculated). (6100)

C. WATER AND SOIL POLLUTION FACTORS

Biodegradation

decomposition period by a soil microflora: >64 d. (176)

Biodegradation

Inoculum	*Method*	*Concentration*	*Duration*	*Elimination results*	
	Closed bottle test	2- 80 mg/L	20 d	0% ThOD	(7666)

D. BIOLOGICAL EFFECTS

Toxicity

Bacteria			
A.S. oxygen consumption inhibition	3 h EC_{50}	>100 mg/L	(7666)
Photobacterium phosphoreum	30 min EC_{50}	20 mg/L	(7666)
Microtox test	15 min EC_{50}	1.1 mg/L	(8899)

Tetrahymena pyriformis: growth inhibition at 27°C, 48 h EC_{50}: 22 mg/L (2704)

Algae			
Scenedesmus subspicatus	72 h EC_{10}	1.8; 3,0 mg/L	
(cell multiplication inhibition test)	72 h EC_{50}	5.2; 10 mg/L	(7667)
Haematococcus pluvialis	4 h EC_{50}	4 mg/L	(7666)
Crustaceans			
Daphnia magna	24 h EC_0	20; 33 mg/L	(7667)
	24 h EC_{50}		
	24 h EC_{100}		(7666)
	21 d NOEC	0.13 mg/L	
	48 h EC_{50}	26 mg/L	(8085)
Fishes			
Bluegill	96 h LC_{50},S	1.2 mg/L	
Pimephales promelas	30- 35 d LC_{50},F	208 mg/L	(2704)
Brachydanio rerio	96 h LC_0	42 mg/L	(7666)
Menidia beryllina	96 h LC_{50}	0.55 mg/L	
Pimephales promelas	96 h LC_{50}	18 mg/L	(7666)
Brown trout, bluegill sunfish, yellow perch, goldfish	24 h LC_0	5 mg/L	(9594)
Guppy	96 h LC_{50}	20 mg/L	(8085)
Fathead minnow	96 h LC_{50}	18 mg/L	(8481)
Mammals			
Mouse	oral LD_{50}	390 mg/kg bw	
Rat	oral LD_{50}	470 mg/kg bw	(9789, 9995)

o-chloronitrobenzene (*o*-nitrochlorobenzene; ONCB)

$ClC_6H_4NO_2$

$C_6H_4ClNO_2$

CAS 88-73-3

USES AND FORMULATIONS

chemical synthesis. Intermediate in dyestuff synthesis.

A. PROPERTIES

needles; molecular weight 157.56; melting point 32.5°C; boiling point 245.7°C; vapor pressure 0.058 hPa at 20°C; 3.0 x 100^{-2} mmHg at 20 °C; density 1.37 at 22/4 °C; solubility < 100 mg/L at 22 °C; 590 mg/L at 20 °C; log P_{oct} 2.2 (measured), 2.24.

B. AIR POLLUTION FACTORS

Photochemical reactions

Photolysis: $t_{1/2}$: 2 d, based on reactions with OH° (calculated). (9259)

C. WATER AND SOIL POLLUTION FACTORS

Water quality

in river Maas (the Netherlands): avg in 1973: n.d.-0.33 µg/L (n.s.i.). (342)

Biodegradation

decomposition period by a soil microflora: >64 d. (176)

Odor threshold

detection: 0.015-0.020 mg/kg (n.s.i.). (903)

Biodegradation

Inoculum	*method*	*concentration*	*duration*	*elimination results*	
	Closed Bottle Test		20 d	0% ThOD	(7672)
facultative aerobic bacteria	in mineral medium		7 d	80% reduction to o-chloroaniline	(7674)
river water, sewage		21 mg/L	175 d	0%	(9988)

Waste water treatment

A reduction of chloronitrobenzenes in water, from 1.9 mg/L to < 0.03 mg/L was achieved by the application of 8 mg/L ozone and 3 mg/L hydrogen peroxide with a 20 min contact time. (9994)

Hydrolysis

Hydrolysis is not expected under environmental conditions.

Biodegradation

Inoculum	method	concentration	duration	elimination	
Sludge miscelaneous	Modified MITI I	30 mg/L	14 days	8.2%	
A.S. industrial	Modified Zahn-Wellens	200 mg/L DOC	5 days	80%	
			10 days	>90%	
A.S.	Respirometer	200 mg/L DOC	10 days	<10%	
Domestic sewage adapted	OECD 301D		20 days	0% ThOD	(10717)

D. BIOLOGICAL EFFECTS

Bioaccumulation

	Concentration µg/L	*exposure period*	*BCF*	*organ/tissue*	
Fishes					
Cyprinus carpio	250	8 weeks	7-21		
	25	8 weeks	7-22		(7674)
Poecilia reticulata	6,000	3 days	11.6-11.9		

Toxicity

Micro organisms			
Pseudomonas putida	30min EC_0	100 mg/L	
	72h EC_{10}	11; 19 mg/L	
(cell multiplication inhibition test)	72h EC_{50}	34; 75 mg/L	(7676)
Activated sludge domestic	24h EC_0	ca 80 mg/L	(10717)
Photobacterium phosphoreum Microtox test	15 min EC_{50}	4.5 mg/L	(9657)
Algae			
Scenedesmus obliquus	96h EC_{50}	18.1 mg/L	
Scenedesmus pannonicus	72h EC_{50}	24 mg/L	(10717)
Scenedesmus subspicatus	72h EC_{10}	11; 19 mg/L	
(cell multiplication inhibition test)	72h EC_{50}	34; 75 mg/L	(7676)
Chlorella pyrenoidosa growth	96h EC_{50}	6.8 mg/L	(7914)
Fungi			
Pythium altimum	88h ED50	157.6 mg/L	
Rhizoctonia solani	88h EC_{50}	48.9 mg/L	
Crustaceae			
Daphnia magna	48h EC_{50}	3.2; 23.9 mg/L	
	48h LC_{50}	49 mg/L	
Reproduction rate	21d NOEC	3 mg/L	
	21d LOEC	9.9 mg/L	
	24h EC_0	5 mg/L	
	24h EC_{50}	12 mg/L	(7675)
	21d NOEC	4 mg/L	(7675)
Daphnia carinata	48h EC_{50}	21.3 mg/L	(10717)
Plants			
Lactuca sativa	14d EC_{50}	3.2-10 mg/kg soil	
Cucumus sativus	6d EC_{50}	18.1 mg/kg sand	
Phaseolus aureus	6d EC_{50}	29.9 mg/kg sand	(10717)
Fish			
Pimephales promelas juveniles	33d NOEC	1.02 mg/L	
	33d LOEC	2.04 mg/L	
Cyprinus carpio	96h LC_{50}	25.5 mg/L	
Poecilia reticulata	14d LC_{50}	30 mg/L	
	96h LC_{50}	30 mg/L	(10717)
Brachydanio rerio	96h LC_{50}	35 mg/L	(7673)
	14d NOEC	2.9 mg/L	

Leuciscus idus	24h LC_0	5 mg/L	(7672)
	96h LC_{100}	10 mg/L	
Oryzias latipes	48h LC_{50}	28 mg/L	(7674)
guppy	14d LC_{50}	30 mg/L	(7902)
Mammals			
Rat	oral LD_{50}	144; 219; 251; 263; 270; 288; 300; 339; 350; 457; 510; 560; 630 mg/kg bw	
Mouse	oral LD_{50}	135; 140; 340; 440 mg/kg bw	
Rabbit	oral LD_{50}	280 mg/kg bw	(10717)

p-chloronitrobenzene (*p*-nitrochlorobenzene)

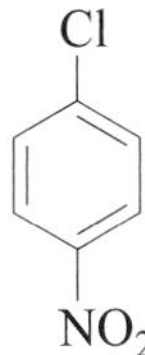

$C_6H_4ClNO_2$

$C_6H_4ClNO_2$

CAS 100-00-5

USES AND FORMULATIONS

Feedstock for chemical synthesis. Used in the manufacture of dyestuffs and insecticides, and in rubber industry.

A. PROPERTIES

molecular weight 157.56; melting point 83.5°C; boiling point 239- 242°C; vapor pressure 0.085 hPa at 20°C, 8 hPa at 110°C; vapor density 5.44; density 1.52 at 18/4°C; solubility <1,000 mg/L; $LogP_{OW}$ 2.6 (measured); 2.39/2.41.

B. AIR POLLUTION FACTORS

1 mg/m^3 = 0.15 ppm, 1 ppm = 6.55 mg/m^3.

C. WATER AND SOIL POLLUTION FACTORS

Biodegradation

period by a soil microflora: >64 d. (176)

Biodegradation

Inoculum	*Method*	*Concentration*	*Duration*	*Elimination results*	
mun. waste water	Closed bottle test	2.4 mg/L	20 d	>50% ThOD	(7661)
mun. waste water	Respirometer test	100 mg/L	28 d	6% O_2 cons.	
A.S.		100 mg/L	14 d	0% ThOD	(7663)

Concentrations of <0.05 mg/L in wastewater are reported to be completely degraded under anaerobic conditions.	(9997)

Inhibition of biodegradation

at 100 mg/L, no inhibition of NH_3 oxidation by *Nitrosomonas* sp.	(390)

D. BIOLOGICAL EFFECTS

Bacteria			
Pseudomonas putida	30 min EC_{10}	59 mg/L	(7661)
(cell multiplication inhibition test)	10 h EC_{10}	47 mg/L	(7661)
Algae			
Scenedesmus subspicatus	7 d EC_3	10 mg/L	(7661)
(cell multiplication inhibition test)	48 h EC_{10}	2.2; 4.9 mg/L	
	48 h EC_{50}	8; 16 mg/L	(7665)
Haematococcus pluvialis	4 h EC_{50}	4 mg/L	(7661)
Crustaceans			
Daphnia magna	24 h EC_0	2.0; 3.3 mg/L	
	24 h EC_{50}	6; 15 mg/L	
	24 h EC_{100}	14 mg/L	(7661, 7664)
	21 d NOEC	0.32 mg/L	(7661)
	48 h LC_{50}	8.9 mg/L	(8085)
Fishes			
Brachydanio rerio	96 h LC_{50}	14 mg/L	(7662)
	14 d NOEC	1.5 mg/L	(7662)
Leuciscus idus	48 h LC_0	1.0 mg/L	
	48 h LC_{50}	2.0 mg/L	
	48 h LC_{100}	2.5 mg/L	(7661)
	48 h LC_{50}	18 mg/L	(7661)
Salmo gairdneri	96 h LC_{50}	6 mg/L	(7661)
Oryzas latipes	48 h LC_{50}	14 mg/L	(7663)
Brown trout, bluegill sunfish, yellow perch, goldfish	24 h LC_0	5 mg/L	(9594)
Guppy	96 h LC_{50}	13 mg/L	(8085)
Carp	48 h LC_{50}	35 mg/L	(9996)
Mammals			
Rat	oral LD_{50}	420 mg/kg bw	
Mouse	oral LD_{50}	650 mg/kg bw	(9998, 9701)

1-chlorooctane (octylchloride)

$CH_2Cl(CH_2)_6CH_3$

$C_8H_{17}Cl$

CAS 111-85-3

USES

alkylation agent in organic synthesis.

A. PROPERTIES

molecular weight 148.7; melting point -58°C; boiling point 182°C; vapor pressure 1.5 hPa at 20°C; density 0.87 g/cm^3; solubility 19 mg/L at 20°C; $LogP_{OW}$ 3.79.

C. WATER AND SOIL POLLUTION FACTORS

Biodegradability

inoculum/methods	*test conc.*	*test duration*	*removed*	
Municip. waste water, OECD screening	10 mg/L	14 d	97% DOC	(5485)
A.S., Sturm test	10 mg/L	28 d	40% DOC	(5485)

D. BIOLOGICAL EFFECTS

Bacteria

Pseudomonas putida	6h EC_{10}	42 µL/L	(5485)
FISHES: *Leuciscus idus*	48h LC_0	8 mg/L	
	48h LC_{50}	12 mg/L	(5485)

Chloroparaffin waxes (chlorohydrocarbon waxes)

CAS 63449-39-8

USES AND FORMULATIONS

Chlorinated paraffins with carbon number C-17 in Cl content 10- 72%. Other chlorinated hydrocarbons have been assigned the following CAS numbers: C_{10-13}: CAS 85535-84-8 and C_{14-17}: CAS 85535-85-9. The products are mainly used as plasticizers for PVC and paints, additives for high-pressure metal working fluids and flame retardants for rubber and plastics.

A. PROPERTIES

molecular weight approx. 1,100 (C 23.5; 72% Cl); melting point - 25/0°C; vapor pressure 2.7 x 10^{-6} hPa at 80°C (42%Cl); density 1.1 (C_{18-20}; 35% Cl), 1.3 (C_{18-20}, 52% Cl); solubility <0.005 mg/L at 20°C (C_{25}, 43% Cl), 0.0059 mg/L at 70°C (C_{25}, 70% Cl); $logP_{OW}$ >6 (measured for C_{18-26}, 42-44% Cl)

B. AIR POLLUTION FACTORS

Photochemical reactions

$t_{1/2}$: 0.9- 1.4 d, based on reactions with OH° (calculated for C_{18-26}, 35- 53% Cl) (8282)

C. WATER AND SOIL POLLUTION FACTORS

ThOD	1.36 g/g

Biodegradation

Inoculum	*Method*	*Concentration*	*Duration*	*Elimination results*	
Adapted soil m.o.	Closed bottle test (C_{20-30}, 42% Cl)		5 d	0% ThOD	
			10 d	0.7% ThOD	
			15 d	10% ThOD	
			20 d	27% ThOD	
			25 d	32% ThOD	(8276)
Nonadapted soil m.o.	Closed bottle test (C_{20-30}, 42% Cl)		5 d	0% ThOD	
			10 d	12% ThOD	
			15 d	13% ThOD	
			20 d	23% ThOD	
			25 d	310% ThOD	(8276)

D. BIOLOGICAL EFFECTS

Toxicity

Bacteria			
A.S. oxygen consumption inhibition	3 h EC_0	>2,000 mg/L (C_{18-20}, 35% Cl)	(8281)
		>2,000 mg/L (C_{18-20}, 44% Cl)	(8281)
Primary municipal sludge	24 h EC_0	1,250 mg/L (C_{18-20}, 52% Cl)	(8278)
		1,250 mg/L (C_{18-20}, 35% Cl)	(8274)
		600 mg/L (C_{18-20}, 44% Cl)	(8278)
		1,250 mg/L (C_{18-20}, 49% Cl)	(8278)

Crustaceans				
Daphnia magna	21 d EC_{50}	41 mg/L	(C_{18-27}, 60% Cl) + emulsifier	
	21 d NOEC	4.2 mg/L	(C_{18-27}, 60% Cl) + emulsifier	(8279)
	24 h EC_{50}	149 mg/L	(C_{18-27}, 60% Cl) + emulsifier	
	24 h NOEC	45 mg/L	(C_{18-27}, 60% Cl) + emulsifier	(8280)
	24 h EC_{50}	1,024 mg/L	(C_{18-27}, 60% Cl) + solvent	
	24 h NOEC	100 mg/L	(C_{18-27}, 60% Cl) + solvent	(8280)
	24 h EC_{50}	102 mg/L	(C_{18-27}, 60% Cl) + emulsifier	
	24 h NOEC	23 mg/L	(C_{18-27}, 60% Cl) + emulsifier	(8280)
	24 h EC_{50}	553 mg/L	(C_{18-27}, 60% Cl) + solvent	
	24 h NOEC	100 mg/L	(C_{18-27}, 60% Cl) + solvent	(8280)

Fishes			
Salmo gairdneri	96 h LC_{10}	>770 mg/L (C_{20-30}, 42% Cl)	(8276)
	96 h LC_{50}	>300 mg/L (C_{20-30}, 40% Cl)	(8272, 8273)
	96 h LC_{50}	>300 mg/L (C-17, 39% Cl)	(8272, 8273)
	96 h LC_{50}	>300 mg/L (C_{22-26}, 50% Cl)	(8272, 8273)
	60 d NOEC	>4 mg/L (C_{22--26}, 43% Cl)	(8275)
		>3.8 mg/L (C-20, 70% Cl)	(8275)
Lepomis macrochirus	96 h LC_{50}	>300 mg/L (C_{20-30}, 40% Cl)	(8272, 8273)
		>300 mg/L (C_{22-26}, 70% Cl)	(8273)
		>300 mg/L (C-17, 39% Cl)	(8272, 8273)
		>300 mg/L (C_{22-26}, 50% Cl)	(8272, 8273)
Leuciscus idus	48 h LC_{10}	>500 mg/L (C_{18-20}, 52% Cl)	(8278)
	48 h LC_{10}	400 mg/L (C_{18-20}, 35% Cl)	(8274)
	48 h LC_{10}	>500 mg/L (C_{18-20}, 49% Cl)	(8274)

	96 h LC_{10}	500 mg/L (C_{18-20}, 44% Cl)	(8274)
	96 h LC_{50}	400 mg/L (C_{18-20}, 35% Cl)	(8274)
Alburnus alburnus	96 h LC_{50}	>5,000 mg/L (C_{22-26}, 42% Cl)	(8277)

chloropentafluoroethane (CFC 115; F 115; Freon 115; pentafluorochloroethane)

C_2ClF_5

CAS 76-15-3

USES

in closed systems, refrigerant, heat transfer agent. Production stopped in Germany in 1993.

A. PROPERTIES

molecular weight 154.47; melting point -106°C; boiling point -38°C; density liquid 1.31 g/cm^3 at 20°C; LogP_{OW} 2.4 at 25°C.

B. AIR POLLUTION FACTORS

Indirect photolysis

reaction with OH°: $t_{1/2}$: 220 years: (3094)

compound	*global warming potential*	*ozone depletion potential*	
CFC 11	1.0	1.0	
CFC 12	3.0	1.0	
CFC 113	1.4	0.8	
CFC 114	3.9	1.0	
CFC 115	7.5	0.6	
HCFC 22	0.36	0.055	(3095)

Estimated total atmospheric lifetime: 550 years. (3096)

C. WATER AND SOIL POLLUTION FACTORS

Biodegradation

in marine model ecosystem: half-life time: >87 d. (3126)

1-chloropentane *see* amylchloride

2-chloropentane (2-pentylchloride)

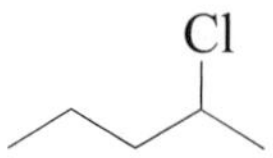

CAS 625-29-6

EINECS 210-885-7

A. PROPERTIES

molecular weight 106.59; boiling point 94-95°C; density 0.87 at 25°C; LogH 0.03 at 25°C.

5-chloropentanoic acid (5-chlorovaleric acid; 5-chlorovalerianic acid)

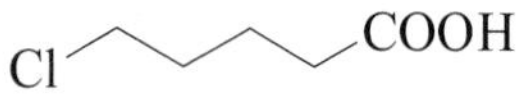

$CH_2Cl(CH_3)_3COOH$

$C_5H_{12}ClO_2$

CAS 1119-46-6

D. BIOLOGICAL EFFECTS

Toxicity to microorganisms

30 min microtox inhib. EC_{50}: 169 mg/L. (2624)

Chlorophenamidine (N'-(4-chloro-2-methylphenyl)-N,N-dimethylmethanimidamide; N'-(2-methyl-4-chlorophenyl)-N,N-dimethylformamidine; Fundex; Galecron; Spanone; Carzol; chlordimeform; N#-(4-chloro-*o*-tolyl)-N,N-dimethylformamidine;)

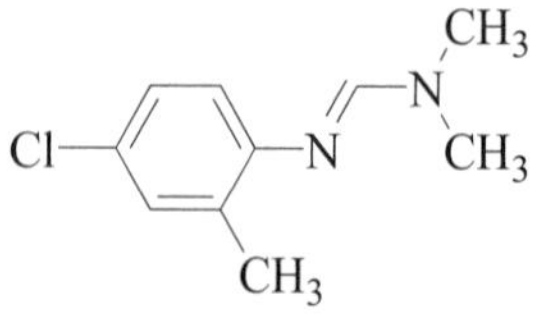

$C_{10}H_{13}ClN_2$

CAS 6164-98-3

USES

broad spectrum insecticide that is effective for all stages of insects and mites, including eggs and adults. Chlorphenamidine hydrochloride has been used for the control of the rice stem borer in Japan. (1547)

A. PROPERTIES

buff-colored crystals; molecular weight 196.70; melting point 32°C; boiling point 156-157 °C at 0.4 mm Hg; density 1.1 at 25 °C ; vapor pressure 3.5 x 10± mm Hg at 20 °C;solubility 250 ppm at 20°C.

C. WATER AND SOIL POLLUTION FACTORS

Residues of chlorphenamidine and three degradation products in rice grains, straws, and soil:

µg/kg	chlorphenamidine	desmethyl chlorphenamidine	N-formyl-4-chloro-o-toluidine	4-chloro-o-toluidine	
rice grains	4-48	0.2-1	10-38	3-61	
straws	260-9,700	10-180	67-500	80-6,900	
soil, 0-5 cm depth	35-2,900	5-15	8-380	2-68	
soil, 5-10 cm depth	33-150	1-4	4-9	1-20	(1547)

Persistence in soils

in non-sterile soil $t_{1/2}$ estimate of 1 month . (10026)

Mobility in soils: adsorbed in soil by a cationic exchange mechanism. (10027)

desmethyl chlorphenamidine

4-chloro-o-toluidine

N-formyl-4-chloro-o-roluidine

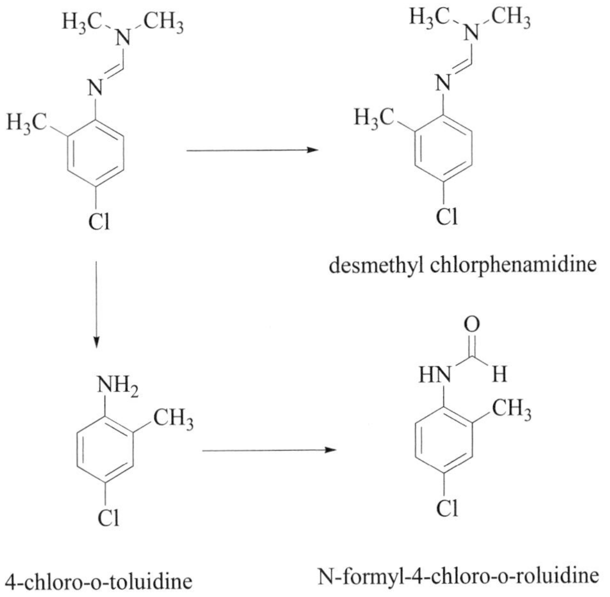

The major metabolites by plants are: N′-(4-chloro-*o*-tolyl)N-methylformamidine (desmethylchlorphenamidine), N-formyl-4-chloro-*o*-toluidine and 4-chloro-*o*-toluidine (1548, 1549, 1550)

Persistence in soils

in non-sterile soil t1/2 estimate of 1 month . (10026)

Mobility in soils

adsorbed in soil by a cationic exchange mechanism. (10027)

D. BIOLOGICAL EFFECTS

Toxicity

Fish			
trout	24h LC_{50}	11.7 mg/L	
bluegill sunfish	24h LC_{50}	1 mg/L	
Japanese killifish	24h LC_{50}	33 mg/L	(9600)
Mammals			
Rat	oral LD_{50}	160 mg/kg bw	(10028, 10029)
Mouse	oral LD_{50}	224mg/kg bw	(10030)
rabbit	oral LD_{50}	625 mg/kg bw	(9775)

2-chlorophenanthrene

$C_{14}H_9Cl$

CAS 24423-11-8

A. PROPERTIES

$LogP_{OW}$ 5.16.

D. BIOLOGICAL EFFECTS

Fishes

log BCF: fathead minnow: 3.63 (2606)

chlorophene (4-chloro-2-(phenylmethyl)phenol; o-benzyl-p-chlorophenol))

$C_{13}H_{11}ClO$

CAS 120-32-1

TRADENAMES

Santophen 1

USES

Antiseptic

A. PROPERTIES

molecular weight 218.68

C. WATER AND SOIL POLLUTION FACTORS

Occurrence

in municpal treatment works in Germany: chlorophen is routinely found in both influents and effluents up to 710 ng/L. (7178)

m-chlorophenol (1-chloro-3-hydroxybenzene; 3-chlorophenol)

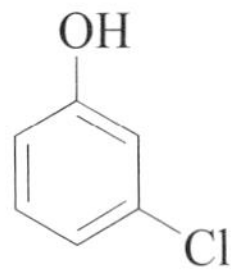

ClC_6H_4OH

C_6H_5ClO

CAS 108-43-0

A. PROPERTIES

needles; molecular weight 128.56; melting point 32.8°C; boiling point 214°C; vapor pressure 5 mm at 72°C, 40 mm at 118°C; density 1.2 at 45°C; solubility 26,000 mg/L at 20°C; $LogP_{OW}$ 2.47/2.50.

B. AIR POLLUTION FACTORS

Control methods

wet scrubber: water at pH 8.5: 45 odor units/scf
$KMnO_4$ at pH 8.5: 25 odor units/scf (115)

Photodegradation

By photocatalytic degradation in UV-illuminated TiO_2 aqueous suspension, 50% and 99% disappearance are obtained after 37 and 110 min at an initial conc. of 20 mg/L. Chorohydroquinone and hydroxyhydroquinone were detected as the major metabolites. The decomposition rate is not much affected by the pH over a wide range. (2371)

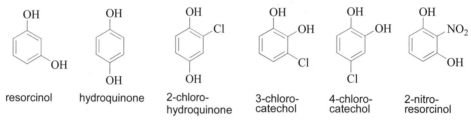

3-chlorophenol + •OH → ... + •OH → ...

C. WATER AND SOIL POLLUTION FACTORS

Phototransformation

Irradiation of a 1 mmol/L aqueous solution of 2-chlorophenol in the presence of 50 mmol of nitrate/L leads to the formation of mainly dihydroxybenzenes. The hydroxylation occurs mainly in the *ortho* and *para* positions with respect to the hydroxy function. Neither nitroso- nor nitroderivatives were formed other than 2-nitroresorcinol. The following compounds were detected:

resorcinol, hydroquinone, 2-chloro-hydroquinone, 3-chloro-catechol, 4-chloro-catechol, 2-nitro-resorcinol

(2675)

Biodegradation

biological degradation in 7-d tests; 0% degradation for original, 1st, 2nd, and 3rd subculture.

decomposition rate in soil suspensions: >72 d for complete disappearance. (175)
decomposition period by a soil microflora: >64 d. (176)

lag period for degradation of 16 mg/L by
waste water at pH 7.3 and 30°C: 14-25 d
soil suspension at pH 7.3 and 30°C: >25 d (1096)

Impact on biodegradation processes

toxicity to microorganisms: 24 h A.S. growth inhib. EC_{50}: 115 mg/L (2624)
toxicity to microorganisms: OECD 209 closed system inhib. EC_{50}: >0.3 mg/L (2624)

Reduction of amenities

odor threshold: 0.100 to 0.200 mg/L; detection: 0.05 mg/L (226)
taste threshold: 0.900 to 1.0 mg/L; 0.0001 mg/L (226, 998)

Waste water treatment

ion exchange: adsorption on Amberlite X AD-4 at 25°C, influent: 350 ppm; solute adsorbed:

zero leakage:	2.40 lb/cu ft	
100 ppm leakage:	2.53 lb/cu ft	(40)

Degradation

by *Pseudomonas:* 200 mg/L at 30°C

parent:	100% ring disruption in 72 h	
mutant:	100% ring disruption in 28 h	(152)

W, unadapted A.S.: at	1 mg/L:	100% removal after 6 h	
	10 mg/L:	40% removal after 6 h	
	100 mg/L:	0% removal after 6 h	(1639)

Removal efficiency from water incubated with 1,000 units of horseradish peroxidase enzyme per liter, at pH 7.0: 66.9%. (2452)

Biodegradation

in anoxic habitats in the dark: % substrate disappearance after 3 months:

in sewage sludge at 37°C:	98
in pond sediment at 20°C:	98
in methanogenic aquifer at 20°C:	100
in sulfate-reducing aquifer at 20°C:	18, loss not above control

Phenol was detected as metabolite in sewage sludge and pond sediment but not in methanogenic aquifer. (2621)

Biodegradation

in upflow anaerobic sludge blanket reactor at 35°C and a hydraulic retention time of 13.2 h at 0.1-0.5 mg/L influent: 0% removal efficiency at steady state during a 4-week experiment. (2650)

D. BIOLOGICAL EFFECTS

Bacteria			
inhibition of nitrifying bacteria	2h EC_{20}	0.2 mg/L	
	2h EC_{50}	0.9 mg/L	
	2h EC_{80}	4.0 mg/L	(7051)
inhibition of luminiscent bacteria	30min EC_{20}	1 mg/L	
Photobacterium phosphoreum	30min EC_{50}	7 mg/L	
	30min EC_{80}	25 mg/L	(7051)
municipal A.S. biomass growth inhibition	4h EC_{20}	50 mg/L	
	4h EC_{50}	92 mg/L	
	4hEC_{80}	>100 mg/L	(7052)
industrial A.S. biomass growth inhibition	4h EC_{20}	32 mg/L	
	4h EC_{50}	95 mg/L	
	4h EC_{80}	200 mg/L	(7052)
industrial A.S. respiration inhibition	30min EC_{20}	3 mg/L	
	30min EC_{50}	25 mg/L	
	30min EC_{80}	160 mg/L	(7052)

Algae			
Chlorella pyrenoidosa	toxic	40 mg/L	(41)
Selenastrum capricornutum	4d EC_{50} growth inhib.	29 mg/L	(2625)

Crustaceans			
Daphnia magna	24h EC_{50}	15,800 μg/L	
	48h EC_{50}	7,900 μg/L	(2708)

Worms

Eisenia andrei: in 10% organic matter soil at 23°C:

14d LC_{50}	162; 213; 220; 271 mg/kg dw	(2708)
14d LC_{50}	152-315 μg/L (estim. conc. in the soil pore water phase)	(2789)

Lumbricus rubellus: in 10% organic matter soil at 15°C:

14d LC_{50}	309; 378; 406; 561 mg/kg dw	(2708)
14d LC_{50}	343-603 μg/L (estim. conc. in the soil pore water phase)	(2789)

Fishes		
Idus idus melanotus	48h LC_0	1 mg/L

	48h LC_{50}	3 mg/L	
	48h LC_{100}	6 mg/L	(998)
Poecilia reticulata	24h LC_{50}	6.5 mg/L at pH 7.3	(1833)
	14d LC_{50}	6.4 mg/L	(2696)
	>7d LC_{50}	6.4 mg/L	(2708)
flounder *Platichthys flesus*	96h LC_{50}	4.0 mg/L	(7085)

o-chlorophenol (1-chloro-2-hydroxybenzene)

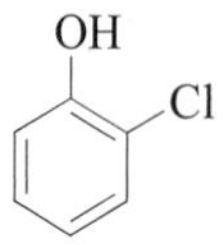

C_6H_5ClO

CAS 95-57-8

USES

organic synthesis.

A. PROPERTIES

colorless liquid; molecular weight 128.56; melting point 0-7°C; boiling point 175.6°C; vapor pressure 40 mm at 82°C, 100 mm at 106°C; density 1.24 at 18/15°C; solubility 28,500 mg/L at 20°C; $LogP_{OW}$ 2.15/2.19.

B. AIR POLLUTION FACTORS

Odor

characteristic: quality: medicinal		
threshold concentrations:	0.019 mg/m³	(710)
	3.6 ppb	(279)
	recognition: 0.0005 mg/m³	(712)

Control methods

wet scrubber:	water at pH 8.5:	outlet: 200 odor units/scf	
	$KMnO_4$ at pH 8.5:	outlet: 1 odor units/scf	(115)

Incinerability

thermal stability ranking of hazardous organic compounds: rank 102 on a scale of 1 (highest stability) to 320 (lowest stability). (2390)

C. WATER AND SOIL POLLUTION FACTORS

Manmade sources

sewage works in agricultural region:	influent: 27 µg/L effluent: 0.1 µg/L; removal efficiency: 99.6%	(7055)
In Canadian municipal sludges and sludge compost: September 1993- February 1994:	mean values of 11 sludges ranged from 0.02 to 0.15 mg/kg dw.; mean: 0.06 mg/kg dw; mean value of sludge compost: 0.01 mg/kg dw.	(7000)

in sewage effluents: 1.7 µg/L. (237)

Reduction of amenities

approx. conc. causing adverse taste in fish:	0.015 mg/L	(41)
faint odor:	0.00018 mg/L	(129)
taste threshold:	0.0001 to 0.006 mg/L	(998, 226)
adult bluegills: taste to the flesh at	2.0 mg/L	(226)

Phototransformation

Irradiation of a 1 mmol/L aqueous solution of 2-chlorophenol in the presence of 50 mmol of nitrate/L leads to the formation of mainly dihydroxybenzenes. The hydroxylation occurs mainly in the *ortho* and *para* positions with respect to the hydroxy function. Neither nitroso- nor nitroderivatives were formed other than 3-nitrocatechol. The following compounds were detected:

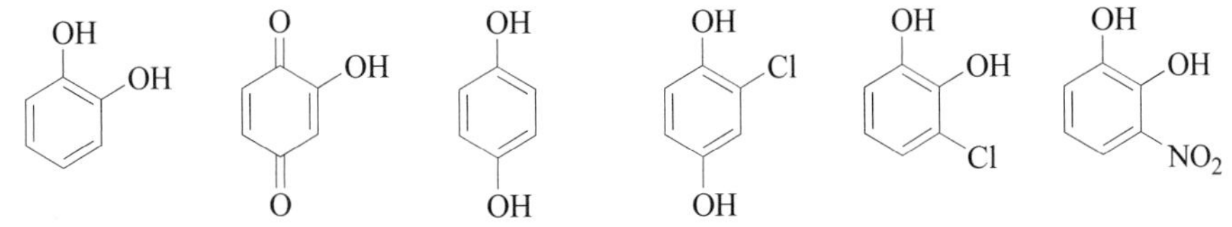

catechol 2-hydroxyquinone hydroquinone 2-chloro-hydroquinone 3-chloro-catechol 3-nitrocatechol

(2675)

By photocatalytic degradation in UV-illuminated TiO_2 aqueous suspension, 50% and 99% disappearance are obtained after 34 and 120 min at an initial conc. of 20 mg/L. Chorohydroquinone, 1,2 dihydroxybenzene, and hydroxyhydroquinone were detected as the major metabolites. The decomposition rate is not much affected by the pH over a wide range.

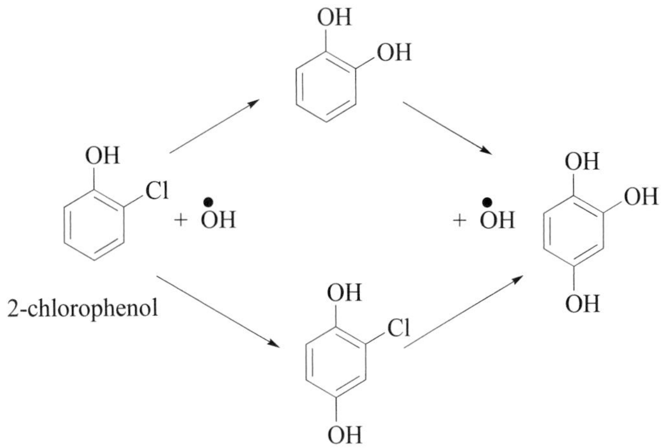

2-chlorophenol

(2371)

Biodegradation

Anaerobic mineralisation at 50 mg C/L at 35 °C in the dark:

after 56 d in digested sewage sludge	0-30% of theoretical mineralization	
after 56 d in freshwater swamp	30-75% of theoretical mineralization	
after 96 d in marine sediment	0-30% of theoretical mineralization	(7037)

Decomposition

rate in soil suspensions: 14 d for complete disappearance. (175)

Decomposition

period by a soil microflora: >64 d. (176)

Adapted A.S. at 20°C-product is sole carbon source: 95% COD removal at 25 mg COD/g dry inoculum/h. (327)

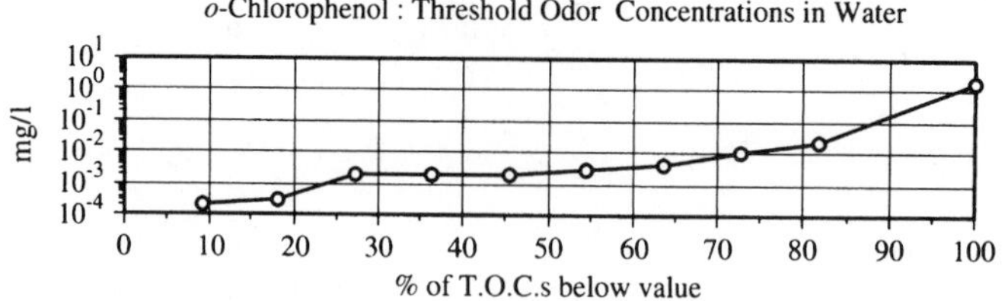

(97, 226, 300, 304, 879, 894, 925, 998)

Waste water treatment

methods	temp, °C	d observed	feed, mg/L	d acclim.	% removed	
SEW, CA	20	25	1	23	4	
RW, CA	20	15	1	6+	100	(93)

degradation by *Pseudomonas:* 200 mg/L at 30°C:
parent: 100% ring disruption in 52 h
mutant: 100% ring disruption in 26 h (152)

lag period for degradation of 16 mg/L by:
waste water at pH 7.3 and 30°C: 14-25 d
soil suspension at pH 7.3 and 30°C: >25 d (1096)

W, unadapted A.S.: at	1 mg/L:	100%	removal after 3 h	
	10 mg/L:	97%	removal after 6 h	
	100 mg/L:	20%	removal after 6 h	(1639)

removal efficiency from water incubated with 1,000 units of horseradish peroxidase enzyme per liter, at pH 7.0: 99.8%. (2452)

Biodegradation

in anoxic habitats in the dark:	% substrate disappearance after 3 months:	
in sewage sludge at 37°C:	57	
in pond sediment at 20°C:	100	
in methanogenic aquifer at 20°C:	100	
in sulfate-reducing aquifer at 20°C:	20, loss not above control	(2621)

Impact on biodegradation processes

inhibition of degradation of glucose by *Pseudomonas fluorescens* at 30 mg/L
inhibition of degradation of glucose by *E. coli* at 400 mg/L (293)

24-h act. sludge growth inhib. EC_{50}: 162-382 mg/L
24-h act. sludge respiration inhib. EC_{50}: 355 mg/L
biodegradation inhib.: 90 mg/L (2624)

Groundwater quality

in groundwater infiltrated through river bank in the Netherlands 1986: max: 0.12 µg/L. (6018)

D. BIOLOGICAL EFFECTS

Bioaccumulation

Lepomis macrochirus at 9.18µg/L at 16°C during 28 d:

BCF: 214; half-life in tissues: <1 d (2602)

log BCF: 2.33 (2606)

Toxicity

Bacteria *Pseudomonas*	toxic: 30 mg/L		(30)

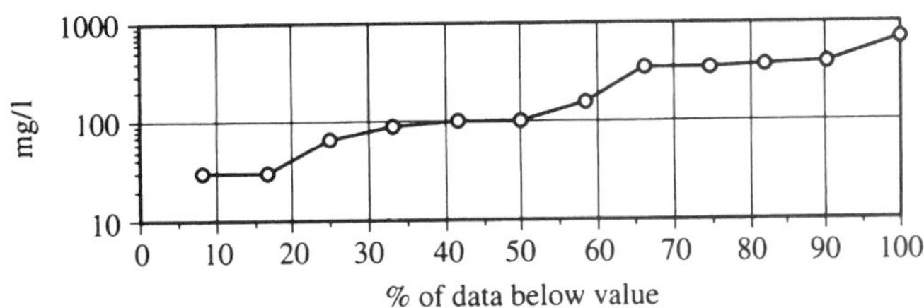

Summary of short-term EC_{50} values for bacteria (n = 12) based on data reported in this and other works (3999)

Algae			
Chlorella pyrenoidosa	toxic	96 mg/L	(41)
Scenedesmus	toxic:	60 mg/L	
Chlorella vulgaris	4d EC_{50}	170 mg/L	(2708)
Selenastrum capricornutum	4d EC_{50}	70 mg/L	(2708)

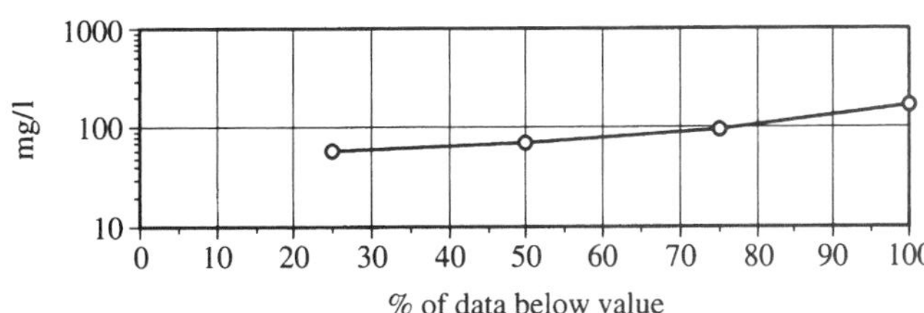

Summary of short-term EC/LC_{50} values for algae (n = 4) based on data reported in this and other works (3999)

Protozoa	*Colpoda*	toxic: 30 mg/L	(30)

Crustaceans			
Daphnia magna	24h EC_{50}	6,300; 17,900 µg/L	
	48h EC_{50}	2,600; 7,400; 8,950 µg/L	(2708)
Daphnia magna	life cycle test:	3 week NOEC: 0.5 mg/L	
		3 week MATC: 0.7 mg/L	(2708)
	7d LC_{50}	3.7 mg/L	(2625)

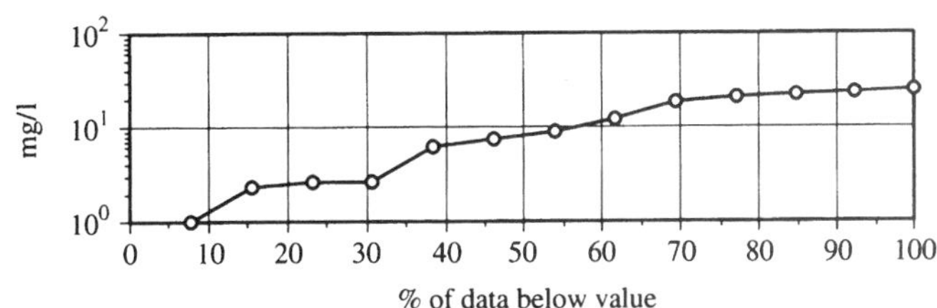

Summary of short-term EC/LC_{50} values for crustaceans (n = 13) based on data reported in this and other works (3999)

Fishes

goldfish: amount found in dead fish at 20 ppm: 128 µg/g
BCF (at 20 ppm) = 6.4 (1850)

Poecilia reticulata	>7d LC_{50}	11 mg/L	(2708)
Pimephales promelas	8d LC_{50}	6.3 mg/L	(2708)
	96h LC_{50}	9.4; 12 mg/L	(2708, 2625)
eggs 4 weeks after hatching: early life stage test:	>4 week NOEC	4 mg/L	
	>4 week MATC	5.69 mg/L	(2708)
Poecilia reticulata	96h LC_{50}	14 mg/L	(2708)
	14d LC_{50}	11 mg/L	(2696)
bluegill fingerlings	96h LC_{50}	8.4 mg/L	(226)
bluegill sunfish	24h LC_{50}	8.2 mg/L	(253)
fatheads	24-96h LC_{50}	11-22 mg/L	
bluegills	24-96h LC_{50}	8-12 mg/L	
goldfish	24-96h LC_{50}	12-15 mg/L	
guppies	24-96h LC_{50}	20-23 mg/L	(158)
goldfish	24h LC_{50}	16 mg/L	
Poecilia reticulata	24h LC_{50}	11 mg/L	(1833)
Idus idus melanotus	48h LC_0	5 mg/L	
	48h LC_{50}	8.5 mg/L	
	48h LC_{100}	10 mg/L	(998)
flounder Platichthys flesus	96h LC_{50}	7.0; 6.3 mg/L	(7085)
sole Solea solea	96h LC_{50}	6.6 mg/L	(7085)

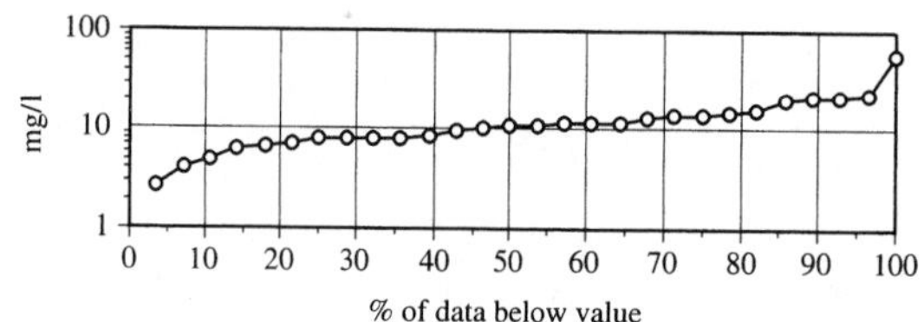

Summary of 24-96h LC_{50} values for fishes (n = 28) based on data reported in this and other works (3999)

p-chlorophenol (1-chloro-4-hydroxybenzene)

Cl—⟨benzene ring⟩—OH

C_6H_5ClO

CAS 106-48-9

OCCURRENCE

phenylurea metabolite.

A. PROPERTIES

molecular weight 128.56; melting point 43°C; boiling point 217°C; vapor pressure 0.10 mm at 20°C, 0.25 mm at 30°C, 10 mm at 92°C, 40 mm at 125°C; vapor density 4.4; density 1.31; solubility

27,100 mg/L at 20°C; saturation concentration in air 0.70 g/m^3 at 20°C, 1.39 g/m^3 at 30°C; $LogP_{OW}$ 2.39/2.44.

B. AIR POLLUTION FACTORS

1 mg/m^3 = 0.19 ppm, 1 ppm = 5.34 mg/m^3.

Odor

recognition:	0.001 mg/m^3	(712)
	1.2 ppm	(279)

Control methods

wet scrubber:	water at pH 8: outlet: 5 odor units/scf $KMnO_4$ at pH 8: outlet: 1 odor unit/scf	(115)

C. WATER AND SOIL POLLUTION FACTORS

Manmade sources

sewage works in agricultural region:	influent: 0.1 μg/L effluent: 0.03 μg/L; removal efficiency: 70%	(7055)

By photocatalytic degradation in UV-illuminated TiO_2 aqueous suspension 50% and 99% disappearance are obtained after 26 and 80 min at an initial conc. of 20 mg/L. 1,4-dihydroxybenzene was detected as the major metabolite. The decomposition rate is not much affected by the pH over a wide range.

Cl–C6H4–OH + °OH ⟶ HO–C6H4–OH

(2371)

Phototransformation

Irradiation of a 1 mmol/L aqueous solution of 2-chlorophenol in the presence of 50 mmol of nitrate/L leads to the formation of mainly dihydroxybenzenes. The hydroxylation occurs mainly in the *ortho* and *para* positions with respect to the hydroxy function. Neither nitroso- nor nitroderivatives were formed other than 2-nitrohydroquinone. The following compounds were detected:

2-hydroxyquinone

hydroquinone

quinone

4-chlorocatechol

4-chlororesorcinol

2-nitrohydroquinone

(2675)

Biodegradation
Decomposition

rate in soil suspensions: 9 d for complete disappearance.	(175)
Decomposition period by a soil microflora: 16 d.	(176)
Adapted A.S. at 20°C-product is sole carbon source: 96% COD removal at 11 mg COD/g dry inoculum/h.	(327)
Removal efficiency from water incubated with 1,000 units of horseradish peroxidase enzyme per liter, at pH 5.5: 98.7%.	(2452)

Biodegradation

in anoxic habitats in the dark: % substrate disappearance after 3 months:

in sewage sludge at 37°C:	11, loss not above control
in pond sediment at 20°C:	100
in methanogenic aquifer at 20°C:	100
in sulfate reducing aquifer at 20°C:	26

Phenol was detected as metabolite in pond sediment and methanogenic aquifer; no products were detected in sulfate reducing aquifer. (2621)

Anaerobic mineralisation

at 50 mg C/L at 35 °C in the dark:

after 56 d in digested sewage sludge	0-30% of theoretical mineralization
after 56 d in freshwater swamp	inhibition of mineralization
after 96 d in marine sediment	inhibition of mineralization

(7037)

Photolysis and microbial degradation

at 25 µg/L and pH 7.6; in winter 10°C, in summer 24°C.

The following half-lives are in hours.

	transformation		*mineralization*	
water	*winter*	*summer*	*winter*	*summer*
distilled	99	63	224	58
poisoned estuary	63	46	334	53
estuary	3	28	95	10
estuary dark	116	11	231	2

(2691)

Impact on biodegradation processes

Inhibition of degradation

of glucose by *Pseudomonas fluorescens* at 20 mg/L.

Inhibition of degradation

of glucose by *E. coli* at 200 mg/L. (293)

24 h act. sludge growth inhib. EC_{50}: 105; 183 mg/L.
24 h act. sludge respiration inhib. EC_{50}: 225 mg/L.
Biodegradation inhib. EC_{50}: 42 mg/L. (2624)

The mineralization of 4-chlorophenol in methanogenic river sediment is affected by the following chemicals:

chemical	EC_{50}
benzene	2,500 mg/kg dw
pentachlorophenol	62 mg/kg dw
1,2-dichloroethane	1,100 mg/kg dw
chloroform	1.5 mg/kg dw

(2961)

Reduction of amenities

approx. conc. causing adverse taste in FISH: 0.05 mg/L. (41)

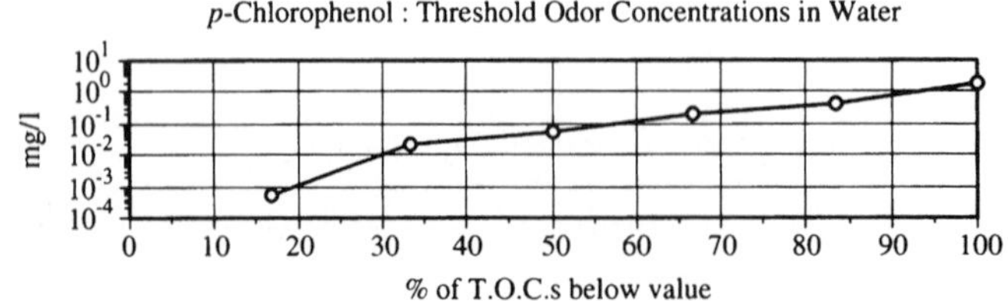

(226, 875, 879, 894, 998)

Taste threshold

average: 1.24 mg/L
range: 0.02-20.4 mg/L (294, 30)
0.25 mg/L (304)
1.0 to 1.35 mg/L; 0.0001 mg/L (226, 998)

Waste water treatment

methods	temp, °C	d observed	feed, mg/L	d acclim.	% removed	
SEW, CA	20	25	1	15	33	
RW, CA	20	36	1	23	100	
RW, CA	20	13	1	6+	100	(93)

Degradation

by *Pseudomonas:* 200 mg/L at 30°C:

parent: 100% ring disruption in 96 h
mutant: 100% ring disruption in 33 h (152)
Adapted culture: 27% removal after 48-h incubation, feed = 200 mg/L (292)

lag period for degradation of 16 mg/L by:
waste water at pH 7.3 and 30°C: 7-14 d
soil suspension at pH 7.3 and 30°C: 14-25 d (1096)

Continuous-flow fluidized-bed reactor using cellite carrier for cell immobilization at 24-29°C and hydraulic retention time of 5 h, under anoxic conditions and 50 mg/L nitrates: 80% removal at approx. 40 mg/L influent, no intermediate degradation products were detected. (2796)

Degradation in soil

biodegradation $t_{1/2}$ in nonadapted aerobic subsoil: sand Oklahoma: 1.47 d. (2695)

D. BIOLOGICAL EFFECTS

Bioaccumulation

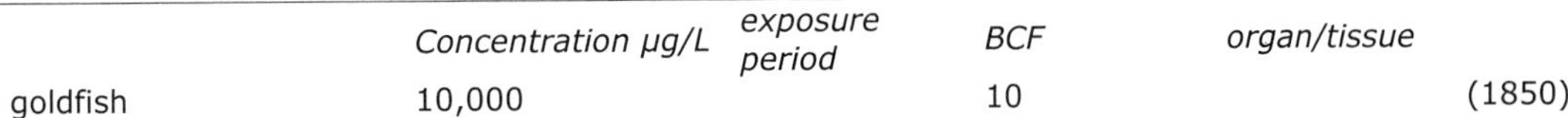

	Concentration µg/L	exposure period	BCF	organ/tissue	
goldfish	10,000		10		(1850)

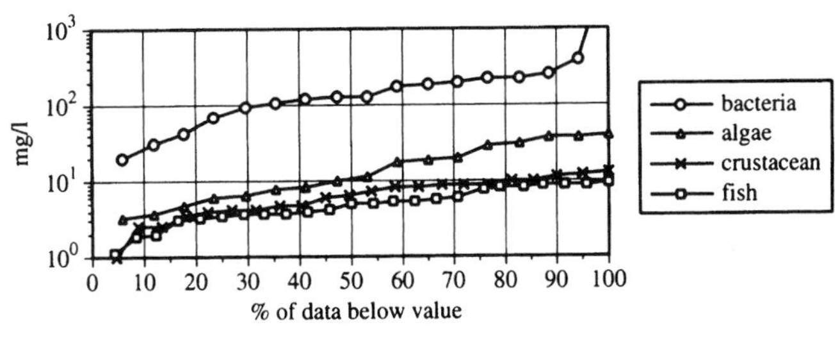

p-chlorophenol

Frequency distributions of short-term values for bacteria (*n* = 17) and algae (*n* = 17), 24-96h EC_{50}/LC_{50} values for crustaceans (*n* = 22) and fishes (*n* = 24), based on data from this and other works (3999)

Toxicity

RubisCo test: inhib. of enzym. activity of Ribulose-P2-carboxylase in protoplasts:	IC_{10}	39 mg/L	
Oxygen test: inhib. of oxygen production of protoplasts:	IC_{10}	643 mg/L	(2698)

Bacteria			
Pseudomonas	toxic	20 mg/L	
Algae			
Scenedesmus	toxic	20 mg/L	
Chlorella pyrenoidosa	toxic	40 mg/L	(41)
Chlorella vulgaris	4d EC_{50}	29 mg/L	
Selenastrum capricornutum	4d EC_{50}	38 mg/L	(2708)
Skeletonema costatum	EC_{50}	3.3 mg/L	(2708)
Mysidopsis bahia	EC_{50}	30 mg/L	(2708)
Scenedesmus subspicatus inhib. of fluorescence	IC_{10}	6.6 mg/L	
Scenedesmus subspicatus growth inhib.	IC_{10}	1.9 mg/L	
Protozoa			
Colpoda:	toxic	5 mg/L	(30)
Crustaceans			
Daphnia magna	24h EC_{50}	3.4; 8.1; 8.6 mg/L	
	48h EC_{50}	2.5; 4.0; 4.1; 4.8 mg/L	(2708)
life cycle test	3w NOEC	0.63 mg/L	
	3w MATC	0.9 mg/L	(2708)
	7d LC_{50}	2.3 mg/L	(2625)
copepod *Tisbe battagliai* copepodid stage	24h LC_{50}	21 mg/L	(7085)
Fishes			
Idus idus melanotus	48h LC_{0}	2 mg/L	
	48h LC_{50}	3.5 mg/L	
	48h LC_{100}	5.5 mg/L	(998)
Pimephales promelas	96h LC_{50}	6.2 mg/L	
Lepomis macrochirus	96h LC_{50}	3.7 mg/L	(2679)
	24h LC_{50},S	4.0 mg/L	
	96h LC_{50},S	3.8 mg/L	(2697)
Pimephales promelas	96h LC_{50}	>3.8 mg/L	(2708)
Poecilia reticulata	96h LC_{50}	8.5 mg/L	(2708)
Cyprinodon variegatus	96h LC_{50}	5.4 mg/L	(2708)
flounder *Platichthys flesus*	96h LC_{50}	5.0 mg/L	(7085)

chlorophenotane *see* DDT

4-chlorophenoxy-ϖ-butyric acid

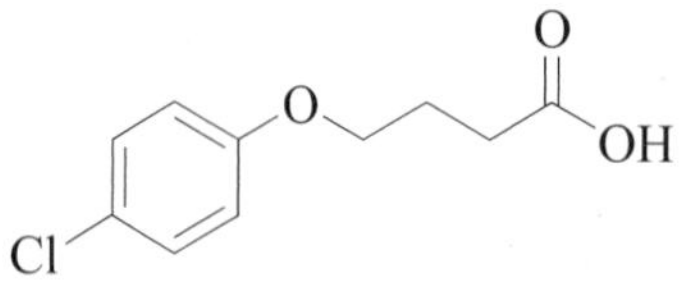

$C_{10}H_{11}ClO_3$

CAS 3547-07-7

C. WATER AND SOIL POLLUTION FACTORS

Biodegradation

53 d for ring cleavage in soil suspension. (1827)

3-(p-(p-chlorophenoxy)phenyl)1,1-dimethylurea *see* chloroxuron

4-chlorophenoxy-α-propionic acid

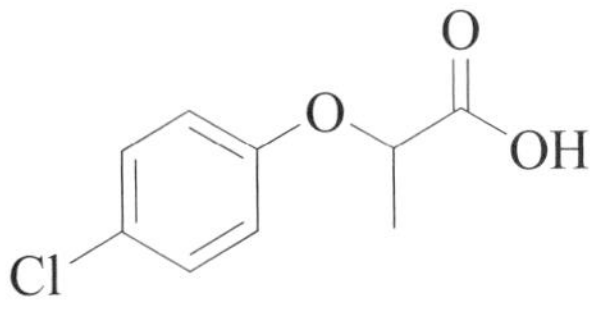

$C_9H_9ClO_3$

CAS 3307-39-9

C. WATER AND SOIL POLLUTION FACTORS

Biodegradation

>205 d for ring cleavage in soil suspension. (1827)

4-chlorophenoxy-ϖ-propionic acid

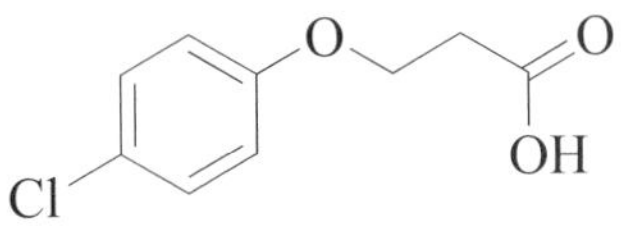

$C_9H_9ClO_3$

CAS 3284-79-5

C. WATER AND SOIL POLLUTION FACTORS

Biodegradation

11 d for ring cleavage in soil suspension. (1827)

4-chlorophenoxy-α-valeric acid

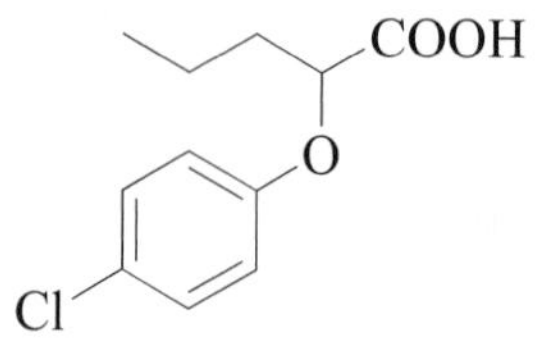

$C_{10}H_{12}ClO_3$

C. WATER AND SOIL POLLUTION FACTORS

Biodegradation

>81 d for ring cleavage in soil suspension. (1827)

4-chlorophenoxyacetic acid (4-CPA; Tomatone-; Tomatotone; 4-CP; PCPA)

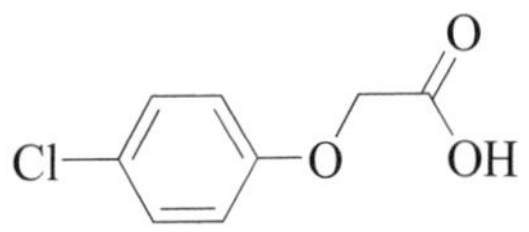

$ClC_6H_4OCH_2COOH$

$C_8H_7ClO_3$

CAS 122-88-3

USES

A plant growth regulator used to improve tomato set.

OCCURRENCE

Degradation product of 2,4-dichlorophenoxyacetic acid (2,4-D).

IMPURITIES, ADDITIVES, COMPOSITION

Technical grade contains 4-chloro-*o*-cresol (<34%). (1606)

A. PROPERTIES

molecular weight 186.59; melting point 157-159°C; solubility miscible; $logP_{OW}$ 1.99

C. WATER AND SOIL POLLUTION FACTORS

Biodegradation

11 d for ring cleavage in soil suspension. (1827)

Waste water treatment

A.C. type BL (Pittsburgh Chem. Co.): % adsorbed by 10 mg A.C.

from 10^{-4} M aqueous solution at	pH 3:	55%	
	pH 7:	18%	
	pH 11:	10%	(1313)

Aquatic reactions

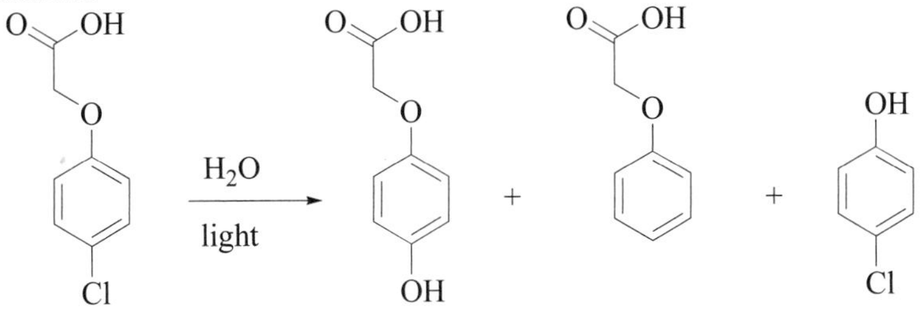

Photolysis of 4-CPA

Photolysis

Decomposed under UV or sunlight to p-chlorophenol, phenol, hydroquinone, p-chlorophenyl formate, phenoxyacetic acid, p-hydroxyphenoxyacetic acid and humic acids. (10007)

Biodegradation

Soil microorganisms metabolise 4-CPA to 2-hydroxy-4-chlorophenoxyacetic acid, further degradation occurs via phenoxyacetic acid to yield the o- and p-hydroxy derivatives. (10004)

The following were identified in culture extracts from a *Pseudomonas species* isolated from soil: 4-chloro-2-hydroxyphenoxyacetate; 4-chlorocatechol; β-chloromuconolactone; γ-carboxyethylene-d-a-b-butenolide. It was found that β-chloromuconolactone was unstable in aqueous solutions and hydrolysed to the corresponding b-hydroxy analogue. (9893)

Azotobacter chrococcum was able to utilise 4-CPA as sole carbon source. (10006)

Persistence in soils

in clay loam soil: $t_{1/2}$: 20 d. (10005)

D. BIOLOGICAL EFFECTS

Bacteria			
Photobacterium phosphoreum Microtox test	5-30 min EC_{50}	98 mg/L	(9657)
Fishes			
sea trout *(Salmo trutta)*	24h LD_{50}	147 mg/L	(1606)
brown trout, bluegill sunfish, yellow perch, goldfish	24 LC_0	5 mg/L	(9594)
Mammals			
rat	oral LD_{50}	850 mg/kg bw	(9763)

1-(4-chlorophenyl)-3-(2,6-difluorobenzoyl)urea *see* diflubenzuron

3-(p-chlorophenyl)-1,1-dimethylurea (N'(4-chlorophenyl)-N,N-dimethylurea; CMU; monuron; chlorfenidim; telvar)

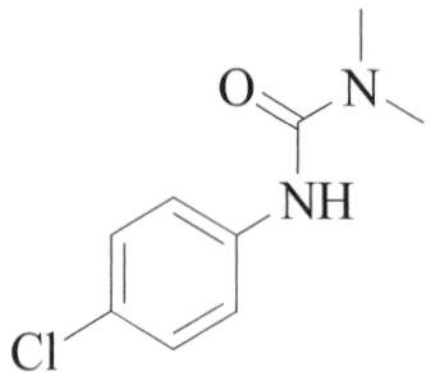

$ClC_6H_4NHCON(CH_3)_2$

$C_9H_{11}ClN_2O$

CAS 150-68-5

USES

herbicide, is an inhibitor of photosynthesis and is absorbed via the roots; sugar-cane flowering suppressant.

A. PROPERTIES

odorless crystalline solid; molecular weight 198.5; melting point 174-175°C; vapor pressure 5 × 10^{-7} mm at 25°C; density 1.27 at 20/20°C; solubility 230 ppm at 25°C; $LogP_{OW}$ 3.05.

C. WATER AND SOIL POLLUTION FACTORS

Aquatic reactions

75-100% disappearance from soils: 10 months. (1815)

Persistence in river water

in a sealed glass jar under sunlight and artificial fluorescent light-initial conc. 10 μg/L:

	% of original compound found					
after	*1 h*	*1 wk*	*2 wk*	*4 wk*	*8 wk*	
	80	40	30	20	0	(1309)

Waste water treatment

Powdered A.C.: Freundlich adsorption parameters: *K*: 0.066; *1/n* = 0.33.

Carbon dose to reduce 5 mg/L to a final conc. of 0.1 mg/L: 160 mg/L.
Carbon dose to reduce 1 mg/L to a final conc. of 0.1 mg/L: 29 mg/L. (594)

Inhibition of biodegradation

in soil 40 mg/L did not inhibit nitrification. (9892)

D. BIOLOGICAL EFFECTS

Bacteria			
Photobacterium phosphoreum Microtox test	5 min EC_{50}	228 mg/L	(9657)
Algae			
Protococcus sp.	10d EC_{10}	1 ppb	(2347)
	10d EC_{100}	20 ppb	(2347)
Chlorella sp.	10d EC_{70}	1 ppb	(2347)
Clorococcum sp.	10d EC_{46}	100 ppb	(2349)
	10d EC_{50}	90-150 ppb	(2348)
Dunaliella euchlora	10d EC_{0}	1 ppb	(2347)
	10d EC_{100}	20 ppb	(2348)
Isochrysis galbana	10d EC_{50}	100-130 ppb	(2348)
Monochrysis lutheri	10d EC_{17}	1 ppb	(2347)
Phaeodactylum tricornutum	10d EC_{50}	90-100 ppb	(2348)
	10d EC_{35}	1 ppb	(2347)
	10d EC_{100}	20 ppb	(2347)
Fishes			
Oncorhynchus kisuth	48h LC_{50}	110,000 µg/L	(2106)
Salmo gairdneri	96h LC_{50}	75,000 µg/L	(2679)
Mammals			
rat	oral LD_{50}	1,053-3,700 mg/kg bw	(9759, 9763)

((3-chlorophenyl)hydrazono)-propanedinitrile

D. BIOLOGICAL EFFECTS

Crustaceans			
Ceriodaphnia dubia	7d LC_{50}	159 mg/L	
Daphnia pulex	7d LC_{50}	159 mg/L	(2625)

1-(4-chlorophenyl)thiourea (1-(p-chlorophenyl)-2-thiourea)

S NH2 NH Cl

$C_7H_7ClN_2S$

CAS 3696-23-9

A. PROPERTIES

molecular weight 186.67.

B. AIR POLLUTION FACTORS

Incinerability

thermal stability ranking of hazardous organic compounds: rank 286 on a scale of 1 (highest stability) to 320 (lowest stability). (2390)

2-chlorophenylamine *see* *o*-chloroaniline

3-chlorophenylamine *see* *m*-chloroaniline

4-chlorophenylamine *see* *p*-chloroaniline

S-((chlorophenylthio)methyl)-O,O-diethylphosphorodithioate *see* carbophenothion

4-chlorophenylurea

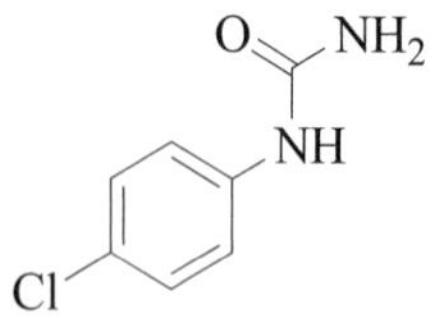

O NH2
NH
Cl

$C_7H_7ClN_2O$

CAS 140-38-5

A. PROPERTIES

$LogP_{OW}$ 1.86.

C. WATER AND SOIL POLLUTION FACTORS

Diflubenzuron hydrolyzes in water to *p*-chlorophenylurea. (1332)

D. BIOLOGICAL EFFECTS

Fishes			
Rainbow trout	96h LC_{50},S	72 mg/L	
fathead minnow:	96h LC_{50},S	>100 mg/L	
channel catfish:	96h LC_{50},S	>100 mg/L	
bluegill:	96h LC_{50},S	>100 mg/L	(1510)
Salmo gairdneri	96h LC_{50}	0.43 mmol/L	(2679)

6-chloropicolinic acid

Cl N COOH

ClC_5H_3NCOOH

$C_6H_4ClNO_2$

CAS 4684-94-0

C. WATER AND SOIL POLLUTION FACTORS

Manmade sources

sole metabolite other than carbon dioxide from degradation of 2-chloro-6-(trichloromethyl)pyridine.

Degradation

in soil at 1.0 ppm initial conc. after 35 d incubation:

soil temperature,°C	*% decomposition*
34-35	48
18-21	36
2-3	17

The most important factor influencing the decomposition rate is soil temperature. (1610)

chloropicrin (trichloronitromethane; nitrochloroform; nitrotrichloromethane; Acquinite; Larvicide 100; Picfume)

CCl_3NO_2

CAS 76-06-2

USES

organic synthetics; dye-stuffs; fumigants; fungicides; insecticides; rat exterminator; poison gas.

A. PROPERTIES

molecular weight 164.39; melting point -64°C; boiling point 112°C; vapor pressure 16.9 mm at 20°C, 33 mm at 30°C; vapor density 5.7; density 1.65 at 20/4°C; saturation concentration in air 170 g/m^3 at 20°C, 286 g/m^3 at 30°C; solubility 2,000 mg/L; logP_{OW} 2.09

B. AIR POLLUTION FACTORS

1 mg/m^3 = 0.149 ppm, 1 ppm = 6.72 mg/m^3.

Odor

T.O.C.: 1.1 ppm (211, 57)

O.I. at 20°C: 22,200 (316)

Photolysis

Photolysis in atmosphere: $t_{1/2}$: 20 d. (9603)

Vapor phase photolysis: In the laboratory, chloropicrin underwent photochemical oxygenation followed by rearrangement and cleavage to produce phosgene as the major product. (2921)

Partitioning coefficients

$K_{air/water}$ 0.10

Cuticular matrix/air partition coefficient* 1900 ± 0 (7077)

* experimental value at 25°C studied in the cuticular membranes from mature tomato fruits (*Lycopersicon esculentum* Mill. cultivar Vendor)

C. WATER AND SOIL POLLUTION FACTORS

Reduction of amenities

faint odor: 0.0073 mg/L. (129)

Photolysis:

in water: $t_{1/2}$: 31 h yielding CO_2, bicarbonate, chloride, nitrate, and nitrite. (9602)

in surface layers of water, $t_{1/2}$: 3 d. (9603)

$CCl_3NO_2 \longrightarrow H_3C{-}NO_2$ (nitromethane)

Biodegradation

In sandy loam soil: $t_{1/2}$: 4.5 d with carbon dioxide being the terminal breakdown product.

In an anaerobic aquatic/soil: $t_{1/2}$: 1.3 h, dehalogenation to nitromethane. (9602)

Partitioning coefficients

Cuticular matrix/water partition coefficient**: 190 (7077)

Waste water treatment

Removed from wastewater by treatment with an alkali metal sulfate at pH 6-14. (10009)

D. BIOLOGICAL EFFECTS

Mammals			
rat	acute oral LD_{50}	250 mg/kg	(9604)

chloroprene (2-chloro-1,3-butadiene; neoprene)

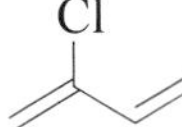

$CH_2CClCHCH_2$

C_4H_5Cl

CAS 126-99-8

USES

Intermediate in the synthesis of polychloroprene. The residual contents of monomeric chloroprene in polymeric products is at maximum 500 ppm (polychloroprene latices).

A. PROPERTIES

Colorless liquid; molecular weight 88.5; melting point -130°C; boiling point 59.4°C; vapor pressure 118 mm at 10°C, 275 mm at 30°C, 200 mm at 20°C; vapor density 3.06; density 0.96 at 20°C; solubility 260; 480 mg/L at 20°C, 250 mg/L at 25°C; saturation concentration in air 964 g/m^3 at 20°C; log P_{oct} 2.2 (calculated).

B. AIR POLLUTION FACTORS

1 mg/m^3 = 0.27 ppm, 1 ppm = 3.68 mg/m^3.

Odor

characteristic: quality: slightly etheric.

T.O.C.: recogn.: 0.40 mg/m^3 = 0.11 ppm (73, 754)

Human odor perception:	Nonperception:	0.25 mg/m^3	
	Perception:	0.4 mg/m^3	
Human reflex response:	Adverse response:	0.4 mg/m^3	
Animal chronic exposure:	no effect:	0.22 mg/m^3	
	Adverse effect:	0.48 mg/m^3	(170)

O.I. at 20°C: 2,390,000 (316)

Incinerability

thermal stability ranking of hazardous organic compounds: rank 69 on a scale of 1 (highest (2390) stability) to 320 (lowest stability).

Photochemical reactions

$t_{1/2}$: 11 d, based on reactions with OH° (calculated). (6100)

Photodegradation

T/2 (OH radicals) = 18.3 hours (10719)

C. WATER AND SOIL POLLUTION FACTORS

Biodegradation

Inoculum	*Method*	*Concentration*	*Duration*	*Elimination results*	
Municipal waste water	Closed bottle test	12.8 mg/L	28 d	10% ThOD	(7957)

D. BIOLOGICAL EFFECTS

Bacteria			
Pseudomonas fluorescens	24 h EC_0	1,000 mg/L	(7957)
Escherichia coli	24h NOEC	1000 mg/L	(10719)
Algae			
Naviculum seminulum	7 d EC_{50}	380 mg/L	(7958)
Crustaceans			
Daphnia magna	24 h EC_0	100 mg/L	
	24 h EC_{50}	348 mg/L	
	24 h EC_{50}	348 mg/L	
reproduction	21 d NOEC	3.2 mg/L	(10719)
Fishes			
Lepomis macrochirus	96 h LC_{50}	245 mg/L	
Leuciscus idus	96 h LC_0	200 mg/L	(7957)
Goldfish	24 h LC_{50}	10 mg/L	(10719)
Mammals			
Rat	oral LD_{50}	50; 146; 251; 260; 384; 450 mg/kg bw	(10719)

1-chloropropane (propylchloride)

Cl

$CH_3CH_2CH_2Cl$

C_3H_7Cl

CAS 540-54-5

USES

Anaesthetic. Antiparasiticide. Blowing agent. Solvent.

A. PROPERTIES

Colorless liquid; molecular weight 78.54; melting point - 122.8°C; boiling point 47. 2°C; vapor pressure 350 mm at 25°C; vapor density 2.71; density 0.89 at 20/4°C; solubility 2,700 mg/L at 20°C; $LogP_{OW}$2.04; log *H* - 0.26 at 25°C.

B. AIR POLLUTION FACTORS

1 ppm = 3.26 mg/m^3, 1 mg/m^3 = 0.306 ppm.

C. WATER AND SOIL POLLUTION FACTORS

Waste water treatment

A.S.: after	6 h:	0.7% of ThOD	
	12 h:	0.8% of ThOD	
	24 h:	1.9% of ThOD	(88)

D. BIOLOGICAL EFFECTS

Bacteria			
Activated sludge oxygen consumption inhibition	5 d EC_{50}	>1,000 mg/L	(7047)
Photobacterium phosphoreum (Microtox test)	15 min EC_{50}	438 mg/L	(7047)
Photobacterium phosphoreum	30 min EC_{50}	4.7 mg/L	(10010)
Algae			
Scenedesmus subspicata inhib. of growth	72 h EC_{50}	1,090 mg/L	(7047)
Crustaceans			
Daphnia magna	24 h EC_{50}	75 mg/L	(7047)
Fishes			
	48 h LC_{50}	160 mg/L	(7047)

2-chloropropane (isopropylchloride)

$CH_3CHClCH_3$

C_3H_7Cl

CAS 75-29-6

USES

Solvent. Blowing agent. Chemical intermediate. Anaesthetic.

A. PROPERTIES

Colorless liquid; molecular weight 78.54; melting point -117°C; boiling point 36.5°C; vapor pressure

523 mm at 25°C; vapor density 2.7; density 0.86 at 20/4°C; solubility 3,440 mg/L at 12.5°C, 3,100 mg/L at 20°C; $LogP_{OW}$1.90; LogH -0.18 at 25°C.

B. AIR POLLUTION FACTORS

1 ppm = 3.26 mg/m^3, 1 mg/m^3 = 0.306 ppm.

D. BIOLOGICAL EFFECTS

Bacteria			
Activated sludge oxygen consumption inhibition	5 d EC_{50}	>1,000 mg/L	(7047)
Photobacterium phosphoreum (Microtox test)	15 min EC_{50}	149 mg/L	(7047)
Photobacterium phosphoreum	30 min EC_{50}	1.6 mg/L	(10010)
Algae			
Scenedesmus subspicata inhib. of growth	72 h EC_{50}	870 mg/L	(7047)
Crustaceans			
Daphnia magna	24 h EC_{50}	103 mg/L	(7047)
Fishes			
	48 h LC_{50}	180 mg/L	(7047)
Mammals			
Guinea pig	oral LD_{100}	10,000 mg/kg bw	(9649)

2-chloropropanoic acid (2-chloropropionic acid)

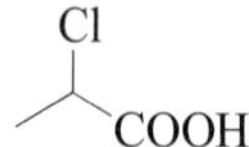

$C_3H_5ClO_2$

CAS 598-78-7

USES

Intermediate for the manufacture of pesticides.

A. PROPERTIES

molecular weight 108.53; melting point -12°C; boiling point 186°C, decomposes >60°C with the emission of HCl; vapor pressure 4 hPa at 60°C, 30 hPa at 100°C; density 1.27 at 20°C; solubility miscible; $LogP_{OW}$0.65 at pH 3.0, <-2.4 at pH 7.2 (measured), 0.85 (calculated).

C. WATER AND SOIL POLLUTION FACTORS

Biodegradation

Inoculum	*Method*	*Concentration*	*Duration*	*Elimination results*	
Adapted A.S.	BOD test	100 mg/L	5 d	100% ThOD	(6310)

Zahn-Wellens test		>70% DOC	(6308)

D. BIOLOGICAL EFFECTS

Bacteria			
Pseudomonas putida (cell multiplication inhibition test)	17 h EC_{10}	135 mg/L at pH 7.5	(6309)
Fishes			
Leuciscus idus	96 h LC_{50}	100-150 mg/L	(6307)

3-chloropropionitrile (3-chloropropanonitrile; β-chloropropionitrile)

Cl—CH₂CH₂—CN

C_3H_4ClN

CAS 542-76-7

USES

In pharmaceutical and polymer synthesis.

A. PROPERTIES

molecular weight 89.53; melting point -51°C; boilking point 176°C decomoses; density 1.14 at 25°C; vapor pressure 6 mm Hg at 50°C; vapor density 3.1; solubility 19 mg/L at 25°C.

B. AIR POLLUTION FACTORS

Incinerability

Temperature for 99% destruction at 2.0-sec residence time under oxygen-starved reaction conditions: 655°C.

Thermal stability ranking of hazardous organic compounds

rank 142 on a scale of 1 (highest stability) to 320 (lowest stability). (2390)

D. BIOLOGICAL EFFECTS

Mammals			
rat	oral LD_{50}	9 mg/kg bw	
mouse	oral LD_{50}	100 mg/kg bw	(9649)

chloropropylate (isopropyl 4,4'-dichlorobenzilate; propyl p,p'-dichlorobenzilate; 4-chloro-α-(4-chlorophenyl)-α-hydroxybenzene acetic acid, 1-methylethyl ester; Acaralate-; Chlormite; Rospan; Rospin)

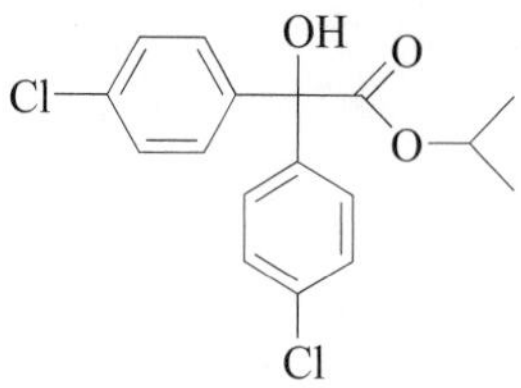

$C_{17}H_{16}Cl_2O_3$

CAS 5836-10-2

USES

Acaricide. Marine anti-fouling agent.

A. PROPERTIES

molecular weight 339.22; melting point 73°C; boiling point 148-150°C; density 1.35; vapor pressure 1.8×10^{-7} mm Hg at 20°C.

D. BIOLOGICAL EFFECTS

Bioaccumulation

biotransfer factor in milk: log B_m: -3.65			(2644)
Fishes			
harlequin fish	24-48 LC_{50}	22-20 mg/L	(10011)
bluegill sunfish, rainbow trout	96h LC_{50}	0.45-0.66 mg/L	(9662)
Mammals			
rat, mouse	oral LD_{50}	5000 mg/kg bw	(9662, 9775)

chloropropylene-oxide *see* epichlorohydrin

2-chloropyrazine

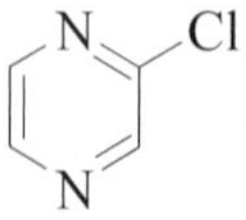

$C_4H_3ClN_2$

CAS 14508-49-7

A. PROPERTIES

molecular weight 14.53; boiling point 153-154°C; density 1.28.

C. WATER AND SOIL POLLUTION FACTORS

Reductive dehalogenation in anaerobic lake sediment suspension at 20°C in the dark

half-life: 78 d (n.s.i.). (2700)

3-chloropyridine

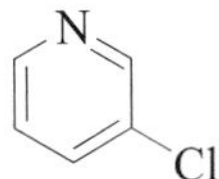

$(C_5H_4N)Cl$

C_5H_4ClN

CAS 626-60-8

USES

Chemical synthesis. Lubricating oil antiwear additive.

A. PROPERTIES

molecular weight 113.55; boiling point 148°C; density 1.19; solubility miscible

C. WATER AND SOIL POLLUTION FACTORS

Reductive dehalogenation in anaerobic lake sediment suspension at 20°C in the dark

half-life: >70 d. (2700)

2-chloropyrimidine

$C_4H_3ClN_2$

CAS 1722-12-9

A. PROPERTIES

molecular weight 114.53; melting point 66-68°C; boiling point 75-76°C at 10 mm.

C. WATER AND SOIL POLLUTION FACTORS

Reductive dehalogenation in anaerobic lake sediment suspension at 20°C in the dark

half-life: 26 d. (2700)

chloroquine

(7-chloro-4-(4-diethylamino-1-methylbutylamino)-quinoline; Aralen; Capquin)

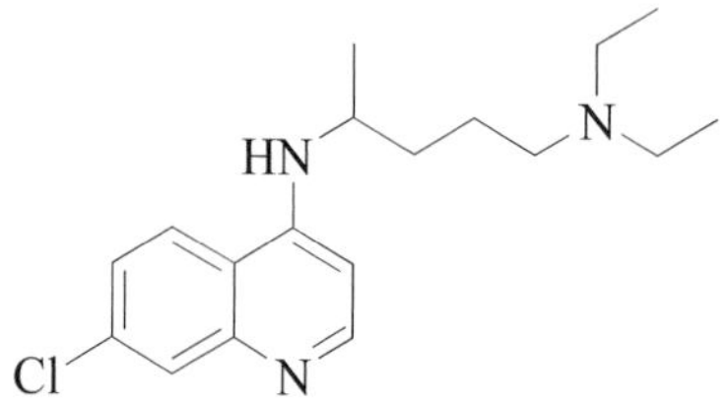

$C_9H_5NClNHCH(CH_3)(CH_2)_3N(C_2$

$C_{18}H_{26}ClN_3$

CAS 54-05-7

USES

medicine (antimalarial); usually dispensed as the phosphate; antiamebic. Antirheumatic, Lupus erythematosus suppressant.

A. PROPERTIES

colorless crystals; bitter taste; molecular weight 319.92; melting point 87°C

D. BIOLOGICAL EFFECTS

Marine tests

Microtox™ *(Photobacterium)* test	5 min EC_{50}	537 mg/L	
Artoxkit M *(Artemia salina)* test	24h LC_{50}	1,267 mg/L	
Freshwater tests			
Streptoxkit F *(Streptocephalus proboscideus)* test	24h LC_{50}	7.4 mg/L	
Daphnia magna test	24h LC_{50}	27 mg/L	
Rotoxkit F *(Brachionus calyciflorus)* test	24h LC_{50}	2.7 mg/L	(2945)

Mammals			
rat	oral LD_{50}	330 mg/kg bw	(10012)
mouse	oral LD_{50}	311 mg/kg bw	(10013)

chloroquine phosphate (7-chloro-4-[[4-(diethylamino)-1-methylbutyl]amino] quinoline phosphate (1:2); 1, 4-Pentanediamine, N4- (7-chloro-4-quinolinyl)-N1,N1-diethyl-, phosphate (1:2))

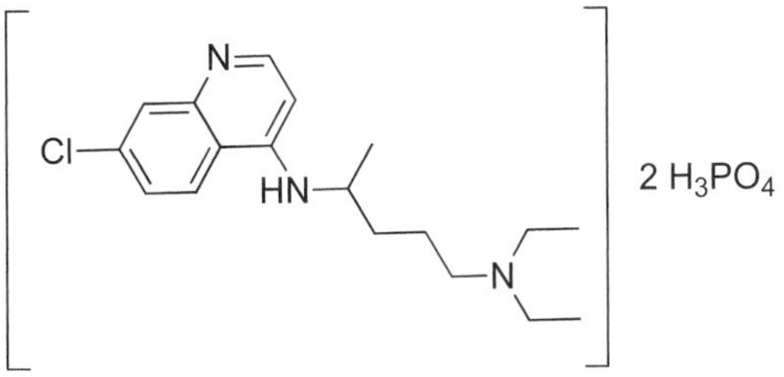

$C_{18}H_{26}ClN_3.2H_3PO_4$

CAS 1446-17-9

TRADENAMES

Aralen®; Arechin ®; Avloclor®; Chingamin®; Resoquine diphosphate®; Sanoquin®; Tanakan®

USES

Chloroquine phosphate is in a class of drugs called antimalarials and amebicides. It is used to prevent and treat malaria. It is also used to treat amoebiasis

A. PROPERTIES

molecular weight 515.87; melting point 194°C; water solubility 50,000 mg/L

C. WATER AND SOIL POLLUTION FACTORS

Hydrolysis

T/2 = stable

D. BIOLOGICAL EFFECTS

Toxicity

Micro organisms			
Vibrio fischeri	5min EC_{50}	856 mg/L	(7237)
Crustaceae			
Artemia salina	24h EC_{50}	2054 mg/L	
Daphnia magna	24h EC_{50}	42.9 mg/L	
Streptocephalus probiscideus	24h LC_{50}	11.3 mg/L	
Brachionus calyciflorus	24h EC_{50}	3.26 mg/L	(7237)
Mammals			
Rat	oral LD_{50}	623 mg/kg bw	
Mouse	oral LD_{50}	500 mg/kg bw	(10792)

Metabolism in animals

Chloroquine is rapidly and almost completely absorbed from the gastrointestinal tract, and only a small proportion of the administered dose is found in the stools. Excretion of chloroquine is quite slow, but is increased by acidification of the urine.

Chloroquine undergoes appreciable degradation in the body. The main metabolite is desethylchloroquine, which accounts for one fourth of the total material appearing in the urine; bisdesethylchloroquine, a carboxylic acid derivative, and other metabolic products as yet uncharacterized are found in small amounts. Slightly more than half of the urinary drug products can be accounted for as unchanged chloroquine.

m-chlorostyrene (Benzene,1-chloro-3-ethenyl; 3-chlorostyrene)

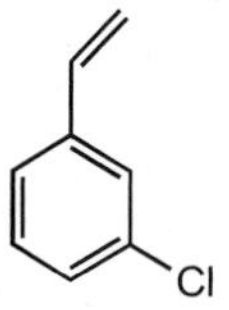

C_8H_7Cl

CAS 2039-85-2

EINECS 218-024-7

USES

Uses of m-chlorostyrene include the manufacture of polystyrene

A. PROPERTIES

molecular weight 138.60; boiling point 62-63°C; density 1.12 at 20°C; log Pow 3.6 (estimated); H 2.4 x 10^{-3} atm.m^3/mol;

C. WATER AND SOIL POLLUTION FACTORS

Mobility in soils

K_{OC}	840; 2100 (estimated)

D. BIOLOGICAL EFFECTS

Bioaccumulation

log BCF : 2.5 (estimated)

o-chlorostyrene (Benzene,1-chloro-2-ethenyl; 2-chlorostyrene)

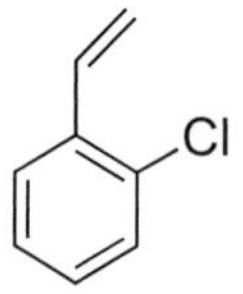

C_8H_7Cl

CAS 2039-87-4

EINECS 218-026-8

USES

o-chlorostyrene is produced commercially as a monomer from o-chloroethylbenzene and is also produced as a mixture with p-chlorostyrene. Uses of o-chlorostyrene include the manufacture of polystyrene, the incorporation into polyesters and the formulation of plastics, rubber and resin.

A. PROPERTIES

molecular weight 138.60; melring point −63.1°C; boiling point 188.7°C; density 1.1 at 20°C; vapour pressure 9.6 x 10^{-1} mm Hg; log Pow 2.5 (estimated); H 2.4 x 10^{-3} atm.m^3/mol

B. AIR POLLUTION FACTORS

Photodegradation

T/2 : 14 hours

C. WATER AND SOIL POLLUTION FACTORS

Mobility in soils

K_{OC} 2100 (estimated)

D. BIOLOGICAL EFFECTS

Bioaccumulation

log BCF : 2.5 (estimated)

p-chlorostyrene (4-chlorostyrene; benzne,1-chloro-4-ethenyl)

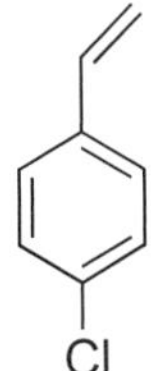

C_8H_7Cl

CAS 1073-67-2

EINECS 214-028-8

USES

p-chlorostyrene is produced commercially as a monomer from o-chloroethylbenzene and is also produced as a mixture with o-chlorostyrene. Uses of p-chlorostyrene include the manufacture of polystyrene, the incorporation into polyesters and the formulation of plastics, rubber and resin.

A. PROPERTIES

molecular weight 138.60; melting point −15.9°C; boiling point 192.0°C; density 1.09 at 20°C; vapour pressure 1.5 mm Hg at 25°C; log Pow 3.6 (estimated); H 2.4 x 10^{-3} atm.m^3/mol

B. AIR POLLUTION FACTORS

Photodegradation

14 hours (estimated)

C. WATER AND SOIL POLLUTION FACTORS

Mobility in soils

K_{OC} 2100 estimated)

D. BIOLOGICAL EFFECTS

Bioaccumulation

log BCF : 2.5 (estimated)

chlorothalonil (bravo; tetrachloroisophthalonitrile; 1,3-benzenedicarbonitrile, 2,4,5,6-tetrachloro-; 1,3-dicyanotetrachlorobenzene; tetrachlorometaphthalodinitrile; m-tetrachlorophthalodinitrile; Bravo; Chloroalonil; Daconil)

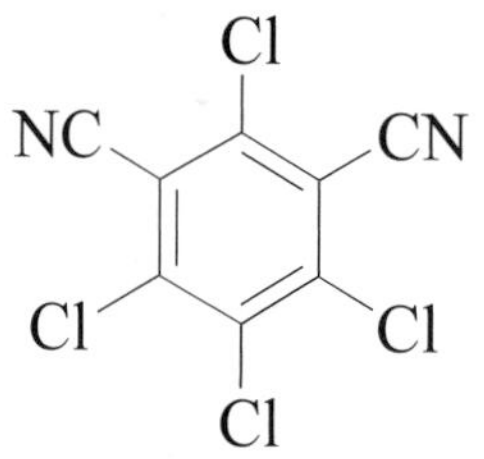

$C_8Cl_4N_2$

CAS 1897-45-6

USES

broad-spectrum fungicide. Preservative in paints and adhesives. Bactericide. Nematocide.

A. PROPERTIES

white, crystalline solid; molecular weight 265.9; melting point 250-251°C; boiling point 350°C at 1,010 mbar; density 1.8 at 25°C; solubility 0.9 mg/L; 0.6 mg/L; log P_{OW} 2.9; vapour pressure 4.3 x 10^{-12}; Henry's constant 1.7 x 10^{-2} $Pa.m^3.mol^{-1}$

C. WATER AND SOIL POLLUTION FACTORS

In groundwater wells

in the U.S.A.:150 investigations up to 1988:

median of the conc.'s of positive detections for all studies	0.02 μg/L	
maximum conc.	13 μg/L	(2944)

Biodegradation

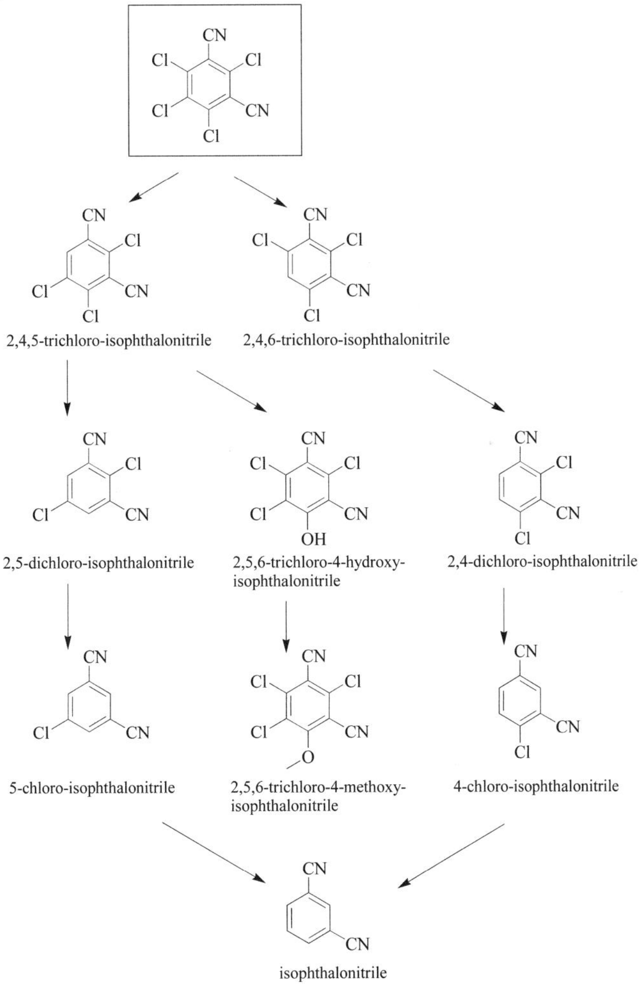

2,4,5-trichloro-isophthalonitrile
2,4,6-trichloro-isophthalonitrile

2,5-dichloro-isophthalonitrile
2,5,6-trichloro-4-hydroxy-isophthalonitrile
2,4-dichloro-isophthalonitrile

5-chloro-isophthalonitrile
2,5,6-trichloro-4-methoxy-isophthalonitrile
4-chloro-isophthalonitrile

isophthalonitrile

Microbial degradation

products identified. (10018)

Persistence

in aerobic and anaerobic soil conditions:	$t_{1/2}$: 5-36 d.	
in aerobic and anaerobic aquatic soil studies:	$t_{1/2}$: a few hours to a few d.	(9600)

Inhibition of biodegradation

Strongly inhibited bacterial degradation of cellulose in soil under aerobic and anaerobic conditions. (10015)

Soil sorption

log K_{oc} 2.9-3.8

D. BIOLOGICAL EFFECTS

Bioaccumulation

	Concentration µg/L	*exposure period*	*BCF*	*organ/tissue*
Fishes				

willow shiner *(Gnathopogon caerulescens)*	18	whole body	
carp *(Cyprinus carpio)*	25	whole body	(10017)
Molluscs: blue mussels	910	whole body	(10016)

SOIL MICROBIAL PROCESSES

Nitrification inhibition	6h EC_{25}	50 mg/kg dw	
	3d EC22	1.6 mg/kg dw	
	3d EC88	20 mg/kg dw	
	3d EC_30	1.8; 19 mg/kg dw	
	1w NOEC	>10 mg/kg dw	
	2w EC44	10 mg/kg dw	
	28d NOEC	4.4 mgkg dw	
Aerobic nitrogen fixation	18h EC47	100 mg/kg dw	
	6h EC43	1.6 mg/kg dw	
	6h EC_{100}	20 mg/kg dw	
	6h EC_55	1.8 mg/kg dw	
	6h $EC_{20}0+$	19 mg/kg dw *(stimulation !)*	
Anaerobic nitrogen fixation	6h $EC_{90}+$	1.6 mg/kg dw *(stimulation !)*	
	6h EC99	20 mg/kg dw	
	6h NOEC	1.8 mg/kg dw	
	6h EC400+	19 mg/kg dw *(stimulation !)*	
Sulfur oxidation	4w EC26	10 mg/kg dw	
	8w NOEC	>10 mg/kg dw	
Respiration	28d NOEC	4.4 mg/kg dw	
Bacteriophyta population growth	7d EC29	10 mg/kg dw	(7257)

ENZYME ACTIVITY

Amylase	1d EC_30	10 mg/kg dw	
Dehydrogenase	4d NOEC	>10 mg/kg dw	
Invertase	1d EC_{25}	10 mg/kg dw	
Phosphatase	2h NOEC	>10 mg/kg dw	
Urease	2d NOEC	>10 mg/kg dw	(7257)
Fungi			
population growth	7d EC_52	10 mg/kg dw soil	(7257)
Algae			
Pseudokirchneriella subsp.	96h NOEC	0.05 mg/L	
Scenedesmus subspicatus	96h NOEC	0.06 mg/L	
Anabaena flos-aquae	120h NOEC	0.02 mg/L	
Navicula pelliculosa	120h NOEC	0.0035 mg/L	(7257)
Plants			
Lemna gibba	72d NOEC	0.29 mg/L	(7257)
Insects			
Chironomus riparius, 1st instar	28d NOEC	0.125 mg/L	(7257)
Crustaceae			
Daphnia magna	48h EC_{50}	0.054; 0.115; 0.117 mg/L	
	24h LC_{50}	0.195 mg/L	
	48h LC_{50}	0.13-0.20 mg/L	(10016)
	21d NOEC	0.0006; 0.019; 0.035; 0.035 mg/L	
Parataya australiensis	96h LC_{50}	0.016 mg/L	
Astacopsis gouldi	96h LC_{50}	0.012 mg/L	
Penaeus duorarum	96h LC_{50}	0.162 mg/L	(7257)
Worms			
Eisenia foetida	14d LC_{50}	537 mg/kg dw	(7257)
Molluscs			
Crassostrea virginica	96h EC_{50}	0.005; 0.026 mg/L	(7257)
Mya arenari	96h LC_{50}	35 mg/L	
blue mussel	96h LC_{50}	5.9 mg/L	(10016)
Fishes			
Cyprinus carpio	96h LC_{50}	0.06 mg/L	
	48h LC_{50}	0.11 mg/L	(2680)

Galaxias auratus	96h LC_{50}	0.029 mg/L	
Galaxias maculatus	96h LC_{50}	0.016 mg/L	
Galaxias truttaceus	96h LC_{50}	0.019 mg/L	
Ictalurus punctatus	96h LC_{50}	0.047; 0.052 mg/L	
Lepomis macrochirus	96h LC_{50}	0.059 mg/L	
Oncorhynchus mykiss	96h LC_{50}	0.017; 0.043; 0.076 mg/L	
	21d NOEC	0.0069 mg/L	
Cyprinodon variegatus	96h LC_{50}	0.033 mg/L	
Pimephales promelas	45w NOEC	0.003 mg/L	(7257)
channel catfish	96h LC_{50}	0.044; 0.052 mg/L	(9600, 10014)
rainbow trout	96h LC_{50}	0.049 mg/L	
bluegill sunfish	96h LC_{50}	0.062 mg/L	(9600)
cytotoxicity to goldfish GF-Scale cells	NR_{50}	1.5 mg/L	(2680)
Salmo gairdneri	4d LC_{50}	7.6-17 µg/L	(2625)
Mammals			
rat	oral LD_{50}	10,000 mg/kg bw	(10019)

chlorothion (chlorthion; O,O-dimethyl-O-(3-chloro-4-nitrophenyl)phosphorothioate)

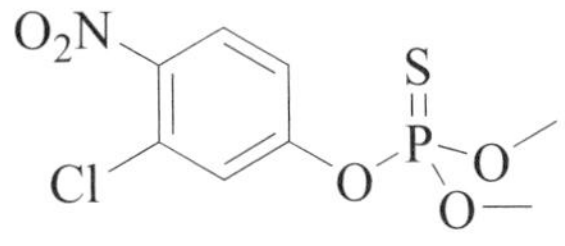

$C_8H_9ClNO_5PS$

CAS 500-28-7

USES

insecticide.

A. PROPERTIES

molecular weight 297.66; melting point 21°C; boiling point 136°C at 0.3 mbar; vapor pressure 5.5 × 10^{-6} mbar at 20°C; density 1.43; solubility 40 mg/L at 20°C.

C. WATER AND SOIL POLLUTION FACTORS

Aquatic reactions

hydrolyzes rapidly in alkaline media, more slowly in the presence of acids, to dimethylthiophosphate and 3-chloro-4-nitrophenol. To a limited extent there is oxidation to the phosphate and hydrolysis to phosphoric acid and 3-chloronitrophenol.

D. BIOLOGICAL EFFECTS

Crustaceans			
Daphnia magna	48h LC_{50}	4.5 µg/L	
Fishes			
Pimephales promelas	96h LC_{50}	2,800 µg/L	(2123)
Lepomis macrochirus	96h LC_{50}	700 µg/L	(2123)

α-chlorotoluene *see* benzylchloride

m-chlorotoluene (1-chloro-3-methylbenzene; 3-chlorotoluene; 3-methylchlorobenzene; m-tolyl chloride)

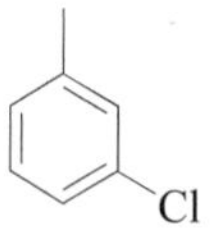

$CH_3C_6H_4Cl$

C_7H_7Cl

CAS 108-41-8

USES

Solvent. Dyestuff intermediate. Intermediate in organic synthesis.

A. PROPERTIES

molecular weight 126.59; melting point -48°C; boiling point 160-162°C; vapor pressure 9.8 mmHg at 43°C; density 1.07; $LogP_{OW}$ 3.28.

C. WATER AND SOIL POLLUTION FACTORS

Biodegradation

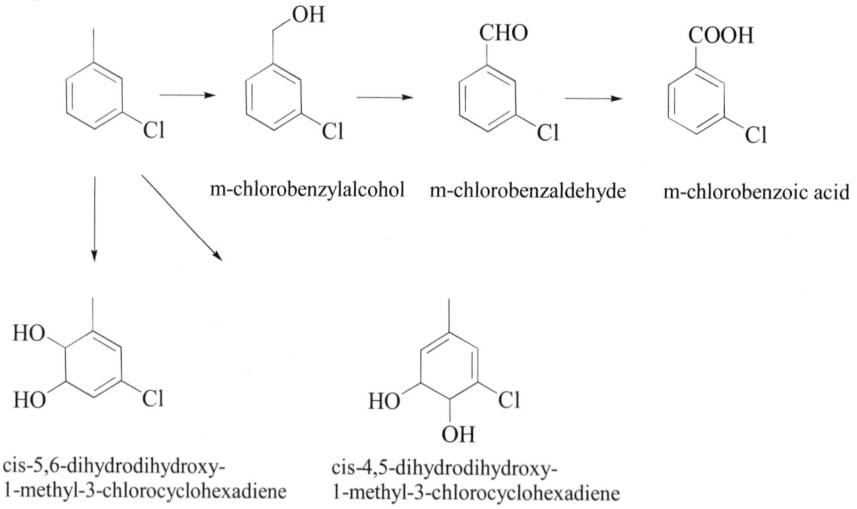

m-chlorobenzylalcohol m-chlorobenzaldehyde m-chlorobenzoic acid

cis-5,6-dihydrodihydroxy-1-methyl-3-chlorocyclohexadiene

cis-4,5-dihydrodihydroxy-1-methyl-3-chlorocyclohexadiene

Proposed transformation of 3-chlorotoluene by *Burkholderia sp.* strain PS12. (7080)

D. BIOLOGICAL EFFECTS

Fishes			
guppy *(Poecilia reticulata)*	7d LD_{50}	18 ppm	(1833)

o-chlorotoluene (2-chloro-1-methylbenzene; o-tolylchloride; HALSO 99)

$CH_3C_6H_4Cl$

C_7H_7Cl

CAS 95-49-8

USES

solvent and intermediate for organic chemicals and dyes. Dyestuff intermediate. Intermediate in organic synthesis. Manufacture of pharmaceuticals and synthetic rubber compounds.

A. PROPERTIES

molecular weight 126.58; melting point -36.5/-34°C; boiling point 159°C; vapor pressure 2.7 mm at 20°C, 5 mm at 30°C; vapor density 4.37; density 1.1 at 20/4°C; saturation concentration in air 18.6 g/m^3 at 20°C, 33.3 g/m^3 at 30°C; solubility 47 mg/L at 20 °C; LogP_{OW} 3.42.

B. AIR POLLUTION FACTORS

1 mg/m^3 = 0.19 ppm, 1 ppm = 5.26 mg/m^3.

Photochemical reactions

$t_{1/2}$: 8.4 d, based on reactions with OH° (calculated) (8481)

C. WATER AND SOIL POLLUTION FACTORS

ThOD: 2.21

Water quality

In river Maas at Eysden (the Netherlands) in 1976: median: n.d.: range: n.d.-0.1 µg/L.
In river Maas at Keizersveer (the Netherlands) in 1976: median: n.d.: range: n.d.-0.1 µg/L. (1368)
In Delaware River (U.S.A.): conc. range (n.s.i.): winter: 3 µg/L; summer: n.d. (1051)

Biodegradation

Inoculum	*method*	*concentration*	*duration*	*elimination results*	
domestic sewage	Closed Bottle Test	8-80 mg/L	28 d	0% ThOD	(8962)
adapted domestic sewage	Modified MITI Test		28 d	0% ThOD	(8963)
domestic seage	Modified Zahn-Wellens Test	23 mg DOC/L	28 d	86% DOC	(8962)
adapted A.S.	Modified MITI Test	100 mg/L	28 d	0% ThOD	(8964)

o-chlorobenzylalcohol o-chlorobenzaldehyde o-chlorobenzoic acid

cis-5,6-dihydrodihydroxy-1-methyl-2-chlorocyclohexadiene

cis-3,4-dihydrodihydroxy-1-methyl-2-chlorocyclohexadiene

Proposed transformation of 3-chlorotoluene by Burkholderia sp. strain PS12. (7080)

D. BIOLOGICAL EFFECTS

Bioaccumulation

	Concentration µg/L	*exposure period*	*BCF*	*organ/tissue*	
Fishes: *Cyprinus carpio*	45	56 d	20-112	total body	
		56 d	42-87	total body	(8964)

Bacteria			
Pseudomonas putida	16h EC_0	15 mg/L	(1900)
	18h EC_0	250 mg/L	(8963)
methanogens	96h IC_{50}	53 mg/L	(8966)
Protozoans			
protozoa *(Entosiphon sulcatum)*:	72h EC_0	>80 mg/L	(1900)
protozoa (*Uronema parduczi* Chatton-Lwoff):	EC_0	>80 mg/L	(1901)
Algae			
Scenedesmus quadricauda	8d EC_0	>100 mg/L	(1900)
Scenedesmus subspicatus	72h EC_{20}	60 mg/L	
(cell multiplication inhibition test)	72h EC_{50}	>100 mg/L	(8965)
Crustaceans			
Daphnia magna	24h EC_0	9 mg/L	
	24h EC_{50}	20 mg/L	
	21d NOEC	0.27 mg/L	(8963)
Fishes			
Salmo gairdneri	96h LC_{50}	2.3 mg/L	(8963)
Brachydanio rerio	96h LC_{50}	70-100 mg/L	(8963)
Oryzias latipes	48h LC_{50}	9.6 mg/L	(8964)
trout, bluegill sunfish, yellow perch, goldfish	24h LC_0	5 mg/L	(9594)
Mammals			
rat	oral LD_{50}	5,700; 3,227 mg/kg body wt	

p-chlorotoluene (1-chloro-4-methylbenzene; p-tolylchloride; 4-chlorotoluene; 4-methylchlorobenzene)

$CH_3C_6H_4Cl$

C_7H_7Cl

CAS 106-43-4

USES

Solvent and intermediate for organic chemicals and dyes.

A. PROPERTIES

molecular weight 126.59; boiling point 162°C; melting point 7.3- 7.6°C; vapor pressure 3.6 hPa at 20°C; 6.5 hPa at 30°C; 16 hPa at 50°C; 152 hPa at 100°C; density 1.07 (25/15°C); solubility 15- 40 mg/L at 20°C, 106 mg/L at 20°C; LogP_{OW}3.33 (measured).

B. AIR POLLUTION FACTORS

1 mg/m^3 = 0.19 ppm; 1 ppm = 5.26 mg/m^3.

Photochemical reactions

$t_{1/2}$: 8.4 d, based on reactions with OH° (calculated). (8359)

C. WATER AND SOIL POLLUTION FACTORS

ThOD: 2.15

Water quality

In river Maas at Eysden (the Netherlands) in 1976: median: 0.1 µg/L; range: n.d.- 0.3 µg/L. In river Maas at Keizersveer (the Netherlands) in 1976: median: 0.1 µg/L; range: n.d.-0.2 µg/L. (1368)

Biodegradation

Inoculum	*Method*	*Concentration*	*Duration*	*Elimination results*	
Domestic sewage adapted	Closed bottle test		20 d	0% ThOD	(9010)
Domestic sewage	Modified Zahn-Wellens test	22 mg DOC/L	28 d	86% DOC	(9009)
Domestic sewage adapted	Modified MITI test		28 d	1% ThOD	(9009)
A.S	Respirometer test	60 mg DOC/L	20 d	<10% $ThCO_2$	(9011, 9012)
A.S.	Modified MITI Test	100 mg/L	14 d	0% ThOD	(9013)

Bacterial metabolism of 4-chlorotoluene. (7081, 7082)

846 p-chlorotoluene

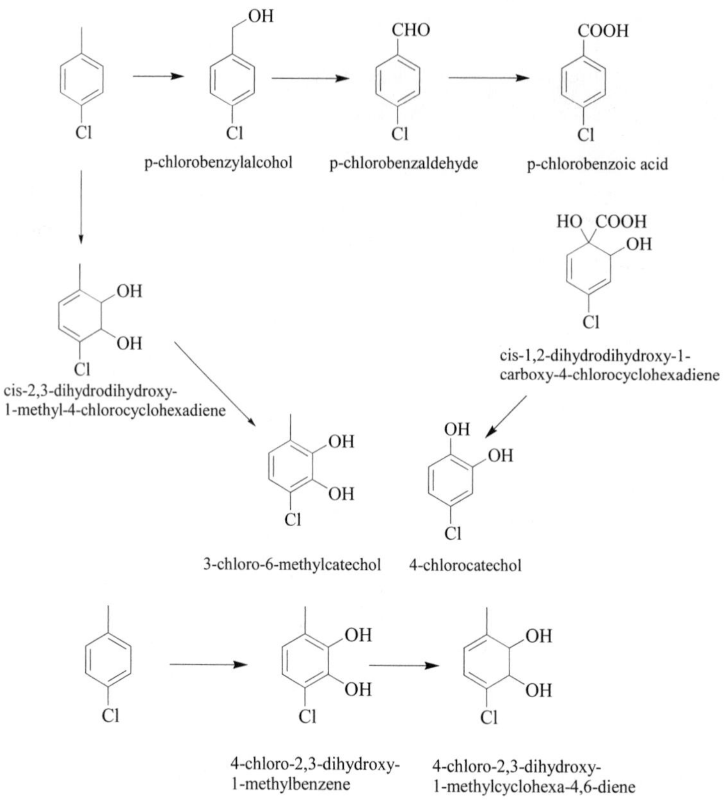

p-chlorobenzylalcohol p-chlorobenzaldehyde p-chlorobenzoic acid

cis-2,3-dihydrodihydroxy-1-methyl-4-chlorocyclohexadiene

cis-1,2-dihydrodihydroxy-1-carboxy-4-chlorocyclohexadiene

3-chloro-6-methylcatechol 4-chlorocatechol

4-chloro-2,3-dihydroxy-1-methylbenzene 4-chloro-2,3-dihydroxy-1-methylcyclohexa-4,6-diene

Metabolism

by *Pseudomonas putida* to yield (+)- cis-4-chloro-2,3-dihydroxy-1-methylcyclohex-4,6-diene and 4-chloro-2, 3-dihydroxy-1-methylbenzene. (10020)

D. BIOLOGICAL EFFECTS

Bioaccumulation

	Concentration µg/L	*Exposure period*	*BCF*	*Organ/tissue*	
Fish: *Cyprinus carpio*	30	56 d	14- 102	total body	
	300	56 d	22- 76	total body	(9013)

Bacteria			
anaerobic municipal sludge	24 h EC_0	12 mg/L	(9011, 9012)
Pseudomonas putida	30 min EC_0	250 mg/L	(9010)
Algae			
Scenedesmus quadricauda	8 d EC_3	>100 mg/L	(9010)
Crustaceans			
Daphnia magna (n.s.i.):	EC_{50} reproduction	1.3 mg/L	
	NOEC reproduction	0.57 mg/L	
	NOEC growth	0.32 mg/L	
	LC_{50}	2.3 mg/L	(2706)
	24 h EC_0	ca. 3 mg/L	
	24 h EC_{50}	6-12 mg/L	(9011, 9012)
	48 h EC_{50}	3.6 mg/L	(9010)
Fishes			
guppy *(Poecilia reticulata)*	14 d LC_{50}	5.9 mg/L	(1833)

Brachydanio rerio	96 h LC_0	15 mg/L	
	96 h LC_{50}	24 mg/L	
	96 h LC_{100}	50 mg/L	(9010)
Oryzias latipes	48 h LC_{50}	5.2 mg/L	(9013)
Mammals			
Rat	oral LD_{50}	2,389; 2,273; 3,600; 2,100; 1,920 mg/kg bw	
Mouse	oral LD_{50}	1,900 mg/kg bw	(9897, 9929)

2-chlorotoluene-4-sodium sulfonate

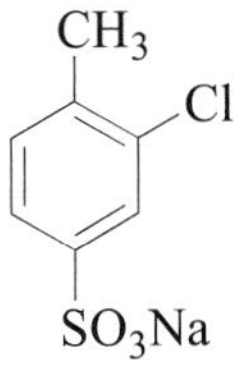

$C_7H_6ClNaO_3S$

USES

synthesis of dyes, intermediates, and drugs.

D. BIOLOGICAL EFFECTS

Fishes			
Lepomis macrochirus	24h LC_{50}	>1,374 mg/L	(1295)

2-chlorotoluene-5-sodium sulfonate

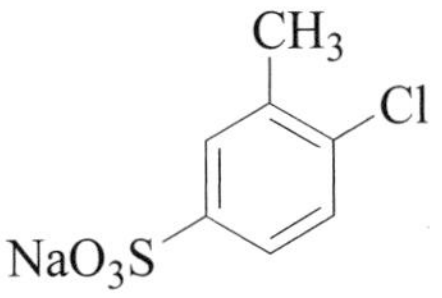

$C_7H_6ClNaO_3S$

D. BIOLOGICAL EFFECTS

Crustaceans			
Daphnia magna		*young*	*adult*
	25h LC_{50}	0.8 mg/L	3.3 mg/L
	50h LC_{50}	0.6 mg/L	1.3 mg/L

	100h LC_{50}	0.4 mg/L	(1295)
Snail eggs			
Lymnaea sp.	25h LC_{50}	30 mg/L	
Fishes			
Mollienesia latipinna	25h LC_{50}	115.2 mg/L	
	50h LC_{50}	66.1 mg/L	(1295)

chlorotoluron (N'-(3-chloro-4-methylphenyl)-N,N-dimethylurea; 3-(3-chloro-4-methylphenyl)-1,1-dimethylurea; Chlortoluron; Dicuran; Dikurin; 3-(3-chloro-*p*-tolyl)-1,1-dimethylurea; N#-(3-chloro-4-methylphenyl)-N,N-dimethylurea)

Cl N H N O

$C_{10}H_{13}ClN_2O$

CAS 15545-48-9

USES

herbicide

A. PROPERTIES

molecular weight 212.69; melting point 147-148°C; vapor pressure 4.8 x 10^{-8} mbar at 20°C; density 1.40 at 20°C; solubility 70 mg/L at 20°C; log P_{OW} 2.53.

C. WATER AND SOIL POLLUTION FACTORS

Soil degradation

half-life outdoors: 30 d in springtime. The metabolites are 3-chloro-*p*-toluidine, 3-(3-chloro-4-methylphenyl)-1-methylurea and 1-(3-chloro-4-methylphenyl)urea. (2962)

Ninety strains of micromycetes isolated from soil were cultivated in liquid synthetic medium with 100 mg/L chlorotoluron. The chlorotoluron concentration was >50% depleted by 4% of the strains in five d. (10031)

Hydrolysis

t1/2: 18-45 months (calculated). (9600)

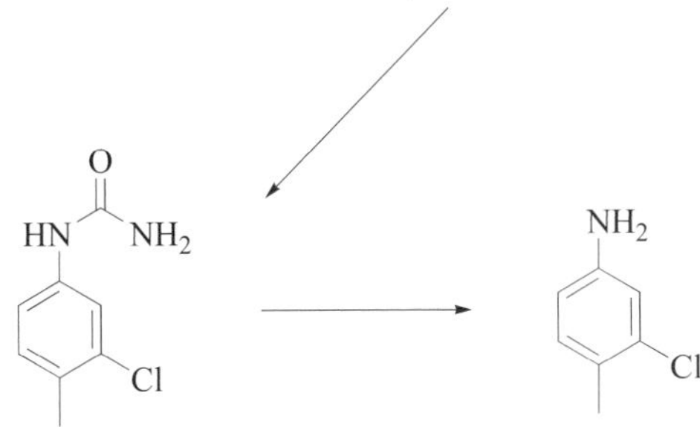

chlortoluron 3-(3-chloro-4-methylphenyl)-1-methylurea

1-(3-chloro-4-methylphenyl)-urea 3-chloro-p-toluidine

Biodegradation

Ninety strains of *micromycetes* isolated from soil were cultivated in liquid synthetic medium with 100 mg/L chlorotoluron. The chlorotoluron concentration was > 50% depleted by 4% of the strains in five days. (10031)

D. BIOLOGICAL EFFECTS

Toxicity

Fishes			
Salmo gairdneri	96h LC_{50}	34 mg/L	
Ictalurus punctatus	96h LC_{50}	60 mg/L	
Perca species	96h LC_{50}	51 mg/L	(2679)
rainbow trout	96h LC_{50}	20-35 mg/L	
bluegill	96h LC_{50}	40-50 mg/L	
crucian carp	96h LC_{50}	>100 mg/L	(2963)
Mammals			
rat	oral LD_{50}	> 10,000 mg/kg bw	(9600)

chlorotrifluoromethane (F 13; trifluoromethylchloride; trifluorochloromethane; Freon-13; CFC-13)

$CClF_3$

CAS 75-72-9

USES

Fire extinguishing agent. Lubricant. Refrigerant. Propellant for aerosol sprays. Etching gas in production of integrated circuits.

OCCURRENCE

Present in landfill gas.

A. PROPERTIES

molecular weight 104.46; melting point -181°C; boiling point -82°C; vapor pressure 400 mm Hg at -93°C; LogH 1.85 at 25°C.

B. AIR POLLUTION FACTORS

Photochemical reactions

Indirect effects from accumulation in stratosphere may lead to substantial effects on human health due to ozone depletion with resulting increase in effects of UVB radiation . (10025)

5-chlorovanillin

5-chlorovanillin

$C_8H_7ClO_3$

CAS 19463-48-0

A. PROPERTIES

molecular weight 186.60

C. WATER AND SOIL POLLUTION FACTORS

Manmade sources

In sediment of rivers and lakes downstream of a bleached kraft pulp and paper mill in New Zealand (1991/1992)

km's downstream		
1.5	<0.1; 3.8; <0.1; <0.1 µg/kg dw	
11	9.8; <0.1; <0.1 µg/kg dw	(7042)

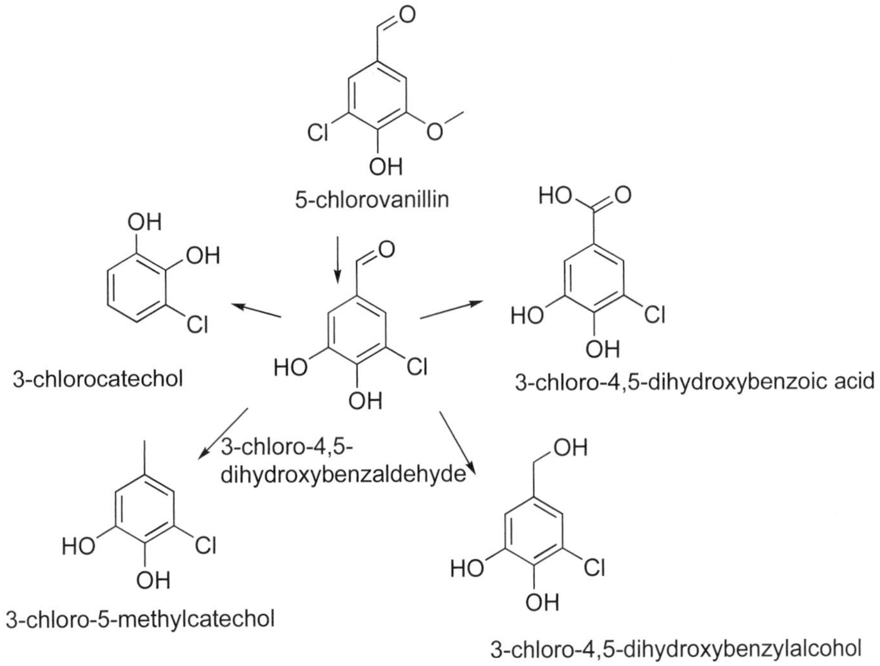

Transformation pathway of 5-chlorovanillin by metabolically stable anaerobic enrichment cultures obtained from sediments from the Gulf of Bothnia and the Baltic sea. (10821)

6-chlorovanillin (2-chloro-4-hydroxy-5-methoxybenzaldehyde)

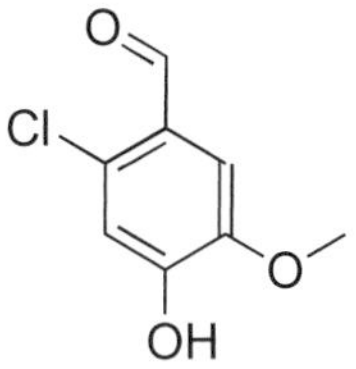

$C_8H_7ClO_3$

CAS 18268-76-3

SOURCES

a degradation product of chloroguaiacols which are derived from wood extractives and are discharged from paper pulp mills. A reaction product of lignin from black spruce (*Picea mariana*) with sodium chlorite in acid solution. (10820)

A. PROPERTIES

melting point 167-168°C; log P_{OW} 1.5-2.0

C. WATER AND SOIL POLLUTION FACTORS

Manmade sources

In sediment of rivers and lakes downstream of a bleached kraft pulp and paper mill in New Zealand

(1991/1992):	
km's downstream	µg/kg dw
1.5	4.5; <0.1; <0.1: <0.1
11	5.7; <0.1; 7.6

Concentrations in Fraser basin, Canada (1992-1996) (10820)

Bed sediment µg/kg dw	Suspended sediment µg/kg dw	Clarified water ng/L
0.23-3.5	<0.65-1100	<0.11-16

Anaerobic degradation of 6-chlorovanillin

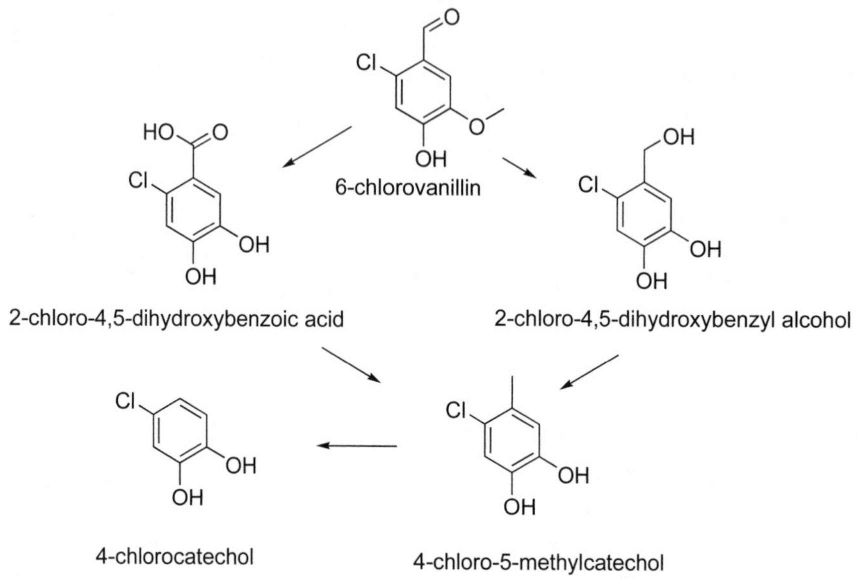

6-chlorovanillin

2-chloro-4,5-dihydroxybenzoic acid

2-chloro-4,5-dihydroxybenzyl alcohol

4-chlorocatechol

4-chloro-5-methylcatechol

Transformation pathway by metabolically stable anaerobic enrichment cultures obtained from sediments from the Gulf of Bothnia and the Baltic sea (10821)

D. BIOLOGICAL EFFECTS

Toxicity

Crustaceae			
Cerio daphnia dubia	7d LC_{50}	13.9 mg/L	(10820)
	7d IC_{50}	6.5 mg/L	
	7d NOEL	2.5 mg/L	(10944)
Insecta			
Chironomus tentans	48h NOEC	>100 mg/L	(10820)
Fishes			
Oncorhynchus mykiss	96h LC_{50}	2.6 mg/L	(10820)
Samo trutta	96h LC_{50}		
Acipenser transmontanus	24h LC_{50}	0.41 mg/L	(10820)
larvae	24h NOEC	0.001 mg/L	
Lepomis macrochirus	96h LC_{50}	13.2 mg/L	(10820)
Pimephales promelas	7d IC_{50}	15.7 mg/L	
	7d NOEL	4.9 mg/L	(10944)

chloroxuron (tenoran; 3-*p*(*p*-(*p*-chlorophenoxy)phenyl)-1,1-dimethylurea)

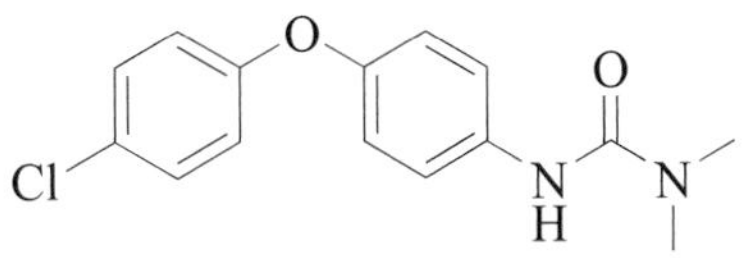

$C_{15}H_{15}ClN_2O_2$

CAS 1982-47-4

USES

herbicide absorbed by both roots and leaves.

A. PROPERTIES

white crystals; molecular weight 290.77; melting point 149-150°C; solubility 3 mg/L; $LogP_{OW}$ 4.0.

D. BIOLOGICAL EFFECTS

Molluscs			
Lymnaea stagnalis:	20d EC_{50}	17.5 mg/L	
	locomotor effect after 0.25 d at	14.5 mg/L	
Physa fontinalis:	locomotor effect after 0.17 d at	14.5 mg/L	(2625)

Fishes			
Salmo gairdneri:	96h LC_{50}	0.44 mg/L	
Ictalurus punctatus:	96h LC_{50}	0.44 mg/L	
Lepomis macrochirus:	24-48h LC_{50}	25 mg/L	
Poecilia reticulata:	96h LC_{50}	7.8 mg/L	(2679)

chlorphoxim (2-chloro-α-(((diethoxyphosphinothioyl)oxy)imino)benzeneacetonitrile; (*o*-chlorophenyl)glyoxylonitrileoxime-O,O-diethylphosphorothioate)

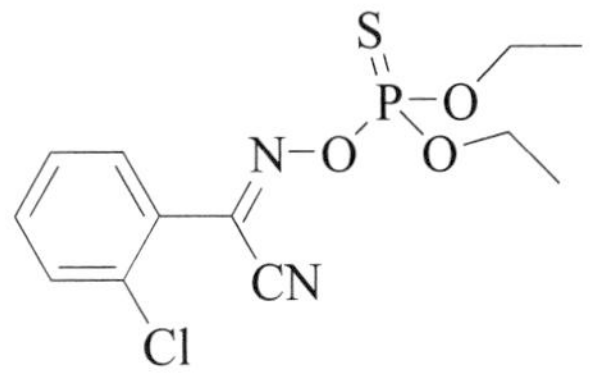

$C_{12}H_{14}ClN_2O_3PS$

CAS 14816-20-7

USES

pesticide; effective against the larval stages of *Simulium damnosum* (blackfly), the vector of human onchocerciasis in Africa (1692); effective against adult mosquitoes (1693) and agricultural insects (1694, 1695).

A. PROPERTIES

molecular weight 332.7; melting point 66.5°C; boiling point not distillable; vapor pressure <1 mPa at 20°C; solubility 1.7 mg/L at 20°C.

C. WATER AND SOIL POLLUTION FACTORS

Method of analysis

High concentrations of chlorphoxim form under gas-chromatographic conditions several trialkylphosphates, among them O,O-diethyl-O-methylphosphorothioate (DEMTP) and O,O-diethyl-S-methylphosphorothioate. (1695)

A GC-method is based on the derivation of chlorphoxim to DEMTP with a 99% efficiency of conversion. (1691)

D. BIOLOGICAL EFFECTS

Fishes

bluegill: at 0.02 mg/L, 24h BCF: 150 (chlorphoxim emulsifiable concentrate): (1691)

rainbow trout	96h LC_{50}	0.1-1.0 mg/L	
carp	48h LC_{50}	8.5 mg/L	(2963)

chlorpropham *see* isopropyl-N-(3-chlorophenyl)carbamate

chlorpyrifos (lorsban; dursban; O,O-diethyl-O-(3,5,6-trichloro-2-pyridyl)phosphorothioate)

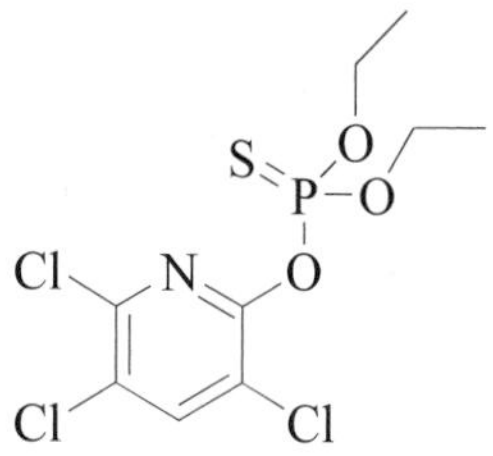

$C_9H_{11}Cl_3NO_3PS$

CAS 2921-88-2

USES

insecticides.

A. PROPERTIES

amber solid cake with amber oil; molecular weight 350.6; melting point 41.5-43.5°C; vapor pressure 1.87×10^{-5} mm at 20°C; 2.5×10^{-5} hPa at 20°C; density (liq.) 1.4 at 43.5°C; decomposition temp.

approx. 160°C; solubility 0.4 mg/L at 23°C; $LogP_{OW}$ 5.11 at 20°C; 4.82; 4.7.

B. AIR POLLUTION FACTORS

Manmade sources

in ambient air of storage and office rooms of commercial pest control buildings in a 4-h period:

storage room:	avg: 220 ng/m^3; range: 83-595 ng/m^3	
office room:	avg: 129 ng/m^3; range: 26-357 ng/m^3	(1868)

Ambient air concentrations

Bloomington, Indiana, U.S.A., 1985-1986:

inside 12 homes:	0.3-150 ng/m^3	
outside:	0.2-0.4 ng/m^3	(2642)

C. WATER AND SOIL POLLUTION FACTORS

Hydrolyzed by strong alkalis. 50% decomposition of an aqueous-alcoholic solution occurs in about 7 d at pH 10. (2962)

Degradation

persistence in soil at 10 ppm initial concentration:

	50% remaining	*5% remaining*	
sterile sandy loam	17 weeks		
sterile organic soil	>24 weeks		
nonsterile sandy loam	<1 week	1 week	
nonsterile organic soil	2.5 weeks	8 weeks	(1433)

Hydrolysis

in buffered distilled water at 25°C:

half-life: 22 d at pH 8.1
35 d at pH 6.9
62 d at pH 4.7

A 16-fold rate enhancement was demonstrated in canal and pond water at 25°C. The following products of hydrolysis were identified:

3,5,6-trichloro-2-pyridinol
O-ethyl,O-hydrogen-O-(3,5,6-trichloro-2-pyridyl)phosphorothioate
O,O-dihydrogen-O-(3,5,6-trichloro-2-pyridyl)phosphorothioate (1652)

Persistence

in vitro, in capped bottles at 21°C in darkness: % remaining after 8 weeks at initial conc. of 5 ppm:

in sterilized natural water:	45%	
in sterilized distilled water:	50%	
in natural water:	40%	
in distilled water:	55%	(2559)

Biodegradation: in soil in greenhouse at 23-26°C at constant light at initial conc. of 5-10 mg/kg soil, experiment time 50 d:

conditions	t/2 d	
in sandy soil	7	
in organic rich orchard soil	12	
in agricultural soil	9	
in soil from volcanic area	15	
in field experiments	30 (dissipation t/2)	(7045)

Mobility

log K_{OC}: 3.8. (7045)

D. BIOLOGICAL EFFECTS

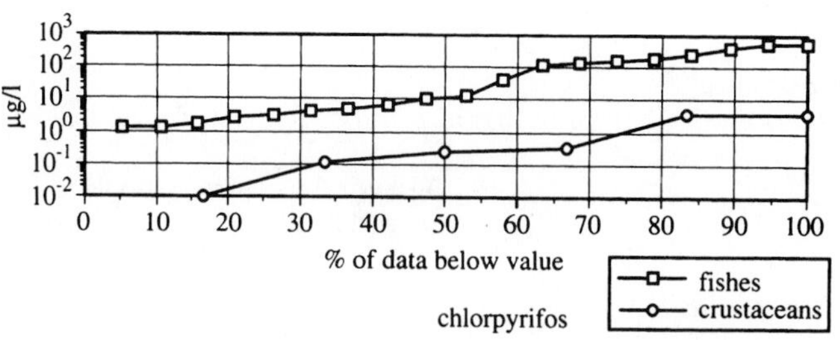

Frequency distributions of 24-96h EC_{50}/LC_{50} values for crustaceans (n = 6) and fishes (n = 19) based on data from this and other works (3999)

Crustaceans			
Gammarus lacustris	96h LC_{50}	0.11 μg/L	(2124)
Gammarus fasciatus	96h LC_{50}	0.32 μg/L	(2126)
shiner perch *(Cymatogaster aggregata)*	96h LC_{50}	3.5; 3.7 μg/L	(2343)
Korean shrimp *(Palaemon macrodactylus)*	96h LC_{50}	0.01; 0.25 μg/L	(2353)
Gammarus pulex	1h $LC_{90\text{-}95}$	0.05-0.1 μg/L	(1653)
Insects			
Pteronarcys californica	96h LC_{50}	10 μg/L	(2128)
Pteronarcella badia	96h LC_{50}	0.38 μg/L	(2128)
Claassenia sabulosa	96h LC_{50}	0.57 μg/L	(2128)
Culex pipiens	LC_{50}	2.6 μg/L	(2625)
Ephimeroptera: *Bactis rhodani*	1h $LC_{90\text{-}95}$	0.01-0.02 ppm	
Trichoptera:			
Brachycentrus subnubilis	1h $LC_{90\text{-}95}$	0.2-0.5 ppm	
Hydropsyche pellucidula	1h $LC_{90\text{-}95}$	>0.5 ppm	
Odonata: *Agrion*	1h $LC_{90\text{-}95}$	0.2 ppm	
Diptera: *Simulium ornatum*	1h $LC_{90\text{-}95}$	0.05-0.1 ppm	(1653)
Fishes			
Lepomis macrochirus	96h LC_{50}	2.6 μg/L	(2137)
Salmo gairdneri	96h LC_{50}	11 μg/L	(2137)
Fundulus heteroclitus	96h LC_{50}	4.7-12.2 μg/L	(1615)
green sunfish	72h LC_{50},S	40 μg/L	
mosquitofish	72h LC_{50},S	260 μg/L	(1203)
Pimephales promelas	30d EC_{50} growth	3.9-7.1 μg/L	
	96h LC_{50}	122; 200-506 μg/L	
	30d LC_{50}	7.1 μg/L	
Tilapia aurea	24h LC_{50}	418 μg/L	
	5d BCF	72-1,060 μg/L	
Tilapia nilotica	72h LC_{50}	151 μg/L	(2625)
mosquitofish	log BCF	2.67	(2606)
	96h LC_{50}, μg/L	*MATC, early life stage μg/L*	
Mysidopsis bahia	1.7	0.37	
Leuresthes tenuis	1.3	1.2	
Menidia beryllina	4.2	0.54	
Menidia peninsulae	1.3	46.0	
Cyprinodon variegatus	-	2.3	
Opsanus beta	560	<3.7	(2643)

cytotoxicity to goldfish GF-Scale cells: NR_{50}: 6.0 mg/L = 0.017 mmol/L (2680)

Cyprinus carpio	48h LC_{50}	0.13 mg/L	(2680)
bluegill (*Lepomis macrochirus)*	24h LC_{50},S	0.007 mg/L	
	96h LC_{50},S	0.003 mg/L	(2697)

Biotransfer factors

Biotransfer factor in beef: log B_b: -3.55

Biotransfer factor in milk: log B_m: -4.73 (2644)

chlorpyrifosmethyl *see* O,O-dimethyl-O-(3,5,6-trichloro-2-pyridyl)phosphorothioate)

chlorsulfuron (2-chloro-N-[(4-methoxy-6-methyl-1,3,5-triazin-2-yl)aminocarbonyl]-benzenesulfonamide)

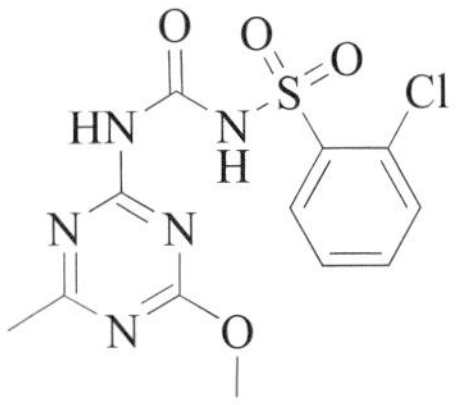

$C_{12}H_{12}ClN_5O_4S$

CAS 64902-72-3

USES

herbicide.

A. PROPERTIES

colorless, odorless crystals; molecular weight 357.8; melting point 174-178°C; decomposes at 192°C; vapor pressure 6.1 × 10^{-6} mbar at 25°C; solubility 100-125 mg/L at 25°C.

C. WATER AND SOIL POLLUTION FACTORS

Soil sorption and degradation

Leaching was studied using field lysimeters containing undisturbed soil monoliths or packed soil. The herbicide was applied at two rates, 4 g and 8 g active ingredient/ha. All lysimeters received supplementary watering in addition to natural rainfall. The compound was not detectable at the lowest dose. At the highest dose, up to 43 ng/L was detected in the leachate. Converted to fluxes over the 7-month period, considerably less than 1% of the applied compound appeared in the leachate. (2941)

Persistence in dry soil: $t_{1/2}$: 6-8 d. (9740)

Aquatic reactions

Hydrolysis occurs with an average half-life of 4-8 weeks at pH 5.7-7.0 at 20°C. Significant (2962) degradation occurs in 24-48 hours in acidic aqueous solutions below pH 5. Polar organic solvents such as methanol and acetone also promote hydrolysis.

D. BIOLOGICAL EFFECTS

Bioaccumulation

	Concentration µg/L	*exposure period*	*BCF*	*organ/tissue*	
Algae: *Chlorella fusca*			< 9 at pH 6.0 53 at pH 5.0.		

It is suggested that chlorsulfuron penetrates the algal cell membranes in its undissociated state and accumulates via an ion-trapping mechanism. (9739)

Fishes			
bluegill sunfish	96h LC_{50}	>250 mg/L	
rainbow trout	96h LC_{50}	>250 mg/L	(2962)

Algae			
Selenastrum capricornutum	72h EC_{50}	0.68-0.81 mg/L	
	72h EC_{100}	10 mg/L	(7024)
Chlorella saccharophila	96h EC_{50}	74.5 mg/L	(9738)
Scenedesmus acutus	96h EC_{50}	0.19 mg/L	
Pseudanabaena galeata	96h EC_{50}	21.1 mg/L	(9738)

Mammals			
rat	oral LD_{50}	5,550-6,290 mg/kg bw	(9662, 9743)

chlortetracycline hydrochloride (aureocycline; aureomycine; auxeomycin; isphamycin)

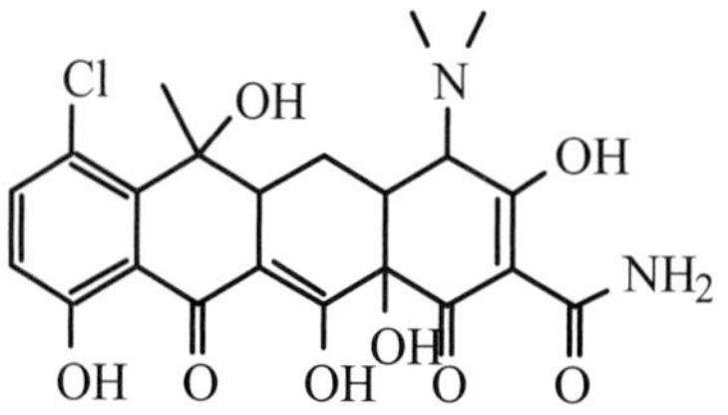

$C_{22}H_{23}ClN_2O_8.HCl$

CAS 64-72-2 (hydrochloride)

CAS 57-62-5

A. PROPERTIES

molecular weigth 460.44; melting point 220 –245 °C

C. WATER AND SOIL POLLUTION FACTORS

Environmental concentrations in soils

Northern Germany: concentrations in soil fertilised in april 2000 with pig slurry containing 4 mg/L tetracycline, samples taken in may 2000: (mean ± SD, n=4) (7214)

soil depth	inside test area	soil depth	outside test area
0 – 10 cm	3.0 ± 0.7 µg/kg	0 - 30 cm	4.0 ± 0.7 µg/kg
10 – 20 cm	2.9 ± 0.2 µg/kg	30 – 60 cm	n.d.
20 – 30 cm	3.0 ± 0.7 µg/kg	60 –90 cm	n.d.

Northern Germany: concentrations in soil under "crusty" pig and cattle slurry, samples taken in may 2000. (7214)

soil depth	"crusty" pig slurry	"crusty" cattle slurry
"crusty" slurry	3.4 – 1001.6	7.6 ± 2.7 µg/kg
0 – 10 cm	1.2 – 41.8 µg/kg	5.0 µg/kg
10 – 20 cm	1.8 – 8.5 µg/kg	4.6 µg/kg
20 – 30 cm	2.0 – 10.4 µg/kg	2.3 µg/kg

Environmental concentrations in waters

(7221, 7224)	Location and year	min ng/L	median ng/L	max ng/L	n	n > d.l.
Surface water	Germany 1999	<dl	<dl	<dl	14	0
Surface water	USA 1999	<150		690	90	2
Effluent MTP	Germany 1999	<dl	<dl	<dl	5	0
Groundwater	Germany 1999	<dl	<dl	<dl	37	0
Groundwater	Germany 1999	<dl		170	8	1
Groundwater	Germany 1999	<dl		220	16	1

D. BIOLOGICAL EFFECTS

Algae			
Micriocystis aeruginosa	7d EC_{50}	0.05 mg/L	
Selenastrum capricornutum	72h EC_{50}	3.1 mg/L	(7237)
Mammals			
Rat	acute oral LD_{50}	10300 mg/kg bw	(7215)

5*β*-cholestan-3*β*-ol *see* coprostanol

5*β*-cholesten-3*β*-ol *see* cholesterol

cholesterin *see* cholesterol

cholesterol (cholesterin; 5β-cholesten-3β-ol; cholestrin; cholesteryl-alcohol; Hydrocerin; Lanol)

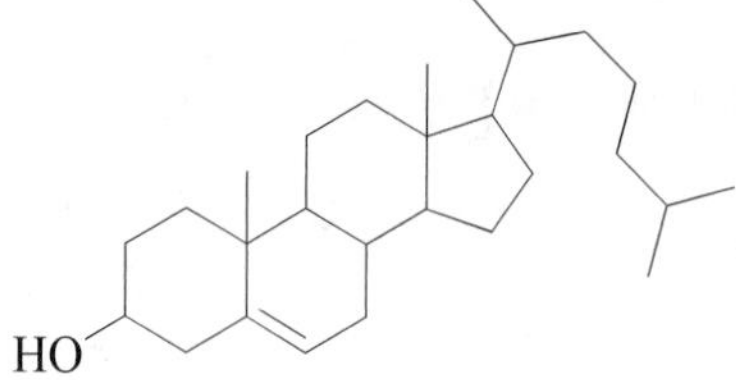

$C_{27}H_{46}O$

CAS 57-88-5

USES

Emulsifying agent.

OCCURRENCE

Found in all body tissues especially in the brain, spinal cord and in animal fats, oils and egg yolk.egg yolk, liver, kidneys, etc.

A. PROPERTIES

white or faintly yellow, almost odorless pearly granules or crystals; molecular weight 386.66; melting point 148.5°C; boiling point 360°C decomposes; density 1.067 at 20/4°C; solubility 2 mg/L.

C. WATER AND SOIL POLLUTION FACTORS

Biodegradation

Mycobacterium strain isolated from marine coastal sediment oxidized cholesterol in Tween 80-cholesterol (1000 mg/L) medium. Products included 4-cholesten-3-one, 4-androsten-13,17-dione, 1,4-androstadien-13,17-dione, testosterone and 1-dehydrotestosterone. All cholesterol disappeared in approx. 4 days. (10032)

Pseudomonas sp. isolated from humus soil was shown to oxidize the C-3 and C-6 positions by introduction of a hydroxyl or ketone group. (10033)

BOD_{25}^{35}: 27% ThOD in seawater, inoculum: enrichment cultures of hydrocarbon- oxidizing bacteria

ThOD: 3.12 (521)

Surface water quality

in Delaware River (U.S.A.):	in winter:	5-10 ppb	
	in summer	3-8 ppb	(1051)

Goteborg (Sweden) sew. works 1989-1991: infl.: 5-300 μg/l; effl.: n.d.-0.8 μg/l (sum of isomers). (2787)

Choline chloride (ethanaminiumchloride,2-hydroxy-N,N,N-trimethyl-; trimethyl(2-hydroxyethyl)ammonium chloride)

HO N+ Cl-

$C_5H_{14}ClNO$

CAS 67-48-1

TRADENAMES

Bilineurin chloride; Biocoline; Hepacholine; Hormocline; Lipotril; Luridin chloride; Neocolina; Paresan

USES

food additive; choline chloride has had wide dispersive use as a food additive for animal husbrandry since the early's 1930s. For this application area almost 100% of the produced choline chloride is either premixed as solid and then directly mixed with animal feed or marketed as a fluid compound to the customers and directly released into special mixing apparatus. A very small amount of the choline chloride production is used for formulations in the field of plant growth regulators.

IMPURITIES, ADDITIVES, COMPOSITION

trimethylamine (max 55 mg/L; ethylene glycol (max 500 mg/L)

A. PROPERTIES

molecular weight 139.63; melting point -18°C; vapor pressure 10 hPa at 20°C, 413 hPa at 100°C; density 1.1 at 20°C; log P_{OW} -3.77; white crystalline solid; water solubility 650,000 mg/L; log K_{OC} 0.37 (calculated); H 2.06×10^{-11} Pa m^3/mol at 25°C

B. AIR POLLUTION FACTORS

Odour

amine like

Photodegradation

indirect photolysis: t/2: 6.9-20.7 hours

C. WATER AND SOIL POLLUTION FACTORS

BOD_5	60% ThOD	(6035)

Hydrolysis

Choline chloride is a quaternary ammonium salt and dissociates in water into the corresponding positively charged quaternary hydroxyl alkylammonium ion and the negatively charged chloride ion.

Biodegradation

Choline chloride is readily biodegradable. (10596)

Inoculum	method	concentration	duration	elimination	
Activated sludge	Modified MITI Test	100 mg/L	14 days	93.5% ThOD	(10596)

D. BIOLOGICAL EFFECTS

Bioaccumulation

based on the low Pow value no bioaccumulation is expected.

Toxicity

Micro organisms			
Pseudomonas putida	17h EC_{10}	113 mg/L	
	17h EC_{50}	133 mg/L	
	17h EC_{90}	278 mg/L	(6033, 10596)
Algae			
Scenedesmus subspicatus	72h Ecb+gr50	>500 mg/L	
	72-96h EC_{10}	>500 mg/L	

	72-96h EC_{50}	>500 mg/L	(6034)
Selenastrum capricornutum	96h EC_{50}		
Pseudokirchneriella subcapitata	72h Ecb+gr50	>1,000 mg/L	
	72h NOEC gr	32 mg/L	(10596)
Crustaceae			
Daphnia magna	48h EC_{50}	349; >500 mg/L	
	48h NOEC	125 mg/L	
	21d NOEC	30.2 mg/L	
Daphnia magna Straus	24h EC_0	250 mg/L	
	48h EC_0	125 mg/L	
	24-48h EC_{50}	>500 mg/L	
	24-48h EC_{50}	>500 mg/L	(6033)
Crangon crangon	48h EC_{50}	>1,000 mg/L	(10596)
Fishes			
Leuciscus idus	96h LC_{50}	>10,000 mg/L	(6032)
Limanda limanda	96h LC_{50}	>1,000 mg/L	
Oryzias latipes	96h LC_{50}	>100 mg/L	(10596)
Mammals			
Rat	oral LD_{50}	3,150; >5,000 mg/kg bw	
Mouse	oral LD_{50}	3,900 – 6,000 mg/kg bw	(10596)

chrysene (1,2-benzophenanthrene)

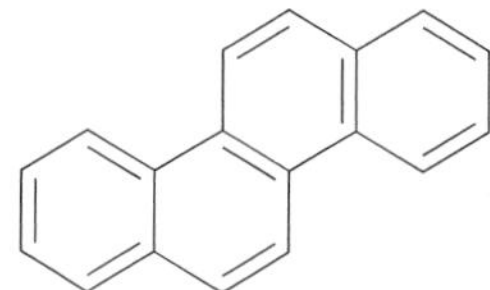

$C_{18}H_{12}$

CAS 218-01-9

MANMADE SOURCES

in gasoline	0.052-2.0 mg/L
in high-octane gasoline	6.7 mg/L (+ cyclopenteno(*cd*)pyrene)
in fresh motor oil	0.56 mg/L (+ cyclopenteno(*cd*)pyrene)
in used motor oil after 5,000 km	86-190 mg/L (+ cyclopenteno(*cd*)pyrene)
in used motor oil after 10,000 km	129-237 mg/L (+ cyclopenteno(*cd*)pyrene)
in motor oils	0.4-2.6 mg/L (n = 6)
in bitumen	1.6-5.1 mg/L
in S. Louisiana crude oil	18 mg/L
in Kuwait crude oil	6.9 mg/L
in no. 2 fuel oil	2.2 mg/L
in bunker C fuel oil	196 mg/L
constituent of coal tar creosote	40,000 mg/L
in coal tar	8,600 mg/L
in carbon black	20 mg/kg
constituent of coals	0.25-1.2 mg/kg
in compost	26 mg/kg dry wt (+ benz(*a*)anthracene)
in horse manure	0.24 mg/kg dry wt (+ benz(*a*)anthracene)

NATURAL SOURCES

coal tar.

A. PROPERTIES

crystals; molecular weight 228.2; melting point 254°C; boiling point 488°C; vapor pressure 6.3 × 10^{-7} mm Hg at 20°C; density 1.27 at 20/4°C; solubility 1.5 µg/L at 15°C, 6 µg/L at 25°C, in seawater 1-50 µg/L, 17 µg/L at 24°C (practical grade), $LogP_{OW}$ 5.61.

B. AIR POLLUTION FACTORS

Manmade sources

in tail gases of gasoline engine	27-318 µg/m^3	(340)
in tail gases of typical European gasoline engine	30-65 µg/L fuel burnt	(1291)
in exhaust condensate of gasoline engine	85-123 µg/L gasoline consumed	(1070)
in cigarette smoke	6 µg/100 cigarettes	(1298)
in emissions from open burning of scrap rubber tires	71-92 mg/kg of tire	(2950)

Ambient air quality

glc's in Budapest 1973			
heavy-traffic area	winter	121 ng/m^3	
(6 a.m.-8 p.m.)	summer	78 ng/m^3	
low-traffic area	winter	30 ng/m^3	
	summer	56 ng/m^3	
glc's in cities in the Netherlands	summer 1970-1971	1-5 ng/m^3	
	winter 1968-1971	4-38 ng/m^3	(1277)
glc's in Birkeness, Norway	Jan.-June 1977	avg 0.81 ng/m^3	
		range n.d.-6.2 ng/m^3	
glc's in Rörvik, Sweden	Dec. 1976-April 1977	avg 1.57 ng/m^3	
(+ triphenylene)		range 0.21-12 ng/m^3	(1236)

comparison of glc's inside and outside houses at suburban sites:

outside	4.6 ng/m^3 or 91 µg/g SPM	
inside	4 ng/m^3 or 72 µg/g SPM	(1234)

Photodecomposition

absorbs solar radiation strongly. (2903)

Incinerability

thermal stability ranking of hazardous organic compounds: rank 10 on a scale of 1 (highest stability) to 320 (lowest stability). (2390)

C. WATER AND SOIL POLLUTION FACTORS

Manmade sources

in suspended fraction of leachates of coal: <0.02-11.2 mg/kg. (2413)

Water and sediment quality

in sediments in Severn estuary (UK): 1.1-5.6 mg/kg dry wt (+ perylene)	(1467)
in sediment of Wilderness Lake, Coline Scott, Ontario (1976): 23 µg/kg dry wt	(932)
river Main at Seligenstadt, Germany: 12-38 ng/L	(530)
in rapid sand filter solids from Lake Constance water: 0.5 mg/kg	
in river Argen water solids: 1.6 mg/kg	(531)

Waste water treatment

Aerobic slurry bioremediation reactor with mixed aeration chamber, treating a petrochemical waste sludge at 3.5 to 7.5% solids containing approx. 25% oil and grease and 0.5 to 2.5 g of waste material per gram of microorganisms, at 22-24°C during batch treatment:

waste residue: infl.: 8.6 mg/kg; effl. after 90 d: <2.5 mg/kg (2800)
Göteborg (Sweden) sew. works 1989-1991: infl.: n.d.; effl.: n.d. (2787)

Degradation in soil

degradation in soil system at initial conc. between 4 and 500 mg/kg soil:
$t_{1/2}$: 5; 10; ∞ d (2903)

degradation in sandy loam in the dark at 20°C at 700 mg/kg:
abiotic $t_{1/2}$: no loss within 16 months
biodegradation $t_{1/2}$: 7 years (2806)

D. BIOLOGICAL EFFECTS

Algae			
Anabaena flos-aquae	2w EC_{35} growth	±0.002 mg/L	
	NOEC growth	±0.001 mg/L	(2917)
Crustaceans			
Daphnia magna	2h LC_{50}	1.9 mg/L	(2917)
Insects			
Aedes aegypti	24h LC_{50}	1.7 mg/L	(2917)

Worms

*Apparent bioconcentration factors in polychaete worms (*Polychaeta *dry wt/sediment* dry wt):

Prionospio cirrifera and *Spiochaetpoterus costarum*	BCF	14.7	
Capitella capitata	BCF	6.2	(2673)
Amphibians			
Rana pipiens	24h LC_{50}	>6.7 mg/L	(2917)
Fishes			
Neanthes arenaceodentata	96h LC_{50}	>1 mg/L	(995)
calculated BCF: 4,230			(2917)

cimetidine

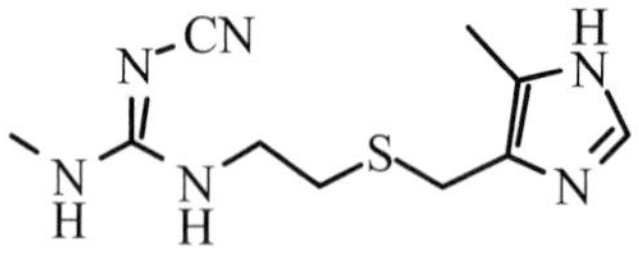

$C_{10}H_{16}N_6S$

CAS 51481-61-9

TRADENAMES

Acibilin; Acinil; Cimal; Cimetag; Cimetum; Edalene; Dyspamet; Eureceptor; Gastromet; Peptol; Stomedine; Tagamet; Tametin; Tratul; Ulcedine; Ulcerfen; Ulcimet; Ulcofalk; Ulcomedina.

USES

Pharmaceutical

A. PROPERTIES

molecular weight 252.35; melting point 141-143°C; water solubility : 11,400 mg/L at 37°C

C. WATER AND SOIL POLLUTION FACTORS

Environmental concentrations

Samples from 139 streams in 30 states in the US thought to be susceptible to contamination from agricultural or urban activities during 1999-2000: detected in 6 out of 88 samples in values ranging from 0.007 – 0.58 µg/L (7221)

D. BIOLOGICAL EFFECTS

Mammals			
Rat	acute oral LD_{50}	2600; 5000 mg/kg bw	(7244)

cineole (eucalyptol; 1,3,3-trimethyl-2-oxabicyclo(2,2,2; 1,8-cineole; 1,8-epoxy-p-menthane; limonene-oxide; Cajeputol; Terpan; Zadoary oil)

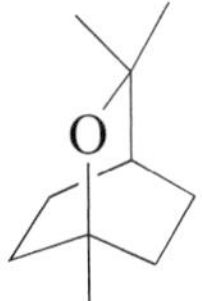

$C_{10}H_{18}O$

CAS 470-82-6

USES

pharmaceuticals (cough syrups, expectorants), flavoring, perfumery.

OCCURRENCE

Major constituent of oil of eucalyptus and oil of cajeput. Constituent of rosemary oil, tea tree oil and lavender oil.

A. PROPERTIES

colorless essential oil; a terpene ether having a camphor-like odor and pungent, cooling, spicy taste; slightly soluble in water; molecular weight 154.25; melting point 1.5°C; boiling point 176-177°C; density 0.92; solubility < 100 mg/L at 20°C.

B. AIR POLLUTION FACTORS

Calculated tropospheric lifetimes

reactant	*lifetime*
OH°	1.4 d
ozone	>110 d

	NO_3	7.8 years	(2451)

C. WATER AND SOIL POLLUTION FACTORS

Biodegradation

Facultatively denitrifying bacteria are able to utilise cineole as sole carbon and energy source. (10034)

D. BIOLOGICAL EFFECTS

Fishes			
Pimephales promelas	96h LC_{50}	102 mg/L	(2625)
Mammals			
rat	oral LD_{50}	2,480 mg/kg bw	(9753)

cinerin I

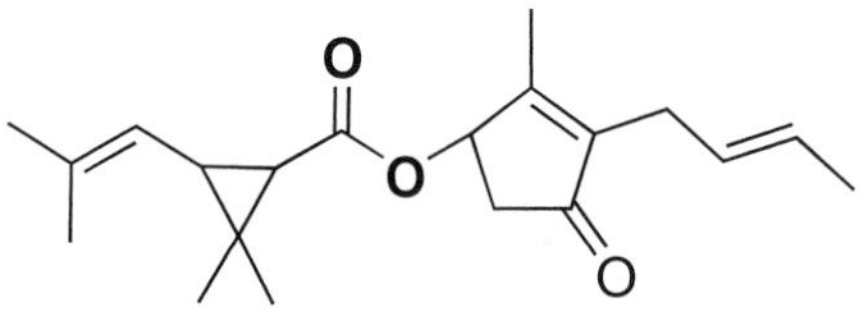

$C_{21}H_{30}O_3$

CAS 25402-06-6

USES

Ingredient of pyrethrins, extract of *Chrystsanthemum* flower (see also pyrethrins)

A. PROPERTIES

molecular weight 316.4; boiling point 136-138°C at 0.008 mm Hg; vapour pressure 1.1 x 10^{-6} mm Hg; water solubility 3.6 mg/L; log Pow 4.8

C. WATER AND SOIL POLLUTION FACTORS

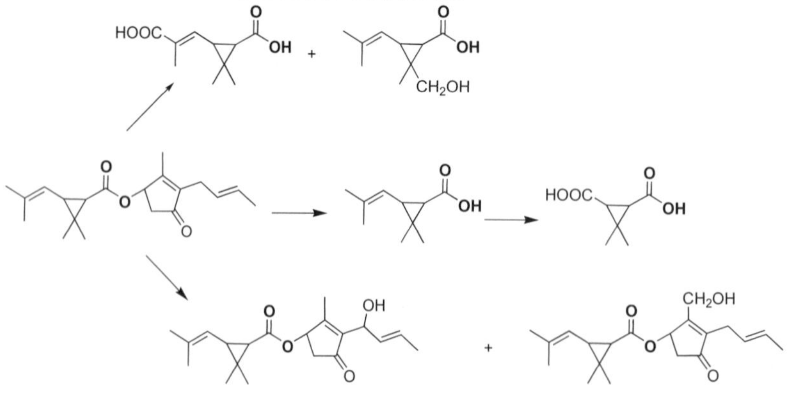

Microbial degradation routes of cinerin I [11188]

Microbial degradation of cinderin I	(11188)

Mobility in soils

Log K_{OC}	4.0	(11188)

D. BIOLOGICAL EFFECTS

Toxicity

Insecta			
Musca domestica	LD_{50}	1.77 µg/fly	(11188)

cinerin II

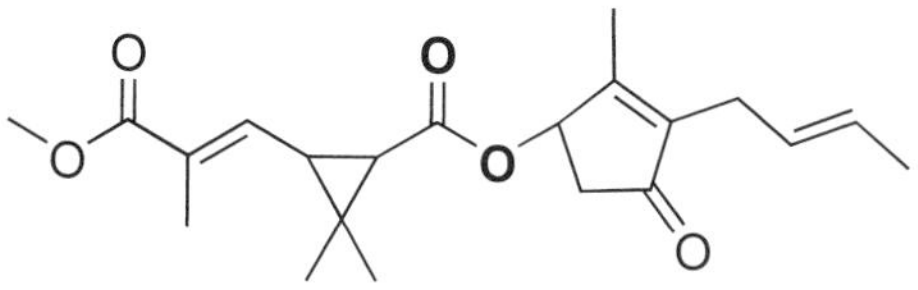

$C_{21}H_{28}O_5$

CAS 121-20-0

USES

Ingredient of pyrethrins, extract of *Chrystsanthemum* flower (see also pyrethrins)

A. PROPERTIES

molecular weight 360.4; boiling point 182-184°C at 0.001 m Hg; vapour pressure 4.6 x 10^{-7} mm Hg; water solubility 1038 mg/L; log Pow 2.7

C. WATER AND SOIL POLLUTION FACTORS

Biodegradation

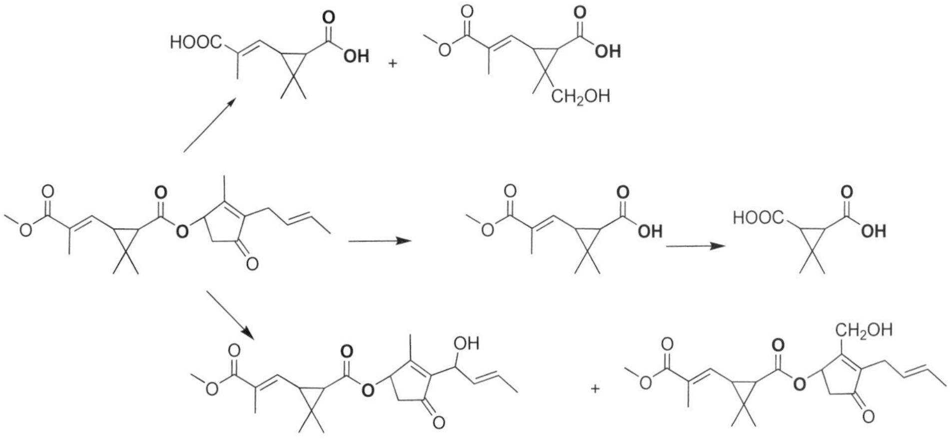

Microbial degradation routes of cinerin II [11188]

Microbial degradation of cinerin II	(11188)

Mobility in soils

Log K_{OC} : 2.9	(11188)

D. BIOLOGICAL EFFECTS

Toxicity

Insecta			
Musca domestica	LD_{50}	0.43 µg/fly	(11188)

Cinidon-ethyl ((Z)-ethyl-2-chloro-3-[2-chloro-5-(cyclohex-1-ene-1,2-dicarboximido)phenyl]acrylate; (Z)-2-chloro-3-[2-chloro-5-(1,3,4,5,6,7-hexahydro-1,3-dioxo-2H-isoindol-2-yl)phenyl]-2-propenoic acid ethyl ester)

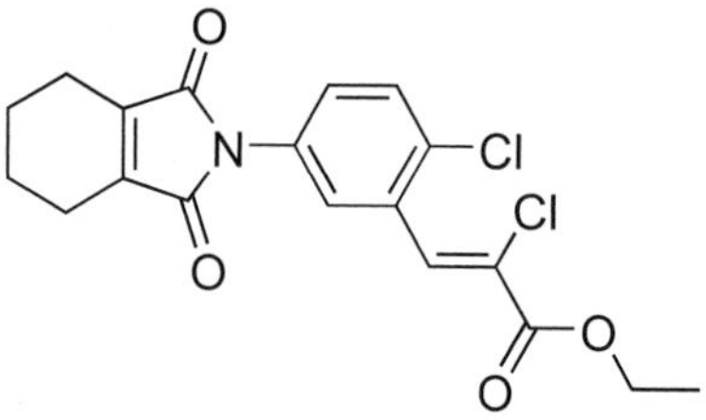

$C_{19}H_{17}Cl_2NO_4$

CAS 142891-20-1

USES

herbicide

A. PROPERTIES

White crystalline solid; molecular weight 394.3; melting point 112.2-112.7°C; relative density 1.4 at 20°C; vapour pressure <1 x 10^{-5} Pa at 20°C; water solubility 0.057 mg./L at 20°C; log Pow 4.5 at 25°C and pH 5.4; H <6.9 x 10^{-2} Pa.m^3/mol at 20°C

B. AIR POLLUTION FACTORS

Photodegradation

T/2 (OH radicals) = <2.5 hours

C. WATER AND SOIL POLLUTION FACTORS

Hydrolysis (10892)

T/2 at 20°C: at pH 5 : 5 days; at pH 7 : 35 hours; at pH 9: 54 minutes
Major metabolites at pH 7: 615M20; 615M07; 615M16; 615M01; 615M03

615M07	Z)-2-chloro-3-[2-chloro-5-(((2-hydroxycarbonyl)cyclohexen-1-yl)carbonylamino)phenyl]-acrylic acid ethyl ester)
615M16	(ethyl (2Z)-3-(3-amino-6-chlorophenyl)-2-chloroprop-2-enoate)
615M20	cyclohexene-1,2-dicarboxylic acid
615M21	cyclohexene-1,2-dicarboxylic anhydride
615M01	(Z)-2-chloro-3-[2-chloro-5-(1,3-dioxo-4,5,6,7-tetrahydroisoindol-2-yl)phenyl]acrylic acid
615M03	(Z)-2-chloro-3-[2-chloro-5-(((2-hydroxycarbonyl)cyclohexen-1-yl)carbonylamino)-phenyl]acrylic acid

615M04	2-N-[3-((1Z)-2-carboxy-2-chlorovinyl)-4-chlorophenyl]carbamoyl-?-hydroxycyclohex-1-enecarboxylic acid. 615M04 is a mixture of structural isomers with different hydroxy substitution positions

Biodegradation

	% mineralisation	% bound-residues	After
in soil aerobic (10892)	11-57%	40-87%	270 days
Major metabolites are 615M01; 615M03; 615M04			
in soil anaerobic	8%	76%	120 days
Major metabolites are 615M01; 615M03; 615M10			

Biodegradation in laboratory studies (10892)

degradation in soils	%	Cinidon-ethyl	615M01	615M03	615M04
Aerobic at 20°C	50	0.6-1.9 days	10-54 days	23-33 days	13 days
Aerobic at 20°C	90	6-22 days	34-180 days	77-110 days	42 days
Aerobic at 10°C	50	2.4 days	40 days	85 days	
Anaerobic at 20°C	50	0.3 days	14 days	18 days	

Biodegradation in water/sediment systems (10892)

DT50 water	1.5-7 hours
DT90 water	2-3 days
DT50 whole system	5 hours
DT90 whole system	2 days

Mobility in soils (10892)

	Cinidon-ethyl	615M01	615M03	615M04
K_{OC}	869-5654	90-435	0->2013	16-28

D. BIOLOGICAL EFFECTS

Bioaccumulation

Species	conc. µg/L	body parts	BCF	
Oncorhynchus mykiss	0.7	Whole fish	24 (parent compound)	
			597-707 (based on radioactivity)	(10892)

Toxicity

Algae			
Pseudokirchneriella subcapitata	ECb_{50}	0.021 mg/L	(10892)
Aquatic plants			
Lemna gibba	EC_{50}	0.174 mg/L	(10892)
Worms			
Eisenia foetida	14d LC_{50}	>1000 mg/kg soil	
	14d NOEC	1000 mg/kg soil	(10892)
Crustaceae			
Daphnia magna	48h EC_{50}	59.2 mg/L	
	48h NOEC	0.11 mg/L	(10892)
Insecta			
Chironomus riparius			
Honeybees	Acute oral LD_{50}	>200 µg/bee	
	Acute contact LD_{50}	>200 µg/bee	(10892)
Fish			
Oncorhynchus mykiss	96h LC_{50}	200 mg/L	
	96h NOEC	1 mg/L	(10892)
Mammals			
Rat	oral LD_{50}	>2,200 mg/kg bw	(10892)

Toxicity of metabolite ***615M10***

Oncorhynchus mykiss	96h LC_{50}	>100 mg/L	
Daphnia magna	48h EC_{50}	>100 mg/L	
Pseudokirchneriella subcapitata	ECb_{50}	>100 mg/L	(10892)

Toxicity of metabolite ***615M01***			
Eisenia foetida	14d LC_{50}	>1000 mg/kg soil	
	14d NOEC	1000 mg/kg soil	(10892)

Toxicity of metabolite ***615M03***			
Eisenia foetida	14d LC_{50}	>1000 mg/kg soil	
	14d NOEC	1000 mg/kg soil	(10892)

cinnamaldehyde (3-phenylpropenal; β-phenylacrolein; cinnamic aldehyde; cinnamylaldehyde; cinnamal)

CHO

$C_6H_5CHCHCHO$

C_9H_8O

CAS 104-55-2 (trans)

USES

Flavour and perfume industries.

OCCURRENCE

Found in Ceylon and Chinese cinnamon oils.

A. PROPERTIES

molecular weight 132.15; melting point -7.5°C; boiling point 251°C; vapor pressure 1 mm at 76.1°C, 40 mm at 152°C; density 1.1 at 15/4°C; $LogP_{OW}$ 1.88.

B. AIR POLLUTION FACTORS

1 mg/m^3 = 0.182 ppm, 1 ppm = 5.439 mg/m^3.

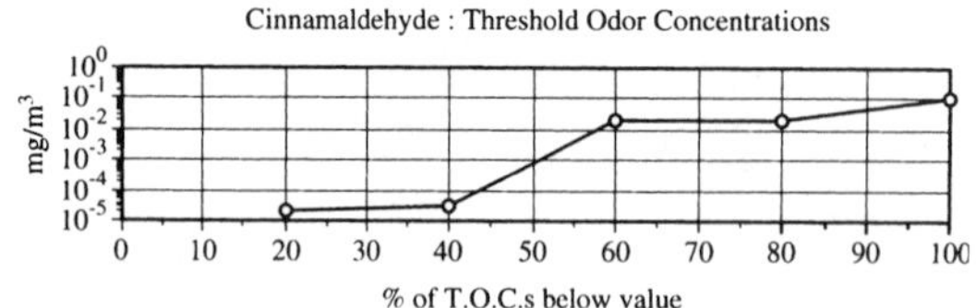

O.I. at 20°C: 53,000 (316)

C. WATER AND SOIL POLLUTION FACTORS

Waste water treatment

RW, Sd, BOD^{20}, 10 d observed, 100 d acclimation: 88% removed. (93)

D. BIOLOGICAL EFFECTS

Mammals			
rat	oral LD_{50}	2,220 mg/kg bw	(9753)
mouse	oral LD_{50}	2,225 mg/kg bw	(10035)
guinea pig	oral LD_{50}	1,160 mg/kg bw	(9753)

cinnamene *see* styrene

cinnamic acid (β-phenylacrylic acid; 3-phenylpropenoic acid; cinnamylic acid)

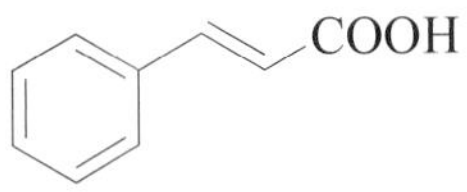

$C_6H_5CHCHCOOH$

$C_9H_8O_2$

CAS 140-10-3

CAS 621-82-9

USES

medicine (anthelmintic), perfumes, intermediate; food additive.

OCCURRENCE

obtained from the styrax tree (Benjamin gum) in Southeast Asia and Sumatra.

A. PROPERTIES

white, crystalline scales; honey floral odor; molecular weight 148.16; melting point 132-135°C; boiling point 300°C; density 0.91; solubility 500 mg/L at 25°C.

C. WATER AND SOIL POLLUTION FACTORS

Biodegradation

half-lives in nonadapted aerobic subsoil: sand Oklahoma: 222 d. (2695)

cinnamic aldehyde *see* cinnamaldehyde

Cinnamidopropyl Trimonium Chloride (Cinnamidopropyl Trimethyl Ammonium Chloride)

$C_{15}H_{23}N_2O.Cl$

TRADENAMES

Incroquat-UV-283

USES

hair conditioning agent in cosmetic preparations up to 4% in the finished product.

A. PROPERTIES

viscous yellow liquid; molecular weight ;> 700,000 mg/L

C. WATER AND SOIL POLLUTION FACTORS

Hydrolysis

significant hydrolysis is unlikely to occur at pH 4 to 9. (7419)

cinnamylalcohol (cinnamic alcohol; 3-phenyl-2-propen-1-ol; phenylallyl alcohol; styrylcarbinol; styrone)

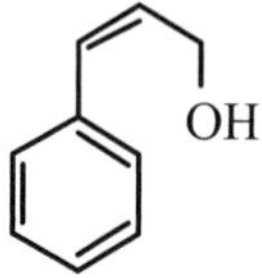

$C_9H_{10}O$

CAS 104-54-1

USES

a fragrance in perfumed consumer products such as dishwash, car shampoo, hand cleaner, toilet paper, nappies in concentrations of 0.002 to 0.007 wt%. (7195)

A. PROPERTIES

pale yellow solid; molecular weight 134.2; melting point 33 °C; density 1.04; $LogP_{OW}$ 1.95; 1.84 (calculated)

cinnamylaldehyde *see* cinnamaldehyde

ciodrin (dimethyl-1-methyl-2-(1-phenylethoxycarbonyl)vinylphosphate; crotoxyphos; α-methylbenzyl 3-hydroxycrotonate, dimethyl phosphate; 2-butenoic acid, 3-[(dimethoxyphosphinyl)oxy]-, 1-phenylethyl ester, (E)-; crotonic acid, 3-hydroxy-,α-methylbenzyl ester, dimethyl phosphate, (E)-; Crotoxypho-; Cyodrin; Volfazol)

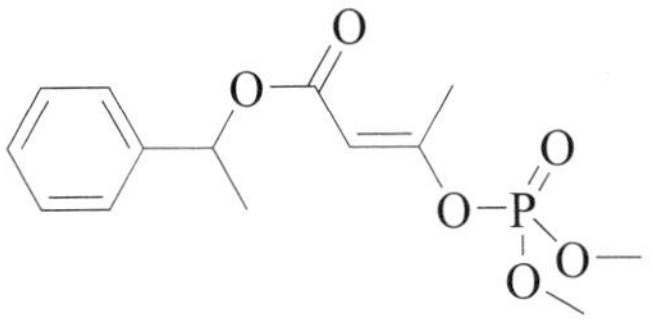

$C_{14}H_{19}O_6P$

CAS 7700-17-6

USES

Livestock insecticide.

A. PROPERTIES

molecular weight 314.28; boiling point 135°C at 0.04 mbar; vapor pressure 1.8 10^{-5}mbar at 20°C; density 1.2 at 15°C; solubility 1,200 mg/L; $\log P_{OW}$ 0.82

C. WATER AND SOIL POLLUTION FACTORS

Hydrolysis

at 38°C and pH 9	$t_{1/2}$:35 hours	
in silty clay loam soil	$t_{1/2}$: 2 hours	
in loamy sand	$t_{1/2}$: 71 hours	(10036)

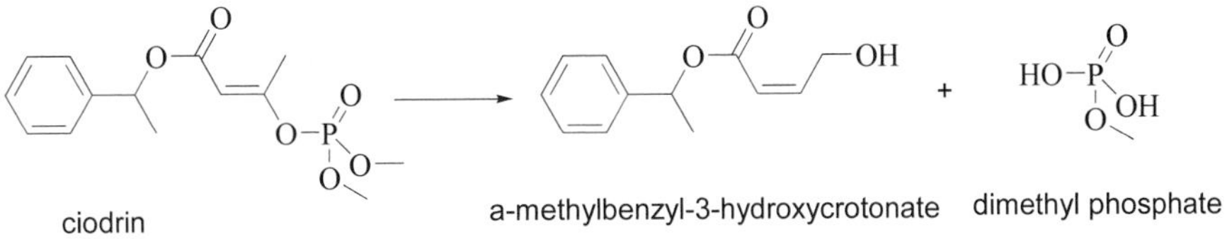

ciodrin → a-methylbenzyl-3-hydroxycrotonate + dimethyl phosphate

An enzyme isolated from clay loam hydrolysed the compound to dimethyl phosphate and α-methylbenzyl 3-hydroxycrotonate in 16 h at 37°C. (10037)

D. BIOLOGICAL EFFECTS

Crustaceans			
Gammarus lacustris	96h LC_{50}	15 µg/L	(2124)
	24h LC_{50}	0.049 mg/L	(9344)
Gammarus fasciatus	96h LC_{50}	11 µg/L	(2126)

Fishes			
Lepomis macrochirus	96h LC_{50}	250 µg/L	(2137)
Micropterus salmoides	96h LC_{50}	1,100 µg/L	(2137)
Salmo gairdneri	96h LC_{50}	55 µg/L	(2137)

Ictalurus punctatus	96h LC_{50}	2,500 µg/L	(2137)
bluegill sunfish	96h LC_{50}	0.15 mg/L	
fathead minnow	96h LC_{50}	12 mg/L	(9344)
Mammals			
rat, mouse	oral LD_{50}	38-40 mg/kg bw	(10038)
male, female rat	oral LD_{50}	74; 110 mg/kg bw	(9827)

CIPC *see* isopropyl-N-(3-chlorophenyl)carbamate

Ciprofloxacin (1-cyclopropyl-6-fluoro-4-oxo-7-piperazin-1-yl-quinoline-3-carboxylic acid)

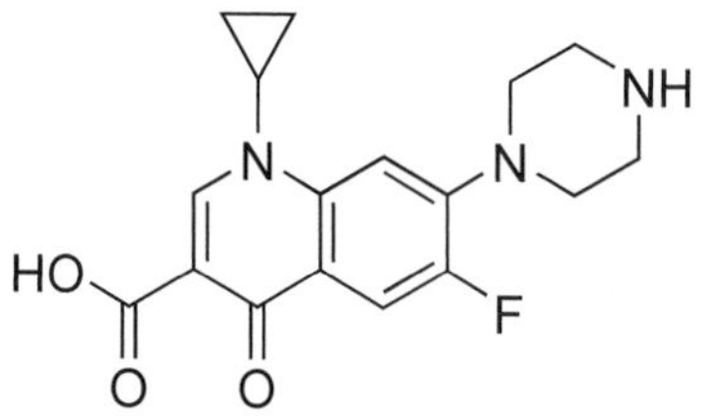

$C_{17}H_{18}FN_3O_3$

CAS 85721-33-1

TRADENAMES

Cipro®,Ciproxin® andCiprobay®

USES

Antibiotic belonging to the fluoroquinolones

A. PROPERTIES

molecular weight 331.35

C. WATER AND SOIL POLLUTION FACTORS

Environmental concentrations

	range µg/L	n	
in effluent hospital	3-87	16	(7237)
	14.5; 2-30	estimated	
influent MWTP	0.6	estimated worst case	(7237)
surface waters (1999-2000)	<0.02-0.03	90	(7221)

D. BIOLOGICAL EFFECTS

Micro organisms			
Pseudomonas putida	16h IC_{50}	0.08 mg/L	(7237)

	16h EC_{100}	0.32 mg/L	
	16h NOEC	0.010 mg/L	
Algae			
Selenastrum capricornutum	72h EC_{50}	2.97 mg/L	(7237)

cis-2-butene (*cis*-β-butylene; dimethylethylene-; pseudobutylene)

$CH_3CHCHCH_3$

C_4H_8

CAS 590-18-1

USES

In the production of gasolines, butadiene and other chemicals.

A. PROPERTIES

molecular weight 56.10; boiling point 4°C; melting point -139°C; boiling point 3.7°C; vapor pressure 760 mm Hg at 37°C; vapor density 1.94; density 0.6 liquefied; $logP_{OW}$ 2.3.

B. AIR POLLUTION FACTORS

1 mg/m^3 = 0.43 ppm, 1 ppm = 2.33 mg/m^3.

Odor threshold

4.8 mg/m^3 (n.s.i.); recognition: 28.5 mg/m^3. (710, 761)

Atmospheric reactions

R.C.R.: 4.83 (49)
Estimated lifetime under photochemical smog conditions in S.E. England: 0.6 hr (1699, 1700)

Atmospheric half-lives

for reactions with OH°: 0.2 days
for reactions with O_3: 0.05 days (2716)

Manmade sources

in gasoline: 0.09-0.35 vol % (312)
evaporation from gasoline fuel tank: 4.2 vol % of total evaporated HCs
evaporation from carburator: 0.2-0.3 vol % of total evaporated HCs (398, 399, 400, 401, 402)

cis-butenedioic acid *see* maleic acid

cis-butenedioic anhydride *see* maleic anhydride

cis-13-docosenoic acid *see* *cis*-erucic acid

cis-erucic acid (*cis*-13-docosenoic acid; Z-13-docosenoic acid)

$CH_3(CH_2)_7CHCH(CH_2BH$

$C_{22}H_{42}O_2$

CAS 112-86-7

A. PROPERTIES

colorless needles; molecular weight 338.56; melting point 33.5°C; boiling point 281°C at 33 mm; density 0.86 at 55/4°C.

C. WATER AND SOIL POLLUTION FACTORS

Waste water treatment

A.S.: after	6 h:	4.2% of ThOD	
	12 h:	5.8% of ThOD	
	24 h:	11% of ThOD	(89)

cis-1,2-ethylenedicarboxylic acid *see* maleic acid

Cis-3-hexen-1-ol (leaf alcohol)

OH

$CH_3CH_2CHCHCH_2CH_2OH$

$C_6H_{12}O$

CAS 928-96-1

EINECS 213-192-8

USES

cis-3-Hexenol is used as refreshing top note of delicate floral fragrances, such as Muguet and Lilac. This material finds considerable use in flavors, for instance, Mint and various fruit complexes. Use level : 0.05-5 ppm as consumed.

OCCURRENCE NATURAL

Occurs in leaves of odourous plants, including shrubs and trees. Isolated from Japanese oil of peppermint.
cis-3-Hexenol has been found in many flowers, fruits, and vegetables, for instance, Carnation, Gardenia, Honeysuckle, Hyacinth, Osmanthus, Apple, Apricot, Cherry, Grape, Kiwi, Lemon, Melon, Mint, Orange, Plum, Strawberry, Corn, Pumpkin, Tomato, Tea, etc.

A. PROPERTIES

colorless to pale yellow liquid;molecular weight 100.16; boiling point 156-157°C; density 0.85.

B. AIR POLLUTION FACTORS

Odour

A powerful, fresh and intensely green, grassy odour.

D. BIOLOGICAL EFFECTS

Toxicity

Fishes			
Pimephales promelas	4d LC_{50}	381 mg/L	(2625)
Mammals			
rat, mouse	oral LD_{50}	4,700-7,000 mg/kg bw	(10328, 10329)

cis-3-hexenal

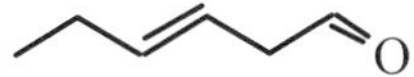

cis-3-hexenal

$C_6H_{10}O$

CAS 6789-80-6

B. AIR POLLUTION FACTORS

Natural sources

emitted from the leaves of *Cupressus sempervirens* : 0.20 % of total terpene emissions which varied between 3-35 μg/g dw.h. The major component was limonene which constituted 83% of total terpene emissions. (7049)

Wounding of plants as a result of cutting, mowing, drying, grinding etc release enzymes which produces a rapid emission of the compounds hexanal, (Z)-3-hexenal and its metabolites including (E)-2-hexenal, hexanol, hexenols and hexenyl acetates (7147)

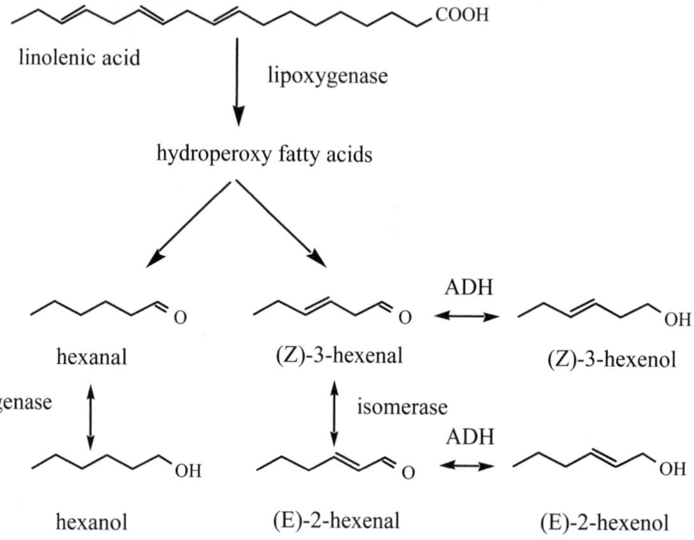

COOH
linolenic acid
lipoxygenase
hydroperoxy fatty acids
ADH
O hexanal
O (Z)-3-hexenal
OH (Z)-3-hexenol
alcoholdehydrogenase (ADH)
isomerase
ADH
OH hexanol
O (E)-2-hexenal
OH (E)-2-hexenol

Wound induced volatile organic carbons emitted by various woody and non-woody plants.

(7147)

cis-nonachlor (1,2,3,4,5,6,7,8,8-nonachloro-3a,4,7,7a-tetrahydro-4,7-methanoindan; <font class="SimpleText"> 1,2,3,4,5,6,7,8,8-nonachloro-2,3,3a,4,7,7a-hexahydro-4,7-methano-1H-indene, (1alpha, 2alpha, 3alpha, 3aalpha, 4beta, 7beta, 7aalpha)-)

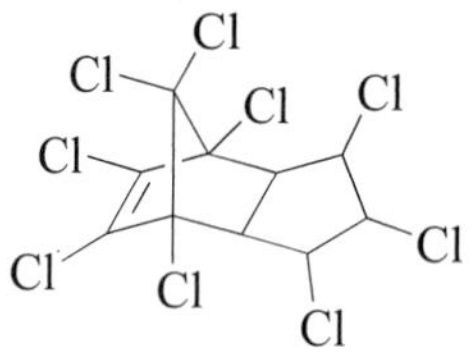

Cl Cl Cl Cl Cl Cl Cl Cl Cl

$C_{10}H_5Cl_9$

CAS 5103-73-1

USES

insecticide.

A. PROPERTIES

molecular weight 444.23

D. BIOLOGICAL EFFECTS

Fishes
goby fish (*Acanthogobius flavimanus*) collected at the seashore of Keihinjima along Tokyo Bay-Aug. 1978: residue level: 8 ppb (1721)

cis-9-octedecenoic acid *see* oleic acid

cis-4-propenylguaiacol (*cis*-isoeugenol; o-methoxy-4-propenylphenol)

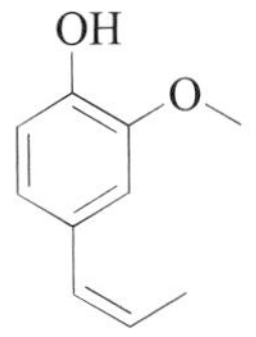

$C_{10}H_{12}O_2$

CAS 5912-86-7

A. PROPERTIES

molecular weight 164.22.

B. AIR POLLUTION FACTORS

avg. conc. in smoke from 28 woodstoves and fireplaces:

burning hardwood:	1.5 g/kg carbon collected on filter	
softwood:	1.0 g/kg carbon collected on filter	(2414)

cis-propenylsyringol

B. AIR POLLUTION FACTORS

avg. conc. in smoke from 28 woodstoves and fireplaces:
burning hardwood: 3.1 g/kg carbon collected on filter

softwood:	0.1 g/kg carbon collected on filter	(2414)

cis-N-((trichloromethyl)thio)-4-cyclohexene-1,2-dicarboximide *see* captan

citral (3,7-dimethyl-2,6-octadienal)

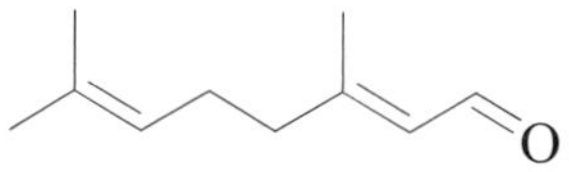

$C_{10}H_{16}O$

CAS 5392-40-5

USES

a fragrance in perfumed consumer products such as dishwash, car shampoo, hand cleaner, toilet paper, nappies in concentrations of up to 0.05 wt%. (7195)

OCCURRENCE

constituent (75-85%) of oil of lemon grass, the volatile oil of the graminae *Cymbopogon citratus* and *Cymbopogon flexuosus*. Also present to a limited extent in oils of verbena, lemon and orange. Found in oils *of Litsea citrata* (90%), *Litsea cubeba Blume* (70%), *Lindera citriodoura* (65%), *Backhousia citriodoura* (95-97%), *Calypranthes particulata* (62%), *Leptospermum liversidgei* (70-90%), *Ocimum gratissimum* (66%).

IMPURITIES, ADDITIVES, COMPOSITION

Citral is a mixture of two geometric isomers, geranial (trans configuration 55-70%) and Neral (cis configuration 35-45%)

A. PROPERTIES

Citral is a pale yellow liquid having a strong lemon like odour.; boiling point 226-228°C;220-240°C decomposes; water solubility 590 mg/L at 25°C;molecular weight 152.24; melting point <-20°C; vapor pressure 0.06 hPa at 20°C; density 0.88-0.9 at 20°C; log P_{oct} 2.8 (measured), 2.7 (calculated)

C. WATER AND SOIL POLLUTION FACTORS

Hydrolysis

t/2 =	pH 4	PH 7	pH 9	
Neral	9.5 days	230 days	30 days	
Geranial	9.8 days	106 days	23 days	(10565)

Citral is hydrolysed into geranic acid (3,7-dimethyl-1,6-octadienoic acid) and 6-methyl-5-heptene-2-on. It is also reported that citral is converted into 2-formylmethyl-2-methyl-5(1-hydroxy-1-methylethyl)-tetrahydrofuran under oxygen atmosphere in aqueous conditions.

H_2O → 3,7-dimethyl-1,6-octadienoic acid + 6-methyl-5-hepten-2-one

Hydrolysis of citral

Biodegradation

Inoculum	method	concentration	duration	elimination	
	MITI	100 mg/L	28 days	88; 93; 94% ThOD 76; 76; 82% TOC 100; 100;100% GC analysis	
	Modified MITI		28 days	70; >70%	(10565)
Activated sludge	MITI	100 mg/L	28 days	88-94% ThOD	(10565)
	Sapromat Test		>70% ThOD		(6436)

D. BIOLOGICAL EFFECTS

Toxicity

Micro organisms			
Domestic activated sludge	30min EC_{50}	ca 200 mg/L	
	30min EC_{20}	ca. 90 mg/L	
	30min EC_{90}	ca. 1000 mg/L	(10565)
A.S. adapted (oxygen consumption inhibition)	30min EC_{20}	>100 mg/L	(6436)
Pseudomonas putida	30min EC_{10}	800 mg/L	(10565)
(oxygen consumption inhibition test)	30min EC_{50}	2,100 mg/L	(10565)
	30min EC_{90}	>10,000 mg/L	(6435)
Algae			
Selenastrum capricornutum	72h ECb_{50}	5.0 mg/L	
	72h NOECb	3.1 mg/L	(10565)
Scenedesmus subspicatus	72h EC_{10}	4.9 mg/L	
(cell multiplication inhibition test)	72h EC_{50}	16 mg/L	(6435)
	96h EC_{10}	1.9 mg/L	
(cell multiplication inhibition test)	96h EC_{50}	19 mg/L	(6435)
Scenedesmus subspicatus	72h EC_{10}	4.9 mg/L	
(cell multiplication inhibition test)	72h EC_{50}	16 mg/L	(6435)
	96h EC_{10}	1.9 mg/L	
(cell multiplication inhibition test)	96h EC_{50}	19 mg/L	(6435)
Crustaceae			
Daphnia magna reproduction	21d EC_{50}	1.6 mg/L	
	21d NOEC	1.0 mg/L	
Daphnia magna Straus	24h EC_0	6.2 mg/L	
	24h EC_{50}	11 mg/L	
	24h EC_{100}	25 mg/L	(6435)
	48h EC_0	3.1 mg/L	
	48h EC_{50}	7 mg/L	
	48h EC_{100}	25 mg/L	(6435)
Fish			
Oryzias latipes	96h LC_{50}	4.1 mg/L	(10565)
Leuciscus idus	96h LC_{50}	4.6-10 mg/L	(6434)
Mammals			
Rat	oral LD_{50}	4,950; 4,960; 6.800 mg/kg bw	

Mouse	Oral LD_{50}	1,440; 6,000 mg/kg bw	(10565)

Mammalian metabolism

Citral is probably metabolized to 1,5-dimethyl-1,5-hexadien-1,6-dicarboxylic acid and 7-carboxy-3-methylocta-6-enoic acid. (10999)

citrazinic acid (2,6-dihydroxyisonicotinic acid)

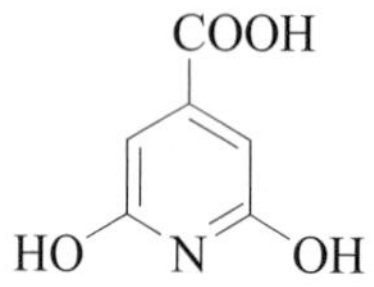

$(HO)_2(C_5H_2N)COOH$

$C_6H_5NO_4$

CAS 99-11-6

USES

competing coupler in color developer solutions.

A. PROPERTIES

molecular weight 155.11; melting point >300°C.

C. WATER AND SOIL POLLUTION FACTORS

Oxidation parameters

theoretical	analytical	
TOD = 1.34	Reflux COD = 100% recovery	
COD = 0.93	TKN = 99.4% recovery	
NOD = 0.41	BOD_5 = 0.086	
	BOD_5/COD = 0.092	
	BOD_5 acclimated = 0.10	(1828)

Impact on conventional biological treatment systems

unacclimated system	NOEC >500 mg/L	
acclimated system	NOEC >1,010 mg/L	(1828)

The compound did not affect the systems, but there was no evidence to indicate that the compound was biodegradable under either condition. (1828)

D. BIOLOGICAL EFFECTS

Algae

Selenastrum capricornutum	NOEC	10 mg/L	
	inhibitory	100 mg/L	(1828)

Crustaceans

Daphnia magna	LC_{50}	32 mg/L	(1828)

Fishes

Pimephales promelas	LC_{50}	>100 mg/L	(1828)

citric acid (2-hydroxy-1,2,3-propanetricarboxylic acid; β-hydroxytricarballylic acid)

HO COOH
HOOC COOH

$COOHCH_2C(OH)(COOH)CH_2COOH$

$C_6H_8O_7$

CAS 77-92-9

USES

preparation of citrates, flavoring extracts, confections, soft drinks, effervescent salts; acidifier; dispersing agent; medicines; acidulant and antioxidant in foods; sequestering agent, water-conditioning agent and detergent builder; cleaning and polishing stainless steel and other metals; alkyd resins; mordant; removal of sulfur dioxide from smelter waste gases, abscission of citrus fruit in harvesting; cultured dairy products.

OCCURRENCE

in lemon, lime, and pineapple juice and molasses.

A. PROPERTIES

molecular weight 192.12; melting point 153°C; $-H_2O$ 70/75°C; boiling point decomposes; density 1.66 at 18/4°C; solubility 1,330,000 mg/L cold; THC 474 kcal/mole; $LogP_{OW}$ -1.72.

C. WATER AND SOIL POLLUTION FACTORS

BOD_{5d}	58; 61% ThOD	(30, 282, 163)
BOD_{20d}	89% ThOD	(30)
ThOD	0.686	(30)

Manmade sources

excreted by humans	in urine 3-17 mg/kg body wt/d in sweat 0.2 mg/100 ml	(203)

Waste water treatment

A.S. after	6 h	1.7% of ThOD	
	12 h	1.0% of ThOD	
	24 h	13% of ThOD	(89)

A.S., Resp., BOD^{20}, 1-15 d observed, feed: 720 mg/L, <1 d acclimation: 30% theoretical oxidation, 98% removed. (93)

Biodegradable dissolved organic carbon test at 21°C in continuously circulating biofilm reactor:

50% removal of DOC at:	225 mg/L DOC: 20 hours 4 mg/L DOC: ±180 hours	(2803)

Impact on biodegradation

at 100 mg/L no inhibition of NH_3 oxidation by *Nitrosomonas* sp. (390)

Methods of analysis: recovery of citric acid by iron coprecipitation:

pH	*% citric acid recovery*[a] *Distilled water*	*Lake Mendota water*
8.0		47
8.5		44
9.0	70	38
9.5	70	33
10.0	75	25

[a]Initial concentrations: 6.5×10^{-7} M of citric acid and 0.01 M $FeCl_3$.

recovery from Lake Mendota water at initial conc. 6.5 × 10⁻⁷ M				
$FeCl_3$, M	*pH 8*	*pH 9*	*pH 10*	
0.005	48	36	20	
0.01	47	38	25	
0.015	49	38		
0.02	49	38	20	(1310)

Automated fluorometric method based on the Furth-Herman reaction, 10 samples per hour without any separation or preconcentration, detection limit: 10 μg/L. (217)

D. BIOLOGICAL EFFECTS

Toxicity threshold (cell multiplication inhibition test)			
bacteria *(Pseudomonas putida)*	16h EC_0	>10,000 mg/L	(1900)
algae *(Microcystis aeruginosa)*	8d EC_0	80 mg/L	(329)
green algae *(Scenedesmus quadricauda)*	7d EC_0	640 mg/L	(1900)
protozoa *(Entosiphon sulcatum)*	72h EC_0	485 mg/L	(1900)
protozoa (*Uronema parduczi* Chatton-Lwoff)	EC_0	622 mg/L	(1901)

Arthropoda			
Daphnia magna	LD_0	80 mg/L, long-time exposure in soft water	(245)
	LD_{100}	120 mg/L long-time exposure in soft water	(245)
Daphnia	toxic	100 mg/L	(30)

Fishes			
goldfish 48-h period of survival	at pH 4.0	894 mg/L	
	at pH 4.5	625 mg/L	(157)
	LD_0	625 mg/L, long-time exposure in hard water	
	LD_{100}	894 mg/L, long-time exposure in hard water	(245)

citronellol (3,7-dimethyl-6-octen-1-ol; cephrol)

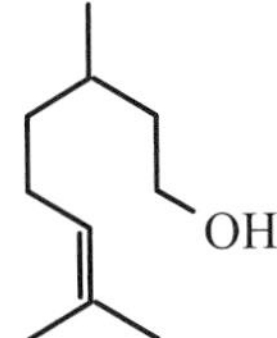

$C_{10}H_{20}O$

CAS 106-22-9

USES

a fragrance in perfumed consumer products such as dishwash, car shampoo, hand cleaner, toilet paper, nappies in concentrations of up to 0.076 wt%. (7195)

A. PROPERTIES

colorless to pale yellow liquid; molecular weight 156.3; boiling point 222 °C; density 0.86

clarithromycin (6-O-methyl-erythromycin; 6-(4-dimethylamino-3-hydroxy- 6-methyl-tetrahydropyran-2-yl) oxy-14-ethyl-12,13-dihydroxy-4-(5-hydroxy-4-methoxy-4,6- dimethyl-tetrahydropyran-2-yl) oxy-7-methoxy-3,5,7,9,11, 13-hexamethyl-1- oxacyclotetradecane-2,10-dione)

cladinose (3-O-methylmycarose) desosamine

$C_{38}H_{69}NO_{13}$

CAS 81103-11-9

TRADENAMES

Biaxin, Klacid, Claripen, Claridar.

USES

Clarithromycin is a macrolide antibiotic

A. PROPERTIES

molecular weight 747.95

C. WATER AND SOIL POLLUTION FACTORS

Environmental concentrations

	range µg/L	n	median µg/L	90 percentile µg/L	
effluent MWTP (1999)	up to 0.24	1			(7237)
surface waters (1999)	up to 0.26	33		0.150	(7237)

Clenbuterol (4-amino-3,5-dichloro-α-[[(1,1-dimethylethyl)amino]-methyl]benzenemethanol)

$C_{12}H_{18}Cl_2N_2O$

CAS 37148-27-9

TRADENAMES

Clenbuterol, Monores

USES

β2-Sympathomimetic, bronchodilator

A. PROPERTIES

molecular weight 277.19

C. WATER AND SOIL POLLUTION FACTORS

Environmental concentrations

	range µg/L	n	median µg/L	90 percentile µg/L	
effluent MWTP (1996-2000)	<0.025-0.181	25	<0.025	0.072	
	<0.050-0.080	29	<0.05	<0.05	(7237)
surface waters (1996-1998)	<0.005	25	<0.005	<0.005	
	<0.010-0.050	45	<0.010	<0.010	(7237)
drinkingwater (1996)	<0.005	16			(7237)

Clodinafop-propargyl (Prop-2-ynyl-(R)-2-[4-(5-chloro-3-fluoro-pyridin-2-yloxy)-phenoxy]propionate; (R)-2-[4-[(5-chloro-3-fluoro-2-pyridinyl)oxy]phenoxy]-propanoic acid-2-propynyl ester)

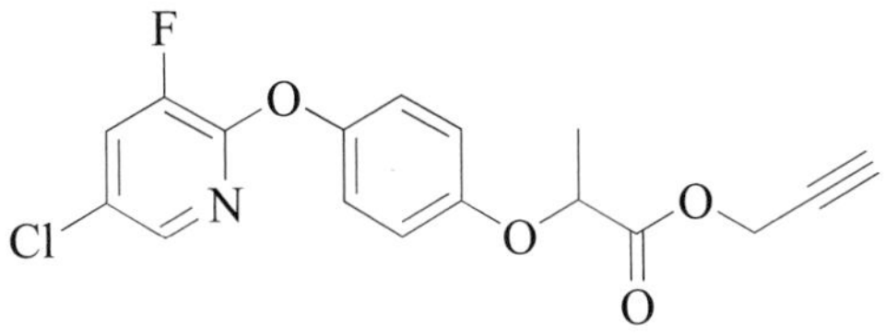

$C_{17}H_{13}ClFNO_4$

CAS 105512-06-9

TRADENAMES

Discover® Herbicide

USES

Herbicide. Clodinafop-propargyl is a member of the Oxyphenoxy acid ester chemical class, which includes the active ingredients fluazifop-butyl, fenoxaprop-ethyl, diclofop methyl, quizalofop-ethyl and haloxyfop-methyl.

A. PROPERTIES

molecular weight 349.8; melting point 59.5°C; themal decomposition starts at about 285°C; density 1.35 kg/L; vapour pressure 3.2 x 10^{-6} Pa at 25°C; water solubility 4.0 mg/L at 25°C; log Pow 3.9 at 25°C; H 2.8 x 10^{-4} Pa.m^3/mol

C. WATER AND SOIL POLLUTION FACTORS

Hydrolysis

25°C	pH 5	T/2 = 27 days
25°C	pH 7	T/2 = 4.8 days
25°C	pH 9	T/2 = 0.07 days

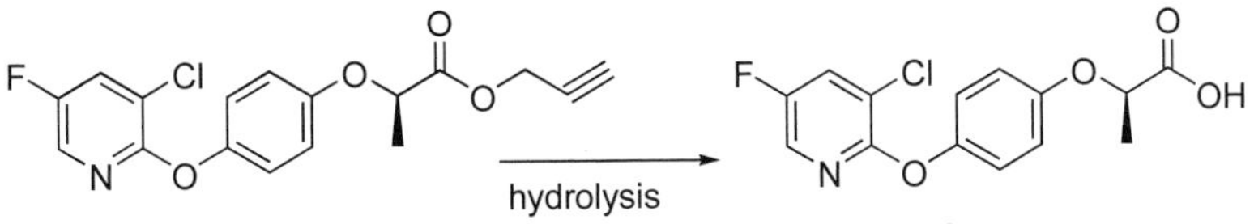

hydrolysis

clodinafop-propargyl

clodinafop
CGA 193469

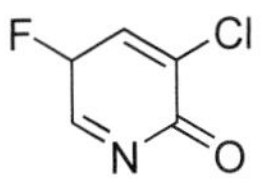

CGA 302371

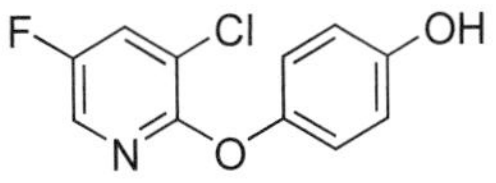

CGA 193468

Metabolites of clodinafop-propargyl

Biodegradation in soil aerobic

% mineralisation	% bound-residues	After
24-42	48-58	84 days

Relevant metabolites: clodinafop (CGA 193469) and CGA 302371
Anaerobic flooded soil: rate of degradation is similar to that in aerobic soil, only metabolite detected clodinafop stabel, limited formation of CO2 and bound residues.(10988)

CGA 302371	2(1H)-Pyridinone, 5-chloro-3-fluoro-
Clodinafop	(R)-2-[4-(5-chloro-3-fluoro-2-pyridyloxy)phenoxy]propionic acid

CGA 193469	
CGA 193468	Phenol, 4-[(5-chloro-3-fluoro-2-pyridinyl)oxy]-

Biodegradation in laboratory studies

degradation in soils	%	Clodinafop propargyl	Clodinafop	CGA 302371
Aerobic at 20°C	50	<0.1-1.5 days	7-13 days	9-12 days
Aerobic at 20°C	90	<0.1-4.9 days	24-61 days	29-41 days
Aerobic at 10°C	50	<0.2-3.3 days	17-40 days	20-27 days
Anaerobic at 20°C	50	0.1 days	stable	

Mobility in soils

	Clodinafop propargyl	Clodinafop	CGA 302371	CGA 193468
K_{OC}	252-2364	25-53	25-82	238-253

D. BIOLOGICAL EFFECTS

Toxicity

Algae			
Scenedesmus subspicatus	72h EC_{50}	>1.6 mg/L	(10988)
Aquatic plants			
Lemna gibba	14d EC_{50}	>1.4 mg/L	(10988)
Crustaceae			
Daphnia magna	21d NOEC	0.23 mg/L	(10988)
Birds			
Bobwhite quail	Acute oral LD_{50}	1363 mg/kg bw	
Mallard duck	Acute oral LD_{50}	>1874 mg/kg bw	(10988)
Mollusca			
Crassostrea virginica	96h EC_{50}	0.77 mg/L	(10988)
Fish			
Oncorhynchus mykiss	21d NOEC	0.10 mg/L	
Lepomis macrochirus	96h LC_{50}	0.21 mg/L	(10988)
Mammals			
Rat	oral LD_{50}	1,392; 2,271 mg/kg bw	(10988)

Toxicity of metabolite ***Clodinafop CGA 193469***

Lepomis macrochirus	96h LC_{50}	>76 mg/L	
Daphnia magna	48h EC_{50}	>10 mg/L	
	22d NOEC	0.16 mg/L	
Mycrocystis aeruginosa	120h EC_{50}	42; 71 mg/L	
Lemna gibba	14d EC_{50}	>4.5 mg/L	(10988)

Toxicity of metabolite ***CGA 193468***

Oncorhynchus mykiss	96h LC_{50}	5.7 mg/L	
Daphnia magna	48h EC_{50}	12 mg/L	
Scenedesmus subspicatus	72h EC_{50}	2.0; 2.4 mg/L	(10988)

Toxicity of metabolite ***CGA 302371***

Oncorhynchus mykiss	96h LC_{50}	>100 mg/L	
Daphnia magna	48h EC_{50}	>100 mg/L	
Selenastrum capricornutum	72h EC_{50}	44; >100 mg/L	(10988)

Clofibrate (2-(4-chlorophenoxy)-2-methylpropanoic acid ethyl ester; clofibric acid ethylester)

Cl

$C_{12}H_{15}ClO_3$

CAS 637-07-0

TRADENAMES

Clofibrate; Bioscleran; Etofylin; Theofibrate; Etofibrate

USES

Lipid regulator.Clofibrate causes a decrease in plasma triacylglycerol levels by increasing the activity of lipoprotein lipase, thereby, increasing the removal of VLDL (very low density lipoprotein) from the plasma.

A. PROPERTIES

colourless oil; molecular weight 242.70; boiling point 148-150 °C @ 20 mm Hg; density 1.14 at 20 °C;

C. WATER AND SOIL POLLUTION FACTORS

Occurrence

Not detected in effluent of municipal water treatment works because of rapid hydrolysis to Clofibric acid upon ingestion. (7178)

Cl —hydrolysis→ Cl OH + CH_3-CH_2OH

Clofibrate Clofibric acid

	range µg/L	n	median µg/L	90 percentile µg/L	
effluent MWTP	<0.10	20	<0.10	<0.10	(7237)
surface waters (1985-1998)	0.040	1			
	<0.050	10			
	<0.030	36			(7237)
groundwater (1997)	<0.0005	3			(7237)
drinkingwater (1996)	<0.0005	3			(7237)
river sediment (1997)	<0.1 µg/kg	10			

D. BIOLOGICAL EFFECTS

Micro organisms			
Vibrio fischeri	30min EC_{10}	14.2 mg/L	
	30min EC_{50}	40.3 mg/L	(7237)
Algae			
Scenedesmus subspicatus	72h EC_{10}	5.4 mg/L	
oo	72h EC_{50}	12 mg/L	(7237)
Crustaceae			
Daphnia magna	21d EC_{10}	0.0084 mg/L	

	21d EC_{50}	0.106 mg/L	
	21d NOEC	0.01 mg/L	
	24h EC_{10}	17.7 mg/L	
	24h EC_{50}	28.2 mg/L	(7237)
Mammals			
Rat	oral LD_{50}	940 mg/kg	(7210)

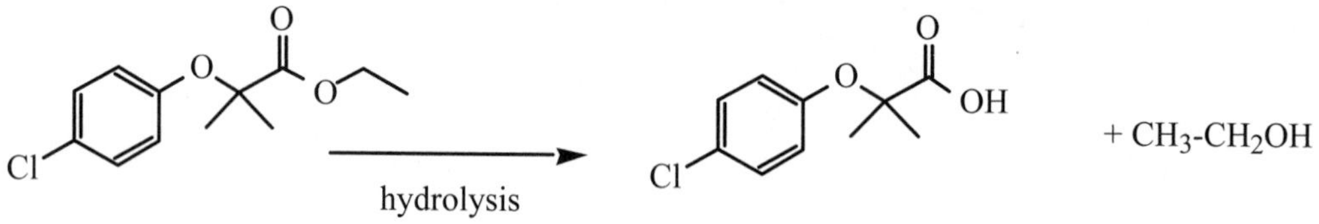

hydrolysis

Clofibrate → Clofibric acid + CH_3-CH_2OH

Metabolism

rapidly hydrolyzed to Clofibric acid upon ingestion which is excreted primarily as glucuronide. (7178)

Clofibric acid (2-(4-chlorophenoxy)-2-methylpropionic acid)

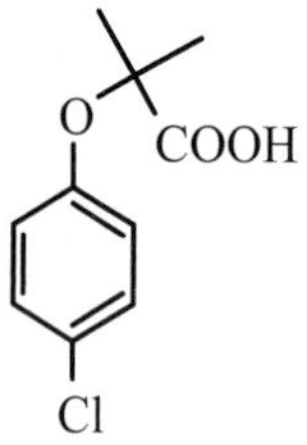

$ClC_6H_4O(CH_3)_2COOH$

$C_{10}H_{11}ClO_3$

CAS 882-09-7

EINECS 212-925-9

TRADENAMES

Atromid-S, Regulipid

USES

Plant growth regulator (antiauxins).Clofibric acid is the intermediate to produce clofibrate, an antihyperlipidemic agent to reduce elevated serum lipids (cholesterol).

A. PROPERTIES

White to off-white crystalline powder; molecular weight 214.65; melting point 120-122°C; vapor pressure 0.00011 mm Hg at 25°C; water solubility 583 mg/L; log Pow 2.57; H 2.19 x 10^{-8}atm.m^3/mol

C. WATER AND SOIL POLLUTION FACTORS

Occurrence

Clofibric acid is the active metabolite of Clofibrate, formed via hydrolysis very soon after ingestion and excreted primarily as glucuronide (very little as free acid). Presence in effluents of sewage works indicates hydrolysis of conjugate.
(7178)

hydrolysis

+ CH_3-CH_2OH

Clofibrate — Clofibric acid

Environmental concentrations

In North sea	1 – 2 ng/L up to 7.8 ng/L	(7178)
In tap water in the city of Berlin, germany	10 – 65 ng/L	(7151)
In tap waters in germany	up to 270 ng/L	(7178)
Swiss rural/urban lakes:	1 – 9 ng/L	
In surface waters	maximum 550 ng/L	(7178)

Influents to sewage works

	removal efficiency	effluents of sewage works
Germany >50 g/day	51 %	max 1600 ng/L
Switzerland 500-1800 ng/L		
Brazil 1000 ng/L	15 - 34 %	
Missouri 800 – 2000 ng/L		800 – 2000 ng/L (2.1 kg/day)
Belgium and The Netherlands 2000		

	ng/L	n	
in effluents of MWTP's	<10-70	2	
in surface waters	<10-30	11	
in drinking water	<10	6	(7237)

	range µg/L	n	median µg/L	90 percentile µg/L	
influent MWTP	0.005-0.014	11			
	1.0	1			(7237)
effluent MWTP (1996-2000)					
	2.54-9.71	7			
	< 0.02-1.6	49	0.36	0.72	
	0.45-0.68	2			
	0.05-1.06				
	0.06-0.42				
	0.46-1.03				
	<0.01-0.072				(7237)
waste waters (1996-1997)	0.46-1.03				
	<0.05-1.56				(7237)
surface waters (1996-2000)	<0.01-0.03	11			
	0.001-0.009	4			
	0.0005-0.0078	6			
	0.027-0.157	?			
	<0.005-0.051	8			
	<0.0005-1.75	?			
	<0.0005-0.22				
	<0.010-0.550	43	0.066	0.210	
	<0.020-0.030	7			
	up to 0.875	27			
	<0.001	10			
	<0.010-0.030	8			(7237)

groundwater (1995-97)	0.001-4.0	1	
	0.070-7.30	17	(7237)
	<0.0005	3	
drinkingwater (1994-97)	<0.001-0.170	14	
	<0.0005	3	
	up to 0.270	48	
	<0.010	6	
	up to 0.165	64	(7237)
river sediment (1997)	<0.1 µg/kg	10	(7237)

Environmental concentrations in Germany 2001 (10923)

	Positive samples	Min ng/L	Max ng/L	Average ng/L	Median ng/L
River Schussen	5/7	<10	24	15	15
River Körsch upstream WWTP	0/8	<10	<10	<10	<10
River Körsch downstream WWTP	8/8	20	346	162	127
Influent WWTP Stuttgart-M	6/6	108	317	223	244
Effluent WWTP Stuttgart-M	7/7	125	351	234	189
Suspended solids influent WWTP Stuttgart-M µg/kg	0/6	n.d.	n.d.	n.d.	n.d.
Influent WWTP Reutlingen-W	5/5	168	634	307	254
Effluent WWTP Reutlingen-W	5/5	40	233	138	128
Suspended solids influent WWTP Reutlingen-W µg/kg	0/5	n.d.	n.d.	n.d.	n.d.
Influent WWTP Steinlach-W	5/5	57	229	108	84
Effluent WWTP Steinlach-W	5/5	61	119	83	79
Suspended solids WWTP Steinlach-W µg/kg	0/5	n.d.	n.d.	n.d.	n.d.
Percolation water municipal dumping ground Reutlingen-S	5/5	2165	4992	3334	2879
Percolation water municipal dumping ground Dusslingen	5/5	1588	5627	3249	2658

D. BIOLOGICAL EFFECTS

Micro organisms			
Vibrio fischeri	30min EC_{50}	100 mg/L	(7237)
Protozoa			
Tetrahymena pyriformis	48h EC_{50}	175 mg/L	(7237)
Algae			
Scenedesmus subspicatus	4d EC_{50}	89 mg/L	(7237)
Crustaceae			
daphnia magna	48h LC_{50}	106 mg/L	(7237)
Fish			
Brachidanio rerio	48h EC_{50}	126 mg/L	(7237)
Worms			
Lumbricus variegatus	96h LC_{50}	>0.4 mg/L	
Insecta			
Chironomus riparius	24h LC_{50}	>0.4 mg/L	

Clomazone (2-(2-chlorobenzyl)-4,4-dimethyl-1,2-oxazolidin-3-one; 2-[(2-chlorophenyl)methyl]-4,4-dimethyl-3-isoxazolidinone)

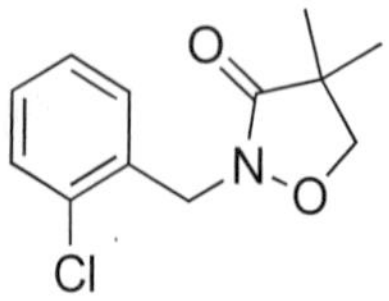

O
N
O
Cl

$C_{12}H_{14}ClNO_2$

CAS 81777-89-1

TRADENAMES

Clomazone

USES

Clomazone is an agrochemical manufactured from OCBC (CAS 611-19-8)

A. PROPERTIES

slight yellow viscous liquid; molecular weight 239.7; boiling point 275.4°C; relative density 1.19; vapour pressure 19.2 mPa at 23°C; water solubility 1100 mg/L; H 4.1 x 10-3 Pa.m^3/mol; LogP_{OW} 2.5; log K_{OC} 2.4

C. WATER AND SOIL POLLUTION FACTORS

Field dissipation

24; 16-36; 28-84 days (10556)

Halflife in soil

in aerobic sandy loam: 36 days; in aerobic silt loam: 87; 137 days (10556)

Biodegradation

OCBC is not formed during any known mechanism of degradation of clomazone in soil. (10551) Under aerobic conditions, clomazone was primarily converted to bound soil residues and CO_2; in flooded soils, the compound was rapidly converted to a reductive metabolite, N-[(2'-chlorophenyl)methyl]-3-hydroxy-2,2-dimethylpropanamide. (10558)

D. BIOLOGICAL EFFECTS

Toxicity

Molluscs			
Crassostrea virginica	96h EC_{50}	5.3 mg/L	(10557)
Algae			
Selenastrum capricornutum	5d EC_{50}	3.5 mg/L	(10557)
Crustaceae			
Americamysis bahia(Opossum shrimp)	96h EC_{50}	0.57 mg/L	
Daphnia magna	48h EC_{50}	5.2 mg/L	(10557)
Fish			
Cyprinodon variegatus	96h LC_{50}	40.6 mg/L	
Lepomis macrochirus	96h LC_{50}	34 mg/L	
Oncorhynchus mykiss	96h LC_{50}	19 mg/L	(10557)
Mammals			
Rat	oral LD_{50}	1360; 1710 mg/kg bw	(10555)

clopidol *see* 3,5-dichloro-2,6-dimethyl-4-pyridinol

clopyralid (3,6-dichoropyridine-2-carboxylic acid; clopyralid; 3,6-dichloro-2-pyridinecarboxylic acid; Cirtoxin; Lontrel; Matrigon)

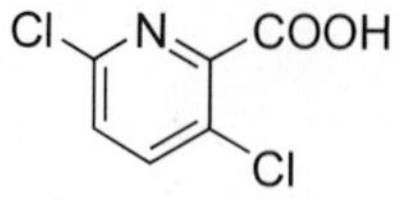

clopyralid

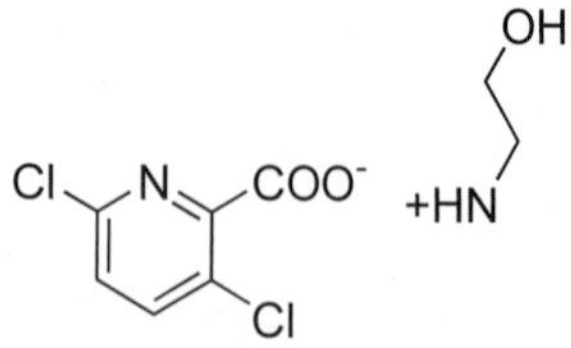

Clopyralid monoethanolamine salt

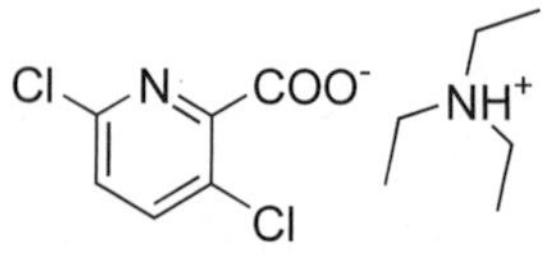

clopyralid triethylamine salt

$C_6H_3Cl_2NO_2$

CAS 1702-17-6

EINECS 216-935-4

TRADENAMES

Clopyralid is sold as an acid, ester, or salt under the trade names Transline, Stinger®, Reclaim®, and Curtail®. Formulations labelled for non-cropland use include Transline® (clopyralid amine salt formulation) and Curtail® (clopyralid amine salt plus 2,4-D amine salt formulation).

USES

Clopyralid is a selective systemic herbicide used to control broadleaf weeds on forests, rights-of-way, rangeland and pastures. Clopyralid, is a chemical that has an auxin-like activity. That is, it disrupts cell growth. Plants affected by clopyralid show twisting of the growing tips that is typical of this class of herbicide. Clopyralid is not metabolised in plants. Tolerant plants just accept the chemical.Systemic post-emergence herbicide

ENVIRONMENTAL FATE

In soil and water, clopyralid is degraded primarily by microbial metabolism. It is resistant to degradation by sunlight, hydrolysis, or other chemical degradation. It is water-soluble, does not bind strongly with soils, and has the potential to be highly mobile in soils, especially sandy soils. Clopyralid is not highly volatile.

A. PROPERTIES

White crystalline solid; molecular weight 191.96; melting point 149.6°C: density 1.76 kg/L at 21°C; vapour pressure 1.36 x 10^{-3} Pa at 25°C;solubility 7850 mg/L at 20°C(99.2% pure compound in distilled water); log P_{ow} -2.6 at pH 7.

pH	Water solubility at 20°C	LogP_{ow} at 20°C
5	118,000 mg/L	-1.81
7	143,000 mg/L	-2.63
9	157,000 mg/L	-2.55

(10989)

C. WATER AND SOIL POLLUTION FACTORS

Biodegradation

under favourable microbiological conditions: $t_{1/2}$: approx. 49 days. (10115)

Persistence

In field dissipation studies: $t_{1/2}$: 8-14 days. (10116)

Hydrolysis

No hydrolysis at pH 4 to 9

Biodegradation in soil aerobic

	% mineralisation	% bound-residues	After
	47-65%	11-35%	92 days

Biodegradation in laboratory studies

degradation in soils	%		
Aerobic at 20°C	50	13-65 days	
Aerobic at 20°C	90	43-217 days	
Aerobic at 10°C	50	73-198 days	
Anaerobic at 20°C	50	> 1 year	(10989)

Mobility in soils

K_{OC}	0.4-12.9	(10989)

D. BIOLOGICAL EFFECTS

Biotransfer factor in beef

log B_b: -5.50. (2644)

Toxicity

Algae			
Selenastrum capricornutum	96h EC_{50}	6.9 mg/L	(10989)
Crustaceae			
Daphnia magna	48h EC_{50}	225 mg/L	(10989)
Insecta			
Chironomus riparius	12h LC_{50}	991 mg/L	
	24h LC_{50}	851 mg/L	
	48h LC_{50}	750 mg/L	(10989)
Birds			
Bobwhite quail	Acute oral LD_{50}	>2,000 mg/kg bw	
Mallard duck	Acute oral LD_{50}	1465 mg/kg bw	(10989)
Fishes			
Oncorhynchus mykiss	96h LC_{50}	103.5 mg/L	
Lepomis macrochirus	96h LC_{50}	125.4 mg/L	(10989)
rainbow trout, bluegill sunfish	96h LC_{50}	103-125 mg/L	(10114, 9662)
Mammals			
Rat	oral LD_{50}	>5,000 mg/kg bw	(10989)
	oral LD_{50}	4,300-5,000 mg/kg bw	(962)

Clothianidin ((*E*)-1-(2-chloro-1,3-thiazol-5-ylmethyl)-3-methyl-2-nitroguanidine; [C(*E*)]-*N*-[(2-chloro-5-thiazolyl)methyl]-*N*'-methyl-*N*''-nitroguanidine)

NH
N NH
NO2 S Cl
N

896 Clothianidin

$C_6H_8ClN_5O_2S$

CAS 210880-92-5

TRADENAMES

Poncho

USES

Nitroguanidine Insecticide.

A. PROPERTIES

Clear and colorless solid powder; molecular weight 249.7; melting point 176.8 °C; vapour pressure 3.8 x 10^{-11} Pa at 20°C; water solubility 327 mg/L at 20°C; log Pow 0.7 at 25°C; H 2.9 x 10^{-11} Pa x m^3/mol

B. AIR POLLUTION FACTORS

Photodegradation

T/2 (OH radicals) = 1.4 hours.

C. WATER AND SOIL POLLUTION FACTORS

Hydrolysis

Nosignificant hydrolysis at temperature of environmental concern.

pH 4	50°C	hydrolytically stable
pH 7	50°C	hydrolytically stable
pH 9	20°C	t/2 = 1401 days
pH 9	50°C	t/2 = 14.4 days

Biodegradation

Inoculum	method	concentration	duration	elimination	
Water/sediment	Aerobic		31-50 days	50%	(10757)

Biodegradation in soil

Clothianidin slowly degraded in the soil under laboratory conditions (in the dark and at 20°C) forming several metabolites, of which N-methyl-N'-nitroguanidine (MNG) and N-(2-chlorothiazol-5-ylmethyl)-N'-nitroguanidine (TZNG) are major metabolites. Other minor degradation compounds are N-(2-chlorothiazol-5-ylmethyl)-N'-methylurea (TZMU) and nitroguanidine (NTG). Besides these metabolites 11 to 17% of the appliedamount was finallly mineralised to CO2 after 120 days and up to 13% of the applied radioactivity was irreversibly bound to the soil (bound residu). Half-lives for 9 soils raanged from 143 tot 1328 days.

The degradation half-lives of the relevant soil metabolites were : 82 to 108 for MNG and 62 to 111 days for TZNG indicating a slight degradability for both metabolites. (10757)

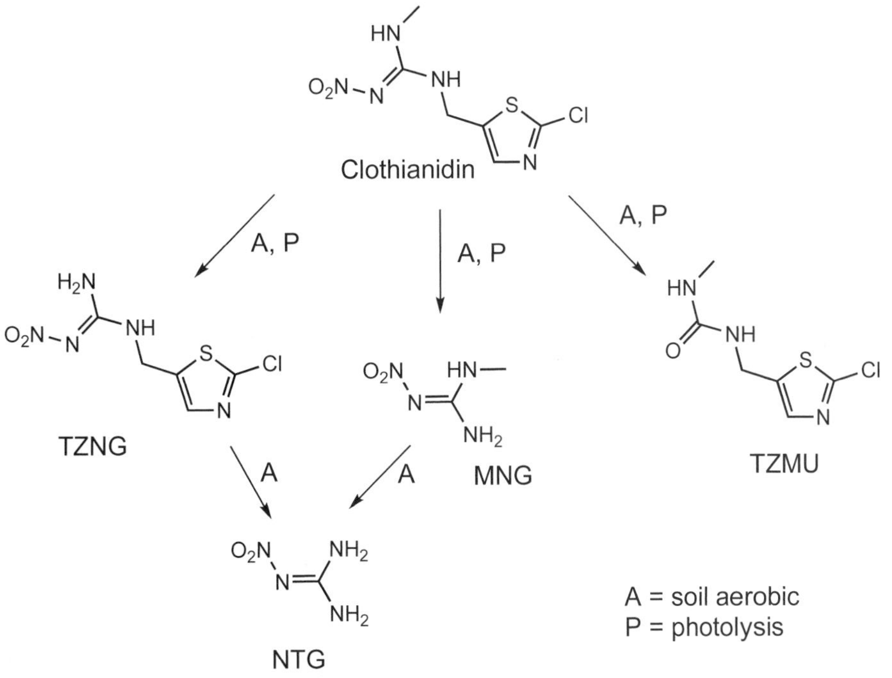

A = soil aerobic
P = photolysis

Proposed metabolic degradation pathway of clothianidin in soil

Aerobic water/sediment systems

The degradation and metabolism of clothianidin was investigated under aerobic laboratory in the dark at 20 °C for 100 days. Clothianidin was moderately degraded, with half-lives of 48 and 65 days for the entire systems. Most of the applied radioactivity was translocated into the sediments. The bound residues in the sediments reached maximum amounts of 30 and 43 %, and the main part of this radioactivity could be assigned to the parent compound and the metabolite TMG (N-(2-Chlorothiazol-5-ylmethyl)-N'-methyl-guanidine). No metabolite exceeded 1.3 % of the applied amount and about 3 to 4 % of the applied amount was mineralised to CO2. (10757)

Anaerobic water/sediment system

The anaerobic metabolism and degradation of clothianidin was investigated using a farm pond system in the dark at 20 °C for 360 days under nitrogen. Clothianidin was degraded rapidly, with a half-life of 21 days calculated for the entire system. Clothianidin is translocated rapidly from the water phase into the sediment, where the major part of the active substance was strongly bound as the study progressed. The amount of the applied radioactivity in the water phase decreased from about 90 % to 1.4 % at the end of the study, while in the sediment the radioactivity increased to more than 90 % within 59 days. After day 90, more than 80 % of the applied radioactivity was still bound to the sediment. In the water phase, only the parent compound was identified. (10757)

Photodegradation on soil surface

The photodegradation of active substances is of relevance in the case of spray applications on crops and/or soil. A photolysis study was performed with radioactively labelled clothianidin applied onto a loamy sand. The samples were exposed to artificial light under laboratory conditions. The active substance was thoroughly decomposed forming a large number of degradation products each of low amounts, e.g. metabolites also detected in the soil degradation study. Besides these metabolites, also mineralisation to CO2 was observed amounting to 5 % of the applied radioactivity. Clothianidin degraded with a half-life of 8.2 days under artificial laboratory conditions, corresponding to 34 days under environmental conditions (sunny days at Phoenix, USA, representing conditions of a desert). It may be concluded that the photolytic degradation of clothianidin will contribute to the dissipation of the substance from the environment.(10757)

Photolytic degradation in the aquatic environment

The photolytic degradation behaviour of clothianidin was investigated in sterile buffer solution adjusted to pH 7 under artificial laboratory conditions as well as in natural water with natural sunlight. The photolytic degradation in sterile buffer solution was very fast. Clothianidin was completely metabolised forming finally formamide and carbon dioxide. The calculated degradation half-life (DT50) was < 4 hours. Additionally, a photolytic degradation study in natural water – river water from USA – was performed under summer sunlight conditions. Again, clothianidin degraded quickly with a half-life of 26.6 days. A great number of degradation compounds were found and identified in both studies, which showed similar metabolisation patterns. Major metabolites under natural conditions were: TMG (*N*-(2-chlorothiazol-5-ylmethyl)-*N'*-methyl-guanidine), MAI (3-methylamino-1*H*-imidazo[1,5-*c*]imidazole), MU (methylurea), MG (methyl guanidine), HMIO (4-hydroxy-2-methylamino-2-imidazoline-5-one), CTCA (2-chlorothiazole-5-carboxylic acid) and urea. In both studies CO2 was the ultimate degradation compound, found at up to 34 % of the applied amount, indicating an extensive photomineralisation of clothianidin.(10757)

Mobility in soil

The mean KOC-value calculated for the adsorption was 160 mL/g, while the mean KOC-value for the desorption was 188 mL/g, indicating that the adsorption is partially irreversible. The KOC-values of clothianidin increased during 99 days ageing by a factor of 2.1 to 3.5, demonstrating the stronger binding of the active substance to the soil when it is adsorbed by the soil under realistic field conditions for a longer time.

The mean KOC-value calculated for the adsorption of MNG was 21 , for TZNG it was 275, for TZMU 62, for TMG 2459 and for TNG 16.(10757)

Degradation of clothianidin in soil under field conditions (first order kinetics) (10757)

Report	Location	Condition	org. C [%]	DT50 [days]	Mean soil tempera-ture [°C]	DT50 [days] at 20 °C	DT90 [days] at 20 °C
Schramel (2000 a)	Burscheid (Germany)	bare soil	0.97	32	8.9	13	44
	Monheim (Germany)		0.89	165	10.6	79	261
	Bury St. Edmunds (UK)		0.86	506	10.5	239	795
	Guiseniers (France)		1.12	240	11.7	125	415
Schramel (2000 b)	Bury St. Edmunds (UK)	cropped	0.86	394	10.5 10.6	186 125	619 416
	Guiseniers (France)		1.12	263			
Schramel (2000 c)	St. Etienne (France)	cropped	0.95	369	17.6 18.5	305 195	1018 645
	Torrebonica (Spain)		0.80	219			
Average DT50/DT90 (geometric)				219		120	399

D. BIOLOGICAL EFFECTS

Toxicity

Algae			
Selenastrum capricornutum	120h EC_{b50}	56 mg a.i./L	
Scenedesmus subspicatus	120h EC_{b50}	228 mg a.i./L	(10758)
Aquatic plants			
Lemna	14d	59 mg a.i./L	(10758)
Birds			
Japanese quails	Acute oral LD_{50}	430 mg a.i./kg bw	
Bobwhite quail	Acute oral LD_{50}	>2,000 mg a.i./kg bw	(10758)
Crustaceae			
Daphnia magna	48h EC_{50}	>120 mg a.i./L	
	21d NOEC	0.12 mg a.i./L	(10758)
Insecta			
Chironomus riparius	48h EC_{50}	0.029 mg a.i./L	
	28d EC_{15}	0.00072 mg a.i./L	(10758)
Fish			
Rainbow trout	96h LC_{50}	>100 mg a.i./L	
Bluegill sunfish	96h LC_{50}	>100 mg a.i./L	
Fathead minnow early life stage	33d NOEC	>20 mg a.i./L	(10758)
Mammals			
Rat	oral LD_{50}	>5,000 mg a.i./kg bw	
mouse	oral LD_{50}	425 mg a.i./kg bw	(10758)

Aquatic toxicity

of the Clothianidin metabolites MNG, TZNG and TMG to aquatic species

Crustaceae		48h LC_{50}
Daphnia	MNG	>100 mg a.i./L
	TZNG	57 mg a.i./L
	TMG	>100 mg a.i./L
Algae		96h LC_{50}
Selenastrum capricornutum	MNG	>100 mg a.i./L
	TZNG	>100 mg a.i./L
	TMG	24.6 mg a.i./L
Insects		48h LC_{50}
Chironomus	MNG	>100 mg a.i./L
	TZNG	0.433 mg a.i./L
	TMG 28d NOEC	>0.1 mg a.i./L
Fish		96h LC_{50}
Rainbow trout	MNG	>100 mg a.i./L
	TZNG	>100 mg a.i./L
	TMG	>100 mg a.i./L (10758)

cloxacillin sodium

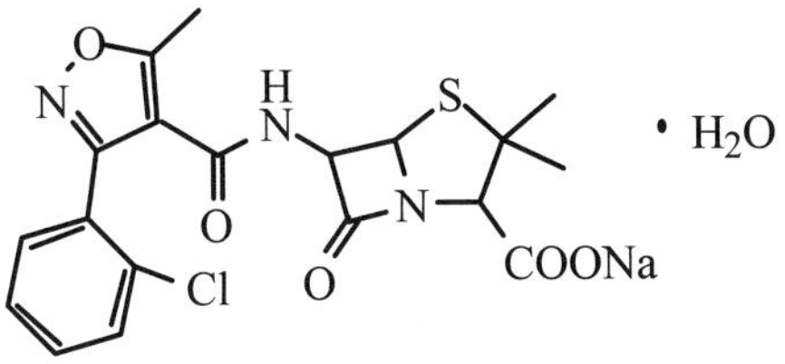

$C_{19}H_{17}ClN_3O_5SNa.H_2O$

CAS 7081-44-9

USES

pharmaceutical; semisynthetic penicillin

A. PROPERTIES

white or almost white crystalline powder; molecular weight 475.9; water solubility freely

C. WATER AND SOIL POLLUTION FACTORS

Environmental concentrations

	range µg/L	n	median µg/L	90 percentile µg/L	
effluent MWTP (1999)	<0.020	4			(7237)
surface waters (1999)	<0.050	14			(7237)

CMME *see* chloromethylmethylether

CMT *see* carboxymethyltartronate

co-op brushkiller 112

USES

herbicide; active ingredient: iso-octylesters of 2,4-D and 2,4,5-T.

C. WATER AND SOIL POLLUTION FACTORS

Persistence

persistence of 2,4-D and 2,4,5-T iso-octylesters in soil (0-10 cm depth).

	mg/kg ester					
application rate:	*7.8 kg/ha*		*15.7 kg/ha*		*31.4 kg/ha*	
residue of ester:	*2,4-D*	*2,4,5-T*	*2, 4-D*	*2,4,5-T*	*2,4-D*	*2,4,5-T*
d after application						
1	n.d.	n.d.	0.01	0.02	0.11	0.30
14	n.d.	n.d.	n.d.	0.01	0.01	0.12
28	n.d.	n.d.	n.d.	n.d.	n.d.	0.03
42	n.d.	n.d.	n.d.	n.d.	n.d.	0.005

(n.d. = <0.005 ppm)

The major residues in the soil following the application of the 2,4-D/2,4,5-T isooctylester formulation were the respective free phenoxy acids. The persistence of these residues is shown in the table below.

	mg/kg acid											
application rate:												
	7.8 kg/ha				*15.7 kg/ha*				*31.4 kg/ha*			
residue of *ester:*	*2,4-D*		*2,4,5-T*		*2,4-D*		*2,4,5-T*		*2,4-D*		*2,4,5-T*	
d after application	*0-10 cm*	*10-20 cm*	*0-10 cm*	*10-20 cm*	*0-10 cm*	*10-20 cm*	*0-10 cm*	*10-20 cm*	*0-10 cm*	*10-20 cm*	*0-10 cm*	*10-20 cm*
1	0.59		0.28		0.82		0.42		3.66		1.16	
14	0.87		1.05		1.47		1.32		12.4		7.1	
28	0.14		0.92		0.28		3.60		0.60		8.9	
42	0.06	0.03	0.26	0.22	0.50	0.14	2.39	0.76	0.39	0.03	10.9	0.22
56	0.01	0.06	0.84	0.18	0.05	0.09	1.20	0.59	0.28	0.06	7.2	0.18
70	0.03	0.02	0.11	0.10	0.11	0.03	0.81	0.19	0.24	0.02	7.5	0.10
265	0.01	n.d.	0.02	0.01	0.05	tr	0.11	0.01	0.1	n.d.	1.2	0.01
385	n.d.	n.d.	tr	tr	n.d.	n.d.	tr	tr	n.d.	n.d.	0.02	n.d.

Coal tar creosote (Creosote oil; coal tar oil; naphthalene oil; tar oil)

CAS 8001-58-9

USES

Coal tar creosote is a wood preservative and water-proofing agent for structures on land and in marine and fresh waters and for railway crossing timbers and sleepers (railroad ties), bridge and pier decking, poles, log homes, fencing, and equipment for childrens's playgrounds. The majority of creosote used in the European Union is for the pressure impregnation of wood. Non-wood uses include anti-fouling applications on concrete marine pilings. Creosote can be a component of roofing pitch, fuel oil, lamp black and a lubricant for die moulds. Other uses reported include animal and bird repellent, insecticide, animal dip and fungicide.

CHEMICAL IDENTITY

A brownish-black/yellowish-dark green oily liquid with a characteristic odour, obtained by the fractional distillation of crude coal tars. The approximate distillation range is 200-400°C. The chemical composition of creosote is influenced by the origin of the coal and also by the nature of the distilling process; as a result, the creosote components are rarely consistent in their type and concentration.

Creosote is a mixture of several hundred chemicals, but only a limited number of them are present in amounts greater than 1%.
There are six major classes of compounds in creosote:
-aromatic hydrocarbons, including polycyclic aromatic hydrocarbons (PAH's) and alkylated PAH's (which can constituted up to 90% of creosote);
-tar acids (1-3%)/phenolics (2-17%);
-tar bases (1-3%)/ nitrogen-containing heterocycles (4.4-8.2%);
-aromatic amines
-sulfur-containing heterocycles (1-3%);
-oxygen-containing heterocycles, including dibenzofurans (5-7.5%.

Creosote used in wood preservation are classified in grades A, B and C with different boiling ranges: Grade A : 200-400°C; Grade B: 235-400°C; Grade C: 300-400°C.(10917)

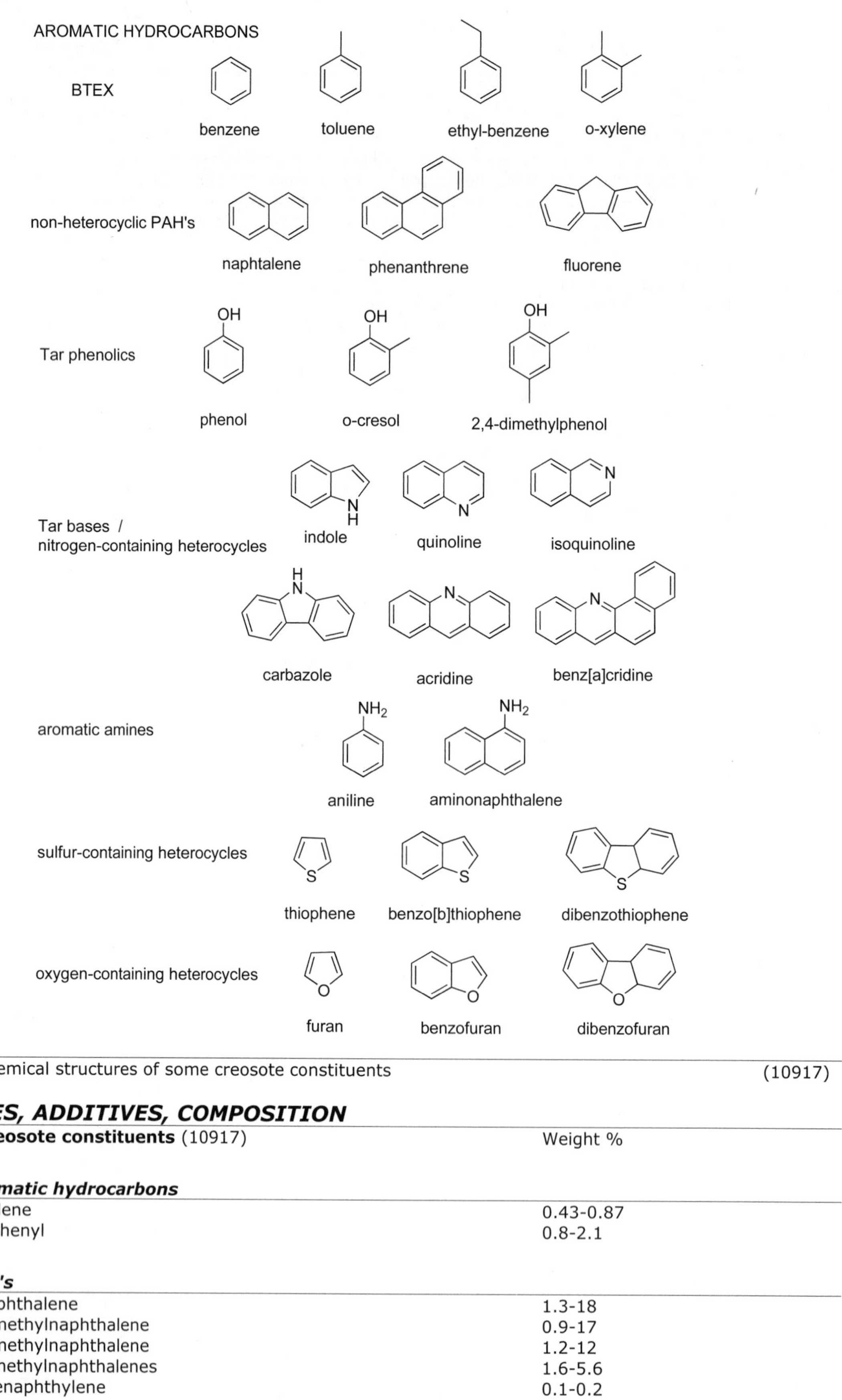

Chemical structures of some creosote constituents (10917)

IMPURITIES, ADDITIVES, COMPOSITION

Creosote constituents (10917)	Weight %
Aromatic hydrocarbons	
Indene	0.43-0.87
Biphenyl	0.8-2.1
Pah's	
Naphthalene	1.3-18
1-methylnaphthalene	0.9-17
2-methylnaphthalene	1.2-12
Dimethylnaphthalenes	1.6-5.6
Acenaphthylene	0.1-0.2

Acenaphthene	3.1-14.7
Fluorene	3.1-10.0
Methylfluorenes	2.3-3.1
Phenanthrene	1-16.9
Methylphenanthrenes	0.45-3.1
Anthracene	0.4-8.2
Methylanthracenes	4.0-5.9
Fluoranthene	0.2-10.0
Pyrene	0.1-8.5
Benzofluorenes	1.0-3.4
Benzo[a]anthracene	0.17-0.5
Benzo[k]fluoranthene	0.16-0.3
Chrysene	<0.05-3.0
Benzo[a]pyrene	0.02-0.43
Benzo[e]pyrene	0.2
Perylene	0.1
Tar acid/phenolics	
Phenol	0.24-0.56
o-cresol	0.1-0.2
m-,p-cresol	0.24-2.3
2,4-dimethylphenol	0.12-0.59
Naphthols	0.12
Tar bases/ nitrogen-containing heterocycles	
Indole	2.0
Quinoline	0.58-2.0
Isoquinoline	0.3-0.7
Benzoquinoline	0.05-4.0
Methylbenzoquinoline	0.3
Carbazole	0.22-3.9
Methylcarbazoles	2.0
Benzocarbazoles	0.1-2.8
Dibenzocarbazoles	3.1
Acridine	0.12-2.0
Aromatic amines	
Aniline	0.05-0.21
Sulfur-containing heterocycles	
Benzothiophene	0.3-0.5
Dibenzothiophene	0.73-1.0
Oxygen-containing heterocycles	
Benzofuran	<0.1
Dibenzofuran	1.1-7.5
Other not specified components	23.1

Concentrations of creosote compounds in old railway sleepers (railroad ties) (10917)

compound	mg/kg shavings Germany (n=5)	Canada (n=27)	
	Range	Range	mean
Acenaphtene	44-973	139-5600	1410
Acenaphthylene		n.d.-42	11
Anthracene		273-5300	1170
Benzo[a]anthracene		167-2110	599
Benzo[a]fluoranthene	22-419		
Benzo[b]fluoranthene	+[j]: 307-2316	82-948	421
Benzo[k]fluoranthene	100-1930	52-811	310
Benzo[ghi]perylene		28-339	142
Benzo[a]pyrene	43.8-1573	86-656	342
Benzo[e]pyrene	30.8-1300		

Cyclopenta[def]phenanthrene	418-3917		
Chrysene	+ trph: 266-12,950	22-2260	681
Dibenzo[a,h]anthracene		n.d.-187	64
Fluoranthene	833-23,067	481-7820	2560
Fluorene	58-1849	178-4910	1420
Indeno[1,2,3-cd]pyrene	322-354	18-389	193
Naphthalene	6.4-392		
Perylene	32-231		
Phenanthrene	+anth: 1005-19,892	654-13,500	3720
phenylnaphthalene	101-2140		
Pyrene	553-11,683	356-5110	1670
Dibenzofuran	23-990		
Dibenzothiophene	22-1420		
Quinoline	7.8-30.5		
Phenols	0.48-37.8		
1-naphthol	0.8-5.1		
4-phenylphenol	0.5-7.7		

A. PROPERTIES

molecular weight variable; boiling range 200-400°C; density 1.0-1.17 kg/L at 25°C; log Pow 1.0

C. WATER AND SOIL POLLUTION FACTORS

Concentrations of PAH's detected at creosote-contaminated sites(10917)

	Groundwater mg/L	Sediment mg/kg dw	Soil mg/kg dw
Acenaphtene	0.8-140	0.02-890	n.d-5300
Acenaphthylene	0.06-15	n.d.-86	<0.05-635
Anthracene	0.07-61	0.1-1650	<0.05-5400
Anthraquinone	4-19		
Benzo[a]anthracene	0.28-39	0.02-140	5.3-1200
Benzo[a]fluorene		0.29-120	
Benzo[b]fluorene	4.9-36	0.28-120	2.2-500
Benzofluoranthenes		0.23-2280	
Benzo[b]fluoranthene	0.12-11.9	n.d.-120	
Benzo[k]fluoranthene		n.d.-69	0.7-550
Benzo[ghi]perylene		0.13-97	0.2-60
Benzo[a]pyrene	0.0003-37.6	0.2-610	<0.05-490
Benzo[e]pyrene		0.08-120	0.94
Biphenyl	0.22-5.7	0.01-0.4	0.79
Chrysene	0.0007-37.6	n.d.-290	n.d.-1119
Dibenzo[a,h]anthracene	<0.0001	n.d.-144	0.3-240
1,4-dimethylnaphthalene			2.5
1,7-dimethylnaphthalene	0.038		1.9
2,3-dimethylnaphthalene	1.8-8.8		
2,6-dimethylnaphthalene		n.d.-438	
2,6-dimethylphenanthrene	6.0-19.4	n.d.-0.6	
Fluoranthene	1.0-230	0.3-6580	<0.05-6600
Fluorene	0.13-141	n.d.-7720	n.d.-5600
9H-fluorenone			28
1-methylfluorene		0.3-75	0.47
Indeno[1,2,3-cd]pyrene	3.2	0.1-453	0.7-2.3
2-methylanthracene	9.7-74.3		
1-methylfluorene	0.35		
Methylnaphthalenes	0.82		
1-methylnaphthalene	0.6-29.1	0.01-1.4	2.1
2-methylnaphthalene	0.56-2.8	0.02-22	3.4-6100
1-methylphenanthrene	11.5	0.01-1.2	
3-methylphenanthrene		2.5	
Naphthalene	1.1-83	0.1-7720	<0.05-23,226
trimethylnaphthalene		n.d.-0.6	
Perylene		0.05-14	
Phenanthrene	0.2-357	0.2-5687	<0.05-16,000
9,10-Phenanthrenedione			17
Pyrene	0.09-171	0.2-1700	n.d.-4600

Concentrations of nitrogen-containing heterocycles at creosote-contaminated sites (10917)

	Groundwater mg/L	Sediment mg/kg dw	Soil mg/kg dw
9-acridinone	0.0005-0.10		
Acridine	0.00001-4.1		
Alkylpyridines, other	0.11		
Alkylquinolines, other	0.086		
Carbazole	0.15-30.4	n.d.-18	
2,4-dimethylpyrimidine	0.008-0.027		
1-hydroxyquinoline	1.1-6.9		
2-hydroxy-4-methylquinoline	0.45-1.1		
2-hydroxyquinoline	0.27-42		
Indole	0.083		
Isoquinoline	0.029-5.4		
Isoquinolinone	4.2		
1-methylpyrrole	n.d.		
2-methylquinoline	0.02-0.3		
4-methylquinoline	0.6-1.6		
Pyrrole	0.0002		
Quinoline	0.04-11.4		
Quinolinone	10		

Concentrations of sulfur-containing heterocycles at creosote-contaminated sites (10917)

	Groundwater mg/L	Sediment mg/kg dw	Soil mg/kg dw
Alkylthiophenes	0.006		
Benzothiophenes	0.1-2.5		
Benzothiophene-2,3-dione	0-0.18		
Dibenzothiophene	0.005-56	0.35	1.18
Dibenzothiophenesulfone	0.0003		
Thiophene	0.009		

Concentrations of oxygen-containing heterocycles at creosote-contaminated sites (10917)

	Groundwater mg/L	Sediment mg/kg dw	Soil mg/kg dw
Benzofuran	0.016		
Dibenzofuran	0.03-84	0.03-6.0	n.d.-4500
Methylbenzofurans	0.011		

Concentrations of monocyclic aromatic compounds at creosote-contaminated sites (10917)

	Groundwater mg/L	Sediment mg/kg dw	Soil mg/kg dw
Benzene	0.03-8.4		
p-dichlorobenzene	0.033		
Ethylbenzene	0.039		
Toluene	0.05-1.2		
1,2,3-trimethylbenzene	0.047		
Xylenes	0.094-1.7		

Concentrations of phenolic compounds at creosote-contaminated sites (10917)

	Groundwater mg/L	Sediment mg/kg dw	Soil mg/kg dw
Cresols	3.2		
o-cresol	0.65-6.6		
m-cresol	0.1-25.2		
p-cresol	n.d.		
2,3-dimethylphenol	0.05-1.1		
2,4-dimethylphenol	0.15-5.7		
2,5-dimethylphenol	0.15-3.0		
2,6-dimethylphenol	0-0.9		
3,4-dimethylphenol	0-2.2		
3,5-dimethylphenol	0.2-9.5		
2-methylphenol	1.3-7.1		

3-methylphenol	13.7
4-methylphenol	3.6-6.2
Naphthol	1.2
Phenol	0.05-11.4
Trimethylphenol	0.02-1.9
Xylenols	0.6-9.4

D. BIOLOGICAL EFFECTS

Crustaceans			
adult lobster *(Homarus americanus)*	at 10°C	96h LC_{50}	1.76 mg/L
larval lobster *(Homarus americanus)*	at 20°C	96h LC_{50}	0.02 mg/L
Crangon	at 10°C	96h LC_{50}	0.13 mg/L
Crangon	at 20°C	96h LC_{50}	0.11 mg/L

The solutions were aerated gently and renewed at 48 h intervals. The concentration of creosote in water decreased exponentially with time according to the equation $C = ae^{-}bt$ (C = relative concentration, t = time in h). The a and b coefficients for creosote at 5 mg/L nominal concentration were 0.511 and -0.022 for the 1-l volume and were 0.841 and -0.067 for 30 l. (1557)

Toxicity

Micro organisms			
Photobacterium phosphorium	15min EC_{50}	0.38; 0.63 mg/L	(10917)
Aquatic plants			
Lemna gibba	EC_{50}	12; 54 mg/L	
Myriophyllum spicatum	EC_{50}	33; 55; 86 mg/L mg/L	(10917)
Crustaceae			
Daphnia pulex L	48h EC_{50}	0.02-4.3 mg/L	
Daphnia magna	48h NOEC	0.52; 3.07 mg/L	
	48h EC_{50}	1.04-4.3 mg/L	
Nitocra spinipes	96h LC_{50}	0.76-1.56 mg/L	
Mysidopsis bahia	7d LC_{50}	0.015-0.021 mg/L	
Penaeus duodarum	96h LC_{50}	0.18-0.34 mg/L	
Homarus americanus	96h LC_{50}	0.02; 1.76 mg/L	
Crangon septemspinosa	96h LC_{50}	0.11; 0.13 mg/L	(10917)
Mollusca			
Crassostrea virginica	96h EC_{50}	0.71; 0.41-1.01 mg/L	(10917)
Fish			
Leuciscus idus	48h LC_{50}	50-100 mg/L	
Lepomis macrochirus	24h LC_{50}	3.7 mg/L	
	96h LC_{50}	0.99 mg/L	
	96h NOEC	0.75 mg/L	
Oncorhynchus mykiss	24h LC_{50}	2.16 ; 4.42 mg/L	
	24h NOEC	0.32 mg/L	
	96h LC_{50}	0.88 mg/L	
	96h NOEC	0.49 mg/L	
Carassius auratus	24h LC_{50}	3.5 mg/L	
	96h LC_{50}	2.6 mg/L	
	96h NOEC	0.25 mg/L	
Cyprinodon variegates	96h LC_{50}	0.72; 3.5 mg/L	
Brachydanio rerio	24h LC_{50}	5.5 mg/L	
	48h LC_{50}	5.2 mg/L	
	96h LC_{50}	4.1 mg/L	
	96h NOEC	2.6; 4.1 mg/L	
Alburnus alburnus	96h LC_{50}	7.9-10.5 mg/L	(10917)
Mammals			
Rat	oral LD_{50}	725; 1700; 2524; 3800;	

		3870; 5430 mg/kg bw	
Mouse	oral LD_{50}	433 mg/kg bw	
sheep	oral LD_{50}	4000 mg/kg bw	
calf	oral LD_{50}	>4000 mg/kg bw	(10917)

coco alkylamines (coconut oil alkylamine; cocoamine)

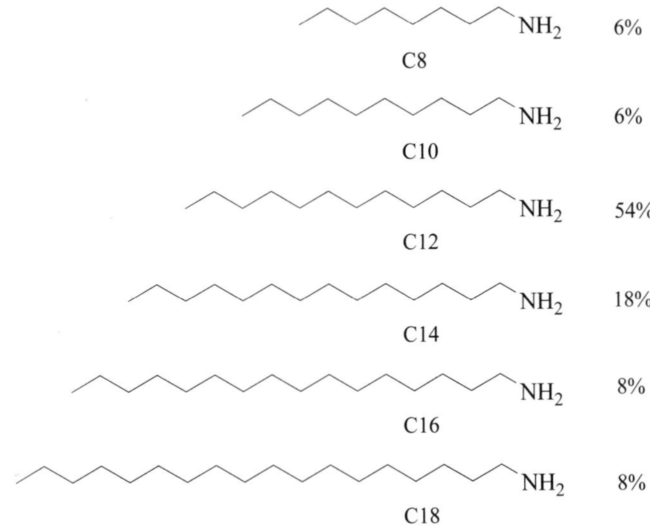

CAS 61788-46-3

USES AND FORMULATIONS

a mixture of alkylamines prepared from coconut oil. A typical composition (in wt.%) is shown above (Genamin CC 100 D). Feedstock for chemical synthesis.

A. PROPERTIES

molecular weight 188-204; melting point 15°C; boiling point 150-340°C; density 0.78 at 60°C; $logP_{OW}$ 5.2 (calculation based on 1-dodecanamine)

B. AIR POLLUTION FACTORS

Photochemical reactions

$t_{1/2}$: 7-10 h, based on reactions with OH° (calculated) (8231)

C. WATER AND SOIL POLLUTION FACTORS

Biodegradation

Inoculum	*method*	*concentration*	*duration*	*elimination results*	
A.S.	Modified MITI Test		4 d	25% COD	(8232)
industrial A.S.	Zahn-Wellens Test	200 mg/L	3 hours	13% DOC	
			10-20 d	0% DOC	
			26 d	50-65% DOC	(8233)
municipal	Closed Bottle test	2 mg/L	5 d	0% ThOD	

effluent					
			15 d	22% ThOD	(8234)
industrial adapted	A.S. Closed Bottle Test	2.5 mg/L	7 d	30% ThOD	
			14 d	30% ThOD	
			21 d	47% ThOD	
			28 d	64% ThOD	(8235)

D. BIOLOGICAL EFFECTS

Bioaccumulation

	conc µg/L	*exposure period*	*BCF*	*organ/tissue*
Fishes	*Lepomis macrochirus*	30 d	98	total body
Poecilia reticulata	48 hours	6918	lipid	

Toxicity

Bacteria

A.S. domestic (respiration inhibition test)	3h EC_{10}	5.5 mg/L	
	3h EC_{20}	7.5 mg/L	
	3h EC_{50}	14 mg/L	
	3h EC_{80}	25 mg/L	(8237)
Anaerobic bacteria from domestic water treatment plant 20h EC_{10} 1,000 mg/L			(8233)

Fishes

Brachydanio rerio	48h LC_0	0.12 mg/L	
	48h LC_{50}	0.3 mg/L	
	48h LC_{100}	0.5 mg/L	(8236)
	96h LC_0	0.12 mg/L	
	96h LC_{50}	0.24 mg/L	
	96h LC_{100}	0.35 mg/L	(8236)
Poecila reticulata	24h LC_{10}	0.2 mg/L	(8232)

cocoamidopropyl betaine (1-propanaminium,3-amino-N-(carboxymethyl)-N,N-dimethyl-,N-cocosacyl-deriv.,hydroxide, internal salt; dimethylcarboxymethyl coconut fatty acid propylamido ammonium betaine; (N-cocoamididopropyl)-N,N-dimethylglycine, hydroxyde, internal salt; N,N-dimethyl-N-(cocosacylamidopropyl)-ammoniumacetobetaine; dimethylcarboxymethyl coconut fatty acid propylamido ammonium betaine)

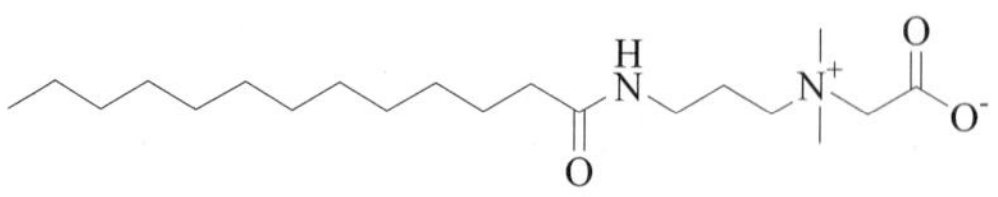

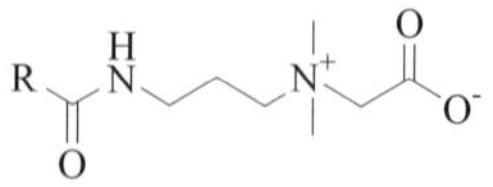

$RCONH(CH_2)_3N^+(CH_3)CH_2COO^-$

CAS 61789-40-0

CAS 4292-10-8

CAS 70851-07-9

EINECS 263-058-8

USES AND FORMULATIONS

an amphoteric surfactant for the manufacture of hair shampoos (15 vol%), foam baths, shower gells (20 vol%) and other body cleansing products (4-10 vol%). The alkylmoiety R consists mainly of C_{12} and C_{14}.The technical grade contains approx. 35% water.

A. PROPERTIES

solid substance; pourpoint 10°C; boiling point 100°C; density 1.05; solubility 350 000 mg/L at 20°C; LogP_{ow}no calculation possible (substance dissociates in aqueous solution and experimental deteremination not feasible due to surfactant properties (enrichment in interface).

B. AIR POLLUTION FACTORS

Photochemical reactions

$t_{1/2}$: 11 d, based on reactions with OH° (calculated) (6100)

C. WATER AND SOIL POLLUTION FACTORS

Biodegradation

Inoculum	*method*	*concentration*	*duration*	*elimination results*	
Municipal effluent	Closed Bottle Test	5 mg/L	30 d	84% COD	(8242)
Municipal effluent	Coupled units test	10 mg/L		97.6% DOC	(8246)
Municipal effluent	Zahn-Welens Test	250 mg/L	7 d	84% COD	
			28 d	99% COD	(8242)
Municipal effluent	Modified OECD Screening Test	10 mg/L	14 d	90% DOC	
			28 d	100% DOC	(8242)
Digester sludge from municipal sewage plant	50 mg/L	42 d	25-61%	$ThCO_2$+CH_4	(8246)

D. BIOLOGICAL EFFECTS

Toxicity

Bacteria			
Pseudomonas putida			
(cell multiplication inhibition test)	16h EC_0	>3,000 mg/L	
(oxygen consumption inhibition)	30min EC_0	>3,000 mg/L	(8245)
Pseudomonas aeruginosa	24h EC_0	>100 mg/L	(8248)
Staphylococcus epidermis	24h EC_{10}	2.5 mg/L	
Bacillus albicans	24h EC_{10}	20 mg/L	
Aspergillus niger	24h EC_{10}	20 mg/L	(8248)
Algae			
Scenedesmus subspicatus	96h EC_0	0.09 mg/L	
	96h EC_{10}	0.14 mg/L	
(cell multiplication inhibition test)	96h EC_{50}	0.55 mg/L	(8244)
Selenastrum capricornutum growth	3w EC_{14}	10 mg/L	(8247)
Crustaceans			
Daphnia magna	48h EC_0	1.6 mg/L	
	48h EC_{50}	6.5 mg/L	

	48h EC_{100}	27 mg/L	(8243)
	21d NOEC	0.9 mg/L	
	21d LOEC	3.6 mg/L	(8243)
Fishes			
Brachydanio rerio	96h LC_0	1.7 mg/L	
	96h LC_{50}	2.0 mg/L	
	96h LC_{100}	2.4 mg/L	(8242)
Leuciscus idus	LC_0	2.5 mg/L	
	LC_{100}	5.0 mg/L	(6000)
Mammals			
Rat: oral LD_{50} >5,000 mg/kg			(6000)
No mutagenicity in Ames Salmonella/microsomes test			(6000)

coconitriles (coconut oil fatty acid nitriles)

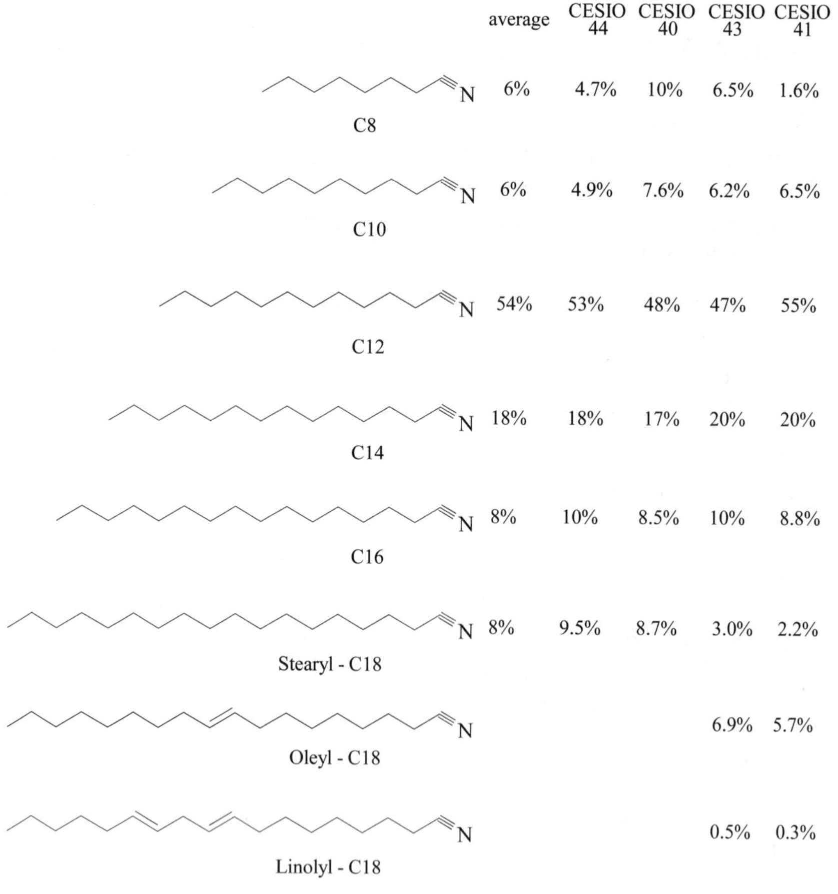

	average	CESIO 44	CESIO 40	CESIO 43	CESIO 41
C8	6%	4.7%	10%	6.5%	1.6%
C10	6%	4.9%	7.6%	6.2%	6.5%
C12	54%	53%	48%	47%	55%
C14	18%	18%	17%	20%	20%
C16	8%	10%	8.5%	10%	8.8%
Stearyl - C18	8%	9.5%	8.7%	3.0%	2.2%
Oleyl - C18				6.9%	5.7%
Linolyl - C18				0.5%	0.3%

CAS 61789-53-5

USES AND FORMULATIONS

intermediate, used in chemical synthesis. A mixture of alkylnitriles derived from coconut oil. Coconitriles is a mixture of C_{8-18} alkylnitriles with approx. 50% dodecanenitrile. A number of typical compositions is presented above. The concentrations of the individual components vary between the following values:

		%
C_8	octanenitrile	1.6-10
C_{10}	decanennitrile	4.9-7.6
C_{12}	dodecanenitrile	47-55
C_{14}	tetradecanenitrile	17-20
C_{16}	hexadecanenitrile	8-10
Stearyl-C_{18}	octadecanenitrile	2.2-9.5
Oleyl-C_{18}	octadecenenitrile	0-6.9
Linolyl-C_{18}	octadecadienenitrile	0-0.5

A. PROPERTIES

melting point -20/4°C; boiling point approx 205°C; density approx. 0.83 at 20°C; $\log P_{OW}$ 5.0/5.12 for CESIO 44;

B. AIR POLLUTION FACTORS

Photochemical reactions

$t_{1/2}$: 2.6-8 d, based on reactions with OH° (calculated for octanenitrile and octadecanenitrile) (8249)

C. WATER AND SOIL POLLUTION FACTORS

Biodegradation

Inoculum	*method*	*concentration*	*duration*	*elimination results*	
	Modified Sturm Test	10 mg/L CESIO 40	28 d	25% $ThCO_2$	
		20 mg/L CESIO 40	28 d	42% $ThCO_2$	(8250)

D. BIOLOGICAL EFFECTS

Bioaccumulation

	conc µg/L	*exposure period*	*BCF*	*organ/tissue*
Fishes				
Lepomis macrochirus		30 d	98	total body
Poecilia reticulata	48 hours	6918	lipid	

Crustaceans				
Daphnia magna	(CESIO 41)	24h EC_{50}	40; 58 mg/L	(8251)
		48h EC_{50}	34; 45 mg/L	
		48h NOEC	7.4; 12 mg/L	(8251)
Fishes				
Brachydanio rerio	(CESIO 43)	96h LC_0	0.9; 5.6 mg/L	
		96h LC_{50}	3.5; 32 mg/L	
		96h LC_{100}	14; 56 mg/L	
		96h NOEC	0.38; 1 mg/L	(8252)
		72h LC_{50}	5.4; 38 mg/L	(8252)
		48h LC_{50}	6.2; 40 mg/L	
		24h LC_{50}	8.1; 45 mg/L	(8252)

Coconut oil (coconut palm oil; cocoanut oil; coconut butter; copra oil; cocos nucifera oil)

6-18
6-18
6-18

CAS 8001-31-8

EINECS 232-282-8

USES

food and foodstuff and personal care additives. Coconut oil is obtained from copra which is dried coconut (Cocos Nucifera) meat. Copra usually contains 65-68% oil. Coconut oil is used as a raw material for the production of C_{8-18} fatty acids and glycerine through splitting at high temperatures (>150°C) and high pressures (>30 kg/cm^2). Coconut oil is also a raw material for the production of C_{8-18} methylesters which are then hydrogenated in an continuous process into fatty alcohols.

IMPURITIES, ADDITIVES, COMPOSITION

a natural mixture of triglycerides with even-numbered, unbranched C-chains. When hydrolysed yealds the follwing range of fatty acids (average)

Common name	Composition	Mean wt %
Caproic acid	C6:0	0.4
Caprylic acid	C8:0	7 - 9
Capric acid	C10:0	6 – 7
Lauric acid	C12:0	49 - 51
Myristic acid	C14:0	18 - 20
Palmitic acid	C16:0	7.5 – 8.5
Stearic acid	C18:0	3
Oleic acid	C18:1	4.5
Linoleic acid	C18:2	0.8
Others	C18:3; C20:0; C24:1	0.3

A. PROPERTIES

melting point 20-27 °C; density 0.92 at 20 °C, 0.90 at 50°C; water solubility poor

C. WATER AND SOIL POLLUTION FACTORS

Environmental fate

Fatty acids can be liberated from the glycerides of fats due to enzymatic activity. Hydrolysis can be catalyzed enzymatically, by moisture, acids basis and heat. Unsaturated fatty acids contained in coconut oil can be subject to autoxidation which involves the formation of a hydroperoxide group adjacent to a double bound. This step proceeds via a free-radical mechanism. The formation of free radicals can be triggered by light. Singlet oxygen, formed under the influence of short-wave radiation and a sensitizer propably plays a key role in this reaction. The intermediate hydroperoxides decompose into a number of different products including epoxides, alcohols, diols, keto compounds, dicarboxylic acids and isomerization products. (7403)

D. BIOLOGICAL EFFECTS

Toxicity

Mammals

Rat	oral LD_{50}	> 5,000; >23,500 mg/kg bw	(7403)

COD *see* 1,5-cyclooctadiene

compound 497 *see* dieldrin

Copper phthalocyanine (copper phthalocyaninato; phthalocyanine copper complex; cyanine blue; phthalocyanine blue)

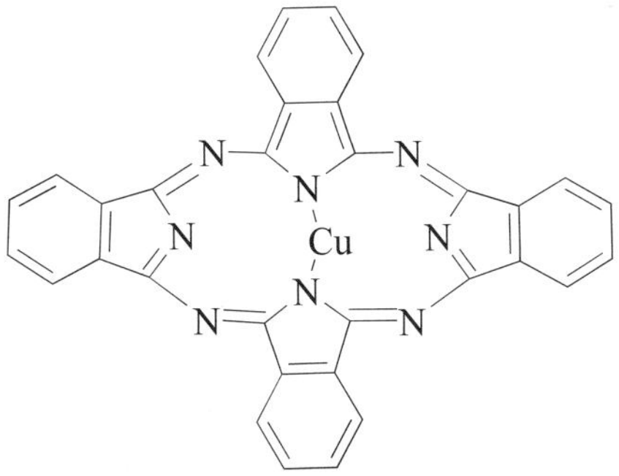

$C_{32}H_{16}CuN_8$

CAS 147-14-8

TRADENAMES

Accosperse cyan blue GT; Bahama blue BC; Blue GLA; Blue pigment; Ceres blue BHR; Chromofine blue 4920; cromofine blue 4950; Dainichi cyanine blue B; Duratint blue 1001; Euvinyl blue 702; Fastolux blue; Fenalac blue B disp; Graphtol blue BL; Helio fast blue B; Hostaperm blue AFN; Irgalite blue LGLD; Irgaplast blue RBP; Aquqline blue; Bermuda blue; Blue toner GTNF; calcotone blue GP; Chromatex blue BN; Cono blue B4; Cromophtal blue 4G; Cyan blue BNC 55-3745; Cyan peacock blue G; Daltolite fast blue B; Fastogen blue 5007; Fastolux peacock blue; Franconia blue A 4431; Helio blue; Isol fast blue B

USES

Pigment in inks, paint and plastics;color additive for coloring polybutester nonabsorbable sutures; dye stuff

A. PROPERTIES

molecular weight 576.08; melting point >250°C decomposes; log P_{ow} 6.6 (calculated); solub < detection limit

C. WATER AND SOIL POLLUTION FACTORS

Impact on biodegradation processes

Pseudomonas respiration test:	30 min EC_{10}	>10 000 mg/L	(7500)
municipal A.S.: inhibition of oxygen consumption	30 min EC_{20}	750 mg/L	(7501)

Biodegradation

Inoculum	method	concentration	duration	elimination	
A.S.	MITI test		14 days	0% ThOD	(10700)
A.S.	modified MITI test	107 mg/L	28 d	<1% ThOD	(7502)

D. BIOLOGICAL EFFECTS

Toxicity

Plants			
Coryza sativa	LC_{50}	>100 mg/L	
Brassica rapa Hikari	LC_{50}	>100 mg/L	
Lettuca sativa	LC_{50}	>100 mg/L	(10700)
Fish			
Oryzias latipes	48h LC_{50}	>100 mg/L	(10700)
Mammals			
Rat	oral LD_{50}	>10,000 mg/kg bw	
Rabbit	oral LD_{50}	>16,000 mg/kg bw	(10700)

Copper pyrithione ((bis(1-hydroxy-1H-pyridine-2-thionato-O,S)copper; Copper-2-pyridinethiol-1-0xide))

$C_{10}H_8N_2O_2S_2Cu$

CAS 14915-37-8

TRADENAMES

Copper Omadine; CleanBio-Cu

USES

Copper pyrithione is a menber of the group of antimicrobial agents classified as cyclic thiohydroxamine acids to be used in antifouling paint on marine vessels.

A. PROPERTIES

molecular weight 315.86

C. WATER AND SOIL POLLUTION FACTORS

Photodegradation

When exposed to sunlight in water : t/2 = 7-8 minutes . In the dark no t/2 could be (10739) determined after 48 hours.

D. BIOLOGICAL EFFECTS

Toxicity

Fish			
Oncorhynchus mykiss	7d LC_{50}	0.0076 mg/L	
	14d LC_{50}	0.0030 mg/L	
	21d LC_{50}	0.0017 mg/L	
	28d LC_{50}	0.0013 mg/L	
suspension cultured fish cell CHSE-sp	24h EC_{50}	0.10 mg/L	(7258)

coprostanol (5β-cholestan-3β-ol; dihydrocholesterol)

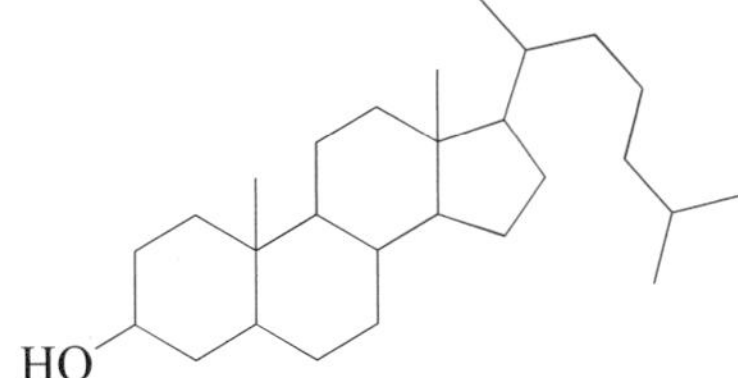

$C_{27}H_{48}O$

CAS 80-97-7

A. PROPERTIES

molecular weight 388.75.

Coprostanol is produced in the intestine of mammals by the microbial reduction of cholesterol, which is the main steroid in the tissue of vertebrates. (1795)

Coprostanol is thus excreted by mammals along with cholesterol and other steroids, although coprostanol is generally the dominant one. (1796)

C. WATER AND SOIL POLLUTION FACTORS

Sediment and water quality

in sediments:			
Veracruz Harbor, Mexico, 1979		0.006-0.44 mg/kg	(1794)
near the city of Mazatlan, Mexico, 1979		0.020-0.20 mg/kg	(1794)
Clyde Estuary near the city of Glasgow, 1977		0.19-14 mg/kg	(1797)
New York, Bight, 1978		0.056-5.2 mg/kg	(1798)
in surface waters:			
Delaware River, U.S.A.	winter	4-9 µg/L	
Delaware River, U.S.A.	summer	1-2 µg/L	(1051)
Old Deerfield River, U.S.A.		1.2 µg/L	
Chicopee River, Ludlow		3.6 µg/L	
Turners Falls		0.9 µg/L	
Connecticut River, Hadley		n.d.	
Connecticut River,		n.d.	(527)

Sunderland		
Mill River, Amherst, U.S.A.	130-180 μg/L	
Mill River, Millers Falls	7 μg/L	
Lake Warner	0.45 μg/L	(527)

Manmade sources

Göteborg (Sweden) sew. works 1989-1991: infl.: 5-30 μg/L; effl.: n.d.-2 μg/L. (2787)

Water treatment methods

Old Deerfield sewage treatment plant: infl.: 170 μg/L; eff.: 4.6-6.8 μg/L. (527)

coronene

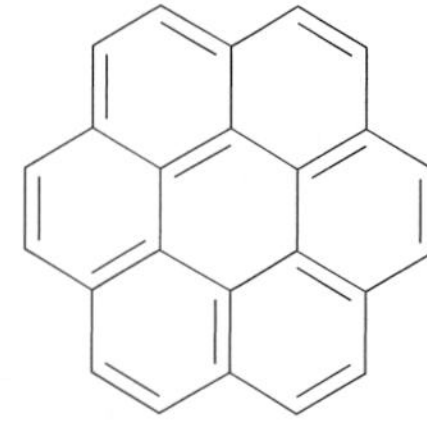

$C_{24}H_{12}$

CAS 191-07-1

MANMADE SOURCES

in gasoline: 630 μg/kg; 165 μg/L; 60-85 μg/kg; 120-1,110 μg/kg; 278 μg/L; 1,900 μg/kg (340, 1070, 385, 380, 1220)

in motor oil:	fresh	0.00 mg/kg	
	after 5,000 km	24-37 mg/kg	
	after 10,000 km	28-63 mg/kg	(1220)

in coal tar: 2.0 g/kg (2600)

A. PROPERTIES

molecular weight 300.36; melting point >360°C; boiling point 525°C.

B. AIR POLLUTION FACTORS

in tail gases of gasoline engine	14-209 μg/m^3	(340)
	1.4-52 μg/L fuel burnt	(1291)
in coke oven emissions	766-865 mg/kg of sample	(960)

emissions from space-heating installation burning:

coal (underfeed stoker)	0.33 mg/10^6 Btu input	
gasoil	2.1 mg/10^6 Btu input	
gas	5.3 mg/10^6 Btu input	(954)

in stack gases of municipal incinerator, after spray tower and electrostatic precipitator: 0.04 mg/1,000 m^3, in residues: <20 μg/kg. (341)

Ambient air quality

glc's in cities in the Netherlands:			
summer 1968-1971	n.d.-2 ng/m^3		
winter 1968-1971	n.d.-4 ng/m^3		(1277)
glc's in Budapest, 1973:			
heavy traffic	winter	0.8 ng/m^3	
	summer	4.7 ng/m^3	
low traffic	winter	1.1 ng/m^3	
	summer	11.1 ng/m^3	(1259)

glc's in the average American urban atmosphere, 1963:

15 μg/g airborne particulates or 2 ng/g m^3 air (1293)

comparison of glc's inside and outside houses at suburban sites:

outside	0.92 ng/m^3 or 21.8 μg/g SPM	
inside	0.43 ng/m^3 or 7.3 μg/g SPM	(1234)

C. WATER AND SOIL POLLUTION FACTORS

Manmade sources

In effluent spray tower of stack gases of municipal incinerator: <0.01 μg/L. (341)

Sediment quality

Wilderness Lake, Colin Scott, Ontario, 1976	6 μg/kg dry wt	(932)
Severn Estuary, UK	traces-1.1 mg/kg dry wt	(1467)

D. BIOLOGICAL EFFECTS

Plants

Carrots *Daucus carota* (UK, 1987):
conc. through a cross section: μg/kg dry wt
peel: 14; inner peel: 3; outer core: 5; core: n.d.
effect of food processing on conc. in whole carrot: μg/kg dry wt
uncooked: 2; cooked: n.d.; tinned: 13; frozen: 9 (2674)

cortison acetate

D. BIOLOGICAL EFFECTS

Algae

Chlorella fusca BCF (wet wt): 40 (2659)

cotinine

MANMADE SOURCES

constituent of environmental tobacco smoke. (2437)

cotinine (1-methyl-5-(3-pyridinyl)2-pyrrolidinone; (2)-1-methyl-5-(3-pyridinyl)-2-pyrrolidinone)

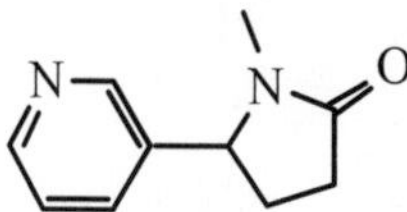

$C_{10}H_{12}N_2O$

CAS 486-56-6

A. PROPERTIES

molecular weight 176.2; melting point 41 °C; boiling point 250°C; vapor pressure 8.6 x 10^{-5} mm Hg at 25°C; water solubility 999,000 mg/L at 25 °C; log Pow 0.07; H 3.33 x 10^{-12} atm-m^3/mole at 25 °C;

B. AIR POLLUTION FACTORS

Occurrence

In tobacco smoke in 30 m^3 teflon chambers from smoking 1 to 4 1R1 Kentucky reference cigarettes

0.6-97; <2-5.2 nmol/m^3 gas-phase sample (7108, 7109)

20±11 μ mol/g particulate sample (7108)

C. WATER AND SOIL POLLUTION FACTORS

Environmental concentrations

Samples from 139 streams in 30 states in the US thought to be susceptible to contamination from agricultural or urban activities during 1999-2000: detected in 15 out of 88 samples in values ranging from 0.004 – 0.90 μg/L (7221)

coumaphos (e coral)

o-coumaric acid lactone *see* coumarin

Coumarin (2H-1-benzopyran-2-one; 1,2-benzopyrone; *o*-coumaric acid lactone; coumarinic lactone; cumarin; tonka bean camphor)

$C_9H_6O_2$

CAS 91-64-5

USES

Flavouring agent. Fixatives and enhancing agent for essential oil odours in perfumes. Tobacco products. Cosmetics. Deodourizing and odour-enhancing agent, pharmaceutical preparations.. (7195)

A fragrance in perfumed consumer products such as dishwash, car shampoo, hand cleaner, toilet paper, nappies in concentrations of 0.003 to 0.03 wt%.

OCCURRENCE

constituent of Angelica, Tonka beans, lavender oil, woodruff (*Asperula spp.*), sweet clover (*Melilotis*), Cassia, Peru Balsam, other plants of Orchidaleae and lavender families, parsnip, fruits, cinnamon oil and other essential oils.

A. PROPERTIES

colorless crystals, flakes, or powder; molecular weight 146.14; melting point 67-68°C; boiling point 290-302°C; vapor pressure 1 mm at 106°C, 40 mm at 189°C; density 0.93 at 20/4°C; solubility 100 mg/L at 25°C, 2,500 mg/L at 20°C; log P_{OW} 1.39.

B. AIR POLLUTION FACTORS

Odor

vanilla, pleasant.

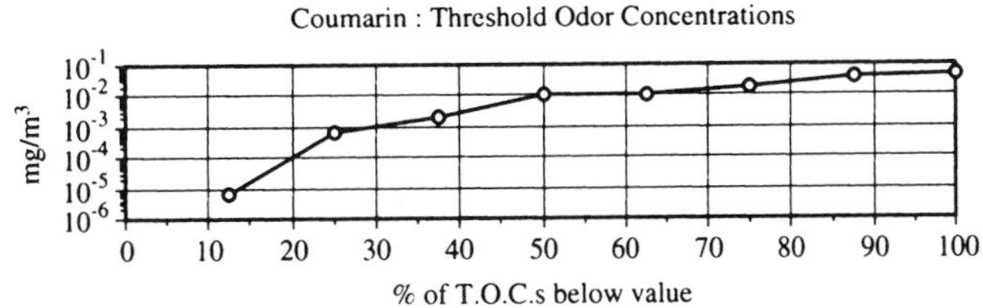

(279, 307, 607, 610, 710, 774, 775, 840)

C. WATER AND SOIL POLLUTION FACTORS

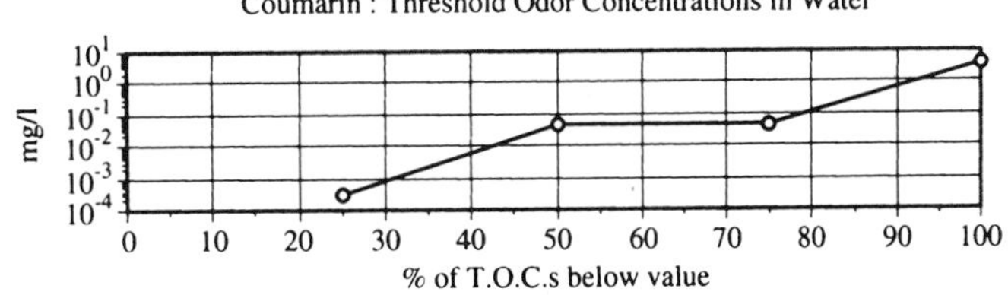

(129, 885, 894)

D. BIOLOGICAL EFFECTS

Bioaccumulation

	Concentration μg/L	*exposure period*	*BCF*	*organ/tissue*	
Algae					
Chlorella fusca			42	Wet weight	(2659)

Bacteria			
Biotox™ test *Photobacterium phosphoreum*	5min EC_{20}	22 mg/L	
	5min EC_{50}	86 mg/L	
Microtox™ test *Photobacterium phosphoreum*	5min EC_{50}	41 mg/L	(7023)
Crustaceans			
Daphnia pulex	48h EC_{50}	30.6 mg/L	(2625)
Fishes			
brown trout, bluegill sunfish, yellow perch, goldfish	24h LC_0	5 mg/L	(9638)
Mammals			
mouse, guinea pig, rat	oral LD_{50}	196-680 mg/kg bw	(9639, 9640)
rat	acute oral LD_{50}	293 mg/kg bw	(7169)

coumarinic lactone *see* coumarin

coumarone *see* 2,3-benzofuran

CPTH *see* 3-chloro-4-methylbenzenaminehydrochloride

m-cresol (*m*-cresylic acid; 3-hydroxytoluene; 3-methylphenol)

C_7H_8O

CAS 108-39-4

MANMADE SOURCES

constitutent of coal tar creosote: 1 wt %. (2386)

A. PROPERTIES

yellowish liquid; molecular weight 108.13; melting point 12°C; boiling point 202°C; vapor pressure 0.04 mm at 20°C, 0.12 mm at 30°C, 5 mm at 76°C; vapor density 3.72; density 1.04 at 20/4°C; solubility 23,500 mg/L at 20°C, 58,000 mg/L at 100°C; THC 880 kcal/mole; saturation concentration in air 0.24 g/m^3 at 20°C, 0.68 g/m^3 at 30°C; log P_{OW} 1.96/2.01.

B. AIR POLLUTION FACTORS

1 mg/m^3 = 0.22 ppm, 1 ppm = 4.50 mg/m^3.

m-Cresol and *p*-Cresol : Threshold Odor Concentrations

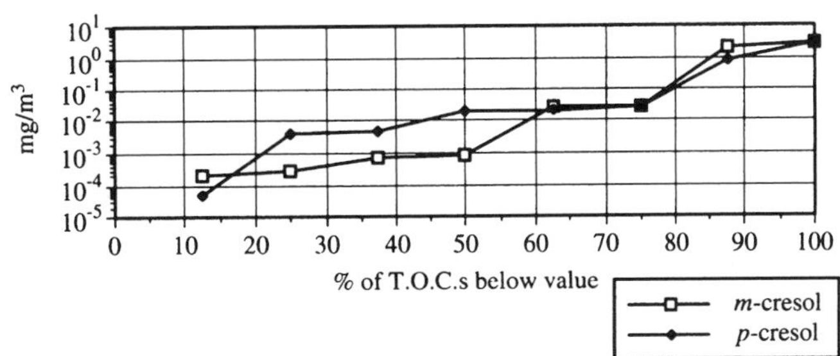

Manmade sources

in exhaust of a 1970 Ford Maverick gasoline engine operated on a chassis dynamometer following the 7-mode California cycle: 0.2-0.4 ppm (1053)

avg conc. in smoke from 28 woodstoves and fireplaces:
burning hardwood: 19.7 g/kg carbon collected on filter
softwood: 15.8 g/kg carbon collected on filter (2414)

Incinerability

thermal stability ranking of hazardous organic compounds: rank 103 on a scale of 1 (highest stability) to 320 (lowest stability). (2390)

C. WATER AND SOIL POLLUTION FACTORS

Impact on biodegradation processes

75% inhibition of nitrification process in nonadapted activated sludge at	11 mg/L	(30)
inhibition of degradation of glucose by *Pseudomonas fluorescens* at	40 mg/L	
inhibition of degradation of glucose by *E. coli* at	600 mg/L	(293)

biodegradation inhib. EC_{50}	500 mg/L	(2624)

Reduction of amenities

causing taste in trout and carp from approx.:	10 mg/L onwards	(41)
causing taste in fish from approx.:	0.2 mg/L onwards	(41)
odor threshold in water:	0.016-4 mg/L; 0.25 mg/L; 0.68 mg/L; 0.8 mg/L (97, 325, 326, 998)	
taste threshold in water	0.002 mg/L	(998)

Biodegradation

decomposition period by a soil microflora: 1 d. (176)

adapted A.S. at 20°C-product is sole carbon source: 95% COD removal at 55 mg COD/g dry inoculum/h. (327)

Activated sludge acclimated to the following aromatics at 250 mg *m*-cresol/L after 30 min:

phenol	37% of ThOD	
o-cresol	41% of ThOD	
m-cresol	38% of ThOD	
p-cresol	2% of ThOD	
mandelic acid	13% of ThOD	(92)

methods	*temp, °C*	*d observed*	*feed, mg/L*	*d acclim.*	*% ThOD*	*% removed*	
RW, CA	20	2	1	nat.		100	
RW, CA	4	7	1	nat.		99	
AS, W	20	0.5	250	25+	38		(93)

W, unadapted A.S. at	1 mg/L	100% removal after 3 hours	
	10 mg/L	100% removal after 3 hours	
	100 mg/L	4% removal after 6 hours	(1639)

Göteborg (Sweden) sew. works 1989-1991: infl.: 0.1-50 µg/L; effl.: n.d. (*m* + *p*). (2787)

Degradation in soil

biodegradation $t_{1/2}$ in nonadapted aerobic subsoil: clay Oklahoma: 0.28; 2.0 d. (2695)

BOD_5	67; 67-75% ThOD acclim. sewage.	(27, 41)
	68% of ThOD.	(274)
$BOD_{24h}30°C$	46% of ThOD (seed water from phenol-degradion plant).	
BOD_{2d}	62% of ThOD (seed water from phenol-degradion plant).	
BOD_{5d}	80% of ThOD (seed water from phenol-degradion plant).	(564)
COD	95% ThOD.	(27, 41)
	100% of ThOD.	(274)
ThOD	2.52	

Biodegradation

Possible reactions for the initial transformation of *m*-cresol under sulfate-reducing conditions by *Desulfotomaculum* sp. *m*-Cresol is hydroxylated on the methyl group to give 3-hydroxybenzyl alcohol, which is oxidized to 3-hydroxybenzaldehyde and then to 3-hydroxybenzoate. 3-Hydroxybenzoate is subsequently dehydroxylated to form benzoate, which is further metabolized (see figure) . (11120)

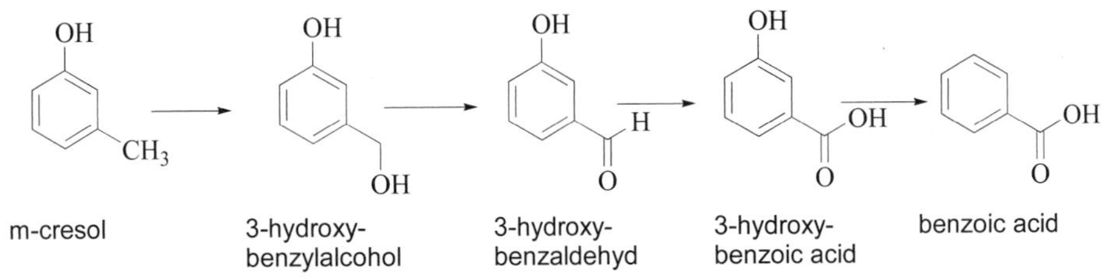

Pathway for the transformation of m-cresol under sulfate-reducing conditions by *Desulfomaculum sp.*	(11120)

D. BIOLOGICAL EFFECTS

Bacteria			
E. coli	LC_0	600 mg/L	
Pseudomonas putida)	16h EC_0	53 mg/L	(1900)
Biotox™ test *Photobacterium phosphoreum*	5min EC_{20}	37 mg/L	
	5min EC_{50}	108 mg/L	
	15min EC_{20}	35 mg/L	
	15min EC_{50}	84 mg/L	
Microtox™ test *Photobacterium phosphoreum*	5min EC_{20}	1.4 mg/L	
	5min EC_{50}	4.8 mg/L	
	15min EC_{20}	1.6 mg/L	
	15min EC_{50}	5.5 mg/L	(7022)
Algae			
Chlorella pyrenoidosa	toxic	148-171 mg/L; 800 mg/L (n.s.i.)	(41)
Scenedesmus	LC_0	40 mg/L	
Microcystis aeruginosa	8d EC_0	13 mg/L	(329)
Scenedesmus quadricauda	7d EC_0	15 mg/L	(1900)
Protozoa			
Vorticella campanula	perturbation level	0.5 mg/L	
Paramecium caudatum	perturbation level	0.9 mg/L	
Tetrahymena pyriformis	24h LC_{100}	378 mg/L	(1662)
Entosiphon sulcatum	72h EC_0	31 mg/L	(1900)
Uronema parduczi Chatton-Lwoff	EC_0	62 mg/L	(1901)
Arthropoda			
Daphnia	LD_0	28 mg/L	
Gammarus pulex	perturbation level	0.7 mg/L	
Mollusc			
Glossosiphonia complanata	perturbation level	1.1 mg/L	(30)
Fishes			
mosquito fish	24-96h LC_{50}	24 mg/L	
bluegill	96h LC_{50}	10-13.6 mg/L	(41)
crucian carp	24h LC_{50}	25 mg/L	
roach	24h LC_{50}	23 mg/L	
tench	24h LC_{50}	21 mg/L	
trout embryos	24h LC_{50}	7 mg/L	(222)

o-cresol (*o*-cresylic acid; 2-hydroxytoluene; 2-methylphenol)

C_7H_8O

CAS 95-48-7

USES AND FORMULATIONS

disinfectant; food antioxidant; perfume mfg.; dye mfg.; plastics and resins mfg.; herbicide mfg. (98%-DNOC, UCPA); tricresylphosphate mfg.; ore flotation, textile scouring agent, organic intermediate; mfg. of salicylaldehyde, coumarin; surfactant; cresylic acid constitutent. (347)

MANUFACTURING SOURCES

coal tar refining; petroleum refining; organic chemical mfg.; wood processing. (347)

SOURCES

automobile exhaust, roadway runoff, runoff from asphalt; general use of plastics, petroleum distillates, fuels, perfumes, oils, lubricants, metal cleaning and scouring compounds; laboratory chemical; constituent of domestic sewage; constituent of coal tar creosote: 1 wt %. (2386, 347)

NATURAL SOURCES (WATER AND AIR)

coal, petroleum, constituent in wood, constituent in natural runoff. (347)

A. PROPERTIES

yellowish liquid; molecular weight 108.13; melting point 31°C; boiling point 191°C; vapor pressure 0.24 mm at 25°C, 5 mm at 64°C, 0.24 hPa at 20°C, 0.55 hPa at 30°C, 2.5 hPa at 50°C; vapor density 3.7; density 1.04; solubility 26,000 mg/L at 25°C, 31,000 mg/L at 40°C, 56,000 mg/L at 100°C; THC 883 kcal/mol, 856 kcal/mol; saturation concentration in air 1.2 g/m^3 at 20°C, 2.8 g/m^3 at 30°C; $LogP_{OW}$ 1.95 (measured); LogH -4.30 at 25°C.

B. AIR POLLUTION FACTORS

1 mg/m^3 = 0.22 ppm, 1 ppm = 4.50 mg/m^3.

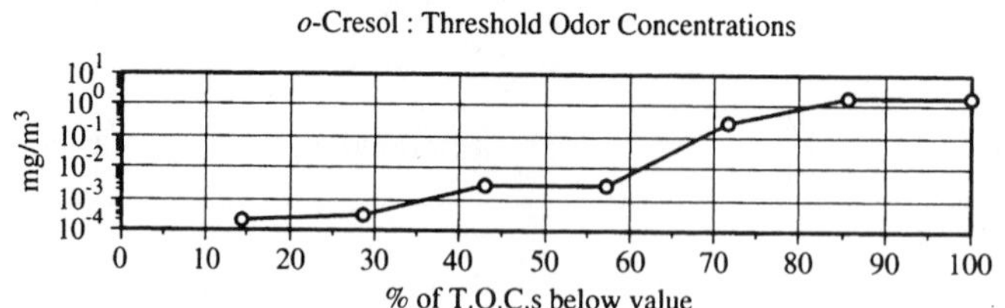

(57, 73, 279, 610, 712, 829)

Manmade sources

in exhaust of a 1970 Ford Maverick gasoline engine 0.5-1.0 ppm (including unknown compound) (1053)

avg conc. in smoke from 28 woodstoves and fireplaces:
burning hardwood: 12.3 g/kg carbon collected on filter

softwood:	8.7 g/kg carbon collected on filter	(2414)

Control methods

wet scrubber: water at pH 8.5: outlet: 20 odor units/scf: $KMnO_4$ at pH 8.5: outlet: 1 odor unit/scf. (115)

two-step catalytic combustion on aluminum oxide (n.s.i.):

influent: 10,000 mg/m^3
after 1st step: 840 mg/m^3
after 2nd step: 23 mg/m^3 (190)

Incinerability

thermal stability ranking of hazardous organic compounds: rank 104 on a scale of 1 (highest stability) to 320 (lowest stability). (2390)

C. WATER AND SOIL POLLUTION FACTORS

BOD_5:	63; 65; 67; 69; 70% ThOD.	(30, 36, 41, 274)
$BOD^{30°C}$:	53% ThOD at 15 mg/L (seed water from phenol-degradation plant).	
BOD_2:	61% ThOD at 15 mg/L (seed water from phenol-degradation plant).	
BOD_5:	77% ThOD at 15 mg/L (seed water from phenol-degradation plant).	(564)
BOD_{20}:	70; 71; 86% ThOD at 10 mg/L, unadapted sew.; lag period: 1 d.	(30, 554)
COD:	94; 95% ThOD.	(36, 41)
	92% of ThOD (0.05 *n* $K_2Cr_2O_7$).	(274)
ThOD:	2.52.	(36)

Reduction of amenities

odor threshold (tentative):	average: 0.65 mg/L range: 0.016-4.1 mg/L	(294, 97)
T.O.C. in water:	0.09 ppm 0.65 ppm	(326)
T.O.C. in water: 0.26 ppm		(325)

Odor threshold

detection: 1.4 mg/L. (998)

Taste threshold conc.

0.003 mg/L. (998)

Photooxidation

$t_{1/2}$ in top meter of Greifensee (Switzerland): 11 d. (2640)

Biodegradability

transformation by adapted A.S. into the following metabolites: 3-methylcatechol, 4-methylresorcinol, and methylhydrochinon. (5544)

% degradation in 7 d tests:			
original culture:	avg: >99.5	range: >99.5	
1st subculture:	>99.5	98.5->99.5	
2nd subculture:	>99.5	99.0->99.5	
3rd subculture:	>99.5	>99.5	(87)

Decomposition

period by a soil microflora: 1 d. (176)
Adapted culture: 98% elimination after 48 h incubation, initial conc. = 500 mg/L. (292)
Adapted A.S. at 20°C-product is sole carbon source: 95% COD removal at 54 mg COD/g dry inoculum/h. (327)

W, unadapted A.S. at	1 mg/L:	100% removal after 3 h	
	10 mg/L:	100% removal after 3 h	
	100 mg/L:	17% removal after 6 h	(1639)

Oxidation

by activated sludges acclimated to the following aromatics: 250 mg/L o-cresol after 30 min:

phenol:	34% ThOD	
o-cresol:	35% ThOD	
m-cresol:	29% ThOD	
p-cresol:	6% ThOD	
mandelic acid:	9% ThOD	(93)

methods	*temp, °C*	*d observed*	*feed, mg/L*	*% removed*	
RW, CA	20	2	1	100	
RW, CA	4	7	1	98	(93)

inoculum/methods	*test conc.*	*test duration*	*removed*	
A.S., adapted	200 mg/L	5 d	95% COD	(5542)
municipal sewage, Closed bottle test	3 mg/L	30 d	80% ThOD	(5543)
adapted culture	500 mg/L	2 d	98%	(5550)

Mixed culture, adapted to pentachlorophenol at initial conc. of 5 mg/L:

	10% degrad.	*50% degrad.*	*90% degrad.*	
aerobic	20 hours	45 hours	50 hours	
anaerobic	20 hours	57 hours	66 hours	(5548)

Impact on biodegradation processes

biodegradation inhib.	EC_{50}	500 mg/L	(2624)

inhibition of nitrification in activated sludge at 11-16 mg/L (75% reduction) (n.s.i.) (43)

inhibition of degradation of glucose by *Pseudomonas fluorescens* at 50 mg/L (293)
inhibition of degradation of glucose by *E. coli* at 600 mg/L (293)
A.S.: EC_{50}: 940 mg/L after 5 d exposure (5540)

Waste water treatment

Photooxidation: An aqueous solution of cresol (n.s.i.) is destroyed by photooxidation using visible light as direct energy source and methyleneblue as a dye-sensitizer. (188)

Ion exchange: adsorption on Amberlite X AD-2: 100% retention, infl.: 0.3 ppm, autoxidation at 25°C: $t_{1/2}$: 11,000 h at pH 9.0. (1908)

Göteborg (Sweden) sew. works 1989-1991: infl.: 0.1-4 µg/L; effl.: n.d. (2787)

Manmade emissions

60 ml/min pure water passed through 25-ft, 1/2-inch I.D. tube of general grade PVC contained 4.6 ppb *o*-cresol, which constituted 43% of total contaminant concentration. (430)

D. BIOLOGICAL EFFECTS

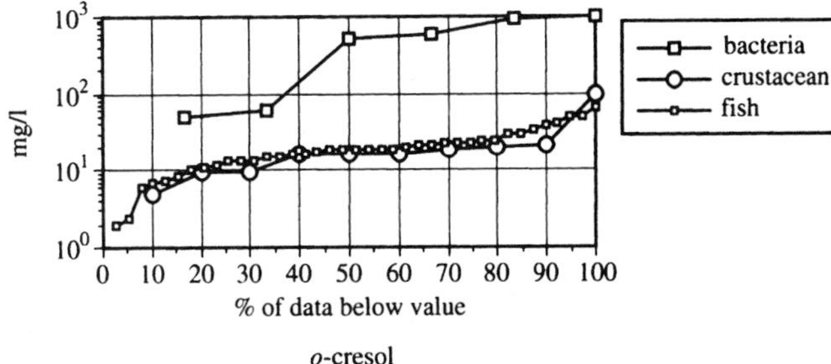

o-cresol

Frequency distribution curves for short-term EC_{50} values for bacteria (*n* = 6) and 24-96h EC_{50}/LC_{50} values for crustaceans (*n* = 10) and fishes (*n* = 39) based on data from this work and from the work of Rippen (3999)

Bacteria			
E. coli	LD_0	60 mg/L	
Pseudomonas putida	16h EC_0	33 mg/L	(1900)
Photobacterium phosphoreum	30 min EC_{50}	32 mg/L	(5541)
Biotox™ test *Photobacterium phosphoreum*	5min EC_{20}	23 mg/L	
	5min EC_{50}	63 mg/L	
	15min EC_{20}	21 mg/L	
	15min EC_{50}	52 mg/L	
Microtox™ test *Photobacterium phosphoreum*	5min EC_{20}	5.8 mg/L	
	5min EC_{50}	15 mg/L	
	15min EC_{20}	7.3 mg/L	
	15min EC_{50}	16 mg/L	(7022)
Microtox *Vibrio fisheri*	15min EC_{50}	15 mg/L	(7025)
Algae			
Scenedesmus	LD_0	40 mg/L	
Microcystis aeruginosa	8d EC_0	6.8 mg/L	(329)
Scenedesmus quadricauda	7d EC_0	11 mg/L	(1900)
Chlorella pyrenoidosa	48h EC_{50}	34 mg/L	(5523)
Scenedesmus pannonicus	8d EC_{50}	11 mg/L	(5523)
Selenastrum capricornutum	96h EC_{50}	65 mg/L	(5523)
Rhaphidocellis subcapitata	72h EC_{50}	65 mg/L	(7025)
Protozoa			
Entosiphon sulcatum	72h EC_0	17 mg/L	(1900)
	72h EC_{50}	34 mg/L	(5523)
Uronema parduczi Chatton-Lwoff	EC_0	31 mg/L	(1901)
Tetrahymena pyriformis	14h LC_{100}	3.7 mmol/L	(1662)
Chilomonas paramecium	48h EC_{50}	132 mg/L	(5523)
Insects			
Aedes aegypti	48h NOLC	65 mg/L	
	48h EC_{50}	80 mg/L	(5523)
Culex pipiens	48h NOLC	31 mg/L	
	48h EC_{50}	46 mg/L	(5523)
Chironomus gr. *thummi*	48h LC_{50}	34 mg/L	(5549)
Asellus aquaticus	48h LC_{50}	23 mg/L	(5549)

Worms			
Tubificidae	48h LC_{50}	165 mg/L	(5549)
Molluscs			
Limnaea stagnalis	48h NOLC	56 mg/L	
	48h EC_{50}	160 mg/L	(5523)
Crustaceans			
Daphnia magna	48h EC_{50}	5; 9.5 mg/L	(5523, 5539)
	48h EC_{50},S	16 mg/L	(5537)
	48h EC_{50},F	>94 mg/L	(5529)
	48h NOLC	2.9 mg/L	(5523)
	24h EC_0	6.3; 9.5 mg/L	
	24h EC_{50}	19; 20 mg/L	
	24h EC_{100}	36; 50 mg/L	(5535, 5536)
Daphnia pulex	48h EC_{50}	9.6 mg/L	
	48h NOLC	5.2 mg/L	(5523)
Daphnia cucullata	48h EC_{50}	16 mg/L	(5523)
Daphnia	48h EC_0	16 mg/L	(5534)
Crangon septemspinosa	59h LC_{50}	14 mg/L	(5545)
Gammarus pulex	48h LC_{50}	21 mg/L	(5549)
Fishes			
goldfish	24-96h LC_{50}	19-49 mg/L soft water	
bluegills	24-96h LC_{50}	21-22 mg/L soft water	
fatheads	24-96h LC_{50}	13-18 mg/L hard water	(158)
Poecilia reticulata	24-96h LC_{50}	18-50 mg/L hard water	(41)
	48h LC_{50}	38 mg/L	
	48h NOLC	32 mg/L	(5523)
	96h LC_{50}	19 mg/L soft water	(5528)
crucian carp	24h LC_{50}	30 mg/L	
roach	24h LC_{50}	16 mg/L	
tench	24h LC_{50}	15 mg/L	
'trout' embryos	24h LC_{50}	2 mg/L	(222)
Leuciscus idus	48h LC_{50}	2; 10; 18 mg/L	(5522, 5523, 5524)
Pimephales promelas	48h LC_{50}	24; 34 mg/L	(5523, 5527)
	48h NOLC	30 mg/L	(5523)
	96h LC_{50}	12.5 mg/L soft water	(5528)
	96h LC_{50}	13.4 mg/L hard water	(5528)
	96h LC_{50},F	18 mg/L	(5529)
Pimephales promoxis	96h LC_{50}	20 mg/L	(5525)
Salmo gairdneri	48h LC_{50}	13 mg/L	
	48h NOLC	3.8 mg/L	(5523)
	96h LC_{50}	7; 17 mg/L	(5525, 5532)
	96h LC_{50},F	8.4 mg/L	(5529)
Salmo trutta	96h LC_{50}	6.2 mg/L	(5532)
Brachydanio rerio	96h LC_0	20 mg/L	
	96h LC_{50}	24 mg/L	
	96h LC_{100}	30 mg/L	(5524)
Oryzias latipes	48h LC_{50}	41 mg/L	
	48h NOLC	32 mg/L	(5523)

Lepomis humilis	1h LC_{100}	55-56 mg/L	(5526)
Lepomis macrochirus	96h LC_{50}	21 mg/L soft water	(5528)
Carassius auratus	96h LC_{50}	23 mg/L hard water	(5528)
Ictalurus punctatus	96h LC_0	4; 15 mg/L	
	96h LC_{50}	11; 67 mg/L	
	96h LC_{100}	16; 100 mg/L	(5530)
Trutta iridea embryos	24h LC_{50}	2.3 mg/L	(5531)
Tinca vulgaris	24h LC_{50}	15 mg/L	(5531)
Leuciscus rutilus	24h LC_{50}	16 mg/L	(5531)
Carassius vulgaris	24h LC_{50}	29 mg/L	(5531)
Salvelinus fontinalis	96h LC_{50}	7.2 mg/L	(5532)
Sarotherodon mossambica	96h LC_{50}	23 mg/L	(5533)
Amphibians			
Mexican axolotl (3-4 w after hatching)	48h LC_{50}	40 mg/L	
clawed toad (3-4 w after hatching)	48h LC_{50}	38 mg/L	(1823)
Erpobdella octoculata	48h LC_{50}	135 mg/L	(5549)
Dugesia cf. lugubris	48h LC_{50}	24 mg/L	(5549)
Hydra oligactis	48h LC_{50}	75 mg/L	(5549)
Corixa punctata	48h LC_{50}	80 mg/L	(5549)
Ischnura elegans	48h LC_{50}	46 mg/L	(5549)
Cloeon dipterum	48h LC_{50}	50 mg/L	(5549)

p-cresol

(*p*-cresylic acid; 4-hydroxytoluene; 4-methylphenol)

C_7H_8O

CAS 106-44-5

MANMADE SOURCES

constituent of coal tar creosote: 1 wt %. (2386)

A. PROPERTIES

yellowish liquid; molecular weight 108.13; melting point 34.8°C; boiling point 202°C; vapor pressure 0.04 mm at 20°C, 0.11 mm at 25°C, 1 mm at 53°C; vapor density 3.72; density 1.04 at 20/4°C; solubility 24,000 mg/L at 40°C, 53,000 mg/L at 100°C; THC 880 kcal/mole; saturation concentration in air 0.24 g/m^3 at 20°C, 0.74 g/m^3 at 30°C; LogP_{OW} 1.92/1.94; LogH -4.49 at 25°C.

B. AIR POLLUTION FACTORS

1 mg/m^3 = 0.22 ppm 1 ppm = 4.50 mg/m^3.

Odor

tarlike, pungent.

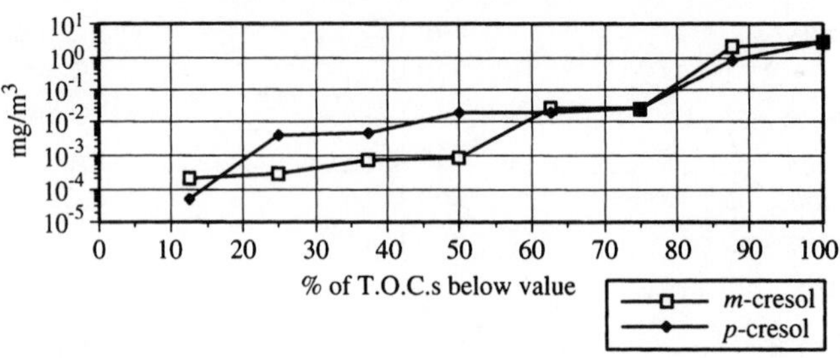

Manmade sources

in exhaust of a 1970 gasoline engine: 0.4-0.7 ppm. (1053)

Incinerability

Temperature for 99% destruction at 2.0-sec residence time under oxygen-starved reaction conditions: 745°C.

Thermal stability ranking of hazardous organic compounds:rank 104 on a scale of 1 (highest stability) to 320 (lowest stability). (2390)

C. WATER AND SOIL POLLUTION FACTORS

BOD_5:	55-70; 57% ThOD (acclim. sewage).	(27, 41)
$BOD_{24h}{}^{30°C}$:	55% of ThOD (seed water from phenol-degradation plant).	
$BOD_{2d}{}^{30°C}$:	61% of ThOD (seed water from phenol-degradation plant).	
$BOD_{5d}{}^{30°C}$:	81% of ThOD (seed water from phenol-degradation plant).	(564)
COD:	95% ThOD.	(41)
ThOD:	2.52.	(27)

Impact on biodegradation processes

75% reduction of the nitrification process in nonadapted activated sludge at 16.5 mg/L. (30)

inhibition of degradation of glucose by *Pseudomonas fluorescens* at 30 mg/L.

inhibition of degradation of glucose by *E. coli* at >1,000 mg/L. (293)

biodegradation inhib. EC_{50} 500 mg/L. (2624)

Photooxidation

$t_{1/2}$ in top meter of Greifensee (Switzerland): 4.4 d. (2640)

Reduction of amenities: T.O.C. in water: 0.055 ppm. (325)

Biodegradation: decomposition period by a soil microflora: 1 d. (176)

Adapted A.S. at 20°C-product is sole carbon source: 96% COD removal at 55 mg COD/g dry inoculum/h. (327)

CH_3 / OH → COOH / OH / OH → Ring Fissior

Accumulation of dihydroxybenzoic acid by cells of *Pseudomonas* sp. (1219)

Manmade sources

in sewage effluents:	0.090 mg/L	(227)
in primary domestic sewage plant effluent:	0.020-0.029 mg/L	
in secondary domestic sewage plant effluent:	0.020-0.090 mg/L	(517)

landfill leachate from lined disposal area (Florida 1987-1990): mean conc. 556 µg/L (2788)

Waste water treatment

activated sludges acclimated to the following aromatics: 250 mg/L p-cresol, *after 30 min:*

phenol:	33% theor. oxidation	
o-cresol:	34% theor. oxidation	
m-cresol:	35% theor. oxidation	
p-cresol:	39% theor. oxidation	
mandelic acid:	20% theor. oxidation	(92)

methods	*temp, °C*	*d observed*	*feed, mg/L*	*d acclim.*	*% theor. oxidation*	*% removed*	
RW, CA	20	6	1	nat.	-	95	
RW, CA	4	19	1	nat.	-	95	
AS, W	20	1/2	250	25+	39	-	(93)

W, unadapted A.S. at	1 mg/L	100% removal after 3 h	
	10 mg/L	100% removal after 3 h	
	100 mg/L	3% removal after 6 h	(1639)

Taste threshold conc.

0.002 mg/L. (998)

Odor threshold: detection: 0.2 mg/L. (998)

Soil sorption

K_{OC}:	for Fullerton soil, 0.06% OC:	3,350	
	for Apison soil, 0.11% OC:	3,420	
	for Dormont soil, 1.2% OC:	115	(2599)

Biodegradation

Anaerobic mineralisation at 50 mg C/L at 35 °C in the dark:

after 56 d in digested sewage sludge	30-75% of theoretical mineralization	
after 56 d in freshwater swamp	>75% of theoretical mineralization	
after 96 d in marine sediment	30-75% of theoretical mineralization	(7037)

$t_{1/2}$ in nonadapted aerobic subsoil: groundwater Florida: 3 d. (2695)

D. BIOLOGICAL EFFECTS

Bacteria			
Biotox™ test *Photobacterium phosphoreum*	5min EC_{20}	2.2 mg/L	
	5min EC_{50}	8.7 mg/L	
	15min EC_{20}	1.7 mg/L	
	15min EC_{50}	7.7 mg/L	
Microtox™ test *Photobacterium phosphoreum*	5min EC_{20}	0.34 mg/L	
	5min EC_{50}	0.92 mg/L	
	15min EC_{20}	0.40 mg/L	
	15min EC_{50}	7.7 mg/L	(7022)

Protozoa			
ciliate *(Tetrahymena pyriformis)*	24h LC_{100}	400 mg/L	(1662)

Algae			
Scenedesmus	LC_0	6 mg/L	
Arthropoda			
Daphnia	LD_0	12 mg/L	(30)
Fishes			
crucian carp	24h LC_{50}	21 mg/L	
roach	24h LC_{50}	17 mg/L	
tench	24h LC_{50}	16 mg/L	
trout' embryos	24h LC_{50}	4 mg/L	(222)
rainbow trout	toxic	5 mg/L	(156)
fathead minnows: static bioassay in Lake Superior water at 18-22°C: LC_{50} (1; 24; 48; 72; 96h): >30; 26; 21; 21; 19 mg/L			(350)

m-cresylic acid *see* *m*-cresol

o-cresylic acid *see* *o*-cresol

p-cresylic acid *see* *p*-cresol

crotamiton (N-ethyl-N-(2-methylphenyl)-2-butenamide; N-ethyl-o-crotonotoluidide; crotonyl-N-ethyl-o-toluidine))

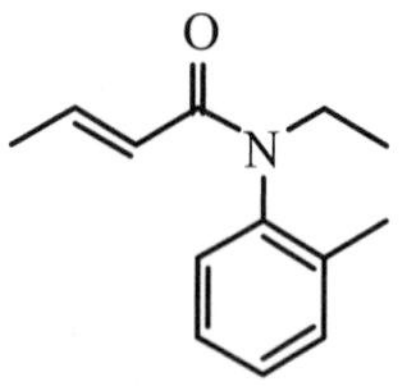

$C_{13}H_{17}NO$

CAS 483-63-6

TRADENAMES

Crotamiton; crotamitex; Eurax; Euraxil; Veteusan

USES

fungicide; insecticide; scabicide; antipruritic

A. PROPERTIES

yellowish oil; molecular weight 203.28; boiling point 153-155 °C

C. WATER AND SOIL POLLUTION FACTORS

Manmade sources

	concentration µg/L	n	
influent MWTP	<0.12-0.13	3	(7237)

crotonaldehyde (β-methylacrolein; 2-butenal; crotonic aldehyde; propylene aldehyde)

$CH_3CHCHCHO$

C_4H_6O

CAS 4170-30-3

USES

intermediate for the synthesis of sorbic acid, trimethylhydroquinone, 3-methoxybutanol, thiophenes, chinaldines, pyridines, dyes, pesticides, pharmaceuticals. Feedstock for the mfg. of alkydresins.

A. PROPERTIES

molecular weight 70.1; melting point -75°C; boiling point 99/104°C; vapor pressure 19 mm at 20°C, <43 hPa at 20°C; vapor density 2.41; density 0.85; solubility 155,000 mg/L; LogP_{OW} 0.63 (calculated).

B. AIR POLLUTION FACTORS

1 mg/m^3 = 0.349 ppm, 1 ppm = 2.8 mg/m^3.

Odor

T.O.C.:	0.6 mg/m^3 = 0.2 ppm	(57)
	0.1 mg/m^3 = 0.035 ppm	(307)
	0.18-0.57 mg/m^3	(710)
	0.420 mg/m^3	(842)

Incinerability

Temperature for 99% destruction at 2.0-sec residence time under oxygen-starved reaction conditions: 710°C.

Thermal stability ranking of hazardous organic compounds

rank 113 on a scale of 1 (highest stability) to 320 (lowest stability). (2390)

Manmade sources

in exhaust of a 1970 Ford Maverick gasoline engine: 0.1-0.9 ppm (1053)

in exhaust of gasoline engine: 0.4-1.4 vol. % of total exhaust aldehydes (395, 396, 397)

Photochemical degradation

reaction with OH°: $t_{1/2}$: 0.5 d (calculated) (5146)

reaction with O_3: $t_{1/2}$: 13 d (5152)

C. WATER AND SOIL POLLUTION FACTORS

BOD_5:	37% of ThOD	
BOD_{10}:	60; 95% ThOD	(256, 5130)
	70% ThOD (adapted)	(5130)
COD:	97% of ThOD std. dil. sew.	
ThOD:	1.37	

Reduction of amenities

faint odor: 0.021 mg/L. (129)

Biodegradability

inoculum/method	*test conc.*	*test duration*	*removed*	
A.S., industrial, nonadapted		3 hours	~20% ThOD	
		5 d	50% ThOD	
		10 d	65% ThOD	
		15 d	75% ThOD	(5124)
A.S., laboratory test unit	600 mg/L	1 d	100%	(5139, 5140)
A.S., industrial, lab aeration unit	400 mg/L	90 min	93%	(5141)
isolated bacteria	200 mg/L	30 min	80-90%	
	500 mg/L	60 min	80-90%	
	800 mg/L	90 min	80-90%	(5142)
A.S., industrial, adapted	1,800 mg/L	10 d	25-28%	(5143)
anaerobic bacteria, adapted	100 mg/L	10 d	95%	(5144)
Pseudomonas fluorescens		1 hour	85%	
		10 d	98%	(5145)
A.S., industrial, nonadapted		5 d	78% ThOD	
		10 d	83% ThOD	
		15 d	94% ThOD	(5128)

Fungi

Beauveria sulfurescens:	700-800 mg/L;	48 hours	90% transformation to 2-butenol; crotonalde- hyde could not be detected among the reaction products	(5147)
Pseudomonas putida mutant	700 mg/L	24 hours	73% transformation to 2-butenol	(5147)

Impact on biodegradation processes

municipal primary sludge	24h EC_0	50 mg/L	
	24h EC_{50}	90 mg/L	(5124)
municipal sewage	16h EC_0	850 mg/L	(5129)
	16h toxic	8,500 mg/L	(5129)
municipal sewage, nonadapted	72h EC_0	14 mg/L	(5130)
municipal sewage, adapted	72h EC_0	36 mg/L	(5130)
A.S., industrial	3h EC_{100}	100 mg/L	(5131)
	24h EC_{100}	50 mg/L	
	3h EC_{80}	1 mg/L	(5131)
A.S., industrial	24h EC_{80}	1 mg/L	(5131)

Waste water treatment

activated carbon: adsorbability: 0.092 g/g, 45% reduction, infl.: 1,000 mg/L; effl.: 544 mg/L. (32)

methods	*temp, °C*	*d observed*	*feed, mg/L*	*d acclim.*	*% theor. oxidation*	*% removed*	
NFG, BOD	20	1-10	1,000	365 + P	-	27	
NFG, BOD	20	1-10	50-250	365 + P	-	inhibition	
NFG, BOD	20	1-10	1,600	365 + P	-	36	
RW, Sd, BOD	20	10	?	100	70	-	(93)

Abiotic degradation

in aqueous solution 1,400-2,100 mg/L at 25°C, after 75 hours: 0.3% decrease in the presence of mineral acids (0.5-1.9N); a rapid reversible hydration to 3-hydroxybutanol, equilibrium is reached at 25°C and 35°C after 47% and 39% of crotonaldehyde have been transformed. (5149)

Photochemical degradation

in aqueous solution at wavelength *>290 after 30 min exposure:*

conc.	*% degradation*	
1.8 mg/L	10	
3.5 mg/L	6	
17	2.4	(5153)

D. BIOLOGICAL EFFECTS

Bacteria			
E. coli	LC_0	50,000 mg/L	
	EC_5, immobilization	15,000 mg/L	
	EC_{100}, immobilization	40,000 mg/L	(5132)
	minimum inhib. conc.	500 mg/L	(5133)
	30 min EC_0	>2,500 mg/L	(5133)
	72h EC_0	10 mg/L	(5137)
	72h EC_{100}	100 mg/L	
	12h EC_0	100 mg/L	
	12h EC_{100}	1,000 mg/L	
	1h EC_0	1,000 mg/L	
	1h EC_{100}	10,000 mg/L	(5137)
Staphylococcus aureus	LC_0	50,000 mg/L	(5132)
	EC_5, immobilization	20,000 mg/L	
	EC_{100}, immobilization	>50,000 mg/L	(5132)
	minimum inhib. conc.	500 mg/L	(5133)
	30 min EC_0	>2,500 mg/L	(5133)

Bacillus subtilis	LC_0	50,000 mg/L	(5132)
	EC_5, immobilization	12,500 mg/L	
	EC_{100}, immobilization	40,000 mg/L	(5132)
Pseudomonas aeruginosa	minimum inhib. conc.	250 mg/L	(5133)
	30 min EC_0	>2,500 mg/L	(5133)
Proteus vulgaris	30 min EC_0	>2,500 mg/L	(5133)
Algae			
Dunaliella bioculata	2h EC_{50}, immobilization	75 mg/L	
	2h EC_0, immobilization	50 mg/L	
	toxic	10 mg/L	(5127)
Protozoans			
Chilomonas paramecium	48h EC_{10}	2 mg/L	(5156)
Paramecium caudatum	48h EC_{50}	20 mg/L	(5126)
Actinosphaerium sp.	30h LC_{50}	15 mg/L	(5126)
Fungi			
Candida albicans	minimum inhib. conc.	250 mg/L	(5133)
	30 min EC_0	>2,500 mg/L	(5133)
Candida utilis	EC_8	50 mg/L	
	EC_{14}	100 mg/L	
	EC_{20}	150 mg/L	
	EC_{25}	200 mg/L	(5155)
	EC_{10}	50 mg/L	
	EC_{100}	500 mg/L	(5134, 5135)
Microsporum gypseum	30 min EC_{100}	2,500 mg/L	(5133)
	30 min EC_0	1,000 mg/L	(5133)
Trichophyton mentagrophytes	30 min EC_0	>2,500 mg/L	(5133)
Crustaceans			
Daphnia magna	26h LC_{50}	10 mg/L	(5126)
Fishes			
Leuciscus idus	48h EC_0	1; 5 mg/L	
(after degradation of crotonaldehyde)	48h LC_0	500 mg/L	(5124)
Cyprinus carpio	2h LC_{100}	32 mg/L	
	4h LC_{100}	17 mg/L	
	8h LC_{100}	6.4-8.5 mg/L	
	30h LC_{100}	4.3 mg/L	
	10d EC_0	0.85 mg/L	
Poecilia reticulata	14d LC_{50},S	556 mg/L	(5124)

Lepomis macrochirus: static bioassay in fresh water at 23°C, mild aeration applied after 24 h (85% aqueous):

material added	*% survival after*				*best fit 96h LC_{50}*
ppm	*24 h*	*48 h*	*72 h*	*96 h*	*ppm*
7.5	0	-	-	-	
5.6	0	-	-	-	
4.2	0	-	-	-	
3.2	90	90	90	70	3.5
1.8	100	100	100	100	

Menidia beryllina: static bioassay in synthetic seawater at 23°C: mild aeration applied after 24 h:

material added	*% survival after*				*best fit 96h LC_{50}*
ppm	*24 h*	*48 h*	*72 h*	*96 h*	*ppm*

3.2	0	-	-	-		
1.8	90	90	20	10	1.3	
1.0	100	100	100	90		(352)

crotonic aldehyde *see* crotonaldehyde

crotoxyphos *see* ciodrin

crotylmercaptan

$CH_3CHCHCH_2SH$

C_4H_8

A. PROPERTIES

molecular weight 88.

B. AIR POLLUTION FACTORS

Odor

characteristic: skunk odor: (51)

T.O.C.: 0.0075 ppb; 56 ppt; 0.00043-0.0014 mg/m^3 (10, 279, 710)

0.0002 mg/m^3 = 0.055 ppb. (307)

C. WATER AND SOIL POLLUTION FACTORS

Reduction of amenities

faint odor: at 0.000029 mg/L. (129)

crystal violet (Basic violet 3; C.I. 42555; Gentian Violet)

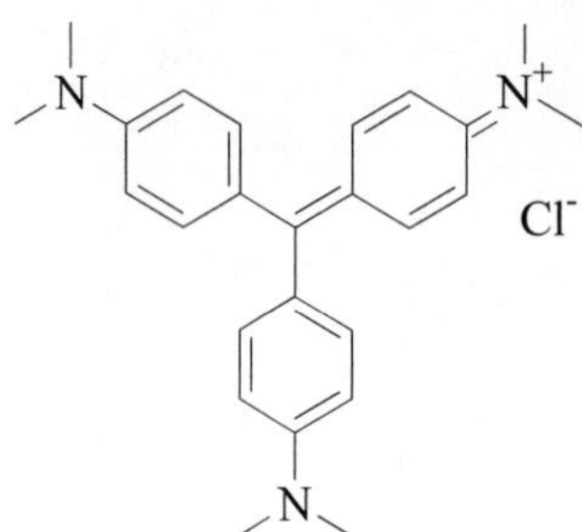

$C_{25}H_{30}ClN_3$

CAS 548-62-9

USES

indicator.

A. PROPERTIES

molecular weight 407.98; melting point 215°C decomposes.

D. BIOLOGICAL EFFECTS

Toxicity to microorganisms

Bacillus subtilis growth inhib. EC_{50}: 0.0013 mmol/L. (2624)

cumarin *see* coumarin

cumene *see* isopropylbenzene

Cumene sulfonate, sodium salt ((1-methylethyl)benzenesulfonic acid, sodium salt)

SO_3Na SO_3Na SO_3Na

ortho- meta- para-

$C_9H_{11}O_3SNa$

CAS 28348-53-0

CAS 32073-22-6

EINECS 248-93-87; 250-91-35

USES

Hydrotrope. Hydrotropes are used as coupling agents to solubilize the water insoluble and often incompatible functional ingredients of household and institutional cleaning products and personal care products. Important application products are household laundry and cleaning products, such as laundry powders and liquids (maximum concentration 0.66%), liquid fabric conditioners (maximum concentration 0.66%), liquid and powder laundry bleach additives, hand dishwashing liquid, machine dishwashing liquid, liquid and gel toilet cleaners (maximum concentration 1.9%), and liquid, powder (maximum concentration 6%), gel and spray surface cleaners (maximum concentration 2%).

IMPURITIES, ADDITIVES, COMPOSITION

Commercial cumene sulfonates consists of mixtures of 3 isomers

A. PROPERTIES

molecular weight 222; melting point 182->300°C; vapour pressure 1.1 x 10^{-9} Pa; water solubility 330,000-400,000 mg/L; log Pow –1.5

B. AIR POLLUTION FACTORS

Atmospheric oxidation

t/2 estimated at 40 – 105 hours (10588)

C. WATER AND SOIL POLLUTION FACTORS

Hydrolysis

Hydrolysis: negligible (10588)

Biodegradation

Inoculum	method	concentration	duration	elimination	
	OECD 301E screening			94%	
	Zahn Wellens		28 days	100% ThOD	
	Pre-OECD			73%	(10588)
Municipal waste water	Modified OECD screening test	20 mg/L	21 days	94% DOC	(6539)
A.S.	Coupled units test	10 mg/L		82– 92% DOC	(6539)
A.S.	Zahn-Wellens test	240 mg/L	19 days	100% DOC	(6539)

D. BIOLOGICAL EFFECTS

Toxicity

Bacteria			
Pseudomonas putida	48h EC_{10}	>16,000 mg/L	(10588)
	48h EC_{10}	40,000 mg/L	(6539)
Algae			
Scenedesmus subspicatus	72h $ECgr_{50}$	>1,000 mg/L	(6539, 10588)
Crustaceae			
Daphnia magna	48h EC_{50}	>450 mg/L	
	24h EC_{50}	>1,000 mg/L	(10588)
	21d NOEC	ca 30 mg/L	(6539)
	21 d EC_{50}	154 mg/L	
Fishes			
Fathead minnow	96h LC_{50}	>450 mg/L	
Leuciscus idus	48h LC_{50}	>1,000 mg/L	(6539, 10588)
Mammals			
Rat	oral LD_{50}	>7,000 mg/kg bw	(10588)

cumenehydroperoxide *see* isopropylbenzenehydroperoxide

p-cumenyl isocyanate (1-isocyanato-4-(1-methylethyl)benzene; p-isopropylphenylisocyanate; 4-(1-methylethyl)phenylisocyanate; isocyanic acid, p-isopropylphenylester)

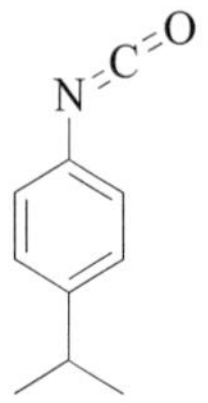

N=C=O

$C_{10}H_{11}NO$

CAS 31027-31-3

USES AND FORMULATIONS

Manufacturing of pesticides.

A. PROPERTIES

molecular weight 161.21; melting point - 50°C; boiling point 222°C; vapor pressure 0.22 hPa at 20°C, 1.7 hPa at 50°C; density 1.0 at 20°C; solubility hydrolyses.

B. AIR POLLUTION FACTORS

Photochemical reactions

$t_{1/2}$: 2.4 d, based on reactions with OH° (calculated) (6543)

C. WATER AND SOIL POLLUTION FACTORS

Biodegradation

Inoculum	*Method*	*Concentration*	*Duration*	*Elimination results*	
Industrial A.S.	Zahn-Wellens test	10,000 mg/L at pH 8	5 d	60% DOC	
			10 d	>95% DOC	(6542)

D. BIOLOGICAL EFFECTS

Toxicity

Bacteria			
Primary municipal sludge	24 h EC_{10}	100- 1,000 mg/L at pH 8	(6542)
Crustaceans			
Daphnia magna	24 h EC_{10}	100- 1,000 mg/L at pH 8	(6542)
Fishes			
Brachydanio rerio	96 h LC_0	125 mg/L	
	96 h LC_{50}	142 mg/L	(6541)

p-cumylphenol (*p*-(2-phenylisopropyl)phenol)

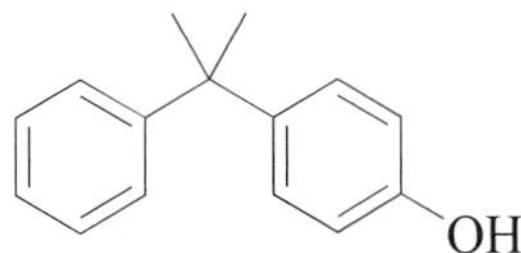

$C_6H_5C(CH_3)_2C_6H_4OH$

$C_{15}H_{16}O$

CAS 599-64-4

USES

intermediate for resins, insecticides, lubricants.

A. PROPERTIES

white to tan crystals; molecular weight 212.29; melting point 70-73°C; boiling point 335°C.

B. AIR POLLUTION FACTORS

Odor

characteristic: phenol odor.

C. WATER AND SOIL POLLUTION FACTORS

COD: 93% ThOD (27)
ThOD: 2.8 (27)

cupric acetate (acetic acid, copper salt; copper diacetate; copper-acetate; crystallized-verdigris; neutralized-verdigris)

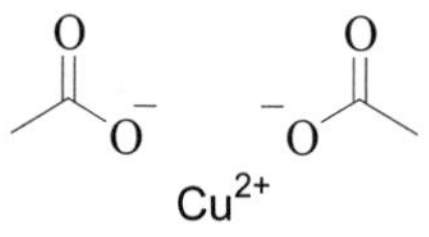

$(CH_3COO)_2Cu$

$C_4H_6CuO_4$

CAS 142-71-2

USES

Fungicide. Intermediate in the manufacture of Paris green. Catalyst. Textile dyestuff. Pigment for ceramics.

A. PROPERTIES

molecular weight 181.64; melting point 115°C; density 1.9 at 20°C; solubility miscible.

C. WATER AND SOIL POLLUTION FACTORS

Copper is more dangerous in soft waters than in hard waters; the ionic form Cu_{2+} being the most toxic. In hard water the final precipitation product is the basic carbonate, malachite, $CuCO_3.Cu(OH)_2$.

D. BIOLOGICAL EFFECTS

Fishes

	24h LC_{50}	*48h LC_{50}*	*96h LC_{50}*	
fathead minnows *(Pimephales promelas)*	0.48 mg/L	0.42 mg/L	0.30 mg/L	
grass shrimp *(Palaemonetes pugio)*			37.0 mg/L	(1904)

Mammals

rat	oral LD_{50}	710 mg/kg bw	(10040)

cutrine

USES

algicide.

IMPURITIES, ADDITIVES, COMPOSITION

copper-triethanolamine complexes. (331)

D. BIOLOGICAL EFFECTS

Molluscs

decapod: *Penaeus californiensis*	96h LC_{50},F	1,000 mg/L	(1134)

Fishes

harlequin fish *(Rasbora heteromorpha)*:

mg/L	*24 h*	*48 h*	*96 h*	*3 m (extrap.)*	
LC_{10},F	0.7	0.26			
LC_{50},F	1.2	0.35	0.24	0.01	(331)

Procambarus clarki	96h LC_{50}	461-2,945 mg/L	(2625)

cyanatryn

(2-(4-ethylamino-6-methylthio-*s*-triazin-2-ylamino)-2-methylpropionitrile; 2-(1-cyano-1-methylethylamino)-4-ethylamino-6- methylthio-1,3,5-triazine)

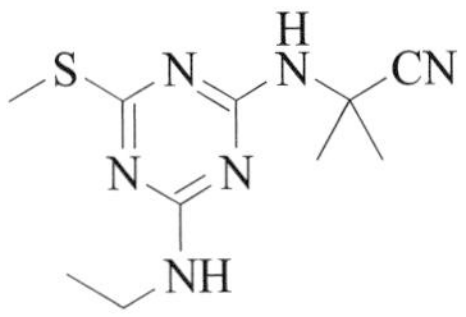

$C_{10}H_{16}N_6S$

CAS 21689-84-9

USES

herbicide. Active ingredient: triazine (50%). (331)

D. BIOLOGICAL EFFECTS

Snail

Lymnaea peregra:	adult:	2d LC_{60}	20 ppm	(1414)
		8d LC_0	10 ppm	(1415)
	eggs:	20 ppm: high incidence of embryonic deformities, and all failed to hatch		(1414)
		NOEC	0.2 ppm	

Amphibians

Rana temporaria tadpoles:	96h LC_{50}	30 ppm	(1415)

Crustaceans			
Daphnia longispina	96h LC_{50}	15.4 ppm	(1415)
Daphnia pulex	3d EC_{25}	2 ppm	
	3d EC_{10}	0.2 ppm	(1414)

Fishes

harlequin fish *(Rasbora heteromorpha):*

mg/L	*24 h*	*48 h*	*96 h*	*3 m (extrap.)*	
LC_{10},F	15	9	9		
LC_{50},F	35	18	15	5	(331)

cyanazine (2-chloro-4-(1-cyano-1-methylethylamino)-6-ethylamino-1,3,5-triazine; bladex; 2-(4-chloro-6-ethylamino)-S-triazine-2-ylamino-2-propionitrile)

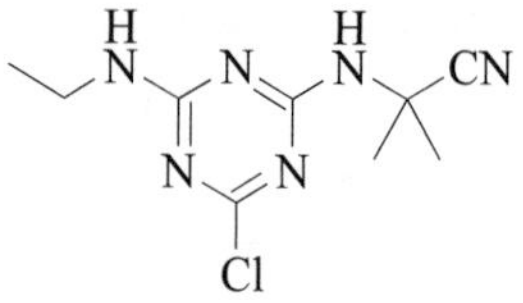

$C_9H_{13}ClN_6$

CAS 21725-46-2

USES

herbicide.

A. PROPERTIES

white, crystalline solids; molecular weight 240.7; melting point 167.5-169°C; vapor pressure 2 × 10^{-7} Pa at 20°C; solub 170 mg/L at 20°C; P_{ow}: 1.8-2.25.

USERS AND FORMULATIONS:

C. WATER AND SOIL POLLUTION FACTORS

in soil: $t_{1/2}$ 1-5 weeks; 12-15 d. (2915, 2962)

The first degradation step is hydrolysis; then bacterial degradation occurs according to the following pathway:

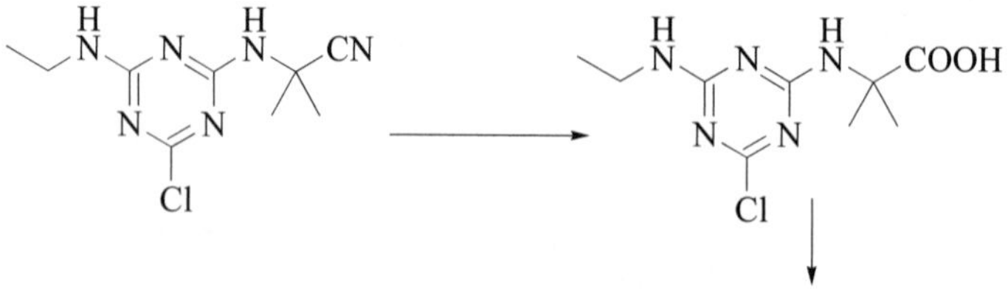

cyanazine → biodegradation

In plants and soil, the nitrile group is hydrolyzed to a carboxylic acid group, and the chlorine atom is replaced by a hydroxyl. (2915)

In surface waters

biodegradation proceeds until the triazine ring is reached, which is relatively stable. In an ecosystem after 35 d incubation, 60% was identified as desethylcyanazine, 18% as cyanazine, and 19% as a nonextractable material. (2915)

Surface and groundwater quality

Rhein, Germany, 1985-86: <0.01-0.36 μg/L (2915)

in groundwater wells in the U.S.A.-150 investigations up to 1988:

median of the conc.'s of positive detections for all studies	0.4 μg/L	
maximum conc.	7 μg/L	(2944)

Water treatment

initial conc., μg/L	*treatment method*	*% removal*	
2	flocc. + sed. + filtr. + chlorination	20	
0.6-0.7	chlorination	2-3	(2915)

D. BIOLOGICAL EFFECTS

Plants

Bioconcentration factor for VEGETATION: log BCF_V: -0.06. (2644)

FRESHWATER ECTOPROCTA

no appreciable effect at 2.5 mg/L for 84 h exposure. (1902)

Fishes

	24h LC_{50}	*96h LC_{50}*	
Sarotherodon mossambica	24 mg/L	64 mg/L	
Cirrhinus mrigala	13 mg/L	40 mg/L	(1902)
harlequin fish	16 mg/L		(2962)

(+)-cyano-3-phenoxybenzyl-(+)-α-(4-chlorophenyl)isovalerate *see* fenvalerate

(+)α-cyano-3-phenoxybenzyl-(+)cis, trans-2,2-dichlorovinyl-2,2-dimethylcyclopropane- carboxylate *see* cypermethrin

5-cyanoacenaphthene

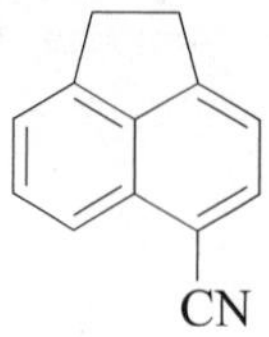

CN

USES

in soots generated by the combustion of aromatic hydrocarbon fuels doped with pyridine. (1723)

1-cyanoacenaphthylene

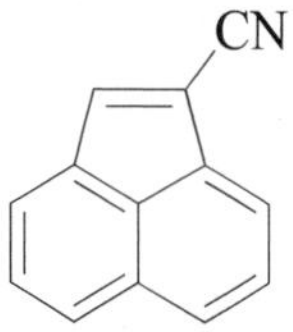

CN

USES

in soots generated by the combustion of aromatic hydrocarbon fuels doped with pyridine. (1723)

5-cyanoacenaphthylene

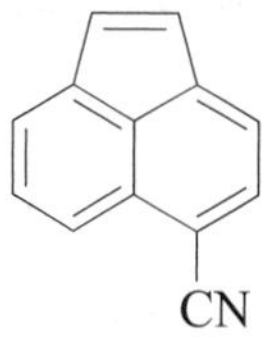

CN

USES

in soots generated by the combustion of aromatic hydrocarbon fuels doped with pyridine. (1723)

1-cyanoallyl acetate (2-acetoxy-3-butennitril; 3-butenenitrile,2-hydroxy-, acetate; acrolein cyanohydrin acetate; acroleincyanhydrin-O-acetate)

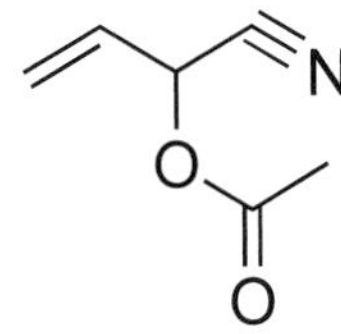

$C_6H_7NO_2$

CAS 15667-63-7

EINECS 239-743-2

A. PROPERTIES

molecular weight 125.13; melting point <-75 °C; boiling point 170 °C; density 1.03 kg/L at 20°C, < 1.0 kg/L at 80°C; vapour pressure 8 hPa at 20°C, 13 hPa at 30°C; water solubility 10,000 mg/L at 25 °C at pH 2.7, 27,000 mg/L at 80 °C; log Pow −0.21 (calculated)

C. WATER AND SOIL POLLUTION FACTORS

Hydrolysis

the following reaction mechanisms occur in water : the second reaction step (decomposition of 2-hydroxy-3-butenenitrile) results at pH 3 in an equilibrium. (7394)

1-cyanoallyl acetate $\xrightarrow{+H_2O}$ 2-hydroxy-3-butenenitrile + acetic acid (OH + HOAc) $\xleftrightarrow{+H_2O}$ 2-propenal + hydrogen cyanide (+ HCN)

Biodegradation

BOD5	9530 mg O2/L at 10,000 mg/L	(7395)
COD	14,500 mg/g substance	(7395)
BOD5/COD	0.64	

D. BIOLOGICAL EFFECTS

Toxicity

Fish			
Leuciscus idus	48h NOEC	0.32 mg/L	
	48h LC_0	0.56 mg/L	
	48h LC_{50}	0.75 mg/L	
	48h LC_{100}	1.0 mg/L	(7395)
Mammals			
Rat	oral LD_{50}	31.6-33.7 mg/kg bw	(7395)

β-cyanoethyl-2,3-dibromopropionate *see* busan 76

cyanoethylene *see* acrylonitrile

cyanogenbromide (bromine cyanide)

$$Br-C\equiv N$$

BrCN

CAS 506-68-3

USES

organic synthesis, brominating agent; parasiticide; fumigating compositions; rat exterminants.

A. PROPERTIES

crystals; penetrating odor; slowly decomposed by cold water; molecular weight 105.93; melting point 49-51°C; boiling point 61-62°C; density 2.01 at 20/4°C; vapor pressure 100 mm at 22.6°C.

B. AIR POLLUTION FACTORS

Incinerability

thermal stability ranking of hazardous organic compounds: rank 23 on a scale of 1 (highest stability) to 320 (lowest stability). (2390)

D. BIOLOGICAL EFFECTS

Fishes

Lepomis macrochirus: static bioassay in fresh water at 23°C, mild aeration applied after 24 h:

material added ppm	*% survival in water 24 h*	*48 h*	*72 h*	*96 h*	*best fit 96h LC_{50} ppm*
0.42	0	-	-	-	
0.32	70	70	70	70	
0.18	100	100	80	70	0.24
0.10	100	100	100	100	

Menidia beryllina: static bioassay in synthetic seawater at 23°C, mild aeration applied after 24 h:

material added ppm	*% survival after 24 h*	*48 h*	*72 h*	*96 h*	*best fit 96h LC_{50} ppm*
0.56	0	-	-	-	
0.42	85	85	85	85	
0.32	100	100	100	100	0.47
0.18	100	100	100	100	

(352)

(chlorinecyanide)

$$Cl-C\equiv N$$

CNCl

CAS 506-77-4

USES

intermediate for the synthesis of chlorocyano-derivatives such as cyanamide, guanidine, cyanate, thiocyanate etc.

A. PROPERTIES

colorless liquid or gas; molecular weight 61.48; melting point -6.5°C; boiling point 12.5/13°C; vapor pressure 1,000 mm at 20°C, 1,324 hPa at 20°C; vapor density 2; density 1.22 at 4/4°C; solubility 30,000 mg/L, 85,000 mg/L at 20°C; $LogP_{OW}$ 0.64 (calculated).

B. AIR POLLUTION FACTORS

1 mg/m^3 = 0.398 ppm, 1 ppm = 2.51 mg/m^3.

Odor

characteristic: quality: bitter almonds

hedonic tone: pungent
T.O.C.: recogn.: 2.5 mg/m^3 = 1 ppm (73)
O.I. at 20°C = 1,300,000 (316)

Incinerability

thermal stability ranking of hazardous organic compounds: rank 17 on a scale of 1 (highest stability) to 320 (lowest stability). (2390)

C. WATER AND SOIL POLLUTION FACTORS

Reduction of amenities

faint odor: 0.0025 mg/L. (129)

D. BIOLOGICAL EFFECTS

Crustaceans

Daphnia magna	age 24 hours	24h LC_{50},S	0.04 mg/L	
		48h LC_{50},S	0.029 mg/L	
	age 5 d	24h LC_{50},S	0.086 mg/L	
		48h LC_{50},S	0.065 mg/L	

Fishes

goldfish:	6-8h LC	1 mg/L	(154)

cyanoguanidine (Dicyandiamide; dicyanodiamide)

NC–N=C(NH$_2$)$_2$

$C_2H_4N_4$

CAS 461-58-5

EINECS :203-615-4

USES

The substance is a basic chemical and is used in industry for electrical/electronic engineering, metal extraction, refining and processing of metals, paper, pulp and board, textile processing, pharmaceuticals and intermediates. The substance is also a food additive, absorbent, adhesive, binding, coloring, electroplating, surface-active agent and fertilizer.The chemical is applied in agricultural and horticultural practice, in order to reduce nitrate losses from soils. (10626)

IMPURITIES, ADDITIVES, COMPOSITION

Melamine (0.7%); thiourea 200 mg/L; heavy metals 10 mg/L

A. PROPERTIES

White crystalline odourless powder; molecular weight; melting point 209.5°C; boiling point solidified at 252°C; density 1.4 kg/L at 25°C; vapour pressure 4.5 x 10^{-3} Pa at 100°C; water solubility 12,700 mg/L at 0°C; 40,000 mg/L at 25°C; log Pow −0.52 at 25°C; H 2.2 x 10^{-10} atm.m^3/mol

C. WATER AND SOIL POLLUTION FACTORS

Hydrolysis

No hydrolysis at 50°C after 5 days. Solutions above 80°C decompose slowly yielding ammonia (10626)

Biodegradation

Inoculum	method	concentration	duration	elimination	
A.S.	Ready biodegradability test	25 mg/L	28 days	0% ThOD	
A.S.	OECD 302 C	30 mg/L	14 days	0% ThOD	
Soil bacteria	Anaerobic	100 mg/L	40 days	40%	
A.S.	Aerobic	30 mg/L	10 days	0%	
A.S.			10 days	30-40%	(10626)

In soil cyanoguanidine seems to be gradually degraded via guanylurea, guanidine and urea. The first step in this degradation has been assumed to be catalyzed by the interaction with metal oxides in soil minerals rather than being due to microbial mineraliziation. (10626)

(H$_2$N)$_2$C=N–CN → HN=C(NH$_2$)–NH–C(=O)–NH$_2$ → [HN=C(NH)–NH]$^+$ → H$_2$N–C(=O)–NH$_2$

cyanoguanidine → guanylurea → guanidine → urea

D. BIOLOGICAL EFFECTS

Bioaccumulation

Species	conc. µg/L	duration	BCF
Cyprinus carpio	2,000	42 days	<0.3

	200	42 days	<3.1	(10626)

Toxicity

Algae			
Selenastrum capricornutum	72h ECb_{50}	935 mg/L	
	72h NOECb	171 mg/L	
	72h $ECgr_{50}$	>1,000 mg/L	
	72h NOECgr	556 mg/L	(10626)
Crustaceae			
Daphnia magna	48h EC_{50}	>1,000 mg/L	
	48h NOEC	1,000 mg/L	
reproduction	21d NOEC	25 mg/L	
	21d LOEC	50 mg/L	
	21d LC_{50}	>100 mg/L	
	21d EC_{50}	69.6 mg/L	(10626)
Fish			
Salmo gairdneri	96h NOEC	3,600 mg/L	
	96h LC_{50}	7,700 mg/L	
Oryzias latipes	24-96h LC_0-LC_{100}	>100 mg/L	
	24-48h LC_{50}	>2,300 mg/L	
	14d LC_0-50	>100 mg/L	
	14d NOEC	>100 mg/L	(10626)
Mammals			
Rat	oral LD_{50}	>30,000 mg/kg bw	
Mouse	oral LD_{50}	>10,000 mg/kg bw	(10626)

1-cyanonaphthalene

$C_{11}H_7N$

CAS 86-53-3

MANMADE SOURCES

in soots generated by the combustion of aromatic hydrocarbon fuels doped with pyridine. (1723)

2-cyanonaphthalene

$C_{11}H_7N$

CAS 613-46-7

MANMADE SOURCES

in soots generated by the combustion of aromatic hydrocarbon fuels doped with pyridine. (1723)

---

cyanophos *see* cyanox

---

cyanopropane *see* butyronitrile

2-cyanopyridine (picolinonitrile; 2-pyridinecarbonitrile; 2-azobenzonitrile)

N CN

$(C_5H_4N)CN$

$C_6H_4N_2$

CAS 100-70-9

USES

Chemical intermediate.

A. PROPERTIES

molecular weight 104.11; melting point 26-28°C; boiling point 212-215°C; solub 51,200 mg/L; $LogP_{OW}$: 0.23; 0.50.

D. BIOLOGICAL EFFECTS

Bacteria			
Photobacterium phosphoreum Microtox test	5-30 min EC_{50}	89 mg/L	(8899)
Algae			
Tetrahymena pyriformis	8 EC_{50}	645 mg/L	(2704)
	60h EC_{50}	647 mg/L	(8903)
Fishes			
Pimephales promelas:	30-35d LC_{50},F	728 mg/L	(2704)
Pimephales promelas	24h LC_{50}	1,050 mg/L	

	96h LC_{50}	726 mg/L	(2709)

cyanox (O-*p*-cyanophenyl-O,O-dimethylphosphorothioate cyanophos)

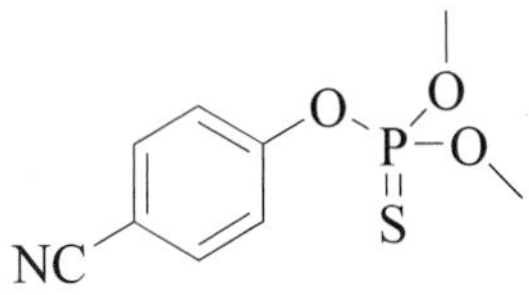

$C_9H_{10}NO_3PS$

CAS 2636-26-3

USES

insecticide.

A. PROPERTIES

molecular weight 243.2; clear amber liquid, melting point 14-15°C.

D. BIOLOGICAL EFFECTS

Fishes

harlequin fish *(Rasbora heteromorpha):*

mg/L	*24 h*	*48 h (40% active ingredient)*	
LC_{10},F	20	6.7	
LC_{50},F	36	14	(331)

Cytotoxicity to goldfish GF-Scale cells: NR_{50}: 45.5 mg/L = 0.19 mmol/L. (2680)

Cyprinus carpio: 48h LC_{50}: 5.2 mg/L. (2680)

cyanurotriamide *see* melamine

cyazofamid (4-chloro-2-cyano-N,N-dimethyl-5-P-tolylimidazole-1-sulfonamide)

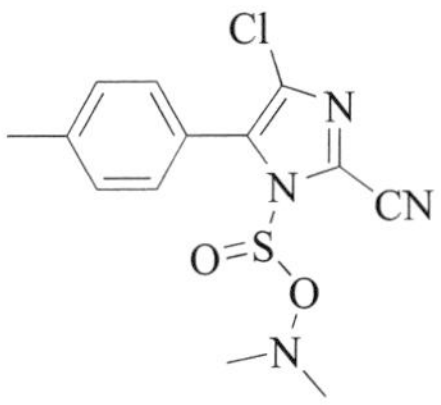

954 cyazofamid

$C_{13}H_{13}ClN_4O_2S$

CAS 120116-88-3

USES

fungicide

A. PROPERTIES

White to ivory powder; molecular weight 324.8; melting point 152.7°C; relative density 1.45 kg/L at 20°C; vapour pressure <$1.3.10^{-5}$ Pat at 35°C; water solubility 0.1-0.12 mg/L at 20°C and pH 5 to 9; log Pow 3.2 at 25°C; H <4 $x10^{-2}$ $Pa.m^3/mol$

C. WATER AND SOIL POLLUTION FACTORS

Hydrolysis

T/2 = at 25°C and pH 5 : 13 days, pH 7 : 12 days, pH 9 : 11 days

Photodegradation

T/2 = 0.5 hour at pH 5; metabolites:

CCIM	t/2 = 23 days
HTID:	t/2 = 44 days
CCTS	t/2 = 2.2 days

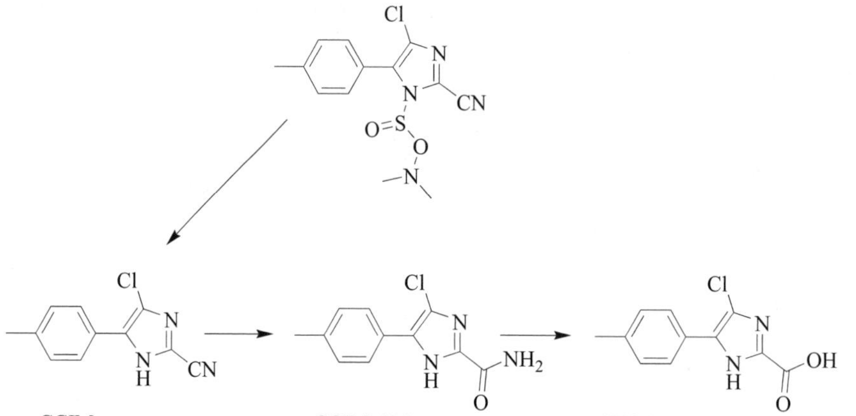

CCIM
4-chloro-5-p-tolylimidazole

CCIM-AM
4-chloro-5-p-tolylimidazole-
-2-carboxamide

CTCA
4-chloro-5-p-tolylimidazole-
-2-carboxylic acid

Biodegradation in laboratory studies (10769)

degradation in soils	%	Cyazofamide	CCIM	CCIM-AM	CTCA
Aerobic at 20°C	50	5.9-15 days	3.8-29 days	1-57 days	18-395 days
Aerobic at 20°C	90	17-50 days	12-71 days	123-187 days	
Aerobic at 10°C	50	37 days	7 days	9-38 days	229 days
Anaerobic at 20°C	50	5.8 days	4.7 days	35 days	slow

Biodegradation in water/sediment systems (10769)

Readily biodegradable	no
DT50 water	4.9-9.9 days
DT90 water	24-36 days
DT whole system	11-16 days
DT whole system	38-58 days

Mobility in soils (10770)

	Cyazofamid	CCIM	CCIM-AM	CTCA

K_{oc}	657–2,900	475-1,158	1,941-3,398	572-1,357

D. BIOLOGICAL EFFECTS

Toxicity

Algae			
Selenastrum capricornutum	72h EC_{50}	0.025 mg a.i./L	(10769)
Skeletonema costatum	72h EC_{50}	0.071 mg a.i./L	
	72h NOEC	0.0036 mg a.i./L	(10770)
Navicula pelliculosa	Acute EC_{50}	>0.12 mg a.i./L	
	Acute NOEC	0.019 mg a.i./L	
Anabaena flos-aquae	Acute EC_{50}	>0.25 mg a.i./L	(10770)
Birds			
Colinus virginianus	Oral LD_{50}	>2,000 mg/kg bw	
Anas platyrhnchus	Oral LD_{50}	>2,000 mg/kg bw	(10769)
Worms			
Eisenia foetida	14d LC_{50}	>1,000 mg a.io./kg bw	(10769)
Crustaceae			
Daphnia magna	48h EC_{50}	0.19 mg/L	(10769)
	48h NOEC	0.107 mg a.i./L	
Americanysis bahia	Acute LC_{50}	0.087 mg a.i./L	
	Acute NOEC	0.0369 mg a.i./L	(10770)
Insecta			
Honeybees	Oral LD_{50}	>151.7 μg/bee	
	Contact LD_{50}	>100 μg/bee	(10769)
Mollusca			
Crassostrea virginica	Acute EC_{50}	0.0147 mg a.i./L	(10770)
Fish			
Lepomis macrochirus	96h LC_{50}	>0.107 mg a.i./L	
	96h NOEC	0.107 mg a.i./L	(10770)
Pimephales promelas	Chronic	0.09 mg a.i./L	(10770)
Oncorhynchus mykiss	96h LC_{50}	0.56 mg/L	(10769)
	96h NOEC	0.107 mg a.i./L	
Cyprinodon variegates	96h LC_{50}	>0.167 mg a.i./L	
	96h NOEC	0.108 mg a.i./L	(10770)
Mammals			
Rat	oral LD_{50}	>5,000 mg/kg bw	(10769)
Mouse	oral LD_{50}	>5,000 mg/kg bw	(10770)

Toxicity of metabolite ***CCIM***

Worms			
Eisenia foetida	14d LC_{50}	56 mg a.io./kg bw	(10769)
Rat	oral LD_{50}	324 mg/kg bw	(10769)

Toxicity of metabolite ***CCIM-AM***

Worms			
Eisenia foetida	14d LC_{50}	>1,000 mg a.io./kg bw	(10769)
Rat	oral LD_{50}	>3,000 mg/kg bw	(10769)

Toxicity of metabolite ***CTCA***

Worms			
Eisenia foetida	14d LC_{50}	>1,000 mg a.io./kg bw	(10769)
Rat	oral LD_{50}	1,839 mg/kg bw	(10769)

Toxicity of metabolite ***DMSA***

Worms			
Eisenia foetida	14d LC_{50}	>1,000 mg a.io./kg bw	(10769)

Toxicity of metabolite ***CCBA***			
Worms			
Eisenia foetida	14d LC_{50}	>1,000 mg a.io./kg bw	(10769)

cycasin

B. AIR POLLUTION FACTORS

Incinerability

thermal stability ranking of hazardous organic compounds: rank 301 on a scale of 1 (highest stability) to 320 (lowest stability). (2390)

cyclanilide (1-(2,4-dichlorophenylaminocarbonyl)-cyclopropane carboxylic acid)

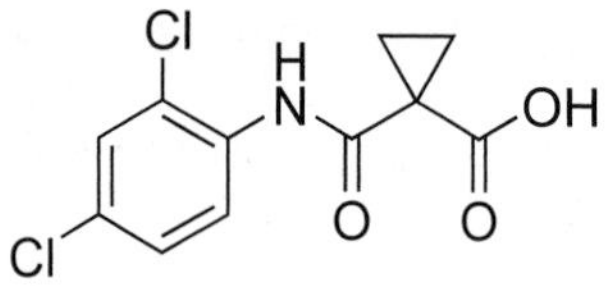

$C_{11}H_9Cl_2NO_3$

CAS 113136-77-9

TRADENAMES

Cyclanilide technical; FINISH Harvest Aid for Cotton

USES

Malonanilate plant growth regulator

A. PROPERTIES

White powdery solid; molecular weight 274.1 melting point 195°C; relative density 1.48 kg/L at 20°C; vapour pressure 0.84 x 10^{-5} Pa at 50°C; water solubility at 20°C and pH 5.2 : 37 mg/L; at pH 7 and 9 : 48 mg/L; log Pow 3.25 at 21°C; H <7.4 x 10^{-5} Pa.m^3/mol

B. AIR POLLUTION FACTORS

Photodegradation

T/2 (OH radicals) = 8 daylight hours. (10893)

C. WATER AND SOIL POLLUTION FACTORS

Hydrolysis

Not hydrolysed at 25 °C and pH 5, 7 and 9. (10893)

Photodegradation in water

T/2 (summer days in Florida) : 50-55 days at pH 5,7 and 9. (10893)

Soil photolysis

T/2 = 151 days at 25°C

Biodegradation in soil aerobic (10893)

% mineralisation	% bound-residues	After
4.3%	30%	120 days

Relevant metabolites : 2,4-dichloroaniline

Biodegradation of Cyclanilide

Biodegradation in laboratory studies (10893)

degradation in soils	%	
Aerobic at 20°C	50	15-62 days
Aerobic at 20°C	90	50-162 days
Anaerobic at 20°C	50	>15 months

Biodegradation in water/sediment systems (10893)

DT50 water	17-18 days
DT90 water	157-201 days
DT50 whole system	56-63 days
DT90 whole system	185-208 days

Mobility in soils (10893)

	Cyclanilide	2,4-dichloroaniline
K_{OC}	194-565	492-883

D. BIOLOGICAL EFFECTS

Toxicity

Algae			
Navicula pelliculosa	120h EC_{50}	>0.17 mg/L	(10894)
Anabaena flos-aquae	120h EC_{50}	0.08 mg/L	(10894)
Skeletonema costatum	120h EC_{50}	>0.27 mg/L	(10894)
Kirchneriellla subcapitata	120h EC_{50}	>0.27 mg/L	(10894)
Aquatic plants			
Lemna gibba	14d EC_{50}	>0.22 mg/L	(10894)
Worms			
Eisenia foetida	15d EC_{50}	469 mg/kg soil	(10894)
Crustaceae			
Daphnia magna	48h EC_{50}	13 mg/L	(10894)
Mysid shrimp	96h LC_{50}	5 mg/L	(10894)
Insecta			
Honeybees	Acute oral LD_{50}	89.5 µg/bee	
	Acute contact LD_{50}	>100 µg/bee	(10894)
Birds			
Bobwhite quail	14d LD_{50}	240 mg/kg bw	
Mollusca			

Crassostrea virginica	96h EC_{50}	19 mg/L	(10894)
Fish			
Oncorhynchus mykiss	96h LC_{50}	>11 mg/L	
Bluegill sunfish	96h LC_{50}	>16 mg/L	
Fathead minnow	NOEC	1.2 mg/L	
Sheepshead minnow	96h LC_{50}	49 mg/L	(10894)
Mammals			
Rat	oral LD_{50}	4208; 315 /kg bw (1)	(10894)
(1) Cyclanilde Technical			

---

cycloate *see* ro-neet

---

cyclobutanate (butanoic acid, 3a,4,5,6,7,7a-hexahydro-4,7-methano-1H-indenyl ester; methanoindene)

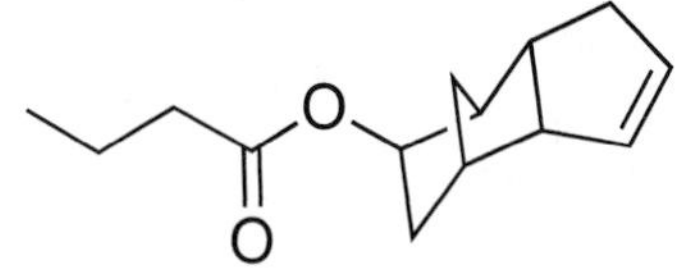

$C_{14}H_{20}O_2$

CAS 113889-23-9

USES

component of fragrance oils to be used in the cosmetic industry for production of toiletries, shampoos, soap, and household cleaning agents and detergents (containing <10 mg/L cyclobutanate) following mixing with other ingredients.

A. PROPERTIES

molecular weight 220.31; freezing point <-20°C; boiling point 274°C; density 1.03 kg/L at 20°C; vapour pressure 1.12 x 10^{-2} kPa at 20°C; water solubility 12 mg/L at 20°C; log Pow 4.48 at 21 °C; log K_{OC} 3.18 at 30 °C

C. WATER AND SOIL POLLUTION FACTORS

Hydrolysis

> 1 year at pH 4 to 7; 13 days at pH 9 (7405)

Biodegradation

Inoculum	method	concentration	duration	elimination	
Non-adapted AS	CO2 evolution test	300 mg/L	28 days	38%	(7405)

The substance is not readily biodegradable, see also graph (7405)

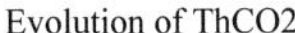

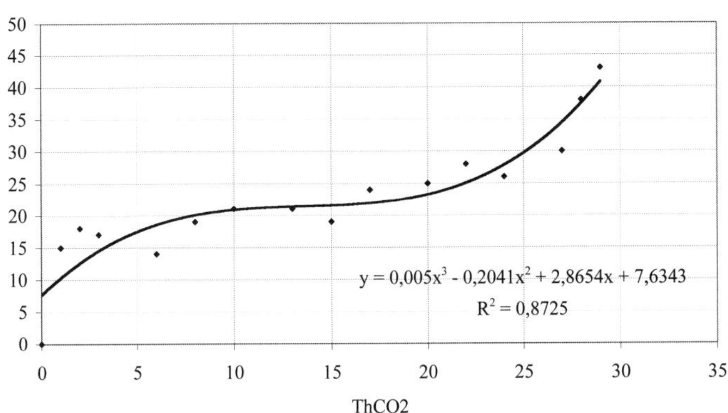

D. BIOLOGICAL EFFECTS

Bioaccumulation

the substance has a strong potential to bioaccumulate in animals. (7405)

Toxicity

Micro organisms			
Activated sludge respiration	30min IC_{50}	>1,000 mg/L	
	3h IC_{50}	>1,000 mg/L	(7405)
Algae			
Scenedesmus subspicatus	72h NOAECb	0.17 mg/L	
	72h ECb_{50}	0.29 mg/L	
	72h NOAECgr	0.17 mg/L	
	72h $ECgr_{50}$	0.39 mg/L	(7405)
Crustaceae			
Daphnia magna	24h EC_{50}	7.1 mg/L	
	48h LC_{50}	4.7 mg/L	
	24h NOAEC	3.8 mg/L	
	48h NOAEC	2.1 mg/L	(7405)
Fish			
Oncorhynchus mykiss	96h LC_{50}	3.6 mg/L	
	96h NOAEC	1.4 mg/L	(7405)
Mammals			
Rat	oral LD_{50}	>5,000 mg/kg bw	(7405)

cyclododecane

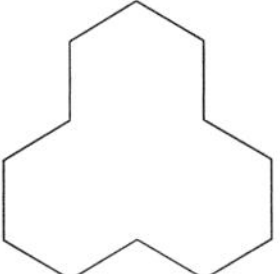

$C_{12}H_{24}$

CAS 294-62-2

USES

Feedstock for the manufacturing of C_{12}-polyamide.

A. PROPERTIES

molecular weight 168.32; melting point 61°C; boiling point 243°C; vapor pressure 0.1 hPa at 20°C; solubility 10 mg/L at 20°C; $LogP_{OW}$6.7 (measured).

C. WATER AND SOIL POLLUTION FACTORS

Biodegradation

Inoculum	*Method*	*Concentration*	*Duration*	*Elimination results*	
Municipal sludge	Closed bottle test	2 mg/L	28 d	3% product	(6255)

D. BIOLOGICAL EFFECTS

Bacteria			
Pseudomonas putida (oxygen consumption inhibition)	6 h EC_{10}	1.7 mg/L	(6255)
Crustaceans			
Daphnia pulex	48 h EC_{50}	21 mg/L	(2625)
Daphnia magna Straus	24 h EC_0	>10 mg/L	(6255)
Fishes			
Leuciscus idus	48 h LC_0	>10 mg/L	(6255)

cyclododecanol

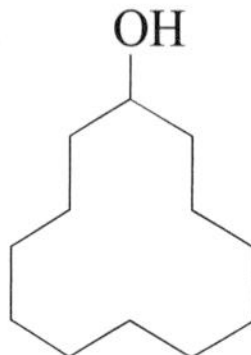

$C_{12}H_{24}O$

CAS 1724-39-6

USES

Feedstock for the manufacturing of laurinlactam, C_{12}-polyamide, and dodecanediodic acid.

A. PROPERTIES

molecular weight 184.3; melting point 75- 77°C; boiling point 278°C; vapor pressure <0.1 hPa at 20°C; solubility 40 mg/L at 20°C; $LogP_{OW}$4.2 (measured), 4.6 (calculated).

C. WATER AND SOIL POLLUTION FACTORS

Biodegradation

Inoculum	*Method*	*Concentration*	*Duration*	*Elimination results*	
	Closed bottle test	2 mg/L	28 d	100% ThOD	(6367)

D. BIOLOGICAL EFFECTS

Bacteria			
Pseudomonas putida (oxygen consumption inhibition test)	6 h EC_{10}	452 mg/L	(6367)
Algae			
Scenedesmus subspicatus (cell multiplication inhibition test)	72 h EC_{10}	1.7 mg/L	
	72 h EC_{50}	6.5 mg/L	
	72 h EC_{90}	25 mg/L	(6367)
Crustaceans			
Daphnia magna	24 EC_{50}	3.5 mg/L	(6367)
Fishes			
Leuciscus idus	48 h LC_{50}	3.9 mg/L	(6367)

cyclododecanone

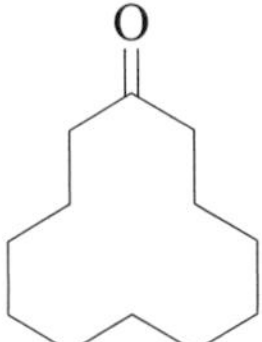

$C_{12}H_{22}O$

CAS 830-13-7

USES

Feedstock for the manufacture of laurinelactam and dodecandioic acid.

A. PROPERTIES

molecular weight 182.3; melting point 61°C; boiling point 277°C; vapor pressure 0.03 hPa at 20°C; density 0.97 at 20°C; solubility 45 mg/L at 20°C; $LogP_{OW}$ 3.8 (measured).

C. WATER AND SOIL POLLUTION FACTORS

Biodegradation

Inoculum	*Method*	*Concentration*	*Duration*	*Elimination results*	
Municipal waste water	Closed bottle test	2 mg/L	28 d	55% ThOD	(6329)

D. BIOLOGICAL EFFECTS

Bacteria			
Pseudomonas putida (oxygen consumption inhibition test)	6 h EC_{10}	>1,700 mg/L	(6329)
Algae			
Scenedesmus subspicatus (cell multiplication inhibition test)	72 h EC_{10}	2 mg/L	
	72 h EC_{50}	3.9 mg/L	
	72 h EC_{90}	7.6 mg/L	(6329)
Crustaceans			
Daphnia magna	24 h EC_{50}	7.5 mg/L	(6329)
Fishes			
Leuciscus idus	48 h LC_0	8.3 mg/L	
	48 h LC_{50}	11 mg/L	
	4 h LC_{100}	13 mg/L	(6329)

1,5,9-cyclododecatriene (CDT)

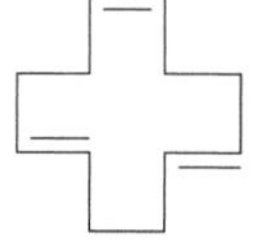

$C_{12}H_{18}$

CAS 4904-61-4

CAS 2765-29-9 (trans, trans, -cis-)

CAS 676-22-2 (trans, trans, -trans-)

USES AND FORMULATIONS

Feedstock for the manufacturing of C_{12}-polyamides, dodecanedioic acid, and flame retardants. The product is stabilized with 30-50 mg/L p-tert-butylcatechol in order to prevent the formation of peroxides.

A. PROPERTIES

molecular weight 162.28; melting point -15°C; boiling point 231°C; vapor pressure 0.1 hPa at 20°C; density 0.89 at 20/20°C; solubility 5 mg/L at 20°C; $LogP_{OW}$ 4.5 (measured).

C. WATER AND SOIL POLLUTION FACTORS

BOD_5:	0.02	(277)
COD:	3.02	(277)

Biodegradation

Inoculum	*Method*	*Concentration*	*Duration*	*Elimination*

			results	
Closed bottle test	2 mg/L	28 d	0% ThOD	(6504)

D. BIOLOGICAL EFFECTS

Bacteria			
Pseudomonas putida cell multiplication	18 h EC_{10}	>5 mg/L	(6504)
Oxygen consumption inhibition	6 h EC_{10}	12 mg/L	(6504)
Crustaceans			
Daphnia magna	24 h EC_{50}	2.9 mg/L	(6504)
Fishes			
Goldfish	24 h LD_{50}	4 mg/L	(277)
Leuciscus idus	48 h LC_{50}	3.2 mg/L	(6504)

1,3,5-cycloheptatriene (tropilidene)

C_7H_8

CAS 544-25-2

A. PROPERTIES

dark yellow liquid; molecular weight 92.14; melting point -79.5°C; boiling point 110-130°C; density 0.89 at 20/4°C.

C. WATER AND SOIL POLLUTION FACTORS

BOD_5:	0.10 = 3% of ThOD	(277)
COD:	2.30 = 74% of ThOD	(277)

D. BIOLOGICAL EFFECTS

Fishes			
goldfish	24h LD_{50}	15 mg/L	(277)

cycloheptene

C_7H_{12}

CAS 628-92-2

A. PROPERTIES

molecular weight 96.17; boiling point 112-113°C; density 0.824.

D. BIOLOGICAL EFFECTS

Threshold conc. of cell multiplication inhibition of the protozoan Uronema parduczi *Chatton-Lwoff:* EC_0 >40 mg/L (1901)

1,3-cyclohexadiene

C_6H_8

CAS 592-57-4

A. PROPERTIES

molecular weight 80.13; boiling point 80°C; density 0.84.

B. AIR POLLUTION FACTORS

Odor threshold

detection: 0.0025-0.0066 mg/m^3. (653)

D. BIOLOGICAL EFFECTS

Fishes

young Coho salmon:	24-96h LC_{33}	50 ppm in artificial seawater at 8°C	(317)

2,5-cyclohexadiene-1,4-dione *see p*-benzoquinone

cyclohexane (hexahydrobenzene; hexamethylene)

C_6H_{12}

CAS 110-82-7

A. PROPERTIES

colorless liquid; molecular weight 84.16; melting point 6.3°C; boiling point 81°C; vapor pressure 77 mm at 20°C, 120 mm at 30°C; vapor density 2.9; density 0.78 at 20/4°C; solubility 55 mg/L at 20°C, 45 mg/L at 15°C; THC 936 kcal/mol, LHC 881 kcal/mol; saturation concentration in air 357 g/m^3 at 20°C, 532 g/m^3 at 30°C; LogH 0.90 at 25°C.

B. AIR POLLUTION FACTORS

1 mg/m^3 = 0.29 ppm, 1 ppm = 3.49 mg/m^3.

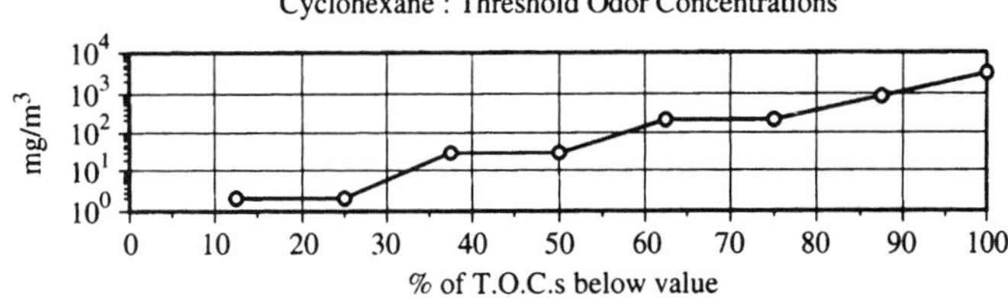

(54, 279, 307, 601, 643, 708, 737, 805, 828)

O.I. at 20°C = 203,000

Indoor/outdoor glc's winter 1981/1982 and 1982/1983 the Netherlands

µg/m^3	*median*	*maximum*	
pre-war homes	2	26	
post-war homes	1	22	
homes <6 years old	1	355	
outdoors	0.4	2	(2668)

C. WATER AND SOIL POLLUTION FACTORS

$BOD^{25°C}{}_{35\ d}$: 70% ThOD in seawater/inoculum: enrichment cultures of hydrocarbon-oxidizing bacteria. (521)

ThOD: 3.42

Manmade sources

In Canadian municipal sludges and sludge compost: September 1993- February 1994: mean values of 11 sludges ranged from 0.035 to 0.12 mg/kg dw.; mean: 0.070 mg/kg dw mean value of sludge compost: ND. (7000)

cyclohexanol → cyclohexanone → 1-oxa-2-oxo-cycloheptane → 6-hydroxy-hexanoate → 6-oxohexanoate → adipic acid → acetyl-CoA + scuccinyl-CoA

Microbial degradation of cyclohexane

(1244, 1245)

Incubation with natural flora in the groundwater-in presence of the other components of high-octane gasoline (100 µl/L) biodegradation: 45% after 192 h at 13°C (initial conc. 0.12 µl/L).	(956)
Rotating disk contact aerator: infl. 231 mg/L, effl. 0.2 mg/L; elimination: >99% or 19,077 $mg/m^2/24$ h or 5,151 $g/m^3/h$.	(406)

Reduction of amenities

T.O.C. = 0.02 mg/L.
Partition coefficients:

Cuticular matrix/air partition coefficient*	170 ± 10

* experimental value at 25 °C studied in the cuticular membranes from mature tomato fruits (*Lycopersicon esculentum* Mill. cultivar Vendor)

D. BIOLOGICAL EFFECTS

Protozoa

Threshold conc. of cell multiplication inhibition of the protozoa:

Uronema parduczi Chatton-Lwoff:	EC_0	>50 mg/L	(1901)

Mussels

mussel larvae *(Mytilus edulis):* 10-20% increase of growth rate at 1 to 100 ppm. (475)

Fishes

guppy *(Poecilia reticulata):* 7d LC_{50}: >84 (1833)

fathead minnows: static bioassay in Lake Superior water at 18-22°C: LC_{50} (1; 24; 48; 72; 96 h): 95; 93; 93; 93; 93 mg/L
fathead minnows: static bioassay in reconstituted water at 18-22°C: LC_{50} (1; 24; 48; 72; 96 h): 126; 117; 117; 117; 117 mg/L (350)

mosquito fish:	24h LC_{50}	15,500 mg/L in turbid Oklahoma water	(244)
fatheads:	24-96h LC_{50}	43-32 mg/L	
bluegills:	24-96h LC_{50}	43-34 mg/L	
goldfish:	24-96h LC_{50}	42.3 mg/L	
guppies:	24-96h LC_{50}	57.7 mg/L	(158)

young Coho salmon:	96h NOLC	100 ppm in artificial seawater at 8°C	(317)

cyclohexanecarboxylic acid (hexahydrobenzoic acid)

$C_6H_{11}COOH$

$C_7H_{12}O_2$

CAS 98-89-5

A. PROPERTIES

molecular weight 128.17; melting point 30-32°C; boiling point 232-233°C; density 1.03.

Manmade sources

landfill leachate from lined disposal area (Florida 1987-'90): mean conc. 0.99; 2.5 mg/L. (2788)

1,2-cyclohexanediol

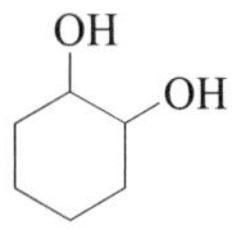

$C_6H_{10}(OH)_2$

$C_6H_{12}O_2$

CAS 1460-57-7 (trans)

CAS 931-17-9 (mixture of cis and trans)

A. PROPERTIES

molecular weight 116.16; melting point 72.5-75°C (mixture of *cis* and *trans*); boiling point 118-120°C/ 10 mm.

C. WATER AND SOIL POLLUTION FACTORS

Biodegradation

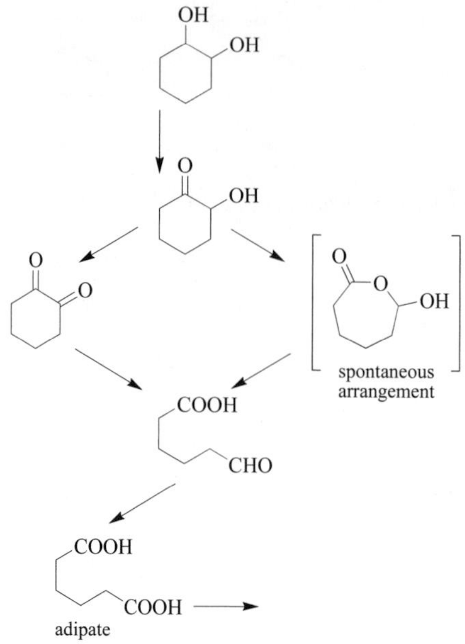

Alternative routes for the conversion of cyclohexan-1,2-diol to adipate	(426)
adapted A.S.-product as sole carbon source: 95% COD removal at 66 mg COD/g dry inoculum/h	(327)

cyclohexanol (hexahydrophenol; cyclohexylalcohol; hexalin; adronol; hydrophenol; hydralin; anol)

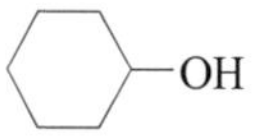

$C_6H_{11}OH$

$C_6H_{12}O$

CAS 108-93-0

USES

Manufacture of adipic acid for the production of nylon 66. Used in production of lacquers, paints, varnishes, degreasing agents, plastics, plasticizers, soaps, detergents, rubber cements, textiles, dyes, and insecticides.

OCCURRENCE

Metabolite of cyclamate.

IMPURITIES, ADDITIVES, COMPOSITION

May contain cyclohexanone.

A. PROPERTIES

colorless liquid; molecular weight 100.16; melting point 24°C; boiling point 161°C; vapor pressure 1 mm at 20°C, 3.5 mm at 34°C, 1.3 hPa at 20°C, 4.7 hPa at 34°C; vapor density 3.45; density 0.95 at 25/4°C; solubility 56,700 mg/L at 15°C, 36,000 mg/L at 20°C; THD 890 kcal/mol, LHC 838 kcal/mol; saturation concentration in air 4.9 g/m^3 at 20°C, 10.0 g/m^3 at 30°C; LogP_{OW} 1.23.

B. AIR POLLUTION FACTORS

1 mg/m^3 = 0.244 ppm, 1 ppm = 4.163 mg/m^3.

Odor

T.O.C.:	0.21 mg/m^3	(639, 57)
Human reflex response: no response:	0.04 mg/m^3	
Animal chronic exposure: no effect:	0.059 mg/m^3	
Adverse effect:	0.61 mg/m^3	(170)
O.I. at 20°C = 26,300		(316)

Photochemical reactions

$t_{1/2}$: 22 h, based on reactions with OH° (calculated). (8720)

C. WATER AND SOIL POLLUTION FACTORS

Volatilization

from a model river:	$t_{1/2}$: 23 h	
from a model pond:	$t_{1/2}$: 11 d	(8721)

BOD_5:	3, 4, 13% ThOD	(30, 41, 27, 220)
BOD_{20}:	69% ThOD	(30)
COD:	76; 96% of ThOD	(220)
TOC:	96% of ThOD	(220)
ThOD:	2.83	(30)

Biodegradation pathways

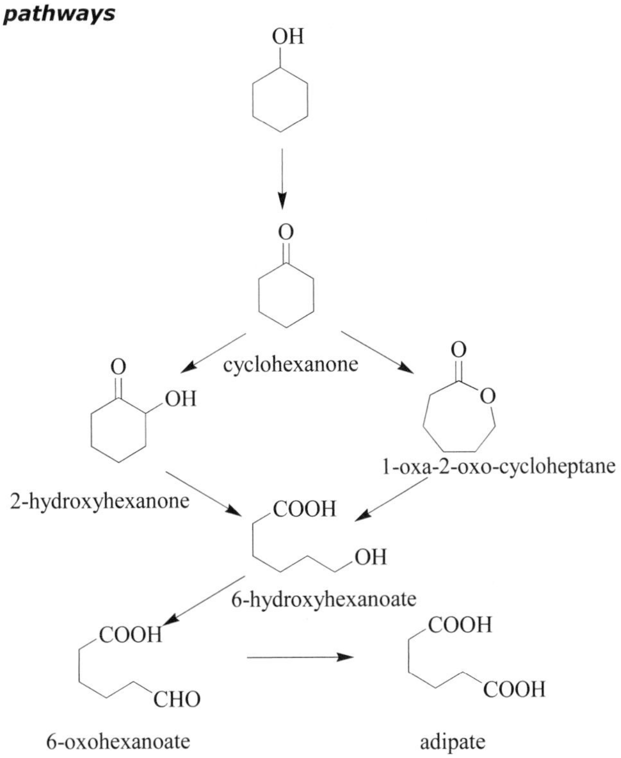

Alternative routes for the metabolic conversion of cyclohexanol to adipate (426)

Biodegradation

Inoculum	*Method*	*Concentration*	*Duration*	*Elimination results*	
Mixed culture adapted	Closed bottle test			55% ThOD	(8723)
Municipal waste water, nonadapted	Laboratory pilot plant	500 mg/L		66% ThOD	
Municipal waste water, adapted	Laboratory pilot plant	500 mg/L		70% ThOD	(8722)
Industrial waste water	Closed bottle test		5 d	63% ThOD	(8724)
Adapted inoculum	Closed bottle test		5 d	74% ThOD	(8725)
	Closed bottle test		5 d	65% ThOD	(8726)
A.S. adapted	Closed bottle test	ca. 100 mg/L		96% COD	(8727)
A.S.	MITI test	100 mg/L	28 d	94- 99% ThOD	(8728)
Municipal effluent, nonadapted			5 d	78% ThOD	(8730)
A.S. industrial		398 mg DOC/L	3 h	11% DOC	
			1 d	45% DOC	
			4 d	98% DOC	(8726)
A.S. industrial nonadapted	Zahn-Wellens test	400 mg/L	7 d	97% DOC	(8731)
Municipal waste water, nonadapted	Respirometer test	500 mg/L	1 d	45% ThOD	
Municipal waste water, adapted	Respirometer test	500 mg/L	1 d	51% ThOD	
Nocardia globerula adapted			2 d	78% ThOD	(8732)
Methanogenic laboratory culture		213 mg/L	23 d	<5% ThCH4	(8733)

$$\text{cyclohexanol (C}_6\text{H}_{11}\text{–OH)} \longrightarrow \text{phenol (C}_6\text{H}_5\text{–OH)} + CO_2$$

Anaerobic denitrifying bacteria isolated from anaerobic municipal sludge metabolized cyclohexanol under aerobic and anaerobic conditions. One of the strains degraded cyclohexanol to phenol (40%) and to CO_2 (60%). (8734)

Odor thresholds

detection:	3.5 mg/kg	(886)
	0.4 mg/kg	(894)

Waste water treatment

reverse osmosis: 68% rejection from 0.01 M solution adapted A.S.-product as sole carbon source-96.0% COD removal at 28 mg COD/g dry inoculum/h. (327)

D. BIOLOGICAL EFFECTS

Bacteria

A.S. oxygen consumption inhibition	3 h EC_{50}	>10,000 mg/L	(6000)
Primary municipal sludge	24 h EC_0	100-1,000 mg/L	(6100)

Pseudomonas putida	17 h EC_{10}	472 mg/L	
(cell multiplication inhibition test)	17 h EC_{50}	955 mg/L	
	17 h EC_{90}	1,989 mg/L	(8741)
A.S. industrial respiration test	30 min EC_{10}	>1,995 mg/L	(8743)
	30 min EC_{20}	>400 mg DOC/L	(8726)
Photobacterium fosforeum	5 min EC_{50}	115 mg/L	(8744)
Microbial mixed culture respiration test	75 min EC_{50}	3,105 mg/L	(8746, 8747)
Algae			
Scenedesmus subspicatus	24 h EC_{20}	414 mg/L	
(*see also graph*)	48 h EC_{20}	32 mg/L	
	48 h EC_{50}	452 mg/L	
	72 h EC_{20}	0.11 mg/L	
(cell multiplication inhibition test)	72 h EC_{50}	29 mg/L	
	72 h EC_{90}	>500 mg/L	
	96 h EC_{20}	0.22 mg/L	
	96 h EC_{50}	29 mg/L	
	96 h EC_{90}	470 mg/L	(8741)

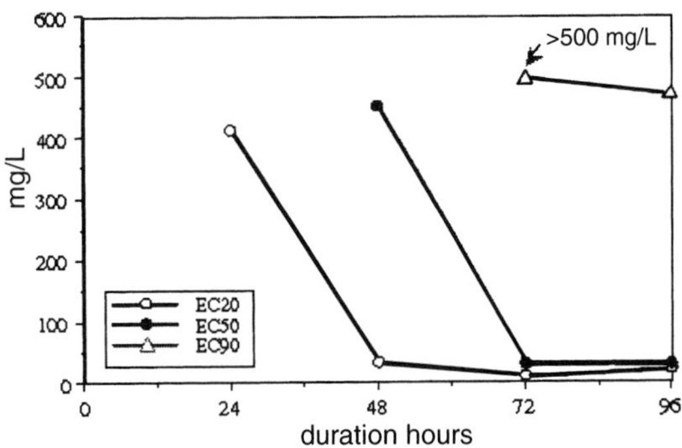

Crustaceans			
Daphnia magna Straus	24 h EC_{0}	250 mg/L	
	24 h EC_{50}	>500 mg/L	
	24 h EC_{100}	>500 mg/L	(8741)
Elminius modestus	15 min EC_{50}	3.7 mg/L	(8742)
Fishes			
Fathead minnows:	1 h LC_{50}	1,550 mg/L	
	24-96 h LC_{50}	704; 705; 1,033 mg/L	(8735, 8737, 8740, 350)
Menidia beryllina	96 h LC_{50}	720 mg/L	(352)
	96 h NOEC	>500 mg/L	(8738)
Lepomis macrochirus	96 h LC_{50}	1,100 mg/L	(352)
	96 h NOEC	>790 mg/L	(8738)
Mammals			
Rabbit, rat	oral LD_{50}	2,060; 2,200 mg/kg body wt	(8690, 8691)

cyclohexanone (ketohexamethylene; pimelic ketone; sextone, anone)

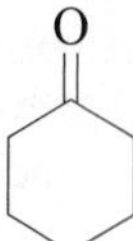

$C_6H_{10}O$

CAS 108-94-1

USES

Cyclohexanone is used in organic synthesis, particularly in the production of adipic acid and caprolactam, PVC and its copolymers, and methacrylate ester polymers. Additional uses include wood stains, paint and varnish removers, spot remover, degreasing of metals, polishes, levelling agents, dyeing and delustering silk, lubricating oil additives, solvent for herbicides, cellulosics, natural and synthetic resins, waxes, fats, etc...

Feedstock for mfg. of caprolactam-polyamide 6 and other plastics; solvent for pesticides, PVC printing inks and varnishes.

A. PROPERTIES

mw. 98.2; melting point -26/-38°C; boiling point 157°C; vapor pressure 4 mm at 20°C, 6.2 mm at 30°C, vapor pressure 4.5 hPa at 20°C; 13 hPa at 40°C; 35.4 hPa at 60°C; vapor density 3.38; density 0.95 at 20/4°C; solubility 23,000 mg/L at 20°C, 24,000 mg/L at 31°C; saturation concentration in air 19 g/m³ at 20°C, 32 g/m³ at 30°C; log P_{ow} 0.81.

B. AIR POLLUTION FACTORS

1 mg/m³ = 0.25 ppm, 1 ppm = 4.08 mg/m³.

Odor

sweet, sharp, pleasant.

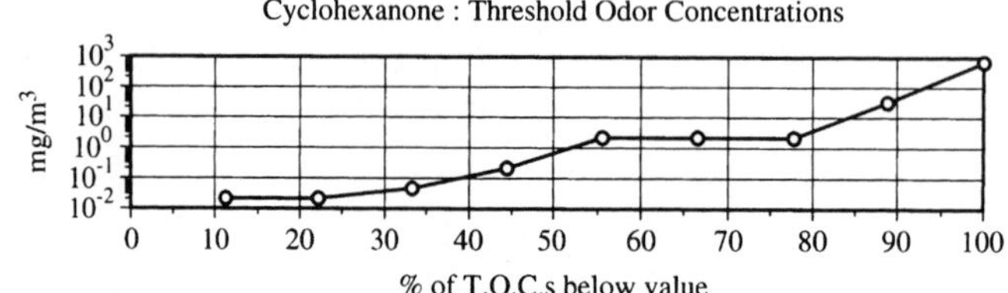

human reflex response:	no response	0.06 mg/m³	
animal chronic exposure:	no effect	0.042 mg/m³	
	adverse effect	0.46 mg/m³	(170)
O.I. at 20°C: 21,900			(316)

Photodegradation

By direct photolyis : t/2 = 4.3 days; reasction with OH radicals: t/2 = 1 day.

C. WATER AND SOIL POLLUTION FACTORS

BOD_5:	32; 47% of ThOD	(30, 220)
BOD_{20}:	77% ThOD	(30)
COD:	100% of ThOD	(220)
ThOD:	2.60	(30)

Biodegradation

adapted A.S. product as sole carbon source: 96% COD removal at 30 mg/COD/g dry inoculum/h.	(327)

Waste water treatment

A.C. adsorbability: 0.13 g/g C, 67% reduction, infl.: 1,000 mg/L.	(32)

Impact on biodegradation processes

OECD 209 closed system inhib. EC_{50} >1,000 mg/L.	(2624)

Partition coefficients

Cuticular matrix/air partition coefficient*	8400 ± 230	(7077)

* experimental value at 25°C studied in the cuticular membranes from mature tomato fruits (*Lycopersicon esculentum* Mill. cultivar Vendor)

D. BIOLOGICAL EFFECTS

Toxicity threshold (cell multiplication inhibition test)

Bacteria			
Pseudomonas putida	16h EC_0	180 mg/L	(1900)
Algae			
Microcystis aeruginosa	8d EC_0	52 mg/L	(329)
green algae			
Scenedesmus quadricauda	7d EC_0	370 mg/L	(1900)
Protozoa			
Entosiphon sulcatum	72h EC_0	545 mg/L	(1900)
Uronema parduczi Chatton-Lwoff	EC_0	280 mg/L	(1901)

Bacteria			
Pseudomonas	toxic	500 mg/L	
Pseudomonas fluorescens	16h EC_0	180 mg/L (pH = 7)	(5267)
Algae			
Scenedesmus	not toxic at 1 g/L		(30)
Protozoans			
Chilomonas paramecium			
Ehrenberg	48h EC_0	573 mg/L	(5272)
Crustaceans			
Daphnia magna Straus	24h EC_0	526 mg/L	
	24h EC_{50}	820 mg/L	
	24h EC_{100}	1,240 mg/L	(5265)
Daphnia magna	24h EC_0	540 mg/L	
	24h EC_{50}	800 mg/L	
	24h EC_{100}	1,540 mg/L	(5266)
Fishes			
fathead minnow	96h LC_{50}	526; 618; 630 mg/L	(5257, 5259)
Leuciscus idus	24h LC_{50}	538 mg/L	(5258)
	96h LC_{50}	536; 539; 752 mg/L	(5260, 5261)

Toxicity

Micro organisms			
Pseudomonas putida	16h NOEC	90 mg/L	(10689)

Algae			
Microcystis	7d NOEC	26 mg/L	
Scenedesmus quadricauda	7d NOEC	185 mg/L	(10689)
Protozoans			
Entosiphon sulcatum	72h NOEC	273 mg/L	(10689)
Mammals			
Rat	oral LD_{50}	1,296; 1,620; 1,800; 1,840; 2,070; 2,110 mg/kg bw	
Mouse	oral LD_{50}	1,400 mg/kg bw	(10689)

Metabolites in human newborns

trans-1,2-cyclohexanediol with small amounts of 1,3- and 1,4-cyclohexanediol, and sometimes, traces of cis-1,2-cyclohexanediol. Glucuronide conjugates were not detected.

Metabolites

trace amounts of hydrocycyclohexylmercapturic acid and cis-2-hydrocycyclohexylmercapturic acid were excreted in the urine
Metabolites in beagle dogs: less than 1% of the dose was excreted as cyclohexanone and cyclohexanol.

Cyclohexanone oxime ((Hydroxyimino)cyclohexane; Antioxidant D)

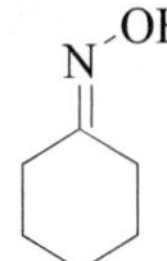

$C_9H_{12}O$

$C_6H_{10}(NOH)$

CAS 100-64-1

USES

Cyclohexanone oxime, a white crystalline solid, is a "closed-system intermediate," and is used primarily as a captive intermediate in the synthesis of caprolactam which, in turn, is polymerized to polycaprolactam (Nylon-6) fibers, resins and plastics. (10801)

A. PROPERTIES

molecular weight 113.18; melting point 89-91°C; boiling point 206-210°C; specific gravity 0.97 mg/L; vapour pressure 0.029 mm Hg at 77°F; water solubility 15,000 mg/L at 68°F; log Pow 0.84 at 77°F; H 8 x 10^{-6} atm.m^3/mole;vapor density 3.91

B. AIR POLLUTION FACTORS

1 mg/m^3 = 0.21 ppm, 1 ppm = 4.70 mg/m^3.

Odour

A pungent to slightly sweet odour

Atmospheric degradation

Atmospheric photo-oxidation may be an important removal process for cyclohexanone oxime. (10801)

C. WATER AND SOIL POLLUTION FACTORS

Hydrolysis

Stable at environmental conditions

BOD_5	0.030	(30)
BOD_{20}	0.130	(30)

D. BIOLOGICAL EFFECTS

Toxicity

Fish			
Pimephales promelas	96h LC_{50}	208 mg/L	(10801)
Mammals			
Mouse	oral LD_{50}	>500 mg/kg bw	(10801)
Bacteria			
Pseudomonas	toxic	30 mg/L	
Algae			
Scenedesmus	toxic	480 mg/L	
Protozoa			
Colpoda	toxic	60 mg/L	
Arthropoda			
Daphnia	toxic	120 mg/L	(30)

Metabolism in rats

Three urinary metabolites were identified: cyclohexylglucuronide and the monoglucuronides of cis- and trans-cyclohexane-1,2-diol. (10801)

N-OH
glucuronic acid
cyclohexylglucuronide
OH
O
glucuronic acid
monoglucuronides of
cis- and trans-cyclohexane-
-1,2-diol
urine

cyclohexanoneisooxime *see* α-caprolactam

cyclohexene (1,2,3,4-tetrahydrobenzene; benzenetetrahydride)

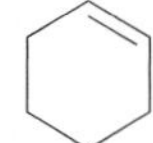

C_6H_{10}

CAS 110-83-8

USES

Alkylation component in the manufacture of adipic acid, maleic acid, hexahydrobenzoic acid and aldehyde. Used in the preparation of butadiene in laboratory. Used in oil extraction. Inorganic synthesis. Catalyst solvent.

OCCURRENCE

: In coal tar.

A. PROPERTIES

molecular weight 82.14; melting point -104°C; boiling point 83°C; vapor pressure 160 mm at 38°C; vapor density 2.9; density 0.81 at 20/4°C; solubility 213 mg/L at 20°C.

B. AIR POLLUTION FACTORS

1 mg/m^3 = 0.29 ppm, 1 ppm = 3.41 mg/m^3.

Odor threshold

detection: 0.6 mg/m^3. (637)

Atmospheric half-lives: for reactions with OH°: 0.1 d; for reactions with O_3: 0.05 d (2716)

Photodegradation: t/2: 0.8 hours (reaction of ozone and hydroxyl radical) (10622)

C. WATER AND SOIL POLLUTION FACTORS

Biodegradation

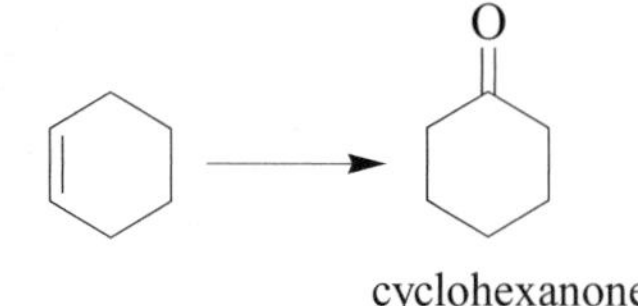

(203)

Waste water treatment

rotating disk contact aerator: infl. 123 mg/L, effl. 0.4 mg/L; elimination: >99% or 10,139 mg/m^2/h or 2,737 g/m^3/h. (406)

Biodegradation

Inoculum	method	concentration	duration	elimination	
A.S.	Modified MITI test	100 mg/L	28 days	0% ThOD	
				0% GC analysis	(10622)

D. BIOLOGICAL EFFECTS

Bacteria

Pseudomonas putida	16h EC_0	17 mg/L	(329)
Algae			
Microcystis aeruginosa	8d EC_0	>160 mg/L	(329)
Protozoa			
Uronema parduczi Chatton-Lwoff cell multiplication inhibition	EC_0	>50 mg/L	(1901)
Crustaceans			
Daphnia magna strauss	48h EC_{50}	720 mg/L	(8763)
Fishes			
young Coho salmon	96h NOLC	100 mg/L,	(317)

Bioaccumulation

Species	conc. µg/L	duration	body parts	BCF	
Cyprinus carpio	100	28 days		12-38	(10622)

Toxicity

Algae			
Chlorella pyrenoidosa	48h NOECb	0.22 mg/L	
	48h ECb_{50}	3.8 mg/L	
Selenastrum capricornutum	72h NOECgr	0.67 mg/L	(10622)
Crustaceae			
Daphnia magna	48h EC_0	1.5; 3.8 mg/L	
	48h EC_{50}	2.1; 5.3 mg/L	
	24h EC_0	563 mg/L	
	24h EC_{50}	720 mg/L	
	24h EC_{100}	750 mg/L	
reproduction	21d NOEC	0.53 mg/L	
	21d LOEC	0.74 mg/L	
	21d EC_{50}	1 mg/L	
	15d NOEC	2.4 mg/L	
	15d LOEC	2.9 mg/L	
	15d EC_{50}	4 mg/L	(10622)
Protozoans			
Chilomonas sp.	48hLOEC	>160 mg/L	(10622)
Fish			
Poecilia reticulata	96h LC_0	3.1 mg/L	
	96h LC_{50}	12.4 mg/L	
Oryzias latipes	96h LC_0	4 mg/L	
	96h LC_{50}	5.8 mg/L	
	96h LC_{100}	17 mg/L	(10622)
Mammals			
Rat	oral LD_{50}	>1,000 mg/kg bw	(10622)

N-cyclohexyl-2-benzothiazolylsulfenamide

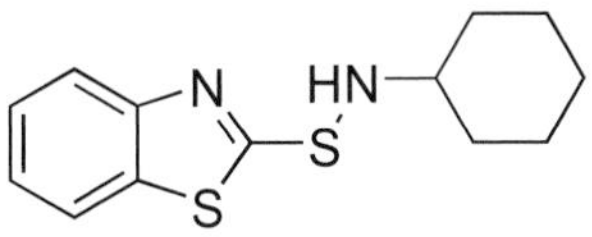

$C_{13}H_{16}N_2S_2$

CAS 95-33-0

TRADENAMES

Accelerator CBS

USES

Rubber auxiliary, vulcanization accelerator

A. PROPERTIES

molecular weight 264.40

D. BIOLOGICAL EFFECTS

Toxicity

Algae	72h EC_{50}	0.10; 0.15 mg/L	(11221)
	72h NOEC	0.0084; 0.016 mg/L	(11221)
Daphnids	48h EC_{50}	0.79 mg/L	(11221)
	21d EC_{50}	0.12 mg/L	(11221)
	21d NOEC	0.058 mg/L	(11221)
Fish	96h LC_{50}	2.1 mg/L	(11221)
	14d LC_{50}	0.78 mg/L	(11221)
	14d NOEC	0.14 mg/L	(11221)

2-cyclohexyl-4,6-dinitrophenol

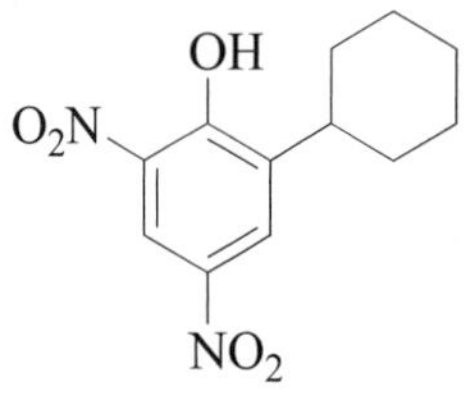

CAS 131-89-5

B. AIR POLLUTION FACTORS

Incinerability

thermal stability ranking of hazardous organic compounds: rank 187 on a scale of 1 (highest stability) to 320 (lowest stability). (2390)

2-cyclohexyl propanal (Pollenal II)

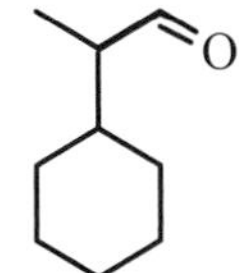

$C_6H_{11}CH(CH_3)CHO$

$C_9H_{16}O$

CAS 2109-22-0

USES

the chemical is a fragrance enhancer in formulated perfumes used in household, toiletry and cosmetic products such as soaps, detergents, fabric softeners and may contain between 0.04 and 0.23 % of the chemical. (7118)

IMPURITIES, ADDITIVES, COMPOSITION

2-cyclohexyl propionic acid (<1.0%) and up to 4% of unidentified impurities; <0.05% of 2,6-di-tert.butyl-4-hydroxytoluene is added as anti-oxidant. (7118)

A. PROPERTIES

clear colourless, non-viscous liquid; molecular weight 140.2; boiling point 196.5-200.0 °C at 1025 mbar; density 0.91; vapour pressure 0.082 kPa at 25 °C; water solubility 423 mg/L at 20 °C; $LogP_{OW}$ 2.95 at 20 °C; H 27.04 Pa/m^3/mole (calculated)

B. AIR POLLUTION FACTORS

Photochemical reactions

The compound is not expected to persist in the atmosphere (7118)

C. WATER AND SOIL POLLUTION FACTORS

Hydrolysis

no significant hydrolysis at pH 4 and 7 (t/2 > 1 year at 25 °C), some hydrolysis at pH 9 (t/2 between 1 day and 1 year at 25 °C). The compound is not expected to exhibit significant hydrolysis under ambient environmental conditions. (7118)

Mobility in soil

log K_{OC} : 2.49 (QSAR calculations) (7118)

Biodegradation

Inoculum	method	concentration	duration	elimination	
	Closed Bottle Test		28 days	11% COD	(7118)

D. BIOLOGICAL EFFECTS

Bioaccumulation

Species	conc. µg/L	duration	body parts	BCF	
Fathead minnow				58 (calculated)	(7118)

Toxicity

Crustaceae

Daphnia magna	48h EC_{50}	1.2 mg/L

	48h NOEC	0.6 mg/L	(7118)
Fish			
Oncorhynchus mykiss	96h LC_{50}	3.2 mg/L	
	96h NOEC	1.5 mg/L	(7118)
Mammals			
Rat	oral LD_{50}	> 5000 mg/kg bw for males	
	oral LD_{50}	> 2000 mg/kg for females	(7118)

cyclohexylacetate (hexaline acetate; cyclohexanol acetate)

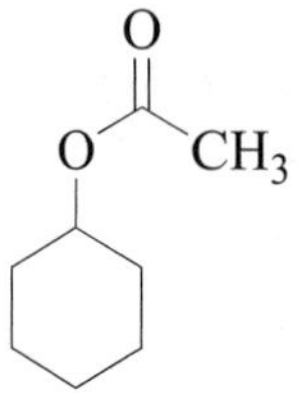

$C_8H_{14}O_2$

CAS 622-45-7

A. PROPERTIES

molecular weight 142.19; melting point 177°C; vapor density 4.9; density 1.0.

D. BIOLOGICAL EFFECTS

Toxicity threshold (cell multiplication inhibition test)

Bacteria			
Pseudomonas putida	16h EC_0	83 mg/L	(1900)
Algae			
Microcystis aeruginosa	8d EC_0	46 mg/L	(329)
green algae			
Scenedesmus quadricauda	7d EC_0	5.3 mg/L	(1900)
Protozoa			
Entosiphon sulcatum	72h EC_0	120 mg/L	(1900)
Uronema parduczi Chatton-Lwoff	EC_0	>400 mg/L	(1901)

cyclohexylacetic acid

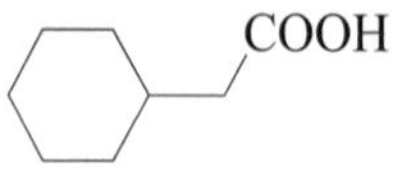

$C_6H_{11}CH_2COOH$

$C_8H_{14}O_2$

CAS 5292-21-7

A. PROPERTIES

molecular weight 142.2; melting point 31-33°C; boiling point 242-244°C; density 1.01.

C. WATER AND SOIL POLLUTION FACTORS

microbial metabolite of *n*-alkyl-substituted cyclohexanes

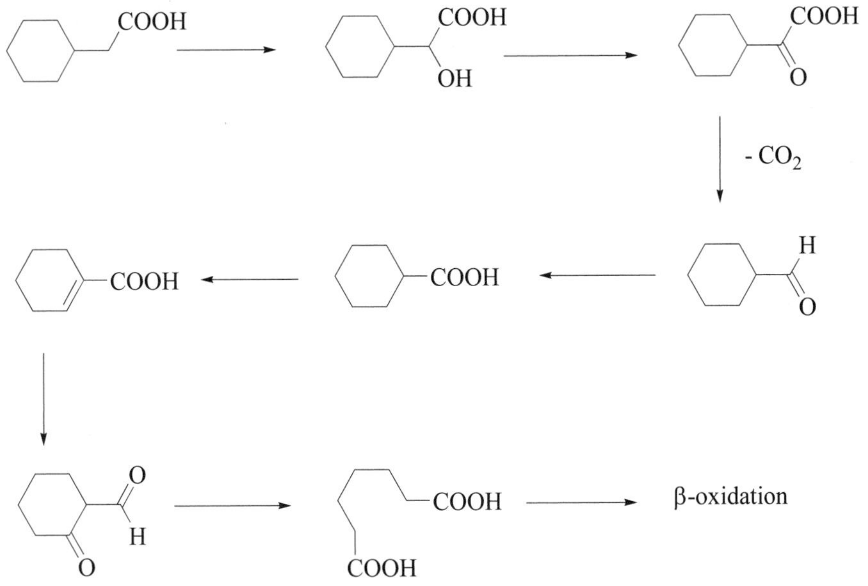

COOH → COOH OH → COOH O

$- CO_2$

COOH ← COOH ← H O

O H → COOH COOH → β-oxidation

Proposed pathway for the utilization of cyclohexylacetic acid by the marine bacteria *Alcaligenes* sp. (2604)

cyclohexylalcohol *see* cyclohexanol

cyclohexylamine (CHAM; hexahydroaniline; aminocyclohexane)

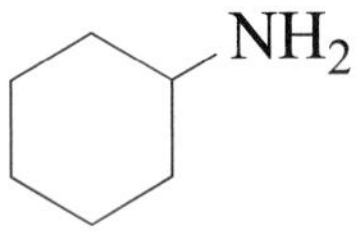

NH_2

$C_6H_{11}NH_2$

$C_6H_{13}N$

982 cyclohexylamine

CAS 108-91-8

USES

Organic synthesis. Manufacture of plasticizers, corrosion inhibitors, rubber chemicals, dyestuffs, emulsifying agents, dry cleaning soaps, acid gas absorbants, and insecticides.

Cyclohexylamine (CHAM) is widely used as an insecticide and antiseptic in various industries

A. PROPERTIES

molecular weight 99.17; melting point – 18°C; boiling point 134°C; vapor pressure 8.4– 14 hPa at 20°C, 24 hPa at 30°C, 59 hPa at 59°C; vapor density 3.42; density 0.87 at 20/4°C; solubility miscible; log P_{OW} 1.2 (measured); 1.4– 1.5 (calculated).

B. AIR POLLUTION FACTORS

1 mg/m^3 = 0.247 ppm, 1 ppm = 4.06 mg/m^3.

C. WATER AND SOIL POLLUTION FACTORS

BOD_5:1.15
COD: 2.4

Biodegradation

Inoculum	*Method*	*Concentration*	*Duration*	*Elimination results*	
Domestic sewage	Closed bottle test	2.4 mg/L	20 d	>90% ThOD	(8710)
River mud, enriched culture	Sapromat	10 mg/L	12 d	82% ThOD	
(*see also graph*)		50 mg/L	12 d	0% ThOD	
		100 mg/L	12 d	0% ThOD	(8711)
A.S. adapted (*see also graph*)	Sapromat	10 mg/L	12 d	62% ThOD	
		50 mg/L	12 d	79% ThOD	
		100 mg/L	12 d	79% ThOD	(8711)
A.S. nonadapted	sapromat	10 mg/L	12 d	92% ThOD	
(*see also graph*)		50 mg/L	12 d	68% ThOD	
		100 mg/L	12 d	0% ThOD	(8711)

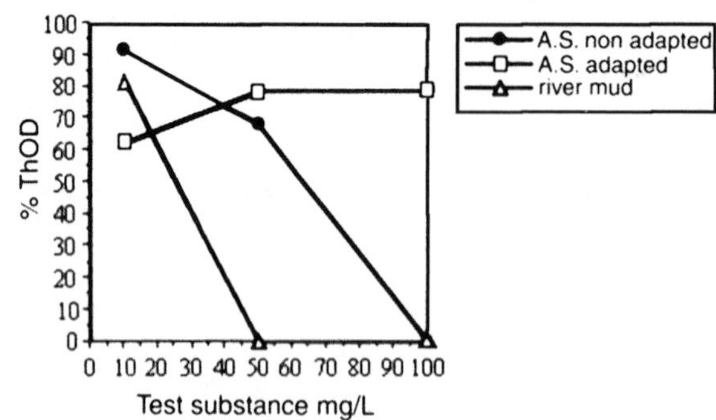

Biodegradation

rate in function of inoculum and concentration of test substance in respirometer test (Sapromat) after 12 d. (8711)

Impact on biodegradation processes

Activated sludge inhibition of NH_3 oxidation	EC_{75}	230 mg/L	(9698)
Activated sludge bacteria growth inhibition	EC_{100}	1,500 mg/L	(10041)
Act. Sludge respiration inhib.	EC_{50}	870 mg/L	(2624)

Biodegradation

A bacterial strain, IH-35A, classified as a member of the genus *Brevibacterium*, is capable of growing on cyclohexylamine (CHAM). The degradation of CHAM proceeds via cyclohexanone (CHnone), 6-hexanolactone, 6-hydroxyhexanoate, and adipate. Further degradation pathways by strains of *Pseudomonas* (see pathway). Alkane-induced cells of *Pseudomonas spp*. catalyzed thew-oxidation of hexanoate or 1,6-hexanediol into oxohexanoate and the further oxidation thereof into adipic acid.β-oxidation of 6-hydroxyhexanoate generates a nonmetabolizable intermediate, namely 2-tetrahydrofuran-acetic acid (see figure). (10799, 10800)

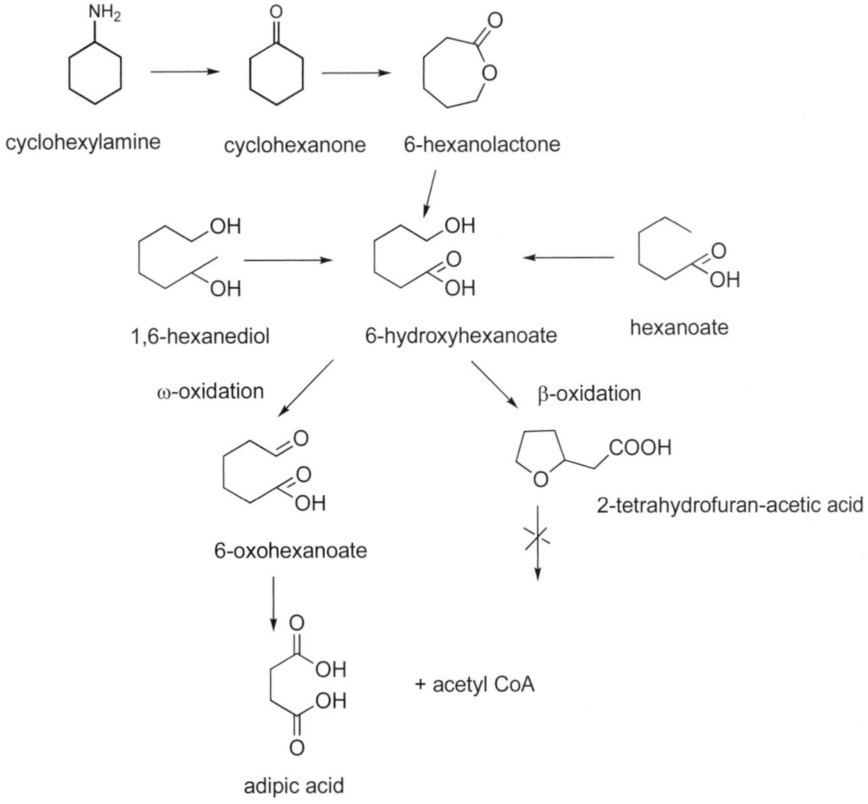

Biodegradation of cyclohexylamine

Metabolism in mammals

CHAM is excreted largely unchanged and only 4 to 5% of CHAM is metabolized to cyclohexanol, *trans*-cyclo-hexane-1,2-diol, and *trans*-3-, *cis*-3-, *trans*-4-, and *cis*-4-amino-cyclohexanols in rats, guinea pigs, and humans. (10799)

cyclohexanol cyclo-hexane-1,2-diol

4-aminocyclohexanol 3-aminocyclohexanol

Metabolism of hexylamine in mammals

Metabolism of hexylamine in mammals	(10799)

D. BIOLOGICAL EFFECTS

Bioaccumulation

	Concentration µg/L	*Exposure period*	*BCF*	*Organ/tissue*
Fishes: *Lepomis macrochirus*		30 d	98	total body
Poecilia reticulata		48 h	6,918	lipid

Bacteria			
A.S. oxygen consumption inhibition	3 h EC_{50}	2,152 mg/L	
	3 h EC_{95}	24,163 mg/L	(8710)
A.S. inhibition of NH_3 oxidation	EC_{75}	230 mg/L	(8718)
Escherichia coli	24 h EC_0	500 mg/L	(8710)
Pseudomonas fluorescens	24 h EC_0	500 mg/L	(8710)
Luminiscent bacteria	15 min EC_0	2,500 mg/L	(8710)
Pseudomonas putida	18 h EC_0	1,000 mg/L	(8710)
Pseudomonas putida	16 h EC_0	420 mg/L	(1900)
Algae			
Microcystis aeruginosa	8 d EC_0	0.002 mg/L	(329)
Scenedesmus quadricauda	7 d EC_0	0.51 mg/L	(1900)
Selenastrum capricornutum	96 h EC_0	5 mg/L	
(cell multiplication inhibition test)	96 h EC_{50}	20 mg/L	(8711)
Protozoans			
Entosiphon sulcatum	72 h EC_0	0.6 mg/L	(1900)
Uronema parduczi Chatton-Lwoff	EC_0	>200 mg/L	(1901)
Chilomonas paramecium	48 h EC_0	>400 mg/L	(8719)
Tetrahymena pyriformis	24 h EC_{50}	210 mg/L	(8715)
Crustaceans			
Daphnia magna	24 h EC_0	22; 50 mg/L	
	24 h EC_{50}	49; 58; 80 mg/L	
	24 h EC_{100}	120; 120mg/L	(8711, 8716, 8717)
Fishes			
Brachydanio rerio	96 h LC_0	200 mg/L	

	96 h LC_{50}	470 mg/L	
	96 h LC_{100}	600 mg/L	(8713)
Leuciuscus idus	48 h LC_0	50; 50; 174 mg/L	
	48 h LC_{50}	58; 195 mg/L	
	48 h LC_{100}	82; 100; 217 mg/L	(8713, 8714)
Oncorhynchus mykiss	96 h LC_{50}	44 mg/L (soft water)	
		90 mg/L (hard water)	(8711)
Oryzias latipes	48 h LC_{50}	54 mg/L	(8715)
Mammals			
Rat	oral LD_{50}	376- 496; 441- 839; 136- 238; 143- 167; 251- 309; 278; 219- 256; 237; 348 mg/kg bw	
Rat	oral LD_{50}	400- mg/kg bw (5% solution)	(9649)

cyclohexylcarboxylic acid

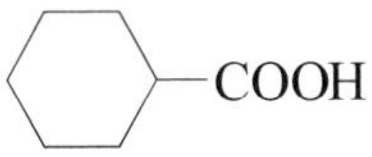

C_6H_5COOH

$C_7H_6O_2$

CAS 98-89-5

C. WATER AND SOIL POLLUTION FACTORS

microbial metabolite of *n*-alkyl-substituted cyclohexanes.

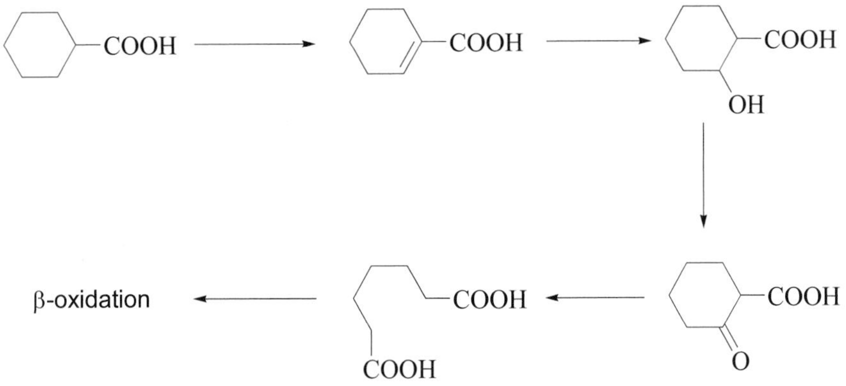

COOH → COOH → COOH OH → COOH O → COOH COOH → β-oxidation

Proposed pathway for the utilization of cyclohexylcarboxylic acid by the marine bacteria *Alcaligenes* sp. (2604)

2-cyclohexylidene-2-phenylacetonitrile (benzeneacetonitrile, α-cyclohexylidene)

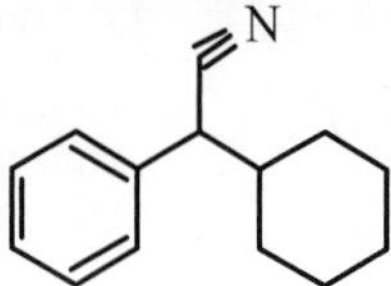

$C_{14}H_{15}N$

CAS 10461-98-0

TRADENAMES

Peonile

USES

The chemical possesses a fresh, floral, geranium-like odour and is mixed with other synthetic and natural components to make fragrances. The chemical is used as an aroma in alcoholic perfumery, cosmetics, toiletries, household products, soaps, detergents and industrial perfumery. The concentration of the chemical in alcohol based perfumes will not eceed 0.2 %. The chemical will typically be present in soaps at <0.1% and in detergents or softeners at <0.05 %. (7181)

IMPURITIES, ADDITIVES, COMPOSITION

citric acid 40 mg/L, tocopherol alpha 200 mg/L

A. PROPERTIES

colourless to pale yellow liquid; molecular weight freezing point <-50°C; boiling point 305 °C at 101.3 kPa; density 1.03 kg/L at 20°C; vapour pressure 0.043 Pa at 20°C; water solubility 7.5 mg/L at 20°C; $LogP_{OW}$ 4 at 30°C (measured)

B. AIR POLLUTION FACTORS

Photochemical reactions

t1/2: 11 d, based on reactions with OH° (calculated)

C. WATER AND SOIL POLLUTION FACTORS

Hydrolysis

t/2 > 1 year. The chemical contains no bonds which are susceptible to hydrolysis under the environmental pH region where 4<pH<9, and so it is expected to be stable.

Biodegradation

Inoculum	method	concentration	duration	elimination	
activated sludge	Ready biodegradability	100mg/l	28 days	0% ThOD	(7181)

Mobility in soil

log K_{OC} : 3.42 (calculated)

D. BIOLOGICAL EFFECTS

Toxicity

Algae

Selenastrum capricornutum	72h EC_{50}	0.86; 1.97 mg/L

	72h NOEC	0.5 mg/L	(7181)
Crustaceae			
Daphnia magna	48h EC_{50}	2.3 mg/L	(7181)
Fish			
Oncorhynchus mykiss	96h LC_{50}	1.4 mg/L	(7181)
Mammals			
Rat	oral LD_{50}	619 mg/kg bw	(7181)

cyclohexylmethane *see* methylcyclohexane

cyclohexylphenoxy dinitrile (Propanedinitrile, [[4-[[2-(4-cyclohexylphenoxy)ethyl]ethylamino]-2-methylphenyl]methylene]-; [[4-[[2-(4-cyclohexylphenoxy)ethyl]ethylamino]-2-methylphenyl] methylene] malononitrile; [4-[[2-(4-Cyclohexylphenoxy)ethyl] ethylamino]-2-methylbenzylidene] malononitrile; *N*-2-[(4-Cyclohexyl) phenoxy]ethyl-N-ethyl-4-(2,2-dicyanoethenyl)-3-methylaniline)

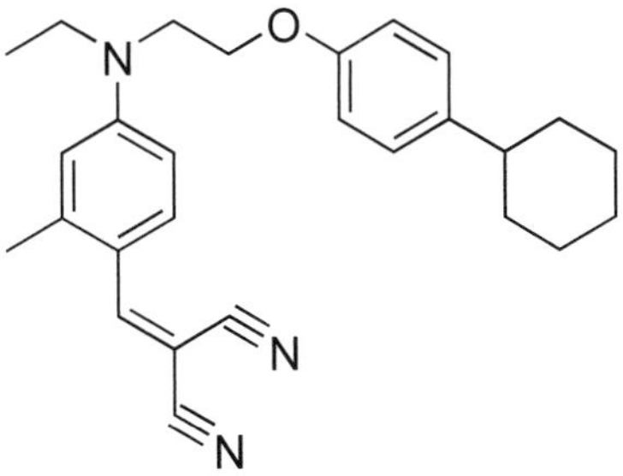

$C_{27}H_{31}N_3O$

CAS 54079-53-7

SMILES N#CC(=Cc1c(cc(N(CCOc2ccc(cc2)C2CCCCC2)CC)cc1)C)C#N

USES

Colourant, pigment, dye and printing ink

A. PROPERTIES

(all modelled) molecular weight 413.55; melting point 244°C; boiling point 567°C; vapour pressure 3 x 10^{-10} Pa; water solubility 0.25 µg/L ; log Pow 7.9; H 3.7 x 10^{-12} atm.m^3/mol

B. AIR POLLUTION FACTORS

Photodegradation

t.2 = 0.04 day. (11164)

C. WATER AND SOIL POLLUTION FACTORS

Mobility in soils

Log K_{OC} 6.3 (11164)

D. BIOLOGICAL EFFECTS

Bioaccumulation

Species	BCF	
Fish	4,000-362,000 (modelled)	(11164)

Toxicity

Fish	96h LC_{50}	0.04-4 µg/L (modelled)	(11164)

1,3-cyclooctadiene

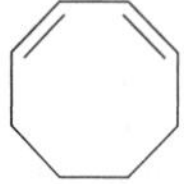

C_8H_{12}

CAS 1700-10-3 (Z, Z-)

CAS 3806-59-5 (cis, cis-)

USES

intermediate.

A. PROPERTIES

molecular weight 108.18.

C. WATER AND SOIL POLLUTION FACTORS

Waste water treatment methods

rotating disk contact aerator: infl. 60 mg/L, effl. 0.7 mg/L; elimination: 99% or 4,906 $mg/m^2/24$ h or 1325 $g/m^3/24$ h. (406)

1,5-cyclooctadiene (COD)

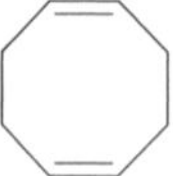

C_8H_{12}

CAS 111-78-4

USES

resin intermediate.

A. PROPERTIES

molecular weight 108.18; melting point -69.5°C; boiling point 150.9°C; vapor pressure 5 mm at 20°C; density 0.88 at 20/20°C.

C. WATER AND SOIL POLLUTION FACTORS

BOD_5:	0.19	(277)
COD:	2.62	(277)

D. BIOLOGICAL EFFECTS

Fishes

goldfish:	24h LD_{50}	14 mg/L	(277)

cyclooctane

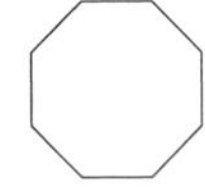

C_8H_{16}

CAS 292-64-8

A. PROPERTIES

molecular weight 112.22; melting point 10-13°C; boiling point 151°C at 740 mm; density 0.83.

B. AIR POLLUTION FACTORS

Odor threshold

detection: 3.6 mg/m^3. (828)

C. WATER AND SOIL POLLUTION FACTORS

Waste water treatment methods

rotating disk contact aerator: inf. 241 mg/L, effl. 0.5 mg/L; elimination: >99% or 19,866 mg/m^2/h or 5,364 g/m^3/h. (406)

cyclooctene

C_8H_{14}

CAS 931-88-4

A. PROPERTIES

molecular weight 110.2; melting point -16°C; boiling point 145-146°C; density 0.85.

C. WATER AND SOIL POLLUTION FACTORS

Waste water treatment methods

rotating disk contact aerator: infl. 61 mg/L, effl. 0.6 mg/L; elimination: 99% or 5,054 mg/m^2/h or 1,365 g/m^3/h. (406)

cyclopenta(cd)pyrene

$C_{18}H_{10}$

CAS 27208-37-3

MANMADE SOURCES

in coal tar:	0.63 g/kg.	(2600)
in carbon black:	70 mg/kg.	(2600)

A. PROPERTIES

molecular weight 226.28.

4H-cyclopenta(def)phenanthrene

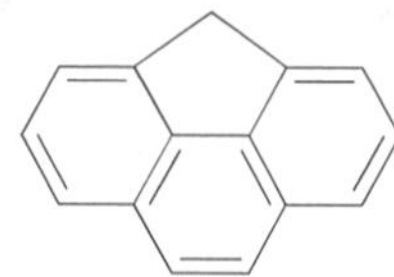

$C_{15}H_{10}$

CAS 203-64-5

MANMADE SOURCES

in coal tar: 6.3 g/kg. (2600)
in exhaust condensate of gasoline engine: 0.47-0.76 mg/L gasoline consumed. (1070)
in carbon black: 270 mg/kg. (2600)

A. PROPERTIES

molecular weight 190.25; melting point 113-115°C; boiling point 353°C; solubility 1.1 mg/L at 24°C.

11,H-cyclopenta(grs)benzo(e)pyrene *see* 8,9-methylenebenzo(e)pyrene

10,H-cyclopenta(mno)benzo(a)pyrene *see* 10,11-methylene-benzo(a)pyrene

cyclopentane (pentamethylene)

C_5H_{10}

CAS 287-92-3

USES

solvent for cellulose ethers.

A. PROPERTIES

colorless liquid; molecular weight 70.14; melting point -94°C; boiling point 50°C; density 0.75; vapor pressure 400 mm at 31°C; log H 0.88 at 25°C.

C. WATER AND SOIL POLLUTION FACTORS

Biodegradation

incubation with natural flora in the groundwater-in presence of the other components of high-octane gasoline (100 µl/L): biodegradation: 0% after 192 h at 13°C (initial conc.: 0.17 µl/L). (956)

D. BIOLOGICAL EFFECTS

Fishes

young Coho salmon	96h NOLC	100 ppm in artificial seawater at 8°C	(317)

cyclopentanol (cyclopentylalcohol)

$C_5H_{10}O$

CAS 96-41-3

USES

perfume and pharmaceutical solvent; intermediate for dyes, pharmaceuticals, and other organics.

A. PROPERTIES

molecular weight 86.13; boiling point 139-140°C; melting point -19°C; density 0.95.

B. AIR POLLUTION FACTORS

Odor threshold

detection: 4,200-8,700 mg/m^3. (727)

C. WATER AND SOIL POLLUTION FACTORS

Biodegradation

adapted A.S.-product as sole carbon source: 97% COD removal at 55 mg COD/g dry inoculum/h. (327)

D. BIOLOGICAL EFFECTS

Toxicity threshold (cell multiplication inhibition test)

Bacteria			
Pseudomonas putida	16h EC_0	250 mg/L	(1900)
Algae			
Microcystis aeruginosa	8d EC_0	28 mg/L	(329)
green algae			
Scenedesmus quadricauda	7d EC_0	255 mg/L	(1900)
Protozoa			
Entosiphon sulcatum	72h EC_0	290 mg/L	(1900)
Uronema parduczi Chatton-Lwoff	EC_0	>800 mg/L	(1901)

cyclopentanone (adipic-ketone)

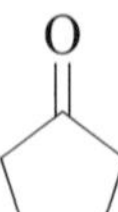

$C_5H_8(O)$

C_5H_8O

CAS 120-92-3

USES

intermediate for pharmaceuticals, insecticides, and rubber chemicals.

A. PROPERTIES

molecular weight 84.12; melting point -51°C; boiling point 125-126°C at 630 mm; density 0.94.

B. AIR POLLUTION FACTORS

Odor threshold

detection: 31-1,120 mg/m^3. (727)

C. WATER AND SOIL POLLUTION FACTORS

Biodegradation

adapted A.S.-product as sole carbon source: 95% COD removal at 57 mg/COD/g dry inoculum/h. (327)

Lactone forming mono-oxygenases metabolic biodegradation (436)

D. BIOLOGICAL EFFECTS

Toxicity threshold (cell multiplication inhibition test)

Bacteria			
Pseudomonas putida	16h EC_0	175 mg/L	(1900)
Algae			
Microcystis aeruginosa	8d EC_0	63 mg/L	(329)
green algae			
Scenedesmus quadricauda	7d EC_0	1,900 mg/L	(1900)
Protozoa			
Entosiphon sulcatum	72h EC_0	232 mg/L	(1900)
Uronema parduczi Chatton-Lwoff	EC_0	1,210 mg/L	(1901)
Crustaceans			
Daphnia magna	EC_{50}	1,440 mg/L	(7504)

cyclopentene

C_5H_8

CAS 142-29-0

USES

organic synthesis.

A. PROPERTIES

molecular weight 68.12; melting point -135°C; boiling point 44°C; density 0.77.

D. BIOLOGICAL EFFECTS

Fishes

young Coho salmon	96h NOLC 100 ppm in artificial seawater at 8°C	(317)

cyclopenteno(cd)pyrene

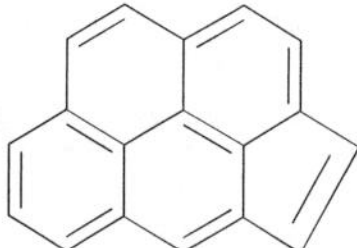

C_9H_{11}

A. PROPERTIES

molecular weight 226.

Manmade sources

in gasoline: >0.2 g/L
in exhaust condensate of gasoline engine: 0.75-0.99 mg/L gasoline consumed (1070)
in exhaust condensate of gasoline engine: 2.0 mg/g (1069)

Cyclophosphamide (N,N-bis(2-chloroethyl)tetrahydro-2H-,1,3,2-oxazaphosphorin-2-amine-2-oxide; cyclofosfamide)

$C_7H_{15}Cl_2N_2O_2P$

CAS 50-18-0

TRADENAMES

Cyclophosphane; Cycloblastin; Cyclophosphamide

USES

Antineoplastic

A. PROPERTIES

molecular weight 261.09

C. WATER AND SOIL POLLUTION FACTORS

Environmental concentrations

	range µg/L	n	median µg/L	90 percentile µg/L	
in effluent hospital	0.019-4.49	7			(7237)
influent MWTP	<0.006-0.14	21			(7237)
effluent MWTP (1996-2000)	0.006-0.017	21			
	<0.010-0.020	16	<0.010	0.018	
	up to 0.060				(7237)
surface waters (1998)	<0.010	26	<0.010	<0.010	(7237)

Environmental concentrations in Germany 2001 (10923)

	Positive samples	Min ng/L	Max ng/L	Average ng/L	Median ng/L
River Schussen	0/7	n.d.	n.d.	n.d.	n.d.
River Körsch upstream WWTP	2/8	<10	42	11	<10
River Körsch downstream WWTP	2/8	<10	10	<10	<10
Influent WWTP Stuttgart-M	6/6	14	90	39	32
Effluent WWTP Stuttgart-M	2/7	<10	10	<10	<10
Suspended solids influent WWTP Stuttgart-M µg/kg	4/6	n.d.	10,559	2621	1266
Influent WWTP Reutlingen-W	5/5	66	176	102	79
Effluent WWTP Reutlingen-W	1/5	<10	11	<10	<10
Suspended solids influent WWTP Reutlingen-W µg/kg	5/5	77	235	133	102
Influent WWTP Steinlach-W	5/5	24	176	76	45
Effluent WWTP Steinlach-W	0/5	<10	9	<10	<10
Suspended solids WWTP Steinlach-W µg/kg dw	5/5	94	664	267	135
Leachate of landfill Reutlingen-S	5/5	66	344	141	97
Leachate of landfill Dusslingen	5/5	103	222	169	192

D. BIOLOGICAL EFFECTS

Toxicity

Algae			
Pseudokirchneriella subcapitata	72h EC_{50}	>100 mg/L	(10928)
Crustaceae			
Daphnia magna	21d NOEC	56 mg/L	
	21d EC_{50}	>100 mg/L	(10928)

cyclophosphamide monohydrate

$C_7H_{15}Cl_2N_2O_2P.H_2O$

CAS 6055-19-2

USES

cancer research tool.

A. PROPERTIES

molecular weight 279.1; melting point 49-51°C.

B. AIR POLLUTION FACTORS

Incinerability

thermal stability ranking of hazardous organic compounds: rank 273 on a scale of 1 (highest stability) to 320 (lowest stability). (2390)

cyclopropanecarboxylic acid, 3-hexenyl ester

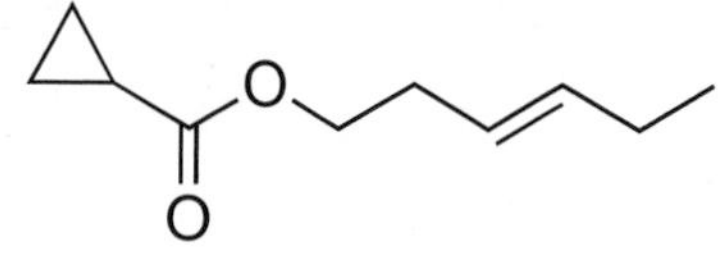

$C_{10}H_{16}O_2$

CAS 188570-78-7

USES

a fragrance ingredient in a variety of cosmetic and domestic products up to 2 mg/L. Typical end use products are body lotions, creams, hairsprays, shampoos, deodourant sprays, soap bars, foam baths, toilet waters, dishwashing liquids, fabric washing liquids, sulface sprays and air freshener sprays.

A. PROPERTIES

clear colorless liquid; molecular weight 168.1; freezing point −25°C; boiling point 214.5-227°; density 0.94 kg/L at 20°C; vapour pressure 17.5 Pa at 25°C; water solubility 200 mg/L at 20°C; log Pow 3.24 at 21 °C; log K_{OC} 2.4 (estimated)

C. WATER AND SOIL POLLUTION FACTORS

Hydrolysis

t/2 estimated at 23 years at pH 7 and 2.3 years at pH 8. (7420)

Biodegradation

the chemical is considered readily biodegradable according to the OECD criteria. (7420)

Inoculum	method	concentration	duration	elimination	
Activated sludge	Closed Bottle test	100 mg/L	28 days	77 to 82%	(7420)

ThCO2

D. BIOLOGICAL EFFECTS

Toxicity

calculation for aquatic organisms based on QSARS using ECOSAR program			(7420)
GREEN ALGAE	96h EC_{50}	0.46 mg/L	
Daphnids	48h LC_{50}	9.9 mg/L	
Fish	96H LC_{50}	5.5 mg/L	(7420)
Rat	oral LD_{50}	>2,000 mg/kg bw	(7420)

cyclosiloxanes

C. WATER AND SOIL POLLUTION FACTORS

Göteborg (Sweden) sew. works 1989-1991: infl.: nd-0 μg/L; effl.: nd. (2787)

β-cyfluthrin (3-(2,2-dichloroethenyl)-2,2-dimethylcyclopropanecarboxylic acid cyano-(4-fluoro-3-phenoxyphenyl)methyl ester; 3-(2,2-dichlorovinyl)-2,2-dimethylcyclopropanecarboxylic acid-α-cyano-(4-fluoro-3-phenoxy-phenyl)methyl ester; cyano(4-fluoro-3-phenoxyphenyl)methyl 3-(2,2-dichloroethyenyl)-2,2-dimethylcyclopropanecarboxylate)

Cl O CN O F

$C_{22}H_{18}Cl_2FNO_3$

CAS 68359-37-5

EINECS 269-855-7

USES

Non-systemic contact insecticide.
Cyfluthrin is a nonsystemic synthetic pyrethroid insecticide used to control chewing and sucking insects. Through contact and stomach poisoning, it attacks the nervous system; resulting in swift debilitation and has a long residual effect.
Its uses are extensive, including: agricultural crops, stored products, public health situations (i.e. cockroaches, mosquitoes, and flies), ornamentals, turf, and domestic pests. Target insects include ants, silverfish, cockroaches, grain beetles, fleas, flies, European corn borer, Colorado potato beetle, and many others

IMPURITIES, ADDITIVES, COMPOSITION

Cyfluthrin comprises a mixture of four diastereoisomeric pairs of enantiomers: I (R)-α-cyano-4-fluoro-3-phenoxybenzyl(1R)-cis- 3-(2,2-dichlorovinyl)-2,2-dimethylcyclopropanecarboxylate + (S)-α, (1S)-cis-; II S-α,(1R)-cis- + (R)-alpha, (1S)-cis-; III (R)-α,(1R)-trans- + (S)-α, (1S)-trans-; IV (S)-α,(1R)-trans- + (R)-α, (1S)-trans. Technical grade cyfluthrin contains: 23-26% diastereoisomer I, 16-19%

diastereoisomer II, 33-36% diastereoisomer III, and 22-25% diastereoisomer IV.

A. PROPERTIES

	melting point°C	Solubility at pH 3	Log P_{ow}
diastereoisomer I	64	0.0022	6.0
diastereoisomer II	81	0.0021	5.9
diastereoisomer III	65	0.0032	6.0
diastereoisomer IV	106	0.0043	5.9
technical grade	60		

molecular weight 434.29; density 1.27 at 20°C(supercooled melt); melting point 80.7-106.2°C ; vapour pressure 1.4-8.5 x 10^{-8} Pa at 20°C; water solubility 1-2 µg/L at 20°C; log Pow 5.9 at 22°C;

C. WATER AND SOIL POLLUTION FACTORS

In rivers in U.K. (1986): 139 samples from 70 river sites yielded no values above the detection limit of 0.005 µg/l. (2711)

Hydrolysis

20°C	pH 4	T/2 = > 1 year
20°C	pH 7	T/2 = 160-270 days
20°C	pH 9	T/2 = 33 to 42 hours

cyfluthrin

4-fluoro-3-phenoxybenzaldehyde

4-fluoro-3-phenoxybenzoic acid

Photolysis in water

Photolysis of b-cyfluthrin in water (11008)

Biodegradation in soil aerobic

% mineralisation	% bound-residues	After
Up to 36%	Up to 42%	190 days

Relevant metabolites: 3-(2,2-dichlorovinyl)-2,2-dimethylcyclopropanecarboxylic acid; 4-fluoro-3-phenoxybenzoic acid and 4-fluoro-3-phenoxybenzaldehyde.(11007)

Biodegradation in laboratory studies

degradation in soils	%	
Aerobic at 20°C	50	48-54 days
Aerobic at 20°C	90	253-1664 days

(11007)

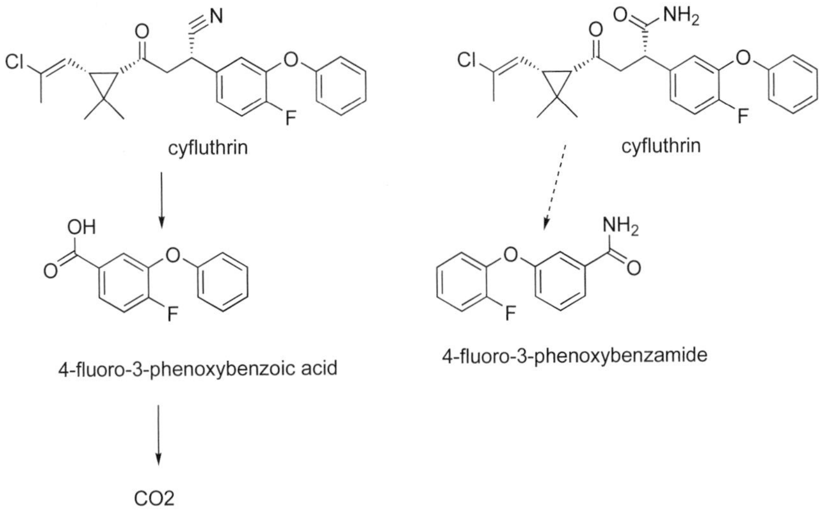

cyfluthrin

cyfluthrin

4-fluoro-3-phenoxybenzoic acid

4-fluoro-3-phenoxybenzamide

CO_2

Aerobic soil degradation of Cyfluthrin

Aaerobic degradation of Cyfluthrin in soil (11008)

cyfluthrin

4-fluoro-3-phenoxybenzoic acid

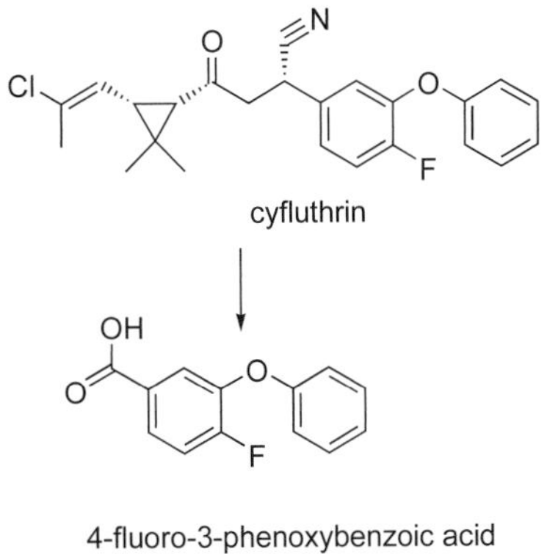

Anaerobic degradation of Cyfluthrin in soil (11008)

Mobility in soils

	cyfluthrin	Metabolite DCVA
K_{OC}	64,000-180,000	14-356

D. BIOLOGICAL EFFECTS

Bioaccumulation

Species	conc. µg/L	duration	body parts	BCF	
Fish				506	(11007)

Toxicity

Algae			
Scenedesmus subspicatus	96h EC_{50}	>10 mg/L	(11007)
Crustaceae			
Daphnia magna	48h EC_{50}	0.29 µg/L	
	21d NOEC	0.02 µg/L	(11007)
Insecta			
Chironomus riparius	28d EC_{50}	0.45 µg/L	(11007)
Fish			
Pimephales promelas	307d NOEC	0.14 µg/L	

Oncorhynchus mykiss	96h LC_{50}	0.068 µg/L	
	58d NOEC	0.01 µg/L	
Lepomis macrochirus	96h LC_{50}	0.28 µg/L	(11007)
rainbow trout, golden orfe, bluegill sunfish, carp	96h LC_{50}	0.0006-0.022 mg/L	(9662)
Mammals			
Rat	oral LD_{50}	77 mg/kg bw	(11007)
	oral LD_{50}	400 mg/kg bw	(9662)

Cyhalofop-butyl (Butyl-(R)-2-[4(4-cyano-2-fluorophenoxy)phenoxy]propionate)

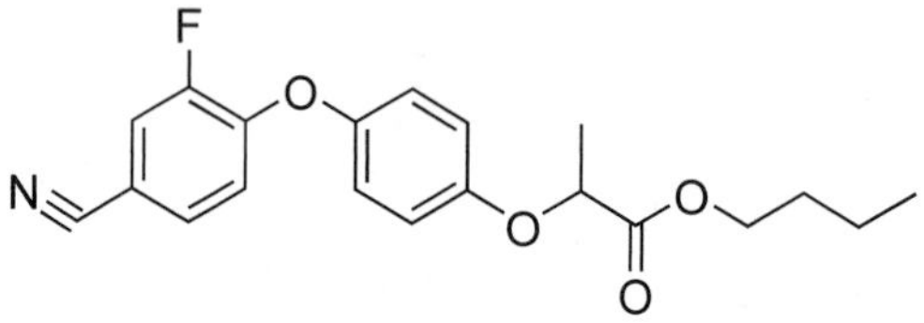

$C_{20}H_{20}FNO_4$

CAS 122008-85-9

TRADENAMES

Clincher; Barnyard Herbicide

USES

Aryloxyphenoxy propionate herbicide

A. PROPERTIES

molecular weight 357.39; melting point 45.5-49.5°C; decomposition at 270°C; density 1.17 kg/L; vapour pressure 5.3 x 10^{-5} Pa at 25°C; water solubility at 20°C at pH 5 : 0.46 mg/L; at pH 7 : 0.44 mg/L; at pH 9: rapid hydrolysis; log Pow 3.32; H 9.5 x 10^{-4} Pa.m^3/mol

B. AIR POLLUTION FACTORS

Photodegradation

T/2 (OH radicals)= 6 hours

C. WATER AND SOIL POLLUTION FACTORS

Hydrolysis the only relevant hydrolysis product is Cyhalofop-acid(10883)

	25°C	37°C
pH 1.2	-	42 hours
pH 4.0	> 1 year	> 1 year
pH 7.0	97 days	31 days
pH 9.0	43 hours	11.5 hours

Photodegradation in water

T/2 = 25-28 days. No relevant photodegradation products.

Biodegradation in soil (10883)

	% mineralisation	%	After

		bound-residues	
aerobic biodegradation in soil	36-46%	34-44%	120 days

The parent compound degraded rapidly to the acid, then the amide and the diacid. Under anaerobic conditions the same sequence of metabolites was observed.

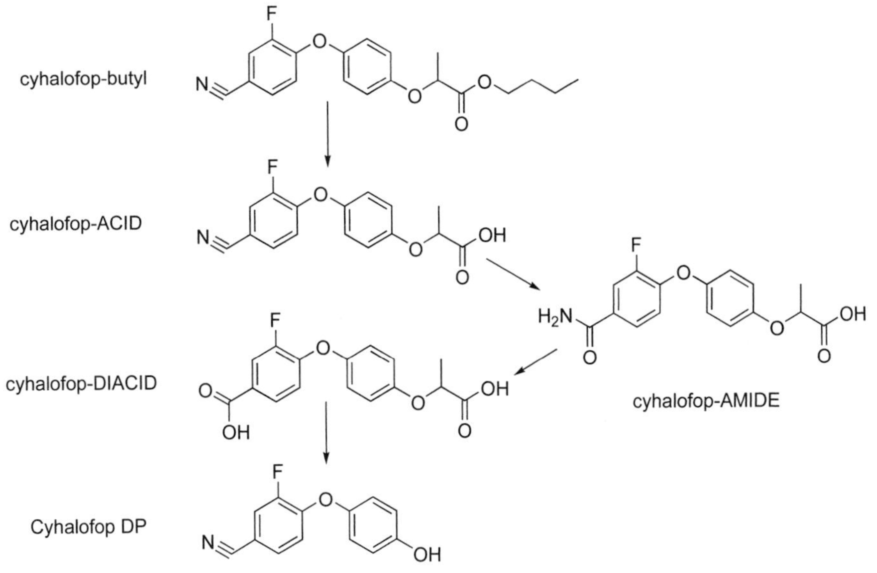

cyhalofop-butyl

cyhalofop-ACID

cyhalofop-AMIDE

cyhalofop-DIACID

Cyhalofop DP

Key metabolic transformations of cyhalofop-butyl

Key metabolic transformations of Cyhalofop-butyl	(10884)

Biodegradation in water/sediment systems (10883)

DT50 water	1.7-4.5 hours
DT90 water	5.6-15 hours
DT50 whole system	1.4-5.3 hours
DT90 whole system	4.7-18 hours

Mobility in soils (10884)

	Cyhalofop-butyl	-acid	-amide	-diacid
K_{OC}	2066-9637	57-195	50-152	79-614

D. BIOLOGICAL EFFECTS

Bioaccumulation

Species	conc. µg/L	duration	body parts	BCF	
Fish				<7-8	(10883)

Toxicity

Algae			
Selenastrum capricornutum	72h EC_{50}	>0.96 mg/L	
Anabaena flos-aquae	72h EC_{50}	>8.4 mg/L	
Navicula pelliculosa	120h EC_{50}	1.33 mg/L	(10883)
Aquatic plants			
Lemna gibba	14d EC_{50}	>5.3 mg/L	(10883)
Worms			
Eisenia foetida	14d LC_{50}	>1000 mg/kg soil	
	14d NOEC	1000 mg/kg soil	(10883)
Crustaceae			
Daphnia magna	48h EC_{50}	>2.7 mg/L	(10883)

Insecta			
Honeybees	Acute oral LD_{50}	>100 µg/bee	
	Acute contact LD_{50}	>100 µg/bee	(10883)
Birds			
Bobwhite quail	Acute oral LD_{50}	>2,250 mg/kg bw	
Mallard duck	Acute oral LD_{50}	>2,250 mg/kg bw	(10883)
Fish			
Lepomis macrochirus	96h LC_{50}	0.79 mg/L	
	96h NOEC	0.31 mg/L	(10883)
Pimephales promelas	28d NOEC	0.13 mg/L	(10883)
Mammals			
Rat	oral LD_{50}	>5,000 mg/kg bw	(10883)

Toxicity of metabolite ***Cyhalofop butyl-DIACID***

Lepomis macrochirus	96h LC_{50}	>100 mg/L	
Pimephales promelas	28d NOEC	10 mg/L	
Daphnia magna	48h EC_{50}	>100 mg/L	
Selenastrum capricornutum	72h EC_{50}	>100 mg/L	(10883)

Toxicity of metabolite ***Cyhalofop butyl-ACID***

Lepomis macrochirus	96h LC_{50}	>100 mg/L	
Oncorhynchus mykiss	96h LC_{50}	>100 mg/L	
Daphnia magna	48h EC_{50}	>100 mg/L	
Selenastrum capricornutum	72h EC_{50}	>100 mg/L	(10883)

Toxicity of metabolite ***Cyhalofop butyl-AMIDE***

Lepomis macrochirus	96h LC_{50}	>100 mg/L	
Daphnia magna	48h EC_{50}	>100 mg/L	
	21d NOEC	100 mg/L	
Selenastrum capricornutum	72h EC_{50}	>50 mg/L	(10883)

cyhalothrin ((RS)-α-cyano-3-phenoxybenzyl(Z)-(1RS,3RS)-(2-chloro-3,3,3-trifluoropropenyl)-2,2-dimethylcyclopropanecarboxylate)

F F Cl F O CN O O

$C_{23}H_{19}ClF_3NO_3$

CAS 68085-85-8

USES

to control animal ectoparasites on cattle.

A. PROPERTIES

molecular weight 449.9; boiling point 187-190°C; vapor pressure 7.5 x 10^{-9} mm Hg at 20°C; solubility 0.003 mg/L at 20°C.

C. WATER AND SOIL POLLUTION FACTORS

Hydrolysis

by water is slow at pH 7-9, more rapid at pH >9. (2962)
In rivers in U.K. (1986): 139 samples from 70 river sites yielded no values above the detection limit of 0.005 μg/l. (2711)
Persistence: in river water in sunlight : $t_{1/2}$: ca. 20 days. (9600)

D. BIOLOGICAL EFFECTS

Fishes			
rainbow trout	96h LC_{50}	0.00054 mg/L	(9600)

Mammals			
female, male rats	oral LD_{50}	144-243 mg/kg bw	(9600)

cyhexatin (tricyclohexyltin hydroxide; tricyclohexylhydroxystannane; DOWCO-213; Plictran; Acarstin; Mitacid)

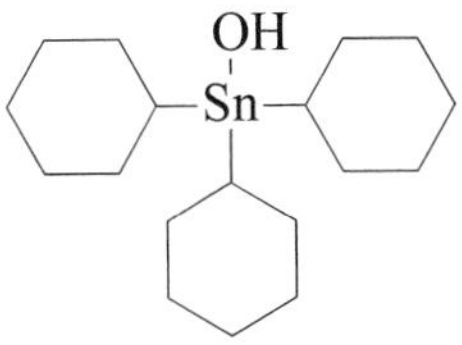

$C_{18}H_{34}OSn$

CAS 13121-70-5

USES

acaricide.

A. PROPERTIES

molecular weight 385.16; melting point 195-198°C; vapor pressure $<10^{-5}$ mbar at 20°C; solubility <1 mg/L.

D. BIOLOGICAL EFFECTS

Biotransfer factor in beef

log B_b: -4.44. (2644)

Fishes			
large-mouth bass	24h LC_{50}	0.060 mg/L	(9662)
goldfish	24h LC_{50}	0.55 mg/L	(9600)

Mammals			
rat, mouse	oral LD_{50}	190-1,070 mg/kg bw	(10042, 10043, 10044, 10045)
rabbit	oral LD_{50}	500 mg/kg bw	(10046)

p-cymene (dolcymene; *p*-isopropyltoluene; *p*-methylisopropylbenzene; Cymol; Camphogen; Paracymol)

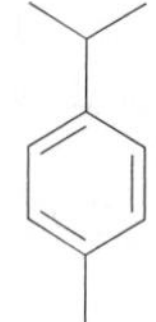

$CH_3C_6H_4CH(CH_3)_2$

$C_{10}H_{14}$

CAS 99-87-6

USES

Synthesis of p-cresol and carvacrol. Used as a diluent for lacquers, varnishes, dyes; in production of synthetic resin; component of fragrances for soap, cream and perfume.

OCCURRENCE

Occurs in essential oils distilled from plants including *Cuminum cyminum, Pinus palustris, Pinus caribaea, Cupressus sempervirens.* In cola: 105 mg/L, in cherry cola: 102 mg/L (7734)

A. PROPERTIES

molecular weight 134.22; melting point -68°C; boiling point 176-178°C; vapor density 4.6; density 0.86;log P_{OW} 4.1; vapor pressure 1 mm Hg at 17°C.

B. AIR POLLUTION FACTORS

Natural sources

emitted from the leaves of *Cupressus sempervirens*: 0.16% of total terpene emissions which varied between 3-35 µg/g dw.h. The major component was limonene which constituted 83% of total terpene emissions. (7049)

Calculated tropospheric lifetimes

reactant	*lifetime*	
OH°	1.0 d	
ozone	>330 d	
NO_3	1.3 years	(2451)

indoor/outdoor glc's winter 1981/1982 and 1982/1983, Netherlands:

	µg/m³ *median*	*maximum*	
pre-war homes`	0.6	11	
post-war homes	0.7	32	
homes <6 years old	1	10	
outdoors	<0.3	<0.3	(2668)

C. WATER AND SOIL POLLUTION FACTORS

Göteborg (Sweden) sew. works 1989-1991: infl.: 0.1-5 µg/L; effl.: n.d. (2787)

D. BIOLOGICAL EFFECTS

Mammals

rat (weanling)	oral LD_{50}	4,750 mg/kg bw	(9753)

3-p-cymenol *see* thymol

Cypermethrin ((+)α-cyano-3-phenoxybenzyl-(+)*cis, trans*-2,2,dichlorovinyl-2,2-dimethylcyclopropanecarboxylate)

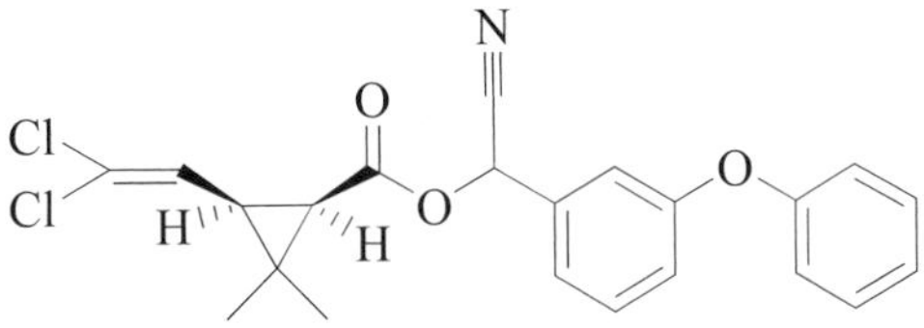

$C_{22}H_{19}Cl_2NO_3$

CAS 52315-07-8

TRADENAMES

Ammo, Arrivo, Barricade, Basathrin, CCN52, Cymbush, Cymperator, Cynoff, Cypercopal, Cyperguard, Cyperhard Tech, Cyperkill, Cypermar, Demon, Flectron, Fligene, Folcord, Kafil super, NRDC 149, Polytrin, PP 383, Ripcord, Siperin, Stockade, Super.

USES

cypermethrin is a synthetic pyrethroid insecticide used to control many pests, including moth pests of cotton, fruit, and vegetable crops. It is also used for crach, crevice, and spot treatment to control pests in stores, warehouses, industrial buildings, houses, apartment buildings, greenhouses, laboratories, and on ships, railcars, buses, trucks and aircraft.
Impurities and additives marketed product: Technical cypermethrin is a mixture of eight different isomers, each of which may have its own chemical and biological properties.

A. PROPERTIES

yellow brown viscous liquid; molecular weight 416.3; melting point 80°C; boiling point 220°C; vapor pressure 5.1 × 10^{-8} mbar at 70°C; density 1.12 at 22°C; solubility 0.041 mg/L at room temp., 0.0004 mg/L at pH 7; log P_{OW} 4.47; 6.3.; vapour pressure 1.3 x 10^{-9} mm Hg at 20°C; H 2.5 x 10-7 atm-m^3/mol at 20 °C

B. AIR POLLUTION FACTORS

Photolysis

Cypermethrin photodegrades rapidly on soil surfaces to many byproducts, with half-lives of 8-16 days. The principal photoproducts are 3-phenoxybenzoic acid (PBA) and 3-(2,2-dichlorovinyl)-2,2-dimethyl cyclopropanecarboxylic acid (DCVA)

C. WATER AND SOIL POLLUTION FACTORS

In rivers in U.K. (1986): 139 samples from 70 river sites yielded 3 positive values (detection limit: 0.005 μg/L): median of positive values: 0.01 μg/L; maximum: 0.015 μg/L. (2711)

Degradation in soil

In soil, hydrolysis (ester cleavage) occurs within about 16 weeks. 3-phenoxybenzoic acid is formed, among other metabolites. Further hydrolytic and oxidative degradation occurs. (2962)

1006 Cypermethrin

Biodegradation

cypermethrin + H_2O → 3-phenoxybenzoic acid

Under water-logged conditions the rate of hydrolysis of cypermethrin was slower than under aerobic conditions and 3-phenoxybenzoic acid accumulated in the anaerobic soil. (10049)

On crops, cypermethrin degradation occurs mainly by hydrolysis of the ester bond followed by hydrolytic and oxidative processes. Mainly unchanged chemical was found 21 d after. application . (10048, 10049)

Hydrolysis

> 50 days at environmental temperatures and pH values. Hydrolysis of the ester linkage is the principal degradation route and leads to the formation of 3-phenoxybenzoic acid (PBA) and cyclopropanecarboxylic acid derivatives, principally, 3-(2,2-dichlorovinyl)-2,2-dimethyl cyclopropanecarboxylic acid (DCVA) (see pathways). (7160)

Soil sorption

log K_{OC}: 4.8 (average of 5 soil types) which indicates that cypermethrin binds strongly to organic matter. (7160)

Biodegradation

The anaerobic half-life is reported at <14 days, similar to the half-life in aerobic soils ranging from 6-20 days,but the major metabolite PBA does not continue to break down anaerobically. The half-life in sterile soils was 20 to 25 weeks. (7160)

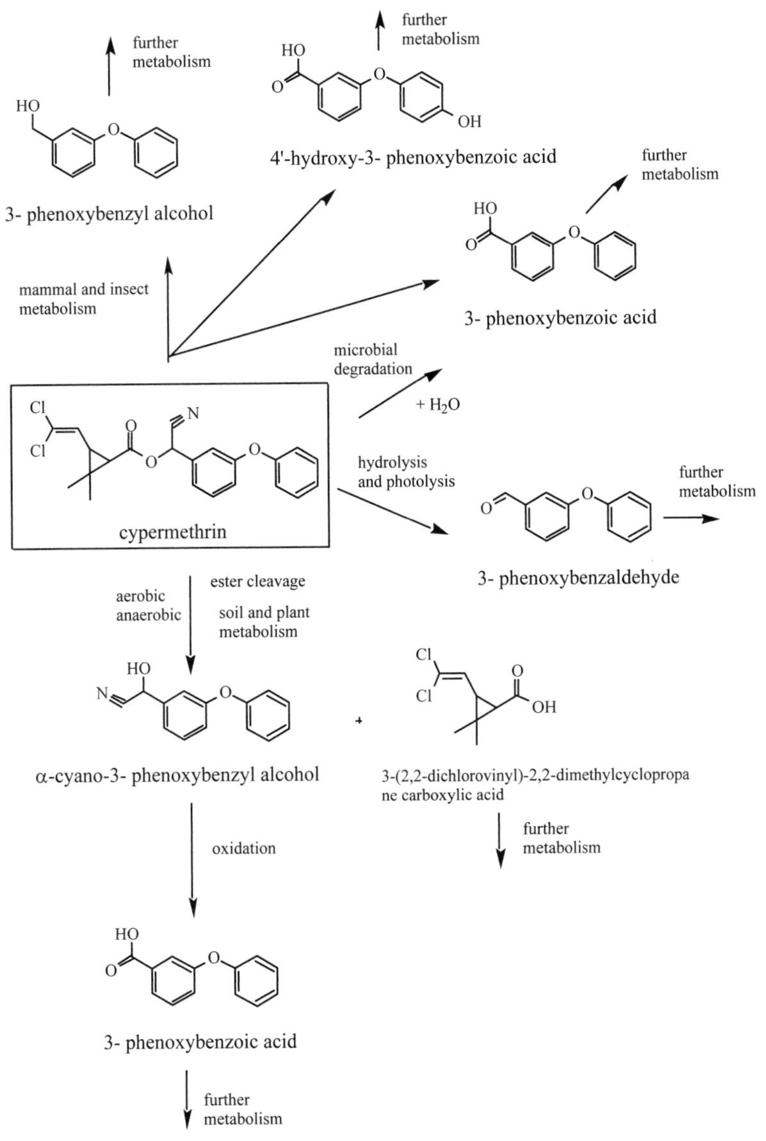

Cypermethrin degradation pathways	(7160)

D. BIOLOGICAL EFFECTS

Bioaccumulation

Species	conc. µg/L	duration	body parts	BCF	
Rainbow trout			total body	180-438	(7160)
Algae: *Chlorella fusca*			wet wt	3,280	(2659)

Toxicity

Birds			
Bobwhite quail	Acute oral LD_{50}	>20,000 mg/kg	
Mallard duck	Acute oral LD_{50}	>20,000 mg/kg	
Chickens	Acute oral LD_{50}	> 2000 mg/kg	(7160)
Algae			
Scenedesmus bijuga	14d EC_{50},growth	0.05 mg/L	(2624)
Insects			
mayfly	24h EC_{50}	0.0006 mg/L	(9600, 9662)

Crustaceans			
Daphnia magna	24h EC_{50}	0.002 mg/L	(9600, 9662)
lobster	24h EC_{50}	0.00004 mg/L	(9600, 9662)
Fishes			
rainbow trout	96h LC_{50}	0.82 µg/L	(7160)
bluegill sunfish	96h LC_{50}	1.78 mg/L	(7160)
rainbow trout: technical grade	24h LC_{50},S	55 ppb active ingredient	
formulated product	24h LC_{50},S	11 ppb active ingredient	(1577)
Heteropneustes fossilis	96h LC_{50}	9.1 µg/L	(2625)
brown trout	96h LC_{50}	0.002-0.0028 mg/L	
Atlantic salmon	96h LC_{50}	0.002-0.0024 mg/L	(9600, 9662)
Mammals			
hamster, rat	oral LD_{50}	200-250 mg/kg bw	
mouse	oral LD_{50}	82 mg/kg bw	
guinea pig	oral LD_{50}	500 mg/kg bw	(10049)

cyprodinil ((4-cyclopropyl-6-methyl-pyrimidin-2-yl)phenylamine; 4-cyclopropyl-6-methyl-N-phenyl-2-pyrimidinamine)

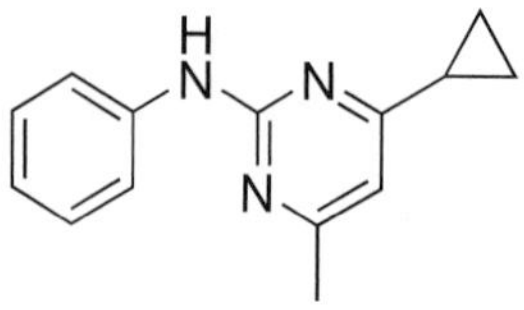

$C_{14}H_{15}N_3$

CAS 121552-61-2

USES

Cyprodinil is a systemic fungicide that acts by inhibiting the biosynthesis of methionine.

A. PROPERTIES

Fine white crystals; molecular weight 225.3; melting point 71-76°C; vapour pressure at 25°C 5 x 10^{-4} Pa; water solubility 13-20 mg/L at 25°C; log Pow 4.0 at 25°C; H 7 x 10^{-3} Pa.m^3/mol

C. WATER AND SOIL POLLUTION FACTORS

Hydrolysis

No hydrolysis at pH 4 to 9 for 5 days at 50°C.

Biodegradation in soil aerobic

% mineralisation	% bound-residues	After
5-14%	37-68%	90 days

Code	metabolites	CAS
CGA 232449	(6-cyclopropyl-2-phenylaminopyrimidin-4-yl)methanol	121552-66-7
CGA 249287	4-cyclopropyl-6-methylpyrimidin-2-ylamine	92238-61-4

CGA 263208	Phenyl-1-guanidine	2002-16-6
CGA 275535	3-(4-cyclopropyl-6-methylpyrimidin-2-ylamino)phenol	
CGA 304075	4-(4-cyclopropyl-6-methylpyrimidin-2-ylamino)phenol	195157-66-5
CGA 304076	4-cyclopropyl-6-methyl-2-phenylaminopyrimidin-5-ol	
CGA 321186	3-[5-(4-cyclopropyl-6-methyl-pyrimidin-2-ylamino)-2-hydroxy-phenylsulfinyl]-2-hydroxypropionic acid	
CGA 321915	4-cyclopropyl-6-methylpyrimidin-2-ol	121553-48-8
NOA-413167	2-(4-cyclopropyl-6-methylpyrimidine-2-ylamino)phenol	
NOA-422054	(2-amino-6-cyclopropylpyrimidin-4yl)methanol	
NOA-436942	4-cyclopropyl-6-methylpyrimidin-2,5-diol	
L1	guanidinophenol	
L2	2-sulfate conjugate of 4-(4-cyclopropyl-6-methylpyrimidin-2-ylamino)benzene-1,2-diol	
L3a	Glucuronic acid caonjugate of 4-cyclopropyl-6-methyl-2-phenylaminopyrimidin-5-ol	
Le3b	6-cyclopropyl-2-(4-hydroxyphenylamino)pyrimidin-4-ylmethanol	
L3c	Glucuronic acid conjugate of (6-cyclpropyl-2-phenylamino-Pyrimidin-4-yl)methanol	
L4	5-glucuronic acid conjugate of 6-cyclopropyl-4-hydroxymethyl-2-phenylaminopyrimidin-5-ol	
1G	Glucuronic acid conjugate of 4-(4-cyclopropyl-6-methylpyrimidin-2-ylamino)phenol	
2G	2-Glucuronic acid conjugate of 4-(4-cyclopropyl-6-methylpyrimidin-2-ylamino)benzene-1,2-diol	
1U	5-sulfate conjugate of 4-cyclopropyl-2-(4-hydroxyphenylamino)-6-methylpyrimidin-5-ol	
2U	Sulfate conjugate of 4-cyclopropyl-6-methyl-2-phenylamino-pyrimidin-5-ol	
3U	Sulfate conjugate of 4-(4-cyclopropyl-6-methylpyrimidin-2-ylamino)phenol	
4U	2-sulfate conjugate of 4-(6-cyclopropyl-4-hydroxymethyl-pyrimidin-2-ylamino)benzene-1,2-diol	
5U	5-glucuronic acid conjugate of 4-cyclopropyl-2-(4-hydroxyphenylamino)-6-methylpyrimidin-5-ol	
6U	Disulfate conjugate of 4-cyclopropyl-2-(4-hydroxyphenylamino)-6-Methylpyrimidin-5-ol	
7U	6-cyclopropyl-4-hydroxymethyl-2-(4-hydroxyphenylamino)-pyrimidin-5-ol	

Biodegradation in laboratory studies (11010)

degradation in soils	%	Cyprodinil	CGA 249287	CGA 275535
Aerobic at 20°C	50	31-41 days	76-153 days	0.4-0.9 days
Aerobic at 20°C	90	103-136 days	252-508 days	1.3-3 days
Aerobic at 10°C	50	85 days		
Anaerobic at 20°C	50	stable		

Biodegradation in soil

Degradation was shown to involve hydroxylation of the phenyl or the pyrimidyl ring. Cleavage of the anilino-pyrimidyl bridge of Cyprodinil and its hydroxyphenyl derivatives gave the major metabolite CGA 249287 and the minor GCA 321915. The phenyl ring and the pyrimidyl ring were poorly mineralized. Under anaerobic conditions it was found that Cyprodinil and its major metabolite CGA 275535 were stable. No novel breakdown products were identified.(11010)

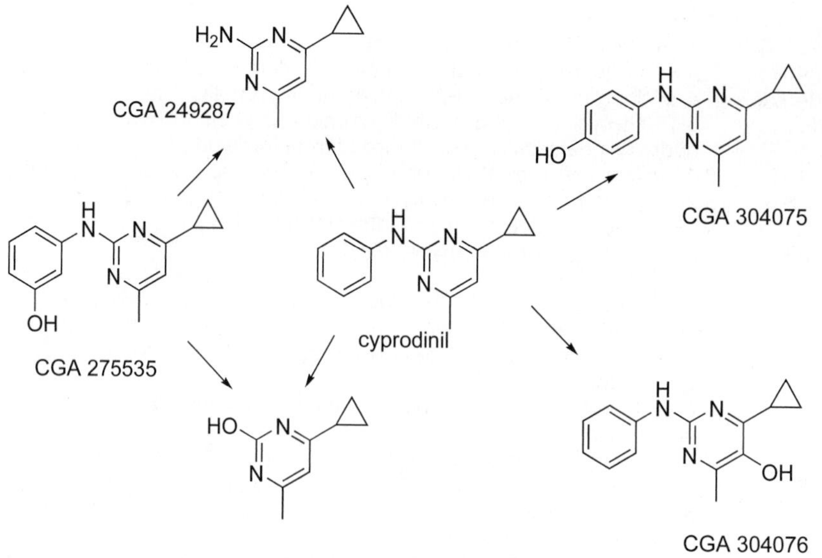

1010 cyprodinil

CGA 249287

CGA 304075

CGA 275535

cyprodinil

CGA 304076

Proposed pathway of cyprodinil metabolism in aerobic soil

Proposed pathway of Cyprodinil metabolism in aerobic soil	(11009)

Metabolism in plants

The metabolism in plants proceeds mainly by hydroxylation of the phenyl and pyrimidin rings followed by sugar conjugation. Cleavage of the amine bridge between the two ring systems represents a minor degradation route. Incorporation of ultimate degradation products in natural plant components has been demonstrated.	(11010)

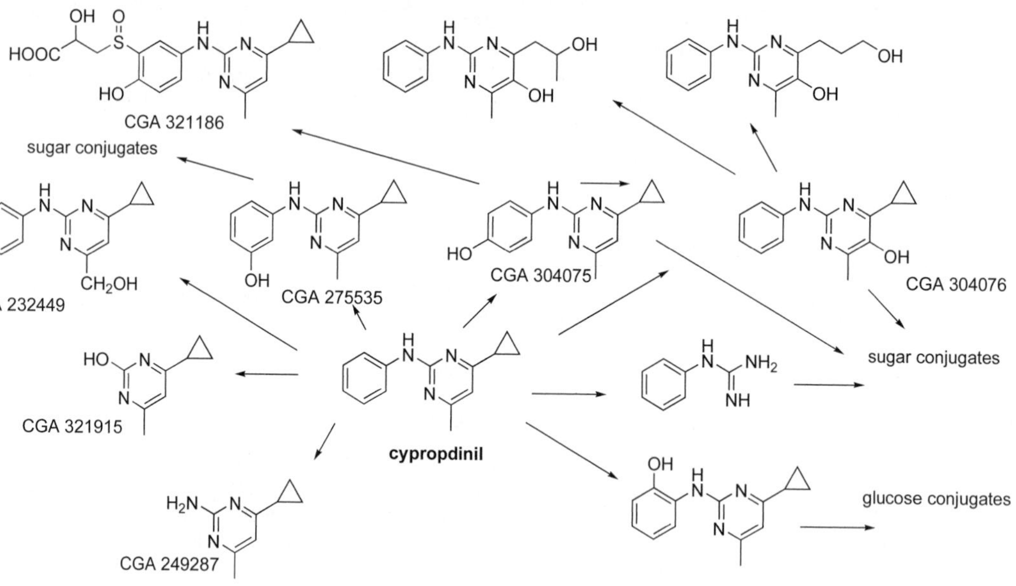

CGA 321186

sugar conjugates

CGA 232449

CGA 275535

CGA 304075

CGA 304076

sugar conjugates

CGA 321915

cypropdinil

glucose conjugates

CGA 249287

Cyprodinil plant metabolism, proposed pathways

Cyprodinil plant metabolism	(11009)

Mobility in soils

	Cyprodinil	CGA 249287	CGA 275535	
K_{OC}	1536-2012	173-867	1810	(11010)

D. BIOLOGICAL EFFECTS

Bioaccumulation

Species	conc. μg/L	duration	body parts	BCF
Fish			Whole fish	393

Toxicity

Algae			
Pseudokirchneriella subcapitata	72h EC_{50}	2.6 mg/L	
Anabaena flos-aquae	72h EC_{50}	3.76 mg/L	
Navicula pelliculosa	72h EC_{50}	2.1 mg/L	(11010)
Aquatic plants			
Lemna gibba	14d EC_{50}	7.7 mg/L	(11010)
Invertebrates			
Daphnia magna	48h EC_{50}	0.033 mg/L	
	21d NOEC	0.0082 mg/L	
Daphnia longispina	48h EC_{50}	0.22 mg/L	
Daphniopsis sp.	24h EC_{50}	0.21 mg/L	
Brachionus calyciflorus	24h EC_{50}	>9.5 mg/L	
Chaoborus sp.	48h EC_{50}	4.0 mg/L	
Cloeon sp.	48h EC_{50}	3.5 mg/L	
Lymnea stagnalis	48h EC_{50}	2.9 mg/L	
Gammarus sp.	48h EC_{50}	1.8 mg/L	
Ostracoda	48h EC_{50}	1.1 mg/L	
Simocephalus vetulus	48h EC_{50}	0.15 mg/L	
Thamnocephalus platyurus	24h EC_{50}	0.12 mg/L	(11010)
Insecta			
Chironomus riparius	27d NOEC	80 mg/kg sediment dw	(11010)
Birds			
Bobwhite quail	Acute oral LD_{50}	>2,000 mg/kg bw	
Mallard duck	Acute oral LD_{50}	>500 mg/kg bw	(11010)
Fish			
Pimephales promelas	36d NOEC	0.231 mg/L	
Oncorhynchus mykiss	96h LC_{50}	3.2 mg/L	
	21d NOEC	0.083 mg/L	(11010)
Mammals			
Rat	oral LD_{50}	>2,000 mg/kg bw	(11010)

Toxicity of metabolite ***CGA 249287***

Oncorhynchus mykiss	96h LC_{50}	55 mg/L	
Daphnia magna	48h EC_{50}	>100 mg/L	
Pseudokirchneriella subcapitata	72h EC_{50}	>100 mg/L	
Chironomus riparius	27d NOEC	25.6 mg/kg sediment dw	(11010)

Toxicity of metabolite ***CGA 275535***

Oncorhynchus mykiss	96h LC_{50}	2.1 mg/L	
Daphnia magna	48h EC_{50}	6.8 mg/L	
Pseudokirchneriella subcapitata	72h EC_{50}	9.4 mg/L	(11010)

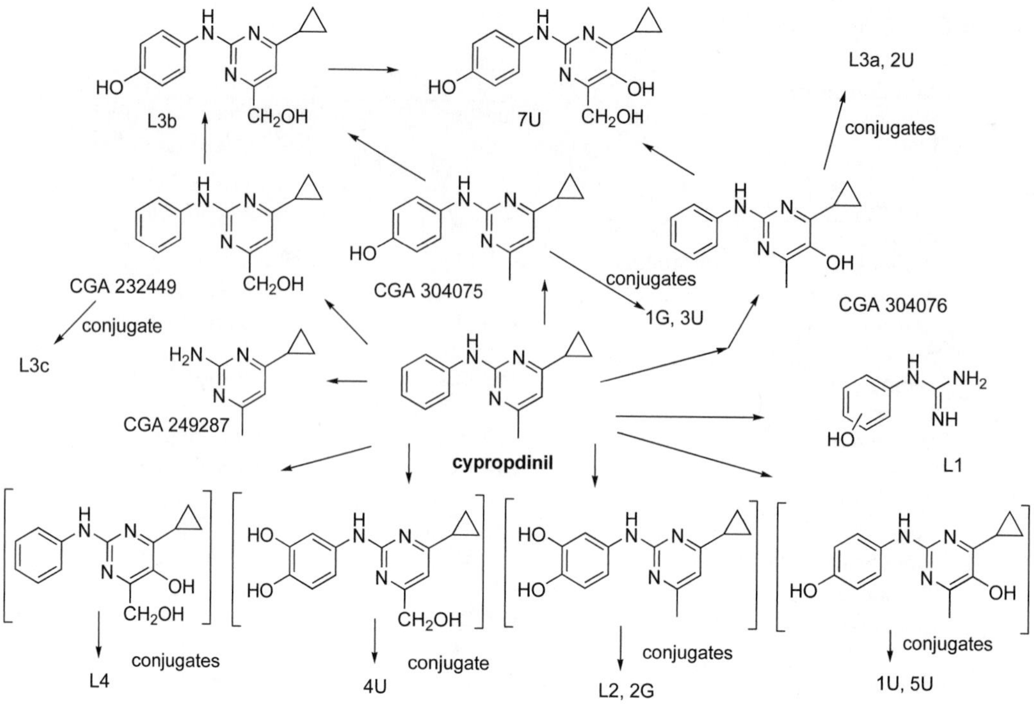

Proposed pathways of Cyprodinil animal metabolism

Proposed pathways of Cyprodinil animal metabolism	(11009)

L-cysteine (L-2-amino-3-mercaptopropanoic acid; L-β-mercaptoalanine)

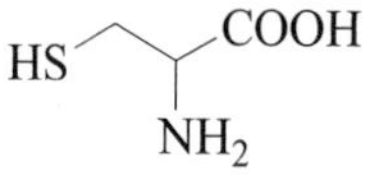

$HSCH_2CH(NH_2)COOH$

$C_3H_7NO_2S$

CAS 52-90-4 (L-(+)-)

CAS 3374-22-9 (DL-)

A. PROPERTIES

molecular weight 121.15; melting point 220°C decomposes.

C. WATER AND SOIL POLLUTION FACTORS

A.S. after	6 h:	7.5% of ThOD	
	12 h:	8.6% of ThOD	
	24 h:	11% of ThOD	(89)

L-cystine (dicysteine; L-β,β'-dithiodialanine; L-3,3'-dithio-*bis*-(2-amino propanoic acid))

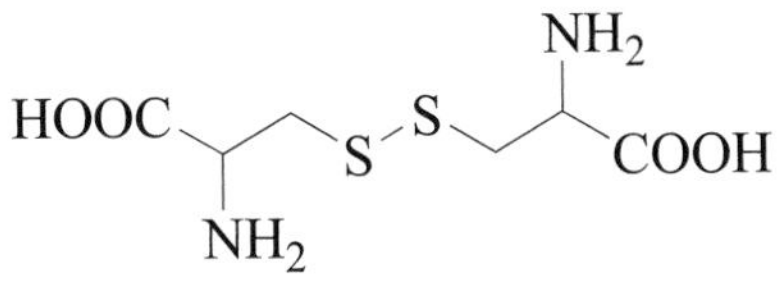

HOOC–CH(NH_2)–CH_2–S–S–CH_2–CH(NH_2)–COOH

$(SCH_2CH(NH_2)COOH)_2$

$C_6H_{12}N_2O_4S_2$

CAS 56-89-3

A. PROPERTIES

molecular weight 240.29; melting point 258/261°C decomposes; solubility 110 mg/L at 25°C, 520 mg/L at 75°C.

C. WATER AND SOIL POLLUTION FACTORS

Manmade sources

excreted by humans: in urine: 1.5-2.4 mg/kg body wt/d. (203)

Waste water treatment

A.S.: after	6 h:	1.5% of ThOD	
	12 h:	2.4% of ThOD	
	24 h:	4.7% of ThOD	(89)
A.S. Resp, BOD, 20°C, 1-5 d observed, feed: 1,000 mg/L,<1 d acclimation:		0% ThOD	
		0% removed	(93)